KIRK-OTHMER

ENCYCLOPEDIA OF
CHEMICAL
TECHNOLOGY

FOURTH EDITION

VOLUME **20**

POWER GENERATION
TO
RECYCLING, GLASS

EXECUTIVE EDITOR
Jacqueline I. Kroschwitz

EDITOR
Mary Howe-Grant

KIRK-OTHMER

ENCYCLOPEDIA OF CHEMICAL TECHNOLOGY

FOURTH EDITION

VOLUME **20**

POWER GENERATION
TO
RECYCLING, GLASS

A Wiley-Interscience Publication
JOHN WILEY & SONS

New York • Chichester • Brisbane • Toronto • Singapore

Library of Congress Cataloging-in-Publication Data

Encyclopedia of chemical technology/executive editor, Jacqueline
    I. Kroschwitz; editor, Mary Howe-Grant.—4th ed.
      p. cm.
    At head of title: Kirk-Othmer.
    "A Wiley-Interscience publication."
    Contents: v. 20, Power Generation to Recycling, Glass
    ISBN 0471-52689-4 (v. 20)
    1. Chemistry, Technical—Encyclopedias. I. Kirk, Raymond E.
    (Raymond Eller), 1890–1957. II. Othmer, Donald F. (Donald
    Frederick), 1904–1995. III. Kroschwitz, Jacqueline I., 1942– .
    IV. Howe-Grant, Mary, 1943– . V. Title: Kirk-Othmer encyclopedia
    of chemical technology.
    TP9.E685 1992   91-16789
    660'.03—dc20

Printed in the United States of America

10 9 8 7 6 5 4 3 2

CONTENTS

EDITORIAL STAFF
FOR VOLUME 20

Executive Editor: **Jacqueline I. Kroschwitz**
Editor: **Mary Howe-Grant**
Associate Managing Editor: **Lindy Humphreys**
Copy Editors: **Lawrence Altieri**
               Jonathan Lee

CONTRIBUTORS
TO VOLUME 20

Lowell Ray Anderson, *International Specialty Products, Wayne, New Jersey*, Pyrrole and pyrrole derivatives

Laszlo Beres, *DuPont-NEN Products, Boston, Massachusetts*, Radioactive tracers

Christopher Boerner, *Washington University, St. Louis, Missouri*, Introduction (under Recycling)

Narasimhan Calamur, *Amoco Corporation, Naperville, Illinois*, Propylene

Martin Carrera, *Amoco Corporation, Naperville, Illinois*, Propylene

vii

Kenneth Chilton, *Washington University, St. Louis, Missouri,* Introduction (under Recycling)

Paul W. Collins, *GD Searle & Company, Skokie, Illinois,* Prostaglandins

Steven Collins, *Empire State Electric Energy Research Corporation, New York, New York,* Power generation

John A. Conkling, *American Pyrotechnic Association, Chestertown, Maryland,* Pyrotechnics

Robert A. Copeland, *DuPont-Merck Pharmaceutical Company, Wilmington, Delaware,* Protein engineering

Paul D. Crane, *DuPont-Merck Pharmaceutical Company, North Billerica, Massachusetts,* Radiopharmaceuticals

June P. Davis, *DuPont-Merck Pharmaceutical Company, Wilmington, Delaware,* Protein engineering

Maurice Dery, *Akzo Nobel Chemicals Inc., Dobbs Ferry, New York,* Quaternary ammonium compounds

D. Scott Edwards, *DuPont-Merck Pharmaceutical Company, North Billerica, Massachusetts,* Radiopharmaceuticals

K. Thomas Finley, *State University of New York, Brockport,* Quinolines and isoquinolines; Quinones

Joseph M. Genco, *University of Maine, Orono,* Pulp

T. Godel, *F. Hoffmann-La Roche Ltd., Basel, Switzerland,* Psychopharmacological agents

Richard G. Helmer, *Idaho National Engineering Laboratories, Idaho Falls,* Radioisotopes

W. Hunkeler, *F. Hoffmann-La Roche Ltd., Basel, Switzerland,* Psychopharmacological agents

F. Jenck, *F. Hoffmann-La Roche Ltd., Basel, Switzerland,* Psychopharmacological agents

Gabe I. Kornis, *Pharmacia & Upjohn Inc., Kalamazoo, Michigan,* Pyrazoles, pyrazolines, and pyrazolones

Joel L. Lazewatsky, *DuPont-Merck Pharmaceutical Company, North Billerica, Massachusetts,* Radiopharmaceuticals

Youlin Lin, *Mallinckrodt Medical, Inc., St. Louis, Missouri,* Radiopaques

Kou-Chang Liu, *International Specialty Products, Wayne, New Jersey,* Pyrrole and pyrrole derivatives

John E. Logsdon, *Union Carbide Corporation, Texas City, Texas,* Isopropyl alcohol (under Propyl alcohols)

Richard A. Loke, *Union Carbide Corporation, Texas City, Texas,* Isopropyl alcohol (under Propyl alcohols)

Harry V. Makar, *Consultant, Ellicott City, Maryland,* Ferrous metals (under Recycling)

James R. Martin, *F. Hoffmann-La Roche Ltd., Basel, Switzerland,* Psychopharmacological agents

William H. McBride, *UCLA Medical Center, Los Angeles, California*, Radioprotective agents

Vincent D. McGinniss, *Battelle Columbus Laboratory, Columbus, Ohio*, Radiation curing

J-L. Moreau, *F. Hoffmann-La Roche Ltd., Basel, Switzerland*, Psychopharmacological agents

David Murray, *University of Alberta, Cross Cancer Institute, Edmonton, Canada*, Radioprotective agents

Ramiah Murugan, *Reilly Industries, Inc., Indianapolis, Indiana*, Pyridine and pyridine derivatives

Raman Nambudripad, *Beth Israel Hospital, Boston, Massachusetts*, Proteins

Lev Nelik, *Roper Pumps Company, Commerce, Georgia*, Pumps

T. A. Orofino, *AstroTurf Industries, Dalton, Georgia*, Recreational surfaces

Dennis Pearson, *Hoechst Celanese Corporation, Corpus Christi, Texas*, n-Propyl alcohol (under Propyl alcohols)

H. Wayne Richardson, *Phibro-Tech, Inc., Sumter, South Carolina*, Nonferrous metals (under Recycling)

C. Philip Ross, *Creative Opportunities, Inc., Laguna Niguel, California*, Glass (under Recycling)

J. L. Ryans, *Eastman Chemical Company, Kingsport, Tennessee*, Pressure measurement

John Paul San Giovanni, *Jockey Hollow Technologies, Boston, Massachusetts*, Process control

Eric F. V. Scriven, *Reilly Industries, Inc., Indianapolis, Indiana*, Pyridine and pyridine derivatives

Irwin Silverstein, *International Specialty Products, Wayne, New Jersey*, Quality assurance

A. J. Sleight, *F. Hoffmann-La Roche Ltd., Basel, Switzerland*, Psychopharmacological agents

Temple F. Smith, *Boston University, Massachusetts*, Proteins

Dan Steinmeyer, *Monsanto Company, St. Louis, Missouri*, Process energy conservation

H. G. Sweenie, *AstroTurf Industries, Dalton, Georgia*, Recreational surfaces

Michael Szycher, *PolyMedica Industries, Inc., Woburn, Massachusetts*, Prosthetic and biomedical devices

Arthur J. Taggi, *DuPont Printing and Publishing, Boothwyn, Pennsylvania*, Printing processes

Barry L. Tarmy, *TBD Technology, Berkeley Heights, New Jersey*, Reactor technology

Joseph E. Toomey Jr., *Reilly Industries, Inc., Indianapolis, Indiana*, Pyridine and pyridine derivatives

David Trent, *The Dow Chemical Company, Freeport, Texas*, Propylene oxide

Aaron Twerski, *Brooklyn Law School, New York*, Product liability

Jerry D. Unruh, *Hoechst Celanese Corporation, Corpus Christi, Texas*, *n*-Propyl alcohol (under Propyl alcohols)

Peter Walker, *DuPont Printing and Publishing, Boothwyn, Pennsylvania*, Printing processes

U. Widmer, *F. Hoffmann-La Roche Ltd., Basel, Switzerland*, Psychopharmacological agents

NOTE ON CHEMICAL ABSTRACTS SERVICE REGISTRY NUMBERS AND NOMENCLATURE

Chemical Abstracts Service (CAS) Registry Numbers are unique numerical identifiers assigned to substances recorded in the CAS Registry System. They appear in brackets in the *Chemical Abstracts* (CA) substance and formula indexes following the names of compounds. A single compound may have synonyms in the chemical literature. A simple compound like phenethylamine can be named β-phenylethylamine or, as in *Chemical Abstracts*, benzeneethanamine. The usefulness of the *Encyclopedia* depends on accessibility through the most common correct name of a substance. Because of this diversity in nomenclature careful attention has been given to the problem in order to assist the reader as much as possible, especially in locating the systematic CA index name by means of the Registry Number. For this purpose, the reader may refer to the CAS Registry Handbook—Number Section which lists in numerical order the Registry Number with the *Chemical Abstracts* index name and the molecular formula; eg, **458-88-8**, Piperidine, 2-propyl-, (*S*)-, $C_8H_{17}N$; in the *Encyclopedia* this compound would be found under its common name, coniine [*458-88-8*]. Alternatively, this information can be retrieved electronically from CAS Online. In many cases molecular formulas have also been provided in the *Encyclopedia* text to facilitate electronic searching. The Registry Number is a valuable link for the reader in retrieving additional published information on substances and also as a point of access for on-line data bases.

In all cases, the CAS Registry Numbers have been given for title compounds in articles and for all compounds in the index. All specific substances indexed in *Chemical Abstracts* since 1965 are included in the CAS Registry System as are a large number of substances derived from a variety of reference works. The CAS Registry System identifies a substance on the basis of an unambiguous computer-language description of its molecular structure including stereochemical detail. The Registry Number is a machine-checkable number (like a Social Security number) assigned in sequential order to each substance as it enters the registry system. The value of the number lies in the fact that it is a concise and unique means of substance identification, which is independent of, and therefore bridges, many systems of chemical nomenclature. For polymers, one Registry Number may be used for the entire family; eg, polyoxyethylene (20) sorbitan monolaurate has the same number as all of its polyoxyethylene homologues.

Cross-references are inserted in the index for many common names and for some systematic names. Trademark names appear in the index. Names that are incorrect, misleading, or ambiguous are avoided. Formulas are given very frequently in the text to help in identifying compounds. The spelling and form used, even for industrial names, follow American chemical usage, but not always the usage of *Chemical Abstracts* (eg, *coniine* is used instead of *(S)-2-propylpiperidine*, *aniline* instead of *benzenamine*, and *acrylic acid* instead of *2-propenoic acid*).

There are variations in representation of rings in different disciplines. The dye industry does not designate aromaticity or double bonds in rings. All double bonds and aromaticity are shown in the *Encyclopedia* as a matter of course. For example, tetralin has an aromatic ring and a saturated ring and its structure

appears in the *Encyclopedia* with its common name, Registry Number enclosed in brackets, and parenthetical CA index name, ie, tetralin [*119-64-2*] (1,2,3,4-tetrahydronaphthalene). With names and structural formulas, and especially with CAS Registry Numbers, the aim is to help the reader have a concise means of substance identification.

CONVERSION FACTORS, ABBREVIATIONS, AND UNIT SYMBOLS

SI Units (Adopted 1960)

The International System of Units (abbreviated SI), is being implemented throughout the world. This measurement system is a modernized version of the MKSA (meter, kilogram, second, ampere) system, and its details are published and controlled by an international treaty organization (The International Bureau of Weights and Measures) (1).

SI units are divided into three classes:

BASE UNITS

length	meter[†] (m)
mass	kilogram (kg)
time	second (s)
electric current	ampere (A)
thermodynamic temperature[‡]	kelvin (K)
amount of substance	mole (mol)
luminous intensity	candela (cd)

SUPPLEMENTARY UNITS

plane angle	radian (rad)
solid angle	steradian (sr)

[†]The spellings "metre" and "litre" are preferred by ASTM; however, "-er" is used in the *Encyclopedia*.

[‡]Wide use is made of Celsius temperature (t) defined by

$$t = T - T_0$$

where T is the thermodynamic temperature, expressed in kelvin, and $T_0 = 273.15$ K by definition. A temperature interval may be expressed in degrees Celsius as well as in kelvin.

DERIVED UNITS AND OTHER ACCEPTABLE UNITS

These units are formed by combining base units, supplementary units, and other derived units (2–4). Those derived units having special names and symbols are marked with an asterisk in the list below.

Quantity	Unit	Symbol	Acceptable equivalent
*absorbed dose	gray	Gy	J/kg
acceleration	meter per second squared	m/s^2	
*activity (of a radionuclide)	becquerel	Bq	1/s
area	square kilometer	km^2	
	square hectometer	hm^2	ha (hectare)
	square meter	m^2	
concentration (of amount of substance)	mole per cubic meter	mol/m^3	
current density	ampere per square meter	$A//m^2$	
density, mass density	kilogram per cubic meter	kg/m^3	g/L; mg/cm^3
dipole moment (quantity)	coulomb meter	C·m	
*dose equivalent	sievert	Sv	J/kg
*electric capacitance	farad	F	C/V
*electric charge, quantity of electricity	coulomb	C	A·s
electric charge density	coulomb per cubic meter	C/m^3	
*electric conductance	siemens	S	A/V
electric field strength	volt per meter	V/m	
electric flux density	coulomb per square meter	C/m^2	
*electric potential, potential difference, electromotive force	volt	V	W/A
*electric resistance	ohm	Ω	V/A
*energy, work, quantity of heat	megajoule	MJ	
	kilojoule	kJ	
	joule	J	N·m
	electronvolt[†]	eV[†]	
	kilowatt-hour[†]	kW·h[†]	
energy density	joule per cubic meter	J/m^3	
*force	kilonewton	kN	
	newton	N	$kg·m/s^2$

[†]This non-SI unit is recognized by the CIPM as having to be retained because of practical importance or use in specialized fields (1).

Quantity	Unit	Symbol	Acceptable equivalent
*frequency	megahertz	MHz	
	hertz	Hz	1/s
heat capacity, entropy	joule per kelvin	J/K	
heat capacity (specific), specific entropy	joule per kilogram kelvin	J/(kg·K)	
heat-transfer coefficient	watt per square meter kelvin	W/(m²·K)	
*illuminance	lux	lx	lm/m²
*inductance	henry	H	Wb/A
linear density	kilogram per meter	kg/m	
luminance	candela per square meter	cd/m²	
*luminous flux	lumen	lm	cd·sr
magnetic field strength	ampere per meter	A/m	
*magnetic flux	weber	Wb	V·s
*magnetic flux density	tesla	T	Wb/m²
molar energy	joule per mole	J/mol	
molar entropy, molar heat capacity	joule per mole kelvin	J/(mol·K)	
moment of force, torque	newton meter	N·m	
momentum	kilogram meter per second	kg·m/s	
permeability	henry per meter	H/m	
permittivity	farad per meter	F/m	
*power, heat flow rate, radiant flux	kilowatt	kW	
	watt	W	J/s
power density, heat flux density, irradiance	watt per square meter	W/m²	
*pressure, stress	megapascal	MPa	
	kilopascal	kPa	
	pascal	Pa	N/m²
sound level	decibel	dB	
specific energy	joule per kilogram	J/kg	
specific volume	cubic meter per kilogram	m³/kg	
surface tension	newton per meter	N/m	
thermal conductivity	watt per meter kelvin	W/(m·K)	
velocity	meter per second	m/s	
	kilometer per hour	km/h	
viscosity, dynamic	pascal second	Pa·s	
	millipascal second	mPa·s	
viscosity, kinematic	square meter per second	m²/s	
	square millimeter per second	mm²/s	

Quantity	Unit	Symbol	Acceptable equivalent
volume	cubic meter	m^3	
	cubic diameter	dm^3	L (liter) (5)
	cubic centimeter	cm^3	mL
wave number	1 per meter	m^{-1}	
	1 per centimeter	cm^{-1}	

In addition, there are 16 prefixes used to indicate order of magnitude, as follows:

Multiplication factor	Prefix	Symbol	Note
10^{18}	exa	E	
10^{15}	peta	P	
10^{12}	tera	T	
10^9	giga	G	
10^6	mega	M	
10^3	kilo	k	
10^2	hecto	h^a	[a] Although hecto, deka, deci, and centi
10	deka	da^a	are SI prefixes, their use should be
10^{-1}	deci	d^a	avoided except for SI unit-multiples
10^{-2}	centi	c^a	for area and volume and nontech-
10^{-3}	milli	m	nical use of centimeter, as for body
10^{-6}	micro	μ	and clothing measurement.
10^{-9}	nano	n	
10^{-12}	pico	p	
10^{-15}	femto	f	
10^{-18}	atto	a	

For a complete description of SI and its use the reader is referred to ASTM E380 (4) and the article UNITS AND CONVERSION FACTORS which appears in Vol. 24.

A representative list of conversion factors from non-SI to SI units is presented herewith. Factors are given to four significant figures. Exact relationships are followed by a dagger. A more complete list is given in the latest editions of ASTM E380 (4) and ANSI Z210.1 (6).

Conversion Factors to SI Units

To convert from	To	Multiply by
acre	square meter (m^2)	4.047×10^3
angstrom	meter (m)	$1.0 \times 10^{-10\dagger}$
are	square meter (m^2)	$1.0 \times 10^{2\dagger}$

†Exact.

To convert from	To	Multiply by
astronomical unit	meter (m)	1.496×10^{11}
atmosphere, standard	pascal (Pa)	1.013×10^{5}
bar	pascal (Pa)	$1.0 \times 10^{5\dagger}$
barn	square meter (m^2)	$1.0 \times 10^{-28\dagger}$
barrel (42 U.S. liquid gallons)	cubic meter (m^3)	0.1590
Bohr magneton (μ_{B})	J/T	9.274×10^{-24}
Btu (International Table)	joule (J)	1.055×10^{3}
Btu (mean)	joule (J)	1.056×10^{3}
Btu (thermochemical)	joule (J)	1.054×10^{3}
bushel	cubic meter (m^3)	3.524×10^{-2}
calorie (International Table)	joule (J)	4.187
calorie (mean)	joule (J)	4.190
calorie (thermochemical)	joule (J)	$4.184^{\dagger}$
centipoise	pascal second (Pa·s)	$1.0 \times 10^{-3\dagger}$
centistokes	square millimeter per second (mm^2/s)	$1.0^{\dagger}$
cfm (cubic foot per minute)	cubic meter per second (m^3/s)	4.72×10^{-4}
cubic inch	cubic meter (m^3)	1.639×10^{-5}
cubic foot	cubic meter (m^3)	2.832×10^{-2}
cubic yard	cubic meter (m^3)	0.7646
curie	becquerel (Bq)	$3.70 \times 10^{10\dagger}$
debye	coulomb meter (C·m)	3.336×10^{-30}
degree (angle)	radian (rad)	1.745×10^{-2}
denier (international)	kilogram per meter (kg/m)	1.111×10^{-7}
	tex‡	0.1111
dram (apothecaries')	kilogram (kg)	3.888×10^{-3}
dram (avoirdupois)	kilogram (kg)	1.772×10^{-3}
dram (U.S. fluid)	cubic meter (m^3)	3.697×10^{-6}
dyne	newton (N)	$1.0 \times 10^{-5\dagger}$
dyne/cm	newton per meter (N/m)	$1.0 \times 10^{-3\dagger}$
electronvolt	joule (J)	1.602×10^{-19}
erg	joule (J)	$1.0 \times 10^{-7\dagger}$
fathom	meter (m)	1.829
fluid ounce (U.S.)	cubic meter (m^3)	2.957×10^{-5}
foot	meter (m)	$0.3048^{\dagger}$
footcandle	lux (lx)	10.76
furlong	meter (m)	2.012×10^{-2}
gal	meter per second squared (m/s^2)	$1.0 \times 10^{-2\dagger}$
gallon (U.S. dry)	cubic meter (m^3)	4.405×10^{-3}
gallon (U.S. liquid)	cubic meter (m^3)	3.785×10^{-3}
gallon per minute (gpm)	cubic meter per second (m^3/s)	6.309×10^{-5}
	cubic meter per hour (m^3/h)	0.2271

†Exact.
‡See footnote on p. xiii.

To convert from	To	Multiply by
gauss	tesla (T)	1.0×10^{-4}
gilbert	ampere (A)	0.7958
gill (U.S.)	cubic meter (m^3)	1.183×10^{-4}
grade	radian	1.571×10^{-2}
grain	kilogram (kg)	6.480×10^{-5}
gram force per denier	newton per tex (N/tex)	8.826×10^{-2}
hectare	square meter (m^2)	$1.0 \times 10^{4\dagger}$
horsepower (550 ft·lbf/s)	watt (W)	7.457×10^{2}
horsepower (boiler)	watt (W)	9.810×10^{3}
horsepower (electric)	watt (W)	$7.46 \times 10^{2\dagger}$
hundredweight (long)	kilogram (kg)	50.80
hundredweight (short)	kilogram (kg)	45.36
inch	meter (m)	$2.54 \times 10^{-2\dagger}$
inch of mercury (32°F)	pascal (Pa)	3.386×10^{3}
inch of water (39.2°F)	pascal (Pa)	2.491×10^{2}
kilogram-force	newton (N)	9.807
kilowatt hour	megajoule (MJ)	$3.6^{\dagger}$
kip	newton (N)	4.448×10^{3}
knot (international)	meter per second (m/S)	0.5144
lambert	candela per square meter (cd/m^3)	3.183×10^{3}
league (British nautical)	meter (m)	5.559×10^{3}
league (statute)	meter (m)	4.828×10^{3}
light year	meter (m)	9.461×10^{15}
liter (for fluids only)	cubic meter (m^3)	$1.0 \times 10^{-3\dagger}$
maxwell	weber (Wb)	$1.0 \times 10^{-8\dagger}$
micron	meter (m)	$1.0 \times 10^{-6\dagger}$
mil	meter (m)	$2.54 \times 10^{-5\dagger}$
mile (statute)	meter (m)	1.609×10^{3}
mile (U.S. nautical)	meter (m)	$1.852 \times 10^{3\dagger}$
mile per hour	meter per second (m/s)	0.4470
millibar	pascal (Pa)	1.0×10^{2}
millimeter of mercury (0°C)	pascal (Pa)	$1.333 \times 10^{2\dagger}$
minute (angular)	radian	2.909×10^{-4}
myriagram	kilogram (kg)	10
myriameter	kilometer (km)	10
oersted	ampere per meter (A/m)	79.58
ounce (avoirdupois)	kilogram (kg)	2.835×10^{-2}
ounce (troy)	kilogram (kg)	3.110×10^{-2}
ounce (U.S. fluid)	cubic meter (m^3)	2.957×10^{-5}
ounce-force	newton (N)	0.2780
peck (U.S.)	cubic meter (m^3)	8.810×10^{-3}
pennyweight	kilogram (kg)	1.555×10^{-3}
pint (U.S. dry)	cubic meter (m^3)	5.506×10^{-4}
pint (U.S. liquid)	cubic meter (m^3)	4.732×10^{-4}

†Exact.

To convert from	To	Multiply by
poise (absolute viscosity)	pascal second (Pa·s)	$0.10^{\dagger}$
pound (avoirdupois)	kilogram (kg)	0.4536
pound (troy)	kilogram (kg)	0.3732
poundal	newton (N)	0.1383
pound-force	newton (N)	4.448
pound force per square inch (psi)	pascal (Pa)	6.895×10^3
quart (U.S. dry)	cubic meter (m^3)	1.101×10^{-3}
quart (U.S. liquid)	cubic meter (m^3)	9.464×10^{-4}
quintal	kilogram (kg)	$1.0 \times 10^{2\dagger}$
rad	gray (Gy)	$1.0 \times 10^{-2\dagger}$
rod	meter (m)	5.029
roentgen	coulomb per kilogram (C/kg)	2.58×10^{-4}
second (angle)	radian (rad)	$4.848 \times 10^{-6\dagger}$
section	square meter (m^2)	2.590×10^6
slug	kilogram (kg)	14.59
spherical candle power	lumen (lm)	12.57
square inch	square meter (m^2)	6.452×10^{-4}
square foot	square meter (m^2)	9.290×10^{-2}
square mile	square meter (m^2)	2.590×10^6
square yard	square meter (m^2)	0.8361
stere	cubic meter (m^3)	$1.0^{\dagger}$
stokes (kinematic viscosity)	square meter per second (m^2/s)	$1.0 \times 10^{-4\dagger}$
tex	kilogram per meter (kg/m)	$1.0 \times 10^{-6\dagger}$
ton (long, 2240 pounds)	kilogram (kg)	1.016×10^3
ton (metric) (tonne)	kilogram (kg)	$1.0 \times 10^{3\dagger}$
ton (short, 2000 pounds)	kilogram (kg)	9.072×10^2
torr	pascal (Pa)	1.333×10^2
unit pole	weber (Wb)	1.257×10^{-7}
yard	meter (m)	$0.9144^{\dagger}$

†Exact.

Abbreviations and Unit Symbols

Following is a list of common abbreviations and unit symbols used in the *Encyclopedia*. In general they agree with those listed in *American National Standard Abbreviations for Use on Drawings and in Text* (*ANSI Y1.1*) (6) and *American National Standard Letter Symbols for Units in Science and Technology* (*ANSI Y10*) (6). Also included is a list of acronyms for a number of private and government organizations as well as common industrial solvents, polymers, and other chemicals.

Rules for Writing Unit Symbols (4):

1. Unit symbols are printed in upright letters (roman) regardless of the type style used in the surrounding text.
2. Unit symbols are unaltered in the plural.
3. Unit symbols are not followed by a period except when used at the end of a sentence.
4. Letter unit symbols are generally printed lower-case (for example, cd for candela) unless the unit name has been derived from a proper name, in which case the first letter of the symbol is capitalized (W, Pa). Prefixes and unit symbols retain their prescribed form regardless of the surrounding typography.
5. In the complete expression for a quantity, a space should be left between the numerical value and the unit symbol. For example, write 2.37 lm, *not* 2.37lm, and 35 mm, *not* 35mm. When the quantity is used in an adjectival sense, a hyphen is often used, for example, 35-mm film. *Exception:* No space is left between the numerical value and the symbols of degree, minute, and second of plane angle, degree Celsius, and the percent sign.
6. No space is used between the prefix and unit symbol (for example, kg).
7. Symbols, not abbreviations, should be used for units. For example, use "A," not "amp," for ampere.
8. When multiplying unit symbols, use a raised dot:

$$N{\cdot}m \quad for \quad newton \ meter$$

In the case of W·h, the dot may be omitted, thus:

$$Wh$$

An exception to this practice is made for computer printouts, automatic typewriter work, etc, where the raised dot is not possible, and a dot on the line may be used.

9. When dividing unit symbols, use one of the following forms:

$$m/s \quad or \quad m{\cdot}s^{-1} \quad or \quad \frac{m}{s}$$

In no case should more than one slash be used in the same expression unless parentheses are inserted to avoid ambiguity. For example, write:

$$J/(mol{\cdot}K) \quad or \quad J{\cdot}mol^{-1}{\cdot}K^{-1} \quad or \quad (J/mol)/K$$

but *not*

$$J/mol/K$$

10. Do not mix symbols and unit names in the same expression. Write:

$$\text{joules per kilogram} \quad or \quad \text{J/kg} \quad or \quad \text{J·kg}^{-1}$$

but *not*

$$\text{joules/kilogram} \quad nor \quad \text{joules/kg} \quad nor \quad \text{joules·kg}^{-1}$$

ABBREVIATIONS AND UNITS

A	ampere	AOAC	Association of Official Analytical Chemists
A	anion (eg, HA)		
A	mass number	AOCS	American Oil Chemists' Society
a	atto (prefix for 10^{-18})		
AATCC	American Association of Textile Chemists and Colorists	APHA	American Public Health Association
		API	American Petroleum Institute
ABS	acrylonitrile–butadiene–styrene	aq	aqueous
abs	absolute	Ar	aryl
ac	alternating current, *n*.	*ar-*	aromatic
a-c	alternating current, *adj*.	*as-*	asymmetric(al)
ac-	alicyclic	ASHRAE	American Society of Heating, Refrigerating, and Air Conditioning Engineers
acac	acetylacetonate		
ACGIH	American Conference of Governmental Industrial Hygienists		
		ASM	American Society for Metals
ACS	American Chemical Society	ASME	American Society of Mechanical Engineers
AGA	American Gas Association	ASTM	American Society for Testing and Materials
Ah	ampere hour		
AIChE	American Institute of Chemical Engineers	at no.	atomic number
		at wt	atomic weight
AIME	American Institute of Mining, Metallurgical, and Petroleum Engineers	av(g)	average
		AWS	American Welding Society
		b	bonding orbital
AIP	American Institute of Physics	bbl	barrel
		bcc	body-centered cubic
AISI	American Iron and Steel Institute	BCT	body-centered tetragonal
		Bé	Baumé
alc	alcohol(ic)	BET	Brunauer-Emmett-Teller (adsorption equation)
Alk	alkyl		
alk	alkaline (not alkali)	bid	twice daily
amt	amount	Boc	*t*-butyloxycarbonyl
amu	atomic mass unit	BOD	biochemical (biological) oxygen demand
ANSI	American National Standards Institute		
		bp	boiling point
AO	atomic orbital	Bq	becquerel

C	coulomb	DIN	Deutsche Industrie Normen
°C	degree Celsius		
C-	denoting attachment to carbon	dl-; DL-	racemic
		DMA	dimethylacetamide
c	centi (prefix for 10^{-2})	DMF	dimethylformamide
c	critical	DMG	dimethyl glyoxime
ca	circa (approximately)	DMSO	dimethyl sulfoxide
cd	candela; current density; circular dichroism	DOD	Department of Defense
		DOE	Department of Energy
CFR	Code of Federal Regulations	DOT	Department of Transportation
cgs	centimeter-gram-second	DP	degree of polymerization
CI	Color Index	dp	dew point
cis-	isomer in which substituted groups are on same side of double bond between C atoms	DPH	diamond pyramid hardness
		dstl(d)	distill(ed)
		dta	differential thermal analysis
cl	carload		
cm	centimeter	(E)-	entgegen; opposed
cmil	circular mil	ε	dielectric constant (unitless number)
cmpd	compound		
CNS	central nervous system	e	electron
CoA	coenzyme A	ECU	electrochemical unit
COD	chemical oxygen demand	ed.	edited, edition, editor
coml	commercial(ly)	ED	effective dose
cp	chemically pure	EDTA	ethylenediaminetetra- acetic acid
cph	close-packed hexagonal		
CPSC	Consumer Product Safety Commission	emf	electromotive force
		emu	electromagnetic unit
cryst	crystalline	en	ethylene diamine
cub	cubic	eng	engineering
D	debye	EPA	Environmental Protection Agency
D-	denoting configurational relationship		
		epr	electron paramagnetic resonance
d	differential operator		
d	day; deci (prefix for 10^{-1})	eq.	equation
d	density	esca	electron spectroscopy for chemical analysis
d-	dextro-, dextrorotatory		
da	deka (prefix for 10^1)	esp	especially
dB	decibel	esr	electron-spin resonance
dc	direct current, n.	est(d)	estimate(d)
d-c	direct current, adj.	estn	estimation
dec	decompose	esu	electrostatic unit
detd	determined	exp	experiment, experimental
detn	determination	ext(d)	extract(ed)
Di	didymium, a mixture of all lanthanons	F	farad (capacitance)
		F	faraday (96,487 C)
dia	diameter	f	femto (prefix for 10^{-15})
dil	dilute		

FAO	Food and Agriculture Organization (United Nations)	hyd	hydrated, hydrous
		hyg	hygroscopic
		Hz	hertz
fcc	face-centered cubic	i (eg, Pr^i)	iso (eg, isopropyl)
FDA	Food and Drug Administration	i-	inactive (eg, i-methionine)
		IACS	International Annealed Copper Standard
FEA	Federal Energy Administration	ibp	initial boiling point
FHSA	Federal Hazardous Substances Act	IC	integrated circuit
		ICC	Interstate Commerce Commission
fob	free on board		
fp	freezing point	ICT	International Critical Table
FPC	Federal Power Commission		
		ID	inside diameter; infective dose
FRB	Federal Reserve Board		
frz	freezing	ip	intraperitoneal
G	giga (prefix for 10^9)	IPS	iron pipe size
G	gravitational constant = 6.67×10^{11} N·m^2/kg^2	ir	infrared
		IRLG	Interagency Regulatory Liaison Group
g	gram		
(g)	gas, only as in H$_2$O(g)	ISO	International Organization Standardization
g	gravitational acceleration		
gc	gas chromatography	ITS-90	International Temperature Scale (NIST)
gem-	geminal		
glc	gas–liquid chromatography	IU	International Unit
		IUPAC	International Union of Pure and Applied Chemistry
g-mol wt; gmw	gram-molecular weight		
GNP	gross national product	IV	iodine value
gpc	gel-permeation chromatography	iv	intravenous
		J	joule
GRAS	Generally Recognized as Safe	K	kelvin
		k	kilo (prefix for 10^3)
grd	ground	kg	kilogram
Gy	gray	L	denoting configurational relationship
H	henry		
h	hour; hecto (prefix for 10^2)	L	liter (for fluids only) (5)
ha	hectare	l-	$levo$-, levorotatory
HB	Brinell hardness number	(l)	liquid, only as in NH$_3$(l)
Hb	hemoglobin	LC$_{50}$	conc lethal to 50% of the animals tested
hcp	hexagonal close-packed		
hex	hexagonal	LCAO	linear combination of atomic orbitals
HK	Knoop hardness number		
hplc	high performance liquid chromatography	lc	liquid chromatography
		LCD	liquid crystal display
HRC	Rockwell hardness (C scale)	lcl	less than carload lots
		LD$_{50}$	dose lethal to 50% of the animals tested
HV	Vickers hardness number		

LED	light-emitting diode	N-	denoting attachment to
liq	liquid		nitrogen
lm	lumen	n (as n_{D}^{20})	index of refraction (for
ln	logarithm (natural)		20°C and sodium light)
LNG	liquefied natural gas	^{n} (as Bu^{n}),	
log	logarithm (common)	n-	normal (straight-chain
LOI	limiting oxygen index		structure)
LPG	liquefied petroleum gas	n	neutron
ltl	less than truckload lots	n	nano (prefix for 10^9)
lx	lux	na	not available
M	mega (prefix for 10^6);	NAS	National Academy of
	metal (as in MA)		Sciences
M	molar; actual mass	NASA	National Aeronautics and
$\overline{M}_w$	weight-average mol wt		Space Administration
$\overline{M}_n$	number-average mol wt	nat	natural
m	meter; milli (prefix for	ndt	nondestructive testing
	10^{-3})	neg	negative
m	molal	NF	*National Formulary*
m-	meta	NIH	National Institutes of
max	maximum		Health
MCA	Chemical Manufacturers'	NIOSH	National Institute of
	Association (was		Occupational Safety and
	Manufacturing Chemists		Health
	Association)	NIST	National Institute of
MEK	methyl ethyl ketone		Standards and
meq	milliequivalent		Technology (formerly
mfd	manufactured		National Bureau of
mfg	manufacturing		Standards)
mfr	manufacturer	nmr	nuclear magnetic
MIBC	methyl isobutyl carbinol		resonance
MIBK	methyl isobutyl ketone	NND	New and Nonofficial Drugs
MIC	minimum inhibiting		(AMA)
	concentration	no.	number
min	minute; minimum	NOI-(BN)	not otherwise indexed (by
mL	milliliter		name)
MLD	minimum lethal dose	NOS	not otherwise specified
MO	molecular orbital	nqr	nuclear quadruple
mo	month		resonance
mol	mole	NRC	Nuclear Regulatory
mol wt	molecular weight		Commission; National
mp	melting point		Research Council
MR	molar refraction	NRI	New Ring Index
ms	mass spectrometry	NSF	National Science
MSDS	material safety data sheet		Foundation
mxt	mixture	NTA	nitrilotriacetic acid
μ	micro (prefix for 10^{-6})	NTP	normal temperature and
N	newton (force)		pressure (25°C and 101.3
N	normal (concentration);		kPa or 1 atm)
	neutron number		

NTSB	National Transportation Safety Board	qv	quod vide (which see)
O-	denoting attachment to oxygen	R	univalent hydrocarbon radical
o-	ortho	(R)-	rectus (clockwise configuration)
OD	outside diameter	r	precision of data
OPEC	Organization of Petroleum Exporting Countries	rad	radian; radius
o-phen	o-phenanthridine	RCRA	Resource Conservation and Recovery Act
OSHA	Occupational Safety and Health Administration	rds	rate-determining step
		ref.	reference
owf	on weight of fiber	rf	radio frequency, n.
Ω	ohm	r-f	radio frequency, adj.
P	peta (prefix for 10^{15})	rh	relative humidity
p	pico (prefix for 10^{-12})	RI	Ring Index
p-	para	rms	root-mean square
p	proton	rpm	rotations per minute
p.	page	rps	revolutions per second
Pa	pascal (pressure)	RT	room temperature
PEL	personal exposure limit based on an 8-h exposure	RTECS	Registry of Toxic Effects of Chemical Substances
		^{s} (eg, Bu^s);	
pd	potential difference	sec-	secondary (eg, secondary butyl)
pH	negative logarithm of the effective hydrogen ion concentration	S	siemens
		(S)-	sinister (counterclockwise configuration)
phr	parts per hundred of resin (rubber)	S-	denoting attachment to sulfur
p-i-n	positive-intrinsic-negative		
pmr	proton magnetic resonance	s-	symmetric(al)
p-n	positive-negative	s	second
po	per os (oral)	(s)	solid, only as in $H_2O(s)$
POP	polyoxypropylene	SAE	Society of Automotive Engineers
pos	positive		
pp.	pages	SAN	styrene-acrylonitrile
ppb	parts per billion (10^9)	sat(d)	saturate(d)
ppm	parts per million (10^6)	satn	saturation
ppmv	parts per million by volume	SBS	styrene–butadiene–styrene
ppmwt	parts per million by weight	sc	subcutaneous
PPO	poly(phenyl oxide)	SCF	self-consistent field; standard cubic feet
ppt(d)	precipitate(d)		
pptn	precipitation	Sch	Schultz number
Pr (no.)	foreign prototype (number)	sem	scanning electron microscope(y)
pt	point; part		
PVC	poly(vinyl chloride)	SFs	Saybolt Furol seconds
pwd	powder	sl sol	slightly soluble
py	pyridine	sol	soluble

soln	solution	*trans-*	isomer in which
soly	solubility		substituted groups are
sp	specific; species		on opposite sides of
sp gr	specific gravity		double bond between C
sr	steradian		atoms
std	standard	TSCA	Toxic Substances Control
STP	standard temperature and		Act
	pressure (0°C and 101.3	TWA	time-weighted average
	kPa)	Twad	Twaddell
sub	sublime(s)	UL	Underwriters' Laboratory
SUs	Saybolt Universal seconds	USDA	United States Department
syn	synthetic		of Agriculture
^{t} (eg, Bu^t),		USP	*United States*
t-, tert-	tertiary (eg, tertiary		*Pharmacopeia*
	butyl)	uv	ultraviolet
T	tera (prefix for 10^{12}); tesla	V	volt (emf)
	(magnetic flux density)	var	variable
t	metric ton (tonne)	*vic-*	vicinal
t	temperature	vol	volume (not volatile)
TAPPI	Technical Association of	vs	versus
	the Pulp and Paper	v sol	very soluble
	Industry	W	watt
TCC	Tagliabue closed cup	Wb	weber
tex	tex (linear density)	Wh	watt hour
T_g	glass-transition	WHO	World Health
	temperature		Organization (United
tga	thermogravimetric		Nations)
	analysis	wk	week
THF	tetrahydrofuran	yr	year
tlc	thin layer chromatography	(*Z*)-	zusammen; together;
TLV	threshold limit value		atomic number

Non-SI (Unacceptable and Obsolete) Units		Use
Å	angstrom	nm
at	atmosphere, technical	Pa
atm	atmosphere, standard	Pa
b	barn	cm^2
bar†	bar	Pa
bbl	barrel	m^3
bhp	brake horsepower	W
Btu	British thermal unit	J
bu	bushel	m^3; L
cal	calorie	J
cfm	cubic foot per minute	m^3/s
Ci	curie	Bq
cSt	centistokes	mm^2/s
c/s	cycle per second	Hz

†Do not use bar (10^5 Pa) or millibar (10^2 Pa) because they are not SI units, and are accepted internationally only for a limited time in special fields because of existing usage.

Non-SI (Unacceptable and Obsolete) Units		Use
cu	cubic	exponential form
D	debye	C·m
den	denier	tex
dr	dram	kg
dyn	dyne	N
dyn/cm	dyne per centimeter	mN/m
erg	erg	J
eu	entropy unit	J/K
°F	degree Fahrenheit	°C; K
fc	footcandle	lx
fl	footlambert	lx
fl oz	fluid ounce	m^3; L
ft	foot	m
ft·lbf	foot pound-force	J
gf den	gram-force per denier	N/tex
G	gauss	T
Gal	gal	m/s^2
gal	gallon	m^3; L
Gb	gilbert	A
gpm	gallon per minute	(m^3/s); (m^3/h)
gr	grain	kg
hp	horsepower	W
ihp	indicated horsepower	W
in.	inch	m
in. Hg	inch of mercury	Pa
in. H_2O	inch of water	Pa
in.-lbf	inch pound-force	J
kcal	kilo-calorie	J
kgf	kilogram-force	N
kilo	for kilogram	kg
L	lambert	lx
lb	pound	kg
lbf	pound-force	N
mho	mho	S
mi	mile	m
MM	million	M
mm Hg	millimeter of mercury	Pa
mμ	millimicron	nm
mph	miles per hour	km/h
μ	micron	μm
Oe	oersted	A/m
oz	ounce	kg
ozf	ounce-force	N
η	poise	Pa·s
P	poise	Pa·s
ph	phot	lx
psi	pounds-force per square inch	Pa
psia	pounds-force per square inch absolute	Pa
psig	pounds-force per square inch gage	Pa
qt	quart	m^3; L
°R	degree Rankine	K
rd	rad	Gy
sb	stilb	lx
SCF	standard cubic foot	m^3
sq	square	exponential form
thm	therm	J
yd	yard	m

BIBLIOGRAPHY

1. The International Bureau of Weights and Measures, BIPM (Parc de Saint-Cloud, France) is described in Appendix X2 of Ref. 4. This bureau operates under the exclusive supervision of the International Committee for Weights and Measures (CIPM).
2. *Metric Editorial Guide (ANMC-78-1)*, latest ed., American National Metric Council, 5410 Grosvenor Lane, Bethesda, Md. 20814, 1981.
3. *SI Units and Recommendations for the Use of Their Multiples and of Certain Other Units (ISO 1000-1981)*, American National Standards Institute, 1430 Broadway, New York, 10018, 1981.
4. Based on *ASTM E380-89a (Standard Practice for Use of the International System of Units (SI))*, American Society for Testing and Materials, 1916 Race Street, Philadelphia, Pa. 19103, 1989.
5. *Fed. Reg.*, Dec. 10, 1976 (41 FR 36414).
6. For ANSI address, see Ref. 3.

R. P. LUKENS
ASTM Committee E-43 on SI Practice

Continued

POWER GENERATION

Power, P, defined as the rate at which work is performed, is expressed in terms of energy divided by time and is most commonly given in units of horsepower, as for the power supplied by mechanical devices such as diesel engines, or in the SI units of watts, especially when measuring electrical power. One horsepower is equivalent to the amount of power needed to lift 33,000 pounds (14,982 kg) one foot (30.5 cm) in one minute. One watt is equivalent to the power required to perform one joule of work per second. In a simple direct-current circuit where potential is represented by E:

$$P = EI = I^2R$$

One watt of power is required to force one ampere of electric current, I, through a length of conductor having a resistance, R, to flow of one ohm. The calculation of power in alternating-current circuits is more complex because of the characteristics of these circuits and their loads. For a single-phase circuit having an effective line voltage, E, and an effective line current, I:

$$P = EI \times PF$$

where PF is the power factor corresponding to the lag between the voltage and current wave forms. For multiphase systems, a factor is introduced on the right-hand side of the equation. The factor is two for a two-phase system; 1.73 for a three-phase system.

History

In the industrial arena, the term power generation most typically refers to the production of electrical or mechanical power via any of several energy conversion processes. Early examples of practical power generation devices include water-wheel-powered mills for grinding grain, which were reportedly used as early as 100 BC in the Balkans and areas of the Middle East, and wind-powered mills, which were widely used as early as the tenth century in the Middle East.

In the early eighteenth century, the advent of large-scale coal mining in England led to the development of a steam-powered pump for evacuating ground water for the deep coal (qv) mines. The first practical steam (qv) engine, on which the steam-powered pump relied, was developed in England in 1690 by Newcomen. The device consisted of a piston that reciprocated in a cylinder by the alternating admittance and condensation of steam generated in a large copper vessel located beneath the cylinder. The engine, similar in principle to steam engines later applied during the industrial revolution to power ships, was successfully used for pumping water at an English mine in 1711.

In the mid-1700s, Watt was commissioned to repair a model of the Newcomen steam engine. Realizing that the device's need to condense steam in its cylinder following the power stroke via a jet of cold water made the engine a voracious consumer of steam and therefore inefficient, Watt built a working model of a new type of steam engine that avoided the need to condense steam within the cylinder. A separate condensing chamber was employed that, by avoiding the need to heat and cool the working cylinder alternately, reduced steam and fuel consumption. Another benefit was the ability to admit steam alternately above and below the piston, which created a double-acting cylinder. This doubled the work that the machine was capable of doing. In addition, the cylinder of Watt's steam engine was insulated to further stem heat loss and improve the efficiency of the device. Watt's steam engine was patented in 1769 and by 1776 it was being developed for commercial use in pumping applications.

A sun-and-planet gear arrangement was then adapted to enable this reciprocating machine to drive rotating equipment. The ability to power rotating machinery using steam engines was an important development during the industrial revolution because factories no longer needed to be located next to rivers or waterfalls where water wheels could be employed to drive the factories' rotating machinery. Instead, factories could be located where it was most economical in terms of available work forces, such as in cities, and means of transport, such as near harbors.

The industrial revolution, often said to have commenced in the 1750s upon the widespread use of mechanized shuttle looms in textile factories, ended in the 1880s upon the advent of the electric lamp. The revolution relied heavily on power supplied by steam engines. In addition to revolutionizing manufacturing, the steam engine was used to power ships and locomotives, greatly improving the ability to deliver goods reliably over long distances and thereby expanding international trade. By the end of the nineteenth century, internal combustion engines, steam- and water-turbine-powered electric generators, and electric motors were being applied commercially.

The first centralized electric generating plant in the United States was Edison's three-unit steam-engine-based station, which supplied electric power to light approximately 5000 electric lamps in a group of homes and businesses in New York City in 1882. Also in 1882, the first hydroelectric power plant went into operation in Appleton, Wisconsin, generating approximately 25 kW of power, enough to power more than 200 100-watt light bulbs.

Early power plants had to be located close to the users of the electricity to minimize power losses associated with the resistance to current flow through the transmission lines. The development of transformers significantly improved the ability to transport electricity efficiently over long distances. By boosting the voltage of the electricity, a given amount of power can be transmitted over long distances at much lower line losses compared to that transmitted at lower voltages. Where loads exist along a line, step-down transformers can be used.

In the early part of the twentieth century, many industrial plants had fossil-fuel-fired boilers that produced steam for both mechanical-drive equipment, such as turbine-driven refrigeration compressors, and steam-turbine-powered electric generators serving the plant's electrical needs, eg, motors. As extensive electric power transmission and distribution systems were erected, industrial power users came to rely more on power supplied by centralized utility-owned power plants. Electric utilities made huge investments in generation and delivery equipment. It was not economically feasible for more than one power company to install transmission and distribution systems in a given geographic area. Thus utilities were considered to be natural monopolies and were subjected to tight government controls to ensure fair business practices and consumer protection.

A significant number of energy-intensive industrial facilities, such as many steel (qv) mills and paper (qv) plants, have maintained on-site power generation capacity to meet their unique electric needs. These and other nonutility generators (NUGs) have formed a growing industry in the late twentieth century. In the United States, this industry has been driven by regulations that allow NUGs to generate power for sale to utilities, and other factors such as the widespread availability of low cost natural gas (see GAS, NATURAL). Internationally, many countries have paved the way for NUGs by privatizing their power generation sectors, which were previously government controlled.

The widespread availability of electrical energy completely transformed modern society and enabled a host of breakthroughs in manufacturing, medical science, communications, construction, education, and transportation. Centralized fossil fuel-powered, steam-turbine-based power plants remain the dominant means of electricity production. However, hydropower facilities such as the 1900-MW Hoover Dam Power Project located on the Arizona–Nevada border, commissioned by the U.S. Bureau of Reclamation during the 1930s, have also made significant contributions.

In 1956, the world's first commercial nuclear power plant started operation in England. By the 1960s, many nuclear power plants were built worldwide. At the end of the twentieth century, nuclear generating plants are used widely by U.S. electric utilities. Since 1984, these plants have provided the second largest share of total U.S. electricity generation, 21% of annual GW·h generated, behind coal-fired power plants (see NUCLEAR REACTORS).

The use of nuclear power has been a topic of debate for many years. Nuclear fuel represents a resource for generating energy well into the future, whereas economically recoverable fossil fuel reserves may become depleted. Worker exposure, injuries, and fatalities in nuclear fuel mining are reportedly far less compared to those associated with recovery and handling of fossil fuels. Potential hazards associated with transporting and storing radioactive wastes do exist, however.

At the same time that utilities began utilizing nuclear power, combustion turbine generators became an important means by which extra power required during times of peak demand were produced. Widely installed during the late 1960s and 1970s in the United States for peaking power service, the combustion turbines, also referred to as gas turbines, were based on a thermodynamic cycle similar to that used in jet aircraft. However, instead of creating thrust for propulsion, the explosive power developed by the combustion of compressed air and fuel, oil, or gas is used to drive a power turbine which in turn drives an electric generator. Combustion turbines proved to be ideal for peaking power applications because of rapid start-up capabilities. Unlike steam-based power plants, which can take hours to come on-line owing to thermal stress constraints, combustion turbines can be started and brought to full load in a matter of minutes. However, frequent and rapid start-ups can shorten a combustion turbine's operating life and necessitate increased maintenance.

Early combustion turbine/generators were fairly inefficient and expensive to operate compared to base-load steam power plants. These were therefore used predominantly for special applications where rapid start-up capability, small size, and ease of operation proved advantageous. Such applications included remote sites in oil and gas fields, off-shore drilling platforms, gas transmission pipelines (qv), emergency power for buildings and plants, and peaking power service. By the 1980s, however, improvements in gas turbines; reduced prices and increased availability of natural gas; the technology's adaptability to efficient cogeneration power plants, which coproduce thermal and electrical energy; and advanced combined-cycle power systems, all converged to make combustion-turbine-based power options competitive with conventional steam-turbine-based cycles.

In the United States, laws passed since the late 1970s have encouraged the development of cogeneration plants and independent power plants developed by nonutility power producers. Combustion-turbine-based facilities have become extremely popular among nonutility power plant developers. Among the many reasons for selecting this technology is the ability to develop compact, modular, combustion-turbine-based power plants within one or two years.

The idea of harnessing renewable energy sources for power generation has long intrigued those in the power industry. Types of renewable energy facilities used in the United States include hydroelectric, geothermal, biomass, photovoltaic, solar thermal, and wind. However, except for hydropower, financial and technological obstacles have prevented widespread utilization of these power options. In 1993, renewable energy sources accounted for 11% of total net electric utility generation and 25% of total nonutility generation in the United States. However, excluding hydropower, these percentages drop to 0.3% and 21%, respectively (see CHEMURGY; FUELS FROM BIOMASS; GEOTHERMAL ENERGY; PHOTOVOLTAIC CELLS; RENEWABLE ENERGY SOURCES; SOLAR ENERGY).

In the United States and elsewhere, governmental agencies have long been involved in the electric power business. Public agencies were formed to build and operate power generation, transmission, and distribution systems. For example, the U.S. Rural Electrification Administration was established primarily to provide power to agricultural regions. Numerous state-run and municipal power systems were also established. These have long provided inexpensive power, benefitting residents and also attracting and/or retaining businesses and industry. However, investor-owned utilities have come to be the main suppliers of electric power in the United States (Table 1). For example, investor-owned utilities supplied approximately 80% of the total power generated by utilities in 1993. Nonutility power producers having a capacity of 1 MW or greater supplied approximately 325×10^9 kWh of electricity in the United States in 1993, accounting for approximately 11% of the power generated that year. The 1993 total installed generating capacity for electric utilities was nearly 700,000 MW; for nonutility power producers, 61,000 MW (Table 2). Consumption of the 2.861×10^{12} kWh of electricity in the United States was 35% to residential consumers, 34% to industrial, 28% to commercial, and the remaining 3% to others (1).

During the 1980s and 1990s, there was an international trend toward privatization of the electric power industry within countries where the industry was previously government controlled. In the United States as of 1995, where the electric power industry has been dominated by investor-owned utilities subject to tight government controls, deregulation was bringing sweeping changes and increased competition among utilities and from nonutility power producers (see REGULATORY AGENCIES, POWER GENERATION).

Table 1. U.S. Power Supply by Ownership for 1993[a]

	Utility ownership			
Parameter	Investors	Public	Federal	Cooperative
generating capability, % of 700 GW	76	11	9	4
net generation, % of 2883 $\times$ 10^9 kWh	79	9	8	4
total sales, % of 2861 $\times$ 10^9 kWh	76	14	2	8

[a]Ref. 1.

Combustion Fundamentals

The combustion of fossil fuels, typically coal, oil (see PETROLEUM), or natural gas, is central to the energy conversion process by which most electric power is generated worldwide, particularly in the United States (Table 3). Thus, an understanding of the principles of fossil fuel combustion is integral to the understanding of modern power plant operation. A fundamental goal of power plant designers and operators is to ensure that the maximum economical amount of energy is extracted for practical use from a given amount of fuel burned (see COMBUSTION TECHNOLOGY).

Combustion is basically an exothermic chemical reaction by which the potential chemical energy stored in a fuel is released during its reaction with oxygen in the presence of heat. Thus, the basic constituents of combustion are fuel, oxygen, and heat. At normal temperatures, the oxygen, usually from air, and fuel can typically mingle without any combustion effect. However, upon the addition of sufficient heat, the fuel and air are excited to a point at which

Table 2. U.S. Electric Power Industry Summary for 1993[a]

Parameter	Electric utilities	Nonutilities
generating capability[b], MW	699,971[c]	60,778
coal	300,795	9,772[d]
petroleum	69,519	2,043[e]
gas	132,495	23,463[f]
petroleum–natural gas[g]		8,505
nuclear	99,041	20[h]
renewable		
hydroelectric, conventional	74,763	2,741
geothermal	1,747	1,318
biomass[i]	459	10,177
wind	1	1,813
solar thermal	0	354
photovoltaic	4	7
other	21,146[j]	566[k]
generation[l], kWh $\times 10^6$	2,882,525	325,226
coal	1,639,151	53,367[d]
petroleum	99,539[m]	13,364[e]
gas	258,915	174,282[f]
nuclear	610,291	78
renewable		
hydroelectric, conventional	269,098	11,511
geothermal	7,571	9,749
biomass[i]	1,990	55,746
wind	0	3,052
solar thermal	0	895
photovoltaic	4	2
other	4,036[j,n]	3,181[k]

[a]Ref. 2. [b]Values for the nonutilities are installed capacity values. [c]Net summer capability based on primary energy source. Waste heat, waste gases, and waste steam are included in the original primary energy source, ie, coal, petroleum, or gas. Historical data have been revised to reflect this change. [d]Includes coal, anthracite culm, and coal waste. [e]Includes petroleum coke, diesel, kerosene, petroleum sludge, and tar. [f]Includes natural gas, butane, ethane, propane, waste heat, and waste gases. [g]Combined fuel. [h]Nuclear reactor and generator at Argonne National Laboratory used primarily for research and development in testing reactor fuels as well as for training. The generation from the unit is used for internal consumption. [i]Includes wood, wood waste, peat, wood liquors, railroad ties, pitch, wood sludge, municipal solid waste, agricultural waste, straw, tires, landfill gases, fish oils, and/or other waste. [j]Hydroelectric pumped storage. [k]Includes hydrogen, sulfur, batteries, chemicals, and spent sulfite liquor. [l]Values for utilities are net; values for nonutilities are gross. [m]Includes petroleum coke. [n]Represents total pumped storage facility production minus energy used for pumping.

high velocity molecular collisions occur. If the velocity of these collisions is high enough, the bonds holding the various molecules together break and release heat.

The point at which enough heat has been added to start combustion is known as the ignition point. Once initiated, external heating sources are typically not required to maintain the combustion process, because most fuels release sufficient heat during the combustion process.

Table 3. 1993 U.S. Power Generation by Energy Source[a]

Source	Quantity consumed	Fraction of power generated[b], %
coal, t	738	57
petroleum, m^{3c}	25.8×10^6	4
natural gas, m^{3d}	75.9×10^6	9
nuclear		21
hydroelectric		9
other		21

[a]Refs. 1 and 2.
[b]Total net generation was 2883×10^9 kWh.
[c]To convert m^3 to barrels, multiply by 6.29.
[d]To convert m^3 to ft^3, multiply by 35.31.

Designers of combustion equipment, including boilers, reciprocating engines, and gas turbines, rely heavily on combustion calculations to determine the rate of fuel and air input required to achieve the desired thermal output. The main fossil fuel constituents enabling combustion are carbon (qv), hydrogen (qv), and various hydrocarbons (qv). The main products resulting from combustion are water (qv) and carbon dioxide (qv). However, a variety of trace compounds and unburned matter may also be released during the combustion of power plant fuels. These materials vary, depending on the specific fuel being burned and whether complete combustion occurs. For example, coal and fuel oil may contain small amounts of sulfur (qv). This is an undesirable constituent, despite its nominal heat content, because sulfur reacts during combustion to form acidic sulfur dioxide.

Common combustion reactions and heat releases for 0.454 kg of reactant under ideal combustion conditions are as follows, where Btu represents British thermal unit:

$$C + O_2 \longrightarrow CO_2 + 14,756 \text{ kJ} \ (14,000 \text{ Btu/lb C})$$

$$2 H_2 + O_2 \longrightarrow 2 H_2O (g) + 64,400 \text{ kJ} \ (61,100 \text{ Btu/lb } H_2)$$

$$2 C + O_2 \longrightarrow 2 CO + 4,216 \text{ kJ} \ (4,000 \text{ Btu/lb C})$$

$$2 CO + O_2 \longrightarrow 2 CO_2 + 4,532 \text{ kJ} \ (4,300 \text{ Btu/lb CO})$$

$$CH_4 + 2 O_2 \longrightarrow CO_2 + H_2O (g) + 25,164 \text{ kJ} \ (23,875 \text{ Btu/lb } CH_4)$$

$$S + O_2 \longrightarrow SO_2 (g) + 4,216 \text{ kJ} \ (4,000 \text{ Btu/lb S})$$

Steam Cycles

Conventional fossil fuel-fired power plants, nuclear power facilities, cogeneration systems, and combined-cycle facilities all have one key feature in common: some type of steam generator is employed to produce steam. Except for simple-cycle cogeneration facilities, the steam is used to drive one or more rotating turbines coupled to rotating electric generators for electricity production. The thermodynamic cycle by which water is boiled in a steam generator (heat addition), forced through a turbine (expansion), and then condensed (heat rejection) before being pumped back to the steam generator is known as the Rankine cycle (Fig. 1).

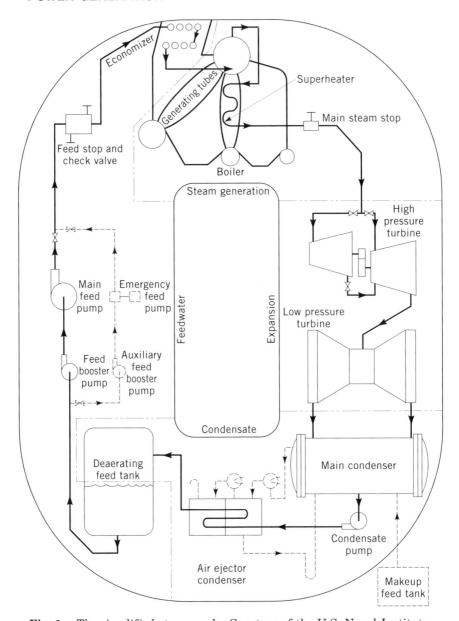

Fig. 1.   The simplified steam cycle. Courtesy of the U.S. Naval Institute.

Because of the simplicity and reliability of the Rankine cycle, facilities employing this method have dominated the power industry in the twentieth century and typically play an important role in most modern combined-cycle facilities. Water is the working fluid of choice in nearly all Rankine cycle power plants because water is nontoxic, abundant, and low cost.

Under standard pressure and temperature conditions of 101.3 kPa and 15°C (14.7 psia and 60°F), the amount of heat (or thermal energy in transition) that must be added to one pound of water to raise the temperature of the pound of water 1°F is known as one British thermal unit (Btu). One Btu, a common unit of measure in the energy field, is equivalent to 1.05435 kJ. The heating content of fuels is often measured in heat energy per unit mass or heat energy per unit volume. In addition, the heat rate of a fossil fuel-fired power plant, a measure of plant efficiency, describes how much heat energy per hour of fuel input is required to generate a continuous 1-kW power output. Because water vapor is a combustion by-product of all hydrogen-bearing fuels, a fuel's actual measured heating value varies based on whether the heat of vaporization is included. In the United States, a fuel's higher heating value (HHV) is most commonly referenced for combustion calculations. The HHV gives a gross heat content for the fuel, including the heat of vaporization. In Europe, a fuel's lower heating value (LHV) is often used, which does not include the heat of vaporization. When calculating thermal efficiency based on a fuel's lower heating value, a given combustion system seems to have a higher efficiency compared to use of the fuel's HHV in the calculation. In the SI system, a fuel's heating content is typically given in kilojoules per unit mass or volume.

The generation of steam (qv) is essentially a two-stage process. Through the addition of heat, the temperature of water is first raised to its boiling point, ie, 100°C at atmospheric pressure. After the boiling point is reached, the temperature at the contact point between the liquid and vapor remains constant as long as the pressure is not permitted to build, such as in a partially closed vessel, until vaporization is completed. When water is boiled in a closed vessel, such as in the steam drum of a utility boiler, vapor generation increases the pressure within the vessel. As the pressure increases, the boiling point (or saturation temperature) increases. Thus, in a boiler operating at 8.27 MPa (1200 psia), water must be heated to nearly 299°C to achieve boiling (Fig. 2).

In most utility boilers, steam pressure regulation is achieved by the throttling of turbine control values where steam generated by the boiler is admitted into the steam turbine. Some modern steam generators have been designed to operate at pressures above the critical point where the phase change between liquid and vapor does not occur.

The visible plume associated with steam leaks, or the vapor that spouts from a tea kettle, is actually caused by fine particles of water entrained in the steam flow. When steam is free of entrained moisture, it is invisible. Because moisture can damage piping, turbines, and other steam path components found in a power plant, the steam is generally superheated to a point where the potential for moisture carryover is minimized. In addition, superheating of steam (adding more heat beyond the boiling point) is also generally desirable to maximize the achievable output from a given size boiler. Designing a plant to operate at higher superheated steam temperatures enables higher plant efficiencies because a smaller percentage of the steam's total heat input remains as it exits the turbine and is condensed.

Rankine Cycle Power Plants

Power plants based on the Rankine thermodynamic cycle have served the majority of the world's electric power generation needs in the twentieth century. The most common heat sources employed by Rankine cycle power plants are

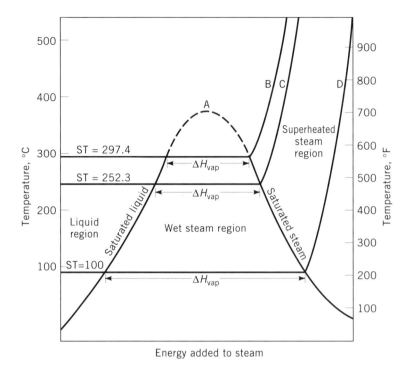

Fig. 2. Properties of steam, where ST = saturation temperature and numbers are in °C. Point A corresponds to 22.1 MPa, and lines B, C, and D to 8.27, 4.13, and 0.101 MPa, respectively. To convert MPa to psia, multiply by 145. Courtesy of the U.S. Naval Institute.

either fossil fuel-fired or nuclear steam generators. The former are the most widely used.

Fossil Fuel-Fired Plants. In modern, fossil fuel-fired power plants, the Rankine cycle typically operates as a closed loop. In describing the steam–water cycle of a modern Rankine cycle plant, it is easiest to start with the condensate system (see Fig. 1). Condensate is the water that remains after the steam employed by the plant's steam turbines exhausts into the plant's condenser, where it is collected for reuse in the cycle. Many modern power plants employ a series of heat exchangers to boost efficiency. As a first step, the condensate is heated in a series of heat exchangers, usually shell-and-tube heat exchangers, by steam extracted from strategic locations on the plant's steam turbines (see HEAT-EXCHANGE TECHNOLOGY).

Another stage of heating occurs in a direct contact, deaerating heater before the water is directed to the boiler feed pumps. The deaerating heater serves a dual role: it not only adds heat to the water, but also improves the water quality by removing dissolved oxygen and other noncondensable gases in the condensate that can cause corrosion within the plant's steam and water piping, steam generator, and steam turbines. Condensate is introduced into the upper portion of the deaerating heater via spray nozzles. There, the fine droplets of condensate are heated to the boiling point by swirling jets of low pressure steam, liberating the entrained, noncondensable components of air, including oxygen and carbon dioxide. The steam literally scrubs the gas molecules off of the water molecules to which they cling. The liberated gas travels up through the top of the deaerater where a vent condenser separates out any vaporized condensate for return to the system before the air is ejected through a vent line. Deaerated water collects in the bottom of the heater vessel.

The deaerator is typically considered the dividing point between the condensate and feedwater systems. Water exiting the deaerator is fed directly to a boiler's high speed centrifugal feedwater pump, which boosts the water's pressure to a level high enough to enable its ultimate introduction into the steam drum where boiling occurs. The steam drum, located at the top of a boiler, is a cylindrical rolled-steel vessel serving multiple purposes. These include (1) receiving point and short-term storage location for feedwater before the water enters the boiler tubes; (2) serving as the start and termination point for heat-transfer tubes in the boiler; and (3) being the chemical dosing and solids blowdown point for water treatment. In a boiler that has a steam drum operating at 4.13 MPa (600 psig), the feedwater pump may be required to boost the feedwater pressure to 5.17 MPa (750 psig) to ensure positive flow to the boiler. Thus, unlike condensate systems that operate at relatively low pressures, the boiler feedwater system operates at a high pressure.

Because boiler feedwater pumps operate at such high pressures and volumetric throughputs, it is critical that the inlet pressure to the pump be maintained at a pressure high enough to prevent the pump from reducing the pressure at its inlet to a point where the heated feedwater would flash into steam. For this reason, the deaerating feedwater tank is typically located at the highest elevation possible in a power plant, with the feedwater pumps in the lowest practical location. In some plants, feedwater booster pumps are employed to ensure sufficient pressure at the inlet to the main feedwater pump. Flashing of steam at a feed pump's suction side can result in overheating and loss of that pump, and possibly even boiler damage if insufficient water is being supplied to generate steam and dissipate the heat produced by the combustion taking place in the refractory-lined furnace (see FURNACES, FUEL-FIRED).

After exiting the feedwater pump and before entering the boiler's steam drum, feedwater usually goes through additional high pressure feedwater heaters (heated by steam turbine extraction steam) as well as another stage of heating in the plant's economizer. The economizer consists of a bank of heat exchanger tubes located in the boiler's backpass section, where heat is transferred from the hot exhaust gas exiting the boiler to the incoming feedwater flowing through the economizer tubes. Most modern fossil fuel-fired power plants share these design features. However, many different types of feedwater heater and boiler arrangements are available, based on the specific steam needs of an application, ie, the required steam temperature and pressure, the type of fuel being burned, space constraints, desired efficiency, and emissions regulations. For example, compact prepackaged boilers have been widely used in industrial and small utility applications. These units can be delivered via rail, barge, or truck fully preassembled or as modular components, which greatly reduces the amount of on-site construction required. Shop assembly of components under controlled conditions can help ensure high quality construction. However, packaged boilers have a limited size range. Thus, field-erected boilers are still required for most large industrial and utility installations.

Boilers can be further categorized by the type of firing system as well as the configuration and type of steam-generating tube employed. Only the basics of furnace heat transfer and steam-generator design are discussed herein. A simplified description of a steam boiler that employs a steam drum is used. However, certain high pressure, once-through supercritical boilers do not employ steam drums. Instead, feedwater is pumped directly into boiler heat-transfer tubes that ultimately terminate at a header where steam is collected for use in the steam turbine.

After exiting the economizer, the feedwater is directed into the boiler's cylindrical steam drum via a common header pipe that penetrates the drum's wall and distributes the water evenly within the drum through holes drilled in the upper side of the distribution pipe. Because the distribution pipe is located axially below the waterline in the lower section of the steam drum, the incoming feedwater mixes thoroughly with the water in the drum and prevents any significantly uneven temperature distributions within the drum.

In modern natural-circulation watertube boilers, the steam drum makes up the uppermost end of a thermal circuit (Fig. 3). Downcomer tubes make up one side of this thermal circuit, bringing hot, high pressure water from the steam drum down to waterwall headers (or manifolds) located at the bottom of the furnace. The downcomers initiate along the length of the steam drum's underside and run vertically downward on the outside of the hot furnace. Waterwall tubes (risers) extend upward from the waterwall headers, forming a box around the furnace, within which combustion takes place (Fig. 4). Heat is released to the waterwall tubes mainly through radiant heating. As the water in the waterwall

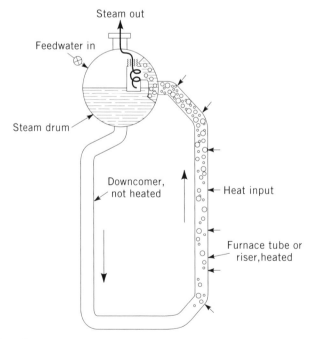

Fig. 3. Simple schematic of a natural or thermal circulation loop. Courtesy of The Babcock and Wilcox Co.

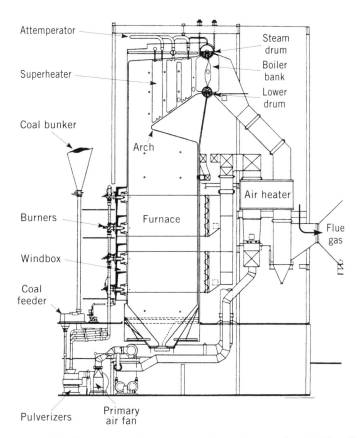

Fig. 4. Two-drum Stirling power boiler system for pulverized coal. The flue gas exits via back-end environmental control devices. Courtesy of The Babcock and Wilcox Co.

riser tubes is heated, its density decreases, causing it to rise upward in the tubes, where it eventually reaches its saturation temperature and steam bubbles form. The saturated steam formed in the waterwalls exits into the upper portion of the steam drum.

It is important that the rate of circulation within the waterwall tubes be great enough to carry heat away from the metal tube walls fast enough to prevent the walls from overheating. Because the circulation is dependent on the difference in density between the cooler water found in the downcomers and the hotter water and steam located in the waterwalls, the rate of circulation increases as this differential pressure increases. Thus, the rate of heat transfer from the combustion zone to waterwalls, the height of the boiler, and its operating pressure all combine to determine the rate of circulation.

Certain boilers employ forced circulation, whereby a pump helps impart the circulation through the downcomer lines to the waterwall header, particularly to improve or control circulation at low loads. Forced-circulation pumps are also required in high pressure and supercritical pressure boilers, because once the pressure within a boiler approaches the critical pressure, 22.1 MPa (3208 psia), the densities of the water and steam become similar, limiting or eliminating the potential for natural circulation.

After the waterwall tubes deliver the saturated steam back into the top of the boiler drum, moisture is separated out by a series of baffles, steam separators, and corrugated screens. The water removed drops down into the hot water contained in the steam drum. The steam travels out through either a dry pipe, which leads to a superheater header, or a series of superheater tubes that connect directly into the top of the steam drum. The superheater tubes wind back into the top of the furnace and/or a hot flue-gas backpass section, next to the economizer, where heat from the combustion gases exiting the furnace superheats the steam traveling through the tubes.

The superheated steam generated in the superheater section is collected in a header pipe that leads to the plant's high pressure steam turbine. The steam turbine's rotor consists of consecutive sets of large, curved, steel alloy disks, each of which anchors a row of precision-cast turbine blades, also called buckets, which protrude tangentially from the shaft and impart rotation to the shaft when impacted by jets of high pressure steam. Rows of stationary blades are anchored to the steam turbine's outer shell and are located between the rows of moving rotor blades.

Superheated steam first enters the high pressure (HP) turbine, after passing through a control valve, via a nozzle box where the steam is allowed to expand through nozzles, creating a controlled, high energy jet of steam directed at the first stage of moving buckets. Each subsequent stage of interstage nozzles further expands and redirects the steam, so that it impacts the succeeding row of moving blades at the optimum angle for imparting maximum rotational force. As the steam travels axially through each row of turbine blades, it loses energy and expands. Thus, subsequent rows of blades are proportionally larger to accommodate the expanding steam volume and to ensure that maximum energy is transferred from the steam to the moving blades.

Some steam may be extracted from the turbine at interstage locations for use in feedwater heating. Optimizing the plant's overall efficiency via such extraction and heat exchange is one of the many complex thermodynamic challenges facing designers of modern high efficiency power plants. Another way that plant efficiency is optimized is through the use of a reheat turbine, often located on the same shaft as the high pressure turbine. Here, steam exiting the last stage of the HP turbine is collected in a header and directed to a reheater tube bank located within the boiler's upper furnace or flue-gas backpass near the superheater tubes. This reheated steam is then used to drive the reheat turbine. A principal benefit of using reheat is that thermal and friction losses associated with condensation in the latter turbine stages can be minimized.

In many large plants, an intermediate pressure (IP) turbine is located on the same shaft as either the HP turbine or the low pressure (LP) turbine. In this arrangement, a common gearbox translates the torque from the high speed turbine shaft to a generator shaft turning at a much lower speed. Unless reheat is used, the steam exiting the HP turbine is typically used directly in the IP turbine.

Steam exiting the IP turbine is used to power a LP turbine, which usually, but not always, drives a separate generator. LP turbines employ larger blades than those of IP and HP turbines to best utilize the expanded steam volume available in this latter stage of the cycle. By the time the steam exits the last stage of the LP turbine, a significant amount of energy has been extracted and condensation begins to occur.

A wide variety of turbine types and arrangements are used in modern power plants. For example, some small industrial plants may use only a single condensing turbine. Prepackaged steam/turbine generator systems can be economical for small-to-midsize installations.

To accommodate the expanding steam exiting the LP turbine, the steam is directed into a condenser, typically a large rectangular shell-and-tube heat exchanger within which thousands of cooling water tubes are located. The incoming steam flows over the cold tubes, condenses, then drops to the bottom of the vessel for recycling into the condensate system. The rapid contraction and condensation of steam that occurs in the condenser causes a partial vacuum to form there. Air ejectors are used to draw out noncondensable gases, eg, air, that can build up in the condenser. A loop seal is located in the line that feeds the condensate back to the feed system to enable the withdrawal of condensate without loss of vacuum within the condenser.

The materials used in the Rankine cycle power plant must be carefully chosen to ensure long service life in the severe temperature, pressure, and cyclic conditions that exist in modern power plants. Steam turbines, piping, and valving must be resistant to high temperature corrosion and erosion by entrained moisture. The exteriors of steam generation tubes located in the path of the hot flue-gas stream must stand up to the erosive forces of uncombusted and partially combusted particulates in those gases. Components must also be designed to allow for significant thermal expansion and contraction.

Water Treatments. As for many other industrial processes, water treatment and monitoring is critical for the integrity of system components. The goal of water treatment is to ensure that the water used for steam generation is pure, such that any contaminants that could lead to corrosion, erosion, or scaling of boiler components are absent. The deaerating feedwater heater is one means by which noncondensable gases are removed from the system. Air and contaminant in-leakage can occur in a number of locations, but particularly through joints, seals, and leaks in the low pressure turbine and condenser, because of the vacuum present in the condenser. Oxygen scavenging chemicals, such as hydrazine, hydroquinine, or erythorbic acid, are also dosed into the feedwater system to eliminate any oxygen not removed by the deaerator or air ejectors.

As the water evaporates into steam and passes on to the superheater, solid matter can concentrate in a boiler's steam drum, particularly on the water's surface, and cause foaming and unwanted moisture carryover from the steam drum. It is therefore necessary either continuously or intermittently to blow down the steam drum. Blowdown refers to the controlled removal of surface water and entrained contaminants through an internal skimmer line in the steam drum. Filtration and coagulation of raw makeup feedwater may also be used to remove coarse suspended solids, particularly organic matter.

Scaling of boiler tubes is a particularly serious problem because it can cause overheating and failure of tubes. Thus, to control scaling, compounds of magnesium and calcium can be removed from the makeup water added to the cycle, or in some cases the condensate, by treatment in a softener or polishing system or a demineralizer. In one type of softener, an ion-exchange (qv) process is used whereby undesirable cations, particularly dissolved magnesium and calcium, are exchanged for comparatively harmless sodium ions present on the surface of small resin balls located in a closed vessel. Periodically, the sodium ions on the resin balls become depleted and it is necessary first to backwash, then to regenerate the resin bed by passing a salt solution through it. Polishing with a sodium-cycle softener can also help filter out suspended metal oxides, which are corrosion by-products. Another type of system, the hot lime zeolite softener, can be used if there is a need to reduce greatly the amount of total dissolved solids in the water. A two-stage zeolite process can be employed for improved control of the pH of the plant's water supply. The goal is to control alkalinity that can cause foaming and accelerated corrosion. The addition of chemical amines is also applied for pH control. The optimal pH for boiler water is in the 8–9.5 range for minimizing the corrosion of steel and copper components.

Demineralizers are often used to treat raw makeup water or condensate where high purity is required, such as in large central station boilers that operate at high steam pressures. Demineralizers employ a combination of cation and anion exchange to remove additional material, including sodium and ammonium cations. Virtually all salt anions, such as bicarbonate, sulfate, and chloride, are removed and replaced by hydroxide ions in the demineralizer.

Fuel System.   Many different types of boilers are available, depending on site-specific requirements. The furnace design and combustion system employed are the primary differences distinguishing boilers based on the fuel burned. However, the main goals of any firing system are the same, ie, to ensure that fuel and air are delivered in such a way as to promote safe, efficient, smooth combustion, and to minimize pollutant emission formation and maximize heat transfer to the waterwall tubes. Properly controlling the air-to-fuel ratio is central to these goals. Insufficient air leads to incomplete combustion, the formation of high levels of carbon monoxide, high particulate emissions, and the potential for hazardous buildup of unburned fuel in the boiler and exhaust stack. Excessive air can significantly decrease boiler efficiency. To maximize both safety and efficiency, boilers are typically operated having 10–20% excess air over the level required to achieve complete or stoichiometric combustion. Oxygen monitors are employed and incorporated into combustion control systems for this reason. Whether a boiler fires liquid fuel, coal, or natural gas, there are typically multiple burners which serve to direct the fuel into the furnace for combustion and which ensure that proper fuel–air mixing occurs so that combustion takes place at the desired rate and in the desired location.

Many liquid fuel-fired boilers burn residual, ie, No. 6, oil, which has a high heating value (HHV) of approximately 44,270–46,600 kJ/kg (19,000–20,000 Btu/lb). Oil fuel is typically delivered to a power plant via barge, rail, or pipeline. The liquid fuel is stored in large tanks featuring explosion-proof gooseneck vents and a surrounding spill-containment basin. Oil is delivered under pressure to the boiler's burners by a positive displacement pump. Strainers upstream from the pump filter out sediment and contaminants. A recirculation line from the pump's outlet to its suction side helps prevent damage to the pump when pressure fluctuations occur, such as in a sudden shutdown.

Oil-firing systems use either pressurized steam, air, or mechanical means of atomizing the fuel oil into small droplets as it is sprayed from the burner. Atomizing maximizes the surface area of fuel oil exposed to air, ensuring thorough mixing of fuel and air, and efficient combustion. Pressurized air is introduced through primary, secondary, and sometimes tertiary ports in and around the burner. Forced-draft fans deliver the air to the burner through windboxes, which are essentially flow-ways built into the walls of the furnace, exterior to the waterwalls. Combustion air is preheated via heat exchangers that extract heat from the boiler's hot exhaust gases. Induced-draft (ID) fans located in the exhaust duct near the stack may be applied to provide further control of gas flow and heat transfer within the boiler. ID fans may also help compensate for the pressure drop that occurs across equipment located in the flue-gas stream, such as cyclone separators, fabric filters, or electrostatic precipitators (ESPs) systems used to capture flyash entrained in the exhaust gas.

Natural gas, which has a heating content of 51,260–55,290 kJ/kg (22,000–24,000 Btu/lb) HHV, is usually delivered to a power plant via pipeline (see PIPELINES). The gas pressure typically must be stepped down from the gas-supply line pressure via pressure reducing valving for use in the plant. Gas distributors maintain a fuel-gas-metering station at the fence line of gas-fired power plants. From that point on, the specific gas-handling equipment requirements vary significantly, depending on site-specific factors. Pipeline-quality gas is generally clean and dry. However, shell-and-tube heat exchangers may be required to keep gas temperatures safely above the dew point and eliminate the potential for condensation of water or liquid hydrocarbons. Piping runs must also be designed to eliminate pockets or low points where condensate might collect to form slugs.

Gas-fired boilers employ burners that are similar in appearance to oil-fired burners. However, these do not require atomizers and are designed predominantly for controlling the mixing of fuel and air in the desired combustion zone. Natural gas-fired plants typically have relatively low air pollutant emissions when compared to oil- or coal-fired facilities, because only minute levels of trace compounds are present in natural gas. For example, gas-fired facilities require no controls for sulfur dioxide, SO_2, the main precursor to the formation of acidified rain. Emissions of oxides of nitrogen, NO_x, although much lower when compared to typical oil- or coal-fired facilities, may require control. NO_x emissions can lead to the formation of low level ozone (qv) or smog and contribute to acid rain, but to a much lower degree compared to SO_2. Many plants feature dual-fuel burners, which allow either gas or oil to be burned.

Coal-fired plants utilize much different fuel-firing systems. The most common firing system is found in pulverized coal-fired (PC-fired) facilities. Central to this system is a pulverizer that crushes the raw coal into a fine powder and dries it by using hot, incoming combustion air. The pulverized coal can then be delivered pneumatically to burners where additional air is added via the furnace's windbox (Figs. 5 and 6). The burners are normally located on the furnace's wall (wall firing) or mounted in every corner (tangential firing).

Although PC-fired units account for a large portion of the coal-fired facilities in the United States and elsewhere, other firing options are available for solid fuel firing. For example, fluidized-bed-fired boilers have become

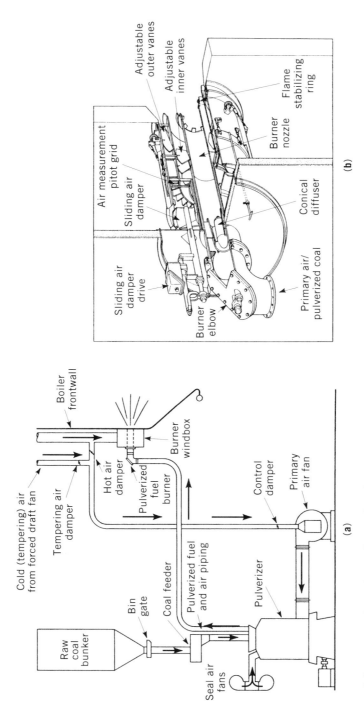

Fig. 5. Schematics of (**a**) direct-firing hot-fan system for pulverized coal; and (**b**) low NO_x burner for coal firing. Courtesy of The Babcock and Wilcox Co.

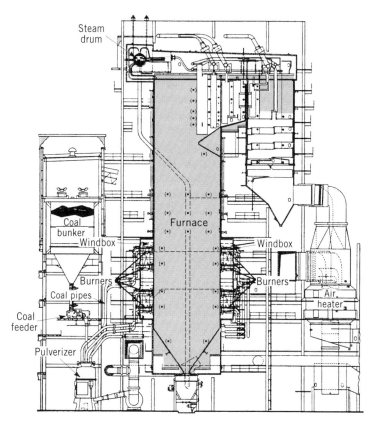

Fig. 6. Conventional pulverized coal-fired system. Courtesy of The Babcock and Wilcox Co.

increasingly popular because these can control pollutant emissions to low levels as well as fire a wide variety of solid fuels, such as low grade coals and coal wastes, petroleum coke from the oil-refining process, wood (qv) and agricultural wastes, and de-inking sludge from newspaper recycling.

Fluidized-bed combustors (FBCs) employ a unique combustion process whereby the fuel is injected into and burned in a highly turbulent bed of combustion air, inert material such as sand, and burning fuel. FBCs can typically be classified as either fixed-bed, bubbling-bed, or circulating fluidized-bed combustors (CFBCs), depending on the velocity of burning solids within the unit (see FLUIDIZATION). Fixed- and bubbling-bed FBCs feature relatively low velocities and the burning fuel bed is isolated to the lower furnace region. In CFBs, a portion of the burning solids leaves the combustion bed and becomes entrained in the hot gases of combustion that rise up out of the bed. The burning particles are recirculated back into the bed, before reaching the boiler's convective backpass, by a cyclone separator–loop seal arrangement.

In most cases, FBCs employ some type of air injection system in the floor of the furnace both to impart turbulence into the burning fuel bed and supply combustion air. Secondary and tertiary air ports may be located above the burning fuel bed.

Other types of solid fuel-firing systems used in modern power plants include traveling grate stokers, ie, fuel injected onto a burning pile on a moving grate; roller grates; inclined grates; cyclonic burners; and suspension burning. Traveling grate systems have been widely used for burning waste fuels, including refuse-derived fuels (RDF), municipal solid waste, and wood chips (see FUELS FROM WASTE).

Emission Control.   In 1993, for a net electric power generation of 3196×10^9 kWh, power industry emissions were as follows:

Air pollutant	Quantity, $t \times 10^6$
SO_2	13.8
NO_x	6.25
CO_2	2124

In oil-, gas-, and coal-fired facilities, low NO_x burners have been successfully applied to reduce NO_x emissions by limiting the peak burner flame temperatures either by rapid premixing of fuel and air, or by staging the introduction of air or fuel to achieve a longer, cooler flame. A key design goal of rich-burn low NO_x systems is to limit the amount of oxygen available in the peak temperature zone where NO_x formation occurs. Post-combustion or back-end NO_x-control treatments, such as selective catalytic reduction (SCR) or selective noncatalytic reduction (SNCR) of NO_x via ammonia injection, can be applied where stringent NO_x-control regulations are in effect.

$$4\,NO + 4\,NH_3 + O_2 \longrightarrow 4\,N_2 + 6\,H_2O$$
$$6\,NO + 8\,NH_3 + 3\,O_2 \longrightarrow 7\,N_2 + 12\,H_2O$$

A schematic of a SCR system is shown in Figure 7. Systems capable of operating at higher temperatures than those shown in Figure 7**b** were under development as of 1995.

Emissions control systems play an important role at most coal-fired power plants. For example, PC-fired plants sited in the United States require some type of sulfur dioxide control system to meet the regulations set forth in the Clean Air Act Amendments of 1990, unless the boiler burns low sulfur coal or benefits from offsets from other highly controlled boilers within a given utility system. Flue-gas desulfurization (FGD) is most commonly accomplished by the application of either dry- or wet-limestone systems. Wet FGD systems, also referred to as wet scrubbers, are the most effective solution for large facilities. Modern scrubbers can typically produce a saleable wallboard-quality gypsum as a by-product of the SO_2 control process (see SULFUR REMOVAL AND RECOVERY).

Similar to oil-fired plants, either low NO_x burners, SCR, or SNCR can be applied for NO_x control at PC-fired plants. Likewise, fabric filter baghouses or electrostatic precipitators can be used to capture flyash (see AIR POLLUTION

21

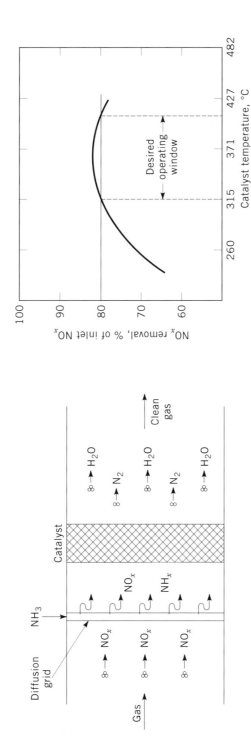

Fig. 7. NO$_x$ reduction using selective catalytic recovery (SCR): (**a**) basic principles of the SCR process where ⊗ represent gas particles; and (**b**) effect of temperature on NO$_x$ removal. Courtesy of General Electric.

CONTROL METHODS). The collection and removal of significant levels of bottom ash, unburned matter that drops to the bottom of the furnace, is a unique challenge associated with coal-fired facilities. Once removed, significant levels of both bottom ash and flyash may require transport for landfilling. Some beneficial reuses of this ash have been identified, such as in the manufacture of Portland cement.

A principal advantage of fluidized-bed combustors for firing high sulfur fuel is that limestone sorbent can be injected directly into the burning bed where the sorbent reacts directly with the SO_2 and other acidic gases requiring control to achieve high capture rates. Another advantage of FBCs relates to NO_x control. Because of the thorough mixing of the burning fuel, temperatures within the burning bed and furnace are relatively uniform and comparatively low. This limits the NO_x production associated with higher combustion region temperatures, ie, thermal NO_x. In addition, ammonia or other reagents can be injected into the upper furnace area to achieve low NO_x levels at the plant's stack. Because the solids entrained in CFBs can be abrasive, these units must be well designed to ensure that the refractory lining protects key components and withstands rapid erosion by the circulating solids.

Nuclear Fuel-Fired Plants.   There are some basic design differences of the nuclear plants steam cycles (see NUCLEAR REACTORS) as compared to conventional fossil fuel plants. In nuclear power plants, thermal energy is released during the fissioning of a nuclear fuel (or fissile), such as uranium-235. Fission, the splitting of a heavy nucleus, is typically initiated when the material is struck by a neutron. When the nucleus of a fissile material's atom is split, the energy associated with the atom is released. This energy, approximately 100 billion times that released during the combustion of one carbon atom in fossil fuels, is significant. In addition, extra neutrons are released that impact other nuclei, creating chain reaction.

Extremely safe means of controlling the nuclear reaction process have been devised by introducing materials that can moderate the production and absorption of neutrons released during fission. The two most popular types of light-water nuclear reactors are boiling-water reactors (BWRs) and pressurized-water reactors (PWRs). These feature similar fuel assemblies consisting of 2–4% enriched uranium dioxide fuel pellets stacked in zirconium alloy cladding tubes. In both designs, the fuel is housed within water-filled pressure vessels. Control rods consisting of a moderating material can be lowered (for PWRs) or raised (for BWRs) into the reactor core to absorb neutrons and temper the reaction.

In BWRs, steam is generated in a single-loop arrangement, whereby water enters through the bottom of the reactor vessel, picks up heat from the reacting nuclear fuel assembly, and exits through the top of the vessel. For safety reasons, the fuel assembly remains immersed in water. However, steam is generated and released in the top of the reactor vessel. A series of mechanical separators and dryers serve to remove water entrained in the steam to protect downstream piping and steam turbine. Overall, steam conditions in nuclear power plants are much less severe than those in conventional fossil-fired stations. For example, steam exiting a BWR has an average temperature below 300°C, compared to 550°C or higher for a similar-sized fossil-fired steam generator.

PWRs operate differently from BWRs. In PWRs, no boiling takes place in the primary heat-transfer loop. Instead, only heating of highly pressurized water occurs. In a separate heat-exchanger vessel, heat is transferred from the pressurized water circuit to a secondary water circuit that operates at a lower pressure and therefore enables boiling. Because of thermal transfer limitations, ultimate steam conditions in PWR power plants are similar to those in BWR plants. For this reason, materials used in nuclear plant steam turbines and piping must be more resistant to erosion and thermal stresses than those used in conventional units.

Cogeneration

In the power industry, cogeneration refers to the simultaneous generation of heat and power. One example of a cogeneration plant is the central utility boiler that provides steam for both electric power generation and supply to a local district heating system where it can be used for facilities or process heating, or cooling, via absorption coolers or steam-turbine-driven refrigeration and/or air conditioning systems. Such a plant can use a noncondensing steam turbine to supply some or all of the steam required by the district heating system. Because of the significant (up to 48%) losses associated with condensing, all or part of the steam exiting the turbine can be eliminated. A well-designed cogeneration facility can convert 80% or more of the fuel energy input for useful purposes, ie, power generation and process heating. In a conventional Rankine cycle power plant, the thermal energy remaining in the steam as it enters the condenser can range up to 50% or more of the fuel energy input, boiler associated losses are ca 15%, and other losses about 2%. Thus, a conventional Rankine cycle plant may convert only 35% of the fuel energy for power generation. In a modern coal-fired cogeneration system, heat losses can be cut to 16%, 15% of which are boiler-associated. Such a system, where the waste heat is recovered from the main power generation cycle for reuse, is often referred to as a topping cycle (see ENERGY MANAGEMENT; PROCESS ENERGY CONSERVATION).

Another example of a topping cycle cogeneration system is one based on recovering the thermal energy exhausting from a gas turbine/generator. Gas turbines are ideal for cogeneration because the high temperature, high flow exhaust gas streams contain a large amount of recoverable heat energy. Heat recovery steam generators (HRSGs) can be used to capture much of the heat in a gas turbine's exhaust (Fig. 8). Unfired-HRSGs are essentially convective steam generators through which a turbine's exhaust gas is directed. Steam generated in the HRSG can be applied in the same fashion as that exhausting from the noncondensing steam turbine. HRSGs can also contain additional burners that require only fuel injection. No combustion air system is needed because gas turbine exhaust streams contain a significant volume of hot, fresh air which is used to prevent overheating of the gas turbine's rotating blades and other hot gas path components.

Gas Turbines and Combined-Cycle Power Plants

Gas turbine engines have undergone substantial improvement in efficiency and have become increasingly popular for power generation and cogeneration. Units

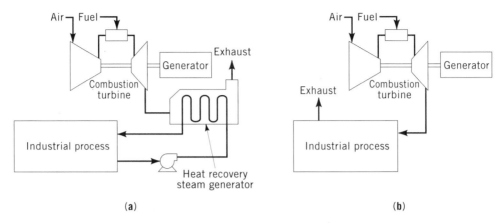

Fig. 8. Combustion turbines with process heat recovery: (**a**) represents direct use of exhaust gas for process heating where industrial process includes refinery, chemicals, food processing, and ethanol production; and (**b**) exhaust-to-water heat exchanger where industrial process includes material drying, water chiller, and CO_2 production. Courtesy of General Electric.

introduced as of the mid-1990s feature simple-cycle efficiencies of more than 40% LHV for just the gas turbine. Combined-cycle power plants based around these units having power conversion efficiencies as high as 60% LHV have been introduced. Combined-cycle facilities recover the waste heat remaining in the combustion turbine's exhaust to drive another power cycle. In this case, the 60% efficiency refers only to energy conversion for electrical output. If steam is also recovered for heating or use in an industrial process, overall plant efficiencies can be significantly higher.

Gas turbines are based on the Brayton thermodynamic cycle (Fig. 9). Most modern units operate in the following manner. Combustion and cooling air is first drawn in and compressed in a multistage axial compressor located on the cold end of the gas turbine's rotor (point A of Fig. 9**b**). The compressed air is injected into a combustor section where it is combined with fuel for combustion, generating hot, expanding gaseous exhaust. The expanding combustion gases are then directed axially along the rotor into a power turbine or turbines. There

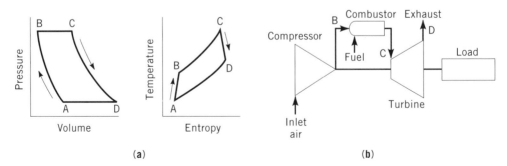

Fig. 9. Brayton cycle, where A = compressor inlet, B = combustor inlet, C = power turbine inlet, and D = exhaust: (**a**) thermodynamic relationships and (**b**) schematic of a simple-cycle, single-shaft gas turbine. Courtesy of General Electric.

the gases impart torque on the tangential turbine blades in a manner similar to a steam turbine, before exhausting to atmosphere or a heat recovery steam generator. Depending on the size of the gas turbine, the unit's rotor may be linked directly to a generator or attached via a speed-reducing gearbox.

Because gas turbine efficiency and power output can be maximized by increasing a unit's mass throughput and combustor-inlet temperature, gas turbines require a huge volume of air, both for combustion and component cooling. For example, one 168-MW unit, introduced in 1990, employs an 18-stage axial compressor to ingest ambient air at a rate of 409 kg/s at full load. The air exits the compressor at 1.4 MPa (200 psig) for injection into the machine's combustors. Thus, up to 50% or more of the power generated in a gas turbine's power turbine section may be required to drive the unit's compressor. This is one of the reasons that many gas turbines, particularly those derived from aircraft engine designs that feature high compression ratios, are designed with two separate shafts. One shaft incorporates the compressor and initial turbine stage or stages used to drive the compressor; the second shaft contains the power turbine stages that drive the unit's generator (Figs. 10 and 11). However, large heavy-duty gas turbines are generally designed to have a single shaft. Because increasing mass throughput can boost plant output, many utility facilities employ inlet-air cooling schemes to maximize plant output, particularly during peak-load hours.

Gas turbine/generators have long been used in remote, and often compact, locations, such as on oil rigs, at gas pipeline pumping stations, and on ships. These systems have evolved to be extremely compact and modular, making them ideal for modern utility and industrial power applications. Site work thus consists mainly of connecting prepackaged modules having a low profile compared to conventional power facilities.

Gas turbine-based power plants, particularly natural gas-fired cogeneration and combined-cycle facilities, have proven to be highly reliable, efficient, and environmentally attractive. Advances in machine design, more efficient plant integration, and optimistic forecasts for the availability of affordable natural gas worldwide have boosted the appeal of these systems for both base-load and peaking service.

Combined-cycle power plants, typically facilities that utilize waste heat from a gas turbine cycle or such technology as reciprocating engines to generate steam for use in a second power generation cycle, have become the most efficient and

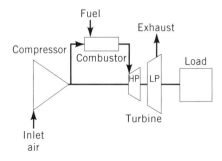

Fig. 10.   Simple-cycle, two-shaft gas turbine where HP and LP are high and low pressure, respectively. Courtesy of General Electric.

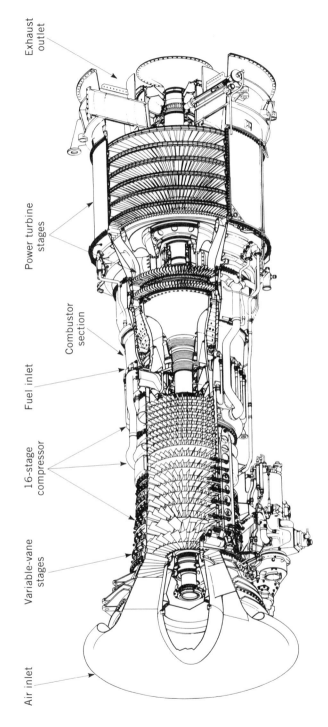

Air inlet

Variable-vane
stages

16-stage
compressor

Fuel inlet

Combustor
section

Power turbine
stages

Exhaust
outlet

Fig. 11. Two-shaft combustion turbine engine. Courtesy of General Electric Marine and Industrial Engines.

26

economical means of generating power in many areas of the world (Figs. 12 and 13). The combined-cycle facility, based on linking the gas turbine or Brayton cycle and the Rankine cycle, is the most common type.

The next generation of gas turbine-based, combined-cycle power plants, under construction in many parts of the world, is to feature net plant efficiencies in the 60% range based on LHV of fuel input. These facilities, scheduled for start-up in the latter 1990s, are anchored by large gas turbines capable of simple-cycle efficiencies >40% LHV in some cases. To develop these machines, manufacturers have scaled up and improved upon designs that have already proved to be highly reliable.

New units can be ordered having dry, low NO_x burners that can reduce NO_x emissions below 25 ppm on gaseous fuels in many cases, without back-end flue-gas cleanup or front-end controls, such as steam or water injection which can reduce efficiency. Similar in concept to low NO_x burners used in boilers, dry low NO_x gas turbine burners aim to reduce peak combustion temperatures through staged combustion and/or improved fuel–air mixing.

One of the principal challenges faced by burner designers is achieving low NO_x levels and stable combustion simultaneously over a machine's entire load range. Thus some low NO_x burners operate in different modes, depending on what works best at a given load range. For example, one two-stage, low NO_x burner design acts more like a conventional burner at low loads, then biases combustion toward lean/premix operation at higher loads (Fig. 14).

Improved materials, coatings, and cooling techniques permit newer machines to operate at higher turbine inlet temperatures, yielding both increased output and efficiency. Further efficiency gains result from improved aerodynamics in the hot gas path, compressor, and turbine sections. Use is also made of variable inlet guide vanes (IGV).

Manufacturers have also strived to improve the reliability, availability, and maintainability (RAM) of the newer units. Many existing designs have achieved availabilities in the 95–99% range. Based on this performance, makers have introduced ever-larger machines, some up to 282 MW. A more recently introduced, integrated steam-cooled combined-cycle gas turbine is even larger. Although untested as of 1995, the newer gas turbines, to judge from existing designs, should operate reliably in both base-load and utility-peaking service. Facilities based around one large gas turbine/generator typically have lower capital requirements and operating costs compared to plants based around multiple gas turbines. In addition, the large units being introduced in the mid-1990s retain much of the start-up and cycling flexibility displayed by smaller machines and are also better suited for future conversion to coal-gas firing (see COAL CONVERSION PROCESSES).

Natural gas-fired combined cycles are usually less expensive to install and maintain than are conventional oil- or coal-fired power plants. The only clear disadvantage of natural gas as a fuel is the high delivered cost in certain areas. Per unit of heating value, gas is about 100 times bulkier than oil. Even when liquefied in a costly process (see LIQUEFIED PETROLEUM GAS), natural gas is still 60% bulkier than oil. Only about 15% of the world's gas supply is traded internationally, compared to half of the world's oil. However, the combination of increasingly stringent emissions regulations, advances in natural gas exploration

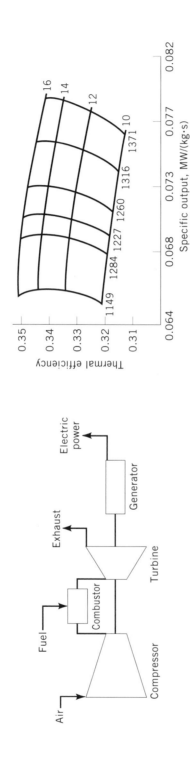

Fig. 12. Combustion turbine engine simple cycle: (**a**) schematic of plant; and (**b**) thermodynamics, where the horizontal lines correspond to the pressure ratio, X_c, given and vertical lines to the combustor temperature, T_F, in °C as indicated. Courtesy of General Electric.

28

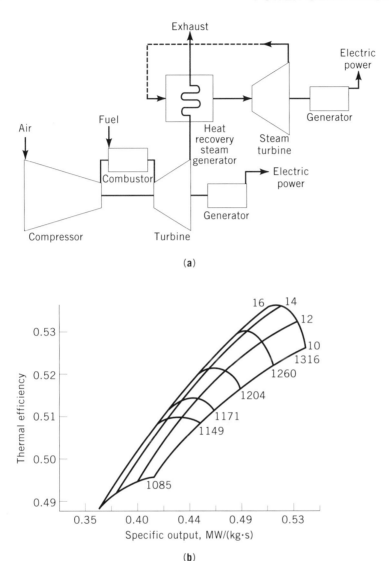

Fig. 13. Combustion turbine engine combined cycle: (**a**) schematic of plant; and (**b**) thermodynamics, where the vertical lines correspond to the pressure ratio, X_c, given and the horizontal lines to the combustor temperature, T_F, in °C as indicated. Courtesy of General Electric.

and recovery techniques, and international cooperation on new gas drilling and pipeline projects, has decreased the delivered cost of gas compared to coal and oil in many regions.

Electric utilities and others are beginning to apply combustion turbines in a number of unique ways. For example, gas turbines can be used to repower aging Rankine cycle power plants. Repowering typically refers to the retirement of a plant's boiler, combined with the installation of a new combustion

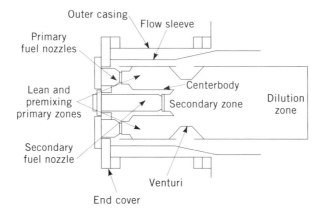

Fig. 14. Dry, low NO_x combustor schematic (3,4). Courtesy of General Electric.

turbine/generator and HRSG capable of generating enough steam to operate the plant's original steam turbine in a combined-cycle configuration (Fig. 15).

Other innovative concepts include the incorporation of coal-gas-fired gas turbine generators into highly efficient combined-cycle arrangements with integrated coal gasification and air separation facilities (Figs. 16–18) (see NITROGEN). Integrated gasification combined-cycle (IGCC) facilities can achieve net efficiencies above 40% while maintaining extremely low emissions. For example, 98–99% of the sulfur can be removed from the coal gas and recovered as a potentially saleable by-product. Some utilities installing large natural gas-fired peaking turbines plan eventually to convert these facilities for combined-cycle operation as required load grows. Others are keeping the option open to add coal gasification equipment at some future date if natural gas prices rise dramatically.

Several utility-scale demonstration facilities having power outputs in the 300-MW class have been constructed in the United States and Europe. These

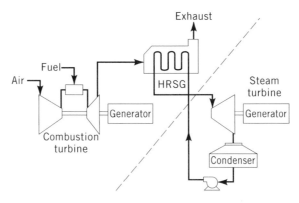

Fig. 15. Repowering schematic where the modules to the left of the dashed line have been added to the existing Rankine cycle plant shown on the right of the dashed line. HRSG = heat recovery steam generator. Courtesy of Black and Veatch.

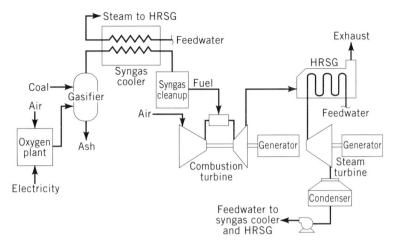

Fig. 16. Coal gasification power plant; HRSG = heat recovery steam generator. Courtesy of Black and Veatch.

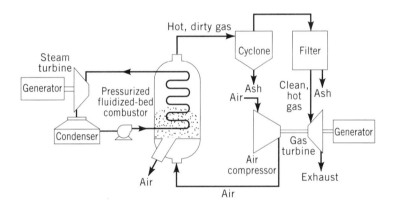

Fig. 17. Pressurized fluidized-bed combustion. Courtesy of Black and Veatch.

started accumulating operating experience in 1995 and 1996. Other IGCC plants have been constructed, including units fueled by petroleum coke and refinery bottoms. Advanced 500-MW class IGCC plants based around the latest heavy-duty combustion turbines are expected to be priced competitively with new pulverized-coal-fired plants utilizing scrubbers.

Advanced Gas Turbine Designs

Scaled-up gas turbine designs typically have higher efficiencies than their predecessors. Power increases by roughly the square of the scaling factor whereas machine losses increase in a linear manner. Output and efficiency are further boosted in some cases because the larger machines operate at slower speeds and therefore do not have the gearing-related losses inherent in small high speed designs.

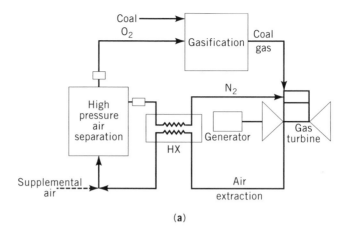

(a)

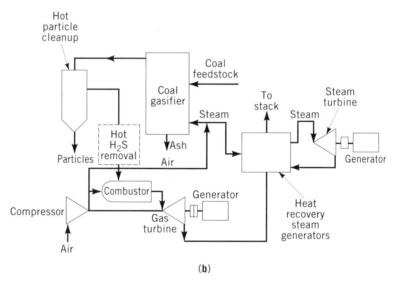

(b)

Fig. 18. Integrated gasification combined-cycle (IGCC) process: (**a**) integrated air separation; and (**b**) simplified IGCC. Courtesy of General Electric.

Gas turbine technology has advanced rapidly in the latter 1990s. The most up-to-date information in combustion turbine technology is available in the *Annual Technology Report* published by the International Gas Turbine Institute (IGTI) in Atlanta, Georgia, part of the American Society of Mechanical Engineers (ASME) International.

Beyond increases directly linked to scaling up, manufacturers can extract higher outputs and efficiencies from new designs by increasing the turbine inlet temperature and the relative mass flow of air, and therefore of combustion gases. One European manufacturer has introduced a 170-MW gas turbine/generator for 60-Hz service and a 240-MW unit for 50-Hz service. These are basically scaled-up versions of what was originally an advanced 60-MW machine. The newer machines reportedly feature simple-cycle efficiencies of 38%. Combined cycles based around two such machines, a single reheat steam turbine and triple-pressure heat recovery steam generators (HRSG) yield plant efficiencies up to 58% or a heat rate of under 5687 kJ/kWh (6000 Btu/kWh).

The 170-MW machine, a popular size for reasons of economics and applicability to phased construction, features a 15-stage compressor and a four-stage power turbine section. Compared to a 106-MW unit developed by the manufacturer during the 1980s, the pressure ratio, X_c, ie, the compressor outlet pressure divided by the air-inlet pressure, has increased from 10.8 to 16.5. Total mass flow through the compressor has increased from 353 to 454 kg/s. To accommodate increased mass flow and elevated pressure, the 170-MW machine features compressor-intake dimensions that are 15% larger. In addition, the first two compressor stages are designed to permit stable supersonic flow velocities there. This is an important feature for maximizing the flow without drastically increasing a unit's size and length compared to earlier models.

The firing temperature for the newest gas turbines being put into service as of the mid-1990s has reached 1310°C. The value was only 1105°C for earlier designs. To accommodate this high firing temperature, newer machines feature a double-shell construction. An inner wall of metallic heat-shield pads keeps the high temperature gas flow away from the turbine's horizontally split outer shell. Impingement air cooling and film cooling are used to minimize hot spots on the metallic heat shields. Both external cooling air and intercooled compressor–extraction air are used to cool various stages of the power turbine. Air is directed to cooling passages in the blades via internal pathways in the rotor and turbine casing.

The unit's annular combustor is also a significant modification. Earlier models featured relatively large, cylindrical silo combustion chambers which, for models designed in the 1980s, were arranged vertically and for machines designed in the early 1990s, horizontally. Newer combustor design is essentially an annular ring that circumscribes the cylindrical engine and houses 24 burners which penetrate the cylindrical ring. This design reportedly reduces the combustion chamber surface area compared to silo-type combustors and thereby decreases cooling air requirements, leaving more air for combustion. In addition, the shorter combustor residence time cuts NO_x production. Other important design features enabling reliable high temperature operation include the use of (*1*) advanced manufacturing techniques, such as directionally solidified and single-crystal cast turbine blades having improved creep rupture strength; (*2*) special high temperature alloys (qv); and (*3*) vacuum-plasma coating techniques refined in the 1990s (see PLASMA TECHNOLOGY; REFRACTORY COATINGS).

The newer turbine/generators also include features designed to improve reliability and maintainability. All moving blades can be replaced without lifting the rotor off its bearings, all stationary blade carriers can be removed by removing only one half of the shell, compressor blade carriers can be aligned from outside the machine's casing, and exhaust bearings without horizontal joints permit axial assembly and disassembly of components, thus easing these processes. Using hybrid, dry, low NO_x burners, full-load NO_x emissions can be held below 25 ppm for gas fuel firing. Units firing No. 2 oil can maintain full-load NO_x emissions below 45 ppm by using water or steam injection to temper thermal NO_x production.

A 165-MW-class gas turbine/generator has been introduced by another manufacturer. This machine, also developed by scaling up a proven design, features a simple-cycle efficiency of 37.5%; a turbine inlet temperature of 1235°C; a pressure ratio of 30:1, up from 16:1 on the previous generation; and an output of 165 MW for gas fuel firing under International Standards Organization (ISO) conditions (101 kPa, 15°C (14.7 psia, 59°F)). A combined-cycle facility based around this machine could achieve efficiencies up to 58% or a heat rate of about 6209 kJ/kWh (5885 Btu/kWh).

The unit features a single-shaft, welded rotor supported by two bearings, a 22-stage subsonic compressor, five turbine stages, and forced-air cooling of the turbine rotor, blade carriers, and early turbine stages. An annular combustor is used, reducing the overall height of the machine by 4 m. There is no increase in length compared to the previous machine in this power range.

The machine's dual annular combustor is a continuous chamber containing 72 double-cone, low NO_x burners. One of the main goals of using an annular design is to optimize the flow of hot gases to the turbine, thereby minimizing thermal losses. The annular arrangement also yields a homogeneous mixture of hot gases having a uniform temperature distribution during start-up, part-load, and full-load conditions, thereby helping to limit thermal NO_x production. The unit's lean-premix burners can maintain NO_x levels below 25 ppm on a continuous basis for natural-gas firing, and 42 ppm for oil firing using water or steam injection. The premix burners consist of an axial split cone, the two halves of which are offset by two constant-width air-inlet slots. For gas firing, fuel is injected through fine holes at the end of the slots. The geometry of the burner causes a high speed vortex flow to develop within the cone, yielding a lean mixture as the fuel and air enter the detached flame. The flame is stabilized by an aerodynamically induced recirculation zone in the free space, eliminating the need for any type of flame holder. For oil firing, fuel is sprayed into the burner via an atomizer at the base of the inner cone that vaporizes the oil and permits it to mix with the air and burn in a similar fashion as the gas fuel. Advanced combined-cycle facilities based on this machine can potentially attain net efficiencies of 58% if a three-pressure reheat cycle is employed. In this case, the gas turbine exhausts through an HRSG that generates steam at three separate pressure and temperature levels.

At least two manufacturers have developed and installed machines rated to produce more than 210 MW of electricity in the simple-cycle mode. In both cases, the machines were designed and manufactured through cooperative ventures between two or more international gas turbine developers. One 50-Hz unit, first installed as a peaking power facility in France, is rated for a gross output of 212 MW and a net simple-cycle efficiency of 34.2% for natural-gas firing. When integrated into an enhanced three-pressure, combined-cycle with reheat, net plant efficiencies in excess of 54% reportedly can be achieved.

The 212-MW unit features a turbine inlet temperature of 1260°C and a pressure ratio of 13.5:1. The manufacturer has subsequently installed a number of larger, more powerful versions of this unit, which produce up to 226.5 MW. Turbine inlet temperature is 1288°C; the pressure ratio is 15:1. Five of these high output machines anchor a 1675-MW facility in the Netherlands. These machines were developed by geometric scaling from a 168-MW, 60-Hz unit. To accommodate the higher firing temperatures and mass flows, these units employ advanced alloys, coatings, and modified cooling schemes. Many features of these units were derived from aircraft engine designs.

A notable difference between the newer large machines and the somewhat smaller units is the use of multiple, reverse-flow can combustors configured annularly. Because the individual cans are relatively small, they reportedly lend themselves well to laboratory experimentation with various fuel types, including reduced-heat value synfuels (see FUELS, SYNTHETIC). A dry, low NO_x version of the can combustors has been developed for both gas and liquid fuel firing. NO_x emissions can reportedly be held below 25 ppm when firing gas fuel. By employing water injection, NO_x emissions can be held below 60 ppm for oil-fired units.

Similar to other newer gas turbines, the 226-MW design has been optimized for integration into a high efficiency combined-cycle plant featuring a triple-pressure, reheat steam cycle. In this case, the gas turbine and steam turbine can be configured on a single power train to drive one large generator.

The principal advantages of the single-shaft design include its compactness and slightly lower capital costs compared to the conventional multishaft configuration where the gas and steam turbines drive separate generators (Fig. 19). When the 226-MW gas turbine/generator is linked with a vertical HRSG and a 118-MW steam turbine, the single-shaft power block requires a ground space smaller than 135×40 m.

The design also has the advantages of component reduction. For example, only one thrust bearing and one main transformer is required. In addition, both the steam and gas turbines can be automated from a single prepackaged control center. Finally, a single, larger, more efficient water-cooled generator can be used. In contrast, principal advantages of using the multishaft arrangement include the possibility of phased construction and start-up, which can reduce the lag time between initial investment and revenue generation, and increased flexibility in the ability to extract steam reliably for district heating and cogeneration.

The most popular HRSG for use with these machines is a vertical, assisted-circulation design featuring a built-in stack and low groundspace requirements.

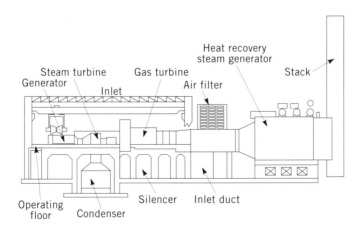

Fig. 19. Single-shaft combined-cycle elevation. Courtesy of General Electric.

More recently, an L-shaped HRSG has been developed to reduce the hot flue-gas temperature abruptly after the gas turbine's exhaust diffuser by locating both the reheater and HP and IP superheaters in the horizontal inlet section to the HRSG. This design reportedly simplifies and reduces the cost of the exhaust duct and expansion joint design. Like most popular designs, the L-shaped HRSG incorporates prefabricated tube bundles packaged and shipped to ease on-site construction. For this design, a typical three-pressure HRSG having reheat capability requires about 30 separate tube bundles.

In May of 1995, one gas turbine manufacturer introduced a new engine series expected to push the achievable net efficiency ratings for base-load combined-cycle power plants to the 60% level. The power ratings for these units are 240 and 280 MW for simple cycle, and 350 and 480 MW for combined cycle using turbine inlet (firing) temperatures of 1430°C. To achieve the 60% thermal efficiency level, either of the machines can be integrated into a single-shaft combined-cycle configuration and employ steam cooling on the four-stage power turbine. The steam cooling, as well as thermal barrier coatings (qv) and single-crystal cast first-stage turbine blades, enables the units to operate at the higher firing temperature at a high combined-cycle efficiency. The intermediate pressure steam used for cooling is extracted from the combined-cycle arrangement's HP steam turbine. Some is directed through the gas turbine's hollow rotor to cool the rotating turbine blades; some is directed through the engine's outer shell to cool the stationary vanes and shrouds. To conserve energy, the steam cooling system is a closed loop. As the steam passes through the gas turbine, it is heated to near-reheat temperature and can therefore be injected into the reheat steam, which drives the unit's LP–IP steam turbine. Also helping to boost efficiency is the unit's 18-stage compressor, a geometrically scaled-up aeroengine design, which achieves pressure ratios of 23:1, compared to 15:1 on the previous generation of machines.

Natural Gas for Power Plant Use

Most gas-fired, heavy-duty gas turbines installed as of 1996 operate at gas pressures between 1.2 and 1.7 MPa (180–250 psig). However, aeroderivative gas turbines and newer heavy-duty units can have such high air-inlet compression ratios as to require booster compressors to raise gas inlet pressures, in some cases as high as 5.2 MPa (750 psig).

Motor-driven, multistage reciprocating compressors have reportedly been the most popular choice for aeroderivatives. Motor-driven, oil-flooded screw compressors are also used in some cases. High horsepower, multistage centrifugal compressors, similar to those used at many pipeline compressor stations, may be required for the newer heavy-duty units if the distribution pipeline pressure is insufficient (see PIPELINES). Gas turbines have more stringent fuel-gas specifications in terms of cleanliness than do gas-fired boilers. Thus oil- and water-knockout systems, coalescing filters, and fine-mesh filters are used.

All gas-fired power plants require oxygen analyzers to ensure that air has not been drawn into the piping system. Oxygen intake can lead to the presence of an explosive mixture in the pipeline before the fuel reaches the burner or combustor zone. When gas-fired units are located in an enclosed area, multiple ultraviolet flame detectors are used to shut down equipment and flood the area with CO_2 or a chemical fire suppressant whenever a spark or flame is detected.

Consumption of natural gas, as of the mid-1990s, was about 2000×10^9 m^3/yr. Using seismic detection equipment, exploration firms search for gas reserves buried deep underground and beneath the sea floor. Advanced computer systems process the seismic data to pinpoint the most likely locations for reserves. These advanced systems have both cut the time required for data analysis, by 80%, and greatly improved the success rate for new drill rigs.

In North America, other technological breakthroughs, such as drilling multiple wells from a single platform, drilling deeper underground, horizontal drilling, and deep-water drilling, have increased yield by over 50% in the 1980s. In addition, tax incentives have enticed producers to recover gas from less conventional sources, including tight-sand formations and coal beds. As of 1995, approximately 272,000 wells were operating in the United States and Canada (see GAS, NATURAL).

Processed gas is compressed for transmission through interstate and intrastate pipelines (qv). These pipelines vary in size according to system needs, but typically have diameters between 0.6–1.1 m and a wall thickness up to 1.3 cm. Average operating pressures for new pipelines range from 5–7 MPa (700–1000 psig). To guard against corrosion, epoxy-coated piping and cathodic protection systems are used.

Modern pipeline companies control gas flow through their systems in response to and anticipation of demand swings via computerized operations centers. These facilities typically use satellite or modem-based telemetry systems to monitor gas flow and automatically regulate valves and compressor stations located along the pipeline and at gas storage facilities. In addition to continuously reviewing data collected from the telemetry system, dispatchers also closely monitor weather conditions in various regions so that gas can be delivered to key areas in anticipation of greater demand. Similarly, operators must maintain close communications with power plant dispatchers and other gas system operators.

On a typical pipeline, compressor stations are located at 81–161-km intervals and may contain up to 15 compressors. These stations may use either gas-turbine, reciprocating-engine, and/or motor-driven centrifugal compressors capable of boosting pipeline pressure and keeping gas moving at an average speed of about 24 km/h. Gas-turbine-driven units are the most popular.

Filtration and water-knockout systems are used to clean up the gas before it enters a compressor. Cooling systems are sometimes required to maintain compressor discharge temperatures below 54°C to avoid damage to the pipeline's protective coatings. Automated compressor stations are typically staffed by maintenance and repair personnel eight hours per day, five days per week. Other stations are staffed on a 24-hour basis because personnel must start, stop, and regulate compressors in response to orders from the dispatch office.

The demand for gas is highly seasonal. Thus pipeline companies economize by sizing production facilities to accommodate less than the system's maximum wintertime demand. Underground storage facilities are used to meet seasonal and daily demand peaks. In North America, gas is stored in three main types of underground formations: depleted oil or gas fields, aquifers that originally contained water, and caverns formed by salt domes or mines.

When full, storage reservoirs may exhibit pressures above 14 MPa (2000 psig). The rate at which a storage facility can deliver gas declines as more volume is withdrawn from it. Thus a certain volume of permanent cushion gas is needed to maintain required rates of delivery. Peaking reservoirs may contain up to 75% cushion gas; base-load storage basins may contain closer to 50%. Specific cushion gas volumes vary greatly, depending on the application and the properties of the reservoir. In aquifer storage systems, for example, excessive withdrawals may allow water to encroach, rendering pockets of gas irretrievable.

In certain areas, the lack of underground storage capacity necessitates the use of liquefying stations. Cooled liquefied natural gas (LNG) occupies only about 1/600 the volume required when it is in the gaseous form. However, above-ground LNG storage facilities reportedly are significantly more expensive to operate than underground storage reservoirs and gaining permits can be difficult. In addition to fuel storage, many power producers improve system reliability and control costs by ensuring that gas-fired units can be rapidly switched over to liquid fuel firing or that a separate unit in the system can be engaged if gas supplies are curtailed for any reason.

Other Generation Options

Reciprocating Engines. Reciprocating engines, particularly medium- and slow-speed diesels similar to those used for shipboard propulsion and power generation, have been used for land-based power generation since the 1940s. Because of their relatively high efficiency (up to 45%), reliability, and quick start-up capability, these units have been popular for peaking, emergency, and base-load power generation. However, reciprocating units have not been nearly as popular as combustion turbines for cogeneration or combined-cycle power generation. The exhaust exiting a reciprocating unit has a comparatively lower volumetric flow, entrained air, and overall energy content compared to combustion turbines. In addition, available machine sizes are smaller for reciprocating engines.

Advantages of reciprocating engines for small power plant applications include the following: in gas-fired applications, reciprocating engines require much lower gas inlet pressures (14–28 kPa (20–40 psig) compared to combustion turbines; reciprocating engines are not significantly impacted by changes in ambient air temperature (achievable gas turbine output falls at higher temperatures); and reciprocating engines have better part-load efficiency characteristics than gas turbines.

Nuclear Reactors. Nuclear power facilities account for about 20% of the power generated in the United States. Although no new plants are planned in the United States, many other countries, particularly those that would otherwise rely heavily on imported fuel, continue to increase their nuclear plant generation capacity. Many industry observers predict that nuclear power may become more attractive in future years as the price of fossil fuels continues to rise and environmental regulations become more stringent. In addition, advanced passive-safety reactor designs may help allay concerns over potential safety issues.

Fuel Cells. Fuel cells (qv) are essentially batteries (qv) that run on fuel and therefore do not run down. Advanced fuel cells feature efficiencies above 40% and consist of a series of porous, conducting electrode layers (a layer corresponding to one anode and one cathode), separated by an electrolyte (an ionic charge carrier). The multiple cells are configured in series to achieve the voltages required for industrial and commercial use.

As of 1995, the cost for fuel cell-based power plants was prohibitive when compared to conventional options. This cost may come down in the latter 1990s if manufacturers of these devices receive enough orders to achieve economies of scale. Fuel cells have many potential advantages compared to conventional power cycles. For example, fuel cells generate very little pollution either in emissions or noise. Thus these cells are ideal for siting in populated areas close to loads where it may be difficult to site conventional power sources because of permitting issues. In addition, small fuel cell power plants can potentially run as unstaffed facilities. Finally, fuel cells do generate significant levels of waste heat that can be captured and utilized to improve overall plant efficiency (see PROCESS ENERGY CONSERVATION).

BIBLIOGRAPHY

"Power Generation" in *ECT* 1st ed., Vol. 11, pp. 65–87, by R. A. Budenholzer, Illinois Institute of Technology; in *ECT* 2nd ed., Vol. 16, pp. 436–469, by R. A. Budenholzer, Illinois Institute of Technology; in *ECT* 3rd ed., Vol. 19, pp. 62–94, by S. K. Batra, GIMRET International, Inc., and R. A. Budenholzer, Illinois Institute of Technology.

1. "Monthly Power Plant Report," Form EIA-759; "Annual Electric Generator Report," Form EIA-860; "Annual Electric Utility Report," EIA-861, Energy Information Administration, U.S. Department of Energy, Washington, D.C., 1993.
2. *Electric Power Annual 1993*, Energy Information Administration, U.S. Department of Energy, Washington, D.C., 1993.
3. *Electric Power Annual 1993*, U.S. Department of Energy, Energy Information Administration, Washington, D.C., 1994.
4. S. Collins, *Power*, **137**(2) (Feb. 1993).

General References

T. Elliott, *Standard Handbook of Powerplant Engineering*, McGraw-Hill Book Co., Inc., New York, 1989.

Steam: Its Generation and Use, 40th ed., The Babcock and Wilcox Co., Barberton, Ohio, 1992.

B. H. Bunch and A. Hellemans, *The Timetables of Technology*, Simon and Schuster, New York, 1993.

J. Makansi, *Managing Steam*, Leslie Controls Inc., Tampa, Fla., 1985.

S. Collins, *Power*, **136**(2) (Feb. 1992).

S. Collins, *Power*, **138**(6) (June 1994).

Proceedings of the 37th GE State-of-the-Art Technology Seminar, General Electric Co., Schenectady, N.Y., July 1993.

R. L. Dickenson and D. R. Simbeck, *JSME-ASME International Conference on Power Engineering*, Sept. 1993.

R. Farmer, *Gas Turbine World* (Sept./Oct. 1993).

R. Farmer, *Gas Turbine World* (Jan./Feb. 1995).
R. Farmer and K. Fulton, *Gas Turbine World* (May/June 1995).
1995 Technology Report, International Gas Turbine Institute, ASME, Atlanta, Ga., 1995.

STEVEN COLLINS
Empire State Electric Energy Research Corporation

PPS FIBERS.   See HIGH PERFORMANCE FIBERS.

PRASEODYMIUM.   See LANTHANIDES.

PRESERVATIVES.   See ANTIOXIDANTS; ANTIOZONANTS; COATINGS; FOOD
ADDITIVES; PAINT; WOOD.

PRESSURE MEASUREMENT

Pressure measurement is important in the chemical process industries (CPI) and in laboratories for a number of reasons: differential pressure is the driving force in fluid dynamics; product quality frequently depends on certain pressures (or vacuums) being reached and accurately maintained for specific lengths of time during a process; and pressure is a crucial safety consideration in the operation of process equipment particularly where boiler or reactor pressures must not exceed certain limits (see FLUID MECHANICS; HIGH PRESSURE TECHNOLOGY; VACUUM TECHNOLOGY).

Units of Measurement

Pressure is defined as force per unit of area. The International System of Units (SI) pressure unit is the pascal (Pa), defined as $1.0 \ N/m^2$. Conversion factors from non-SI units to pascal are given in Table 1 (see also UNITS AND CONVERSION FACTORS; front matter). An asterisk after the sixth decimal place indicates that the conversion factor is exact and all subsequent digits are zero. Relationships that are not followed by an asterisk are either the results of physical measurements or are only approximate. The factors are written as numbers greater than 1 and less than 10, with 6 or fewer decimal places (1).

Table 1. Conversion of Pressure Units to the SI Unit [a]

To convert from	To	Muliply by[b]
atmosphere		
normal = 760 torr	pascal (Pa)	$1.013\ 25 \times 10^5$
technical = 1 kgf/cm^2	pascal (Pa)	$9.806\ 650^* \times 10^4$
bar	pascal (Pa)	$1.000\ 000^* \times 10^5$
centimeter of mercury (0°C)	pascal (Pa)	$1.333\ 22 \times 10^3$
centimeter of water (4°C)	pascal (Pa)	$9.806\ 38 \times 10$
decibar	pascal (Pa)	$1.000\ 000^* \times 10^4$
dyne per centimeter^2	pascal (Pa)	$1.000\ 000^* \times 10^4$
foot of water (39.2°F)	pascal (Pa)	$2.988\ 98 \times 10^3$
gram-force per centimeter^2	pascal (Pa)	$9.806\ 650^* \times 10$
inch of mercury		
32°F	pascal (Pa)	$3.386\ 389 \times 10^3$
60°F	pascal (Pa)	$3.376\ 85 \times 10^3$
inch of water		
39.2°F	pascal (Pa)	$2.490\ 82 \times 10^2$
60°F	pascal (Pa)	$2.488\ 4 \times 10^2$
kilogram-force per centimeter^2	pascal (Pa)	$9.806\ 650^* \times 10^4$
kilogram-force per meter^2	pascal (Pa)	$9.806\ 650^*$
kilogram-force per millimeter^2	pascal (Pa)	$9.806\ 650^* \times 10^6$
millibar	pascal (Pa)	$1.000\ 000^* \times 10^2$
millimeter of mercury (0°C)	pascal (Pa)	$1.333\ 224 \times 10^2$
poundal per foot^2	pascal (Pa)	$1.488\ 164$
poundal-force per foot^2	pascal (Pa)	$4.788\ 026 \times 10$
pound-force per inch^2 (psi)	pascal (Pa)	$6.894\ 757 \times 10^3$
torr (mm Hg absolute, 0°C)	pascal (Pa)	$1.333\ 22 \times 10^2$

[a] Ref. 1.
[b] Asterisk means conversion is exact.

Definition of Terms

Absolute pressure is pressure measured relative to a perfect vacuum, an absolute zero of pressure (2). Like the absolute zero of temperature, perfect vacuum is never realized in a real world system but provides a convenient reference for pressure measurement. The acceptance of strain gauge technology in the fabrication of pressure sensors is resulting in the increased use of absolute pressure measurement in the CPI (see SENSORS). The pressure reference for most of the pressure gauges used in the CPI as of the mid-1990s is atmospheric or local barometric pressure. Barometric pressure varies with elevation and weather. These variables have been eliminated by establishing a standard atmospheric pressure of 101,325 Pa (14.696 psi) as a basis for correcting gauge indication for variations in barometric pressure. One standard atmosphere is equal to the pressure exerted by a column of mercury 760 mm high at a temperature of 0°C where the acceleration owing to gravity is 9.80665 m/s^2. The Pa is an absolute pressure unit. SI conventions make no provisions for differentiating between absolute and gauge pressure. Absolute and gauge pressure are often differentiated in psi by the notation psia and psig, respectively.

Gauge pressure is equal to absolute pressure minus barometric pressure. Absolute pressure is gauge pressure plus barometric pressure. If a gauge indicates, for example, that the pressure is 15 psig and the local barometric pressure is 14.7 psia, then the absolute pressure is 29.7 psia (205 kPa). Gauge pressure can be either positive or negative. When the term pressure gauge is used, the reference is almost always to a gauge that is used to measure positive pressures, ie, pressures that exceed local barometric pressure. A vacuum gauge is used to measure negative pressures. A compound gauge is designed to measure both positive and negative pressures and indicates gauge pressure and vacuum on the same scale. A negative gauge pressure indicates that the system is operating under a vacuum, ie, the absolute pressure is less than barometric. For systems that operate under negative pressures, ie, under vacuum, absolute pressure is equal to the barometric pressure minus the vacuum. If a gauge, for example, indicates −25 in. of mercury (−25 in. Hg), the local barometric pressure is 29.9 in. Hg, and the ambient temperature is 60°F, then the absolute pressure is 4.9 in. Hg (16.5 kPa).

Mechanical Gauges

Pressure, particularly where pressure is monitored as opposed to being controlled, is generally measured by directly actuated mechanical elements. Mechanical gauges, reliable and inexpensive, dominate process applications in older plants. Moreover, the sensing elements for many of the more modern and sophisticated electronic transmitters are simple mechanical elements such as Bourdon tubes or diaphragms. Mechanical gauges may be divided into two groups. The first, which includes liquid manometers, bell gauges, and slack diaphragm gauges, measure pressure by balancing an unknown force against a known force. The second, Bourdon gauges, diaphragm gauges, and bellows elements, rely on elastic deformation of a sensing element for pressure measurement (3).

Liquid Manometers. Liquid manometers were used extensively by scientists in the seventeenth through the early twentieth centuries to measure pressure. The practice of expressing pressure as a certain height of liquid evolved from this usage. The typical liquid manometer consists of a cyclindrical glass U-tube partially filled with liquid. One end is connected to the process; the other can be either open or closed (Fig. 1).

An open manometer is normally used to measure pressure relative to local barometric pressure or to measure differential pressure, ie, the difference between two pressures. Open manometers are used, for example, for measuring pressure differentials across fans, heat exchangers, and distillation column trays. Almost any liquid of known density can be used. Water (qv), mercury (qv), and heavy oils are most common. Closed manometers are used to measure absolute pressures. The mercury barometer is a familiar example of the closed manometer. Liquid manometers are increasingly regarded as too fragile for general use in the CPI, and restrictions placed on the use of mercury in most processes frequently precludes consideration of mercury manometers for measuring process pressures. The principal use of liquid column manometers as of this writing (1996) is as a primary standard for calibrating other gauges.

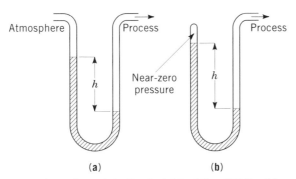

Fig. 1. U-Tube manometer where h is the height of the (▨) liquid employed: (**a**) open; (**b**) closed.

Inverted Bell-Type Pressure Element.   An inverted bell manometer, illustrated in Figure 2, consists of two inverted bells immersed in oil. The oil provides a liquid seal. The bells are suspended from opposite ends of a balance beam and are arranged so that pressure, P, can be introduced under each bell. One of the lines is usually open to atmospheric pressure, the other to the pressure to be measured. The bell subjected to the higher pressure rises in the oil, tilting the beam which moves a pointer on a scale. This instrument responds to a pressure difference, ΔP, as small as 0.1 Pa (0.0004 in. H_2O). The gauge ranges available range from 0–0.05 kPa (0–0.2 in. H_2O) up to 0–3.7 kPa (0–15 in. H_2O) pressure or vacuum. Inverted bell manometers are used for measuring very low positive pressures, such as those found in furnace kiln drafts and conveyor dryers (see DRYING; FURNACES, FUEL-FIRED).

Bourdon Tube.   A Bourdon tube is made from a flattened or elliptical tube, where one end is sealed, the other open to the process. Figure 3 illustrates the three basic designs of this sensing element. All Bourdon tubes are based on the simple principle that a closed-end, flattened or elliptical coiled tube tends to straighten out when a gas or a liquid under pressure is allowed to enter the

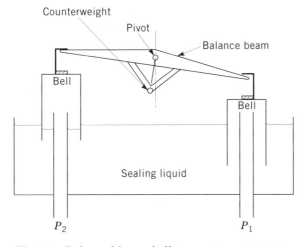

Fig. 2.   Balanced-beam bell-type pressure gauge.

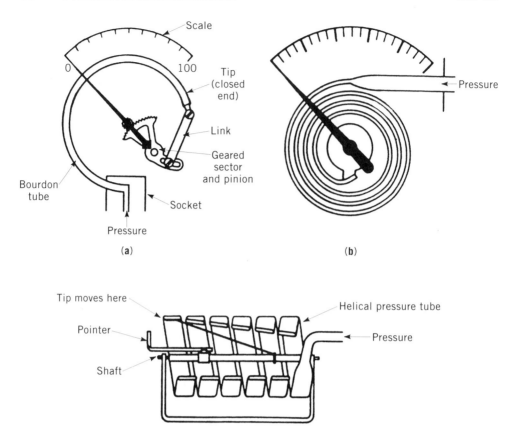

Fig. 3. Bourdon pressure elements: (**a**) C-Bourdon tube; (**b**) spiral Bourdon tube; (**c**) helical Bourdon tube.

tube. The Bourdon tube responds to the pressure difference between the inside and the outside of the tube. If a Bourdon tube is connected to a system under vacuum, atmospheric pressure causes the tube to curl inward. Bourdon tubes are, therefore, used extensively in pressure gauges, in vacuum gauges, and in compound gauges.

The C-Bourdon tube (Fig. 3**a**) usually has an arc of 250°. The open end of the tube is fixed. The closed end of the tube, ie, the tip of the tube, is connected through a mechanical linkage to a pointer or a pen. The principal limitation of the C-design is tip travel. The degree of movement per unit pressure change is small, and the design of the mechanical linkages required for amplification can be quite complex. The spiral Bourdon (Fig. 3**b**) is made by winding the tube in a spiral of several turns, instead of the relatively short 250° arc of the C-design. This gives the spiral a much higher degree of movement per unit of pressure change. The helical Bourdon tube (Fig. 3**c**) has the same advantage and even more tip travel than the spiral (4). The C-Bourdon tube is more often used as the sensing element for a pressure gauge, a vacuum gauge, or a transmitter. Spiral or helical Bourdon tubes are more likely to be used in receivers and recorders.

The advantages of Bourdon tube gauges include low cost, simple construction, availability of instruments for measuring both high and low pressures, and many years of application experience. Bourdon tube limitations include loss of precision below 345 kPa (50 psi) because the Bourdon tube has a very low spring gradient; mechanical linkages usually required for amplification being a source of hysteresis; accumulation of process materials in the Bourdon tube compromising accuracy; and gauges that are susceptible to damage resulting from shock and vibration. Instruments using C-Bourdon elements span the range from 1.3 kPa to 689×10^5 MPa (10 torr to 100,000 psi). The minimum span is about 69 kPa (10 psi). Spiral elements span the range from 0–138 kPa to 0–27.6 MPa (0–20 to 0–4,000 psi), and helical elements are used for 0–689 kPa to 0–689 MPa (0–100 to 0–100,000 psi). Bourdon tubes are fabricated from phosphor bronze, beryllium copper, steel, stainless steel, Monel, Ni-Span C, and special alloys.

Diaphragm Gauges. The sensing element for a diaphragm gauge is a flexible disk, either flat or having concentric corrugations, made of sheet metal. Some gauges use the diaphragm itself as the pressure sensor; others use it as the basic component for a capsule, manufactured by fusion-welding two diaphragms together at their peripheries. Figure 4 shows a flat diaphragm, a corrugated diaphragm, and two basic types of capsules. Most diaphragms have concentric corrugations which make possible deflection-to-pressure ratios many times greater than those of flat plates. Capsules are of two types: convex (Fig. 4**d**), in which

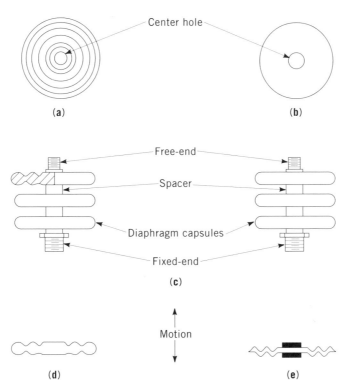

Fig. 4. Diaphragm pressure elements: (**a**) corrugated diaphragm; (**b**) flat diaphragm; (**c**) multielement diaphragm capsules; (**d**) convex capsule; and (**e**) nested capsule.

the orientation of the corrugations of the two diaphragms is opposed; and nested (Fig. 4e), in which the corrugations match.

Diaphragm elements are sensitive to small pressure changes and are therefore particularly useful in the measurement of low pressures. Absolute pressures approaching 13 Pa (0.1 torr) can be accurately measured by using thin-walled beryllium copper diaphragms and designing the gauge for a narrow pressure range. The diaphragm capsule behaves much like a simple bellows element, but the capsule is more accurate and more durable, because the movement of the diaphragm is small, ie, well within the elastic limits of the metal. Diaphragm capsules are used for the pressure ranges 0–0.5 kPa (0–2 in. H_2O) to 0–689 kPa (0–100 psi). Diaphragm and diaphragm capsules are fabricated from a wide variety of materials, including phosphor bronze, beryllium copper, stainless steel, Monel, Ni-Span C, Hastelloy, and Inconel.

Slack diaphragm gauges use diaphragms made from an elastomer such as silicone rubber (see ELASTOMERS, SYNTHETIC). A slack diaphragm gauge does not rely on elastic deformation of the diaphragm for pressure measurement. The pressure-sensing element is a calibrated spring. The spring gradient determines the deflection of the discharge for an applied pressure. The gauge measures pressure by balancing an unknown force against a known force. The Magnehelic gauge is an example. The diaphragm is made of silicone rubber and is balanced by a spring. The motion of the spring is transmitted through a magnetic linkage to the pointer. Slack diaphragm gauges, which cover the range 0–405 kPa (0–120 in. H_2O), are used extensively for measuring pressure drop across filters used in ventilating systems and to measure pressure drop in pneumatic conveying systems (see CONVEYING). In this and a number of similar applications, these gauges are both more accurate and more reliable than conventional Bourdon tube gauges.

Bellows Elements. A cross-section of a spring-and-bellows pressure element is shown in Figure 5. The bellows is enclosed in a metal housing connected by piping to the process and restrained at the top by a form-fitted nut. A rod resting on the bottom of the bellows transmits any vertical motion of the bellows through a suitable linkage to a pointer or pen. As the pressure inside the bellows increases, the bellows compresses the spring. The stiffness of the bellows is small compared to the stiffness of the spring, and therefore the pressure range is primarily a function of the stiffness of the spring. A spring-and-bellows pressure element can be used at pressures from approximately 0–1.24 kPa (0–5 in. H_2O) to 0–345 kPa (0–50 psi).

The bellows is formed from a length of thin-walled tubing by extrusion in a die. The metals used in the construction of the bellows must be ductile enough for reasonably easy fabrication and have a high resistance to fatigue failure. Materials commonly used are brass, bronze, beryllium copper, alloys of nickel and copper, steel, and Monel (5).

Meters. Diaphragms, diaphragm capsules, and bellows elements are used extensively in meter bodies designed to measure differential pressure. These units can be used to measure the differences in pressure between two water lines, two steam headers, two stills, etc, from 25 Pa (0.1 in. H_2O) to 0–4826 kPa (0–700 psi) and are designed to operate at pressures as great as 69 MPa (10,000 psi). A bellows-actuated meter body is illustrated in Figure 6. The high

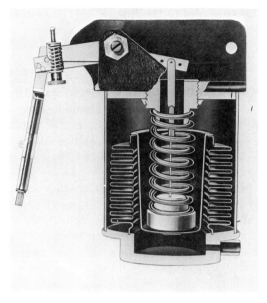

Fig. 5. Cross section of a spring-and-bellows pressure element.

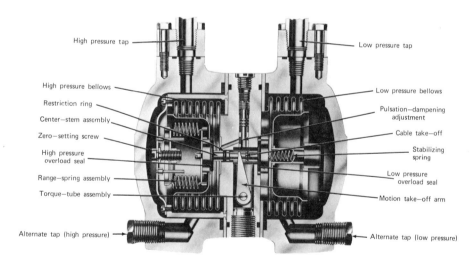

Fig. 6. Bellows-actuated differential-pressure element.

pressure and low pressure bellows are joined by the center-stem assembly. The entire volume inside the bellows is filled with liquid, and the bellows is sealed. When the pressure at the high pressure tap exceeds the pressure at the low pressure tap, the high pressure bellows moves to the right and, by means of the center stem and the liquid fill, forces the low pressure bellows to the right. Motion stops when the force on the stabilizing spring equals the differential pressure, ie, the difference between the high and low pressures. The cable and motion take-off arm translate the center-stem movement to the torque tube assembly, and this is connected to the linkage mechanism for positioning of the pointer or pen arm.

Electronic Sensors

Electronic sensors and electronic control systems have displaced many of the mechanical sensors and pneumatic control systems in the CPI. This change, occurring in the 1980s and 1990s, is the result of the superiority of electronic sensor technology, as well as the superiority of electronic control systems. Mechanical gauges rely on a mechanical linkage to convey movement of the sensor to a pointer or a pen. This mechanical linkage has a high degree of hysteresis and deadband that limits accuracy and repeatability of the movement. Electronic sensing of element movement eliminates many such problems, and is inherently more accurate. Electronic transmitters have the additional advantages, compared to pneumatic transmitters, of built-in calibration checks, temperature compensation, self-diagnostics, signal conditioning, and other features that may be derived from the use of microprocessors (6).

It is especially difficult in discussing electronic instruments to distinguish between a pressure sensor, a pressure transducer, and a pressure transmitter. Sensor and transducer are synonymous. The distinction between transducer and transmitter is, however, fundamental. All transmitters are transducers; not all transducers are transmitters. Most electronic transducers incorporate a mechanical element as the primary sensing element. An electronic transducer converts movement of a mechanical element, or the output from an electronic sensing element, to an electrical quantity such as resistance, capacitance, or voltage. In its simplest form, a pressure transmitter is a combination of a pressure transducer and a signal-conditioning circuit that outputs a current proportional to the process pressure. An electronic transmitter converts the movement of a mechanical element or the output from an electronic sensing element to a 4–20-mA signal, or to digital information, for transmission to an indicator, a recorder, a controller, or to a distributive control system (DCS).

Piezoelectric Elements. Designs of piezoelectric pressure elements are based on the principle that certain crystalline insulators, such as quartz (see SILICA, SYNTHETIC QUARTZ SILICA), when properly cut and oriented with respect to their crystallographic axes, generate a small electric charge when stressed. In practice, a stack of quartz plates is mounted in a housing, which has a thin diaphragm at one end. The housing is usually designed to be mounted in the wall of a pressure vessel. The diaphragm is exposed to the pressure and deflects, thereby applying a compressive force to the stack, which in turn generates a charge that is directly proportional to the force. A typical design is shown in Figure 7. Piezoelectric sensors cannot measure static or absolute pressures for more than a few seconds, but this automatic elimination of static signals allows drift-free operation. Piezoelectric sensors are very rugged. Consequently these sensors are used extensively in demanding applications to measure the dynamic pressures associated with the operation of shock tubes, rocket motors, internal-combustion engines, mufflers, pumps (qv), compressors, pipelines (qv), and oil exploration imploders (7). Such devices are available in pressure ranges between 0–345 kPa (0–50 psi) and 0–69 MPa (0–10,000 psi). Some are available in ranges as low

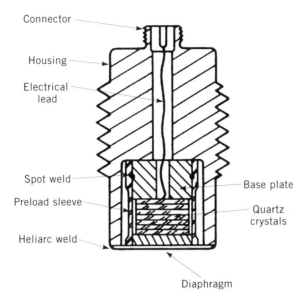

Connector

Housing

Electrical
lead

Spot weld

Preload sleeve

Heliarc weld

Base plate

Quartz
crystals

Diaphragm

Fig. 7. Piezoelectric pressure sensor.

as 0–69 Pa (0–0.01 psi) and as high as 0–827 MPa (0–120,000 psi). Response time is generally very fast. Frequency response can be as high as 500 kHz.

Linear-Variable-Differential-Transformer and Reluctive Pressure Transducers. In a linear-variable-differential-transformer (LVDT) pressure transducer, the pressure to be measured is fed to a Bourdon tube or diaphragm. The motion of this element is transferred to the magnetic core of a transformer. Using a-c excitation of the primary coil of the transformer, a varying voltage is produced in the secondary coil as the core is moved. The two secondary coils are wound in opposite directions. Therefore, when the core is centered between them, the voltages cancel each other. When the core is moved from center, a differential voltage is produced. This voltage is proportional to the movement of the Bourdon tube or diaphragm, which is proportional to the pressure.

Measurement by a reluctive transducer is based on the ratio of the reluctance of the magnetic flux path of two coils. Reluctance is the resistance to magnetic flow offered by a magnetic substance to magnetic flux. Reluctance is the equivalent of electrical resistance in a magnetic circuit. In the sensor system shown in Figure 8, a diaphragm of magnetically permeable material is supported between two symmetrical E-core inductance assemblies and completes a magnetic circuit with each one. Application of pressure to the diaphragm causes it to deflect, thereby increasing the gap in the magnetic flux path of one core and decreasing the gap in the other by an equal amount. Because the magnetic reluctance varies with the gap, the inductance ratio changes with the position of the diaphragm. This can be measured in a bridge circuit, which produces an output voltage proportional to the pressure. Pressure ranges from 0–86 Pa (0–0.0125 psi) to 0–862 MPa (0–125,000 psi) are available. The pressure range can be changed in the field by simply replacing the diaphragm. The LVDT and reluctive pressure transducers are particularly well suited for low pressure

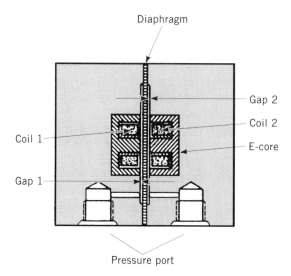

Fig. 8. Reluctance-type pressure sensor. The E-core corresponds to the area where the magnetic fields are produced. Courtesy of Validyne Engineering Corp.

measurement. This technology has been used extensively for absolute pressure measurements in the range 13–1333 Pa (0.1–10.0 mm Hg absolute).

Strain Gauges.   The simplest version of a strain gauge uses the change in the electrical resistance of a metal wire under strain to measure pressure. A bonded strain gauge is constructed by using an adhesive to bond a metal wire (or a metal foil) to an elastic sensing element, usually a Bourdon tube, diaphragm, or a bellows element. Electrical insulation is provided by the adhesive or by backing material on the strain gauge. The gauge changes movement of the sensor to an electrical signal. When the length of the wire is changed by tension or compression, the result is a change in the diameter of the wire and, hence, a change in electrical resistance. The change in resistance is a measure of the pressure.

The construction of a bonded strain gauge is illustrated in Figure 9. The ideal strain gauge would change resistance in response to deformation of the surface to which it is bonded, and for no other reason (8). The measurement is obtained, however, by transferring strain from the sensing element through

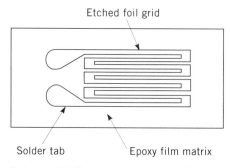

Fig. 9. Bonded metal foil strain gauge showing etched foil grid.

the adhesive and the backing material to the strain gauge. The accuracy of the measurement is, therefore, limited by the characteristics of the adhesive and the backing material. The adhesives (qv) and backing materials are usually epoxies.

The development of thin-film and diffused semiconductor strain gauges represents significant advances in strain-gauge technology. A thin-film strain gauge is produced by depositing a thin layer of metal on a metal diaphragm by either vacuum deposition or sputtering (see THIN FILMS). To produce thin-film strain-gauge transducers, first an electrical insulator such as a ceramic is deposited on the diaphragm (see CERAMICS AS ELECTRICAL MATERIALS). The strain-gauge alloy is deposited on the insulator. This technique produces a strain gauge that is molecularly bonded to the element, thus eliminating temperature effects and stress creep associated with organic adhesives. The result is long-term stability, which is the principal advantage of thin-film strain-gauge technology.

In the construction of a diffused semiconductor strain gauge, the gauge is diffused directly into the surface of a silicon diaphragm, using photolithographic masking techniques and solid-state diffusion of an impurity element, such as boron. Bonding does not involve an adhesive. Creep and hysteresis are, therefore, eliminated. The diffusion process does, however, require that the diaphragm be made from silicon.

All strain gauges convert a pressure change to a change in electrical resistance. The resistors are usually arrayed in the four arms of a Wheatstone bridge. The advantages of the strain-gauge transducer are fast response, practically infinite resolution, minimum movement of elastic elements, high accuracy, comparative ease of compensation for temperature effects, low source impedance, and relative freedom from acceleration effects. The disadvantages include the difficulty of obtaining zero output at zero pressure, ie, bridge imbalance, low output levels, difficulties associated with isolating the excitation ground from the output ground, and signal conditioning requirements that include zero nulling and calibration. Generally, strain-gauge accuracies range from ± 0.1 to $\pm 2\%$ of full scale. Strain-gauge pressure transducers are available for measurement of 0–3.3 kPa (0–25 mm Hg) to 0–1379 MPa (0–200,000 psi).

Piezoresistive Sensors. The distinction between strain-gauge sensors and piezoresistive (integrated-circuit) sensors is minor. Both function by measuring the strain on an elastic element as it is subjected to pressure. A piezoresistive transducer is a variation of the strain gauge that uses bonded single-crystal semiconductor wafers. Pressure applied to one side of a silicon wafer strains resistors diffused into the wafer. The change in applied pressure causes a linear change in the resistance value, which can then be converted and amplified to a usable output signal. In some designs, the transducer consists of a single silicon crystal sensor and its supporting electronics mounted in a pressure-tight housing on a ceramic substrate. If the transducer is to be used in an application that involves a corrosive or a conductive process fluid, the housing is filled with silicone oil. An elastic diaphragm isolates the oil from the process fluid, as it allows pressure to be applied to the oil. Because of high natural frequencies (greater than 50 kHz), single silicon crystal sensors are very reliable for use under conditions of severe shock and vibration. These sensors can be used to measure pressure from 69–34,000 kPa (10–5,000 psi). Operating temperatures are −1 to 85°C.

In other designs, a diffused silicon sensor is mounted in a meter body that is designed to permit calibration, convenient installation in pressure systems and electrical circuits, protection against overload, protection from weather, isolation from corrosive or conductive process fluids, and in some cases to meet standards requirements, eg, of Factory Mutual. A typical process pressure meter body is shown in Figure 10. Pressure measurement from $0-746$ Pa ($0-3$ in. H_2O) to $0-69$ MPa ($0-10,000$ psi) is available for process temperatures in the range -40 to $125°C$. Differential pressure- and absolute pressure-measuring meter bodies are also available. As transmitters, the output of these devices is typically $4-20$ mA dc with 25-V-dc supply voltage.

Capacitive Pressure Transducers. In all capacitive pressure detectors, the basic operating principle is that a change in capacitance occurs owing to the movement of an elastic element. In a conventional capacitance-type pressure transducer, the sensing element is a diaphragm. A cutaway view of a transducer designed to measure differential, absolute, or gauge pressure is shown in Figure 11. The process comes in contact with an isolating diaphragm, a metal diaphragm used to isolate the sensing element from the process. Process pressure is hydraulically coupled to the sensing element. A change in process pressure is hydraulically transmitted by the silicone oil fill fluid from the isolating diaphragm to the sensing element. Deflection of the sensing diaphragm is detected by capacitor plates positioned on either side of it. Diaphragm movement changes the capacitance between the plates and the diaphragm; one side increases while the other decreases. Signal conditioning converts the capacitance change to a stable direct current or voltage signal. Standard output for a two-wire transmitter is $4-20$ mA dc.

The transducer is very rugged. The most sensitive components in the transducer are isolated from the process, and movement of the sensing diaphragm is restricted to give over-pressure protection as high as 31 MPa (4500 psi). Fabrication of wetted parts from stainless steel, Hastelloy C-276, or Monel ensures that exposure to the process does not drastically reduce transducer life. Capaci-

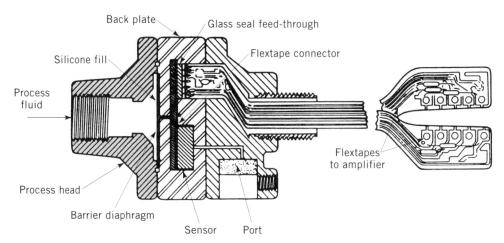

Fig. 10. Cross-section of meter body having a piezoresistive pressure sensor. Courtesy of Honeywell, Inc.

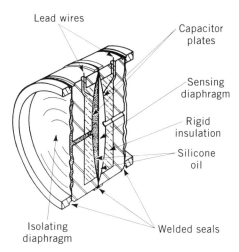

Fig. 11. Capacitive pressure sensor in the differential measurement configuration. Courtesy of Rosemount, Inc.

tive pressure transducers are available for measurement of gauge pressures in the range 0–1.24 kPa (0–5 in. H_2O) to 41.4 MPa (0–6000 psi). The lower end of the range for transducers designed to measure differential pressure is 0–124 Pa (0–0.5 in H_2O). Absolute pressure transducers have been designed to measure pressures as low as 0–6.8 kPa (0–2 in. Hg absolute), but the conventional capacitive pressure transducer is significantly less accurate than a capacitance manometer for absolute pressures less than approximately 1.3 kPa (10 torr).

Vacuum Measurement

Vacuum measurement spans the range from 10^5 Pa (atmospheric pressure) to pressures that are less than 10^{-10} Pa (7.5×10^{-13} torr), ie, more than 16 orders of magnitude. Subatmospheric pressures are divided into six regions for convenience (9):

Vacuum region	Pressure, Pa	Torr
low	10^5–3.3×10^3	750–25
medium	3.3×10^3–10^{-1}	25–7.5×10^{-4}
high	10^{-1}–10^{-4}	7.5×10^{-4}–7.5×10^{-7}
very high	10^{-4}–10^{-7}	10^{-6}–10^{-10}
ultrahigh	10^{-7}–10^{-10}	7.5×10^{-10}–7.5×10^{-13}
extreme ultrahigh	$\leq 10^{-10}$	7.5×10^{-13}

The pressure sensors discussed herein are reasonably accurate for measurement into the low vacuum region. Upon some modification, these sensors can be extended into the lower end of the range, to approximately 1.3×10^{-1} Pa (10^{-3} torr). Vacuum gauges are required for accurate measurement of lower pressures.

Vacuum gauges may be broadly classified as either direct or indirect (10). Direct gauges measure pressure as force per unit area. Indirect gauges measure a physical property, such as thermal conductivity or ionization potential, known to change in a predictable manner with the molecular density of the gas.

The definition of vacuum advanced herein assumes that pressure in the gaseous phase adequately characterizes vacuum environments. This assumption is normally good for processing applications in the range $10^5-1.3 \times 10^3$ Pa (760 torr to 100 μm Hg). Medium vacuum operations, and in particular high vacuum and ultrahigh vacuum operations, cannot be characterized by this simple, single-parameter approach. Molecular concentrations and the chemical identity of the molecules present in the vapor space and on the walls of the process vessel are important parameters in describing the vacuum environment in those pressure regions.

Capacitance Manometers.   Capacitance manometers were first used in research laboratories in the early 1950s. The development of capacitance manometers designed specifically for the harsh environments that characterize chemical processes is, however, a more recent development. The near-phenomenal accuracy of capacitance manometers, typically 0.10% of a reading $\geq$13.3 Pa ($\geq$0.1 torr) to 3% of a reading at 10^{-2} Pa (10^{-4} torr), excellent linearity over a wide range of pressures, and the development of sensors that are rugged and reliable, increasingly makes this the technology of choice for vacuum measurements in the range $10^{-3}-10^3$ Pa ($10^{-5}-10$ torr). Capacitance manometers are rugged for two reasons. Gauge electronics never come in contact with the process, and the sensor body and metal diaphragm are fabricated from stainless steel, Monel, Inconel, or other high nickel alloys.

A pressure sensor for a capacitance manometer (Fig. 12) measures absolute pressure independent of gas composition. A permanently sealed cavity evacuated to a very low pressure, $10^{-5}-10^{-6}$ Pa ($10^{-7}-10^{-8}$ torr), and maintained by a getter material, ie, an adsorbent, provides an absolute pressure reference. The device is remarkably sensitive. Diaphragm deflections as small as 10^{-11} m (10^{-9} in.) can be detected. Using the thinnest diaphragms, this corresponds to a pressure of approximately 10^{-4} Pa (10^{-6} torr). The lower limit for accurate measurement, however, is approximately 10^{-3} Pa (10^{-5} torr). Nonlinearity is the most important single source of error, representing approximately 80% of all accumulated errors. Temperature effects, typically 0.06% of a reading for a 1°C

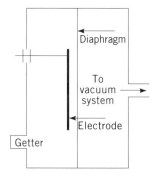

Fig. 12.   Schematic of a capacitance manometer. Courtesy of MKS Instruments.

change in temperature, are also important (11). A change in the temperature of the sensor causes it to undergo dimensional changes that affect sensor output. The effects of changes in ambient temperature on process gas temperature can be minimized by using sensor heaters to maintain the sensor at a constant elevated temperature (typically 40–80°C).

The capacitance manometer is in essence an electronic diaphragm gauge. It differs from the mechanical diaphragm gauge in the way diaphragm deflections are translated into pressure readings. The mechanical gauge employs a mechanical linkage. The capacitance manometer uses capacitance changes to measure diaphragm deflections. Capacitance manometers are very accurate and exhibit excellent retention of calibration. They have, therefore, been almost universally adopted as secondary calibration standards for pressure measurement in the range $10^{-2}-10^5$ Pa ($10^{-4}-760$ torr). Capacitance manometers exhibit excellent linearity over a wide range of pressures, and sensors are routinely designed to cover three or four decades, ie, three or four orders of magnitude. Sensors that cover six decades have also been developed.

Thermal Conductivity Gauges. Thermal conductivity gauges measure thermal conductivity and are thus indirect gauges. Pressure is measured indirectly. The thermal conductivity of a gas is essentially independent of pressure for pressures greater than approximately 1.33 kPa (10 torr). As the pressure is reduced below 1.33 kPa (10 torr), the thermal conductivity of a gas decreases, first by a logarithmic relation, then linearly as the pressure approaches 10^{-1} Pa (10^{-3} torr). The range of thermal conductivity gauges spans almost eight decades, from atmospheric to 10^{-3} Pa, but application is limited almost exclusively to the pressure range 133–0.1 Pa ($1-10^{-3}$ torr).

The sensor for a thermal conductivity gauge is a thin metal wire or heat-sensitive element enclosed in a metal cyclinder. The element is electrically heated. If a constant current is maintained across the element, the temperature of the element increases as the pressure in the metal cylinder is reduced below 1.33 kPa (10 torr). The rate at which heat is transferred from the element to the cyclinder wall decreases, because the thermal conductivity of the gases surrounding the element decreases with pressure. Convection is insignificant and radiation and conduction through the connecting wires can usually be ignored at pressures $>10^{-1}$ Pa (10^{-3} torr). The thermal conductivity of the gases surrounding the element, and indirectly the pressure, can therefore be measured by measuring the temperature of the element. The different means used to measure the temperature of the heated element distinguish the two main types of thermal conductivity gauges, the Pirani gauge, shown schematically in Figure 13, and the thermocouple gauge (12).

The Pirani gauge is so constructed that the resistance of the heated element is directly proportional to the element temperature. System pressure is measured by measuring the resistance of the element. The heated element comprises one arm of a Wheatstone bridge. An identical element, sealed in an evacuated tube, makes up an adjacent arm of the network. The sealed element, or compensating element, partially compensates for small changes in bridge voltage and ambient temperature. In the operation of a thermocouple gauge, the temperature of the heated element is measured directly by a thermocouple. The thermocouple, usually iron or copper–constantan, is spot welded to the element. The current

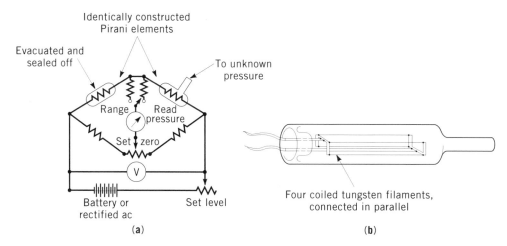

Fig. 13. Pirani gauge circuit: (**a**) gauge in a fixed-voltage Wheatstone bridge; (**b**) sensing element (6).

used to heat the element is kept constant, and the pressure is indicated by the output of the thermocouple. The current produced by the thermocouple is measured by a microammeter that is calibrated in pressure units. The pressure indicated by a thermal conductivity gauge is composition-sensitive. The thermal conductivity of a gas depends on the gas composition. Dry air or nitrogen are used in calibrating the gauge. Gauge readings must, therefore, be corrected if an absolute indication of pressure is required and gases other than those used for calibration are present in the system.

Thermal conductivity gauges are simple and robust, but not very accurate. Accuracy quoted in instrument specifications is typically $\pm 20\%$ of reading across the range $0.1-133$ Pa ($10^{-3}-1$ torr). Thermal conductivity gauges are generally used as pressure indicators, to monitor rather than measure system pressure. The advantages of these gauges include low cost, simplicity, and interchangeability of the sensing elements. They are well adapted for applications in which a single power supply and measuring circuit is used, with sensing elements located at different parts of the same vacuum system or on several different systems.

Hot-Cathode Ionization Gauges. For pressures below approximately 10^{-4} Pa, it is not possible, except under carefully controlled conditions, to detect the minute forces that result from the collision of gas molecules with a solid wall. The operation of the ion gauge is based on ionization of gas molecules as a result of collisions with electrons. These ions are then subsequently collected by an ion collector. Ionization gauges, used almost exclusively for pressure measurement in high, very high, ultrahigh, and extreme ultrahigh vacuums, measure molecular density or particle flux, not pressure itself.

The earliest form of ion gauge, the triode gauge, looks much like a triode vacuum tube (Fig. 14**a**). The gauge consists of three electrodes in a hermetically sealed tube: a filament surrounded by a grid wire helix and a large-diameter, solid cylinder. The filament, typically tungsten, serves as the cathode. It is heated by an electric current, and electrons are released from the filament surface into the surrounding vacuum. Emission of electrons is controlled by

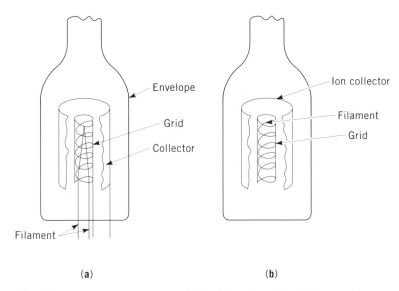

Fig. 14. Ionization gauges: (**a**) triode head and (**b**) Bayard-Alpert.

controlling the electric current to the cathode. The grid serves as the anode and is set at a positive potential of $100-300$ V with respect to the cathode. The grid attracts the electrons emitted by the cathode. The third electrode, the plate, is set at a potential of from -2 to -25 V with respect to the cathode. The plate, the ion collector, attracts positive ions generated by collisions between the electrons emitted by the cathode and molecules of gas. The usual practice using commercial hot-cathode ionization gauges is to precisely control the emission current to the cathode and measure pressure by measuring the ion current at the plate.

The operating range of conventional triode ionization gauges is $10^{-1}-10^{-6}$ Pa ($10^{-3}-10^{-8}$ torr). Burnout caused by positive ions reaching the cathode dramatically reduces filament life for gauges operating above approximately 10^{-1} Pa (10^{-3} torr). The lower limit for the operating range is established by a phenomenon known as the x-ray effect and an associated phenomenon, the photoemission current. Electrons striking the anode, or grid, produce low energy x-rays. Photoemission of electrons occurs when x-rays produced at the anode strike the ion collector. Gauge electronics cannot distinguish between the current produced by positive ions striking the plate and the photoemission current, the current produced by the loss of electrons from the plate. The photoemission current is independent of pressure, and the minimum plate current, a result of the x-ray effect, corresponds to a pressure of approximately 10^{-6} Pa (10^{-8} torr).

The discovery of the x-ray effect spurred the development of a new generation of hot-cathode gauges designed to minimize this effect. One of the earliest, and commercially the most successful, the Bayard-Alpert gauge shown in Figure 14**b**, was developed in 1950 (13). A fine wire, the ion collector, is suspended along the axis of a cylindrical grid. The two-cathode design illustrated is typical. When one filament burns out, the other can be put into service, thus doubling the life of the gauge head. The most important feature of the Bayard-Alpert gauge is the use of a fine wire for the ion collector. The oncoming x-rays see a cross section that is at least two orders of magnitude less than the cross-section of a plate collector. X-ray effects are proportionately less and lower pressure readings are therefore possible. The lower limit for Bayard-Alpert gauges is 10^{-10}–10^{-11} Pa (10^{-12}–10^{-13} torr).

The accuracy of indirect-reading gauges is always a source of concern. This is especially true of ionization gauges, because the ionization gauge is a particle density gauge. The ion current measured at the plate is a function of particle density and ionization probability. Particle density is proportional to pressure if, and only if, temperature is constant. Dry air or nitrogen is used in calibrating ionization gauges. These gauges must be recalibrated when used to measure the pressure of other gases. The ionization probability depends on the gas species (14). Hot-cathode ionization gauges are built to very demanding tolerances, and carbon contamination of electrodes over time compromises accuracy. In spite of these problems and under the less-than-ideal conditions that characterize industrial application of these gauges, hot-cathode ionization gauges are generally accurate to within ±25% for measurements in high and very high vacuum. The various ultrahigh versions of the triode and Bayard-Alpert gauges are accurate to within an order of magnitude.

Cold-Cathode Ionization Gauges. The cold-cathode gauge, the Penning gauge shown schematically in Figure 15, is more rugged, but less accurate, than the hot-filament ionization gauge (15). A potential of 2–10 kV is applied between the cathodes and the anode. The cathodes are constructed of zirconium, thorium, or another surface-active material. Free electrons are formed by the collision of positive ions with the cathodes. Electrons ejected from the cathodes collide with gas molecules to produce additional positive ions, setting up a Townsend discharge between the cathode and the anode. An essential element of this gauge is the permanent magnet positioned outside the gauge tube. Without it, the discharge cannot be maintained at pressures below approximately 1 Pa (10^{-2} torr). The magnetic field prevents electrons from traveling directly from cathode to anode. The electrons are constrained to move in helical paths along the lines of magnetic flux.

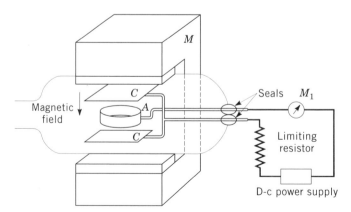

Fig. 15. Cold-cathode ionization gauge, where A = anode, C = cathode, M = horseshoe magnet, and M_1 = microammeter (15).

Electrons emitted by the cathode pass through the anode and continue until repelled by the other cathode. The electrons oscillate between the cathodes and eventually are captured by the anode. Because the electrons are forced to travel several meters before reaching the anode, the probability of colliding with gas molecules and forming positive ions is greatly enhanced. This allows a discharge to be maintained at pressures as low as 10^{-4} Pa (10^{-6} torr). Sensor output, the sum of the positive ion current to the cathode and the electron current leaving the cathode, is proportional to pressure over the range $1–10^{-4}$ Pa ($10^{-2}–10^{-6}$ torr), the nominal pressure range for Penning cold-cathode ionization gauges.

Cold-cathode gauges are simpler, less expensive, and more rugged than hot-cathode gauges. One advantage of the cold-cathode gauges, as compared to hot-cathode gauges, is that cold-cathode gauges withstand accidental exposure to atmospheric pressure without damage. Because of safety concerns, the requirement for maintaining a potential of several thousand volts between the cathodes and the anode can be a significant disadvantage. The cold-cathode gauge is less accurate than the hot-cathode gauge. Under ideal conditions the Penning gauge is accurate to within $\pm 25\%$ over the range $1–10^{-4}$ Pa ($10^{-2}–10^{-6}$ torr), but manufacturers' specifications that the accuracy of the gauge is within a factor of two is a more realistic estimate (16). The various ultrahigh versions of the gauge, the magnetron and the inverted magnetron, for example, are accurate to within an order of magnitude.

Smart Pressure Transmitters

In the mid-1980s, microprocessors were mated to pressure transmitters, and the first smart transmitters were the result (17). Conventional electronic pressure transmitters are strictly analogue in nature, converting the motion or changes in the electrical resistance of a sensor to standard 20–100-kPa (3–15-psig) pneumatic signals or to 4–20-mA d-c electrical signals. Smart transmitters convert the response of the sensor to a high resolution digital signal (Fig. 16). This signal is then linearized and compensated for temperature and, in the case of differential pressure transmitters, for static pressure effects. Once configured, the accuracy of the smart transmitter is dependent only on the accuracy of the sensor.

Intelligent or smart transmitters feature digital electronics, remote communication and configuration, and high turndown. The digital electronics that characterize the transmitter virtually eliminate errors introduced by rearranging. A smart transmitter has the ability to continuously monitor its operating conditions and correct itself for potential errors such as nonlinearity, ambient temperature effects, static pressure effects, etc. A smart transmitter performs continuous diagnostics of its sensing element, ie, the meter body, and electronics and of the loop power supply and wiring. Smart transmitters have the advantage, as compared to conventional analogue electronic transmitters, of greater accuracy and stability, and much greater range, ie, order one model, install it anywhere, and respan (without recalibration) at any time. Smart transmitters are easier to specify and use, inherently more reliable, and have lower maintenance costs. The advantages of smart transmitters are increasingly understood by the process control (qv) engineers who specify the control systems and the

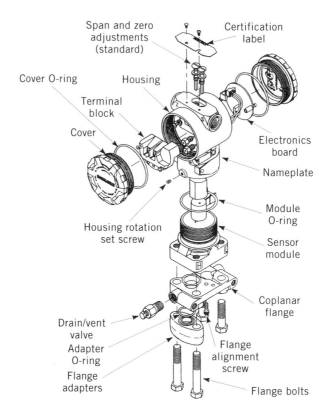

Fig. 16. Exploded view of a smart transmitter based on variable capacitance sensor technology. Courtesy of Rosemount, Inc.

associated transmitters and by the instrument mechanics who maintain the system. Higher price tags notwithstanding, smart transmitters are rapidly replacing conventional transmitters.

The advent of the smart transmitter revolutionized the CPI's approach to process instrumentation. However, the widespread acceptance of smart transmitters, as of 1996, has been less than that expected. In applications where smart transmitters are being used, they have, for the most part, simply replaced conventional analogue transmitters. Full communication capabilities have not been utilized because of the need to maintain existing control systems that rely on 4–20-mA signals. Because all smart transmitters use digital electronics, the digital signal must pass through a digital-to-analogue (D/A) converter in order to transmit the 4–20-mA analogue signal. The D/A converter can be a significant source of error, and the time lag associated with the D/A conversion can be an issue in some applications. For example, for many years, 20–100 kPa (3–15 psig) and 4–20 mA have functioned as open protocols for conventional analogue systems universally adopted by instrument manufacturers and users worldwide. There is, however, no open, comprehensive, universally accepted digital communications protocol, ie, no fieldbus standard.

At first, each smart transmitter manufacturer had a unique and proprietary digital protocol which greatly reduced the user's flexibility in configuring control systems. Products from different vendors were simply not compatible. The ISA Standards and Practices Committee 50 (SP50) began working on a universal fieldbus standard in 1985. SP50 was given the mission of developing a nonproprietary (open), comprehensive, universal digital communications protocol. As of this writing (ca 1996), SP50 and a number of trade organizations that are pushing adoption of a fieldbus standard, are slowly making progress. Addition of fieldbus capability is an issue that can no longer be postponed by manufacturers. The next generation of smart transmitters, multivariable smart transmitters that can, for example, simultaneously measure absolute pressure, differential pressure, and temperature, are already in production. Adoption of a fieldbus standard will ensure compatibility of digital instrumentation, and the enhanced digital integration that is expected to follow should result in significant improvements in the CPI's process control systems (see PROCESS CONTROL).

BIBLIOGRAPHY

"Pressure Measurement" in *ECT* 2nd ed., Vol. 16, pp. 470–481, by C. F. Cusick, Honeywell, Inc.; in *ECT* 3rd ed., Vol. 19, pp. 95–110, by C. R. Brandt, Honeywell, Inc.

1. *Metric Practice Guide*, ASTM Bulletin E380-19, American Society for Testing and Materials, Philadelphia, Pa., 1979.
2. D. L. Roper and J. L. Ryans, *Chem. Eng.* **96**(3), 125 (1989).
3. H. E. Soisson, *Instrumentation in Industry*, John Wiley & Sons, Inc., New York, 1975, p. 61.
4. W. J. Demorest, *Chem. Eng.* **92**(20), 56 (1985).
5. Ref. 3, p. 76.
6. D. M. Considine, ed., *Process/Industrial Instruments & Controls Handbook*, 4th ed., McGraw-Hill Book Co., Inc., New York, 1993, p. 4.54.
7. Ref. 6, p. 4.74.
8. W. G. Andrew and H. B. Williams, *Applied Instrumentation in the Process Industries*, 2nd ed., Vol. 1, Gulf Publishing, Houston, Tex., 1979, pp. 153–155.
9. M. S. Kaminsky and J. M. Lafferty, *Dictionary of Terms for Vacuum Science and Technology*, American Vacuum Society, Thorofare, N.J., 1980, pp. 70–71.
10. Z. C. Dobrowolski, in Ref. 6, p. 4.78.
11. J. J. Sullivan, *J. Vac. Sci. Technol. A* **3**(3), 1721 (1985).
12. A. Berman, *Total Pressure Measurements in Vacuum Technology*, Academic Press, Inc., Orlando, Fla., 1985, pp. 140–146.
13. R. J. Bayard and D. Alpert, *Rev. Sci. Instr.* **21**, 70–71 (1950).
14. Ref. 12, pp. 159–161.
15. A. Guthrie, *Vacuum Technology*, John Wiley & Sons, Inc., New York, 1963, pp. 176–181.
16. W. Jitschin, *J. Vac. Sci. Technol. A* **8**(2), 951 (1990).
17. G. A. Orrison IV, *Control Eng.* **42**(1), 73–76 (1995).

General References

References 3, 6, 8, and 12 are general references.

B. G. Liptak, *Instrument Engineers' Handbook—Process Measurement and Analysis*, 3rd ed., Chilton Book Co., Radnor, Pa., 1995, pp. 523–601.

B. G. Liptak, *Instrument Engineers' Handbook—Process Control*, 3rd ed., Chilton Book Co., Radnor, Pa., 1995, pp. 252–284.

S. Dushman, in J. M. Lafferty, ed., *Scientific Foundations of Vacuum Techniques*, 2nd ed., John Wiley & Sons, Inc., New York, 1962, pp. 258–370.

J. F. O'Hanlon, *A Users's Guide to Vacuum Technology*, 2nd ed., John Wiley & Sons, Inc., New York, 1989, pp. 75–100.

J. L. RYANS
Eastman Chemical Company

PRESSURE SWING ADSORPTION.   See MEMBRANE TECHNOLOGY.

PRESSURE VESSELS.   See HIGH PRESSURE TECHNOLOGY; TANKS AND PRESSURE VESSELS.

PRIMING COMPOSITIONS.   See EXPLOSIVES AND PROPELLANTS.

PRINTING INK.   See INKS.

PRINTING PROCESSES

Significant change has taken place in printing processes in the latter part of the twentieth century. This change, which embraces electronic technology, involves a shift in the printing and publishing industry to electronic prepress, electronic printing, and alternative media such as compact disk–read only memory (CD-ROM) and multimedia. Meeting the needs of the Electronic Age, these new developments could conceivably rival the cultural and commercial impact of the introduction of printing to the Western world. Just as the fast, economical printing production processes to which Johannes Gutenberg contributed around 1450 served to spread knowledge and advance civilization in the Renaissance, in the 1990s, the tools of electronic printing are helping to enhance global education and change the ways in which information is communicated. Spurred on by the impact of television, printing is demanding more color; shorter press runs, ie, more information flexibility; and a change in workflow integration. Printing and mass distribution are being replaced by electronic distribution of information and local printing for specialized, sometimes even individual, interests.

The role of chemical technology in printing is also changing. Whereas a need exists for hard copy, whether for visual, legal, or historical reasons, the hard copy must meet new standards of performance, including visual and environmental. Many traditional printing processes have become unacceptable in the workplace, and are being replaced by processes that are water-based, dry, desktop, or in some other ways more convenient.

There are four main printing processes: planography or lithography, intaglio or gravure, porous or screen, and relief (flexography or letterpress). Use of letterpress, which once excelled in the reproduction of text and pictures, has rapidly waned as first photolithography, and then electronic prepress systems, have increased in ability to provide text material without the setting of metal type. Although letterpress printing has diminished in importance, the relief image has been revived in the form of flexography using photopolymer elastomeric relief plates. These plates offer significant advantages in process and performance over rubber plates.

In general, the process of printing involves generating two physically different areas: the printing or image area, and the nonprinting or nonimage area. In relief printing, whether flexographic or letterpress, the image or printing area is raised above the nonprinting area. Ink is applied to the raised surface, which is brought into direct contact with the paper (qv) or other surface upon which the print is to appear. The relief printing process is used to print on a variety of paper and plastic packaging (qv) materials as well as for some magazines and newspapers, labels, and business forms. Water-based or solvent inks are used. Letterset describes the use of relatively thin relief plates for printing by the offset principle (see INKS).

In the intaglio process, the nonprinting area is at a common surface level but the printing area is recessed, consisting of wells etched or engraved, usually to different depths. The most typical method of intaglio printing is the gravure process. Solvent inks with the consistency of light cream are transferred to the whole surface and a metal doctor blade is used to remove excess ink from the nonprinting surface. Ink is transferred directly to the substrate, usually with an electrostatic assist. Gravure printing is used to print long-run magazines, mail-order catalogs, newspaper supplements, preprints for newspapers, plastic laminates, floor coverings, etc.

In the planographic or lithographic process, the image and nonimage areas are on the same plane, and the difference between image and nonimage areas is maintained by the physicochemical principle that oil and water do not mix. The image area is oil-receptive and water-repellent; the nonimage area is water-receptive and oil-repellent. Therefore, the ink adheres only to the image areas, from which it is transferred to the surface to be printed, usually by the offset method. This process is used for printing general commercial literature, books, catalogs, greeting cards, letterheads, business forms, checks, maps, art reproductions, labels, packages, etc.

In the stencil or screen printing process, a stencil representing the non-printing areas is applied to a silk (qv), nylon, or stainless-steel fine-mesh screen to which ink having the consistency of paint is applied and transferred to the surface to be printed by scraping with a rubber squeegee. This process is used for printing displays, posters, signs, instrument dials, wallpaper, textiles, etc.

Direct printing is the transfer of the image directly from the image carrier to the paper. Most letterpress and gravure and all screen printing are done by this method. In indirect or offset printing, the image is transferred from the image carrier to an intermediate rubber-covered blanket cylinder, from which it is transferred to the paper (Fig. 1). Because most lithography is printed in this way, lithography is usually referred to as offset printing. Letterpress and gravure can also be printed by the offset method.

Images are defined for these printing processes in a number of different ways. Letterpress uses cast-metal type for printing. The other processes produce images on a support by manual, chemical, mechanical, or increasingly by electronic imaging means. As of this writing (ca 1995), the greatest number of plates and images are made by photomechanical methods. These systems are characterized by photographic images and light-sensitive coatings that, by using chemical etching or other treatments, lead to the formation of a printing surface. Increasingly, this printing surface is produced directly by electronic imaging without the traditional photographic intermediates. In some cases the final creation of a printing surface is electronic. These processes are termed computer-to-plate or direct-to-press.

A typical workflow involves image creation, capture, assembly, storage, approval, duplication, output, delivery, and distribution. A printing process workflow is shown in Figure 2.

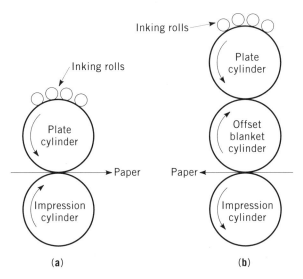

Fig. 1. Printing cycle for (**a**) direct and (**b**) offset printing.

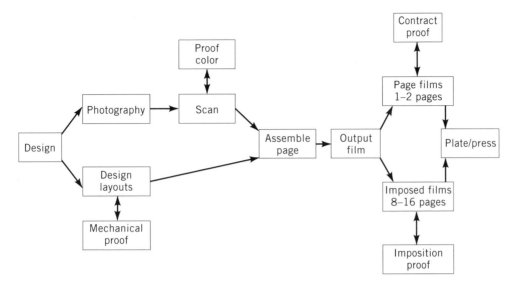

Fig. 2. Printing industry workflow.

Image Creation

Photography. Most press-printed illustrations are reproductions of photographic originals (1,2). These originals may include paintings, drawings, or digital image files created using drawing or illustration software. Press-printed photographic images fall into three main photography (qv) categories based on intended use: commercial, editorial, and fine art.

Commercial photography typically illustrates a product or other salable item for a package, advertisement, catalog, or brochure. Commercial originals must be technically excellent and are often retouched to improve or enhance the image, or to remove blemishes. Editorial photography usually illustrates a magazine or newspaper story, thus pictorial content is more important than technical excellence. Fine-art photography is regarded as attractive, collectible, or salable as works of art, and may be press-printed in the form of posters, books, or prints.

Both the camera and the media play a role. The photographer chooses a film and camera, adjusting the camera variables to achieve particular effects and meet the requirements of the project.

Media Types. Black-and-White Negative Film. Black-and-white photographs are usually reproduced from photographic prints. To preserve maximum details, eg, in fine-art reproductions, an original black-and-white negative may be scanned directly.

Color Film. Most color photographic images reproduced on press are made from positive color transparencies known as slides. Transparencies are preferred over color prints for reproduction because of superior sharpness, tonal contrast, and color saturation (see COLOR PHOTOGRAPHY).

Photographs made from color negative film are usually reproduced from a color print. These can also be reproduced directly from the negative either to extract the full sharpness and tonal range of the negative or, because negative films develop faster than transparencies, to meet a publication deadline. Color negative film has wider exposure latitude and greater tonal sensitivity range, making it ideal in rapidly changing or high contrast lighting, such as news or sports photography. About 2% of all press-printed images are produced from color prints made from an original color negative or transparency.

Common Film Sizes. The most popular photographic film format uses rolls of film 35-mm wide, producing a full-frame image measuring 24 × 36 mm (1 × 1.5 in.). A wide array of 35-mm films are available. Owing to economy, quality, and versatility, about 80% of all press-printed color transparencies are 35 mm. However, the small image size and high enlargements required can challenge lens performance, emulsion technology, and cleanliness. Dust and scratches are also enlarged. Larger originals tend to produce sharper, smoother, and cleaner results.

The next larger film sizes, considered a medium format, are 120-, 220-, and 70-mm roll films. The first two are about 62-mm wide and unperforated; the last is bulk motion-picture stock perforated along both edges. These are also available in a range of emulsion types.

When finest grain structure and maximum image sharpness are required, the original is photographed in a large-format camera using a separate sheet of film for each exposure. These cameras are heavier than roll-film cameras and usually require a tripod and/or a controlled studio environment. Common sheet film sizes include 102 × 127 mm (4 × 5 in.) and 203 × 254 mm (8 × 10 in.).

Camera. Choice of camera and camera settings affects the final appearance of the image. Variables such as the lens, which affects sharpness, magnification, and field of view of the image; lens aperture, which determines the lens opening through which light can reach the film plane; depth of the image field; and shutter speed, which controls the maximum amount of light reaching the film, all combine to produce the desired artistic effect. The other parts of the print process chain must then maintain that effect into the final printed page.

Filmless Photography. A rapidly emerging type of photography is the direct acquisition of monochrome and red–green–blue (RGB) images electronically using digital cameras (3). Digital cameras fall into two categories: those that can stop moving subjects with short shutter speeds or electronic flash, and those that require relatively long, tripod-based exposures and continuous illumination. The latter typically produce superior image quality but can only work with static, ie, nonmoving, subjects.

Digital cameras use two-dimensional charge-coupled device (CCD) arrays to generate instantaneously a digital image, thus substantially shortening the color production process. Both the traditional generation of film-based original artwork and the entire scanning process are eliminated. This technology is in its infancy as of the mid-1990s, but promises to improve rapidly, driven by savings in cost and time.

One digital camera technique uses a high density pixel device consisting of three interwoven arrays representing the primary colors red, green, and blue. Over each of these charge-coupled sensors is deposited a filter that determines its color. This system has the advantage of capturing all three colors simultaneously. Such arrays are difficult to manufacture and therefore expensive for high resolution image capture. An alternative system uses a single array, ie, either a scanned linear device or an area array, and movable filters whereby the system captures the color planes sequentially. This has the advantage of high resolution at a lower cost, but exposure times are longer. A unique variant of area array technology uses a low resolution color video CCD array mounted on a piezoelectric platform the location of which can be incrementally positioned so as to increase the resolution of the video sensor sixfold.

Disadvantages of these technologies include short depth of focus and intense lighting requirements. Moreover, they are not suitable for live models. Alternatively, high density television or vidicon tube technology can be used. This latter technology has the advantage of using standard flash lighting, which increases depth of focus while delivering only moderate resolution.

Creative Processes. Use of computer technology (qv) by designers, artists, illustrators, and photographers involved in the design and layout of printed materials has propelled the printing industry into a new digital era. Work processes have been radically transformed, giving the creators more prepress control of the printing process. More design alternatives can be provided to the client earlier in the process, the final solution can be developed faster and at less cost than previous conventional methods, and a fairly accurate representation of the color layout can be approved prior to release for prepress and printing.

Creative processes (Fig. 3) have four distinct workflow steps: conceptualization, electronic design and initial layout, preparation of graphic and photographic elements, and finalization of electronic mechanical. During conceptualization, one or more creative solutions are generated. Known as loose comprehensives or comps, these solutions representing text, headlines, and graphic elements can be prepared either conventionally, ie, by pen and pencil or color marker sketches, or electronically, by electronic sketches. After the sketches, the creator begins to develop a tight comprehensive of the design. Text and headlines are typeset and placed, along with existing graphics, illustrations, or photographs. This first layout is output in either black and white or representative color for review and approval of the design and layout by the client. In the third step, supporting graphic elements used within the design are created using drawing software applications. Photographs are taken and provided for scanning and placement. In the last step, all text and graphic elements are finalized and placed in position, and an electronic file is prepared for output.

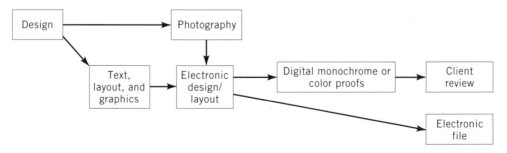

Fig. 3.   Creative process workflow steps.

Whereas each step has a unique objective, electronic tools allow the steps to be accomplished totally by the creator, thus changing the roles and responsibilities across the printing supply chain.

Synthetic Image Creation. Two types of synthetic image creation programs are available: bit map-based and vector-based (object oriented). Ultimately, all digital images are converted into bit-mapped (raster) images for display or output. The distinction between bit-mapped and vector is the form of the image in the creating application or program. Digital image creation can best be understood from the concept of bit-mapped images (4).

Bit-Mapped Images. A bit map is a grid pattern composed of tiny cells or picture elements called pixels. Each pixel has two attributes: a location and a value or set of values. Location is defined as the address of the cell in a Cartesian, ie, x and y coordinate, system. Value is defined as the color of the pixel in a specified color system. Geometric qualities of images are a function of the location attribute, ie, the finer the grid pattern, the more precisely can the geometric qualities be controlled. Color qualities are a function of the value attribute, ie, the more bytes of computer memory assigned to describe each pixel, the more precisely can the color qualities be controlled.

The fineness of the grid pattern, called resolution, is quantified in terms of pixels per linear distance. Common practice for commercial quality images assumes that photographic images can be reproduced accurately using about 300 pixels per inch (120 pixels/cm).

The number of bytes, each of which contains eight binary bits, assigned to describe each pixel dictates the number of colors that can be represented by the pixel. Assigning one bit to each cell allows two color values, eg, 0 or 1 (black or white). Assigning two bits allows four (2^2) color values, three bits allows eight (2^3), eight bits allows 256, and so on. Whereas 256 is not enough colors for commercially acceptable color photographic images, it is more than enough for commercially acceptable black-and-white photographic images, providing 256 shades of gray.

Assigning three eight-bit bytes, one byte for each of the additive primary colors, red, green, and blue, to a single pixel provides 16,777,216 colors (256^3). This arrangement, called 24-bit color, provides commercially acceptable color photographic images. Hence, depending on the number of bytes assigned to the value, images can be represented from monochrome through true color.

Creation. Users create bit-mapped images by accessing each cell in the bit map and assigning a color value to it. Software programs provide users with the tools to accomplish this, making the task easier, faster, and more intuitive. In effect, the user draws on the computer screen, using the electronic equivalent of conventional drawing tools. Because there are only two attributes associated with each cell, ie, location and value, the software tools allow the user to manipulate either the location attribute, the value attribute, or both. So many different tools have been developed that the effects available to the user are limited only by the user's imagination. For example, a program may have a fill tool that allows the user to fill a specified area with a desired color.

Vector Images. Conceptually, a vector image is not an image at all, but a set of equations defined by a user that directs the way a computer creates or alters a raster image. By storing only the equations and not the pixel information, vector images take up much less space than bit-mapped images. Moreover, they are also resolution-independent, because resolution is not specified until the vector image is converted to a raster image. Resolution independence is an advantage when a user desires to enlarge an image, whereas bit-mapped images lose information, ie, become lower resolution, when enlarged.

Vector-based programs provide tools for the user to define the equations in simple, intuitive ways. In fact, most users are unaware that their instructions to the computer are actually being used to define equations. For example, if a user wants a line across the top of the page, the line shows up on the screen as the user expects. However, the software has done more than simply display the line. The vector-based program in effect creates and stores an equation that defines the line. Complex definitions are, of course, required for more complex images. Typographic or typesetting programs are common examples of vector-based graphics.

Software tools are used to modify images in just about any way imaginable. For example, cloning allows a user to copy pixels from one portion of an image and place these in a systematic way in another portion. The principal application of cloning is to remove parts of images. For example, by cloning a background over an undesired image, the subject can be removed from the picture while retaining a natural appearance. Color correction functions allow a user to alter the color of the image. Typically, the user can make the image lighter or darker, or alter the red, blue, and green channels to change the color in a desired way. Sharpen and blur functions allow the user to alter an image to emphasize or de-emphasize portions of it. Images can also be resized, cropped, or inserted into other images.

Natural Images. Natural images typically originate as photographs and are invariably represented in some type of bit-mapped form. They are similar to synthetically created bit-mapped images except they are defined by a scanner that samples the light reflected from a photographic print.

Merger of Natural and Synthetic Images. Natural images and synthetic bit-mapped images can be merged directly to form a single image. A common software tool, for example, allows cutting of images from one file and pasting of the same image into another file. Such tools allow the user to define both the location and the amount to cut and the location to paste.

For a long time, the printing industry struggled to combine text (vector) and graphics (bit-mapped or natural images) capabilities in the same prepress system. This problem was solved with the advent of the PostScript (Adobe Systems, Inc.) page description language, which is capable of handling both bit-mapped and vector images, eg, a page that contains images scanned or created in a bit-mapped program and text or vector-based linework created in a page layout program. The two image types are not merged into a single file, but instead are processed separately as two separate layers until the image is converted to a raster image for final output. At that point, all layers are converted to bit-mapped format and merged together.

Creative Tools. Software applications, designed to perform specific functions, have changed the methods of producing a layout from conventional cut-and-paste techniques to electronic ones. Image creators are likely to have available powerful desktop publishing capabilities, an array of sophisticated software applications, and digital monochrome or color printers for digital proofing and review. Software applications allow the creator to produce illustrations and graphics, retouch photographs, compose and create page layouts, and perform prepress techniques such as placing high resolution images and other graphic elements into a page design or layout. Photographs and illustrations provided as hard copy can also be scanned and placed within the page layout as low resolution images and output to a digital printer.

Image Capture

Color and Color Separation. In 1860, James Clerk Maxwell discovered that all visible colors could be matched by appropriate combinations of three primary colors, red, green, and blue (RGB). His experiment involved mixtures of colored lights added together to produce other colors or white light. This additive color is well represented by the primaries RGB. Indeed, human color vision is trichromatic, ie, human visual response approximates receptors for the colors recognized as red, green, and blue (see COLOR).

Printers use colored materials, eg, inks (qv), that absorb or subtract regions of the visible spectrum from white light. Subtractive color is usually represented by the three printer's primaries: cyan, magenta, and yellow (CMY). Cyan absorbs red light, magenta absorbs green, and yellow absorbs blue light.

The nature of printing is such that each color of the primary set must be printed sequentially, one on top of the other, building up a color image one color at a time. The basic task of color prepress work is converting a color image into color separations that represent each of the CMY color components of an image. In addition, the color planes must be adjusted for the characteristics of the printing system. Inks used in printing do not represent the ideal CMY primaries. The inks are made from real pigments and must satisfy many other requirements, not the least of which is cost. The colors show overlap in their visible spectra, referred to as color contamination. Thus areas that are printed by using equal amounts of CMY, instead of appearing neutral, ie, grey or black, are actually colored. Also, the maximum amount of ink that can be printed plays a role. Ink quantity does not allow for a true deep black. In addition, it restricts the color gamut that can be achieved on a press.

Some of the color deficiencies can be overcome by using a fourth ink, black, which allows printing neutral tones and dark blacks and colors. Black also improves the contrast of an image and its apparent sharpness. Black, usually referred to as K to distinguish it from blue, makes up the fourth member of the printer's primaries, CMYK.

Although adding black increases the color gamut somewhat in the darker colors, it does not help in other regions of the spectrum. Compensation is made in the color-separation process by adjusting the tone of an image to provide a pleasing rendition, but not necessarily matching the original color.

Color contamination is another factor that needs to be considered. For example, a small amount of cyan commonly contaminates magenta. Thus, adding magenta to color-correct an image also adds cyan, which must be removed for compensation.

The color-separation process, then, becomes one of taking an image, usually analyzed in the RGB primary color space, and converting that image to the

CMYK primary color space, while accommodating the deficiencies of the printing process. This is a highly nonlinear problem, which, because of the four channels, does not have a unique solution. Much effort has been expended, both in the traditional photomechanical process of color separation, and in the newer electronic systems, to achieve a goal of press-ready separations. Much of the workflow in a prepress shop is centered around separating color and checking the accuracy and suitability of the separations for printing. Reviews of color reproduction in printing (5) and electronic color separation (6,7) are available.

Color Scanning. Historically, the process of color separation was done by using a process camera. A colored original was placed in a copy frame and photographed three times, once through each of a red, a green, and a blue filter. The red filter image produced the cyan plate, the green produced the magenta, and the blue the yellow. An additional image is created to make a black plate using white light (8,9).

The color-separation process has been automated through the use of electronic color scanners. These devices convert color images to electronic data, which are used either to image directly monochrome photographic film that represents the cyan, magenta, yellow, and black content, or to provide CMYK or RGB image data files to computer-based pagination and retouching systems.

Color scanning technology can be grouped into two principal categories: photomultiplier or linear charged-coupled device (CCD) array. These are the actual components within the scanners that convert light to voltage (see PHOTODETECTORS). White light is passed through a spot or line on a transparency or reflected from a photographic print. The resulting light, which has been altered, taking on the color of the spot through which it has passed, is then passed through red, green, and blue filters and onto either photomultipliers or a CCD array where a voltage, proportional to the amount of color in the pixel being analyzed, is generated. These voltages are converted to dot percentages that range from 0 to 100%, where 0% represents no color, ie, white, and 100% represents a solid color. The spot or line of light is moved (scanned) across the entire image. After scanning, the image ends up as a rasterized array of numerical values representing the color value of each pixel in a file typically containing 32 bits (eight bits per color) of data per pixel (10–12).

One of the benefits of color scanning equipment is that is allows corrections to be made to an image during the image capture process. As each pixel is analyzed it is passed through a series of look-up tables or mathematical algorithms that alter the amount of each color (CMYK) based on operator instructions. The benefit of this process is that color can be adjusted as the image is captured, thus dispensing with the need for post-scan processing.

Process Camera. Although replaced by the digital scanner and imagesetter to a large extent, a process camera is still used in many shops to convert originals to film. The camera consists of a movable glass-covered copyboard to hold the original, a movable lens board, and a stationary vacuum back to hold the film. A bellows connects the lens board to the vacuum back. The degree of enlargement or reduction is governed by the lens and the distance from copyboard to lens and lens to film plane. The camera may be set up vertically or horizontally. The vertical camera's optical axis is perpendicular to the floor and has the advantage of compactness over the horizontal camera, in which the optical axis is parallel. Owing to a larger range of copyboard movement, however, the horizontal camera has a greater range of enlargement and reduction and is usually designed for larger original and film sizes (13,14).

Image Assembly

Page Creation. *Traditional Methods.* In the preparatory stages of printing, the traditional process starts with one or more originals in the form of written matter, line art (diagrams, drawings), and photographs (1,14). These originals are photographed to make films that, in a final step, are used to expose a printing plate. The conversion process often includes changing the size of the original to the size required on the page. Because most printing processes cannot print continuous-tone photographs directly, these photographs are converted to halftones in which the intermediate tones of the original are represented by solid dots on the film, which are spaced equally but vary in area (Fig. 4).

Halftone Screens. Halftone screens are used in cameras to convert a continuous-tone photograph to halftone dots (Fig. 5). Most screens, sometimes called contact screens, are made from film and consist of evenly spaced rows of vignetted dots. The screen ruling, usually expressed in lines per centimeter, governs the level of detail of the printed halftone. Finer rulings give greater detail.

(a)   (b)

Fig. 4. Example of a halftone illustration. The circle in (**a**) indicates the enlarged area shown in (**b**), which reveals the image to be composed of halftone dots.

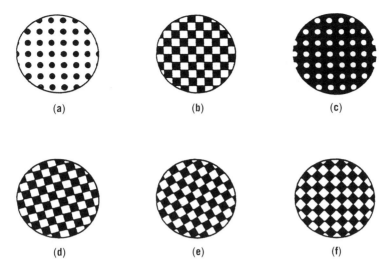

Fig. 5. Halftone dots magnified: (**a**), (**b**), and (**c**) are 20, 50, and 80% dots at 0°, respectively; (**d**), (**e**), and (**f**) are 50% dots at 15, 30, and 45°, respectively.

Newspaper halftones are generally produced by using a screen ruling between 25 and 50 lines per centimeter. Magazine halftones are produced with screen rulings between 50 and 60 lines per centimeter. Screen rulings up to 120 lines per centimeter may be used for special purposes.

Besides changing size, additional photographic steps may be required to achieve the desired effect on the printed page. For halftone images, the color balance may have to be changed to achieve a visual match with original art work. Line art and text images may have to be manipulated to allow for registration variance on the printing press.

If the line art and text originals are sized correctly in relation to one another, these can be assembled together onto a stiff paper or acetate base to form a paste-up or mechanical and photographed as a unit. Line art or text that is not sized correctly is enlarged or reduced on the camera and assembly is done at a later stage. Continuous-tone originals are treated separately and assembled later.

To create the halftone image, the screen is contacted against the film in the camera. Exposure of the original through this screen creates a pattern of dots on the film that correspond in size to the different tonal areas in the original. The lightest tones of the original reflect the highest percentage of light, which penetrates most of the screen dot's vignette. In this area, a large dot is formed on the film. Darker tones reflect less light and penetrate less of the screen dot's vignette and result in a smaller dot being formed on the film.

Typesetting. Typesetting or typography is the process of arranging, composing, and placing type onto the printed page. Typesetting has evolved through three important stages: hot metal, phototypesetting, and electronic (8,13). The roots of typesetting began in hot metal with Gutenberg, who cast the earliest metal type used in Western civilization. Prior to metal, type for each individual printed piece was carved in wood, a slow and precise craft. Wooden type deteriorated after brief use.

Gutenberg developed a hand-held mold for casting individual type characters. These cast or blocks were then arranged, composed, and placed in the printing press (a letterpress). These metal casts made more impressions than wood before deteriorating, and new casts were quickly made. Lead alloys were the easiest and cheapest metal for use, and could be reused. Because each block resembled a single typeface and size, each different size and typeface required a mold and casting. Most of the terminology used in typesetting was derived from the hot-metal process, including face, feet, shoulders, body, and character.

The basic hot-metal process remained in use from Gutenberg's time to the late 1960s, but over the centuries became automated with the advent of equipment such as the Mergenthaler Linotype. This equipment allowed an operator to enter characters through a typewriter-like keyboard, which caused type to be cast one line at a time. These lines, or slugs, were then placed together to form a page.

Phototypesetting represented an easier way to compose type. Early phototypesetters used an optical process, whereby a disk of characters, in different sizes and typefaces, was spun under computer control. Each character was projected in turn onto photosensitive film or paper. This was followed by systems where characters drawn on a cathode ray tube (CRT) exposed the photosensitive material. In each case, the operator interacted with the system at a video screen that only showed the characters of the text (the information content) and codes that indicated how the characters were to look on paper. An experienced operator was required to obtain high quality results.

Conventional typesetting has been largely superseded by electronic typesetting using the PostScript page description language. Fonts and typefaces in PostScript, as well as in a similar system, TrueType (Microsoft Corporation), actually exist as mathematical descriptions of the characters. This allows each character to be rendered at arbitrary sizes and resolutions for different output devices. Electronic typographic programs are usually implemented in such a way that the operator sees a close approximation of the final output on a computer screen, ie, what you see is what you get (WYSIWYG). The operator can visually compose a page, making best use of the myriad fonts at his disposal.

The page layout program, which is used to prepare the text, generates a representation of the printed page, including text in the PostScript language. A PostScript raster image processor (RIP) interprets the PostScript commands and renders the image into a bit map, which can then be output on a printer or imagesetter.

Digital Page Creation.   Historically, designers created the page, a layout artist recreated the page to exact specification as a mechanical, and then the prepress house, referred to as a trade shop, created the films for producing the printing plates. Individual page elements, type, and photographic images were transferred to color-separated film by the use of cameras. Strippers in the trade shop gathered the pieces of film for a given page and recomposed them for plate creation, one for each color printing plate that needed to be generated. As of the mid-1990s, many shops still use this technique, but usually the film has been electronically generated (8).

Using color scanners, photographs are digitized directly onto a computer system. These digital images are then positioned in an electronic page layout document that contains type and other page elements, created by the designer or production artist. Finally, the page is imaged to color-separation film or directly to the printing plate (4).

The pioneers of electronic page layout, ie, Du Pont-Crosfield, Linotype-Hell, Scitex, and Dai Nippon Screen, developed proprietary pagination systems to compose pages for color separation. Initially, only the largest commercial trade shops and printers used these complex, expensive systems. The conventional mechanical is sent to the prepress house, along with any photographic and type elements used in composing the page. Scanner operators input text, photographs, and other design elements into the pagination system. The operator digitizes the page by using the mechanical as an exact reference. Photographic images are scanned and added to the layout. Once the page layout has been established, it is imaged to film using an output recorder or imagesetter that exposes the individual colors to separation film or digital printing plates. The real electronic boom came as designers began generating final electronic pages using desktop personal computers.

Many of the tasks once performed only in a high end electronic pagination system, often called color electronic prepress systems (CEPS), are possible by using personal computers and work stations. Pages are composed on screen, and the results are output typically as PostScript files. The data are transferred over a network to the final output device.

Image Storage and Movement

Electronic Data Storage.   Images in the prepress industry range from one megabyte to hundreds of megabytes, thus data storage capacity is a primary concern. Modern color electronic prepress systems rely on high capacity magnetic hard disks for quick data storage and retrievel. These drives reach capacities of several gigabytes (see INFORMATION STORAGE MATERIALS, MAGNETIC).

Many different types of removable storage media are also used. These have the advantage of utilizing a relatively inexpensive storage medium used in the more expensive drive equipment. Removable media storage systems also have the advantage of providing a simple means of moving large data files from place to place. These systems typically have capacities of tens to low hundreds of megabytes for magnetically based media, and close to a gigabyte for optically based media.

Compact disk–read only memory (CD-ROM), employing technology similar to that of consumer audio compact disks, has become a commonly used medium for the distribution of large image files because of their large data capacity and low manufacturing costs. Magnetic tape is still used for inexpensive long-term bulk storage and backup of files stored on other media (see MAGNETIC MATERIALS). Computer storage technologies are changing rapidly. A good source for information to remain current is any of a number of computer oriented magazines (15).

Networks.   Image files are typically transferred among many different stages in the prepress process. This is most often done over a network that connects the various computer work stations used in the process. Although the subject of networks is beyond the scope of this article, some discussion is useful in understanding prepress workflow. The literature contains a more complete description (16).

At the lowest level, the network is the physical medium that connects the various pieces of equipment. This can be copper wire, often known as Ethernet, or optical fiber, ie, fiber-distributed data interface (FDDI). Networks allow transmission of data at nominal speeds of 10 to 100 megabits per second, depending on the physical medium used.

The next level is the protocol that governs how the data are transmitted over the wire. Many protocols are in use. A typical installation may have multiple protocols running simultaneously on the same physical network. Vendors of network hardware and software develop protocols that are optimized for the type of application for which their product is targeted. Among the protocols commonly seen in a prepress network are Apple Computer's AppleTalk and EtherTalk, and TCP/IP used by many UNIX work station vendors. The application software a user employs automatically uses whatever protocol is necessary to move information over the network.

The same concepts of a network as the local area network (LAN) in an office can be extended over a broader area, eg, wide area network (WAN), effectively to extend the network around the world. This type of network is becoming more important as prepress operations try to work more cooperatively with their clients, by accepting digital files over the network, or passing finished work back for approval.

The physical medium that links the LAN at each end of the network is typically supplied by a telecommunications provider. This can range from private networks, where the telecommunications company maintains physical links among the desired locations, to switched services such as switched multimegabit data services (SMDS), where high speed data links are made and broken as needed, much like conventional telephone connections. These services can range in speed from a few kilobits per second up to about 45 megabits per second. Again, these represent the physical link layer, and may support multiple transmission protocols such as frame relay or asynchronous transfer mode (ATM). In any case, to the user the result is an extended network that appears to be directly connected to the local network. The user is able to move images among locations as easily as moving them on a personal computer.

Image Compression.   Despite the increasing capacity of storage media and speed of LANs and WANs, storing and transmitting large image files represent a significant expense. Many schemes to compress image files have been developed. Data compression encodes the information contained so that the resulting file is smaller. Two general types of compression are used.

Lossless compression preserves all of the information in a file, so the original file can be reconstructed bit for bit. This is usually accomplished by replacing data redundancies by tokens that represent the data. An example would be an image where much of the background is white. Most of the white pixels could be substituted by tokens representing the number of white pixels replaced. Because the tokens are much smaller than the runs of white pixels, the file is compressed. Lossless compression, represented by run length encoding (17) or Lempel-Ziv-Welch methods (18), are capable of moderate levels of compression, often about two or three to one. These do not provide much compression of natural images, which often do not contain much redundant information.

Lossy methods, on the other hand, do not preserve all the information. The original file cannot be reconstructed bit for bit. Lossy compression takes advantage of the insensitivity of the human eye to rapid changes in color in an image. These methods deliberately remove high frequency color information, but maintain all of the luminance (lightness) information where the eye is most sensitive. These methods, the best known of which is called JPEG, for the Joint Photographic Experts Group that developed the standard, are capable of compression ratios of twenty to one or more, depending on the amount of image quality degradation that can be tolerated (19).

Image Files.   Only the physical problem of moving or storing files has been discussed herein. The format in which the data is actually stored or transmitted is another part of the process. There are a number of file formats, many of which are proprietary, developed to suit the needs of a particular computer platform or software application. There has been some effort to standardize to a few file types, but many others are in use. A thorough discussion of graphics file formats is available (20).

Image Manipulation

After creating the set of halftones, line art, and type for a given page, the craftsperson may have to manipulate the images to produce the page as designed. These manipulations, which depend on the type of printing plate and press to be used, include changing negatives to positives or vice versa, combining images, duplicating images, changing image orientation from right reading to wrong reading or vice versa, correcting color balance, and making line or text images thicker or thinner to provide overlap between colors during printing (14).

Most often the manipulations require making contacts by placing unexposed film with the film to be contacted into a vacuum frame. Contact is maintained between the two films by evacuating the air between a plate covered with a corrugated rubber blanket and a sheet of plate glass (qv). Originally, orthochromatic films exposed with low power tungsten lamps under red safelight conditions were used. To enhance productivity and worker comfort, films are now formulated to work under bright yellow or white light in normal room conditions. Metal halide or quartz halogen lamps having high ultraviolet output are used as exposure sources. Negative working contact films and positive working duplicating films are available.

Once films corresponding to all of the separate components of a page, eg, text, line art, and halftone images, have been made in their final forms, they are assembled onto a carrier sheet to create a flat. More than one flat are necessary for complex or multicolor pages. This stage of image assembly is commonly known as stripping, because this process at one time involved stripping film emulsions from a glass plate (8). The flat is used to expose the printing plate to form a printable image. For negative working printing plates, negative film images are set into windows of a plastic carrier sheet to create the negative flat. The carrier sheet is colored to block actinic light so that only the clear areas of the film appear as images on the plate. For positive working printing plates, positive film images are positioned onto a clear plastic base to create the positive flat.

Positioning the various pieces of film is critical. Stripping tables are used to provide an illuminated surface to position films on the carrier sheet. The working surface is usually a sheet of plate glass frosted to diffuse light from fluorescent tubes underneath the glass. Often the edges of the stripping table are machined to provide straight edges for T-squares. For more critical work, precision layout tables having micrometer-adjustable straight edges are used. Flat-to-flat registration is provided by pin punch systems.

A special form of image assembly, used especially in label making, involves creating one image as a flat and using that image to make multiple identical images on different places of either a film or the printing plate itself. Step-and-repeat or photocomposing machines automate this task (13).

Electronic Film Output.   Imagesetters are precision electromechanical devices that image monochromatic color separations onto photographic film or paper. Each separation is used to expose a printing plate for a single ink color (4).

Imagesetters.   All modern imagesetters use a laser light source to expose the media. Imagesetters should not be confused with color film recorders, which use similar technology to expose full-color images onto a single piece of film. Many different laser technologies are in use, each requiring different media sensitivities (see LASERS). Helium neon, HeNe, lasers produce a visible red light (633 nm) and use media that must be handled under green safelight conditions. Argon ion, Ar^+, lasers produce blue light (488 nm) and allow more convenient amber safelights to be used. Other laser sources include far red laser diodes (785 nm), frequency-doubled yttrium aluminum garnet (YAG) lasers (532 nm), and infrared laser diodes (826 nm). The light produced by these lasers is focused to a tiny spot and modulated to control the placement of marks on the medium. In general, laser diodes can be modulated directly by controlling current flow to the diode, whereas gas lasers, eg, argon ion, require external beam modulation through an acousto-optic modulator (AOM) or similar device (see LIGHT GENERATION).

Three principal imaging technologies are in use as of the mid-1990s. External drum imagesetters use vacuum or mechanical means to affix a sheet of media to a cylinder. The cylinder rotates at a high speed, typically 300–1000 rpm, while the laser head slowly traverses the width of the drum. An advantage of external drum imagesetters is that multiple laser beams can easily be fitted to the imaging head, thus increasing the film exposure speed.

Internal drum imagesetters affix the film to the inside of a semicircular shape. The film typically covers between a 180 and 270° arc. Both the drum and the film remain stationary during the imaging process. A single laser beam is aimed down the axis of the drum at right angles to a spinning mirror, which deflects the beam onto the medium. The spinning mirror slowly traverses the length of the cylinder to allow exposure of the complete sheet of film or paper.

Capstan imagesetters, the third common technology, slowly advance the film or paper across a flat surface. A single laser beam is scanned across the width of the medium by a rotating polygon or resonant mirror.

All of these technologies require careful attention to the film transport mechanism to avoid imaging defects. The sizes of the features produced on the film, typically resolution of 1000 spots/cm or greater, and the accuracy of spot placement both require high precision equipment. Even minute misplacements of spots can produce objectionable visible artifacts.

PostScript. Designers use electronic page composition systems to create pages, but imagesetters plot only monochromatic film separations. Imagesetter output requires that the data be converted from full-color pictorial information to its individual color components. Additionally, the data must undergo a significant format conversion. Electronic page composition systems represent data in a relatively low resolution format, typically 120 pixels per centimeter for desktop printing and 28 pixels per centimeter for video monitors, using eight bits each of cyan, magenta, yellow, and black data for each pixel. Because most imagesetter lasers are binary devices, where the laser beam is either on or off, the data must be converted to monochromatic format with only one bit representing each pixel. Imagesetters compensate for the loss of binary levels by operating at a much higher resolution, up to 2000 spots per centimeter, and using area coverage algorithms to represent different shades of gray as dots of varying area. These rows of dots, made up of the much smaller imagesetter spots, are referred to as screens for historical reasons. Screens typically consist of rows of dots having a resolution of 33 to 120 lines per centimeter.

The PostScript page description language was developed by Adobe Systems, Inc. in the 1980s to provide a common format for describing page contents (21). It has been adopted almost universally by desktop publishing vendors, and many of the high end proprietary page makeup systems (Du Pont-Crosfield, Scitex, Linotype-Hell, Dai Nippon Screen) also support PostScript.

PostScript, an intermediate language, is converted into the appropriate imagesetter data stream by a device called the raster image processor (RIP). The RIP, which can be either a dedicated hardware device or a software program that runs on a standard computing platform, typically operates in two stages. First, it processes the entire PostScript file and builds a display list of all images, text, and linework elements that comprise the page. This display list is independent of the characteristics of the imagesetter. Then, the display list is screened at the resolution required for the imagesetter, and a binary data stream (bit map) is created for each color separation.

Conventional Screening. The primary objective of screening is to use area coverage algorithms to represent different shades of color, despite the fact that the imagesetter laser is typically a binary (on/off) device. The biggest difficulty in screening is the prevention of moiré patterns, which are interference patterns that occur when two or more color separations overlap.

It can be mathematically shown that no moiré patterns occur when the rows of dots in each color separation are rotated exactly 30° apart. Color printing, however, is a four-color process. Only three colors can be assigned screens that are exactly 30° apart, because a screen at 120° is the same as a screen at 30°. One color must be assigned an angle that is only 15° apart from two other screens. This introduces a moiré pattern. The yellow plate is usually assigned this angle because its lighter moiré patterns are less objectionable than moiré patterns in the cyan, magenta, or black plates (22). A typical set of screen angles is therefore 0, 15, 45, and 75° (see Fig. 5). Another problem is that these angles do not line up exactly with the rectangular grid of spots that is exposed by the imagesetter.

Complex screening algorithms are used to minimize artifacts from the mismatch between the recorder grid and the desired angles (4).

Stochastic Screening. Stochastic screening (23), also called frequency-modulated (FM) screening, is another way to eliminate problems with moiré. Instead of using a fixed angle for each screen and varying the size of each dot, a random distribution of small, fixed-size dots is used. Different amounts of color are represented by more or fewer dots in a particular region.

Stochastic screening can produce images that rival photographs in their quality level. However, the small dots used by stochastic screening can cause problems in proofing, plate-making, and printing. It remains to be seen whether stochastic screening is to become a permanent part of the printing process.

Image Approval

At this stage of the printing process, it is desirable to check the position and appearance of the text and color images. The process check can also serve as the standard by which customer approval of the completed print order is defined. This step in the overall process is called proofing and can be carried out by analogue or digital methods, depending on the overall printing process steps in use.

For both analogue and digital prepress processes it is frequently necessary to check the appearance of an image and then to gain customer approval to proceed with the expensive step of image duplication by printing. For this purpose, proofs are made at several steps of the workflow (see Fig. 2).

Analogue Proofing. Analogue proofing can be described as the process of making an image, either monochrome or color, by photomechanical methods for the following purposes. (*1*) Production or quality control within the shop. Production procedures prior to plate-making, such as scanning and stripping, are areas of frequent change or opportunities for error in the printing process. A proof can be used to evaluate these procedures and errors can be corrected prior to the expensive process of making final plates for the press. Internal uses within the printing operation for analogue proofs include scanner setup, color correction, check on stripping procedures, press guides, quality assurance, and bindery operation setup. (*2*) Communications medium between the trade shop, printer, and customer who desires the print job. The parties producing and desiring a printed image often need a means of visualizing and understanding what is said. A proof is a tool that can be used to bridge the communications gap and ensure that the printer is producing what the customer wants. (*3*) Contract or contract proof between the parties producing the print and the customer. The proof, a representation of the appearance of the final image off-press, can serve as a contract to protect all parties should a significant discrepancy occur in the final image from the press. Photomechanical off-press proofing systems yielding halftone proofs have enjoyed wide usage as a result of their consistency and accuracy in predicting the color obtained from the press and the high cost of press proofing in terms of materials, labor, and press ready-time to make color images on a press. In addition, the ready acceptance and confidence of the customer base, in viewing and evaluating off-press proofs as an integral feature in the image approval process, has steadily progressed over time.

Color-Proofing Methods. There are three principal analogue color-proofing methods commonly used in the industry. (1) Monochrome proofing. A monochrome proof is a single-color image on paper or polyester film that depicts image placement, imposition, limited registration detail, and production errors. Some commercially available monochrome systems are Dylux (Du Pont), CopiArt CP-3 (Fuji), and Dry Silver (3M). (2) Overlay proofing. The proofs are an assembly of individual monochrome images on individual polyester film sheets, typically yellow, magenta, cyan, and black colors, which can be overlaid in register to produce a four-color image representing information from each color separation generated from the scanner. The proofs can be used as progressives by the printer in the form of one-, two-, or three-color overlays; checks on registration by the stripper; and for lower quality color approval work. Distortions in color perception arising from light reflections off the polyester image-bearing sheet detract from overlay proofing as a tool for accurate color prediction. Some commercially available overlay systems include Color Key (3M), Cromacheck (Du Pont), and NAPS/PAPS (Hoechst-Celanese). (3) Surprint proofing. These proofs are the highest quality four-color images available for off-press proofing uses. The proofs are assembled in such a way that the individual colored images lie directly in contact with one another and, unlike overlay proofs, are not capable of being leafed through or pulled apart into individual images after the assembly process. Surprint proofs provide the best print prediction or simulation of color obtainable from the press and are used extensively in the color approval process as the contract proof between the trade shop, printer, and customer. Some representative, commercially available surprint proofing systems are MatchPrint (3M), Cromalin (Du Pont), WaterProof (Du Pont), Color Art (Fuji), and PressMatch (Hoechst-Celanese). A proof can be made in the printing operation anytime a question arises as to the appearance or quality of the image to be produced by the press. The proof approval process can be complicated because of the many iterations for color acceptance by the customer. The separations used to make the proof can be color-corrected many times before it is shown to a customer; even then the customer can request more changes, which may necessitate the remake of the separations. In any event, the product of the process is a set of final film separations to make the printing plates and a final proof to be used as the guide and contract for the printing process.

Monochrome Proofing. Several monochrome photoimaging systems can be used to produce a single color for proofing applications. The photoimaging ingredients are coated on paper and/or film and can be used for proofing and/or as a registration master. The chemistries used for color formation are complex reaction schemes involving the oxidation of leuco dyes by a biimidazole dimer or the photolysis of polyhalogenated compounds upon exposure to uv light (360 nm). Du Pont's Dylux is typical of the oxidation of a leuco dye and the key reactions can be represented by the following, where L_2 is a biimidazole dimer (**1**); $L^•$, the biimidazole radical; RH, a leuco dye (**2**); and R^+, a dye.

$$L_2 \xrightarrow{h\nu} 2\, L^•$$

$$L^• + RH \longrightarrow RH^{+•} + L^-$$

$$2\, RH^{+•} \longrightarrow R^+ + RH + H^+$$

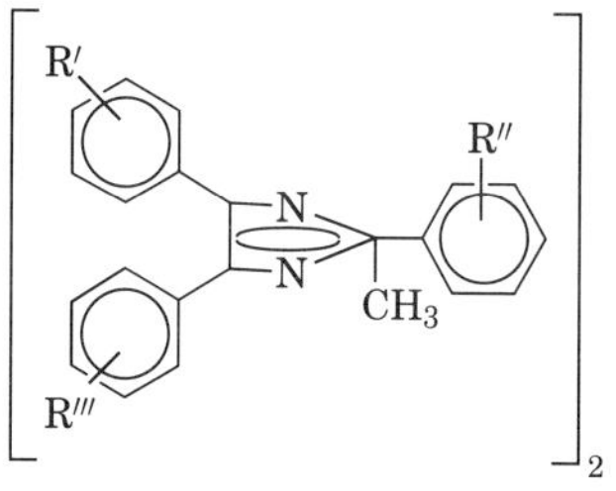

(1) 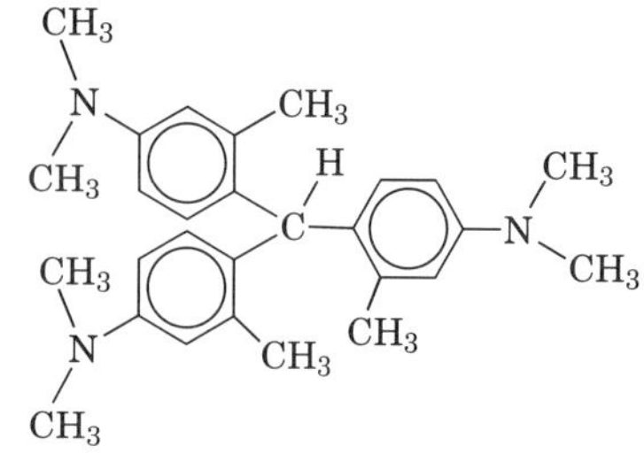　　　　　　　　　　　　　　　(2)

After imagewise exposure to uv light through a negative separation, a deactivation (or fixing) exposure using only visible light is recommended to stabilize the image by the photoreduction of quinone. The photo-oxidation of the leuco dye and photoreduction schemes have been described (24–28). A unique feature of Dylux is the dual response, which can be used to generate both positive and negative images in either a deep blue or gray color. The intensity of the color image is controlled by the amount of exposure. Thus color breaks, ie, simulation of two or three density levels in a multicolor job, can be performed by the customer. Some other important characteristics of this monochrome proofing system include an intense, blue image to aid register and alignment; an image on paper that can be folded and marked; and room-light handling.

The photolysis of polyhalogenated compounds forms the basis for another monochrome system. Iodoform can undergo photolysis to produce hydrogen iodide, which subsequently reacts with a di(2-furfuryl) derivative (3) and aromatic amines to produce a colored dye adduct (4) (29). The photolysis scheme and subsequent reactions can be shown by the following:

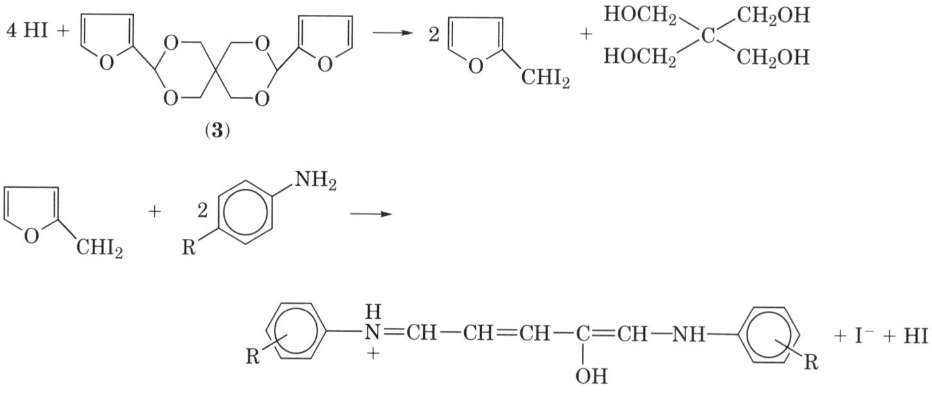

(3)

(4)

Another halogenated photolysis (30), using carbon tetrabromide to produce hydrogen bromide and subsequent reaction with spiropyran (5), produces a highly colored spiropyrilium bromide salt.

$$CBr_4 \xrightarrow{h\nu} CBr_3^{\bullet} + Br^{\bullet} \xrightarrow{[H]} HBr + HCBr_3$$

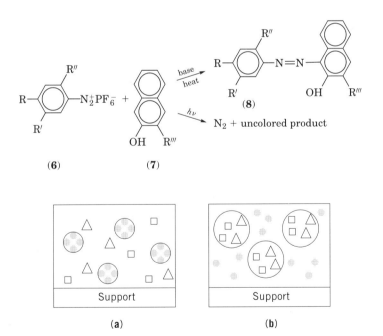

(5)

Thermally processed silver systems can also be used as monochrome proofing products. Dry Silver papers and films, as developed by 3M, are exposed to light to form a stable, invisible latent image, which upon heat processing at 126–138°C for about 20 seconds produces a high resolution black image. The chemistry involves the reduction of silver ion in the latent image by hydroquinone or a similar substance. Several references are available that detail the chemistry (31–34). Dry Silver coatings can be applied to paper and polyester films to provide opaque or translucent images.

The Fuji CopiArt monochrome proofing system is based on the photogeneration of color from leuco dyes or diazo-coupling (35). CopiArt includes both positive and negative working systems (Fig. 6). For the positive working system, a diazo compound (**6**) reacts with a coupler (**7**) as shown.

Fig. 6. CopiArt proofing system: (**a**) structure of positive working system, where △ = organic base, □ = coupler, ○ = microcapsule, and ◉ = diazo compound; and (**b**) structure of negative working system, where △ = photoinitiator, □ = leuco dye, ○ = microcapsule, and ◉ = radical quencher.

Compound (**8**) is an azo dye (see AZO DYES). The color-forming ingredients in the positive imaging system are microencapsulated and diazo-based. During imagewise exposure through a positive separation, nitrogen is released from the diazo compound so that coupling, which gives rise to color, cannot occur in the areas struck by light. Further heat processing allows the coupling agent, which is outside the microcapsules, to interact in unexposed areas to form a color image.

The negative system incorporates a coating containing microencapsulated image-forming ingredients applied to paper. The colored image is formed by the oxidation of a leuco dye within the microcapsules upon imagewise exposure through a negative separation. The image is further processed by heating to stabilize (fix) the system. Heating allows the nonencapsulated fixing agents to stabilize the image.

Chemical reactions for the negative systems in CopiArt can be generalized by the reaction of the leuco form (**9**) of crystal violet [548-62-9] to produce the colored form (**10**).

Overlay Proofing.   Overlay proofing systems can be categorized as wet- or dry-processed systems. The negative working wet-processed systems are generally composed of polymeric diazo resin salts (halides or heavy metal), which after photolysis form an insoluble adduct.

Wet-processed systems (36–38), such as Color Key (3M) and NAPS (negative-acting proofing system) (Hoechst-Celanese), are comprised of films containing pigments and/or dyes that are dispersed in nonreactive, film-forming polymer systems along with the diazo polymer resin. After exposure to uv light, through a negative separation, the film sample undergoes a wash-off process of the unexposed material using an aqueous solvent or aqueous alkaline developer, yielding the desired negative image. Each process color, ie, yellow, magenta, cyan, or black, on a polyester support is processed in the manner described and overlaid in register to form the overlay proof.

The negative working dry-processed Cromacheck overlay proofing system from Du Pont (39) is based on adhesion balance. After exposure and peel-apart development, the exposed areas of the pigmented photopolymer layer become selectively attached to the polyester coversheet by cohesively breaking away from the integral pigmented photopolymer layer. After peeling apart, the unexposed pigmented photopolymer layer remains on the tacky adhesive. Figure 7 illustrates the image development process.

The right-reading, pigmented photopolymer image from the negative separation remains attached to the polyester coversheet. A simple exposure, through the appropriate separations followed by peel development and overlay of the polyester sheets in register, gives rise to the overlay proof. The dry-processed system is convenient, easy to use, and environmentally friendly. Disposal of waste developer is not an issue.

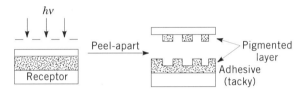

Fig. 7. Image development process.

Surprint Proofing. The high quality surprint proofing systems used by the printing industry generally give rise to a laminated structure on a paper or plastic receptor bearing the colored image. The laminated structures are composed of four individual laminations, one for each process color, plus, in some instances, a protective topcoat. Surprint proofing systems can be either positive or negative working and wet- or dry-processed to yield the desired high quality four-color image. Du Pont pioneered the development of dry-processed photopolymer surprint proofing systems, Cromalin, using colored, powdered toners to represent the printed image. The positive system relies on a difference of tack between the exposed and unexposed photopolymer areas and the acceptance of toner by the tacky areas to yield a positive colored image (40) as shown in Figure 8**a**. Sequential laminations, exposures through the appropriate separations, and application of the process-colored toners provide a positive proof. The photopolymer layer contains a complex mixture of monomers, plasticizers (qv), initiators, sensitizers, and polymeric binders to achieve the necessary tack/adhesion balance.

The photopolymerization process taking place within a representative mixture of sensitizer, initiator, chain-transfer agent, and monomer, typical of positive Cromalin, has been studied in detail (41,42). The exact mechanism is still controversial, but a generalized reaction scheme can be postulated as follows, where L_2 = biimidazole dimer, S = sensitizer, RH = chain-transfer agent, L_2^* = excited biimidazole dimer, $L^\bullet$ = biimidazole radical, S^* = excited sensitizer,

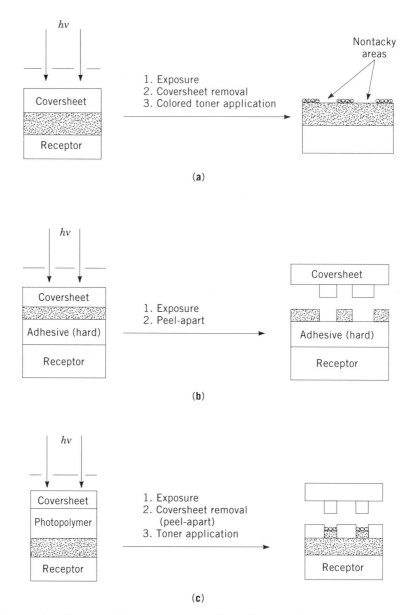

Fig. 8. (**a**) Positive proofing image formation; (**b**) positive working dry peel-apart process; and (**c**) peel-apart process.

$R^\bullet$ = chain-transfer agent radical, M = monomer, $M^\bullet$ = monomer radical, and P = polymer.

$$S \xrightarrow{h\nu} S^*$$

$$L_2 + S^* \longrightarrow S + L_2^*$$

$$L^\bullet + RH \longrightarrow R^\bullet + LH$$

$$L_2^* \longrightarrow 2L^\bullet$$
$$L^\bullet + RH \longrightarrow L^- + RH^{\bullet +}$$
$$RH^{\bullet +} \longrightarrow R^\bullet + H^+$$
$$R^\bullet + M \longrightarrow RH + M^\bullet$$
$$M^\bullet \longrightarrow polymer$$

The dry, peel-developable process has been extended to positive working, precolored surprint proofing systems, eg, Eurosprint (Du Pont) (43). The system contains two basic layers: a pigmented photopolymer or pigmented diazo layer, and a hard thermoplastic adhesive composed of mostly poly(vinyl acetate). The proof-making process is to laminate, expose (through a positive), and peel-develop successively the films corresponding to the appropriate colors desired for the proof. The unexposed pigmented layer remains behind on the adhesive and the exposed pigmented material remains on the coversheet when exposed with a positive separation.

The dry-processed, peel-apart system (Fig. 8**b**) used for negative surprint applications (39,44) is analogous to the peel-apart system described for the overlay proofing application (see Fig. 7) except that the photopolymer layer does not contain added colorant. The same steps are required to produce the image. The peel-apart system relies on the adhesion balance that results after each exposure and coversheet removal of the sequentially laminated layer. Each peel step is followed by the application of the appropriate process-colored toners on a tacky adhesive to produce the image from the negative separations. The mechanism of the peel-apart process has been described in a viscoelastic model (45–51) and is shown in Figure 8**c**.

Negative-working, peel-apart films incorporating pigments in the photopolymer layer can also be used for proofing applications (52). The exposure process causes a reversal of the image in the sense that the unexposed pigmented material is removed with the coversheet upon peeling. The exposure causes a photo release of the exposed material from the coversheet and the nonexposed material is removed during the peel step as shown in Figure 9**a**.

The process used to make the proof involves the same lamination, exposure, and peel-apart steps shown before. Different photoactive systems having variants on the peel-developable system, ie, dual exposures to achieve a right-reading image on receptor or transfer processes, are reported in the patent literature (53,54).

The wet-processed negative surprint systems (55,56) introduced by 3M, Fuji, and others are generally composed of various types of diazo chemistry in a thin pigmented film. The negative systems rely on the solubility of the diazo salt resin in the unexposed state to effect image development. After sequential laminations, exposures, and aqueous base wash-off steps with the individual colored films, the proof consists of exposed diazo and film-forming polymers. The wash-out systems used to process the products are generally aqueous, alkaline developers (Fig. 9**b**).

Newer systems have appeared on the market, in which tap water is used as the developer, thus making disposal of spent developer easy for the customer. The WaterProof surprint proofing system from Du Pont is

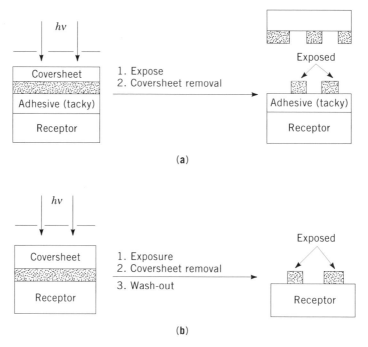

Fig. 9. (**a**) Reversal image formation and (**b**) wet-processed negative surprint.

characteristic of these environmentally friendly developer systems. Some tap water developer-based systems utilize the photocross-linking of poly(vinyl alcohol) (PVA) acetylized with a pyridinium salt (SbQ–PVA) (57,58). Pigments are inserted into the SbQ–PVA polymer to provide the necessary colors for a surprint proofing system. SbQ–PVA photocross-links to the point of providing excellent distinction between the exposed and the unexposed material, thereby providing the basis for superior resolution.

The positive pigmented film systems are generally composed of various types of phenol–formaldehyde, cresol–formaldehyde, or other polymer derivatives containing the diazoquinone structure at various points along the polymer backbone (59). In the unexposed state, the polymer system containing the diazoquinone is insoluble. However, upon exposure to uv light, a photosolubilization reaction ensues when the diazoquinone moiety is converted to a carboxyl group. The reaction proceeds with the loss of nitrogen to form a ketene, which then readily reacts with ambient water to yield the solubilizing carboxylic acid group. The general reaction scheme can be depicted as follows, where R represents a polymer chain such as a phenol–formaldehyde derivative.

The right-reading positive image resides on the receptor film after lamination, exposure through a positive, and wash-out of the unexposed areas with a solvent or aqueous developer. The four-color proof is built up by the sequential lamination, exposure, and development steps.

Digital Proofing. In a modern electronic prepress environment, much of the page makeup is done using computer systems. Images separated on a scanner may be stored electronically; images created using an electronic drawing system may exist only as a digital file. It is necessary to proof these elements, as well as those that exist as film separations. One approach is to output the files on an imagesetter, ie, make films of the color separation, and proof by using an analogue system. This, however, may add extra expense and time to the process, and, as printers move closer to the goal of computer-to-plate or computer-to-press, film may never be needed. Digital proofing technologies then become necessary.

An ideal digital proofer would take digital files and output a hard copy that looks exactly like the printed page, including accurate color to judge the adequacy of the color-separation process. Several digital proofing technologies are being used. Brief descriptions follow. In all cases these technologies are also used as printing technologies.

Thermal Printing. Dye sublimation and thermal wax-transfer printers are commonly used as moderate quality digital proofers. Dye sublimation printers are the more acceptable, although these do not produce the halftone dots that a printer sees on the plate. The images produced have a photographic quality and the color accuracy is acceptable using modern color-matching technology. The typical dye-sublimation printer is capable of printing one or two standard size (8.5 × 11 in. (21.6 × 27.9 cm)) pages on a single sheet of paper. These printers are frequently used in design studios, where design comps are printed for internal use or customer approval, and where color accuracy is less important. Although the price of the printer is moderate, material costs tend to be high.

A high quality version of a dye-sublimation printer has been developed specifically for color proofing. This device uses a laser writing head, rather than the typical thermal printhead, to produce higher resolution images. The device is capable of true halftones, providing an accurate rendition of a printed page. It is, however, expensive both in equipment and materials cost.

Electrophotographic Printing. Two different categories of printer are used for electrophotographic printing. Common office color copiers or printers are often used for a low to moderate quality proof, where color accuracy is not required. These printers typically do not have high enough resolution to produce true halftone dots, so dithering schemes, ie, dot patterns used to simulate continuous tone, are used. The machines use dry powder toners. Monochrome office-type printers may also be used for proofing text-only documents or making layout proofs of color documents.

A higher quality category of electrophotographic printer has been developed strictly for proofing. This device has a much higher resolution, using liquid toners to produce true halftones. Thus the proofer can produce images that are close representations of a printed color piece. This proofer must be color-accurate for customer approval. It is also an expensive device, because of the accuracy required to achieve quality halftones.

Ink-Jet Printing. Continuous-flow ink-jet printers have made strong inroads as digital proofers in the 1990s. These printers, which do not have the resolution required for halftones, are capable of achieving multiple ink-density levels by striking the same spot on paper with multiple drops of ink. This results in a continuous-tone appearance. Color can be adjusted to achieve accurate matches to press, making ink-jet printing a satisfactory and moderately priced choice for proofing.

Monitors. Cathode ray tube (CRT) monitors are a key element of electronic prepress systems, providing an electronic canvas for the operator. They may also be used to judge general adequacy of color in a process called soft proofing.

For many years manufacturers of CRT monitors have sought to provide accurate soft proofing on their equipment. The process is difficult in that all CRTs work in a light-emissive RGB color space, whereas the printer uses subtractive color (CMYK). Black is therefore the most difficult color to reproduce, because a CRT cannot be any darker than it appears when current flow is off. The displayed color is also operator-dependent unless a calibration scheme is used to measure the output of the CRT. Soft proofing using a calibrated monitor is accurate for position and color when used by a trained operator. In general, for the customer who is primarily interested in seeing color as it is to appear on the printed page, CRT soft proofing is inadequate.

Image Duplication and Output

The principal processes for making many printed copies from a prepress image include lithography, gravure, flexography, letterpress, and screen processes, as well as the newer technologies of thermal printing, electrophotography, and ink-jet printing (see IMAGING TECHNOLOGY).

Lithography. Of the principal printing processes, lithography is by far the most widely used. But in spite of usage and the numerous studies (60) to which lithography has been subject over many years, the mechanism of the process is not well understood. This lack of knowledge reflects the complexity of the various interactions of ink, plate, and water that come into play whenever a lithographic plate runs on press.

Lithography is a planographic process. Image, or printing areas, and non-image, or nonprinting areas, reside in the same plane and are differentiated by the extent to which these areas accept printing ink. Nonprinting areas are hydrophilic, accepting water and repelling ink; printing areas are oleophilic, repelling water and accepting ink. During printing, water is applied to the surface of the plate as an aqueous solution of surface-active agents, generally referred to as a fountain solution, and the ink is applied to the plate surface through a roller train.

Offset plates usually comprise a support, which may be fabricated from paper or plastic but most frequently from metal, and one or more layers of a radiation-sensitive composition. Aluminum has gained general acceptance as support material for several reasons. It is light in weight and has a well-adhered surface oxide layer that is resistant to corrosion under normal press-room conditions. Unlike paper or plastic, aluminum is not stretched significantly when mounted on the press cylinder, hence it is easily capable of maintaining good image registration, an important requirement when color work is being printed. The role of the aluminum in a conventional offset plate is not just simply one of acting as support for the radiation-sensitive coating and subsequently the printing image; it also participates directly in the printing process by providing the nonprinting regions of the final plate.

The first step in creating a printing image is usually to contact the surface of the radiation-sensitive coating of the printing plate with a photographic film positive or negative, and to expose it to a light source emitting between 380 and 420 nm, the spectral region to which most commercial offset plates are sensitized. So as to obtain acceptable image definition, the plate surface and overlying film must be in intimate contact during the exposure stage. This is best achieved by the use of a vacuum printing frame. For particularly rapid drawdown, ie, evacuation of air, a matte layer is applied to the surface of the radiation-sensitive coating. The next step is to treat the imagewise exposed plate with a specially designed aqueous or organic solvent-based developer solution in order to remove selectively the now comparatively more soluble portions of the coating, thus revealing the support surface in the underlying areas.

This surface must be such that the radiation-sensitive coating can be quickly and completely removed during the development process. Any residual material can cause ink to be picked up in the nonimage areas and transferred to the paper during the printing operation. The support surface in the revealed areas must be strongly hydrophilic, possessing a high affinity for water, and repelling oil-based printing inks. In the case of a positive working plate, it is the unexposed areas of coating that provide the final printing image; in the case of a negative working plate, the exposed areas. The image in either case must be strongly bonded to the support in order for it to resist the powerful abrasion forces which come into play during printing.

Substrate.   To provide a surface that exhibits these properties, a multi-stage process is used. The surface of the aluminum sheet is first roughened or grained, either mechanically, using oscillating and rotating wire brushes, or by electrolysis in solutions of acid, typically hydrochloric and nitric acids, or possibly in mixtures of mineral and organic acids (61). The electrochemical process is the more expensive of the two, using a significant amount of electrical power and generating waste solutions, containing aluminum ions, that accumulate as the process continues. The surface structure obtained, however, is more uniform and consistent and is generally preferred. Mechanically grained plates are frequently used when high print quality is not a requirement, for example in the printing of newspapers. A grained surface is important for two reasons. This surface improves adhesion of the coating and of the printing image, and it also allows, through a considerably increased surface area, more water to be held in the nonprinting regions of the plate surface.

In addition to being grained, aluminum substrate used for presensitized printing plates is almost invariably anodized. The oxide film, weighing from $1-8$ g/m^2 and produced by anodizing in a bath of acid, usually phosphoric or sulfuric, serves two purposes. It acts in concert with the rough surface structure to increase coating adhesion and, because it possesses the hardness, scratch resistance, and general ease of hydrophilization required for an effective lithographic surface, plays an essential role in providing the printer with a consistently acceptable level of press performance.

The anodized surface is often subjected to additional treatment before the radiation-sensitive coating is applied. The use of aqueous sodium silicate is well known and is claimed to improve the adhesion of diazo-based compositions in particular (62), to reduce aluminum metal-catalyzed degradation of the coating, and to assist in release after exposure and on development. Poly(vinyl phosphonic acid) (63) and copolymers (64) are also used. Silicate is normally employed for negative-working coatings but rarely for positive ones. The latter are reported (65) to benefit from the use of potassium fluorozirconate.

Some printing environments are aggressive and demand ultratough images and abrasion-resistant backgrounds. Conventional plates having polymer images tend to fail under these conditions. As a result there is a small market for plates in which both image and nonimage areas are metal and which comprise hydrophilic electroplated chromium on an ink-receptive copper (bimetal) or copperized steel (trimetal) base. The chromium layer has a higher pore density than anodized aluminum, providing an increased affinity for water, hence a reduced potential for dirty backgrounds. The plate is imaged using photographic masking followed by etching.

Radiation-Sensitive Coatings. There are two basic types of presensitized offset plates, negative and positive working. Plates are classified according to how their coatings respond to exposure to actinic radiation. Coatings that become less soluble in specific solvents, either through polymerization, cross-linking, or loss of certain functional groups, are employed in negative plate systems. In contrast, a coating that degrades on exposure to radiation, and hence becomes more soluble, is used in positive systems.

The differences in performance, which determine the printer's choice for one type of plate over the other, are no longer clear cut. Advances in electronic prepress have made it possible for the two systems to be more or less interchangeable for many types of work. The choice of negative or positive determines, possibly more than any other factor, the degree by which a halftone dot increases in size in going from film image to printed copy. When exposing a plate through a piece of film, light passes through the clear areas of film, then through the radiation-sensitive coating on the plate surface. Any remaining unabsorbed light it reflected from the support surface. As a result, the area of coating exposed is greater than the image size of film, leading to a developed dot size on a negative plate being 10% or so greater than that of the corresponding dot on the film; on a positive plate, the size is 10% or so smaller. This cause of dot gain can be corrected at the film-making stage, and is becoming less of a serious problem as scanners come into widespread use.

Positive plates are more expensive to make than negative, and also tend to be less resistant to the chemicals used on press. In order to increase their useful life, positive plates are often baked for a few minutes at temperatures of 220°C or more in order to harden, or insolubilize, the image material. Their use offers advantages, however. Apart from the reduced dot gain, which leads to better printed image quality, fewer problems are experienced at the film image assembly stage. The presence of dust on film and contact frame glass is not as serious a problem as it is in using negative systems.

Negative Plate Coatings. The bulk of negative plates have a diazo-based coating. This often comprises an *N*-aryl- or alkyl aminobenzenediazonium salt condensed with formaldehyde (66) or a methylol derivative (67) to form a low molecular weight polymer such as the following:

$$\left[\begin{array}{c} \text{—} \underset{\substack{| \\ \text{HN} \\ | \\ \bigcirc \\ | \\ N_2^+}}{\bigcirc}\text{—CH}_2\text{—} \\ \\ SO_3^-(CH_2)_{11}CH_3 \end{array} \right]_{4-6}$$

The hydrochloric and sulfuric acid-derived salts, although widely used, tend to give products having poor shelf life. Significantly improved coating quality and stability is possible, however, if the diazo resin is substantially water insoluble and coated from organic solvents. This can be achieved by the use of counterions derived from halogenated Lewis acids such as hexafluorophosphoric or tetrafluoroboric acids. Alternatively, sulfonic acids substituted with long-chain alkyl groups can be employed (68). High molecular weight polymers are often added to the coating formulation to boost the strength of the final image. Photolysis of the diazo compound with loss of the highly polar diazonium group provides sufficient solubility differential in aqueous surfactant solution for easy and rapid development.

Photopolymerizable compositions based on monomeric acrylic or other ethylenically unsaturated acid derivatives are becoming increasingly popular.

When multifunctional derivatives are employed, three-dimensional networks having high strength and abrasion resistance are possible on exposure to light. A typical composition may contain an ethoxylated trimethylolpropane triacrylate monomer, a perester phenacylidene initiator (69), and an acrylic acid–alkyl methacrylate copolymer as binder.

Unlike diazo-based compositions, such materials have widespread use in other fields, eg, can coatings, paint (qv), varnishes, and inks. Their attractiveness stems mainly from the potential for amplification, hence high photographic speed, and from the relative ease of tailoring photoresponse to a particular region of the uv–vis spectrum. Free-radical polymerization suffers from oxygen inhibition effects, however, and lithographic or litho plate formulations using this technology are not free of the problem. Possible solutions include the use of oxygen scavengers (70) or of photoinitiating systems more resistant to oxygen quenching (71). The most effective approach is to apply an oxygen-impermeable overlayer to the radiation-sensitive coating (72).

Positive Plate Coatings.   1,2-Naphthoquinone diazide sulfonic acid esters are used extensively in the formulation of both photoresists and litho plate coatings (see LITHOGRAPHIC RESISTS (SUPPLEMENT)). As in the case of negative plates, a polymer binder is normally used to provide additional strength. In positive plate coatings this is almost invariably a phenol or cresol novolak resin. Di- and trihydroxy benzophenone esters and sulfonaphthoquinone diazides are frequently found in commercial formulations, as are esters of cresol–formaldehyde and pyrogallol acetone condensation polymers (73).

The process by which a solubility differential between exposed and unexposed areas occurs is well known (74). Photodegradation products of the naphthoquinone diazide sensitizer, eg, a 1,2-naphthoquinonediazide-5-sulfonic acid ester (**11**), where Ar is an aryl group, to an indene carboxylic acid confers much increased solubility in aqueous alkaline developer solutions.

(**11**)

An interesting development allows a relatively conventionally formulated positive plate to be used in the negative mode (37). Heating a plate, which has been exposed through a film negative, to about 135°C for two minutes causes the indene carboxylic acid photodegradation products to decarboxylate and become alkali-insoluble. Cooling the plate and blanketwise exposing to light permit the originally unexposed material to be washed away on development.

Waterless Lithography Printing.   Waterless printing processes have been actively researched since the late 1950s. The first commercial waterless plate was introduced in the 1960s. Waterless printing, or driography, is a method of printing that provides high quality reproduction without recourse to a dampening system, or fountain solution, on press. Without the problem of water-induced ink emulsification, prints exhibit sharper dots and good tonal gradation with little variation in density throughout the run.

The plate, available in both positive and negative versions, comprises a support, usually of aluminum, coated with a primer, a radiation-sensitive layer, and a top silicone layer. The silicone layer is ink-repellent and the underlying radiation-sensitive layer is ink-attractive. When the exposed plate is developed, the silicone layer is selectively removed according to the image pattern, revealing the underlying surface. This surface, being oleophilic, now constitutes the printing image. Whether the system is negative or positive working, ie, whether the silicone film is released from or bonded to the underlying surface following exposure and development, is determined by the nature of the radiation-sensitive composition. Positive systems result by using photopolymerizable or cross-linkable material that hardens on exposure to light, bonding to the silicone at the interface and causing it to be retained on treatment with a wash-off solution (75). Negative systems employ quinonediazides (76), which undergo photodegradation and release the overlying silicone.

In comparison to the organic photopolymer where the surface has been selectively revealed during wash-off treatment and which now forms the printing image, the silicone exhibits a low surface free energy and shows relatively little affinity for the specially formulated printing ink. This is borne out in practice so long as the ink viscosity on press is carefully controlled. Inks for waterless printing are designed to demonstrate optimum performance within a 5°C temperature range and therefore require ink-cooling systems in presses. A disadvantage of waterless plates in the 1990s is the relatively poor scratch and abrasion resistance of the nonimage surface compared to anodized aluminum.

Computer-to-Plate. The desire for direct digital output to litho plate goes back to the early 1970s when the first attempts were made to eliminate the film intermediate. At that time the drive came from a growing opportunity for facsimile transmission for newspapers and similar systems (77). Since then the digital revolution has changed the viability of direct digital to litho plate from a curiosity to a financial necessity for the lithographic industry as alternative direct-to-press (78) and direct-to-paper (79) approaches come closer to commercial reality. The litho plate technology avenues open to the system designer are high speed photopolymer, electrophotography, silver halide, and thermal imaging. These have not changed because laser imaging technology has been unable to accommodate the low sensitivity in the uv of diazo coatings. Early attempts provided plates of 1–10 mJ/cm^2 at 360 nm using a water-cooled argon ion laser. For a laser tube life of 100–500 hours, at \$10,000 per tube, and plates having poor (1–100 days) shelf life and inferior plate resolution, the expected convergence between laser technology and photopolymer speed did not happen. Printers have continued to use conventional presensitized negative and positive plates. Plate chemistry has to match the capabilities of the commercial graphic arts lasers, ie, the air-cooled gas lasers, argon ion and helium neon, and gallium arsenide diode lasers.

The energy in mJ required to expose the various plate chemistries is given below. To convert J to cal, divide by 4.184.

Plate chemistry	Energy required, mJ/cm^2
diazo	
positive	100–600
negative	100–600
photopolymer	0.1–500
organic photoconductor	0.001–0.05
silver halide	0.0001–0.05

At 1000 dots per centimeter, the standard imagesetter output resolution for color litho printing, the data rate to the laser is 28 MHz, a rate only just possible in the mid-1990s. Printers, however, require access to plates within five minutes to keep expensive presses rolling. As can be seen from the list above, only the organic photoconductor (OPC) and the silver halide plates can deliver this photographic speed. The diazo chemistry is entirely too slow.

Photopolymer. The best fit of commercial lasers and plate technology for speed is that of organic photoconductor and silver halide. Photopolymer technology is close to being viable, using improved laser power, more efficient optics, or high speed photopolymers (see PHOTOCONDUCTIVE POLYMERS). Optimization of photopolymer speed, spectral sensitization to the visible wavelengths, and the emergence of new lasers, such as the diode-pumped frequency-doubled Nd^{3+} yttrium aluminum garnet (YAG) laser, promise convergence of laser technology and photopolymer speed. Figure 10 shows the photopolymer plate structure and process of use.

Typical photopolymer plates comprise a photosensitive layer containing an acrylate monomer; a photoinitiation system, eg, a perester; and a sensitizing dye, eg, a coumarin (qv). The photosensitive coating is usually covered by a 2-μm overcoat, generally a water-soluble polymer such as poly(vinyl alcohol), which substantially prevents polymer-chain termination by oxygen. In order to ensure shelf stability at high photographic speeds, laser exposure is used, which provides a free radical that is relatively stable at room temperatures. Only when the plate is heated to over 100°C does the polymerization proceed to completion. Apart from this heating step the processing chemistry is familiar to all plate-

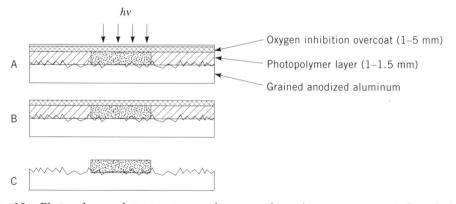

Fig. 10. Photopolymer plate structure and process of use: A, upon exposure to laser light, $h\nu$; B, upon heating to 120°C; and C, after development, rinsing, and gum treatment.

making: alkaline development followed by application of a hydrophilic polymer layer common to each of the processes described herein.

Electrophotography. Electrophotographic plate-making systems using organic photoconductor materials have been used successfully in cameras, eliminating film, by newspapers since the early 1980s and have been tried in the computer-to-plate application (80). The coating composition is typically an OPC, such as an oxadiazole, a sensitizing dye such as a rhodamine, and an alkali-soluble support resin.

The structure of an electrophotographic plate is shown in Figure 11. Latent image instability and problems in the control of the complex imaging system of charging, toning, fusing, and developing together make the application of OPC to quality four-color work difficult. These problems limit OPC application to simple camera systems (see ELECTROPHOTOGRAPHY).

Silver Halide on Diazo Plates. Silver halide has high photographic speed and a highly developed sensitizing and stabilizing technology. The principal problem is stabilizing the silver halide materials on an aluminum substrate and providing a durable printing image. One system (81) takes a standard positive working diazo plate overcoated with an appropriately sensitized silver halide emulsion. The structure and process of use are shown in Figure 12. This system has one significant advantage over other silver halide-coated aluminum plates (82). The silver halide emulsion is separated from the aluminum by the layer of diazo, thus preventing interaction between the highly sensitized emulsion layer and the aluminum metal, a notorious source of fog for photographic film.

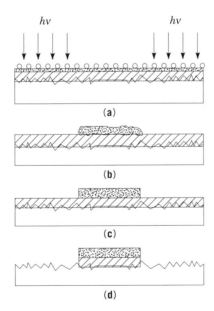

Fig. 11. Electrophotographic plate structure and process of use: (**a**) laser exposure of the negatively charged photoconductor, ○, sitting on the plate surface; (**b**) application of oppositely charged toner particles from a magnetic brush or an aprotic liquid carrier; (**c**) fusing of the toner particles on the coating; (**d**) decoating the unprotected alkali-soluble coating and washing and gumming the plate.

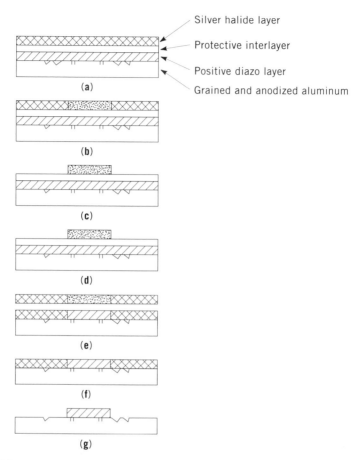

Fig. 12. Silver halide on diazo plate: (**a**) structure upon imagewise laser exposure; (**b**) development of silver image; (**c**) fixation of unexposed areas; (**d**) after rinsing; (**e**) upon blanket uv exposure; (**f**) after washing off; and (**g**) after developing, rinsing, and finishing.

Although the processing cycle is complicated, a conventional, well-understood plate results for use on the press.

Single-Sheet Diffusion Transfer. Another solution to the interaction between silver halide and aluminum is that of single-sheet diffusion transfer (83,84). By carefully specifying the graining and anodizing conditions, a stable and substantially continuous anodic layer of aluminum oxide, which minimizes the fogging and associated interactions, can be obtained. This litho plate on aluminum is produced using the well-known diffusion transfer process (85). Du Pont Silverlith is a computer-to-plate product based on this technology. The aluminum plate is coated first with a receptor layer, then with a conventional silver halide emulsion. On development the developer monobath first chemically develops the exposed silver halide areas, then solubilizes the exposed silver grains as complexes, which are reduced at the nucleating receptor layer by hydroquinone. Because these sequential processes are taking place simultaneously, controlling the kinetics of such a process is crucial. It is therefore of prime importance that the chemical development process is much faster than the solubilization.

The silver image (83) is made ink-receptive by reaction, ie, chemisorption, with an oleophilic molecule such as phenyl mercaptotetrazole, which has an active sulfur group (thiol-thiolate) and a phenyl group to impart oleophilicity. The system has the advantage of a simple processing scheme at the expense of an unconventional printing image. The structure and process of use are shown in Figure 13.

Thermal Imaging.   Advances in near-infrared (nir) laser diode technology have led to the promise of economical thermal imaging at 800–1100 nm (see LIGHT GENERATION, LIGHT-EMITTING DIODES). State-of-the-art plate materials, as of 1995, had a sensitivity of 200 mJ/cm^2 at 850 nm, 1000 times less sensitive than the fastest available visible sensitive photopolymer plates. A number of different thermal imaging mechanisms are being evaluated, including ablation (79), thermal coalescence, and photopolymerization. Each of these approaches is interesting but, as of this writing, laser technology is expected to dictate which technology is to be employed.

Direct-to-Press.   A novel approach, pioneered by Heidelberg Druckmaschinen AG and Presstek, is to expose printing plates directly on the printing press, using a laser-based system. In this case, a special waterless printing plate is mounted on all four units of a special printing press. Each unit also has a laser head that scans the width of the plate cylinder as the latter slowly rotates. The laser ablates a surface layer, exposing the ink-receptive surface of the plate. Because the plates are exposed directly on the press, color-to-color registration is not a problem. Printing can begin as soon as the plates are written. This type of technology serves a market between the short-run color electrophotographic printers and a long-run conventional plate and press system.

Gravure.   The gravure printing process, sometimes called intaglio or rotogravure, utilizes a recessed image plate cylinder to transfer the image to the substrate. The plate cylinder can be either chemically or mechanically etched or

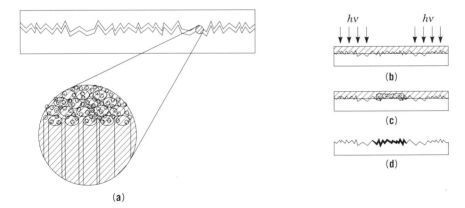

Fig. 13.   Single-sheet diffusion transfer plate: (**a**) structure; (**b**) upon exposure to light; (**c**) development; and (**d**) washing off and finish. In (**a**) the plate is first coated with a receiver layer of small (<5 nm) catalytic sites. The photographic layer is a spectrally sensitized silver halide emulsion. In (**c**) the exposed areas develop as silver metal. Unexposed areas diffuse down to the receiver layer and form the printing image. In (**d**) the emulsion is washed off, revealing the silver printing image on the anodic layer (〰〰).

engraved to generate the image cells. The volume of these cells determines the darkness or lightness of the image. If an area is darker, the cells are larger; if the area is lighter, the cells are smaller.

Traditionally, a gravure cylinder was prepared chemically by using an etch process. Since the early 1970s, electromechanical engraving, in which a diamond stylus cuts the cells into cylinders, has been the preferred approach. Of particular significance over the years has been Hell's Helioklischograph electromechanical engraving system. Later alternative approaches have included laser and electron-beam engraving.

In gravure, all elements within the image are screened. This is in contrast to flexographic and lithographic plates, which can contain true solids as well as halftones.

The gravure printing process is based around an inking system that is extremely simple, giving the process a high degree of consistency, particularly with regard to color printing. This consistency is difficult to match using other printing techniques. The system, shown in Figure 14, utilizes a liquid ink that has traditionally been solvent-based, although environmental pressures have resulted also in the development of aqueous-based inks.

The gravure cylinder sits in the ink fountain and is squeegeed off with a doctor blade as it rotates. The impression cylinder is covered with a resilient rubber composition, which presses the paper into contact with the ink in the tiny cells of the printing surface. The image is thus transferred directly from the gravure cylinder to the substrate. Frequently an electrostatic assist is used to help the ink transfer from the gravure cylinder to the substrate. Gravure inks are comprised of pigment, resin binder, and, most frequently, a volatile solvent. The ink is quite fluid and dries entirely by evaporation. In multicolor printing, where two or more gravure units operate in tandem, each color dries before the next is printed. This is seen as a particular advantage for gravure over, for example, offset, where the placing of wet ink on wet ink can lead to inferior print quality. The wet ink on dry and the simple ink train, together with a long run-length capability, have led to the belief that gravure has traditionally set the standard for high quality printing.

The gravure market can be considered to comprise three approximately equal segments: publications, packaging, and specialty printing. In publications,

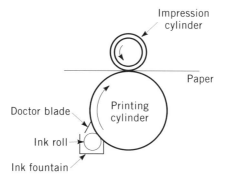

Fig. 14.   Gravure printing system.

gravure retains a significant proportion of the long-run magazine market; however, web offset is rapidly eating into this market as litho plates become capable of printing one million and more impressions. In packaging printing, where paperboard and repeat-run cartons are encountered, gravure is the ideal process. The cylinder lasts virtually forever and color consistency is high. The final third of the gravure market is specialty printing of such items as wallpaper, gift wrap, and floor coverings. Shifts between these gravure market segments are anticipated but the total share, about 10% for the U.S. commercial printing market, is expected to remain unchanged throughout the 1990s.

The fundamental strengths of gravure are that the process provides consistent color throughout long print runs, and, because of its ability to apply heavy ink coverage, can be used to print high quality work or to print on a lower grade of paper than offset and still maintain acceptable print quality. Additionally, gravure generates overall less waste than offset.

In contrast, the primary process disadvantages of gravure are long lead time, high cost of manufacture for gravure cylinders, generally long press make-ready times, and environmental hazards associated with the use of solvent-based inks. These need to be eliminated if the technique is to remain competitive. Developments in the gravure process have therefore been directed at addressing these shortcomings. Work has continued on the assessment of laser engraving to speed up the engraving process. Lasers are capable of cutting between 25,000 and 30,000 cells a second, significantly faster than electromechanical systems. As of this writing, the Max Daetwyler Corporation is developing a direct digital laser engraving system. Linotype-Hell is also working on developments to transfer digital data to the engraver faster. Of particular interest is Gypsy, Linotype-Hell's digital cylinder preparation system that outputs a full cylinder of pages digitally for instant engraving on the Helioklischograph. No film intermediate is required.

Additionally, attempts are being made to streamline the gravure process by improving make-ready times. Press manufacturers Cerutti and Albert-Frankenthal are working on cassette systems where jobs are prepared outside the press and subsequently loaded with virtually zero stop time. Computer control of press functions such as compensators, angle bars, and folders also help reduce press make-ready time.

The environmental concerns associated with the use of toluene, a toxic and flammable aromatic hydrocarbon, as a gravure ink solvent must be addressed. Whereas ink manufacturers are working on the development of water-based inks, the slow drying times and poor printing qualities of the prototype products have impeded commercialization. Furthermore, the high cost of these materials is seen as a barrier to their introduction.

Flexography. Flexography is a variation of letterpress printing used mainly for packaging applications. It is characterized by the use of an elastomeric printing plate, fast drying inks, and an anilox roll ink-metering system. A principal advantage of flexography is its ability to print on a wide range of substrates, including plastic films, foils, coated and uncoated paper, paperboard, and corrugated board. Other advantages include low cost and short cycle time, ability to change cylinder diameters to reduce stock waste, precise ink transfer with minimum on-press adjustments, ability to print one layer and laminate another layer over it, and ability to print continuous patterns such as wallpaper and gift wrap.

Limitations of flexography include higher highlight dot gain and lower solid density as compared to gravure and offset. The increased dot gain results from a combination of printing plate deformation and ink spread on press. Another limitation is the inability to print uniform solids and halftones using the same plate without substantial make-ready. The increased pressure required for uniform solids increases dot gain in highlights.

Flexographic Printing Press.   A typical print station is shown in Figure 15. The three basic types of flexographic presses are shown in Figure 16. Dryers generally separate individual print stations. In the stack press (Fig. 16**a**), individual print stations are in sequence one on top of the other. This press is used primarily for paper and laminated films. Advantages include accessibility to print stations and the ability to reverse the web, thus allowing printing both sides in one pass. The central impression press (Fig. 16**b**) has its print stations distributed around a large central impression cylinder, which is precisely geared to each print station and thus improves registration. This press is used mainly to print high quality wide-web films. The in-line press (Fig. 16**c**) is used mainly for printing corrugated and folding cartons, as well as for narrow-web tags and labels.

Anilox Rolls.   The heart of the flexographic printing system is the anilox roll, a steel cylinder optionally coated with ceramic and engraved with a pattern of pits or cells. The function of the anilox roll is to meter a uniform film of ink from the ink fountain to the printing plate without the need for continuous adjustment. There are many types of anilox rolls, distinguished by the mode of engraving, the materials of construction, the pattern of the cells, and the cell geometry. Two main families are mechanically engraved chrome rolls and laser-engraved ceramic rolls.

Mechanically engraved chrome rolls are steel cylinders engraved by a precision tool. Following engraving the rolls are plated with copper, which acts as a bonding layer, and chrome, which hardens the surface. Mechanically engraved chrome rolls have been the mainstay of the industry for many years. However,

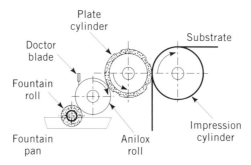

Fig. 15.   Flexographic printing station, where the fountain pan supplies ink to the rubber fountain roll, which in turn supplies ink to the anilox roll. The doctor blade removes excess ink from the surface of the anilox roll so that it transfers a uniform layer of ink to the printing plate. The printing plate then transfers this layer of ink to the substrate, which is supported by the impression roll.

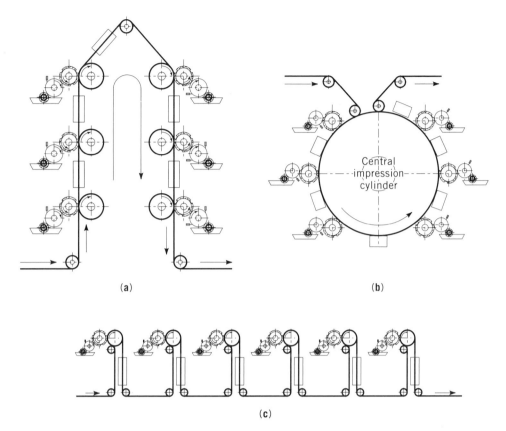

Fig. 16. Schematics of flexographic presses, where the arrow designates the direction of the web flow and ▭ represents an interstation dryer. (**a**) Stack press; (**b**) central impression press; and (**c**) in-line press.

upon the introduction of the reverse-angle doctor blade, which has better control of the metering of the ink, and the subsequent wear problems, the industry has been steadily shifting to laser-engraved ceramic rolls.

Ceramic rolls are steel cylinders that have been coated with a ceramic, usually chromium oxide, layer and then engraved with the beam from a CO_2 laser. By controlling the energy and timing of the laser pulse, the depth and diameter of the anilox cell can be tightly controlled. Anilox roll variables include the screen count and screen angle as well as cell volume. Unlike mechanically engraved rollers, the volume of laser-engraved rolls is independent of screen count, allowing a variety of rollers having a wider range of volumes and screen counts. The ceramic coating reduces doctor blade wear by an order of magnitude, thus prolonging the life of the roll.

Flexographic Printing Plates. There are three primary types of flexographic printing plates: molded rubber, solid-sheet photopolymer, and liquid photopolymer.

Molded Rubber Plates. Initially, flexographic printing plates were made of hand-cut or molded rubber. The basic steps of rubber plate-making include

(*1*) preparing an engraving by exposing a photoresist-coated magnesium plate to uv light through a photographic negative, washing away the unexposed resist with chlorinated solvent, and acid-etching the unprotected magnesium; (*2*) molding a phenolic matrix board using the magnesium engraving; and (*3*) molding the rubber plate using a matrix mold.

Molded rubber plates transfer ink well and, when large numbers of identical designs are needed, are inexpensive to make. However, environmental concerns over the use of acid etching solutions to prepare the magnesium engraving, poor thickness uniformity, and poor dimensional stability of the molded plates are the disadvantages. Photopolymer molding plates, both solid and liquid, have replaced most magnesium engravings.

Solid-Sheet Photopolymer Plates.   Solid photopolymer plates consist of an elastomeric, photosensitive layer bonded to a polyester support. The plate formulation usually contains an elastomer that provides the required properties of flexibility and resilience (see ELASTOMERS). Acrylic or methacrylic monomers, which, in the presence of sufficient energy, polymerize to reduce the solubility of the material, are employed along with a photoinitiator, which absorbs uv radiation and initiates polymerization. Various additives such as thermal stabilizers, plasticizers (qv), and dyes are also present. These photopolymer plates are made by back exposure to uv to consume stabilizers and define relief height, uv exposure through a photographic negative to form the image, removal of unexposed areas with a solvent to form the relief image, drying to remove solvent, overall exposure to eliminate surface tack, and post-exposure to further cross-link and toughen the plate. Advantages of solid-sheet photopolymer plates over rubber include improved dimensional stability derived from the polyester support, longer print run length, improved print quality including more predictable rendition of four-color process images, and better thickness uniformity.

Newer technology involves aqueous-processible photopolymer plates. Many plate-makers and printers are eager to switch to water processing in order to eliminate volatile organic solvents. The chemistry and process of use are similar to that of the solvent-processible plate except that in the aqueous plate, the elastomer has pendent carboxyl, hydroxyl, or other water-soluble groups to allow aqueous processing.

Liquid Photopolymer Plates.   Another form of photopolymer plate is the liquid system, which is used successfully in corrugated and newspaper markets. The plate-making process is similar to the solid-sheet photopolymer plate except that before exposure to uv, the liquid photopolymer, which consists of liquid rubber oligomers and cross-linkers, is sandwiched between thick glass plates to provide defined and uniform thickness during exposure. The main advantage over solid plates is the use of lower cost materials. Disadvantages include inferior thickness uniformity and limited-run life.

Laser-Engraved Rubber.   Relief images can also be made by laser engraving a rubber-coated cylinder. The rubber layer is selectively ablated by a CO_2 laser beam, leaving a relief printing image. This method eliminates the need for a photographic negative, and is ideal for printing continuous tone. Laser imaging cannot reproduce the fine screen rulings obtainable with solid plates and is limited to ca 34 lines/cm (85 lines/in.).

Flexographic Printing Inks. Conventional flexography uses low viscosity fast-drying inks. These are typically solutions having 25–35% solids content, which consists mainly of nearly equal parts of a pigment or pigments dispersed in polymer resins (see PIGMENT DISPERSIONS). These resins form the ink film and attach the pigment to the substrate. About 5% of the ink may consist of additives such as slip agents, surfactants, plasticizers, and antifoams. The solvent portion is usually a blend of alcohols and acetate esters. Drying rate can be varied by changing the ratio of acetate to alcohol in the solvent.

For aqueous inks, the resins are water- or alkali-soluble or dispersible and the solvent is mostly water containing sufficient alcohol (as much as 25%) to help solubilize the resin. To keep the alkali-soluble resin in solution, pH must be maintained at the correct level. Advances include the development of uv inks. These are high viscosity inks that require no drying but are photocurable by uv radiation. In these formulations, the solvent is replaced by monomers and photoinitiators that can be cross-linked by exposure to uv radiation. The advantage of this system is the complete elimination of volatile organic compounds (VOC) as components of the system and better halftone print quality. Aqueous and uv inks are becoming more popular as environmental pressure to reduce VOC increases.

Letterpress. Letterpress is the oldest and, until the mid-1960s, the dominant and most versatile printing process. As of the 1990s, it is used mainly in high quality design work, fine books, and quality stationery. Unlike lithography, which prints best on coated papers, letterpress can print on any paper, provided that the paper is of even thickness, flat, and reasonably rigid.

Letterpress is printed directly by the relief method from cast metal or plates on which the image or printing areas are raised above the nonprinting areas. Ink rollers apply ink to the surface of the raised areas, which transfer it directly to paper. Flat-bed cylinder presses are available but most letterpress is printed on rotary presses.

Make-Ready. Because pressure is needed for ink wetting and transfer and because the image elements in letterpress vary in size, the same amount of printing pressure or squeeze exerts more pressure on highlight dots than on shadow dots. This necessitates considerable make-ready to even the impression so that the highlights print correctly and do not puncture the paper. Precision plates and premake-ready systems have helped reduce make-ready time, but it is still appreciable for quality printing and is a reason letterpress has been largely replaced by other processes.

Screen Printing and Stencil Processes. *Image Carriers for Stencil Processes.* There are two stencil processes in general use: screen printing and stencil duplicating. Screen printing used for art reproduction is called seriography.

Screen-printing image carriers can be produced manually or by photomechanical means. The screens can consist of silk cloth having mesh counts of 40–80 openings per lineal centimeter. Nylon screens are used for textile printing and metal screens of phosphor bronze and stainless steel are used for fine-detail printing in meshes as fine as 120 openings per lineal centimeter. The screen material is attached to a rigid frame and stretched tightly so that it is level and smooth. The stencil is applied to the bottom side of the screen, ie, the side in contact with the surface to be printed. Ink having a consistency similar to thick paint is used in the screen. Ink is transferred by rubbing on the screen surface using a rubber squeegee. The screens can be used for ca 10^5 impressions.

Screen printing is also practiced by using rotary screens, made by plating a metal cylinder electrolytically onto a steel cylinder, removing the cylinder after plating, applying a photopolymer coating to the cylinder, exposing it through a positive and a screen, developing the image, and etching it. The result is a cylinder having solid metal in the nonimage areas and pores in the image area. On rotary screen presses, the ink is pumped into the cylinder, and the squeegee, which is inside the cylinder, controls the flow of ink through the image pores to the substrate.

Manual stencils are made by knife-cutting special film stencil materials. These consist of two plastic layers. The image to be printed is cut through one layer, and this part of the stencil is placed in contact with the underside of the screen. A solvent, which is insoluble in the ink but attaches the cut stencil to the screen, is applied and then the backing layer is removed. Manual stencils can also be produced by drawing directly on the screens using special materials.

Photomechanical stencils are of two types: direct coatings and transfer films. Direct coatings are either bichromated gelatin or bichromated poly(vinyl alcohol) (PVA). The coated screens are exposed through a positive, washed, and inspected. These screens are used for printing electronic components. They are not practical for commercial work because of the difficulty of reclaiming the screen after use.

There are four transfer-film methods for making screens: carbon tissue, unsensitized film, presensitized film, and photographic transfer film.

Screen Printing. The process of screen printing has been reviewed (86). In general, printing is done by feeding the paper into the press in sheets, ie, sheet-fed, or from rolls, ie, web-fed. Greeting cards, maps, and some textiles (qv) are printed on sheet-fed presses. Web-fed presses are also used in printing textiles. Many presses are multicolor, ie, they can print a number of colors in succession. Usually each color requires a separate complete unit, ie, inking, plate, and impression mechanism, on the press. Some presses can print both sides of the sheet in one pass through the press.

Manufacture by Screen Printing. A large use of screen printing is in the manufacture of electrical circuits having high volumetric efficiency (small size) compared to standard breadboard circuits. The manufacturing process involves the sequential printing of a circuit layer followed by a drying step. The circuit is usually fired at high temperature ($>600°C$) to sinter the inorganic phases. For some applications the firing step is omitted or replaced by other types of processing, eg, etching. Screen printing uses a screen with a pattern formed by polymer, and an electrically functional ink. The printing is done on a substrate, usually made of ceramic, eg, alumina, but plastic and other materials are also used. These substrates vary in size, but are normally no larger than 15×10 cm and less than 1-mm thick.

Thermal Printing. Thermal printing is a generic name for methods that mark paper or other media with text and pictures by imagewise heating of special-purpose consumable media. Common technologies are direct thermal; thermal, ie, wax, transfer; and dye-sublimation, ie, diffusion, transfer. Properties and preferred applications are diverse, but apparatus and processes are similar (87–89).

Thermal printing usually involves passing materials over a full-width array of electronically controlled heaters (a thermal printhead). This marks thousands of spots simultaneously, so pages print relatively quickly. Image data to control the printhead usually come from computer systems. Black-and-white and full-color systems are both practical. Color is slower and more costly to purchase and use, primarily because this involves three or four successive printing operations, one for each color used.

Printhead Technology. The printhead is common to the various thermal printing technologies. Printhead design and control are important. The heads are expensive page-wide printed circuit-like linear arrays of resistors, which must be fabricated uniformly (90–93). Electronic controls must maintain functional uniformity in use. Self-heating, abrasion, and chemicals evolved during printing reduce both image quality and head life. High power lasers can replace printheads for certain precise thermal printing applications, especially in the area of digital printing. Laser-addressed printers are capable of writing halftones, making such printers useful as digital proofers (94,95).

Direct Thermal Printing. Inexpensive telephone-facsimile printing commonly involves the early direct thermal marking method. Another use is wide-format computer-aided drafting output. Direct thermal involves imagewise heating of a single-component consumable. For example, a coated paper can change color or density as it passes over a printhead. Early processes melted wax overlayers to expose a colored base. However, most systems employ chemical means: a coating of two or more colorless but reactive components is fused to generate a high contrast mark, for instance by using leuco dye chemistry (96). The printing layer may contain a flouran leuco dye precursor, and an acid developer such as bisphenol A in a kaolin and poly(vinyl alcohol) binder (97,98). Direct thermal prints are usually monochrome, but are available in various colors. Marking is usually binary (full density or nothing), but time and temperature adjustment can achieve intermediate values. Maximum density (~ 1.3 optical density) requires 1 to 2 J/cm^2 (0.24–0.48 cal/cm^2) dissipation to heat the coating above $\sim 100°C$ for 1–2 milliseconds. Systems usually write another line of marks about every 5 ms. Most coatings and printhead technology work best at 8 to 24 marks per millimeter spacing (200–600 dots/in.). Prints, which can be a meter or more wide and any length, are relatively unattractive. The media coating feels slippery and is hard to write on. Images also have low contrast. However, printing is fast, uses simple and inexpensive apparatus, and is cheap, running around $0.01 per page.

Thermal Transfer Printing. Wax transfer is another name for thermal transfer printing. Some typewriters and many reliable, high quality ticket and label makers use black-only thermal transfer. Computer-display capture, presentation transparencies, publication proofing, business chart and document printing, and sign-making are uses for color thermal transfer printing.

Thermal transfer involves imagewise heating of a donor ribbon facing a receptor page (99). The donor and receptor meet at the thermal printhead. Donor ribbons print an equivalent size page. These ribbons are typically page-wide rolls of color-coated plastic film, consisting of 3–12-μm thick biaxially oriented polyester or polycarbonate. The printhead side of the film is treated to reduce friction. The marking side has several micrometers of waxes (qv) and binders, pigments and dyes, dispersants, and softening agents coated over a release layer. A typical formulation may be 40% ester wax, 20% microcrystalline binder wax, 20% pigment and dye colorants, 10% oils and other softening agents, and 10% dispersants and other components. Coatings have carefully tailored melting and rehardening temperatures (100).

Color ribbons hold sequentially coated page-sized regions of marking material: magenta, cyan, yellow, and sometimes black. Donor ribbon and receptor pages move together between the thermal printhead and a pressure roller. About 1 J/cm^2 (0.25 cal/cm^2) raises the donor coating temperature above 50 to 80°C to induce melting and adhesion to the cool receptor. Proper peel angle and delay after imaging are important for good image quality.

Full-color printing requires pages be reprinted three or four times in registration. Receptors pass over the printhead several times at exactly the same speed and location, but contact a differently colored section of donor. A fourth pure black color can improve dark picture regions, sharpen text and line art, and mask color misregistration.

Thermal transfer marks are usually shiny, thick, soft and scratchable, and relatively high in optical density and color saturation. The color coating either transfers entirely or not at all. This binary marking process requires picture regions to be halftoned for simulation of intermediate densities and colors. Marked-spot size and shape are usually about the same size as the printhead heating elements. Donor properties and printhead technology limit addressability to about 12 marks per millimeter (300 dots/in.), limiting the quality of halftone reproduction.

Thermal transfer printers are relatively complex, both mechanically and electronically. In addition to precise mechanisms they require sophisticated electronics for networking and PostScript interpretation. Cost of equipment and consumables ($0.50–$1.00 per page) is relatively high.

Receptors range from standard size (A4) to tabloid size (A3) pages, ie, 8.5 × 11 in. (21.6 × 27.9 cm) to 11 × 17 in. (30 × 43 cm), but media-handling mechanisms often prevent marking close to page edges. Special papers are usually required for good print quality. Most color pages physically print in less than one minute, which is often faster than image data acquisition and interpretation.

Dye-Sublimation Thermal Printing.   Dye-diffusion thermal transfer was developed primarily for reproduction of natural pictures. A combination of modest resolution and precise control of individual spot density gives good results for electronic photography and publication proofing. Printing engines are superficially similar to thermal transfer apparatus. Some operate either way. Media handling is about the same, and full color also employs overprinting a page three or four times. However, the actual marking processes are quite different.

Transfer occurs by sublimation, condensation, and diffusion (101). Printhead thermal dissipation causes donor dye to travel to the surface of the donor ribbon and convert directly to a gas. Colorant puffs immediately strike the nearby receptor and soak in, assisted by residual printhead heat.

The donor coatings are thin ($\sim$1 μm) solid solutions of dye ($\sim$6%) in a thermally stable binder that remains attached during printing. This coating may include tiny beads to establish precise donor–receptor spacing, reduce heat loss, and prevent sticking. The coatings are typically solid solutions of volatile dyes in a thermally stable binder such as ethyl cellulose. Receptors must be smooth, and are polyester–silicone coated to aid dye diffusion and stabilization. Finished prints are glossy.

Dye sublimation requires more heat dissipation and a longer ($>$10 ms) heating period to make a dark mark than does thermal transfer. Careful manipulation of heating time and temperature can proportion mark size and dye content to cover a wide density range (0 to ca 2 optical density).

Dye-sublimation printheads usually have 8 to 12 elements per millimeter (200–300 dots/in.). However, colorant dye evolution, transfer, and diffusion cause relatively big, round, and fuzzy marks that usually significantly overlap their neighbors. Thus adjacent dye-sublimation marks mix quite smoothly. Because the amount of dye transferred can be controlled by thermal input from the printhead, the process is actually analogue. Halftoning is not normally required. The resulting pictures can be attractive and resemble photographs. Unfortunately the large fuzzy marks do a poor job of reproducing sharp-edged features such as text.

Print sizes range from tiny (2 $\times$ 3 in. (5 $\times$ 7.6 cm)) to fairly large (A3) (11 $\times$ 17 in. (30 $\times$ 43 cm)). Printing time varies accordingly, but is seldom faster than a few minutes per page. Equipment cost depends on image size, quality, and speed. Cost is usually higher than a similar thermal transfer printer. Consumable costs are high, averaging from $2 to $8 per page.

Electrophotography. The importance of electrophotography (qv) cannot be overstated. This well-named, complex technology describes both the product, ie, an optical copy or photograph of an original, and the key to its process, ie, electrostatics.

The electrophotographic system (102,103) involves two key physicochemical elements: a photoreceptor and a toner. The minimum requirements of the process are (*1*) to charge a photoconductive photoreceptor uniformly; (*2*) to illuminate selectively the photoreceptor to form a latent electrostatic image; and (*3*) to develop the image by applying charged toner. These steps are illustrated in Figure 17.

Some photoreceptors are used with a positive surface charge; others are used with a negative charge. By electrically biasing the toner development module with respect to the photoreceptor and using toner of the proper charge, toner can be attracted to either charged or discharged portions of the latent image. In an office copier it is common to illuminate an original document and relay reflected light to the photoreceptor in order to discharge areas that do not receive toner, eg, the paper background, while attracting toner to the remaining charged areas, eg, text on the original. In electronic printers, however, it is common to illuminate and thus discharge areas that are intended to hold toner, eg, text, while repelling toner from the still charged background. Various models that optimize system performance, as well as competitive and patent positions, have resulted in literally thousands of variants in systems and materials choices.

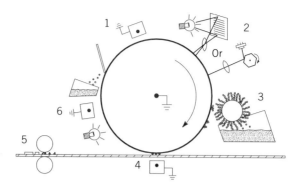

Fig. 17. Schematic photocopier or digital printer. The numbers correspond to the six process steps described in text. The arrow indicates the direction of the drum movement.

Surrounding the three minimum-requirement steps of the electrophotographic system are many additional steps that depend on system design, but by far the most common implementation of electrophotography adds (4) electrostatic transfer of the toned image to paper, and (5) thermal fusing of toner to surface paper fibers and to neighboring toner particles. This fusing often involves contacting the toned image with heated, coated rubber rolls that must not remove significant amounts of toner or disturb the fragile image while heating the image to >100°C.

Once toned, the latent image on the photoreceptor can be used only once; hence, after the transfer step (6) the residual toner is cleaned from the photoreceptor and residual charge is optically erased, after which the process repeats from step (1). This process can be repeated from as slowly as four times per minute in equipment designed for personal use, to more than 200 times per minute in high speed copying or printing equipment. The photoreceptor must survive this series of process steps until it wears out, after 2,000–1,000,000 copies, depending on the design of the equipment. In low volume equipment the photoreceptor, toner supply, and toning, charging, erasing, and cleaning systems are contained in a replaceable toner cartridge that can be discarded or recycled. In high volume applications each subsystem is individually replaced at the end of its service life.

The photoreceptor must hold an electrostatic charge from the time it is charged by a corona device until the image is toned, ie, from 0.25–10 s. The photoreceptor must also be photoconductive so that charge is dissipated where light strikes it. Inorganic photoreceptors can be based on amorphous Se and Se–Te alloys, CdS or CdS–polyester laminates, amorphous Si:H, or ZnO. A huge range of organic photoconductors (OPCs) have been developed to reduce manufacturing costs and eliminate heavy-metal waste, particularly in toner cartridge-based systems. These OPCs involve a thin lower layer designed to absorb light and generate electron–hole pairs, and a thicker dielectric layer above the first, which is designed to resist mechanical wear while transporting photogenerated holes toward the negative surface charge.

Toner particles are always composite materials (qv), sufficiently large to contain the colorant, magnetic material, and polymers, yet small enough to have an average particle size of 3–20 μm. Toner must be electrostatically chargeable, most often by tribocharging, and have a polarity such that the toner is attracted to the appropriate areas of the photoconductor. The toner must fuse to form a durable surface upon mild heating, yet be low cost and easily ground or milled to the appropriate particle size. A wide range of thermoplastics is used, including poly(methyl methacrylate), polystyrene, polyesters, and polyamides. Black toner contains carbon black, and may also contain iron oxide to make it magnetic. Toners can be conductive or insulating, and may have other important properties for a final application, such as magnetic readability and pigment to impart spot or process color. In dual-component development systems, common in high volume systems, the toner particles are used in conjunction with large magnetic carrier beads which, together with a magnetic roller system, brush toner onto the latent image. In monocomponent systems, common in low copy volume systems, the toner particles are themselves made magnetic by inclusion of magnetic materials (qv).

Liquid toners are suspensions of toner particles in a fluid carrier. The carrier is typically a hydrocarbon. Dielectric, chemical, and mechanical properties of the liquid must be compatible with the photoreceptor, the suspended toner particles, and the materials of the development equipment. Liquid toners are capable of producing higher resolution than dry toners because of the smaller (3–5 μm) particle size achievable. Development of the latent image occurs as it passes through a bath of toner and the charged particles are attracted to the oppositely charged surface.

Monochrome image quality has become nearly impeccable in the 1990s, and the difference between lithographic and electrophotographic printing of text is more economic than aesthetic. Halftone images, however, are clearly below newspaper quality in all but the most expensive high resolution digital printers. In inexpensive machines, however, there may still be limitations such as background toning, lack of edge sharpness of lines and curves, image quality limited by the copy lens, banding in solid areas, poor solid area development, mottle, low density, and limited selection of compatible printing substrates.

Electrophotography is especially valuable in those applications where access to one or a small number of copies is required instantly from desktop computers or in a small office, and where a per-copy cost of \$0.07–\$0.10 is acceptable. In walk-up copier/finishers where 1–50 collated and stapled, single- or two-sided documents are required, per-copy cost of \$0.03–\$0.05 is expected. At the highest end of electrophotographic volume where a single machine might produce one-half to a million copies per month, possibly as a copier/digital printer capable of electronic collation, individualization of copy sets, binding with covers, and management of data from a network, cost is \$0.015–\$0.035 per copy. In emerging short-run color printing where spot-color or full-color sheet-fed or web-fed electrophotographic presses push the limits of the technology, cost may be from \$0.25 to as high as \$1 per copy.

Two manufacturers have developed electrophotographic printers for the short-fun fast-response color printing market (104). The advantage to the user is that there is no need to print a large number of an item to reach an economical point for a printing press. Print jobs do not need to be inventoried or discarded when obsolete. Single or a few copies can be rapidly produced with no extra expense. Reprinting, and changing information in the job, is a simple task without the need to prepare film and printing plates because the entire job is handled electronically until the sheets are actually printed.

These machines show markedly different design philosophies. Indigo, Ltd., of Israel, has developed a liquid toner-based sheet-fed printer. It can print on both sides of the page in turn. With appropriate binding options, it is capable of providing fully finished color books. The Xeikon machine (Belgium) is a dry-toner web printer that prints on both sides of the paper in a single pass. It is also capable of in-line binding to produce finished books. These machines are aimed at the large and growing market for a few hundred copies of a color piece at fast turnaround.

Despite the many advantages of electrophotographic printing processes, a principal economic disadvantage remains: each print costs the same. There is no economy of scale as more prints are made, as is the case for conventional printing. This limits electrophotography to the short-run market.

Ink-Jet Printing.   Ink jet is a digital printing process in which small drops of ink are propelled from a nozzle to a receiving surface without contact between the ink source and the surface. There are two type of ink jets: continuous and impulse. The former generate a continuous stream of ink drops from each nozzle. The size and frequency of drops in a continuous ink-jet printer are determined by pumping pressure, ink viscosity, and nozzle size. Drops not needed for printing are electrostatically charged and deflected into a sump. Impulse printers generate drops only in response to a computer signal.

There are two modes of printing using continuous ink jets. These differ in whether deflection is used to adjust the trajectory of a drop to a spot on the paper other than the one it would strike without deflection. Each drop in the binary, or undeflected drop, printer follows one of two possible trajectories. It either flies straight to the target spot on the paper or is deflected to a sump. The operation of a continuous ink-jet system is shown in Figure 18. Binary printers provide the greatest accuracy in drop placement and best image quality because deflection is used only on nonprinting drops. Deflected drop printers are used for low resolution printing applications. For example, a single stationary jet can print lines of readable product codes onto products or containers passing by in a factory.

There are two types of impulse printers (Fig. 19). A piezoelectric ink jet propels a drop by flexing one or more walls of the firing chamber to decrease rapidly the volume of the firing chamber. This causes a pressure pulse and forces out a drop of ink. The flexing wall is either a piezoelectric crystal or a diaphragm driven by a piezoelectric incorporated into the firing chamber (Fig. 19a) (see PIEZOELECTRICS (SUPPLEMENT)). Thermal impulse ink jets also propel one drop at a time, but these use rapid bubble formation to force part of the ink in a firing chamber out the orifice (Fig. 19b).

Continuous Ink Jet is the oldest, most mature of the ink-jet technologies. Development as a computer printer technology began long before the age of digital computers. In 1867 Lord Kelvin conducted experiments on continuous jets and in 1878 Lord Rayleigh published his work on the basic physics of drop formation. Since that time much work has been directed toward understanding and controlling the process, first for printing output from analogue devices

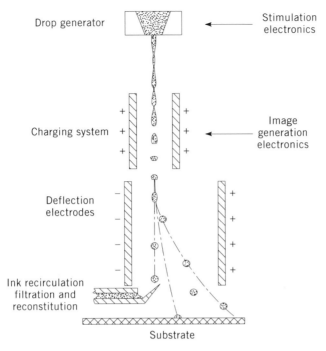

Fig. 18. Schematic of a continuous ink-jet system.

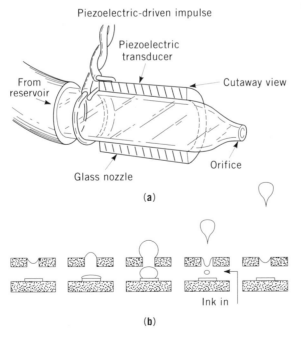

Fig. 19. Ink-jet system: (**a**) piezoelectric ink-jet firing chamber and (**b**) bubble formation in thermal ink-jet technology. Courtesy of Trident Inc.

and later from digital computers. Milestone products include an oscillograph recorder in 1951; the first commercial computer printer, for industrial marking of packages, in 1958; and a large-format four-color printer in 1985. Although continuous ink-jet technology is relatively mature, continued innovations in drop charging technology have enabled high quality printing at print speeds up to 305 m/min. Further advances in equipment are expected to come from improvements in the design and engineering of new printers for specific market applications.

There was a logical progression of technology development from continuous to piezoelectric ink jet. Designers of continuous ink-jet systems ensure that the ink stream breaks into drops of constant size and frequency by applying vibrational energy with piezoelectric crystals at the natural frequency of drop formation. This overcomes the effects of any random forces from noise, vibrations, or air currents.

By changing the role of the piezoelectric crystal from regulating drop formation to propelling drops, the need for high pressure ink-pumping systems, drop charging and deflection systems, and waste ink plumbing systems were eliminated.

The earliest significant technical work on piezoelectric ink jet began in the 1930s and the first true commercial activity was begun in the late 1960s. This early development effort, aimed at office printing applications, had limited commercial success. The first successful piezoelectric ink-jet printer was introduced in 1977. It printed a relatively crude character set using an array of 12 jets in its printhead.

Canon and Hewlett-Packard independently began work on thermal impulse ink-jet technology for a brief period before introducing successful products in 1984. Since that time, both companies have introduced a steady stream of improved products, primarily for office use. Initially, thermal ink-jet products were thought of as improved versions of dot matrix printers. The newest products offer text quality comparable to high quality office printers, and incorporate color capability, even in the lowest cost offerings. Near-photographic color pictures can be created by using software available on standard personal computers. Several million of these devices are sold annually for office and home use.

Inks. The main components of the inks are typically water, colorants, and humectants. Additives are used to control drying time, waterfastness, lightfastness, and consistency of drop formation. Water is an excellent vehicle for ink jet because of its high surface tension and safety in all environments.

The principal physical properties influencing ink performance are surface tension and viscosity. High surface tension is desired for good droplet formation and capillary refill in drop-on-demand ink jet. Low viscosity is desired because less energy is required to pump and eject ink. Conductivity is also an important parameter. Continuous ink-jet inks must have some conductivity to allow for charging. Low conductivity is generally preferred for impulse, particularly thermal ink jet, because excess ions can cause corrosion of the printhead.

Inks for continuous ink-jet printers typically comprise dyes dissolved in water or solvent having salts added to make the ink conductive for electrostatic charging. Whenever waterproof printing is required, low boiling solvent inks are used. For printers that are used in office environments, water is used as the ink solvent. Using water-based inks, humectants may be added to inhibit drying of ink in the sump and surfactants are added to wet the printing surface.

Piezoelectric impulse ink-jet inks may be aqueous or solvent-based. They may also be solid at room temperature. Phase-change inks consist of colorant dissolved or dispersed in waxy polymer. When the printer is turned on, the temperature of the ink is raised and maintained above its melting temperature. The ink solidifies immediately after printing. The first phase-change ink-jet printer was introduced to the market in 1988, setting a new standard for print quality on various types of paper and other substrates. Several companies, eg, Tetranix, have developed products based on the phase-change ink-jet technology.

Piezoelectric impulse ink-jet printers are especially sensitive to bubbles in the ink. A bubble in the firing chamber absorbs some of the compressional force from the flexing of the chamber wall and reduces drop volume and drop velocity, thereby affecting print quality. Because of the limited range of motion of the crystal, bubbles are not readily ejected, and the loss of print quality owing to their presence is persistent.

Thermal impulse ink-jet systems impose extreme requirements on stability of ink ingredients to high temperatures. The temperature of the resistive heater in the firing chamber rises from its steady-state temperature to about 400°C within several microseconds. When the bubble of superheated steam forms at the heater surface and expands to propel some of the ink out of the nozzle, dissolved solids precipitate onto the resistor surface. If these are not immediately redissolved after firing and refilling, then repeated firing causes an accumulation of solids on the resistor that interferes with heat transfer and firing. Even trace quantities of materials that are only sparingly soluble in ink tend to accumulate on the heater and damage the jet over time.

There are several conflicting demands on the ink to achieve high quality prints, fast print speed, and high reliability. For print speed, fast drying time is desired, but the ink must not dry in the ink-jet nozzles in order to avoid plugging. High quality text is achieved by holding the colorant on the surface of the paper, but this slows drying time and leads to intermingling or bleed between adjacent colors. Also, improved waterfastness and durability of the printed images is needed for many applications.

The colorant in ink for ink jet has traditionally been water-soluble dyes. High solubility is needed to prevent the dyes from precipitating and clogging the nozzles. The dye itself, its counterions, and impurities in the dyes can cause crusting. Pigment dispersions are gaining importance as colorants in ink jet because of the generally greater lightfastness and image durability compared to dyes. Using pigmented inks, ink-jet performance is dependent on dispersion quality. The first commercial thermal ink-jet printers containing pigment dispersions were introduced in 1993 by Hewlett-Packard using Du Pont inks.

Humectants and low vapor pressure cosolvents are added to inhibit drying of ink in the nozzles. Surfactants or cosolvents that lower surface tension are added to promote absorption of ink vehicle by the paper and to prevent bleed. For improvements in durability, additional materials such as film-forming polymers have been added. Ink developments are providing ink-jet prints with improved lightfastness, waterfastness, and durability. As a result, such prints are beginning to rival the quality of electrophotographic prints.

Applications. *Industrial Marking.* Ink jets are routinely used to print product identification codes directly onto products or product packages as part of the manufacturing process. The continuous ink-jet system suppliers have made small-character industrial printing systems used for component marking and unit product identification a fertile market area. The continuous ink-jet system's capability of printing at high speeds without contact on a variety of packages has led to a large market opportunity in printing "best used by" messages, lot numbers, and identifying dates and manufacturers on a variety of products from beer cans to wire and cables.

Personalization of Commercial Printing. Ink jet is used for personalizing magazines and bulk-mail advertisements printed on traditional offset printing presses. Scitex Digital Printing sells continuous ink-jet printing stations that may be installed in-line with traditional web-fed offset printing presses. The system includes a data system, a database of variable information, and the ink-jet imager. The variable information consists of names, addresses, etc. The supplier information consists of product names, logos, maps to local outlets, etc.

The improved quality of binary continuous arrays is beginning to enable printing of both the fixed and variable information by the ink-jet imager. This trend is expected to accelerate with improvements in image quality, because of the cost advantage of single-process printing. Scitex Digital Printing has demonstrated a 61 m/min printing press that uses the same technology in a page-wide array. Equipment advances such as these are breathing new life into what had been a mature technology.

Office. Various segments of the office market have the largest population of computers, and thus the greatest demand for computer printers. In 1994, however, more computers were purchased for home use than for the office. The fastest developing computer printer technology is thermal ink jet, whether measured by rates of performance increase, price decrease, or purchase.

Computer-Aided Design. Computer-aided design (CAD) program users are a significant growing market for large-format thermal ink-jet printers. These printers are rapidly replacing pen plotters as the preferred output device (see COMPUTER-AIDED DESIGN AND MANUFACTURING (CAD/CAM)).

Graphic Arts. Continuous ink-jet printers were introduced to the graphic arts in the mid-1980s for proofing of digital image files prior to lithographic printing. Their quality and functionality have been improved and by the mid-1990s they were widely accepted, in spite of the differences in dot structure between ink-jet images and litho images. Because of the slow printing speed, their use has been limited to proofing.

Future Possibilities. Ink jet has the potential to become the predominant digital printing technology because of its cost advantage and simplicity. Growth in image quality of ink-jet technologies is shown in Figure 20. Printing by ink jet requires only one critical process step, ie, the jetting of ink, and can apply all four colors to the page at the same time. By contrast, electrophotography has six critical process steps for each color: clean, charge, expose, tone, transfer, and fuse.

Originally conceived as a universal printing technology suitable for any application, ink jet satisfies a wider range of user requirements than any other printing technology and has become one of the most important methods for office computer printing, industrial marking, graphic arts, and other applications. The

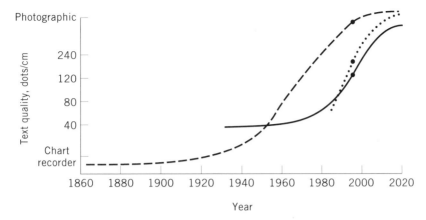

Fig. 20. Growth in image quality of ink-jet technology, where • corresponds to quality in 1995 for (– – –) continuous flow, (—) piezoelectric, and (· · ·) thermal drop-on-demand.

direct, noncontact features of ink jet have been extremely important in opening up applications of ink-jet printing, especially in industrial settings, where the surfaces to be printed are often rough or uneven.

Much of the future performance of ink jet is dependent on advances in ink formulation (see INKS). Early ink-jet printers used special media for printing, but there is a demand for printing on a wide variety of paper stocks using a given ink. A strong driving force in the commercial success of ink jet has been the simplicity and low cost of the printer. Thus there is market resistance to achieving improvements through costly printer enhancements.

Image Distribution

Media and Inks for Printing. Some of the complex interactions in printing processes are exemplified in offset lithographic printing, where several components act together to produce an image on the printed sheet. The offset lithographic printing process is illustrated in Figure 21, where the following steps take place. (*1*) The fountain system delivers water, via the dampening roll, at a fixed temperature, pH, and conductivity, to the surface of the rotating plate; (*2*) the inking system delivers a controlled amount of ink to the plate; (*3*) the fountain solution interacts with the nonimage part of the plate surface; at the same time, the ink adheres to the image areas (the inked image area is composed of halftone dots and/or solids); (*4*) the plate cylinder rotates in contact with the rubber-covered offset (or blanket) cylinder; (*5*) the inked areas of the plate are transferred to the blanket; (*6*) the printing paper, which has its own set of properties, including texture, surface hardness, thickness, strength, and absorbency, passes between the blanket cylinder and the impression cylinder, and the inked image area is transferred to the paper; (*7*) the ink dots settle onto the surface of the printed sheet; some mechanical spreading of the dots occurs because of the pressure between the cylinders and from diffusion; part of the ink is absorbed into the paper; (*8*) solvents in the ink begin to evaporate as soon as the printed dot is exposed to air; assisted in some presses by heat and/or uv radiation, the ink dots begin to dry and the tack of the ink changes accordingly; and (*9*) the printing stock passes into the next color section of the press, where the process

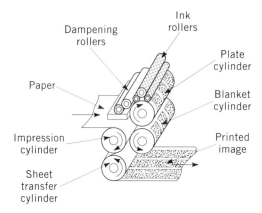

Fig. 21. Offset lithographic printing process. See text.

begins again. These interactions can be better controlled by understanding the various printing media (13).

Printing Ink Formulation.   Inks typically contain three main components: pigments, ie, the colored, solid ingredients; vehicles, ie, the fluid ingredients; and additives (qv), such as driers and extenders.

Pigment Color.   Pigments, the solid coloring substance in ink, are the most noticeable on a printed sheet. Ink pigments typically fall into a few classes of compound: cyans (and greens) are usually copper phthalocyanines, whereas yellows, reds, and oranges are azo pigments or metal salts of azo dyes. Carbon black is usually used for black. Pigment choice is dependent on a number of properties in addition to color. Cost is extremely important in this competitive industry.

Vehicles.   The solid pigments are dispersed into the ink vehicle, which consists of a combination of resin, oil, and solvent. The solvent is absorbed by the paper, leaving a partially dry ink film of resin and oil that binds the pigment to the paper. This film then hardens by oxidation. Oxidation of the vehicle is aided by varnish driers, ie, metallic salts. Cobalt driers are considered the most effective (see DRIERS AND METALLIC SOAPS).

The oils can be either vegetable oils (qv) or hydrocarbon oils. The choice depends on the desired ink properties and cost. Soybean oil, although commonly used, is only one of a number of vegetable oils used successfully. Linseed, corn, canola, and tung oil are also common. A drive toward increased use of soy and other vegetable oil-based inks is motivated in large part by public and governmental concern for the environment. Petroleum-based inks can contribute to air pollution (qv) because of the volatile organic components, principally hydrocarbons, that such inks release as they dry.

A primary source of environmental pollution from printing ink comes from the metal-based pigments used, as well as various resins, waxes, and drying agents that are also part of the inks. These materials are added to inks regardless of the source of the oil. As a result, petroleum inks are just as suitable for landfill disposal under U.S. EPA regulations as are vegetable oil inks.

Ink Transparency.   Four-color lithographic printing depends on the subtractive properties of colored light. When viewing a printed piece, the source of illumination light must pass through several layers of ink before it is reflected back from the paper surface, and finally passes back through the ink layers. The inks act like selective color filters, only permitting certain colors to pass through.

Complex secondary and tertiary colors are formed when the light passes through multiple layers of these ink filters. In order for overprinted inks to act as subtractive color filters, the ink layers must be transparent enough to allow light through in the first place. Completely opaque inks would show only the color reflected from the surface of the top color. Transparency is largely a function of pigment particle size, and how well the pigment is dispersed in the vehicle. Poor dispersion, or large particle size, results in inks that scatter light, and which are thus not transparent. The apparent hue of an ink may also be affected by these properties.

Ink Tack and Flow. Tack is a measure of the cohesion of an ink film, which gives the ink resistance to splitting between two rapidly separating surfaces, such as the plate and the blanket. Ink flow must be balanced with tack. High tack affects the ease with which paper runs through the press. Excessive tack can cause dot loss (picking), and may even damage the surface of the plate. Ink flow refers to the ink property that causes it to level out the way true liquids do. On press, an ink should not exhibit markedly long or short flow properties. Long (low viscosity) inks, such as used with newsprint, have a tendency to fly or mist on press. Short inks tend to cake up on the rollers, plate, or blanket.

Density and Ink Film Thickness. Because ink film thickness directly impacts color, it is important to control film thickness on press. The amount of ink transferred to paper, along with a related factor, dot gain, which refers to how the halftone dots spread under the pressure of the printing process, is the tool by which a press operator monitors color on press.

Paper. A wide variety of substrates, such as paper (qv), polyester or other plastic films, or even metal, can be used as substrates for printing (13). By far, however, most printing is done onto paper. Printing paper is manufactured in an enormous variety of sizes, shapes, colors, and surface finishes. Many of these variables have a significant impact on the appearance of the printed sheet. Others may affect how the paper is handled by the press, how quickly ink dries after printing, and how fast the paper can be moved through the press. Different paper types are suited to the needs and requirements of different types of printing. Categories of paper include ground wood, in which the wood pulp is retained, and wood-free or free sheets, in which the wood pulp is removed and the remaining fibers are bleached. Examples of paper types are book papers, business papers, envelope papers, label papers, and news print (see PULP).

Paper Production and Properties. During the paper production process, the wood pulp is mixed into an aqueous slurry, which is poured onto a continuously moving wire mesh and drained by gravity. Additional water is then squeezed out of the paper by a felt web pressed from above. This pressed side, referred to as the felt side, has smaller, finer fibers and is therefore smoother. It is also the preferred printing side.

The bottom or wire side of the paper exhibits a more pronounced grain and a greater openness, which affects ink appearance on paper, because the wire side loses short fibers, paper sizing, and filler through the wires. Although the wire side is usually not the preferred printing side, if both sides are to be printed, the characteristics of both sides need to be considered.

As the paper moves through the production process, the fibers tend to align themselves parallel to the course in which they are moving. This general orientation of the paper fibers is referred to as the grain or grain direction. Grain is mainly significant when the paper is to be folded. Folding cross-grain weakens the paper along the fold.

Paper absorbency is important for proper ink drying. A paper's surface should allow the ink vehicle to penetrate at the proper rate to achieve proper setting of the ink. If the surface is too absorbent, it causes low ink holdout and loss of gloss. If it is not absorbent enough, it causes ink to transfer to other sheets in a stack, or sheets to stick together.

Many papers are coated to add smoothness, gloss, and brightness. Coatings include binders (such as latex), starches, and clays. Fluorescent brightening agents may also be added. Coated papers are typically used for high quality printing because colors print better on those stocks. Some paper stocks are calendered as an alternative to coating. Calender rollers are sets of steel rollers having polished surfaces, located at the dry end of a paper machine. As paper runs through the rollers under high pressure, it becomes progressively smoother, more compact, and more glossy.

Recycled Paper. Recycling paper involves repulping and chemical de-inking (see RECYCLING, PAPER). Repulping the paper further shortens the wood fibers, lowering bonding strength. Because fiber strength is also reduced by bleaching, it is often not practical to bleach recycled papers to the same extent as virgin paper. As a result, recycled papers do not have the same purity, brightness, and strength of virgin papers. Recycled paper also accepts ink on press much differently from virgin paper.

Paper runability, literally its ability to run through the press, is affected by (*1*) its surface strength, ie, papers used in offset printing require greater surface strength than other papers because of the force applied by viscous inks; (*2*) shear and tensile strength, ie, papers need to be able to withstand the stresses of passing between multiple sets of rollers, often under high pressure; and (*3*) dimensional stability, ie, in offset lithography, a paper must be able to endure stretching in different directions as it passes between rollers and has ink and moisture transferred to it, and when heat is applied to dry ink. Any permanent stretch can lead to registration problems, which affect the ability of the press to print color consistently.

Paper printability, its ability to be ink-receptive, is determined by smoothness, absorbency, porosity, and ink holdout capability.

The Printing Process. True integration of all facets of the printing process is limited, in most cases, to sporadic measurements of the individual elements and processes that contribute to the final product. There is only a limited effort underway as of the mid-1990s to establish close links between electronic prepress and conventional processes such as proofing, plate-making, and printing.

Variables. The basic operations of printing use consumable products that affect print quality. Photographic film used in conjunction with a laser film recorder or imagesetter accepts the digital data as a halftone representation of a scanned image. This film is then used for analogue proofing applications and ultimately for making plates for color printing.

Exposure and development must be predictable and repeatable because multiple pieces of film may be used for a single plate exposure. The proper exposure source and the intensity of that source must be established. The imagesetter must be properly linearized to produce the desired precise halftone dots. This film must be handled under correct safelight conditions and machine-processed in chemical developers at the correct speed and temperature.

Proofing materials have many of the same sources of variability as film, plus the added problem of registering, ie, accurately overlaying, the different colored layers. The printing plate must be exposed precisely to hold all that is discernible in the films. The accuracy of the exposure is critical because the plate must retain all the information contained in the films for faithful reproduction on the press.

Instrumentation Used for Process Control. Although measurement tools have been used for some time in the printing industry, instruments such as densitometers have only begun to gain wider acceptance in the 1990s, especially in the production of final films from scanning. Rather than trained personnel, densitometers are widely used at the printing press to control color. As print buyers become more sophisticated, more accurate color control is needed to ensure that the job is produced according to specifications.

Usually specifications are dictated by readings of the solid density numbers (100% dot or solid area), uniform density across the sheet, and dot gain (apparent dot growth from film to printed sheet) values in the quartertone (25%), midtone (50%), and three-quartertone (75%) dot regions, that typify the high quality printing expected when using new four- to eight-color printing presses. The proof is often used as the quality standard which the printed piece must meet.

Instruments like colorimeters and spectrophotometers are used less often. These are used primarily for manufacturing control of printing inks. Frequently, however, inks other than yellow, magenta, and cyan are used for spot-color applications, and in those instances a spectrophotometer ensures the correct match of an ink blend to standard.

Process control (qv) of a lithographic printing press is a difficult chore even for an experienced operator. Largely, control of the press has been left to experienced operators who manipulate a mixture of water and ink to control an ink film emulsion that eventually transfers to the paper to create the printed piece. Among the variables that may need to be adjusted are the following: ink tack, which measures the cohesion of an ink film; ink–water balance, because too much water in ink can cause emulsification and destroy the necessary relationship needed to print sharply; pH and temperature of both the fountain solution, which comprises the water, alcohol, or alcohol substitutes, and the buffering agents, which prevent inking the nonimaged areas of the plate; hardness and squeeze of the various rollers that transport the ink/water and paper through the printing press; amount of ink delivered to each zone across the width of the press; condition of the blankets that pick up the ink from the plate and transfer it to paper, ie, if blankets are worn or misused, image deterioration may result, and if packed incorrectly, image quality may be affected by higher or lower dot gain.

Guidelines for uniform density and dot gain limits have been established. These criteria are the standards used to control the press. A densitometer is used to measure these standards and measurements are made on sheets removed as the press is running. The operator then uses these measurements to guide press adjustments, primarily ink flows, to correct the printed result to standard (8).

Finishing.   Some printing, such as stationery or small posters, can be delivered as printed, but most printing must be converted from printed press sheets to a finished piece through various finishing and bindery processes. Finishing is a general term that includes a number of different and often specialized operations that transform printed press sheets into their final form (8,13). Some of the most common finishing operations include cutting, folding, stitching, collating, binding, scoring, perforating, round cornering, drilling or punching, die-cutting, embossing, laminating, and padding.

Most finishing work cannot be done in-line with a press. However, in some instances, such as web-fed presses, some folding, collating, and binding can take place. This is confined to printing where large quantities are required, such as newspapers, catalogs, and magazines. When preparing press sheet layouts, print planners must consider the type of finishing equipment they use because the printer's imposition, ie, the arrangement of pages in a press layout to ensure the correct order after the printed sheet is folded, must suit the finishers' equipment.

Binding.   The work required to convert press sheets into finished books, magazines, and catalogs is called binding. Binding includes processes such as saddle and side stitching, perfect binding, mechanical or coil binding, and book binding. All binding methods usually begin by folding a single press sheet printed on both sides down to a signature. Signatures vary from 4–64 pages on a single sheet to maximize press sheet usage, but are usually 16 or 32 pages.

The simplest and least expensive method of binding is the saddle stitch method, in which signatures are collated together and placed on a saddle beneath a mechanical stitching head. Staples are forced through the spine of the book. Most magazines are assembled using the saddle stitch method because of cost and the reader's need to open and lay the magazine flat. When too many signatures or pages are required, side stitching is used as an alternative to saddle stitching. Side stitching allows for bulk assembly. Wire staples are forced through the side of a group of pages about 0.25 in. from the edge. A cover is generally glued around the piece to finish it off. A disadvantage of side stitching is that the finished book cannot lay flat. *National Geographic* magazine is an example of side stitch binding.

Perfect binding or adhesive binding has become a popular alternative to stitched binding. However, this kind of binding is not as durable as stitched pages. The process starts with the collation of pages into a set. The set moves to a roughing station where the back or folded edge is ground off, leaving a rough surface for the adhesive to grip. Heated adhesive is applied and then a cover is attached. Perfect binding allows for books to lay flat and is commonly used for telephone and paperback books.

Mechanical or coil bindings are used for notebooks and calendars. The pages are collated and special drilling equipment is used to punch holes in the paper. Metal or plastic coils are inserted. This durable binding allows finished books to lay flat.

Edition binding or case binding is the most common method used for school textbooks and other hardbound books, including this *Encyclopedia*. This process uses collated signatures that are sewn on special machines which pass thread through the folds of each signature and are knotted at the back. Glue is applied to the spine and hard covers are attached. Finished books are then dried in special hydraulic presses.

Throughout the binding process, many finishing operations may be used to enhance the final appearance of the book. Round cornering of pages, covers, and inserts, along with trimming the final product, all add to the successful assembly of the final piece.

Environmental Aspects

Printing processes generate waste. Many environmental regulations apply, as do several alternative choices for action. With the common goal of protecting the environment, printers must balance and carefully analyze all waste minimization, treatment, and recycling programs to minimize costs in a competitive industry (see RECYCLING; WASTE REDUCTION; WASTES, INDUSTRIAL).

The environmental restrictions under which printers must operate continue to increase. Federal, state, and local governments all have a say in controlling discharges to the air, land, and water. Improving environmental performance means moving up the waste management scale. On the lower end of the scale the strategies of waste disposal and waste minimization are found. On the higher end are the superior strategies of waste recycle and the ultimate strategy of no waste.

Recycling. Many printing wastes can easily be directed to well-established recycling outlets, ie, corrugated packaging material, scrap paper, and aluminum-based printing plates. There is also an existing market for recovery of silver and polyester from films used in the printing process.

Photoprocessing chemicals, solvents, and inks are much more difficult to recycle, but printers who partner with their suppliers and promote the use of environmental products have made progress in this area of recycling as well.

Waste Materials and Emissions. In 1990, there were significant changes in the Clean Air Act, which resulted in provisions for better control of volatile organic compounds (VOCs) that directly impact air quality. Printing presses and associated operations usually generate VOCs through the use of solvents, press washes, fountain solutions, and inks. Minor contributions also come from film and plate production. Various approaches to reducing VOCs include refrigeration of fountain solutions; add-on controls, ie, combustion of ink/solvent residue; press enclosures; and substitution of low VOC products.

Sewer Disposal. Photoprocessing and printing wastes tend to be aqueous solutions that are combined with other plant effluents and sent to the local sewer plant for treatment. The parameters of concern include silver, pH, and biological oxygen demand (BOD). BOD is a measure of how well a waste material degrades in the environment. Lower values are preferred. Silver-bearing waste streams are typically treated on-site, and the treated effluent is released to the drain. The printer usually receives a small cash credit for silver recovered.

As of 1995, the trend was to reduce the impact on sewer systems and the printer's environmental liability. This can be accomplished through aqueous systems where quality can be maintained, through recycling, or through waste minimization programs. Quite often, printers avoid local sewer systems in treating wastes, for fear of violating permit conditions. The newest technologies involve fully recyclable chemistries, thus eliminating sewer waste.

Solid and Hazardous Wastes. Among the most stringent environmental regulations impacting the printing industry are the hazardous waste control standards issued under the Resource Conservation and Recovery Act (RCRA). These rules focus on hazardous wastes such as chlorinated solvents, flammable waste inks, and wastes containing heavy metals. Printers must segregate these materials on-site and ship them via an approved hazardous waste hauler to a permitted facility. Many solid wastes generated from the printing process are not hazardous and often may be recycled locally, eg, corrugated paperboard, aluminum printing plates, and paper from scrap product.

Employee Safety and Health Issues. Printers, like other industrial employers, must be concerned with employee safety and occupational health regulations. Increased public awareness and a trend toward healthier life-styles have resulted in a push for reducing hazardous chemicals, especially carcinogens, in the workplace. Among other things, printers should have thorough training programs in the areas of employee safety and training.

Economic Aspects

The printing and publishing industry, though large and growing, is highly fragmented. The value of shipments from the industry worldwide exceeded $400 billion in 1993, $176 billion in the United States alone. The industry is growing at a real growth rate of 3–8%/yr, ~2–4%/yr in North America and Western Europe, and 6–8%/yr in developing countries of the world. The printing industry consumed approximately $82 billion in materials in 1993 to produce its products, of which approximately 90% was paper and ink. The outlook for the printing industry is reviewed annually (105).

On a worldwide basis the printing industry ranks as number four in manufacturing employment. Over 4.5 million people are employed at more than 350 thousand sites. In the United States, 1993 employment exceeded 1.5 million, an increase of over 180 thousand jobs in the period of 1982–1993. Similar growth in employment is being experienced worldwide. Table 1 shows industry trends in the United States, which are typical of those worldwide. These do not include revenues from packaging or office copying.

The printing industry is undergoing unprecedented change that is expected to continue into the twenty-first century as digital technology, economic restructuring, and global competition combine to change the operating environment. The printers' role in this changing industry is still evolving but is certainly expected to involve more than just producing ink on paper. It is estimated that before the year 2050, 50% of sales revenues will come from products not produced in 1994 (106). Traditional print, however, should continue to be the primary communication vehicle for information well into the twenty-first century. Significant restructuring and capital investments are being made to optimize production processes and to position for future growth. This may result in short-term excess capacity. Over the long-term, however, the printing industry, as it becomes a more integral part of the multimedia communications industry, should continue to grow and prosper.

Table 1. 1987–1993 Printing and Publishing Industries Value of Shipment, $ \times 10^9$

Industry segment	1987	1990	1993	CAGR[a], %
newspapers	31.90	34.60	35.80	1.9
periodicals	17.30	20.40	22.70	4.6
books				
publishing	12.60	15.30	18.20	6.3
printing	3.30	4.10	4.90	6.8
miscellaneous publishing	7.80	8.90	10.80	5.8
commercial printing	44.80	52.90	61.60	5.5
printed packaging	54.00	58.00	60.00	1.7
business forms	7.40	7.80	7.10	−0.7
greeting cards	2.90	3.80	4.80	8.8
blank books	2.90	3.10	3.70	4.1
bookbinding	1.20	1.40	1.60	4.9
typesetting	1.80	1.90	2.00	1.8
plate-making services	2.40	2.80	3.10	4.4
Total	*190.30*	*215.00*	*236.30*	*4.0*

[a]CAGR = compound annual growth rate.

BIBLIOGRAPHY

"Printing and Reproducing Processes" in *ECT* 1st ed., Vol. 11, pp. 126–149, by R. F. Reed and M. H. Bruno, Lithographic Technical Foundation; "Printing Processes" in *ECT* 2nd ed., Vol. 16, pp. 494–546, by M. H. Bruno, International Paper Co.; in *ECT* 3rd ed., Vol. 19, pp. 110–163, by M. H. Bruno, Consultant.

1. J. N. Fields, *Graphic Arts Manual*, Arno Press Inc./Musarts Publishing Corp., New York, 1980.
2. L. Stroebel and co-workers, *Photographic Materials and Processes*, Butterworth & Co. (Publishers) Ltd., Stoneham, Mass., 1986.
3. P. Dyson and R. Rossello, *Digital Cameras for Studio Photography*, Seybold Publications, Media, Pa., 1995.
4. *Digital Color Prepress*, Agfa Prepress Education Resources, Mt. Prospect, Ill., 1993.
5. R. W. G. Hunt, *The Reproduction of Color in Photography, Printing and Television*, Fountain Press, Tolworth, U.K., 1987.
6. R. K. Molla, *Electronic Color Separation*, Chapman Printing Co., Parkersburg, W. Va., 1988.
7. J. A. C. Yule, *Principles of Color Reproduction*, John Wiley & Sons, Inc., New York, 1967.
8. J. M. Adams, D. D. Faux, and L. J. Rieber, *Printing Technology*, 3rd ed., Delmar Publishers Inc., Albany, N.Y., 1988.
9. F. Wentzel, R. Blair, and T. Destree, *Graphic Arts Photography: Color*, Graphic Arts Technical Foundation, Pittsburgh, Pa., 1987.
10. *A Guide to Color Separation*, 3rd ed., Agfa Prepress Education Resources, Mt. Prospect, Ill., 1995.
11. *An Introduction to Digital Scanning*, Agfa-Gevaert NV, Belgium, Brussels, 1994.
12. M. Southworth and D. Southworth, *Color Separation on the Desktop*, Graphic Arts Publishing, Livonia, N.Y., 1993.
13. R. Blair and T. M. Destree, *The Lithographers Manual*, 8th ed., Graphic Arts Technical Foundation, Inc., Pittsburgh, Pa., 1988.
14. J. Cogoli and co-workers, *Graphic Arts Photography: Black and White*, Graphic Arts Technical Foundation, Pittsburgh, Pa., 1981.

15. *Mac World* and *PC World*, International Data Group, San Francisco, Calif.
16. M. G. Naugle, *The Illustrated Network Book*, Van Nostrand Rheinhold, Co., Inc., New York, 1994; R. W. Klessig and K. Tesink, *SMDS: Wide Area Data Networking with Switched Multi-megabit Data Service*, Prentice Hall, Inc., Englewood Cliffs, N.J., 1995; R. P. Davidson, *Broadband Networking ABCs for Managers*, John Wiley & Sons, Inc., New York, 1994.
17. M. R. Nelson, *The Data Compression Book*, M & T Books, Redwood City, Calif., 1991.
18. T. A. Welch, *IEEE Computer*, **17**, 6, (1984).
19. G. K. Wallace, *Commun. ACM*, **34**, 30 (1991).
20. J. D. Murray and W. van Ryper, *Encyclopedia of Graphics File Formats*, O'Reilly and Associates, Sebastopol, Calif., 1994.
21. *PostScript Language Reference Manual*, 2nd ed., Adobe Systems, 1990.
22. M. H. Bruno, *Pocket Pal*, 13th ed., International Paper Co., New York, 1983.
23. R. Ulichney, *Digital Halftoning*, MIT Press, Cambridge, Mass., 1987.
24. R. Dessauer, *Image Technology*, **12**(2), 27 (1970).
25. C. E. Looney and co-workers, *Photogr. Sci. Eng.* **16**(6) 433, 1972.
26. U.S. Pat. 3,445,234 (May 20, 1969), L. A. Cescon and R. Dessauer (to E. I. du Pont de Nemours & Co.).
27. U.S. Pat. 3,615,454 (Oct. 26, 1971), L. A. Cescon, R. Dessauer, and R. L. Cohen (to E. I. du Pont de Nemours & Co.).
28. U.S. Pat. 3,390,996 (July 2, 1968), A. MacLachlan (to E. I. du Pont de Nemours & Co.).
29. J. Mattor and co-workers, *Symposium on Novel Imaging Systems*, Boston, Mass., 1969; U.S. Pats. 3,394,391-5 (July 23, 1968); 3,410,687 (Nov. 12, 1968); 3,413,121 (Nov. 26, 1968) (to S. D. Warren Co.).
30. E. Brinckman and co-workers, *Unconventional Imaging Processes*, Focal Press, New York, 1978.
31. J. Shepard, *J. Appl. Photogr. Sci.* **52**, 3486 (1982).
32. U.S. Pats. 3,152,904 (Oct. 13, 1964) and Reissue 26,719 (1969), D. F. Sorensen and J. Shepard (to 3M Co.).
33. U.S. Pat. 3,392,020 (July 9, 1968), H. C. Yutzy (to Kodak Co.).
34. U.S. Pat. 3,457,075 (July 22, 1969), D. Morgan and B. Shely (to 3M Co.).
35. T. Tanaka, *Advances in Non-Impact Printing Technology*, IS & T, Portland, Oreg., 1991.
36. U.S. Pat. 3,136,637 (June 9, 1964), G. W. Larson (to 3M Co.).
37. U.S. Pat. 3,671,236 (June 20, 1972), P. C. Van Beusekom (to 3M Co.).
38. Eur. Pat. 0048160 (July 24, 1981), R. Hallman and co-workers, (to NAPP Systems).
39. U.S. Pat. 4,282,308 (Aug. 4, 1981), A. B. Cohen and R. N. Fan (to E. I. du Pont de Nemours & Co.).
40. U.S. Pat. 3,649,268 (Mar. 14, 1972), V. F. H. Chu and A. B. Cohen (to E. I. du Pont de Nemours & Co.).
41. D. F. Eaton, A. G. Horton, and J. P. Horgan, *J. Photochem. Photobiol. A: Chem.* **58**, 373 (1991).
42. B. M. Monroe and G. C. Weed, *Chem. Rev.* **93**, 435 (1993).
43. U.S. Pats. 4,895,787 (Jan. 23, 1990) and 5,049,476 (Sept. 17, 1991), S. Platzer (to Hoechst-Celanese).
44. U.S. Pats. 4,174,216 (Nov. 13, 1979); 4,247,619 (Jan. 27, 1981); 4,304,839 (Dec. 8, 1981), A. B. Cohen and R. N. Fan (to E. I. du Pont de Nemours & Co.).
45. J. H. Choi, *J. Img. Tech.* **15**, 190 (1989).
46. U.S. Pat. 4,396,700 (Aug. 2, 1983), M. Kitajima, H. Tachikawa, and T. Ikeda (to Fuji Photo Film Co.).
47. S. Honma and E. Inoue, *Photogr. Sci. Eng.* **27**, 47 (1983).

48. Y. Yamamura, S. Hayashi, and T. Kaneko, *J. App. Photo. Eng.* **9**, 143 (1983).
49. R. W. Woodruff, W. Jeffers, and R. A. Snedeker, *Photogr. Sci. Eng.* **11**, 93 (1967).
50. I. J. Berkower, J. Buetel, and P. Walker, *Photogr. Sci. Eng.* **12**, 283 (1968).
51. T. Nakayama and K. Shimazu, *Photogr. Sci. Eng.* **22**, 138 (1978).
52. U.S. Pats. 5,001,036 (Mar. 19, 1991); 5,028,511 (Mar. 17, 1991), J. H. Choi (to E. I. du Pont de Nemours & Co.).
53. U.S. Pats. 4,923,780 (May 8, 1990); 4,489,154 (Dec. 18, 1984), H. W. Taylor, Jr. (to E. I. du Pont de Nemours & Co.).
54. U.S. Pats. 5,108,868 (Apr. 28, 1992); 5,053,310 (Oct. 1, 1991), S. Platzer (to Hoechst Celanese).
55. U.S. Pat. 4,482,625 (Nov. 13, 1984), A. Namiki and co-workers (to Fuji Photo Film Co.).
56. U.S. Pats. 4,666,817 (May 19, 1987), L. W. Sachi; 4,656,114 (Apr. 7, 1987), B. M. Cederberg and A. K. Mussur (to 3M Co.).
57. E. Cockburn and co-workers, *Eur. Polym. J.* **24**, 1015 (1988).
58. K. Ichimura and T. Komatsu, *J. Poly. Sci.* **25**, 1475 (1987).
59. U.S. Pats. 4,260,673 (Apr. 7, 1981), R. I. Krech; 4,571,373 (Feb. 18, 1986) A. K. Mussur and P. M. Koelsch (to 3M Co.).
60. K. F. Hird, *Offset Lithographic Technology*, Goodheart, New York, 1991.
61. U.S. Pat. 4,172,772 (Oct. 30, 1979), M. Ould and G. N. Stevens (to Vickers Ltd.).
62. U.S. Pat. 2,714,066 (July 26, 1955), J. M. Case and C. L. Jewett (to 3M Co.).
63. U.S. Pat. 3,276,868 (Oct. 4, 1966), F. Uhlig (to Azoplate Corp.).
64. U.S. Pat. 4,153,461 (May 8, 1979), G. Berghauser and F. Uhlig (to Hoechst AG).
65. Jpn. Pat. K 85,123,846 (Dec. 9, 1983), K. Shimura and S. Gotou (to Konishiroku Photo KK).
66. U.S. Pat. 4,631,245 (Feb. 4, 1985), G. Pawlowski (to Hoechst AG).
67. U.S. Pat. 3,867,147 (Feb. 18, 1975), J. A. Teuscher (to American Hoechst Co.).
68. Eur. Pat. 547,578 (Dec. 16, 1992), T. Sekiya (to Fuji Photo Film Co.).
69. U.S. Pat. 5,286,603 (Feb. 15, 1994), J. R. Wade, R. M. Potts, and M. J. Pratt (to Vickers Ltd.).
70. U.S. Pat. 4,113,893 (Sep. 12, 1978), E. A. Hahn (to PPG Industries Inc.).
71. H. A. Gaur and co-workers, *Makromol. Chem.* **185**(9) 1795–1808 (1984).
72. Eur. Pat. 403,096 (May 25, 1990), M. Ali (to 3M Co.).
73. U.S. Pat. 3,635,709 (June 5, 1969), K. Kobayashi (to Polychrome Corp.).
74. J. Kosar, *Light-Sensitive Systems*, John Wiley and Sons, Inc., New York, 1965, p. 342.
75. U.S. Pat. 3,894,873 (July 15, 1975), M. Kobayashi (to Toray Industries).
76. Brit. Pat. GB 2,034,911 (Oct. 24, 1979), T. Fujita and M. Iwamoto (to Toray Industries).
77. L. G. Larson, *TAGA Proceedings*, 334–344 (1975).
78. R. Demmerle, *Proceedings of IS&T 3rd Technical Symposium on Prepress, Proofing and Printing*, 1993, pp. 215–218.
79. R. Demmerle, *Ibid.*, pp. 219–231.
80. E. Atkins, *Lasers Graphics*, **2**, 36–59 (1981).
81. A. Katsuhiko, *J. Soc. Photogr. Sci. Technol. Japan*, **54**, 668–673 (1992).
82. L. A. DeSchamphelaere, *Research Disclosures*, **18805**, 708–710 (1979).
83. U.S. Pat. 4,859,290 (Aug. 22, 1989), P. J. Watkiss (to E. I. du Pont de Nemours & Co.).
84. S. B. Doyle and P. J. Watkiss, *TAGA Proceedings*, 173–193 (1989).
85. A. Rott and E. Weyde, *1972 Photographic Silver Halide Diffusion Transfer*, Focal Press, Boston, Mass., 1972.
86. W. Appleton, *Screen Printing: A Literature Review*, Pira International, Leatherhead, Surrey, U.K., 1984.

87. J. Sturge, V. Walworth, and A. Shepp, *Imaging Processes and Materials*, 8th ed., Van Nostrand Reinhold Co., Inc., New York, 1989, pp. 9, 274–278, 280–281, 384–388.

88. J. F. Komerska, in A. S. Diamond, ed., *Handbook of Imaging Materials*, Marcel Decker Inc., New York, 1991, pp. 487–526.

89. G. A. Nothmann, *Nonimpact Printing*, Graphic Arts Technical Foundation, Pittsburgh, Pa., 1989, pp. 93–98.

90. P. A. McLaughlin and co-workers, *Proceedings of Microelectronics Symposium*, San Francisco, Calif., 1973.

91. P. Barnwell and co-workers, *Electro. Comp. Sci. Tech.* **6**, 3–4 (1980).

92. A. Buttler, *Proceedings of European Hybrid Microelectronics Conference*, Avignon, France, 1981.

93. H. Schofield, *5th Annual European ICG Conference on Thermal Printing Technology*, Amsterdam, Netherlands, 1989.

94. C. DeBoer, *IS&T Proceeding of 7th International Congress on Advances in Non-Impact Printing Technologies*, Portland, Oreg., 1991, p. 449.

95. W. A. Tolbert and co-workers, *J. Imaging Sci. Tech.* **37**, 411–421 (1993).

96. A. Igarashia and T. Ikeda, in J. Gaynor, ed., *Advances in Non-Impact Printing Technologies for Computer and Office Applications*, Van Nostrand Reinhold Co., Inc., New York, 1982, pp. 886–892.

97. U.S. Pat. 4,536,220 (Aug. 29, 1985), M. Kondo and co-workers (to Kanzaki Paper Co.).

98. U.S. Pat. 4,539,225 (Sept. 3, 1985), K. Kojima, H. Yamahira, and Y. Oeda (to Kanzaki Paper Co.).

99. Y. Tokunaga and K. Sugiyama, *IEEE Trans. Elect. Dev.* **27**, 218–222 (1980).

100. U.S. Pat. 4,503,095 (Mar. 5, 1989), T. Seto and Y. Shimazaki (to Fuji Photo Film Co.).

101. R. A. Hahn and N. A. Beck, *IS&T Proceedings of 5th International Congress on Advances in Non-Impact Printing Technologies*, San Diego, Calif., 1989, pp. 441–448.

102. R. M. Schaffert, *Electrophotography*, Focal Press, Boston, Mass., 1980.

103. L. B. Schein, *Electrophotography and Development Physics*, Springer-Verlag, Berlin, 1993.

104. A. E. Karsh, "Agfa Shows System for Short Run Color," *Seybold Report on Publishing Systems*, **23**, 13 (Oct. 1993).

105. Technical data, Graphic Arts Technical Foundation, Pittsburgh, Pa., 1994.

106. Technical data, Printing Industry of America, Alexandria, Va., 1994.

General References

M. H. Bruno, *Principles of Color Proofing*, GAMA Communications, Salem, N.H., 1986.

J. Kosar, *Light-Sensitive Systems*, John Wiley & Sons, Inc., New York, 1965.

J. Sturge, V. Walworth, and A. Shepp, *Imaging Processes and Materials*, 8th ed., Van Nostrand Reinhold Co. Inc., New York, 1989.

J. M. Adams, D. D. Faux, and L. J. Rieber, *Printing Technology*, 3rd ed., Delmar Publishers Inc., Albany, N.Y., 1988.

J. N. Fields, *Graphic Arts Manual*, Arno Press Inc./Musarts Publishing Corp., New York, 1980.

R. Blair and T. M. Destree, *The Lithographers Manual*, 8th ed., Graphic Arts Technical Foundation, Inc., Pittsburgh, Pa., 1988.

ARTHUR J. TAGGI
PETER WALKER
Du Pont Printing and Publishing

PROCESS CONTROL

Basic Elements and Equipment

In order to operate a process facility in a safe and efficient manner, it is essential to be able to control the process at a desired state or sequence of states. This goal is usually achieved by implementing control strategies on a broad array of hardware and software. The state of a process is characterized by specific values for a relevant set of variables, eg, temperatures, flows, pressures, compositions, etc. Both external and internal conditions, classified as uncontrollable or controllable, affect the state. Controllable conditions may be further classified as controlled, manipulated, or not controlled. Excellent overviews of the basic concepts of process control are available (1–6).

Process Systems. Because of the large number of variables required to characterize the state, a process is often conceptually broken down into a number of subsystems which may or may not be based on the physical boundaries of equipment. Generally, the definition of a system requires both definition of the system's boundaries, ie, what is part of the system and what is part of the system's surroundings; and knowledge of the interactions between the system and its environment, including other systems and subsystems. The system's state is governed by a set of applicable laws supplemented by empirical relationships. These laws and relationships characterize how the system's state is affected by external and internal conditions. Because conditions vary with time, the control of a process system involves the consideration of the system's transient behavior.

Process systems are broadly categorized as self-regulatory and nonself-regulatory. The former is one in which a change in an external condition can cause the system to move from an initial steady state to another steady state without additional external intervention. The latter, a nonself-regulatory process system, does not achieve another steady state without additional control action once the first external change occurs.

Controlled Conditions, Correcting Conditions, and Control Algorithm. The basic elements of process control are the conceptual definition of the process system; the selection of the controlled conditions, the correcting conditions, and the disturbance sources to be addressed; and the selection of the control algorithm. The goal of process control is achieved by adjusting the values of an appropriate subset of process variables, ie, the correcting conditions or manipulated variables, so as to change the values of other process variables, ie, the controlled conditions or variables, to compensate for variations and disturbances in the process system. The controlled variables are selected so that their values characterize the state of the process system and the process and operating objectives. The manipulated variables are selected so that these can easily be manipulated to affect the controlled variables. The control algorithm defines how the manipulated variables are to be adjusted to bring the controlled variables to their desired values, ie, to bring the process system to its desired state.

Generic Control Strategies. The two generic strategies for process control are feedback and feedforward control. Most process control strategies are based on one or a combination of these strategies (1–3).

The conceptual structure of the feedback control strategy is shown in Figure 1a. A feedback control strategy measures the controlled variable after it has been affected and, therefore, after a deviation between the controlled variable measurement and its desired setpoint value may have occurred. The deviation is compensated for by adjusting the manipulated variable(s). In this strategy, the flow of information is from the output of the process, namely, the measured controlled variable, to the process's manipulated variable inputs. By continuing to measure the output response as well as the manipulated variable compensating changes, the feedback control strategy in essence continues to seek, by trial and error, the values of the manipulated variables to bring the process into balance. This strategy is based on deviations from the desired controlled variable setpoint. It continually compensates for disturbances regardless of their source.

Some of the inherent advantages of the feedback control strategy are as follows: regardless of the source or nature of the disturbance, the manipulated variable(s) adjusts to correct for the deviation from the setpoint when the deviation is detected; the proper values of the manipulated variables are continually sought to balance the system by a trial-and-error approach; no mathematical model of the process is required; and the most often used feedback control algorithm (some form of proportional–integral–derivative control) is both robust and versatile.

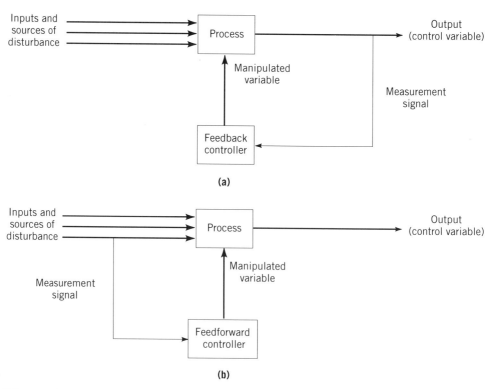

Fig. 1.   Control structures, where (——) is the process and (——) the measurement and control, for (**a**) feedback and (**b**) feedforward.

The main disadvantage of the feedback control strategy is that no corrective action is taken until after a deviation between the measured controlled variable and its setpoint is detected. In processes having large, frequent disturbances and large dead times, large lag times, or both, this characteristic can significantly degrade control performance. Pure feedback control in such situations leads to continual oscillations, large time period of deviation from the setpoint, or instability.

The feedforward control strategy (Fig. 1b) addresses the disadvantages of the feedback control strategy. The feedforward control strategy measures the disturbance before it affects the output of the process. A model of the process determines the adjustment in the manipulated variables(s) to compensate for the disturbance. The information flow is therefore forward from the disturbances, before the process is affected, to the manipulated variable inputs.

The primary advantage of the feedforward over the feedback control strategy is that corrective action is initiated before the controlled variable is upset. Feedforward control, however, has its own drawbacks, ie, variables used to characterize the disturbances must be measurable; a model of the response of the controlled variable to the disturbance must be available (when the feedforward strategy is used alone, the control performance depends on the accuracy of the model); and the feedforward control strategy does not compensate for any disturbance not measured or modeled.

In most process plant situations where feedforward control is appropriate, a combination of the feedforward and feedback control is usually used. The feedforward portion reduces the impact of measured disturbances on the controlled variable while the feedback portion compensates for model inaccuracies and unmeasured disturbances. This control strategy is referred to as feedforward control with feedback trim.

Control Equipment. In the early 1960s, electrical analogue control hardware replaced much of the pneumatic analogue control hardware in many process industries. Certain control elements, ie, control valve actuators, remain pneumatic as of this writing (1996). Electrical analogue controllers of the 1960s were single-loop controllers in which each input was brought from the measurement point in the process to the control house where most of the controllers were located. The output from the controller was then sent from the control house to the final control element. The operator interface consisted of a control panel having a combination of display faceplates, chart recorders, etc, for the single-loop controllers and indicators. Control strategies were primarily feedback.

During the late 1960s and early 1970s, a few firms introduced process control computers (PCC) to perform direct digital control (DDC) and supervisory process control (SPC). In cases where the system made extensive use of DDC, the DDC loops often had close to 100% analogue control backup making the systems costly. Other early systems primarily used the PCC for SPC. Regulatory control was provided by analogue controllers, which did not require backup, but the operator's attention was split between the control panel and the computer screens. The terminal displays provided the operator interface when DDC/SPC was being used, but the control panels were still located in the control house for the times when the analogue backup was necessary. Within the PCC environment, some firms began to broaden the use of advanced control techniques such as feedforward control, model-based predictive control, interaction decoupling control, multilevel cascades, and dead-time compensation (7).

Distributed Control Systems. From the mid-1970s to early 1980s, control vendors introduced microprocessor-based distributed control systems (DCSs) and programmable logic controllers (PLCs) (Fig. 2). The local control units (LCUs) containing microprocessors implemented a number of controller blocks using a proportional–integral–derivative (PID) or other algorithm, and performed limited internal logic. The process interface unit (PIU), also containing microprocessors, provided a means of bringing in and sending many analogue and digital signals to field hardware. Cathode ray tubes (CRTs) provided the operator interface as in the case of the PCCs. However, DCS displays tended to be more graphical in nature. Communication between the components of the DCS utilized digital communications through a data highway. These communications, controller block functions and capabilities including alarming, and interactions between controllers, inputs, and outputs were all defined by software, not by hardwiring. DCSs, therefore, revolutionized many aspects of process control, from the appearance of the control room to the widespread use of advanced control strategies.

Since the early 1980s, the capability and power of DCSs have increased. There has been a general increase in the use of digital communications technology within process control. Some advanced control strategies, previously only implemented on a PCC, are, as of the mid-1990s, implemented within the DCS. Most local control units perform their own analogue-to-digital (A/D) and digital-to-analogue (D/A) conversion. The LCUs can be located in equipment rooms closer to the process. Digital communications via a coaxial or fiber optic cable send information back to the control room, thus saving on wiring costs. With this

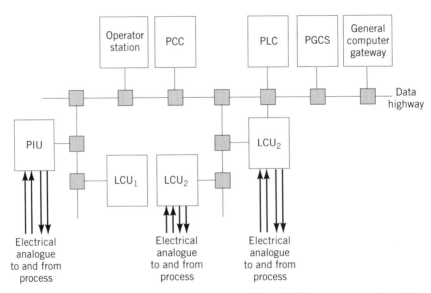

Fig. 2. Distributed control system structure, where PIU = process interface unit; LCU_1 = local control unit, model 1; LCU_2 = local control unit, model 2; PCC = process control computer; PLC = programmable logic controller; and PGCS = process gas chromatograph system.

trend toward increased use of digital communications technology, smart transmitters and smart actuators are also gaining in popularity (see SMART MATERIALS (SUPPLEMENT)). These devices, equipped with their own microprocessor, perform tasks such as autoranging, autocalibration, characterization, signal conditioning, and self-diagnosis at the device location. Thus tasks required of the local control unit or the data acquisition unit are reduced.

The features that have made DCSs popular are (1) reduction in wiring and installation costs through the use of data highways and remotely located local control units; (2) reduction in the space requirements for panel in the control room, leading to less area needed for blast resistance; (3) improved operator interface, most operators prefer customized CRT displays to control panels; (4) ease of expansion, modularity of DCSs eases system expansion and integration; (5) increased flexibility in control configuration, control strategies may often be modified without any need for rewiring; and (6) improved reliability for control systems as a whole owing to redundancy, modularity, etc.

Programmable Logic Controllers. Initially, programmable logic controllers (PLCs) were dedicated, stand-alone, microprocessor-based devices executing straightforward binary logic for sequencing and interlocks. These were originally intended for applications which, prior to that time, had been implemented with hardwired electromechanical and electrical relays, switches, pushbuttons, timers, etc (8). PLCs significantly improved the ease with which modifications and changes could be implemented to such logic. Although many of the early applications were in the discrete manufacturing industries, the use of PLC quickly spread to the process-related industries. From the time of their introduction in the late 1960s, PLCs have become increasingly more powerful in terms of calculational capabilities, eg, PID algorithms; data highways to connect multiple PLCs; improved operator interfaces; and interfaces with personal computers and DCSs. In processes where control is dominated by logic-type controls, PLCs are a preferred alternative to a DCS. Because of the availability of relatively smooth integrated interfaces between DCSs and PLCs, practice as of the 1990s is generally to use an integrated combination of a DCS and PLCs.

Safety and Shutdown Systems. Safety interlocking, shutdown, and sequencing logic is used to protect personnel and equipment from potentially dangerous situations. These systems are an important part of the overall process control system. Prior to the 1970s, this logic was implemented by hardwiring combinations of discrete input and output devices which were electromechanical or electrical. Whereas PLCs have become a popular means of implementing this process logic, often portions are still implemented in hardware for reasons of reliability, cost, etc. Figure 3 shows a schematic where the contact has been hardwired and the PLC is used for purposes of lockout, ie, the pushbutton must be pressed once the switch has tripped. This type of setup is used in especially hazardous situations where the operator must verify that everything is all right before the unit can be taken out of its safe state. Usually a bypass switch is also included to allow the output device to be returned if the PLC fails.

PLCs allow the connection of combinations of software relays, switches, etc, and do not require the hardwiring of physical devices. The more complex and interconnected the logic and the greater the number of devices, the greater the benefits of the PLCs in terms of flexibility in implementing, changing the

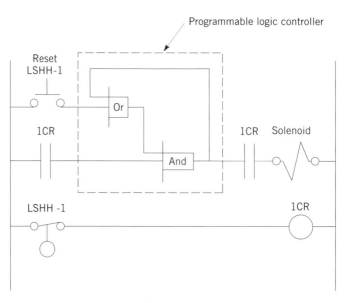

Fig. 3. Hardwired system having a PLC backup with lockout where the reset button must be pushed once the switch trips. CR = contact relay and LSHH = level switch high–high alarm; "or" and "and" gates are designated as boxes.

logic, and saving physical space. As digital technology has improved, the use of PLCs in these applications has also increased. Various approaches have been devised to increase reliability (8–10). Dynamically redundant PLCs may be hooked up in series (Fig. 4**a**) or in parallel (Fig. 4**b**). The series configuration minimizes the chance that a triac failing to turn on causes the solenoid to energize when it should not, because now the triacs in both systems would have to fail. However, the series hookup also increases the possibility of shutdown owing to the failure of a single triac failing open. The parallel hookup reduces the possibility of a shutdown owing to a single triac failing open but, in turn, increases the probability of an unsafe condition owing to a triac failing closed. Whereas redundant PLCs increase reliability, they still require regular checking.

Special high reliability systems have been developed for safety shutdowns, eg, triple modular redundant systems (TMRS), which are designed to be fault-tolerant (11,12), ie, the system can have an internal failure and still perform its basic function. Basically a TMRS (Fig. 5) consists of three identical subsystems actively performing identical functions simultaneously. The results of the three subsystems are compared in a two-of-three voting network prior to sending the signals to the output devices. If any one of the subsystems experiences a failure, the overall system can still function properly as long as two of the subsystems are working. This setup allows the identification of components suspected of failure. To further increase reliability, multiple sensor and output devices may be used (12,13). When multiple sensors are used, the system is often implemented along with a two-of-three voting network.

Smart Transmitters, Valves, and Field Bus.   There is a clearly defined trend in process control technology toward increased use of digital technology. The trend, which started with digital controllers, has increasingly spread from that

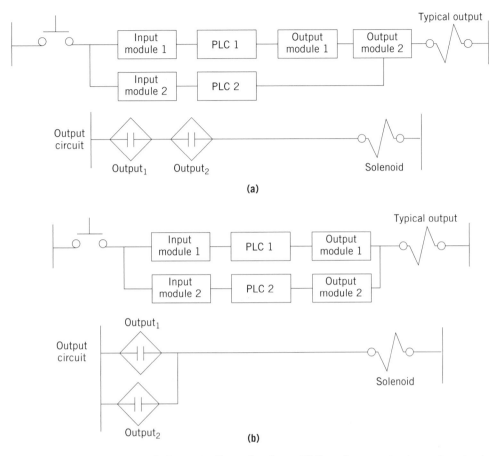

Fig. 4. Configuration of dynamically redundant PLCs where output$_1$ and output$_2$ correspond to the outputs from output modules 1 and 2, respectively: (**a**) series, and (**b**) parallel.

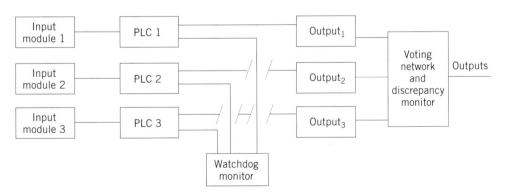

Fig. 5. Structure of a triple modular redundant system (TMRS). See text.

portion of the overall control system outward toward field elements such as smart transmitters and smart control valves. Digital communication occurs over a field bus, ie, a coaxial or fiber optic cable, to which intelligent devices are directly connected and transmitted to and from the control room or remote equipment rooms as a digital signal. The field bus approach reduces the need for twisted pairs and associated wiring (Fig. 6).

Facility Control Hierarchy.   The hierarchy of control levels in a process-related plant is outlined in Table 1. Advanced higher level controls are a means to achieve and maintain consistent and improved quality. Advances in the

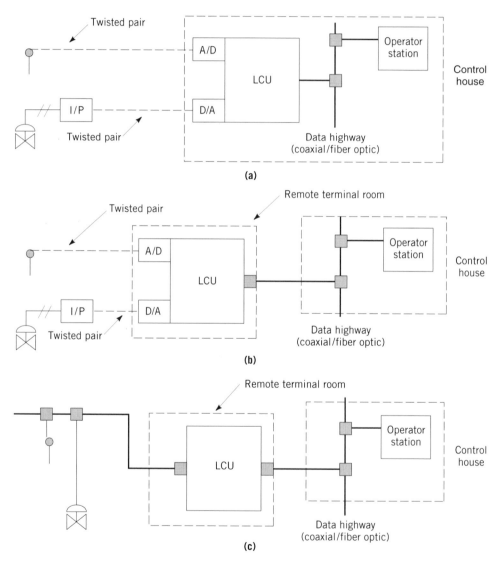

Fig. 6. Impact of field bus on wiring costs, where A/D and D/A correspond to analogue-to-digital and digital-to-analogue converters, respectively; LCU = local control unit; and I/P = current-to-pressure transducer. (**a**) DCS without remote terminal rooms; (**b**) DCS with remote terminal rooms; and (**c**) DCS with remote terminal rooms and field bus.

Table 1. Conceptual Levels in a Process Facility Control Hierarchy

Conceptual control level[a]	Class of plant functions	Operation	Communication time	Data level	Class of disturbances[b]	Frequency Action	Frequency Scan
plant-wide optimization		directions for local optimization and supervisory level	hours		low or intermittent frequency[c]	low or event-driven	low
	stewardship			CIP[d]			
local optimization and supervisory		optimum settings on local basis, eg, unit operation	minutes to hours		low to medium frequency	low	medium
	analysis and coordination			PCC[e] simulation and database management			
advanced regulatory		assist in achieving better regulatory control of operations	seconds to minutes		medium to high frequency	medium	medium
	operations			DCS[f]/PCC and other transaction or event-driven system			
basic regulatory		stabilize operations at F, P, L, T[g] setpoints	0.1 s to seconds		high to medium frequency	high	high

[a] Control increases from top to bottom; information from bottom to top.
[b] Amplitudes vary unless otherwise noted.
[c] Amplitude is usually large.
[d] CIP = computer-integrated processing.
[e] PCC = process control computer.
[f] DCS = distributed control system.
[g] F = flow, P = pressure, L = level, and T = temperature.

137

capabilities of DCSs make the development of advanced higher level controls within the DCS feasible. Modern process facilities are often designed with a relatively high degree of process integration in order to minimize the theoretical cost of producing the product. From an operation's standpoint, however, this integration gives rise to relatively complex interactions between the operating variables. Thus, it can be relatively complex to determine how to adjust the plant to optimize the operation.

Each of the four conceptual control levels has its own requirements and needs in terms of hardware, software, techniques, and customization. Because information flows up in the hierarchy and control decisions flow down, effective control at a particular level occurs only if all the levels beneath the level of concern are working well. Thus, good instrumentation is the foundation of good hierarchial control. Generally, the advanced higher level control applications are aimed at one or more of the following: (*1*) determining and maintaining the plant at a practical optimal operating point given the current conditions and economics; (*2*) maintaining safe operation for the protection of personnel and equipment; (*3*) minimizing the need for operator attention and intervention; and (*4*) minimizing the number, extent, and propagation of upsets and disturbances. The control applications in the upper control levels must emphasize reliability and take into account the person–machine interface as well as the control aspects. For example, particular control applications are designed to ease the operator's effort in commissioning and monitoring the multilevels of cascades along with the control applications themselves. Part of the application is designed to monitor and manage the suitability of the cascades and to break them to a safe operating level when appropriate. Other important factors are proper training and documentation.

The highest level of control is the plant-wide optimization level. The primary goal at this level is to determine the optimal operating point of the plant's mass and energy balance and to adjust the relevant setpoints in an appropriate manner. There are many possible process interactions and combinations of constraints involved, thus these control applications help identify the optimal operating point. The appropriate optimization objective is dependent on the market situation, so that, at times, the optimization objective is to maximize production; at other times it is to minimize operating costs for a fixed production rate. The control applications at this level often utilize steady-state mathematical models for the process or portions thereof. These models must be tuned to a specific plant's operation and constraints in order to ensure that the key aspects of the overall plant operation are characterized. Execution frequency of the control applications at this level are therefore from hours to days or months, depending on the frequency of relevant variations. Standard mathematical techniques and heuristics based on experience are used to determine the operating point that best accomplishes the selected optimization objective. Once the optimum operating point has been determined, the relevant setpoints are passed down to the lower control levels. Because of the time constants and dynamics associated with the top level's control and manipulated variables, setpoints are usually ramped incrementally to their new values in a manner such that the process is not disturbed and the proximity to constraints can be periodically checked before the next increment is made.

The plant optimization control level applications determine the values of key variables that optimize the overall plant material and energy balance. The control applications at the local optimization and supervisory control level, on the other hand, focus on subsystems within the overall plant. These subsystems usually consist of a single or, at most, a few highly interactive pieces of equipment. Most of the applications at this level are aimed at optimizing the subsystem within an operating window defined by soft constraints, eg, values determined by the plant optimization level applications, and hard constraints, eg, equipment material limits. Often the optimal operating point of the subsystems is against one of the constraints of the operating window. Hence, many of these control applications employ a constraint control strategy, ie, a strategy that pushes the subsystem against the closest active constraint. Typically the closest currently active constraint changes with time and situations, eg, between day and night, different weather conditions, different operating states of upstream equipment, etc. The constraint control strategies continually make minor adjustments to keep the substantials along the active constraint, ie, near optimum. As such adjustments are made continually and with vigilance by the advanced control application, these applications can generate significant benefits over the course of a year, although these benefits may appear minor when viewed at a single point in time. In addition to local optimization applications, the control level also includes multivariable, predictive, and model-based control strategies such as dynamic matrix control, inferential controls, closed-loop analyzer controls, and generally control strategies involving extensive measurement validity checking and programming logic and computation. The control applications at the local optimization and supervisory level typically provide setpoints for the controls at the advanced regulatory and basic regulatory control levels.

The general objective of the advanced regulatory control level applications is to improve the performance of basic regulatory control level controllers. The execution frequency of applications at this level is typically in the range of seconds to minutes. Controls are often implemented in the DCS. They are focused on portions of the subsystems of the local optimization and supervisory control level, and differ from the controls at the basic regulatory level in that the former controls are often multivariable and anticipatory in nature. These control applications often fall into one of the following categories: (*1*) complete or partial interaction decoupling; (*2*) feedforward ratio or feedforward additive with feedback trim; and (*3*) compensated control or calculated control variable. The level of control closest to the process is the basic regulatory control level. Good performance of this level is crucial for the benefits of higher levels of control.

Instrumentation

Components of a Control Loop. Instrumentation, which provides the direct interface between the process and the control hierarchy, serves as the fundamental source of information about the process state and the ultimate means by which corrective actions are transmitted to the process. Figure 7 illustrates the hardware components of a typical modern control loop. The function of the process measurement device is to sense the value, or changes in value, of process variables. The choice of a specific device typically requires considerations of the

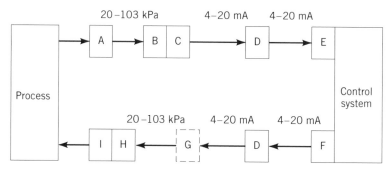

Fig. 7. Instrument components of a control loop, where A = process measurement devices, in this case, pressure measurement; B = transducer; C = transmitter; D = signal transmission; E = analogue-to-digital converter (A/D); F = digital-to-analogue converter (D/A); G = current-to-pressure transducer (I/P); H = actuator; and I = final control element. Some devices combine A, B, and C, and (– – –) represents the line to and from the control system. Examples of pressure measurements and the corresponding current values are given. To convert kPa to psia, multiply by 0.145.

specific application, economics, and reliability requirements. The actual sensing device may generate a physical movement, pressure signal, millivolt signal, etc. The function of the transducer is to transform the measurement signal from one physical or chemical quantity to another; in this instance, pressure to milliamps. The transduced signal is then transmitted to a remote location through the transmission line. The transmitter is therefore a signal generator and a line driver. Often the transducer and the transmitter are contained in the same device. Most modern control equipment requires a digital signal for displays and control algorithms, thus the analogue-to-digital converter transforms the transmitter analogue signal to a digital format. Because the A/D converters may be relatively expensive if adequate digital resolution is required, the incoming digital signals are usually multiplexed. Prior to sending the desired control action, which is often in a digital format, to the final control element in the field, the desired control action is usually transformed by a D/A converter to an analogue signal for transmission. The D/A converters are relatively inexpensive and are not normally multiplexed. Widespread use of digital control technologies has made A/D and D/A converters part of the control system.

Once the desired control action has been transformed to an analogue signal, it is transmitted to the final control element over the transmission lines. However, the final control element's actuator may require a different type of signal and thus another transducer may be necessary. Many control valve actuators utilize a pressure signal so a current-to-pressure (I/P) transducer is used to provide a pressure signal to the actuator.

Process Measurements.   The most commonly measured process variables are pressures, flows, levels, and temperatures (see FLOW MEASUREMENT; LIQUID-LEVEL MEASUREMENT; PRESSURE MEASUREMENT; TEMPERATURE MEASUREMENT). When appropriate, other physical properties, chemical properties, and chemical compositions are also measured. The selection of the proper instrumentation for a particular application is dependent on factors such as the type and nature of the fluid or solid involved; relevant process conditions; rangeability, accuracy, and repeatability required; response time; installed cost; and maintainability and reliability. Various handbooks are available that can assist in selecting sensors (qv) for particular applications (14–16).

Pressure. Most pressure measurements are based on the concept of translating the process pressure into a physical movement of a diaphragm, bellows, or a Bourdon element. For electronic transmission, these basic elements are coupled with an electronic device for transforming a physical movement associated with the element into an electronic signal proportional to the process pressure, eg, a strain gauge or a linear differential variable transformer (LDVT).

To avoid maintenance problems, the location of pressure measurement devices must be carefully considered to protect against vibration, freezing, corrosion, temperature, overpressure, etc. For example, in the case of a hard-to-handle fluid, an inert gas is sometimes used to isolate the sensing device from direct contact with the fluid.

Flow. The principal types of flow rate sensors are differential pressure, electromagnetic, vortex, and turbine. Of these, the first is the most popular. Orifice plates and Venturi-type flow tubes are the most popular differential pressure flow rate sensors. In these, the pressure differential measured across the sensor is proportional to the square of the volumetric flow rate.

Orifice plates are relatively inexpensive and are available in many materials to suit particular applications. This sensor is generally preferred for measuring gas and liquid flows. However, orifice plates have a relatively high unrecoverable pressure drop and limited range. Investment cost for a Venturi-type flow tube is generally higher than for an orifice plate for the same application. However, the higher unrecoverable pressure drop of the orifice plate sometimes dictates the use of a Venturi-type flow tube because of the overall cost in particular instances, ie, investment plus operating, or process considerations.

The proper installation of both orifice plates and Venturi-type flow tubes requires a length of straight pipe upstream and downstream of the sensor, ie, a meter run. The pressure taps and connections for the differential pressure transmitter should be located so as to prevent the accumulation of vapor when measuring a liquid and the accumulation of liquid when measuring a vapor. For example, for a liquid flow measurement in a horizontal pipe, the taps are located in the horizontal plane so that the differential pressure transmitter is either close-coupled or connected through downward sloping connections to allow any trapped vapor to escape. For a vapor measurement in a horizontal pipe, the taps should be located on the top of the pipe and have upward sloping connections to allow trapped liquid to drain.

Magnetic flow meters are sometimes utilized in corrosive liquid streams or slurries where a low unrecoverable pressure drop and high rangeability is required. The fluid is required to be electrically conductive. Magnetic flow meters, which use Faraday's law to measure the velocity of the electrically conductive liquid, are relatively expensive. Their use is therefore reserved for special situations where less expensive meters are not appropriate. Installation recommendations usually specify an upstream straight run of five pipe diameters, keeping the electrodes in continuous contact with the liquid.

Vortex meters have gained popularity where the unrecoverable pressure drop of an orifice meter is a concern. The vortex shedding meter is based on the principle that fluid flow about a bluff body causes vortices to be shed from alternating sides of the body (von Karman vortex street) at a frequency proportional to the fluid velocity. The vortex shedding meter can be designed to produce either a linear analogue or a digital signal. The meter range, depending on process conditions, is relatively large (10:1). Because of the lack of moving parts and the absence of needs for additional manifolds or valves, the reliability and safety of this meter are relatively good. The main restrictions on its use are (1) it should not be used for dirty or very viscous fluids; (2) the Reynolds number should be greater than 10,000 but less than a process condition-dependent maximum set by cavitation, compressibility, and unrecoverable pressure drop; and (3) meters >20 cm (8 in.) tend to have limited applicability because of relatively high cost and limited resolution. Similar to other flow meters, the vortex shedding meter requires a fully developed flow profile, and therefore a meter run.

Liquid Level. The most widely used devices for measuring liquid levels involve detecting the buoyant force on an object or the pressure differential created by the height of liquid between two taps on the vessel. Consequently, care is required in locating the tap. Other less widely used techniques utilize concepts such as the attenuation of radiation; changes in electrical properties, eg, capacitance and impedance; and ultrasonic wave attenuation (see ULTRASONICS).

Temperature. Temperature sensor selection and installation should be based on the process-related requirements of a particular situation, ie, temperature level and range, process environment, accuracy, and repeatability. Accuracy and repeatability are affected by the inherent characteristics of the device and its location and installation. For example, if the average temperature of a flowing fluid is to be measured, mounting the device nearly flush with the internal wall may cause the measured temperature to be affected by the wall temperature and the fluid boundary layer. Thermocouples are the most widely used means of measuring process temperatures. These are based on the Seebeck effect. The emf developed by the hot junction is compared to the emf of a reference or cold junction, which either is held at a constant reference temperature or has compensation circuitry. The difference between the hot junction and the reference junction temperature is thus determined. Depending on the temperature range, temperature level, etc, various combinations of metals are used by the thermocouple, eg, Chromel/Alumel (Type K), iron/Constantan (Type J), and platinum–10% rhodium/platinum (Type S). Because thermocouple emfs are low level signals, it is important to prevent the contamination of the signal by stray currents, ie, noise, resulting from the proximity to other electrical devices and wiring.

Thermocouples are placed within protecting tubes, called thermowells, for protection against mechanical damage, vibration, corrosion, stresses owing to flowing fluids, etc. These thermowells impact the speed of response of the thermocouple by placing an additional lag in the control loop. Special thermowell designs do exist, which minimize this added lag. In hostile process environments where the reliability of a temperature measurement is a concern, multiple temperature sensors are sometimes used in conjunction with a majority voting system which can be implemented in software or hardware. Where exceptional accuracy and repeatability are required, resistance thermometry detectors

(RTDs) are sometimes used although these are more expensive than thermocouples. RTDs are based on the principle that as the temperature increases, the electrical resistance of conductors also increases. RTDs can experience many of the same problems as thermocouples, so considerations such as thermowells and protection from electrical noise contamination are also appropriate in the case of RTDs.

Analyzers. Onstream analyzers measure various physical and chemical properties as well as component compositions. Compared to most other instrumentation, analyzers are relatively expensive, more complex and sensitive, and require more regular maintenance by trained personnel. Therefore, the expense for onstream analyzers needs to be justified by the benefits generated through use. Improvements in analyzer technology, digital control systems, and process control technology have led to increasing use of analyzers in closed-loop automatic control applications. The most common physical and chemical properties measured by analyzers include density, viscosity, vapor pressure, boiling point, flash point, cloud point, moisture, heating value, thermal conductivity, refractive index, and pH. Some of these analyzers are continuous; others are discrete. More detailed discussions are available (14,15,17). Compositional analysis is usually performed by various types of spectrographic or chromatographic analyzers. Some are designed for specific components, eg, oxygen, carbon monoxide, H_2S, and total sulfur; others, eg, process gas chromatographs, are customized for particular applications. Reference 17 contains a discussion of the theory and practice associated with compositional analyzers.

Key considerations in using an analyzer for closed-loop control are repeatable, reliable analyzer measurement and appropriate analyzer system response time. In control applications at the supervisory control level and above, accuracy as well as repeatability is often required. On-line process analyzers may be placed into three categories. In the first, *in situ*, the analysis is continuous and the probe mounted directly in the process stream. In this category, the measurement can be treated similarly to other process measurements, as long as some additional care is taken owing to the reliability issues. In the second category, the analysis is continuous but the sample is not naturally in a form required by the analyzer. Thus, the sample must be conditioned. In the third category, the analyzer takes a period of time to analyze a discrete sample which must usually be conditioned. Also, a sample-and-hold circuitry is required to keep an output signal at its last value between analyses. In the latter two categories, the analysis introduces dead time into the control loop. Therefore, to achieve good closed-loop control in these instances, the additional dead time introduced by the measurement should be minimized and the control algorithm should contain some form of dead-time compensation.

A sketch of the setup for an analyzer in the third category is shown in Figure 8. Most of these analyzers require sample takeoff, sample transport, sample conditioning and preparation, analysis, and sample return or disposal. These subsystems need to be carefully designed to ensure that the analyzer meets its intended purpose and is reliable. The sample transport subsystem is required because analyzers are often placed in the protected, controlled environment of an analyzer shelter remote from the sample takeoff. The sample location should be selected so that the sample is representative, the complexity of the

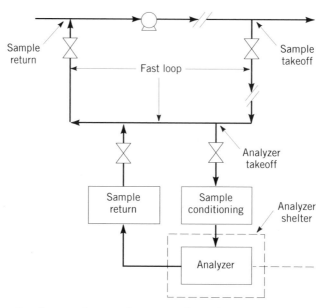

Fig. 8.   Schematic of an analyzer subsystem where (−−−) represents the line to and from the control system. The sample conditioner is usually located on the outside wall of the analyzer shelter for maintenance ease.

sample conditioning subsystem is minimized, and the equipment is accessible. It is therefore preferable that the sample is a single phase, relatively clean, and that the takeoff location does not add significant dead time to the control loop. Furthermore, the pressure at the sample takeoff should be such that the pressure differential between the sample capture and sample return point is adequate to drive the sample through the fast loop at a sufficient velocity to avoid the need for a sample pump. The purpose of the fast loop is to bring the sample to the vicinity of the analyzer takeoff at a high velocity.

Many factors need to be considered for the proper selection, design, and specification of an analyzer, its sample handling subsystems, and its intended use within a process control strategy and hierarchy, including the interface to the control system. As a rule of thumb, the measurement effective dead time plus lag time should be no greater than one-sixth of the time constant of the process and other elements in the control loop. Consequently, most analyzers are utilized at the supervisory control level and above. Thus, for example, a process gas chromatograph system having a cycle time, including the sample handling subsystems, of five minutes is required for a supervisory or local optimization control loop where the effective time constant is thirty minutes. Therefore, only *in situ* analyzers should be considered at the regulatory control level. Furthermore, there is usually a direct relationship between reduced analyzer system cycle time and increased analyzer system cost.

Signal Transmission and Conditioning.   A wide variety of physical and chemical phenomena are used to measure the many process variables required to characterize the state of a process. Because most processes are operated from a control house, these values must be available there. Hence, the measurements are usually transduced to an electronic form, most often 4 to 20 mA, and then transmitted to the control house or to a remote terminal unit and then to the control house (see Fig. 6). Wherever transmission of these signals takes place in twisted pairs, it is especially important that proper care is taken so that these measurement signals are not corrupted owing to ground currents, interference from other electrical equipment and distribution, and other sources of noise. Most instrument and control system vendors publish manuals giving advice and instructions for installation and engineering practices for the proper grounding, shielding, cable selection, cable routing, and wiring for control systems. The importance of these considerations should not be underestimated (14,18).

Control Valves and Other Final Control Elements. Good control at any hierarchial level requires good performance by the final control elements in the next lower level. At the higher control levels, the final control element may be a control application at the next lower control level. However, the control command must ultimately affect the process through the final control elements at the regulatory control level, eg, control valves.

Control Valves. Material and energy flow rates are the most commonly selected manipulated variables for control schemes. Thus, good control valve performance is an essential ingredient for achieving good control performance. A control valve consists of two principal assemblies, a valve body and an actuator (Fig. 9**a**). Good control valve performance requires a consideration of the process characteristics and requirements, eg, fluid characteristics, range, shutoff, and safety; and control requirements, eg, installed control valve characteristics and response time. The proper selection and sizing of control valves and actuators is an extensive topic in its own right (14–16,19). Many control valve vendors provide computer programs for the sizing and selection of valves.

The valve body, the portion that contains the process fluid, consists of the internal valve trim, packing, and bonnet. The internal trim determines the relationship between the flow area and the stem position, which is usually proportional to the air signal. This relationship, referred to as the inherent valve characteristic, is often classified as linear, equal percentage, or quick opening.

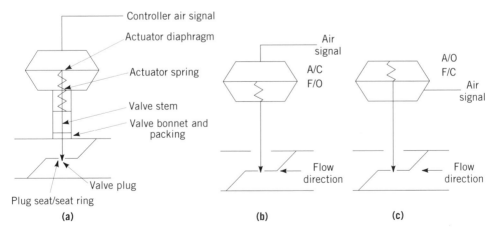

Fig. 9. Control valve and actuator: (**a**) nomenclature; (**b**) fail-open (F/O) or air-to-close (A/C) action; and (**c**) fail-closed (F/C) or air-to-open (A/O) action.

Actual valve characteristics generally fall within these three classifications. For a globe-type valve, the internal trim consists of plug, seat ring, plug stem, plug guide, and in some instances a cage. In rotary style valves, such as a ball- or butterfly-type, the internal trim consists of a ball or vane, seal ring, rotary shaft, and bushings. Although the internal trim fixes the relationship between flow area and stem position, the relationship between the control air signal and the flow area may be modified by the use of different cams in the case of rotary style valves, or via software in the case of smart valves or control systems having software capabilities.

The actuator provides the force to move the stem or rotary shaft in response to changes in the controller output signal. Actuators must provide sufficient motive force to overcome the forces developed by the process fluid and the valve assembly; be responsive in quick and accurate positioning corresponding to changes in the control signal; and be responsive in automatically positioning the valve in a safe position when a failure (eg, instrument air or electrical, a safety interlock, or a shutdown) has occurred. Most actuators are either of the spring–diaphragm-, piston-, or motor-type.

The general approach to specifying a control valve involves selecting a valve body type, trim characteristics, size, and material based on the process fluid characteristics, desired installed valve characteristics, process conditions, and process requirements. The actuator is then specified based on the valve selected, process flow conditions, required speed of response, etc. Meeting the requirements of a safe fail-position involves considering both the valve and the actuator. Figure 9**b** and 9**c** are schematics of how a fail-open (F/O) or a fail-closed (F/C) requirement may be met by using a spring–diaphragm actuator (direct acting configuration) and the preferred flow-to-open plug configuration. Figure 10 illustrates how a three-way solenoid valve may be used to implement a safety interlock or shutdown condition where the fail-safe condition is open. Analogous setups may be devised for other conditions based on the action of the solenoid with and without power and the control valve failure action.

Various accessories can be supplied along with the control valves for special situations. Positioners ensure that the valve stem is accurately positioned

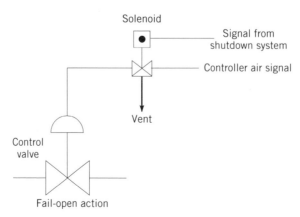

Fig. 10.   Use of solenoid valve for shutdown. Upon signal to shutdown, the solenoid closes a controller air signal port and opens a vent port.

following small or slowly changing control signals or where unbalanced valve forces exist. Boosters, which are actually pneumatic amplifiers, can increase the speed of response or provide adequate force in high pressure applications. Limit switches are sometimes included to provide remote verification that the valve stem has actually moved to a particular position.

In addition to control valves which regulate the flow of one stream, process facilities sometimes use modulating three-way control valves to adjust simultaneously the flows of two streams, either to divert one stream into two streams or to combine two streams into one. The stable operation of these three-way control valves requires that the flow tends to open the plugs. These valves are of either a diverting or a mixing design (Fig. 11). The most common uses of three-way control valves are in controlling the heat transferred by regulating the flow through and around a piece of heat-transfer equipment and in controlling the blending of two streams. The choice of using a diverting or mixing service in a heat-transfer application is determined by pressure and temperature consideration. The upstream valve location (diverting service) is preferred if there is no overriding consideration. If there is a change of phase involved in the heat-transfer equipment, then a diverting valve should be used. Three-way control valves should not be used in services if the temperature is high (>260°C), if there is a high temperature differential (>150°C), or if there is a high pressure or pressure differential. If any of these conditions exist, two single-seated two-way valves should be used to implement the bypass control strategy even though this is a more expensive initial investment.

Other Types of Final Control Elements. Devices other than control valves are also used as final control elements. Dampers are used to control the flow of gases and vapors. Louvers are also used to control the flow of gases, eg, flow of air in air fin coolers. Feeders such as screw feeders, belt feeders, and gravimetric feeders are often used to control the flow of solids. Metering pumps and certain feeders combine the functions of the measurement and final control element in some control loops. Pumps (qv) having variable speed drives are also used as final control elements. In these cases, the pump speed adjustment is achieved by adjusting the speed of the prime mover, eg, electric motor and steam turbine, or through the transmission linkage.

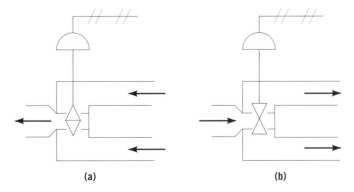

(a)   (b)

Fig. 11. Three-way control valves: (**a**) mixing and (**b**) diverting.

Generic Control Techniques and Strategies

Classical Feedback Control. The majority of controllers in a continuous process plant is of the linear feedback controller type. These controllers utilize one or more of three basic modes of control: proportional (P), integral (I), and derivative (D) action (1,2,6,7). In the days of pneumatic or electrical analogue controllers, these modes were implemented in the controller by hardware devices. These controllers implemented all or parts of the following control algorithm:

$$m(t) = K_C \left[e(t) + \left(\frac{1}{T_I}\right) \int_0^t e(t^*)\,dt^* + T_D \frac{de}{dt} \right]$$

where $m(t)$ = controller output at time t; $e(t)$ = deviation between control variable and its setpoint at time t; K_C = controller gain; T_I = integral time constant; and T_D = derivative time constant. Upon the advent of digital control devices, this basic control algorithm was implemented as a digital approximation:

$$M_n = K_C \left[E_n + \left(\frac{T_C}{T_I}\right) \sum_0^n E_k + \left(\frac{T_D}{T_C}\right)(E_n - E_{n-1}) \right]$$

where $T_C = t_i - t_{i-1}$ = control interval; $M_n = m(t_n)$; and $E_i = e(t_i)$, in the full position form, or

$$\Delta M_n = M_n - M_{n-1} = K_C \left[\Delta E_n + \left(\frac{T_C}{T_I}\right) E_n + \left(\frac{T_D}{T_C}\right)(E_n - 2\,E_{n-1} + E_{n-2}) \right]$$

in the incremental or velocity form. DCS and computer vendors often utilize a slight variation of these basic forms particular to their systems. In each form, the proportional, the integral, and the derivative actions which make up the PID algorithm can be distinguished. These are, however, almost always used in combinations in actual process control situations.

The proportional action term is given by

$$m(t) = K_C e(t) \qquad \text{or} \qquad \Delta M_n = K_C(E_n - E_{n-1})$$

The proportional action computes the magnitude of the output, which is proportional to the deviation of the controlled variable from the setpoint. It acts as soon as a deviation is detected. The larger the deviation, the larger the amount of immediate control action. Some systems replace the change in the deviation in the digital form with the change in control variable to avoid what is referred to as proportional kick when a rapid change in the setpoint is made. This is done assuming other controller actions can take care of the change in setpoint. The sensitivity of the proportional action is adjusted by adjusting the magnitude of the proportional gain, K_C. Where the controller has other control actions, eg, integral and derivative, the adjustment of K_C usually affects these actions also. One of the disadvantages of proportional-only control is that the controller stops changing the output once the deviation stops changing. Consequently, a steady-state deviation, referred to as offset, can persist.

The integral action of a PID algorithm is designed to remove this offset. The integral action term is given by

$$m(t) = \frac{K_C}{T_I} \int_0^t e(t^*)\, dt^* \quad \text{or} \quad \Delta M_n = K_C\left(\frac{T_C}{T_I}\right)E_n$$

The integral action continues to change the output as long as a deviation persists. The longer the time that the deviation exists in one direction, the greater the change in output. Not only does the integral action act to eliminate the offset, it also tends to smooth the effect of process noise. However, in the continuous and the full-position digital form, the integral action can also lead to a potential problem because the same feature that eliminates offset continues to act as long as there is a deviation. This difficulty, called reset windup, can arise if the final control element is constrained but the deviation from setpoint continues to exist. In this case, the integral action continues to increase and hence does not react quickly in the opposite direction when it should. The incremental digital form does not suffer from this difficulty. Also, most DCSs and some analogue controllers have antireset windup features to prevent reset windup. In most actual process control situations, integral action is combined with proportional action. This combination often accounts for over 80% of the PID controllers in a process plant. In general, the stronger the integral action, ie, the smaller the T_I, the more oscillatory the response.

The derivative action assists with settling out these oscillations faster. The derivative action term is as follows:

$$m(t) = K_C T_D \frac{de}{dt} \quad \text{or} \quad \Delta M_n = K_C\left(\frac{T_D}{T_C}\right)(E_n - 2E_{n-1} + E_{n-2})$$

The derivative term anticipates future deviations by considering the rate of the deviation change. The greater the rate of deviation change, the stronger the derivative action. The relative strength of the derivative action, for a given rate of change of deviation, is adjusted by the value of T_D. Although derivative action tends to settle oscillations faster, it also tends to amplify the effect of process and measurement noise. As a result, noisy signals must be filtered if derivative action is present. When derivative action is used in process plants, it is usually combined with proportional and integral action. Because a sudden change in setpoint can cause a large increase in the derivative action, the de/dt term is often replaced with $-dc/dt$, that is, the rate of change of the control variable. Using this modification, the controller responds identically for a load disturbance but with only the proportional and integral action to a setpoint change. This generally leads to a smoother response to a setpoint change.

PID controller is a flexible, effective, and reliable controller for the process industries. A considerable range of controller actions is possible by selecting tuning parameters to provide different weights to the present (proportional), the past (integral), and the projected future (derivative). References related to the tuning of PID controllers are available (20–24).

Cascade Control.   Cascade control (2–4) is a control strategy where the control output of a primary (master) controller is used to set the setpoint of the secondary (slave) controller. Generally, the controlled variable of the primary controller is the primary controlled variable of interest. The secondary controlled variable is selected to recognize the effect of particular disturbances to the process faster than the primary controller, and to compensate before these have had a chance to disturb significantly the primary controlled variable. Therefore, the secondary control loop improves the performance of the primary control loop by lessening the magnitude and type of disturbances for which the primary controller must compensate. Figure 12 gives the general structure of a cascade control strategy.

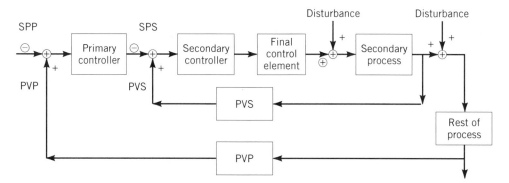

Fig. 12.   Cascade control signal flow diagram, where SPP = primary control variable setpoint; PVP = primary control variable measurement; SPS = secondary control variable setpoint; and PVS = secondary control variable measurement. The + and − indicate to multiply the signal by +1 or −1 before adding signals.

Schemes to control the outlet temperature of a process furnace by adjusting the fuel gas flow are shown in Figure 13. In the scheme without cascade control (Fig. 13**a**), if a disturbance has occurred in the fuel gas supply pressure, a disturbance occurs in the fuel gas flow rate, hence, in the energy transferred to the process fluid and eventually to the process fluid furnace outlet temperature. At that point, the outlet temperature controller senses the deviation from setpoint and adjusts the valve in the fuel gas line. In the meantime, other disturbances may have occurred in the fuel gas pressure, etc. In the cascade control strategy (Fig. 13**b**), when the fuel gas pressure is disturbed, it causes the fuel gas flow rate to be disturbed. The secondary controller, ie, the fuel gas flow controller, immediately senses the deviation and adjusts the valve in the fuel gas line to maintain the set fuel gas rate. If the fuel gas flow controller is well tuned, the furnace outlet temperature experiences only a small disturbance owing to a fuel gas supply pressure disturbance.

Both control schemes react in a similar manner to disturbances in process fluid feed rate, feed temperature, feed composition, fuel gas heating value, etc. In

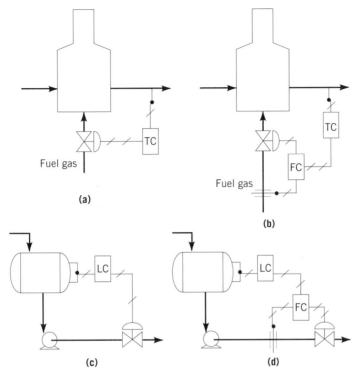

Fig. 13. Cascade control schemes, where TC = temperature controller; FC = fuel gas flow controller; and LC = liquid level controller. (**a**) Simple circuit having no cascade control; (**b**) the same circuit employing cascade control; and (**c**) and (**d**) liquid level control circuits with and without cascade control, respectively.

fact, if the secondary controller is not properly tuned, the cascade control strategy can actually worsen control performance. Therefore, the key to an effective cascade control strategy is the proper selection of the secondary controlled variable considering the source and impact of particular disturbances and the associated process dynamics.

In the drum level control scheme (Fig. 13**c**), if the inflow changes, the level changes accordingly and the level controller increases or decreases the valve opening to balance the in-flow and the out-flow. However, if the upstream or the downstream pressure varies, the flow through the control valve changes, which in turn causes the level to change. The level controller reacts by increasing or decreasing the valve opening and, hence, the flow. In the latter case, however, the pressure variations give rise to disturbances in the level. If a cascade control strategy (Fig. 13**d**) is used, the secondary controller compensates for these pressure variations before they significantly affect the level. In this cascade control strategy, the primary controller, ie, the level controller, needs to compensate primarily for in-flow variations.

The dynamics of the secondary control loop should be approximately two to four times as fast as the dynamics of the primary control loop in order to achieve stable control. The secondary controller is actually part of the primary controller's process system. Hence, changes in the secondary controller tuning constants change the process system of the primary controller. Therefore, cascade control loops should always be tuned by first tuning the secondary controller and then the primary controller. If the secondary controller tuning is changed for any reason, the primary controller may need to be retuned also.

Many misconceptions exist about cascade control loops and their purpose. For example, many engineers specify a level-flow cascade for every level control situation. However, if the level controller is tightly tuned, the out-flow bounces around as does the level, regardless of whether the level controller output goes directly to a valve or to the setpoint of a flow controller. The secondary controller does not, in itself, smooth the outflow. In fact, the flow controller may actually cause control difficulties because it adds another time constant to the primary control loop, makes the proper functioning of the primary control loop dependent on two process variables rather than one, and requires two properly tuned controllers rather than one to function properly. However, as pointed out previously, the flow controller compensates for the effect of the upstream and downstream pressure variations and, in that respect, improves the performance of the primary control loop. Therefore, such a level-flow cascade may often be justified, but not for the smoothing of out-flow.

Cascade control strategies are among the most popular and useful process control strategies. Modern control systems have made their implementation and operation both easier from the standpoint of operations personnel, and cost effective as they are implemented in software rather than hardwiring the connections.

Feedforward Control.   A feedforward control strategy is one in which a variable characterizing a particular disturbance, referred to as the disturbance variable, is measured or computed. This variable is then used in a model to determine the adjustment appropriate for the manipulated variable to mitigate the impact of the disturbance on the controlled variable (1,2,7). In other words, the measurement of a source of disturbance is used to anticipate the need for a corrective control action. Generally, a feedforward control strategy can be classified either as dynamic or steady state, or as additive or multiplicative. Examples of additive and multiplicative feedforward control strategies are shown in Figure 14. Most feedforward control applications are combined with an appropriate feedback controller. The feedback controller (trim) compensates for deviations in the controlled variable, whether or not it resulted from the measured disturbance variable and corrected for errors in the feedforward control adjustments.

The decision to implement a feedforward control strategy should be based on the quality of control required, the nature and frequency of the disturbance, the impact of the disturbance on the process, and the availability of a measured variable which can indicate the disturbance. Feedforward controls should not be installed unless justified by one or more of these factors, because of unnecessary complications in the overall control. Feedforward control may actually cause self-induced disturbances if improperly tuned.

Steady-State Feedforward.   The simplest form of feedforward (FF) control utilizes a steady-state energy or mass balance to determine the appropriate manipulated variable adjustment. This form of feedforward control does not account for the process dynamics of the disturbance or manipulated variables on the controlled variable. Consider the steam heater shown in Figure 15. If a steady-state

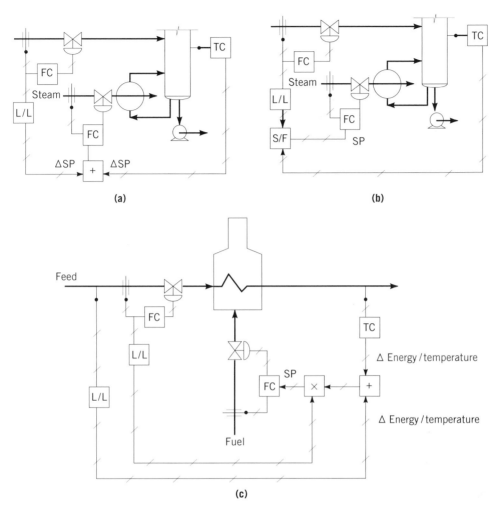

Fig. 14. Examples of feedforward (FF) controls having feedback (FB) trim, where L/L = lead/lag element; FC and TC = flow and temperature controllers, respectively; SP = setpoint; S/F = steam/feed ratio; × = multiplication of signals; and + = sum of signals. (**a**) Additive; (**b**) multiplicative; and (**c**) combined additive and multiplicative.

feedforward control is designed to compensate for feed rate disturbances, then a steady-state energy balance around the heater yields:

$$FC_P(T_2 - T_1) = F_S H_S$$

where C_P = heat capacity; F = flow rate; H = enthalpy of condensation; T = temperature; and the subscript S indicates steam. Because the feedforward scheme is to address only variations in the feed rate and not the inlet temperature, heat capacity, or steam enthalpy, the following difference equation can be derived for the variation in feed rate, where the subscript SP = setpoint:

$$C_P(T_{2,\mathrm{SP}} - T_1)\Delta F = H_S \Delta F_S$$

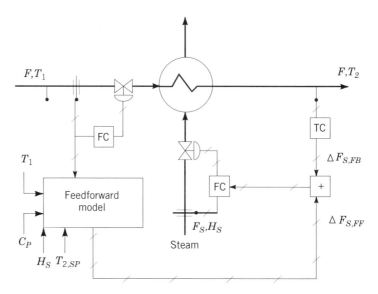

Fig. 15.   Example of steady-state feedforward controls, where + indicates the summation of signals. Terms are defined in text.

or, solving for the feedforward adjustment, the following equation is derived:

$$\Delta F_{S,FF} = \frac{C_P}{H_S}(T_{2,SP} - T_1)\Delta F$$

As shown in Figure 15, this incremental feedforward adjustment in the steam flow controller setpoint is trimmed by an incremental feedback control signal, $\Delta F_{S,FB}$. The feedforward controller adjusts the steam rate for variations in the feed rate and for changes in the outlet temperature setpoint, assuming the current value of the setpoint is accessed by the application. However, the feedforward controller does not adjust the steam rate for variations in the inlet temperature, heat capacity, steam enthalpy, or outlet temperature, owing to causes other than setpoint changes. These other disturbances and errors in the feedforward model are handled by the feedback controller in this particular scheme. If there is a significant error in the feedforward model, the feedforward scheme can actually degrade the overall control performance as compared to a simple feedback controller. If inlet temperature variations are a concern, the feedforward control model can be based on the heat energy and compensate for both temperature and feed rate variations. Therefore, the model used should be based on the types of disturbances of primary concern. When the disturbance variable has a controller associated with it (in this case, the feed rate controller), the feedforward control uses the feed flow controller setpoint as the feedforward signal when the feed controller is in auto mode, and only uses the measurement signal when the feed controller is in manual mode. In this way, the feedforward controller does not adjust the steam flow controller based on noise or minor variations when the feed controller is in auto mode. Often the dynamics of the process are important for good feedforward control (1,3,5,6).

Ratio and Multiplicative Feedforward Control. In many physical and chemical processes and portions thereof, it is important to maintain a desired ratio between certain input (independent) variables in order to control certain output (dependent) variables (1,3,6). For example, it is important to maintain the ratio of reactants in certain chemical reactors to control conversion and selectivity; the ratio of energy input to material input in a distillation column to control separation; the ratio of energy input to material flow in a process heater to control the outlet temperature; the fuel–air ratio to ensure proper combustion in a furnace; and the ratio of blending components in a blending process. Indeed, the value of maintaining the ratio of independent variables in order more easily to control an output variable occurs in virtually every class of unit operation.

There are two fundamental ways to implement ratio controls. In Figure 16a, the disturbance (or load) variable and the variable to be manipulated to maintain the ratio are measured, the ratio is calculated and compared with the desired value, and, based on the deviation, the manipulated variable is adjusted. In this approach, the ratio is used as the controlled variable for a feedback controller,

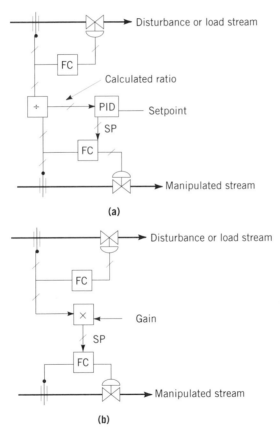

Fig. 16. Methods of ratio control implementation, where ÷ and × indicate the division and multiplication of signals, respectively: (**a**) use of a divider and ratio controller and (**b**) use of a multiplier. Calculated ratio = actual value of control variable; both setpoint and gain are equivalent to the desired ratio. Terms are defined in text.

usually containing proportional and integral action, as shown. This approach has a number of potential problems, eg, a variable loop gain because the control variable involves division. The approach shown in Figure 16**b** is preferred for most ratio controls. In this approach, the disturbance (or load) variable, which is also the denominator in the ratio, is measured and multiplied by the value of the desired ratio and the result is used as a setpoint for the controller for the variable being manipulated. This second approach was used in the multiplicative feedforward control examples of Figure 14. The desired ratio, which is the gain of the multiplier element, is usually adjusted by the feedback trim controller at a somewhat slower rate than the ratio control application being executed.

Ratio control and multiplicative feedforward control, in general, are subject to the same considerations. Ratio control can be of a steady-state or a dynamic form. It is often implemented using a setpoint as the load variable when the load variable has a controller associated with it and the controller is in auto mode.

Selective Control. Selective control strategies can be used to implement overrides, prioritized control actions, and constraint control applications (1,3,4). In selective control, signal select functions, or their equivalent in custom code, pick a signal or group of signals from a larger group of signals, eg, the highest, the lowest, or the median. Selective control strategies can be implemented by using hardware devices having appropriate wiring, preprogrammed software blocks, or custom-coded software programs. The use of selective control strategies has increased with the availability of software for implementation. Selective control strategies can choose a signal from either a number of controlled variables or manipulated variable signals. Discrete logic control functions can also be used to implement many selective control applications.

Figure 17 shows four simple selective control schemes. Figure 17**a** illustrates the use of a select function to control a reactor hot-spot temperature by selecting from a number of possible controlled variable measurements. Owing to phenomena such as channeling, the location of a reactor hot spot in a fixed-bed reactor may vary. The $>$ block selects the highest signal. Figure 17**b** illustrates the use of selective control strategy on controller outputs to implement an override. For example, if the tower level is maintained below 80% by manipulating the bottoms flow, then the tower feed is throttled until the tower level falls below 80%. The $<$ block selects the lowest signal and the $\times$ block subtracts the current value of the level, $h\%$, from 100% and multiplies the result by 5. In Figure 17**c**, the control scheme implements a strategy that utilizes a limited but less costly fuel supply up to its high limit, before using the supplementary (and more costly) fuel. The $<$ block again selects the lowest of its input signals and the $-$ block subtracts its input signals. Computing functions are often used together with the select functions to achieve a particular operating objective. Additional computing elements can be added to this scheme to account for differences in heating value between the fuels. Selective control strategies are also useful in implementing constraint control strategies, as shown in Figure 17**d**. Oil is heated in a fired heater and then circulated to reboilers for a number of distillation towers. A differential pressure controller (dPC) recirculates any of the hot oil not needed by the reboilers. A control signal from control schemes in the individual distillation towers adjusts control valves located in the inlet to their respective reboilers. The most economical operation of

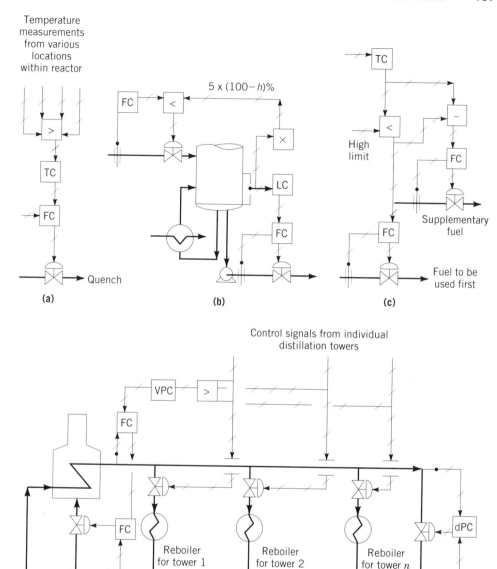

Fig. 17. Examples of selective control strategy: (**a**) reactor hot spot; (**b**) level override; (**c**) prioritized; and (**d**) constraint controls, where VPC = value position controller. See text.

the hot-oil circuit occurs when the oil flow through the reboilers is at a maximum and the heater outlet temperature is at a minimum, subject to the constraint that the heat demands of all the reboilers are satisfied. Such an operating point minimizes fuel use, stack losses, and bypass flow. In Figure 17**d**, the high selector (>) picks the maximum of all the distillation controller outputs to the individual reboiler control valves. That control signal is used as the process measurement for a valve position controller (VPC) having a setpoint at a value close to the full open position, eg, 90%. The VPC therefore adjusts the setpoint of the heater outlet temperature controller to keep at least one of the reboiler valves close to the full open position. The VPC algorithm and relevant control parameters should be selected to adjust the temperature setpoint and bring the system toward the optimized condition in the long term. The reboiler controls and control valves should have enough flexibility to respond to disturbances in the distillation towers in the short term without reaching a constrained condition, ie, the fully opened reboiler valve.

Multivariable Control. A situation where the adjustment of one input variable affects the values of a number of output variables is common in processes and process equipment. Most of these process interactions are the natural consequences of underlying physical and chemical phenomena and particular designs. Using the proper pairing of controlled and manipulated variables, a great majority of the controlled variables can be adequately controlled by using single input–single output (SISO) controllers (3,25–27). The interactions between the controllers resulting from the process interactions are handled by the normal feedback action of the controllers. However, there are situations where the strength of the interactions between a group of variables is such that the process and control objectives cannot be met by using SISO controllers alone. In such situations, the SISO controllers can cause instability or need continual retuning as these interact with one another. Multivariable control strategies, which explicitly address these interactions, can maintain good control performance in these instances (28–30).

Multivariable control strategies utilize multiple input–multiple output (MIMO) controllers that group the interacting manipulated and controlled variables as an entity. Using a matrix representation, the relationship between the deviations in the n controlled variable setpoints and their current values, e_i, and the n controller outputs, m_i, is

$$
\begin{pmatrix} m_1 \\ m_2 \\ \cdot \\ \cdot \\ m_n \end{pmatrix} = \begin{pmatrix} G_{C11} & G_{C12} & \cdot & \cdot & G_{C1n} \\ G_{C21} & & \cdot & \cdot & \cdot \\ \cdot & \cdot & \cdot & \cdot & \cdot \\ \cdot & & \cdot & \cdot & \cdot \\ G_{Cn1} & \cdot & \cdot & \cdot & G_{nm} \end{pmatrix} \begin{pmatrix} e_1 \\ e_2 \\ \cdot \\ \cdot \\ e_n \end{pmatrix}
$$

which in matrix shorthand is

$$
m = G_C e
$$

Similarly, the relationship between the n outputs, m_i, and the current values of the controlled variables, c_i, can be written as a matrix comparable to the inputs, which in matrix shorthand is

$$
c = G_P m
$$

From these two relations, the following equation is derived:

$$c = G_P m = G_P G_C e$$

This equation relates the deviations from setpoints to the observed current values of the controlled variables.

From a theoretical standpoint, if a multivariable controller is to compensate for the process interactions, then the elements of the controller, G_C, should be chosen so that the matrix product, $G_P G_C$, results in a diagonal matrix. This design methodology requires a knowledge of the elements of the process dynamics transfer function, G_P. For a two-by-two controller, the resulting setup is usually manageable. A three-by-three, although considerably more complex, is also manageable. For dynamic interaction decoupling controllers of levels greater than three-by-three, the setup becomes relatively complex and other approaches, such as model-based control strategies, should be considered. More details about these decoupling control schemes are available (25,30).

When faced with a situation where strong process interactions exist, the following steps are recommended. If some of the process and control objectives can be compromised, then the approach of tuning the controllers associated with the less important objectives loosely (sluggish response), while tuning the controller associated with the most important objective tightly, is the easiest approach. If such a compromise of quality in control is not acceptable, different pairings of the controlled and manipulated variables should be considered. At times, a different pairing or simple functional relationships such as a sum, a difference, or a ratio can reduce the interactions between the controlled variables significantly. The relative gain array technique (25,26) and singular value decomposition (27), as well as a good understanding of the process and related physical and chemical phenomena, can assist such an analysis. If the relative gain array approach is used, consideration of the process dynamics may need to be added, as some of the modifications of the method have proposed (31,32). If neither of these straightforward approaches has satisfactorily reduced the interactions in light of the process and control objectives, decoupling controls or model-based controls, eg, dynamic matrix control (33) and model algorithmic control (34), should be considered. If the decoupling control strategy is used, a number of practical aspects of design and implementation must be taken into account. In actual implementations, it has been found that full decoupling controllers, ie, where G_C has been designed to make the product $G_P G_C$ a diagonal matrix, results in a multiple input–multiple output (MIMO) controller in actual plant application that tends to be extremely sensitive both to the accuracy of the process dynamic models in G_P, and to how well the interacting controllers are tuned and remain tuned. On the other hand, it has been found that a MIMO controller which has been designed only for partial decoupling, eg, compensating only in the direction of the stronger interaction between two variables and resulting in a G_C that is a triangular matrix, leads to a MIMO controller that is more robust when actually implemented (28).

Model-Based Control. A number of the control strategies discussed utilize models as part of the strategy, eg, feedforward control and adaptive control. In the model-based predictive control strategies, the use of models is the central theme of the strategies. The fundamental concept underlying these model-based predictive control strategies is a feedback control strategy with a feedback signal of some measure of the deviation between the actual controlled variables and the predicted values over some time horizon. Examples of model-based control strategies are dead-time compensation (1,7) and techniques based on a discrete convolution model. References 1, 6, and 35 should be consulted for the theoretical basis and development of these models.

Dead-Time Compensation.   Dead time within a control loop can greatly increase the difficulty of close control using a PID controller. Consider a classical feedback control loop (Fig. 18a) where the process has a dead time of T_{DT}. If the setpoint is suddenly increased at time t, the controller immediately senses the deviation and adjusts its output. However, because of the dead time in the loop, the controller does not begin to see the impact of that change in its feedback signal, that is, a reduction in the deviation from setpoint, until the time $t + T_{DT}$. Because the deviation does not change until $t + T_{DT}$, the controller continues to change its output in the same direction until the deviation begins to change. If the controller is tightly tuned (relatively large controller gain), it has a tendency to overshoot the required change by a significant amount. This situation then occurs in the opposite direction, leading to cycling and possibly instability. In order to avoid cycling and instability, the controller gain must be set relatively low. Thus, the controller does not react quickly to disturbances or setpoint changes.

The Smith dead-time compensator is designed to allow the controller to be tuned as tightly as it would be if there were no dead time, without the concern for

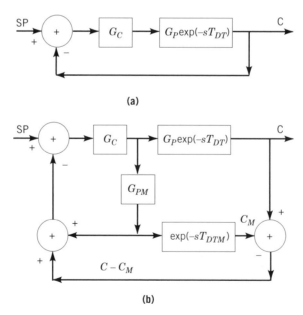

Fig. 18.   Dead-time compensation: (**a**) classical feedback and (**b**) Smith dead-time compensator. SP = setpoint; C = controlled variable; and (+) and (−) indicate multiplication by +1 or −1, respectively, before summation.

cycling and stability. Therefore, the controller can exert more reactive control. The dead-time compensator utilizes a two-part model of the process, ie, G_{PM}, which models the portion of the process without dead time, and $\exp(-sT_{DTM})$, which models the dead time. As seen from Figure 18**b**, the feedback signal is composed of the sum of the model (without dead time) and the error in the overall model $(G_{PM}\exp(-sT_{DTM}))$, ie, $C - C_M$. Using the same type of setpoint change discussed for Figure 18**a** and assuming the model is perfect, that is, when the setpoint change is made, the controller again immediately senses the deviation and adjusts its output:

$$G_P \equiv G_{PM}\exp(-sT_{DTM})$$
$$T_{DT} \equiv T_{DTM}$$

where the subscript *DTM* corresponds to dead-time model. However, output in Figure 18**b** is sent not only to the process but also to the portion of the model without dead time, ie, G_{PM}. The output from this portion of the model, ie, prediction of the response of the process as if there were no dead time, is summed with the error in the overall model, $C - C_M$, which is zero because a perfect model is assumed for purposes of explanation. The controller, seeing only the model output from G_{PM} between t and $t + T_{DT}$, therefore believes that the deviation has been reduced and begins to cut back on its output. After the time period T_{DT} has passed, the actual controlled variable begins to respond to the first control move that was made by the controller. Because the model was assumed to be perfect, the output of the dead-time model block, $\exp(-sT_{DTM})$, begins to output at the same time. The overall modeling error, $C - C_M$, remains zero during the entire period of the actual response. Hence, the controller may be tuned as if there were no dead time. When the model is not perfect, the modeling error, $C - C_M$, is not zero and the controller acts to reduce it to zero. It is more important to have an accurate estimate for the dead time for the Smith dead-time compensator than for a well-tuned classical feedback controller. Other variations have been developed for the use of an analytic model for predictive control, including the analytical predictor (36), the generalized analytical predictor (37), and a general methodology known as internal model control (35,38).

Discrete Convolution Model. At times the process cannot be adequately modeled by using a preselected analytical form and determining a set of model parameters as required by the above techniques. A discrete convolution model (39), as used within control techniques such as dynamic matrix control (33) and model algorithmic control (34,40), avoids the difficulty of having to preselect an analytical form at the cost of additional computation. A discrete convolution model is capable of modeling complex and unusual process responses in a relatively straightforward manner based on experimental data. Similar to the analytical predictor and the generalized analytical predictor, the discrete convolution model is designed around digital technology. A discrete convolution model for the response of a particular variable, $c(t)$, to a change in the variable, $m(t)$, at a set of evenly spaced discrete instants of time, $(t_0, t_1,$ etc to $t_I; t_i = t_{i-1} + \Delta t)$, where c^*_{n+1} = predicted value of $c(t_{n+1})$, is given by the general form

$$c^*_{n+1} = c_0 + \sum_{j=1}^{J} h_j m_{n+1-j}$$

where $h_j = a_j - a_{j-1}$; $j = 1, 2$, etc to J; and a_j = step response coefficient. The upper limit of the summation, J, referred to as the model horizon, should be selected so that $J\Delta t$ is the time for the particular response to settle out. The discrete convolution model is sometimes expressed as a step response model, that is,

$$c^*_{n+1} = c_0 + \sum_{j=1}^{J} a_j \Delta m_j$$

where $\Delta m_j = m_j - m_{j-1}$ for a series of step changes Δm_j. This step response model may be expressed in matrix form:

$$\begin{pmatrix} c^*_1 - c_0 \\ c^*_2 - c_0 \\ . \\ . \\ c^*_J - c_0 \end{pmatrix} = \begin{pmatrix} a_1 & 0 & . & . & 0 \\ a_2 & a_1 & 0 & . & 0 \\ . & . & . & . & . \\ . & . & . & . & . \\ a_J & a_{J-1} & . & . & a_1 \end{pmatrix} \begin{pmatrix} \Delta m_0 \\ \Delta m_1 \\ . \\ . \\ \Delta m_J \end{pmatrix}$$

or $c = A\Delta m$, where A is the triangular matrix given. The number of coefficients depends on the process characteristics as well as the sampling period, Δt. The fundamental concept of this technique is that the next control move, Δm, is determined by solving a set of equations based on the condition that the deviation between the corrected prediction of the response, determined by the discrete convolution model, and the desired trajectory to the set of setpoints is optimized. For the next time period, the whole process is repeated.

The corrected prediction for the next time step's control move is determined by adjusting the prediction from the discrete convolution model based on the error between the actual observed and the predicted values for the past time steps and shifting it forward one time step. The size of the set of equations to be solved (the matrix) is determined by two parameters: the prediction horizon, which sets the number of rows, and the control horizon, which sets the number of columns. The prediction horizon specifies the number of predicted values of the control variable. The larger the prediction horizon, the more conservative the control action but the greater the amount of computation required. The control horizon is the number of future control actions. If the control horizon is too large, an excessive control action may result; if too small, the resulting controller may not be sensitive enough to modeling errors.

Various practical modifications have been added to commercial versions of these discrete convolution model-based techniques. For example, weighing matrices have been added to the theoretical algorithm so the calculated control moves are of a reasonable magnitude; constraints on controlled and manipulated variables are handled by utilizing either a linear programming (33,41) or a quadratic programming (42) method to solve the constrained optimization. Most of the commercially available versions also come with an identification package to assist in determining the discrete convolution model coefficients. Most of the applications of the discrete convolution model-based predictive control are of the multivariable control type (43). Considering the number of possible responses to be modeled, as well as all the variables and interactions between them and the possible number of coefficients to be determined in each response, the experimental and calculational effort associated with identification can be formidable. However, based on documented applications of these techniques, good model identification is a key factor in developing a robust, well-performing predictive controller. A summary of the theory of the discrete convolution model-based controls is available (33).

Adaptive Control. An adaptive control strategy is one in which the controller characteristics, ie, the algorithm or the control parameters within it, are automatically adjusted for changes in the dynamic characteristics of the process itself (34). The incentives for an adaptive control strategy generally arise from two factors common in many process plants: (1) the process and portions thereof are really nonlinear; and (2) the process state, environment, and equipment's performance all vary over time. Because of these factors, the process gain and process time constants vary with process conditions, eg, flow rates and temperatures, and over time. Often such variations do not cause an unacceptable problem. In some instances, however, these variations do cause deterioration in control performance, and the controllers need to be retuned for the different conditions.

The dual-fired heater shown in Figure 19 illustrates some of the practical situations that can arise. This heater fires both fuel oil and fuel gas. Depending on the overall plant fuel balance during a particular time period, one of the fuels is baseloaded and the other is modulated to control the heater's outlet temperature. Depending on which fuel is used as the temperature controller's manipulated variable, the process gain, that is, the ratio of the percentage of change in outlet temperature for a percentage of change in the manipulated variable setpoint, can be drastically different. The process dead time and lag times are likely to be about the same, however. In addition to the changes in gain owing to different fuels as the manipulated variable, the process dynamics also change significantly if the heater is run at significantly different feed rates. For example, if the feed rate is reduced 50%, the process gain doubles and the effective dead time also increases. Although the temperature controller may be performing well when fuel oil is the manipulated variable or when it is being run at 100% feed rate, its performance deteriorates when the manipulated variable is changed to fuel gas or if the feed rate is reduced to 50%, unless the controller is retuned. One approach to addressing this situation is to tune the controller initially so as to strike a compromise between its performance in each situation or over a large range of flows. It is not always possible, however, to reach a compromise that is acceptable for both process and control objectives. For example, such a compromise is not likely to be acceptable for the fired heater shown (Fig. 19) if its outlet was the feed to a reactor that required close control of its inlet temperature, ie, the heater outlet temperature.

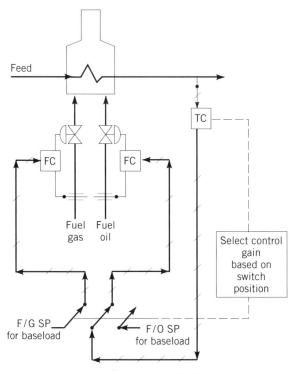

Fig. 19. Example of programmed adaptation. F/O SP indicates the fuel oil setpoint and F/G SP indicates the fuel gas setpoint.

Two general classes of adaptive control strategies (44,45) are programmed adaptation and self-adaptation. If the adaptation occurs primarily as a result of the measurement of a factor that has caused the need for adaptation, the strategy is referred to as programmed adaptation. If, on the other hand, the adaptation that occurs is based on a measure of the controller's performance, and which, therefore, requires a measurement of the controlled variable, then the strategy is referred to as self-adaptive. The programmed adaptive control strategy in Figure 19 compensates for the use of different fuels as the manipulated variable. The position of the switch is sensed and, based on its position, the temperature controller's gain is adjusted to a value appropriate for that particular fuel. Although the scheme adapts the temperature controller for switches in its manipulated variable, it is not designed, as outlined, to adapt the controller to significant changes in the feed rate. If significant feed rate changes occurred often enough to be an important disturbance, one of the following schemes would be appropriate: (*1*) use of a more sophisticated programmed adaptation based on not only the fuel type but also feed rate range; (*2*) implementation of a self-adaptive scheme requiring sets of data on feed rate, fuel type, and outlet temperature deviation from setpoint; or (*3*) employment of a combination of feedforward with feedback trim and adaptive control.

Programmed adaptation can be straightforward to implement. At times, only relatively simple logic is required. Self-adaptation, on the other hand, often requires more sophisticated logic and algorithms. At the basic regulatory control level, however, some DCS vendors offer self-tuning controllers (45,46). Where the cause of the changes in the control performance is unknown, uncontrollable, or unmeasurable, programmed adaptation may not be possible. Hence, if an adaptive control strategy is required, a self-adaptive approach will be necessary. The self-tuning regulatory controllers mentioned above utilize some means of estimating the values of the parameters that characterize the process dynamics as well as relationships between these and the tuning parameters. Another type of self-adaptive control is model reference adaptive control (MRAC) (47), which utilizes a reference model to define the desired response and an adaptation mechanism to adjust the control parameters.

Constraint and Optimizing Control. Operating objectives for process facilities are set by economics, product orders, availability of raw materials and utilities, etc. At different points in time it may be advantageous or necessary to operate a process in different ways to meet a particular operating objective. A process plant, however, is a dynamic, integrated environment where external and internal conditions can cause the optimal operating point for each operating objective to vary from time to time. Therefore, the two higher levels of control, ie, the local and the plant-wide optimization levels (see Table 1), can generate significant credits from following the optimal operating point.

Often, in operating facilities, the optimal operating points at the local optimization level lie along a constraint. When such is the case, a constraint control strategy is a straightforward way to achieve and maintain operation close to the optimal operating point. A constraint control strategy monitors the proximity to relevant constraints, identifies which are active, and incrementally adjusts the manipulated variables to control and maintain operation close to a constrained condition. Credits are achieved whenever opportunity presents itself. Consider the example of maximizing the feed to a particular distillation column while meeting the separation specifications and optimizing the energy consumption. Figure 20 is a representation of the operating window for the particular distillation column. The various constraints for the column, condenser, reboiler, and flooding are shown. The relative position of some of these constraints are dependent on uncontrolled external conditions, eg, the condenser constraint position depends on the ambient air temperature. Also shown are the lines of constant separation at various feed rates. From the relative slopes of the flooding constraint line and these contour lines, the maximum feed rate is achieved by operating in the upper left-hand corner of the operating window, ie, the intersection of the flooding constraint and the condenser constraint. The constraint control strategy can thus operate as follows: (1) at a constant column pressure, increase the column feed rate incrementally until the flooding constraint is approached (assuming that lower level controls are automatically adjusting the boilup); (2) once the proximity of the flooding constraint is reached, lower the pressure controller constraint incrementally until the proximity of the condenser constraint is reached (again assuming that the lower level controls are automatically adjusting the boilup). If the constraints never changed position, this operating point would maximize the feed while meeting the separation specifications and optimizing the energy consumption. Indeed, if operators could concentrate on a particular column's operation, they could also perform these steps. However, a process plant and its environment are dynamic and some of the constraints that

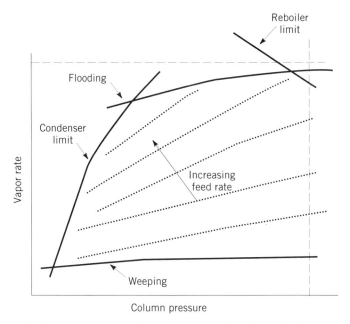

Fig. 20. Example of constraint control: operating window for a distillation column, where (− − −) is the vessel pressure limit, and (···) represents constant separation at different fixed feed rates.

define the operating window move as these conditions change. For example, if the ambient air temperature significantly drops in the evening, the condenser constraint line moves to the left. This movement, in turn, creates an opportunity to further increase the feed, which the constraint control application does incrementally until the unit is again in the proximity of an active constraint. In a like manner, as daylight returns and the ambient temperature begins to rise, the constraint control application incrementally decreases the feed rate to keep the operating point within the operating window.

Constraint control strategies can be classified as steady-state or dynamic. In the steady-state approach, the process dynamics are assumed to be much faster than the frequency with which the constraint control application makes its control adjustments. The variables characterizing the proximity to the constraints, called the constraint variables, are usually monitored on a more frequent basis than actual control actions are made. A steady-state constraint application increases (or decreases) a manipulated variable by a fixed amount, the value of which is determined to be safe based on an analysis of the proximity to relevant constraints. Once the application has taken the control action toward or away from the constraint, it waits for the effect of the control action to work through the lower control levels and the process before taking another control step. Usually these steady-state constraint controls are implemented to move away from the active constraint at a faster rate than they do toward the constraint. The main advantage of the steady-state approach is that it is predictable and relatively straightforward to implement. Its major drawback is that, because it does not account for the dynamics of the constraint and manipulated variables, a conservative estimate must be taken in how close and how quickly the operation is moved toward the active constraints.

Considerable effort has been devoted to the application of linear and non-linear optimization methods to improve the operation of process facilities (48,49). Most of the applications that have been implemented have been off-line us-ing steady-state models. These optimization methods have also been embedded within control algorithms (33,42,50) for optimal control rather than process op-timization. The increased computing capability in control systems, the improve-ments in software packages, and the increase in process integration in process plants have rekindled interest in real-time process optimization. However, the problems associated with its practical application are many, eg, the large num-ber of variables involved in applications; accounting for the reliability, accuracy, and availability of all of the associated measurements; accounting for the many practical constraints in a quantitative manner; accounting for the accuracy of the models, physical and chemical property data methods; accounting for associ-ated dynamics of the many subsystems in interpreting measurements and taking control actions, especially when steady-state models are being used; and the re-quirement for good control performance at all hierarchial control levels below this level to ensure the proper implementation of control actions. Thus the relative credits of on-line optimization must be compared to a regularly performed off-line optimization. In spite of potential complications, a practical technology for real-time optimization is evolving. Successful applications have been reported in the literature (1,6,48).

Discrete Logic and Control. Discrete logic and control functions are used to implement overrides, interlocks, sequencing, signal selection, etc, based on inputs that have a discrete number of states. There are many ways to represent process logic (51), including relay ladder logic, flow charts, and Boolean logic. Examples of simple physical systems are available in the literature (14).

Batch and Sequence Control. In batch processes the product is made in discrete batches by sequentially performing a number of processing steps on the raw materials and intermediate products. For example, fixed amounts of reac-tants may be charged to a vessel, mixed and heated to a reaction temperature, reacted for a fixed period of time, drained from the vessel, separated, dried, and packaged. In a batch process, the path followed in state space is often im-portant. Consequently, compared to a continuous process, batch process control requires a greater percentage of discrete logic and sequential control than regu-latory control loops. Batch control applications must control the timing and se-quencing of the process steps based on discrete input and outputs as well as analogue outputs. A hierarchial model for batch control is given in Table 2. The safety interlocking and shutdown applications, designed to protect personnel, environment, and equipment from damage, are the lowest level of process logic. Typically, applications of this type are much more extensive in a batch plant than in a continuous process plant. Applications at the discrete control level use discrete inputs, outputs, as well as normal and exception logic to place the batch in a desired fixed state depending on prevalent conditions and the step in the batch process. Applications at the sequential control level transition a batch sequentially through a series of discrete steps. Typically, each step of the process has a number of states such as start-up, normal, hold, normal shutdown,

Table 2. Conceptual Levels in a Batch Control Hierarchy

Conceptual control level[a]	Operation
production planning	develops production plan based on customer orders, resource availability, constraints, etc
production scheduling	develops production schedule to satisfy the production plan
recipe management	stores, retrieves, and modifies recipes
batch management	controls selection of proper recipe; initiates and supervises production of a batch
sequential control	controls and directs process through a sequence of distinct states
regulatory and discrete control	adjusts specific control elements that interface directly with process equipment
safety interlocking	controls functions that ensure safety of personnel, equipment, etc

[a]Control increases from top to bottom; information from bottom to top.

emergency shutdown, and restart. The batch management level applications select the proper recipe for a batch and provide the overall supervision of the production of the batch. Most batch process facilities make numerous products and grades, each having its own recipe. The recipe management level provides for the modification, storage, and retrieval of recipes by the batch management level. Applications at the production scheduling level provide a cost-effective, short range production schedule, producing a wide variety of products and grades, scheduling of batch runs based on resource and raw material constraints, various product and grade transition costs, production plan demands, and delivery time requirements. Applications at the production planning level include product and quantity determination, resource determination, and timing. To provide effective batch control, there must be close coordination between the three bottom control levels. For optimal or near optimal batch plant operation, the higher control levels are becoming increasingly important.

The complexity of the interactive logic within and between the various control levels, the required interactions with operators, and the need for ongoing application modification and maintenance are reasons why organization, functional design, and clear documentation are so important to the successful use of batch control applications. In order to describe what must be done, structural models are commonly used to represent the required batch processing actions, the batch equipment, and the combination of components. Figure 21 outlines popular structural models for representing the batch processing actions and the batch equipment. Various formats have been proposed for describing the batch control applications, eg, how the batch processing steps are carried out with the batch equipment and instrumentation, interfaces between the various levels of control, interfaces between the batch control and the operator actions and responses, and interactions and coordination with the safety interlocks. The formats proposed include flow charts, state charts, decision tables, structured pseudocode, state transition diagrams, petri nets, and sequential function charts. These and other formats and their individual advantages and disadvantages have been discussed in the literature (52–56).

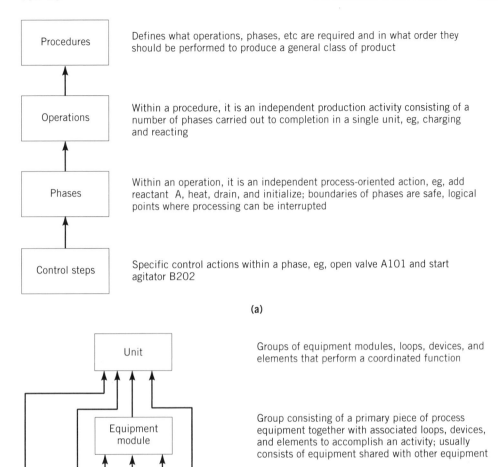

Procedures — Defines what operations, phases, etc are required and in what order they should be performed to produce a general class of product

Operations — Within a procedure, it is an independent production activity consisting of a number of phases carried out to completion in a single unit, eg, charging and reacting

Phases — Within an operation, it is an independent process-oriented action, eg, add reactant A, heat, drain, and initialize; boundaries of phases are safe, logical points where processing can be interrupted

Control steps — Specific control actions within a phase, eg, open valve A101 and start agitator B202

(a)

Unit — Groups of equipment modules, loops, devices, and elements that perform a coordinated function

Equipment module — Group consisting of a primary piece of process equipment together with associated loops, devices, and elements to accomplish an activity; usually consists of equipment shared with other equipment

Loop / Device — Loop: two or more elements where a signal passes between them for measurement and/or control, eg, sensor + transmitter + control + valve; device: two or more elements combined for a discrete state condition, eg, block valve + solenoid valve + actuator + limit switches

Element — Piece of equipment that has an active function, eg, sensor, transmitter, and switch

(b)

Fig. 21. Batch structural models: (**a**) processing and (**b**) equipment.

Since 1989, the Instrument Society of America (ISA) has been working on developing a standard, referred to as ISA-S88, for standardizing batch control concepts, structures, terminology, data structures, and batch control configuration and programming languages. As of this writing (1996), the first portion is available (57).

Economic Aspects

Process Instrumentation and Control Systems. Investment for instrumentation and control systems and their installation typically range between 3 to 10% of the total installed cost for a grassroots continuous process facility. Instrumentation and control systems also represent a substantial percentage of the overall facility maintenance (qv) costs. Investment costs may be placed in one of two categories, ie, nondiscretionary and discretionary.

The nondiscretionary category contains items required for the adequate protection of personnel and equipment, for protection of the environment, for satisfying governmental and industry codes and regulations, and for maintaining specific company standards. The discretionary category, on the other hand, contains items that are either directly or indirectly related to improved profitability. The discretionary portion can be appropriately analyzed by a benefit–cost ratio study. Such a justification establishes control objectives, identifies important aspects for increased profitability, and, in general, improves the quality of decision-making with regard to the instrumentation and control systems. Associated benefits should be related to the overall facility's economics. For example, a control application that requires certain instrumentation and control system features, and that increases the average production rate, normally has little economic value in a market-limited environment, but may have significant value within a production-limited environment. A list of the benefits of improved control follows:

Raw material costs and utilization
    yield improvement
    reduced losses
    improved product and raw material recovery
    ability to operate closer to constraints and specifications consistently
      and safely
    reduced quality giveaway
Operating and inventory cost
    reduction in utility requirements
    decreased need for conservative operation
    reduced need for building inventory and associated provisions
    consistent operation closer to optimum operation
    reduced reprocessing of off-spec material
Other
    improvement in service factor
    reduction of off-grade product when shifting operations and control
      strategies for different situations such as grade
      changes and feedstock changes
    improved data for operations and control performance engineering
      analysis

With the advent of DCSs, PLCs, smart transmitters, smart valves, and digital communications, it is becoming more and more imperative to take an overall system approach in analyzing the benefit–cost ratios of alternative systems to meet control objectives. It is no longer appropriate to view benefits or costs on a loop-by-loop basis. For example, a minimal DCS configuration may consist of one operator station, a data highway, and one control unit. Thus, for the controls up to the capacity of the control unit, the per-loop cost includes the cost of the operator station and the data highway. However, it is usual to find that the capacity of both the operator station and the data highway is much greater than that of a single control unit. Therefore, adding another control unit, ie, doubling the control capacity, can be done without investing in another operator station or data highway. Much modern control equipment consists of prepackaged modularized components configured into a system. Thus similar types of overall considerations are required in arriving at a cost-effective control design for a particular facility.

The classic systems design approach is becoming more common in the 1990s for the overall instrumentation and control systems design. First, a control philosophy is developed that sets some general guidelines and objectives. Next, a functional definition of the instrumentation and control system is developed. The functional definition, which should include a clear statement of specific control objectives, provides the basis for determining the relative benefits of alternatives and for the system selection process. Using the functional definition, alternative instrumentation and control systems can be roughly configured to satisfy the functional definitions to different degrees. Rough cost estimates and rough benefit–cost ratios can then be developed.

In roughing out feasible alternatives, rough cost estimates are often informally developed based on previous designs and factored estimates. As a check on the reasonableness of such estimates, comparison can be made to more general factors related to overall cost estimates. Generally, the installed cost of instrumentation and control systems is typically in the range of 3 to 10% of the total installed cost of the facility. The engineering and detailed design costs are typically about 35% of the purchased equipment cost of the process-related equipment or approximately 13% of the total installed cost. Of these engineering and detailed design costs, those associated with instrumentation and control systems are typically around 11% in a full engineering and procurement project for a grassroots facility. It is often possible, once a rough design has been developed, to work with a number of vendors to obtain preliminary costing for their systems. Table 3 contains some rough prices for various types of instrument measurement devices and other miscellaneous control components. These prices can vary significantly based on materials of construction, services conditions, etc. These prices do not reflect installed cost, unless so noted.

Surveys by many instrumentation and control system vendors as well as others indicate that as of the mid-1990s technology advances have far outpaced the ability of many organizations to utilize effectively the enhanced capabilities of the instrumentation and control systems equipment. These surveys have indicated (1) that approximately 80% of the installed DCSs were being used for little more, from a control standpoint, than a replacement of a large control panel with CRT custom displays; (2) that most of the actual control strategies

Table 3. Costs for Different Classes of Instrumentation

Type of device	Cost range, $	Comments
Flow measurement		
orifice	500–3,500	5–30 cm, flanged run and transmitters
mass coriolis	5,000–21,000	2–15 cm
Venturi/flow tube	3,000–10,000	
vortex shedding	2,000–6,000	
magnetic	3,000–10,000	2–20 cm, flanged
turbine	3,000–9,000	5–20 cm, flanged
Level measurement		
differential pressure	200–1,500	
displacer	2,000	81 cm, 1.5 m length
Temperature measurement		
resistance thermometry detector	750–2,000	sensor, thermowell, and transmitter
thermocouple	500–2,000	sensor, thermowell, and transmitter
Pressure measurement		
diaphragm	800–2,000	
differential pressure	750–2,000	
electronic pressure sensor	750–2,000	
Concentration measurement		
gas chromatograph	60,000–120,000[a]	
liquid chromatograph	50,000–120,000[a]	
mass spectrometer	50,000–120,000[a]	
O_2 in gases	5,000–10,000	
Density measurement		
displacement	2,000–4,000	
oscillating coriolis	4,000–21,000	
vibrating densitometers	5,000–21,000	
Control valves		
ball	1,500–20,000	5–30 cm
butterfly	2,000–23,000	10–56 cm
globe	2,000–25,000	5–25 cm
Other measurements		
conductivity	700–3,000	
viscosity	3,000–20,000	

[a] Cost is that of an installed device.

employed were no different from those in the days of analogue controllers; and (3) that the standard DCS features which make it relatively easy to improve control strategies, eg, improved features of PID algorithms, simple feedforward controls, and conditional controls, were not being utilized. Thus considerable benefits were not being reaped from equipment already installed. It is expected that the better the process, the control equipment, and the control concepts are understood, the more likely it is that benefits can be reaped. Achieving ongoing benefits from improved control requires a coordinated team effort.

BIBLIOGRAPHY

"Instrumentation" in *ECT* 1st ed., Vol. 7, pp. 908–926, by J. Procopi, Minneapolis-Honeywell Regulator Co.; in *ECT* 2nd ed., Vol. 11, pp. 739–774, by N. A. Fiorino, Honeywell, Inc.; "Instrumentation and Control" in *ECT* 3rd ed., Vol. 13, pp. 485–512, by T. J. Williams, Purdue University.

1. D. E. Seborg, T. F. Edgar, and D. A. Mellichamp, *Process Dynamics and Control*, John Wiley & Sons, Inc., New York, 1989.
2. P. Harriott, *Process Control*, McGraw-Hill Book Co., Inc., New York, 1964.
3. F. G. Shinskey, *Process Control Systems*, McGraw-Hill Book Co., Inc., New York, 1967.
4. W. L. Luyben, *Process Modeling, Simulation, and Control for Chemical Engineers*, McGraw-Hill, Book Co., Inc., New York, 1973.
5. C. A. Smith and A. B. Corripio, *Principles and Practice of Automatic Process Control*, John Wiley & Sons, Inc., New York, 1985.
6. G. Stephanopoulos, *Chemical Process Control*, Prentice-Hall, Inc., Englewood Cliffs, N.J., 1983.
7. C. L. Smith, *Digital Computer Process Control*, International Textbook Co., Scranton, Pa., 1964.
8. N. L. Conger, "Designing Safety Shutdown Systems: A Systematic Approach," *Proceeding of Instrument Society of America Conference*, ISA/73, Paper #73-756, Houston, Tex., 1973.
9. J. A. Benedetto, *Control Engineering*, **28**(7), 136 (1981).
10. J. A. Wilkinson and B. W. Balls, "Microprocessor-Based Safety Systems Designed for Fire and Gas and Emergency Shutdown Applications," *Proceedings of Instrument Society of America Conference*, ISA/85, Paper #85-0841, Philadelphia, Pa., 1985.
11. J. A. Humphrey, *Instr. Cont. Syst.* **60**(10), 74 (1987).
12. J. Machulda, *InTech*, **32**(9), 105 (1985).
13. A. Krigman, *InTech*, **32**(7), 41 (1985).
14. B. G. Liptak, ed., *Instrument Engineer's Handbook*, Vols. 1 and 2, Chilton Publishing Co., Philadelphia, Pa., 1969.
15. W. G. Andrew, *Applied Instrumentation in the Process Industries*, Gulf Publishing Co., Houston, Tex., 1974.
16. D. M. Considine, ed., *Process Instruments and Controls Handbook*, 2nd ed., McGraw-Hill Book Co., Inc., New York, 1974.
17. K. J. Clevett, *Process Analyzer Technology*, John Wiley & Sons, Inc., New York, 1986.
18. D. A. Mellichamp, ed., *Real-Time Computing with Applications to Data Acquisition and Control*, Van Nostrand Reinhold, Co., Inc., New York, 1983.
19. J. W. Hutchinson, ed., *ISA Handbook of Control Valves*, 2nd ed., Instrument Society of America, Research Triangle Park, N.C., 1976.
20. H. A. Fertig, *ISA Trans.* **14**(3), 292 (1975).
21. A. M. Lopez, J. A. Miller, and P. W. Murrill, *Instr. Tech.* **14**(11), 57 (1967).
22. C. L. Smith, A. B. Corripio, and J. Martin Jr., *Instr. Tech.* **22**(12), 39 (1975).
23. M. Yuwanda and D. E. Seborg, *AIChE J.*, **28**(3), 434 (1982).
24. A. Jutan and E. S. Rodriguez II, *Can. J. Chem. Eng.* **62**, 802 (1984).
25. T. J. McAvoy, *Interaction Analysis Theory and Application*, Instrument Society of America, Research Triangle Park, N.C., 1983.
26. E. H. Bristol, *IEEE Trans. Auto. Control*, **AC-11**, 133 (1966).
27. C. F. Moore, "Application of Singular Value Decomposition to the Design, Analysis, and Control of Industrial Processes," *Proceeding of American Control Conference*, Boston, Mass., 1986, p. 643.
28. J. P. Gagnepain and D. E. Seborg, *IEC Des. Develop.* **21**, 5 (1982).
29. W. H. Ray, *Computers Chem. Eng.* **7**, 367 (1983).

30. W. H. Ray, *Advanced Control*, McGraw-Hill, Book Co., Inc., New York, 1981.
31. K. A. McDonald and T. J. McAvoy, *AIChE J.* **37**, 583 (1983).
32. T. S. Chang and D. E. Seborg, *Int. J. Control*, **37**, 583 (1983).
33. C. R. Cutler and B. L. Ramaker, "Dynamic Matrix Control: A Computer Control Algorithm," *Proceedings of Joint Auto. Control Conference*, Paper WP5-B, San Francisco, Calif., 1980.
34. J. Richalet and co-workers, *Automatica*, **14**, 413 (1978).
35. M. Morari and E. Zafiriou, *Robust Process Control*, Prentice-Hall, Inc., Englewood Cliffs, N.J., 1989.
36. J. E. Doss and C. F. Moore, *ISA Trans.* **20**(4), 77 (1982).
37. K. P. Wong and D. E. Seborg, *AIChEJ.* **32**, 1597 (1986).
38. C. A. Garcia and M. Morari, *IEC Proc. Des. Develop.* **21**, 308 (1982).
39. G. E. P. Box and G. M. Jenkins, *Time Series Analysis: Forecasting and Control*, Holden-Day, San Francisco, Calif., 1970.
40. R. K. Mehra, R. Rouhani, and J. Eterno, in T. F. Edgar and D. E. Seborg, eds., *Chemical Process Control 2*, Engineering Foundation, New York, 1982.
41. A. M. Morshedi, C. R. Cutler, and T. A. Skrovanek, "Optimal Solution of Dynamic Matrix Control with Linear Programming Techniques," *Proceedings of American Control Conference*, Boston, Mass., 1986, p. 199.
42. C. E. Garcia and A. M. Morshedi, *Chem. Eng. Comm.* **46**, 73 (1986).
43. C. R. Cutler and R. B. Hawkins, "Application of a Large Model Predictive Controller to a Hydrocracker Second Stage Reactor," *Proceedings of American Control Conference*, Atlanta, Ga., 1988, p. 284.
44. K. J. Astrom and B. Wittenmark, *Adaptive Control Systems*, Addison-Wesley, Publishing Co., Inc., Reading, Mass., 1988.
45. K. J. Astrom and T. Hagglund, *Automatic Tuning of PID Controllers*, Instrument Society of America, Research Triangle Park, N.C., 1987.
46. E. H. Bristol, *ISA Trans.* **22**(3), 17 (1983).
47. D. E. Seborg, T. F. Edgar, and S. L. Shah, *AIChE J.* **32**, 881 (1986).
48. T. F. Edgar and D. M. Himmelblau, *Optimization of Chemical Processes*, McGraw-Hill Book Co., Inc., New York, 1988.
49. G. S Beveridge and R. S. Schechter, *Optimization: Theory and Practice*, McGraw-Hill Book Co., Inc., New York, 1968.
50. J. L. Marchetti, D. A. Mellichamp, and D. E. Seborg, *IEC Proc. Des. Devel.* **22**, 488 (1983).
51. *Binary Logic Diagrams for Process Operations*, ISA-S5.2, Instrument Society of America, Research Triangle Park, N.C., 1992.
52. T. G. Fisher, *Batch Control Systems*, Instrument Society of America, Research Triangle Park, N.C., 1990.
53. H. P. Rosenof and A. Ghosh, *Batch Process Automation: Theory and Practice*, Van Nostrand Reinhold, Co., Inc., New York, 1987.
54. H. P. Rosenof, *Control Eng.*, **28**(3), 73 (1981).
55. E. H. Bristol, *InTech*, **32**(10), 47 (1985).
56. B. D. Beatty, *Control*, **1**(1), 238 (1988).
57. *Batch Control I: Models and Terminology*, ISA-dS88.01, Instrument Society of America, Research Triangle Park, N.C., 1995.

JOHN PAUL SAN GIOVANNI
Jockey Hollow Technologies

PROCESS ENERGY CONSERVATION

The main driving force for increased energy conservation, which continues in times of both rising and falling energy prices, is broadscale technological process. Advances in technology are responsible for the historical rise in energy efficiency of 1–3% per year achieved by process industries. A wide range of big and little steps have contributed to these advances, such as improved gas turbine efficiency, structured packing in distillation (qv), computer control (see PROCESS CONTROL), variable speed drives, computer design tools, and improved catalysts (see CATALYSIS) and synthetic processes for a variety of materials, eg, low density polyethylene (see OLEFIN POLYMERS), acrylonitrile (qv), ammonia (qv), and acetic acid (see ACETIC ACID AND DERIVATIVES).

The second force that has driven increased energy conservation is the trade of capital for energy. This trade is optimized within an existing technology and nets large increases when energy prices rise rapidly compared to capital price as in the 1975–1985 time period. The effect of energy usage on total cost is shown in Figure 1. If proper design is used, total costs are relatively tolerant of large deviations from the optimum design. For example, in Figure 1a, if the piping pressure drop is anywhere between one-third and three times the optimal, the penalty in total cost is ≤10%. In piping systems (qv), capital costs dominate over energy costs (1).

Energy Balance

Historically, an energy balance has been prepared for the components of a process primarily to ensure that heat exchangers and utility supply are adequate (see ENERGY MANAGEMENT; HEAT-EXCHANGE TECHNOLOGY; POWER GENERATION; PROCESS CONTROL). Often, an overall process energy balance was not developed. However, beginning in the mid-1980s, the energy balance for the overall process has become a document almost as important as the material balance. The overall energy balance serves as an evergreen framework during design to highlight the areas having greatest potential for improvement. Moreover, this document serves as a tool for plant-operating personnel after start-up, to aid optimization of energy use.

The energy balance should analyze the energy flows by type and amount, ie, present summaries of electricity, fuel gas, steam level, heat rejected to cooling water, etc. It should include realistic loss values for turbine inefficiencies and heat losses through insulation.

Exergy, Lost Work, and Second-Law Analysis. When energy is critically important to process economics, the simple energy balance is sometimes carried into an analysis of lost work. This compares the actual design against the theoretical ideal at each step and defines where the true energy use, or lost work, is occurring. In the discussions herein of reaction, separation, heat exchange, compression, refrigeration, and steam systems, the importance of this concept is illustrated. A few terms are defined below.

Exergy, E, is the potential to do work. It is also sometimes called availability or work potential. Thermodynamically, this is the maximum work a stream

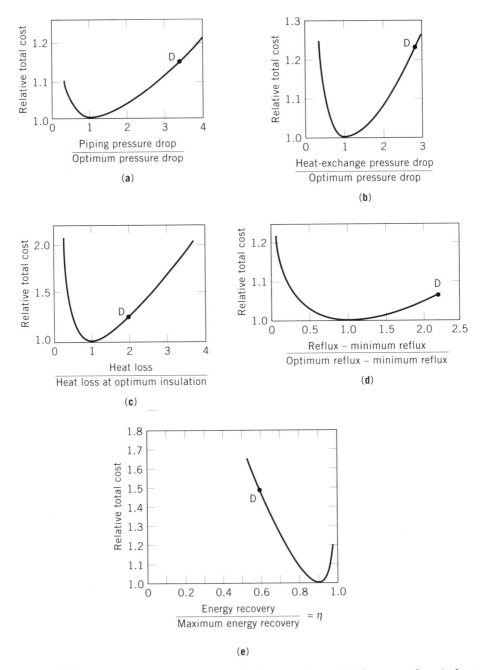

Fig. 1. Effect of energy use on total cost where total cost is the sum of capital and energy costs for the lifetime of the plant, discounted to present value. Point D corresponds to the design point if the designer uses an energy price that is low by a factor of four in projected energy price. Effects on costs of (**a**) pressure drop in piping, (**b**) pressure drop in exchangers, (**c**) heat loss through insulation, (**d**) reflux use, and (**e**) energy recovery through waste-heat boiler use.

can deliver by coming into equilibrium with its surroundings:

$$E = (H - H_0) - T_0(S - S_0)$$

where E = the maximum theoretical work potential; H and S = enthalpy and entropy of the stream at its original conditions; H_0 and S_0 = enthalpy and entropy of the same stream at equilibrium with the surroundings; and T_0 = temperature of the surroundings (sink).

Free energy, G, is a related thermodynamic property. It is most commonly used to define the condition for equilibrium in a processing step. It is identical to ΔE if the processing step occurs at T_0.

$$\Delta G = \Delta H - T \Delta S$$

Lost work, LW, is the irreversible loss in exergy that occurs because a process operates with driving forces or mixes material at different temperatures or compositions.

$$LW = E_{\text{in}} - E_{\text{out}}$$

Second-law analysis looks at the individual components of an overall process to define the causes of lost work. Sometimes it focuses on the efficiency of a step and ratios the theoretical work needed to accomplish a change, eg, a separation, to that actually used.

Sometimes it is more cost-effective simply to compare the design against a second-law violation checklist covering items such as mixing streams at different temperatures and compositions, high pressure drops in control valves, reactions running far from equilibrium, high temperature differentials, and pump-discharge recirculation (2).

Reactor Design for Energy Conservation

How closely a design approaches minimum energy is largely determined by the raw materials and catalyst system chosen. However, if reaction temperature, residence time, and diluent are the only variables, there is still a tremendous opportunity to influence energy use via the effect on yield. Even given none of these, there is still wide freedom to optimize the heat interchange system (see REACTOR TECHNOLOGY).

Design Variables. *Maximizing Yield.* Often the greatest single contribution to reduced energy cost is increased yield. High yield reduces the amount of material to be pumped, heated, and cooled while also simplifying downstream separation. This says nothing about the indirect energy reduction achieved through reduced raw material use. On average, the chemical industry uses almost as much energy in its raw materials as it does in direct purchases of fuel.

Minimizing Diluent. The case concerning diluent is less clear. A careful balance must be made of the benefits a diluent gives in higher yield against the costs needed for mass handling and separation.

Optimizing Temperature. Temperature is usually dictated by yield considerations. The choice of temperature for yield often overrides any desire to choose a temperature that minimizes the energy bill.

Heat Recovery and Feed Preheating. The objective is to bring the reactants to and from reaction temperature at the least utility cost, and to recover maximum waste heat at maximum temperature. The impact of feed preheating merits a more careful look. In an exothermic reaction, preheated feed permits the reactor to act as a heat pump, ie, to buy low and sell high. The most common example is combustion-air preheating for a furnace.

Batch vs Continuous Reactors. Usually, continuous reactors yield much lower energy use because of increased opportunities for heat interchange. Sometimes the savings are even greater in downstream separation units than in the reaction step itself. Especially for batch reactors, any use of refrigeration to remove heat should be critically reviewed. Batch processes often evolve little from the laboratory-scale glassware setups where refrigeration is a convenience.

Separation

About one-third of the chemical industry's energy is used for separation. A correlation exists between selling price and feed concentration (Fig. 2) as well as between selling price and product purity.

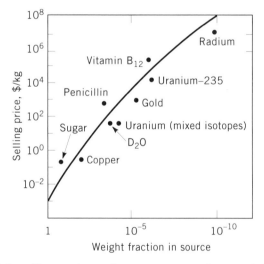

Fig. 2. Commercial selling prices of some separated materials (3). Courtesy of McGraw-Hill Book, Co., Inc.

Concentration and purity can both be traced to the minimum work of separation, W, where T_0 is the sink temperature; N_i is the number of moles of a

$$W = RT_0 \Sigma N_i \ln(x_i \gamma_i)$$

species present in the feed; $x_i = N_i/\Sigma N_i$; and γ_i is an activity coefficient. The value of W provides a target that is easily calculated and approachable in practice. For example, work calculated from this expression closely approaches the performance of a real-world distillation after inefficiencies for driving forces are taken into account.

For ideal solutions ($\gamma_i = 1$) of a binary mixture, the equation simplifies to the following, which applies whether the separation is by distillation or by any other technique.

$$W = RT_0[x_1 \ln x_1 + (1 - x_1) \ln(1 - x_1)]$$

When a separation is not completed, less work is required. For x_1 equal to 0.5,

Product purity, %	Relative work
100	1
99.9	0.99
99.0	0.92
90.0	0.53

This relative work is an important consideration when comparing separation techniques. Some leave much of the work undone, as, for example, in crystallization (qv) involving an unseparated eutectic mixture.

Distillation.   Distillation (qv) is by far the most common separation technique because of its inherent advantages. Its phase separation is clean, its equilibrium is closely approached in each stage, and its multistage countercurrent device is relatively easy to build.

Minimum work for an ideal separation at first glance appears unrelated to the slender vertical vessel having a condenser at the top and a reboiler at the bottom. The connection becomes evident when one calculates the work embedded in the heat flow that enters the reboiler and leaves at the condenser. An ideal engine can extract work from this heat.

$$E = QT_0\left(\frac{1}{T_{\text{condenser}}} - \frac{1}{T_{\text{reboiler}}}\right)$$

Comparison of actual use of work potential against the minimum allows calculation of an efficiency relative to the best possible separation:

$$\eta = \frac{RT_0[x_1 \ln x_1 + (1 - x_1) \ln(1 - x_1)]}{QT_0\left(\frac{1}{T_{\text{condenser}}} - \frac{1}{T_{\text{reboiler}}}\right)}$$

There is still no obvious reason to believe that the efficiency of separating a mixture and an α (relative volatility) of 1.1 is related to that for an α of 2; however, it is known that when α is small, the required reflux and Q are large, but ($T_{\text{condenser}} - T_{\text{reboiler}}$) is small (see DISTILLATION).

The two effects almost cancel one another to yield an approximation for the minimum work potential used in a distillation (3,4).

$$E = RT_0(1 + [\alpha - 1]x_1)$$

When this is combined with the definition of minimum separation work, an approximation for distillation efficiency for an ideal binary can be obtained:

$$\eta = \frac{x_1 \ln x_1 + (1 - x_1) \ln(1 - x_1)}{1 + (\alpha - 1)x_1}$$

This efficiency is high and shows only minor dependence on α over a broad range of α. For $x_1 = 0.5$:

α	η
1.1	0.66
1.5	0.55
2.0	0.46

The dependence on x_1 is greater:

	$\alpha = 1.05$	$\alpha = 2$
η for $x_1 = 0.1$	0.32	0.30
η for $x_1 = 0.01$	0.056	0.053

These values, which match experience, suggest that distillation should be the preferred separation method for feed concentrations of 10–90%, but is probably a poor choice for feed concentrations of less than 1%. Techniques such as adsorption (qv), chemical reaction, and ion exchange (qv) are chiefly used to remove impurity concentrations of <1%.

The high η values above conflict with the common belief that distillation is always inherently inefficient. This belief arises mainly because past distillation practices utilized such high driving forces for pressure drop, reflux ratio, and temperature differentials in reboilers and condensers. A real example utilizing an ethane–ethylene splitter follows, in which the relative number for the theoretical work of separation is 1.0, and that for the net work potential used before considering driving forces is 1.4.

$$\eta = \frac{\text{theoretical work}}{\text{net work potential used}} = \frac{1.0}{1.4} = 0.7$$

losses for driving forces:

reflux above the minimum	0.1
exchanger ΔT	2.1
ΔP in tower	0.5
ΔP in condenser and tower	0.8
Total	*3.5*

$$\eta_{\text{including losses}} = \frac{1.0}{1.4 + 3.5} = 0.2$$

These numbers show that, first, the theoretical work can be closely approached by actual work after known inefficiencies are identified and, second, the dominant driving force losses are in pressure drop and temperature difference. This is a characteristic of towers having low relative volatilities.

Optimum Design. *Condenser and Reboiler* ΔT. The losses for ΔT are typically far greater than those for reflux beyond the minimum. The economic optimum for temperature differential is usually under 15°C, in contrast to the values of over 50°C often used in the past. This is probably the biggest opportunity for improvement in the practice of distillation. A specific example is the replacement of direct-fired reboilers with steam (qv) heat.

Adjusting Process to Optimize ΔT. At first glance, there appear to be only three or four utility levels (temperatures), and these can be 50°C apart. Different ways to increase the options include using multieffect distillation, which spreads the ΔT across two or three towers; using waste heat for reboil; and recovering energy from the condenser. To make these options possible, the pressure in a column may have to be raised or lowered.

Reflux Ratio. Generally, the optimum reflux ratio is below 1.15 and often below 1.05 minimum. At this point, excess reflux is a minor contributor to column inefficiency. When designing for this tolerance, correct vapor–liquid equilibrium (VLE) and adequate controls are essential.

State-of-the-Art Control. Computer control using feed-forward capability can save 2–20% of a unit's utilities by reducing the margin of safety (5). Unless the discipline of a controller forces the reduction of the safety margin, operators typically opt for increased safety. Operators are probably correct to do so when a proper set of analyzers and controllers has not been provided and maintained.

Right Feed Enthalpy. Often it is possible to heat the feed with a utility considerably less costly than that used for bottom reboiling. Sometimes the preheating can be directly integrated into the column-heat balance by exchange against the condensing overhead or against the net bottoms from the column. Simulation and careful examination of the overall process are required to assess the value of feed preheating.

A vapor feed is favored when the stream leaves the upstream unit as a vapor or when most of the column feed leaves the tower as overhead product. The use of a vapor feed was a key component in the high efficiency cited previously for the C_2 splitter, where most of the feed goes overhead.

Low Column Pressure Drop. The penalty for column pressure drop is an increase in temperature differential:

$$\Delta T = \left(\frac{dT}{dP} \right) \Delta P$$

$$\frac{dT}{dP} = \frac{R}{\Delta H} \frac{T^2}{P}$$

As this suggests, the penalty becomes large for low vapor pressure materials, ie, for components that are distilled at or below atmospheric pressure. The work penalty associated with this ΔT is approximately defined by the following ratio.

$$\frac{\Delta T_{\text{pressure drop}}}{T_{\text{reboiler}} - T_{\text{condenser}}} = \text{fraction of } W \text{ for } \Delta P$$

This penalty is greatest for close-boiling mixtures. A powerful technique for cutting ΔP is the use of packing. Conventional packings such as 5-cm (2-in.) pall rings can achieve a factor of four reductions over trays, and structured packing can achieve a factor of 10 reduction. Structured packing is more vulnerable to mistakes in detailed engineering and much less tolerant of fouling than trays. Almost 50% of the installations have encountered serious performance problems (6). It is also 2 to 10 times as expensive as the trays it typically replaces. However, despite these obstacles, structured packing is the biggest innovation in energy-saving hardware in the chemical processing industries. The overhead line and condenser pressure drop should be considered as well. (Note the high loss in the C_2 splitter example.)

Intermediate Condenser.   As shown in Figure 3, an intermediate condenser forces the operating line closer to the equilibrium line, thus reducing the inherent inefficiencies in the tower. Using intermediate condensers and reboilers, it is possible to raise the efficiency above that for a simple reboiler–condenser system, particularly when the feed composition is far from 50:50 in a binary mixture.

	Maximum efficiency of heavy component in feed	
	50%	95%
one condenser, one reboiler	67	20
two condensers, one reboiler	73	47
three condensers, one reboiler	77	62

The intermediate condenser is most effective when a less costly coolant can be substituted for refrigeration.

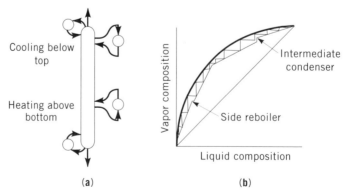

Fig. 3. (a) Schematic of an intermediate condenser and reboiler; (b) the corresponding vapor–liquid equilibrium.

Intermediate Reboiler. Inclusion of an intermediate reboiler moves the heat-input location up the column to a slightly colder point. It can permit the use of waste heat for reboil when the bottoms temperature is too hot for the waste heat.

Heat Pumps. Because of added capital and complexity, heat pumps are rarely economical, although they were formerly commonly used in ethylene/ethane and propylene/propane splitters. Generally, the former splitters are integrated into the refrigeration system; the latter are driven by low level waste heat, cascading to cooling water.

Lower Pressure. Usually, relative volatility increases as pressure drops. For some systems, a 1% drop in absolute pressure cuts the required reflux by 0.5%. Again, if operating at reduced pressure looks promising, the process can be evaluated by simulation. In a complete study of distillation processes, other questions that need to be asked include, Is the separation necessary? Is the purity necessary? Are there any recycles that could be eliminated? Can the products be sent directly to downstream units, thereby eliminating intermediate heating and cooling?

Other Separation Techniques. Under some circumstances, distillation is not the best method of separation. Among these instances are the following: when relative volatility is <1.05; when <1% of a stream is removed, as in gas drying (adsorption or absorption) or C_2H_2 removal (reaction or absorption); when thermodynamic efficiency of distillation is <5%; and when a high boiling point pushes thermal stability limits. A variety of other techniques may be more applicable in these cases.

Reaction. Purification by reaction is relatively common when concentrations are low (ppm) and a high energy but low value molecule is present. Some examples are the hydrogenation of acetylene and the oxidation of waste hydrocarbons:

$$C_2H_2 + H_2 \longrightarrow C_2H_4$$

$$\text{waste hydrocarbon} + O_2 \longrightarrow H_2O + CO_2$$

Absorption. As a separation technique, absorption (qv), also called extractive distillation, starts with an energy deficit because the process mixes in a pure material (solvent) and then separates it again. This process is nevertheless quite common because it shares most of the advantages of distillation. Additionally, because it separates by molecular type, it can be tailored to obtain a high α. The following ratios are suggested for equal costs (7):

$\alpha_{\text{distillation}}$	$\alpha_{\text{extraction distillation}}$	$\alpha_{\text{extraction}}$
1.2	1.4	2.5
1.4	1.9	5
1.6	2.3	8

In practice, most of the applications have come where a small part (<5%) of the feed is removed. Examples include H_2S/CO_2 removal and gas drying with a glycol (see DISTILLATION, AZEOTROPIC AND EXTRACTIVE).

Extraction. The advantage of extraction is that a liquid is purified rather than a vapor, allowing operation at lower temperatures and the removal of a series of similar molecules at the same time, even though these molecules differ widely in boiling point. An example is the extraction of aromatics from hydrocarbon streams (see EXTRACTION, LIQUID–LIQUID).

The disadvantage of extraction relative to extractive distillation is the greater difficulty of getting high efficiency countercurrent processing.

Adsorption. Adsorbents can achieve even more finely tuned selectivity than extraction. The most common application is the fixed bed with thermal regeneration, which is simple, attains essentially 100% removal, and carries little penalty for low feed concentration. An example is gas drying. A variant is pressure-swing adsorption. Here, regeneration is attained by a drop in pressure. By using multiple stages, high impurity rejection can be achieved, but at the expense of losing part of the desired product (see ADSORPTION).

Another approach is the simulated moving-bed system, which has large-volume applications in normal-paraffin separation and *para*-xylene separation. Since its introduction in 1970, the simulated moving-bed system has largely displaced crystallization in xylene separations. The unique feature of the system is that, although the bed is fixed, the feed point shifts to simulate a moving bed (see ADSORPTION, LIQUID SEPARATION).

Melt Crystallization. Crystallization (qv) from a melt is inherently more attractive than distillation because the heat of fusion is much lower than that of evaporation. It also benefits from lower operating temperature. In addition, organic crystals are virtually insoluble in each other so that a pure product is possible in a one-stage operation.

However, crystallization has unique disadvantages that outweigh its virtues and have sharply limited its application. Industry practice suggests the use of a workable alternative, if one exists. The disadvantages of melt crystallization include the following. (*1*) Difficulty of physical separation. Impure liquid is trapped as occlusion, and wets all crystal surfaces. (*2*) Requirement of a second separating process for eutectic mixture. The process thus resembles formation of two liquid phases. Although little energy is required to get the two phases, a great amount of it is required to finish the purification. (*3*) Difficulty of adding or removing heat on account of the thermal resistance of the crystal. (*4*) Difficulty of moving the liquid countercurrent to the crystals.

Thermodynamic efficiency is hurt by the large ΔT between the temperatures of melting and freezing. In an analogy to distillation, the high α comes at the expense of a big spread in reboiler and condenser temperature. From a theoretical standpoint, this penalty is smallest when freezing a high concentration (ca 90%) material.

One process, shown in Figure 4, is a semibatch operation in which liquid falls down the walls of long tubes. This permits both staged operation and sweating of crystals. Sweating is the removal of impurities by melting a small portion of crystals after mother liquor is first drained. The sweating operation washes residual mother liquor off the remaining crystals, and also removes some impurities from within the crystals. Typically, the sweating and staged operations require melting 5 kg of material for each kg of product (8).

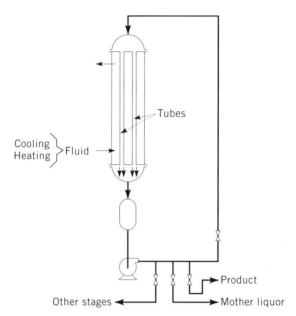

Fig. 4. Falling-film crystallizer, semibatch. Tube length is 12.2 m; tube diameter is 7.6 cm.

Membranes. Liquid separation via membranes, ie, reverse osmosis (qv), is used in production of pure water from seawater. The chief limit to broader use of reverse osmosis is the high pressure required as the concentration of reject rises.

Mole fraction of reject	Minimum ΔP, MPa (psi)
0.05	7.6 (1100)
0.10	15.2 (2200)
0.20	31.8 (4600)

As a result, most systems are limited to achieving a mole fraction reject of 0.1 or less (see MEMBRANE TECHNOLOGY).

Membranes are also used to separate gases, for example, the production of N_2 and O_2 from air and the recovery of hydrogen from ammonia plant purge gas. The working principle is a membrane that is chemically tuned to pass a molecular type.

Heat Exchange

Most processing is thermal. Reaction systems and separation systems are typically dominated by the associated heat exchange. Optimization of this heat exchange has tremendous leverage on the ultimate process efficiency (see HEAT-EXCHANGE TECHNOLOGY).

Heat exchangers use energy two ways: as frictional pressure drop, and as the loss in ability to do work when heat is degraded.

$$LW = QT_{\text{sink}}\left(\frac{1}{T_{\text{cold}}} - \frac{1}{T_{\text{hot}}} \right) + \text{frictional work for } \Delta P$$

In an optimized system, the lifetime value of the lost work associated with ΔT typically exceeds the cost of the heat exchanger. The lifetime value of the ΔP lost work in an optimized system is typically one-third as great as the heat exchanger capital (9). This means that when the costs for pumping power to overcome the heat exchanger ΔP (for the lifetime of the heat exchanger) are discounted to the time of heat exchanger purchase, their sum approximates one-third the heat exchanger cost. (This assumes a large heat exchanger designed for optimum pressure drop.)

The selection of design numbers for ΔP and ΔT is frequently the most important decision the process designer makes. The designer commonly becomes lost in the detail of tube length and baffle cut in an effort to optimize the hardware to meet a target, and spends far too little time on choosing that target.

Heat-Exchange Networks. A basic theme of energy conservation is to look at a process broadly, ie, to look at how best to combine process elements. The heat-exchange network analysis can be a useful part of this optimization. Figure 5 illustrates the basic concept of what the network analysis does. The analysis builds cumulative heating and cooling curves, pinching them together until a minimum ΔT is reached. This is discussed in greater detail elsewhere (see HEAT-EXCHANGE TECHNOLOGY).

Network analysis, or pinch technology, has become an increasingly powerful approach to process design that includes most of the virtues of second-law analysis. For example, pinch technology can be broadened to include process revisions such as changing the temperature and pressure of distillation columns to fit into the natural cascade of high level heat dropping down to ambient (10). Other extensions of the concept include analysis of distillation column profiles, total site integration, and batch processing (11). The approach yields a quantitative estimate of readily achievable improvement. For example, in Figure 5,

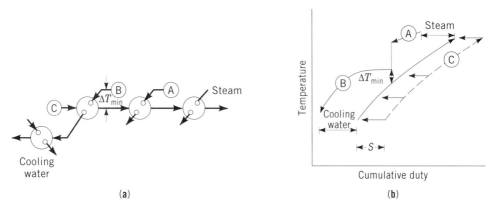

Fig. 5. Simple heat-exchange network where stream C is heated, and streams A and B are cooled: (**a**) schematic; (**b**) temperature–cumulative duty curve where $S =$ savings in system and cooling water owing to driving to ΔT_{min}.

note the reduction in steam and cooling water obtained by driving the design to the pinch, $\Delta T_{\min}$.

Overdesign. Overdesign has a great impact on the cost of heat exchange and sometimes is confused with energy conservation, through lower ΔT and ΔP. The best approach is to define clearly what the objective of overdesign is and then to specify it explicitly. If the main concern is a match to other units in the system, a multiplier is applied to flows. If the concern is with the heat balance or transfer correlation, the multiplier is applied to area. If the concern is fouling, a fouling factor is called for. If low ΔT or ΔP is the principal concern, however, that should be specified. Adding extra surface saves energy only if the surface is configured to do so. Doubling the area may do nothing more than double the ΔP, unless it is configured properly.

ΔT and ΔP Optimization. Ideally, ΔT and ΔP are optimized by trying several values, making preliminary designs, and finding the point where savings in utility costs just balance the incremental surface costs. Where the sums at stake are large, this should be done. However, for many cases the simple guidelines given below are adequate. The primary focus is the impact of surface and utility prices; a secondary focus is the impact of fluid properties on heat-transfer coefficient (9).

Optimum ΔT. There are three general cases of high importance: the waste-heat boiler, in which only one fluid involves sensible heat transfer, ie, a temperature change; the feed–effluent exchanger, in which both fluids involve sensible heat transfer and are roughly balanced, ie, undergo essentially the same temperature change; and the reboiler, in which neither fluid involves a temperature change, ie, one fluid condenses and the other boils.

Waste-Heat Boiler. In a waste-heat boiler (Fig. 6), the approach ΔT sets both the amount of the unrecovered energy and the amount of heat-exchange surface. When terms are added for energy value, K_v, and surface cost, K_l, the optimum occurs when

$$\Delta T_{\text{approach}} = \frac{K_l}{K_v} \frac{1.33}{U}$$

where K_l = annualized cost per unit of surface, $\$/(m^2 \cdot yr)$; K_v = annualized cost per unit of utility saved, $\$/(W \cdot yr)$; and U = heat-transfer coefficient, $W/(m^2 \cdot K)$. The factor 1.33 includes the value of the pressure drop for the added surface.

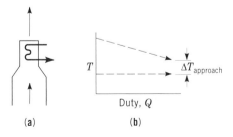

(a) (b)

Fig. 6. ΔT in a waste-heat boiler: (**a**) schematic; (**b**) corresponding graphic representation.

For example, the optimum $\Delta T_{\text{approach}}$ is computed as follows:

$$K_l = \frac{\$215/\text{m}^2}{2 \text{ yr}} = \frac{\$107.5}{(\text{m}^2 \cdot \text{yr})}$$

$$K_v = \frac{0.017}{\text{kW} \cdot \text{h}} \cdot 8322 \text{ h/yr} = \frac{\$142}{(\text{kW} \cdot \text{yr})} = \frac{\$0.142}{(\text{W} \cdot \text{yr})}$$

$$\Delta T_{\text{approach}} = \frac{107.5/(\text{m}^2 \cdot \text{yr})}{0.142/(\text{W} \cdot \text{yr})} \frac{1.33}{56.8 \text{ W}/(\text{m}^2 \cdot \text{K})} = 17.7 \text{ K}$$

where $U = 56.8$ W/(m^2·K) (10 Btu/(h·ft^2·°F)); surface cost = $215/m^2 ($20/ft^2); payout time = 2 yr; energy price = $0.017/(kW·h) ($5/10^6 Btu); and onstream time = 8322 h/yr. This case underlines a dramatic change in process design. Note that $\Delta T_{\text{approach}}$ varies with the first power of the ratio of surface price to energy price. The most visible result has been a change in typical fired heater design efficiency from 65–75% to 92–94%. A secondary result has been the appearance of waste-heat recovery units in many processes at the point where air coolers were once used.

Feed–Effluent Exchanger.   The detailed solution for the optimum ΔT in a feed–effluent exchanger (Fig. 7) involves a quadratic equation for $\Delta T_{\text{approach}}$, but within the following restrictions.

$$0.8 < \frac{\Delta T_{\text{hot}}}{\Delta T_{\text{cold}}} < 1.25$$

$$\frac{T_{\text{hot}_{\text{in}}} - T_{\text{cold}_{\text{in}}}}{\Delta T_{\text{log mean}}} < 10$$

An excellent approximation is given by Reference 9.

$$\Delta T_{\text{log mean}} = \left[\frac{K_l}{K_v} \frac{1.33}{U} \left(T_{\text{hot}_{\text{in}}} - T_{\text{cold}_{\text{in}}} \right) \right]^{1/2}$$

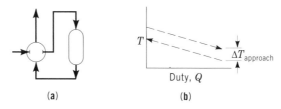

Fig. 7. ΔT in a feed–effluent exchanger: (**a**) schematic; (**b**) corresponding graphic representation.

For example, the optimum $\Delta T_{\text{log mean}}$ for a feed–effluent exchanger is computed as follows:

$$K_l = \frac{\$107.5/\text{m}^2}{2 \text{ yr}} = \$53.8/(\text{m}^2 \cdot \text{yr})$$

$$K_v = \frac{\$0.027}{(\text{kW}\cdot\text{h})} \cdot 8322 \frac{\text{h}}{\text{yr}} \cdot 1000 \frac{\text{kW}\cdot\text{h}}{\text{W}} = \frac{\$0.227}{(\text{W}\cdot\text{yr})}$$

$$\Delta T_{\text{log mean}} = \left(\frac{53.8}{0.227} \frac{1.33}{284} (200 - 100) \right)^{1/2} = 10.5°\text{C}$$

where $T_{\text{hot}_{\text{in}}} = 200°\text{C}$; $T_{\text{cold}_{\text{in}}} = 100°\text{C}$; $U = 284 \text{ W}/(\text{m}^2\cdot\text{K})$ (50 Btu/(h·ft²·°F)); surface cost $= \$107.5/\text{m}^2$ ($10/ft²); payout time $= 2$ yr; energy price $= \$0.027/(\text{kW}\cdot\text{h})$ ($8/10^6$ Btu); onstream time $= 8322$ h/yr; and $\Delta T_{\text{hot}}/\Delta T_{\text{cold}} = 1.20$.

Reboiler. The case shown in Figure 8 is common for reboilers and condensers on distillation towers. Typically, this ΔT has a greater impact on excess energy use in distillation than does reflux beyond the minimum. The capital cost of the reboiler and condenser is often equivalent to the cost of the column they serve.

The concept of an optimum reboiler or condenser ΔT relates to the fact that the value of energy changes with temperature. As the gap between supply and rejection widens, the real work in a distillation increases. The optimum ΔT is found by balancing this work penalty against the capital cost of bigger heat exchangers.

If the Carnot cycle is used to calculate the work embedded in the thermal flows with the assumption that the heat-transfer coefficient, U, is constant and the process temperature is much greater than ΔT, a simple derivation yields the following:

$$\Delta T_{\text{optimum}} = T_p \left[\frac{K_l}{K_v U T_{\text{sink}}} \right]^{1/2}$$

where $T_{\text{sink}} =$ temperature (absolute) at which heat is rejected; $T_p =$ process temperature (absolute); $K_l =$ annualized cost per unit of surface; and $K_v =$ annualized value of power.

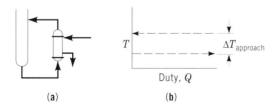

(a)   (b)

Fig. 8. ΔT in the reboiler: (**a**) schematic; (**b**) corresponding graphic representation.

For utilities above ambient temperature,

$$K_v = K_p \text{ (turbine efficiency)}$$

where K_p is the annualized cost of purchased power. The above relations typically give ΔT values in the 10–20°C range.

One strong caution is that the assumption of a constant U is usually inaccurate for boiling applications. Simulation is generally needed to fix ΔT accurately, particularly at ΔT values below 15°C.

Optimum Pressure Drop. For most heat exchangers there is an optimum pressure drop. This results from the balance of capital costs against the pumping (or compression) costs. A common prejudice is that the power costs are trivial compared to the capital costs. The total cost curve is fairly flat within ±50% of the optimum (see Fig. 1b), but the incremental costs of power are roughly one third of those for capital on an annualized basis. This simple relationship can be extremely useful in quick design checks.

The best approach is to have a computer program check a series of pressure drops and see how energy requirements decrease as surface increases. If this option is not available, the following simple method can be used to obtain specification sheet values. Start with a pressure drop of 6.9 kPa (1 psi), and apply three correction factors, $F_{\Delta T}$, F_{cost}, and F_{prop}, as follows.

$$\Delta P_{\text{optimum}} = 6.9(F_{\Delta T})(F_{\text{cost}})(F_{\text{prop}})$$

The correction for temperature difference is given by the following.

$$F_{\Delta T} = \frac{T_{\text{in}} - T_{\text{out}}}{(T_{\text{hot}} - T_{\text{cold}})_{\text{mean}}}$$

This term is a measure of the unit's length. Sometimes it is referred to as the number of transfer units (see MASS TRANSFER (SUPPLEMENT)). This simply says that the optimum pressure drop increases as the heat exchanger gets longer, ie, has more transfer units. The forms of F_{cost} and F_{prop} both follow from the fact that in turbulent flow the heat-transfer coefficient varies approximately with (power dissipated/volume)$^{0.25}$ (9). The preexponential terms are empirical. They net a 35% higher $\Delta P_{\text{optimum}}$ than the (power dissipated/volume) correlation for heat transfer in a 1.5-cm diameter tube. The correction for costs is

$$F_{\text{cost}} = 0.017 \left(\frac{\$/(\text{m}^2 \cdot \text{yr})}{\$/\text{kW} \cdot \text{h}} \right)^{0.75}$$

The correction for physical properties is

$$F_{\text{prop}} = \left(\frac{c}{c_w} \right)^{0.6} \left(\frac{k_w}{k} \right)^{0.6} \left(\frac{\mu}{\mu_w} \right)^{0.35} \left(\frac{\rho}{\rho_w} \right)^{0.5}$$

where c = specific heat, k = thermal conductivity, μ = viscosity, ρ = density, and c_w, k_w, μ_w, and ρ_w are the same properties for water at 25°C.

From these equations, the optimum ΔP for a feed–effluent exchanger, where the fluid has the physical properties of water and the following values:

$$T_{in} - T_{out} = 20°C$$

$$\Delta T = (T_{hot} - T_{cold})_{mean} = 10°C$$

$$\left.\begin{array}{l} \text{surface cost} = \$215/m^2 \\ \text{payout time} = 2 \text{ yr} \end{array}\right\} \$107.5/(m^2 \cdot yr)$$

$$\text{power cost} = \$0.03/(kW \cdot h)$$

is calculated by

$$F_{\Delta T} = \frac{20}{10} = 2$$

$$F_{cost} = 0.017\left(\frac{107.5}{0.03}\right)^{0.75} = 7.8$$

$$F_{prop} = 1$$

$$\Delta P_{optimum} = 6.9(2)(7.8)(1) = 107.67 \text{ kPa (15.6 psi)}$$

If all else remains the same as above, except that the process fluid is a gas with $\mu = 0.02 \ \mu_w$, $\rho = 0.00081 \ \rho_w$, $c = 0.25 \ c_w$, and $k = 0.066 \ k_w$, then

$$F_{prop} = (0.25)^{0.6}\left(\frac{1}{0.066}\right)^{0.6}(0.02)^{0.35}(0.00081)^{0.5} = 0.016$$

$$\Delta P_{optimum} = 6.9(2)(7.8)(0.016) = 1.7 \text{ kPa (0.25 psi)}$$

The great impact of density in this example and in Table 1 should be noted. Probably the most common specification error is to use the large ΔP values characteristic of liquids in low density gas systems.

Table 1. Impact of Fluid Density on Optimum ΔP

Fluid	$\dfrac{\rho}{\rho_w}$	$\left(\dfrac{\rho}{\rho_w}\right)^{0.5}$	$=$	Relative optimum ΔP
water	1	$(1)^{0.5}$		1
oil	0.8	$(0.8)^{0.5}$		0.9
gas				
high pressure	0.05	$(0.05)^{0.5}$		0.22
atm pressure	0.001	$(0.001)^{0.5}$		0.03
vacuum	0.0002	$(0.0002)^{0.5}$		0.014

Fired Heaters. The fired heater is first a reactor and second a heat exchanger. Often, in reality, it is a network of heat exchangers.

Fired Heater as a Reactor.   When viewed as a reactor, the fired heater adds a unique set of energy considerations, such as, Can the heater be designed to operate with less air by O_2 and CO analyzers? How does air preheating affect fuel use and efficiency? How can a lower cost fuel (coal) be used? Can the high energy potential of the fuel be used upstream in a gas turbine?

Air Preheating.   Use of unpreheated air in the combustion step is probably the biggest waste of thermodynamic potential in industry (Table 2).

Air preheating has the unique benefit of giving a direct cut in fuel consumed. It also can increase the heat-input capability of the firebox because of the hotter flame temperature. The drawback is that it tends to increase NO_x formation.

The most common type of air preheater on new units is the rotating wheel. On retrofits, heat pipes or hot-water loops are often more cost-effective because of ductwork costs or space limits.

Limitations in the material of construction make it difficult to use the high temperature potential of fuel fully. This restriction has led to the insertion of gas turbines into power generation steam cycles and even to the use of gas turbines in preheating air for ethylene-cracking furnaces.

Fired Heater as a Heat-Exchange System.   Improved efficiency in fired heaters has tended to focus on heat lost with the stack gases. When stack temperatures exceed 150°C, such attention is proper, but other losses can be much bigger when viewed from a lost-work perspective. For example, a reformer lost-work analysis by Monsanto gave the breakdown shown in Table 2.

Table 2. Lost-Work Analysis for a Fired Heater

Parameter	Lost-work potential, %
combustion step	54
radiant section ΔT	7
convection section ΔT	24
stack losses (exit temp 225°C)	13
wall losses	2

Losses for ΔT in the convection section are almost twice those for the hot exit flue gas. Furnace optimization is the clearest illustration of the benefits of lost-work analysis. If losses from a stack are nearly transparent, the losses embedded in an excessive ΔT in a convection section are even harder to identify. They do not show up on the energy balance that highlights the hot stack. These losses can be cut by adding surface to the convection section and shifting load from the radiant section, as well as by looking at the overall process (including steam generation) for streams to match the cooling curve of the flue gases.

Concern over corrosion from sulfuric acid when burning sulfur-bearing fuels often governs the temperature of the exit stack gas. However, the economics of heat recovery is so strong that flue gases are sometimes designed into the condensing range of weak sulfuric acid, recognizing that tube replacement will be required in the future.

Simple heat losses through the furnace walls are also significant. This follows from the high temperatures and large size of fired heaters, but these losses are not inevitable. In an optimized system, losses through insulation (1) are roughly proportional to

$$\left(\frac{\text{refractory price}}{\text{energy price}} \right)^{1/2}$$

This means that if the price ratio has decreased by a factor of four, then losses should be down by a factor of two. If the old optimum allowed a 2% loss, the new optimum would be closer to 1%.

Dryers. A drying (qv) operation needs to be viewed as both a separation and a heat-exchange step. When it is seen as a separation, the obvious perspective is to cut down the required work. This is accomplished by mechanically squeezing out the water. The objective is to cut the moisture in the feed to the thermal operation to less than 10%. In terms of hardware, this requires centrifuges and filters, and may involve mechanical expression or a compressed air blow. In terms of process, it means big crystals.

When the dryer is seen as a heat exchanger, the obvious perspective is to cut down on the enthalpy of the air purged with the evaporated water. Minimum enthalpy is achieved by using the minimum amount of air and cooling as low as possible. A simple heat balance shows that for a given heat input, minimum air means a high inlet temperature. However, this often presents problems with heat-sensitive material and sometimes with materials of construction, heat source, or other process needs. All can be countered somewhat by exhaust-air recirculation.

Minimum exhaust-air enthalpy also means minimum temperature. If this cannot be attained by heat exchange within the dryer, preheating the inlet air is an option. The temperature differential guidelines of the feed–effluent interchange apply.

Like the fired heater, the dryer is physically large, and proper insulation of the dryer and its allied ductwork is critical. It is not uncommon to find 10% of the energy input lost through the walls in old systems.

Optimum Design of Pumping, Compression, and Vacuum Systems

Pumping. Many companies have optimum pipe-sizing programs, but in the absence of one, a good rule of thumb is that, in an optimized system, the annualized cost for pumping power should be one-seventh the annualized cost of piping (1). Piping is always a significant cost component and should be optimized (see PIPING SYSTEMS). Similarly, for an optimized heat exchanger, the annualized cost for pumping should be one-third the annualized cost of the surface for the thermal resistance connected with that stream.

The pump should be specified for the right flow. As Figure 9 shows, a 50% overdesign factor increases power by 35% in a combination of higher head and lower efficiency (see PUMPS).

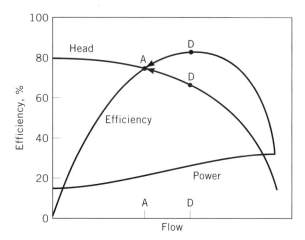

Fig. 9. Impact of excess design capacity on pump energy use, where A corresponds to actual operating flow, and D to the design point if the designer calls for 50% more flow than the actual operating flow value.

If the allowance for control can be reduced, it should be. One option is the use of variable-speed drives. This eliminates the control valve and its pressure drop and piping. Its best application is where a large share of the head is required for friction and where process demands cause the required flow to vary.

Compression. The work of compression is typically compared against the isentropic–adiabatic case.

$$\eta_{\text{comp}} = \frac{W_{\min}}{E_{\text{out}} - E_{\text{in}}}$$

For an ideal gas, this can be expressed in terms of temperatures

$$\eta_{\text{comp}} = \frac{W_{\min}}{W_{\text{actual}}} = \frac{T_{\text{in}}\left(\left(\frac{P_{\text{out}}}{P_{\text{in}}}\right)^{R/c} - 1\right)}{T_{\text{out}} - T_{\text{in}}}$$

where R/c is the ratio of gas constant to molar specific heat. Minimum work is directly proportional to suction temperature. This means that close temperature approaches are justified on suction coolers.

Sometimes $W_{\min}$ for compression is expressed for the isothermal case, which is always lower than that for the adiabatic case. The difference defines the maximum benefit from interstage cooling.

The measuring of temperature rise permits monitoring efficiency for a fixed pressure ratio and suction temperature. Efficiencies should always exceed 0.6, and 1.00 is approachable in reciprocating devices. Their better efficiency needs to be balanced against their greater cost, greater maintenance, and lower capacity.

Thermocompressors. A thermocompressor is a single-stage jet using a high pressure gas stream to supply the work of compression. One application is in boosting waste-heat-generated steam to a useful level. An example is shown in Figure 10. Thermocompressors can also be used to boost a waste combustible gas into a fuel system by using high pressure natural gas. The mixing of the high energy motive stream with the low energy suction stream inherently involves lost work, but as long as the pressures are fairly close, the net efficiency for the device can be respectable (25–30%). Here, efficiency is defined as the ratio of isentropic work done on the suction gas to the isentropic work of expansion that could have been obtained from the motive gas. The thermocompressor has the advantage of no moving parts and low capital cost.

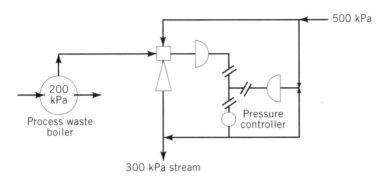

Fig. 10. A thermocompressor. To convert kPa to psi, multiply by 0.145.

Vacuum Systems.   The most common vacuum system is the vacuum jet. Because of the high ratio of motive pressure to suction pressure, the efficiency of vacuum systems is generally only 10–20%. The optimum system often employs several stages with intercondensers. Steam use in this range varies roughly as $(1/P)^{0.3}$, where P is absolute suction pressure. Like the thermocompressor, the steam jet has the advantage of no moving parts. However, the velocity in the steam nozzle (the part through which the steam discharges) is extremely high, which makes it subject to erosion and replacement every few years to maintain efficiency (see VACUUM TECHNOLOGY).

Because of the low efficiency of steam-ejector vacuum systems, there is a range of vacuum above 13 kPa (100 mm Hg) where mechanical vacuum pumps are usually more economical. The capital cost of the vacuum pump goes up roughly as (suction volume)$^{0.6}$ or $(1/P)^{0.6}$. This means that as pressure falls, the capital cost of the vacuum pump rises more swiftly than the energy cost of the steam ejector, which increases as $(1/P)^{0.3}$. Usually below 1.3 kPa (10 mm Hg), the steam ejector is more cost-effective.

Other factors that favor the choice of the steam ejector are the presence of process materials that can form solids or require high alloy materials of construction. Factors that favor the vacuum pump are credits for pollution abatement and high cost steam. The mechanical systems require more maintenance and some form of backup vacuum system, but these can be designed with adequate reliability.

Refrigeration

Refrigeration is a high value utility (see REFRIGERATION AND REFRIGERANTS). The value of heat in a hot stream is the amount of work it can surrender:

$$\frac{W}{Q} = \left(\frac{T - T_{\text{sink}}}{T} \right) \eta_{\text{turbine}}$$

and the value of refrigeration is the work required to heat-pump it to the sink temperature:

$$\frac{W}{Q} = \left(\frac{T_{\text{sink}} - T}{T} \right) \frac{1}{\eta_{\text{compressor}}} \frac{1}{\eta_{\text{fluid}}}$$

The value of refrigeration is compared to heating in Figure 11 for $\eta_{\text{turbine}} = \eta_{\text{compressor}} = 0.7$ and for $\eta_{\text{fluid}} = 0.8$. In Figure 12, η_{fluid} accounts for cycle inefficiencies such as the letdown valve.

Because of its value, refrigeration justifies thicker insulation, lower ΔT values in heat exchange, and generally much more care in engineering (12). The designer should ensure that the capital cost of the refrigeration users has

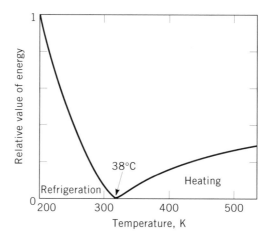

Fig. 11.   Relative value of energy at various temperatures.

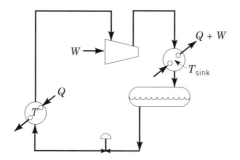

Fig. 12.   Schematic of compression refrigeration.

been optimized and integrated with the cost of the refrigeration system, and that the cost of supplying power to the refrigeration machine driver has been integrated with the refrigeration system optimization. It is also good to ask, Is refrigeration really necessary? Can river water or cooling-tower water be used directly for part of the year? Can part of the refrigeration be replaced? Can the refrigerant-condensing temperature be reduced during part of the year? Can the system be designed to operate without the compressor during cold weather? Is a central system more efficient than scattered independent systems? Does the control system cut required power for part-load operations? Are enough gauges and meters provided to monitor operation? Is there an abundance of waste heat (above 90°C) available from the plant? If so, refrigeration can be supplied by an absorption system.

Absorption chiller units (Fig. 13) need 1.6–1.8 J (0.38–0.43 cal) of waste heat per joule (0.24 cal) of refrigeration. Commercially available LiBr absorption units are suitable for refrigeration down to 4.5°C. For low level waste heat (90–120°C), absorption chillers utilize waste heat as efficiently as steam turbines using mechanical refrigeration units. Absorption refrigeration using 120°C saturated steam delivers 4.5°C refrigeration, having an efficiency, with respect to the work potential in the steam of 35%.

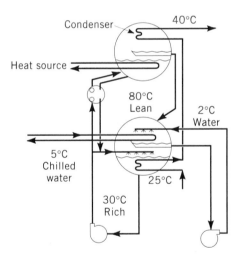

Fig. 13. Absorption refrigeration.

Steam and Condensate Systems

In the process industry, steam (qv) serves much the same role as money does in an economy, ie, it is the medium of exchange. If its pricing fails to follow common sense or thermodynamics, strange design practices are reinforced. For example, many process plants employ accounting systems where all steam carries the same price regardless of temperature or pressure. This may be appropriate in a polymer or textile unit where there is no special use for the high temperature; it is wrong in a petrochemical plant.

Some results of the constant-value pricing system are as follow: generation in a central unit at relatively low pressure, <4.24 MPa (600 psig); tremendous economic pressure to use turbines rather than motors for drives; lack of incentive for high efficiency turbines; excessively high temperature differentials in steam users; tremendous incentive to recover waste heat as low pressure steam; and a large plume of excess low pressure steam vented to the atmosphere.

A number of alternative pricing systems have been proposed that hinge on turbine efficiency and the relative pricing of fuel and electricity. All pricing systems have problems, which suggests that the best system is a simpler one that relates the value of steam to that at the generation pressure by its work potential (exergy content).

$$\frac{\text{value at pressure}}{\text{value at highest generation pressure}} = \frac{\text{exergy at pressure}}{\text{exergy at highest generation pressure}}$$

Design of a central power steam system is beyond the scope of this discussion, but the interaction between the steam system and the process must be considered at all stages of design. There is a long list of factors to consider in designing a steam system (see also ENERGY MANAGEMENT):

Is there a computer program that monitors the system and advises on what turbines to operate and how to minimize steam venting?

Can a gas turbine be utilized for power generation upstream of the boiler?

Can steam-use pressure be lowered? (If ΔT in the heater is above 20°C, the steam pressure is probably above the economic optimum.)

Are there any turbines under 65% efficiency? (Turbines are often limited to sizes above 500 kW, where good efficiencies can be obtained; they are used for smaller drives only where they are essential to the safe shutdown of the unit.)

Are there waste streams with unutilized fuel value that can be burned in the boilers?

Is there a program to monitor turbine efficiency by checking temperatures in and out?

Is condensate recovered?

Is the flash steam from condensate recovered?

Is feedwater heating optimized?

Is there any pressure letdown without power recovery?

Has enough flexibility been built into the overall condensing turbine system? (The balance changes over the history of a unit as a process evolves, generally in the direction of less condensing demand.)

Is steam superheat maintained at the maximum level permitted by mechanical design?

Can a thermocompressor be used to increase steam pressures from waste heat?

Are all users metered?

Is low level process heat used to preheat deaerator makeup?

Are ambient sensing valves used to turn off steam tracer systems?

Cooling-Water Systems

Cooling water is a surprisingly costly utility. On the basis of price per unit energy removed, it can cost one-fifth as much as the primary fuel. Roughly half of this cost is in delivery (pump, piping, and power). This fact has several important implications for design. Heat exchangers should be designed to use the available pressure drop. A heat exchanger that is designed for 10 kPa (1.45 psi) when 250 kPa (36 psi) is available will have five times the design flow. If an exchanger cannot be economically designed to use available ΔP, orifices should be provided to balance the system. This can be done without compromising the guidelines that no unit should be designed for less than 0.6 m/s on tubeside or less than 0.3 m/s on shellside. If temperature requirements permit, the system will cost less to operate with exchangers in series. An installed flow-measuring element is usually justified. If only part of the system requires a high head, this could be supplied by a booster pump. The whole system need not be designed for the high head.

Other energy considerations for cooling towers include the use of two-speed or variable-speed drives on cooling-tower fans, and proper cooling-water chemistry to prevent fouling in users (see WATER, INDUSTRIAL WATER TREATMENT). Air coolers can be a cost-effective alternative to cooling towers at 50–90°C, just below the level where heat recovery is economical.

Special Systems

Heat Pumps. A heat pump is a refrigeration system that raises heat to a useful level. The most common application is the vapor recompression system for evaporation (qv) (Fig. 14). Its application hinges primarily on low cost power relative to the alternative heating media. If electricity price per unit energy is less than 1.5 times the cost of the heating medium, it merits a close look. This tends to occur when electricity is generated from a cheaper fuel (coal) or when hydroelectric power is available.

Use in distillation systems are rare. The reason is the recognition that almost the same benefits can be achieved by integrating the reboiling–condensing via either steam system (above ambient) or refrigeration system (below ambient).

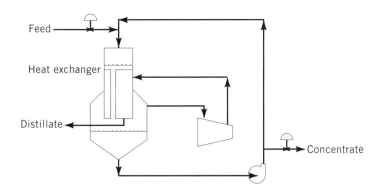

Fig. 14. Vapor recompression evaporator system.

In an optimized system, where T_{hot} and T_{cold} are in absolute units, K, the following is true.

$$\frac{Q}{W} = \frac{T_{\text{hot}}\,\eta_{\text{compressor}}}{T_{\text{hot}} - T_{\text{cold}}}$$

This provides another criterion for testing whether a heat-pump system may be cost-effective. A power plant takes three units of Q to yield one unit of W. Therefore, to provide any incentive for less overall energy use, Q/W must be far in excess of 3.

Energy Management Systems. The reduction in computing costs has made it possible to do a wide range of routine monitoring and controlling. For example, a distillation system can be monitored continuously, the energy use can be compared against an optimum, and the cost-per-hour deviation from the optimum setpoint can be displayed.

Existing Plants

Good design ideas for new plants are also good for existing plants, but there are three basic differences. (*1*) Because a plant already exists, the capital–operating cost curve differs. Usually, this makes it more difficult to reduce utility costs to as low a level as in a new plant. (*2*) The real economic justification for change is more likely to be obscured by the plant accounting system and other nontechnical inputs. (*3*) The real process needs are measurable and better defined.

An example in support of the first point is the case of optimum insulation thickness. A tank, optimally insulated when first installed, can fall below optimal if the value of heat is quadrupled. This change can justify twice the old insulation thickness on a new tank. However, the old tank may have to function with its old insulation. The reason is that there are large costs associated with preparation to insulate. This means that the cost of an added increment of insulation is much greater than assumed in the optimum insulation thickness formulas (Fig. 15).

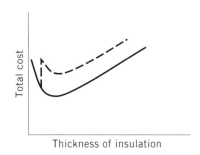

Fig. 15. Tank insulation costs: existing (---) vs new (——). Total cost represents the sum of heat loss and insulation.

An example of the second difference is that many things appear to be strongly justified by savings in low pressure steam if the steam is valued artificially high. A designer of a new plant has the advantage of focusing attention on savings in the primary budget items, ie, fuel and electricity at the plant gate, rather than on cost-sheet items such as steam at battery limits.

The third difference is that many process details are relatively uncertain when a plant is designed. For example, inert loading for vacuum jets is rarely known to within 50%. Although the first two differences are negative, the third provides a unique opportunity to measure the true need and revise the system accordingly.

Energy Audit

The energy audit has seven components: as-it-is balance, field survey, equipment tests, checking against optimum design, idea-generation meeting, evaluation, and follow-up.

As-It-Is Balance.　This is a mandatory first step for the energy audit. It permits the targeting of principal potentials; checking of use against design; checking of use against optimum, ie, how a new plant would be designed; definition of possible hot or cold interchanges; definition of unexpected uses, eg, the large steam purge to process or high pressure drop exchanger; and contribution from specialists not familiar with the unit.

Field Survey.　This is often done by a team of two: one who knows what to look for and one who knows the process. They should look and listen for things such as air leaks, high pressure drops across values, frost on piping, lights of the wrong type or lights on at the wrong time, steam plumes (a reason to climb to the top of the unit), and minimum-flow bypasses in use. They should also talk to the operators and ask such questions as, What runs when the unit is down? What happens when the reflux is cut? Where are the guidelines for steam–feed ratio? How close is the unit run to these guidelines? The field survey should develop detailed repair lists for leaking traps, uninsulated metal, lighting, and steam leaks.

Equipment Tests.　Procedures for rigorous, detailed efficiency determination are available (ASME Test Codes) but are rarely used. For the objective of defining conservation potentials, relatively simple measurements are adequate. For fired heaters, stack temperature and excess O_2 in stack should be measured; for turbines, pressures (in and out) and temperatures (in and out) are needed.

Checking Against Optimum Design.　This attempts to answer the question whether a balance needs to be as it is. The first thing to compare against is the best current practice. Information is available in the literature (13) for large-volume chemicals such as NH_3, CH_3OH, urea, and ethylene. The second step is to look for obvious violations of good practice on individual pieces of equipment. Examples of violations are stack temperatures $>150°C$; process streams $>120°C$, cooled by air or water; process streams $>65°C$, heated by steam; $\eta_{turbine} < 65\%$; reflux ratio >1.15 times minimum; and excess air $>10\%$ on clean fuels.

Idea-Generation Meeting.　This meeting has the following guidelines: gather people with expertise and experience, discuss the as-it-is balance for each area, record all ideas, and assign follow-up responsibilities.

Evaluation.   The evaluation of each idea should include a technical description as well as its economic impact and technical risk. Ideas should be ranked for implementation. A report should provide a five-year framework for energy projects.

Follow-Up.   If no savings result, the effort has been wasted. Thus the audit leader must ensure that the potential of every good design idea is recognized by management and the project-generation channels of the company.

Perspective.   Often, what seems like negligence can be a tried-and-proven practice. Lists offering huge savings are frequently not accurate. The process has to work, and safety cannot be compromised for energy savings. If a change to save energy is justified, the control and hardware revisions that permit it to be implemented safely are also justified. Similarly, present utility savings may or may not compensate for future repair bills or lost products. For example, an idling turbine may be necessary to permit a safe plant shutdown if a power failure occurs; a cooling-water flow that is throttled to below 0.6 m/s in winter is likely to require a heat-exchanger cleaning in late spring; a furnace that runs too low on excess air may run into after-burning; and a column run too close to the minimum reflux ratio without adequate controls runs a risk of off-specification product.

The plant accounting system often needs questioning. All energy is not created equal. The energy that is recovered from flashed steam or that is shaved off a reboiler's duty may not be worth its cost-sheet value. The meters that matter are the primary meters at the plant gate. Only if the recovered energy reduces the plant gate meters does it save the plant money. Solutions to an energy waste problem must fit into the over plant–energy balance.

BIBLIOGRAPHY

"Process Energy Conservation" in *ECT* 3rd ed., Suppl. Vol., pp. 669–697, by D. Steinmeyer, Monsanto Co.

1. D. E. Steinmeyer, *Chemtech*, **12**(3), 188 (1982).
2. W. F. Kenney, *Proceedings of Industrial Energy Conservation Technology Conference*, Energy Systems Laboratory, Texas A and M University, College Station, Tex., 1981, p. 247.
3. C. J. King, *Separation Processes*, 2nd ed., McGraw-Hill Book Co., Inc., New York, 1980.
4. C. S. Robinson and E. R. Gilliland, *Elements of Fractional Distillation*, 4th ed., McGraw-Hill Book Co., Inc., New York, 1950.
5. T. L. Tolliver, *Chem. Eng.* **93**(22), 99 (1986).
6. A. Sloley, personal communication, Process Consulting Services, Houston, Tex., Aug. 25, 1995.
7. M. Souders, *Chem. Eng. Prog.* **60**(2), 75 (1964).
8. Technical data, D. Carter, Monsanto Corp., St. Louis, Mo., 1982.
9. D. Steinmeyer, *Hydrocarbon Process.* **71**(4), 53 (1992).
10. R. Smith and B. Linnhoff, *Chem. Eng. Res. Des.* **66**, 195 (1988).
11. B. Linnhoff, *Chem. Eng. Prog.* **90**(8), 33 (1994).
12. W. F. Furgerson, *Conserving Energy in Refrigeration*, Massachusetts Institute of Technology Press, Cambridge, Mass., 1982.
13. *Hydrocarbon Process.* **74**(3), 89 (1995).

General References

W. F. Kenney, *Energy Conservation in the Process Industries*, Academic Press, Inc., New York, 1984.

B. Linnhoff and co-workers, *A User's Guide on Process Integration for the Efficient Use of Energy*, Institution of Chemical Engineers, Warwickshire, U.K., 1994.

R. Smith, *Chemical Process Design*, McGraw-Hill Book Co., Inc., New York, 1995.

DAN STEINMEYER
Monsanto Company

PROCESSING, MODELING. See SUPPLEMENT.

PROCESS RESEARCH AND DEVELOPMENT. See RESEARCH/TECHNOLOGY MANAGEMENT

PRO DRUGS. See PHARMACEUTICALS; PHARMACODYNAMICS.

PRODUCER GAS. See FUELS, SYNTHETIC.

PRODUCT LIABILITY

In the early 1960s, the law of American products liability underwent significant change. Until that time, the ability of claimants to recover for personal injury resulting from defective and dangerous products was severely restricted by doctrines that either barred or limited the prosecution of successful tort actions. Since the early 1960s until the mid-1990s, the most dynamic and explosive area of tort law has been the field of products liability (1). However, even the most dynamic areas of the law tend, after a period of expansion, to settle down doctrinally. Thus, though products liability cases continue to be brought across a broad range of products, it has become increasingly clear since the early 1980s that the period of doctrinal expansion that eased recovery for claimants has slowed down considerably. Courts throughout the United States have entered a period of consolidation.

Historical Overview

Prior to 1960, an injured plaintiff seeking recovery in a products liability action could bring a case under either of two theories. The plaintiff could allege that the product seller was negligent, or that the seller breached a warranty that attended every product sale provided for under the Uniform Sales Act and, later, under Section 2-314 of the Uniform Commercial Code (UCC), which stipulated that for a product to be merchantable it must be "reasonably fit for the ordinary purposes for which such goods are used." Each of these theories came with a distinct disadvantage.

Negligence.   Early in the development of the law of negligence-based liability for defective products, the courts almost universally held that the negligent supplier of a defective product could only be held liable to an injured person with whom the supplier had directly contracted. This rule, which limited a product supplier's tort liability, is generally referred to as the privity rule. Its origins have been traced to the opinion of Lord Abinger in Winterbottom vs Wright (2), in which the plaintiff sought to recover in negligence for injuries suffered when a horse-drawn mail coach collapsed while the plaintiff was driving it. The defendant had supplied the coach in question to the Postmaster General, pursuant to a contract that called for the defendant to keep the coach in good repair. The plaintiff alleged that the defendant negligently failed to fulfill his contractual promise to keep the coach in repair, thus causing the coach to collapse and injure the plaintiff.

In granting judgment for the defendant, the English court concluded that, given the absence of any contractual relationship between the defendant and the plaintiff, no recovery could be had in negligence. Refusing to permit the contract between the defendant and the Postmaster General to be turned into a tort, Lord Abinger observed:

> There is no privity of contract between [the plaintiff and the defendant]; and if the plaintiff can sue, every passenger, or even any person passing along the road, who was injured by the upsetting of the coach, might bring a similar action. Unless we confine the operation of such contracts as this to the parties who entered into them, the most absurd and outrageous consequences...would ensure.

Although the privity rule in this case generally came to be recognized by the American courts by the end of the nineteenth century, a number of exceptions were developed judicially. In an early New York case, Thomas vs Winchester (4), a vendor of vegetable extracts falsely labeled a poison and sold it to a druggist, who in turn sold it to a customer. The customer, seriously injured as a result, recovered damages from the seller who affixed the erroneous label even though the injured plaintiff had no direct privity relationship with him. The defendant's negligence, the court said, "put human life in imminent danger." Once the breach of the privity barrier had been established for products that created "imminent danger," claimants barraged the courts with the contention that, indeed, their injuries were brought about by products that had high danger levels and which thus met the threshold test for bypassing the privity rule.

In MacPherson vs Buick Motor Company, Justice Cardozo put an end to the temporizing and effectively eliminated privity as an obstacle to recovery against negligent manufacturers (5):

> We hold...that the principle of *Thomas vs Winchester* is not limited to poisons, explosives, and things of like nature, to things which in their normal operation are implements of destruction. If the nature of a thing is such that it is reasonably

certain to place life and limb in peril when negligently made, it is then a thing of danger. Its nature gives warning of the consequences to be expected. If to the element of danger there is added knowledge that the thing will be used by persons other than the purchaser, and used without new tests, then, irrespective of contract, the manufacturer of this thing of danger is under a duty to make it carefully.... There must be knowledge of a danger, not merely possible, but probable. It is possible to use almost anything in a way that will make it dangerous if defective. That is not enough to charge the manufacturer with a duty independent of his contract. Whether a given thing is dangerous may be sometimes a question for the court and sometimes a question for the jury. There must also be knowledge that in the usual course of events the danger will be shared by others than the buyer. Such knowledge may often be inferred from the nature of the transaction.... We have put aside the notion that the duty to safeguard life and limb, when the consequences of negligence may be foreseen, grows out of contract and nothing else. We have put the source of the obligation where it ought to be. We have put its source in the law....

The MacPherson ruling abolishing privity in negligence actions was widely accepted, and in the 1990s is the law in all jurisdictions. Though privity was abolished and a plaintiff was free to sue a manufacturer directly for negligence, a substantial obstacle to recovery remained. A plaintiff was still required to prove that the defendant was negligent in bringing about the defective product. When a product contained a manufacturing defect, defendants would argue that they had instituted reasonable quality control procedures and hence the products were not negligently manufactured (see QUALITY ASSURANCE). In order to assist plaintiffs in establishing their cases, the court increasingly relied on the doctrine of *res ipsa loquitur*. The evidentiary doctrine of *res ipsa loquitur* (the thing speaks for itself) allows an inference of negligence to be drawn from the occurrence of an accident involving an instrumentality (in a products liability case, the product itself) within the defendant's control, under circumstances where such an accident would not ordinarily occur in the absence of negligence. Thus, from the fact that a defective product failed in use and caused an accident, courts allowed triers of fact to infer that the manufacturer of the product negligently caused the defect to occur.

It was clear, however, that this was a stopgap device and that courts were becoming less concerned with manufacturer fault and more concerned with whether the product marketed was, in fact, defective. In the famous case of Escola vs Coca-Cola Bottling Company (6), the plaintiff was injured when a bottle of Coca-Cola beverage broke in her hand. She alleged negligence on the part of the bottler and relied on *res ipsa loquitur* to support her claim. The defendant presented compelling evidence to show that the bottler had exercised considerable precaution by carefully regulating and checking the pressure in the bottles and in making visual inspections for defects in the glass at several stages in the bottling process. The court nonetheless held that the jury was entitled to find the defendant negligent. In a landmark concurring opinion, Justice Traynor made the following observation:

The injury from a defective product does not become a matter of indifference because the defect arises from causes other than the negligence of the manufacturer, such as negligence of a submanufacturer, of a component part whose defects could not be revealed by inspection or unknown causes that even by the device of *res ipsa loquitur* cannot be classified as negligence of the manufacturer. The inference of negligence may be dispelled by an affirmative showing the proper care....An injured person, however, is not ordinarily in a position to refute such evidence or identify the cause of the defect, for he can hardly be familiar with the manufacturing process as the manufacturer himself is. In leaving it to the jury to decide whether the inference has been dispelled, regardless of the evidence against it, the negligence rule approaches the rule of strict liability. It is needlessly circuitous to make negligence the basis of recovery and impose what is in reality liability without negligence. If public policy demands that a manufacturer of goods be responsible for their quality regardless of negligence there is no reason not to fix that responsibility openly....

As handicrafts have been replaced by mass production with its great markets and transportation facilities, the close relationship between the producer and consumer of a product has been altered. Manufacturing processes, frequently valuable secrets, are ordinarily either inaccessible to or beyond the ken of the general public. The customer no longer has means or skill enough to investigate for himself the soundness of a product, even when it is not contained in a sealed package, and his erstwhile vigilance has been lulled by the steady efforts of manufacturers to build up confidence by advertising and marketing devices such as trademarks. Consumers no longer approach products warily but accept them on faith, relying on the reputation of the manufacturer or the trademark. Manufacturers have sought to justify that faith by increasingly high standards of inspection and a readiness to make good on defective products by way of replacements and refunds. The manufacturer's obligation to the consumer must keep pace with the changing relationship between them. . . .

It would take two decades before Justice Traynor's prophetic opinion became a reality. In the interim, plaintiffs were formally saddled with the responsibility of proving that the defendant–seller of a product was negligent. The implications of retaining negligence as the operative theory were of considerable consequence. Plaintiffs seeking to sue retailers or wholesalers had difficulty doing so. These parties were rarely negligent, in that they could not ordinarily be expected to discover flaws and defects in the products sold. Thus, if a manufacturer were not subject to the jurisdiction of the courts of a given state, or if the manufacturer were unavailable for suit (eg, insolvency), the plaintiff would be left with no recourse.

In short, the abolition of privity opened manufacturers to liability for negligence. Plaintiffs, however, could not establish claims merely by proving that they were harmed by defective products from a manufacturer. The requirement that classic fault be established often stood as a formal barrier to a successful tort action.

Implied Warranty of Merchantability.   An alternative method of recovery that required the plaintiff to establish only defect (no fault) was available. The plaintiff could bring an action for breach of the implied warranty of merchantability. Under the Uniform Sales Act and later under the UCC Section 2-314, a warranty accompanied every sale of goods (unless disclaimed) stating that the product is "reasonably fit for the ordinary purposes for which such goods are used." Defective products fail to meet the statutory definition and provide a predicate for a cause of action. The implied warranty of merchantability can be characterized as strict liability in contract. Disadvantages of the Code warranty were quite substantial, however. First, because the cause of action was contractual, the requirement that the parties to the suit be in privity with each other was a necessary requisite for standing to sue. The MacPherson ruling dealt the death blow to privity in causes of action based on negligence; in contract, however, the privity doctrine remained very much alive. Thus, direct actions by injured consumers against product manufacturers were barred. Plaintiffs could sue immediate sellers for breach of implied warranty; however, this right was often misleading. Retail sellers were often judgement-proof. In some cases they disclaimed liability as allowed by the UCC Section 2-316. Furthermore, the UCC statute of limitations, Section 2-715, provides for a maximum of four years from tender of delivery (sale). A tort statute of limitations generally runs from the time of injury, thus in many cases providing a longer time to bring suit.

Combining Tort and Contract Advantages. Two methods were available to allow plaintiffs an easier road to recovery. Courts either stripped the tort action of the necessity for establishing fault, or interpreted the UCC in such a way that privity was not necessary and the other Code defenses were not applicable to cases involving personal injury or property damage. Either way a manufacturer would be open to direct suit without the need to prove fault. The decision that overturned the privity requirement was an action for breach of the implied warranty of merchantability. In Henningsen vs Bloomfield Motors, Inc. (7), a plaintiff brought suit for injuries sustained when a newly purchased car went out of control. The plaintiff was driving when she heard a loud noise from the bottom by the hood. The steering wheel then spun in her hands; the car veered sharply and crashed into a wall. The trial judge dismissed negligence counts against the manufacturer of the car. The judge submitted the issue of breach of implied warranty of merchantability to the jury, which found against both the retailer and the manufacturer. The auto manufacturer subsequently appealed the verdict against it on the grounds that the company was not in privity with the injured plaintiff. In striking down the privity defense the court said:

> Under modern condition the ordinary layman...has neither the opportunity nor the capacity to inspect or to determine the fitness of an automobile for use; he must rely on the manufacturer who has control of its construction, and to some degree on the dealer who, to the limited extent called for by the manufacturer's instructions, inspects and services it before delivery. In such a marketing milieu his remedies and those of persons who properly claim through him should not depend "upon the intricacies of the law of sales. The obligation of the manufacturer should not be based alone on privity of contract. It should rest, as was once said, upon 'the demands of social justice.'"....

> Accordingly, we hold that under modern marketing conditions, when a manufacturer puts a new automobile in the stream of trade and promotes its purchase by the public, an implied warranty that it is reasonably suitable for use as such accompanies it into the hands of the ultimate purchaser. Absence of agency between the manufacturer and the dealer who makes the ultimate sale is immaterial (85).

The Henningsen decision sought to impose strict liability against manufacturers within the framework of the Uniform Commercial Code. Only a short time elapsed before the courts recognized that the language used by the UCC to address liability provided a clumsy tool for prosecuting personal injury cases, and that strict liability was a purely tort doctrine.

In Greenman vs Yuba Power Products Company (9), a plaintiff brought an action for damages he suffered on account of a power tool. The plaintiff was working on the power lathe when a piece of wood suddenly flew out of the machine and struck him in the forehead. About 10 months later, the plaintiff gave the retailer and the manufacturer written notice of claimed breaches of warranties. Under a provision of the Uniform Commercial Code applicable in California (Section 2-607), a buyer, in order to maintain an action under the Code, was required to give the seller notice of the breach of warranty within a "reasonable time after the buyer knows or ought to know" of the breach. The defendant claimed that the buyer had not done so in this case and was thus barred from proceeding under the Code cause of action. Justice Traynor rejected the defence, holding that a valid claim lies in strict liability in tort without the necessity of implicating the UCC:

> Although in these cases strict liability has usually been based on the theory of an express of implied warranty running from the manufacturer to the plaintiff, the abandonment of the requirement of a contract between them, the recognition that the liability is not assumed by agreement but imposed by law ... and the refusal to permit the manufacturer to define the scope of its own responsibility for defective products ... make clear that the liability is not one governed by the law of contract warranties but by the law of strict liability in tort. Accordingly, rules defining and governing warranties that were developed to meet the needs of commercial transactions cannot properly be invoked to govern the manufacturer's liability to those injured by its defective products unless those rules also serve the purposes for which such liability is imposed.

The Greenman decision was a watershed, and privity-free strict liability in tort swept the country as a tidal wave. In 1965, the American Law Institute embraced the concept in Section 402A, and thousands of decisions cited to the Restatement. Within a decade the decision became the majority rule in the United States; in the 1990s all but a tiny minority of states ascribe to it.

The Second Restatement

All discussion of the post-1965 era must begin with the Restatement (Second) of Torts, Section 402A. The person responsible for this section was Dean William Prosser, who became the reporter for American Law Institute's Restatement (Second) of Torts at a time when change was on the horizon in the field of products liability. His initiative in drafting Section 402A provided the courts with a ready-made formulation for the adoption of strict tort liability. Entitled "Special Liability of Seller of Product for Physical Harm to the User or Consumer," Section 402A reads as follows:

(*1*) One who sells any product in a defective condition unreasonably dangerous to the user or consumer or to his property is subject to liability for physical harm thereby caused to the ultimate user or consumer, or to his property, if (a) the seller is engaged in the business of selling such a product, and (b) it is expected to and does reach the user or consumer without substantial change in the condition in which it is sold.

(*2*) The rule stated [above] applies although (a) the seller has exercised all possible care in the preparation and sale of his product, and (b) the user or consumer has not bought the product from or entered into any contractual relation with the seller.

Two comments sought to give content to the terms "defective condition" and "unreasonably dangerous" that appear in this rule:

(g) Defective condition. The rule stated in this Section applies only where the product is, at the time it leaves the seller's hands, in a condition not contemplated by the ultimate consumer, which will be unreasonably dangerous to him. The seller is not liable when he delivers the product in a sale condition, and subsequent mishandling or other causes make it harmful by the time it is consumed. The burden of proof that the product was in a defective condition at the time that it left the hands of the particular seller is upon the injured plaintiff; and unless evidence can be produced which will support the conclusion that it was then defective, the burden is not sustained.

Safe condition at the time of delivery by the seller will, however, include proper packaging, necessary sterilization, and other precautions required to permit the product to remain safe for a normal length of time when handled in a normal manner.

(i) Unreasonably dangerous. The rule stated in this Section applied only where the defective condition of the product makes it unreasonably dangerous to the user or consumer. Many products cannot possibly be made entirely safe for all consumption, and any food or drug necessarily involves some risk of harm, if only from overconsumption.... The article sold must be dangerous to an extent beyond that which would be contemplated by the ordinary consumer who purchases it, with the ordinary knowledge common to the community as to its characteristics. Good whiskey is not unreasonably dangerous merely because it will make some people drunk, and is especially dangerous to alcoholics; but bad whiskey, containing a dangerous amount of fusel oil, is unreasonably dangerous....

Prima Facie Case and Affirmative Defenses.   The imposition of strict tort liability did not mean that a plaintiff was entitled to automatic recovery when injured by a product. The following decisions set forth the basic elements that a plaintiff must establish in order to make out a prima facie case of products liability and the affirmative defenses that either reduce or bar plaintiff's recovery.

Defect.   A plaintiff must establish that the product which injured him was defective. A product can be defective in three ways: it may be defectively manufactured, defectively designed, or sold with inadequate warnings.

Manufacturing defects are those that arise during the production process of the product. Quality control should eliminate most manufacturing defects from reaching the market. However, no quality control system is foolproof and occasionally flawed products do reach the consumer.

Unlike manufacturing defects that are idiosyncratic and random, design defects arise on the drawing board and are generic. If the product is defectively designed, every unit that is in the marketplace suffers from the same defect. A classic example of a design defect is McCormack vs Hankscraft Company (10). In that case, the plaintiff, a three-year old child, suffered serious burns when she tripped over the cord of a hot water vaporizer. The top of the vaporizer dislodged and scalding hot water poured out on her. The contention of the plaintiff was that the design of the vaporizer was defective. Plaintiff's experts testified that the vaporizer could have been designed so that the cap of the vaporizer was secured to the jar, which would have prevented the water from spilling when the vaporizer tipped over.

A product may also be defective because it was sold with inadequate warnings. Burch vs Amsterdam Corporation (11) is an early example of a failure-to-warn case. A plaintiff was badly burned in an explosion and flash fire that occurred while he was applying a floor tile adhesive sold by the defendant. The label on the can of mastic adhesive warned that the product was extremely flammable and should not be used near a fire or flame. Although the plaintiff checked his surrounding to be sure there were no flames he failed to notice the gas stove pilot light in the kitchen where he was working. After plaintiff applied a coat of adhesive, the vapors from the adhesive reached the pilot light and exploded. The court held that whether the warning given was adequate should be decided by a jury. It noted that an ordinary user might not have realized that "near fire or flame" included nearby pilot lights or that fumes and vapors, as well as the adhesive itself, were extremely flammable.

Cause-In-Fact.   It is not sufficient that a product is found to be defective. In order to be successful in a products liability action, a plaintiff must establish that the product defect was causally related to the harm the plaintiff suffered. If the selfsame harm would have occurred even if the product had not been defective, then plaintiff's harm cannot be linked to the product defect and recovery is barred. In a 1994 case, O'Bryan vs Volkswagenwerk, AG (12), the plaintiff was injured following a serious collision when he was thrown from the car and was rendered a quadriplegic. The plaintiff alleged that the door lock of the car should have been designed to withstand greater pressure in the event of a collision. The court entered verdict for the defendant holding that the plaintiff had not established that even a better-designed door lock would have withstood the violent impact of the collision which caused the plaintiff's harm. Thus, even if the design of the door lock were hypothetically defective, there was no causal nexus between the alleged defect and the harm.

Proximate Cause.   Even when there is a causal connection between the defect and the harm, it is still necessary for a plaintiff to establish that the nature of the plaintiff's harm, together with the circumstances of its occurrence, were reasonably foreseeable. The law, in general, does not protect against remote unforeseeable risks. For example, in Buckley vs Bell (13), the plaintiff ordered gasoline from the defendant and the defendant delivered diesel fuel by mistake, thus making the product "defective." The plaintiff filled his hay baler with the diesel fuel and, when it would not start, discovered the error in delivery. He emptied the fuel tank onto the ground, refilled it with gasoline, and attempted to start the baler. A backfire-induced fire ensued, involving the spilled diesel fuel, and the hay baler was destroyed. The plaintiff sought to recover for the value of the machine. The trial court sitting as trier of fact decided that there was no causal connection between the delivery of the defective fuel and the plaintiff's loss, because the consequence could not be foreseen. The Wyoming Supreme Court affirmed, concluding that the but-for connection is not, by itself, enough to support liability.

Another example of the proximate cause requirement is presented by Ritter vs Narragansett Electric Company (14). A suit was brought against both the retailer and the manufacturer of a stove for injuries sustained by a minor while playing in the kitchen of her home. The minor attempted to look into a pot atop the stove in which water was boiling. She opened the oven door, which was a drop-type door, and placed her foot on the edge of the door with the intention of standing on it to look into the pot. As she put her weight on the door, the range toppled over, and the pot of boiling water scalded her.

The evidence against the manufacturer of the stove revealed that when weight of approximately thirty pounds or more were placed on the door, the range would tip forward. The court decided that a jury could conclude that as a result of the design of the range, the danger in the use of the oven door as a shelf was foreseeable. The defendant would thus be negligent in either failing to warn about the danger, or in not designing the stove with a better center of gravity.

Having decided the defect question, the court still had to face the question of abnormal use or misuse. To be sure, if a 30-lb turkey had been placed on the oven door and the stove had tipped, there would be little question that injury was within the scope of the risk created by the defect. But the injury did not occur in that manner. Instead, a minor decided to use the oven door as a step-stool for peering over the top of the range. Was the manner of the occurrence so unforeseeable that it is not fair to assign this harm to the product defect? Was this injury less a result of product defect and more a result of children not being properly supervised? No formula has been devised to resolve this kind of proximate cause question. In this instance the court sent the issue of abnormal use back to the trial court for a jury determination as to whether the injury had been the product of an abnormal or improper use of the range.

The Role of Plaintiff's Conduct. Until the early 1970s, the majority rule in the United States was that if a plaintiff's fault was a contributing cause to the injury, recovery was barred. Since that time a revolution of sorts has taken place. Almost all states follow a rule that reduces a plantiff's recovery by the percentage of fault attributed to the plaintiff. Thus juries are asked to compare the conduct of the plaintiff and the defendant and then to assign a percentage of fault to each. Plaintiff's recovery is then reduced by a given percentage. Many jurisdictions take the position that if the fault assigned to the plaintiff is over 50% then recovery is barred.

In cases where the plaintiff has suffered harm as a result of a defective product, the claim is often made that the plaintiff contributed to the harm by negligent conduct. For instance, the steering mechanism of an automobile may be defective, but the accident may have occurred because plaintiff was driving over the speed limit. The injury may thus have been caused by the joint fault of a bad product and faulty plaintiff behavior. A small minority of courts has taken the position that, in products liability cases, a plaintiff's fault should not serve to reduce recovery. The argument is that the product should function well even when a plaintiff is acting negligently. To allow the defendant to reduce liability compromises the role of product sellers in guaranteeing the integrity of the products being placed on the market. However, the overwhelming majority of courts applies principles of comparative responsibility and allows for the reduction of the plaintiff's recovery based on the percentage of fault assigned by a jury to the conduct of the plaintiff.

The Third Restatement

New Definitions of Defect.   The concept espoused by Section 402A of the Restatement (Second) of Torts, namely, that a manufacturer was strictly liable for selling a defective product and that proof of fault was unnecessary, worked well with regard to products that contained manufacturing defects.

If the plaintiff can demonstrate that a particular product unit came off the assembly line with a manufacturing defect which made the product significantly more dangerous than other similar units, then the conclusion can be drawn that the defect rendered the product legally unacceptable. In effect, the manufacturer is the author of the standard (the intended design) against which the allegedly defective unit is measured. In a design defect case, the individual unit is exactly what the manufacturer wanted it to be. If the product is defective it is because the design standard set by the manufacturer is found wanting. In most cases, this means that the manufacturer has failed to include design features that would have provided greater safety. In essence, the plaintiff's argument is that the design fails to measure up to some hitherto adoptable, but as yet unadopted, standard. In practical terms, the plaintiff must hypothesize a reasonable, safer alternative design and the court must find the hypothetical alternative a preferred substitute for the offending product.

In examining whether a design was defective, most courts found it necessary to revert back to a test analogous to negligence. The question that had to be confronted was whether the manufacturer had acted reasonably in designing the product. Some courts have sought to bypass this inquiry by adopting the consumer expectation test, which allows for the imposition of liability when the plaintiff establishes that the product failed to perform as a reasonable consumer would have expected. Although this rule has some currency, it is on the wane in the 1990s. This is because for most product designs of any complexity it is difficult to determine consumer expectations with regard to product performance. Consumers generally expect well-designed products, but such an expectation merely brings back the question of whether the product was reasonably designed.

In June 1992, the American Law Institute undertook the task of drafting the Restatement (Third) of Torts: "Products Liability." Tentative Draft No. 2 takes the position that different liability rules must apply for manufacturing defects and defects based on inadequate design or failure to warn:

(a) a product contains a manufacturing defect when the product departs from its intended design even though all possible care was exercised in the preparation and marketing of the product;

(b) a product is defective in design when the foreseeable risks of harm posed by the product could have been reduced or avoided by the adoption of a reasonable alternative design by the seller or other distributor, or a predecessor in the commercial chain of distribution, and the omission of the alternative design renders the product not reasonably safe;

(c) a product is defective because of inadequate instructions or warnings when the foreseeable risks of harm posed by the product could have been reduced or avoided by the provision of reasonable instructions or warnings by the seller or other distributor, or a predecessor in the commercial chain of distribution, and the omission of the instructions or warnings renders the product not reasonably safe.

A large number of U.S. states are in agreement with this new formulation. In cases where the claim is based on design defect the need for plaintiff to establish the availability of a reasonable alternative design gives content to the liability rule. In cases based on inadequate warning, the rule requiring the plaintiff to establish that the harm could have been avoided by the provision of a reasonable instruction or warning is not an easy one to administer. Claimants can easily allege that the product should have provided a warning of the very risk that caused the injury. Such a position pushed to its extreme would lead to a plethora of warnings of remote risks. Overwarning can lead to the debasement of warnings and to their becoming trivialized by the consumers. Careful balance is necessary to make certain that claims of inadequate warnings are reality-based and will actually render products safer.

The Role of Foreseeability. In cases based on design defect and failure to warn, defect is established by showing the presence of foreseeable risks of harm which warranted either a better design or a warning. The requirement that foreseeability be established is at odds with the concept of strict liability. However, a overwhelming majority of courts has refused to impose liability for risks that were unforeseeable at the time of product sale. In most cases dealing with mechanical defects, it is rare that the risk of harm from foreseeable product use is not known. However, in cases dealing with chemicals, drugs, or other toxic agents, it is not uncommon for risks of harm to be discovered only after a long period of product use. The foreseeability requirement thus provides significant protection for manufacturers when they can legitimately claim that the risk of harm was neither known nor knowable by the application of reasonable scientific inquiry available at the time of sale.

Defendant Identification. It is normally incumbent on a plaintiff to establish which defendant's product caused harm. In a small class of toxic tort cases, courts have relaxed this requirement. A problem developed in the 1980s in connection with which plaintiffs had great difficulties identifying the defendant that caused the relevant harm. The problem centered on the distribution and sale of the drug diethylstilbesterol [56-53-1] (DES) from 1941 to 1971. DES was prescribed to help prevent miscarriage. Daughters of mothers who consumed DES during pregnancy claimed that as a result they suffered various forms of cancerous and precancerous conditions and required surgery to prevent the spread of the disease. They alleged that the drug companies that marketed the drug were negligent in failing to discover that DES was a carcinogen and failing to test for the efficacy and safety of the drug, and hence breached a duty to warn about such dangers. The diseases complained of did not manifest themselves until at least 10 to 12 years after birth. There was a substantial latency period between the time of ingestion and the time the disease became known to its victims.

DES was a generic drug without any clearly identifiable shape, color, or markings, and was produced with a single chemical formula. Over three hundred companies at one time or another produced or marketed DES. Because of the long latency period between exposure and injury, it was often impossible to locate records with which to identity which manufacturers' drugs were dispensed to the mothers of particular plaintiffs. Several leading courts recognized the exceptions to the normal rule requiring the plaintiff to bear the burden of establishing the defendant's identity. Instead, they imposed liability in proportion to the defendant's share of the relevant DES market.

Other courts addressing this problem have refused to adopt a rule of proportional recovery. Some have asserted that such a fundamental change of a basic tort principle is more appropriately a legislative function. Others have expressed concern that the long latency period renders any reconstruction of market shares highly speculative. The Restatement notes the opposing views but takes no position on this controversial issue.

After a period of turbulence, the field of products liability has begun to settle down in the 1990s. The most explosive aspects of the field, ie, litigation based on design defect and failure to warn, are recognized to be closely analogous to traditional rules of negligence. Liability will attach only if there was a reasonable alternative design or warning that would have prevented the plaintiff's harm. Over the years numerous bills have been introduced in Congress to federalize aspects of products liability law. It appears that the basic structure of the law will not be affected by the legislation pending in U.S. Congress in 1996. The law of products liability is part of the great common law tradition administered by state courts throughout the United States. It is likely to remain so for the foreseeable future.

BIBLIOGRAPHY

"Product Liability" in *ECT* 3rd ed., Suppl. Vol., pp. 698–710, by A. Twerski, Hofstra Law School.

1. A. W. Weinstein and co-workers, *Products Liability and the Reasonably Safe Product*, John Wiley & Sons, Inc., New York, 1978.
2. *Winterbottom vs Wright*, Exch. 10 M. & W. 109 (1842).
3. *Ibid.*, 114.
4. *Thomas vs Winchester*, 6 N.Y. 307 (1852).
5. *MacPherson vs Buick Motor Co.*, 217 N.Y. 382, 111 Northeastern 1050 (1916).
6. *Escola vs Coca-Cola Bottling Co.*, 24 Calif. 2d 453, 150 P.2d 436 (1944).
7. *Henningsen vs Bloomfield Motors, Inc.*, 32 N.J. 358, 161 A.2d 69 (1960).
8. *Ibid.*, 32 N.J. 384, 161 A.2d 83–84.

9. *Greenman vs Yuba Power Products Co.*, 59 Calif. 2d 57, 377, P.2d 987 (1962).
10. *McCormack vs Hankscraft Co.*, 258 Minn. 322, 158 N.W.2d 488 (1967).
11. *Burch vs Amsterdam Corp.*, D.C. Cir. 366 A.2d 1079 (1976).
12. *O'Bryan vs Volkswagenwerk, AG*, unpublished opinion, 6th Cir. (1994).
13. *Buckley vs Bell*, Wyo. 703 P.2d 1089 (1985).
14. *Ritter vs Narragansett Electric Co.*, 109 R.I. 176, 283, A.2d 255 (1971).

AARON TWERSKI
Brooklyn Law School

PROLINE. See AMINO ACIDS.

PROMETHIUM. See RARE-EARTH ELEMENTS.

PROPANE. See HYDROCARBONS.

PROPANOLAMINES. See ALKANOLAMINES.

PROPANOLS. See PROPYL ALCOHOLS.

PROPELLANTS. See EXPLOSIVES AND PROPELLANTS.

PROPENE. See PROPYLENE.

PROPIONALDEHYDE. See ALDEHYDES.

PROPIONIC ACID. See CARBOXYLIC ACIDS, SURVEY.

PROPYL ALCOHOLS

Isopropyl alcohol, **216**
n-Propyl alcohol, **241**

ISOPROPYL ALCOHOL

Isopropyl alcohol [*67-63-0*], also known as isopropanol, 2-propanol, dimethyl-carbinol, and *sec*-propyl alcohol, is a colorless, volatile, and flammable liquid, having a molecular weight of 60.09 and a slight odor resembling a mixture of ethyl alcohol [*64-17-5*] and acetone [*67-64-1*]. Isopropyl, the lowest member of the class of secondary alcohols, is generally known as the first petrochemical. Of the lower (C$_1$–C$_5$) alcohols, isopropyl alcohol is third in commercial production, behind methanol (qv) and ethyl alcohol (qv). The 1993, U.S. production was 5.4×10^5 metric tons (1). Production of isopropyl alcohol has been declining at an average annual rate of 3% since 1980, mostly because of the declining use of isopropyl alcohol as an acetone (qv) feedstock. An estimated 50% of isopropyl alcohol was used in solvent applications in 1992 (see SOLVENTS, INDUSTRIAL). Isopropyl alcohol is used for the manufacture of agricultural chemicals, pharmaceuticals (qv), process catalysts, and solvents. Properties, preparation, and uses of isopropyl alcohol have been discussed (2).

Physical Properties

Physical properties of isopropyl alcohol are characteristic of polar compounds because of the presence of the polar hydroxyl, −OH, group. Isopropyl alcohol is completely miscible in water and readily soluble in a number of common organic solvents such as acids, esters, and ketones. It has solubility properties similar to those of ethyl alcohol (qv). There is a competition between these two products for many solvent applications. Isopropyl alcohol has a slight, pleasant odor resembling a mixture of ethyl alcohol and acetone, but unlike ethyl alcohol, isopropyl alcohol has a bitter, unpotable taste.

Physical and chemical properties of isopropyl alcohol reflect its secondary hydroxyl functionality. For example, its boiling and flash points are lower than *n*-propyl alcohol [*71-23-8*], whereas its vapor pressure and freezing point are significantly higher. Isopropyl alcohol boils only 4°C higher than ethyl alcohol.

Anhydous and 91 vol % alcohol, the two main grades of isopropyl alcohol marketed in the United States, differ mainly in water content. The latter represents the azeotrope with water and is usually referred to as constant boiling mixture (CBM) isopropyl alcohol. A listing of some important physical constants of anhydrous and CBM isopropyl alcohol is given in Table 1. Because of its tendency to associate in solution, isopropyl alcohol forms azeotropes with compounds from a variety of classes, including hydrocarbons, esters, halocarbons, amines, ketones (qv), and aromatics. Examples of some binary azeotropes are given in Table 2. Isopropyl alcohol does not form binary azeotropes with acetone, ethyl alcohol, ethylbenzene [*100-41-4*], hexylamine [*111-26-2*], or methyl isobutyl ketone [*108-10-1*]. It does, however, form ternary systems with many of these and other compounds (2).

Table 1. Physical Properties of Isopropyl Alcohol[a]

Property	Anhydrous	91 Vol %
molecular weight	60.10	
boiling point, at 101.3 kPa[b], °C	82.3	80.4
freezing point, °C	−88.5	−50.0
specific gravity, 20/20	0.7864	0.8183
density, at 20°C, g/cm³	0.7854	0.8173
surface tension, at 20°C, mN/m(=dyn/cm)	21.32	21.40[c]
specific heat, liquid at 20°C, J/(kg·K)[d]	2510.4	
refractive index, n_D^{20}	1.3772	1.3769
heat of combustion, at 25°C, kJ/mol[d]	2005.8	
latent heat of vaporization, at 101.3 kPa[b], kJ/mol[d]	39.8	
vapor pressure, at 20°C, kPa[b]	4.4	4.5
critical temperature, °C	235.2	
critical pressure, at 20°C, kPa[b]	4764	
viscosity, mPa·s(=cP)		
at 0°C	4.6	
20°C	2.4	2.1[c]
40°C	1.4	
coefficient of expansion[e]	$V_t = V_o[1 + (1.0743 \times 10^{-3})t + (3.28 \times 10^{-7})t^2]$	
flammability limit in air, vol %[f]		
lower	2.5	
upper	12	
flash point, °C		
Tag open cup	17.2	21.7
closed cup	11.7	18.3
autoignition temperature, °C[g]	399	

[a]Refs. 3–5, except where noted.
[b]To convert kPa to mm Hg, multiply by 7.50.
[c]At 25°C.
[d]To convert J to cal, divide by 4.184.
[e]Ref. 6; t in °C.
[f]Ref. 7.
[g]Ref. 8.

Chemical Properties

Chemical properties of isopropyl alcohol are determined by its functional hydroxyl group in the secondary position. Except for the production of acetone, most isopropyl alcohol chemistry involves the introduction of the isopropyl or isopropoxy group into other organic molecules by the breaking of the C—OH or the O—H bond in the isopropyl alcohol molecule.

Isopropyl alcohol undergoes reactions typical of an active secondary alcohol. It can be dehydrogenated, oxidized, esterified, etherified, aminated, halogenated, or otherwise modified at the OH moiety more readily than primary alcohols such as n-propyl or ethyl alcohol. Manufacture of the commercially important aluminum isopropoxide [555-31-7] and isopropyl halides illustrates this reactivity. The aluminum isopropoxide reaction involves the aluminum replacement of the hydroxyl hydrogen atom and concomitant hydrogen evolution; the

Table 2. Azeotropes of Isopropyl Alcohol[a]

Component	CAS Registry Number	Bp[b], °C	Azeotrope bp[b], °C	Isopropyl alcohol composition, wt %
water		100	80.3	87.4
			120.5[c]	88.3
toluene	[108-88-3]	110.6	80.6	69
methyl propionate	[554-12–1]	79.6	77	28
methyl ethyl ketone	[78-93-3]	79.6	77.9	32
ethyl acetate	[141-78-6]	77.05	75.9	25
2-chlorobutane	[78-86-4]	68.25	64	18
hexane	[110-54-3]	68.9	62.7	23
cyclohexane	[110-82-7]	80.8	68.6	33
butylamine	[109-73-9]	77.8	84.7	60
diisopropyl ether	[108-20-3]	69	66.2	16.3

[a]Ref. 9.
[b]Determined at 101.3 kPa (1 atm), except where noted.
[c]At 411.5 kPa (4.06 atm).

isopropyl halides reaction involves hydroxyl group displacement. Aluminum isopropoxide is produced in quantitative yield by refluxing isopropyl alcohol with aluminum turnings (10,11).

$$6 \, (CH_3)_2CHOH + 2 \, Al^0 \longrightarrow 2 \, ((CH_3)_2CHO)_3Al + 3 \, H_2$$

Catalytic amounts of mercuric chloride are usually employed in this preparation. Aluminum isopropoxide is a useful Meerwein-Ponndorf-Verley reducing agent in certain ester-exchange reactions and is a precursor for aluminum glycinate, a buffering agent (see ALKOXIDES, METAL).

Displacement of the hydroxyl group is exemplified by the production of isopropyl halides, eg, isopropyl bromide [75-26-3], by refluxing isopropyl alcohol with a halogen acid, eg, hydrobromic acid [10035-10-6] (12).

$$(CH_3)_2CHOH + HBr \longrightarrow (CH_3)_2CHBr + H_2O$$

The order of reactivity with acid is HI > HBr > HCl. Reaction with hydrochloric acid [7647-01-0] to form isopropyl chloride [75-29-6] is facilitated by a zinc chloride catalyst.

Dehydrogenation.   Before the large-scale availability of acetone as a co-product of phenol (qv) in some processes, dehydrogenation of isopropyl alcohol to acetone (qv) was the most widely practiced production method. A wide variety of catalysts can be used in this endothermic (66.5 kJ/mol (15.9 kcal/mol) at 327°C), vapor-phase process to achieve high (75–95 mol %) conversions. Operation at 300–500°C and moderate pressures (207 kPa (2.04 atm)) provides acetone in yields up to 90 mol %. The most useful catalysts contain Cu, Cr, Zn, and Ni, either alone, as oxides, or in combinations on inert supports (see CATALYSTS, SUPPORTED) (13–16).

$$(CH_3)_2CHOH \xrightarrow[\text{ZnO catalyst}]{380°C, \; 100\text{-}200 \, kPa \; (1\text{-}2 \, atm)} CH_3COCH_3 + H_2$$

Although the selectivity is high, minor amounts of by-products can form by dehydration, condensation, and oxidation, eg, propylene [115-07-1], diisopropyl ether, mesityl oxide [141-79-7], acetaldehyde [75-07-0], and propionaldehyde [123-38-6]. Hydrotalcites having different Al/(Al + Mg) ratios have been used to describe a complete reaction network for dehydrogenation (17). This reaction can also be carried out in the liquid phase.

Dehydrogenation processes for acetone, methyl isobutyl ketone [108-10-1], and higher ketones (qv) utilizing, in one process, a copper-based catalyst have been disclosed (18,19). Dehydrogenation reaction is used to study the acid–base character of catalytic sites on a series of oxides (20,21).

Oxidation. Isopropyl alcohol can be catalytically oxidized using air or oxygen at high temperatures to give acetone and water.

$$(CH_3)_2CHOH + 0.5\,O_2 \xrightarrow[\text{catalyst}]{400\text{-}600°C} CH_3COCH_3 + H_2O$$

The catalysts are of the same general type as those used for dehydrogenation processes. In contrast to dehydrogenation, oxidation is highly exothermic (180 kJ/mol (43 kcal/mol) at 295°C). Therefore, careful control of processing conditions is critical in order to minimize the formation of by-products, especially those of dehydration (22,23). It is possible to run this oxidation and the dehydrogenation reactions simultaneously by proper choice of catalysts and conditions. Use of this technology for conversion of isopropyl alcohol to acetone is minimal compared to the dehydrogenation route.

Isopropyl alcohol can be partially oxidized by a noncatalytic, liquid-phase process at low temperatures and pressure to produce hydrogen peroxide [7722-84-1] and acetone (24–26).

$$(CH_3)_2CHOH + O_2 \xrightarrow[253\,\text{kPa (2.5 atm)}]{\text{peroxide } 90\text{-}140°C} CH_3COCH_3 + H_2O_2$$

Oxygen or air can be used with a peroxide initiator, eg, hydrogen peroxide (qv). The oxidation rate is sensitive to the quantity of by-product acetic acid [64-19-7] generated. The theoretical yield ratio of acetone to hydrogen peroxide produced is 1.7 by weight. This process is normally employed when hydrogen peroxide is desired, in which case acetone and unreacted isopropyl alcohol are recycled. The process is used by Shell, in which hydrogen peroxide is used for oxidation of allyl alcohol [107-18-6] to acrolein [107-02-8], and by Burmah Oil Company, in which hydrogen peroxide is converted to peracetic acid (26,27).

Isopropyl alcohol can be oxidized by reaction of an α,β-unsaturated aldehyde or ketone at high temperature over metal oxide catalysts (28). In one Shell process for the manufacture of allyl alcohol, a vapor mixture of isopropyl alcohol and acrolein, which contains two to three moles of alcohol per mole of aldehyde, is passed over a bed of uncalcined magnesium oxide [1309-48-4] and zinc oxide [1314-13-2] at 400°C. The process yields about 77% allyl alcohol based on acrolein.

$$(CH_3)_2CHOH + CH_2{=}CHCHO \longrightarrow CH_3COCH_3 + CH_2{=}CHCH_2OH$$

Esterification. Isopropyl alcohol is esterified readily by treatment of carboxylic acids in the presence of an acidic catalyst, eg, p-toluenesulfonic acid [6192-52-5]. An equilibrium is established in the reaction typically carried out at 100–160°C and atmospheric pressure, using an excess of alcohol.

$$RCOOH + (CH_3)_2CHOH \overset{acid}{\rightleftharpoons} RCOOCH(CH_3)_2 + H_2O$$

Energy is supplied to remove the water as an azeotrope, thus forcing the reaction in the desired direction. Excess alcohol is distilled and recycled, and yields of ester are nearly quantitative. For example, isopropyl acetate [108-21-4] can be prepared by the reaction of isopropyl alcohol and acetic acid in the presence of sulfuric acid catalyst and using toluene as the azeotroping agent. The kinetics of the esterification of isopropyl alcohol with some organic acids by sulfonated cation-exchange resins have been reported (29). Esterification of isopropyl alcohol with myristic acid [544-63-8] forms isopropyl myristate [110-27-0], an emollient and lubricant in various cosmetic products and topical medicinals (11,30) (see COSMETICS). A jellied product is marketed as Estergel (30).

Xanthate esters are prepared by reaction of isopropyl alcohol and carbon disulfide [75-15-0]. Isopropyl xanthates have wide use in mineral flotation (qv) processes, and sodium isopropyl xanthate [140-93-2], $C_4H_7OS_2Na$, is a useful herbicide for bean and pea fields (see HERBICIDES) (30).

Phosphite esters are formed readily by the reaction of phosphorus halides and isopropyl alcohol. For example, triisopropyl phosphite [116-17-6] is prepared from phosphorus trichloride [7719-12-2] and isopropyl alcohol at low temperatures in the presence of an acid scavenger, eg, pyridine [110-86-1].

$$3 (CH_3)_2CHOH + PCl_3 \longrightarrow ((CH_3)_2CHO)_3P + 3 HCl$$

Similarly, another important esterification reaction of isopropyl alcohol involves the production of tetraisopropyl titanate [546-68-9], a commercial polymerization catalyst, from titanium tetrachloride [7550-45-0] and isopropyl alcohol.

Isopropyl nitrate [1712-64-7] can be prepared by the reaction of isopropyl alcohol with nitric acid [52583-42-3].

$$(CH_3)_2CHOH + HNO_3 \longrightarrow (CH_3)_2CHONO_2 + H_2O$$

The reactants are fed separately into a still, from which the product is continuously removed by distillation (qv) (31). Isopropyl nitrate is a valuable engine-starter fuel and can be used in explosives (see EXPLOSIVES AND PROPELLANTS) (32). The nitrite ester, isopropyl nitrite, can be prepared from the reaction of isopropyl alcohol and either nitrosyl chloride or nitrous acid at ambient temperature (33). The ester is used as a jet engine propellant (30).

Etherification. Isopropyl alcohol can be dehydrated in either the liquid phase over acidic catalysts, eg, sulfuric acid, or in the vapor phase over acidic aluminas to give diisopropyl ether (DIPE) and propylene (qv).

$$(CH_3)_2CHOH \longrightarrow CH_3CH{=}CH_2 + H_2O$$

$$2 (CH_3)_2CHOH \longrightarrow ((CH_3)_2CH)_2O + H_2O$$

Either product can be favored over the other by proper selection of catalyst and reaction conditions. However, the principal source of DIPE is as a by-product from isopropyl alcohol production. Typically, excess DIPE is recycled over acidic catalysts in the alcohol process where it is hydrated to isopropyl alcohol. DIPE is used to a minor extent in industrial extraction and as a solvent.

The 1990 Clean Air Act mandates for blended oxygenates in gasoline created a potentially large new use for DIPE as a fuel oxygenate. Isopropyl alcohol can react with propylene over acidic ion-exchange (qv) catalysts at low temperatures, which favor high equilibrium conversions per pass to produce DIPE (34).

Glycol ethers can be prepared from isopropyl alcohol by reaction of olefin oxides, eg, ethylene oxide [75-21-8] (qv) or propylene oxide [75-56-9] (qv). Reactions such as that to produce 2-isoproxyethanol [109-59-1] (isopropyl Cellosolve) are generally catalyzed by an alkali hydroxide.

$$(CH_3)_2CHOH \ + \ \overset{\displaystyle O}{\overset{\displaystyle /\backslash}{CH_2CH_2}} \ \xrightarrow{\text{KOH}} \ (CH_3)_2CHOCH_2CH_2OH$$

Higher alkoxylated products, ie, oligomers, are formed by secondary reaction of oxide and the hydroxy group of the previous product.

$$(CH_3)_2CHOCH_2CH_2OH \ + \ \overset{\displaystyle O}{\overset{\displaystyle /\backslash}{CH_2CH_2}} \ \xrightarrow{\text{KOH}} \ (CH_3)_2CHO(CH_2CH_2O)_xH$$

This is a particularly troublesome competing reaction when the olefin oxide, eg, ethylene oxide, produces the more reactive terminal primary hydroxy group. Glycol ethers are used as solvents in lacquers, enamels, and waterborne coatings to improve gloss and flow.

Amination. Isopropyl alcohol can be aminated by either ammonolysis in the presence of dehydration catalysts or reductive ammonolysis using hydrogenation catalysts. Either method produces two amines: isopropylamine [75-31-0] and diisopropylamine [108-18-9]. Virtually no trisubstituted amine, ie, triisopropylamine [122-20-3], is produced. The ratio of mono- to diisopropylamine produced depends on the molar ratio of isopropyl alcohol and ammonia [7664-41-7] employed. Molar ratios of ammonia and hydrogen to alcohol range from 2:1–5:1 (35,36).

$$(CH_3)_2CHOH \ + \ NH_3 \ \xrightarrow[\Delta, \text{ pressure}]{\text{catalyst}} \ (CH_3)_2CHNH_2 \ + \ H_2O$$

$$(CH_3)_2CHOH \ + \ (CH_3)_2CHNH_2 \ \xrightarrow[\Delta, \text{ pressure}]{\text{catalyst}} \ [(CH_3)_2CH]_2NH \ + \ H_2O$$

In the reductive ammonolysis process, there is virtually no consumption of added hydrogen, which is used to increase catalyst life by inhibiting coking and tar formation. In this reaction, a gaseous mixture of the alcohol, ammonia, and hydrogen is fed to a fixed-bed reactor that contains a dehydrogenation catalyst, eg, Cu, Cr, or Ni, supported on alumina. Operating conditions are 150–250°C and 790–2860 kPa (100–400 psig) (35–40). The liquid hourly space velocity (LHSV) of isopropyl alcohol is about 0.5/h. Isopropyl alcohol conversions per pass are in excess of 85% and yields are >90%. By-products, eg, nitrile (from dehydrogenation of the amine) and amide (from hydrolysis of the nitrile) are produced, but these are recycled to increase productivity.

Direct ammonolysis involving dehydration catalysts is generally run at higher temperatures (300–500°C) and at about the same pressure as reductive ammonolysis. Many catalysts are active, including aluminas, silica, titanium dioxide [13463-67-7], and aluminum phosphate [7784-30-7] (41–43). Yields are acceptable (>80%), and coking and nitrile formation are negligible. However, little control is possible over the composition of the mixture of primary and secondary amines that can be obtained.

Isopropylamine is the most widely used of the propylamines. Most of it is consumed in herbicide manufacture, primarily in production of 2-chloro-4-ethyl-6-isopropylamino-*sym*-triazine. A smaller quantity is used for pesticide manufacture (40,44). Diisopropylamine is used chiefly in pesticides (qv) and as a corrosion inhibitor, eg, diisopropylammonium nitrate (see CORROSION AND CORROSION CONTROL) (44,45).

Halogenation.   Normally, 2-halopropane derivatives are prepared from isopropyl alcohol most economically by reaction with the corresponding acid halide. However, under appropriate conditions, other reagents, eg, phosphorus halides and elemental halogen, also react by replacement of the hydroxyl group to give the halide (46).

$$3 \ (CH_3)_2CHOH + PBr_3 \longrightarrow 3 \ (CH_3)_2CHBr + H_3PO_3$$

Halogenation of isopropyl alcohol in aqueous solution results in concomitant oxidation. Thus, chlorination at 65°C produces a mixture of chloroacetone derivatives, chiefly 1,3-dichloroacetone [534-07-6] and 1,1,3-trichloroacetone [921-03-9] (47,48). Further chlorination at 70–100°C provides nearly complete conversion of lower chloroacetones into 1,1,1,3,3-pentachloroacetone [1768-31-6] and hexachloroacetone [116-16-5] (48–52). Chlorination of isopropyl alcohol reportedly can be conducted to give 1,1,1,3-tetrachloroacetone [16995-35-0], which is converted to 1,1,1-trichloro-2,3-epoxypropane [3083-23-6], a useful intermediate for synthesis of agricultural and pharmaceutical chemicals and for the preparation of plastics having low flammability (53,54). However, commercial processes to chloroacetones are believed to be based on chlorination of acetone rather than isopropyl alcohol, because of undesired by-products produced in the latter route. In addition to use as chemical intermediates, chloroacetones have excellent solvent properties, especially for plastics. However, toxicity may limit use.

Halogenated 2-propanol derivatives, eg, 1,3-dichloro-2-propanol [96-23-1], are generally prepared from glycerol [56-81-5] (qv). These materials are used in the preparation of halogen-containing phosphates to plasticize and lower the flammability of plastics, eg, polyurethanes and cellulosics.

Miscellaneous Reactions.   Reactions of potential commercial significance include acylation by ketene [463-51-4]:

$$(CH_3)_2CHOH + CH_2{=\!=}C{=\!=}O \longrightarrow CH_3COOCH(CH_3)_2$$

and the Ritter reaction to prepare *N*-isopropylacrylamide [2210-25-5] from acrylonitrile [107-13-1] and isopropyl alcohol:

$$CH_2 =\!\!= CHCN + (CH_3)_2CHOH \longrightarrow CH_2 =\!\!= CHCONHCH(CH_3)_2$$

Manufacture

The first industrial quantities of isopropyl alcohol were produced in 1920 in the world's first petrochemical plant, owned by Standard Oil (Exxon) Company (Bayway, New Jersey). This was followed in 1921, by the start-up of isopropyl alcohol production in Clendenin, West Virginia, by the Carbide and Carbon Chemicals (Union Carbide) Corporation. The Shell Oil Company began production in the 1930s at Dominguez, California (55). These three companies are the principal domestic manufacturers as of the mid-1990s.

The indirect hydration, also called the sulfuric acid process, practiced by the three U.S. domestic producers, was the only process used worldwide until ICI started up the first commercial direct hydration process in 1951. Both processes use propylene and water as raw materials. Early problems of high corrosion, high energy costs, and air pollution using the indirect process led to the development of the direct hydration process in Europe. However, a high purity propylene feedstock is required. In the indirect hydration process, C_3-feedstock streams from refinery off-gases containing only 40–60 wt % propylene are often used in the United States.

Other potential synthetic methods include fermentation (qv) of certain carbohydrates (qv), oxidation of propane, hydrogenation of acetone, and hydrolysis of isopropyl acetate. The hydrogenation of by-product acetone is the only method practiced commercially.

Indirect Hydration. Indirect hydration is based on a two-step reaction of propylene and sulfuric acid. In the first step, mixed sulfate esters, primarily isopropyl hydrogen sulfate, but also diisopropyl sulfate, form. These are then hydrolyzed, forming the alcohol and sulfuric acid.

Step 1. Esterification:

$$CH_3CH =\!\!= CH_2 + H_2SO_4 \rightleftharpoons (CH_3)_2CHOSO_3H$$

$$(CH_3)_2CHOSO_3H + CH_3CH =\!\!= CH_2 \rightleftharpoons ((CH_3)_2CHO)_2SO_2$$

Step 2. Hydrolysis:

$$(CH_3)_2CHOSO_3H + H_2O \rightleftharpoons (CH_3)_2CHOH + H_2SO_4$$

$$((CH_3)_2CHO)_2SO_2 + 2\,H_2O \rightleftharpoons 2\,(CH_3)_2CHOH + H_2SO_4$$

By-Products. Diisopropyl ether is the principal by-product formed by reaction of the intermediate sulfate esters with isopropyl alcohol.

$$(CH_3)_2CHOSO_3H + (CH_3)_2CHOH \rightleftharpoons ((CH_3)_2CH)_2O + H_2SO_4$$

$$((CH_3)_2CHO)_2SO_2 + (CH_3)_2CHOH \rightleftharpoons (CH_3)_2CHOSO_3H + ((CH_3)_2CH)_2O$$

The principal reactions are reversible and a mixture of products and reactants is found in the crude sulfate. High propylene pressure, high sulfuric acid concentration, and low temperature shift the reaction toward diisopropyl sulfate. However, the reaction rate slows as products are formed, and practical reactors operate by using excess sulfuric acid. As the water content in the sulfuric acid feed is increased, more of the hydrolysis reaction (Step 2) occurs in the main reactor. At water concentrations near 20%, diisopropyl sulfate is not found in the reaction mixture. However, efforts to separate the isopropyl alcohol from the sulfuric acid suggest that it may be partially present in an ionic form (56,57).

$$(CH_3)_2CHOH + H_2SO_4 \rightleftharpoons (CH_3)_2CHOH_2^+ + HSO_3^-$$

Other by-products include acetone, carbonaceous material, and polymers of propylene. Minor contaminants arise from impurities in the feed. Ethylene and butylenes can form traces of ethyl alcohol and 2-butanol. Small amounts of *n*-propyl alcohol carried through into the refined isopropyl alcohol can originate from cyclopropane [75-19-4] in the propylene feed. Acetone, an oxidation product, also forms from thermal decomposition of the intermediate sulfate esters, eg,

$$(CH_3)_2CHOSO_3H \xrightarrow{heat} CH_3COCH_3 + SO_2 + H_2O$$

In addition to generating malodorous sulfur dioxide [7446-09-5], the acetone formed can undergo further condensation in the acidic medium to generate mesityl oxide [141-79-7], $(CH_3)_2C{=}CHCOCH_3$, and higher products.

High propylene concentrations in the presence of acids can form dimers, trimers, and higher homologues, which can polymerize or hydrate to C_6, C_9, or higher alcohols, and olefins. These derivatives can emit musty, woody, and camphoraceous odors, and their reaction products with sulfur-containing compounds can give a cat-like odor to the product (58). Odor can be improved by employing appropriate reaction conditions and by contacting isopropyl alcohol with various metals, eg, copper and nickel, or certain partially reduced metal oxides (59,60).

Process. A typical indirect hydration process is presented in Figure 1. In the process, propylene reacts with sulfuric acid (>60 wt %) in agitated reactors or absorbers at moderate (0.7–2.8 MPa (100–400 psig)) pressure. The isopropyl sulfate esters form and are maintained in the liquid state at 20–80°C. Low propylene concentrations, ie, 50 wt %, can be tolerated, but concentrations of 65 wt % or higher are preferred to achieve high alcohol yields. Because the

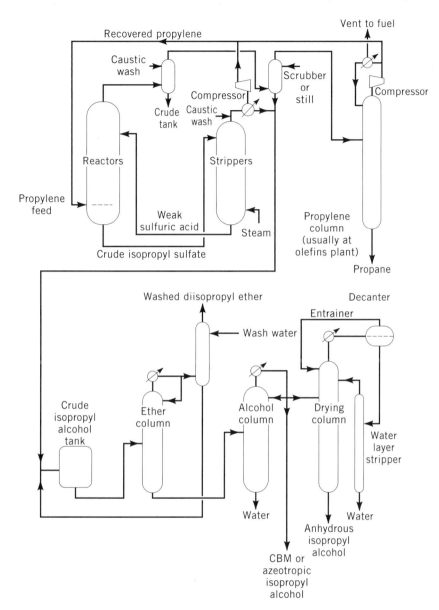

Fig. 1. Indirect hydration process for the manufacture of isopropyl alcohol; CBM = constant boiling mixture (61,62).

reaction is exothermic, internal cooling coils or external heat exchangers are used to control the temperature.

There are two general operational modes for conducting the reaction. In the two-step strong acid process, separate reactors are used for the propylene absorption and sulfate ester hydrolysis stages. The reaction occurs at high sulfuric acid concentration (>80 wt %), at 1–1.2 MPa (130–160 psig) pressure, and low (eg, 20–30°C) temperature. The weak acid process is conducted in a single stage at low acid (60–80 wt %) concentration and at higher (2.5 MPa (350 psig)) pressure and (60–65°C) temperature. Chemical selectivity to isopropyl alcohol and diisopropyl ether are above 98% for each process.

The sulfate ester hydrolysate is stripped to give a mixture of isopropyl alcohol, isopropyl ether, and water overhead, and dilute sulfuric acid bottoms. The overhead is neutralized using sodium hydroxide and refined in a two-column distillation system. Diisopropyl ether is taken overhead in the first, ie, ether, column. This stream is generally recycled to the reactors to produce additional isopropyl alcohol by the following equilibrium reaction:

$$((CH_3)_2CH)_2O + H_2SO_4 \rightleftharpoons (CH_3)_2CHOSO_3H + (CH_3)_2CHOH$$

Wet isopropyl alcohol (87 wt % and 91 vol %) is taken overhead in the second still. More than 93% of the charged propylene is converted to isopropyl alcohol in this system. If available, a propylene column may recover unreacted feedstock.

The bottoms from the stripper (40–60 wt % acid) are sent to an acid reconcentration unit for upgrading to the proper acid strength and recycling to the reactor. Because of the associated high energy requirements, reconcentration of the diluted sulfuric acid is a costly operation. However, a propylene gas stripping process, which utilizes only a small amount of added water for hydrolysis, has been described (63). In this modification, the equilibrium quantity of isopropyl alcohol is stripped so that acid is recycled without reconcentration. Equilibrium is attained rapidly at 50°C and isopropyl alcohol is removed from the hydrolysis mixture. Similarly, the weak sulfuric acid process minimizes the reconcentration of the acid and its associated corrosion and pollution problems.

The 91 vol % alcohol is sold as such or is dehydrated by azeotropic distillation using either diisopropyl ether or cyclohexane to produce an anhydrous product (see Table 2) (see DISTILLATION, AZEOTROPIC AND EXTRACTIVE) (64). Wet isopropyl alcohol is fed at about the center of a dehydrating column, and the azeotroping agent is fed near the top. As the ternary azeotrope forms, it is taken overhead, condensed, and the layers are separated. The upper layer, which is mainly azeotroping agent and alcohol, is returned to the top of the column as reflux. Anhydrous isopropyl alcohol is removed from the base of the column. The lower layer is mostly water. It is fed to a stripping column for recovery of isopropyl alcohol and azeotroping agent. A purge may be taken to remove by-products such as acetone.

Acid corrosion presents a problem in isopropyl alcohol factories. Steel (qv) is a satisfactory material of construction for tanks, lines, and columns where concentrated (>65 wt %) acid and moderate (<60°C) temperatures are employed. For dilute acid and higher temperatures, however, stainless steel, tantalum, Hastelloy, and the like are required for corrosion resistance and to ensure product purity (65).

The extent of purification depends on the use requirements. Generally, either intense aqueous extractive distillation, or post-treatment by fixed-bed absorption (qv) using activated carbon, molecular sieves (qv), and certain metals on carriers, is employed to improve odor and to remove minor impurities. Essence grade is produced by final distillation in nonferrous, eg, copper, equipment (66).

Manufacturing plants in the United States are believed to use solely indirect propylene hydration. Several European companies, eg, British Petroleum, Shell, and Deutsche Texaco, also employ this older technology in plants in Europe and Japan (67).

Direct Hydration. The acid-catalyzed direct hydration of propylene is exothermic and resembles the preparation of ethyl alcohol (qv) from ethylene (qv).

$$CH_3CH = CH_2 + H_2O \xrightleftharpoons{catalyst} (CH_3)_2CHOH \qquad \Delta H = -50 \text{ kJ/mol } (-12 \text{ kcal/mol})$$

The equilibrium can be controlled to favor product alcohol if high pressures and low temperatures are applied. The advantage of low temperature is difficult to utilize, however, because most known catalysts require high or moderate temperatures to be effective.

Process. There are three basic processes in commercial operation: (*1*) vapor-phase hydration over a fixed-bed catalyst of supported phosphoric acid (Veba-Chemie) (68–71), or silica-supported tungsten oxide with zinc oxide promoter (ICI) (66,67,72); (*2*) mixed vapor–liquid-phase hydration at low (150°C) temperature and high (10.13 MPa (100 atm)) pressure using a strongly acidic cation-exchange resin catalyst (Deutsche Texaco AG) (73–77); and (*3*) liquid-phase hydration at high (270°C) temperature and high (20.3 MPa (200 atm)) pressure in the presence of a soluble tungsten catalyst (Tokuyama Soda) (78–80).

A typical process scheme for the direct hydration of propylene is shown in Figure 2. Turnkey plants based on this technology are available (71,81). The principal difference between the direct and indirect processes is the much higher pressures needed to react propylene directly with water. Products and by-products are also similar, and refining systems are essentially the same. Under some conditions, the high pressures of the direct process can increase the production of propylene polymers.

The first direct hydration plant (ICI, 1951) used a WO_3–ZnO catalyst supported on SiO_2, high (230–290°C) temperature, and high (20.3–25.3 MPa (200–250 atm)) pressure. Similarly, in the Veba-Chemie process (see Fig. 2), a vaporized stream of propylene and water is passed through an acidic catalyst bed (H_3PO_4 supported on SiO_2) at 240–260°C and 2.5–6.6 MPa (25–65 atm) (69). The gas stream from the reactor is cooled and fed to a scrubber where the remaining isopropyl alcohol is removed. Isopropyl alcohol selectivity is ca 96% for the gas-phase process. Owing to equilibrium limitations in the gas phase at high temperature and low pressure, a low propylene conversion (5–6%) results and thus a large amount of unreacted propylene is recycled. Both processes involve high plant costs owing to high pressure requirements, gas recycles, and the requirement for high purity propylene (ca 99 wt %).

Deutsche Texaco developed a trickle-bed process to avoid the disadvantages of the gas-phase process. In the trickle-bed process, a mixture of liquid water and propylene gas in a molar ratio of 12 to 15:1 is introduced at the top of a fixed-bed reactor and allowed to trickle down over a sulfonic acid ion-exchange resin. Reaction between the liquid and gas phases takes place at 130–160°C and 8–10 MPa (80–100 atm), forming aqueous isopropyl alcohol. Propylene conversions per pass are greater than 75%, and isopropyl alcohol selectivity is 93%.

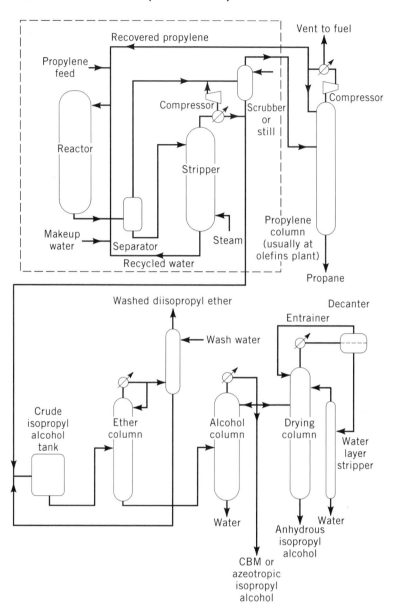

Fig. 2. Direct hydration process for the manufacture of isopropyl alcohol. The steps within the dashed box differentiate the direct from the indirect processes (see Fig. 1).

Only 92 wt % propylene purity is needed for this process. Approximately 5% di-isopropyl ether and some alcohols of the higher oligomers form as by-products. The life of the cation-exchange resin is at least eight months.

A liquid-phase variation of the direct hydration was developed by Tokuyama Soda (78). The disadvantages of the gas-phase processes are largely avoided by employing a weakly acidic aqueous catalyst solution of a silico-tungstate (82). Preheated propylene, water, and recycled aqueous catalyst solution are pressurized and fed into a reaction chamber where they react in the liquid state at 270°C and 20.3 MPa (200 atm) and form aqueous isopropyl alcohol. Propylene conversions of 60–70% per pass are obtained, and selectivity to isopropyl alcohol is 98–99 mol % of converted propylene. The catalyst is recycled and requires little replenishment compared to other processes. Corrosion and environmental problems are also minimized because the catalyst is a weak acid and because the system is completely closed. On account of the low gas recycle ratio, regular commercial propylene of 95% purity can be used as feedstock.

After flashing the propylene, the aqueous solution from the separator is sent to the purification section where the catalyst is separated by azeotropic distillation; 88 wt % isopropyl alcohol is obtained overhead. The bottoms containing aqueous catalyst solution are recycled to the reactor, and the light ends are stripped of low boiling impurities, eg, diisopropyl ether and acetone. Azeotropic distillation yields dry isopropyl alcohol, and the final distillation column yields a product of more than 99.99% purity.

Catalysts. Because low temperature and high pressure favor product formation in propylene hydration, one of the earliest efforts in developing the isopropyl alcohol process technology was a search for a catalyst that maximizes alcohol productivity at low temperatures within a reasonable time. Patents issued in the early 1950s involved various tungsten compounds (83–87). Studies of acidic supports and of acids adsorbed on various porous supports were conducted at about the same time (88–99). Since that time, many other acidic materials have been claimed, including cation-exchange resins (100–105), molybdophosphoric acid [51429-74-4] (106), titanium and zinc oxides (107), tungsten and zirconium oxides (108), silicotungstates (82,109), molybdenum oxalate [24958-46-1] (110), and zeolites (111).

Improved versions of acid ion-exchange resin catalysts have been offered by Universal Oil Products (112–114), where increased activity allowing lower temperature operation and higher equilibrium conversions per pass is claimed. When used in a variation of the Deutsche Texaco process with reduced water feed, the trickle-bed process becomes a single liquid-phase process that simplifies reactor design. By refeeding the isopropyl alcohol as shown in Figure 3, the final product is diisopropyl ether. The ether process is offered as a means of making oxygenates for reformulated gasoline.

Reaction Mechanism. Propylene hydration in dilute acid solution probably proceeds according to the rate-determining formation of propyl carbonium ion (115).

$$CH_3CH{=}CH_2 + H_3O^+ \xrightleftharpoons{fast} CH_3CH\underset{\underset{H}{+}}{-}CH_2 + H_2O$$

$$CH_3CH\underset{\underset{H}{+}}{-}CH_2 \xrightleftharpoons{slow} (CH_3\overset{+}{C}HCH_3)$$

$$(CH_3\overset{+}{C}HCH_3) + 2\,H_2O \xrightleftharpoons{fast} CH_3\overset{\overset{OH}{|}}{C}HCH_3 + H_3O^+$$

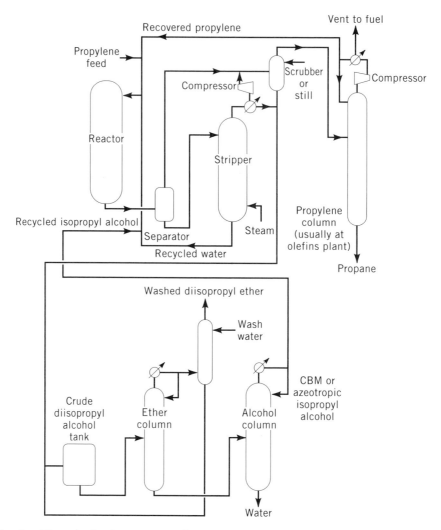

Fig. 3. Direct hydration process where the product is isolated as diisopropyl ether.

According to this mechanism, the reaction rate is proportional to the concentration of hydronium ion and is independent of the associated anion, ie, rate = $k[CH_3H_6][H_3O^+]$. However, the acid anion may play a marked role in hydration rate, eg, phosphomolybdate and phosphotungstate anions exhibit hydration rates two or three times that of sulfate or phosphate (78). Association of the polyacid anion with the propyl carbonium ion is suggested. Protonation of propylene occurs more readily than that of ethylene as a result of the formation of a more stable secondary carbonium ion. Thus higher conversions are achieved in propylene hydration.

Thermochemical Data. Equilibrium considerations significantly limit alcohol yield at low pressures in the vapor-phase process (116). Consequently, conditions controlling equilibrium constants have been determined and give the following relation, where T is in K (116,117):

$$\log K = 2624/T - 7.584$$

Likewise, hydration in the liquid phase can be expressed by the free-energy ($\Delta F°$) equation:

$$\Delta F° = 23.25T - 9352$$

The effects of pressure and temperature on the equilibrium concentration of alcohol in both phases of hydration of propylene when both liquid and vapor phases are present have been calculated and are presented in Table 3. Low temperature reduces by-product diisopropyl ether.

Other Processes. Isopropyl alcohol can be prepared by the liquid-phase oxidation of propane (118). It is produced incidentally by the reductive condensation of acetone, and is partly recovered from fermentation (119). Large-scale commercial biological production of isopropyl alcohol from carbohydrate raw materials has also been studied (120–123).

Table 3. Calculated Vapor–Liquid Equilibrium Composition for Propylene Hydration[a]

Pressure, MPa[b]	Alcohol concentration in liquid phase, mol %	Concentrations in vapor phase, mol %			
		Isopropyl alcohol	Water	Propylene	Isopropyl ether
Temperature of 200°C					
10.13	3.8	17.3	23.3	54.8	4.6
20.26	5.5	24.0	18.9	48.8	8.3
30.39	6.2	28.3	18.2	43.6	9.9
40.52	6.3	31.2	18.8	37.6	12.4
50.65	5.7	35.3	19.8	31.9	13.0
Temperature of 250°C					
10.13	1.2	7.9	49.3	42.5	0.3
20.26	2.1	13.0	37.6	48.6	0.8
30.39	2.6	14.5	36.1	48.5	0.9
40.52	3.2	15.3	36.1	47.5	1.1
50.65	3.7	15.2	37.8	45.7	1.3
Temperature of 275°C					
10.13	0.9	3.9	69.7	26.3	0.1
20.26	1.6	8.1	52.0	39.7	0.2
30.39	2.0	9.5	49.2	41.0	0.3
40.52	2.3	9.8	48.5	41.3	0.4
50.65	2.4	9.7	51.0	38.9	0.4

[a]Ref. 116.
[b]To convert MPa to psi, multiply by 145.

Storage and Shipping

Anhydrous isopropyl alcohol is typically shipped in plain steel railroad tank cars, tank trucks, 3.8-L (1-gal) glass jugs, and 19- and 208-L (5- and 55-gal) DOT 17E phenolic-lined steel pails and drums, respectively. Plain steel is not suitable for isopropyl alcohol containing water because rusting can result. Instead, baked phenolic-lined steel tanks are used. Aluminum is also unsuitable. It is attacked by isopropyl alcohol, especially the anhydrous grade, resulting in the formation of aluminum isopropoxide. Containers must comply with DOT specifications. Tanks, piping, and equipment can be made of similar material.

The U.S. domestic shipping name of isopropyl alcohol is UN No. 1219 Isopropanol. Anhydrous as well as water solutions to 91 vol % alcohol are considered flammable liquid materials by the DOT. Both have flash points below 37.8°C by the Tag closed-cup method. Acceptable modes of transportation include air, rail, road, and water (124). For international air and water shipments, the International Maritime Organization (IMO) class is 3.2, the packaging group is II, and the primary hazard label required is "Flammable Liquid."

Economic Aspects

Economic comparisons of the indirect and direct hydration processes show a cost savings for use of the latter technology (79,125). The largest savings are in capital investment, processing, and maintenance (qv), all of which are necessitated by the troublesome sulfuric acid reconcentration of the indirect process. Greater corrosion and pollution problems are also associated with indirect hydration. However, drawbacks of direct hydration are its high energy usage and the need for highly concentrated propylene feedstock. Economics of the various direct hydration processes are roughly comparable (79).

Price and Demand. In terms of production volume, isopropyl alcohol is about the fourth largest chemical produced from propylene (66). Total 1993 U.S. nameplate capacity for isopropyl alcohol production was 8.48×10^5 metric tons. The total world capacity is about 2.0×10^9 metric tons (Table 4) (126–128). The 1995 U.S. prices were \$0.55/L (\$2.10/gal) for refined 91 vol % and \$0.62/L (\$2.36/gal) for anhydrous alcohol (129), an increase from the \$0.18/L (\$0.70/gal) average price of 1977. The price of isopropyl alcohol is driven by the price of propylene, the primary feedstock, and by the price of ethyl alcohol, a competing solvent.

U.S. production of isopropyl alcohol declined by 3%/yr from 1983 through the mid-1990s (130). 1993 demand was 5.5×10^5 metric tons, including exports, which have been growing at a 3–5% annual rate. The biggest factor contributing to the decline in demand is the decrease in use for acetone production. Only 8% of the isopropyl alcohol produced in 1993 was used for acetone feedstock, compared to 38% in 1977. Most acetone is produced as a co-product of cumene [98-82-8] oxidation to phenol (qv).

The future demand for isopropyl alcohol is expected either to remain flat or to grow slightly. Its main use as a chemical intermediate is growing, and this should offset the pressure on use as a solvent from tighter volatile organic chemicals (VOC) regulations.

Table 4. Isopropyl Alcohol Capacities

Producer	Plant location	Capacity, t × 10³
United States[a]		
ARCO Chemical Co.	Channelview, Tex.	30
Exxon Chemical Co.	Baton Rouge, La.	295
Shell Chemical Co.	Deer Park, Tex.	273
Union Carbide Corp.	Texas City, Tex.	250
Total		*848*
Europe[b]		
Shell Chimie, SA	Berre-l'Etang, France	120
Hüls Aktiengesellschaft	Herne, Germany	75
RWE–DEA AG für Mineralöl und Chemie	Moers, Germany	140
Shell Nederland Chemie BV	Rotterdam-Pernis, the Netherlands	250
Industrias Quimicas Associadas, SA	Tarragona, Spain	36
BP Chemicals, Ltd.	Port Talbot, U.K.	85
Shell Chemicals U.K., Ltd.	Ellesmere Port, U.K.	105
Total		*811*
Asia[c]		
Jinzou Petrochemical Corp.	China	60
Mitsui Toatsu Chemicals Inc.	Takaishi, Japan	33
Nippon Petrochemicals Co., Ltd.	Kawasaki, Japan	60
Tokuyama Soda Co., Ltd.	Tokuyama, Japan	38
Shell Eastern Chemicals (Pte.) Ltd.	Pulau Ular, Singapore	70
Lucky Ltd.	Yeochon, S. Korea	30
Yukong Ltd.	Ulsan, S. Korea	35
Lee Chang Yung Chemical Industry Corp.	Linyuan City, Taiwan	45
Total		*371*

[a]Ref. 126.　[b]Ref. 127.　[c]Ref. 128.

Specifications

Typical specifications for the two basic grades of isopropyl alcohol are shown in Table 5. Other grades that are marketed include a cosmetic grade (91 vol % and anhydrous) containing perfume, and an electronic grade of low conductivity. Other more restrictive specifications for special grades are shown in Table 6.

Analytical and Test Methods

Purity of commercial aqueous isopropyl alcohol mixtures is most simply determined by specific gravity measurement. However, this technique, based on the assumption that impurities are not present in significant quantities, does not provide a positive identification of isopropyl alcohol. Gas chromatography, which overcomes the disadvantages of the specific gravity method, is an excellent technique for determining isopropyl alcohol in the presence of other organic substances, eg, ethyl alcohol, methyl alcohol, and acetones (131). Colorimetric methods can be used to determine trace amounts of ethyl alcohol upon the addition of specific compounds known to form complexes with alcohols (132). In the

Table 5. Typical Specifications for Isopropyl Alcohol

Test	Anhydrous	Azeotropic, 91–93 vol %
isopropyl alcohol, wt %	99.8, min	85, min
acidity, wt %[a]	0.002	0.005
dilution	clear	clear
alkalinity, meq/mL	0.0001	
water, wt %	0.1, max	
nonvolatile, g/100 mL	0.0008, max	
color, Pt–Co units	5, max	10, max
odor	nonresidual	
sulfur/chlorides, ppm		3
specific gravity, 20/20	0.7861–0.7866	0.8246, max
refractive index, at 20°C	1.3763–1.3780	
appearance	no visible/settled matter	some suspended matter
distillation, °C	0.5, max[b]	
permanganate time test, minutes	45, min	
copper corrosion test	no pitting or black stain	

[a]As acetic acid. [b]Including 82.3 ± 0.1.

Table 6. Special Specifications for Isopropyl Alcohol

Test	ACS reagent 1987	1990 USP XXII Anhydrous[a]	ASTM D770-85
isopropyl alcohol, %[b]		99.0	
acidity, meq/mL	0.0001[c]	0.00028	0.002[d]
dilution	clear		no turbidity
alkalinity, meq/g	0.0001		
water, wt %	0.2, max		0.2, max
nonvolatile, mg/100 mL	1, max	2.5, max[e]	5, max
color, Pt–Co units	10, max		10, max
odor			nonresidual
specific gravity[f]			
20/20	0.7883		0.785–0.787
25/25		0.783–0.787	
refractive index, at 20°C		1.376–1.378	
infrared	identification		
distillation, °C[g]	1.0, max		1.5, max
ultraviolet scan absorbance, max			
330–400	0.01		
300	0.02		
275	0.03		
260	0.04		
245	0.08		
230	0.2		
220	0.4		
210	1		

[a]Values of acidity, nonvolatility, and refractive index for 1990 *U.S. Pharmacopeia* XXII azeotropic (91–93 vol %) isopropyl alcohol are equal to those of 1990 USP XXII anhydrous. [b]By gas chromatography, area %. For USP XXII azeotropic, % of organics. [c]Value is meq/g. [d]Value is percent as acetic acid. [e]Per 50 mL. [f]For ASTM D310-86 (91 vol %) at 20°C, value is 0.8180 ± 0.0005; for USP XXII azeotropic at 25°C, value is 0.815–0.810. [g]Including 82.3; for ASTM D1310-86, value is 1.0°C, max, including 80.4.

case of isopropyl alcohol, some of the colorimetric methods make use of its facile oxidation to acetone, which can provide highly colored complexes. Trace amounts of isopropyl alcohol can also be determined photometrically in the presence of acetone and acetic acid (133). A conductometric method has been developed for isopropyl alcohol–water mixtures (134).

Health and Safety Factors

Alcohols as a class have low toxicity. Isopropyl alcohol, however, is about twice as toxic as ethyl alcohol (135), but less toxic than methyl alcohol. There is no known systematic investigation of the effects of inhalation, eg, from aerosols, of isopropyl alcohol in humans. The known human toxicity is based on numerous cases of accidental ingestion or topical application. Toxic doses of ingested isopropyl alcohol, usually as rubbing alcohol, may produce narcosis, anesthesia, coma, and death. The single lethal dose for humans is about 250 mL, although as little as 100 mL can be fatal. Death occurs from paralysis of the central nervous system. Approximately 70–90% of ingested isopropyl alcohol is oxidized to acetone in the body, thus it is not a cumulative poison (136). Acetone appears on the breath within 15 minutes after ingestion. There is no fixed relationship between blood and urine isopropyl alcohol concentrations. NIOSH has published an extensive review of the toxicology of isopropyl alcohol (137).

Use of isopropyl alcohol in industrial applications does not present a health hazard. The alcohol produces anesthetic effects in high vapor concentration. Consequently, the OSHA permissible exposure limit (PEL) and the ACGIH threshold limit value (TLV) have been established at 400 ppm (0.098 mg/L) for an 8-h exposure (TWA) (138). This level causes a mild irritation of the eyes, nose, and throat (139). However, the TLV level does not produce symptoms of anesthesia (140). The OSHA and ACGIH short-term exposure limits (STELs) are 500 ppm. The odor threshold for isopropyl alcohol ranges from 3 to 200 ppm, which is the minimum concentration having identifiable odor (141).

An EPA–TSCA Section 4 Test Rule (142) requires manufacturers of isopropyl alcohol to test the alcohol for health and environmental effects. The Chemical Manufacturers Association (CMA) formed a panel to oversee the required testing. Rodent studies were designed to assess hazards for systemic effects (rats and mice) as evaluated by subchronic (90 d) inhalation exposure; adult neurotoxicity (rats) evaluated in a subchronic inhalation exposure; developmental toxicity (rats and rabbits) evaluated in an oral, gavage study; developmental neurotoxicity (rats) evaluated in an oral, gavage study; reproductive toxicity (rats) evaluated in a two-generation, oral, gavage study; and oncogenicity (rats and mice) evaluated in a two year inhalation study. Testing was completed in 1994. No adverse effects were found. As of this writing (1996), the test results are under review by the U.S. EPA.

Uses

Uses of isopropyl alcohol are chemical, solvent, and medical. Estimated U.S. uses in 1993 were as chemical intermediates, 34%; personal care and household products, 24%; coatings and ink solvent, 15%; processing solvent, 12%; pharmaceuticals, 10%; and miscellaneous uses, 5% (143).

Chemical.   The use of isopropyl alcohol as a feedstock for the production of acetone is expected to remain stable, as the dominant process for acetone is cumene oxidation. Isopropyl alcohol is also consumed in the production of other chemicals such as methyl isobutyl ketone, methyl isobutyl carbinol [108-11-2], isopropylamine, and isopropyl acetate. The use of diisopropyl ether as a fuel ether may become a significant outlet for isopropyl alcohol.

Solvent.   Because of its balance between alcohol, water, and hydrocarbon-like characteristics, isopropyl alcohol is an excellent, low cost solvent free from government regulations and taxes that apply to ethyl alcohol. The lower toxicity of isopropyl alcohol favors its use over methyl alcohol, even though the former is somewhat higher in cost. Consequently, isopropyl alcohol is used as a solvent in many consumer products as well as industrial products and procedures, eg, gasification and extractions. It is a good solvent for a variety of oils, gums, waxes (qv), resins, and alkaloids, and consequently is used for preparing cements, primers, varnishes, paints, printing inks, etc.

Isopropyl alcohol is also employed widely as a solvent for cosmetics (qv), eg, lotions, perfumes, shampoos, skin cleansers, nail polishes, makeup removers, deodorants, body oils, and skin lotions. In cosmetic applications, the acetone-like odor of isopropyl alcohol is masked by the addition of fragrance (144).

Over 68 aerosol products containing isopropyl alcohol solvent have been reported (145). Aerosol formulations include hair sprays (146), floor detergents (147), shoe polishes (148), insecticides (149,150), burn ointments (151), window cleaners, waxes and polishes, paints, automotive products (eg, windshield deicer), insect repellents, flea and tick spray, air refreshers, disinfectants, veterinary wound and pinkeye spray, first-aid spray, foot fungicide, and fabric-wrinkle remover (152) (see AEROSOLS).

Medical Usage.   Isopropyl alcohol is also used as an antiseptic and disinfectant for home, hospital, and industry (see DISINFECTANTS AND ANTISEPTICS). It is about twice as effective as ethyl alcohol in these applications (153,154). Rubbing alcohol, a popular 70 vol % isopropyl alcohol-in-water mixture, exemplifies the medicinal use of isopropyl alcohol. Other examples include 30 vol % isopropyl alcohol solutions for medicinal liniments, tinctures of green soap, scalp tonics, and tincture of mercurophen. It is contained in pharmaceuticals, eg, local anesthetics, tincture of iodine, and bathing solutions for surgical sutures and dressings. Over 200 uses of isopropyl alcohol have been tabulated (2).

BIBLIOGRAPHY

"Isopropyl Alcohol" in *ECT* 1st ed., Vol. 11, pp. 182–190, by J. G. Park and C. M. Beamer, Enjay Co., Inc.; in *ECT* 2nd ed., Vol. 16, pp. 564–578, by E. J. Wickson, Enjay Chemical Laboratory; "Propyl Alcohols (Isopropyl)" in *ECT* 3rd ed., Vol. 19, pp. 198–220, by A. J. Papa, Union Carbide Corp.

1. *Chem. Mark. Rep.* **244**(6), 45 (Aug. 9, 1993).

2. L. F. Hatch and W. R. Fenwick, *Isopropyl Alcohol*, Enjay Chemical Co., New York, 1966.

3. Technical data, Union Carbide Corp., New York, Sept. 14, 1976.

4. *UCAR Alcohols for Coatings Applications*, Brochure F-48588A, Union Carbide Corp., Danbury, Conn., July 1989.

5. *UCAR Performance Soluents Selection Guide for Coatings*, Bulletin F-7465, Union Carbide Corp., Danbury, Conn., Sep. 1990.

6. R. F. Brunel, J. L. Crenshaw, and R. Tobin, *J. Am. Chem. Soc.* **43**, 561 (1921).

7. Louis and Entezam, *Ann. Combust. Liquids*, **14**, 21 (1939).

8. *Design Institute for Physical Property Data*, DIPPR File, American Institute of Chemical Engineers, University Park, Pa., 1989.

9. L. H. Horsley, "Azeotropic Data-III," *Advances in Chemistry Series 116*, American Chemical Society, Washington, D.C., 1973.

10. U.S. Pat. 2,394,848 (Feb. 12, 1946), T. F. Doumani (to Union Oil Co.).

11. C. F. Brown, *Proceedings of the 116th Meeting of the American Chemical Society*, Atlantic City, N.J., 1949.

12. J. F. Norris and co-workers, *Rec. Trav. Chim.* **48**, 885 (1929).

13. Jpn. Pat. 42,11351 (June 26, 1967), (to Mitsubishi Chemical).

14. Brit. Pat. 868,023 (May 17, 1961), L. E. Addy (to British Hydrocarbon Chemicals).

15. Brit. Pat. 1,097,819 (Jan. 3, 1968), (to Usines de Melle).

16. U.S. Pat. 2,586,694 (Feb. 19, 1952), H. O. Mottern (to Standard Oil Development).

17. A. Corma and co-workers, *J. Catal.*, **148**(1), 205–212 (1994).

18. Czech. Pat. 241 425 (Feb. 1, 1988), Z. Hejda, R. Zidek, and J. Kozuch.

19. Czech. Pat. 234 604 (1987), J. Pasek, V. Pexidr, and R. Zidek, J. Hajek.

20. A. Gervasini and A. Auroux, *J. Catal.* **131**(1), 190–198 (1991).

21. C. Lahousse and co-workers, *J. Mol. Catal.* **87**(2,3), 392–432 (1994).

22. P. W. Sherwood, *Pet. Refiner*, **33**(12), 147 (1954).

23. S. S. Lokras, P. K. Deshpande, and N. R. Kuloor, *Ind. Eng. Chem. Prod. Des. Devel.* **9**(2), 293 (1970).

24. U.S. Pat 2,871,104 (Jan. 27, 1959), F. F. Rust (to Shell Development Co.).

25. U.S. Pat. 2,871,101 (Jan. 27, 1959), F. F. Rust and M. L. Porter (to Shell Development Co.).

26. *Hydrocarbon Process. Pet. Refiner*, **40**(11), 249 (1961).

27. *Eur. Chem. News*, **15**(400), 30 (1969).

28. Brit. Pat. 619,014 (Mar. 2, 1949), S. A. Ballard, H. de V. Finch, and E. A. Peterson (to NV Bataafsche Petroleum Maatschappy).

29. F. H. Kamal and co-workers, *Asian J. Chem.* **3**(1), 92–98 (1991).

30. M. Windholz, ed., *The Merck Index*, 9th ed., Merck and Co., Inc., Rahway, N.J., 1976.

31. U.S. Pat. 2,647,914 (Aug. 4, 1953), W. G. Allan and T. J. Tobin (to Imperial Chemical Industries, Ltd.).

32. Ger. Pat. 2,019,808 (Nov. 5, 1970), W. A. Craig and O. A. Gurten (to Imperial Chemical Industries, Ltd.).

33. M. Arvis and L. Gilles, *J. Chim. Phys. Phys.-Chim. Biol.* **67**(9), 1538 (1970).

34. T. L. Marker and co-workers, *Proceedings of AICHE National Meeting, Mar. 29–31, Houston, Tex.*, 1993.

35. U.S. Pat. 2,636,902 (Apr. 28, 1953), A. W. C. Taylor, P. Davies, and P. W. Reynolds (to Imperial Chemical Industries, Ltd.).

36. U.S. Pat 2,609,394 (Sept. 2, 1952), P. Davies and co-workers (to Imperial Chemical Industries, Ltd.).

37. P. H. Groggins, *Unit Processes in Organic Synthesis*, McGraw-Hill Book Co., Inc., New York, 1958, pp. 407, 434.

38. U.S. Pat. 2,349,222 (May 16, 1944), R. H. Goshorn (to Sharples Chemicals).

39. U.S. Pat. 2,365,721 (Dec. 26, 1944), J. Olin and J. McKenna (to Sharples Chemicals).

40. P. Richter and J. Pasek, *Chemicky Prumysl*, **17**(7), 353 (1967).

41. Brit. Pat. 649,980 (Feb. 7, 1951), W. Whitehead (to Imperial Chemical Industries, Ltd.).
42. S. Coffey, ed., *Rodd's Chemistry of Carbon Compounds*, 2nd ed., Elsevier Science Publishing Co., Inc., New York, 1965, pp. 114, 115.
43. M. R. A. Rao, *J. Indian Inst. Sci.* **39**, 138 (1957).
44. C. Matasa and E. Tonca, *Basic Nitrogen Compounds*, 3rd ed., Chemical Publishing Co., Inc., New York, 1973, p. 282.
45. G. T. Austin, *Chem. Eng.*, 101 (May 27, 1974).
46. Z. E. Zolles, *Bromine and Its Compounds*, Academic Press, Inc., New York, 1966, pp. 65, 387.
47. E. H. Huntress, *Organic Chlorine Compounds*, John Wiley & Sons, Inc., New York, 1948, p. 774.
48. U.S. Pat. 1,391,757 (Sept. 27, 1921), H. E. Buc (to Standard Oil Co.).
49. Rus. Pat. 211,528 (Feb. 18, 1968), E. V. Sergeev, T. V. Uvarova, and A. N. Smirnova.
50. Fr. Pat. 816,956 (Aug. 21, 1937), (to I. G. Farbenind AG).
51. Fr. Pat. 818,131 (May 26, 1937), (to I. G. Farbenind AG).
52. U.S. Pat. 3,325,545 (June 13, 1967), W. W. Levis, Jr. (to BASF Wyandotte Chemicals Corp.).
53. Ref. 41, p. 25.
54. U.S. Pat. 3,361,657 (Jan. 2, 1968), M. Kokorudz (to BASF Wyandotte Chemicals Corp.).
55. E. G. Hancock, *Propylene and Its Industrial Derivatives*, John Wiley & Sons, Inc., New York, 1973, p. 8.
56. A. H. Pelofsky, *Ind. Eng. Chem. Prod. Res. Develop.* **11**(2), (1972).
57. R. F. Robey, *Ind. Eng. Chem.* **33**(8), (1941).
58. H. Maarse and M. C. Ten Noever de Brauw, *Chem. Ind. (London)*, **1**, 36 (1974).
59. Brit. Pat. 2,004,538 (Apr. 4, 1979), C. Savini (to Exxon Research Engineering Co.).
60. U.S. Pat. 4,219,685 (Aug. 26, 1980), C. Savini (to Exxon Research Engineering Co.).
61. *Pet. Refiner*, **38**(11), 264 (1959).
62. *Hydrocarbon Process. Pet. Refiner*, **40**(11), 260 (1961).
63. T. Horie, M. Imaizumi, and Y. Fujiwara, *Hydrocarbon Process.* **49**(3), 119 (1970).
64. Jpn. Pat. 7 7012-166 (Apr. 5, 1977), T. Sato, R. Ohuji, and H. Yamanovchi (to Tokuyama Soda).
65. F. C. Fetter, *Chem. Eng.* **55**, 235 (Oct. 1948).
66. J. C. Fielding, in E. C. Hancock, ed., *Propylene and Its Industrial Derivatives*, John Wiley & Sons, Inc., New York, 1973.
67. K. Weissermel and H. J. Arpe, *Industrial Organic Chemistry*, Springer-Verlag, Weinheim, Austria, 1978.
68. Belg. Pat. 683,923 (Dec. 16, 1966), (to Hibernia-Chemie GmbH).
69. *Hydrocarbon Process.* **46**(11), 195 (1967).
70. U.S. Pat. 3,955,939 (May 11, 1976), A. Sommer and M. Urban (to Veba-Chemie AG).
71. Eur. Chem. News, **32** (July 24, 1970).
72. *Petroleum (London)*, **16**, 19 (1953).
73. W. Neier and J. Woellner, *Chemtech*, 95 (Feb. 1973).
74. *Hydrocarbon Process.* **58**(11), 181 (1979).
75. W. Neier and J. Woellner, *Erdoel Kohle*, **28**(1), 19 (1975).
76. W. Neier and J. Woellner, *Hydrocarbon Process.* **5**(11), 113 (1972).
77. *Hydrocarbon Process.* **52**(11), 141 (1973).
78. Y. Onoue and co-workers, *Chemtech*, 432 (July 1978).
79. Y. Onoue and Y. Izumi, *Chem. Econ. Eng. Rev.* **6**(7), 48 (1974).
80. U.S. Pat. 3,758,615 (Sep. 11, 1973), Y. Izumi, Y. Kawasaki, and M. Tani (to Tokuyama Soda).

81. *Eur. Chem. News*, 14 (July 25, 1975).

82. U.S. Pat. 3,758,615 (Sept. 11, 1973), Y. Izumi, Y. Kawasaki, and M. Tani (to Tokuyama Soda).

83. Brit. Pat. 622,937 (May 10, 1949), P. W. Reynolds and co-workers (to ICI).

84. Brit. Pat. 718,723 (Nov. 7, 1954), D. A. Dowden (to ICI).

85. U.S. Pat. 2,683,753 (July 13, 1954), N. Levy and co-workers (to ICI).

86. U.S. Pat. 2,725,403 (Nov. 29, 1955), M. A. E. Hodgson (to ICI).

87. U.S. Pat. 2,755,309 (July 17, 1956), P. W. Reynolds and co-workers (to ICI).

88. U.S. Pat. 2,504,618 (Apr. 18, 1950), R. C. Archibald and co-workers (to Shell Development Co.).

89. U.S. Pat. 2,579,601 (Dec. 25, 1951), R. C. Nelson and co-workers (to Shell Development Co.).

90. U.S. Pat. 2,658,924 (Nov. 10, 1953), S. J. Lukasiewicz and co-workers (to Socony-Vacuum Oil).

91. U.S. Pat. 2,663,744 (Dec. 22, 1953), S. J. Lukasiewicz (to Socony-Vacuum Oil).

92. Ger. Pat. 963,238 (May 2, 1957), C. Wagner (to Bayer).

93. U.S. Pat. 2,825,704 (Mar. 4, 1958), H. R. Arnold and co-workers (to E. I. du Pont de Nemours & Co., Inc.).

94. Brit. Pat. 750,176 (June 13, 1956), H. Newby (to Huels).

95. Brit. Pat. 996,917 (June 30, 1965), (to Gulf Research & Development).

96. Brit. Pat. 1,159,666 (July 30, 1969), (to Hibernia Chemie).

97. Brit. Pat. 1,159,667 (July 30, 1969), (to Hibernia Chemie).

98. Fr. Pat. 1,531,086 (July 17, 1968), (to Scholven-Chemie).

99. Jpn. Pat. 47-23524 (June 30, 1972), R. Ono, T. Sugirua, and K. Takemori (to Mitsui Toatsu Chemicals).

100. U.S. Pat. 3,256,250 (June 14, 1966), V. J. Frilette (to Socony Mobil Oil).

101. Brit. Pat. 1,238,556 (July 7, 1971), R. H. Scott and D. L. Gaulding (to Celanese Corp.).

102. Ger. Pat. 2,147,737 (Mar. 29, 1973), G. Brands and co-workers (to Deutsche Texaco).

103. Ger. Pat. 2,147,739 (Apr. 5, 1973), G. Brands and co-workers (to Deutsche Texaco).

104. Ger. Pat. 2,147,740 (Apr. 5, 1973), G. Brands and co-workers (to Deutsche Texaco).

105. Ger. Pat. 2,147,738 (Mar. 29, 1973), G. Brands and co-workers (to Deutsche Texaco).

106. U.S. Pat. 3,644,497 (Feb. 22, 1972), F. G. Mesick (to Celanese Corp.).

107. Jpn. Pat. 47-23523 (June 30, 1972), K. Tabe and I. Matsuzaki (to Mitsui Toatsu Chemicals).

108. U.S. Pat. 3,450,777 (June 17, 1969), Y. Mitzutani (to Tokuyama Soda).

109. Brit. Pat. 1,281,120 (July 12, 1972), Y. Izumi, M. Tani, and Y. Kawasaki (to Tokuyama Soda).

110. U.S. Pat. 3,705,912 (Dec. 12, 1972), S. N. Massie (to Universal Oil Products).

111. U.S. Pat. 4,214,107 (July 22, 1980), C. D. Chang and N. J. Morgan (to Mobil Oil Corp.).

112. H. U. Hammershaimb and co-workers, *Proceedings of National Conference on Refinery Process and Reformulated Gasoline*, San Antonio, Tex., 1993.

113. T. L. Marker and co-workers, *Proceedings of AIChE National Meeting*, Houston, Tex., 1993.

114. T. L. Marker and co-workers, *1994 Conference on Clean Air Act and Reformulated Gasolines*, Washington, D.C., 1994.

115. R. W. Taft, Jr., *J. Am. Chem. Soc.* **74**, 5372 (1952).

116. C. S. Cope, *J. Chem. Eng. Data*, **11**(3), 379 (1966).

117. F. M. Majewski and L. F. Marek, *Ind. Eng. Chem.* **30**, 203 (1938).

118. Can. Pat. 1,058,637 (July 17, 1979), B. W. Kiff and J. B. Saunby (to Union Carbide Corp.).

119. G. T. Austin, *Chem. Eng.* 101 (May 27, 1974).

120. *Chem. Age*, 15 (July 21, 1978).
121. *Eur. Chem. News*, 30 (July 21, 1978).
122. *Chem. Eng. News*, 28 (July 24, 1978).
123. *Chem. Eng. Prog.* 70 (Apr. 1978).
124. *Code of Federal Regulations*, Title 49, paragraph 172.101, U.S. Printing Office, Washington, D.C., Oct. 1, 1987.
125. *Eur. Chem. News*, (Mar. 10, 1972).
126. *Directory of Chemical Producers: United States*, SRI International, Menlo Park, Calif., 1993, p. 708.
127. *Directory of Chemical Producers: Western Europe*, SRI International, Menlo Park, Calif., 1993, p. 1165.
128. *Directory of Chemical Producers: East Asia*, SRI International, Menlo Park, Calif., 1993, pp. 625–625.
129. *Chem. Mark. Rep.* **248**(18), 29 (Oct. 30, 1995).
130. *Chem. Mark. Rep.* **244**(6), 45 (Aug. 9, 1993).
131. D. W. Hessel and F. R. Modglin, *J. Forensic Sci.* **9**, 255 (1964).
132. W. H. Simmons, *Perfum. Essen. Oil Rec.* **18**, 168 (1927).
133. M. Mantel and M. Anbar, *Anal. Chem.* **36**(4), 936 (1964).
134. A. M. Arjuna, *Anales Real Soc. Espan. Fis. Quim. (Madrid) Ser. B*, **61**(3), 591 (1965).
135. L. T. Fairhall, *Industrial Toxicology*, Williams & Wilkins, Co., Baltimore, Md., 1949, p. 248.
136. S. Zakhori and co-workers, in L. Golberg, ed., *Isopropanol and Ketones in the Environment*, CRC Press, Inc., Cleveland, Ohio, 1977.
137. *Criteria for a Recommended Standard: Occupational Exposure to Isopropyl Alcohol*, DEHW Pub. No., NIOSH, Washington, D.C., 1976, pp. 76–142.
138. *American Conference of Governmental Industrial Hygienists, Documentation of the Threshold Limit Values*, 5th ed., Cincinnati, Ohio, 1986, p. 337.
139. K. W. Nelson and co-workers, *J. Ind. Hyg. Toxicol.* **25**, 282 (1943).
140. G. D. Clayton and F. E. Clayton, *Patty's Industrial Hygiene and Toxicology*, Vol. II, John Wiley & Sons, Inc., New York, 1979, p. 853.
141. J. H. Ruth, Am. *Ind. Hyg. Assoc. J.* **47**(A), 142–151 (1986).
142. *Fed. Reg.* **54**(203), 43252–43264 (1989).
143. Technical data, Union Carbide Corp., Danbury, Conn.
144. V. Lechnitz, *Kosmet. Aerosole*, **44**, 65 (1971).
145. M. N. Gleason and co-workers, *Clinical Toxicology of Commercial Products*, Williams & Wilkins, Co., Baltimore, Md., 1969.
146. Ger. Pat. 2,239,690 (Feb. 22, 1973), D. Y. Hsiung (to Gillette Co.).
147. U.S. Pat. 3,650,956 (Mar. 21, 1972), D. L. Strand and R. L. Abler (to Minnesota Mining and Manufacturing Co.).
148. U.S. Pat. 3,231,397 (Jan. 25, 1966), A. Kessler, G. L. Layne, and C. L. Spector (to Proctor and Gamble Co.).
149. V. M. Tsetlin, I. V. Bessanova, and E. B. Zhuk, *Khim Promst. (Moscow)*, **47**, 31 (1971).
150. U.S. Pat. 3,244,502 (Apr. 5, 1966), S. M. Woogerd (to Hercules Glue Co.).
151. Ger. Pat. 1,935,939 (Feb. 4, 1971), H. Augart (to Goedecke).
152. U.S. Pat. 3,600,325 (Aug. 17, 1971), K. L. Kaufman, D. N. Martin, W. J. Brown (to CPC International Inc.).
153. M. John, *Hosp. Manage.* **57**, 86 (1944).
154. W. R. Straughn, *Mod. Hosp.* **66**, 90 (1946).

JOHN E. LOGSDON
RICHARD A. LOKE
Union Carbide Corporation

n-PROPYL ALCOHOL

n-Propyl alcohol [71-23-8], 1-propanol, $CH_3CH_2CH_2OH$, mol wt 60.09, is a clear, colorless liquid having a typical alcohol odor; it is miscible in water, ethyl ether, and alcohols. 1-Propanol occurs in nature in fusel oils and forms from fermentation and spoilage of vegetable matter (1).

Properties

A number of physical and chemical properties of 1-propanol are listed in Table 1 (2,3). The chemistry of 1-propanol is typical of low molecular weight primary

Table 1. Physical and Chemical Properties of 1-Propanol[a]

Property	Value
freezing point, °C	−126.2
boiling point, °C	97.20
vapor pressure, kPa[b]	
at 20°C	1.987
40°C	6.986
60°C	20.292
80°C	50.756
Antoine equation,[c] 2–120°C, *t* in °C	$\log P_{kPa} = 6.97257 - 1499.21/(204.64 - t)$
vapor density (air = 1)[d]	2.07
density, at 20°C, g/cm^3	0.80375
Francis equation, −21 to 180°C, *t* in °C	$density = 0.8813 + (5.448\,t \times 10^{-4}) -21.536/(313.09 - t)$
refractive index, n_D^{20}	1.38556
viscosity, at 20°C, mPa·s(=cP) [d]	2.256
surface tension, at 20°C, mN/m(=dyn/cm)	23.75
critical temperature, °C	263.65
critical pressure, kPa[b]	5169.60
critical density, g/cm^3	0.275
heat capacity, liquid at 25°C, J/(mol·K)[e]	141
heat of vaporization, kJ/mol[e]	
at 25°C	47.53
97.20°C	41.78
heat of combustion, liquid at 25°C, kJ/mol[d,e]	2033
heat of formation, vapor at 25°C, kJ/mol[d,e]	−254.7
flash point, Tag open cup, °C[d]	28.9
autoignition temperature, °C[d]	371.1
explosive limit, in air, vol %[d]	
lower	2.2
upper	14.0
electrical conductivity at 25°C, S (=mho)[d]	2×10^{-8}

[a]Ref. 2, except where noted.
[b]To convert kPa to mm Hg, multiply by 7.5.
[c]To convert $\log P_{kPa}$ to $\log P_{mm\,Hg}$, add 0.8751 to the constant.
[d]Ref. 3.
[e]To convert J to cal, divide by 4.184.

alcohols (see ALCOHOLS, HIGHER ALIPHATIC). Biologically, 1-propanol is easily degraded by activated sludge and is the easiest alcohol to degrade (4).

Manufacture

1-Propanol has been manufactured by hydroformylation of ethylene (qv) (see OXO PROCESS) followed by hydrogenation of propionaldehyde or propanal and as a by-product of vapor-phase oxidation of propane (see HYDROCARBON OXIDATION). Celanese operated the only commercial vapor-phase oxidation facility at Bishop, Texas. Since this facility was shut down in 1973 (5,6), hydroformylation or oxo technology has been the principal process for commercial manufacture of 1-propanol in the United States and Europe. Sasol in South Africa makes 1-propanol by Fischer-Tropsch chemistry (7). Some attempts have been made to hydrate propylene in an anti-Markovnikoff fashion to produce 1-propanol (8–10). However, these attempts have not been commercially successful.

Hydroformylation and Hydrogenation. The production of 1-propanol by hydroformylation or oxo technology is a two-step process in which ethylene is first hydroformylated to produce propanal. The resulting propanal is hydrogenated to 1-propanol (eqs. 1 and 2).

$$CH_2 = CH_2 + CO + H_2 \xrightarrow[\Delta, \text{ pressure}]{\text{catalyst}} CH_3CH_2CHO \qquad (1)$$

$$CH_3CH_2CHO + H_2 \xrightarrow[\Delta, \text{ pressure}]{\text{catalyst}} CH_3CH_2CH_2OH \qquad (2)$$

Propane, 1-propanol, and heavy ends (the last are made by aldol condensation) are minor by-products of the hydroformylation step. A number of transition-metal carbonyls (qv), eg, Co, Fe, Ni, Rh, and Ir, have been used to catalyze the oxo reaction, but cobalt and rhodium are the only economically practical choices. In the United States, Texas Eastman, Union Carbide, and Hoechst Celanese make 1-propanol by oxo technology (11). Texas Eastman, which had used conventional cobalt oxo technology with an $HCo(CO)_4$ catalyst, switched to a phosphine-modified Rh catalyst in 1989 (11) (see OXO PROCESS). In Europe, 1-propanol is made by Hoechst AG and BASF AG (12).

The rhodium–triphenylphosphine catalyst system is generally used instead of cobalt in oxo processes because the former gives higher reaction rates, greater stability, lower operation pressure, and lower by-product production. The use of rhodium carbonyl catalysts is a possible choice for companies that have idle high pressure cobalt facilities. Rhodium carbonyls require high pressure but are more active than cobalt carbonyls and result in less high boiling by-products. In Germany, Hoechst AG (Werk Ruhrchemie) uses rhodium carbonyls to make propanal, which is then hydrogenated to 1-propanol (13) (see HIGH PRESSURE TECHNOLOGY).

Simple rhodium carbonyls are ca 100–1000 times more reactive than cobalt carbonyls (14). When triphenylphosphine [603-35-0] is added, the reaction rate increases monotonically until the triphenylphosphine:rhodium mole ratio reaches 20:1 to 50:1 (15). Above this ratio, the rate of reaction decreases. A substantial amount of research on the rhodium oxo process has been directed toward improving the linear aldehyde selectivity and linear:branched (l:b) aldehyde ratio. Conditions which increase l:b-aldehyde ratios also increase oxo catalyst stability and thus decrease the ligand decomposition. In general, increasing triphenylphosphine concentration or decreasing CO partial pressure leads to higher selectivity to linear aldehyde and increases catalyst stability, but decreases the rate of hydroformylation. Because ethylene can lead neither to internal olefin nor to branched aldehyde, the triphenylphosphine:rhodium ratio that maximizes the rate would seem best for n-propyl alcohol production. However, catalyst stability problems dictate operation at higher triphenylphosphine:rhodium ratios. A rhodium-catalyzed, ethylene hydroformylation process operates at 90–120°C, 2.17–3.55 MPa (300–500 psig), 1:1–3:1 H_2:CO, 1–10 mM rhodium, and 0.1–0.4 M triphenylphosphine. The chemical efficiency to propanal under these conditions is 98–99%. Additionally, there is 0.5–1.0% efficiency to ethane, as well as 0.5–1.0% efficiency to heavy ends. The rhodium can be obtained from almost any source except those that contain halogen. Halogens form complexes that have very low hydroformation activity (16).

Rhodium-catalyzed hydroformylation has been studied extensively (16–29). The most active catalyst source is hydridocarbonyltris(triphenylphosphine)rhodium, $HRhCO[P(C_6H_5)_3]_3$ (30). However, a molecule of triphenylphosphine is presumed to dissociate to form the active species (21,28). Further dissociation could occur as shown in equation 3.

$$HRhCO[P(C_6H_5)_3]_3 \underset{P(C_6H_5)_3}{\overset{CO}{\rightleftharpoons}} HRh(CO)_2[P(C_6H_5)_3]_2 \underset{+P(C_6H_5)_3}{\overset{-P(C_6H_5)_3}{\rightleftharpoons}} HRh(CO)_2P(C_6H_5)_3 \quad (3)$$

Catalyst and ligand stability and linear aldehyde formation are favored in catalytic cycles which have more triphenylphosphine bound to Rh, eg, $HRh(CO)_2[P(C_6H_5)_3]_2$. It is probable that the first step in catalyst deactivation and ligand decomposition is phenyl migration from phosphorus to Rh (31,32). The resulting Rh–phosphide complex can either react with olefin to form alkyldiphenylphosphine (32), or with another Rh complex to form inactive Rh–phosphide clusters (30,31). Figure 1 illustrates possible pathways. Apparently phenyl migration is inhibited by large excesses of triphenylphosphine, but not by CO; ie, phenyl migration competes better with CO than with triphenylphosphine for coordination sites on Rh.

The technology for hydrogenating the propanal to 1-propanol is well established. Commercially, nickel-based catalysts, eg, Raney nickel or supported nickel, and copper chromium oxide catalysts, are used (33). Vapor-phase or liquid-phase hydrogenation can be carried out. The reactor can be a fixed-bed, slurry-bed, or trickle-bed system (34,35). Conditions for liquid-phase hydrogenation are 2.17–4.24 MPa (300–600 psig) and 100–170°C. Vapor-phase reactors generally operate below 790 kPa (100 psig) (36). Efficiencies to 1-propanol of >95% can be expected, and the principal by-products are acetals, ethers, esters, and diols (35). Both CO and triphenylphosphine poison the hydrogenation catalysts and should be removed prior to hydrogenation of propanal.

Figure 2 is a composite flow diagram for the production of 1-propanol by a two-step oxo/hydrogenation process where the oxo catalyst is rhodium–triphenylphosphine (37,38). The propanal is removed by vapor stripping using excess synthesis gas, CO + H_2 (15,38). After condensation, the propanal is sent to a CO-stripping column to remove traces of CO prior to hydrogenation. The crude 1-propanol (after hydrogenation) is purified in a standard two-tower

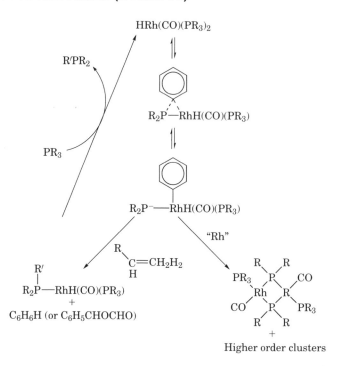

Fig. 1. Pathways for catalyst deactivation and ligand decomposition, where R = C_6H_5.

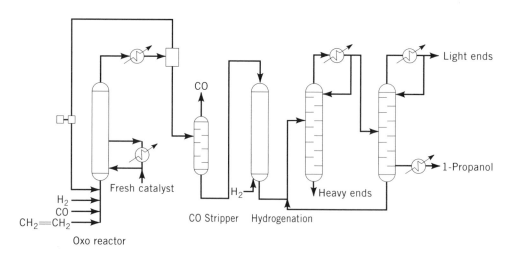

Fig. 2. Oxo/hydrogenation process for 1-propanol.

purification system. If $HCo(CO)_4$ or a rhodium carbonyl is the oxo catalyst, the propanal must be removed as a liquid stream because of the higher reactor pressure. Liquid recycle is necessary. Also, provisions for catalyst removal and handling must be installed.

Sasol Fischer-Tropsch Process. 1-Propanol is one of the products from Sasol's Fischer-Tropsch process (7). Coal (qv) is gasified in Lurgi reactors to produce synthesis gas (H_2/CO). After separation from gas liquids and purification, the synthesis gas is fed into the Sasol Synthol plant where it is entrained with a powdered iron-based catalyst within the fluid-bed reactors. The exothermic Fischer-Tropsch reaction produces a mixture of hydrocarbons (qv) and oxygenates. The condensation products from the process consist of hydrocarbon liquids and an aqueous stream that contains a mixture of ketones (qv) and alcohols. The ketones and alcohols are recovered and most of the alcohols are used for the blending of high octane gasoline. Some of the alcohol streams are further purified by distillation to yield pure 1-propanol and ethanol in a multiunit plant, which has a total capacity of 25,000–30,000 t/yr (see COAL CONVERSION PROCESSES, GASIFICATION).

Economic Aspects

There are six 1-propanol producers in the world, ie, Hoechst Celanese, Texas Eastman, and Union Carbide in the United States; BASF AG and Hoechst AG in Western Europe; and Sasol in South Africa. In 1993, worldwide production capacity for 1-propanol was in excess of 180,000 t/yr; actual production was approximately 110,000 t/yr (39). The December 1995 average delivered price for 1-propanol in the United States was $1144/t (40).

1-Propanol economics are sensitive to the raw material costs of ethylene (qv) and the feedstock for synthesis gas, ie, natural gas or liquid petroleum feedstocks (qv). Natural gas-based technology is slightly more economical.

Analytical and Test Methods, and Purity Specifications

The separation and analysis of 1-propanol are straightforward. Gas chromatography is the principal method employed. Other instrumental techniques, eg, nmr, ir, and classical organic qualitative analysis, are useful. Molecular sieves (qv) have been used to separate 1-propanol from ethanol and methanol. Commercial purification is accomplished by distillation (qv).

1-Propanol is a commodity chemical sold on specification (Table 2). Whereas the specification requires at least 99.8% 1-propanol, purity is generally in excess of 99.9%.

Table 2. Sales Specifications for 1-Propanol[a]

Specification	Limits[b]
1-propanol, wt %	99.8[c]
water, wt %	0.1
color, Pt–Co units	5
acidity, as acetic acid, wt %	0.003
2-methylpentanal, ppm	10
solubility in water	complete
odor	nonresidual

[a]Ref. 41.
[b]Values given are maximum unless otherwise noted.
[c]Value given is minimum.

Health and Safety Factors

1-Propanol is defined as a flammable liquid. Its flash point is below 38°C. OSHA defines 1-propanol as hazardous according to 29 CFR 1910.1200 for the purposes of communication to employees and employers (42). However, this alcohol is only slightly toxic to animals, as can be seen by the toxicity data given in Table 3 (43). 1-Propanol gives negative results in the Ames test and in the Mouse Lymphoma Forward Mutatation Assay (43–46). The National Toxicology Program (NTP) and International Agency for Research on Cancer (IARC) do not list 1-propanol as a carcinogen (43).

Table 3. 1-Propanol Toxicity[a]

Adminstration method	Value
oral dose, rats, LD_{50}, g/kg	1.9
dermal, rabbits, LD_{50}, g/kg	5.4
inhalation, rats, LC_{50}, ppm	24,000
HCGIH TLV, TWA	200
hazard ratings[b]	
NFPA[c]	1, 3, 0
HMIS[d]	2, 3, 0

[a]Ref. 43.
[b]Numbers are for health, flammability, and reactivity, respectively, with a range of 0 to 4.
[c]National Fire Protection Association.
[d]Hazardous Materials Identification System.

Eye contact can cause irritation or burns. Repeated skin contact can result in dermatitis. Exposure to excessive vapor concentrations irritates the eyes and respiratory tract. Very high concentrations have a narcotic effect (43).

In the United States, the reportable quantity of 1-propanol for spills under CERCLA "Superfund" is 100 lb/d (45.4 kg/d). However, no reportable quantity is assigned for transport (43). The substance is on the list for atmospheric standards, as defined in 40 CFR 60.489 (47). The intent of these standards is to require all newly constructed, modified, and reconstructed manufacturing units to use the best demonstrated system of continuous emission reduction for equipment leaks of volatile organic compounds (47). 1-Propanol is also on the right-to-know regulations of the states of Connecticut, Florida, Illinois, Louisiana, Massachusetts, New Jersey, Pennsylvania, and Rhode Island (43). 1-Propanol is allowed as a flavoring substance and adjuvants according to 21 CFR 172.515 (48), and is exempted from the requirement of tolerance when used as a solvent or cosolvent in pesticide formulations (49) (see FLAVORS AND SPICES; PESTICIDES).

Uses

1-Propanol is used mainly as a solvent and as a chemical intermediate. The largest uses are as a specialty solvent in flexographic printing inks (40%), particularly for printing on polyolefin and polyamide film (39). n-Propyl acetate

[*109-60-4*] (33%) is the predominant derivative of propanol. The ester is also used as a solvent in flexographic printing inks as well as a solvent in nitrocellulose lacquers, cellulose esters and ethers, waxes, and insecticide formulations. 1-Propanol improves drying rates, reduces foaming, and controls viscosity at constant solids levels in water-based printing inks. It is also used to produce *n*-propylamines that are herbicide intermediates (18%), glycol ethers (12%), and as a ruminant feed supplement (see FEEDS AND FEED ADDITIVES). 1-Propanol has minor uses as a flavor and fragrance in foods and as a solvent or cosolvent in pesticide formulations. Demand for 1-propanol is expected to grow by 1.8% between 1993 and 1998 (50).

BIBLIOGRAPHY

"*n*-Propyl Alcohol" in *ECT* 1st ed., Vol. 11, pp. 178–182, by M. J. Curry, Celanese Corp. of America; in *ECT* 2nd ed., Vol. 16, pp. 559–564, by J. J. Wocasek, Celanese Chemical Co.; "*n*-Propyl Alcohol" under "Propyl Alcohols" in *ECT* 3rd ed., Vol. 19, pp. 221–227, by J. D. Unruh and L. Spinicelli, Celanese Chemical Co., Inc.

1. J. A. Monick, *Alcohols, Their Chemistry, Properties and Manufacture*, Van Nostrand Reinhold, New York, 1968, pp. 117–119.
2. R. C. Wilhoit and B. J. Zwolinski, *J. Phys. Chem. Ref. Data 2*, (Suppl. 1), 1–66 (1973).
3. *n-Propanol Product Bulletin*, Hoechst Celanese Chemical Group, Dallas, Tex.
4. J. C. Buzzell, Jr. and co-workers, *Behavior of Organic Chemicals in the Aquatic Environment, Part III—Behavior in Aerobic Treatment Systems (Activated Sludge)*, Association of Manufacturing Chemists, Washington, D.C., 1969, pp. 26–31.
5. *Chemical Products Synopsis*, Mannsville Chemical Products, Mannsville, N.Y., Oct. 1977.
6. *Chem. Mark. Rep.* **3**,15 (Oct. 6, 1976).
7. B. Bryant, Sasolchem Ltd., private communication, June 7, 1995.
8. U.S. Pat. 2,830,091 (Apr. 8, 1958), B. Freedman and F. Marritz (to Sinclair Refining Co.).
9. Ger. Pat. 1,041,938 (Oct. 30, 1958), O. Bankowski and G. Hoffman (to VEB Leuna-Werke "Walter Ulbricht").
10. U.S. Pat. 2,873,290 (Feb. 10, 1959), D. Esmay and C. Johnson (to Standard Oil Co.).
11. *Chemical Economics Handbook*, Stanford Research Institute, Stanford, Calif., Feb. 1995, pp. 682.7000J–682.7000T.
12. *Ibid*, pp. 682.7002M–682.7002P.
13. Personal communication, Mar. 1994.
14. H. Wakamotsu, *Nippon Kagaku Zasshi* **85**, 257 (1964).
15. R. L. Olivier and F. B. Booth, *Hydrocarbon Process.*, 112 (1970).
16. J. A. Osborn, G. Wilkinson, and J. F. Young, *Chem. Commun.*, 17 (1965).
17. F. H. Jardine, J. A. Osborn, G. Wilkinson, and J. F. Young, *Chem. Ind. (London)*, 560 (1965).
18. J. A. Osborn, F. H. Jarine, J. F. Young, and G. Wilkinson, *J. Chem. Soc. A*, 1711 (1966).
19. P. S. Hallman, D. Evans, J. A. Osborn, and G. Wilkinson, *Chem Commun.*, 305 (1967).
20. D. Evans, G. Yagupsky, and G. Wilkinson, *J. Chem. Soc. A*, 2660 (1968).
21. D. Evans, J. A. Osborn, and G. Wilkinson, *J. Chem. Soc. A*, 3133 (1968).
22. M. C. Baird, J. T. Mague, J. A. Osborn, and G. Wilkinson, *J. Chem. Soc. A*, 1347 (1967).
23. M. C. Baird, C. J. Nyman, and G. Wilkinson, *J. Chem. Soc. A*, 348 (1968).
24. G. Wilkinson, *Bull. Soc. Chim. Fr.* **12**, 5055 (1968).

25. C. K. Brown and G. Wilkinson, *Tetrahedron Lett.* **22**, 1725 (1969).
26. M. Yagupsky, C. K. Brown, G. Yagupsky, and G. Wilkinson, *J. Chem. Soc. A*, 937 (1970).
27. G. Yagupsky, C. K. Brown, and G. Wilkinson, *J. Chem. Soc. A*, 1392 (1970).
28. C. K. Brown and G. Wilkinson, *J. Chem. Soc. A*, 2753 (1970).
29. Ger. Pat. 1,939,322 (Feb. 12, 1970), G. Wilkinson (to Johnson, Matthey and Co.).
30. W. R. Moser and C. J. Papile, *J. Organometal. Chem.* **41**, 293 (1987).
31. E. Billig and R. L. Pruett, *J. Organometal. Chem.* **192**, C49 (1980); J. D. Jamerson and co-workers, *J. Organometal. Chem.* **193**, C43 (1980).
32. A. G. Abatjoglou and co-workers, *Organometallics* **3**, 923 (1984).
33. H. Adkins, *Reactions of Hydrogen with Organic Compounds over Chromium Oxide and Nickel Catalysts*, University of Wisconsin Press, Madison, 1946.
34. U.S. Pat. 3,491,159 (Jan. 20, 1970), M. Reich and K. Schneider (to Chemische Werke Hüls).
35. Brit. Pat. 1,182,797 (Mar. 4, 1970), M. W. Fewlass and T. M. B. Wilson (to B. P. Chemicals Ltd.).
36. J. J. McKetta, *Encyclopedia of Chemical Processing and Design*, Vol. 5, Marcel Dekker, Inc., New York, 1977, pp. 393–394.
37. O. A. Hershmann and co-workers, *Ind. Eng. Chem. Prod. Res. Dev.* **8**, 372 (1969).
38. U.S. Pat. 4,247,486 (Jan. 27, 1981), E. A. V. Brewster and R. L. Pruett (to Union Carbide).
39. D. Pearson, personal research, July 1994.
40. *Chem. Mark. Rep.*, 31 (Jan. 1, 1996).
41. *n-Propanol Sales Specifications*, Hoechst Celanese Chemical Group, Dallas, Tex., June 1988.
42. *Code of Federal Regulations*, Title 29, Part 1910.1200, OSHA, Washington, D.C., rev. July 1, 1993.
43. *n-Propanol Material Safety Data Sheet*, Hoechst Celanese Chemical Group, Dallas Tex., Dec. 1994.
44. Celanese Chemical Co., Inc., *Mutagenicity Evaluation of n-Propyl Alcohol in the Ames Salmonella / Microsome Plate Test*, Litton Bionetics, Inc., Kensington, Md., Mar. 1978.
45. Celanese Chemical Co., Inc., *Mutagenicity Evaluation of n-Propyl Alcohol*, in *The Mouse Lymphoma Forward Mutilation Assay*, Litton Bionetics, Inc., Kensington, Md., Oct. 1978.
46. Celanese Corp., *Range Finding Toxicity Test on n-Propanol*, Industrial Hygiene Foundation of America, Inc., Pittsburgh, Pa., July 1962.
47. *Code of Federal Regulations*, Title 40, Part 60.489, U.S. EPA, Washington, D.C., July 1, 1989.
48. *Code of Federal Regulations*, Title 21, Part 172.5.5, U.S. FDA, Washington D.C., Apr. 1, 1988.
49. Ref. 46, Part 180.1001 (c and e), July 1, 1988.
50. *Chemical Economics Handbook*, Stanford Research Institute, Stanford, Calif., Feb. 1995, pp. 682.7001C–682.7001D.

J. D. UNRUH
D. PEARSON
Hoechst Celanese Corporation

PROPYLAMINES. See AMINES, LOWER ALIPHATIC AMINES.

PROPYLENE

Propylene [115-07-1], $CH_3CH{=}CH_2$, is perhaps the oldest petrochemical feedstock and is one of the principal light olefins (1) (see FEEDSTOCKS). It is used widely as an alkylation (qv) or polymer–gasoline feedstock for octane improvement (see GASOLINE AND OTHER MOTOR FUELS; OCTANE IMPROVERS (SUPPLEMENT)). In addition, large quantities of propylene are used in plastics as polypropylene, and in chemicals, eg, acrylonitrile (qv), propylene oxide (qv), 2-propanol, and cumene (qv) (see OLEFIN POLYMERS, POLYPROPYLENE; PROPYL ALCOHOLS). Propylene is produced primarily as a by-product of petroleum (qv) refining and of ethylene (qv) production by steam pyrolysis.

Physical Properties

Physical properties of propylene are listed in Table 1 (1).

Parameters for the van der Waals equation of state per mole of propylene are $a = 6.373$ and $b = -0.08272$ when P is in kPa, V in L, and $R = 8.314$

Table 1. Propylene Physical Properties[a]

Property	Value
mol wt	42.081
freezing temperature, K	87.9
bp, K	225.4
critical temperature, K	365.0
critical pressure, MPa[b]	4.6
critical volume, cm^3/mol	181.0
critical compressibility	0.275
Pitzer's acentric factor	0.148
liquid density, at 223 K, g/cm^3	0.612
dipole moment, 10^{-30} C·m[c]	1.3
standard enthalpy of formation, kJ/mol[d]	20.42
standard Gibbs energy of formation for ideal gas, at 101.3 kPa (=1 atm), kJ/mol[d]	62.72
heat of vaporization at bp, kJ/mol[d]	18.41
Lennard-Jones potential[e]	
T, nm	0.4678
ϵ_0/K, K	298.9
solubility, at 20°C, 101.3 kPa (1 atm), mL gas/100 mL solvent	
in water	44.6
in ethanol	1250
in acetic acid	524.5
refractive index, n_D	1.3567

[a]Ref. 2, unless otherwise noted.
[b]To convert MPa to atm, divide by 0.1013.
[c]To convert C·m to debye, divide by 3.336×10^{-30}.
[d]To convert J to cal, divide by 4.184.
[e]Ref. 3.

J/(mol·K); $a = 8.379$ when P is in atm.

$$(P + a/V^2)(V - b) = RT$$

Other pressure–volume–temperature (PVT) relationships may be found in the literature; ie, Benedict, Webb, Rubin equations of state (4–7); the Benedict, Webb, Rubin, Starling equation of state (8); the Redlich equation of state (9); and the Redlich-Kwong equation of state (10).

For the virial equation of state (11),

$$PV/RT = 1 + B(T)/V,$$

the following parameters are valid.

T, K	$B(T)$
280	-392 ± 5
300	-342 ± 5
320	-297 ± 5
340	-260 ± 5
380	-204 ± 5
420	-162 ± 5
460	-132 ± 5
500	-105 ± 5

The following relationship exists for liquid density ρ_l in g/cm^3 on the saturation line when temperature, T, is in K and is valid at 87.85–365.05 K:

$$\rho_l = AB^{-(1-T/T_c)^{2/7}} \tag{1}$$

where $A = 0.2252$, $B = 0.2686$, and T_c = critical temperature, 365.05 K (12).

Propylene is usually transported in the Gulf Coast as compressed liquid at pressures in excess of 6.9 MPa (1000 psi) and ambient temperatures. Compressed liquid propylene densities for metering purposes may be found in the *API Technical Data Book* (13). Another method (14–17) predicts densities within 0.25% and has a maximum error on average of only 0.83%.

For the vapor pressure, P_v, of propylene, equations 2 and 3 apply. Equation 2 is the Antoine equation where P_v is in kPa, T is in K, and $A = 5.94327$,

$$\log_{10} P_v = A + \frac{B}{T + C} \tag{2}$$

$B = 784.86$, and $C = 26.15$. If P_v is given in mm Hg, $A = 6.81837$ and B and C remain unchanged. The Antoine equation is valid for $T = 160–240$ K. For temperatures of 123–365 K, equation 3 should be used:

$$\log_{10} P_v = A + \frac{B}{T} + C \log T + DT \tag{3}$$

where P_v is in kPa and T is in K, $A = 34.752$, $B = -1725.5$, $C = -12.057$, and $D = 8.9948 \times 10^{-3}$ (12). If P_v is given in mm Hg, $A = 36.877$.

Ideal gas properties and other useful thermal properties of propylene are reported in Table 2. Experimental solubility data may be found in References 18 and 19. Extensive data on propylene solubility in water are available (20). Vapor–liquid–equilibrium (VLE) data for propylene are given in References 21–35 and correlations of VLE data are discussed in References 36–42. Henry's law constants are given in References 43–46. Equations for the transport properties of propylene are given in Table 3.

Chemistry

The chemistry of propylene is characterized both by the double bond and by the allylic hydrogen atoms. Propylene is the smallest stable unsaturated hydrocarbon molecule that exhibits low order symmetry, ie, only reflection along the main plane. This loss of symmetry, which implies the possibility of different types of chemical reactions, is also responsible for the existence of the propylene dipole moment of 0.35 D. Carbon atoms 1 and 2 have trigonal planar geometry identical

$$
\begin{array}{c}
\text{H} \\
| \\
{}^{2}\text{C} \quad \text{H} \\
\text{H} \diagdown \!\!\!\diagup\!\!\diagup \diagdown | \\
\text{C}^{1} \; {}^{3}\text{C}\!-\!\text{H} \\
| \quad\;\; | \\
\text{H} \quad \text{H}
\end{array}
$$

to that of ethylene. Generally, these carbons are not free to rotate, because of the double bond. Carbon atom 3 is tetrahedral, like methane, and is free to rotate. The hydrogen atoms attached to this carbon are allylic.

The propylene double bond consists of a σ-bond formed by two overlapping sp^2 orbitals, and a π-bond formed above and below the plane by the side overlap of two p orbitals. The π-bond is responsible for many of the reactions that are characteristic of alkenes. It serves as a source of electrons for electrophilic reactions such as addition reactions. Simple examples are the addition of hydrogen or a halogen, eg, chlorine:

$$\text{CH}_3\text{CH}\!=\!\text{CH}_2 + \text{H}_2 \xrightarrow{\text{catalyst}} \text{CH}_3\text{CH}_2\text{CH}_3 \qquad (4)$$

$$\text{CH}_3\text{CH}\!=\!\text{CH}_2 + \text{Cl}_2 \xrightarrow{\text{catalyst}} \text{CH}_3\text{CHClCH}_2\text{Cl} \qquad (5)$$

Reactions of alkenes are described in References 47 and 48 (see also OLEFINS, HIGHER; BUTYLENES).

The presence of allylic hydrogens in propylene often serves to distinguish its chemistry from that of ethylene (qv). For example, these hydrogens cause cross-linked, gummy materials to form when propylene polymerizes in the presence of peroxide initiators (49). The effect of the allyl hydrogens on propylene reactions can be explained by the stability of allyl radicals and allyl carbocations. When an allylic hydrogen is abstracted from propylene, the sp^3 hybridized carbon of

Table 2. Thermal Properties of Propylene[a]

Property	Equation[b]	Value of constants			
		A	B	C	D
Ideal gas					
heat capacity,[c] J/mol[d]	$C_p = A + BT + CT^2 + DT^3$	2.85	0.238	-1.2×10^{-4}	2.3×10^{-8}
heat of formation,[c] kJ/mol[d]	$\Delta H_f = A + BT + CT^2$	35.3	-5.77×10^{-2}	2.22×10^{-5}	
free energy of formation,[c] kJ/mol[d]	$\Delta G_f = A + BT$	75.3	1.75		
Liquid					
heat of vaporization,[e] kJ/kg[d]	$\Delta H_v = \Delta H_{v_i}((T_c - T)/(T_c - T_i))^n$	437.6[f]	225.45[g]	365.06[h]	0.38[i]
heat capacity,[j] J/(kg·K)[d]	$C_p = A + BT + CT^2 + DT^3$	1969	7.04	-7.04×10^{-2}	1.84×10^{-4}

[a] Ref. 12.
[b] Temperature in K.
[c] Equation valid for range 298–1500 K.
[d] To convert J to cal, divide by 4.184.
[e] Valid for range 87.85–365.05 K.
[f] Value is for ΔH_{v_i} in kJ/kg. To convert J to cal, divide by 4.184.
[g] Value is for T_i in K.
[h] Value is for T_c in K.
[i] Value is for n.
[j] Valid for range 88–373 K.

Table 3. Transport Properties of Propylene[a]

Property	Equation	Value at 25°C	A	B	C	D
		Gas, from −100 to 1000°C				
thermal conductivity, W/(m·K)[b]	$k_G = A + BT = CT^2 + DT^3$	1.84×10^{-2}	-7.577×10^{-3}	6.096	9.96×10^{-8}	3.94×10^{-11}
viscosity, μPa·s[c]	$\mu_G = A + BT + CT^2$	8.39×10^{-6}	-5.601×10^{-7}	31.88×10^{-9}	-62.91×10^{-13}	
		Liquid				
thermal conductivity,[d] W/(m·K)[b]	$k_L = A + BT + CT$	0.115^e	2.9×10^5	-6.05×10^{-2}	1.26×10^{-2}	
viscosity, mPa·s[c] (=cP)	$\log \mu_L = A + B/T + CT + DT^2$					
from −185.3 to 160°C			-27.8	1.096	2602	-863.5×10^6
from −160 to 91.9°C			-5.009	413.2	1.771	-30.92×10^6
surface tension,[f] mN/m (=dyn/cm)	$\sigma = n \cdot \sigma_l((T_c T)/(T_c - T_l))$		19.98^g	203^h	365.05^i	1.1797^j

[a]Ref. 11.
[b]To convert W/(m·K) to cal/(s·cm·K), divide by 418.4.
[c]To convert Pa·s to centipoise, multiply by 10^3.
[d]Range, −185 to 70°C.
[e]At 20°C.
[f]Range, −185.3 to 91.9°C.
[g]Value is for σ_l in mN/m.
[h]Value is for T_l in K.
[i]Value is for T_c in K.
[j]Value is for n.

253

the methyl group changes to sp^2. The p-orbital of this carbon can then overlap with the p-orbital involved in the π-bond, forming a new π-bond, where the bonding overlap is over all three carbon atoms. The electrons from the alkene π-bond and the free-radical electron are delocalized over the entire molecule. In molecular-orbital terms, the allyl radical can be represented as stabilized by overlap between the three adjacent atomic p-orbitals, forming a molecular orbital that extends over three atoms.

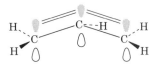

Resonance theory can also account for the stability of the allyl radical. For example, to form an ethylene radical from ethylene requires a bond dissociation energy of 410 kJ/mol (98 kcal/mol), whereas the bond dissociation energy to form an allyl radical from propylene requires 368 kJ/mol (88 kcal/mol). This difference results entirely from resonance stabilization. The electron spin resonance spectrum of the allyl radical shows three, not four, types of hydrogen signals. The infrared spectrum shows one type, not two, of carbon–carbon bonds. These data imply the existence, at least on the time scale probed, of a symmetric molecule. The two equivalent resonance structures for the allyl radical are as follows:

$$\overset{\bullet}{C}H_2 - CH = CH_2 \longleftrightarrow CH_2 = CH - \overset{\bullet}{C}H_2$$

The radical is much more stable if both structures exist. Quantum mechanical theory implies that the radical exists in both states separated by a small potential. Moreover, both molecular orbital theory and resonance theory show that the allyl carbocation is relatively stable.

$$\overset{+}{C}H_2 - CH = CH_2 \longleftrightarrow CH_2 = CH - \overset{+}{C}H_2$$

Polymerization Reactions.   Polymerization addition reactions are commercially the most important class of reactions for the propylene molecule and are covered in detail elsewhere (see OLEFIN POLYMERS, POLYPROPYLENE). Many types of gas- or liquid-phase catalysts are used for this purpose. Most recently, metallocene catalysts have been commercially employed (see METALLOCENES (SUPPLEMENT)). These latter catalysts require higher levels of propylene purity.

Electrophile Addition Reactions.   The addition of electrophilic (acidic) reagents HZ to propylene involves two steps. The first is the slow transfer of the hydrogen ion (proton) from one base to another, ie, from Z^- to the propylene double bond, to form a carbocation. The second is a rapid combination of the carbocation with the base, Z^-. The electrophile is not necessarily limited to a Lowry-Brønsted acid, which has a proton to transfer, but can be any electron-deficient molecule (Lewis acid).

Electrophilic addition requires an acidic reagent. The orientation of the addition is regioselective (Markovnikov) (47). This rule is followed even in the two-step halohydrin (halogen followed by water) formation reaction that goes through the halonium ion intermediate yielding halogen and hydroxyl on adjacent carbon atoms. The Markovnikov rule also explains why hydration of propylene yields isopropyl alcohol rather than *n*-propyl alcohol. The *n*-propyl alcohol is normally obtained through the oxo process (qv), ie, hydroformylation, which goes through an aldehyde intermediate, utilizing a catalyst to add carbon monoxide and hydrogen to ethylene (see PROPYL ALCOHOLS).

The presence of free radicals can invert this rule, to form anti-Markovnikov products. Free-radical addition in this fashion produces a radical on the central carbon, C-2, which is more stable than the allyl radical. This carbon can then experience further addition. For example, acid-catalyzed addition of hydrogen bromide to propylene yields 2-propylbromide; free-radical, Br•-catalyzed addition yields 1-propylbromide.

Substitution Reactions. Substitution reactions can occur on the methyl group by free-radical attack. The abstraction of an allylic hydrogen is the most favored reaction, followed by addition to that position.

Control of addition vs substitution by free radicals can be effected by the reaction conditions, ie, radical concentration, temperature, and phase. Using halogens as propylene reactants, high temperatures and the gas phase favor high radical concentrations and substitution reactions; cold, liquid-phase conditions favor addition reactions.

Manufacture

Steam Cracking. In steam cracking, a mixture of hydrocarbons (qv) and steam (qv) is preheated to ca 870 K in the convective section of a pyrolysis furnace. Then it is further heated in the radiant section to as much as 1170 K (see ETHYLENE; PETROLEUM, REFINERY PROCESSES, SURVEY). Steam reduces the hydrocarbon partial pressure in the reactor. The steam-to-hydrocarbon weight ratio is generally a function of the feedstock and ranges from ca 0.2 for ethane to ≥2.0 for gas oils (50). The amount of steam used is probably a compromise between yield structure (olefin selectivity), energy consumption, and furnace run length, which is limited by coking. The residence time in the radiant section varies from ca 1 s in older plants to as low as 0.1 s in some newer furnaces. The residence time influences olefin selectivity. Generally, selectivity to ethylene improves as residence time decreases. However, a given furnace is relatively inflexible to gross residence time changes for a specific feedstock because of hydrodynamic and heat-flux limitations.

In the radiant section, the hydrocarbon mixture undergoes reactions involving free radicals (51). These mechanisms have been generalized to include the molecular reactions shown below:

Chain initiation reactions : $\qquad$ R — R′ ⟶ R• + R′•

Hydrogen-abstraction reactions : $\qquad$ R• + R′H ⟶ RH + R′•

Radical-decomposition reactions : $\qquad$ R• ⟶ RH + R′•

Radical-addition reactions to unsaturated molecules : $\quad$ RH + R′• ⟶ R″•

Chain-termination reactions : $\qquad$ R• + R′• ⟶ R — R′

Molecular reactions : $RH + R'H \longrightarrow R''H + R'''H$

Radical-isomerization reactions : $R' \cdot \longrightarrow R'' \cdot$

The total number of reactions depends on the number of constituents present in the hydrocarbon feedstock. As many as 2000 reactions can occur simultaneously.

The constituents in the furnace effluent are the same for all hydrocarbon feedstocks. These include all hydrocarbons lighter than pentane plus heavier material, eg, gasoline and fuel oil. The proportion of these components in the effluent depends on the feedstock. For example, the furnace effluent of a plant designed for pure ethane cracking contains large amounts of hydrogen (qv), methane, unconverted ethane, and ethylene. Only small amounts of other constituents are produced. Conversely, a furnace cracking gas oil produces large amounts of gasoline and fuel oil in addition to significant quantities of useful olefins, eg, propylene (see GASOLINE AND OTHER MOTOR FUELS).

The separation train of the plant is designed to recover important constituents present in the furnace effluent. The modern olefin plant must be designed to accommodate various feedstocks, ie, it usually is designed for feedstock flexibility in both the pyrolysis furnaces and the separation system (52). For example, a plant may crack feedstocks ranging from ethane to naphtha or naphtha to gas oils.

The yield of propylene produced in a pyrolysis furnace is a function of the feedstock and the operating severity of the pyrolysis. Typical yields of propylene for various feedstock are available (see ETHYLENE). Under practical operating conditions, ethylene yield increases with increasing severity of feedstock conversion. Propylene yield passes through a maximum, as shown in Figure 1 (53). The economic optimum effluent composition for a furnace usually is beyond the propylene maximum. The furnace operation usually is dictated by computer optimization, ie, linear programming, of the whole plant, where an economic optimum for the plant is based on feedstock price, yield structures, energy considerations, and market conditions for the multitude of products obtained from the furnace. Thus, propylene produced by steam cracking varies according to economic conditions.

In an olefins plant separation train, propylene is obtained by distillation of a mixed C_3 stream, ie, propane, propylene, and minor components, in a C_3-splitter tower. Propylene is produced as the overhead distillation product, and

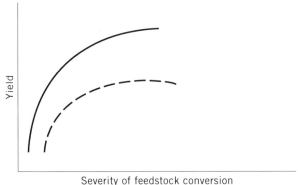

Fig. 1.   Ethylene (———) and propylene (— — —) yields.

the bottoms are a propane-rich stream. The size of the C_3-splitter depends on the purity of the propylene product. Two grades of propylene are commonly produced: a chemical grade, which consists of 92–94 wt % propylene, and a >99 wt %, polymer-grade propylene. Specifications for these two grades are listed in Tables 4 and 5 (54,55). A theoretically larger number of distillation trays is required to produce polymer-grade propylene than to produce the chemical grade, because of the close relative volatilities of propane and propylene.

Refinery Production. Refinery propylene is formed as a by-product of fluid catalytic cracking of gas oils and, to a far lesser extent, of thermal processes,

Table 4. Product Specification for Chemical-Grade Propylene

Component	Specification, wt %	ASTM Test method
	Constituents	
propylene	92–94	D2163
ethane and lighter	<0.4	D2163
ethylene	<0.02	D2723
hydrocarbons		
$\quad$ C_4 and above	<0.2	D2712
$\quad$ C_5 and above	<0.005	D2162
propane	<8.0	D2163
	Impurities[a]	
acetylene + methyl acetylene + propadiene	<100	D2712
total H_2, O_2, CO, CO_2, and N_2	<100	D2504
sulfur	<10	D3120
water	<50	D2713
halides	<10	
alcohols	<50	
amines	<5	
butadiene	<20	D2712
butenes	<125	
dimethyl formamide	0.3	

[a]Values are in ppm wt.

Table 5. Product Specifications for Polymer-Grade Propylene

Component, ppm wt	Specification	ASTM Test method
propylene	99.5[a]	D2163
ethane	<1000	D2163
total acetylenes, dienes, and other unsaturates	<10	D2712
hydrogen	<2	D2504
oxygen	<8	D2504
carbon monoxide	<4	D2504
water	<10	
sulfur	<5	D3120
total nitrogen, aldehydes, ketones, and alcohols	<10	D2504

[a]Value is wt %.

eg, coking. The total amount of propylene produced depends on the mix of these processes and the specific refinery product slate. For example, in the United States, refiners have maximized gasoline production. This results in a higher level of propylene production than in Europe, where proportionally more heating oil is produced.

In fluid catalytic cracking, a partially vaporized gas oil is contacted with zeolite catalyst (see FLUIDIZATION). Contact time varies from 5 s–2 min; pressure usually is in the range of 250–400 kPa (2.5–4 atm), depending on the design of the unit; reaction temperatures are 720–850 K (see BUTYLENES).

Converted feedstock forms gasoline-boiling-range hydrocarbons, C_4 and lighter gas, and coke. Propylene yield varies, depending on reaction conditions, but yields of 2–5% based on feedstock are common (56,57).

Two thermal-cracking processes, ie, delayed coking (58) and Exxon's proprietary process, Flexicoking or fluid coking (59), are used to convert residuum into more valuable products. In delayed coking, residuum and steam are heated in a furnace and then fed into an insulated drum where the free-radical decomposition of the feedstock takes place. Coke eventually fills the drum and must be removed. In fluid coking, a residuum feed is injected into a reactor, where it cracks thermally. Coke formed during the process deposits on other fluidized coke particles and is either removed or gasified. Both fluid and delayed coking occur at 300–600 kPa (3–6 atm). Delayed coking is a lower temperature process (720 vs 820 K) and thus should have lower total olefin yields than fluid coking.

Refinery propylene is recovered at the vapor-recovery unit. Refinery wet gas is passed through an absorber, where it contacts a hydrocarbon liquid, usually a heavy naphtha. The heavier molecular weight dissolves in the liquid, and the lighter material, eg, hydrogen and methane, passes through. The absorbent or rich oil is then passed to a stripper, where dissolved hydrocarbons are removed. The lean oil is recycled to the absorber, and the absorbate is passed to a depropanizer where a propane–propylene stream is taken as the overhead. This refinery-grade stream may require further treatment to remove acid gases, such as hydrogen sulfide, carbonyl sulfide, and carbon dioxide. A chemical- or polymer-grade propylene can be made by further distillation in a propylene concentration unit.

Advanced Cracking Techniques.   Technologies were developed to pyrolyze whole crude oil or various heavy petroleum fractions. These methods have in common very high temperatures, ultrashort residence times in the reactor zone, and rapid quench of the reaction products to minimize undesirable by-products. Among these techniques are the advanced cracking reactor from Union Carbide Corporation and thermal regenerative cracking from Gulf Oil, and Stone and Webster Engineering Corporation (60). These technologies are capital intensive, however, and have never been commercialized.

Synthetic Fuels.   Hydrocarbon liquids made from nonpetroleum sources can be used in steam crackers to produce olefins. Fischer-Tropsch liquids, oil-shale liquids, and coal-liquefaction products are examples (61) (see FUELS, SYNTHETIC). Work using Fischer-Tropsch catalysts indicates that olefins can be made directly from synthesis gas–carbon monoxide and hydrogen (62,63). Shape-selective molecular sieves (qv) also are being evaluated (64).

Catalytic Processes. Commercial interest in propane dehydrogenation has been increasing and numerous plants are being built outside the United States as of the mid-1990s. Several plants are being operated for various strategic reasons. There are four technologies that can be licensed for propane dehydrogenation. These are CATOFIN from ABB Lummus, OLEFLEX from UOP, Fluidized Bed Dehydrogenation (FBD) from Snamprogetti, and Steam Active Reforming (STAR) from Phillips Petroleum. These routes differ primarily in the type of catalyst, the reactor design, and the methods used to increase the conversion, eg, the operating pressure, use of diluents, and reaction temperatures.

The CATOFIN process uses a relatively inexpensive and durable chromium oxide–alumina as catalyst. This catalyst can be easily and rapidly regenerated under severe conditions without loss in activity. To improve propylene selectivity and to increase the propane conversion, this technology uses a vacuum. Dehydrogenation is carried out in the gas phase over fixed beds. Because the catalyst cokes up rapidly, five reactors are typically used. Two are on-stream, while two are being regenerated and one is being purged. The reactors are cycled between the reaction and the reheat/regeneration modes, and the thermal inertia of the catalyst controls the cycle time, which is typically less than 10 minutes (65). This technology was first commercialized in 1986 for the production of isobutylene. The first plant to use this technology for propylene production having a capacity of 245,000 metric tons was built at Antwerp by North Sea Petrochemicals, a joint venture between Himont and Statoil (66,67). As of this writing, Pemex is building a 340,000-t/yr plant in Morelos, Mexico (68,69).

The OLEFLEX process uses multiple side-by-side, radial flow, moving-bed reactors connected in series. The heat of reaction is supplied by preheated feed and interstage heaters. The gas-phase reaction is carried out over a catalyst, platinum supported over alumina, under very near isothermal conditions. The first commercial installation of this technology, having an annual capacity of 100,000 t, was made in 1990 by the National Petrochemical Corporation in Thailand. A second unit, at 245,000 t capacity, has been built in South Korea by the ISU Chemical Company (70).

Fluidized bed dehydrogenation technology is more prevalent in the former Soviet Union. A continuous fluidized-bed reactor system is used with a chromium on alumina catalyst. The catalyst is recirculated from the reactor to the regeneration section on a 30–60-min cycle. This process resembles conventional fluidized catalytic cracking technology used in the oil industry. The process operates under low pressure and has a low pressure drop and uniform temperature profile (70).

Steam-activated reforming technology uses a noble metal-promoted zinc aluminate spinel catalyst in a fixed-bed reactor. The reaction is carried out in the gas phase and uses multiple fired tube reactors in parallel operating under isothermal conditions and at pressures of 344–415 kPa (50–60 psig). Steam is added to the hydrocarbon feed to provide heat to the endothermic reaction, to suppress coke formation, and to increase the equilibrium conversion by lowering partial pressures of hydrogen and propane. Although Phillips has operated a prototype unit at their research and development center, there are no commercial licenses for this technology.

Because propane dehydrogenation is equilibrium-limited and per-pass propylene yield is low, the effluent compression and product purification sections account for nearly 85% of total capital required. Therefore, improvements in the separation section represent the greatest potential for cost reduction. Research efforts are being directed at developing a low cost route to olefins–paraffins separation (71). Concurrent research is being carried out to remove hydrogen *in situ* in a permeable ceramic reactor. Efforts are also directed at developing high temperature catalytic membrane reactors containing palladium and its alloys in the pores (72,73).

Economic Aspects

Production estimates for propylene can only be approximated. Refinery propylene may be diverted captively to fuel or gasoline uses whenever recovery is uneconomic. Steam-cracker propylene production varies with feedstock and operating conditions. Moreover, because propylene is a by-product, production rates depend on gasoline and ethylene demand.

Worldwide propylene production and capacity utilization for 1992 are given in Table 6 (74). The world capacity to produce propylene reached 41.5×10^6 t in 1992; the demand for propylene amounted to 32.3×10^6 t. About 80% of propylene produced worldwide was derived from steam crackers; the balance came from refinery operations and propylene dehydrogenation. The manufacture of polypropylene, a thermoplastic resin, accounted for about 45% of the total demand. Demand for other uses included manufacture of acrylonitrile (qv), oxochemicals, propylene oxide (qv), cumene (qv), isopropyl alcohol (see PROPYL ALCOHOLS), and polygas chemicals. Each of these markets accounted for about 5–15% of the propylene demand in 1992 (Table 7).

Table 6. Worldwide Propylene Capacity and Production in 1992

Geographic location	Nameplate capacity, %	Capacity used, %
North America	33.5	77
western Europe	29.4	82
Japan	11.7	94
others	25.4	67

Table 7. Worldwide Consumption of Propylene in 1992, 10^3 t[a]

Commodity	North America	Western Europe	Japan	Other areas
polypropylene	5,561	5,856	2,622	6,008
acrylonitrile	2,085	1,464	777	1,054
oxo chemicals	973	1,220	388	1,160
propylene oxide	1,390	1,220	291	422
cumene	1,112	854	243	738
isopropyl alcohol	695	610	97	210
polygas chemicals	973	366	97	316
others[b]	1,112	610	340	632
Total	*13,901*	*12,200*	*4,855*	*10,540*

[a]Ref. 74.
[b]Includes acrylic acid, allyl chloride, ethylene–propylene elastomers, and miscellaneous uses.

Although propylene price and ethylene price usually fluctuate together, temporary supply/demand imbalance could cause the price of propylene to rise well above that of ethylene. In early 1995 propylene sold for $800/t on the spot market vs $600/t for ethylene. Thus the propylene-to-ethylene price ratio is important. If the two prices are similar, steam-cracker operators in the United States tend to choose feedstocks that produce more propylene. When propylene price is significantly lower than ethylene price, propylene-derived products are substituted for ethylene derivatives in the areas where applications overlap. Producers have the option of diverting propylene to fuel use or cracking feed. The market price of propylene for chemical uses and gasoline blending must exceed its value as fuel or feedstock for steam cracking.

World propylene demand was expected to increase 4.5%/yr during the 1992–1997 period. The most dramatic growth was expected to be in the Middle East, where construction of several world-scale polypropylene plants was expected to increase propylene consumption 30-fold from the 1992 level by 1997. Other areas, notably Africa, Japan, South America, and eastern Europe, were expected to show vigorous (7–12% range) growth (74). Propylene growth in North America, western Europe, and Japan, however, was expected to be sluggish, at a 2–3% rate. Projected worldwide capacity to produce propylene is expected to reach 49.4×10^6 t by 1997. Construction of splitters in refineries and propane dehydrogenation facilities should bring about some of the propylene capacity growth. The bulk of the additional capacity should come from the expansion of steam crackers and from grassroots ethylene plants.

Storage and Handling

Precautions must be taken to avoid health and fire hazards wherever propylene is handled (75). Equipment capable of causing ignition should be shut down while connecting, disconnecting, loading, and unloading equipment (76). Electrical installations in unloading areas should be classified under Division II requirements of the National Electrical Code (77). No part of any cylinder containing propylene should be subjected to temperatures above 325 K.

Propylene is very volatile and is usually stored as a liquid under pressure. However, it can be stored safely at ambient temperature in approved containers. Storage tanks should be of welded-steel construction in accordance with the ASME Code for Unfired Pressure Vessels (78). All piping and related equipment should be steel and conform to the piping codes of ANSI (79). Waste mixtures containing propylene should not be allowed to enter drains or sewers where the danger of vapor ignition exists. Commercial quantities of propylene are shipped by tanker or pipeline and are stored aboveground in pressure vessels or underground in brine caverns.

Health and Safety Factors

Propylene is a colorless gas under normal conditions, has anesthetic properties at high concentrations, and can cause asphyxiation. It does not irritate the eyes and its odor is characteristic of olefins. Propylene is a flammable gas under normal atmospheric conditions. Vapor-cloud formation from liquid or vapor leaks is the main hazard that can lead to explosion. The autoignition temperature is 731 K in air and 696 K in oxygen (80). Evaporation of liquid propylene can cause skin burns. Propylene also reacts vigorously with oxidizing materials. Under unusual conditions, eg, 96.8 MPa (995 atm) and 600 K, it explodes. It reacts violently with NO_2, N_2O_4, and N_2O (81). Explosions have been reported when liquid propylene contacts water at 315–348 K (82). Table 8 shows the ratio T_w/T_{sl}, where T_w is the initial water temperature, and T_{sl} is the superheat limit temperature of the hydrocarbon.

Table 8. T_w/T_{sl} Ratios for Liquid Propylene Spills in Water[a]

Initial water temperature, T_w, K	T_w/T_{sl}[b]	Result
311–314	0.95–0.96	ice formation
315–348	0.97–1.07	explosion
353–358	1.08–1.10	rapid pops

[a]Ref. 82.
[b]T_{sl} = superheat limit temperature of propylene.

Uses

Propylene has many commercial and potential uses. The actual utilization of a particular propylene supply depends not only on the relative economics of the petrochemicals and the value of propylene in various uses, but also on the location of the supply and the form in which the propylene is available. For example, economics dictate that recovery of high purity propylene for polymerization from a small-volume, dilute off-gas stream is not feasible, whereas polymer-grade propylene is routinely recovered from large refineries and olefins steam crackers. A synthetic fuels project located in the western United States might use propylene as fuel rather than recover it for petrochemical use; a plant on the Gulf Coast would recover it (see FUELS, SYNTHETIC).

The uses of propylene may be loosely categorized as refinery or chemical purpose. In the refinery, propylene occurs in varying concentrations in fuel-gas streams. As a refinery feedstock, propylene is alkylated by isobutane or dimerized to produce polymer gasoline for gasoline blending. Commercial chemical derivatives include polypropylene, acrylonitrile, propylene oxide, isopropyl alcohol, and others. In 1992, ca 64% of U.S. propylene supplies were consumed in the production of chemicals (74). Polypropylene has been the largest consumer of propylene since the early 1970s and is likely to dominate propylene utilization for some time.

Refinery. *Fuel.* Propylene has a net heating value of 45.8 MJ/kg (19,700 Btu/lb) and is often contained in refinery fuel-gas streams. However, propylene is diverted from streams for refinery fuel use in large quantities only when economics for other uses are unfavorable, or equipment for propylene recovery does not exist or is limited in capacity. Propylene is also contained in liquid petroleum gas (LPG), but is limited to a maximum concentration of 5 vol % in certain grades (83) (see LIQUEFIED PETROLEUM GAS).

Alkylate. In petroleum refining, alkylation (qv) is the reaction of light olefins with isoparaffins, principally isobutane, to produce isoparaffins of higher molecular weight for gasoline blending. Alkylation produces material of higher octane rating than polymerization does, and alkylation predominated as gasoline-pool octane requirements increased after World War II. Although butenes are the preferred olefins for alkylation, propylene is also used, depending on the availability of butenes and alkylate demand. The basic alkylation reactions are acid-catalyzed (84). Propylene alkylation requires higher temperatures and acid strengths than alkylation of butenes, and butenes are good promoters of propylene alkylation (85). The principal isomer of propylene alkylation by isobutane is 2,3-dimethylpentane as compared to the higher octane quality 2,2,4-, 2,3,3-, and 2,3,4-trimethylpentanes obtained from butenes (see ALKYLATION). Commercial alkylation units are described in Reference 86.

Polymer Gasoline. Polymerization of butenes and propylene in the refinery, first through thermal, ie, free-radical, reactions and later by acid-catalyzed mechanisms, was the first successful process for upgrading light olefinic gases, which are coproduced in petroleum-cracking processes. Although alkylation gained in importance as gasoline-pool octane requirements increased, polymerization has been the subject of new interest as isobutane supplies have decreased (87) (see GASOLINE AND OTHER MOTOR FUELS).

One of the newest forms of polymerization for gasoline is the Dimersol process (72). Refinery propylene, ie, 67 wt % propylene–33 wt % propane, reacts in the liquid phase with a nickel coordination complex and aluminum alkyl catalyst at ca 330 K and 1725 kPa (250 psi) (88). Propylene conversions are 90–97%. The heat of reaction is removed by circulation through an external cooler. Ammonia and water are injected into the reactor effluent to neutralize the catalyst, hydrocarbon and aqueous phases are separated, and the hydrocarbon is fractionated to produce LPG, which contains unreacted propane and propylene, as well as dimerization products consisting mainly of isohexenes. The process can also be based on butane–butene feeds. As a gasoline blending component, the Dimersol product may have clear octane ratings of 89, which is somewhat higher than that of traditional propylene polymer gasoline, ie, 87 (83,84).

Chemical. *Propylene Oligomers.* Through acid-catalyzed Friedel-Crafts processes similar to those in refinery alkylation and polymerization, propylene forms oligomers, eg, nonenes, dodecenes, and higher molecular weight olefins and viscous polypropenes (89) (see FRIEDEL-CRAFTS REACTIONS). These materials are used in the production of alkyl–aryl sulfonate detergents and motor oil additives (90,91).

Polypropylene. One of the most important applications of propylene is as a monomer for the production of polypropylene. Propylene is polymerized by Ziegler-Natta coordination catalysts (92,93). Polymerization is carried out either in the liquid phase where the polymer forms a slurry of particles, or in the gas phase where the polymer forms dry solid particles. Propylene polymerization is an exothermic reaction (94).

$$n\ C_3H_6\ (g) \longrightarrow (C_3H_6)_n \quad \Delta H_{298} = -104\ \text{kJ/mol}\ (-24.89\ \text{kcal/mol}) \qquad (6)$$

$$n\ C_3H_6\ (l) \longrightarrow (C_3H_6)_n \quad \Delta H_{298} = -89.1\ \text{kJ/mol}\ (-21.30\ \text{kcal/mol}) \qquad (7)$$

The focus of commercial research as of the mid-1990s is on catalysts that give desired and tailored polymer properties for improved processing. Development of metallocene catalyst systems is an example. Exxon, Dow, and Union Carbide are carrying out extensive research on this catalyst system for the production of polyethylene and polypropylene.

Most commercial processes produce polypropylene by a liquid-phase slurry process. Hexane or heptane are the most commonly used diluents. However, there are a few examples in which liquid propylene is used as the diluent. The leading companies involved in propylene processes are Amoco Chemicals (Standard Oil, Indiana), El Paso (formerly Dart Industries), Exxon Chemical, Hercules, Hoechst, ICI, Mitsubishi Chemical Industries, Mitsubishi Petrochemical, Mitsui Petrochemical, Mitsui Toatsu, Montedison, Phillips Petroleum, Shell, Solvay, and Sumimoto Chemical. Eastman Kodak has developed and commercialized a liquid-phase solution process. BASF has developed and commercialized a gas-phase process, and Amoco has developed a vapor-phase polymerization process that has been in commercial operation since early 1980.

As of 1995, Union Carbide was the leading licensor of gas-phase technology, acquiring Shell's polypropylene business. Montel, a joint venture between Royal Dutch-Shell and Himont, has been formed.

Polypropylene is used in battery cases and in the replacement of metal parts in automobiles. It is also widely used in consumer products, eg, kitchen wares, trays, toys, and packaging materials. Its future applications are expected to include an increased portion of the fibers and filaments markets, especially for continued growth in carpet backing and carpet face yarns. Film, both oriented and unoriented, is also expected to be a significant growth market for polypropylene.

Acrylonitrile.   Catalytic oxidation of propylene in the presence of ammonia (qv) yields acrylonitrile (95).

$$CH_2{=}CHCH_3 + NH_3 + \tfrac{3}{2} O_2 \longrightarrow CH_2{=}CHCN + 3 H_2O \tag{8}$$

Yields based on propylene are 50–75%, and the main by-products are acetonitrile and hydrogen cyanide (96).

Propylene Oxide.   Propylene oxide is produced from propylene by two main processes. The first is chlorohydrination of propylene at ca 310 K, followed by epoxidation (qv) of the chlorohydrin by calcium hydroxide.

$$CH_3CH{=}CH_2 + HOCl \longrightarrow CH_3CHOHCH_2Cl \tag{9}$$

$$2\,CH_3CHOHCH_2Cl + Ca(OH)_2 \longrightarrow 2\,CH_3CH\overset{O}{\overset{}{-}}CH_2 + CaCl_2 + 2\,H_2O \tag{10}$$

The second process involves reaction of propylene with peroxides, as in the Oxirane process (97), in which either isobutane or ethylbenzene is oxidized to form a hydroperoxide.

$$\begin{array}{c}\underset{\underset{CH_3}{|}}{\overset{CH_3}{|}}{CH} + O_2 \longrightarrow \underset{H_3C}{\overset{CH_3}{>}}C\underset{CH_3}{\overset{OOH}{<}} \text{ or } \bigcirc\!\!\!\!\overset{CH_2^{\nearrow CH_3}}{} + O_2 \longrightarrow \underset{HOO}{\overset{CH_3}{>}}CH\bigcirc \end{array} \qquad (11)$$

The hydroperoxide reacts with propylene, forming propylene oxide and an alcohol.

$$\underset{H_3C}{\overset{CH_3}{>}}C\underset{CH_3}{\overset{OOH}{<}} + H_3CCH{=}CH_2 \longrightarrow H_2C\overset{}{\underset{O}{\triangle}}CHCH_3 + \underset{H_3C}{\overset{CH_3}{>}}C\underset{CH_3}{\overset{OH}{<}}$$

$$\underset{HOO}{\overset{CH_3}{>}}CH\bigcirc + H_3CCH{=}CH_2 \longrightarrow H_2C\overset{}{\underset{O}{\triangle}}CHCH_3 + \underset{HO}{\overset{CH_3}{>}}CH\bigcirc \qquad (12)$$

tert-Butyl alcohol can be dehydrated to form isobutylene, and methylphenyl carbinol can be dehydrated to form styrene (qv). Thus, in either reaction sequence a by-product of significant value is obtained. Hydroperoxide formation occurs under mild conditions (400 K, 400–3500 kPa (4.3–34.5 atm)) to minimize decomposition to the alcohol (98). Epoxidation of the propylene occurs in the liquid phase under similar conditions in the presence of a catalyst containing Mo, V, or Ti. Conversions are ca 10–20% for propylene and greater than 95% for the hydroperoxide. Both Shell Netherlands and Atlantic Richfield/Halcon have proprietary technology for this process. One other commercialized process involving hydroperoxides is the Daicel process (99). Peracetic acid is produced by oxidation of acetaldehyde in solution in the presence of metal ion catalyst at 300–320 K and 2.53–4.05 MPa (25–40 atm) (see PEROXIDES AND PEROXIDE COMPOUNDS, ORGANIC).

$$\underset{}{\overset{O}{\underset{\|}{CH_3CHH}}} + O_2 \longrightarrow \underset{}{\overset{O}{\underset{\|}{CH_3COOH}}} \qquad (13)$$

Peracetic acid then reacts with propylene at 320–350 K and 1.00–1.34 MPa (10–13.2 atm), forming propylene oxide and acetic acid.

$$\underset{}{\overset{O}{\underset{\|}{CH_3COOH}}} + H_3CCH{=}CH_2 \longrightarrow H_2C\overset{}{\underset{O}{\triangle}}CHCH_3 + \underset{}{\overset{O}{\underset{\|}{CH_3COH}}} \qquad (14)$$

Propylene oxide (qv) uses include manufacture of polyurethanes, unsaturated polyester, propylene glycols (qv) and polyethers, and propanolamines (see

ALKANOLAMINES; GLYCOLS; POLYETHERS; POLYESTERS, UNSATURATED; URETHANE POLYMERS).

Isopropyl Alcohol.   Propylene may be easily hydrolyzed to isopropyl alcohol. Early commercial processes involved the use of sulfuric acid in an indirect process (100). The disadvantage was the need to reconcentrate the sulfuric acid after hydrolysis. Direct catalytic hydration of propylene to 2-propanol followed commercialization of the sulfuric acid process and eliminated the need for acid reconcentration, thus reducing corrosion problems, energy use, and air pollution by SO_2 and organic sulfur compounds. Gas-phase hydration takes place over supported oxides of tungsten at 540 K and 25 MPa (247 atm) or over supported phosphoric acid at 450–540 K and 2.5–6.5 MPa (25–64 atm) (100).

$$CH_3CH{=}CH_2 + H_2O \xrightleftharpoons{\text{catalyst}} \overset{\overset{\displaystyle OH}{|}}{CH_3CHCH_3} \qquad (15)$$

At conditions of high temperature and low pressure, for sufficient catalyst activity and acceptable reaction rates, equilibrium conversions may be as low as 5%, necessitating recycle of large amounts of unreacted propylene (101).

Conversions of ca 75% are obtained for propylene hydration over cation-exchange resins in a trickle-bed reactor (102). Excess liquid water and gaseous propylene are fed concurrently into a downflow, fixed-bed reactor at 400 K and 3.0–10.0 MPa (30–100 atm). Selectivity to isopropanol is ca 92%, and the product alcohol is recovered by azeotropic distillation with benzene.

A third catalytic route to isopropyl alcohol from propylene involves the use of polytungsten compounds in solution in a liquid-phase reactor (101). Propylene is hydrated at ca 540 K under pressure. Conversions are 60–70% and selectivity to 2-propanol is 99%.

Cumene.   Cumene (qv) is produced by Friedel-Crafts alkylation of benzene by propylene (103,104). The main application of cumene is the production of phenol (qv) and by-product acetone (qv). Minor amounts are used in gasoline blending (105).

Butyraldehydes.   Normal and isobutyraldehydes are produced from propylene by the oxo or hydroformylation process (see OXO PROCESS).

$$CH_3CH{=}CH_2 \xrightarrow[\text{catalyst}]{H_2/CO} \overset{\overset{\displaystyle O}{\|}}{CH_3CH_2CH_2CH} + \underset{\underset{\displaystyle CH_3}{|}}{\overset{\overset{\displaystyle O}{\|}}{CH_3CHCH}} \qquad (16)$$

The two main industrial processes that are employed are described in Reference 106. Normal butyraldehyde (qv) is the product of primary interest (103,107).

Other.   Ethylene can be produced by steam cocracking of propylene with ethane and propane. Ethylene and butenes can also be produced by catalytic disproportionation of propylene (108).

$$2\,CH_2{=}CHCH_3 \rightleftharpoons CH_2{=}CH_2 + CH_3CH{=}CHCH_3 \qquad (17)$$

2-Butene is the main C-4 olefin isomer.

Acrolein can be obtained by propylene oxidation in a process similar to ammoxidation (109) (see ACROLEIN AND DERIVATIVES).

$$CH_2=CHCH_3 \xrightarrow{O_2} CH_2=CHCHO + H_2O \tag{18}$$

High temperature chlorination of propylene yields allyl chloride, which is used in glycerol (qv) production (110):

$$CH_2=CHCH_3 + Cl_2 \xrightarrow[770\ K]{} CH_2=CHCH_2Cl + HCl \tag{19}$$

Allyl acetate can be obtained by the vapor-phase reaction of propylene and acetic acid over a supported Pd catalyst (eq. 20) (110). Reaction of acrylic acid and propylene yields isopropyl acrylate (eq. 21), and catalytic reaction with acetic acid produces isopropyl acetate (eq. 22) (110).

$$CH_2=CHCH_3 + CH_3COOH \xrightarrow{O_2} CH_2=CHCH_2O\overset{\displaystyle O}{\overset{\|}{C}}CH_3 \tag{20}$$

$$CH_2=CHCH_3 + CH_2=CHCOOH \xrightarrow{H^+} CH_2=CH\overset{\displaystyle O}{\overset{\|}{C}}OCH(CH_3)_2 \tag{21}$$

$$CH_2=CHCH_3 + CH_3COOH \xrightarrow{catalyst} CH_3\overset{\displaystyle O}{\overset{\|}{C}}OCH(CH_3)_2 \tag{22}$$

Cresols can be made from propylene by reaction with toluene to produce cumene (111).

$$\tag{23}$$

The cumene is oxidized to cymene hydroperoxide, which decomposes to cresols and acetone. The process is similar to phenol (qv) production from cumene.

$$\tag{24}$$

$$\tag{25}$$

BIBLIOGRAPHY

"Propylene" in *ECT* 1st ed., Vol. 11, pp. 193–197, by J. Happel and W. H. Kapfer, New York University; in *ECT* 2nd ed., Vol. 16, pp. 579–584, by W. H. Davis, Texas National Bank of Commerce, and L. K. Beach, Enjay Chemical Laboratory; in *ECT* 3rd ed., Vol. 19, pp. 228–246, by M. R. Schoenberg, J. W. Blieszner, and C. G. Papadopoulos, Amoco Chemicals Corp.

1. D. S. Sanders, D. M. Allen, and W. T. Sappenfield, *Chem. Eng. Prog.* **73**(7), 40 (1977).
2. R. C. Reid, T. M. Prausnitz, and T. K. Sherwood, *The Properties of Gases and Liquids*, 3rd ed., McGraw-Hill Book Co., Inc., New York, 1977.
3. Technical data, NASA Technical Department R-132, Lewis Research Center, Cleveland, Ohio, 1962.
4. M. Benedict, G. B. Webb, and L. C. Rubin, *Chem. Eng. Prog.* **17**(8), 419 (1951).
5. H. W. Cooper and T. C. Goldfrank, *Hydrocarbon Process.* **46**(12), 141 (1967).
6. S. K. Sood and G. A. Haselden, *AIChE J.* **16**, 891 (1970).
7. E. Bender, *Cryogenics* **15**, 667 (Nov. 1975).
8. I. E. Starling and Y. C. Kwok, *Hydrocarbon Process.* **50**(10), 90 (1971).
9. Otto Redlich, *Ind. Eng. Chem. Fundam.* **14**(3), 257 (1975).
10. B. D. Djordjevic and co-workers, *Chem. Eng.* **32**, 1103 (1977).
11. J. H. Dymond and E. B. Smith, *The Virial Coefficients of Gasses, A Critical Compilation*, Clarendon Press, London, 1969, p. 72.
12. C. L. Yaws, *Physical Properties*, McGraw-Hill Book Co., Inc., New York, 1977, pp. 167, 197, and 217.
13. *API Technical Data Book—Petroleum Refining*, Vol. I, American Petroleum Institute.
14. J. R. Tomlinson, *Technical Publication TP-1*, Natural Gas Processors Association, Tulsa, Okla., 1971.
15. R. W. Hankinson and G. H. Thomson, *AIChE J.* **25**, 653 (1979).
16. G. H. Tomson, K. R. Brobst, and R. W. Hankinson, *AIChE J.* **28**, 671 (1982).
17. R. W. Hankinson, T. A. Coker, and G. H. Thomson, *Hydrocarbon Process.* **82**(4), 207 (Apr. 1982).
18. R. Battino, *Chem. Rev.* **66**, 395 (1966).
19. E. Wilhelan, R. Battino, and R. T. Wilcock, *Chem. Rev.* **77**(2), 219 (1977).
20. A. Azarnoosh and J. J. Ketta, *J. Chem. Eng. Data* **4**, 211 (1957).
21. I. Wichterle, J. Linek, and E. Hale, *Vapor–Liquid Equilibrium Data Biography*, Elsevier Science, Inc., New York, 1973, p. 221; Suppl., 1977.
22. H. H. Reamer and B. H. Sage, *Ind. Eng. Chem.* **43**, 1628 (1951).
23. K. Ishii, S. Hayami, T. Shirai, and K. Ishida, *J. Chem. Eng. Data* **11**(3), 288 (1966).
24. D. D. Li and J. J. McKetta, *J. Chem. Eng. Data* **8**, 271 (1963).
25. R. B. Williams and D. L. Katz, *Ind. Eng. Chem.* **46**, 2512 (1954).
26. S. L. McCurdy and D. L. Katz, *Oil Gas J.* **43**(45), 102 (1945).
27. W. G. Schneider and O. Maass, *Can. J. Res.* **19**(10), 231 (1941).
28. G. G. Haselden, F. A. Holland, M. B. King, and R. F. Strickland-Constable, *Proc. Roy. Soc. London Ser. A* **240**, 1 (1957).
29. H. Lee, D. M. Newitt and M. Rubemann, *Proc. Roy. Soc. London Ser. A* **178**, 506 (1941).
30. R. A. McKay, H. H. Reamer, B. H. Sage, and W. N. Lacey, *Ind. Eng. Chem.* **43**, 2112 (1951).
31. G. H. Hanson and co-workers, *Ind. Eng. Chem.* **42**, 735 (1950).
32. A. N. Mann and co-workers, *J. Chem. Eng. Data* **8**(4), 499 (1963).
33. D. B. Manley and G. W. Swift, *J. Chem. Eng. Data* **16**(3), 301 (1971).

34. D. R. Laurence and G. W. Swift, *J. Chem. Eng. Data* **17**(3), 333 (1972).
35. G. H. Goff and co-workers, *Ind. Eng. Chem.* 42, 735 (1950).
36. M. Hirata, S. Ohe, and K. Nagahama, *Computer Aided Data Book of Vapor–Liquid Equilibria*, Kadausha Ltd., Elsevier, Tokyo, Japan, 1975.
37. M. Sagara, Y. Arai, and S. Saito, *J. Chem. Eng. Jpn.* **5**(4), 418 (1972).
38. H. K. Bae, K. Nagahama, and M. Hirata, *J. Jpn. Petrol. Inst.* **21**(4), 249 (1978).
39. M. L. McWilliams, *Chem. Eng.* **80**(25), 138 (Oct. 29, 1973).
40. C. S. Howat and G. W. Swift, *Ind. Eng. Chem. Process Des. Develop.* **19**, 318 (1980).
41. E. W. Funk and J. M. Prausnitz, *AIChE J.* **17**(1), 254 (1971).
42. S. E. M. Haman, W. K. Chung, I. M. Epshayal, and B. C. Y. Lu, *Ind. Eng. Chem. Process Des. Develop.* **16**(1), 1977.
43. J. Y. Lenior and co-workers, *J. Chem. Eng. Data* **17**(3), 340 (1971).
44. S. Ng, H. G. Harris, and J. M. Prausnitz, *J. Chem. Eng. Data* **17**(4), 482 (1969).
45. H. Sayara and co-workers, *J. Chem. Eng. Jpn.* **8**(2), 98 (1975).
46. G. T. Preston and J. M. Prausnitz, *Ind. Eng. Chem. Fundam.* **10**(3), 384 (1971).
47. R. T. Morrison and R. N. Boyd, *Organic Chemistry*, 3rd ed., Allyn and Bacon, Inc., Boston, Mass., 1977, p. 143.
48. T. G. W. Solomons, *Organic Chemistry*, John Wiley & Sons, Inc., New York, 1977.
49. H. Wittcoff, *J. Chem. Educ.* **57**, 707 (1980).
50. S. B. Zdonik, E. J. Green, and L. P. Hallee, *Manufacturing Ethylene*, The Petroleum Publishing Co., Tulsa, Okla., 1970.
51. A. G. Goosens, M. Dente, and E. Ranzi, *Hydrocarbon Process.* **57**(9), 227 (1978).
52. S. B. Zdonik, E. J. Bassler, and L. P. Hallee, *Hydrocarbon Process.* **53**(2), 73 (Feb. 1974).
53. L. E. Chambers and W. S. Potter, *Hydrocarbon Process.* **53**(1), 121 (1974).
54. Amoco Chemicals Manufacturing Specifications, Chicago, Ill., 1975.
55. A. Hahn, A. Chaptal, and J. Sialelli, *Hydrocarbon Process.* **54**, 1989 (1975).
56. E. G. Wollaston, W. J. Haflin, and W. D. Ford, *Hydrocarbon Process.* **54**(9), 93 (Sept. 1975).
57. R. B. Ewell and G. Gadmer, *Hydrocarbon Process.* 57(4), 125 (1978).
58. J. H. Gary and G. E. Handwerk, *Petroleum Refining*, Marcel Dekker, Inc., New York, 1975, p. 52.
59. *Oil Gas J.*, 53 (Mar. 10, 1975).
60. H. G. Davis and R. G. Keister, *ACS Symp. Ser.*, 32 (1976); C. Bowen, paper presented at *The National Meeting of the American Institute of Chemical Engineers*, Houston, Tex., Apr. 5–9, 1981, Series 43J, p. 157.
61. H. J. Glidden and C. F. King, *Chem. Eng. Prog.* **76**(12), 47 (1980).
62. V. U. S. Rao and R. J. Gormley, *Hydrocarbon Process.* **59**(11), 139 (1980).
63. B. Bussemeir, C. D. Frohning, and B. Cornils, *Hydrocarbon Process.* **55**(11), 105 (1976).
64. P. D. Caesar, J. A. Brennan, W. E. Garwood, and J. Cirik, *J. Catal.* **56**, 274 (1979).
65. S. Gussow, D. C. Spence, and E. A. White, *Oil Gas J.* (Dec. 8, 1980).
66. *Chem. Week*, 5 (Sept. 2, 1992).
67. *Chem. Week*, 15 (Nov. 4, 1992).
68. *Chem. Week*, 8 (Jan. 1993).
69. *Chem. Mktg. Reporter*, **240**(15), 3, 14, 15 (1991).
70. E. Chang, *Alkane Dehydrogenation and Aromatization*, Process Economics Program, Report No. 203, SRI International, Menlo Park, Calif., 1992.
71. *Facilitated Transport Process for Low-Cost Olefin–Paraffin Separation*, Advanced Technology Program No. 70NANB4H1528 National Institute of Science and Technology, 1994.

72. Z. D. Ziaka, R. G. Minet, and T. T. Tsotsis, *AIChE J.* **39**(3), 526 (1993).
73. J. Shu, B. P. A. Grandjean, A. Van Nete, and S. Kaliaguine, *Can. J. Chem. Eng.* **69**, 1036 (1991).
74. *Chemical Economics Handbook*, SRI International, Menlo Park, Calif., Jan. 1995.
75. *Chemical Safety Data Sheet SD-59, Propylene*, Manufacturing Chemists Association, Washington, D.C., rev. 1974.
76. Standards for Storage Handling of Liquefied Petroleum Gases, *National Fire Code*, Sect. 5, No. 58, National Fire Protection Assn., Boston, Mass., 1977.
77. *National Electric Code*, No. 70, National Fire Protection Assn., Boston, Mass., 1971.
78. *ASME Boiler and Pressure-Vessel Code*, Section 8, American Society of Mechanical Engineers, New York, 1980.
79. *ANSI B31 Series*, B31.3., American Society of Mechanical Engineers, New York, 1980.
80. H. F. Coward and co-workers, *Limits of Flammability of Gases and Vapors*, Bulletin 503, U.S. Bureau of Mines, Washington, D.C., 1952.
81. N. I. Sax, *Dangerous Properties of Industrial Materials*, 5th ed., Van Nostrand Reinhold Co., Inc., New York, 1979.
82. W. M. Porteous and R. C. Reid, *Chem. Eng. Prog.* **72**, 93 (1976).
83. J. W. Andrews, P. Bonnifay, B. Cha, D. Douillet, and J. Raimbault, paper presented at *The NPRA Annual Meeting*, AM-76-25, San Antonio, Tex., Mar. 28, 1976.
84. R. J. Hengstebeck, *Petroleum Processing, Principles and Applications*, McGraw-Hill Book Co., Inc., New York, 1959.
85. W. S. Knoble and F. E. Herbert, *Petr. Refiner*, 101 (Dec. 1959).
86. C. R. Cupit, J. E. Cwyu, and E. C. Jernigan, *Petro/Chem. Eng.*, 203 (Dec. 1961); *Petro/Chem. Eng.*, 207 (Jan. 1962).
87. G. E. Weisingntel, *Chem. Eng.*, 77 (June 16, 1980).
88. P. M. Kohn, *Chem. Eng.*, 114 (May 23, 1977).
89. E. K. Jones, *Adv. Catal.* **8**, 219 (1956).
90. A. Schrieshein and I. Kirschenbaum, *Chemtech.*, 310 (May 1978).
91. C. M. Fontana, R. J. Herold, E. J. Kinney, and R. C. Miller, *Ind. Eng. Chem.* **44**, 2944 (1952).
92. P. Pin and R. Mulhempt, *Angew. Chem. Int. Ed. Engl.* **19**, 857 (1980).
93. J. Boor, Jr., *Ziepler-Natta Catalysts and Polymerizations*, Academic Press, Inc., New York, 1979.
94. G. S. Parks and H. P. Mosher, *J. Polym. Sci. Part A* **1**, 1979 (1963).
95. D. J. Hadley, R. E. Saunders, and P. T. Mapp, in E. G. Hancock, ed., *Propylene and its Industrial Derivatives*, Earnest Benn Ltd., London, 1973, pp. 416 ff.
96. P. R. Pujado, B. V. Vora, and A. P. Krueding, *Hydrocarbon Process.* **56**(5), 169 (1977).
97. R. Landau, G. A. Sullican, and D. Brown, *Chemtech.*, 602 (Oct. 1979).
98. K. Weissermel and H. J. Arpe, *Industrial Organic Chemistry*, Verlag-Chemie, Weinheim, Austria, 1978, p. 239.
99. 1979 Petrochemical Handbook, *Hydrocarbon Process.* **59**(11), 240 (1979).
100. Ref. 19, p. 175.
101. Y. Onove, Y. Mizutani, S. Akiyama, and Y. Izumi, *Chemtech.*, 432 (July 1978).
102. W. Nier and J. Woellner, *Hydrocarbon Process.* **51**(11), 113 (1972).
103. W. C. Fernelius, H. Wittcoff, and R. E. Varneria, *J. Chem. Educ.* **57**, 707 (1980).
104. J. C. Fielding, in Ref. 95, pp. 244 ff.
105. R. A. Persale, E. L. Pollitzer, D. J. Ward, and P. R. Pujado, *Chem. Econ. Eng. Rev.* **10**(7), 25 (1978).
106. R. Fowler, H. Conner, and R. A. Gaehl, *Hydrocarbon Process.* **55**(9), 247 (1976).
107. H. Weber, W. Demmeling, and A. M. Desia, *Hydrocarbon Process.* **55**(4), 129 (1976).
108. Ref. 19, pp. 77–80.

109. G. E. Schaal, *Hydrocarbon Process.* **52**(9), 218 (1973).
110. L. F. Hatch and S. Matar, *Hydrocarbon Process.* **57**(6), 149 (1978).
111. K. Sto, *Hydrocarbon Process.* **52**(8), 89 (1973).

Narasimhan Calamur
Martin Carrera
Amoco Corporation

PROPYLENE OXIDE

Propylene oxide [75-56-9] (methyloxirane, 1,2-epoxypropane) is a significant organic chemical used primarily as a reaction intermediate for production of polyether polyols, propylene glycol, alkanolamines (qv), glycol ethers, and many other useful products (see Glycols). Propylene oxide was first prepared in 1861 by Oser and first polymerized by Levene and Walti in 1927 (1). Propylene oxide is manufactured by two basic processes: the traditional chlorohydrin process (see Chlorohydrins) and the hydroperoxide process, where either *tert*-butanol (see Butyl Alcohols) or styrene (qv) is a co-product. Research continues in an effort to develop a direct oxidation process to be used commercially.

Physical Properties

Propylene oxide is a colorless, low boiling (34.2°C) liquid. Table 1 lists general physical properties; Table 2 provides equations for temperature variation on some thermodynamic functions. Vapor–liquid equilibrium data for binary mixtures of propylene oxide and other chemicals of commercial importance are available. References for binary mixtures include 1,2-propanediol (14), water (7,8,15), 1,2-dichloropropane [78-87-5] (16), 2-propanol [67-63-0] (17), 2-methyl-2-pentene [625-27-4] (18), methyl formate [107-31-3] (19), acetaldehyde [75-07-0] (17), methanol [67-56-1] (20), propanal [123-38-6] (16), 1-phenylethanol [60-12-8] (21), and *tert*-butanol [75-65-0] (22,23).

Chemical Properties

Propylene oxide is highly reactive owing to the strained three-membered oxirane ring. The ring C–C and C–O bond lengths have been reported as 147 and 144 pm, respectively, whereas the C–C bond for the substituted methyl group is 152 pm (24). Although some reactions, such as those with hydrogen halides or ammonia (qv), proceed at adequate rates without a catalyst, most reactions of industrial importance employ the use of either acidic or basic catalysts.

Ring Opening. The epoxide ring of propylene oxide may open at either of the C–O bonds. In anionic (basic) catalysis, the bond preferentially opens at the least sterically hindered position, resulting in mostly (95%) secondary

Table 1. Physical Property Data for Propylene Oxide

Property	Value	Reference
molecular weight	58.08	
boiling point at 101.3 kPa^a, °C	34.2	2
freezing point, °C	−111.93	2
critical pressure, MPa^a	4.92	3–5
critical temperature, °C	209.1	3–5
critical volume, cm^3/mol	186	4,6
critical compressibility factor, Z_c	0.228	3–5
accentric factor	0.269	5
dipole moment, C·m^b	6.61 × 10^{-30}	6
explosive limits in air, vol %		
upper	36	2
lower	2.3	2
flash point, Tag closed cup, °C	−37	3,7
heat of fusion, kJ/mol^c	6.531	3
heat of vaporization, 101.3 kPa^a, kJ/mol^c	27.8947	2,8,9
heat of combustion, kJ/mol^c	1915.6	2
specific heat, at 20°C, J/(mol·K)^c	122.19	8,10
autoignition temperature, at 101.3 kPa^a, °C	465	2,3
index of refraction, at 25°C	1.36335	2

^{a}To convert kPa to psi, multiply by 0.145.
^{b}To convert C·m to debye, divide by 3.336 × 10^{-30}.
^{c}To convert J to cal, divide by 4.184.

alcohol products (25). Cationic (acidic) catalysts provide a mixture of secondary and primary alcohol products. Weak cationic catalysts, such as zeolites, give up to 30% primary alcohol product when reacting with alcohols. Stronger cationic catalysts, such as H$_2$SO$_4$, give up to 55% primary alcohol product (26).

The ring-opening reactions of epoxides take place by nucleophilic substitution, ie, a S_{N^2} mechanism, on one of the epoxide carbon atoms with displacement of the epoxide oxygen atom. The orientation of ring opening in propylene oxide is determined primarily by the steric hindrance of the substituent methyl group and secondarily by the electron-releasing effect of the methyl group. Thus, acid catalysis increases substitution on the secondary carbon by increasing the positive charge on this carbon (24).

Base-catalyzed reaction

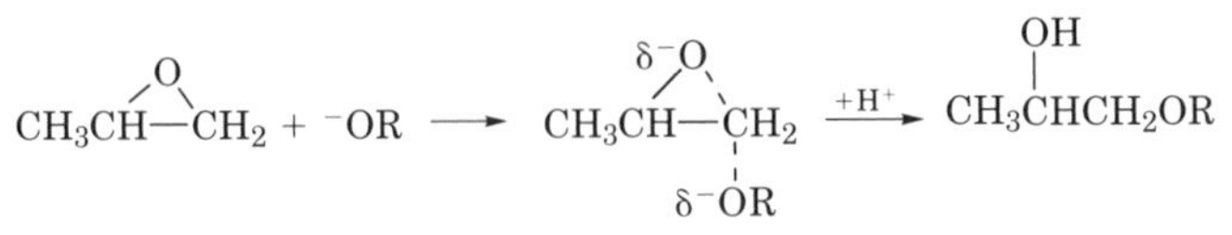

$$CH_3CH-CH_2 + {}^-OR \longrightarrow CH_3CH-CH_2 \xrightarrow{+H^+} CH_3CHCH_2OR$$

Acid-catalyzed reaction

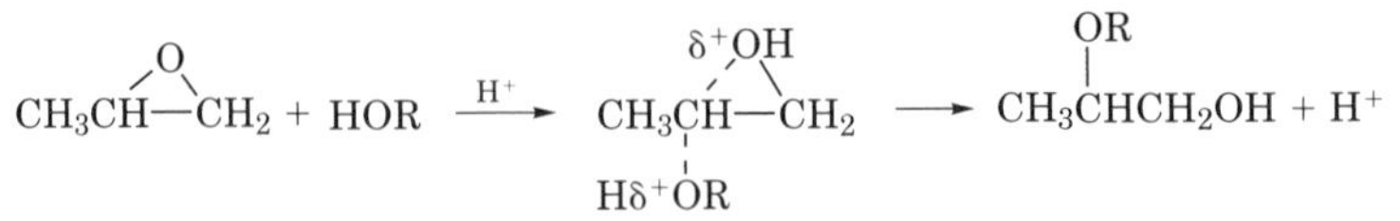

$$CH_3CH-CH_2 + HOR \xrightarrow{H^+} CH_3CH-CH_2 \longrightarrow CH_3CHCH_2OH + H^+$$

Table 2. Propylene Oxide Physical Property Data as a Function of Temperature

Property	Equation	Coefficients				Reference
		A	B	C	D	
		For T in Kelvin				
heat of formation, kJ/mol[a]	$A + BT + CT^2$	-74.450	-7.2182×10^{-2}	3.4979×10^{-5}		11
heat capacity, J/(mol·K)[a]						
vapor	$A + BT + CT^2 + DT^3$	-7.868	0.32282	-1.9498×10^{-4}	4.6455×10^{-8}	12
liquid	$A + BT + CT^2$	113.08	-0.15085	6.728385×10^{-4}		13
		For T in °C				
liquid density, g/mL	$A + BT$	0.8556	-0.00122			2
viscosity, mPa·s(=cP)						
saturated liquid	$A + BT$	0.413	-0.0047			2
vapor	$A + BT$	7.96075×10^{-3}	3.101×10^{-5}			3
heat of vaporization, J/g[a]	$A + BT$	528.23	-1.2552			3
surface tension, N/m	$A + BT$	0.02501	-1.343×10^{-4}			3
thermal conductivity, J/(h·cm·C)[a]						
liquid	$A + BT$	6.5249	-0.01582			3
vapor	$A + BT$	0.12238	0.001025			3
vapor pressure, kPa[b]	$\log_{10} P = A - B/(T + C)$	6.09689	1066.19	226.38		2

[a]To convert J to cal, divide by 4.184.
[b]To convert kPa to mm Hg, multiply by 7.5.

273

Polymerization to Polyether Polyols. The addition polymerization of propylene oxide to form polyether polyols is very important commercially. Polyols are made by addition of epoxides to initiators, ie, compounds that contain an active hydrogen, such as alcohols or amines. The polymerization occurs with either anionic (base) or cationic (acidic) catalysis. The base catalysis is preferred commercially (25,27).

Some of the simplest polyols are produced from reaction of propylene oxide and propylene glycol and glycerol initiators. Polyether diols and polyether triols are produced, respectively (27) (see GLYCOLS).

$$
\underset{\substack{| \\ \text{HOCH}_2\text{CHCH}_3}}{\overset{\text{OH}}{}} + (x + y)\ \text{CH}_3\text{CH}\overset{\text{O}}{\diagdown}\text{CH}_2 \xrightarrow{\text{KOH, H}_2\text{O}} \underset{\substack{\text{CH}_2\text{O(CH}_2\text{CHO)}_{y-1}\text{CH}_2\text{CHCH}_3 \\ | \qquad\qquad\qquad\quad | \\ \text{CH}_3 \qquad\qquad\qquad \text{OH}}}{\overset{\substack{\text{CH}_3 \qquad\qquad\qquad \text{OH} \\ | \qquad\qquad\qquad\quad | \\ \text{CH}_3\text{CHO(CH}_2\text{CHO)}_{x-1}\text{CH}_2\text{CHCH}_3}}{}}
$$

A variety of initiators can be used to produce polyols of varying functionality. Polymerization of propylene oxide on alcohols or mercaptans results in polymers having a single terminal hydroxyl group, whereas use of propylene glycol yields a product having two hydroxyls (diol). Triols result from initiators such as glycerol (qv), trimethylolpropane, 1,2,6-hexanetriol, and triethanolamine. Higher functionality is achieved with initiators such as sorbitol, sucrose, pentaerythritol, sorbitans, and ethylenediamine (28,29). Mixtures of initiators give polymer mixtures of intermediate functionality, eg, glycerol and sucrose.

Propylene oxide can be copolymerized with other epoxides, such as ethylene oxide (qv) (25,29,30) or tetrahydrofuran (31,32) to produce copolymer polyols. Copolymerization with anhydrides (33) or CO_2 (34) results in polyesters and polycarbonates (qv), respectively.

Polyols are typically prepared by base-catalysis using sodium or potassium hydroxide, aqueous ammonia, or trimethylamine (25,27,35). Potassium hydroxide is generally preferred for preparation of polyols for polyurethane applications, whereas sodium hydroxide may be used to prepare polyols used as surfactants (qv). Propylene oxide can also rearrange to allyl alcohol, which reacts with propylene oxide to form a polyether having one hydroxyl and one carbon–carbon double bond end group (27). Use of catalysts such as N-methyltetraphenylporphyrin complexes produces polymers having very narrow molecular weight distribution (36–38). Stereoselective polymerization (reaction rate ratio of 1.05 (R)-propylene oxide to (S)-propylene oxide) of prochiral propylene oxide has been achieved using a chiral aluminum salt of (R)-($-$)-3,3-dimethyl-1,2-butanediol/ZnCl$_2$ (39). Polyols produced by the above methods have molecular weights of ~200–7000 (27).

Molecular weights of poly(propylene oxide) polymers of greater than 100,000 are prepared from catalysts containing $FeCl_3$ (40,41). The molecular weight of these polymers is greatly increased by the addition of small amounts of organic isocyanates (42). Homopolymers of propylene oxide are also prepared by catalysis using diethylzinc–water (43), diphenylzinc–water (44), and trialkylaluminum (45,46) systems.

Reactions. *Water.* Propylene oxide reacts with water to produce propylene glycol [57-55-6], dipropylene glycol, tripropylene glycol, and higher molecular weight polyglycols. This commercial process is typically run using an excess of water (12–20 mol water/mol propylene oxide) to maximize the production of the monopropylene glycol (47).

$$CH_3CH\overset{O}{\overline{\diagup\diagdown}}CH_2 + H_2O \xrightarrow{<200°C} HOCH_2\overset{OH}{\underset{|}{C}}HCH_3 + CH_3\overset{OH}{\underset{|}{C}}HCH_2OCH_2\overset{OH}{\underset{|}{C}}HCH_3$$

$$+ CH_3\overset{OH}{\underset{|}{C}}HCH_2OCH_2\overset{CH_3}{\underset{|}{C}}HOCH_2\overset{OH}{\underset{|}{C}}HCH_3 + \text{isomers} + \text{higher oligomers}$$

Although the commercial process normally uses heat and pressure without a catalyst, acid or base catalysts can be used to enhance reaction rates or product selectivity. Homogeneous catalyst systems, such as carbon dioxide (qv) and a quaternary phosphonium salt, reduce the excess water required (48), whereas heterogeneous hydrotalcite-type catalysts improve selectivity to monopropylene glycol (49).

Ammonia and Amines. Isopropanolamine is the product of propylene oxide and ammonia in the presence of water (see ALKANOLAMINES). Propylene oxide reacts with isopropanolamine or other primary or secondary amines to produce *N*- and *N,N*-disubstituted isopropanolamines. Propylene oxide further reacts with the hydroxyl group of the alkanolamines to form polyether polyol derivatives of tertiary amines (50), or of secondary amines in the presence of a strong base catalyst (51).

Carbon Dioxide and Carbon Disulfide. Propylene oxide and carbon dioxide react in the presence of tertiary amine, quaternary ammonium halides, or calcium or magnesium halide catalysts to produce propylene carbonate (52). Use of catalysts derived from diethylzinc results in polycarbonates (53).

$$CH_3CH\overset{O}{\overline{\diagup\diagdown}}CH_2 + CO_2 \longrightarrow \underset{O}{\overset{CH_3}{\underset{\underset{O}{\|}}{O\diagdown\diagup O}}} \longrightarrow \overset{CH_3\ O}{\underset{|\quad \|}{(CH_2CHOCO)_n}}$$

Similarly, carbon disulfide and propylene oxide reactions are catalyzed by magnesium oxide to yield episulfides (54), and by derivatives of diethylzinc to yield low molecular weight copolymers (55). Use of tertiary amines as catalysts under pressure produces propylene trithiocarbonate (56).

$$CH_3CH\overset{O}{\overline{\diagup\diagdown}}CH_2 + CS_2 \longrightarrow \underset{S}{\overset{CH_3}{\underset{\underset{S}{\|}}{O\diagdown\diagup S}}} \longrightarrow \overset{CH_3\ S}{\underset{|\quad \|}{(CH_2CHOCS)_n}}$$

Hydroxy-Containing Organics. Propylene oxide reacts with the hydroxyl group of alcohols and phenols to produce monoethers of propylene glycol. Use of basic catalysts results in mostly secondary alcohol products; use of acidic catalysts gives a mixture of primary and secondary alcohols. Suitable catalysts include sodium hydroxide, potassium hydroxide, tertiary amines, potassium carbonate, sodium acetate, boron trifluoride, and acid clays. Further addition of propylene oxide yields the di-, tri-, and poly(propylene glycol) ethers. Multiple hydroxyls (glycol, glycerol, glucose, etc) on the organic reactant lead to the polyether polyols discussed earlier (57–59).

Propylene oxide and carboxylic acids in equimolar ratios produce monoesters of propylene glycol. Higher ratios of oxide to acid produce polypropylene glycol monoesters. In the presence of basic catalysts these monoesters can undergo transesterification reactions that yield a product mixture of propylene glycols, monoesters, and diesters (57,60).

$$2\ CH_3CH\overset{O}{\diagup\!\!\diagdown}CH_2 + 2\ RCOH \longrightarrow 2\ \underset{\text{O}}{R}COCH_2\underset{CH_3}{CHOH} \xrightarrow{OH^-} RCOCH_2CHOCR + HOCH_2CHCH_3$$

Natural Products. Many natural products, eg, sugars, starches, and cellulose, contain hydroxyl groups that react with propylene oxide. Base-catalyzed reactions yield propylene glycol monoethers and poly(propylene glycol) ethers (61–64). Reaction with fatty acids results in a mixture of mono- and diesters (65). Cellulose fibers, eg, cotton (qv), have been treated with propylene oxide (66–68).

Hydrogen Sulfide and Mercaptans. Hydrogen sulfide and propylene oxide react to produce 1-mercapto-2-propanol and bis(2-hydroxypropyl) sulfide (69,70). Reaction of the epoxide with mercaptans yields 1-alkylthio- or 1-arylthio-2-propanol when basic catalysis is used (71). Acid catalysts produce a mixture of primary and secondary hydroxy products, but in low yield (72). Suitable catalysts include sodium hydroxide, sodium salts of the mercaptan, tetraalkylammonium hydroxide, acidic zeolites, and sodium salts of an alkoxylated alcohol or mercaptan (26,69,70,73,74).

Friedel-Crafts. 2-Phenylpropanol results from the catalytic (AlCl₃, FeCl₃, or TiCl₄) reaction of benzene and propylene oxide at low temperature and under anhydrous conditions (see FRIEDEL-CRAFTS REACTIONS). Epoxide reaction with toluene gives a mixture of *o*-, *m*-, and *p*-isomers (75,76).

Grignard Reagents. Grignard reagents, RMgX, produce a mixture of secondary alcohols, RCH₂CHOHCH₃, and propylene halohydrin, CH₃CHOHCH₂X, upon reaction with propylene oxide (21,77,78). Use of dialkylmagnesium eliminates the halohydrin formation and rearrangement (79) (see GRIGNARD REACTIONS).

Isomerization and Hydrogenolysis. Isomerization of propylene oxide to propionaldehyde and acetone occurs over a variety of catalysts, eg, pumice, silica gel, sodium or potassium alum, and zeolites (80,81). Stronger acid catalysts favor acetone over propionaldehyde (81). Allyl alcohol yields of 90% are obtained from use of a supported lithium phosphate catalyst (82).

Hydrogenolysis of propylene oxide yields primary and secondary alcohols as well as the isomerization products of acetone and propionaldehyde. Pd and Pt catalysts favor acetone and 2-propanol formation (83–85). Ni and Cu catalysts favor propionaldehyde and 1-propanol formation (86,87).

Reduction of propylene oxide to propylene is accomplished by use of metallocenes, such as $Ti(C_5H_5)_2Cl_2$, and sodium amalgam (88).

Carbonyl Compounds. Cyclic ketals and acetals (dioxolanes) are produced from reaction of propylene oxide with ketones and aldehydes, respectively. Suitable catalysts include stannic chloride, quaternary ammonium salts, glycol sulphites, and molybdenum acetyl acetonate or naphthenate (89–91). Lactones come from Ph_4SbI-catalyzed reaction with ketenes (92).

Other Inorganic Reagents. Propylene oxide reacts with hydrogen halides to give the corresponding isomeric halohydrins (93); with sodium bisulfite to give the sodium salt of 2-hydroxypropanesulfonic acid (94,95); with nitric acid to produce isomeric nitrate esters (96); with hydrogen cyanide to give 1-cyano-2-propanol (97); and with boric acid, boron trichloride, or diborane to give a variety of substituted boranes and borates (98–101). Hydrolysis with water of the reaction products of propylene oxide and PCl_3 or $TiCl_4$ gives 2-chloropropanol; hydrolysis of the product of $AlCl_3$ gives 1,2-dichloropropane (102).

Other Reactions. 2-Dioxolanimines, 2-oxathiolanimines, and 2-oxazolidinimines result from the reaction of propylene oxide with isocyanates, isothiocyanates, and carbodiimides, respectively (103,104).

Trimethylaluminum and propylene oxide form a mixture of 2-methyl-1-propanol and 2-butanol (105). Triethylaluminum yields products of 2-methyl-1-butanol and 2-pentanol (106). The ratio of products is determined by the ratio of reactants. Hydrolysis of the products of methylaluminum dichloride and propylene oxide results in 2-methylpropene and 2-butene, with elimination of methane (105). Numerous other nucleophilic (107) and electrophilic (108) reactions of propylene oxide have been described in the literature.

Manufacture

Propylene oxide is produced by one of two commercial processes: the chlorohydrin process or the hydroperoxide process. The 1995 global propylene oxide capacity was estimated at about 4.36×10^6 t/yr. About half came from each of the two processes. Table 3 summarizes the global production capacities for each of the processes.

The chlorohydrin process involves reaction of propylene and chlorine in the presence of water to produce the two isomers of propylene chlorohydrin. This is followed by dehydrochlorination using caustic or lime to propylene oxide and salt. The Dow Chemical Company is the only practitioner of the chlorohydrin process in North America. However, several companies practice the chlorohydrin process at more than 20 locations in Germany, Italy, Brazil, Japan, Eastern Europe, and Asia.

The hydroperoxide process involves oxidation of propylene (qv) to propylene oxide by an organic hydroperoxide. An alcohol is produced as a coproduct. Two different hydroperoxides are used commercially that result in *tert*-butanol or 1-phenylethanol as the coproduct. The *tert*-butanol (TBA) has been used as a

Table 3. Global Propylene Oxide Production Capacities, t × 10³/yr

Geographical location	Chlorohydrin		Hydroperoxide				Total, production capacity
			Coproduct–TBA[a]		Coproduct–styrene		
	Capacity	Number of sites	Capacity	Number of sites	Capacity	Number of sites	
United States	720	2	732	2	499	1	1951
Latin America	145	1					145
Western Europe	814	5	445	2	225	2	1484
Eastern Europe	125	6					125
Pacific	389	14			270	2	659
Total	*2193*	*28*	*1177*	*4*	*994*	*5*	*4364*
percent of total	50.2		27.0		22.8		

[a]TBA = *tert*-butyl alcohol.

gasoline additive, dehydrated to isobutylene, and used as feedstock to produce methyl *tert*-butyl ether (MTBE), a gasoline additive. The 1-phenyl ethanol is dehydrated to styrene. ARCO Chemical has plants producing the TBA coproduct in the United States, France, and the Netherlands. Texaco has a TBA coproduct plant in the United States. Styrene coproduct plants are operated by ARCO Chemical in the United States and Japan, Shell in the Netherlands, Repsol in Spain, and Yukong in South Korea.

Process flow sheets and process descriptions given herein are estimates of the various commercial processes. There are also several potential commercial processes, including variations on the chlorohydrin process, variations on the hydroperoxide process, and direct oxidation of propylene.

Chlorohydrin Process.   The chlorohydrin process illustrated in Figure 1 is fairly simple, requiring only two reaction steps, chlorohydrination and epoxidation, followed by product purification (109,110,112). Propylene gas and chlorine gas in about equimolar amounts are mixed with an excess of water to generate propylene chlorohydrin and a small amount of chlorinated organic coproducts, chiefly 1,2-dichloropropane (109). Epoxidation, also called saponification or dehydrochlorination, is accomplished by treatment of the chlorohydrin solution with caustic soda or milk of lime (aqueous calcium hydroxide). Propylene oxide and other organics are steam-stripped from the resulting sodium chloride or calcium chloride brine. The brine is treated, usually by biological oxidation, to reduce organic content prior to discharge. The propylene oxide is further purified to sales specifications by removal of lights and heavies via distillation.

Chlorohydrination.   The mechanism for the formation of propylene chlorohydrin is generally believed to be through the chloronium ion intermediate (109,112).

$$CH_3CH{=}CH_2 + Cl_2 \longrightarrow CH_3CH{-}CH_2$$
$$Cl^+Cl^-$$

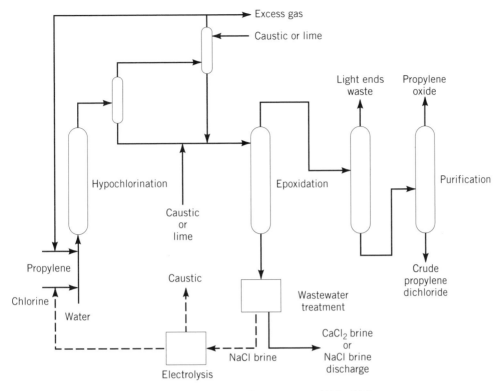

Fig. 1. The chlorohydrin process (109–111).

The chloronium ion intermediate can react with water to produce the desired propylene chlorohydrin, with chloride ion to produce 1,2-dichloropropane, or with propylene chlorohydrin to produce isomers of dichloro-dipropyl ether.

$$CH_3CH{-}CH_2 \;\;\xrightarrow{H_2O}\;\; CH_3\overset{\text{OH}}{\underset{}{C}}HCH_2Cl + CH_3\overset{\text{Cl}}{\underset{}{C}}HCH_2OH + HCl$$

$$Cl^+Cl^-$$

$$CH_3\overset{\text{Cl}}{\underset{}{C}}HCH_2Cl$$

$$CH_3\overset{\text{CH}_2Cl}{\underset{\text{CH}_2Cl}{C}}HOCHCH_3 + HCl$$

chlorohydrin

The 1-chloro-2-propanol isomer represents about 85% of the chlorohydrin produced. In order to minimize the formation of dichloride coproduct and ether, the reactant compositions are chosen such that the effluent liquid contains 4–5 wt % propylene chlorohydrin. Under these conditions, the yield of chlorohydrin, dichloride, and ether from the reactants is reported to be 87–90, 6–9, and 2%, respectively (109,110,112).

In two proposed alternative processes, the chlorine is replaced in the hypochlorination reaction by hypochlorous acid [*7790-92-3*], HOCl, or *tert*-butyl hypochlorite. In the first, a concentrated (>10% by weight) aqueous solution of hypochlorous acid, substantially free of chloride, chlorate, and alkali metal ions, is contacted with propylene to produce propylene chlorohydrin (113). The likely mechanism of reaction is the same as that for chlorine, as chlorine is generated *in situ* through the equilibrium of chlorine and hypochlorous acid (109).

$$HOCl + HCl \rightleftharpoons Cl_2 + H_2O$$

In the second proposed alternative process, *tert*-butyl hypochlorite, formed from the reaction of chlorine and *tert*-butyl alcohol, reacts with propylene and water to produce the chlorohydrin. The alcohol is a coproduct and is recycled to generate the hypochlorite (114–116). No commercialization of the hypochlorous acid and *tert*-butyl hypochlorite processes for chlorohydrin production is known.

$$CH_3C(CH_3)_2OCl + CH_3CH=CH_2 + H_2O \longrightarrow CH_3\overset{\underset{|}{OH}}{C}HCH_2Cl + CH_3C(CH_3)_2OH$$

Epoxidation. Epoxidation, also referred to as saponification or dehydrochlorination, of propylene chlorohydrin (both isomers) to propylene oxide is accomplished using a base, usually aqueous sodium hydroxide or calcium hydroxide.

$$CH_3\overset{\underset{|}{OH}}{C}HCH_2Cl + HCl + Ca(OH)_2 \longrightarrow CH_3\overset{O}{C}H\!-\!CH_2 + CaCl_2 + 2\,H_2O$$

About 10% excess base is used for the epoxidation and neutralization of the acid. Because propylene oxide undergoes hydrolysis to propylene glycol (117,118) in the presence of the base, the oxide is quickly steam-stripped from the brine solution. This operation is carefully controlled to avoid stripping unreacted chlorohydrin. The exit gas, containing about 70 wt % water, 26 wt % propylene oxide, and 4 wt % miscellaneous organic compounds, is further processed to purify the propylene oxide. The brine effluent contains salt, eg, 5–6 wt % calcium chloride; base, 0.1 wt % calcium hydroxide; and 0.05–0.1 wt % propylene glycol (109,110,112). Catalytic dehydrochlorination, using a CsCl/SiO$_2$ catalyst at 150°C, has been reported (119). Propylene oxide is produced in 100% selectivity and HCl is recoverable upon regeneration of the catalyst.

Purification. Purification of propylene oxide from the chlorohydrin process is accomplished by standard distillation (qv) in one or more packed or trayed columns. Lighter components, such as acetaldehyde (qv), are removed overhead. Heavier components, such as water and propylene dichloride, are taken off the bottom of the column (109,110). Glycol formation can be reduced by controlling the bottoms stream at pH between 4 and 7 by adding NaCl or Na$_2$SO$_4$ (120). The bottoms stream is decanted into an aqueous layer and an organic layer.

The aqueous layer can be recycled for additional product recovery. The organic layer contains coproducts that can be further processed for recovery and use (109,110,120). Removal of low levels of close-boiling impurities or poly(propylene oxide) is accomplished by treatment of the oxide using activated carbon, diatomaceous earth, or membranes (121–124).

Effluent Wastewater Treatment.   The volume of water effluent is about 40 times the volume of propylene oxide produced in the chlorohydrin process, representing a significant concern for proper disposal or reuse (112). The options for treatment of the water effluent from the epoxidation process depend on the alkali used. Use of lime in epoxidation results in a calcium chloride brine (4–6 wt %) that has little commercial value and is, therefore, discharged. Use of caustic soda in epoxidation results in a sodium chloride brine that can either be discharged or recycled to a chloralkali electrolysis unit to generate chlorine and caustic (111). The effluent is treated biologically to reduce organic content prior to discharge or recycle (125,126).

Chlorohydrin via tert-Butyl Hypochlorite.   Figure 2 (114,115,127–129,134) shows an integrated chlorohydrin process based on chlorination of a tertiary alcohol to the hypochlorite, reaction with propylene to yield the chlorohydrin and the alcohol, epoxidation with chloralkali cell liquor, recycle of the alcohol, and recycle of brine to the chloralkali cell. With the recycle of brine this process has a lower volume of aqueous effluent. Chlorine, caustic soda (cell liquor), and *tert*-butanol react together to generate *tert*-butyl hypochlorite, which is phase-separated (114) or solvent extracted (128) from the concentrated brine. The *tert*-butyl hypochlorite reacts with propylene and water to make propylene chlorohydrin, regenerating *tert*-butanol (116). Chlorohydrin is recovered by azeotropic distillation (114) or solvent extraction (127,130), and epoxidized to propylene oxide using caustic soda (131). The *tert*-butanol, recovered by distilla-

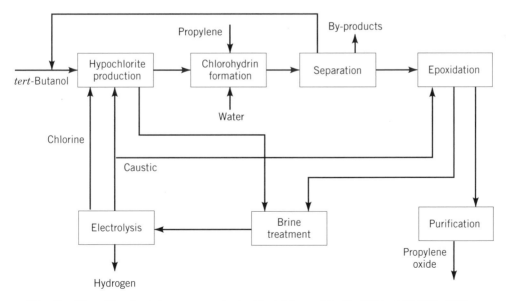

Fig. 2.   The chlorohydrin process via *tert*-butyl hypochlorite (114–116,127–133).

tion or solvent extraction, is recycled to make hypochlorite. Propylene oxide is purified by distillation. Brine from hypochlorite formation and from epoxidation is treated (132,133) and recycled to the electrolytic cell to generate chlorine and caustic. The process can, alternatively, utilize calcium hydroxide, producing a concentrated calcium chloride brine (135).

Hydroperoxide Process.   The hydroperoxide process to propylene oxide involves the basic steps of oxidation of an organic to its hydroperoxide, epoxidation of propylene with the hydroperoxide, purification of the propylene oxide, and conversion of the coproduct alcohol to a useful product for sale. Incorporated into the process are various purification, concentration, and recycle methods to maximize product yields and minimize operating expenses. Commercially, two processes are used. The coproducts are *tert*-butanol, which is converted to methyl *tert*-butyl ether [*1634-04-4*] (MTBE), and 1-phenyl ethanol, converted to styrene [*100-42-5*]. The coproducts are produced in a weight ratio of 3–4:1 *tert*-butanol/propylene oxide and 2.4:1 styrene/propylene oxide, respectively. These processes use isobutane (see HYDROCARBONS) and ethylbenzene (qv), respectively, to produce the hydroperoxide. Other processes have been proposed based on cyclohexane where aniline is the final coproduct, or on cumene (qv) where α-methyl styrene is the final coproduct.

tert-Butyl Hydroperoxide Process.   Figure 3 provides a simplified flow sheet of the propylene oxide and *tert*-butyl alcohol coproduct process. The first step of the process is the liquid-phase air oxidation of isobutane [*75-28-5*] to *tert*-butyl hydroperoxide (TBHP) in the presence of 10–30 wt % *tert*-butyl alcohol (TBA). Temperature is 95–150°C and pressure is 2075–5535 kPa (300–800 psi), resulting in a conversion of 20–30% of the isobutane and a selectivity to TBHP

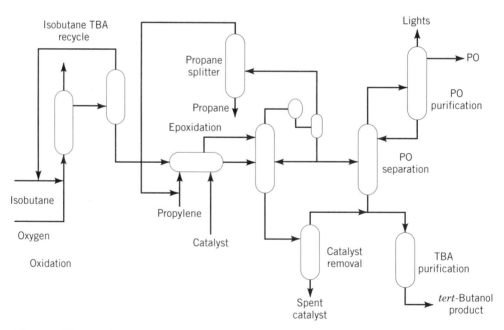

Fig. 3.   The *tert*-butyl hydroperoxide process to propylene oxide (PO) and *tert*-butanol (TBA) (136–156).

of 60–80% and to TBA of 20–40%. Conversion can be increased at the expense of selectivity of TBHP by increasing temperature and reaction time (136). Unreacted isobutane and a portion of the TBA are separated from the product and recycled back to the hydroperoxide-forming reactor.

$$CH_3CHCH_3 + O_2 \longrightarrow CH_3C(CH_3)_2OOH$$
$$\overset{\displaystyle CH_3}{\underset{\displaystyle |}{}}$$

The *tert*-butyl hydroperoxide is then mixed with a catalyst solution to react with propylene. Some TBHP decomposes to TBA during this process step. The catalyst is typically an organometallic that is soluble in the reaction mixture. The metal can be tungsten, vanadium, or molybdenum. Molybdenum complexes with naphthenates or carboxylates provide the best combination of selectivity and reactivity. Catalyst concentrations of 200–500 ppm in a solution of 55% TBHP and 45% TBA are typically used when water content is less than 0.5 wt %. The homogeneous metal catalyst must be removed from solution for disposal or recycle (137,157). Although heterogeneous catalysts can be employed, elution of some of the metal, particularly molybdenum, from the support surface occurs (158). References 159 and 160 discuss possible mechanisms for the catalytic epoxidation of olefins by hydroperoxides.

$$CH_3C(CH_3)_2OOH + CH_2{=}CHCH_3 \longrightarrow H_2C{-}CHCH_3 + CH_3C(CH_3)_2$$

An excess of 2–10 mol propylene to hydroperoxide is used to maximize conversion of hydroperoxide and selectivity to propylene oxide. Temperature is 100–130°C, pressure is 1480–3550 kPa (215–515 psi), and residence time is sufficient (about 2 h) for >95% conversion of the hydroperoxide. An organic solvent such as benzene, chlorobenzene, or *t*-butanol may be employed. Selectivity to propylene oxide is 95–98% based on TBHP and 97–98% based on propylene. The principal by-products are propylene glycol, methyl formate, and a propylene dimer. Some of these by-products are difficult to remove from the product propylene oxide (138–140). The presence of acids, such as carboxylic acids, results in lower product selectivity (142). The reactor design is described as a horizontal tank having partitioned reaction zones, a thermosyphon reactor, or a two-stage reactor having different operating conditions in each stage (142,157).

After epoxidation a distillation is performed to remove the propylene, propylene oxide, and a portion of the TBHP and TBA overhead. The bottoms of the distillation contains TBA, TBHP, some impurities such as formic and acetic acid, and the catalyst residue. Concentration of this catalyst residue for recycle or disposal is accomplished by evaporation of the majority of the TBA and other organics (141,143,144), addition of various compounds to yield a metal precipitate that is filtered from the organics (145–148), or liquid extraction with water (149). Low (<500 ppm) levels of soluble catalyst can be removed by adsorption on solid magnesium silicate (150). The recovered catalyst can be treated for recycle to the epoxidation reaction (151).

Methyl formate and propylene oxide have close boiling points, making separation by distillation difficult. Methyl formate is removed from propylene oxide by hydrolysis with an aqueous base and glycerol, followed by phase separation and distillation (152,153). Methyl formate may be hydrolyzed to methanol and formic acid by contacting the propylene oxide stream with a basic ion-exchange resin. Methanol and formic acid are removed by extractive distillation (154).

Final purification of propylene oxide is accomplished by a series of conventional and extractive distillations. Impurities in the crude product include water, methyl formate, acetone, methanol, formaldehyde, acetaldehyde, propionaldehyde, and some heavier hydrocarbons. Conventional distillation in one or two columns separates some of the lower boiling components overhead, while taking some of the higher boilers out the bottom of the column. The reduced level of impurities are then extractively distilled in one or more columns to provide a purified propylene oxide product. The solvent used for extractive distillation is distilled in a conventional column to remove the impurities and then recycled (155,156). A variety of extractive solvents have been demonstrated to be effective in purifying propylene oxide, as shown in Table 4.

The *tert*-butanol (TBA) coproduct is purified for further use as a gasoline additive. Upon reaction with methanol, methyl *tert*-butyl ether (MTBE) is produced. Alternatively the TBA is dehydrated to isobutylene which is further hydrogenated to isobutane for recycle in the propylene oxide process.

Ethylbenzene Hydroperoxide Process. Figure 4 shows the process flow sheet for production of propylene oxide and styrene via the use of ethylbenzene hydroperoxide (EBHP). Liquid-phase oxidation of ethylbenzene with air or oxygen occurs at 206–275 kPa (30–40 psia) and 140–150°C, and 2–2.5 h are required for a 10–15% conversion to the hydroperoxide. Recycle of an inert gas, such as nitrogen, is used to control reactor temperature. Impurities in the ethylbenzene, such as water, are controlled to minimize decomposition of the hydroperoxide product and are sometimes added to enhance product formation. Selectivity to by-products include 8–10% acetophenone, 5–7% 1-phenylethanol, and <1% organic acids. EBHP is concentrated to 30–35% by distillation. The overhead ethylbenzene is recycled back to the oxidation reactor (170–172).

Table 4. Solvents for Purification of Propylene Oxide From Epoxidation Using *tert*-Butyl Hydroperoxide

Impurity to be removed	Extractive distillation solvent	Reference
water	sulfonate	161
	monohydroxy alkoxyalkanol	162
water–methanol–acetone	1-methyl-2-pyrrolidine	163
	triethyleneglycol	164
	dipropyleneglycol	165
methanol	acetone–water	166
methanol–acetone	water	167
methanol–water–methyl formate	octane or propylene glycol	156
C_4–C_7 hydrocarbons	*tert*-butanol–water	168
	paraffinic hydrocarbon	169
	C_7–C_{10} alkane hydrocarbon	155

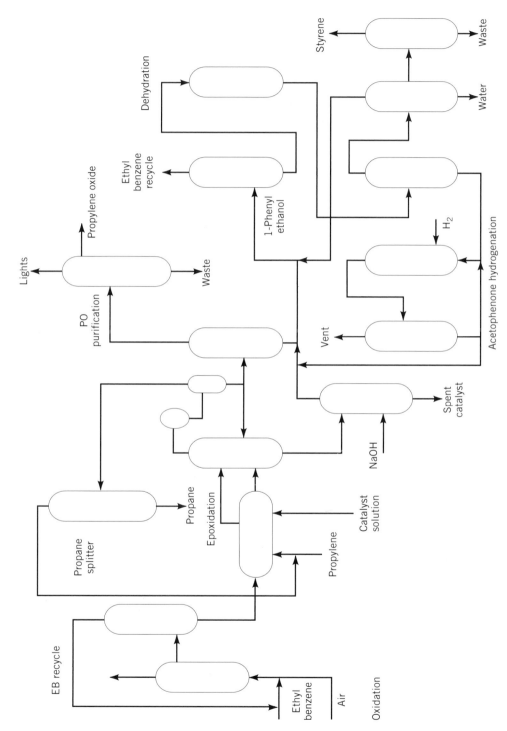

Fig. 4. The ethylbenzene (EB) hydroperoxide process to propylene oxide (PO) and styrene (170–187).

285

$$\text{C}_6\text{H}_5-\text{CH}_2\text{CH}_3 + \text{O}_2 \longrightarrow \text{C}_6\text{H}_5-\overset{\text{OOH}}{\underset{}{\text{CHCH}_3}} + \text{C}_6\text{H}_5-\overset{\text{O}}{\overset{\|}{\text{CCH}_3}}$$

EBHP is mixed with a catalyst solution and fed to a horizontal compart-mentalized reactor where propylene is introduced into each compartment. The reactor operates at 95–130°C and 2500–4000 kPa (360–580 psi) for 1–2 h, and 5–7 mol propylene/1 mol EBHP are used for a 95–99% conversion of EBHP and a 92–96% selectivity to propylene oxide. The homogeneous catalyst is made from molybdenum, tungsten, or titanium and an organic acid, such as acetate, naphthenate, stearate, etc (170,173). Heterogeneous catalysts consist of titanium oxides on a silica support (174–176).

$$\text{C}_6\text{H}_5-\overset{\text{OOH}}{\underset{}{\text{CHCH}_3}} + \text{CH}_2{=}\text{CHCH}_3 \longrightarrow \text{H}_2\text{C}-\text{CHCH}_3 + \text{C}_6\text{H}_5-\overset{\text{OH}}{\underset{}{\text{CHCH}_3}}$$

After epoxidation, propylene oxide, excess propylene, and propane are dis-tilled overhead. Propane is purged from the process; propylene is recycled to the epoxidation reactor. The bottoms liquid is treated with a base, such as sodium hydroxide, to neutralize the acids. Acids in this stream cause dehydration of the 1-phenylethanol to styrene. The styrene readily polymerizes under these conditions (177–179). Neutralization, along with water washing, allows phase separation such that the salts and molybdenum catalyst remain in the aqueous phase (179). Dissolved organics in the aqueous phase are further recovered by treatment with sulfuric acid and phase separation. The organic phase is then distilled to recover 1-phenylethanol overhead. The heavy bottoms are burned for fuel (180,181).

Crude propylene oxide separated from the epoxidation reactor effluent is further purified by a series of conventional and extractive distillations to reduce the content of aldehydes, ethylbenzene, water, and acetone (182,183).

The coproduct 1-phenylethanol from the epoxidation reactor, along with ace-tophenone from the hydroperoxide reactor, is dehydrated to styrene in a vapor-phase reaction over a catalyst of silica gel (184) or titanium dioxide (170,185) at 250–280°C and atmospheric pressure. This product is then distilled to recover purified styrene and to separate water and high boiling organics for disposal. Unreacted 1-phenylethanol is recycled to the dehydrator.

$$\text{C}_6\text{H}_5-\overset{\text{OH}}{\underset{}{\text{CHCH}_3}} \xrightarrow{-\text{H}_2\text{O}} \text{C}_6\text{H}_5-\text{CH}{=}\text{CH}_2$$

Acetophenone is separated for hydrogenation to 1-phenylethanol, which is sent to the dehydrator to produce styrene. Hydrogenation is done over a fixed-bed copper-containing catalyst at 115–120°C and pressure of 8100 kPa (80 atm), a

3:1 hydrogen-to-acetophenone ratio, and using a solvent such as ethylbenzene, to give 95% conversion of the acetophenone and 95% selectivity to 1-phenylethanol (186,187).

Hydrogen Peroxide Processes. A titanium silicalite catalyst (TS-1) is used to produce propylene oxide from propylene and hydrogen peroxide. Methanol or methanol–water mixtures are used as solvents. Yields on peroxide are quantitative and propylene selectivity is high (95%). TS-1 is a molecular sieve having an average pore diameter of 0.55 nm and a TiO_2 content of 2.6 wt %. Although TS-1 does not catalyze *tert*-butylhydroperoxide epoxidation of propylene, it has the advantage of not requiring that alcohols and water be excluded from the reactor, as the molybdenum-based catalysts for *tert*-butylhydroperoxide epoxidation, in contrast, do require. Moreover, TS-1 can catalyze the reaction with as much as 97% water (188,189).

Hydrogen peroxide (70–100%) and propylene react in the presence of a boron-containing catalyst with continuous removal of water to produce propylene oxide (190,191). Another process involves the epoxidation of propylene using hydrogen peroxide in the presence of a fluorinated alcoholic solvent, a transition-metal compound, and a nitrogen-containing organic compound (192). Propylene is epoxidized by hydrogen peroxide in the presence of an organorhenium oxide catalyst and an alkyl aryl secondary alcohol solvent resulting in about 90% peroxide conversion and 90% oxide selectivity based on peroxide (193).

Peracid Processes. Peracids, derived from hydrogen peroxide reaction with the corresponding carboxylic acids in the presence of sulfuric acid and water, react with propylene in the presence of a chlorinated organic solvent to yield propylene oxide and carboxylic acid (194–196).

$$H_2O_2 + CH_3CH_2COOH \longrightarrow CH_3CH_2COOOH + H_2O$$

Although this process has not been commercialized, Daicel operated a 12,000-t/yr propylene oxide plant based on a peracetic acid [79-21-0] process during the 1970s. The Daicel process involved metal ion-catalyzed air oxidation of acetaldehyde in ethyl acetate solvent resulting in a 30% peracetic acid solution in ethyl acetate. Epoxidation of propylene followed by purification gives propylene oxide and acetic acid as products (197). As of this writing (ca 1995), this process is not in operation.

Direct Oxidation of Propylene to Propylene Oxide. Comparison of ethylene (qv) and propylene gas-phase oxidation on supported silver and silver–gold catalysts shows propylene oxide formation to be 17 times slower than ethylene oxide (qv) formation and the CO_2 formation in the propylene system to be six times faster, accounting for the lower selectivity to propylene oxide than for ethylene oxide. Increasing gold content in the catalyst results in increasing acrolein selectivity (198). In propylene oxidation a polymer forms on the catalyst surface that is oxidized to CO_2 (199–201). Studies of propylene oxide oxidation to CO_2 on a silver catalyst showed a rate oscillation, presumably owing to polymerization on the catalyst surface upon subsequent oxidation (202).

Propylene selectivity to propylene oxide in these systems was less than 15% at less than 15% conversion. To enhance selectivity, halocarbon cosolvents such as trichlorotrifluoroethane and chlorobenzene are used in the presence of a silver on alumina catalyst. Propylene selectivity as high as 59% is reported. The principal by-products are carbon dioxide, acetaldehyde, and acetone (203). In gas-phase oxidation using an unsupported silver catalyst promoted with 30–1300 ppm by weight magnesium, selectivities of about 40% at 1.5% conversion are reported (204). A silver catalyst modified with potassium nitrate and supported on alumina having less than 50 ppm by weight leachable sodium contacted with a gas containing propylene, oxygen, ethyl chloride, nitric oxide, and nitrogen gives about 47% selectivity to propylene oxide (205).

Gas-phase oxidation of propylene using oxygen in the presence of a molten nitrate salt such as sodium nitrate, potassium nitrate, or lithium nitrate and a cocatalyst such as sodium hydroxide results in propylene oxide selectivities greater than 50%. The principal by-products are acetaldehyde, carbon monoxide, carbon dioxide, and acrolein (206–207). This same catalyst system oxidizes propane to propylene oxide and a host of other by-products (208).

Propylene oxide is also produced in liquid-phase homogeneous oxidation reactions using various molybdenum-containing catalysts (209,210), cuprous oxide (211), rhenium compounds (212), or an organomonovalent gold(I) complex (213). Whereas gas-phase oxidation of propylene on silver catalysts results primarily in propylene oxide, water, and carbon dioxide as products, the liquid-phase oxidation of propylene results in an array of oxidation products, such as propylene oxide, acrolein, propylene glycol, acetone, acetaldehyde, and others.

Noncatalytic oxidation of propylene to propylene oxide is also possible. Use of a small amount of aldehyde in the gas-phase oxidation of propylene at 200–350°C and up to 6900 kPa (1000 psi) results in about 44% selectivity to propylene oxide. About 10% conversion of propylene results (214–215). Photochemical oxidation of propylene with oxygen to propylene oxide has been demonstrated in the presence of α-diketone sensitizers and an aprotic solvent (216).

Electrochemical Process. Applying an electrical current to a brine solution containing propylene results in oxidation of propylene to propylene oxide. The chemistry is essentially the same as for the halohydrin process. All of the chemistry takes place in one reactor. Most of the reported work uses sodium or potassium bromide as the electrolyte. Bromine, generated from bromide ions at the anode, reacts with propylene and water to form propylene bromohydrin. Hydroxide generated at the cathode then reacts with the bromohydrin to yield propylene oxide (217–219). The net reaction involves transfer of two electrons:

$$CH_2{=}CHCH_3 + H_2O \longrightarrow \overset{\displaystyle O}{CH_2{-}CHCH_3} + H_2$$

By-products include propylene dibromide, bis-(bromopropyl) ether, propylene glycol, and propionic acid. Bromide losses are to the brominated organics and bromate formation. Current efficiency is a function of cell design and losses to bromate. Energy consumption decreases with an increase in electrolyte concentration and a decrease in current density. Space–time yield increases with current density. See Table 5 for performance data (see ELECTROCHEMICAL PROCESSING).

Selectivity of propylene oxide from propylene has been reported as high as 97% (222). Use of a gas cathode where oxygen is the gas, reduces required voltage and eliminates the formation of hydrogen (223). Addition of carbonate and bicarbonate salts to the electrolyte enhances cell performance and product selectivity (224). Reference 225 shows that use of alternating current results in reduced current efficiencies, especially as the frequency is increased. Electrochemical epoxidation of propylene is also accomplished by using anolyte-containing silver–pyridine complexes (226) or thallium acetate complexes (227,228).

Table 5. Electrochemical Production of Propylene Oxide

Reactor	Current efficiency, %	Current density, A/m^2	Space–time yield, kmol/(h·m^3)	Energy, kWh/kg	Reference
sieve plate	60	166			220
	90	600	2	3	219
bipolar graphite rod bed			2.7	8.6	217
bipolar trickle bed	58	2000	1.67		218
	92	65	0.1	3.2	221

Economic Aspects

Production of propylene oxide in the United States in 1993 was estimated at 1,240,000 metric tons, and as having a 10-yr average annual growth rate of 3.9% (229). Projections were for continued growth at about 4%/yr. Producers include Dow Chemical's chlorohydrin plants in Freeport, Texas, and Plaquemine, Louisiana, and ARCO Chemical's hydroperoxide plants in Bayport and Channelview, Texas. Texaco started up a 180,000-t/yr plant in Port Neches, Texas, in late 1994 where MTBE was the coproduct (230). Dow announced a 230,000-t/yr global expansion program, to be achieved through incremental gains at each of four producing sites (231). Globally, about half of the 4,000,000-t capacity is in the chlorohydrin process, and the other half evenly split between the hydroperoxide processes producing styrene and MTBE as coproducts (see Table 3). More than 95% of the propylene oxide produced is converted to derivatives. Polyether polyols (ca 60%) and propylene glycol (ca 25%) are the principal products.

Cost of manufacture is based on capital required to build the production facilities and the operating costs (raw materials, utilities, maintenance, etc). Owing to the high pressure and large volume of recycle streams in the coproduct hydroperoxide processes, the capital required for these processes is considerably higher than for the chlorohydrin process. However, the raw material cost for chlorohydrin is higher than for the hydroperoxide routes. Because the coproducts of styrene and *tert*-butanol are produced in higher volume than propylene oxide, ie, 2.4 styrene/propylene oxide and 3-4 *tert*-butanol/propylene oxide, the profitability of the coproduct is also tied to the market conditions for sale.

Storage and Materials of Construction

The National Fire Protection Association (NFPA) has given propylene oxide a flammability rating of 4 and a reactivity rating of 2, indicative of the low flash point of $-37°C$, flammability range of 2.3–36% by volume in air, and reactivity with water. OSHA classifies propylene oxide as a Class 1A flammable liquid (232). Storage should be in steel or stainless steel vessels in an inert atmosphere such as nitrogen. Pumps (qv) and other electrical devices involved in the transfer of propylene oxide should have explosion-proof motors and housings meeting the electrical requirements of the National Electric Code (NEC-NFPA 70) of Class 1, Division 1, Group B. In addition, all pumps should use mechanical seals to minimize the potential for leaks. For pressures of 103 kPa (15 psig) or less, vessels should be designed according to API-620 code. Higher pressure design should be according to the Code for Unfired Pressure Vessels, Section VIII of the ASME Boiler and Pressure Vessel Code. Propylene oxide feed should be entered below the liquid surface, avoiding free-falling liquid through the vapor space which could cause a spark. All vessels are required to be properly grounded. Storage in drums should be done in accordance with NFPA Codes for storage facilities for flammable and combustible liquids, eg, NFPA 30 and NFPA 80 (2,3).

The reportable quantity (RQ) of propylene oxide under the Comprehensive Environmental Response, Compensation, and Liability Act of 1980 (CERCLA) is 100 lbs (45.4 kg). The potential for leaks from propylene oxide storage and transfer facilities should be minimized by use of mechanical seals on pumps, all-welded tanks and pipes, a minimum of flanged connections and vessel nozzles, sprial-wound stainless steel with Teflon resin fill gaskets, Kalrez 1050 or Teflon O-rings, graphite valve packing, and Teflon paste or tape for pipe threads. Instrumentation on tanks should include high level alarms or shutdowns, temperature indicators, pressure gauges, an automatic pad/depad system using nitrogen with less than 200 ppm oxygen, and combustible gas detectors. All storage and handling facilities should have provision for spill containment such as diking and remote containment. Vapor containment from tank vents can be accomplished through use of scrubbers, flares, or closed pad/depad systems. Flame arrestors may be required to prevent flame propagation into the vessel (2,3).

Transportation

Propylene oxide is classified as a flammable liquid and hazardous substance in the U.S. Department of Transportation (DOT) Hazardous Materials Table. The DOT shipping requirement is Propylene Oxide, 3, UN1280, PG 1. The red flammable liquid label and red flammable placard are required on all shipments (2,3,233).

Shipment by rail is done in DOT 105A 100W tank cars equipped with a 517-kPa (75-psig) safety relief valve and having a capacity of 25,000 gal (95,000 L). Such tank cars are welded carbon steel having 10 cm of fiber glass insulation and a steel jacket. Loading and unloading are done through the top using a dip pipe that extends to the bottom of the car into a shallow sump. Shipment by truck is done in DOT MC-331 steel tanks of about 5,500-gal (20,800-L) capacity, having a pressure rating of 15–35 psig (205–340 kPa). ISO tanks are specified by DOT 51 and IM 101 and have a capacity of about 6,000 gal (22,700 L). All shipments of propylene oxide employ a nitrogen pad of 15–35 psig (205–340 kPa). Procedures for loading and unloading are found in References 2, 3, and 8.

Marine transportation is done by ship or barge in permanent containers on board or by ISO containers. Propylene oxide can be shipped by air freight, but specific regulations for domestic and international transport must be followed. No shipment of propylene oxide can be made on passenger ships or airlines (3,233).

UN Standard 1A1 steel nonremovable head drums or DOT 17C steel drums are used to transport propylene oxide. These drums must be tested to 36 psia (250 kPa) and a stacking height of 10 ft (3 m) (3,8,233).

Specifications and Analysis

Propylene oxide is a high purity product. Thus only the impurities are analyzed and reported. Table 6 lists typical sales specifications (8). The sales specification may vary depending on the application.

Detection of propylene oxide has practical applications in the manufacturing process, in quality control of reaction products, and in environmental monitoring. Propylene oxide content in manufacturing streams is determined by chemical methods such as hydrochlorination (234), gas chromatography, infrared spectrophotometry (235), Raman spectrometry (236), and chemical ionization mass spectrometry (237). These methods can also be applied to analyze unreacted propylene oxide in applications. Reaction with 4-(p-nitrobenzyl) pyridine followed by absorbance measurement at 570 nm is another method (238). In addition, the chemical method involving the use of periodate has been found useful in determining nanomole quantities of propylene oxide (239). Determination of

Table 6. Specifications for Propylene Oxide[a]

Description	Value[b]
acidity, as acetic acid, wt %	0.005
aldehyde, as propionaldehyde, wt %	0.040
color, Pt–Co scale	10.0
nonvolatile residue, g/100 mL	0.002
residual odor	none
specific gravity, 20/20°C	0.829[c]
	0.831
water, wt %	0.050

[a]Ref. 8.
[b]Values given are maximum unless otherwise indicated.
[c]Value given is minimum.

propylene oxide in ambient air can be accomplished by adsorption, followed by thermal desorption in a gas chromatograph and quantification by flame ionization detection (240) or a quadrupole ion trap mass spectrometer (241). Analysis of impurities and physical properties of propylene oxide is accomplished by a variety of methods described in Reference 8.

Health and Safety Factors

Propylene oxide has a variety of toxic effects on humans. Thus, exposure to propylene oxide during manufacture, storage and handling, and use should be minimized. Potential for high exposure to propylene oxide can occur during such routine activities as sampling, analysis, and maintenance, and in disconnecting hoses used for product transfer. Exposure is first minimized through proper design of sampling devices and the handling equipment and use of job procedures for both routine work and maintenance activity. A last resort is the use of personal protective equipment such as respirators, breathing air, gloves, and chemical suits (242,243). Reference 2 provides some recommended methods for safe sampling of propylene oxide.

Propylene oxide released to the atmosphere is relatively stable, having a half-life of 3–10 d. Propylene oxide reacts with hydroxide, producing formaldehyde and acetaldehyde. Reference 244 discusses possible reaction mechanisms for atmospheric reactions. In aqueous systems, propylene oxide is hydrolyzed by water to propylene glycol, or reacts with halides to form halohydrins. Some microorganisms degrade propylene oxide to propylene glycol and other products. Persistence in the environment and bioaccumulation are not likely owing to the high reactivity of propylene oxide (245). Spills and releases to the atmosphere of greater than 100 lbs (45.4 kg) are regulated and reportable to various governmental agencies (246).

Design of processes handling propylene oxide should avoid explosive mixtures (2.3–36% by volume). Reference 247 provides literature references on documented explosions and uncontrolled reactions involving propylene oxide. Reaction vessels and storage tanks must be designed to handle operating pressures and be provided with adequate relief devices to prevent overpressure. Overpressure can result from thermal expansion caused by heating or heat of reaction, runaway reactions, or vaporization (248). Firefighting equipment should include use of Class B dry chemical or foam extinguishers for small fires and deluge systems or 6% foam systems with a capacity of at least a 0.25 gal/min/ft^2 (10.2 L/min/m^2) of storage area (3).

Physiological Effects. Propylene oxide has been studied extensively for its effects on humans and animals. Accordingly, it is regulated under several U.S. Federal statutes and agencies. OSHA has established a time-weighted average 8-h permissable exposure limit of 20 ppm (50 mg/m^3). Because the odor threshold is about 200 ppm, the sweet penetrating odor of propylene oxide is not an adequate warning to prevent overexposure. Exposure to vapors above the permissible exposure limit can be irritating to the eyes and respiratory tract. Low concentrations can cause nausea; high concentrations can cause pulmonary edema (2,3). Although there are no epidemiologic data for long-term exposure to humans, studies on animals suggest that propylene oxide is a pos-

sible human carcinogen and it is classified as such by NIOSH, IARC, and NTP
(3). References 245 and 249–251 provide a summary of the published findings
on propylene oxide carcinogenicity, mutagenicity, and teratogenicity in animals.
Propylene oxide has been shown to cause central nervous system effects such as
ataxia, incoordination, and depression in rats (252–254).

Skin contact can result in irritation, blistering, or burns if confined to the
skin by clothing or shoes, but is not injurious if it evaporates readily. Liquid
propylene oxide exposure to the eyes causes injury in rabbits. Ingestion of
aqueous solutions of 5 and 10% propylene oxide showed LD_{50} values of 1.14 g/kg
for rats and 0.69 g/kg for guinea pigs (253).

Uses

Propylene oxide is a useful chemical intermediate. Additionally, it has found
use for etherification of wood (qv) to provide dimensional stability (255,256),
for purification of mixtures of organosilicon compounds (257), for disinfection of
crude oil and petroleum products (258), for sterilization of medical equipment
and disinfection of foods (259,260), and for stabilization of halogenated organics
(261–263).

Propylene oxide has found use in the preparation of polyether polyols from
recycled poly(ethylene terephthalate) (264), halide removal from amine salts via
halohydrin formation (265), preparation of flame retardants (266), alkoxylation
of amines (267,268), modification of catalysts (269), and preparation of cellulose
ethers (270,271).

Derivatives. *Polyether Polyols.* Polyether polyols produced by polymerization of propylene oxide on polyhydric alcohols account for the largest use of
propylene oxide. The starting polyhydric alcohols have from two to eight hydroxyl
groups and can be mixtures of two or more alcohols. Molecular weights of the
products range from about 400 to about 8000. Some of the polyether polyols may
be made with copolymerization of ethylene oxide. The ethylene oxide can either
be in blocks or randomly distributed in the polymer. The products are useful for
making flexible and rigid urethane foams, adhesives (qv), coatings (qv), sealants
(qv), and reaction moldable products (272).

Propylene Glycol. Propylene glycol, the second largest use of propylene
oxide, is produced by hydrolysis of the oxide with water. Propylene glycol has
very low toxicity and is, therefore, used directly in foods, pharmaceuticals (qv),
and cosmetics, and indirectly in packaging (qv). Propylene glycol also finds use
as an intermediate for numerous chemicals, in hydraulic fluids (qv), in heat-
transfer fluids (antifreeze), and in many other applications (273).

Dipropylene glycol is produced in the manufacture of propylene glycol and
finds utility as an indirect food additive, in brake-fluid formulations, cutting oils,
soaps, and solvents. Tripropylene glycol also finds use as a solvent, as textile
soaps, and as lubricants (273).

Poly(propylene glycol). Polymers of propylene oxide based on reaction with
water or propylene glycol are liquids of 400 to about 4000 molecular weight.
Viscosity increases and water solubility decreases with increasing molecular
weight. Poly(propylene glycol)s find use in cosmetics, as synthetic lubricants,

metalworking fluids, antifoam agents, heat-transfer fluids, nonionic surfactants, and chemical intermediates (274,275).

Glycol Ethers. Glycol ethers are produced by reaction of propylene oxide with various alcohols such as methanol, ethanol, butanol, and phenol. The products are the mono-, di-, and tripropylene glycol ethers. These products are used in protective coatings, inks, textile dyeing, cleaners, antiicing additives for jet fuel, and as chemical intermediates (276).

Isopropanolamines. Reaction of propylene oxide with ammonia yields mono-, di-, and triisopropanolamines. These products find use as soluble oils and solvents, emulsifiers, waterless hand cleaners, cosmetics, cleaners, and detergents. In industrial applications isopropanolamines are used in adhesives, agricultural products, corrosion inhibitors, coatings, epoxy resins, metalworking, and others (51).

BIBLIOGRAPHY

"Propylene Oxide" under "Ethylene Oxide," in *ECT* 1st ed., Vol. 5, pp. 922–923, by R. S. Aries, Consulting Chemical Engineer, and H. Schneider, R. S. Aries & Associates; "Propylene Oxide" in *ECT* 2nd ed., Vol. 16, pp. 595–609, by L. H. Horsley, The Dow Chemical Co.; in *ECT* 3rd ed., Vol. 19, pp. 246–274, by R. O. Kirk and T. J. Dempsey, The Dow Chemical Co.

1. J. H. Saunders and K. C. Frisch, *Polyurethanes Chemistry and Technology*, Part I: *Chemistry*, Robert E. Krieger Publishing Co., Malabar, Fla., 1962, pp. 32–35.
2. *Safe Handling and Storage of DOW Propylene Oxide, Product Bulletin*, Form 109-609-788 SMG, The Dow Chemical Co., Midland, Mich., 1988.
3. *Propylene Oxide, Product Safety Bulletin*, ARCO Chemical Co., Newton Square, Pa., Mar. 24, 1992.
4. C. L. Yaws and M. P. Rackley, *Chem. Eng.* **83**(8), 129–137 (1976).
5. C. L. Yaws, D. Chen, H. C. Yang, L. Tan, and D. Nico, *Hydrocarbon Process.* **68**(7), 61–64 (1989).
6. M. T. Rogers, *J. Am. Chem. Soc.* **69**, 2544–2548 (1947).
7. G. O. Curme, Jr., and F. Johnston, *Glycols*, ACS Monograph No. 114, Reinhold Publishing Corp., New York, 1952, Chapt. 11.
8. *Propylene Oxide, Product Bulletin*, Texaco Chemical Co., Austin, Tex., 1986.
9. G. C. Sinke and D. L. Hildenbrand, *J. Chem. Eng. Data* **1**(1), 74 (1962).
10. E. S. Domalski and E. D. Hearing, *J. Phys. Chem. Ref. Data* **19**(4), 881 (1990).
11. C. L. Yaws and P. Y. Chiang, *Chem. Eng.* **95**(13), 81–88 (1988).
12. C. L. Yaws, H. M. Ni, and P. Y. Chiang, *Chem. Eng.* **95**(7), 91–98 (1988).
13. C. L. Yaws and X. Pan, *Chem. Eng.* **99**(4), 130–135 (1992).
14. L. C. Wilson, W. V. Wilding, and G. M. Wilson, in T. T. Shih and D. K. Jones, eds., *AIChE Symposium Series*, Number 271, Vol. 85, pp. 25–41, New York, 1989.
15. J. Gmehling and U. Onken, *Vapor–Liquid Equilibrium Data Collection, Aqueous Organic Systems*, DECHEMA, Frankfurt, Germany, 1977.
16. S. A. Newman, ed., *Chemical Engineering Thermodynamics*, Ann Arbor Science, The Butterworth Group, Ann Arbor, Mich., 1983, Chapt. 8.
17. V. M. Muhlenbruch and G. Figurski, *Z. Phys. Chemie* **270**(2), 305–314 (1989).
18. L. C. Wilson, W. V. Wilding, and G. M. Wilson, in Ref. 13, pp. 51–72.
19. C. Yokoyama, G. Chen, H.-M. Lin, L.-S. Lee, and K.-C. Chao, in Ref. 13, pp. 79–83.

20. J. Gmehling, U. Onken, and W. Arlt, *Vapor–Liquid Equilibrium Data Collection, Organic Hydroxy Compounds*: *Alcohols*, Suppl. 1, DECHEMA, Frankfurt, Germany, 1982.

21. J. Gmehling, U. Onken, and U. Weidlich, in Ref. 20, *Alcohols and Phenols*, Suppl. 2.

22. J. Gmehling, U. Onken, and J. R. Rarey, in Ref. 20, *Alcohols and Phenols*, Suppl. 4, 1990.

23. J. Gmehling, U. Onken, and W. Arlt, in Ref. 20, *Alcohols and Phenols*, 1978.

24. R. E. Parker and N. S. Isaacs, *Chem. Rev.* **59**, 737–799 (1959).

25. G. Woods, *The ICI Polyurethanes Book*, 2nd ed., ICI Polyurethanes and John Wiley & Sons, Inc., New York, 1990, p. 36.

26. H. Takeuchi, K. Kitajima, Y. Yamamoto, and K. Mizuno, *J. Chem. Soc. Perkin Trans. 2*, 199–203 (1993).

27. J. H. Saunders and K. C. Frisch, *Polyurethanes Chemistry and Technology*, Part I, *Chemistry*, Robert E. Krieger Publishing Co., Malabar, Fla., 1962, p. 32–35.

28. U.S. Pat. 3,370,056 (Feb. 20, 1968), M. Yotsuzuka, K. Kodama, and K. Ogino (to Takeda Chemical Industries, Ltd.).

29. U.S. Pat. 3,433,751 (Mar. 18, 1969), M. Yotsuzuka, A. Keshi, N. Hashimoto, K. Kodama, and S. Nakahara (to Taketa Chemical Industries, Ltd.).

30. U.S. Pat. 3,865,806 (Feb. 11, 1975), L. R. Knodel (to The Dow Chemical Co.).

31. L. A. Dickinson, *J. Polym. Sci.* **58**, 857–868 (1962).

32. L. P. Blanchard, J. Singh, and M. D. Baijal, *Can. J. Chem.* **44**, 2679–2689 (1966).

33. S. Inoue, K. Kitamura, and T. Tsuruta, *Die Makromol. Chem.* **126**, 250–265 (1969).

34. K. Soga, K. Hyakkoku, and S. Ikeda, *J. Polym. Sci. Poly. Chem. Ed.* **17**, 2173–2180 (1979).

35. U.S. Pat. 4,166,172 (Aug. 28, 1979), H. P. Klein (to Texas Development Corp.).

36. S. Inoue, T. Aida, Y. Watanabe, and K. Kawaguchi, *Makromol. Chem., Macromol. Symp.* **42/43**, 365–371 (1991).

37. M. Kuroki, T. Aida, and S. Inoue, *Makromol. Chem.* **189**(6), 1305–1313 (1988).

38. Y. Watanabe, T. Aida, and S. Inoue, *Macromolecules* **23**(10), 2612–2617 (1990).

39. H. Haubenstock, V. Panchalingam, and G. Odian, *Makromol. Chem.* **188**(12), 2789–2799 (1987).

40. S. Aksoy, H. Altinok, H. Tumturk, and K. Alyuruk, *Polymer* **31**(6), 1142–1148 (1990).

41. G. Gee, W. C. E. Higginson, and J. B. Jackson, *Polymer* **3**, 231–242 (1962).

42. U.S. Pat. 3,338,873 (Aug. 29, 1967), A. E. Gurgiolo (to The Dow Chemical Co.).

43. M. Nakaniwa, K. Ozaki, and J. Furukawa, *Die Makromol. Chem.* **138**, 197–208 (1970).

44. F. M. Rabagliati and F. Lopez-Carrasquero, *Polymer Bull.* **14**(3–4), 331–337 (1985).

45. U.S. Pat. 3,468,817 (Sept. 23, 1969), H. L. Hsieh (to Phillips Petroleum Co.).

46. Z. Shen, J. Wu, and G. Wang, *J. Polym. Sci.: Part A: Polym. Chem.* **28**(7), 1965–1971 (1990).

47. *A Guide To Glycols*, Form No. 117-00991-89, Dow Chemical U.S.A., Midland, Mich., 1981.

48. U.S. Pat. 4,160,116 (July 3, 1979), M. Mieno, H. Mori, J. Naknaishi, and J. Kasai (to Showa Denko KK).

49. U.S. Pat. 5,260,495 (Nov. 9, 1993), M. W. Forkner (to Union Carbide Chemicals & Plastics Technology Corp.).

50. K. Slipko and J. Chlebicki, *Pol. J. Chem.* **53**, 2331–2337 (1979).

51. *The Alkanolamines Handbook*, Form No. 111-1159-88R-SAI, The Dow Chemical Co., Midland, Mich., 1988.

52. W. J. Peppel, *Ind. Eng. Chem.* **50**(5), 767–770 (1958).

53. W. Kuran and T. Listos, *Makromol. Chem.* **193**(4), 945–956 (1992).

54. U.S. Pat. 3,542,808 (Nov. 24, 1970), R. C. Vander Linden, J. M. Salva, and P. A. C. Smith (to Esso Research and Engineering Co.).
55. N. Adachi, Y. Kida, and K. Shikata, *J. Polym. Sci. Poly. Chem. Ed.* **15**, 937–944 (1977).
56. J. A. Durden, Jr., H. A. Stansbury, Jr., and W. H. Catlette, *J. Am. Oil Chem. Soc.* **82**, 3082–3084 (1960).
57. J. D. Malkemus, *J. Am. Oil Chem. Soc.* **33**, 571–574 (1956).
58. J. Chlebicki, *Roczniki Chemii* **48**, 1241 (1974).
59. A. Rosowsky, in A. Weissberger, ed., *Heterocyclic Compounds with Three- and Four-Membered Rings*, Part One, Interscience Publishers, a division of John Wiley & Sons, Inc., New York, 1964, pp. 289, 308.
60. F. Scholnick, H. A. Monroe, Jr., E. J. Saffese, and A. N. Wrigley, *J. Am. Oil Chem. Soc.* **44**, 40–42 (1967).
61. U.S. Pat. 3,890,300 (June 17, 1975), M. Huchette and G. Fleche (to Roquette Freres).
62. G. Lammers, E. J. Stamhuis, and A. A. C. M. Beenackers, *Ind. Eng. Chem. Res.* **32**(5), 835–842 (1993).
63. T. Viswanathan, A. Toland, R. Liu, and N. R. Jagannathan, *J. Polym. Sci.: Part C: Polym. Lett.* **28**(3), 95–100 (1990).
64. J. S. Ayers, M. J. Petersen, B. E. Sheerin, and G. S. Bethell, *J. Chromatog.* **294**, 195–205 (1984).
65. A. N. Wrigley, F. D. Smith, and A. J. Stirton, *J. Am. Oil Chem. Soc.* **36**, 34–36 (1959).
66. M. A. Rousselle, M. L. Nelson, H. H. Ramey, Jr., and G. L. Barker, *Ind. Eng. Chem. Prod. Res. Dev.* **19**, 654–659 (1980).
67. U.S. Pat. 4,609,729 (Sept. 2, 1986), J. L. Garner (to The Dow Chemical Co.).
68. U.S. Pat. 4,661,589 (Apr. 28, 1987), G. L. Adams and C. D. Messelt (to The Dow Chemical Co.).
69. U.S. Pat. 5,283,368 (Feb. 1, 1994), J. E. Shaw (to Phillips Petroleum Co.).
70. Ref. 59, p. 328.
71. Ref. 59, p. 321.
72. Ref. 59, p. 337.
73. U.S. Pat. 5,218,147 (June 8, 1993), J. E. Shaw (to Phillips Petroleum Co.).
74. M. Prochazka and P. Durdovic, *Coll. Czech. Chem. Comm.* **42**, 2401–2407 (1977).
75. Ref. 59, p. 433.
76. M. Inoue, K. Chano, O. Itoh, T. Sugita, and K. Ichikawa, *Bull. Chem. Soc. Japan* **53**(2), 458–463 (1980).
77. Ref. 59, pp. 399–400.
78. N. G. Gaylord and E. I. Becker, *Chem. Rev.* **49**, 413–533 (1951).
79. S. Winstein and R. B. Henderson, in R. C. Elderfield, ed., *Heterocyclic Compounds*, Vol. 1, John Wiley & Sons, Inc., New York, 1950, p. 57.
80. M. J. Astle, *The Chemistry of Petrochemicals*, Reinhold Publishing Corp., New York, 1956, pp. 184–185.
81. T. Imanaka, Y. Okamoto, and S. Teranishi, *Bull. Chem. Soc. Japan*, **45**, 3251–3254 (1972).
82. U.S. Pat. 5,262,371 (Nov. 16, 1993), M. K. Faraj (to ARCO Chemical Technology, LP).
83. D. Ostgard, F. Notheisz, A. G. Zsigmond, G. V. Smith, and M. Bartok, *J. Catal.* **129**(2), 519–523 (1991).
84. I. Palinko, F. Notheisz, and M. Bartok, *J. Mol. Catal.* **63**(1), 43–54 (1990).
85. M. Bartok, F. Notheisz, A. G. Zsigmond, and G. V. Smith, *J. Catal.* **100**(1), 39–44 (1986).
86. M. Bartok, F. Notheisz, and A. G. Zsigmond, *J. Catal.* **63**(2), 364–371 (1980).
87. F. Notheisz, A. Molnar, A. G. Zsigmond, and M. Bartok, *J. Catal.* **98**(1), 131–137 (1986).

88. M. Berry, S. G. Davies, and M. L. H. Green, *J. Chem. Soc. Chem. Comm.*, 99 (1978).
89. J. L. E. Erickson and F. E. Collins, Jr., *J. Org. Chem.* **30**, 1050–1052 (1965).
90. U.S. Pat. 3,725,438 (Apr. 3, 1973), B. J. Barone and W. F. Brill (to PetroTex Chemical Corp.).
91. J. F. W. McOmie, *Protective Groups in Organic Chemistry*, Plenum Press, New York, 1973, p. 326.
92. M. Fujiwara, M. Imada, A. Baba, and H. Matauda, *J. Org. Chem.* **53**(25), 5974–5977 (1988).
93. Ref. 59, pp. 350–351.
94. G. S. Yoneda, M. T. Griffin, and D. W. Carlyle, *J. Org. Chem.* **40**(3), 375–377 (1975).
95. Ref. 59, p. 346.
96. Ref. 59, p. 365.
97. Ref. 59, p. 384.
98. H. Steinberg and R. J. Brotherton, *Organoboron Chemistry*, John Wiley & Sons, Inc., New York, 1966, pp. 82, 520, and 549.
99. Ref. 98, p. 487.
100. J. D. Edwards, W. Gerrard, and M. F. Lappert, *J. Chem. Soc.*, 348 (1957).
101. F. G. A. Stone and H. J. Emeleus, *J. Chem. Soc.*, 2755–2759 (1950).
102. N. I. Shuikin and I. F. Bel'skii, *J. Gen. Chem. USSR* **29**, 2936–2938 (1959).
103. I. Shibata, A. Baba, H. Iwasaki, and H. Matsuda, *J. Org. Chem.* **51**, 2177–2184 (1986).
104. N. Watanabe, S. Uemura, and M. Okano, *Bull. Chem. Soc. Japan* **52**(12), 3611–3614 (1979).
105. W. Kuran, S. Pasynkiewicz, and J. Serzyko, *J. Organomet. Chem.* **73**, 187–192 (1974).
106. A. J. Lundeen and A. C. Oehlschlager, *J. Organomet. Chem.* **25**, 337–344 (1970).
107. Ref. 59, pp. 340–342, 421, 428, 430, and 432.
108. Ref. 59, pp. 440, 442–443, 451, and 456.
109. A. C. Fyvie, *Chem. Ind. London* (10), 384–388 (1964).
110. R. B. Stobaugh, V. A. Calarco, R. A. Morris, and L. W. Stroud, *Hydrocarbon Process.* **52**(1), 99–108 (1973).
111. K. H. Simmrock, *Hydrocarbon Process.* **57**(11), 105–113 (1978).
112. A. J. Gait, in E. G. Hancock, ed., *Propylene and Its Industrial Derivatives*, Halsted Press, a division of John Wiley & Sons, Inc., New York, 1973, pp. 273–297.
113. U.S. Pat. 5,146,011 (Sept. 8, 1992), M. Shen and J. A. Wojtowicz (to Olin Corp.).
114. U.S. Pat. 4,008,133 (Feb. 15, 1977), A. P. Gelbein and J. T. Kwon (to The Lummus Co.).
115. *Hydrocarbon Process.* **58**(11), 239 (1979).
116. U.S. Pat. 4,496,777 (Jan. 29, 1985), G. D. Suciu, J. T. Kwon, and A. M. Shaban (to The Lummus Co.).
117. S. Carra, E. Santacesaria, M. Morbidelli, and L. Cavalli, *Chem. Eng. Sci.* **34**(9), 1123–1132 (1979).
118. S. Carra, E. Santacesaria, M. Morbidelli, and G. Buzzi, *Chem. Eng. Sci.* **34**(9), 1133–1140 (1979).
119. I. Mochida, T. Miyazaki, and H. Fujitsu, *Applied Catalysis* **32**, 37–44 (1987).
120. U.S. Pat. 4,243,492 (Jan. 6, 1981), T. Yamamura and co-workers (to Showa Denko KK).
121. U.S. Pat. 5,187,287 (Feb. 16, 1993), T. T. Shih (to ARCO Chemical Technology, LP).
122. U.S. Pat. 4,692,535 (Sept. 8, 1987), H. V. Larson and H. D. Gillman (to Atlantic Richfield Co.).
123. U.S. Pat. 5,235,075 (Aug. 10, 1993), G. W. Bachman and R. K. Brown (to The Dow Chemical Co.).

124. U.S. Pat. 5,248,794 (Sept. 28, 1993), M. L. Chappell and C. T. Costain (to The Dow Chemical Co.).
125. M. A. Zeitoun and W. F. McIlhenny, *Treatment of Wastewater from the Production of Polyhydric Organics*, EPA Project, 12020-EEQ, U.S. EPA, Washington, D.C., 1971, pp. 1–24.
126. L. M. V. Raja, G. Elamvaluthy, R. Palaniappan, and R. M. Krishnan, *Appl. Biochem. Biotechnol.* **28–29**, 827–841 (1991).
127. U.S. Pat. 4,277,405 (July 7, 1981), G. J. Apanel (to The Lummus Co.).
128. U.S. Pat. 4,496,752 (Jan. 29, 1985), A. P. Gelbein and J. T. Kwon (to The Lummus Co.).
129. U.S. Pat. 4,126,526 (Nov. 21, 1978), J. T. Kwon and A. P. Gelbein (to The Lummus Co.).
130. U.S. Pat. 4,376,865 (Mar. 15, 1983), G. J. Apanel (to The Lummus Co.).
131. U.S. Pat. 4,496,753 (Jan. 29, 1985), J. T. Kwon and G. D. Suciu (to Lummus Crest, Inc.).
132. U.S. Pat. 4,415,460 (Nov. 15, 1983), G. D. Suciu and J. E. Paustian (to The Lummus Co.).
133. U.S. Pat. 4,240,885 (Dec. 23, 1980), G. D. Suciu and J. E. Paustian (to The Lummus Co.).
134. U.S. Pat. 4,126,526 (Nov. 21, 1978), J. T. Kwon and A. P. Gelbein (to The Lummus Co.).
135. U.S. Pat. 4,410,714 (Oct. 18, 1983), G. J. Apanel (to The Lummus Co.).
136. U.S. Pat. 4,128,587 (Dec. 5, 1978), J. C. Jubin (to Atlantic Richfield Co.).
137. J. Sobczak and J. J. Ziolkowski, *J. Molecular Catal.* **13**, 11–42 (1981).
138. U.S. Pat. 5,107,067 (Apr. 21, 1992), E. T. Marquis, K. P. Keating, J. R. Sanderson, and W. A. Smith (to Texaco Inc.).
139. U.S. Pat. 4,703,027 (Oct. 27, 1987), E. T. Marquis, J. R. Sanderson, and K. P. Keating (to Texaco Chemical Co.).
140. Eur. Pat. 0188912 (July 30, 1986), E. T. Marquis and co-workers (to Texaco Development Corp.).
141. U.S. Pat. 4,977,285 (Dec. 11, 1990), E. T. Marquis, K. P. Keating, R. A. Meyer, and J. R. Sanderson (to Texaco Chemical Co.).
142. U.S. Pat. 5,274,138 (Dec. 28, 1993), K. P. Keating, E. T. Marquis, and M. A. Mueller (to Texaco Chemical Co.).
143. U.S. Pat. 4,992,566 (Feb. 12, 1991), E. T. Marquis, K. P. Keating, J. R. Sanderson, and R. A. Meyer (to Texaco Chemical Co.).
144. U.S. Pat. 4,455,283 (June 19, 1984), N. H. Sweed (to Atlantic Richfield Co.).
145. U.S. Pat. 4,485,074 (Nov. 27, 1984), R. B. Poenisch (to Atlantic Richfield Co.).
146. U.S. Pat. 5,101,052 (Mar. 31, 1992), R. A. Meyer and E. T. Marquis (to Texaco Chemical Co.).
147. U.S. Pat. 5,128,492 (July 7, 1992), W. A. Smith, R. A. Meyer, and E. T. Marquis (to Texaco Chemical Co.).
148. U.S. Pat. 4,939,281 (July 3, 1990), E. T. Marquis, J. R. Sanderson, and K. P. Keating (to Texaco Chemical Co.).
149. U.S. Pat. 4,315,896 (Feb. 16, 1982), P. D. Taylor and M. T. Mocella (to Atlantic Richfield Co.).
150. U.S. Pat. 5,093,509 (Mar. 3, 1992), R. A. Meyer and E. T. Marquis (to Texaco Chemical Co.).
151. U.S. Pat. 4,598,057 (July 1, 1986), B. H. Isaacs (to Atlantic Richfield Co.).
152. U.S. Pat. 4,691,034 (Sept. 1, 1987), J. R. Sanderson, W. A. Smith, E. T. Marquis, and K. P. Keating (to Texaco, Inc.).
153. U.S. Pat. 4,691,035 (Sept. 1, 1987), J. R. Sanderson, E. T. Marquis, W. A. Smith, and K. P. Keating (to Texaco, Inc.).

154. U.S. Pat. 5,106,458 (Apr. 21, 1992), R. A. Meyer, E. T. Nguyen, and A. Smith (to Texaco Chemical Co.).
155. U.S. Pat. 5,262,017 (Nov. 16, 1993), R. A. Meyer, W. A. Smith, M. A. Mueller, and G. B. Demoll (to Texaco Chemical Co.).
156. U.S. Pat. 5,133,839 (July 28, 1992), T. T. Shih (to ARCO Chemical Technology, LP).
157. U.S. Pat. 5,286,884 (Feb. 15, 1994), R. S. Cowley and D. D. Kinzler (to Texaco Chemical Co.).
158. M. B. Ward, K. Mizuno, and J. H. Lunsford, *J. Molecular Catal.* **27**, 1–10 (1984).
159. H. Mimoun, *J. Molecular Catalysis* **7**, 1–29 (1980).
160. R. A. Sheldon, *J. Molecular Catalysis* **7**, 107–126 (1980).
161. U.S. Pat. 5,116,467 (May 26, 1992), E. T. Marquis, G. P. Speranza, Y. E. Sheu, W. K. Culbreth, III, and D. G. Pattratz (to Texaco Chemical Co.).
162. U.S. Pat. 5,116,465 (May 26, 1992), E. L. Yeakey and E. T. Marquis (to Texaco Chemical Co.).
163. U.S. Pat. 5,116,466 (May 26, 1992), E. T. Marquis, G. P. Speranza, Y. E. Sheu, W. K. Culbreth, III, and D. G. Pattratz (to Texaco Chemical Co.).
164. U.S. Pat. 5,139,622 (Aug. 18, 1992), E. T. Marquis, G. P. Speranza, Y. E. Sheu, W. K. Culbreth, III, and D. G. Pattratz (to Texaco Chemical Co.).
165. U.S. Pat. 5,160,587 (Nov. 3, 1992), E. T. Marquis, G. P. Speranza, Y. E. Sheu, W. K. Culbreth, III, and D. G. Pattratz (to Texaco Chemical Co.).
166. U.S. Pat. 4,971,661 (Nov. 20, 1990), R. A. Meyer, K. P. Keating, W. A. Smith, and R. M. Steinberg (to Texaco Chemical Co.).
167. U.S. Pat. 4,140,588 (Feb. 20, 1979), J. P. Schmidt (to Halcon Research and Development Corp.).
168. U.S. Pat. 5,006,206 (Apr. 9, 1991), T. T. Shih and W. J. Sim (to ARCO Chemical Technology, Inc.).
169. U.S. Pat. 5,127,997 (July 7, 1992), W. A. Smith, R. A. Meyer, and E. T. Nguyen (to Texaco Chemical Co.).
170. U.S. Pat. 3,351,635 (Nov. 7, 1967), J. Kollar (to Halcon International, Inc.).
171. U.S. Pat. 3,459,810 (Aug. 5, 1969), C. Y. Choo and R. L. Golden (to Halcon International, Inc.).
172. U.S. Pat. 4,066,706 (Jan. 3, 1978), J. P. Schmidt (to Halcon International, Inc.).
173. U.S. Pat. 3,849,451 (Nov. 19, 1974), T. W. Stein, H. Gilman, and R. L. Bobeck (to Halcon International, Inc.).
174. U.S. Pat. 3,642,833 (Feb. 15, 1972), H. Wulff and P. Haynes (to Shell Oil Co.).
175. U.S. Pat. 3,702,855 (Nov. 14, 1972), C. S. Bell and H. P. Wulff (to Shell Oil Co.).
176. U.S. Pat. 3,829,392 (Aug. 13, 1974), H. P. Wulff (to Shell Oil Co.).
177. U.S. Pat. 3,860,662 (Jan. 14, 1975), J. Kollar (to Halcon International, Inc.).
178. U.S. Pat. 3,947,500 (Mar. 30, 1976), J. Kollar (to Halcon International, Inc.).
179. U.S. Pat. 5,171,868 (Dec. 15, 1992), R. S. Albal, R. N. Cochran, and T. B. Hsu (to ARCO Chemical Technology, LP).
180. U.S. Pat. 5,276,235 (Jan. 4, 1994), W. S. Dubner (to ARCO Chemical Technology, LP).
181. U.S. Pat. 5,210,354 (May 11, 1993), W. S. Dubner and R. N. Cochran (to ARCO Chemical Technology, LP).
182. U.S. Pat. 3,881,996 (May 6, 1975), J. P. Schmidt (to Oxirane Corp.).
183. U.S. Pat. 3,632,482 (Jan. 4, 1972), S. E. Heary and S. F. Newmann (to Shell Oil Co.).
184. U.S. Pat. 4,049,736 (Sept. 20, 1977), J. J. Lamson, R. H. Hall, E. Stroiwas, and L. D. Yats (to The Dow Chemical Co.).
185. U.S. Pat. 3,442,963 (May 6, 1969), E. I. Korchak (to Halcon International, Inc.).
186. U.S. Pat. 3,927,120 (Dec. 16, 1975), H. R. Grane and T. S. Zak (to Atlantic Richfield Co.).
187. U.S. Pat. 3,927,121 (Dec. 16, 1975), H. R. Grane and T. S. Zak (to Atlantic Richfield Co.).

188. M. G. Clerici, G. Bellussi, and U. Romano, *J. Catal.* **129**, 159–167 (1991).

189. R. L. Burwell, Jr., *Chemtracts Inorg. Chem.* **3**, 344–346 (1991); see Ref. 188.

190. U.S. Pat. 4,303,587 (Dec. 1, 1981), J. P. Schirmann and S. Y. Delavarenne (to Produits Chimiques Ugine Kuhlmann).

191. U.S. Pat. 4,303,586 (Dec. 1, 1981), J. P. Schirmann and S. Y. Delavarenne (to Produits Chimiques Ugine Kuhlmann).

192. U.S. Pat. 4,024,165 (May 17, 1977), T. M. Shryne and L. Kim (to Shell Oil Co.).

193. U.S. Pat. 5,166,372 (Nov. 24, 1992), G. L. Crocco, W. F. Shum, J. G. Zajacek, and H. S. Kesling, Jr. (to ARCO Chemical Technology, LP).

194. U.S. Pat. 4,177,196 (Dec. 4, 1979), A. M. Hildon and P. F. Greenhalgh (to Interox Chemicals Limited).

195. *Hydrocarbon Process.* **64**(11), 166 (1985).

196. U.S. Pat. 4,424,391 (Jan. 3, 1984), R. Walraevens and L. Lerot (to Solvay & Cie).

197. K. Yamagishi, O. Kageyama, H. Haruki, and Y. Numa, *Hydrocarbon Process.* **55**(11), 102–104 (1976).

198. P. V. Greenen, H. J. Boss, and G. T. Pott, *J. Catal.* **77**, 499–510 (1982).

199. N. W. Cant and W. K. Hall, *J. Catal.* **52**, 81–94 (1978).

200. I. L. C. Freriks, R. Bouwman, and P. V. Geenen, *J. Catal.* **65**, 311–317 (1980).

201. M. Stoukides and C. G. Vayenas, *J. Catal.* **82**, 45–55 (1983).

202. M. Stoukides and C. G. Vayenas, *J. Catal.* **74**, 266–274 (1982).

203. U.S. Pat. 4,474,974 (Oct. 2, 1984), J. R. Sanderson, S. B. Cavitt, and E. T. Marquis (to Texaco Inc.).

204. U.S. Pat. 4,859,786 (Aug. 22, 1989), G. E. Vrieland (to The Dow Chemical Co.).

205. U.S. Pat. 4,994,587 (Feb. 19, 1991), T. M. Notermann and E. M. Thorsteinson (to Union Carbide Chemicals and Plastics Co., Inc.).

206. U.S. Pat. 4,883,889 (Nov. 28, 1989), B. T. Pennington (to Olin Corp.).

207. U.S. Pat. 4,943,643 (July 24, 1990), B. T. Pennington and M. C. Fullington (to Olin Corp.).

208. U.S. Pat. 4,885,374 (Dec. 5, 1989). B. T. Pennington (to Olin Corp.).

209. C. Daniel and G. W. Keulks, *J. Catal.* **24**, 529–535 (1972).

210. U.S. Pat. 4,046,783 (Sept. 6, 1977), S. B. Cavitt (to Texaco Development Corp.).

211. C. Daniel, J. R. Monnier, and G. W. Keulks, *J. Catal.* **31**, 360–368 (1973).

212. U.S. Pat. 3,316,279 (Apr. 25, 1967), D. M. Fenton (to Union Oil Co.).

213. U.S. Pat. 4,391,756 (July 5, 1983), P. L. Kuch, D. R. Herrington, and J. M. Eggett (to The Standard Oil Co.).

214. U.S. Pat. 5,241,088 (Aug. 31, 1993), J. L. Meyer, B. T. Pennington, and M. C. Fullington (to Olin Corp.).

215. U.S. Pat. 5,117,011 (May 26, 1992), B. T. Pennington and M. C. Fullington (to Olin Corp.).

216. U.S. Pat. 4,481,092 (Nov. 6, 1984), K. T. Mansfield (to Ciba-Geigy Corp.).

217. T. Bejerano, S. Germain, F. Goodridge, and A. R. Wright, *Trans. Inst. Chem. Eng.* **58**(1), 28–32 (1980).

218. A. Manji and C. W. Oloman, *J. Appl. Electrochem.* **17**, 532–544 (1987).

219. F. Goodridge, S. Harrison, and R. E. Plimley, *J. Electroanal. Chem.* **214**, 283–293 (1986).

220. K. Scott, C. Odouza, and W. Hui, *Chem. Eng. Sci.* **47**(9–11), 2957–2962 (1992).

221. K. G. Ellis and R. E. W. Jansson, *J. Appl. Electrochem.* **11**, 531–535 (1981).

222. K. G. Ellis and R. E. W. Jansson, *J. Appl. Electrochem.* **13**, 651–656 (1983).

223. U.S. Pat. 4,726,887 (Feb. 23, 1988), J. M. McIntyre (to The Dow Chemical Co.).

224. U.S. Pat. 4,634,506 (Jan. 6, 1987), L. R. Novak and D. J. Milligan (to The Dow Chemical Co.).

225. R. C. Alkire and J. E. Tsai, *J. Electrochem. Soc.* **129**(5), 1157–1158 (1982).

226. J. M. Van Der Eijk, T. J. Peters, N. De Wit, and H. A. Colijn, *Catal. Today* **3**, 259–266 (1988).
227. J. A. Switzer, E. L. Moorehead, and D. M. Dalesandro, *J. Electrochem. Soc.: Electrochem. Sci. Tech.* **129**(10), 2232–2237 (1982).
228. U.S. Pat. 4,290,959 (Sept. 22, 1981), R. S. Barker (to Halcon Research and Development Corp.).
229. *Chem. Eng. News* **72**(15), 12–15 (1994).
230. *Chem. Week* **155**(3), 16 (1994).
231. *Chem. Week* **155**(15), 14 (1994).
232. *U.S. Code of Federal Regulations*, Title 29, Paragraph 1910.106, U.S. Government Printing Office, Washington, D.C., 1994.
233. *U.S. Code of Federal Regulations*, Title 49, Parts 172, 174, 176, and 178–180, U.S. Government Printing Office, Washington, D.C., 1994.
234. J. L. Jungnickel, E. D. Peters, A. Polgar, and F. T. Weiss, in J. Mitchell, Jr., and co-workers, eds., *Organic Analysis*, Vol. I, Interscience Publishers, Inc., New York, 1953, pp. 127–154.
235. O. D. Shreve, M. R. Heether, H. B. Knight, and D. Swern, *Anal. Chem.* **23**(2), 277–282 (1951).
236. J. R. Villarreal and J. Laane, *J. Chem. Phys.* **62**(1), 303–304 (1975).
237. S. Suzuki, Y. Hori, R. C. Das, and O. Koga, *Bull. Chem. Soc. Jpn.* **53**, 1451–1452 (1980).
238. S. C. Agarwal, B. L. Van Duuren, and T. J. Kneip, *Bull. Environm. Contam. Toxicol.* **23**, 825–829 (1979).
239. H. E. Mishmash and C. E. Meloan, *Anal. Chem.* **44**(4), 835–836 (1972).
240. J. W. Russell, *Env. Sci. Tech.* **9**(13), 1175–1178 (1975).
241. T. J. Kelly, P. J. Callahan, J. Plell, and G. F. Evans, *Environ. Sci. Technol.* **27**(6), 1146–1153 (1993).
242. G. H. Flores, *Chem. Eng. Prog.* **79**(3), 39–43 (1983).
243. *U.S. Code of Federal Regulations*, Title 29, paragraphs 1910.134 and 1910.1000, U.S. Government Printing Office, Washington, D.C.
244. D. Grosjean, *J. Air Waste Manage. Assoc.* **40**, 1522–1531 (1990).
245. W. Meylan, L. Papa, C. T. De Rosa, and J. F. Stara, *Toxicol. Ind. Health* **2**(3), 219–260 (1986).
246. *U.S. Code of Federal Regulations*, Title 40, Parts 302, 304, 370, and 403, U.S. Government Printing Office, Washington, D.C.
247. L. Bretherick, *Bretherick's Handbook of Reactive Chemical Hazards*, 4th ed., Butterworth & Co., Boston, Mass., 1990, pp. 379–380.
248. G. A. Pogany, *Chem. Ind.* 16–21 (1979).
249. J. Bootman, D. C. Lodge, and H. E. Whalley, *Mutat. Res.* **67**, 101–112 (1979).
250. A. K. Giri, *Mutat. Res.* **277**, 1–9 (1992).
251. R. Nilsson, B. Molholt, and E. V. Sargent, *Reg. Toxicol. Pharmacol.* **14**, 229–244 (1991).
252. A. Ohnishi and Y. Murai, *Environ. Res.* **60**, 242–247 (1993).
253. C. Hine, V. K. Rowe, E. R. White, K. I. Darmer, Jr., and G. T. Youngblood, in G. D. Clayton and F. E. Clayton, ed., *Patty's Industrial Hygiene and Toxicology*, 3rd rev. ed., Vol. 2A, *Toxicology*, Wiley-Interscience, John Wiley & Sons, Inc., New York, 1978, pp. 2148, 2157, and 2186–2191.
254. A. Ohnishi and co-workers, *Arch. Environ. Health* **43**(5), 353–356 (1988).
255. Jpn. Pat. 01200903 (Aug. 14, 1989), S. Ando, T. Teraika, and T. Nishizumi (to Sanyo-Kokusaku Pulp Co.).
256. R. K. Jain and W. G. Glasser, *Holzforschung* **47**(4), 325–332 (1993).
257. Jpn. Pat. 61254594 (Nov. 12, 1986), M. Takamizawa, T. Ishihara, and M. Endo (to Shin-Etsu Chemical Industry Co., Ltd.).

258. Eur. Pat. 185612 (June 25, 1986), B. Mebes (to Sanitized Verwertungs A-G).

259. H. Sato, T. Kidaka, and M. Hori, *Int. J. Artif. Organs* **8**(2), 109–114 (1985).

260. U.S. Pat. 3,919,189 (Nov. 11, 1975), R. A. Empey and D. J. Pettitt (to Kelco Co.).

261. Jpn. Pat. 05301832 (Nov. 16, 1993), M. Fukushima, K. Ooishi, and Y. Ootoshi (to Asahi Glass Co.).

262. Jpn. Pat. 04173749 (June 22, 1992), M. Fukushima, U. Tazawa, K. Ooishi (to Asahi Glass Co.).

263. U.S. Pat. 4,108,910 (Aug. 22, 1978), M. Godfrold and R. Gerkens (to Solvay & Cie.).

264. U.S. Pat. 4,485,196 (Nov. 27, 1984), G. P. Speranza, R. A. Grigsby, Jr., and M. E. Brennan (to Texaco Inc.).

265. U.S. Pat. 4,532,304 (July 30, 1985), M. J. Fazio (to The Dow Chemical Co.).

266. U.S. Pat. 4,631,148 (Dec. 23, 1986), D. P. Braksmayer, F.-T. H. Lee, F. R. Scholer (to FMC Corp.).

267. U.S. Pat. 4,479,010 (Oct. 23, 1984), M. Cuscurida and W. P. Krause (to Texaco Inc.).

268. U.S. Pat. 4,670,557 (June 2, 1987), W.-Y. Su (to Texaco Inc.).

269. L. I. Simandi, E. Zahonyi-Budo, and J. Bodnar, *Inorganica Chimica Acta*, **65**(5), L181–L183 (1982).

270. U.S. Pat. 4,661,589 (Apr. 28, 1987), G. L. Adams and C. D. Messelt (to The Dow Chemical Co.).

271. U.S. Pat. 4,609,729 (Sept. 2, 1986), J. L. Garner (to The Dow Chemical Co.).

272. *621 Ways to Succeed, 1993–1994 Materials Selection Guide*, Form No. 304-00286-1292X SMG, The Dow Chemical Co., Midland, Mich., 1993.

273. *A Guide to Glycols*, Form No. 117-00991-89, The Dow Chemical Co., Midland, Mich., 1981.

274. *The Polyglycol Handbook*, Form No. 118-1026-889-AMS, The Dow Chemical Co., 1988.

275. *Overview Guide to Dow Polyglycols*, Form No. 118-01163-592 AMS, The Dow Chemical Co., 1992.

276. *The Glycol Ethers Handbook*, Form No. 110-00363-290X AMS, Dow Chemical USA, 1990.

David L. Trent
The Dow Chemical Company

PROSTAGLANDINS

Prostaglandins, typified by prostaglandin E_2 (**1**), are a family of naturally occurring substances found in animals and humans. They play important and diverse roles in both health and disease. Some plant species also contain prostaglandins and the coral species *Plexaura homomalla* is one of the richest known sources (1.5–3% of dry weight) (1). Prostaglandins are biosynthesized from 20 carbon polyunsaturated fatty acids. In humans, the predominant precursor of prostaglandins (PGs) is arachidonic acid [*506-32-1*] (**2**) which is available either from the diet or by anabolic conversion of linoleic acid, an essential fatty acid.

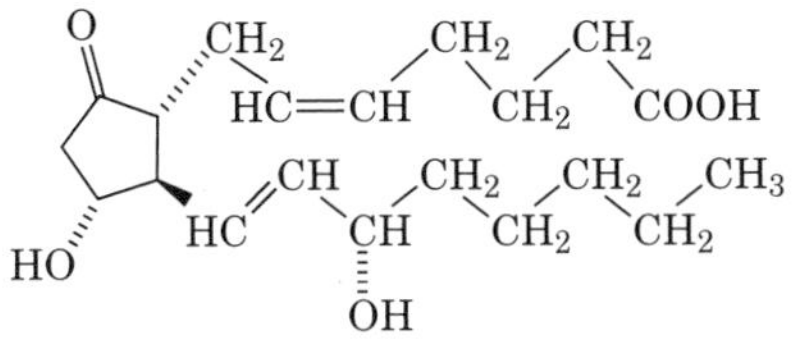

(1)

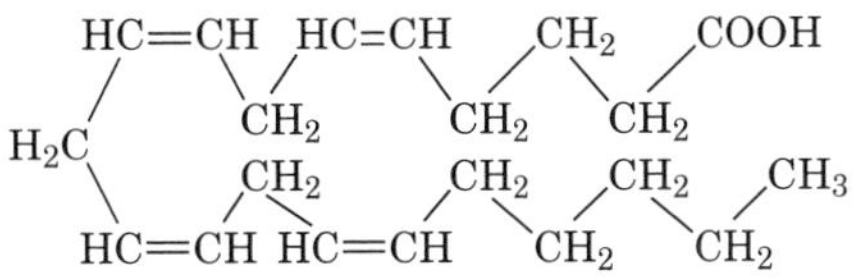

(2)

The enzyme system responsible for the biosynthesis of PGs is widely distributed in mammalian tissues and has been extensively studied (2). It is referred to as prostaglandin H synthase (PGHS) and exhibits both cyclooxygenase and peroxidase activity. In addition to the classical PGs two other prostanoid products, thromboxane A_2 [57576-52-0] (TxA_2) (3) and prostacyclin [35121-78-9] (PGI_2) (4) are also derived from the action of the enzyme system on arachidonic acid (Fig. 1).

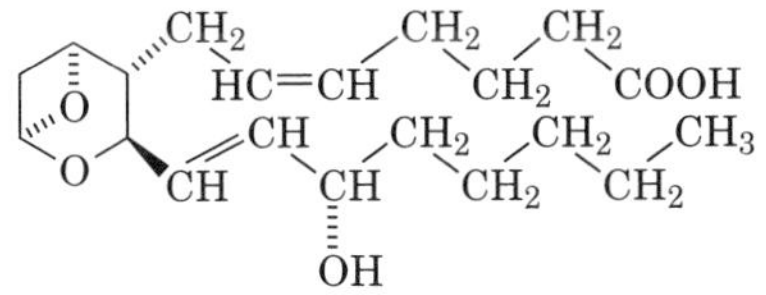

(3)

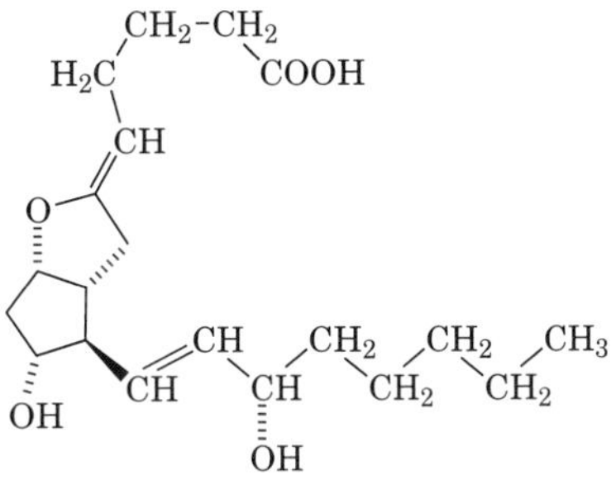

(4)

Prostaglandins were discovered in the 1930s when it was noted (3) that fresh human semen caused strips of human uterine tissue to either relax or contract depending on whether or not the tissue donor had borne children. In 1935 extracts of human seminal fluid were observed to cause a fall in blood pressure in laboratory animals, and contraction of a variety of smooth muscle tissues (4,5). This factor was differentiated from known agents having similar activities, eg, adrenaline and acetylcholine, and named prostaglandin in the belief that it originated from prostate glands (5). The name is somewhat of a misnomer because PGs occur much more ubiquitously. Later work (6) led to the isolation and structural elucidation of prostaglandin E_1 [745-65-3] (PGE_1) (9) and prostaglandin $F_{1\alpha}$ ($PGF_{1\alpha}$) (10).

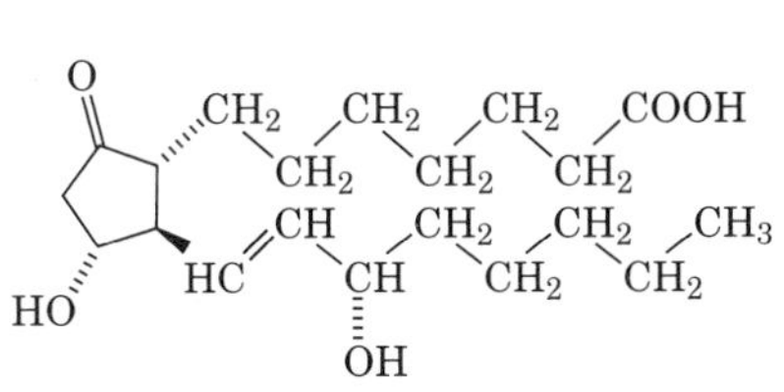

(9)

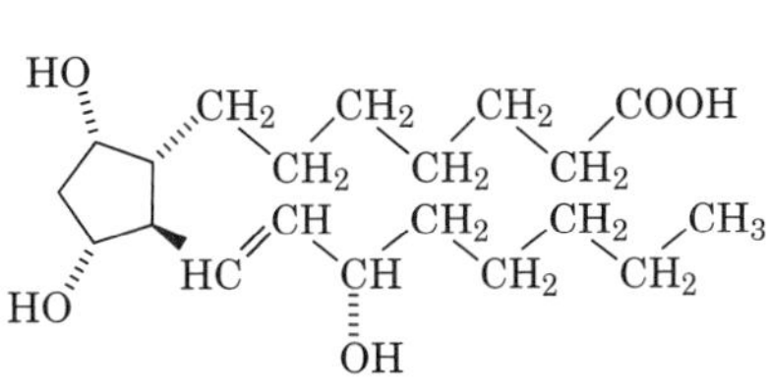

(10)

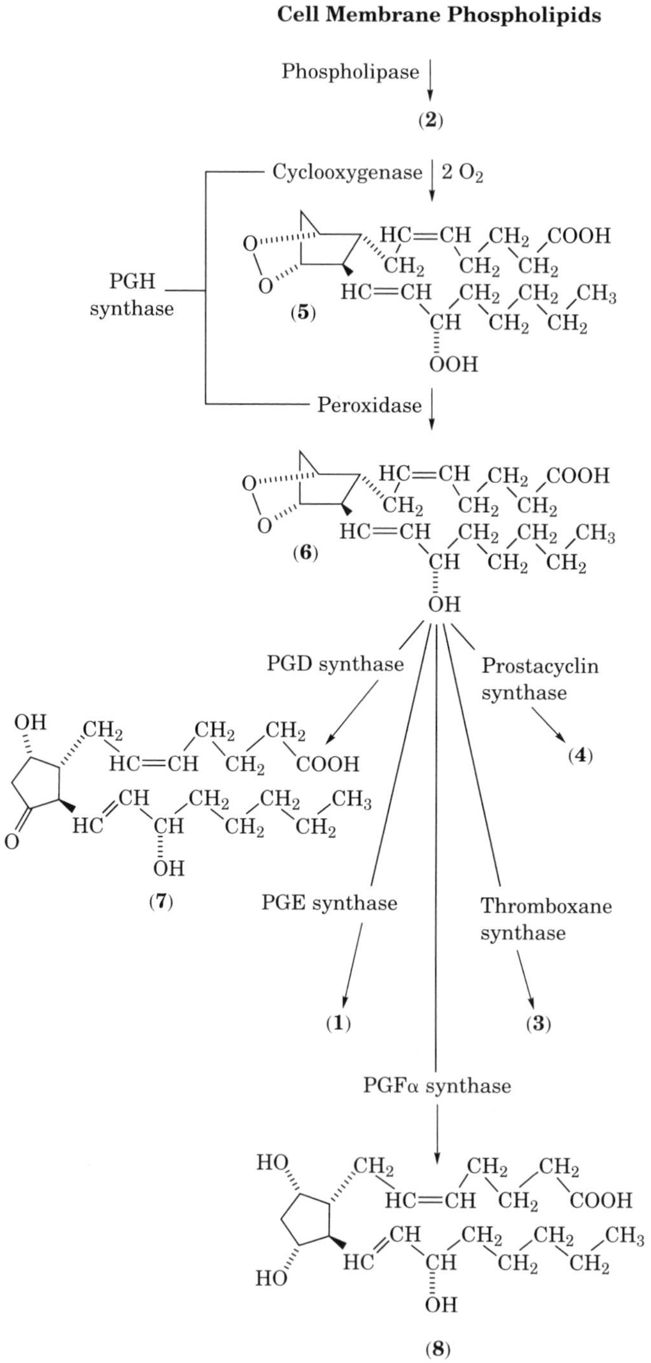

Cell Membrane Phospholipids

Phospholipase

(**2**)

Cyclooxygenase | 2 O$_2$

PGH
synthase

$O^{\cdots}$ HC=CH CH$_2$ COOH
CH$_2$ CH$_2$ CH$_2$
$O^{\cdots}$ HC=CH CH$_2$ CH$_2$ CH$_3$
(**5**) CH CH$_2$ CH$_2$
OOH

Peroxidase

$O^{\cdots}$ HC=CH CH$_2$ COOH
CH$_2$ CH$_2$ CH$_2$
$O^{\cdots}$ HC=CH CH$_2$ CH$_2$ CH$_3$
(**6**) CH CH$_2$ CH$_2$
OH

PGD synthase

Prostacyclin
synthase

(**4**)

OH CH$_2$ CH$_2$ CH$_2$
HC=CH CH$_2$ COOH
CH CH$_2$ CH$_2$ CH$_3$
O HC CH CH$_2$ CH$_2$
OH

(**7**)

PGE synthase

Thromboxane
synthase

(**1**)

(**3**)

PGFα synthase

HO CH$_2$ CH$_2$ CH$_2$
HC=CH CH$_2$ COOH
CH CH$_2$ CH$_2$ CH$_3$
HO HC CH CH$_2$ CH$_2$
OH

(**8**)

Fig. 1. Biosynthesis of prostanoids, where structures (**5**)–(**8**) are PGG$_2$, PGH$_2$, PGD$_2$, and PGF$_{2\alpha}$, respectively.

Additional compounds having similar biological activities and structural components were isolated resulting in the recognition of PGs as a family of closely related compounds. Structural and stereochemical assignments of PGE_1 and $PGF_{1\alpha}$ were confirmed by x-ray crystallographic analysis of their bromo- and iodobenzoates (7,8) (see X-RAY TECHNOLOGY). The absolute stereochemical configuration of PGs is based on the configuration of L-2-hydroxyheptanoic acid, obtained by oxidative ozonolysis of acetylated PGE_1 methyl ester (9). In 1983 Bergström and Samuelsson shared (with John Vane) the Nobel Prize for their contributions to this field (10). Thromboxane A_2 (**3**) and prostacyclin (**4**) were discovered in the mid-1970s (11). Both substances are highly unstable making identification and structural determinations elusive and difficult.

The PGs TxA_2 and PGI_2, commonly referred to as prostanoids, may be thought of as hormone-like substances which are produced on demand rather than stored and which modulate cellular functions at or near their site of generation (see HORMONES). Unlike typical circulating hormones which are released from one principal tissue site, prostanoids are synthesized and released by virtually all tissues. They are extremely potent and exert regulatory influences on the endocrine, reproductive, nervous, gastrointestinal, cardiovascular, renal, and immunological systems. They are also short lived because of rapid metabolism to inactive species. Their effects are generally considered to be beneficial and homeostatic in nature. In some cases, however, their biological effects can be detrimental and disease producing; for example, PGs can exert inflammatory actions and produce fever, and TxA_2 is a pro-aggregatory factor for platelets.

The nomenclature of prostaglandins and prostacyclins is based on the basic prostane skeleton (**11**), whereas thromboxane (**12**) is the parent for the thromboxanes.

(**11**)   (**12**)

Implicit in the base names are the absolute configurations at carbons 8 and 12 and the indicated numbering systems. Derivatives of these parent structures are named according to terpene and steroid nomenclature rules (see STEROIDS; TERPENOIDS). The lengthy and awkward nature of the chemical abstract systematic nomenclature (12) for these compounds has resulted in the development (13) and use of simplified nomenclature based on common names.

The PGs are grouped into several basic families which differ from each other in the nature of the five-membered ring functionalities (Fig. 2). The principal families are designated as PGA through PGJ. The letters E and F originated from the finding that PGE and PGF compounds partition differently. The compound that was more soluble in ethyl ether was called prostaglandin E and the

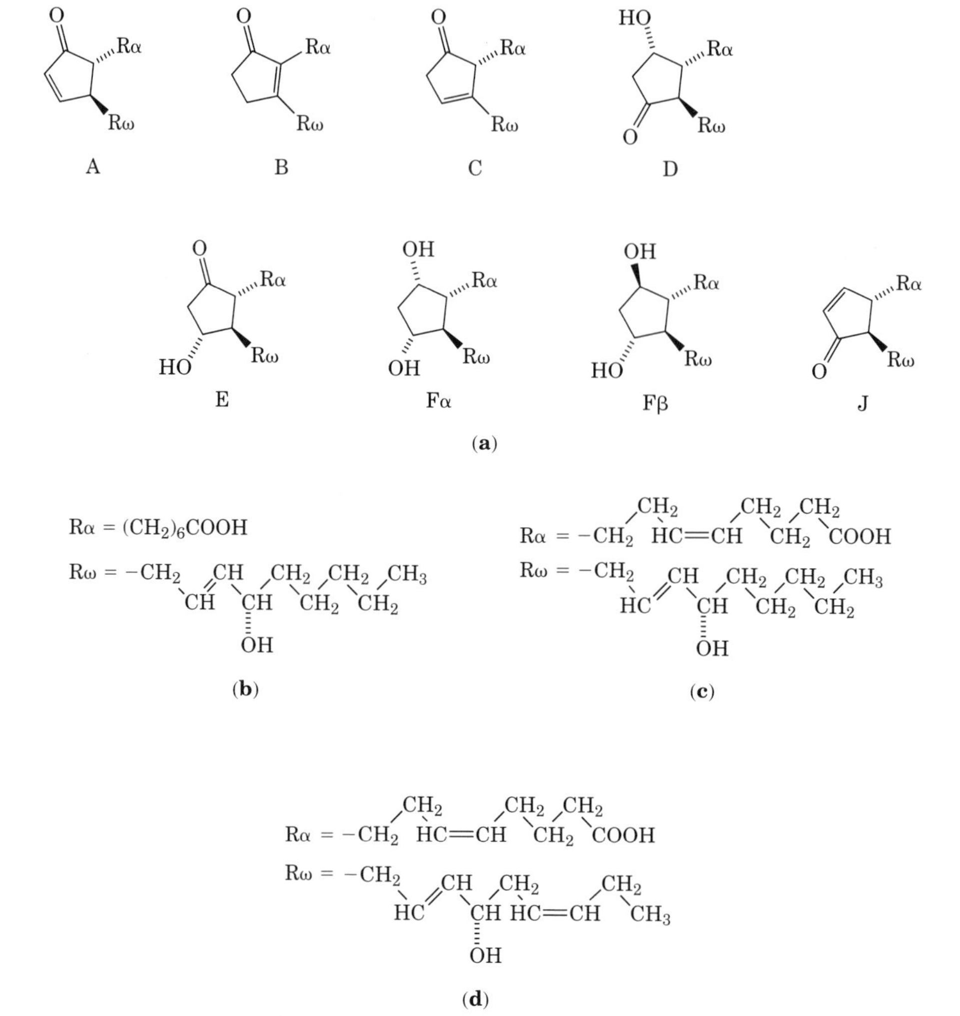

$R\alpha = (CH_2)_6COOH$

$R\omega = -CH_2$ ... CH_3 ... ÖH

(b)

$R\alpha = -CH_2$ ... CH_2 CH_2 CH_2 $COOH$

$R\omega = -CH_2$ ... CH_3 ... ÖH

(c)

$R\alpha = -CH_2$ ... CH_2 CH_2 CH_2 $COOH$

$R\omega = -CH_2$ ... CH_3 ... ÖH

(d)

Fig. 2. (**a**) The basis for prostaglandin nomenclature, where the letters A–F and J define principal families; (**b**) defines the side chains for PG$_1$ derived from dihomo-γ-linolenic acid; (**c**) PG$_2$ derived from arachidonic acid; and (**d**), PG$_3$ derived from eicosapentaenoic acid.

one that was more soluble in phosphate buffer (*fosfate* in Swedish) was termed prostaglandin F (**10**). The letters A and B refer to the formation of these derivatives from PGE compounds by treatment with acid and base, respectively. As indicated in the general structures of Figure 2, the carboxylic acid side chain is referred to as the alpha chain R$_\alpha$, and the hydroxy-bearing chain as the omega chain, R$_\omega$. The number of double bonds in the molecule are denoted by subscript numerals appearing after the name; for example, PGE$_1$ contains one double bond at C$_{13}$; PGE$_2$ has an additional double bond at C$_5$. The stereochemistry of substituents on the cyclopentane ring is designated α- or β- depending on whether the substituent is above or below the plane of the paper.

Biosynthesis and Metabolism

Detailed accounts of the biosynthesis of the prostanoids have been published (14–17). Under normal circumstances arachidonic acid (AA) is the most abundant C-20 fatty acid *in vivo* (18–21) which accounts for the predominance of the prostanoids containing two double bonds eg, PGE_2 (see Fig. 1). Prostanoids of the one and three series are biosynthesized from dihomo-8-linolenic and eicosapentaenoic acids, respectively. Concentrations in human tissue of the one-series precursor, dihomo-8-linolenic acid, are about one-fourth those of AA (22) and the presence of PGE_1 has been noted in a variety of tissues (23). The biosynthesis of the two-series prostaglandins from AA is shown in Figure 1. These reactions make up a portion of what is known as the arachidonic acid cascade. Other lipid products of the cascade include the leukotrienes, lipoxins, and the hydroxyeicosatetraenoic acids (HETEs). Collectively, these substances are termed eicosanoids.

Prostanoid synthesis is initiated by the interaction of a stimulus with the cell surface which results in the activation of cellular phospholipases. Arachidonic acid is generally present in phospholipids esterified at position two. It is released by the action of phospholipase A_2 (PLA_2) which is specific for fatty acid deesterification at that position of the phospholipid molecule. Other pathways for AA mobilization also exist but are less prevalent. Once AA is released, it can be acted upon by prostaglandin H synthase (EC 1.14.99.1) (PGHS) (2), a membrane-bound enzyme which is present in an active form, primarily in the endoplasmic reticulum. This enzyme exhibits two different catalytic activities

(Fig. 1): a cyclooxygenase (bis-oxygenase) which catalyzes the formation of prostaglandin G_2 [*51982-36-6*] (PGG_2) (**5**) from AA, and a peroxidase (or hydroperoxidase) which facilitates the reduction of PGG_2 to prostaglandin H_2 [*42935-17-1*] (PGH_2) (**6**).

The initial step in the action of the cyclooxygenase is the stereospecific removal of the 13-pro-(*S*)-hydrogen from AA. As shown in Figure 3 (2,24), the enzyme is thought to orient an AA molecule by inducing a kink in the carbon chain at C-10. Abstraction of the 13-pro-(*S*)-hydrogen and subsequent isomerization leads to a carbon-centered radical at C-11 which is attacked by molecular

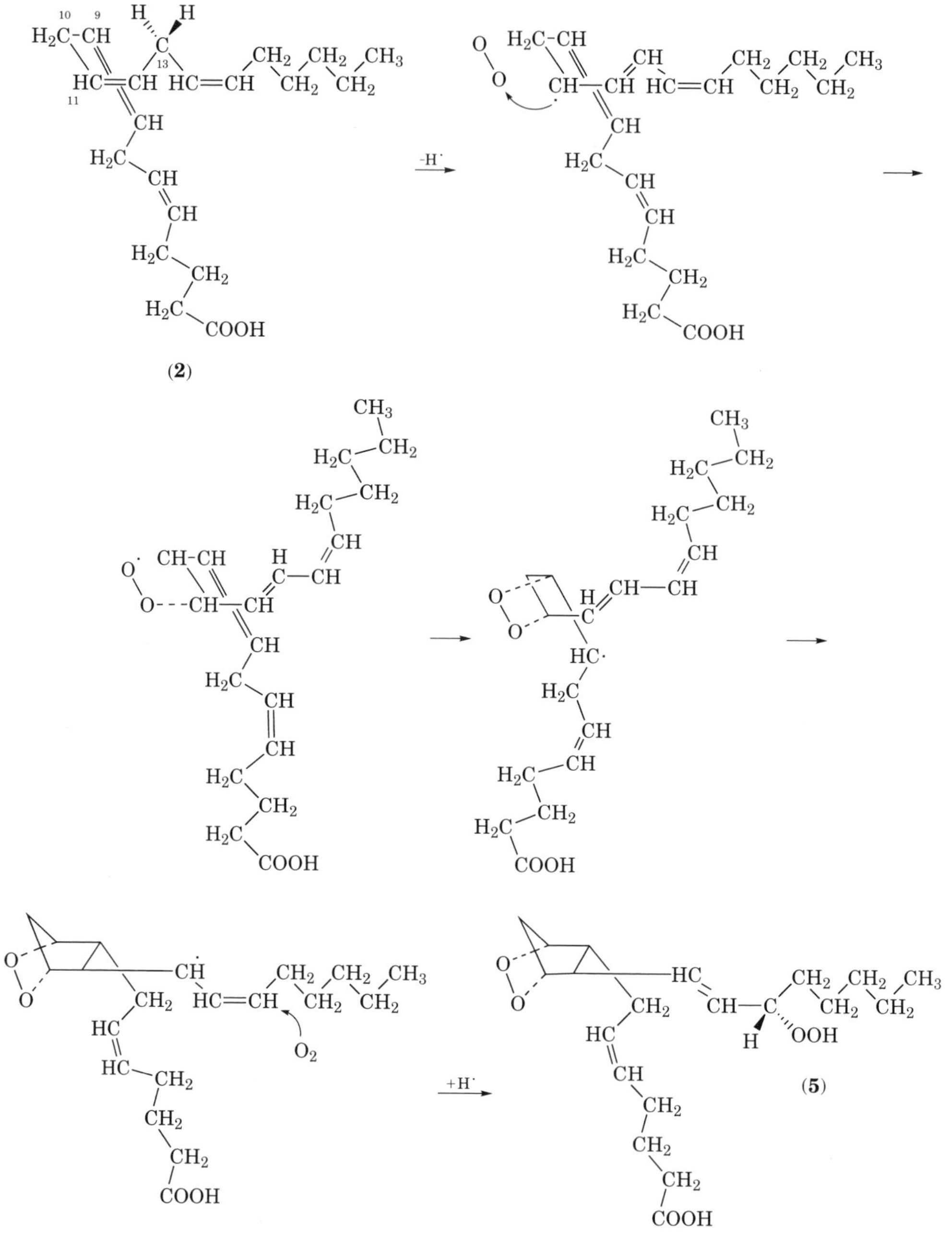

Fig. 3. Putative mechanism of PGH synthase action on arachidonic acid.

oxygen from the solvent side. The resulting 11-hydroperoxyl radical adds to the double bond at C-9; intramolecular rearrangement yields another carbon-centered radical. Reaction of this radical with another molecule of oxygen at C-15 yields PGG$_2$. Newly formed PGG$_2$ can undergo a two-electron reduction to PGH$_2$ catalyzed by the peroxidase activity of PGH synthase. In order for the cyclooxygenase to function, a source of hydroperoxide $(R-O-O-H)$ appears to be required. The hydroperoxide oxidizes a heme prosthetic group at the peroxidase active site of PGH synthase. This in turn leads to the oxidation of a tyrosine residue producing a tyrosine radical which is apparently involved in the abstraction of the 13-pro-(S)-hydrogen of AA (25). The cyclooxygenase is inactivated during catalysis by the nonproductive breakdown of an active enzyme intermediate. This suicide inactivation occurs, on average, every 1400 catalytic turnovers.

The endoperoxides PGG_2 and PGH_2 are extremely important intermediates in the metabolism of AA. In addition to having pronounced biological activity, eg, aggregation of blood platelets and constriction of vascular tissue, they serve as intermediates in the biosynthesis of a variety of prostanoids. PGH_2 is a substrate for a number of enzymes in the cascade including PGE synthase (EC 5.3.99.3), PGD synthase (EC 5.3.99.2), and PGF_α synthase (EC 1.1.1.188) which catalyze the synthesis of the classical prostaglandins, PGE_2, PGD_2, and $PGF_{2\alpha}$. The formation of PGI_2 and TxA_2 from PGH_2 is catalyzed by PGI synthase (EC 5.3.99.4), and TxA synthase (EC 5.3.99.5).

PGE synthases are unique in that each requires reduced glutathione (GSH) as a cofactor. GSH appears to facilitate cleavage of the endoperoxide group and formation of the C-9 keto group (2). Synthesis of $PGF_{2\alpha}$ involves a net two-electron reduction of PGH_2; a PGF_α synthase utilizing NADPH catalyzes this reaction. All other prostanoids are formed via isomerization reactions involving no net change in the oxidation state of PGH_2. PGI synthase and TxA synthase are hemoproteins having molecular weights of 50,000–55,000. Both enzymes, like PGH synthase, undergo suicide inactivation during catalysis. Although all the principal prostanoids are depicted in Figure 1 as being formed by a single cell, prostanoid synthesis appears to be cell specific (2). For example, platelets form mainly TxA_2, endothelial cells form PGI_2 as their primary prostanoid, and PGE_2 is the primary prostanoid produced by renal collecting tubule cells. TxA synthase is found in abundance in platelets and lung. PGI synthase is localized to endothelial cells and vascular and nonvascular smooth muscle. PGE synthase activities are present in several different tissues, but there are differences among these proteins from different tissues.

The primary prostaglandins are believed to be mediators of inflammation and their synthesis can be inhibited by both nonsteroidal antiinflammatory drugs (NSAIDS) and antiinflammatory steroids. The best known of the NSAIDS is aspirin (see ANALGESICS, ANTIPYRETICS, AND ANTIINFLAMMATORY AGENTS; SALICYLIC ACID AND RELATED COMPOUNDS). Aspirin competes with arachidonic acid (AA) for binding to the cyclooxygenase active site, but the binding of AA is about 10,000 times more efficient than that of aspirin. However, once bound, aspirin can acetylate a specific serine residue of PGH synthase: Ser[530]. Acetylation of Ser[530] causes irreversible cyclooxygenase inactivation (see ENZYME INHIBITORS). It was initially thought that the hydroxyl group of Ser[530] was required for catalysis. However, studies using site-directed mutagenesis have shown that replacement of Ser[530] with an alanine residue yields an active enzyme, whereas replacement with an asparagine, which is about the same size as an acetylated serine, yields an inactive enzyme. Apparently, acetylation of Ser[530] by aspirin results in steric hindrance at this position: the bulky acetyl group protrudes into the cyclooxygenase active site and prevents AA from binding (2).

There are many nonsteroidal antiinflammatory drugs. In fact, this is by far the largest portion of the pharmaceutical market, with 1990s worldwide sales exceeding $4 billion. Most other NSAIDS also act by inhibiting the cyclooxygenase activity of PGH synthase. However, unlike aspirin, most of these drugs cause reversible enzyme inhibition by competing with AA for binding. Well-known examples of reversible NSAIDS are ibuprofen (Advil, Whitehall), indomethacin (Indocin, Merck) and naproxen (Naprosyn, Syntex). Antiinflammatory steroids were once thought to attenuate prostanoid synthesis by inhibiting stimulus-induced AA release. However, research in the 1990s has suggested that antiinflammatory steroids may function principally by inhibiting transcription of the PGH synthase gene. Presumably, these steroids interact with a receptor which binds to a negative regulatory element in the promoter region of the PGH synthase gene (26).

There are two isozymes of PGH synthase (27). PGHS-1 (or COX-1) is constitutively expressed in most tissues (28), and is responsible for the production of PGs involved in cellular housekeeping functions such as coordinating the actions of circulating hormones (29) and regulating vascular, gastric, intestinal, and renal homeostasis. PGHS-2 (COX-2), which shares about 59% amino acid homology with PGHS-1, only is expressed following cell activation. Its expression is stimulated by inflammatory mediators (30) and inhibited by antiinflammatory steroids (31). All of the available nonsteroidal antiinflammatory drugs inhibit both forms of the enzyme. As a result they produce side effects in tissues which are dependent on homeostatic levels of PGs. The gastrointestinal tract and kidney are particularly sensitive to PG synthesis inhibition. NSAIDS can cause damage and ulceration in the stomach and intestinal tract and can compromise kidney function especially in people with pre-existing renal impairment. Many pharmaceutical companies are involved in research to identify compounds which would selectively inhibit only the PGHS-2 enzyme and thereby be potentially free of the undesirable side effects of conventional NSAIDS. The x-ray structure of PGHS-1 has been reported (32).

Acetylation of PGH synthase by aspirin has important pharmacological consequences. Besides the analgesic, antipyretic, and antiinflammatory actions of aspirin, low dose aspirin treatment is a useful antiplatelet cardiovascular therapy (33) (see CARDIOVASCULAR AGENTS). TxA_2 is a potent stimulator of platelet aggregation; low dosage aspirin treatment leads to selective inhibition of platelet TxA_2 formation without appreciably affecting the synthesis of prostanoids in other cells. Aspirin causes irreversible inactivation PGH synthase. Since platelets, unlike most other cells, are unable to synthesize new enzyme, new PGH synthase activity must come from new platelets. The replacement time for platelets is 5–10 days, thus considerable time is required for the circulating platelet pool to regain its original complement of active PGH synthase. PGH synthase inactivation also occurs in other cell types, but these cell types can resynthesize PGH synthase relatively quickly.

Once prostanoids are formed they exit the cell, probably via carrier-mediated transport. They act very near their sites of synthesis, then are rapidly inactivated by metabolic enzymes and excreted. The prostanoids are subject to four principal metabolic transformations, as illustrated in Figure 4 for PGE_2: oxidation catalyzed by C-15 prostaglandin dehydrogenase (15-PGDH)

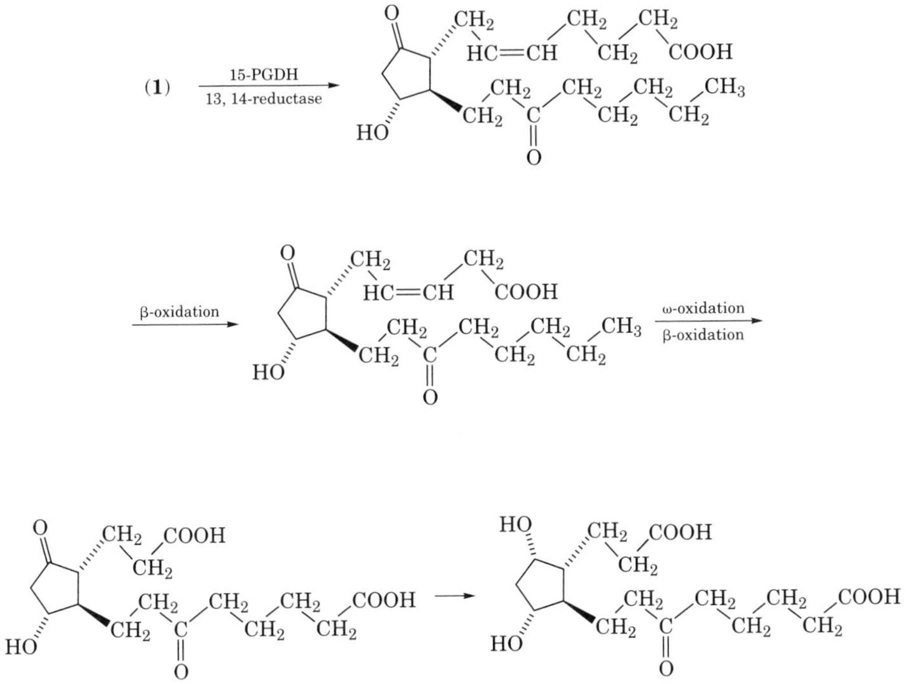

Fig. 4.   Metabolism of natural prostaglandins.

(EC 1.1.1.196); reduction by 13,14-reductase; β-oxidation of the carboxylic acid side chain; and ω- and $(\omega-1)$-oxidation of the aliphatic side chain (34). The 15-PGDH step occurs most rapidly in humans and most other mammals, and it converts the parent PG molecule into its corresponding C-15 ketone structure. The lung is especially enriched in 15-PGDH, and this enzymatic transformation is so rapid that 90–95% of a circulating PG is transformed into its biologically inactive C-15 ketone metabolite on a single pass through the lungs. Reduction of the C-15 ketone metabolite by the 13,14-reductase enzyme produces the saturated metabolite shown. β-Oxidation, a reaction common to fatty acids in general, involves a sequence of dehydrogenation at C-2 and C-3, followed by oxidation to the 3-ketone and finally cleavage to produce the dinor (-2 carbon atoms) metabolite and acetic acid. A second oxidation sequence usually occurs to generate the tetranor (-4 carbon atoms) metabolite. Another point of attack is the ω-chain terminus. Oxidation occurs either at C-20 to give the alcohol and subsequently the acid (**13**) or at C-19 to produce the 19-hydroxy metabolite. The primary urinary metabolite of natural PGs is (**14**) which is the final keto product of all of these processes plus the reduction of the C-9 carbonyl group.

Physical and Chemical Properties

The melting points, optical rotations, and uv spectral data for selected prostanoids are provided in Table 1. Additional physical properties for the

Table 1. Physical Properties of Selected Prostanoids Derived from Arachidonic Acid

Compound	CAS Registry Number	Mp, °C	$[\alpha]_D$ (Solvent), deg	Refs.
PGA$_2$[a]	[13345-50-1]	pale yellow oil	145 (chloroform)	35
PGB$_2$[b]	[13367-85-6]	30–34	16–18	36
PGC$_2$[c]	[49825-91-4]	oil		37
PGC$_2$, methyl ester[d]	[51172-26-0]	oil		38
PGD$_2$	[41598-07-6]	62.8–63.3	13 (chloroform)	39
PGE$_2$	[363-24-6]	65–66	−61 (tetrahydrofuran)	40
		62–64	−52 (tetrahydrofuran)	41
		65.0–67.5		35
PGF$_{2\alpha}$	[551-11-1]	30–35	26 (ethanol)	35,42
			23.8 (tetrahydrofuran)	40
PGI$_2$				
sodium salt	[61849-14-7]	166–168	88 (chloroform)	43
		116–124	97 (ethanol)	43
methyl ester	[61799-74-4]	30–33	78 (chloroform)	43
6-oxo-PGF$_{1\alpha}$	[58962-34-8]	75–78		43
TXB$_2$	[54397-85-2]	91–93	57.4 (ethyl acetate)	44
		92–94		45
		89–90		46

[a] In C_2H_5OH, λ_{max} is 217 nm and $\epsilon = 10,300\ (M \cdot cm)^{-1}$.
[b] In C_2H_5OH, λ_{max} is 278 nm and $\epsilon = 26,000\ (M \cdot cm)^{-1}$.
[c] In CH_3OH, λ_{max} is 234 nm and $\epsilon = 17,000\ (M \cdot cm)^{-1}$.
[d] In CH_3OH, λ_{max} is 229 (sh) and 234 nm.

primary PGs have been summarized in the literature and the physical methods have been reviewed (47). The molecular conformations of PGE$_2$ and PGA$_1$ have been determined in the solid state by x-ray diffraction, and special ^{1}H and ^{13}C nuclear magnetic resonance (nmr) spectral studies of several PGs have been reported (11,48–53). Mass spectral data have also been compiled (54) (see MASS SPECTROMETRY; SPECTROSCOPY).

The E- and D-type PGs are inherently unstable compounds. Their instability is primarily due to the lability of the β-hydroxyketone system in the cyclopentane ring. Under acidic or alkaline conditions (<pH 3 and >pH 7) there is a strong driving force for elimination of the 11-hydroxy group of PGE structures to give the more stable α,β-unsaturated ketone of PGA. In an analogous manner, PGD structures give rise to PGJ compounds. The A-form can isomerize to the PGB derivative under the same conditions. In general, esters or similar derivatives are more stable than their acids which are sufficiently acidic to catalyze their own dehydration. E-type PGs also are susceptible to epimerization of the C-8 side chain under alkaline or thermal conditions. In more acidic media three other processes become significant: C-15-epimerization, allyl transposition of the C-15 hydroxyl group, and dehydration of the C-15 hydroxyl group.

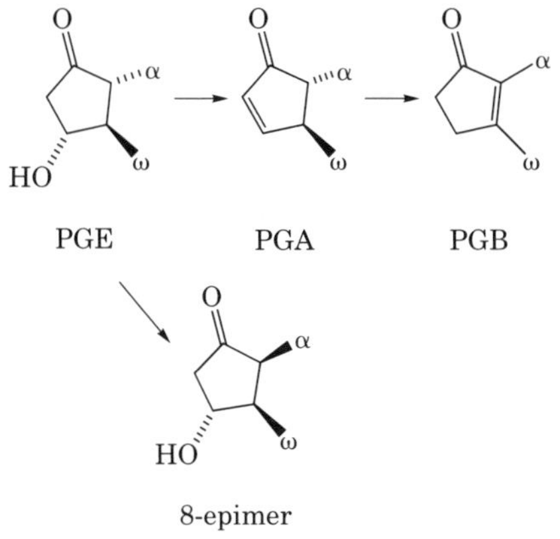

PGE   PGA   PGB

8-epimer

All the bis- and tri-unsaturated prostanoids display sensitivity to atmospheric oxygen similar to that of polyunsaturated fatty acids and lipids. As a result, exposure to the air causes gradual decomposition although the crystalline prostanoids are less prone to oxygenation reactions than PG oils or solutions.

Both thromboxane A_2 (TxA$_2$) and prostacyclin (PGI$_2$) are extremely unstable compounds. The instability of TxA$_2$ (**3**) is due to the strained bicyclic acetal system. Hydrolysis of the acetal to give TxB$_2$ (**15**) releases the strain.

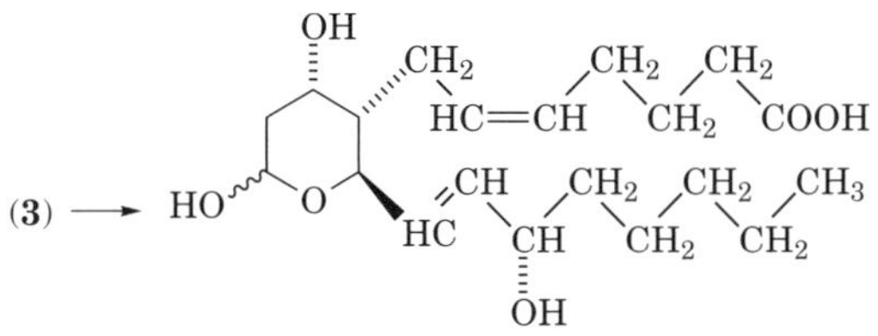

(**15**)

TxB$_2$ (**15**) is a stable but biologically inactive compound and its isolation and characterization were essential to the discovery of TxA$_2$. Measurement of the half-life of TxA$_2$ has been based on the decay of biological activity and the rate of appearance of O-methyl-TxB$_2$ from methanolysis of TxA$_2$. All methods gave values of 30–41 seconds at 37°C in aqueous media free of proteins (55,56). TxA$_2$ has never been isolated and characterized directly. Its structural assignment was based on TxB$_2$ and its more stable synthetic analogues. However, confirmation of the structure of TxA$_2$ by chemical synthesis has been achieved (57).

PGI$_2$ (**4**) contains an acid-labile enol ether which is readily hydrolyzed to generate 6-keto-PGF$_{1\alpha}$ (**16**).

(16)

As with TxA$_2$, the reactivity of PGI$_2$ ($t_{1/2}$ = 3 min at pH 7.6 and 37°C) made isolation of the natural substance difficult, and a pure chemical sample was obtained only through chemical synthesis. PGI$_2$ is stable under more alkaline conditions and can be isolated and stored as a salt. Additional information on the chemistry and stability of TxA$_2$ and PGI$_2$ has been summarized (58).

Biological Properties

The PGs, PGI$_2$ and TxA$_2$ collectively exhibit a wide variety of biochemical and pharmacological activities and are involved in both physiological and pathophysiological processes. However, the individual compounds show different overall activity profiles sometimes in opposing directions. Excellent reviews are available (59–64). A survey of some of the more important biological actions of the prostanoids follow.

Cardiovascular System. In most species and vascular beds PGEs and PGAs are potent vasodilators, whereas responses to PGF$_{2\alpha}$ vary. Prostacyclin, PGI$_2$, is a potent vasodilator with five times the potency of PGE$_2$ and causes prominent hypotension in animals and humans following intravenous administration. Its hydrolysis product, 6-keto-PGF$_{1\alpha}$, is essentially inactive. In contrast, TxA$_2$ is a powerful vasoconstrictor. PGE$_1$ and D$_2$ are inhibitors of platelet aggregation, whereas PGE$_2$ has variable effects depending on its concentration. PGI$_2$ is 30–50-fold more potent than PGE$_1$, inhibiting aggregation at concentrations between 1 and 10 nM. TxA$_2$ is a powerful aggregatory substance. The opposing effects of PGI$_2$ and TxA$_2$ on vascular tone and platelet aggregatory state are believed to be important in vascular homeostasis, and disruption of this balance can lead to cardiovascular disease (see CARDIOVASCULAR AGENTS). Aspirin is used prophylactically to prevent heart attacks because it can selectively reduce TxA$_2$ levels relative to PGI$_2$ levels. TxA$_2$ is produced by platelets which cannot synthesize more PGHS, whereas vascular endothelial cells, which generate PGI$_2$, can rapidly produce fresh enzyme. The PG endoperoxides are vasoconstrictory and platelet aggregators, although less active than TxA$_2$.

Pulmonary System. In general PGFs contract and PGEs relax bronchial and tracheal muscle. Asthmatics are particularly sensitive to PGs and PGF$_{2\alpha}$ can cause severe bronchospasm in these patients. In contrast, PGE$_1$ and -E$_2$ are potent bronchodilators when given by aerosol to asthmatic patients (65) (see ANTIASTHMATIC AGENTS). Prostaglandin endoperoxides and TxA$_2$ are constrictors whereas PGI$_2$ induces slight bronchodilation and antagonizes bronchoconstriction produced by other agents in asthmatics (66).

Reproductive System. The primary PGs are intimately involved in reproductive physiology (67). PGE_2 and $PGF_{2\alpha}$ are potent contractors of the pregnant uterus and intravenous infusion of either of these compounds to pregnant humans produces a dose-dependent increase in frequency and force of uterine contraction. PGI_2 and TxA_2 have mild relaxant and stimulatory effects, respectively, on uterine tissue. The primary PGs also play a role in parturition, ovulation, luteolysis, and lactation and have been implicated in male infertility.

Gastrointestinal System. PGEs, PGAs, and PGI_2 inhibit gastric acid secretion stimulated by feeding, histamine, or gastrin. The volume of secretion, acidity, and content of pepsin are all reduced, probably by an action exerted directly on the secretory cells. In addition, these PGs are vasodilators in the gastric mucosa, and are likely to be involved in the local regulation of blood flow. PGs cause increased mucus secretion in the stomach and small intestine and substantial movement of water and electrolytes into the intestinal lumen. Such effects contribute to the diarrhea noted in animals and humans following the oral or parenteral administration of PGs. In contrast, PGI_2 does not induce diarrhea in animals or humans; it prevents that provoked by other PGs, and inhibits the toxin-induced accumulation of intestinal fluid in experimental models.

Nervous System. PGEs produce sedation and catatonia when injected into the cerebral ventricles of cats (67) (see HYPNOTICS, SEDATIVES, ANTICONVULSANTS, AND ANXIOLYTICS; PSYCHOPHARMACOLOGICAL AGENTS). A number of studies using microiontophoretic techniques, ie, the electrolytic injection of substances into tissues, have been carried out to study the effects of PGs on brain function (68); perhaps the most interesting effect observed is fever production following intraventricular injection of PGEs. This finding prompted the hypothesis that pyrogen-induced fever is due to release of PGE_2 in the brain. The antipyretic activity of aspirin and other NSAIDS is thus hypothesized to be due to inhibition of PG synthesis in the brain. Although this is a satisfying explanation, there is considerable contradictory evidence (69). In humans, PGEs cause pain when injected intradermally or applied to facial skin. They also potentiate the pain-producing effects of bradykinin and histamine (see HISTAMINE AND HISTAMINE ANTAGONISTS).

Kidney Function. Prostanoids influence a variety of kidney functions including renal blood flow, secretion of renin, glomerular filtration rate, and salt and water excretion. They do not have a critical role in modulating normal kidney function but play an important role when the kidney is under stress. For example, PGE_2 and $-I_2$ are renal vasodilators (70,71) and both are released as a result of various vasoconstrictor stimuli. They thus counterbalance the vasoconstrictor effects of the stimulus and prevent renal ischemia. The renal side effects of NSAIDS are primarily observed when normal kidney function is compromised.

Metabolic and Endocrine Effects. The role of PGs in these systems is complex and generally modulatory in nature. PGE_2 is synthesized by adipocytes and is a potent inhibitor of lipolysis. It is also a potent inducer of bone resorption and of calcium release from bone. Under certain circumstances, however, PGEs can stimulate lipolysis and bone growth (72). PGE_2 increases circulating levels of adrenocorticotropic hormone (ACTH), growth hormone, prolactin, and the gonadotropin hormones (see GROWTH REGULATORS; HORMONES, HUMAN GROWTH HORMONE). Other endocrine effects have been reviewed (73). The effect of PGEs on insulin and glucose levels is also complex and regulatory in nature (74) (see INSULIN AND OTHER ANTIDIABETIC DRUGS).

Inflammatory and Immune Responses.   The role of prostanoids in inflammation is controversial and good evidence exists on both sides of the argument as to whether they are pro- or antiinflammatory (75). PGE_2 and PGI_2 are present in inflamed tissues in sufficient concentrations to account for the erythema and increased sensitivity characteristic of acute inflammation (76,77). PGEs are vasodilatory and hyperalgesic, ie, increase sensitivity to pain, and in concert with other mediators such as bradykinin and histamine, PGEs increase vascular permeability. It is also generally accepted that NSAIDS act, at least in part, by inhibiting prostanoid production (78). However, PGEs suppress the secretion of inflammatory mediators by mast cells and the release of lysosomal enzymes from human neutrophils. Possibly the state of the tissue governs the nature of the response to prostanoids (79). Prostanoids also moderate the humoral and cellular immune systems, but their effects are complicated and involve both inhibition and stimulation (80). Thus PGE_2 can inhibit T-cell function and proliferation, but under some circumstances PGEs also stimulate the development of mature T-cells from immature thymocytes and stimulate mitogenic activity of low density T-cells. These dual excitatory/suppressive effects are also observed on B-cells, natural killer cells, and macrophages.

Cytoprotection.   One of the most intriguing properties of prostanoids is their ability to protect cells and tissues from various damaging agents. This phenomenon was first discovered by the observation that natural PGs would prevent damage of the gastric mucosa when administered prior to various chemical irritants such as ethanol, acid, and alkali (81,82). Exogenously administered PGs were also found to prevent the gastric and duodenal damage caused by NSAIDS. The mechanisms underlying the protective properties of prostanoids remain unknown. In the stomach the prostanoids exert actions such as stimulation of mucus and bicarbonate, prevention of mucosal barrier disruption, enhancement of mucosal blood flow, and acceleration of mucosal repair following damage, but none of these are adequate explanations (83). Rather it appears that additional, unidentified cellular protective mechanisms exist. The fact that PGs are protective to tissues other than the gastric mucosa supports this hypothesis. Protective effects of PGs have been documented in the colon (84), liver (85), kidney (86), and pancreas (87), and for injury caused by radiation and chemotherapeutic agents (88,89) (see CHEMOTHERAPEUTICS, ANTICANCER; RADIOPROTECTIVE AGENTS). A comprehensive review of the biological protective properties of prostanoids has been published (90).

Health and Safety.   The prostanoids are extremely potent substances with a wide variety of biological effects. Therefore utmost caution should be used in their handling to avoid adverse effects. As an example, PGE_1, if ingested, may cause fever, diarrhea, abdominal pain, low blood pressure, nausea, vomiting, headache, and dizziness. If inhaled, bronchodilatation and respiratory tract irritation may occur. Skin exposure, especially with esters of PGE_1, can result in reddening and increased pain sensitivity of the skin, particularly the face. Prostaglandins of the E- and F-types have uterine stimulating properties and therefore can endanger pregnancy. The LD_{50} (the dose lethal to 50% of the treated population) for PGE_1 in rats is 228 mg/kg of body weight by oral administration and 19.2 mg/kg intravenously. Material Safety Data Sheets (MSDS) are available for some of the natural compounds, for example, PGE_1, PGD_2, and TxB_2, from MDL Information Systems, Inc. (San Leandro, California). Other sources of MSDS are Sigma-Aldrich Corp. (Milwaukee, Wisconsin) or the individual manufacturers of marketed prostanoids. Prostanoids that are marketed as drugs, for example misoprostol (Cytotec), are regulated by the U.S. FDA and corresponding agencies in other countries.

Prostanoid Receptors

Characterization of prostanoid receptors and quantification of the action of ligands at these receptors have been hampered by lack of selectivity of the natural prostanoids and most synthetic agonists and antagonists for specific receptors, the ability of the prostanoids to induce opposing actions in the same tissue, and the multiplicity of prostanoid receptor subtypes in most tissues. The development of a classification system of prostanoid receptors (91) (Table 2) provides a working framework toward understanding these interactions. PGE receptors are pharmacologically divided into three subtypes, EP_1, EP_2, and EP_3, which differ in their mode of signal transduction; binding at these receptors is believed to lead to elevation of intracellular calcium levels, stimulation of adenylate cyclase, and inhibition of adenylate cyclase, respectively (92). PGE_2 increases cyclic adenosine monophosphate (cAMP) levels in many tissues suggesting that the EP_2 subtype is ubiquitously distributed and mediates various PGE_2 actions in many tissues and cells. EP_2 receptors mediate relaxation of tracheal and ileum circular muscle, vasodilation of various blood vessels, and stimulation of sodium and water reabsorption in kidney tubules. EP_1 receptors, in contrast, mediate contraction of tracheal and gastrointestinal smooth muscle. EP_3 receptors are involved in modulation of neurotransmitter release (see NEUROREGULATORS), inhibition of gastric acid secretion, stimulation and relaxation of smooth muscle, and stimulation of lipolysis in adipose tissue. Two excellent reviews on the prostanoid receptors and their biological actions have been published (93). The cloning and expression of complementary deoxyribonucleic acids (cDNAs) for various animal and human prostanoid receptors have been described (94–99) (see BIOTECHNOLOGY; GENETIC ENGINEERING).

The paucity of selective ligands to activate and antagonize the EP-receptors continues to make characterizations extremely difficult. Sulprostone [60325-46-4] (17) is reported to activate EP_1 and EP_3 but not EP_2 receptors, whereas

Table 2. Prostaglandin Receptor Subtypes

Receptor subtype	Most potent natural PG agonist	Usual response on smooth muscle
EP_1	PGE_2	contraction
EP_2	PGE_2	relaxation
EP_3	PGE_2	inhibition
DP	PGD_2	relaxation
FP	$PGF_{2\alpha}$	contraction
IP	PGI_2	relaxation
TP	TxA_2	contraction

butaprost (**18**) selectively but weakly activates EP_2 receptors (100). A potent and highly selective agonist at EP_3 receptors, SC-46275 (**19**), has been described (101). Only two selective EP_1 antagonists, SC-19220 (**20**) and AH-6809 (**21**), are known and no selective antagonists of the EP_2 and EP_3 receptors have been reported. A number of structurally diverse TxA_2 antagonists have been described (102).

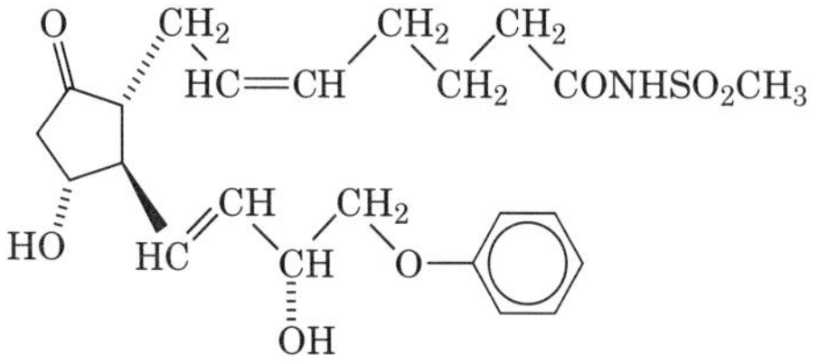

(**17**)

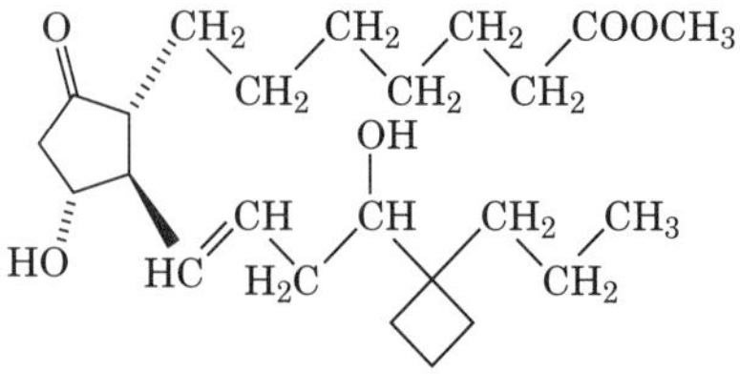

(**18**)

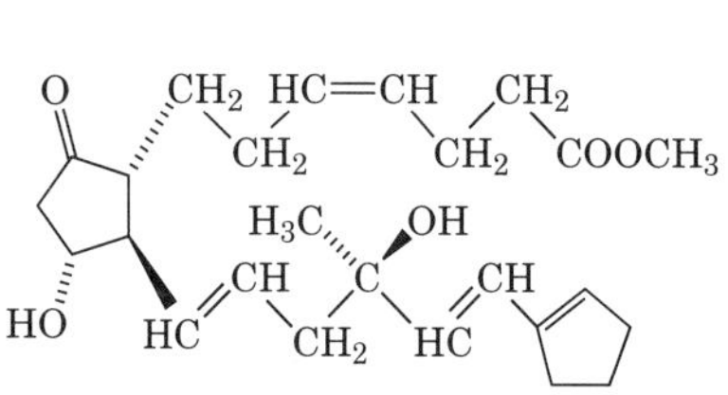

(**19**)

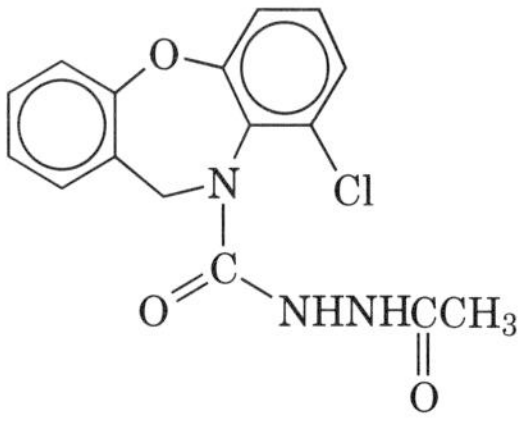

(**20**)

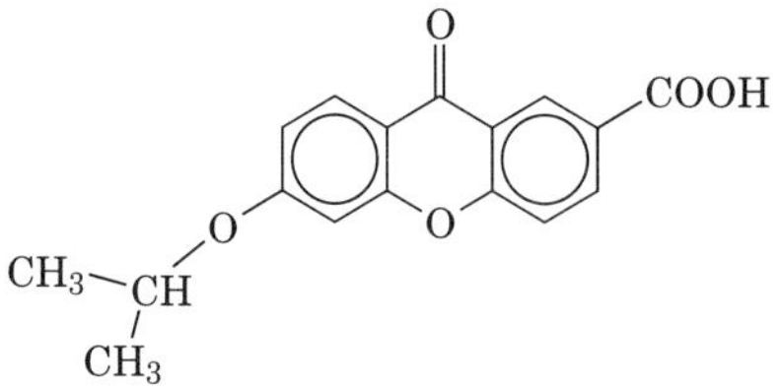

(**21**)

Synthesis of Naturally Occurring Prostanoids and Analogues

Following the rebirth of prostanoid research in the early 1960s, intensive efforts were made in numerous academic and industrial laboratories to develop total chemical syntheses of the natural compounds and, subsequently, their analogues. The driving forces behind this plethora of research were the challenges presented by the stereochemical and functional group complexities of the substances, the wide array of their biological activities, and their extraordinary therapeutic potential in a host of diseases. Likewise, the discovery of prostacyclin and thromboxane in the late 1970s also created a burst of activity. Comprehensive reviews of the total synthesis of the natural prostanoids and their structural analogues are given (6,103–107) (see PHARMACEUTICALS, CHIRAL).

   Classical Prostaglandins.   Prostanoids of the A to F series are known as the classical prostaglandins to distinguish them from those discovered and characterized at later dates. Although the bulk of the synthetic activity occurred in the 1970s, various improvements and new approaches are still appearing in the literature in the 1990s. The numerous syntheses of the classical prostaglandins may be grouped into three basic strategies: cleavage of polycyclic intermediates, the conjugate addition approach, and the cyclization of aliphatic precursors (108). The bulk of the efforts have been devoted to the biologically more important E and F compounds, but direct synthesis of PGAs, -Cs, and -Ds have been reported.

Cleavage of Polycyclic Intermediates. Polycyclic intermediates are constructed so that subsequent cleavage generates cyclopentane derivatives containing the proper functional groups and relative configurations. The synthesis of PGE_2 and $PGF_{2\alpha}$ (Fig. 5) developed by Corey is an elegant example of this approach and is one of the most widely utilized procedures for both laboratory synthesis and commercial production (6). In addition to providing the natural products PGE_2 and $PGF_{2\alpha}$, various intermediates in this process, eg, lactone aldehyde (**29**) and lactone diol (**33**), are extensively used in the synthesis of PG analogues. This stereocontrolled synthesis starts from thallous cyclopentadienide (**22**), which is alkylated with benzyl chloromethyl ether to give the substituted cyclopentadiene (**23**). This reactive and somewhat unstable diene is subjected to a Diels-Alder cycloaddition with α-chloroacrylonitrile, and the resulting cycloadduct is treated with base to generate the ketone function in (**24**). The rigid bicyclic structure of (**24**) has the desired trans relationship between the groups which become the side chains at carbons 8 and 12 of the prostaglandins. Baeyer-Villiger oxidation of ketone (**24**) using MCPBA gives the lactone (**25**), which is hydrolyzed with base to generate a racemic acid. This acid is resolved via its (+)-amphetamine salt to provide the enantiomer (**26**) which has the proper absolute configurations at C-8, -11, and -12. Reaction of (**26**) with potassium iodide and iodine results in the formation of iodolactone (**27**), which has the absolute configurations of $PGF_{2\alpha}$ at C-8, -9, -11, and -12. The hydroxy function at C-11 is acylated with p-phenylbenzoyl chloride, the iodine atom at C-10 is removed by reaction with tributyltin hydride, and the benzyl ether is reductively cleaved to afford lactone alcohol (**28**). This alcohol is oxidized, by a modified Collins oxidation, to produce the lactone aldehyde (**29**) which is commonly referred to as the Corey aldehyde. The lower side chain is stereospecifically attached using the Wadsworth-Emmons modification of the Wittig reaction to give the trans-enone (**30**); the ketone group of this side chain is stereoselectively reduced with lithium diisopinocamphenyl-*tert*-butylborohydride to a 2:1 mixture of the 15(S):15(R) alcohols (**31**) and (**32**). The (R)-isomer (**32**), which is separated from (**31**) by silica gel chromatography (qv), is efficiently recycled to (**30**) by manganese dioxide oxidation; the natural (S)-isomer (**31**) is hydrolyzed to the lactone diol (**33**). The hydroxyl functions of (**33**) are protected as tetrahydropyranyl (THP) ethers, the lactone is reduced to the corresponding lactol, and Wittig reaction of this lactol with the ylid derived from (4-carboxybutyl)triphenylphosphonium bromide affords bis-protected $PGF_{2\alpha}$ (**34**). Deprotection of (**34**) with aqueous acetic acid generates $PGF_{2\alpha}$ (**8**), whereas oxidation of (**34**) with chromic acid followed by aqueous acetic acid deprotection leads to PGE_2 (**1**).

Because the Corey synthesis has been extensively used in prostaglandin research, improvements on the various steps in the procedure have been made. These variations include improved procedures for the preparation of

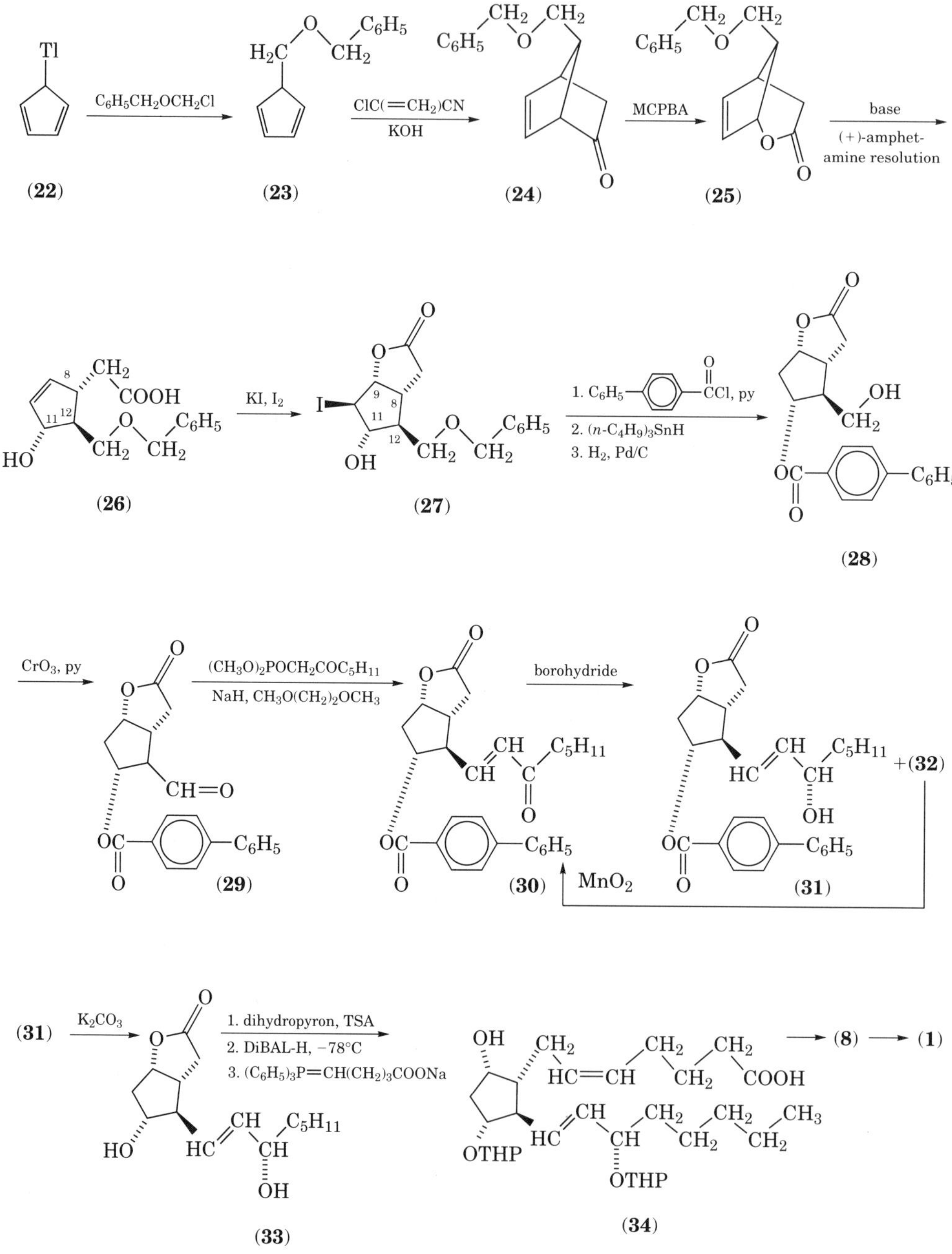

Fig. 5. Corey synthesis of PGE$_2$ and PGF$_{2\alpha}$ where DiBAL-H = diisobutylaluminumhydride; MCPBA = m-chloroperbenzoic acid; py = pyridine; THP = tetrahydropyran; and TSA = p-toluenesulfonic acid (6).

norbornenone (**24**), alternative methods for the resolution of acid (**26**), stereoselective preparations of (**26**), improved procedures for the deiodination of iodolactone (**27**), alternative methods for the synthesis of Corey aldehyde (**29**) or its equivalent, and improved procedures for the stereoselective reduction of enone (**30**) (108–168). For example, a catalytic enantioselective Diels-Alder reaction has been used in a highly efficient synthesis of key intermediate (**24**) in 92% ee (169).

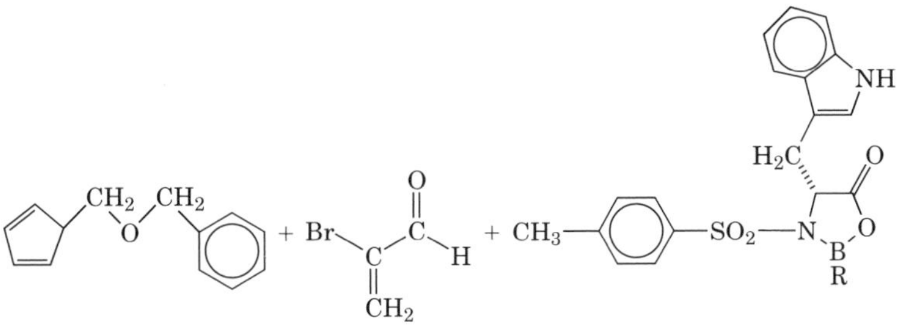

(**35**) R = H or C_4H_9

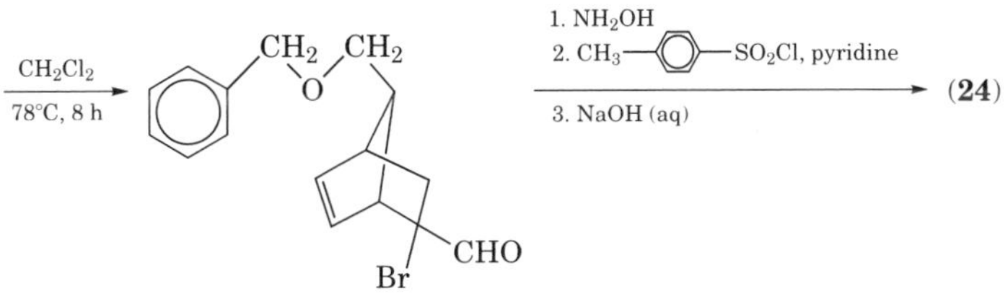

(**36**)

Diels-Alder reaction of 2-bromoacrolein and 5-[(benzyloxy)methyl]cyclopentadiene in the presence of 5 mol % of the catalyst (**35**) afforded the adduct (**36**) in 83–85% yield, 95:5 exo/endo ratio, and greater than 96:4 enantioselectivity. Treatment of the aldehyde (**36**) with aqueous hydroxylamine, led to oxime formation and bromide solvolysis. Tosylation and elimination to the cyanohydrin followed by basic hydrolysis gave (**24**).

The Corey process is also useful for the synthesis of PGs of the 1 and 3 series. Catalytic hydrogenation of (**34**) (see Fig. 5) with 5% Pd/C at −15–20°C results in selective reduction of the 5,6-double bond. Subsequent transformations analogous to those in Figure 5 lead to PGE_1 (**9**) and $PGF_{1\alpha}$ (**10**). The key step for synthesis of the PG_3 series is the Wittig reaction of (**29**) (see Fig. 5) with the appropriate unsaturated ω-chain ylide (170).

Another commercially important total synthesis of PGs is based on the stereoselective cleavage of a bicyclo[3.1.0]hexane structure (Fig. 6). This approach was used in the first published synthesis of a classical PG (171), and was subsequently improved (172,173). The synthesis started with norbornadiene (**37**), which was monoepoxidized, rearranged under acidic conditions, and treated with 2,2-dimethyl-1,3-propanediol to afford the bicyclo[3.1.0]hexene (**38**). This intermediate was converted to cyclobutanone (**39**) by means of a ketene–cycloaddition

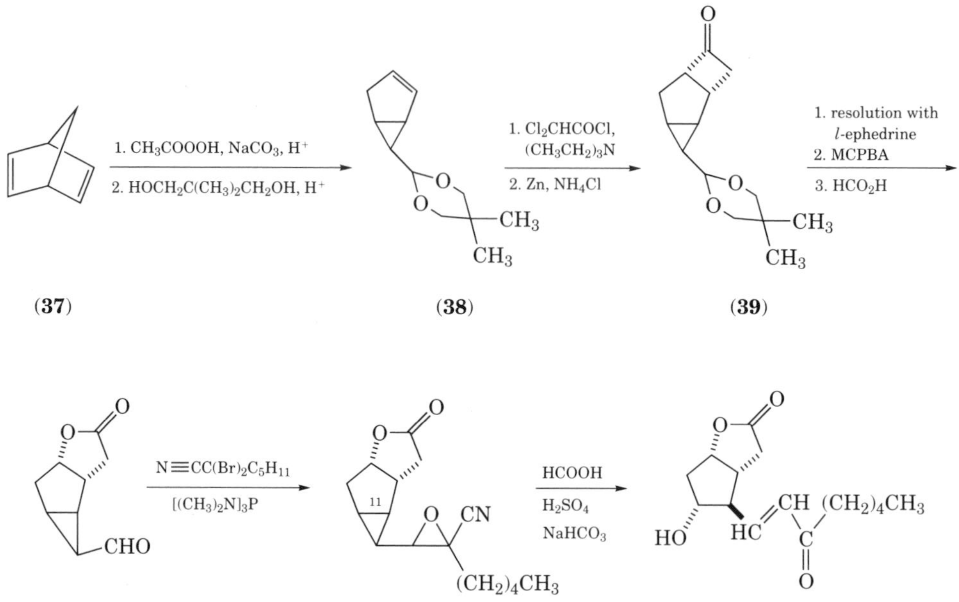

Fig. 6. Synthesis of prostaglandins via a bicyclo[3.1.0]hexane intermediate where MCPBA = m-chloroperbenzoic acid.

reaction, followed by dechlorination with Zn dust. Compound (**39**) was resolved by diastereomeric oxazolidine formation with l-ephedrine. Treatment of (**39**) with m-chloroperbenzoic acid (MCPBA) and hydrolysis of the acetal group gave (**40**); condensation of (**40**) with 1-cyano-1,1-dibromohexane afforded glycidonitrile (**41**). Solvolysis of (**41**) generated enone (**42**) with the desired C-11 hydroxyl and ω-chain allylic alcohol functions. In the solvolysis of (**41**), the hydroxyl ion is sterically directed to the α-face of C-11 and formation of the trans double bond is favored by the spacial disposition of the substituents during the ring-opening sequence. The enone (**42**) was acylated at C-11 to produce the lactone enone (**30**) from the Corey process (see Fig. 5). This synthesis has produced the equivalent of more than 50 kg/yr of PGF$_{2\alpha}$ and has effectively reduced the cost of synthesizing PGs to less than one-hundredth of the cost of bioconversion of arachidonic acid (174).

A third procedure involving cleavage of polycyclic intermediates has been reported (175). This synthesis exploits the stereochemical properties of the bicyclo[3.2.0]heptan-6-one ring system (Fig. 7). Compound (**43**) was reduced with Baker's yeast to give a mixture of the chiral endo and exo (S)-alcohols (**44**) and (**45**) which were readily separated by distillation (qv) or chromatography. Treatment of these intermediates with N-bromoacetamide provided the corresponding bromohydrins (**46**) and (**47**) with concomitant oxidation of the cyclobutane alcohol group. Both isomers were converted to the lactone diol intermediate (**33**) by different processes. With (**46**), protection of the C-11 hydroxyl group as a silyl ether followed by treatment with potassium t-butoxide generated the isolatable intermediate (**48**) (176,177). Regiospecific cleavage of the cyclopropyl ring of (**48**)

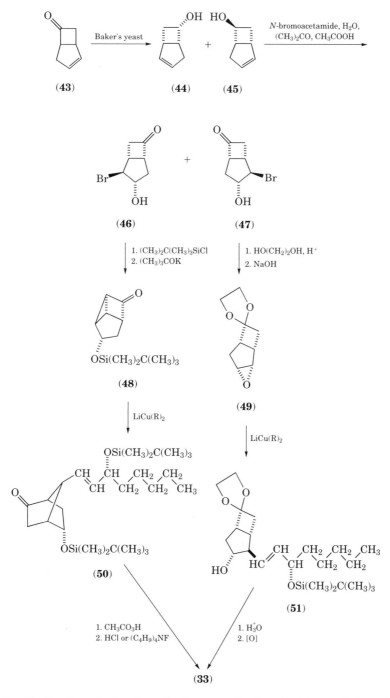

Fig. 7. Synthesis of prostaglandins via cleavage of polycyclic intermediates where R = $HC{=}CHCH(OSi(CH_3)_2C(CH_3)_3)C_5H_{11}$.

with an ω-chain organocuprate reagent provided (**50**) which was subjected to Baeyer-Villiger oxidation, hydrolysis of the silyl ethers, and reverse lactonization to give (**33**). With the other isomer (**47**), an epoxide cleavage route was employed. Thus (**47**) was ketalized and an epoxide formed by base treatment to give (**49**). Cleavage of the strained ketal–epoxide of (**49**) with the ω-chain cuprate reagent gave a 4:1 ratio of (**51**) and its undesired regioisomer. The Corey intermediate (**33**) was generated by hydrolysis and Baeyer-Villiger oxidation of (**51**). A number of other synthetic approaches to Corey intermediates have been developed and reviewed (107,108).

    Conjugate Addition Approach. The use of conjugate addition for prostaglandin synthesis has been widely researched (6,108,178–184). It has been applied to both classical PGs and their analogues and is the basis for the commercial production of several marketed PG analogues (107,185). The approach involves the conjugate addition of an organometallic derivative of the omega chain, $M(HC=CH\text{-}\omega)$, where M is a metal, to a protected hydroxycyclopentenone which generally already contains the appropriate R_α, wherein $R' =$ protecting group.

    An alternative three-component coupling procedure, in which the α-chain is incorporated by reaction with the enolate arising from the initial conjugate addition reaction, has been elegantly researched and refined (186). The advantage of this overall approach is its convergency, versatility for preparing analogues, and the stereoselectivity of the addition reaction. The stereoselectivity arises from enolate quenching under thermodynamic conditions to generate the more stable all-trans product. Thus, fixation of the single asymmetric center in the enone dictates the eventual stereochemical outcome at C-8 and C-12. Although a variety of organometallic reagents have been studied, organocuprates have been the most extensively applied to PG synthesis. The first applications of organocuprate chemistry to PG synthesis were reported in 1972 (187–189). The original synthesis of PGE_1 (**9**) using conjugate addition is shown in Figure 8 (189). The diastereomeric product (**54**) was removed by chromatography. The cuprate species (**53**) was prepared as shown. The enone (**60**), a precursor of (**52**), was prepared from cyclopentadiene. Quantitative alkylation of (**57**) using ethyl 7-bromoheptanoate gave the diene (**58**). Cycloaddition with chemically generated singlet oxygen afforded a mixture of hydroxycyclopentenones (**59**) and (**60**), in which the desired isomer (**60**) was a minor proportion of the product. However, oxidation of the isomeric mixture with Jones reagent to the dione (**61**), followed by reduction of the sterically more accessible ketone function, improved the isomer ratio to 2:1 in favor of the 4-hydroxycyclopentenone (**60**). In a later synthesis, using a resolved enone, PGE_1 was produced exclusively (190).

    Since this original synthesis, a great number of improvements (191–201) have been made in the stereoselective preparation and derivatization of the ω-chain precursor, in cuprate reagent composition and preparation, in protect-

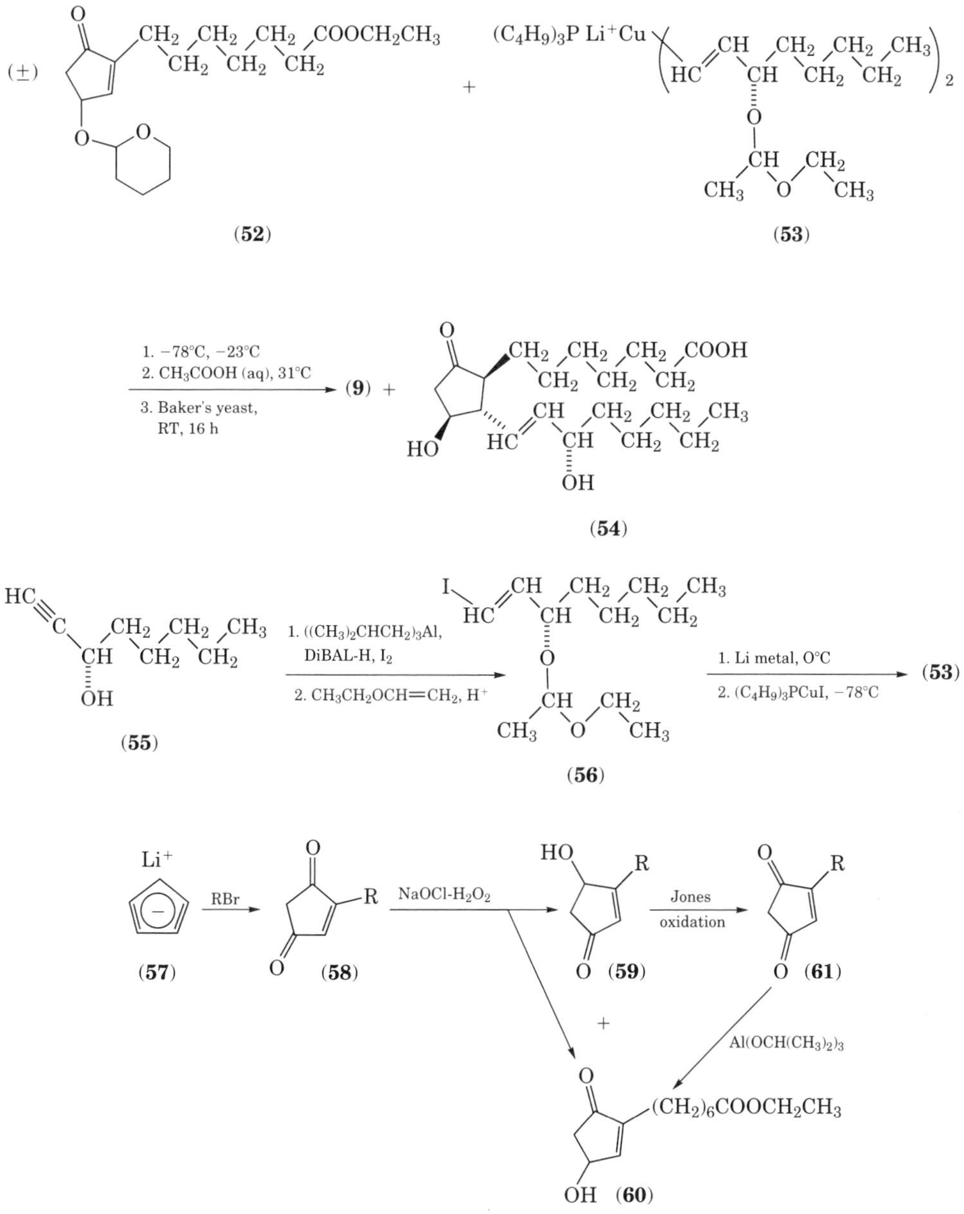

Fig. 8. Synthesis of PGE$_1$ using an organocuprate reagent where DiBAL-H is diisobutyl-aluminumhydride and R = (CH$_2$)$_6$COOCH$_2$CH$_3$.

ing group utilization, and in the preparation and resolution of hydroxycyclopen-tenones. Illustration of some of the many improvements are seen in a synthesis (202) of enisoprost, a PGE$_1$ analogue. The improvements consist of a much more efficient route to the enone as well as modifications in the cuprate reactions. Preparation of the racemic enone is as follows:

(Z,Z)-1,5-Cyclooctadiene (**62**) was ozonolyzed to about 65–70% of completion and quenched with triethylamine and acetic anhydride to give the aldehyde ester (**63**) in 40–50% yield along with about 2–5% of the corresponding dialdehyde. Reaction of crude (**63**) with 2-furanylmagnesium chloride provided the furanyl-carbinol (**64**) which was treated with zinc chloride to produce (**65**). Treatment of (**65**) with a catalytic amount of anhydrous chloral in the presence of triethyl-amine gave the desired enone (**66**). For preparation of the cuprate reagent, zir-conocene chloride hydride followed by iodine was used to generate exclusively the (E)-vinyl iodide (**68**) from (**67**). Treatment of (**68**) with n-C_4H_9Li generated the vinyllithium species which was then converted to the dilithiocyanocuprate reagent (**69**) by addition of lithium methylcyanocuprate, prepared freshly from methyllithium and copper cyanide.

A further improvement in the cuprate-based methodology for producing PGs utilizes a one-pot procedure (203). The ω-chain precursor (**67**) was first functionalized with zirconocene chloride hydride in THF. The vinylzirconium intermediate was transmetalated directly by treatment with two equivalents of n-butyllithium or methyllithium at -30 to $-70°C$. Sequential addition of copper cyanide and methyllithium elicited the *in situ* generation of the higher order cyanocuprate which was then reacted with the protected enone to give the PG.

The primary disadvantage of the conjugate addition approach is the necessity of performing two chiral operations (resolution or asymmetric synthesis) in order to obtain exclusively the stereochemically desired end product. However, the advent of enzymatic resolutions and stereoselective reducing agents has resulted in new methods to efficiently produce chiral enones and ω-chain synthons, respectively (see ENZYME APPLICATIONS, INDUSTRIAL; ENZYMES IN ORGANIC SYNTHESIS). For example, treatment of the racemic hydroxy enone (**70**) with commercially available porcine pancreatic lipase (PPL) in vinyl acetate gave a separable mixture of (S)-hydroxyenone (**71**) and (R)-acetate (**72**) with enantiomeric excess (ee) of 90% or better (204).

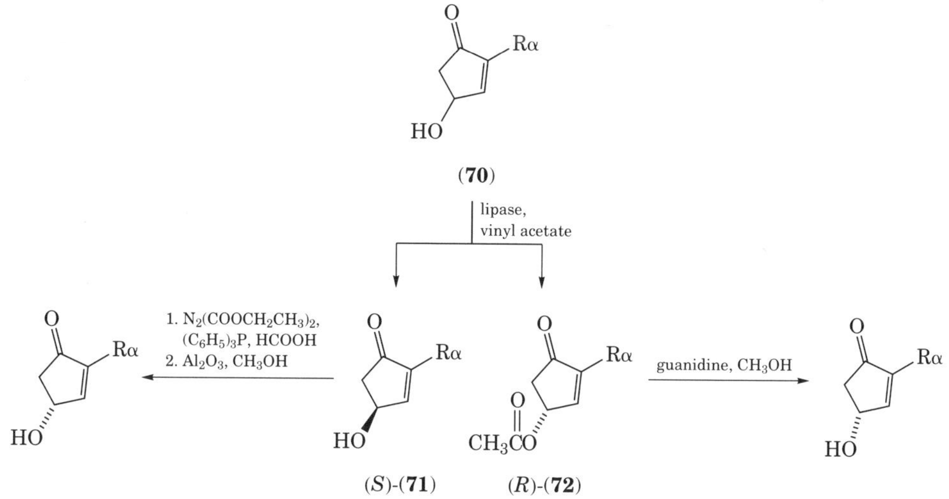

(**70**)

lipase,
vinyl acetate

1. $N_2(COOCH_2CH_3)_2$, $(C_6H_5)_3P$, HCOOH
2. Al_2O_3, CH_3OH

guanidine, CH_3OH

(S)-(**71**)   (R)-(**72**)

The (S)-(**71**) material could be inverted via Mitsunobu chemistry to the desired (R)-isomer without loss of stereochemical integrity, whereas the (R)-(**72**) product was readily cleaved to the (R)-alcohol with guanidine in methanol. The (R)-alcohol could also be recycled to improve its enantiomeric purity. Chiral ω-chain alcohols can be produced with a high ee from the corresponding ketones using stereoselective reducing agents such as (S)-bi-2-naphthol aluminum hydride ((S)-BINAL-H) (191) and alpine-borane (192). A three-component coupling procedure has been devised in which conjugate addition to the chiral cyclopentenone (**74**) is followed by *in situ* quenching of the enolate with an α-chain derivative. A typical synthesis using this methodology is as follows:

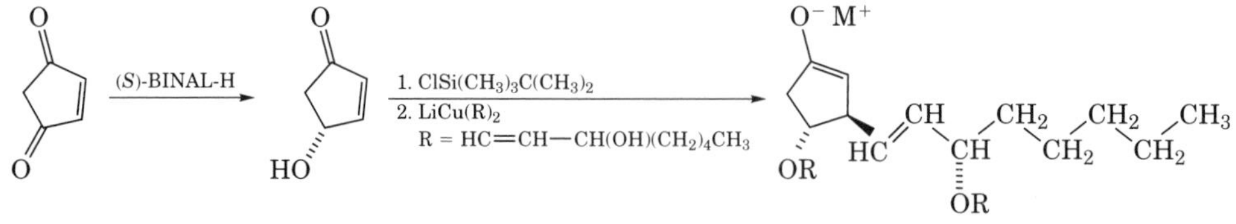

(73)   (74)   (75, R = –Si(CH₃)₂C(CH₃)₃)

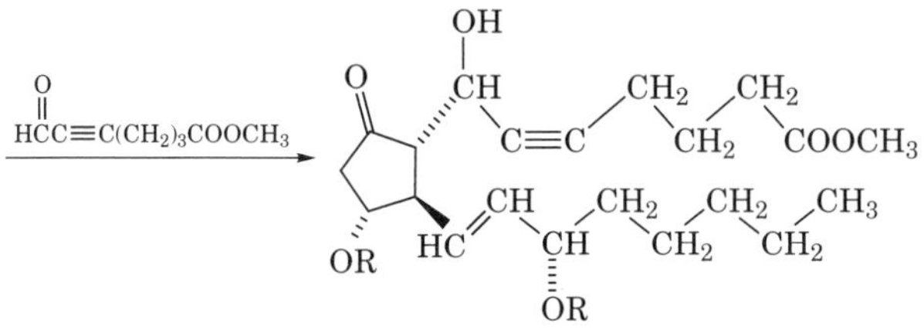

(76)

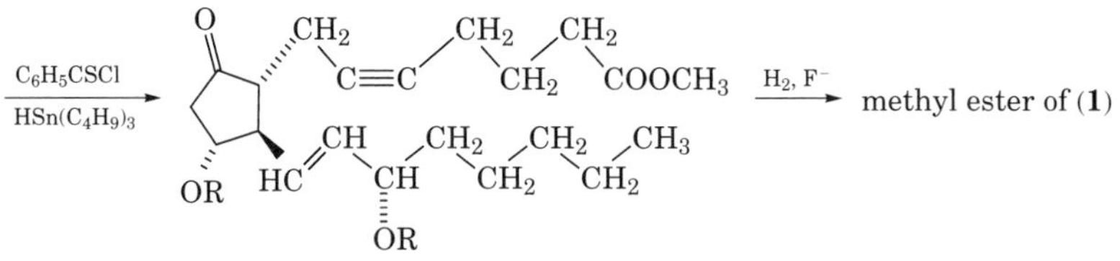

(77)

PGE₂ methyl ester was obtained by reduction of the triple bond to the (Z)-olefin and removal of silyl-protecting groups using fluoride. The primary drawback to this approach is the instability of the enolate leading to elimination of the 11-hydroxy group. Although reactive trapping species such as aldehydes, alkyl halides, Michael acceptors, and allylic halides provide reasonable yields of product, simply alkyl halides fail. In addition, the use of reactive α-chain derivatives often requires further manipulations to generate the desired PG structure. A number of strategies (205–211) have been implemented to avoid or minimize this problem. With propargyl halides as trapping agents, the addition of triphenyltin chloride to the enolate aids the alkylation presumably by enolate metal exchange (186). One improvement (205) of this procedure utilized a vinyl zincate rather than a cuprate.

Cyclization of Aliphatic Precursors. This strategy consists of assembling the key functional groups in an aliphatic format, cyclizing to a cyclopentane intermediate, and completing the synthesis by further elaboration of the side chains. One application of this strategy is as follows:

(78)   (79)   (80)

(81)   (82)   (83)

Reaction of (S)-(−)-2-acetoxysuccinyl chloride (**78**), prepared from (S)-malic acid, using the magnesiobromide salt of monomethyl malonate afforded the dioxosuberate (**79**) which was cyclized with magnesium carbonate to a 4:1 mixture of cyclopentenone (**80**) and the 5-acetoxy isomer. Catalytic hydrogenation of (**80**) gave (**81**) having the thermodynamically favored all-trans stereochemistry. Ketone reduction and hydrolysis produced the bicyclic lactone acid (**82**) which was converted to the Corey aldehyde equivalent (**83**). A number of other approaches have been described (108).

The discovery (1) that the "unnatural" isomer (15R)-PGA$_2$ [*23602-72-4*] and its 15-acetate, methyl ester (**84**) were present in unusually large quantities (1.5–3.0% of dry weight) in the coral species *Plexaura homomalla* provided a stimulus to utilize these materials as sources for synthesis of PGE$_2$ (**1**) and PGF$_{2\alpha}$ (**8**). (15S)-PGA$_2$ was also shown to be present in certain types of *P. homomalla* coral (212,213), making access to PGE$_2$ and PGF$_{2\alpha}$ more direct and efficient. However, modern day total synthesis methods have rendered this source unnecessary.

(84)

Synthesis of Other Prostanoids. The other classical prostaglandins, PGAs, Bs, Cs, Ds, and Js, can be prepared from the PGEs and Fs. Dehydration of PGE$_2$ under acidic conditions (CH$_3$COOH/H$_2$O, heat) generates PGA$_2$, whereas alkaline conditions (NaOH, CH$_3$OH) produce PGB$_2$. PGJ$_2$ can be formed by dehydration of PGD$_2$. Total synthetic procedures for these PGs have been given (6).

Prostaglandin Endoperoxides. The naturally occurring endoperoxides, PGG_1, PGG_2, PGH_1, and PGH_2, as well as a number of their analogues, having variations in the α- and ω-chains, have been prepared by biosynthesis using seminal vesicles (sheep) as the enzyme source, from the corresponding fatty acids (58). The isolation of these products, demonstrating that they have an adequate degree of chemical stability, prompted efforts to prepare them by total synthesis. A practical and efficient synthesis of PGH_2 (**6**) and PGG_2 (**5**) is depicted in Figure 9 (214,215).

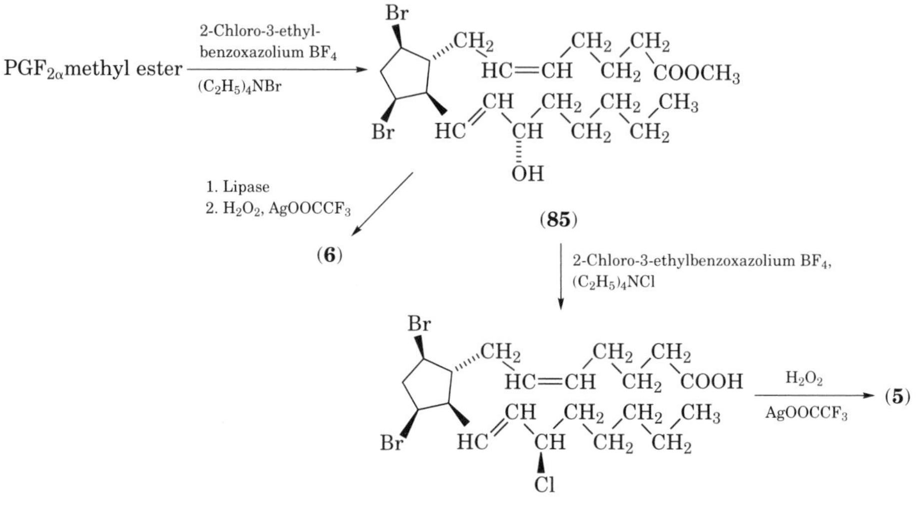

Fig. 9. Synthesis of prostaglandin endoperoxides.

Prostacyclin. The total syntheses of PGI_2 (**4**) have been extensively reviewed (58,103). The first synthesis of PGI_2 as its methyl ester and sodium salt (epoprostenol) (216–218) was pivotal in the chemical characterization of this unstable molecule (58). The key feature of this synthesis was the iodocyclization reaction to produce a pair of diastereomers (**87**) and (**88**).

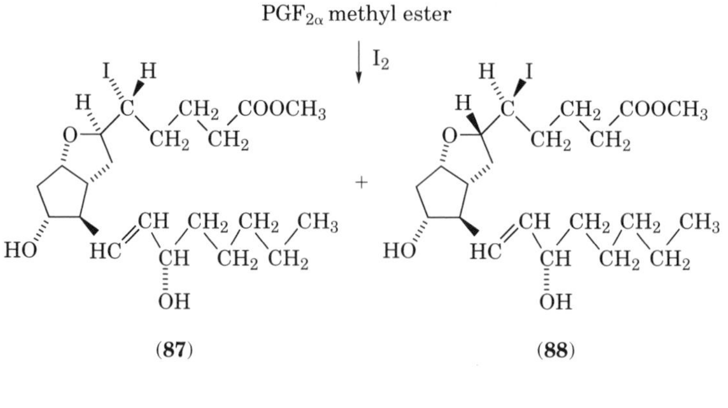

Subsequent dehydrohalogenation afforded exclusively the desired (Z)-olefin of the PGI_2 methyl ester. Conversion to the sodium salt was achieved by treatment with sodium hydroxide. The sodium salt is crystalline and, when protected from atmospheric moisture and carbon dioxide, is indefinitely stable. A variation of this synthesis started with a C-5 acetylenic PGF derivative and used a mercury salt catalyzed cyclization reaction (219). Although natural PGI_1 has not been identified, the syntheses of both (6R)- and (6S)-PGI_1, [62777-90-6] and [62770-50-7], respectively, have been described, as has that of PGI_3 (104,216).

Thromboxanes. Because of its highly unstable nature, TxA_2 (**3**) eluded total synthesis until 1985. However, following the disclosure in 1975 of its proposed structure and pharmacological importance, a great deal of effort was expended to develop syntheses of its stable metabolite, TxB_2 (**15**). A practical, short synthesis (220) from the 9,15-diacetate of $PGF_{2\alpha}$ (**89**) follows.

Treatment of (**89**) with lead tetraacetate generates the unstable open-ring aldehyde (**90**) which is quickly converted to a dimethylacetal (**91**). Following basic hydrolysis of the methyl ester and acetates, the acetal is cleaved with aqueous acid to produce TxB_2. A number of other approaches, including one starting from the Corey aldehyde, have been described (58).

The definitive synthesis of TxA_2 (**3**) (57) is outlined in Figure 10. A lactone (**91**) was formed from TxB_2. Dehydration to the enol ether followed by bromohydrin formation gave intermediate (**92**) which was cyclized to the bromooxetane (**93**) by use of a modified Mitsunobu reaction. Debromination with tri-*n*-butyltin

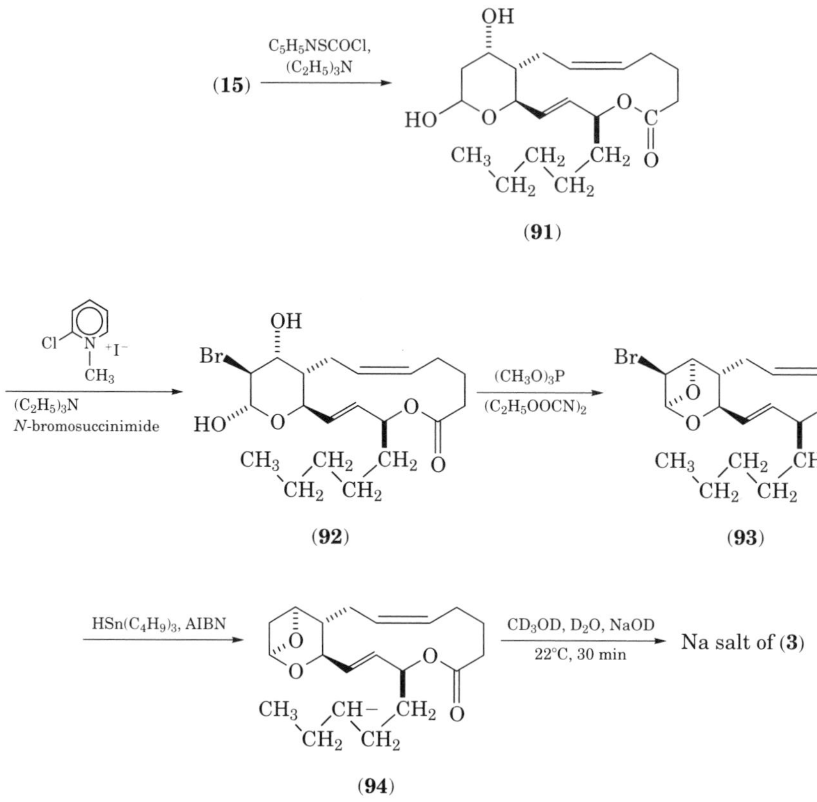

(91)

(92)   (93)

(94)

Fig. 10.   Synthesis of TxA$_2$ (**3**), where AIBN = azobisisobutyronitrile.

hydride cleanly yielded the desired 1,15-lactone form of TxA$_2$, (**94**). Lactone (**94**) was unstable to chromatographic purification but could be obtained free of tin by-products by use of a polymer-bound tin hydride. X-ray crystallography confirmed the structure of (**94**). The macrolactone was saponified in a 1:1 mixture of CD$_3$OD–D$_2$O containing 10 equivalents of NaOD to give the sodium salt of TxA$_2$. This material possessed the appropriate H^1-nmr features of the bicyclic oxetane nucleus and reproduced the biological activities of natural platelet-derived TxA$_2$ in a variety of assays. The lactone (**94**) was inactive in these assays.

Design of Synthetic Analogues

The natural prostanoids have myriad biological effects and held great promise as potential therapeutic agents in numerous diseases. The natural prostanoids, however, also have three notable drawbacks which medicinal chemists have tried to overcome by molecular modification in order to produce acceptable drug candidates. These drawbacks are rapid metabolism which results in lack of activity if taken orally and a short duration of action, numerous side effects due to their multiplicity of biological activities, and poor chemical stability, a characteristic especially pronounced in PGE, -D, and -I structures.

Analogues of the Classical Prostaglandins. A list of the more prominent prostaglandin analogues which are either marketed or have reached some stage of development is given in Table 3. Two reviews (107,221) detail the syntheses of these compounds and the strategies associated with their design. One significant strategy in analogue conception has been directed at impeding the rapid metabolic degradation of the natural compounds. Placement of methyl groups at C-15 or C-16 to block C-15 dehydrogenation has been effective and is exemplified by arbaprostil and trimoprostil. Blockage of β-oxidation of the α-chain has been attempted by placement of double bonds at C-2,3 and C-4,5 and heteroatoms at C-3. Enisoprost, enprostil, and limaprost are examples. Strategies to prevent ω-chain oxidation have also been employed and are seen in enprostil, dimoxaprost, remiprostol, and sulprostone, among others. To improve selectivity, a number of strategies have been attempted, but the task has proven difficult and side effects remain a principal hindrance to the therapeutic use of prostaglandins. Perhaps the most successful modification has been the translocation of the 15-hydroxy group of natural PGEs to the adjacent C-16 position. Analogues such as misoprostol, enisoprost, rioprostol, and remiprostol are examples of this approach. Misoprostol, for example, shows improved separation of therapeutic action from undesired effects such as diarrhea production and cardiovascular effects in comparison with its 15-hydroxy counterparts (222). Replacement of the C-13,14 double bond with a heteroatom as in GR-63779X and tiprostanide has also provided improved selectivity. Modifications to improve chemical stability are evident in trimoprostol, nocloprost, and metenoprost. The propensity for β-elimination of the 11-hydroxy group has been eliminated in each of these analogues. The use of bulky aromatic esters, as found in tiprostanide and GR-63779X, to induce or enhance crystallinity has also been a strategy to improve stability.

Prostacyclin Analogues. Problems of poor chemical stability, rapid metabolic disposition, and lack of selectivity coupled with its high therapeutic potential for treatment of cardiovascular diseases led to an extensive exploration of structural variants of prostacyclin. A list of prominent analogues is given in Table 4, and the synthetic details for these compounds are available (107,223). One of the obvious strategies has been to alleviate the chemical instability inherent in the cyclic enol ether of PGI$_2$. The replacement of oxygen with carbon to give carbacyclic analogues has been a favorite and successful endeavor. Other modifications have been made to the α- and ω-chains to reduce metabolic susceptibility.

Endoperoxide Analogues. The endoperoxide analogues that have been synthesized and biologically evaluated have been summarized (103,104,224). In general, these analogues were designed to be more chemically stable than the naturally occurring substances PGG$_2$ and PGH$_2$, which have labile 2,3-dioxabicyclo[2.2.1]heptane ring systems. The following compounds are notable among the many analogues that have been synthesized. The azo analogue (**95**) and the epoxymethano analogues (**97**) and (**98**) appear to mimic the biological

Table 3. Therapeutically Useful Prostaglandin Analogues

Compound generic name (Trade name)[a]	Structure			CAS Registry Number	Company
	Rα	(ring)	Rω		
arbaprostil (Arbacet)	CH₂ / HC=CH / CH₂ CH₂ / CH₂ CH₂ COOH	cyclopentanone ring with Rα, Rω, HO (O=)	HC=CH–C(HO)(CH₃)– CH₂ CH₂ CH₂ CH₃ / CH₂ CH₃	[55028-70-1]	Upjohn
HR-260 dimoxaprost	same		H₃C CH₃ / HC=CH–C–CH–O–CH₂ CH₂ CH₃ / OH	[90243-98-4]	Hoechst-Roussel
SC-29333 misoprostol (Cytotec)	CH₂ CH₂ CH₂ COOCH₃ / CH₂ CH₂		CH₃ / HC=CH–CH₂–C–CH₂ CH₂ CH₃ / HO (CH₂ CH₃)	[59122-46-2]	Searle
SC-34301 enisoprost	CH₂ / HC=CH / CH₂ / CH₂ COOCH₃		same	[81026-63-3]	Searle
TR-4698 rioprostol	CH₂ CH₂ CH₂ / CH₂ CH₂ CH₂ OH		same	[77287-05-9]	Miles/Ortho and Bayer AG
RS-84,135 enprostil (Gardrin)	CH / HC=C=CH / CH₂ COOCH₃ / CH₂		HC=CH–CH–CH₂–O–C₆H₅ / OH	[73121-56-9]	Syntex
MDL-646 mexiprostil	CH₂ CH₂ CH₂ COOCH₃ / CH₂ CH₂		H₃C OCH₃ / HC=CH–CH–C–CH₂ CH₂ CH₃ / OH (CH₂ CH₃)	[88980-20-5]	Lepetit

334

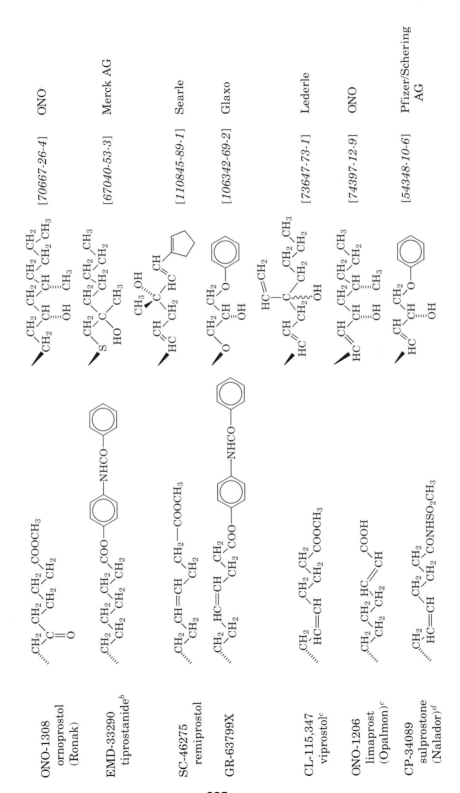

ONO-1308
ornoprostol
(Ronak)

EMD-33290
tiprostanide[b]

SC-46275
remiprostol

GR-63799X

CL-115,347
viprostol[c]

ONO-1206
limaprost
(Opalmon)[c]

CP-34089
sulprostone
(Nalador)[d]

[70667-26-4]   ONO

[67040-53-3]   Merck AG

[110845-89-1]   Searle

[106342-69-2]   Glaxo

[73647-73-1]   Lederle

[74397-12-9]   ONO

[54348-10-6]   Pfizer/Schering
                 AG

Table 3. (Continued)

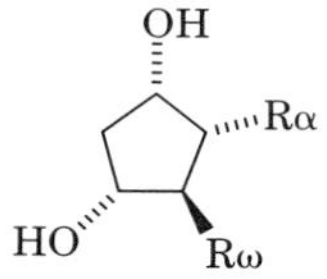

Compound generic name (Trade name)[a]	Structure Rα	Structure Rω	CAS Registry Number	Company
ONO-802 gemeprost (Cervagem)[d]	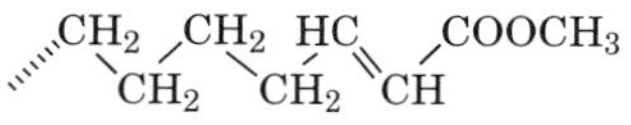 CH_2–CH_2–HC=CH–CH_2–$COOCH_3$	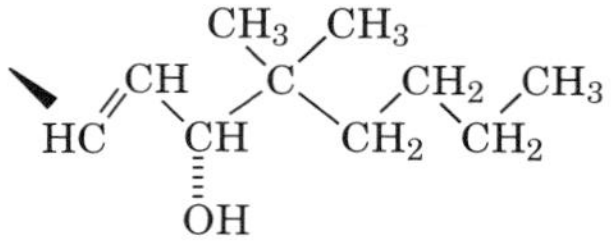 HC≡CH–CH(OH)–C(CH_3)(CH_3)–CH_2–CH_2–CH_2–CH_3	[64318-79-2]	ONO
Carboprost (Prostin 15M)[d]	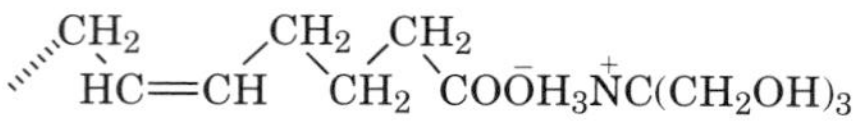 HC=CH–CH_2–CH_2–CH_2–$CO\bar{O}H_3\overset{+}{N}C(CH_2OH)_3$	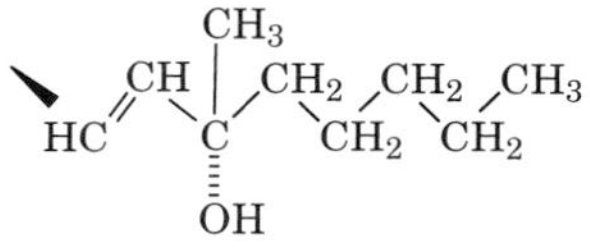 HC=CH–C(CH_3)(OH)–CH_2–CH_2–CH_2–CH_2–CH_3	[58551-69-2]	Upjohn
ICI 80,996 cloprostenol (Estrumate)[e]	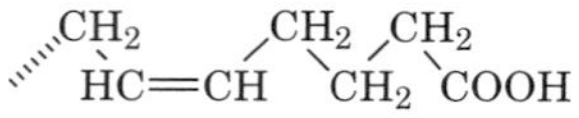 HC=CH–CH_2–CH_2–CH_2–$COOH$	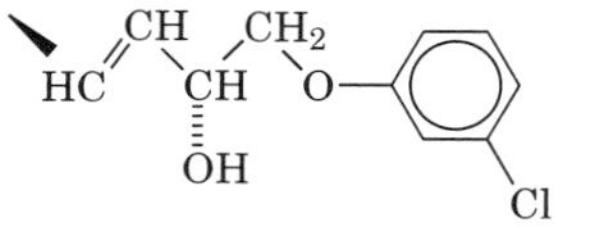 HC=CH–CH(OH)–CH_2–O–(3-Cl-C_6H_4)	[62561-03-9]	ICI
ICI 80,008 fluprostenol (Equimate)[e]	same	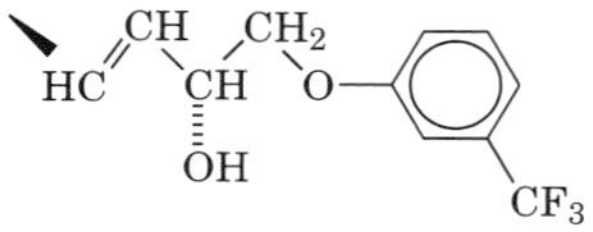 HC=CH–CH(OH)–CH_2–O–(3-CF_3-C_6H_4)	[59685-93-7]	ICI
RS-84043 fenprostalene[e]	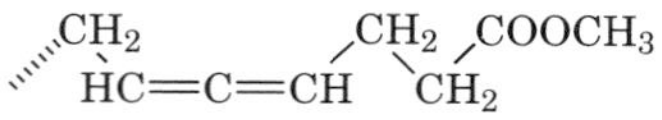 HC=C=CH–CH_2–CH_2–$COOCH_3$	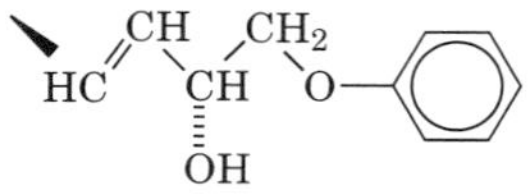 HC=CH–CH(OH)–CH_2–O–C_6H_5	[69381-94-8], [73175-12-9]	Syntex
RS-9390 prostalene (Synchrocept)[e]	same	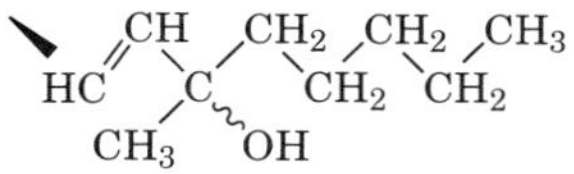 HC=CH–C(CH_3)(OH)–CH_2–CH_2–CH_2–CH_3	[54120-61-5]	Syntex

Central ring structure (common to all):
cyclopentane ring with OH, HO substituents, bearing Rα and Rω side chains.

336

Compound	Structure	CAS Number	Company

PhXA41
latanoprost[f]   [130209-82-4]   Kabi Pharmacia

Structure as given

$HC = CH$ / CH_2 CH_2 CH_2 CH_3 ... $COO-CH$ CH_3

Structure as given

(phenyl ring structure) — HC=CH, CH, CH, OH, CH$_2$, CH$_2$—(phenyl)

RO-21,6937
trimoprostil   [69900-72-7]   Roche

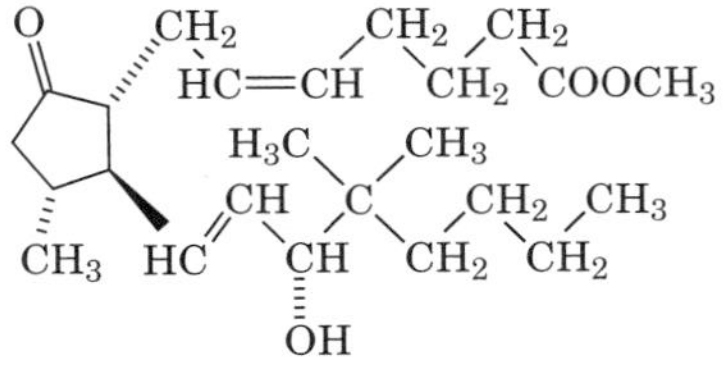

CH_2 CH_2 CH_2 CH_2 $COOCH_3$; $HC=CH$; H_3C CH_3 ; CH_2 CH_3 ; CH_2 CH_2 ; CH, C, OH ; HC ; CH_3 ; O=(cyclopentanone)

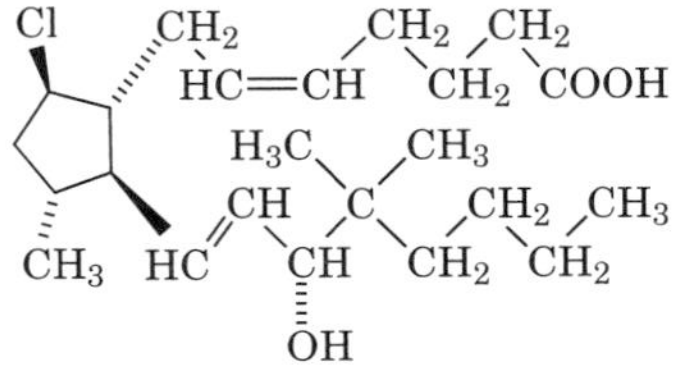

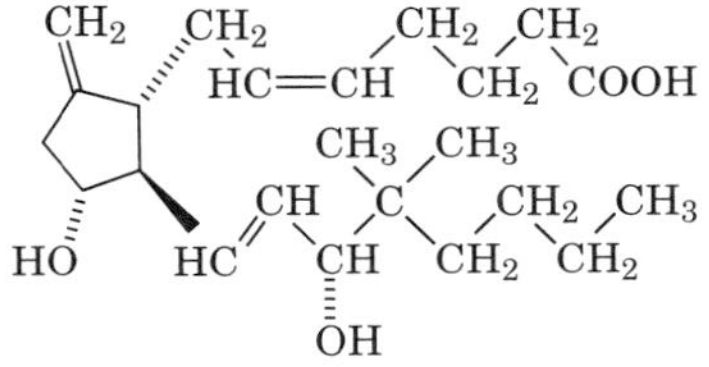

ZK-94726
nocloprost   [79360-43-3]   Schering AG

CH_2 CH_2 CH_2 CH_2 $COOH$; $HC=CH$; H_3C CH_3 ; CH_2 CH_3 ; CH_2 CH_2 ; CH, C, OH ; HC ; CH_3 ; Cl-(cyclopentane)

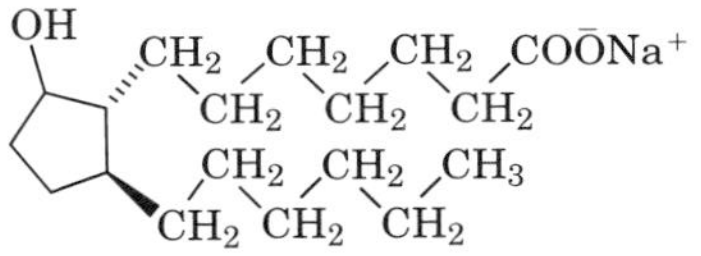

meteneprost[d]   [61263-35-2]   Upjohn

CH_2 CH_2 CH_2 CH_2 $COOH$; $HC=CH$; CH_3 CH_3 ; CH_2 CH_3 ; CH_2 CH_2 ; CH, C, OH ; HC ; CH_2= ; HO-(cyclopentane)

rosaprostol   [56695-65-9]   IBI

OH ; CH_2 CH_2 CH_2 CH_2 $COONa^+$; CH_2 CH_2 CH_2 ; CH_2 CH_2 CH_3 ; CH_2 CH_2 CH_2 (cyclopentane)

---

[a]The therapeutic indication (TI) is antiulcer, unless otherwise noted.
[b]TI is antiulcer and antihypertensive.
[c]TI is antihypertensive.
[d]TI is fertility control and labor induction.
[e]TI is veterinary use and synchronization of estrus.
[f]TI is antiglaucoma.

Table 4. Therapeutically Useful Prostacyclin Analogues

Compound generic name (Trade name)	Structure R	Structure R'	X	CAS Registry Number	Company[a]
epoprostenol (Flolan)			O	[73873-87-7]	Upjohn/Wellcome
ZK-34798 taprostene[b]			O	[108945-35-3]	Gruenenthal
carbacyclin[c]			CH$_2$	[69552-46-1]	Upjohn/Wellcome
ZK-97951 iloprost	same		CH$_2$	[78919-13-8]	Schering AG
OP41483 ataprost	same		CH$_2$	[83997-19-7]	Ono/Dainippon

338

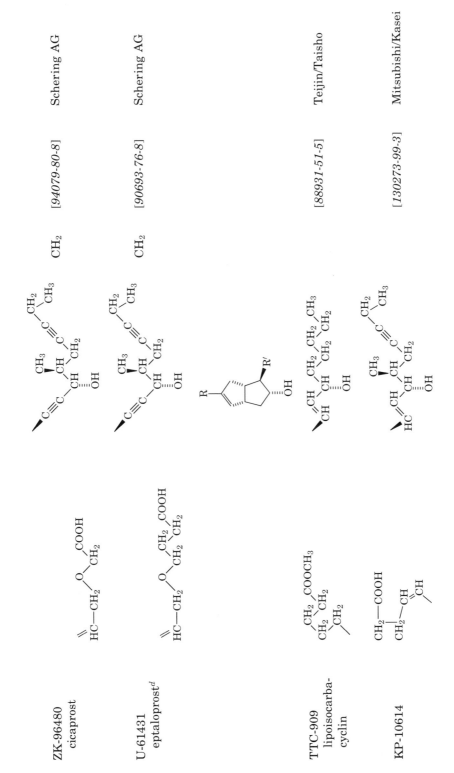

ZK-96480 cicaprost		CH₂	Schering AG
U-61431 eptaloprost[d]		CH₂	Schering AG
TTC-909 lipoisocarba-cyclin			Teijin/Taisho
KP-10614			Mitsubishi/Kasei

Table 4. (*Continued*)

Compound generic name (Trade name)	Structure R	Structure R′	Structure X	CAS Registry Number	Company[a]
	Structures as given				
CG-4203 ciprostene	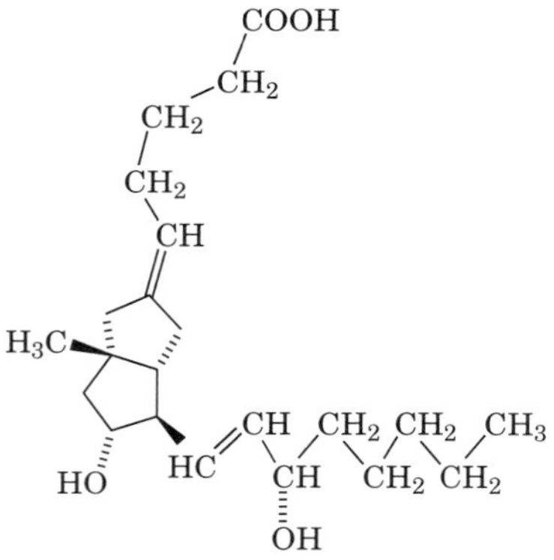			[*81845-44-5*]	Upjohn/Wellcome
beraprost	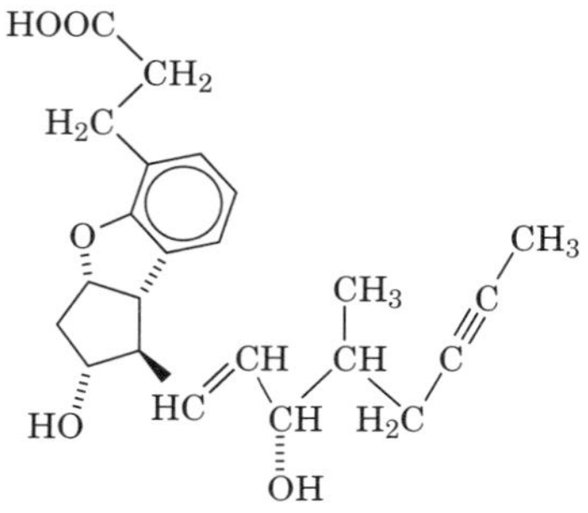			[*88430-50-6*]	Toray/Kaken

ZK-34798
nileprost[e]

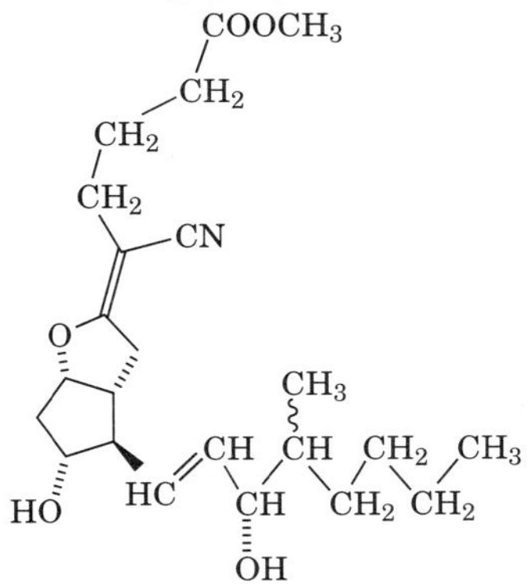

[71097-83-1]   Schering AG

OP-2507[f]

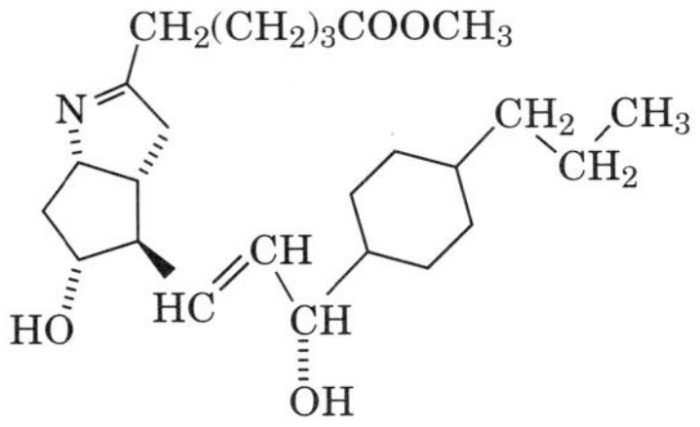

[101758-79-6]   Ono

CH-5084

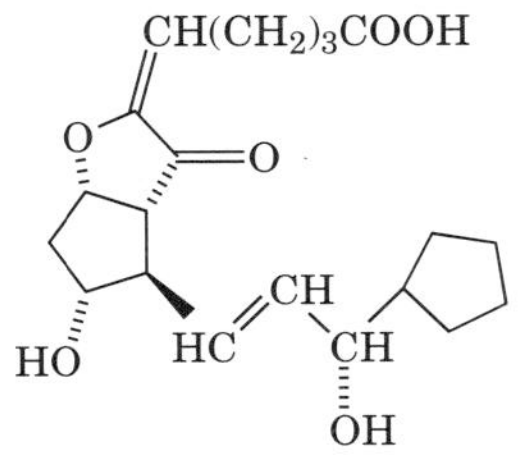

[111319-88-1]   Chinoin

SC-43350

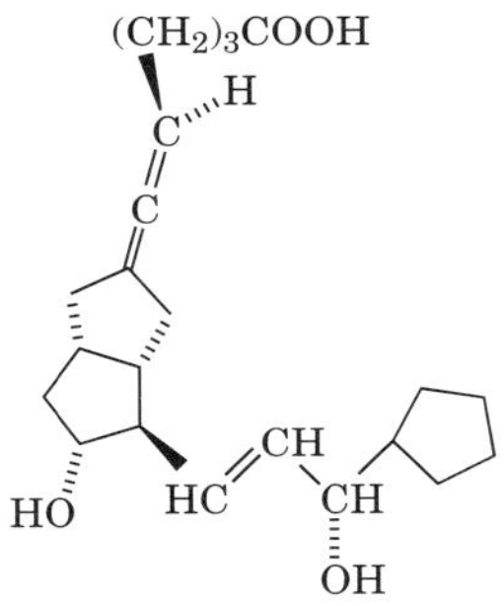

[106138-20-9]   Searle

Table 4. (Continued)

Compound generic name (Trade name)	Structure			CAS Registry Number	Company[a]
	R	R'	X		
RS-93427	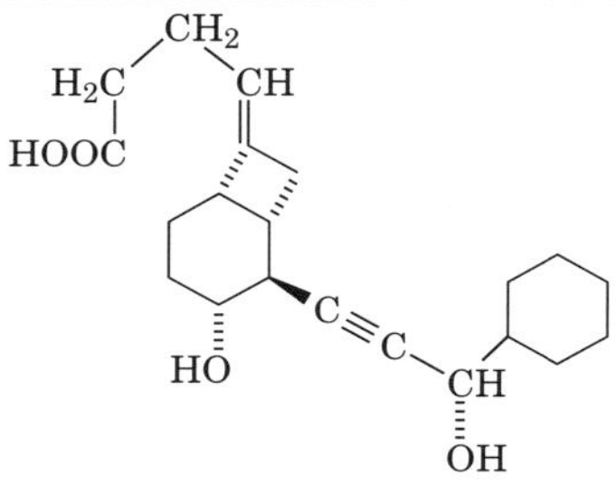			[105284-21-7]	Syntex
U-68,215[e]	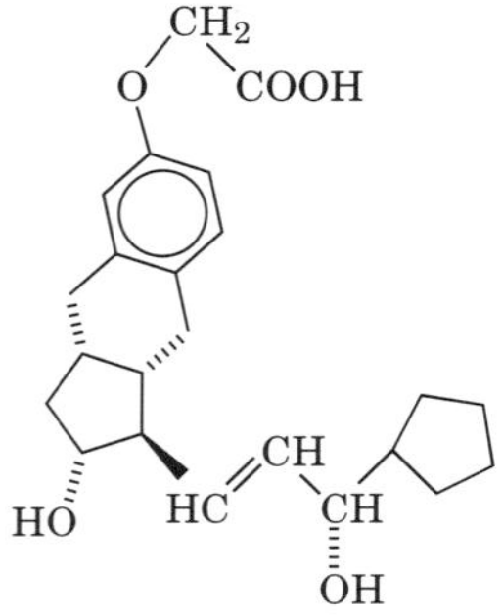			[99570-57-7]	Upjohn

[a]The therapeutic indication (TI) is antithrombotic, unless otherwise noted.
[b]TI is antithrombotic and sudden hearing loss.
[c]TI is peripheral vascular disease.
[d]TI is antithrombotic and antimetastatic.
[e]TI is antiulcer.
[f]TI is antihypoxic.

342

actions of the native compounds, whereas the 15-deoxy analogues (**96**), (**99**), and (**100**) inhibit PGH$_2$-induced human platelet aggregation (225,226).

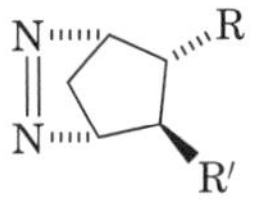

(**95**) 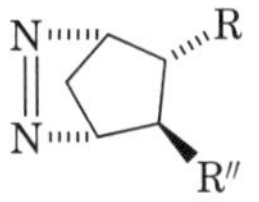 (**96**) 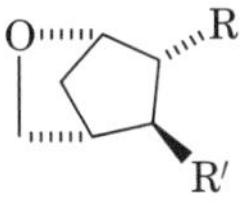 (**97**) 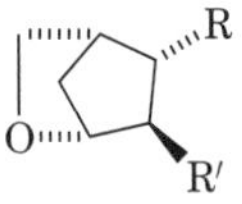 (**98**)

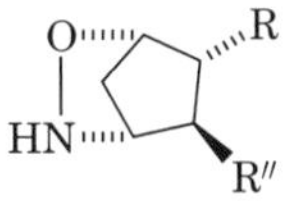

(**99**) 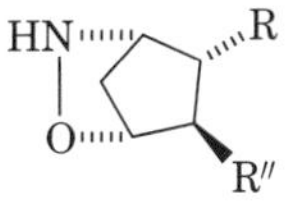 (**100**)

R = CH$_2$CH=CH(CH$_2$)$_3$COOH

R′ = HC=CH·CH(OH)(CH$_2$)$_4$CH$_3$

R″ = HC=CH(CH$_2$)$_5$CH$_3$

Thromboxane A$_2$ Analogues. The thrust of molecular modification in this arena has been to make chemically stable mimics of TxA$_2$ for biological studies and, secondly, to identify antagonists of TxA$_2$ as potential therapeutic agents. The synthesis and biological properties of some of these analogues, as well as those of numerous other analogues have been described (224).

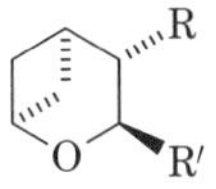

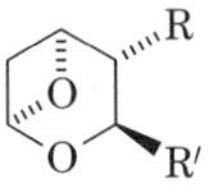

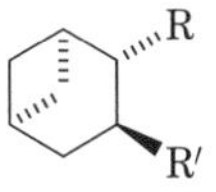

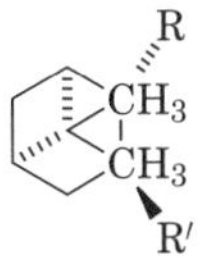

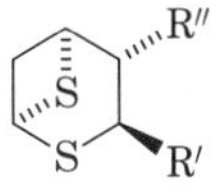

R = CH$_2$CH=CH(CH$_2$)$_3$COOH

R′ = HC=CHCH(OH)(CH$_2$)$_4$CH$_3$

R″ = CH$_2$CH=CH(CH$_2$)$_3$COO$^-$Na$^+$

Therapeutic Role of Prostanoids

The early promise and anticipation of a large and varied therapeutic role for the prostanoids and their analogues has gone, for the most part, unfulfilled. In spite of the potential for therapeutic usefulness in many diseases and a massive effort by many pharmaceutical companies to identify appropriate analogues, only a few compounds have actually been marketed, and the therapeutic applications have, thus far, been limited to the treatment of peptic ulcer disease, gynecological needs (labor induction and fertility control), synchronization of estrus in farm animals, cardiovascular indications (antihypertension and ischemic conditions), and likely, with latanoprost, treatment of glaucoma (see CONTRACEPTIVE DRUGS; GASTROINTESTINAL AGENTS; PHARMACEUTICALS). Even the use of prostanoids (PGEs and PGIs) in peptic ulcer disease, an application intensely pursued with many analogues (see Tables 3 and 4) has been only modestly successful. While these perform about as well as the more widely used histamine blocker agents (cimetidine, etc) in curing ulcers, they are not competitive because of their side effects, most notably diarrhea and abdominal discomfort, and their contraindication in pregnant patients. The most widely approved analogue, misoprostol, has found a niche market in the prevention and treatment of gastric and duodenal ulcers caused by nonsteroidal antiinflammatory drugs. This rather disappointing overall scenario for prostanoid use in therapy may be changing, however. As the tools and knowledge of science have advanced, so have approaches and techniques for studying prostaglandins, resulting in the discovery of potential applications as well as the revisitation of possible use in previously studied diseases. Of special note for PGE analogues has been the examination of their cytoprotective properties in cancer treatment and their role in inflammatory and allergic diseases. Several PG analogues have been found to protect a number of tissues against the effects of radiation and chemotherapeutic agents (88,89,227); these analogues may be useful in preventing side effects, such as mucositis, hair loss, and skin damage, of cancer treatments. The effectiveness of PGEs in animal models of inflammatory diseases (228,229) has been noted and there are anecdotal reports of therapeutic benefit in human patients. Inhaled PGE_2 has been reported to be effective in preventing late-phase response to allergen-induced asthma (230) and misoprostol has prevented the late-phase response to antigens in the skin (231). Thus this renewal of interest and increased breadth of investigation may signal a renaissance for the participation of prostanoids in drug therapy.

The potential of thromboxane-based drugs in cardiovascular disease is also promising. The early attempts to produce effective therapeutics based on inhibition of TxA_2 synthesis failed because of poor efficacy. The lack of effectiveness of these agents was attributed to the accumulation of the TxA_2 precursor, PGH_2, which is itself a thromboxane receptor activator (232). However, a more recent and effective approach has been to combine within one molecule both synthesis-inhibiting properties and receptor antagonism (233,234) and molecules of this type are being pursued (235).

Commercial Sources

Many of the natural prostanoids as well as some of their analogues and related products are available in research quantities from several companies. These include Biomol Research Laboratories (Plymouth Meeting, Pennsylvania), Cayman Chemical (Ann Arbor, Michigan), and Cascade Biochem, Ltd. (Berkshire, England).

BIBLIOGRAPHY

"Prostaglandins" in *ECT* 3rd ed., Suppl. Vol., pp. 711–752, by D. R. Morton, Jr., The Upjohn Co.

1. A. J. Weinheimer and R. L. Spraggins, *Tetrahedron Lett.*, 5185 (1969); W. P. Schneider, R. D. Hamilton, and L. E. Rhuland, *J. Am. Chem. Soc.* **94**, 2122 (1972).

2. W. L. Smith and L. J. Marnett, *Biochem. Biophysic. Acta* **1083**, 1 (1991); W. L. Smith, L. J. Marnett, and D. L. Dewitt, *Pharm. Ther.* **49**, 153 (1991).
3. R. Kurzrok and C. Lieb, *Proc. Soc. Exp. Biol. N.Y.* **28**, 268 (1930).
4. M. W. Goldblatt, *J. Soc. Chem. Ind., London* **52**, 1056 (1933).
5. U.S. Von Euler, *Naunyn-Schmiedeberg's Arch. Exp. Path. Pharmak.* **175**, 78 (1934).
6. J. S. Bindra and R. Bindra, *Prostaglandin Synthesis*, Academic Press, Inc., New York, 1977, p. 7; S. Bergström, *Prog. Lipid Res.* **20**, 7 (1982).
7. S. Abrahamsson, *Acta Crystallogr.* **16**, 409 (1963).
8. S. Abrahamsson, S. Bergström, and B. Samuelsson, *Proc. Chem. Soc.*, 332 (1962).
9. D. H. Nugteren, D. A. van Dorp, S. Bergström, M. Hamberg, and B. Samuelsson, *Nature* **212**, 38 (1966).
10. *Chem. Brit.*, 847 (Dec. 1982).
11. N. H. Anderson, C. J. Hartzell, and B. De, in J. E. Pike and D. R. Morton, eds., *Advances in Prostaglandin, Thromboxane and Leukotriene Research*, Vol. 14, Raven Press, New York, 1985, p. 1; S. Moncada and J. R. Vane, in J. R. Vane and S. Bergström, eds., *Prostacyclin*, Raven Press, New York, 1979, p. 5.
12. J. H. Fletcher, O. C. Dermer, and R. B. Fox, *Nomenclature of Organic Compounds*, American Chemical Society Advances in Chemistry Series, Vol. 126, Washington, D.C., 1974; *Selection of Index Names for Chemical Substances*, *Chem. Abstracts 82*, Index Guide, American Chemical Society, Washington, D.C., 1982.
13. N. A. Nelson, *J. Med. Chem.* **17**, 911 (1974).
14. S. Moncada and J. R. Vane, *J. Med. Chem.* **23**, 591 (1980).
15. B. Samuelsson, M. Goldyne, E. Granström, M. Hamberg, S. Hammarström, and C. Malmsten, *Annu. Rev. Biochem.* **47**, 997 (1978).
16. B. Samuelsson, G. Folco, E. Granström, H. Kindahl, and C. Malmsten, in F. Coceani and P. M. Olley, eds., *Advances in Prostaglandin and Thromboxane Research*, Vol. 4, Raven Press, New York, 1978, pp. 1–25.
17. S. Hammarström, *Arch. Biochem. Biophys.* **214**, 431 (1982).
18. A. J. Marcus, H. L. Ullman, and L. B. Safier, *J. Lipid Res.* **10**, 108 (1969).
19. T. K. Bills and M. J. Silver, *Fed. Proc. Fed. Am. Soc. Exp. Biol.* **34**, 322 (1975).
20. E. J. Christ and D. H. Nugteren, *Biochim. Biophys. Acta* **218**, 296 (1970).
21. P. Cohen and A. Derksen, *Br. J. Haematol.* **17**, 359 (1969).
22. D. F. Horrobin, *Prostaglandins, Leukot. Essent. Fatty Acids* **31**, 181 (1989).
23. R. T. Holman, L. Smythe, and S. Johnson, *Am. J. Clin. Nutr.* **32**, 2390 (1979).
24. M. Hamberg and B. Samuelsson, *J. Biol. Chem.* **242**, 5344 (1967).
25. W. L. Smith and co-workers, in B. Samuelsson, S.-E. Dahlén, J. Fritsch, and P. Hedqvist, eds., *Advances in Prostaglandin, Thromboxane and Leukotriene Research*, Vol. 20, Raven Press, New York, 1990, p. 14.
26. D. L. DeWitt, *Biochem. Biophys. Acta* **1083**, 121 (1991).
27. W. Xie, D. L. Robertson, and D. L. Simmons, *Drug Dev. Res.* **25**, 249 (1992).
28. D. L. Simmons, W. Xie, J. G. Chipman, and G. E. Evett in J. M. Bailey, ed., *Prostaglandins, Leukotrienes, Lipoxins and PAF*, Plenum Press, New York, 1991, p. 67.
29. W. L. Smith, *Biochem. J.* **259**, 315 (1989).
30. J. A. Mitchell, P. Akarasereenont, C. Thiemermann, R. J. Flower, and J. R. Vane, *Proc. Natl. Acad. Sci., USA* **90**, 11693 (1993).
31. J. L. Masferrer, K. Seibert, B. Zweifel, and P. Needleman, *Proc. Natl. Acad. Sci., USA* **89**, 3917 (1992).
32. D. Picot, P. J. Loll, and R. M. Garavito, *Nature (London)* **367**, 243 (1994).
33. J. R. Vane, R. J. Flower, and R. M. Botting, *Stroke* **21** (Suppl. IV), 12 (1990).
34. L. R. Roberts, A. R. Brash, and J. A. Oates, in J. A. Oates, ed., *Advances in Prostaglandin, Thromboxane, and Leukotriene Research*, Vol. 10, Raven Press, New York, 1982, p. 211.

35. W. P. Schneider, G. L. Bundy, F. H. Lincoln, E. G. Daniels, and J. E. Pike, *J. Am. Chem. Soc.* **99** 1222 (1977).

36. W. P. Schneider, *J. Chem. Soc. Chem. Commun.*, 304 (1969).

37. E. J. Corey and C. R. Cyr, *Tetrahedron Lett.*, 1761 (1974).

38. R. C. Kelly, I. Schletter, and R. L. Jones, *Prostaglandins* **4**, 653 (1973).

39. E. E. Nishizawa and co-workers, *Prostaglandins* **9**, 109 (1975).

40. E. J. Corey, T. K. Schaaf, W. Huber, U. Koelliker, and N. M. Weinshenker, *J. Am. Chem. Soc.* **92**, 397 (1970).

41. C. J. Sih and co-workers, *J. Am. Chem. Soc.* **97**, 865 (1975).

42. J. E. Pike, F. H. Lincoln, and W. P. Schneider, *J. Org. Chem.* **34**, 3552 (1969).

43. R. A. Johnson and co-workers, *J. Am. Chem. Soc.* **100**, 7690 (1978).

44. S. Hanessian and P. Lavalle, *Can. J. Chem.* **55**, 562 (1977).

45. W. P. Schneider and R. A. Morge, *Tetrahedron Lett.*, 3283 (1976).

46. N. A. Nelson and R. W. Jackson, *Tetrahedron Lett.*, 3275 (1976).

47. P. W. Ramwell and co-workers, *Prog. Chem. Fats Other Lipids* **9**, 231 (1968); C. Hensby, in P. Crabbe, ed., *Prostaglandin Research*, Academic Press, Inc., New York, 1977, p. 89.

48. W. L. Duax and J. W. Edmonds, *Prostaglandins* **3**, 201 (1973).

49. J. W. Edmonds and W. L. Duax, *Prostaglandins* **5**, 275 (1974).

50. G. Kotovych, G. H. M. Aarts, T. T. Nakashima, and G. Bigam, *Can. J. Chem.* **58**, 974 (1980).

51. G. Kotovych and G. H. M. Aarts, *Org. Magn. Reson.* **18**, 77 (1982).

52. C. Chachaty, Z. Wolkowski, F. Piriou, and G. Lukacs, *J. Chem. Soc. Chem. Commun.*, 951 (1973).

53. G. F. Cooper and J. Fried, *Proc. Natl. Acad. Sci., USA* **70**, 1579 (1973); G. F. Cooper and J. Fried, *Proceedings of the First International Conference on Stable Isotopes in Chemistry, Biology and Medicine*, 1973, pp. 72–83; S. A. Mizsak and G. Slomp, *Prostaglandins* **10**, 807 (1975).

54. C. R. Pace-Asciak, *Advances in Prostaglandins, Thromboxane and Leukotriene Research*, Vol. 18, Raven Press, New York, 1989.

55. M. Hamberg, J. Svensson, and B. Samuelsson, *Proc. Natl. Acad. Sci.* **72**, 2994 (1975).

56. M. Hamberg, J. Svensson, and B. Samuelsson, *Adv. Prostaglandin Thromboxane Res.* **1**, 19 (1976).

57. S. S. Bhagwat, P. R. Hamann, and C. Still, *J. Am. Chem. Soc.* **107**, 6372 (1985).

58. R. A. Johnson, in Ref. 11, p. 131.

59. B. Samuelsson, *The Harvey Lectures, 1979–1980*, Academic Press, Inc., New York, 1981, pp. 1–40.

60. R. P. Robertson, ed., *Med. Clin. North Am.* **65**, 711 (1981).

61. P. W. Ramwell and co-workers, *Prog. Chem. Fats Other Lipids* **9**, 231 (1968).

62. B. Samuelsson and R. Paoletti, eds., *Advances in Prostaglandin, Thromboxane and Leukotriene Research*, Vol. 9, Raven Press, New York, 1982.

63. N. H. Andersen and P. W. Ramwell, *Arch. Intern. Med.* **133**, 30 (1974).

64. E. W. Horton, in S. M. Roberts and F. Scheinmann, eds., *Chemistry, Biochemistry, and Pharmacological Activity of Prostanoids*, Pergamon Press, Oxford, U.K., 1979, p. 1.

65. M. F. Cuthbert, in M. F. Cuthbert, ed., *The Prostaglandins: Pharmacological and Therapeutic Advances*, J. B. Lippincott Co., Philadelphia, Pa., 1973, p. 253.

66. S. Bianco, M. Robuschi, R. Ceserani, and C. Gandolfi, *Int. Res. Commun. Syst. Med. Sci.* **6**, 256 (1978).

67. E. W. Horton, *Brit. J. Pharmacol. Chemother.* **22**, 189 (1964).

68. E. W. Horton, *Prostaglandins: Monographs on Endocrinology*, Springer-Verlag, Berlin, 1972, p. 141.

69. L. S. Wolfe, *J. Neurochem.* **38**, 1 (1982).
70. D. J. Levenson, C. E. Simmons, and B. M. Brenner, *Am. J. Med.* **72**, 354 (1982).
71. J. Tannenbaum, J. A. Splawinski, J. A. Oates, and A. S. Nies, *Circ. Res.* **36**, 197 (1975).
72. K. Ueda and co-workers, *J. Pediatrics* **97**, 834 (1980).
73. S. C. Willey and B. Chernow, in W. D. Watkins, M. B. Peterson, and J. R. Fletcher, eds., *Prostaglandins in Clinical Practice*, Raven Press, New York, 1989, p. 227.
74. R. P. Robertson, *Annu. Rev. Med.* **34**, 1 (1983).
75. J. S. Goodwin, *J. Rheumatol.* **18**(Suppl 28), 26 (1991).
76. L. M. Solomon, L. Juhlin, and M. B. Kirschenbaum, *J. Invest. Derm.* **51**, 280 (1968).
77. G. L. Larsen and P. M. Henson, *Annu. Rev. Immunol.* **1**, 335 (1983).
78. J. R. Vane, *Nature* **231**, 232 (1971).
79. P. Hedqvist, J. Raud, and S.-E. Dahlén, in B. Samuelsson, P. Y.-K. Wong, and F. F. Sun, eds., *Advances in Prostaglandin, Thromboxane and Leukotriene Research*, Vol. 19, Raven Press, New York, 1989, p. 539.
80. J. S. Goodwin and J. L. Ceuppens, *J. Clin. Immunol.* **3**, 295 (1983); J. S. Goodwin and D. R. Webb, in J. S. Goodwin, ed., *Suppressor Cells in Human Disease*, Marcel Dekker, New York, 1981, p. 99.
81. A. Robert, *Gastroenterology* **69**, 1045 (1975).
82. A. Robert, J. E. Nezamis, C. Lancaster, and A. J. Hanchar, *Gastroenterology* **77**, 433 (1979).
83. T. A. Miller, *Am. J. Physiol.* **245**, 6601 (1983); J. L. Wallace, *Gastroenterol. Clin. N. Am.* **21**, 631 (1992).
84. J. L. Wallace, B. J. R. Whittle, and N. K. Boughton-Smith, *Dig. Dis. Sci.* **30**, 866 (1985).
85. J. Stachura, A. Tarnawski, and J. Szezudrawa, *Folia Histochem. Cytochem.* **18**, 311 (1980).
86. M. S. Paller, *Transplan. P.* **20**, 634 (1988).
87. T. Manabe and M. L. Steer, *Gastroenterology* **78**, 777 (1980).
88. W. R. Hanson and K. DeLaurentis, *Prostaglandins* **33**(Suppl.), 93 (1987).
89. F. D. Malkinson, L. Geng, and W. R. Hanson, *J. Invest. Dermatol.* **101**, 1355 (1993).
90. M. M. Cohen, *Biological Protection with Prostaglandins*, Vols. 1 and 2, CRC Press, Boca Raton, Fla., 1985.
91. I. Kennedy, R. A. Coleman, P. P. A. Humphrey, G. P. Levy, and P. Lumley, *Prostaglandins* **24**, 667 (1982).
92. R. A. Coleman, I. Kennedy, P. P. A. Humphrey, K. Bunce, and P. Lumley, in C. Hansch, P. G. Sammes, J. B. Taylor, and J. C. Emmeth, eds., *Comprehensive Medicinal Chemistry*, Vol. 3, Pergamon Press, Oxford, U.K., 1989, p. 643.
93. M. Negishi, Y. Sugimoto, and A. Ichikawa, *Prog. Lipid Res.* **32**, 417 (1993); R. A. Coleman, W. L. Smith, and S. Narumiya, *Biol. Rev.* **46**, 205 (1994).
94. A. Honda and co-workers, *J. Biol. Chem.* **268**, 7759 (1993).
95. Y. Sugimoto and co-workers, *J. Biol. Chem.* **267**, 6463 (1992).
96. S. An, J. Yang, M. Xia, and E. J. Goetzl, *Biochem. Biophys. Res. Comm.* **197**, 263 (1993).
97. M. Hirata and co-workers, *Nature* **349**, 617 (1991).
98. R. M. Breyer and co-workers, *J. Biol. Chem.* **269**, 6163 (1994).
99. M. Adam and co-workers, *FEBS Lett.* **338**, 170 (1994).
100. P. J. Gardiner, in Ref. 25, p. 110.
101. M. A. Savage, C. Moummi, P. J. Karabatsos, and T. H. Lanthorn, *Prostaglandins, Leukot. Essent. Fatty Acids* **49**, 939 (1993).
102. J. E. Pike and D. R. Morton, eds., *Advances in Prostaglandin, Thromboxane and Leukotriene Research*, Vol. 14, Raven Press, New York, 1985.

103. S. M. Roberts and F. Scheinmann, eds., *New Synthetic Routes to Prostaglandins and Thromboxanes*, Academic Press, Inc., New York, 1982.
104. K. C. Nicolaou, G. P. Gasic, and W. E. Barnette, *Angew. Chem. Int. Ed. Engl.* **17**, 293 (1978).
105. M. P. L. Caton and K. Crowshaw, in G. P. Ellis and G. B. West, eds., *Progress in Medicinal Chemistry*, Vol. 15, Elsevier/North-Holland, Inc., New York, 1978, pp. 357–423.
106. P. R. Marcham, in F. D. Gunstone, ed., *Aliphatic and Related Natural Product Chemistry*, Vol. 1, The Chemical Society, London, 1979, pp. 170–235.
107. P. W. Collins and S. W. Djuric, *Chem. Rev.* **93**, 1533 (1993).
108. M. P. L. Caton and T. W. Hart, in Ref. 106, p. 73.
109. E. J. Corey, T. Ravindranathan, and S. Terashima, *J. Am. Chem. Soc.* **93**, 4326 (1971).
110. E. J. Corey and P. L. Fuchs, *J. Am. Chem. Soc.* **94**, 4014 (1972).
111. N. M. Weinshenker, *Prostaglandins* **3**, 219 (1973).
112. S. Ranganathan, D. Ranganathan, and A. K. Mehrotra, *Tetrahedron Lett.*, 1215 (1975).
113. B. M. Trost and Y. Tamaru, *J. Am. Chem. Soc.* **97**, 3528 (1975).
114. E. J. Corey and H. E. Ensley, *J. Am. Chem. Soc.* **97**, 6908 (1975).
115. P. A. Bartlett, F. R. Green, and T. R. Webb, *Tetrahedron Lett.*, 331 (1977).
116. H. E. Ensley, C. A. Parnell, and E. J. Corey, *J. Org. Chem.* **43**, 1610 (1978).
117. J. S. Bindra and R. Bindra, *Prostaglandin Synthesis*, Academic Press, Inc., New York, 1977, pp. 187–245.
118. N. Inukai and co-workers, *Chem. Pharm. Bull.* **24**, 2566 (1976).
119. N. M. Weinshenker, G. A. Crosby, and J. Y. Wong, *J. Org. Chem.* **40**, 1966 (1975).
120. E. J. Corey and J. W. Suggs, *J. Org. Chem.* **40**, 2554 (1975).
121. M. Vandewalle, V. Sipido, and H. DeWilde, *Bull. Soc. Chim. Belg.* **79**, 403 (1970).
122. E. J. Corey, Z. Arnold, and J. Hutton, *Tetrahedron Lett.*, 307 (1970).
123. E. J. Corey and T. Ravindranathan, *Tetrahedron Lett.*, 4753 (1971).
124. D. Brewster and co-workers, *J. Chem. Soc. Chem. Commun.*, 1235 (1972).
125. G. Jones, R. A. Raphael, and S. Wright, *J. Chem. Soc. Chem. Commun.*, 609 (1972).
126. F. Kienzle, G. W. Holland, J. L. Jernow, S. Kwok, and P. Rosen, *J. Org. Chem.* **38**, 3440 (1973).
127. E. J. Corey and B. B. Snider, *Tetrahedron Lett.*, 3091 (1973).
128. D. Brewster and co-workers, *J. Chem. Soc. Perkin Trans. 1*, 2796 (1973).
129. R. B. Woodward and co-workers, *J. Am. Chem. Soc.* **95**, 6853 (1973).
130. J. S. Bindra, A. Grodski, and T. K. Schaaf, *J. Am. Chem. Soc.* **95**, 7522 (1973).
131. E. J. Corey and C. U. Kim, *J. Org. Chem.* **38**, 1233 (1973).
132. J. Van Hooland, P. De Clercq, and M. Vanderwalle, *Tetrahedron Lett.*, 4343 (1974).
133. P. de Clercq and M. Vanderwalle, *Bull. Soc. Chim. Belg.* **83**, 305 (1974).
134. P. de Clercq, D. Van Haver, D. Tavernier, and M. Vandewalle, *Tetrahedron* **30**, 55 (1974).
135. G. Jones, R. A. Raphael, and S. Wright, *J. Chem. Soc. Perkin Trans. 1*, 1676 (1974).
136. E. J. Corey and B. B. Snider, *J. Org. Chem.* **39**, 256 (1974).
137. E. D. Brown, R. Clarkson, T. J. Leeney, and G. E. Robinson, *J. Chem. Soc. Chem. Commun.*, 642 (1974).
138. E. J. Corey, K. C. Nicolaou, and D. J. Beames, *Tetrahedron Lett.*, 2439 (1974).
139. R. Peel and J. K. Sutherland, *J. Chem. Soc. Chem. Commun.*, 151 (1974).
140. R. Coen, P. De Clercq, D. Van Haver, and M. Vandewalle, *Bull. Soc. Chim. Belg.* **84**, 203 (1975).
141. W. Van Brussel, J. Van Hooland, P. De Clercq, and M. Vandewalle, *Bull. Soc. Chim. Belg.* **84**, 813 (1975).

142. M. Samson, P. De Clercq, and M. Vandewalle, *Tetrahedron* **31**, 1233 (1975).
143. H. Shimomura, J. Katsube, and M. Matsui, *Agric. Biol. Chem.* **39**, 657 (1975).
144. A. Fischli, M. Klau, H. Mayer, P. Schonholzer, and R. Ruegg, *Helv. Chem. Acta* **58**, 564 (1975).
145. G. A. Crosby, N. M. Weinshenker, and H.-S. Uh, *J. Am. Chem. Soc.* **97**, 2232 (1975).
146. S. Ranganathan, D. Ranganathan, and A. K. Mehrotra, *Tetrahedron Lett.*, 1215 (1975).
147. P. De Clercq, M. De Smet, K. Legein, F. Vanhulle, and M. Vandewalle, *Bull. Soc. Chim. Belg.* **85**, 503 (1976).
148. P. De Clercq, R. Coen, E. Van Hoff, and M. Vandewalle, *Tetrahedron* **32**, 2747 (1976).
149. I. Tomoskozi, L. Gruber, G. Kovacs, I. Szekely, and V. Simonidesz, *Tetrahedron Lett.*, 4639 (1976).
150. K. G. Paul, F. Johnson, and D. Favara, *J. Am. Chem. Soc.* **98**, 1285 (1976).
151. I. Ernest, *Angew Chem. Int. Ed. Engl.* **15**, 207 (1976).
152. S. Takano, N. Kubodera, and K. Ogasawara, *J. Org. Chem.* **42**, 786 (1977).
153. E. D. Brown, R. Clarkson, T. J. Leeney, and G. E. Robinson, *J. Chem. Soc. Perkin Trans. 1*, 1507 (1978).
154. M. Naruto, K. Ohno, and N. Naruse, *Chem. Lett.*, 1419 (1978).
155. L. A. Paquette, G. D. Crouse, and A. K. Sharma, *J. Am. Chem. Soc.* **102**, 3972 (1980).
156. S. Goldstein and co-workers, *J. Am. Chem. Soc.* **103**, 4616 (1981).
157. L. A. Paquette and G. D. Crouse, *Tetrahedron* **37**(Suppl. 1), 281 (1981).
158. I. Fleming and B.-W. Au-Yeung, *Tetrahedron* **37**(Suppl. 1), 13 (1981).
159. E. J. Corey, K. B. Becker, and R. K. Varma, *J. Am. Chem. Soc.* **94**, 8616 (1972).
160. J. Bowler, K. B. Mallion, and R. A., Raphael, *Synth. Commun.* **4**, 211 (1974).
161. E. J. Corey, K. C. Nicolaou, M. Shibasaki, Y. Machida, and C. S. Shiner, *Tetrahedron Lett.*, 3183 (1975).
162. J. Hutton, M. Senior, and N. C. A. Wright, *Synth. Commun.* **9**, 799 (1979).
163. K. B. Mallion and E. R. H. Walker, *Synth. Commun.* **5**, 221 (1975).
164. R. Noyori, *Pure Appl. Chem.* **53**, 2315 (1981).
165. S. Iguchi, H. Nakai, M. Hayashi, and H. Yamamoto, *J. Org. Chem.* **44**, 1363 (1979).
166. S. Iguchi, H. Nakai, M. Hayashi, H. Yamamoto, and K. Maruoka, *Bull. Chem. Soc. Jpn.* **54**, 3033 (1981).
167. A. L. Gemal and J.-L. Luche, *J. Am. Chem. Soc.* **103**, 5454 (1981).
168. E. J. Corey, N. Imai, and S. Pikul, *Tetrahedron Lett.* **32**, 7517 (1991).
169. E. J. Corey and T.-P. Loh, *J. Am. Chem. Soc.* **113**, 8966 (1991).
170. E. J. Corey and co-workers, *J. Am. Chem. Soc.* **93**, 1490 (1970).
171. G. Just and C. Simonovitch, *Tetrahedron Lett.*, 2093 (1967).
172. R. C. Kelly, V. Van Rheenen, I. Schletter, and M. D. Pillai, *J. Am. Chem. Soc.* **95**, 2746 (1973), and references cited therein.
173. D. R. White, *Tetrahedron Lett.*, 1753 (1976).
174. N. A. Nelson, R. C. Kelly, and R. A. Johnson, *Chem. Eng. News*, 30 (Aug. 16, 1982).
175. R. F. Newton and S. M. Roberts, *Tetrahedron* **36**, 2163 (1980), and references cited therein.
176. S. M. Roberts, *J. Chem. Soc. Chem. Commun.*, 948 (1974).
177. T. V. Lee, S. M. Roberts, and R. F. Newton, *J. Chem. Soc. [Perkins I]*, 1179 (1978).
178. M. P. L. Caton, in S. M. Roberts and F. Scheinmann eds., *New Synthetic Routes to Prostaglandin and Thromboxanes*, Academic Press, London, 1982, p. 105.
179. F. Scheinmann, in S. M. Roberts and R. F. Newton, eds., *Prostaglandins and Thromboxanes*, Butterworth, London, 1982, p. 62.
180. J. P. Marino, R. F. de la Pradilla, and E. Laborde, *J. Org. Chem.* **52**, 4898 (1987).
181. C. R. Johnson and T. D. Penning, *J. Am. Chem. Soc.* **110**, 4726 (1988).
182. E. J. Corey, K. Niimura, Y. Konishi, S. Hashimoto, and Y. Hamada, *Tetrahedron Lett.* **27**, 2199 (1986).

183. R. K. Haynes, D. E. Lambert, P. A. Schober, and S. G. Turner, *Aust. J. Chem.* **40**, 1211 (1987).

184. T. Toru, Y. Yamada, T. Ueno, E. Mackawa, and Y. Ueno, *J. Am. Chem. Soc.* **110**, 4815 (1988).

185. P. W. Collins, *J. Med. Chem.* **29**, 437 (1986).

186. R. Noyori and M. Suzuki, *Angew Chem., Int. Ed. Engl.* **23**, 847 (1984); T. Tanaka and co-workers, *Tetrahedron* **43**, 813 (1987).

187. C. J. Sih and co-workers, *J. Am. Chem. Soc.* **94**, 3643 (1972).

188. A. F. Kluge, K. G. Untch, and J. H. Fried, *J. Am. Chem. Soc.* **94**, 7827 (1972); F. S. Alvarez, D. Wren, and A. Prince, *J. Am. Chem. Soc.* **94**, 7823 (1972).

189. R. Pappo and P. W. Collins, *Tetrahedron Lett.*, 2627 (1972).

190. C. J. Sih and co-workers, *J. Am. Chem. Soc.* **97**, 865 (1975).

191. R. Noyori, L. Tomino, and M. Nishizawa, *J. Am. Chem. Soc.* **101**, 5843 (1979).

192. M. Midland and A. Kasubski, *J. Org. Chem.* **47**, 2814 (1982).

193. E. J. Corey and R. H. Wollenberg, *J. Org. Chem.* **40**, 2265 (1975).

194. P. W. Collins, "Misoprostol," in D. Lednicer, ed., *Chronicles of Drug Discovery*, American Chemical Society, Washington, D.C., 1993.

195. J. R. Behling, P. W. Collins, and J. S. Ng, in L. S. Liebeskind, ed., *Advances in Metal-Organic Chemistry*, Vol. 4, JAI Press, Greenwich, Conn., 1994.

196. P. W. Collins and co-workers, *J. Med. Chem.* **29**, 1195 (1986).

197. M. B. Floyd, in S. M. Roberts and F. Scheinmann, eds., *Chemistry, Biochemistry and Pharmacological Activity of Prostanoids*, Pergamon Press, Oxford, U.K., 1979, p. 161.

198. G. Piancatelli and A. Scettri, *Synthesis*, 116 (1977).

199. R. Pappo, P. Collins, and C. Jung, *Tetrahedron Lett.*, 943 (1973).

200. M. Gill, P. Bainton, and R. W. Rickards, *Tetrahedron Lett.* **22**, 1437 (1981).

201. G. Stork and T. Takahashi, *J. Am. Chem. Soc.* **99**, 1275 (1977).

202. J. H. Dygos and co-workers, *J. Org. Chem.* **56**, 2549 (1991).

203. K. A. Babiak and co-workers, *J. Am. Chem. Soc.* **112**, 7441 (1990); B. H. Lipschutz and E. L. Ellsworth, *J. Am. Chem. Soc.* **112**, 7440 (1990).

204. K. A. Babiak and co-workers, *J. Organic Chem.* **55**, 3377 (1990).

205. Y. Morita, M. Suzuki, and R. Noyori, *J. Org. Chem.* **54**, 1784 (1989).

206. T. Takahashi, M. Nakazawa, M. Kanoh, and K. Yamamoto, *Tetrahedron Lett.* **31**, 7349 (1990).

207. S. J. Danishefsky, M. P. Cabal, and K. Chow, *J. Am. Chem. Soc.* **111**, 3456 (1989); K. Chow and S. J. Danishefsky, *J. Org. Chem.* **54**, 6016 (1989).

208. C. R. Johnson and Y.-F. Chen, *J. Org. Chem.* **56**, 3344, 3352 (1991).

209. G. Stork and M. Isobe, *J. Am. Chem. Soc.* **97**, 4765, 6260 (1975).

210. T. Yoshino, S. Okamoto, and F. Sato, *J. Org. Chem.* **56**, 3205 (1991).

211. R. E. Donaldson, J. C. Saddler, S. Byrn, A. T. McKenzie, and P. L. Fuchs, *J. Org. Chem.* **48**, 2167 (1983).

212. G. L. Bundy, W. P. Schneider, F. H. Lincoln, and J. E. Pike, *J. Am. Chem. Soc.* **94**, 2123 (1972).

213. W. P. Schneider, G. L. Bundy, F. H. Lincoln, E. G. Daniels, and J. E. Pike, *J. Am. Chem. Soc.* **99**, 1222 (1977).

214. N. A. Porter, J. D. Byers, A. E. Ali, and T. E. Eling, *J. Am. Chem. Soc.* **102**, 1183 (1980).

215. N. A. Porter, J. D. Byers, K. M. Holden, and D. B. Menzel, *J. Am. Chem. Soc.* **101**, 4319 (1979).

216. R. A. Johnson and co-workers, *J. Am. Chem. Soc.* **100**, 7690 (1978).

217. R. A. Johnson and co-workers, *J. Am. Chem. Soc.* **99**, 4182 (1977).

218. N. Wittaker, *Tetrahedron Lett.* **32**, 2805 (1977).

219. M. Suzuki, A. Yanagisawa, and R. Noyori, *Tetrahedron Lett.* **24**, 1187 (1983).

220. W. P. Schneider and R. A. Morge, *Tetrahedron Lett.*, 3283 (1976).
221. B. Radüchel and H. Vorbrüggen, in Ref. 102, p. 263.
222. P. W. Collins, *Med. Res. Rev.* **10**, 149 (1990).
223. P. A. Aristoff, in Ref. 102, p. 309.
224. N. H. Wilson and R. L. Jones, in Ref. 102, p. 393.
225. E. J. Corey, K. C. Nicolaou, Y. Machida, C. L. Malmsten, and B. Samuelsson, *Proc. Natl. Acad. Sci. USA* **72**, 3355 (1975).
226. G. L. Bundy, *Tetrahedron Lett.*, 1957 (1975).
227. W. R. Hanson and E. J. Ainsworth, *Radiat. Res.* **103**, 196 (1985).
228. R. N. Fedorak, L. R. Empey, C. MacArthur, and L. D. Jewell, *Gastroenterology* **98**, 149 (1990).
229. H. Allgayer, K. Deschryver, and W. F. Stenson, *Gastroenterology* **96**, 1290 (1989).
230. I. D. Pavord, C. S. Wong, J. Williams, and A. E. Tattersfield, *Am. Rev. Respir. Dis.* **148**, 87 (1993).
231. R. Alam and co-workers, *Am. Rev. Respir. Dis.* **148**, 1066 (1993).
232. T. A. Morinelli and P. V. Halushka, *Trends Cardiovasc. Med.* **1**, 157 (1991).
233. P. Gresele, E. Van Houtte, J. Arnout, H. Deckmyn, and J. Vermylen, *Thromb. Haemostasis* **52**, 364 (1984).
234. P. Gresele and co-workers, *J. Clin. Invest.* **80**, 1435 (1987).
235. R. Soyka and co-workers, *J. Med. Chem.* **37**, 26 (1994).

PAUL W. COLLINS
GD Searle & Company

PROSTHETIC AND BIOMEDICAL DEVICES

Prosthetics or biomedical devices are objects which serve as body replacement parts for humans and other animals or as tools for implantation of such parts. An implanted prosthetic or biomedical device is fabricated from a biomaterial and surgically inserted into the living body by a physician or other health care provider. Such implants are intended to function in the body for some period of time in order to perform a specific task. Medical devices may replace a damaged part of anatomy, eg, total joint replacement; simulate a missing part, eg, mammary prosthesis; correct a deformity, eg, spinal plates; aid in tissue healing, eg, burn dressings; rectify the mode of operation of a diseased organ, eg, cardiac pacemakers; or aid in diagnosis, eg, insulin electrodes.

Prosthetics and biomedical devices are composed of biocompatible materials, often called biomaterials. In the early 1930s the only biomaterials were wood (qv), glass (qv), and metals. These were used mostly in surgical instruments, paracorporeal devices, and disposable products. The advent of synthetic polymers and biocompatible metals in the latter part of the twentieth century has changed the entire character of health care delivery. Polymers, metals, and ceramics (qv) originally designed for commercial applications have been adapted for prostheses, opening the way for implantable pacemakers, vascular grafts, diagnostic/therapeutic catheters, and a variety of other orthopedic devices. The term prosthesis encompases both external and internal devices. This article concentrates on implantable prostheses.

Biomaterials

A biomaterial is defined as a systemic, pharmacologically inert substance designed for implantation or incorporation within the human body (1). A biomaterial must be mechanically adaptable for its designated function and have the required shear, stress, strain, Young's modulus, compliance, tensile strength, and temperature-related properties for the application. Moreover, biomaterials ideally should be nontoxic, ie, neither teratogenic, carcinogenic, or mutagenic; nonimmunogenic; biocompatible; biodurable, unless designed as bioresorbable; sterilizable; readily available; and possess characteristics allowing easy fabrication. The traditional areas for biomaterials are plastic and reconstructive surgery, dentistry, and bone and tissue repair. A widening variety of materials are being used in these areas. Artificial organs play an important role in preventive medicine, especially in the early prevention of organ failure.

To be biocompatible is to interact with all tissues and organs of the body in a nontoxic manner, not destroying the cellular constituents of the body fluids with which the material interfaces. In some applications, interaction of an implant with the body is both desirable and necessary, as, for example, when a fibrous capsule forms and prevents implant movement (2).

Polymers, metals, ceramics, and glasses may be utilized as biomaterials. Polymers (qv), an important class of biomaterials, vary greatly in structure and properties. The fundamental structure may be one of a carbon chain, eg, in polyethylene or Teflon, or one having ester, ether, sulfide, or amide bond linkages. Polysilicones, having a $-Si-O-Si-$ backbone, may contain no carbon.

Plastics are found in implants and components for reconstructive surgery, as components in medical instruments, equipment, packaging materials, and in a wide array of medical disposables. Plastics have assumed many of the roles once restricted to metals and ceramics.

Metals are used when mechanical strength or electrical conductivity is required of a device. For example, as of 1995 the femoral component of a hip replacement device was metal, as were the conductors of cardiac pacemaker leads. Titanium and titanium alloys (qv) are well tolerated in the body. This is partly the result of the strongly adhering oxide layer that forms over the metal surface, making the interface between the body and biomaterial effectively a ceramic rather than a metal. Titanium finds wide use as the femoral component of the artificial hip, where it exhibits great strength, comparatively light weight (the density of titanium is 4.5 g/cm^3), and excellent fatigue resistance. Another area in which titanium has replaced all other metals and alloys is as the casing material for cardiac pacemakers, neural stimulators, and implantable defibrillators.

Stainless steel alloys are also useful in orthopedic applications (see STEEL). Stainless steel alloys are used in the manufacture of staples, screws, pins, etc.

These alloys are used primarily in applications requiring great tensile strength. Elgiloy, an interesting cobalt-based alloy, was originally developed for the mainspring of mechanical watches. This is used essentially as the conductor of neural stimulator leads, which require excellent flexibility and fatigue resistance. Nitinol, an unusual alloy of nickel and titanium, exhibits shape memory (see SHAPE-MEMORY ALLOYS). Its main application has been in dentistry, where its resilience rather than its shape-memory characteristic is of value.

Ceramics (qv) include a large number of inorganic nonmetallic solids that feature high compressive strength and relative chemical inertness. Low temperature isotropic (LTI) carbon has excellent thromboresistance and has found use in heart valves and percutaneous connectors. LTI carbon, known as LTI, was originally developed for encapsulating nuclear reactor fuel. This material was adapted for biomedical applications in the 1970s. LTI is formed by pyrolysis of hydrocarbons at temperatures between 1000 and 2400°C. Aluminum oxide [1344-28-1], Al_2O_3, forms the basis of dental implants (see DENTAL MATERIALS). In the polycrystalline form this ceramic is suitable for load-bearing hip prostheses.

Bioglasses are surface-active ceramics that can induce a direct chemical bond between an implant and the surrounding tissue. One example is 45S5 bioglass, which consists of 45% SiO_2, 6% P_2O_5, 24.5% CaO, and 24.5% Na_2O. The various calcium phosphates have excellent compatibility with bone and are remodeled by the body when used for filling osseous defects.

Definitions

A graft from one biological species to another, historically known as a heterograft, is called a xenograft. In general, it can take two forms. The first is when a living organ is transplanted, eg, the transplant of a porcine heart into another species. The second is when some part is treated chemically and then implanted. Glutaraldehyde-treated porcine heart valves have been used for many years as a replacement for failing heart valves in humans. Specially treated blood vessels can be transplanted also. For example, one technique consists of implanting a woven Dacron tube into the back of a sheep, whereupon a collagen layer grows over the tube. The tube is explanted, treated, and subsequently implanted into a human. The Dacron tube provides the basic strength and the collagenous layer aids the ingrowth and formation of a smooth neo-intima that has thromboresistant properties similar to those of natural intima. Xenograft porcine skin is useful in severe burn cases when large areas of the body require temporary covering.

The terminology of implantation has changed over the years. The historical term autogenic, referring to transplants of any tissue within the same patient, eg, skin, is called an autograft. Procedures that were called allograft or allogeneic, isograft or isogeneic, are considered as homografts. These involve transplantation of compatible organs from one human being to another, eg, hearts, liver, and kidneys.

Medical devices are officially classified into one of three classes. Class I devices are general controls that are primarily intended as devices that pose no potential risk to health, and thus can be adequately regulated without imposing standards or the need for premarket review. Manufacturers of these devices must register with the United States Food and Drug Administration (FDA), provide a listing of products, maintain adequate reports, and comply with good manufacturing practices. Examples are stethoscopes, periodontic syringes, nebulizers, vaginal insufflators, etc.

Class II devices have performance standards and are applicable when general controls are not adequate to assure the safety and effectiveness of a device, based on the potential risk to health posed by the device. To classify a device in the Class II category, the FDA must find that enough data are available on which to base adequate performance standards that would control the safety and effectiveness of the device. Examples are diagnostic catheters, electrocardiographs, wound dressings, percutaneous catheters, gastrointestinal irrigation systems, etc.

Class III devices require premarket approval. When a device is critical, ie, life-supporting and/or life-sustaining, unless adequate justification is given for classifying it in another category, it is a Class III device. Class III also contains devices after 1976 that are not sufficiently similar to pre-1976 devices, and devices that were regulated as new drugs before 1976. Examples are bronchial tubes, ventilators, vascular grafts, pacemakers, cardiopulmonary bypass, surgical meshes, etc.

History

The search for replacements of natural body parts is ancient. Dental implants are traceable to the early Egyptians and other cultures (3). Sutures were used, as were gold plates, for skull repair before 1000 BC. Artificial limbs are mentioned in Roman writing and in the Middle Ages. The mid-1900s saw real advances in prosthetics and biomedical devices. In the mid-1940s the first artificial kidney was developed in Holland by Kolff. The first successful use of a heart–lung machine followed. The pacemaker was invented in the 1950s, the first artificial heart valve was implanted in 1952, and in 1954 the first artificial hip replacement was performed (4). The membrane oxygenator began clinical trials during the mid-1950s. In 1969, the first artificial heart was used in a patient having irreversible myocardial hypertrophy.

All implantable medical devices are complex in design, materials, and implementation procedures. The biocompatibility, biodurability, and efficacy of medical devices are the subject of extensive research by biomaterials scientists, device manufacturers, and health care professionals.

The Federal Food, Drug and Cosmetic Act of 1938 expanded the FDA's regulatory control over food and drugs extending the agency's authority to include medical devices and cosmetics (qv). The Medical Device Amendment to the Federal Food, Drug and Cosmetic Act was enacted in 1976. This established an intricate statutory framework to enable the FDA to regulate nearly every aspect of medical devices, from testing through marketing. To implement this law, the FDA has promulgated regulations published in the nine volumes of Title 21 of the *Code of Federal Regulations* (CFR). The majority of regulations concerning medical devices are contained in Parts 800–1299. If the FDA deletes, adds, or amends a regulation, it is published in the *Federal Register*, which serves as the official public notice system. When a change in a given regulation is envisioned, the FDA publishes a notice, called a proposed rule, which after allowing time for public comment, may become a regulation. The final regulation is then codified in Title 21 of the CFR.

Two statutory provisions of Title 21 govern the introduction of new medical devices into the marketplace. Section 515 establishes a premarket approval application (PMA) containing data and information demonstrating the safety and effectiveness of a device. Section 510(k) establishes a premarket notification process. Under this process, a manufacturer is required to file with the FDA, 90 days before a new device is to be marketed, a premarket notification demonstrating that the device in question is substantially equivalent to a device that was on the market before enactment of the 1976 Amendment and therefore marketable without formal FDA approval.

In enacting Section 510(k), Congress effectively divided medical devices into two broad categories: (1) preamendment, those introduced into commercial distribution prior to the enactment date of the 1976 Amendment, and (2) post-amendment, those introduced into commercial distribution after May 28, 1976.

Post-amendment devices are automatically classified as Class III devices. However, a post-amendment device can be brought to market under the 510(k) process, if the FDA determines that the device is "substantially equivalent" to a preamendment device. If the post-amendment device is identical to a preamendment device, it is substantially equivalent. Then the 510(k) is accepted by the FDA and the post-amendment device is placed in the same class as the preamendment device to which it is substantially equivalent. For example, a wound dressing identical to a preamendment Class I wound dressing would be found substantially equivalent to the preamendment wound dressing and classified in Class I.

All medical devices marketed before May 28, 1976, are distinguished from devices developed after this date, ie, pre-1976 devices were grandfathered. Pre-1976 devices, and those devices similar to them are placed into one of three categories based on the amount of regulation necessary to provide a reasonable assurance of safety and effectiveness.

As of 1992, Title 21 of the CFR listed roughly 1800 generic devices that have been officially classified as medical devices. The term medical device includes such technologically simple articles as hypodermic syringes and blood bags. At the other end of the spectrum, highly sophisticated articles such as pacemakers, surgical lasers, implantable pumps, and vascular grafts are also medical devices. This diversity coupled with the large number of different devices presents a unique challenge for the U.S. Congress and the FDA.

Cardiovascular Devices

Cardiovascular Disease. Treatment of cardiovascular diseases in the United States is a vast and growing industry approaching $100 billion annually (see CARDIOVASCULAR AGENTS). Cardiovascular disease is a progressive condition which can eventually block the flow of blood through the coronary arteries to the heart muscle, thereby causing heart attacks and other life-threatening situations. The same plaque deposits occur in the peripheral arteries, leading to gangrene, amputations, aneurysms, and strokes. Despite enormous progress in cardiovascular medicine since World War II, challenges and unmet needs abound. Mortality rates have declined significantly, millions of people have been helped to lead normal lives, but the prevalence and incidence of cardiovascular diseases remain high (Table 1). Open-heart surgery, cardiac pacing, heart transplants, implantable valves, and, more recently, coronary angioplasty and clot busters, have been developed, but none of the great advances in cardiovascular medicine is preventive. There has been no "Salk vaccine" to preclude the buildup of fatty deposits or plaque in the arteries.

Cardiovascular disease is the leading cause of premature death in the United States. The leading immediate cause of death, referred to as sudden death, in most cases is a massive electrical failure of the heart within the first hour following onset of symptoms. Over 300,000 of the 513,700 coronary heart disease deaths in 1987 were a result of sudden death. Over ten million Americans, including over five million previous heart attack victims and angina patients, have identifiable coronary artery disease (CAD). An estimated two million people in the United States, including pacemaker patients, have severe heart arrhythmia or electrical problems and over eight million suffer from peripheral vascular disease (PVD). Most patients afflicted with PVD already have heart disease, owing to the same underlying causes, namely coronary artery disease. Countless millions of others are at risk of PVD and CAD because of high blood pressure (see Table 1), elevated cholesterol levels, smoking, and other known and identified risk factors.

Cardiovascular devices are identified with the heart, the lungs, and the circulatory road map of the vascular system. This last is comprised of the 96,000 km of arteries, veins, and capillaries which transport blood continuously throughout the body (Fig. 1). Blood serves as the vehicle to carry oxygen, which is exchanged for carbon dioxide in the capillaries, bringing other vital substances to the body's cells, as well as removing waste products.

The heart, a double pump consisting of four chambers (Fig. 2), drives the circulatory system. The atria are the upper two chambers which act as receiving chambers; the much larger lower two ventricles serve as propulsion chambers. After the blood delivers oxygen, it returns to the heart via the venous side of the system which ends up in the venae cavae prior to entering the right atrium. Continuing its flow progression, the blood moves into the right ventricle which pumps it out to the pulmonary arteries and on to the lungs. After the blood enters

Table 1. Cardiovascular Diseases in the United States

Type of heart disease	Prevalence	Incidence	Mortality
congenital	540,000	25,000	5,600
rheumatic	2,180,000		6,100
congestive heart failure	2,000,000	400,000	34,000
coronary	5,000,000	1,500,000	513,700
stroke	2,060,000	500,000	149,200
high blood pressure	60,990,000		30,900
other[a]			237,206
Total			*976,706*

[a]Other cardiovascular diseases cover a long list of circulatory problems, including heart failure, peripheral vascular disease, cardiomyopathy, and arrhythmias.

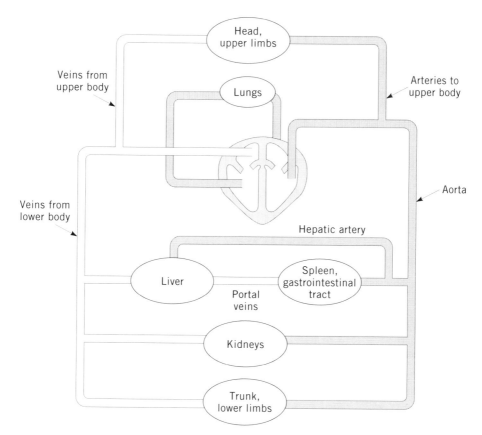

Fig. 1.   An overall view of the cardiovascular system.

the lungs, disposes of carbon dioxide, and picks up fresh oxygen, it returns via the pulmonary veins to the left atrium of the heart and hence to the left ventricle which is the largest of the heart's pumping chambers. The blood starts its journey over again when it is propelled from the left ventricle to the aorta and from there throughout the body via the peripheral arteries.

One of the remarkable features of the heart, which is about the size of a fist, is that the pumping action or contractions (~70 times/min) of atria and ventricles are simultaneous. Deoxygenated blood is pumped out to the lungs from one side of the heart at the same time oxygenated blood is pumped out from the other side to the aorta and onward through the body. The entire process of blood flow between the atria, ventricles, and the principal vessels is in unison, controlled in part by four one-way valves. The atrioventricular valves, tricuspid on the right and mitral on the left, prevent blood from flowing backward into the atria when the ventricles are filled. The pulmonary and aortic semilunar valves prevent blood from returning to the ventricles after they have contracted.

In addition to its internal blood flow operation, the heart has its own system of blood vessels to keep the muscle wall of the heart, the myocardium, supplied with oxygenated blood (Fig. 3a). The coronary arteries, which branch from the aorta to the right and left sides of the heart, are vital to maintaining that supply.

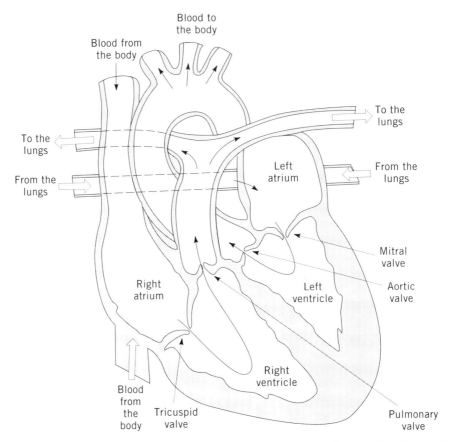

Fig. 2.   Schematic of the human heart, showing the four chambers and various values. The atria act as receiving chambers and the ventricles act as propulsion chambers. The arrows indicate the direction of blood flow.

The heart is an extraordinary electromechanical muscle that can be trained to increase blood flow to the body sixfold. It can range from 5 to 30 L/min during exertion.

An intricate wiring system of muscle fibers conducts electrical impulses which cause the heart to contract and pump blood (Fig. 3**b**). The sinoatrial (SA) node, located in the right atrium, is the heart's natural pacemaker and regulates the heart rate as it receives signals from the autonomic nervous system. A second node, the atrioventricular (AV) node, serves as a junction between the upper and lower chambers. The AV node picks up and sends on the signals from the SA node spreading them throughout the heart to cause first the atria and then the ventricles to contract. The number and frequency of the impulses determine the heart rate which is automatically adjusted to speed up during exertion or, conversely, to slow down during sleep.

Economic Aspects.   The cardiovascular devices market is estimated to be approximately \$2.9 billion annually on a worldwide basis. This market can be further segmented as follows: angiography and angioplasty, \$644 $\times$ 10^6; arrhythmia control, \$1500 $\times$ 10^6; cardiovascular surgery, \$700 $\times$ 10^6; cardiac

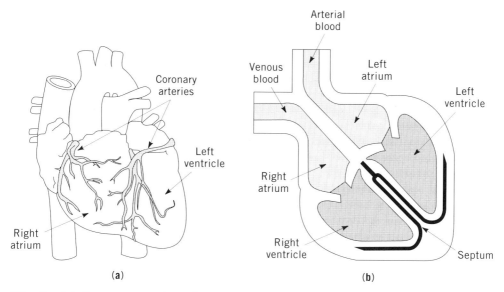

Fig. 3. (**a**) Coronary arteries which form the heart's own blood supply; (**b**) electrical conduction system which powers the human heart.

assist (intra-aortic balloon pump), 80×10^6; and artificial hearts, which are experimental.

The cardiovascular device industry is characterized by high growth and the emergence of numerous companies having products which create additional markets. In the first half of the 1990s, the most notable device developed was used in coronary angioplasty. Introduction of this balloon catheter, used to open up the blocked segment of an artery, was somewhat slow. The barriers to entry were high owing to the technologies involved, the rigorous regulatory process, and the need to gain medical acceptance prior to widespread market introduction.

Device Markets.   New markets for cardiovascular devices are driven by a least five interrelated factors. Technology, competition, economics, consumer demand, and symbiosis of drugs and devices all play roles.

Ideally, new technology results in a close collaboration between the industrial sector and the medical proponent. Because new technologies often come from techniques perfected by specialists, these devices often enjoy instant credibility and rapid market acceptance within the broader medical community. Invariably, the indications, ie, how and when a device can be used, for new device technologies are initially limited. However, indications increase with experience and follow-on improvements. An example is the use of pacemakers. Succeeding generations of devices usually provide additional patient benefits, expanding the market further by commanding higher prices, if not also adding to the number of procedures.

The cardiovascular health care marketplace is highly competitive. This works to the advantage of medical device companies. Hospitals are engaged in intense competition for those who might be candidates for cardiovascular therapies. Therefore, hospitals are anxious to offer the best treatment possible, having the newest devices. Medical specialties are also competing for their share of the same patient population.

In the cardiovascular arena, hundreds of thousands of patients are treated by interventional cardiologists, whereas in the past these people would have been treated by surgeons. The surgeons, to offset the loss of some patients, take on more difficult procedures which over time become routine in nature. Thus, more patients can be successfully treated and the marketplace expands.

Cardiovascular Problems. Despite its durability and resilience, different aspects of the cardiovascular system can malfunction. Some problems are congenital; many are inherited. Diseases can also be caused by infection such as damaged heart valves owing to rheumatic fever. Cardiomyopathy, a diseased heart muscle which may become enlarged, can result from infection or an unknown cause. Other problems may be a function of age. Pacemaker patients often have conduction systems that have simply started to wear out. Lifestyle also plays a role. Although poor diet and smoking cause or contribute to multiple problems such as hypertension and lung disease, when it comes to cardiovascular problems, the main culprit, regardless of its origin, is atherosclerosis. Atherosclerosis is a disease of the arteries resulting from the deposit of fatty plaque on the inner walls.

Plaque. A heart attack, or myocardial infarction, results from insufficient delivery of oxygen to parts of the heart muscle owing to restricted blood flow in the coronary arteries. If heart muscle tissue is deprived of oxygen long enough, it may infarct or die (Fig. 4). The heart attack is often precipitated by a clot, or thrombus, which forms on a severely narrowed portion of a coronary artery. Silent ischemia is somewhat reduced blood supply from narrowing of the arteries. As the name implies, the disease provides no symptomatic warning of an impending problem. When coronary arteries are blocked to the degree that they cannot meet the heart's temporary demand for more oxygenated blood, angina pectoris, or sharp pain, may result. Further progression of the blockage then brings on the myocardial infarction. Atheroma is the medical term used to describe what plaque, the fatty deposits, does to the walls of the arteries.

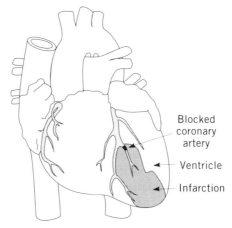

Fig. 4. Myocardial infarction occurs during insufficient delivery of oxygen to a portion of the heart muscle.

Plaque also causes other problems such as strokes and aneurysms, as well as complications of peripheral vascular disease.

Lethal Arrhythmias.   Arrhythmias are a second significant source of cardiovascular problems. An arrhythmia is an abnormal or irregular heart rhythm. Bradyarrhythmias result in heart rates that are too slow; tachyarrhythmias cause abnormally fast rates. A bradyarrhythmia can be debilitating, causing a person to be short of breath, unable to climb stairs, black out, or even to go into cardiac arrest. Tachyarrhythmias can be unsettling and painful at best, life-threatening at worst.

Arrhythmias are caused by disturbances of the normal electrical conduction patterns synchronizing and controlling heartbeats. The wiring leading to the ventricles might, in effect, break or become frayed, causing a slowdown in the signals getting through, or perhaps result in intermittent electrical impulses. If damage to heart muscle tissue occurs, for example, from a myocardial infarction, this could create new electrical pathways. These in turn set up a separate focus of electrical activity (like another natural pacemaker) generating extra beats which can be highly disruptive. If a tachyrhythmia (tachycardia) occurs in the ventricles, the pumping chambers of the heart, the problem can be severely uncomfortable or even cause death if it deteriorates into ventricular fibrillation. Fibrillation is uncontrolled electrical activity. In this chaotic situation, cells become uncoordinated so that the heart muscle only quivers or twitches and no longer contracts rhythmically. Approximately three-fourths of the more than 500,000 deaths per year in the United States from coronary heart disease are sudden deaths.

There is a close correlation between myocardial infarctions and tachyarrhythmias, illustrated by the presence of complex ventricular arrhythmias among heart attack victims which are estimated to affect one-third of the survivors each year. Frequently, the immediate cause of sudden death is ventricular fibrillation, an extreme arrhythmia that is difficult to detect or treat. In the majority of cases, victims have no prior indication of coronary heart disease.

Valvular Disease.   Valve problems severely limit the efficiency of the heart's pumping action bringing forth definitive symptoms. There are two types of conditions, both of which may be present in the same valve. The first is narrowing, or stenosis, of the valve. The second condition is inability of the valve to close completely. Narrowing of the mitral valve, for example, can result in less blood flowing into the left ventricle and subsequently less blood being pumped into the body. If the same valve does not close completely, blood may also back up or regurgitate into the left atrium when the ventricle contracts, preventing even more blood from properly flowing. The backward pressure which results can cause a reduction in the efficiency of the lungs.

Cardiomyopathy.   Cardiomyopathy, or diseased heart muscle, may reach a point at which the heart can no longer function. It arises from a combination of factors, including hypertension, arrhythmias, and valve disease. Other problems, such as congestive heart failure, cause the interrelated heart–lung system to break down. Because the heart can no longer adequately pump, fluid builds up in the lungs and other areas.

Device Solutions.   The first big step in cardiovascular devices was the development of a heart–lung machine in 1953. The ability to shut down the operation of the heart and lungs and still maintain circulation of oxygenated blood throughout the body made open-heart surgery possible. Open heart, really a misnomer, actually refers to opening up the chest to expose the heart, not opening up the heart itself. The principal components of the heart–lung machine, the oxygenator and pump, take over the functions of the lungs and heart.

Atherosclerosis. The first solution to the problem of atherosclerosis was the coronary artery bypass graft (CABG) procedure, first performed in 1964. In a coronary bypass procedure, a graft is taken from the patient's own saphenous vein. The graft is attached to the aorta (Fig. 5) where the coronary arteries originate and the opposite end is connected to the artery below the blocked segment. Blood can then bypass the obstructed area and reach the surrounding tissue below. Extensions of this useful surgery are CABG procedures which utilize mammary arteries of the patient instead of saphenous veins.

The second step toward solving cardiovascular disease from atherosclerosis, ie, angioplasty, was preceded by the diagnostic tool of angiocardiography by nearly 20 years. Angiocardiography, or angiography, permits x-ray diagnosis using a fluoroscope. A radiopaque contrast medium is introduced into the arteries through a catheter (see RADIOPAQUES), and angiography allows accurate location of the plaque blockage. Percutaneous transluminal coronary angioplasty (PTCA), a nonsurgical procedure, emerged in the 1980s as a viable method for opening up blocked arteries. A PTCA catheter has a balloon at its tip which is inflated after it is positioned across the blocked segment of the artery. Plaque is then compressed against the arterial walls, permitting blood flow to be restored. The same solutions of bypass surgery and angioplasty have been applied to atherosclerosis in the peripheral arteries.

Arrhythmias. The first solution to cardiovascular problems arising from arrhythmias came about as a result of a complication caused by open-heart surgery. During procedures to correct congenital defects in children's hearts,

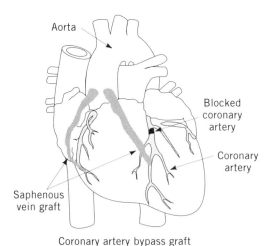

Coronary artery bypass graft

Fig. 5. In coronary bypass, an autologous saphenous vein is used to provide critical blood to the heart muscle, bypassing a blockage in the coronary artery.

the electrical conduction system often became impaired, and until it healed, the heart could not contract sufficiently without outside electrical stimulation. A system that plugged into a wall outlet was considered adequate until an electrical storm knocked out power, leading to the development of the first battery-powered external pacemaker.

The first implantable pacemaker, introduced in 1960, provided a permanent solution to a chronic bradyarrhythmia condition. This invention had a profound impact on the future of medical devices. The pacemaker was the first implantable device which became intrinsic to the body, enabling the patient to lead a normal life.

Early pacemakers paced the heart continuously at a fixed rate, were larger than a hockey puck, and had to be replaced frequently owing to power source technology limitations. Advances in electronics, materials, and knowledge have yielded pacemakers about the size of a U.S. 50-cent piece that last five years or more. More importantly, pacemakers restore the heart to functioning in a completely natural way. The pacemaker senses the electrical activity of the heart and kicks in only when something is wrong. If the impulses initiated by the SA node cannot get all the way through to the lower part of the ventricles, the pacemaker takes over completing the electrical process at the same rate indicated by the heart's natural pacemaker. If the SA node is dysfunctional and cannot put out an appropriate signal, sensors (qv) in rate-responsive pacemakers can correlate other data such as sound waves from body activity, body temperature, or the respiratory rate to compute the proper heart rate.

The first automatic implantable cardioverter defibrillator (AICD) was implanted in 1980. As for pacemakers, early generations of AICDs were bulky and cumbersome, did not last very long, and required open-heart surgery. However, these kept people alive by automatically shocking the heart out of its chaotic electric state whenever it went into ventricular fibrillation. Future devices are being designed to provide the full spectrum of arrhythmia control, including pacing, cardioversion, and defibrillation. Techniques are also being developed to map, ie, locate, the source of certain tachyarrhythmias (an ectopic focus or scar tissue) and remove it without open-heart surgery.

External defibrillation was first performed in 1952 and continues as a routine procedure in hospitals and ambulances. The problem of external defibrillation has not been a technological one, but rather a legal one. Only in the 1990s have laws been passed to permit people other than doctors and paramedics to operate semiautomatic defibrillators to provide help when it is needed. New and better defibrillation devices continue to come to market and are easier and safer to use.

Valve Problems. The primary solution to valve problems has been implantable replacement valves. The introduction of these devices necessitates open-heart surgery. There are two types of valves available: tissue (porcine and bovine) and mechanical. The disadvantage of tissue valves is that these have a limited life of about seven years before they calcify, stiffen, and have to be replaced. The mechanical valves can last a lifetime, but require anticoagulant therapy. In some patients, anticoagulants may not be feasible or may be contraindicated. Of the valves which require replacement, 99% are mitral and aortic valves. The valves on the left side of the heart are under much greater pressure because the left ventricle is pumping blood out to the entire body, instead of only to the lungs. Occasionally, two valves are replaced in the same procedure.

Cardiomyopathy. The best available solution to cardiomyopathy may be one that is less sophisticated than transplant surgery or the artificial heart. The cardiomyoplasty-assist system combines earlier electrical stimulation technology with a new surgical technique of utilizing muscle from another part of the body to assist the heart.

Efforts to develop an artificial heart have resulted in a number of advancements in the assist area. The centrifugal pump for open-heart surgery, the product of such an effort, has frequently been used to support patients after heart surgery (post-cardiotomy), or as a bridge to life prior to transplant. Other efforts have led to the development of ventricular assist devices to support the heart for several months and intra-aortic balloon pumps (IABPs) which are widely used to unload and stabilize the heart.

Interventional Procedures. The emergence of angioplasty created a specialty called interventional cardiology. Interventional cardiologists not only implant pacemakers and clear arteries using balloon catheters, but they also use balloons to stretch valves (valvuloplasty). In addition, they work with various approaches and technologies to attack plaque, including laser (qv) energy, mechanical cutters and shavers, stents to shore up arterial walls and deliver drugs, and ultrasound to break up plaque or to visualize the inside of the artery (see MEDICAL IMAGING TECHNOLOGY; ULTRASONICS).

Typically, procedures have become less invasive as technology evolves. Early pacemaker procedures involved open-heart surgery to attach pacemaker leads (wires) to the outside of the heart. Later, leads could be inserted in veins and pushed through to the interior of the heart, no longer necessitating opening a patient's chest. Using fluoroscopy, the physician can visualize the process, so that the only surgery needed is to create a pocket under the skin for the implantable generator to which the leads are connected.

Clinical evaluation is underway to test transvenous electrodes. Transvenous leads permit pacemakers to be implanted under local anesthesia while the patient is awake, greatly reducing recovery time and risk. As of 1996, the generation of implantable defibrillators requires a thoracotomy, a surgical opening of the chest, in order to attach electrodes to the outside of the heart. Transvenous electrodes would allow cardiologists to perform pacemaker procedures without a hospital or the use of general anesthesia.

Coronary bypass surgery and angioplasty are vastly different procedures, but both procedures seek to revascularize and restore adequate blood flow to coronary arteries. Balloon angioplasty, which looks much like a pacemaker lead except that it has a tiny balloon at the end instead of an electrode, involves positioning a catheter inside a coronary artery under fluoroscopy. The balloon is inflated to compress the offending plaque. Angioplasty is far less invasive than bypass surgery and patients are awake during the procedure. For many patients, angioplasty may not be indicated or appropriate.

Interventional cardiology is but one specialty that has arisen in cardiovascular medicine. Another is interventional radiology for similar procedures in the peripheral arteries, in addition to conventional bypass graft surgery. Competition has been intense among surgeons, cardiologists, and radiologists. Because coronary artery disease is progressive, many patients who are candidates for peripheral and/or coronary angioplasty may be future candidates for bypass surgery.

Cardiologists may be described in terms of three overlapping specialties: interventional, who perform most angioplasty; invasive, who implant about 70% of the pacemakers in the United States; and diagnostic. A subspecialty of diagnostic cardiology, electrophysiology, has grown in importance because it is critical to the treatment of tachrhythmia patients, especially those who are prone to ventricular fibrillation. The further development of implantable devices in this last area depends on close cooperation between companies and electrophysiologists.

Cardiovascular devices are being employed by a wider diversity of specialists and are thus finding applications in other medical areas. This has been particularly true for devices developed to support open-heart surgery. Oxygenators and centrifugal pumps, which take over the functions of the lungs and heart, are used in applications such as support of angioplasty and placing a trauma or heart attack victim on portable bypass in the emergency room. Some devices are finding utility by improving surgical techniques. For example, cardiac surgeons are working with balloon catheters and laser angioplasty systems as an augmentation to regular bypass surgery.

Other cardiovascular devices developed initially for use in open-heart surgery are used extensively in other parts of the hospital and, in many cases, outside the hospital. Patients have been maintained for prolonged periods of time on portable cardiopulmonary support systems while being transported to another hospital or waiting for a donor heart. Blood pumps and oxygenators may take over the functions of the heart and lungs in the catheterization lab during angioplasty, in extracorporeal membrane oxygenation (ECMO) to support a premature baby with severe respiratory problems, or in the emergency room to assist a heart attack victim. It is possible that future patients could be put on portable bypass at the site of the heart attack or accident. The market for cardiac assist devices and oxygenators plus related products such as specialized cannulae and blood monitoring devices is expected to expand rapidly into these areas.

Biomaterials for Cardiovascular Devices. Perhaps the most advanced field of biomaterials is that for cardiovascular devices. For several decades bodily parts have been replaced or repaired by direct substitution using natural tissue or selected synthetic materials. The development of implantable-grade synthetic polymers, such as silicones and polyurethanes, has made possible the development of advanced cardiac assist devices (see SILICON COMPOUNDS, SILICONES; URETHANE POLYMERS).

Devices for the 1990s. The 1990s may turn out to be the decade of active arrhythmia-control devices. Implantable devices to pace, cardiovert, and defibrillate the heart without the need for open-heart surgery should become widely accepted before the year 2000. Dramatic developments and growth are also antipicated in other areas such as the use of laser systems intended to ablate significant amounts of plaque. These have already been approved for peripheral applications and are undergoing clinical evaluation in coronary arteries.

Laser ablation systems hold considerable promise if restenosis (reblocking of the arteries) rates are reduced. The rate as of 1995 is 30%, typically within six months. Mechanical or atherectomy devices to cut, shave, or pulverize plaque have been tested extensively in coronary arteries. Some of these have also been approved for peripheral use. The future of angioplasty, beyond the tremendous success of conventional balloon catheters, depends on approaches that can reduce restenosis rates. For example, if application of a drug to the lesion site turns out to be the solution to restenosis, balloon catheters would be used for both dilating the vessel and delivering the drug. An understanding of what happens to the arterial walls, at the cellular level, when these walls are subjected to the various types of angioplasty may need to come first.

A primary aspect of cardiovascular devices through the twenty-first century is expected to involve the incorporation of diagnostic and visualization capabilities. A separate ultrasound system has been approved for this purpose. Laser angioplasty systems under development include visualization capabilities to distinguish plaque from the arterial wall. Future pacemakers, which already utilize sensors (qv) to determine an appropriate heart rate, are expected to incorporate various other sensors for diagnostic purposes. The biggest challenge in averting sudden death is not so much to perfect a life-sustaining device, but to gain the ability to identify the susceptible patient. Appropriate screening and diagnoses for patients having silent ischemia must be developed. If the presence and extent of coronary artery disease can be identified early, intervention could save thousands of people from an untimely death and help others to live a fuller life. Sensors and specific diagnostic devices are expected to play a large role at about the same time as effective implantable defibrillators.

One of the more intriguing cardiovascular developments is cardiomyoplasty where implantable technologies are blended with another part of the body to take over for a diseased heart. One company, Medtronic, in close collaboration with surgeons, has developed a cardiomyoplasty system to accompany a technique of wrapping back muscle around a diseased heart which can no longer adequately pump. A combination pacemaker and neurological device senses the electrical activity of the heart and correspondingly trains and stimulates the dorsal muscle to cause the defective heart to contract and pump blood. Over 50 implants have been performed to date.

Cardiomyoplasty could greatly reduce the overwhelming need for heart transplants. It might also eliminate the need for immunosuppressive drugs. A broad-based cardiomyoplasty market should emerge by the late 1990s. Successful development of a small-diameter graft to use in coronary bypass surgery instead of the patient's saphenous vein or mammary artery seems likely to occur by the year 2000. Development of appropriate materials and manufacturing methods are needed to maintain patency without damaging blood in grafts below 4 mm in diameter.

Pacemakers.   The implantable cardiac pacemaker (Fig. 6) has been a phenomenal technological and marketing success. In the early 1980s, however, many critics were predicting the demise of these devices and the industry was the subject of congressional investigations over sales practices, alleged overuse, and excessive prices. Critics advocated low priced generic pacemakers, and pacemaker unit volume and prices declined about 10% on average. However, costs have been reduced by curtailing the length of time patients need to stay in the hospital following the implantation procedure and by selection of the correct pacemaker for

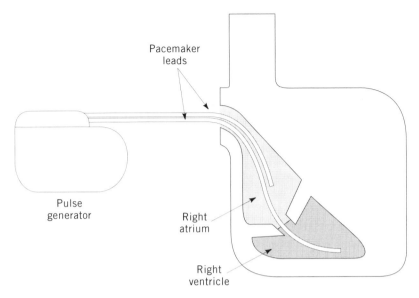

Fig. 6. A pacemaker provides electrical impulses to the heart in an effort to correct potentially fatal arrhythmias.

each patient. Significantly lower cost is attached to a single-chamber device having limited longevity than to the far more expensive dual-chamber device which may be indicated for a young and active patient.

As of the mid-1990s, the market for bradyarrhythmia devices is fully penetrated in Western countries. Some growth is expected to result from an aging population but, by and large, the market is mature. The market for tachyrhythmia devices, in contrast, is only beginning.

Implantable tachyrhythmia devices, available for some years, address far less dangerous atrial tachyarrhythmias and fibrillation. The technical barriers to counteracting ventricular tachyarrhythmias and fibrillation using massive shocks have been formidable and are compounded by the possibility of causing the very problem the shock is designed to overcome. Newer tachyrhythmia devices are being readied that can safely regulate arrhythmias across the full spectrum.

Surgical Devices. Surgical devices comprise the equipment and disposables to support surgery and to position implantable valves and a variety of vascular grafts. Central to open-heart surgery is the heart–lung machine and a supporting cast of disposable products. Two devices, the oxygenator and more recently the centrifugal pump, amount to significant market segments in their own right. Other disposables include cardiotomy reservoirs, filters, tubing packs, and cardioplegia products to cool the heart. In 1993, there were an estimated 250,000–300,000 open-heart surgeries in the United States, including 45,000 valve replacements. The oxygenator market has been driven more recently by the conversion from bubbler to membrane devices which account for about 80% of the oxygenators used in the United States.

Centrifugal pumps are increasingly being used as a safer and more effective alternative to the traditional roller pump in open-heart surgery and liver transplants. As of the mid-1990s, about 45% of open-heart procedures use a centrifugal pump. In the latter 1980s, that number was less than 10%. The market would likely change over more rapidly to centrifugal pumps except for the higher price ($125–$175) of each disposable centrifugal pump compared to the lower cost for roller pump tubing. A strong case can be made that the centrifugal pump is less expensive in the long run because the pump causes less damage to blood and may reduce recovery time. Use of the pump is also a safer method and thus may reduce the liability risk of this procedure.

Implantable valves, particularly mechanical valves which continue to encroach on tissue valves, are unique. Methods such as valvuloplasty, mitral valve repair, or use of ultrasound are unlikely to reduce the number of valve replacements into the twenty-first century. Valve selection remains in the hands of the surgeon because of the critical nature of the procedure. If anything goes wrong, the result can be catastrophic to the patient. Cost of a valve, from $3000–$4000, is a relatively small part of the cost of open-heart surgery which can run as high as $30,000. Growth of the cardiovascular valve market has slowed in the United States with the decline of the threat of rheumatic fever.

Vascular grafts are tubular devices implanted throughout the body to replace blood vessels which have become obstructed by plaque, atherosclerosis, or otherwise weakened by an aneurysm. Grafts are used most often in peripheral bypass surgery to restore arterial blood flow in the legs. Grafts are also frequently employed in the upper part of the body to reconstruct damaged portions of the aorta and carotid arteries. In addition, grafts are used to access the vascular system, such as in hemodialysis to avoid damage of vessels from repeated needle punctures. Most grafts are synthetic and made from materials such as Dacron or Teflon. Less than 5% of grafts utilized are made from biological materials.

Cardiac-Assist Devices. The principal cardiac-assist device, the intra-aortic balloon pump (IABP), is used primarily to support patients before or after open-heart surgery, or patients who go into cardiogenic shock. As of the mid-1990s, the IABP was being used more often to stabilize heart attack victims, especially in community hospitals which do not provide open-heart surgery. The procedure consists of a balloon catheter inserted into the aorta which expands and contracts to assist blood flow into the circulatory system and to reduce the heart's workload by about 20%. The disposable balloon is powered by an external pump console.

Other devices, which can completely take over the heart's pumping function, are the ventricular assist devices (VADs), supporting one or both ventricles. Some patients require this total support for a period of time following surgery (post-cardiotomy); others require the support while being transported from one hospital to another, or while waiting for a donor heart (bridge-to-transplant). Several external and implantable devices are being evaluated for short-term and long-term applications. Temporary support of the ventricles is used in less than 1000 cases a year in the United States.

Considerable interest has emerged in devices providing cardiopulmonary support (CPS), ie, taking over the functions of both the heart and lungs without having to open up the chest. There are several applications for other portable bypass systems or mini-heart–lung machines. Thus far, CPS has been used most frequently in support of anigoplasty prophylactically in difficult cases which could not be otherwise undertaken. The greatest potential is in the emergency room to rest the heart and lungs of heart attack and trauma victims.

Other specialized applications of cardiac arrest devices include extracorporeal membrane oxygenation (ECMO) which occurs when the lungs of a premature infant cannot function properly. The market segments for cardiopulmonary support devices are potentially significant.

Artificial Hearts. Congestive heart failure (CHF) is a common cause of disability and death. It is estimated that three to four million Americans suffer from this condition. Medical therapy in the form of inotropic agents, diuretics (qv), and vasofilators is commonly used to treat this disorder (see CARDIOVASCULAR AGENTS). Cardiac transplantation has become the treatment of choice for medically intractable CHF. Although the results of heart transplantation are impressive, the number of patients who might benefit far exceeds the number of potential donors. Long-term circulatory support systems may become an alternative to transplantation (5).

In 1980, the National Heart, Lung and Blood Institute of NIH established goals and criteria for developing heart devices and support techniques in an effort to improve the treatment of heart disease. This research culminated in the development of both temporary and permanent left ventricular-assist devices that are tether-free, reliable over two years, and electrically powered. The assist devices support the failing heart and systemic circulation to decrease cardiac work, increase blood flow to vital organs, and increase oxygen supply to the myocardium. The newer ventricular assists are required to have no external venting, have a five-year operation with 90% reliability, pump blood at a rate of 3–7 L/min into the aorta at a mean arterial pressure of 90 mm Hg (12 kPa) when assisting the human left ventricle, and have a specific gravity of 1.0 for the implantable ventricular assist device.

In contrast, the total artificial heart (TAH) is designed to overtake the function of the diseased natural heart. While the patient is on heart–lung bypass, the natural ventricles are surgically removed. Polyurethane cuffs are then sutured to the remaining atria and to two other blood vessels that connect with the heart.

One successful total artificial heart is ABIOMED's electric TAH. This artificial heart consists of two seamless blood pumps which assume the roles of the natural heart's two ventricles (Fig. 7). The pumps and valves are fabricated from a polyurethane, Angioflex. Small enough to fit the majority of the adult population, the heart's principal components are implanted in the cavity left by the removal of the diseased natural heart. A modest sized battery pack carried by the patient supplies power to the drive system. Miniaturized electronics control the artificial heart which runs as smoothly and quietly as the natural heart. Once implanted, the total artificial heart performs the critical function of pumping blood to the entire body (6).

Heart Valves. Since the early 1960s nearly 50 different heart valves have been developed. The most commonly used valves as of the mid-1990s include mechanical prostheses and tissue valves. Nearly 75,000 of these prosthetic valves are implanted annually worldwide, and about 30,000 in the United States alone. Caged-ball, caged disk, and tilting-disk heart valves are the types most widely used.

Fig. 7. ABIOMED total heart.

Blood Salvage. In a growing awareness that a patient's own blood is the best to use when blood is needed, newer techniques are reducing the volume of donor blood used in many cardiovascular and orthopedic surgeries. Surgical centers have a device, called the Cell Saver (Haemonetics), that allows blood lost during surgery to be reused within a matter of minutes, instead of being discarded. This device collects blood from the wound, runs it through a filter that catches pieces of tissue and bone, then mixes the blood with a salt solution and an anticoagulant. The device then cleanses the blood of harmful bacteria. Subsequently the blood is reinfused back to the same patient through catheters inserted in a vein in the arm or neck, eliminating the worry of cross-contamination from the HIV or hepatitis viruses (see BLOOD, COAGULANTS AND ANTICOAGULANTS; FRACTIONATION, BLOOD).

Use of intraoperative autotransfusion (IAT) eliminates disease transmission, compatibility testing, and immunosuppression that may result from the use of homologous blood products, reduces net blood loss of the patient, and conserves the blood supply. During vascular surgery, the principal indications for the use of the Cell Saver are ruptured spleen, ruptured liver, aneurysms, and vascular trauma. During orthopedic surgery the principal indications are total hip arthroplasty, spinal fusions, total knee, and any procedure that has wound drains (7). The self-contained apparatus that houses the entire device is shown in Figure 8.

Blood Access Devices. An investigational device called the Osteoport system allows repeated access to the vascular system via an intraosseous infusion directly into the bone marrow. The port is implanted subcutaneously and secured into a bone, such as the iliac crest. Medications are administered as in any conventional port, but are taken up by the venous sinusoids in the marrow cavity, and from there enter the peripheral circulation (8).

Blood Oxygenators. The basic construction of an oxygenator involves any one of several types of units employing a bubble-type, membrane film-type, or hollow-fiber-type design. The most important advance in oxygenator devel-

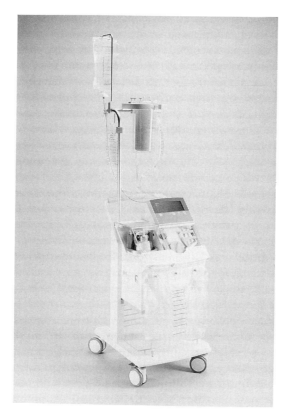

Fig. 8. The Cell Saver.

opment was the introduction of the membrane-type oxygenator. These employ conditions very close to the normal physiological conditions in which gas contacts occur indirectly via a gas-permeable membrane. Blood trauma is minimized by the use of specialized biomaterials such as PTFE, PVC, and cellophane, although lately silicone rubber and cellulose acetate have predominated. A silicone–polycarbonate copolymer, ethylcellulose perfluorobutyrate, and poly(alkyl sulfone) were introduced in the mid-1980s, and tend to dominate this field.

Polyurethanes as Biomaterials. Much of the progress in cardiovascular devices can be attributed to advances in preparing biostable polyurethanes. Biostable polycarbonate-based polyurethane materials such as Corethane (9) and ChronoFlex (10) offer far-reaching capabilities to cardiovascular products. These and other polyurethane materials offer significant advantages for important long-term products, such as implantable ports, hemodialysis, and peripheral catheters; pacemaker interfaces and leads; and vascular grafts.

Implantable Ports. The safest method of accessing the vascular system is by means of a vascular access device (VAD) or port. Older VAD designs protruded through the skin. The totally implanted ports are designed for convenience, near absence of infection, and ease of implantation. Ports allow drugs and fluids to be delivered directly into the bloodstream without repeated insertion of needles into a vein. The primary recipients of totally implanted ports are patients receiving chemotherapy, bolus infusions of vesicants, parenteral nutrition, antibiotics, analgesics, and acquired immune disease syndrome (AIDS) medications.

Vascular access ports typically consist of a self-sealing silicone septum within a rigid housing which is attached to a radiopaque catheter (see RADIO-PAQUES). The catheter must be fabricated from a low modulus elastomeric polymer capable of interfacing with both soft tissue and the cardiovascular environment. A low modulus polyurethane-based elastomer is preferred to ensure minimal trauma to the fragile vein.

Placement of vascular access ports is similar to that of a long-term indwelling arterial catheter. A small incision is made over the selected vein and a second incision is made lower in the anterior chest to create a pocket to house the port. The catheter is tunneled subcutaneously from its entry point into the vein with the tip inside the right atrium. The final position of the catheter is verified by fluoroscopy, secured with sutures, and the subcutaneous pocket is closed. The port septum is easily palpable transcutaneously, and the system may be used immediately. A surgeon typically inserts the vascular access port in an outpatient setting.

To use the port, the overlying skin is prepared using conventional techniques. A local anesthetic is sometimes used to decrease pain of needle insertion, though this is usually not necessary using techniques which utilize small-bore needles. A special point needle is used to puncture the implanted ports as the point of these needles is deflected so it tears the septum rather than coring it, allowing multiple entries. The septum reseals when the needle is removed.

The primary advantages of implantable ports are no maintenance between uses other than periodic flushing with heparinized saline every 28 days to ensure patency, lower incidence of clotting and thrombosis, no dressing changes, insignificant infection incidence, unobtrusive cosmetic appearance, and no restriction on physical activity.

Pacemaker Interfaces and Leads.   Problems of existing pacemaker interfaces and pacemaker lead materials made from silicones and standard polyurethanes are environmental stress cracking, rigidity, insulation properties, and size.

Technical advances in programmable pacemakers that assist both the tachycardia and bradycardia have led to the requirement of implanting a two-lead system. Owing to the ridigity and size of silicones, the only material that fulfills this possibility without significantly impeding blood flow to the heart is polyurethane. The primary needs in this medical area are reduction in making frequent changes and in failure rate, and the ability to have multiple conductors to handle advanced pacemaker technology.

Vascular Grafts.   Although the use of vascular grafts in cardiovascular bypass surgery is widely accepted and routine, numerous problems exist in these surgeries for the materials available. Biocompatibility is often a problem for vascular grafts which also tend to leak and lead to scarring of the anastomosis. The materials are not useful for small-bore (≤ 6 mm) grafts. The primary needs that materials can be developed to address are matching compliance to native vessels, having a lesser diameter for small-bore grafts which would serve as a replacement for the saphenous vein in coronary bypass, thinner walls, biostability, controlled porosity, and greater hemocompatibility for reduced thrombosis.

The advent of newer polyurethane materials is expected to lead to a new generation of cardiovascular devices. The characteristics of polyurethanes, combined with newer manufacturing techniques, should translate into direct medical benefits for the physician, the hospital, and the patient. This field offers exciting growth opportunities.

Orthopedic Devices

Bone, or osseous tissue, is composed of osteocytes and osteoclasts embedded in a calcified matrix. Hard tissue consists of about 50% water and 50% solids. The solids are composed of cartilaginous material hardened with inorganic salts, such as calcium carbonate and phosphate of lime.

Bone is formed through a highly complex process that begins with the creation of embryonic mesenchymal cells. These cells, only found in the mesoderm of the embryo, migrate throughout the human body to form all the types of skeletal tissues including bone, cartilage, muscle, tendon, and ligament. Mesenchymal cells differentiate into the various types of progenitor cells: osteoblasts, chondroblasts, fibroblasts, and myoblasts. Bone tissue begins to form when osteoblasts and chondroblasts synthesize cartilage-like tissue by secreting at least one potent bone cell growth factor (protein hormone), referred to as IGF-II. Bone growth factor is theorized to stimulate an increased expression of IGF-II receptors on the cell walls of other bone cells. This growth factor helps to initiate a cascade of other subcellular activity, leading to the formation of cartilage. The cartilage then hardens into bone when osteoblasts become lodged within the cartilage matrix, and cease to function. These types of mature cells are then known as osteocytes.

A bone is classified according to shape as flat, long, short, or irregular. A living bone consists of three layers: the periosteum, the hard cortical bone, and the bone marrow or cancellous bone. The periosteum is a thin collagenous layer, filled with nerves and blood vessels, that supplies nutrients and removes cell wastes. Because of the extensive nerve supply, normal periosteum is very sensitive. When a bone is broken, the injured nerves send electrochemical neural messages relaying pain to the brain.

Next is a dense, rigid bone tissue, referred to as the hard compact or cortical bone. It is cylindrical in shape and very hard. This dense layer supports the weight of the body and consists mostly of calcium and minerals. Because it is devoid of nerves, it experiences no pain. The innermost layer, known as cancellous bone or spongy bone marrow, is honeycombed with thousands of tiny holes and passageways. Through these passageways run nerves and blood vessels that supply oxygen and nutrients. This material has a texture similar to gelatin. The marrow produces either red blood cells, white blood cells, or platelets.

Rigid bones are needed for kinetic motion, support of internal organs, and muscle strength. The bones that compose the human thigh are pound for pound stronger than steel. Nature meets these needs by separating the skeleton into several bones and bone systems, creating joints where the bones intersect.

Joints are structurally unique. They permit bodily movement and are bound together by fibrous tissues known as ligaments. Most larger joints are encapsulated in a bursa sac and surrounded by synovial fluid which lubricates the joint continuously to reduce friction. The skeleton is constructed of various types of moveable joints. Some joints allow for no movement, such as those connecting the bones of the skull. Other joints permit only limited movement. For example, the joints of the spine allow limited movement in several directions. Most joints have a greater range of motion than the joints of the skull and spine.

The bearing surface of each joint is cushioned by cartilage. This tissue minimizes friction. The cartilage also reduces force on the bone by absorbing shock. The joint area is a narrow space known as an articular cavity which allows freedom of movement.

Ligaments are composed of bands of strong collagenous fibrous connective tissue. This tissue is originally formed by the mesenchymal cells which differentiate into fibroblast cells. These fibroblast cells then further differentiate into specialized cells known as fibrocytes. When fibrocytes mature, they are inactive and compose the ligaments. Ligaments function to tie two bones together at a joint, maintain joints in position preventing dislocations, and restrain the joint's movements. Ligaments may be reattached to bone by the use of an orthopedic anchor (Fig. 9).

Tendons are composed of fibrous connective tissue. Tendon tissue is also formed by the fibroblast cells, similar to the way ligaments are formed. These fibroblast cells then further differentiate into other specialized cells known as fibrocytes. Mature fibrocytes are inactive and compose the cellular portion of tendons. The function of the tendon is to attach muscles to bones and other parts.

The meniscus is skeletal system fibrocartilage-like tissue. It is a type of cartilage found in selected joints, which are subjected to high levels of force. Meniscal tissue originates from mesenchymal cells which differentiate into chondroblast cells. These chondroblast cells then further differentiate into specialized cells known as chondrocytes. When chondrocytes mature, they are inactive and comprise the menisci and other forms of cartilage. The function of the meniscus is to absorb shock by cushioning and distributing forces evenly throughout a joint, and provide a smooth articulating surface for the cartilages of the adjoining bones.

In the knee, the menisci form an interarticular fibrocartilage base for femoral and tibial articulation. The menisci form a crescent shape in the knee. The lateral meniscus is located on the outer side of the knee, and the medial

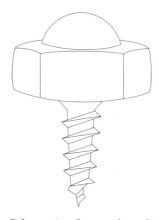

Fig. 9.   Schematic of an orthopedic anchor.

meniscus is located on the inside of the knee. If the knee bends and twists the menisci can overstretch and tear. Menisci tears occur frequently and the knee can sustain more than one tear at a time. If not treated appropriately, however, a menisci tear can roughen the cartilage and lead to arthritis. A meniscus tear acts like grit in the ball bearings of a machine. The longer the torn tissue remains affected, the more irritation it causes.

Meniscus surgery can repair or remove the torn cartilage, depending on the nature of the tear. Arthroscopy, a procedure through small skin incisions to visualize and repair the affected joint, is frequently employed. This procedure is sometimes performed on an out-patient basis. If repair of the menisci is not possible the surgeon removes as little of the meniscus as possible.

The body's frame or skeleton is constructed as a set of levers powered or operated by muscle tissue. A typical muscle consists of a central fibrous tissue portion, and tendons at either end. One end of the muscle, known as the head, is attached to tendon tissue, which is attached to bone that is fixed, and known as the point of origin. The other end of the muscle is attached to a tendon. This tendon is attached to bone that is the moving part of the joint. This end of the muscle is known as the insertion end. An example is the bicep muscle which is connected to the humerus bone of the upper arm at its head or origin. The insertion end of the muscle is connected to the radius bone of the forearm, otherwise known as the moving part of the elbow joint.

Muscle tissue is unique in its ability to shorten or contract. The human body has three basic types of muscle tissue histologically classified into smooth, striated, and cardiac muscle tissues. Only the striated muscle tissue is found in all skeletal muscles. The type of cells which compose the muscle tissue are known as contractile cells. They originate from mesenchymal cells which differentiate into myoblasts. Myoblasts are embryonic cells which later differentiate into contractile fiber cells.

The human body has more than 600 muscles. The body's movement is performed by muscle contractions, which are stimulated by the nervous system. This system links muscle tissue to the spinal cord and brain. The network of nerve cells which carries the brain's signals directs the flow of muscular energy. Most muscular activity occurs beyond the range of the conscious mind. The body, working through the neuromuscular network, manages its own motion.

Typically, in order for motion to occur, several muscle sets must work together to perform even the simplest movements. The bicep is a two-muscle set; the tricep is a three-muscle set. Each set works in tandem. Within each muscle group, muscle fibers obey the all or none principle, ie, all muscle fibers contract or none contract. Therefore, if the muscle fibers of a muscle group are stimulated enough by nerve impulses to contract, they contract to the maximum.

Bones function as levers; joints function as fulcrums; muscle tissue, attached to the bones via tendons, exert force by converting electrochemical energy (nerve impulses) into tension and contraction, thereby facilitating motion. Muscle tissue works only by becoming shorter. It shortens and then rests, ie, a muscle can only pull, it cannot push. Muscles produce large amounts of heat as they perform work. Involuntary contraction of muscle tissue releases chemical energy. This energy produces heat which warms the body, an action known as shivering.

Soft Tissue Injuries.   Some of the more common soft tissue injuries are sprains, strains, contusions, tendonitis, bursitis, and stress injuries, caused by damaged tendons, muscles, and ligaments. A sprain is a soft tissue injury to the ligaments. Certain sprains are often associated with small fractures. This type of injury is normally associated with a localized trauma event. The severity of the sprain depends on how much of the ligament is torn and to what extent the ligament is detached from the bone. The areas of the human body that are most vulnerable to sprains are ankles, knees, and wrists. A sprained ankle is the most frequent injury. The recommended treatment for a simple sprain is usually rest, ice, compression, and elevation (RICE). If a ligament is torn, however, surgery may be required to repair the injury.

A strain is the result of an injury to either the muscle or a tendon, usually in the foot or leg. Strain is a soft tissue injury resulting from excessive use, violent contraction, or excessive forcible stretch. The biomechanical description is of material failure resulting from force being applied to an area causing excessive tension, compression, or shear stress loading, leading to structural tissue distortion and the constant release of energy. The strain may be a simple stretch in muscle or tendon tissue, or it may be a partial or complete tear in the muscle and tendon combination. The recommended treatment for a strain is also RICE, usually to be followed by simple exercise to relieve pain and restore mobility. A serious tear may need surgical repair.

A contusion is an injury to soft tissue in which the skin is not penetrated, but swelling of broken blood vessels causes a bruise. The bruise is caused by a blow of excessive force to muscle, tendon, or ligament tissue. A bruise, also known as a hematoma, is caused when blood coagulates around the injury causing swelling and discoloring skin. Most contusions are mild and respond well to rest, ice, compression, and elevation of the injured area.

Tendonitis, an inflammation in the tendon or in the tendon covering, is usually caused by a series of small stresses that repeatedly aggravate the tendon, preventing it from healing properly, rather than from a single injury. Orthopedic surgeons treat tendonitis by prescribing rest to eliminate the biomechanical tissue stress, and possibly by prescribing antiinflammatory medications, such as steroids (qv). Specially chosen exercises correct muscle imbalances and help to restore flexibility. Continuous stress on an inflamed tendon occasionally causes it to rupture. This usually necessitates casting or even surgery to reattach the ruptured tendon.

A bursa, a sac filled with fluid located around a principal joint, is lined with a synovial membrane and contains synovial fluid. This fluid minimizes friction between the tendon and the bone, or between tendon and ligament. Repeated small stresses and overuse can cause the bursa in the shoulder, hip, knee, or ankle to swell. This swelling and irritation is referred to as bursitis. Some patients experience bursitis in association with tendonitis. Bursitis can usually be relieved by rest and in some cases by using antiinflammatory medications. Some orthopedic surgeons also inject the bursa with additional medication to reduce the inflammation.

Bone Fractures.   A dislocation occurs when sudden pressure or force pulls a bone out of its socket at the joint. This is also known as subluxation. Bone fractures are classified into two categories: simple fractures and compound, complex, or open fractures. In the latter the skin is pierced and the flesh and bone are exposed to infection. A bone fracture begins to heal nearly as soon as it occurs. Therefore, it is important for a bone fracture to be set accurately as soon as possible.

In certain diseases, such as osteomalacia, syphilis, and osteomyelitis, bones break spontaneously and without a trauma. The severity of the fracture usually depends on the force that caused the fracture. If a bone's breaking point was exceeded only slightly, then the bone may crack rather than break all the way through. If the force is extreme, such as in an automobile collision or a gunshot, the bone may shatter. An open or compound fracture is particularly serious because infection is possible in both the wound and the bone. A serious bone infection can result in amputation.

Stress fractures occur when microfractures accumulate because muscle tissue becomes fatigued and no longer protects from shock or impact. These heal, however, if given adequate rest. Without proper rest, unprotected bone beomes fatigued from absorbing the stress which is normally absorbed by muscle. Isolated microfractures become larger and then join together, forming a continuous stress fracture. These are often referred to as fatigue fractures. Stress fractures were first referred to as march fractures.

Stress or fatigue fractures are very painful. Most often symptoms occur after athletic activity or physical exertion. Gradually pain worsens and becomes more constant. Stress fractures do not show up on standard x-rays. A bone scan may be used to confirm the diagnosis. Stress fractures usually occur in the weight-bearing bones of the lower leg and foot. Stress fractures of the tibia account for half of all stress fractures, resulting mostly from athletic activity. These stress fractures are often mistaken for shin splints. In addition to the tibia, the fibula and other small bones of the foot are prone to stress fractures.

Fracture Treatment.   The movement of a broken bone must be controlled because moving a broken or dislocated bone causes additional damage to the bone, nearby blood vessels, and nerves or other tissues surrounding the bone. Indeed, emergency treatment requires splinting or bracing a fracture injury before further medical treatment is given. Typically, x-rays determine whether there is a fracture, and if so, of what type (see MEDICAL IMAGING TECHNOLOGY). If there is a fracture, a doctor reduces it by restoring the parts of the broken bone to their original positions. All treatment forms for fractures follow one basic rule: the broken pieces must be repositioned and prevented from moving out of place until healed. Broken bone ends heal by growing back together, ie, new bone cells form around the edge of the broken pieces. Specific bone fracture treatment depends on the severity of the break and the bone involved, ie, a broken bone in the spine is treated differently from a broken rib or a bone in the arm.

Treatments used for various types of fractures are cast immobilization, traction, and internal fixation. A plaster or fiber glass cast is the most commonly used device for fracture treatment. Most broken bones heal successfully once properly repositioned, ie, fixed in place via a cast. This type of cast or brace is known as an orthosis. It allows limited or controlled movement of nearby joints. This treatment is desirable for certain fractures.

Traction is typically used to align a bone by a gentle, constant pulling action. The pulling force may be transmitted to the bone through skin tapes or a metal pin through a bone. Traction may be used as a preliminary treatment, before other forms of treatment or after cast immobilization.

In internal fixation, an orthopedist performs surgery on the bone. During this procedure, the bone fragments are repositioned (reduced) into their normal alignment and then held together with special screws or by attaching metal plates to the outer surface of the bone. The fragments may also be held together by inserting rods (intramedullary rods) down through the marrow space into the center of the bone. These methods of treatment can reposition the fracture fragments very exactly. A common internal fixation procedure is to surgically fix the femoral neck (broken hip), as shown in Figure 10.

Joint Replacement.   The most frequent reason for performing a total joint replacement is to relieve the pain and disability caused by severe arthritis. The surface of the joint may be damaged by osteoarthritis, ie, a wearing away of the cartilage in a joint. The joint may also be damaged by rheumatoid arthritis, an autoimmune disease, in which the synovium produces chemical substances that attack the joint surface and destroy the cartilage. The swelling, heat, and stiffness that occur in an arthritic joint cause inflammation, the body's natural reaction to disease or injury. Inflammation is usually temporary, but in arthritic joints it is long-lasting and causes disability. When arthritis has caused severe damage to a joint, a total joint replacement may allow the person to return to normal everyday activities.

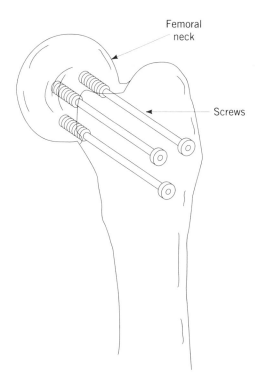

Fig. 10.   Internal femur fixation.

A total joint replacement is a radical surgical procedure performed under general anesthesia, in which the surgeon replaces the damaged parts of the joint with artificial materials. For example, in the knee joint the damaged ends of the bone that meet at the knee are replaced, along with the underside of the kneecap. In the hip joint, the damaged femoral head is replaced by a metal ball having a stem that fits down into the femur. A new plastic socket is implanted into the pelvis to replace the old damaged socket. This is shown schematically in Figure 11. Whereas hips and knees are the joints most frequently replaced, because the scientific understanding of these is best, total joint replacement can be performed on other joints as well, including the ankle, shoulder, fingers, and elbow.

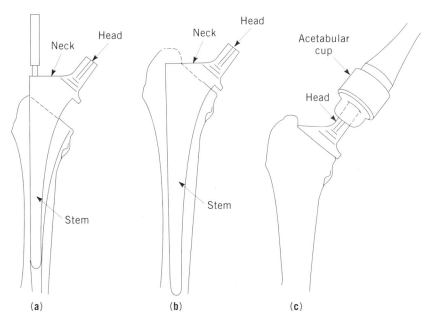

Fig. 11. Three views of a hip implant for joint replacement: (**a**) insertion of the implant into the femur; (**b**) implant in place; and (**c**) femur and implant connected to plastic socket fitted into the pelvis.

The materials used in a total joint replacement are designed to enable the joint to function normally. The artificial components are generally composed of a metal piece that fits closely into bone tissue. The metals are varied and include stainless steel or alloys of cobalt, chrome, and titanium. The plastic material used in implants is a polyethylene that is extremely durable and wear-resistant. Also, a bone cement, a methacrylate, is often used to anchor the artificial joint materials into the bone. Cementless joint replacements have more recently been developed. In these replacements, the prosthesis and the bone are made to fit together without the need for bone cement. The implants are press-fit into the bone.

The recovery period following total joint arthropathy depends on both the patient and the affected joint. In general the patient is encouraged to use the joint soon after replacement. In the case of a hip or knee replacement the patient should be standing and beginning to walk within several days. If the shoulder, elbow, or wrist joint is replaced, use of the new joint can begin very soon after surgery, as these are not weight-bearing joints. The patient generally performs appropriate exercises to strengthen and move the joint during the recovery period.

The main benefits to the patient after total joint replacement are pain relief, which often is quite dramatic, and increased muscle power, which was lost because the painful arthritic joint was not used and usually returns with exercise once pain is relieved. Motion of the joint generally improves as well. The extent of movement depends on how stiff the joint was before the joint was replaced. An extremely stiff joint continues to be stiff for some period of time after replacement.

The principal complication for total joint replacement is infection, which may occur just in the area of the incision or more seriously deep around the prosthesis. Infections in the wound area, which may even occur years after the procedure has been performed, are usually treated with antibiotics (qv). Deep infections may require further surgery, prosthesis removal, and replacement.

Loosening of the prosthesis is the most common biomechanical problem occurring after total joint replacement surgery. Loosening causes pain. If loosening is significant, a second or revision total joint replacement may be necessary. Another complication which sometimes occurs after total joint replacement, generally right after the operation, is dislocation, the result of weakened ligaments. In most cases the dislocation can be relocated manually by the orthopedic surgeon. Very rarely is another operation necessary. A brace may be worn after dislocation occurs for a short time. Although some wear can be measured in artificial joints, wear occurs slowly. Whereas wear may contribute to looseness, it is rarely necessary to do corrective surgery because of wear alone.

Breakage of an implanted joint is rare. Breakage occurs when the bone flexes and the metal implant does not flex as much, thereby exceeding its mechanical fatigue point causing the implant to break or crack. A revision joint replacement operation is necessary if breakage occurs.

Nerves are rarely damaged during the total joint replacement surgery. However, nerve damage can occur if considerable joint deformity must be corrected in order to implant the prosthesis. With time these nerves sometimes return to normal function.

Osteoarthritis, the most common arthritic disorder, affects some 30 million Americans each year. Caused by daily wear and tear on joints or injury, osteoarthritis is painful and restricts daily activity. It can affect the basal joint of the thumb, as well as the knee, hip, and other joints.

Hip Joints. Successful hip joint replacement surgery was introduced in the late 1950s. Since that time design and scientific advances have brought increasingly better clinical results. In excess of 200,000 patients in the United States seek pain relief annually through hip joint replacement. About 18–20%

are revision hip systems, ie, second replacement implants. A hip usually becomes painful when the cartilage that lines the hip socket starts to wear out. As total hip systems evolved, designers attempted to eliminate features which led to failure. Modifications were made to each element of the hip system including the femoral stem and acetabular cup. Wear and tear arthritis may be the result of a genetic defect that prevents the body from manufacturing cartilage rugged enough to last a lifetime. Increased life-expectancy, stresses owing to certain occupations, and prior injury that places abnormal stress on cartilage over a long period may also contribute to the development of osteoarthritis. Often arthritic pain can be controlled through the use of antiinflammatory medication. However, if hip pain becomes intolerable, hip replacement surgery may be elected.

In 1974 a prosthesis introduced by Howmedica combined a biomechanically high strength material, Vitallium, with a professionally engineered geometry. This prosthesis marked the first design departure from the diamond-shaped cross-sectional geometry previously used. Sharp corners were eliminated and replaced by broad, rounded medial and lateral borders. The total sectional area was much greater than any of the previous hip joint implant stems. The result of these combined factors was decreased unit stresses on the cement mantle. This system also marked the first time surgeons could choose components from a selection large enough to provide fit for most primary and revision total hip replacement patients.

The next advance in total hip arthroplasty came with the development of various porous surface treatments which allow bone tissue to grow into the metal porous coating on the femoral stem of the hip implant and on the acetabular component of the total joint replacement. These developments arose because of patients who were not able to tolerate cemented implants because of allergies to the cement, methylmethacrylate. More youthful patients are better served by a press-fit implant as well. Figure 12 shows the difference between textured and beaded surface-treated orthopedic prostheses.

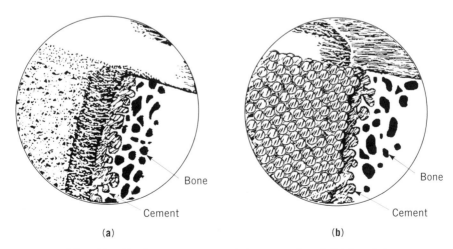

(a)   (b)

Fig. 12. Surface treatments: (**a**) textured and (**b**) beaded.

Hydroxyapatite (HA) coating on the surface of the hip stem and the acetabular cup is the most recent advancement in artificial hip joint implant technology. This substance is a form of calcium phosphate, which is sprayed onto the hip implant. It is a material found in combination with calcium carbonate in bone tissue, and bones can easily adapt to it. When bone tissue does grow into HA, the tissue then fixes the hip joint implant permanently in position. These HA coatings are only used in press-fit, noncemented implants.

The acetabular component is as integral to successful total hip arthroplasty as is the femoral hip stem component. The life of the acetabular component depends on proper placement and bone preparation in the acetabular region of the hip girdle, proper use of bone cement, and superior component design.

The history of the development of the acetabular component parallels that of the femoral component. In 1951 an acetabular prosthesis based on a chromium–cobalt alloy having screw-in sockets was successfully introduced. Beginning in 1955, methylmethacrylate was used as a cementing agent, and work was undertaken to find suitable materials for use as an articulation surface in the acetabular. Teflon appeared to provide a good lubricating, articulating surface. However, over time Teflon exhibited poor wear conditions when it contacted the metal femoral head, and as the body developed systemic reaction to Teflon particles, failures occurred. Since then ultrahigh density polyethylene has proven successful as acetabular cups.

In 1971 a metal-backed polyethylene acetabular cup was introduced. This cup provided an eccentric socket which was replaceable, leaving the metal and replacing only the polyethylene. Because of the success of this component, metal-backed high density polyethylene (HDPE) liner is standard for prosthetic acetabular components. Research confirms that metal-backing reduces the peak stresses in the bone cement, and that HDPE forms a successful articulating surface for the prosthetic joint.

Over time a large variety of materials have been used, including ivory, stainless steel, chromium–cobalt, and ceramics for the acetabular component. None proved sufficient. The implant material composition must provide a smooth surface for joint articulation, withstand hip joint stresses from normal loads, and the substance must disperse stress evenly to the cement and surrounding bone.

The material in use as of the mid-1990s in these components is HDPE, a linear polymer which is tough, resilient, ductile, wear resistant, and has low friction (see POLYOLEFINS, POLYETHYLENE). Polymers are prone to both creep and fatigue (stress) cracking. Moreover, HDPE has a modulus of elasticity that is only one-tenth that of the bone, thus it increases the level of stress transmitted to the cement, thereby increasing the potential for cement mantle failure. When the acetabular HDPE cup is backed by metal, it stiffens the HDPE cup. This results in function similar to that of natural subchondral bone. Metal backing has become standard on acetabular cups.

The femoral component is composed of the head, neck, collar, and stem (see Fig. 11). The head, or ball, is the surface component which articulates with the acetabular cup of the total hip implant. This is an important element in the implant design because this surface absorbs the greatest stress and has the most force applied to it. Consequently the head gets the greatest wear. The diameter of the femoral head affects the distribution of forces in both the femoral and acetabular components. This variable also influences the range of motion that the implant permits, therefore affecting the stability of the ball-in-socket prosthetic joint. The most common head diameters range from 22 to 32 mm. Each size offers advantages and disadvantages. No general consensus exists as to which size is better. As a result, many manufacturers offer more than one head diameter to suit surgeon preference and patient requirements.

The 22-mm diameter is preferred by some doctors who believe that a small-diameter head encourages mechanical fixation of the socket without cement fixation. Other benefits associated with the 22-mm head size include its suitability for use in a patient having a small acetabulum, and the fact that it allows for a thicker acetabular component, which permits more wear and absorbs more energy than do thinner walled components. The larger (32-mm) head diameter range is recommended by other surgeons and has been incorporated into most hip systems. Using this large diameter the surface of the head has a greater area, resulting in decreased stress per unit area. Other advantages include less chance of subluxation (joint dislocation) at the extremes of motion and therefore improved prosthetic joint stability. However, some doctors express concern that the increased articulating surface contact area promotes increased frictional torque.

The stem/neck length and cross-sectional geometry of the neck affects the forces acting on both the neck and the stem. The neck must be large enough to prevent failure, but not so large as to limit range of motion. Further, the neck length should be consistent with the anatomy of the patient.

The most important design consideration in the neck of the femoral component is that it support body weight without breaking. This requires that the head/neck ratio be appropriate. Neck length is measured from the center of the femoral head to the collar of the stem. Variations to neck length, combined with offset neck/stem angle, and head diameter, permit surgeons to adjust leg length of the total hip implant leg to be consistent with that of the opposite leg. It was for this reason that surgeons introduced the concept of providing different neck lengths.

The stem/neck offset, ie, distance from the center of the head to the center of the stem, changes upon a change in the neck length. Increased neck length, and therefore increased offset, raises the bending moment of the stem, thereby increasing the chances for prosthetic failure. The various neck geometries used by designers of prostheses represent attempts to find a satisfactory combination of shape, bulk, and material which can withstand cyclic loading and joint forces without breaking. Most neck/stem angles are neutral (135 deg) in order to estimate average human anatomy and equalize the moment arms.

An important issue of stem design is length. Increased stem length means more stem area for improved stress distribution. Another benefit to a longer implant stem is engagement of the isthmus, the most narrow portion of the femur. Expanding the stem into this area of contact increases prosthetic stability, helps prevent the stem from shifting position, decreases the amount of micromotion, and achieves better alignment along the neutral axis of the femur. Stems are available in varied lengths to match human anatomy and improve isthmic engagement. The most advanced hip implants on the market are totally modular, so that they are nearly custom made to fit into the femur and the acetabulum of the pelvic girdle.

Starting from the bottom of the hip implant, a modular implant begins with a press-fit, distal, high density, ultrahigh molecular weight, polyethylene

(HDPE) plug tip, at the bottom of the femur stem. Then a machined and polished titanium, chromium–cobalt–molybdenum, or vanadium–aluminum metallic alloy diaphyseal–endosteal grooved stem segment is added, followed by a machined, polished, and hydroxyapatite-coated metaphyseal metallic alloy stem segment, the upper portion of the hip implant stem. A custom-fit metallic alloy collar plate, which rests upon the resected (cut-away) head of the femur, comes next, followed by a sized modular metallic alloy neck upon which a chromium–cobalt–molybdenum or zirconium ceramic head, ie, ball that sets upon the hip stem neck, rests. The head then articulates with the acetabular cup liner (see Fig. 11).

The head of the femoral component then articulates with an ion-bombarded, HDPE, high walled, acetabular liner which fits into a screwed in, machined, titanium, chromium–cobalt–molybdenum or vanadium–aluminum metallic alloy hydroxyapatite-coated acetabular shell/cup. Each of the separate parts of the modular system for total hip arthroplasty is manufactured in several different sizes.

Total hip implants of the nature described have hospital list prices in the range of $5000–$8000. Fully custom-made implants cost approximately $10,000. The low end basic total hip implant is forged or cast stainless steel, cemented in place, one size fits all, and costs $1000.

Prosthesis Design. The challenge in prosthesis design is to create an implant that mimics the material characteristics and the exact anatomical functions of the joint. Stress and loading forces on the hip joint and femur are extraordinary. Stresses on the hip joint exceed 8.3 MPa (1200 psi). Standing on one leg produces a loading force on the hip joint of 250% of the total body weight. Running increases these forces to five times the total body weight. The hip joint is surrounded by the most powerful muscle structure in the body enabling movement while supporting sufficient structural force and loads. Proper surgical technique is as critical to the success of an implant procedure as is the design of the device itself. Therefore, matching surgeon skill, an appropriate implant design for the patient, and the correct tools generally forms the best solution for a successful procedure.

A significant aspect of hip joint biomechanics is that the structural components are not normally subjected to constant loads. Rather, this joint is subject to unique compressive, torsion, tensile, and shear stress, sometimes simultaneously. Maximum loading occurs when the heel strikes down and the toe pushes off in walking. When an implant is in place its ability to withstand this repetitive loading is called its fatigue strength. If an implant is placed properly, its load is shared in an anatomically correct fashion with the bone.

Design variables introduced on various prostheses represent efforts to share the stresses and normal loading characteristics of human locomotion. The size, shape, and tissue structure of bone are most commonly affected in the healing of fractures. Bone remodeling was first described in 1892 by the German physician Julius Wolff in *The Law of Bone Transformation*. In terms of force loading and stresses, Wolff's law states that bone responds to mechanical demends by changing its size, shape, and structure.

Resorption of bone tissue occurs in total hip joint replacement patients if sufficient stresses are not adequately transmitted to the remaining bone in exactly the same way that the bone transmitted those stresses originally.

Therefore, the design and proper placement of the neck collar and hip stem must be effective in recreating anatomical structure.

Bone remodeling is the ability of bone to change its size, shape, and structure by adapting to mechanical demands that are placed on it. Bone grows where it is needed, and resorbs where it is not needed. The type of bone tissue that grows depends on the stress level it sustains. Someone who performs strenuous exercise undergoes cortical bone changes resulting in bigger, denser bones. On the other hand, someone who performs minimal physical activity loses bone density through resorption. This is a problem for those who must stay in bed for prolonged periods. The problem of bone density loss owing to minimal physical stress has also been a unique concern for NASA astronauts. Special exercises have been designed for the astronauts to counteract the effects of weightlessness and to slow down bone resorption during long orbital flights.

The process of aging reduces bone size and strength. Thinning and resorption occur in the cancellous bone. Also, cortical bone resorbs and bone shrinks in diameter and thickness. The older the person, the more fragile the bone.

Research based on Wolff's law of bone transformation has resulted in some other important observations. Fluctuating loads, such as those that occur in walking, are better for bone than consistently applied loads, such as weight gain. However, if the effective applied load becomes extreme, pressure necrosis, ie, bone death, occurs. Pressure necrosis is a significant concern in hip arthroplasty. Necrosis means the localized death of living tissue. Undue pressure on living cells causes death. Some total hip replacement failures are the direct result of pressure necrosis.

Some of the early design hip prostheses, created without a complete understanding of stress forces and anatomical loading characteristics of normal activity, had sharp points at the distal end and along the medial and lateral sides of the stem. Improperly seated, or merely subjected to normal forces, these stems directed concentrated stresses into the interfacing cement. This point loading resulted in cement fracturing into fragments and bone tissue suffering pressure necrosis, resulting in implant failure. More rounded prosthetic designs, and the tools and instrumentation to properly seat them, distribute the load over the widest area. This distribution mimics that of natural bone and prevents pressure necrosis.

Biomaterials.   Just as stem designs have evolved in an effort to develop an optimal combination of specifications, so have the types of metals and alloys employed in the construction of total joint implants. Pure metals are usually too soft to be used in prosthesis. Therefore, alloys which exhibit improved characteristics of fatigue strength, tensile strength, ductility, modulus of elasticity, hardness, resistance to corrosion, and biocompatibility are used.

Titanium alloy, composed of titanium, aluminum, and vanadium, is preferred by some orthopedic surgeons primarily for its low modulus of elasticity, which allows for transfer of more stress to the proximal femur. This alloy also exhibits good mechanical strength and biocompatibility (11). The stem flexibility optimizes the transfer of stress directly to the bone, and offers adequate calcar loading to minimize femoral resorption.

Vitallium FHS alloy is a cobalt–chromium–molybdenum alloy having a high modulus of elasticity. This alloy is also a preferred material. When combined with a properly designed stem, the properties of this alloy provide protection for the cement mantle by decreasing proximal cement stress. This alloy also exhibits high yields and tensile strength, is corrosion resistant, and biocompatible. Composites used in orthopedics include carbon–carbon, carbon–epoxy, hydroxyapatite, ceramics, etc.

Tools and Procedures. Arthroscopy is a surgical procedure used to visualize, diagnose, and treat injuries within joints. The term arthroscopy literally means to look inside the joint. During this procedure the orthopedic surgeon makes an incision into the patients skin and inserts a pencil-shaped arthroscope. An arthroscope is a miniature lens and lighting systems that magnifies and illuminates the structures inside the joint. A television screen which is attached to the arthroscope displays the image of the joint on screen.

This is a minimally invasive procedure (MIP) resulting in a shorter hospital stay, faster recovery, and less evident scar in comparison to other types of surgery. Arthroscopic surgery gives the surgeon a precise, direct view of the affected bones and soft tissues. This procedure allows the surgeon to see areas of the joint that are difficult to see on x-rays and more of the joint than is possible even after making a large incision during open surgery. Arthroscopy can be performed under local, general, or spinal anesthesia. The area surrounding the joint is sterilized, and then the joint is expanded to make room for the arthroscope by injecting a sterile solution into the joint. The surgeon makes a small incision into the skin through which the arthroscope is inserted. A surgical instrument probes various parts of the joints to determine the injury. Surgical repair, if needed, is performed using specially designed surgical instruments which are inserted into the joint through the small incisions. This surgery can be viewed on a television screen. The small incisions that were made during surgery are closed usually using only one or two sutures.

Patients' immediate post-operative pain is lower compared to a standard operation and healing and rehabilitation more rapid. Patients can resume near-normal activities in just days. In some cases athletes, who are in prime physical condition, can return to challenging athletic activities within a few weeks. Complications are rare, but do occur on occasion. Most complications associated with this surgery are infection, phlebitis, excessive swelling or bleeding, blood clots, or damage to blood vessels or nerves.

Bioresorbable Polymers

Biomaterials scientists have worked diligently to synthesize polymeric structures which exhibit biocompatibility and long-term biostability. Devices made from these polymers are intended to be implanted in the body for years, and in some cases decades.

The concept of using biodegradable materials for implants which serve a temporary function is a relatively new one. This concept has gained acceptance as it has been realized that an implanted material does not have to be inert, but can be degraded and/or metabolized *in vivo* once its function has been accomplished (12). Resorbable polymers have been utilized successfully in the manufacture of sutures, small bone fixation devices (13), and drug delivery systems (qv) (14).

Several groups have experimented with bioresorbable polymers that have a predictable degree of bioresorbability when exposed to the physiological environment. By the judicious choice of bioresorbability rate it is hoped that as the polymer is resorbed it will leave surface voids where natural tissue would grow, resulting in autologous organ regeneration. The temporary nature of the device will impart initial mechanical functionality to the implant, but after time will be resorbed as the natural tissue regenerates. This concept has been experimentally applied to the regeneration of tissue such as in the liver (15), skeletal tissue (16), cartilage (17), and the vascular wall (18).

One area in which predictable biodegradation is used is the area of degradable surgical sutures. An incision wound, when held together with sutures, heals to about 80% of initial strength within four weeks. Surgical suture is one of the earliest clinical implants in recorded history. Catgut suture, obtained from ovine or bovine intestinal submucosa, was known in 150 AD in the time of Galen, who built his reputation by treating wounded gladiators (19).

Catgut is infection-resistant. The biodegradation of catgut results in elimination of foreign material that otherwise could serve as a nidus for infection or, in the urinary tract, calcification. As a result, chromic catgut, which uses chromic acid as a cross-linking agent, is still preferred in some procedures. Chromic catgut is considered by some to be the most suitable suture material for vaginal hysterectomy owing to its extensibility and rapid absorption. Treatment of natural catgut with synthetic polymers exemplifies the merging of old and new technology. Coating catgut with a polyurethane resin allows catgut to retain its initial tensile strength longer (20).

The first synthetic polyglycolic acid suture was introduced in 1970 with great success (21). This is because synthetic polymers are preferable to natural polymers since greater control over uniformity and mechanical properties are obtainable. The foreign body response to synthetic polymer absorption generally is quite predictable whereas catgut absorption is variable and usually produces a more intense inflammatory reaction (22). This greater tissue compatibility is crucial when the implant must serve as an inert, mechanical device prior to bioresorption.

Polylactic Acid.   Polylactic acid (PLA) was introduced in 1966 for degradable surgical implants. Hydrolysis yields lactic acid, a normal intermediate of carbohydrate metabolism (23). Polyglycolic acid sutures have a predictable degradation rate which coincides with the healing sequence of natural tissues.

Polylactic acid, also known as polylactide, is prepared from the cyclic diester of lactic acid (lactide) by ring-opening addition polymerization, as shown below:

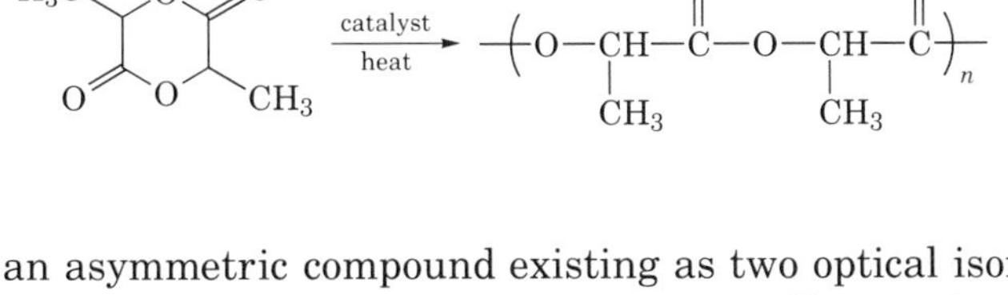

$$\text{(lactide)} \xrightarrow[\text{heat}]{\text{catalyst}} \left(\!\!\begin{array}{c}\text{O}\\\parallel\\\text{O}-\text{CH}-\text{C}-\text{O}-\text{CH}-\text{C}\\\\\text{CH}_3\quad\quad\text{CH}_3\end{array}\!\!\right)_{\!n}$$

Lactic acid is an asymmetric compound existing as two optical isomers or enantiomers. The L-enantiomer occurs in nature; an optically inactive racemic mixture of D- and L-enantiomers results during synthesis of lactic acid. Using these two types of lactic acid, the corresponding L-lactide (mp 96°C) and DL-lactide (mp 126°C) have been used for polymer synthesis. Fibers spun from poly-L-lactide (mp 170°C) have high crystallinity when drawn, whereas poly-DL-lactide (mp 60°C) fibers display molecular alignment on drawing but remain amorphous. The crystalline poly-L-lactide is more resistant to hydrolytic degradation than the amorphous DL form of the same homopolymer. Therefore, pure DL-lactide displays greater bioresorbability, whereas pure poly-L-lactide is more hydrolytically resistant.

The actual time required for poly-L-lactide implants to be completely absorbed is relatively long, and depends on polymer purity, processing conditions, implant site, and physical dimensions of the implant. For instance, 50–90 mg samples of radiolabeled poly-DL-lactide implanted in the abdominal walls of rats had an absorption time of 1.5 years with metabolism resulting primarily from respiratory excretion (24). In contrast, pure poly-L-lactide bone plates attached to sheep femora showed mechanical deterioration, but little evidence of significant mass loss even after four years (25).

Improved techniques for polylactide synthesis have resulted in preparation of exceptionally high molecular weight polymer. Fiber processing research has resulted in fiber samples having tensile breaking strength approaching 1.2 GPa (174,000 psi). This strength value was obtained by hot-drawing filaments spun from good solvents (26).

Polyglycolic Acid. Polyglycolic acid (PGA), also known as polyglycolide, was first reported in 1893, but it wasn't until 1967 that the first commercially successful patent was granted for sutures (27). Like polylactide, polyglycolide is synthesized from the cyclic diester as shown below:

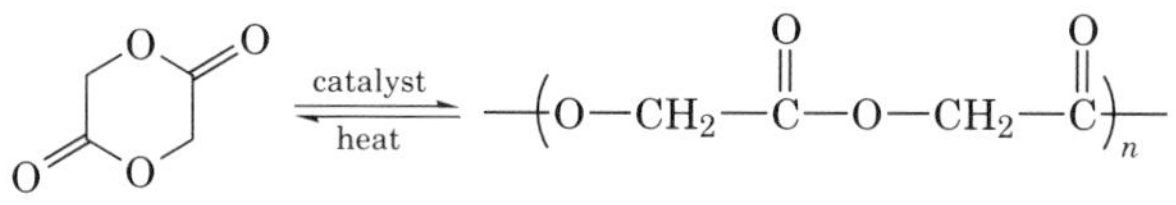

An important difference between polylactide and polyglycolide, is that polyglycolide (mp 220°C) is higher melting than poly-L-lactide (mp 170°C). Although the polymerization reaction in both cases is reversible at high temperature, melt processing of polyglycolide is more difficult because the melting temperature is close to its decomposition temperature.

Unlike poly-L-lactide which is absorbed slowly, polyglycolide is absorbed within a few months post-implantation owing to greater hydrolytic susceptibility. *In vitro* experiments have shown the effect on degradation by enzymes (28), pH, annealing treatments (29), and gamma irradiation (30). Braided polyglycolide sutures undergo surprisingly rapid hydrolysis *in vivo* owing to cellular enzymes released during the acute inflammatory response following implantation (31).

Low humidity ethylene oxide gas sterilization procedures and moisture-proof packaging for polyglycolic acid products are necessary because of the susceptibility to degradation resulting from exposure to moisture and gamma sterilization.

Poly(lactide-*co*-glycolide). Mixtures of lactide and glycolide monomers have been copolymerized in an effort to extend the range of polymer properties and rates of *in vivo* absorption. Poly(lactide-*co*-glycolide) polymers undergo a simple hydrolysis degradation mechanism, which is sensitive to both pH and the presence of enzymes (32).

A 90% glycolide, 10% L-lactide copolymer was the first successful clinical material of this type. Braided absorbable suture made from this copolymer is similar to pure polyglycolide suture. Both were absorbed between 90 and 120 days post-implantation but the copolymer retained strength slightly longer and was absorbed sooner than polyglycolide (33). These differences in absorption rate result from differences in polymer morphology. The amorphous regions of poly(lactide-*co*-glycolide) are more susceptible to hydrolytic attack than the crystalline regions (34).

Similar to pure polyglycolic acid and pure polylactic acid, the 90:10 glycolide:lactide copolymer is also weakened by gamma irradiation. The normal *in vivo* absorption time of about 70 days for fibrous material can be decreased to less than about 28 days by simple exposure to gamma radiation in excess of 50 kGy (5 Mrads) (35).

The crystallinity of poly(lactide-*co*-glycolide) samples has been studied (36). These copolymers are amorphous between the compositional range of 25–70 mol % glycolide. Pure polyglycolide was found to be about 50% crystalline whereas pure poly-L-lactide was about 37% crystalline. An amorphous poly(L-lactide-*co*-glycolide) copolymer is used in surgical clips and staples (37). The preferred composition chosen for manufacture of clips and staples is the 70/30 L-lactide/glycolide copolymer.

Polydioxanone. Fibers made from polymers containing a high percentage of polyglycolide are considered too stiff for monofilament suture and thus are available only in braided form above the microsuture size range. The first clinically tested monofilament synthetic absorbable suture was made from polydioxanone (38). This polymer is another example of a ring-opening polymerization reaction. The monomer, *p*-dioxanone, is analogous to glycolide but yields a poly-(ether–ester) as shown below:

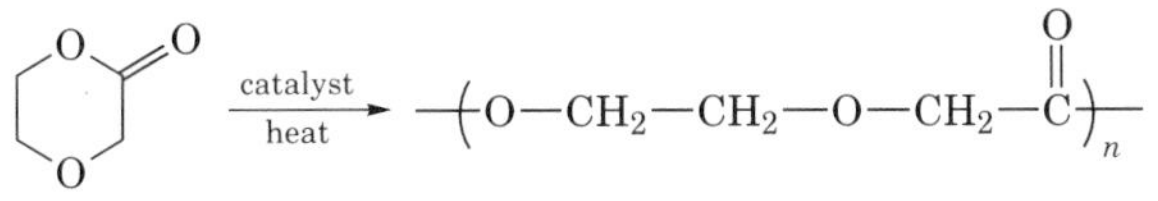

Polydioxanone (PDS) is completely eliminated from the body upon absorption. The mechanism of polydioxanone degradation is similar to that observed for other synthetic bioabsorbable polymers. Polydioxanone degradation *in vitro* was affected by gamma irradiation dosage but not substantially by the presence of enzymes (39). The strength loss and absorption of braided PDS, but not monofilament PDS, implanted in infected wounds, however, was significantly greater than in noninfected wounds.

Poly(ethylene oxide)–Poly(ethylene terephthalate) Copolymers. The poly(ethylene oxide)–poly(ethylene terephthalate) (PEO/PET) copolymers were first described in 1954 (40). This group of polymers was developed in an attempt to simultaneously reduce the crystallinity of PET, and increase its hydrophilicity to improve dyeability. PEO/PET copolymers with increased PEO contents produce surfaces that approach zero interfacial energy between the implant and the adjacent biological tissue. The collagenous capsule formed around the implant is thinner as the PEO contents increase. The structure of a PEO/PET copolymer is shown below:

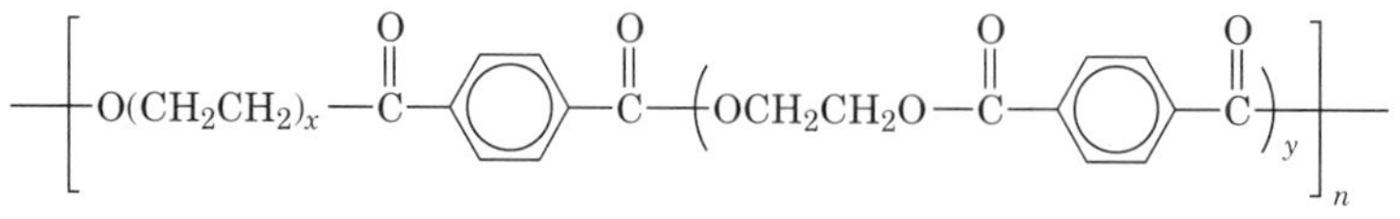

A family of PEO/PET copolymers has been synthesized and the characterized structures found to be close to those expected in theory (41). A wide degradation envelope has been achieved by adjusting the PEO-to-PET ratio. Mechanical properties prove useful for medical applications, and the 60/40 PEO/PET composition is reported as optimal.

Poly(glycolide-*co*-trimethylene carbonate).   Another successful approach to obtaining an absorbable polymer capable of producing flexible monofilaments has involved finding a new type of monomer for copolymerization with glycolide (42). Trimethylene carbonate polymerized with glycolide is shown below:

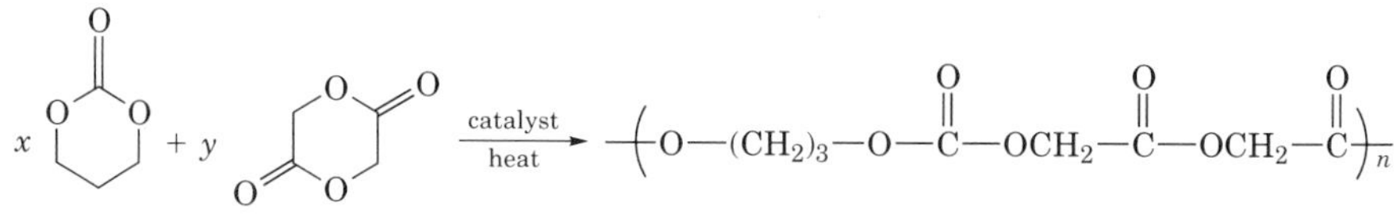

In order to achieve the desired fiber properties, the two monomers were copolymerized so the final product was a block copolymer of the ABA type, where A was pure polyglycolide and B, a random copolymer of mostly poly(trimethylene carbonate). The selected composition was about 30–40% poly(trimethylene carbonate). This suture reportedly has excellent flexibility and superior *in vivo* tensile strength retention compared to polyglycolide. It has been absorbed without adverse reaction in about seven months (43). Metabolism studies show that the route of excretion for the trimethylene carbonate moiety is somewhat different from the glycolate moiety. Most of the glycolate is excreted by urine whereas most of the carbonate is excreted by expired CO_2 and urine.

Poly(ethylene carbonate).   Like polyesters, polycarbonates (qv) are bioabsorbable only if the hydrolyzable linkages are accessible to enzymes and/or water molecules. Thus pellets of poly(ethylene carbonate), $+OCOOCH_2CH_2+_{\overline{n}}$, weighing 200 mg implanted in the peritoneal cavity of rats, were bioabsorbed in only two weeks, whereas similar pellets of poly(propylene carbonate), $+OCOOCH(CH_3)CH_2+_{\overline{n}}$, showed no evidence of bioabsorption after two months (44). Because poly(ethylene carbonate) hydrolyzes more rapidly *in vivo* than *in vitro*, enzyme-catalyzed hydrolysis is postulated as a contributing factor in polymer absorption. Copolymers of polyethylene and polypropylene carbonate have been developed as an approach to achieving the desired physical and pharmacological properties of microsphere drug delivery systems.

Polycaprolactone. Polycaprolactone is synthesized from epsilon-capro-lactone as shown below:

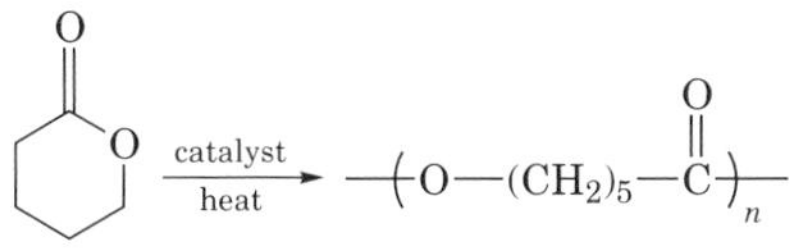

$$\text{catalyst, heat} \longrightarrow -\!\!\left(\!\!O\!-\!(CH_2)_5\!-\!\overset{\displaystyle O}{\overset{\|}{C}}\!\right)_{\!\!n}\!\!-$$

This semicrystalline polymer is absorbed very slowly *in vivo*, releasing ε-hydroxycaproic acid as the sole metabolite. Degradation occurs in two phases: nonenzymatic bulk hydrolysis of ester linkages followed by fragmentation, and release of oligomeric species. Polycaprolactone fragments ultimately are de-graded in the phagosomes of macrophages and giant cells, a process that in-volves lysosome-derived enzymes (45). *In vitro*, polycaprolactone degradation is enhanced by microbial and enzymatic activity. Predictably, amorphous regions of the polymer are degraded prior to breakdown of the crystalline regions (46).

Copolymers of ε-caprolactone and L-lactide are elastomeric when prepared from 25% ε-caprolactone and 75% L-lactide, and rigid when prepared from 10% ε-caprolactone and 90% L-lactide (47). Blends of poly-DL-lactide and polycapro-lactone polymers are another way to achieve unique elastomeric properties. Copolymers of ε-caprolactone and glycolide have been evaluated in fiber form as potential absorbable sutures. Strong, flexible monofilaments have been produced which maintain 11–37% of initial tensile strength after two weeks *in vivo* (48).

Poly(ester–amides). Another approach to obtaining improvements in the properties of synthetic absorbable polymers is the synthesis of poly-mers containing both ester and amide linkages. The rationale for designing poly(ester–amide) materials is to combine the absorbability of polyesters (qv) with the high performance of polyamides (qv). Two types have been reported. Both involve the polyesterification of diols that contain preformed amide link-ages. Poly(ester–amides) obtained from bis-oxamidodiols have been reported to be absorbable only when oxalic acid is used to form the ester linkages (9). Poly(ester–amides) obtained from bis-hydroxyacetamides are absorbable regard-less of the diacid employed, although succinic acid is preferred (50).

The absorption rate has been examined *in vivo* for a series of poly-(ester–amides) having the following formula:

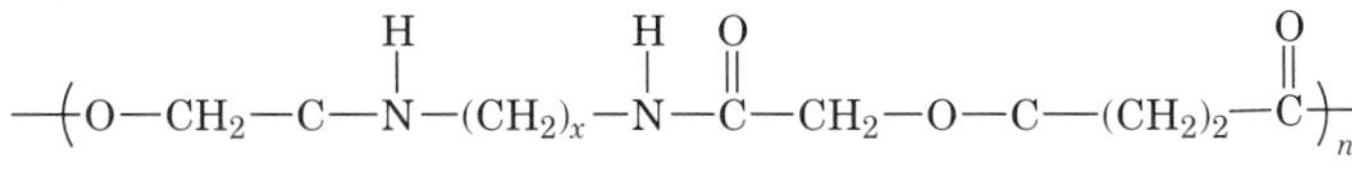

$$-\!\!\left(\!\!O\!-\!CH_2\!-\!C\!-\!\overset{\displaystyle H}{\overset{|}{N}}\!-\!(CH_2)_x\!-\!\overset{\displaystyle H}{\overset{|}{N}}\!-\!\overset{\displaystyle O}{\overset{\|}{C}}\!-\!CH_2\!-\!O\!-\!C\!-\!(CH_2)_2\!-\!\overset{\displaystyle O}{\overset{\|}{C}}\!\right)_{\!\!n}\!\!-$$

Polymers, where $x = 6, 8,$ and 10, are absorbed within six months; the poly-mer where $x = 12$ requires over 19 months for complete absorption. Absorption correlates with the water solubility of the starting amidediol monomers. All are at least sparingly soluble except for the $x = 12$ amidediol which is virtually in-soluble. *In vivo* strength retention of poly(ester–amides) in fiber form is greatest for the $x = 12$ polymer. This material loses very little strength for four weeks then slowly decreases to 50% strength at 8–10 weeks depending on molecular weight and fiber processing conditions.

The metabolic rate of poly(ester–amide) where $x = 6$ has been studied in rats using carbon-14 labeled polymer. This study indicates that polymer degradation occurs as a result of hydrolysis of the ester linkages whereas the amide linkages remain relatively stable *in vivo*. Most of the radioactivity is excreted by urine in the form of unchanged amidediol monomer, the polymer hydrolysis product (51).

Poly(orthoesters). The degradation of a bioresorbable polymer occurs in four stages: hydration, loss of strength, loss of integrity, and loss of mass. This typical behavior limits most of the previously mentioned polymers for use as matrices for slow release drug delivery implants because incorporated drugs that are water soluble have been found simply to leach out at a first-order rate. Thus bioabsorbable polymers which are extremely hydrophobic have been developed to prevent hydration yet still possess hydrolytically unstable linkages. This results in degradation of polymer on the exposed surfaces only thereby releasing the drug content at a more uniform rate. Such polymers have been termed bioerodible.

Poly(orthoesters) represent the first class of bioerodible polymers designed specifically for drug delivery applications (52). *In vivo* degradation of the polyorthoester shown, known as the Alzamer degradation, yields 1,4-cyclohexanedimethanol and 4-hydroxybutyric acid as hydrolysis products (53).

Poly(anhydrides). Poly(anhydrides) are another class of synthetic polymers used for bioerodible matrix, drug delivery implant experiments. An example is poly(bis(p-carboxyphenoxy)propane) (PCPP) which has been prepared as a copolymer with various levels of sebacic anhydride (SA). Injection molded samples of poly(anhydride)/drug mixtures display zero-order kinetics in both polymer erosion and drug release. Degradation of these polymers simply releases the dicarboxylic acid monomers (54). Preliminary toxicological evaluations showed that the polymers and degradation products had acceptable biocompatibility and did not exhibit cytotoxicity or mutagenicity (55).

Economic Aspects

Americans spent \$838.5 billion on health care in 1992, representing an 11.5% increase over the previous year (56). Medical devices and medical disposables contributed significantly to the quality and effectiveness of health care. Worldwise sales for the United States medical device industry was estimated to be \$42 billion in 1993, an increase of 5.8% over 1992. This growth placed medical devices among the fastest growing segments of the United States economy. Medical devices range from wound dressings to artificial hearts, designed to support life in many end-stage cardiac patients. The United States Food and Drug Administration estimates that some 2700 medical devices and over 1500 medical disposables are in use. Medical devices are one of the nation's export champions. The total output of U.S. based medical technology represents 46% of world production, which includes 17% of the EC market, 8% of Japan, and 87% of the United States market, which is the largest in the world.

In 1993, the United States medical device industry employed 282,000 people, averaging a growth rate of about 4% since 1975. Research and development spending in this industry outpaced that in virtually every other industry. Over 7% of sales were spent on research and development in 1993, amounting to a little less than $3 billion. This rate reflects both the rapid rate of innovation and short product life-cycle. Medical devices become obsolete far more rapidly than pharmaceuticals (qv), forcing companies continuously to be innovative.

Worldwide, more than 250,000 cardiac pacemakers were implanted in 1995, accounting for well in excess of $1 billion in sales. It is estimated that over 500,000 people in the United States alone have pacemakers. Millions of people have been able to lead normal lives thanks to this remarkable device.

The worldwide market for vascular grafts was approximately $150 million and growing at about 5% annually as of 1995. The vascular graft area has tremendous market potential in development of small-diameter grafts of 3–4 mm for coronary bypass surgery. The total market for the intra-aortic balloon pump, a cardiac-assist device, is ca $80 million worldwide. About 75,000 patients were supported by these balloon pumps in 1994. This market is thought to have peaked.

BIBLIOGRAPHY

"Prosthetic and Biomedical Devices" in *ECT* 3rd ed., Vol. 19, pp. 275–313, by C. G. Gebelein, Youngstown State University.

1. *Szycher's Dictionary of Biomaterials and Medical Devices*, Technomic Publishing Co., Inc., Lancaster, Pa., 1992.
2. R. L. Whalen, "Connective Tissue Response to Movement at the Prosthesis/Tissue Interface," in *Biocompatible Polymers, Metals and Composites*, Technomic Publishing Co., Lancaster, Pa., 1983.
3. B. E. Balkin, *J. Dent. Ed.* **52**(12), 683–685 (1988).
4. A. Madison, *Transplanted and Artificial Body Organs*, Beaufort Books, Inc., New York, 1981.
5. J. R. Hogness and M. VanAntwerp, eds., *The Artificial Heart: Prototypes, Policies, and Patients*, National Academy Press, Washington, D.C., 1991.
6. *ABIOMED 1994 Annual Report*, Danvers, Mass.
7. R. Turner, *Clin. Orth.* **256**, 299–305 (1990).
8. Technical data, Lifequest Medical, Inc., San Antonio, Tex., 1994.
9. U.S. Pat. 5,229,431 (1993), L. Pinchuk (to Corvita Corp.).
10. U.S. Pat. 5,254,662 (1993), M. Szycher (to PolyMedica Industries, Inc.).
11. J. Lemons, K. M. Nieman, and A. B. Wiess, *J. Biomed. Mater. Res.* **7**, 549–553 (1976).
12. S. I. Ertel and J. Kohn, *J. Biomed. Mater. Res.* **28**(8), 919 (1994).
13. *J. Bone Joint Surg.* **73**, 148–153 (1991).
14. R. Langer, *Science*, Sept. 28, 1527–1533 (1990).

15. L. G. Cima, D. E. Ingber, J. P. Vacanti, and R. Langer, *Biotech. Bioeng.* **38**, 145–158 (1991).
16. C. T. Laurencin, M. E. Norman, H. M. Elgendy, and S. F. El-Amin, *J. Biomed. Mat. Res.* **27**, 963–973 (1993).
17. C. A. Vacanti, R. Langer, B. Schloo, and J. P. Vacanti, *Plast. Reconstr. Surg.* **88**, 753–759 (1991).
18. H. P. Greisler, D. Petsikas, T. M. Lam, and co-workers, *J. Biomed. Mat. Res.* **27**, 955–961 (1993).
19. T. H. Barrows, *Clin. Mat.* **1**(4), 233 (1986).
20. D. Borloz, W. Bichon, and A. L. Cassano-Zoppi, *Biomaterials* **5**, 255–268 (1984).
21. W. H. McCarthy, *Aust NZ J. Surg.* **39**, 422–424 (1970).
22. A. Pavan, M. Bosio, and T. Longo, *J. Biomed. Mater. Res.* **13**, 477–496 (1979).
23. R. K. Kulkarni, K. C. Pani, C. Neuman, and F. Leonard, *Arch. Surg.* **93**, 839–843 (1966).
24. J. M. Brady, D. E. Cutwright, R. A. Miller, G. C. Battistone, and E. E. Hunsuck, *J. Biomed. Mater. Res.* **8**, 218 (1973).
25. M. Vert, H. Garreau, M. Audion, F. Chabot, and P. Christel, *Trans. Soc. Biomater.* **8**, 218 (1985).
26. S. Gogolewski, A. J. Pennings, *J. Appl. Polym. Sci.* **28**, 1045–1061 (1983).
27. U.S. Pat. 3,297,033 (1967), E. E. Schmitt (to Ethicon, Inc.).
28. M. Persson, K. Bilgrav, L. Jensen, and F. Gottrup, *Eur. Surg. Res.* **18**, 122–128 (1986).
29. A. Browning and C. C. Chu, *J. Biomed. Mater. Res.* **20**, 613–632 (1986).
30. C. C. Chu and N. D. Campbell, *J. Biomed. Mater. Res.* **16**, 417–430 (1982).
31. D. F. Williams, in Syrett and Acharya, eds., *Corrosion and Degradation of Implant Materials*, ASTM STP684, American Society for Testing and Materials, Philadelphia, Pa., 1979, pp. 61–75.
32. A. M. Reed and D. K. Gilding, *Polymer* **22**, 459–504 (1981).
33. P. H. Craig, J. A. Williams, and K. W. Davide, *Surg. Gynecol. Obstet.* **141**, 1–10 (1975).
34. R. J. Fredericks, A. J. Melveger, and L. J. Dolegiewitz, *J. Poly. Sci.* **22**, 57–66 (1984).
35. E. Pines and T. J. Cunningham, *Eur. Pat. Appl.* **109**, 197A (1984).
36. D. K. Gilding and A. M. Reed, *Polymer* **20**, 1459–1464 (1979).
37. U.S. Pat. 4,523,591.
38. J. A. Ray, N. Doddi, D. Regula, J. A. Williams, and A. Melveger, *Surg. Gynecol. Obstet.* **153**, 497–507 (1981).
39. D. F. Williams, C. C. Chu, J. Dwyer, *J. Appl. Poly. Sci.* **29**, 1865–1877 (1984).
40. Brit. Pat. 682,866 (1952), D. Coleman (to ICI).
41. D. K. Gilding and A. M. Reed, *Polymer* **20**, 1454–1458 (1979).
42. M. S. Roby, D. J. Casey, R. D. Cody, *Trans. Soc. Biomater.* **8**, 216 (1985).
43. A. R. Katz, D. P. Mukherjee, A. L. Kaganov, and S. Gordon, *Surg. Gynecol. Obstet.* **161**, 213–222 (1985).
44. T. Kawaguchi, M. Nakano, K. Juni, S. Inoue, Y. Yoshida, *Chem. Pharm. Bull.* **31**, 1400–1403 (1983).
45. S. C. Woodward, P. S. Brewer, F. Moatamed, A. Schindler, C. G. Pitt, *J. Biomed. Mater. Res.* **19**, 437–444 (1985).
46. P. Jarrett, C. Benedict, J. P. Bell, J. A. Cameron, and S. J. Huang, *Polym. Prepr.* **24**(1), 32–33 (1983).
47. U.S. Pat. 3,057,537.
48. T. E. Lawler, J. P. English, A. J. Tipton, and R. L. Dunn, *Trans. Soc. Biomater.* **8**, 209 (1985).
49. U.S. Pat. 4,209,607 (June 24, 1980), W. Shalaby (to Ethicon, Inc.).
50. T. H. Barrows, D. M. Grussing, and D. W. Hegdahl, *Trans. Soc. Biomater.* **6**, 109 (1983).

51. T. H. Barrows, S. J. Gibson, and J. D. Johnson, *Trans. Soc. Biomater.* **7**, 210 (1984).
52. U.S. Pat. 4,093,709 (June 6, 1978), G. Heller and N. S. Choi (to Alza Corp.).
53. S. L. Sendelbeck and C. L. Girdin, *Drug Metab. Dispos.* **13**, 29–95 (1985).
54. K. W. Leong, B. C. Brott, R. Langer, *J. Biomed. Mater. Res.* **19**, 941–964 (1985).
55. K. W. Leong, P. D. Amore, M. Marletta, and R. Langer, *J. Biomed. Mater. Res.* **20**, 51–64 (1986).
56. M. Szycher, in Swarbrick and Boylan, eds., *Encyclopedia of Pharmaceutical Technology*, Vol. 9, Marcel Dekker, Inc., New York, 1994, p. 263.

General References

M. Szycher, *Szycher's Dictionary of Biomaterials and Medical Devices*, Technomic Publishing Co., Inc., Lancaster, Pa., 1992.
C. P. Sharma and M. Szycher, *Blood Compatible Materials and Devices*, Technomic Publishing Co., Inc., Lancaster, Pa., 1991.
M. Szycher, *Biocompatible Polymers, Metals and Composites*, Technomic Publishing Co., Inc., Lancaster, Pa., 1983.
M. Szycher, *High Performance Biomaterials*, Technomic Publishing Co., Inc., Lancaster, Pa., 1991.
M. Szycher, *Introduction to Biomedical Polymers*, ACS Audio Courses, American Chemical Society, Washington, D.C., 1989.
I. Wickelgren, *Science* **272**, 668–670 (1996).

MICHAEL SZYCHER
PolyMedica Industries, Inc.

PROTACTINIUM. See ACTINIDES AND TRANSACTINIDES.

PROTEIN ENGINEERING

Protein engineering encompasses a wide variety of techniques, ranging from the rational modification of existing proteins to the de novo design of novel proteins. Protein engineering is most commonly carried out by manipulation of the protein at the genetic level; that is, by mutagenesis of the gene (deoxyribonucleic acid (DNA)) which codes for the protein (see GENETIC ENGINEERING; NUCLEIC ACIDS). Other methods, such as chemical modification and chemical synthesis of proteins (qv), are also included within the scope of protein engineering.

One of the principal goals of protein engineering is to provide insight into the complex relationship between protein structure and function. Many protein engineering studies have involved changing specific amino acid residues within proteins of known crystal structures in order to test predictions as to how such changes may affect the structure or function of the protein. Even in the absence of a well-defined three-dimensional structure, mutagenesis studies can provide a wealth of information on the importance of a particular amino acid residue to various physicochemical properties of the protein, such as stability, enzyme catalysis, ligand binding, and protein–protein interactions.

Protein engineering encompasses a vast amount and wide variety of research. At least two textbooks (1,2) have been devoted exclusively to this topic, and several excellent reviews have been published (3,4). Herein, an overview of principles, an introduction to basic techniques, and a summary of results of representative experiments on protein engineering are provided.

Protein Structure

Much of protein engineering concerns attempts to explore the relationship between protein structure and function. Proteins are polymers of amino acids (qv), which have general structure $^+H_3N — CHR — COO^-$, where R, the amino acid side chain, determines the unique identity and hence the structure and reactivity of the amino acid (Fig. 1, Table 1). Formation of a polypeptide or protein from the constituent amino acids involves the condensation of the amino-nitrogen of one residue to the carboxylate-carbon of another residue to form an amide, also called peptide, bond and water. The linear order in which amino acids are linked in the protein is called the primary structure of the protein or, more commonly, the amino acid sequence. Only 20 amino acid structures are used commonly in the cellular biosynthesis of proteins (qv).

Discrete segments of the polypeptide chain can fold into regular, repeating structural motifs, known as secondary structures. It is the identity and sequence of amino acids within a polypeptide chain that dictate its secondary structure. As of 1996, however, understanding of the factors involved in protein folding is not sufficient to predict accurately the pattern of secondary structure from the amino acid sequence of a protein. The two most common structural motifs in proteins, the α-helix and β-pleated sheet, are shown in Figure 2. Short segments of sequence that allow the contiguous polypeptide to change directions (β-turns), comprise another important secondary structure element in proteins. Regions of the protein that do not contain regular, ie, ordered, secondary structure are referred to as random coil.

The way in which the elements of secondary structure fold upon each other to form compact globular structures is referred to as tertiary structure. Noncovalent forces such as hydrogen bonds, electrostatic interactions, and hydrophobic interactions play a dominant role in determining protein tertiary structure. In some proteins, covalent bonds formed by the oxidation of two cysteine sulfhydryl groups provide additional stability to the folded protein. These cross-links are referred to as disulfide bonds. It is the tertiary structure that gives a protein its overall shape and dimensions. Tertiary structure also provides a means of bringing into spatial proximity amino acids that may be distant in the linear sequence. Often the active form of a protein is a complex of polypeptides held together by noncovalent or covalent (disulfide) interactions. The arrangement of these polypeptide subunits relative to

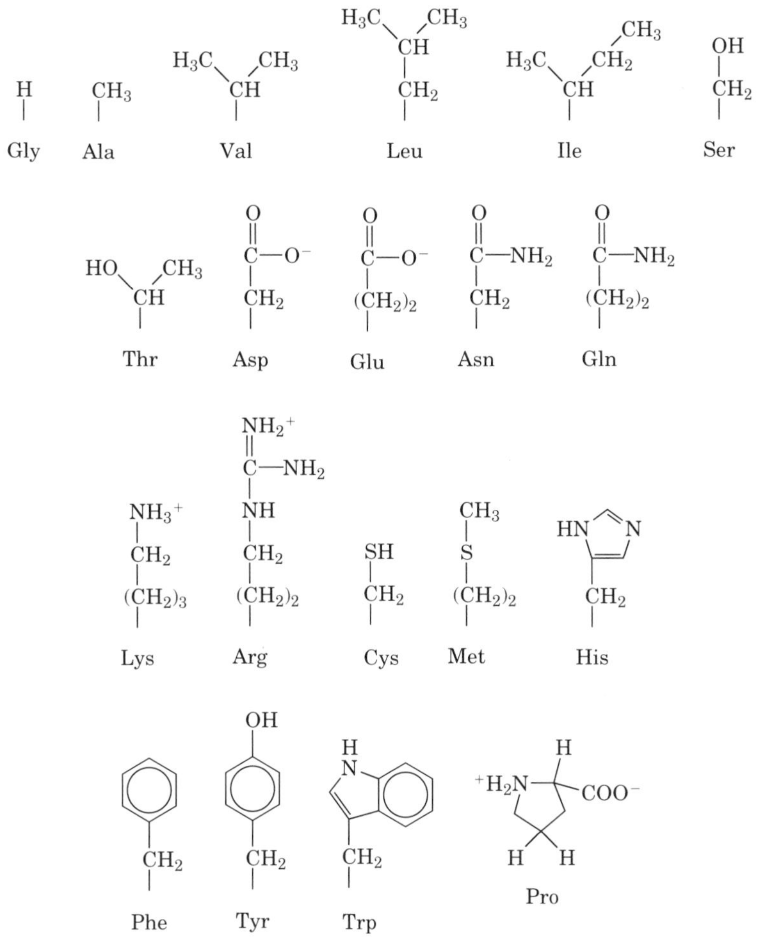

Fig. 1. The side chain R of the 20 standard amino acids $^+H_3N - CHR - COO^-$ at pH 7. For proline, the complete structure is shown. Amino acid side chains can be categorized as aliphatic (Gly, Ala, Val, Leu, and Ile), hydrophilic (Ser, Thr, Asp, Glu, Asn, Gln, Lys, and Arg), sulfur-containing (Cys and Met), aromatic (His, Phe, Tyr, and Trp), and the imino acid proline.

one another defines the quaternary structure of the protein. These different levels of protein structure, ie, primary, secondary, tertiary, and quaternary structure, are summarized in Figure 3.

Noncovalent Forces Stabilizing Protein Structure. Much of protein engineering concerns attempts to alter the structure or function of a protein in a predefined way. An understanding of the underlying physicochemical forces that participate in protein folding and structural stabilization is thus important.

Through combined effects of noncovalent forces, proteins fold into secondary structures, and hence a tertiary structure that defines the native state or conformation of a protein. The native state is then that three-dimensional arrangement

Table 1. Physicochemical Properties of the Natural Amino Acids[a]

Amino acid	Three-letter code	One-letter code	Mass of residue in proteins[b]	Accessible surface area[c], nm²	Hydrophobicity index[d]	pKa of ionizable side chain	Occurrence in proteins, %[e]	Relative mutability[f]
alanine	Ala	A	71.08	1.15	+1.8		9.0	100
arginine	Arg	R	156.20	2.25	−4.5	12.5	4.7	65
asparagine	Asn	N	114.11	1.60	−3.5		4.4	134
aspartate	Asp	D	115.09	1.50	−3.5	3.9	5.5	106
cysteine	Cys	C	103.14	1.35	+2.5	8.4	2.8	20
glutamate	Glu	E	128.14	1.80	−3.5	4.1	3.9	102
glutamine	Gln	Q	129.12	1.90	−3.5		6.2	93
glycine	Gly	G	57.06	0.75	−0.4		7.5	49
histidine	His	H	137.15	1.95	−3.2	6.0	2.1	66
isoleucine	Ile	I	113.17	1.75	+4.5		4.6	96
leucine	Leu	L	113.17	1.70	+3.8		7.5	40
lysine	Lys	K	128.18	2.00	−3.9	10.8	7.0	56
methionine	Met	M	131.21	1.85	+1.9		1.7	94
phenylalanine	Phe	F	147.18	2.10	+2.8		3.5	41
proline	Pro	P	97.12	1.45	−1.6		4.6	56
serine	Ser	S	87.08	1.15	−0.8		7.1	120
threonine	Thr	T	101.11	1.40	−0.7		6.0	97
tryptophan	Trp	W	186.21	2.55	−0.9		1.1	18
tyrosine	Tyr	Y	163.18	2.30	−1.3	10.1	3.5	41
valine	Val	V	99.14	1.55	+4.2		6.9	74

[a]Courtesy of VCH Publishers, Inc. (5).
[b]Values reflect the molecular weights of the amino acids minus that of water.
[c]For residues as part of a polypeptide chain (6).
[d]Ref. 7.
[e]Based on the frequency of occurrence for each residue in the sequence of 207 unrelated proteins (8).
[f]Relative mutability represents the likelihood that a residue will mutate within a specified time period during evolution. Units are arbitrary, with alanine assigned a value of 100 (9).

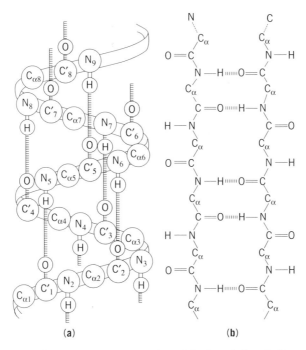

(a)   (b)

Fig. 2. Protein secondary structure: (**a**) the right-handed α-helix, stabilized by intrasegmental hydrogen-bonding between the backbone CO of residue i and the NH of residue $i + 4$ along the polypeptide chain. Each turn of the helix requires 3.6 residues. Translation along the helical axis is 0.15 nm per residue, or 0.54 nm per turn; and (**b**) the β-pleated sheet where the polypeptide is in an extended conformation and backbone hydrogen-bonding occurs between residues on adjacent strands. Here, the backbone CO and NH atoms are in the plane of the page and the amino acid side chains extend from C_α alternating above and below the plane of the page (10).

of the polypeptide chain and amino acid side chains that best facilitates the biological activity of a protein, at the same time providing structural stability. Through protein engineering subtle adjustments in the structure of the protein can be made that can dramatically alter its function or stability.

Electrostatics. Electrostatic interactions, such as salt bridges, result from the electrostatic attraction that occurs between oppositely charged molecules. These usually involve a single cation, eg, the side chain of Lys or Arg, or the amino terminus, etc, interacting with a single anion, eg, the side chain of Glu or Asp, or the carboxyl terminus, etc. This attractive force is inversely proportional to the distance between the charges and the dielectric constant of the solvent, as described by Coulomb's law.

Hydrogen Bonds. An attractive force which involves the sharing of a hydrogen atom between two electronegative atoms, the hydrogen-bond (H-bond), consists of a hydrogen donor to which the H-atom is covalently bound and an acceptor containing either a partial or full negative charge that attracts the electropositive H-atom. Eleven of the 20 amino acids can participate in H-bonds, Arg, Asp, Asn, Cys, Glu, Gln, His, Lys, Ser, Thr, and Tyr. Stronger than van der Waals bonds but much weaker than covalent bonds, hydrogen bonds constitute

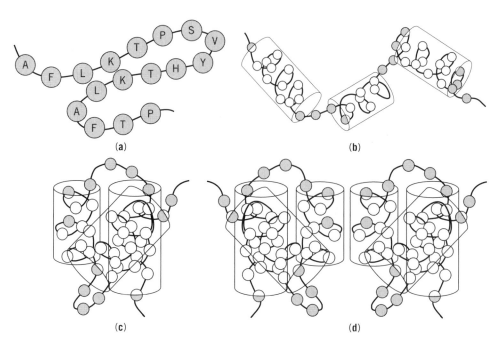

Fig. 3. The hierarchy of protein structures: (**a**) primary structure (see Table 1 for amino acid code); (**b**) secondary structure; (**c**) tertiary structure; and (**d**) quaternary structure (10).

a primary determinant of biological specificity. These bonds are sufficiently strong to direct molecular interactions such as the attraction between an enzyme and its substrate, but sufficiently weak to be reversibly made and broken, as in enzyme catalysis (11).

Van der Waals Interactions. Van der Waals interactions result from the asymmetric distribution of electronic charge surrounding an atom, which induces a complementary dipole in a neighboring atom, resulting in an attractive force. In general, the attractive force of van der Waals interactions is very weak (<4.2 kJ/mol (1 kcal/mol)) but may become significant if steric complementarity creates an opportunity to form a large number of van der Waals attractions.

The Hydrophobic Effect. Noncovalent interactions can occur not only between different parts of the polypeptide chain but also between the polypeptide and the solvent water. Water very effectively solvates polar molecules, weakening electrostatic forces and H-bonds by competing for the elements of those attractions. Nonpolar amino acids cannot, however, participate in such favorable interactions with water. This absence of interaction, coupled with the high affinity of water for itself, forces the nonpolar molecules to associate, forming a sequestered, hydrophobic core in the interior of the protein. This process is termed the hydrophobic effect. It is generally accepted that hydrophobic attractions are the principal driving force in protein folding, primarily owing to the much more favorable entropy of the water resulting from exclusion of nonpolar molecules.

Techniques Used to Study Protein Structure.   The structural features of proteins can be experimentally assessed by a variety of biophysical methods. Spectroscopic methods can assess aspects of secondary and tertiary structure (see SPECTROSCOPY) (10); chromatographic and hydrodynamic methods can provide information on subunit assembly (see CHROMATOGRAPHY). Protein structure at the atomic level can be assessed by x-ray crystallography (12) (see X-RAY TECHNOLOGY) and by multidimensional nmr spectroscopy (see MAGNETIC RESONANCE) (13). These powerful structural methods have played critical roles in protein engineering. Discussions of these techniques may also be found in various journals, such as *Nature, Structural Biology (London)*, and *Current Opinion in Structural Biology*.

The Protein Engineering Process

Identification and Characterization of the Target Protein.   Prior to commencement of the process of protein engineering, a target protein must be selected and some structural information on the natural form of that protein determined. The motivation for selecting a particular target protein varies according to the researcher and the problem to be solved. Once the target protein has been selected, a concerted effort must be made to obtain the natural protein in pure form in order to determine the needed structural information. At a minimum, some amino acid sequence information must be known in order to design the oligonucleotide probes that are used to identify the gene for the target protein. Additionally, information on other properties of the target protein, such as molecular weight and isoelectric point are useful to confirming the identity of the cloned protein at a later date. Knowing what, if any, post-translational modifications (14) occur for the natural protein can guide the choice of an appropriate host organism for expression, and thus it is important to obtain this information prior to attempting cloning and expression.

Whereas most proteins are soluble in the cell cytosol, many proteins are associated with the various membranes found in cells. Membrane proteins present additional challenges and one must therefore know in advance whether or not the target protein is membrane-associated. Finally, it is important to know whether the active form of the target protein is a single, monomeric polypeptide, or if it consists of multiple subunits. Which it is greatly affects the options available for cloning and expression strategies. Methods of protein analysis have been established that can help to answer these and other questions about the nature of the target protein. Descriptions of these various methods can be found in any of a number of texts devoted to protein analysis (10,15).

Cellular Protein Biosynthesis.   The process of cellular protein biosynthesis is virtually the same in all organisms. The information which defines the amino acid sequence of a protein is encoded by its corresponding sequence of DNA (the gene). The DNA is composed of two strands of polynucleotides, each comprising some arrangement (sequence) of the four nucleotide building blocks of the nucleic acids: adenine (A), thymine (T), guanine (G), and cytosine (C). In ribonucleic acid (RNA), T is replaced by uracil (U). The two strands of DNA are held together by hydrogen-bonding patterns between specific nucleotide bases, ie, base pairing. The structures of the bases are such that A always pairs with T, G with C. Hence, the two strands of DNA are complementary; that is, the nucleotide sequence of one strand defines exactly the sequence of the other strand. For example, if one strand had the sequence ATCGA, the complementary strand would have to have the sequence TAGCT. This complementarity provides a convenient means for living organisms to replicate their genetic information and forms the technical basis for protein engineering.

The relationship between the nucleotide sequence of the DNA and the amino acid sequence of the protein is known as the genetic code (Table 2). In this code, a sequence of three nucleotides (a codon) specifies a particular amino acid. The genetic code consists of 64 codons, of which 61 code for amino acids and three are translation stop signals (stop codons). Because the genetic code is nearly the same in all organisms, it is possible to insert genetic information from one organism into a different host organism. The latter can then use its biosynthetic machinery to produce the foreign protein. This process is referred to as heterologous protein expression.

Cellular protein biosynthesis involves the following steps. One strand of double-stranded DNA serves as a template strand for the synthesis of a complementary single-stranded messenger ribonucleic acid (mRNA) in a process called transcription. This mRNA in turn serves as a template to direct the synthesis of the protein in a process called translation. The codons of the mRNA are read sequentially by transfer RNA (tRNA) molecules, which bind specifically to the

Table 2. The Genetic Code

First position (5′-end)	Second position				Third position (3′-end)
	U	C	A	G	
U	Phe	Ser	Tyr	Cys	U
	Phe	Ser	Tyr	Cys	C
	Leu	Ser	STOP	STOP	A
	Leu	Ser	STOP	Trp	G
C	Leu	Pro	His	Arg	U
	Leu	Pro	His	Arg	C
	Leu	Pro	Gln	Arg	A
	Leu	Pro	Gln	Arg	G
A	Ile	Thr	Asn	Ser	U
	Ile	Thr	Asn	Ser	C
	Ile	Thr	Lys	Arg	A
	Met (START)	Thr	Lys	Arg	G
G	Val	Ala	Asp	Gly	U
	Val	Ala	Asp	Gly	C
	Val	Ala	Glu	Gly	A
	Val	Ala	Glu	Gly	G

mRNA via triplets of nucleotides that are complementary to the particular codon, called an anticodon. Protein synthesis occurs on a ribosome, a complex consisting of more than 50 different proteins and several structural RNA molecules, which moves along the mRNA and mediates the binding of the tRNA molecules and the formation of the nascent peptide chain. The tRNA molecule carries an activated form of the specific amino acid to the ribosome where it is added to the end of the growing peptide chain. There is at least one tRNA for each amino acid.

Cloning of the Target Protein. Cloning procedures have been made possible by the availability of several key types of enzymes, including restriction endonucleases, enzymes that cleave double-stranded DNA at specific sites; DNA polymerases (especially thermostable forms), which replicate DNA templates; and DNA ligases, which covalently link (ligate) fragments of DNA. In general, fragments of DNA are ligated into vectors that can autonomously replicate their DNA in the host organism. The two most common vectors used in bacteria are plasmids and phage. Plasmids are naturally occurring circular DNA molecules that can act as accessory chromosomes. Phage, such as lambda phage, are viruses which can both stably integrate into the chromosome of the host (lysogenic pathway) or use the host machinery to produce more viral particles, which in turn lyse the host cell and destroy it (lytic pathway). The M13 phage are single-stranded circular molecules of DNA that do not kill the host when these are packaged and secreted. A double-stranded replicative form of this phage also exists and can be manipulated like plasmids. An excellent introduction to recombinant DNA technology is available (16). Detailed protocols on methodology can be found in References 17 and 18.

Making a cDNA Library. Within the genes of eukaryotes, extraneous stretches of nucleotide sequences that do not code for amino acid residues in the protein (introns) are often interspersed among the coding sequences (exons). When DNA is transcribed, these introns are excised from the mRNA to form a contiguous sequence of exons that translate into the contiguous amino acid sequence of the protein. Eliminating the introns from the DNA directly would provide a convenient tool for cloning purposes. Therefore, complementary DNA (cDNA) libraries are constructed from cellular mRNA (Fig. 4). This form of RNA has already been processed by the cell and therefore the DNA corresponds to the actual nucleotide sequence that codes for the protein without the interrupting introns.

Most eukaryotic genes contain a signal sequence after the stop codon that causes further processing of the mRNA by the enzyme poly-A polymerase, adding about 250 A residues to the $3'$ end of the newly transcribed mRNA. The procedure for making cDNA from mRNA takes advantage of this unique marker of the $3'$-end of the gene by using a poly-T oligonucleotide primer to base-pair to the poly-A tail. This primer is then extended by the enzyme reverse transcriptase which, recognizing the nucleotide sequence of the mRNA, makes a complementary strand of DNA to produce a DNA–RNA hybrid. The RNA part of this duplex is removed by treatment with sodium hydroxide, which hydrolyzes the RNA but leaves the DNA intact. The DNA polymerase is then used to make a complementary copy of the single-stranded DNA. After removal of the single-stranded hairpin loop from the duplex DNA by treatment with S1 nuclease (see step 4 of Fig. 4), the linear cDNA can be ligated to an appropriate plasmid vector

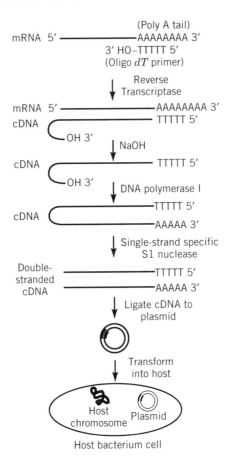

Fig. 4. Steps in making a cDNA library. Cellular mRNA is used as a template to make a complementary DNA. This cDNA is then ligated to a plasmid, which can be transformed into an appropriate host for propagation. See text.

and transformed into bacteria, where it is propagated. Because the cDNA was generated from a mixture of mRNAs, the cDNA is also a mixture of molecules. The transformation of the recombinant cDNA plasmids into the bacteria is designed so that each bacterial cell receives a unique cDNA recombinant.

Screening of a cDNA Library. The bacteria containing individual cDNA recombinant plasmids are grown on an agar plate where individual colonies can be visualized, and these colonies are screened for the presence of the desired DNA, using a probe specific to the DNA of interest. The probe is generally an oligonucleotide where the sequence complements a segment of the nucleotide sequence of the cDNA of interest. The probe is usually radiolabeled to allow easy identification of its location. The bacterial cells on an agar plate are replica-plated onto nitrocellulose filters, where the cells are immobilized. This process involves touching the nitrocellulose to the colonies on the agar plate so that some cells from each colony are transferred to the filter. The filter thus becomes an exact replica of the arrangement of colonies on the agar plate. The cells are lysed and the DNA inside is denatured into single-stranded forms. The probe, which is present in excess, can then hybridize (base-pair) to complementary DNA on the nitrocellulose. The location of the hybridizing probe is then identified by autoradiography, and the colony corresponding to the positive signal is isolated and further tested.

The probe can be virtually any convenient length oligonucleotide, but longer probes are more likely to identify only the specific DNA of interest. If partial amino acid sequence information is used to design a probe, then because the genetic code is degenerate, multiple codons may specify a particular amino acid. Because a positive identification in the screening procedure requires a high degree of base complementarity, the use of highly degenerate nucleotides such as those coding for Leu, Arg, and Ser, which each have six codons, should be minimized. The frequency of codon usage is not necessarily random in a particular species, and tables are available that list the frequency of codon usage for various organisms (18). Thus, peptide sequences containing tryptophan or methionine, which each have only one unique codon, are excellent choices for inclusion in a probe. Where it is not possible to eliminate a highly degenerate residue from the probe, a mixture of degenerate oligonucleotides can be synthesized simultaneously and used in screening.

Characterizing the Clone. After a plasmid containing the cDNA is identified by hybridization techniques, the DNA sequence is determined. If only partial protein sequence was used for screening, the complete coding sequence can be deduced by identifying the translation start signal, ATG, which codes for an initiating methionine (see Table 2), and a translation stop signal. The size or amino acid composition of the gene-encoded protein can then be compared to the size or composition of the purified protein to determine if the results are consistent.

Manipulating the Recombinant DNA Clone. Cloned DNA can be expressed or the gene can be altered by deletions, insertions, or substitutions in the cloned DNA. Deletion and insertion generally involve the use of restriction endonucleases to cut the DNA at a specific site or sites and then remove the desired segment or ligate an additional segment to the gene. Single site-specific substitution is a more common practice for the study of protein structure and function. The general procedure for substituting one amino acid residue for another involves a technique termed oligonucleotide directed mutagenesis. In this procedure, a mutagenic oligonucleotide primer of 20–30 nucleotides, containing usually one to three nucleotide mismatches to change a codon, that specify the desired amino acid substitution, is incorporated into the DNA and specifically selected for further replication. Two particularly popular procedures for this replication are use of single-stranded circular phage DNA (19), and use of the polymerase chain reaction (PCR) using double-stranded DNA (20). These methods are summarized in Figures 5 and 6, respectively.

PCR (Fig. 6) is performed in a thermocycler, which allows various reactions to occur by systematically varying the temperature. For example, denaturation of the duplex DNA to single strands (ss) is carried out at 94°C, followed by annealing of primers to the ssDNA at 45°C and extension of the primed DNA by a thermostable polymerase at 72°C. The process is then repeated 20–30 times to amplify the product. The sequences of primers 2 and 3 (Fig. 6) are complementary to the template sequence, with the exception of the base-pair mismatch(es)

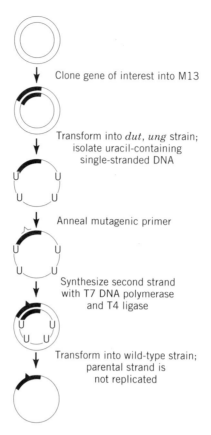

Fig. 5. Generation of mutants using single-stranded DNA. After cloning the target gene into M13, the phage is propagated in the *E. coli dut, ung* strain of *E. coli*, which results in a large percentage of uracil bases incorporated in place of thymine bases. A mutagenic primer is annealed to this DNA, and the remainder of the strand is synthesized *in vitro* with deoxyribonucleotides and DNA polymerase. The circular double-stranded DNA is then propagated in a strain of *E. coli*, which does not replicate the uracil-containing strand and therefore produces only the mutagenic DNA (21).

required to incorporate the desired mutation. Two steps are required to produce the full-length mutant gene. The first step involves two separate reaction tubes: one tube contains the DNA template and primers 1 and 2; the other tube contains the DNA template and primers 3 and 4. In the second step, the two products from the first reaction are combined into one tube, where their overlapping complementary regions (defined by primers 2 and 3 from the first reaction) base-pair to form a template for extension to the full-length DNA by the polymerase. This DNA is then replicated using only the outside primers.

Gene Expression. *Initial Considerations.* Once the cDNA for the natural protein (wild type) or a mutant thereof is cloned, the cDNA is inserted into an appropriate vector containing a ribosome binding site and a promoter that can direct synthesis of the desired protein. The choice of vector depends on the choice of host organism in which to express the protein. Several factors must be considered. Is the gene to be expressed eukaryotic or prokaryotic?

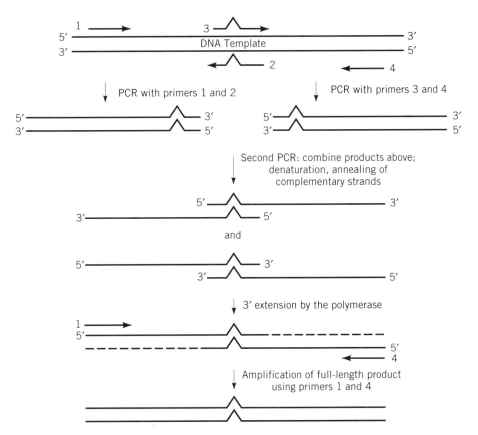

Fig. 6. Polymerase chain reaction (PCR) mediated site-directed mutagenesis. The 5′ and 3′ ends of the nucleotide strands are indicated. The four arrows surrounding the DNA template represent oligonucleotide primers 1–4. See text for discussion.

Is the protein post-translationally modified, eg, glycosylation, phosphorylation, etc, and are the modifications necessary for protein function? The original gene-encoded sequence may not adequately describe the mature protein owing to the occurrence of post-translational modifications. Is the natural protein cytoplasmic, membrane-bound, or secreted? If denatured, can the protein be easily renatured? How much protein is needed? The most common expression systems include bacteria, such as *E. coli*; mammalian cells, such as Chinese hamster ovary (CHO) and monkey kidney (COS) cells; baculovirus-infected insect cells; and yeast. Some advantages and disadvantages of these common hosts are given (22,23).

E. coli. This bacterium has been extensively utilized as a host for protein expression (24). The organism's genome is highly flexible, accepting a great deal of manipulation. Advantages of this expression system are that it is generally fast and inexpensive, has easy scale-up, and high level protein production is possible. A disadvantage is that the overexpressed proteins may be denatured, but this may be turned into an advantage if the denatured protein forms inclusion bodies that can be easily renatured. The cytosol of *E. coli* is such that oxidation of cysteines to form disulfide bonds does not occur. Upon lysis of the cells and exposure of the protein to the oxidizing atmosphere, disulfide bond formation can occur rapidly, sometimes with the formation of misfolded protein. Thus, this host is not always optimal for proteins containing multiple disulfide bonds. Many post-translational modifications, such as specific cleavage of propeptides and attachment of carbohydrate units (glycosylation) cannot be carried out in *E. coli*. If these post-translational modifications are important to protein function, bacterial host systems are generally a poor choice. Finally, the DNA to be transcribed cannot have introns within its sequence, because these are not processed by the bacterium.

Yeast.   The advantages of expression in yeast include potentially high level production of proteins, the ability to have expressed proteins secreted into the media for ease of purification, and relatively low cost, easy scale-up. A disadvantage is that plasmid instability may be a problem which can lead to low product yield. Whereas post-translational modification occurs in yeast, proteins are quite often hyperglycosylated. This is generally a problem with expression in *Saccharomyces cerevisiae* but not for the more recently used yeast host *Pichia pastoris* (25) (see YEASTS).

Insect Cells.   In this system the cDNA is inserted into the genome of an insect virus, baculovirus. Insect cells, or live insect larvae, are then infected with the virus. In this way advantage is taken of the virus's natural machinery for replication utilizing the insect cell. This is one of the best systems available for high level production of native protein having post-translational modifications similar to those seen in mammalian cells. Disadvantages of this system include lytic–batch variations, comparatively slow growth, and costly scale-up.

Mammalian.   For mammalian proteins, mammalian cells offer the most natural host for expression. Problems of incorrect processing and post-translational modification are avoided using these cells. Mammalian cells are usually grown in continuous cell culture, reducing the variability in results (see CELL CULTURE TECHNOLOGY). Moderate-level production of native protein is possible. The procedure, however, is slow and very costly, and the level of protein expression is low. Thus large-scale production of proteins in mammalian cells is not practical. When low quantities of protein are sufficient, this system offers the several advantages described.

Purification of Expressed Proteins.   Once an appropriate host has been selected and the protein has been expressed, isolation of that protein in pure form is needed. Methods for general protein purification have been described in detail in several excellent texts (26–29). These methods typically involve a combination of chromatographic separations based on physicochemical properties of the protein, such as molecular weight (size exclusion chromatography), electrostatic charge (ion exchange (qv) and chromatofocusing), or specificity of interaction with another protein or small molecule (affinity-based chromatography). Such methods are generally applicable to all proteins, whether expressed recombinantly or naturally occurring. For engineered proteins, additional purification strategies are available that take advantage of the ability to manipulate the cDNA of the target protein. Two examples are the use of polyhistidine extensions (30) and formation of fusion proteins (31). In the former, one extends the target protein cDNA to include the coding sequence for several (usually six) histidine residues. Histidines are good transition-metal chelators in their depro-tonated form, ie, above pH 6. After expression and cell lysis, the lysate is applied to a transition-metal, usually nickel, affinity column. Owing to the presence of the polyhistidine extension, the target protein adheres specifically to this column whereas the majority of cellular proteins are not retained. The target protein can then be eluted from the column by lowering the pH or adding excess histidine to the mobile phase to compete for the interaction between the protein and the transition metal.

The second example involves fusing the target protein cDNA to the cDNA of another protein, making a fusion protein. The other protein is chosen because of some unique feature that makes it easy to purify. For example, the maltose binding protein (MBP) binds tightly to an amylose-affinity column, whereas other proteins generally do not. A common strategy is to fuse the cDNA of MBP to that of a target protein in order to use amylose-affinity chromatography as a one-step purification method for the fusion protein (31). Typically, researchers also engineer in a specific protease cleavage site at the interface between the MBP sequence and that of the target protein. After amylose-affinity chromatography, the fusion protein is treated with protease to liberate the free target protein. This protein is then separated from MBP by some conventional chromatographic method, eg, size exclusion or ion-exchange chromatography.

Applications

Studies of Protein Stability. An understanding of the forces which stabi-lize and destabilize proteins is essential both to the de novo design of proteins and rational modification of existing proteins. Forces contributing to the stability of the folded protein can be covalent, such as disulfide bonds, and/or noncovalent, such as hydrophobic or electrostatic interactions. Stabilizing forces are counter-balanced by destabilizing forces such as conformational entropy and hydration, which tend to favor the unfolded form of the protein. As a result, the folded state of naturally occurring proteins has a marginal stability of only 20–60 kJ/mol (5–15 kcal/mol) (32).

Protein engineering has provided a means of assessing the contributions of various noncovalent interactions to protein stability (33–38). In general, one amino acid is replaced by another differing from the original by only one func-tional aspect such as size, charge, or H-bonding ability. By making a series of systematic mutations and determining the effect on protein stability, the im-portance of the deleted functional group can be assessed. Whereas results vary from protein to protein and depend on the details of the local environment experi-enced by the amino acid residue, similar results from studies of several different proteins have provided some encouraging generalizations regarding the contri-butions of noncovalent interactions to protein stability.

Measuring Protein Stability. Protein stability is usually measured quantita-tively as the difference in free energy between the folded and unfolded states of the protein. These states are most commonly measured using spectroscopic tech-niques, such as circular dichroic spectroscopy, fluorescence (generally trypto-phan fluorescence) spectroscopy, nmr spectroscopy, and absorbance spectroscopy (10). For most monomeric proteins, the two-state model of protein folding can be invoked. This model states that under equilibrium conditions, the vast majority of the protein molecules in a solution exist in either the folded (native) or unfolded (denatured) state. Any kinetic intermediates that might exist on the pathway between folded and unfolded states do not accumulate to any significant extent under equilibrium conditions (39). In other words, under any set of solution conditions, at equilibrium the entire population of protein molecules can be accounted for by the mole fraction of denatured protein, f_d, and the mole fraction of native protein, f_n, ie,

$$f_d + f_n = 1.0$$

Folded proteins can be caused to spontaneously unfold upon being exposed to chaotropic agents, such as urea or guanidine hydrochloride (Gdn), or to elevated temperature (thermal denaturation). As solution conditions are changed by addition of denaturant, the mole fraction of denatured protein increases from a minimum of zero to a maximum of 1.0 in a characteristic unfolding isotherm (Fig. 7**a**). From a plot such as Figure 7**a** one can determine the concentration of denaturant, or the temperature in the case of thermal denaturation, required to achieve half maximal unfolding, ie, where $f_d = f_n = 0.5$.

The more stable a protein is, the higher the concentration of denaturant needed to achieve half-maximal unfolding. Again invoking the two-state model, one can calculate the equilibrium constant, K, for unfolding at any point in the unfolding isotherm, and hence the Gibbs free energy of unfolding:

$$K = f_d/(1 - f_d)$$
$$\Delta G = -RT \ln(f_d/(1 - f_d))$$

In Figure 7**b**, the data are plotted as ΔG yielding a linear function. Extrapolation to zero denaturant provides a quantitative estimate of the intrinsic stability of

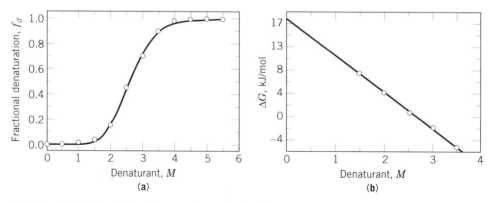

Fig. 7. Unfolding (**a**) isotherm, where the half-maximal unfolding for this protein occurs at 2.6 M denaturant; and (**b**) free energy where in the absence of denaturant, the protein has an extrapolated stability, ΔG_u^0, of 17.6 kJ/mol (4.2 kcal/mol) as shown. To convert J to cal, divide by 4.184.

the protein, ΔG_u^0, which in principle is the free energy of unfolding for the protein in the absence of denaturant. Comparison of the ΔG_u^0 values between mutant and wild-type proteins provides a quantitative means of assessing the effects of point mutations on the stability of a protein.

The Hydrophobic Effect. The importance of the hydrophobic core to protein stability is probed by substituting a nonpolar residue within the core with a smaller nonpolar residue. The resultant difference in free energy of unfolding can then be measured. For example, a replacement of isoleucine by valine yields information on the effect of the δ-methyl group. Further substitution by alanine or glycine studies the importance of the γ- and β-methyl groups, respectively. Studies on staphyloccocal nuclease, gene5 protein of bacteriophage $f1$, barnase, and T4 lysozyme, showed that the average effect on stability of a fully buried hydrophobic residue is 5.0 ± 1.7 kJ/mol (1.2 ± 0.4 kcal/mol) per methylene group (35,40). There is considerable variation within these data, especially from one protein to the next. A comparison of stability data and crystal structures for several T4 lysozyme mutants (41) showed that loss of protein stability tracts linearly with the size of the cavity created by the mutation. Thus the contribution of a residue to hydrophobic stabilization depends not only on the identity of the particular substitution, but also on the size of the cavity created by that substitution.

Hydrogen Bonding and Salt Bridges. Historically, the importance of H-bonds to the stability of folded protein had been thought of as minor, because H-bonding interactions can also occur with water in the unfolded protein. Mutational analysis, aimed at systematically eliminating H-bonding interactions in proteins such as barnase (42), ribonuclease T1 (43), ribonuclease A, lysozyme, cytochrome c, and myoglobin (44), has been carried out by several laboratories. The results suggest that the sum of individual intramolecular H-bonds energies can contribute a significant stabilization energy, approaching that of the hydrophobic effect.

Similarly, electrostatic interactions between amino acids were thought to contribute only minimally to protein stability. Protein engineering studies have shown that this is indeed the case for salt bridges on the surface of proteins, where the high dielectric constant of the solvent water greatly reduces the electrostatic attraction (45). Disruption of buried salt bridges, on the other hand, can have a significant destabilizing effect on protein stability. For example, mutation of either partner in the buried salt bridge between His31 and Asp70 pair results in a destabilization of $13-21$ kJ/mol ($3-5$ kcal/mol) for the protein T4 lysozyme (46).

Disulfides. The introduction of disulfide bonds can have various effects on protein stability. In T4 lysozyme, for example, the incorporation of some disulfides increases thermal stability; others reduce stability (47–49). Stabilization is thought to result from reduction of the conformational entropy of the unfolded state, whereas in most cases the cause of destabilization is the introduction of dihedral angle stress. In natural proteins, placement of a disulfide bond at most positions within the polypeptide chain would result in unacceptable constraint of the α-carbon chain.

Side-Chain Effects on Stability. The nature of the amino acid side chain can affect protein stability, not only by effecting changes in cavity sizes and hydrophobic or electrostatic interactions, but also because some side chains are not tolerable in certain secondary structure forms. For example, proline residues are known to be disruptive to α-helices. Thus, replacement of a residue within a helical region with a proline causes kinking of the helix and an overall destabilization of the folded state of the protein. Not only can a protein be destabilized by eliminating specific side-chain interactions, but replacements that cause interactions not found in the natural protein can likewise be destabilizing. For example, placement of a cysteine on the surface of a protein could lead to intermolecular disulfide bond formation between two molecules of the protein, leading to destabilization.

De Novo Designed Proteins.   To further elucidate the forces which direct the protein folding pathway and stabilize the final native state, several laboratories are studying de novo (from first principles) designed proteins, that is, design of a particular protein structural motif using the level of understanding of forces believed to promote and stabilize a desired structural element. The method is then tested by evaluating the extent to which the designed protein matches the expected structure. This type of protein design is very much an iterative process, involving evaluation and subsequent modification of the design. De novo protein design has been successfully used to create proteins having all α-helix, all β-sheet, and mixed α/β secondary structures. Several excellent reviews are available (49–53).

α-Helical Bundles.   The α-helix is the most extensively studied protein structural motif. Because α-helices form internal hydrogen bonds between the C=O of residue i and the N–H of residue $i + 4$ (see Fig. 2), the individual helix is stabilized and can exist in isolation. Individual helices can be manipulated as independent structural modules designed to associate in some predetermined manner. Often, a minimalist approach to the design of α-helices has been taken. In this approach the goal is to obtain the desired structural motif using the simplest possible construction.

The first de novo designed, all-helical protein was an antiparallel four-helix bundle formed from a simple 16-residue sequence, α_1B, predicted to form an amphiphilic α-helix and associate as a tetramer (54). The designed sequence contains only one type of hydrophobic residue, leucine. The only charged residues are glutamate and lysine. A dimer, α_2, was then formed by linking two identical α_1B sequences by a helical hairpin to form a helix–loop–helix motif. As shown in Figure 8, the structure of the linker plays a critical role in the final structure. The short linker of a single Pro residue promoted the formation of an elongated trimer of dimers, ie, coiled coils, whereas a longer linker of Pro-Arg-Arg permitted the formation of a helix bundle. This feature of negative design was then incorporated into the final sequence, which contained the four identical helix sequences connected by the three-residue linker to form the 73-residue α_4-helix-bundle protein (55). In the α_4-helix-bundle, each of the four helices is oriented antiparallel to its two nearest neighbors and parallel to its diagonal, more-distant neighbor.

The α_4-protein is a very stable, compact protein that exhibits properties of both the native and the molten globule state (50,56). A molten globule is a conformation having a native-like secondary structure, but less compact. This state is thought to represent an intermediate form that is accessed during protein folding. Similar to native proteins, α_4 exhibits a highly cooperative

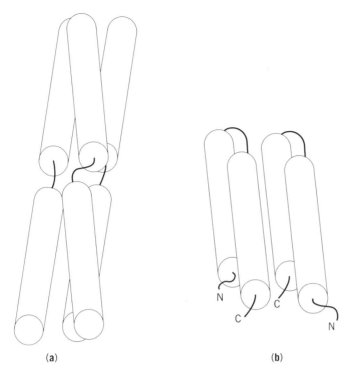

(a)   (b)

Fig. 8. De novo designed α-helical proteins. Dimers of the amphiphilic helix-forming peptide α_1B, GELEELLKKLKELLKG (see Table 1), in which the nature of the linker connecting the individual helices plays a critical role in the structure of the final protein. (**a**) Using a Pro residue as the linker, ie, α_1B-Pro-α_1B, three molecules aggregated to form a trimeric coiled-coil. (**b**) Using Pro-Arg-Arg as the linker, ie, α_1B-Pro-Arg-Arg-α_1B, resulted in the formation of an antiparallel four-helix bundle. Courtesy of S. F. Betz.

guanidinium chloride denaturation curve; it is extremely resistant to unfolding by guanidinium chloride, and its stability is sensitive to substitutions in the putative hydrophobic core of the protein. Several other experiments, however, suggest that α_4 has not achieved a complete native-like structure. Binding of the fluorescent hydrophobic dye 8-anilino-1-naphthalenesulfonic acid (ANS), which binds to the molten globule state, but much less to the native or denatured states of proteins, reveals a low order parameter, suggesting a dynamic molten globule-like hydrophobic interior. The nmr measurements of H/D exchange rates also are intermediate between molten globule and native proteins. Whereas denaturant-induced unfolding is highly cooperative, the thermal unfolding of α_4 is, again, intermediate between native and molten globule states. Thus, as predicted from the design, α_4 folds into a stable, compact state owing to hydrophobic collapse (stabilized by hydrophobic forces), but does not adopt a true native-like structure.

To produce a more native-like protein, modifications which provide specific tertiary interactions have been introduced into α_4. These include substitution of the leucine residues with amino acids possessing aromatic or β-branched side chains in order to promote specific side-chain interactions between helices (57). Metal-binding sites have also been engineered into α_4. Two variants were characterized: H3α_4 and H6α_4, which contain one and two tridentate Zn^{2+} binding sites, respectively (56). The replacement of a partially buried Leu by a His causes a less favorable free energy of folding for the two proteins in the absence of zinc, but zinc binding increases the stability significantly. The nmr spectra have shown that the compactness of the structure increases upon zinc binding for $H6\alpha_4$. However, both the apo- and the metal-bound protein still bind ANS, although the order parameter for ANS bound to the Zn^{2+}-containing protein is similar to that for the binding of ANS to the natural protein, apomyoglobin.

Other approaches to de novo four-helix bundle proteins have emphasized nonrepetitive designs. One such example is the four-helix bundle protein Felix (53), a 79-residue protein which uses 19 of the 20 naturally occurring amino acids: MPEVAENFQQCLERWAKLSVGGELAHMANQAAEAILKGGNEAQLKNAQAL-MHEAMKTRKYSEQLAQEFAHCAYKARASQ (see Table 1). Felix also contains a successfully designed disulfide, which links the first and fourth helices. Although Felix has no homology with any natural protein, its architecture was based on the four helix clusters of hemerythrin and cytochrome b_{562}. The designed protein has been shown to contain a very high α-helix content and is readily soluble in water.

Another approach to identifying structural determinants of four-helix bundles has employed cloning and *in vivo* expression of degenerate synthetic genes, followed by identification of proteins that form stable four-helix bundles. Based on the α-helical periodicity of 3.6 residues per turn, a sequence pattern of polar and nonpolar amino acids required to generate an amphiphilic α-helix, in which one face is predominantly hydrophobic and the other hydrophilic, was designed (58). Using two degenerate codons, one for polar residues and the other for nonpolar residues, a large number of synthetic genes were constructed, and the corresponding proteins were expressed. Out of 108 clones sequenced, 48 contained the desired amphipathic sequence pattern. The others contained insertions, deletions, or aberrant ligations. Twenty-nine of the 48 produced soluble proteins resistant to intracellular degradation, suggestive of compact, stable three-dimensional structures. Further characterization of three of the clones using size-exclusion chromatography and circular dichroic (cd) spectroscopy revealed proper monomeric size and helical content similar to the natural four-helix bundle protein cytochrome b_{562}. Urea denaturation curves revealed that two of the proteins possessed cooperative unfolding transitions and stabilities similar to those of natural proteins. Thus, the initial characterization of some of these clones suggests that this technique has potential value for the study of α-helical bundles.

β-Sheet Proteins.   Successful de novo design of β-sheet proteins has been more difficult than was true of α-helical proteins (59). Because β-strands have very different H-bonding patterns compared to α-helices, several additional variables must be factored into the design. Whereas backbone H-bonding is intrasegmental within an α-helix, the backbone hydrogen bonding in a β-strand occurs between C=O and N–H groups on neighboring strands. In fact, the β-strand can form favorable interactions with neighbors in four directions, ie, left, right, up, and down. This neighborliness frequently causes protein aggregation and precipitation. Thus, negative design is a critical feature of β-sheet design for ensuring unfavorable contacts, where appropriate, to prevent aggregation.

Betabellin (Fig. 9) is an example of a de novo designed β-sheet protein. Betabellin contains two identical four-stranded sheets that dimerize via interac-

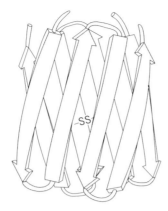

Fig. 9. A de novo designed β-sheet protein, betabellin, formed by the dimerization of two identical four-stranded β-sheets and a disulfide linking the two sheets. This model is for betabellins 9 and later progenies; the earlier betabellins contained a two-armed cross-linker connecting the sheets (51).

tion of their hydrophobic faces to form an antiparallel β-sandwich, ie, a β-sheet, bell-shaped protein. Unlike the modular, minimalist design approach used in helical protein designs, one of the initial design criteria for the betabellins was the construction of a protein-like sequence containing normal amino acid composition, but avoiding direct homology with any natural protein. Other design criteria included secondary structure prediction, statistical preference of residues for positions in β-strands and β-turns, a pattern of alternating hydrophobic and hydrophilic residues to form two opposite faces of the sheet, and consideration of internal packing interactions (53).

Since the first betabellin was designed in the early 1980s, several progeny structures have been designed and tested, each possessing increased amounts of β-structure and improved solubility. The sequences of several of the betabellins are shown in Figure 10. This series of proteins clearly illustrates the iterative nature of de novo protein design. Betabellins 1 through 8 were formed by

Betabellin 1:	STVTARQPNVTYSISPNTATVRLPNfTLSIG
Betabellin 2:	STLTASIPNLTYSISPNTATVKVPNYTLSIG
Betabellin 9:	HTLTASIPDLTYSIDPNTATCKVPDfTLSIGB
Betabellin 12:	HTLTASIpdLTYSINpdTATCKVpdFTLSIGA
Betabellin 14:	HSLTASIkaLTIHVQakTATCQVkaYTVHISE
Betadoublet:	TKLTATQDGLQITINDGTAKCTVDGYQVTIRS
pattern:	epnpnpnttnpnpnprrpnpnpnttnpnpnpe

Fig. 10. Sequences (see Table 1) of betabellins. In each case, only one-half of the β-sandwich is shown. The dimer is formed from identical monomeric sets of four β-strands. In the pattern sequence, e is for end, p is for polar residue, n is for nonpolar residue, and t and r are for turn residues. Lower case f is iodophenylalanine; lower case a, d, k, and p are the D-amino acid forms of alanine, aspartic acid, lysine, and proline, respectively; B is β-alanine (2,53,60,61).

simultaneous synthesis of the two identical 32-residue sequences which were coupled to a two-armed cross-linker on the resin (53). Initial modifications to the design included elimination of consecutive β-branched residues that were believed to couple poorly in the peptide synthesis, and placement of a cysteine residue at position 21 to produce a disulfide across the inside of the barrel. Because these early betabellins were not very water-soluble, betabellins 7 and 8 incorporated eight additional charged residues per molecule, which slightly improved water solubility although secondary structure analysis still had to be performed on the precipitate. For betabellin 9, a significant improvement in solubility was achieved by eliminating the cross-linker and instead synthesizing the single-chain, 32-residue sequence. The cd spectroscopy performed under native conditions confirmed the β-structure, and Raman spectroscopy indicated 60–80% antiparallel β-sheet. Betabellins 12 and 14 (60) incorporated D-amino acids at the turns to promote the formation of tight β-hairpin turns. The more recently characterized β-sandwich proteins betadoublet (61) and betabellin 14D possess significantly more native-like characteristics than their predecessors. In betabellin 14, at least 17 of 32 residues are identical or conserved compared to the earlier sequences. Betadoublet has at least 15 of 32 conserved residues. Whereas the earlier species were insoluble aggregates or only marginally soluble, both of these latter structures are soluble at 10 mg/mL in aqueous buffers.

As for the α-helical proteins, all of the β-designed proteins still possess some characteristics of the molten globule state. Thus, whereas the hydrophobic force plays a dominant role in the folding of a protein, it is not one that is sufficient to define the unique tertiary structure of native proteins. To design a truly native-like protein de novo, specific side-chain interactions which adopt unique packing characteristics must be part of the design process. Design efforts aimed toward reaching this goal are ongoing in several laboratories.

Studies of Protein Function.   For enzymes to catalyze reactions, and receptors to transport molecules and transmit signals across cellular membranes, both classes of proteins require a ligand-binding pocket, the structure of which is highly complementary to that of the ligand. It is this complementarity of structure that provides the high degree of ligand specificity characteristic of these two classes of proteins. The binding-pocket structure depends on the overall tertiary structure of the protein as well as the specific amino acid side chains that line the binding pocket. In the case of enzymes, changes in the residues within the binding pocket can greatly affect the rate or nature of the reaction catalyzed. For both enzymes and receptors, changes of amino acid residues within the binding pocket can also dramatically alter the ligand specificity. Understanding these structure–function relationships would allow the rational design of a binding pocket to alter the ligand specificity of these proteins for specific purposes, such as altered catalysis by an enzyme, or transport of unnatural molecules into cells by an engineered receptor.

Enzymes.   Protein engineering has been used both to understand enzyme mechanism and to selectively modify enzyme function (4,5,62–67). Much as in protein stability studies, the role of a particular amino acid can be assessed by replacement of a residue incapable of performing the same function. An understanding of how the enzyme catalyzes a given reaction provides the basis for manipulating the activity or specificity.

Many enzymes have been the subject of protein engineering studies, including several that are important in medicine and industry, eg, lysozyme, trypsin, and cytochrome P450. Subtilisin, a bacterial serine protease used in detergents, foods, and the manufacture of leather goods, has been particularly well studied (68). This emphasis is in part owing to the wealth of structural and mechanistic information that is available for this enzyme.

Catalysis. The active site of subtilisin BPN$'$ contains the catalytic triad of amino acids common to the serine proteases, ie, Ser221, His64, and Asp32, as well as an oxyanion binding site, Asn155 and the main chain amide of Ser221 (Fig. 11). The catalytic turnover number, k_{cat}, is $10^9 - 10^{10}$ times greater than the first-order spontaneous hydrolytic rate for amide substrates (69). The contribution of the residues of the catalytic triad to enzyme rate enhancement was assessed by mutating individual residues to alanine. Mutation of Ser221, His64, and Asp32 reduced k_{cat} by a factor of 2×10^6-, 2×10^6-, and 3×10^4-fold, respectively (69,70). The fact that only small changes in K_m were observed for these mutants indicates that the reduced catalytic activity is not a result of impaired substrate binding. Even when all three residues in the catalytic triad are replaced simultaneously by alanine, the mutant retains a catalytic activity about 10^3-fold above the uncatalyzed rate. Various studies have suggested other binding interactions contribute to rate enhancement by stabilizing the transition state of the reaction. For example, elimination of the hydrogen bond to Asn155 by substitution with threonine, which is too short to form an optimal H-bond, reduces k_{cat} another 10^3-fold without significantly altering K_m (71,72). Hence, in these mutants substrate binding is not hindered, but the rate of catalysis is reduced.

Engineering Substrate Specificity. Although the serine proteases use a common catalytic mechanism, the enzymes have a wide variety of substrate specificities. For example, the natural variant subtilisins of *B. amyloliquefaciens* (subtilisin BPN$'$) and *B. licheniformis* (subtilisin Carlsberg) possess very similar structures and sequences where 86 of 275 amino acids are identical, but have different catalytic efficiencies, k_{cat}/K_m, toward tetraamino acid *p*-nitroanilide substrates (67). Subtilisin Carlsberg is more active on low molecular weight synthetic substrates. The primary basis for this difference has been attributed to three residues in subtilisin BPN$'$ important for substrate binding: Tyr217, Glu156, and Gly169. When these residues are replaced by the corresponding residues of subtilisin Carlsberg, the substrate specificities become very similar to that of the latter species (73). Because of their efficiency in breaking down proteins, subtilisins have found use commercially in laundry cleaners. Subtilisin BPN$'$ is two times more efficient than subtilisin Carlsberg in laundry applications. The BPN$'$ Tyr217Leu mutant is 10 times more efficient than BPN$'$ for the hydrolysis of low molecular weight synthetic substrates, and its overall performance in laundry detergent is two times better than that of BPN$'$. These studies illustrate the feasibility of designing hybrid enzymes which combine the desired properties of two or more natural proteins.

Complementary steric and hydrophobic interactions have also been engineered into subtilisin BPN$'$ to enhance substrate specificity. In general, large hydrophobic residues are preferred in the P1 substrate position, ie, P1 refers to the first residue on the N-terminal side of the scissle bond of the substrate, 68.

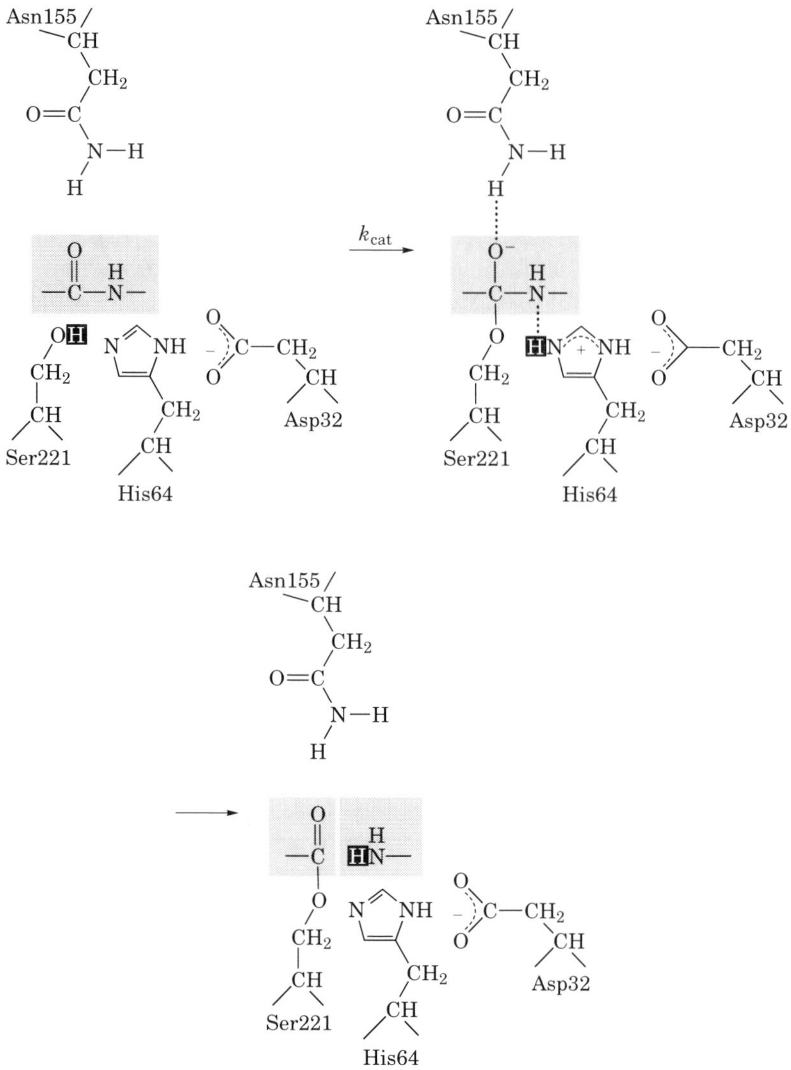

Asn155
CH
CH₂
O=C
N—H
H

O
‖ H
—C—N—

OH N NH   —C—CH₂
CH₂   O   CH
CH   CH₂
Ser221   CH
        His64   Asp32

k_{cat} →

Asn155
CH
CH₂
O=C
N—H
H

O⁻
| H
—C—N—

O   HN⁺ NH
CH₂   CH₂
CH   CH
Ser221   His64

—C—CH₂
O   CH
        Asp32

↓

Asn155
CH
CH₂
O=C
N—H
H

O
‖ H
—C HN—

O   N NH   —C—CH₂
CH₂   O   CH
CH   CH₂
Ser221   CH
        His64   Asp32

Fig. 11. Active site of the serine protease subtilisin BPN′. The amide bond of the peptide substrate is shown in gray. In the rate-limiting acylation step for peptide hydrolysis, k_{cat}, the oxygen atom on the hydroxyl of Ser221 performs a nucleophilic attack on the carbonyl of the substrate, forming a tetrahedral intermediate. His64 facilitates this nucleophilic attack of Ser221 by accepting the hydrogen atom on the hydroxyl. The positively charged His64 is in turn stabilized by the negatively charged Asp32. The substrate peptide bond is cleaved when His64 donates a proton to the substrate amide nitrogen. Deacylation (not shown) occurs by hydrolysis of the acyl enzyme intermediate in essentially the reverse of the steps above (69). Reprinted with permission from Macmillan Magazines Limited.

If, however, Gly166 in the binding pocket of the enzyme is replaced by larger residues, the volume of the binding pocket is reduced, sterically excluding large P1 side chains and enhancing the catalysis of smaller substrates (72,74).

Specificity for a particular charged substrate can be engineered into an enzyme by replacement of residues within the enzyme-active site to achieve electrostatic complementarity between the enzyme and substrate (75). Protein engineering, when coupled with detailed structural information, is a powerful technique that can be used to alter the catalytic activity of an enzyme in a predictable fashion.

Engineering the pH Profile of Subtilisin. The activity of subtilisin BPN′ increases between pH 6 and 8 as His64 ($pK_a \sim 7.2$) is deprotonated (68). Changes in the surface charge of subtilisin have been used to shift the pH activity profile by altering the pK_a of the active site histidine. Two surface acidic residues, Asp99 and Glu156, each located more than 1 nm from the catalytic His64, were replaced with Ser and then Lys. These individual mutants and double mutants exhibited lower His64 pK_a values. The combined effects of the single mutants were cumulative in the double mutants, and in the most extreme case, the double mutant Asp99Lys/Glu156Lys had a pK_a shifted down a full pH unit (68 and references therein). The activity of this double mutant was found to be twofold higher than the wild type enzyme at pH 8, and 10 times greater at pH 6.

Altered Enzyme Function. Manipulation of subtilisin BPN′ has been taken beyond the usual realm of protein engineering studies to include switching the primary enzyme function from peptide hydrolysis to peptide ligation. The general mechanism of serine proteases involves formation of an acyl enzyme intermediate followed by hydrolysis of this intermediate to release the peptide containing a free carboxy terminus. However, an alternative is also possible. In mixed or pure organic solvents, aminolysis, in which a primary amine attacks the acyl enzyme intermediate, is favored over hydrolysis (76). Thus, ligation of the original peptide to another peptide is possible. However, the fact that enzymes are generally not stable in organic solvents precludes the practical use of this synthetic process. A derivative of subtilisin BPN′, in which Ser221 was chemically converted to a cysteine, was shown to favor aminolysis over hydrolysis by >1000-fold (77). The Ser221Cys mutant, termed thiolsubtilisin, was shown to have greater ligase activity than wild-type subtilisin, but still retained significant amidase activity, making its use impractical for synthetic purposes. To improve the ligase activity of thiolsubtilisin, a second mutation, Pro225Ala, was incorporated. The resulting double mutant, termed subtiligase, had a 100-fold reduced amidase activity and 10-fold greater ligase activity than the Ser221Cys mutant, yielding over 95% aminolysis without peptide hydrolysis (76).

Subtiligase has been used to sequentially ligate (couple) synthetic peptide fragments corresponding to RNase A. A full-length protein having higher yield and purity than previously reported for other synthetic processes was generated (78). Characterization of the final product showed it to be identical to commercial RNase A. The use of subtiligase offers advantages over *in vitro* transcription and translation methods, including the ability to incorporate multiple, and different, substitutions and higher yield of purified protein product. The principal disadvantage of its use is that the technology involved is limited to ligation of proteins that can be folded *in vitro*.

Receptors. Integral membrane receptor proteins serve critical roles in the cellular functions of transport and signal transduction. The essential and diverse roles of these receptor proteins make them excellent targets for protein engineering studies. Protein engineering techniques have been used to map receptor binding sites for the various molecules that interact with the receptor, such as agonist, antagonist, and intracellular mediators, and to identify regulatory regions of the receptor, such as phosphorylation sites. Protein domains, as well as specific amino acid residues, involved in receptor–ligand interactions have been identified by deletion mutagenesis, site-directed mutagenesis, and the construction of chimeric proteins (6,79–81).

The β-Adrenergic Receptor. An excellent example of the application of protein engineering techniques to an integral membrane receptor protein is found in the β-adrenergic receptor (βAR). The β-adrenergic receptor belongs to the family of G-protein coupled receptors (GPCRs). Upon agonist binding, the receptor activates a G-protein by catalyzing the exchange of guanosine diphosphate (GDP) for guanosine triphosphate (GTP) to give the active GTP-bound form, which then initiates various intracellular events. Initial ligand specificity is determined by the receptor. Different types of GPCRs couple to specific G-proteins, which in turn stimulate distinct intracellular effector systems. In the case of the β-adrenergic receptor, binding of a catecholamine agonist, such as the biogenic amines epinephrine and norepinephrine (qv), activates the G-protein G_s, which then stimulates the enzyme adenylate cyclase, resulting in production of cyclic adenosine monophosphate (cAMP), the increased concentration of which affects other cellular events, etc.

Models of GPCRs. The extensive biochemical and pharmacological characterization of many GPCRs has been accomplished in the absence of three-dimensional structure information. This fact stands in contrast to that relating to the various other examples described above, in which the rational modification of several soluble proteins, especially enzymes, has been guided by the known three-dimensional structure. Because membrane proteins are generally difficult to crystallize, structural information has been deduced from studies other than direct crystallization. Advances in genetic engineering (qv) techniques have led to the cloning and identification of several GPCRs. The availability of a large database of protein sequences has permitted identification of common structural motifs and conserved residues that may serve essential functional or structural roles. Several models of GPCR structure have been developed (82), all having seven highly conserved hydrophobic regions, proposed to be membrane-spanning α-helices (Fig. 12). These are connected by more divergent hydrophilic regions that form alternating extracellular and intracellular loops. The seven helices are believed to form a transmembrane cluster which creates the ligand binding pocket.

Site-Directed Mutagenesis Studies. The results of several initial protein engineering studies on the beta-2 adrenergic receptor (β2AR) that involved the construction and analysis of deletion mutants suggested that ligand binding occurs within the hydrophobic core. Therefore, several site-directed mutagenesis studies of transmembrane residues were undertaken. Adrenergic agonists and antagonists are biogenic amines having protonatable nitrogens that are critical for receptor binding. Thus all acidic residues in the transmembrane region of the receptor were mutated (83,84). Two important aspartic acid residues, Asp113 and Asp79, were identified as a result of these studies. Ligand binding affinity, K_d, for a radiolabeled antagonist was determined by saturation binding isotherms. The mutant receptor in which Asp79 was replaced by an alanine,

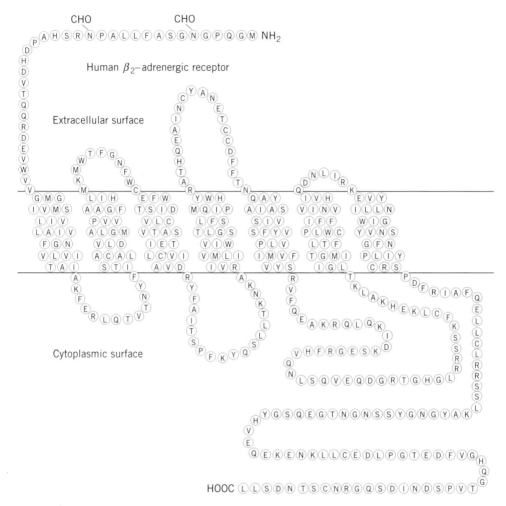

Fig. 12. Model for the β₂-adenergic receptor. It is proposed that the receptor possesses seven hydrophobic helices that span the plasma membrane and are connected by alternating extracellular and intracellular loops (79). The site of glycosylation is represented as CHO. Reprinted with permission from Elsevier Science Ltd., Kidlington, U.K.

Asp79Ala-β_2AR, had a K_d for the antagonist of 60 pM, identical to the wild-type receptor. Based on the fact that agonist and antagonist binding is both mutually exclusive and competitive, the binding affinity of nonradioactive agonist was determined by a competition assay in which the concentration of agonist required to obtain 50% maximal binding (EC$_{50}$) to the receptor in the presence of the antagonist was determined. These experiments showed that the mutant Asp79Ala-β_2AR exhibited a 10-fold increase in K_d for agonist (isoproterenol) binding.

The radioligand binding assays could not be used to assess antagonist or agonist binding to Asp113 mutants, owing to undetectable ligand binding. Therefore, a functional assay was used in which stimulation of the enzyme adenylate cyclase was measured. Using this assay, even very low ligand-binding affinities can be detected by measuring the functional consequences of ligand binding that result in activation of adenylate cyclase. The ability of the isoproterenol to activate adenylate cyclase was assessed by determining a dose response curve in which the rate of production of cAMP as a function of the dose of agonist was measured. The Asp79Ala-β_2AR mutant exhibited a 10-fold increase in the activation constant, K_{act}, for adenylate cyclase stimulation. Substitution of Asp113 with Glu or Asn reduced agonist affinity such that a $10^2 - 10^4$-fold higher concentration of agonist was required for stimulation of adenylate cyclase. These same substitutions reduced the binding affinity of the antagonist propranolol by a factor of 10^4, as assessed by inhibition of isoproterenol-stimulated adenylate cyclase activation. The fact that substitution at position 113 gives increased K_{act} for agonists and increased K_i values for antagonists suggests a similar role for Asp113 in binding these ligands. Based on these results it was concluded that Asp113, within transmembrane helix 3, is essential for high affinity agonist and antagonist binding to the receptor. The observation that binding of isoproterenol still elicited agonist activity suggests that Asp113 is essential for ligand binding, but not for functional activation of the receptor (84). Compared to the Asn mutant, substitution of Asp113 by Glu resulted in affinities closer to the wild-type values, suggesting that the carboxylate moiety may contribute some binding interaction. Although the K_{act} value increased by 2–4 orders of magnitude for Asp113Glu and Asp113Asn, the β_2 subtype ligand-specificity was maintained. Hence, the factors that determine binding selectivity for substituents on the amino groups of isoproterenol (isopropyl), epinephrine (methyl), and norepinephrine (H) do not appear to involve the Asp113 residue. Asp79 may directly interact with agonist or, more likely, maintain the conformation of the agonist-bound receptor. The existence of overlapping but distinct binding sites is supported by the observation that Asp79 mutations affect agonist but not antagonist binding. Asp79 is widely conserved in GPCRs, but Asp113 is present only in GPCRs that bind cationic amines as ligands. The role of Asp113 in determining the specificity of ligand binding was further emphasized by the demonstration (85) that replacement of the residue by a Ser converts the ligand specificity from catecholamines to catechol ketones and esters. Serine contains a hydroxyl side chain capable of hydrogen bonding to the carbonyls of ketones and esters.

Further elucidation of binding interactions which differentiate agonists from antagonists was accomplished by replacing putative hydrogen-bonding residues in the transmembrane region with alanines. These would not productively interact with the hydroxyls of the catechol (86). Pharmacophore mapping studies performed in the late 1970s suggested that hydrogen bonding interactions with the catechol hydroxyl groups were essential for agonist activation of β_2AR. Two serines, Ser204 and Ser207, which are conserved among catecholamine-binding GPCRs, were identified as important for hydrogen-bonding to the hydroxyls of the catecholamine. Mutants of both these residues exhibited normal specificity of antagonist binding but reduced antagonist affinity, as well as reduced efficacy of agonist-induced activation. Stimulation of the mutant receptors by various ligands suggested that Ser204 hydrogen bonds to the *meta*-hydroxyl of the catechol, whereas Ser207 hydrogen bonds to the *para*-hydroxyl group of the ligand. Thus, the serines orient the catechol ring in the binding pocket by hydrogen bonding to the hydroxyl groups of the ligand (Fig. 13). The localization of agonist-specific interactions with the serine residues

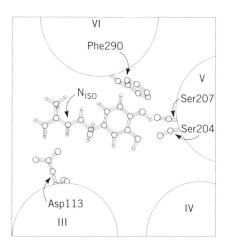

Fig. 13. Model for the ligand binding site of the β_2-adrenergic receptor (87). The view of the receptor is from the extracellular face of the plasma membrane. Four of the seven transmembrane helices are designated by Roman numerals. Arrows indicate those residues that are thought to interact with the agonist isoproterenol, on the basis of site-directed mutagenesis studies. The structure of isoproterenol in the proposed binding pocket is illustrated. The position of the amino nitrogen (N_{ISO}) is highlighted.

in transmembrane helix 5 suggested a mechanism for agonist activation of the receptor. Investigations of the receptor/G-protein coupling have revealed that the third intracellular loop, connecting transmembrane helices 5 and 6, is the region of G-protein binding (88–92). Binding of agonist might cause conformational changes in helix 5 which are transmitted to the third intracellular loop to facilitate activation of G_s, and hence overall signal transduction.

Expanded Genetic Code. The protein engineering techniques described have employed site-directed mutagenesis to elucidate the role of a particular residue in protein structure or function. A technique that uses natural protein synthesis machinery site-specifically to incorporate unnatural amino acids into the protein has also been adapted (93,94). Thus, instead of being limited to the 20 naturally occurring amino acids specified by the genetic code, an amino acid incorporating novel steric, electronic, or spectroscopic properties can be employed permitting a more systematic variation in the properties of the mutated residue. Size, shape, acidity, nucleophilicity, hydrogen-bonding strength, and hydrophobicity can all be probed. Additionally, biophysical probes such as spin and isotopic labels can be incorporated into a protein at the site of interest. This technique has been applied to the study of protein stability, enzymatic catalysis, and signal transduction.

The procedure for incorporating a site-specific unnatural amino acid involves the following steps. The codon for the amino acid of interest is replaced with the stop codon UAG, using standard oligonucleotide-directed mutagenesis. A suppressor tRNA that recognizes this codon is then chemically acylated using the desired unnatural amino acid. Transcription and translation of the mutagenized mutated gene is performed *in vitro* using an *E. coli* or rabbit reticulocyte extract to synthesize the mutant protein containing the unnatural amino acid at the specified position (93).

Examples.   Protein Stability.   Because it is generally accepted that the hydrophobic effect is the principal force in stabilizing globular proteins, amino acids which increase the bulk of buried hydrophobic surface area (packing density) should increase the stability of the protein. Many studies have been performed on the role of cavities in T4 lysozyme. Using natural amino acid replacements, some mutations designed to increase protein stability by filling the largest hydrophobic cavities had a slight destabilizing effect, owing to disruptive interactions with neighboring residues and the adoption of unfavorable dihedral angles (95). In contrast, the unnatural amino acids, *S,S*-2-amino-4-methylhexanoic acid and *S*-2-amino-3-cyclopentylpropanoic acid, designed to fill the cavity with minimal strain, increased the thermal stability of the enzyme (96). Other studies, using unnatural amino acids have probed the importance of H-bonding, and β-branching residues in the stabilization of α-helices (93,97).

Enzyme Mechanism.   Staphylococcal nuclease (SNase) accelerates the hydrolysis of phosphodiester bonds in nucleic acids (qv) some 10^{16}-fold over the uncatalyzed rate (93 and references therein). Mutagenesis studies in which Glu43 has been replaced by Asp or Gln have shown Glu to be important for high catalytic activity. The enzyme mechanism is thought to involve base catalysis in which Glu43 acts as a general base and activates a water molecule that attacks the phosphodiester backbone of DNA. To study this mechanistic possibility further, Glu was replaced by two unnatural amino acids, homoglutamate, which is one methylene group longer than Glu, and *S*-4-nitro-2-aminobutyric acid, which is isoelectronic and isosteric to glutamate but a much poorer base (98). Both mutants had kinetic constants similar to the wild-type enzyme. The conclusion from these studies is that Glu43 may not act as a general base but instead may serve a structural role as a bidentate H-bond acceptor, a role both unnatural substitutions could fill, to fix the conformation of the neighboring loop. Results using deletion mutagenesis of the loop (99) have suggested that this loop may be important in the product dissociation step of the catalytic mechanism.

Cellular Signal Transduction.   Ras p21 is a low molecular weight GTP-binding protein that is important in cell growth and differentiation. When GTP is bound to the enzyme it is in the active signaling state and the signaling is turned off as the GTP is hydrolyzed to GDP. Point mutations that decrease the intrinsic GTPase activity of Ras or the GTPase activity stimulated by the GTPase-activating protein (GAP) are associated with approximately 30% of human cancers. Several regions of Ras, including loop L4 (switch II region), loop L2 (switch I region), and loop L1 (phosphate binding loop) have been studied by replacement with natural and unnatural amino acids (93 and references therein). For example, the conformation of loop L2 is different in the GTP and GDP bound forms. It has been proposed that a cis–trans isomerization of Pro34 may be responsible for the differences in the loop conformation. However, when Pro34 was replaced with a proline analogue that is conformationally locked in the trans state (2,4-methanoproline), no differences were observed in GTPase activity (100,101). Therefore, the cis–trans isomerization of Pro is not essential for signal transduction.

BIBLIOGRAPHY

1. A. R. Rees, M. J. E. Sternberg, and R. Wetzel, *Protein Engineering: A Practical Approach*, IRL Press at Oxford University Press, New York, 1992.
2. D. L. Oxender and C. F. Fox, eds., *Protein Engineering*, Alan R. Liss, Inc., New York, 1987.
3. C. Robinson, ed., *Trends Biotechnol.* **12**, 145–211 (1994). The entire May issue is devoted to the topic of Protein Engineering.
4. A. Fersht and G. Winter, *Trends Biochem. Sci.* **17**, 292–294 (1992); A. Recktenwald, D. Schomburg, and R. D. Schmid, *J. Biotechnol.* **28**, 1–23 (1993).
5. R. A. Copeland, *Enzymes: A Practical Introduction to Structure, Mechanism, and Data Analysis*, VCH, New York, 1996.
6. C. Chothia, *J. Mol. Biol.* **105**, 1–14 (1976).
7. J. Kyte and R. F. Doolittle, *J. Mol. Biol.* **157**, 105–132 (1982).
8. M. H. Klapper, *Biochem. Biophys. Res. Commun.* **78**, 1018–1024 (1977).
9. M. O. Dayhoff, R. M. Schwartz, and B. C. Orcutt, in M. Dayhoff, ed., *Atlas of Protein Sequence and Structure*, Vol. 5, National Biomedical Research Foundation, Washington, D.C., 1978, pp. 345–352.
10. R. A. Copeland, *Methods for Protein Analysis: A Practical Guide to Laboratory Protocols*, Chapman & Hall, New York, 1994.
11. W. H. J. Ward, D. Timms, and A. R. Fersht, *Trends Pharmacol. Sci.* **11**, 280–284 (1990).
12. D. E. McRee, *Practical Protein Crystallography*, Academic Press, Inc., New York, 1993.
13. O. Jardetzky and G. C. K. Roberts, *NMR in Molecular Biology*, Academic Press, Inc., Orlando, Fla., 1981.
14. R. G. Krishna and F. Wold, in A. Meister, ed., *Advances in Enzymology and Related Areas of Molecular Biology*, Vol. 67, John Wiley & Sons, Inc., New York, 1993, pp. 265–298.
15. T. E. Hugli, *Techniques in Protein Chemistry*, Academic Press, Inc., San Diego, Calif., 1989.
16. J. D. Watson, M. Gilman, J. Witkowski, and M. Zoller, *Recombinant DNA*, 2nd ed., Scientific American Books, distributed by W. H. Freeman and Co., New York, 1992.
17. J. Sambrook, E. F. Fritsch, and T. Maniatis, *Molecular Cloning*, 2nd ed., Cold Spring Harbor Press, Cold Spring Harbor, N.Y., 1989.
18. F. M. Ausubel and co-workers, eds., *Current Protocols in Molecular Biology*, John Wiley & Sons, Inc., New York, 1992.
19. T. A. Kunkel, K. Bebenek, and J. McClary, *Methods Enzymol.* **204**, 125–139 (1991).
20. M. A. Innis, D. H. Gelfand, J. J. Sninsky, and T. J. White, *PCR Protocols: A Guide To Methods and Applications*, Academic Press, Inc., San Diego, Calif., 1990.
21. BioRad product bulletin, Lit-245 0890 90-0910a, *Muta-Gene M13 In Vitro Mutagenesis Kit*, Cat. No. 170-3580, BioRad Laboratories, Hercules, Ca., 1990.
22. *Methods in Enzymology* **185** (1990). The entire volume covers gene expression.
23. J. Hodgson, *Bio/Technol.* **11**, 887–893 (1993).
24. R. C. Hockney, *Trends Biotechnol.* **12**, 456–463 (1994).
25. J. M. Cregg, T. S. Vedvick, and W. C. Raschke, *Bio/Technol.* **11**, 905–910 (1993).
26. R. K. Scopes, *Protein Purification: Principles and Practice*, 3rd ed., Springer-Verlag, New York, 1994.
27. M. P. Deutscher, ed., *Methods in Enzymology*, Vol.182, *Guide to Protein Purification*, Academic Press, Inc., San Diego, Calif., 1990.

28. E. L. V. Harris and S. Angal, eds., *Protein Purification Methods: A Practical Approach*, IRL Press at Oxford University Press, Oxford, U.K., 1989.

29. E. L. V. Harris and S. Angal, eds., *Protein Purification Applications: A Practical Approach*, IRL Press at Oxford University Press, Oxford, U.K., 1990.

30. M. Uhlen and T. Miks, *Methods Enzymol.* **185**, 129–143 (1990).

31. C. V. Maina and co-workers, *Gene* **74**, 365–373 (1988).

32. P. L. Privalov, *Advan. Protein Chem.* **33**, 167–241 (1979).

33. A. R. Fersht and L. Serrano, *Curr. Opin. Struct. Biol.* **3**, 75–83 (1993).

34. A. R. Fersht, S. E. Jackson, and L. Serrano, *Phil. Trans. R. Soc. Lond. A.* **345**, 141–151 (1993).

35. B. W. Matthews, *Ann. Rev. Biochem.* **62**, 139–160 (1993).

36. B. W. Matthews, *Curr. Opin. Struct. Biol.* **3**, 589–593 (1993).

37. D. Shortle, *J. Biol. Chem.* **264**, 5315–5318 (1989).

38. P. W. Goodenough and J. A. Jenkins, *Biochem. Soc. Trans.* **19**, 655–662 (1991).

39. C. N. Pace, *CRC Crit. Rev. Biochem.* **3**, 1–43 (1975).

40. C. N. Pace, *J. Mol. Biol.* **226**, 29–35 (1992).

41. A. E. Eriksson and co-workers, *Science* **255**, 178–183 (1992).

42. L. Serrano, J. T. Kellis, P. Cann, A. Matouschek, and A. R. Fersht, *J. Mol. Biol.* **224**, 783–804 (1992).

43. B. A. Shirley, P. Stanssens, U. Hahn, and C. N. Pace, *Biochemistry* **31**, 725–732 (1992).

44. P. L. Privilov and G. I. Makhatadze, *J. Mol. Biol.* **232**, 660–679 (1993).

45. S. Dao-Pin, U. Sauer, H. Nicholson, and B. W. Matthews, *Biochemistry* **30**, 7142–7153 (1991).

46. D. E. Anderson, W. J. Becktel, and F. W. Dahlquist, *Biochemistry* **29**, 2403–2408 (1990).

47. R. Wetzel, L. J. Perry, W. A. Baase, and W. J. Becktel, *Proc. Natl. Acad. Sci.* **85**, 401–405 (1988).

48. M. Matsumara and B. W. Matthews, *Methods Enzymol.* **202**, 336–356 (1991).

49. C. Sander, *Trends Biotechnol.* **12**, 163–167 (1994).

50. S. F. Betz, D. P. Raleigh, and W. F. DeGrado, *Curr. Opin. Struct. Biol.* **3**, 601–610 (1993).

51. J. S. Richardson and co-workers, *Biophys. J.* **63**, 1186–1209 (1992).

52. W. F. DeGrado, D. P. Raleigh, and T. Handel, *Curr. Opin. Struct. Biol.* **1**, 984–993 (1991).

53. J. S. Richardson and D. C. Richardson, *Trends Biochem. Sci.* **14**, 304–309 (1989).

54. S. P. Ho and W. F. DeGrado, *J. Am. Chem. Soc.* **109**, 6751–6758 (1987).

55. W. F. DeGrado, Z. R. Wasserman, and J. D. Lear, *Science* **24**, 622–628 (1989).

56. T. M. Handel, S. A. Williams, and W. F. DeGrado, *Science* **261**, 879–885 (1993).

57. D. P. Raleigh and W. F. DeGrado, *J. Am. Chem. Soc.* **114**, 10079–10081 (1992).

58. S. Kamtekar, J. M. Schiffer, H. Xiong, J. M. Babik, and M. H. Hecht, *Science* **262**, 1680–1685 (1993).

59. M. H. Hecht, *Proc. Natl. Acad. Sci. USA* **91**, 8729–8730 (1994).

60. Y. Yan and B. W. Erickson, *Protein Science* **3**, 1069–1073 (1994).

61. T. P. Quinn, N. B. Tweedy, R. W. Williams, J. S. Richardson, and D. C. Richardson, *Proc. Natl. Acad. Sci. USA* **91**, 8747–8751 (1994).

62. L. Hedstrom, *Curr. Opin. Struct. Biol.* **4**, 608–611 (1994).

63. J. A. Gerlt, *Curr. Opin. Struct. Biol.* **4**, 593–600 (1994).

64. A. S. El Hawrani, K. M. Moreton, R. B. Sessions, A. R. Clarke, and J. J. Holbrook, *Trends Biotechnol.* **12**, 207–211 (1994).

65. C. Wilson and D. A. Agard, *Curr. Opin. Struct. Biol.* **1**, 617–623 (1991).

66. W. M. Atkins and S. G. Sligar, *Curr. Opin. Struct. Biol.* **1**, 611–616 (1991).

67. D. A. Estell, *J. Biotechnol.* **28**, 25–30 (1993).

68. J. A. Wells and D. A. Estell, *Trends Biochem. Sci.* **13**, 291–297 (1988).

69. P. Carter and J. A. Wells, *Nature* **332**, 564–568 (1988).

70. P. Carter and J. A. Wells, *Science* **237**, 394–399 (1987).

71. J. A. Wells, B. C. Cunningham, T. P. Graycar, and D. A. Estell, *Philos. Trans. R. Soc. London A* **317**, 415–423 (1986).

72. P. Bryan, M. W. Pantoliano, S. G. Quill, H. Y. Hsiao, and T. Poulos, *Proc. Natl. Acad. Sci. USA* **83**, 3743–3745 (1986).

73. J. A. Wells, B. C. Cunningham, T. P. Graycar, and D. A. Estell, *Proc. Natl. Acad. Sci.* **84**, 5167–5171 (1987).

74. D. A. Estell and co-workers, *Science* **233**, 659–663 (1986).

75. J. A. Wells, D. B. Powers, R. R. Bott, T. P. Graycar, and D. A. Estell, *Proc. Natl. Acad. Sci. USA* **84**, 1219–1223 (1987).

76. L. Abrahmsen and co-workers, *Biochemistry* **30**, 4151–4159 (1991).

77. K. Nakatsuka, T. Sasaki, and E. T. Kaiser, *J. Am. Chem. Soc.* **109**, 3808–3810 (1987).

78. D. Y. Jackson and co-workers, *Science* **266**, 243–247 (1994).

79. R. J. Lefkowitz, B. K. Kolbika, and M. G. Caron, *Biochem. Pharmacol.* **38**, 2941–2948 (1989).

80. T. M. Fong and C. D. Strader, *Med. Res. Rev.* **14**, 387–399 (1994).

81. T. M. Savarese and C. M. Fraser, *Biochem. J.* **283**, 1–19 (1992).

82. D. Donnelly and J. C. Findlay, *Curr. Opin. Struct. Biol.* **4**, 582–589 (1994).

83. C. D. Strader and co-workers, *Proc. Natl. Acad. Sci. USA* **84**, 4384–4388 (1987).

84. C. D. Strader and co-workers, *J. Biol. Chem.* **263**, 10267–10271 (1988).

85. C. D. Strader and co-workers, *J. Biol. Chem.* **266**, 5–8 (1991).

86. C. D. Strader, M. R. Candelore, W. S. Hill, I. S. Sigal, and R. A. F. Dixon, *J. Biol. Chem.* **264**, 13572–13578 (1989).

87. C. D. Strader, I. S. Sigal, and R. A. Dixon, *FASEB J.* **3**, 1825–1832 (1989).

88. C. D. Strader and co-workers, *J. Biol. Chem.* **262**, 16439–16443 (1987).

89. R. A. F. Dixon and co-workers, *Nature* **326**, 73–77 (1987).

90. R. A. F. Dixon and co-workers, *EMBO J.* **6**, 3269–3275 (1987).

91. B. F. O'Dowd and co-workers, *J. Biol. Chem.* **263**, 15985–15992 (1988).

92. K. A. Kjelsberg, S. Cotecchia, J. Ostrowski, M. G. Caron, and R. J. Lefkowitz, *J. Biol. Chem.* **267**, 1430 (1992).

93. V. W. Cornish and P. G. Schultz, *Curr. Opin. Struct. Biol.* **4**, 601–607 (1994).

94. J. Ellman, D. Mendel, S. Anthony-Cahill, C. J. Noren, and P. G. Schultz, *Methods Enzymol.* **202**, 310–336 (1991).

95. M. Karpusas, W. A. Baase, M. Matsumura, and B. W. Matthews, *Proc. Natl. Acad. Sci. USA* **86**, 8237–8241 (1989).

96. D. Mendel and co-workers, *Science* **256**, 1798–1802 (1992).

97. P. C. Lyu, J. C. Sherman, A. Chen, and N. R. Kallenbach, *Proc. Natl. Acad. Sci. USA* **88**, 5317–5320 (1991).

98. J. K. Judice, T. R. Gamble, E. C. Murphy, A. M. de Vos, and P. G. Schultz, *Science* **261**, 1578–1581 (1993).

99. S. P. Hale, L. B. Poole, and J. A. Gerlt, *Biochemistry* **32**, 7479–7487 (1993).

100. H.-H. Chung, D. R. Benson, V. W. Cornish, and P. G. Schultz, *Proc. Natl. Acad. Sci. USA* **90**, 10145–10149 (1993).

101. H.-H. Chung, D. R. Benson, and P. G. Schultz, *Science* **259**, 806–809 (1993).

JUNE P. DAVIS
ROBERT A. COPELAND
The Du Pont Merck Research Laboratories

PROTEINS

Proteins, ubiquitous to all living systems, are biopolymers (qv) built up of various combinations of 20 different naturally occurring amino acids (qv). The number of proteins in an organism may be as small as half a dozen, as in the case of the simple bacterial virus M13, or as large as 50,000, as in the human system. Proteins are encoded by the deoxyribonucleic acid (DNA) that is present in all living cells.

Protein uses are myriad. The large number of biochemical reactions within the living cell are catalyzed by enzyme proteins (see ENZYME APPLICATIONS; HORMONES; PHARMACEUTICALS). Traditional food processes such as baking (see BAKERY PROCESSES AND LEAVENING AGENTS), brewing (see BEER), and cheese-making (see MILK AND MILK PRODUCTS) involve the action of enzymes found in different microorganisms. Other proteins are involved in the transport of electrons, ions, and small molecules. Proteins are also key components of the immune system (see IMMUNOTHERAPEUTIC AGENTS) and control the genetic expression of other proteins (see GENETIC ENGINEERING). The smallest proteins have a molecular weight of only about 400; the largest protein molecule discovered to date has a molecular weight of 2,700,000.

The study of proteins has a long history. First isolated from both animal and plant sources in the late 1700s, their chemical composition was studied in the 1820s and 1830s. Glycine was the first constituent amino acid to be isolated from gelatin in 1820. The name protein, meaning primary, was coined in 1839. The concept of the linear linkage of amino acids each having the same backbone structure to form the protein was established by Emil Fischer in 1901. Great advances have taken place in protein research since the 1950s through biochemistry and molecular biology (see PROTEIN ENGINEERING).

Properties

The different combinations of amino acid side chains in a protein molecule give the protein its unique structure and function. Although the size and chemical properties of the amino acid side chains vary greatly, these building blocks can be broadly classified into three categories: hydrophobic, charged, and uncharged polar. A set of classical experiments in the 1960s showed that the structure (conformation) of ribonuclease (RNase) is determined by the amino acid sequence under normal physiological conditions (1). This has proved to be true for most proteins.

Purification and Sequencing.   The properties of a protein such as size, net electric charge, and solubility can be exploited in protein purification. Centrifugation is used to separate proteins based on density and shape. Electrophoresis exploits the net charge on a protein as well as its mass and shape (see ELECTROSEPARATIONS). Electrophoretic separation is usually carried out by passing the mixture through a gel medium in an electric field. In chromatography (qv), separation of proteins is carried out by passing them through a column packed with some porous material. The mixture is separated according to size in size-exclusion chromatography, ionic charge in ion-exchange chromatography, or selective adsorption in affinity-labeled chromatography (2).

The method for sequencing proteins was developed in the 1950s (3). A protein was first cleaved into smaller fragments and then hydrolyzed into the constituent amino acids. The individual amino acids were separated on the basis of their charge or hydrophobicity and quantified. The complete sequence was then generated from the fragments. In the 1990s, however, this process has become passé and protein sequences are inferred from their corresponding nucleotide sequences. Large databases (qv) exist for the sequences of proteins from different organisms (4).

Biosynthesis

The whole of molecular biology is based on the central dogma that DNA → ribonucleic acid (RNA) → protein. Each amino acid residue is encoded by a triplet of nucleotides which form the genetic material of chromosomes. The DNA and, in some cases the RNA, is transcribed onto a strand of messenger-RNA (mRNA) and its translation to the protein occurs on ribosomes, specialized organelles that are present in the cytoplasm of the cell. This is a complex process, the main events of which occur in the following series of steps: (1) the ribosome binds to one end of the mRNA strand; (2) the first triplet of the gene coding for a polypeptide, which always corresponds to the amino acid methionine (Met), binds to its specific transfer-RNA (tRNA) molecule, activating methionine; (3) the chain is elongated when the previous amino acid comes off its tRNA and combines with the next amino acid to yield a peptide bond; and (4) elongation continues until a specific termination triplet is encountered, when the polypeptide chain dissociates from the last tRNA and the ribosome.

Post-Translational Modifications. Proteins are often synthesized having 15–26 residue long signal peptides, usually at the amino terminus. The signal peptide serves to translocate the protein across a membrane when it is required in a location other than the cytoplasm. Once the protein has arrived, the signal peptide is cleaved. Certain enzymes and hormones (qv) undergo further cleavage to attain a specific functional form. Many viruses synthesize polyproteins which are then cleaved into individual polypeptide chains.

Proteins also undergo other covalent modifications after synthesis. These include (1) acetylation of the amino-terminal and amidation of the carboxy-terminal residue; (2) addition of fatty acid groups (myristoyl) or lipids, eg, farnesyl, for anchoring proteins to membranes; (3) glycosylation, one of the most common modifications, which serves many purposes such as cell–cell recognition through interaction with other molecules on the surface of cells; (4) phosphorylation, resulting in the addition of a negative charge, which serves as a switch to control the affinity for activators and inhibitors particularly in signal transduction; and (5) disulfide bond formation, in which cysteine (Cys) residues separated in sequence are brought spatially close together through covalent linkage.

Principles of Protein Structure

To understand the function of a protein at the molecular level, it is important to know its three-dimensional structure. The diversity in protein structure, as in many other macromolecules, results from the flexibility of rotation about single bonds between atoms. Each peptide unit is planar, ie, $\omega = 180°$, and has two rotational degrees of freedom, specified by the torsion angles ϕ and ψ, along the polypeptide backbone. The number of torsion angles associated with the side chains, R, varies from residue to residue. The allowed conformations of a protein are those that avoid atomic collisions between nonbonded atoms.

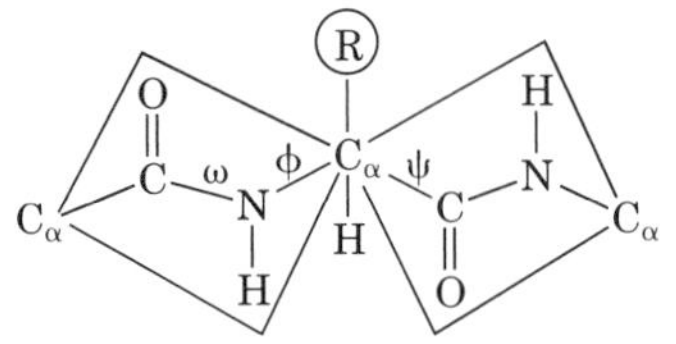

For any given protein, the number of possible conformations that it could adopt is astronomical. Yet each protein folds into a unique structure totally determined by its sequence. The basic assumption is that the protein is at a free energy minimum; however, calometric studies have shown that a native protein is more stable than its unfolded state by only 20–80 kJ/mol (5–20 kcal/mol) (5). This small difference can be accounted for by the favorable forces of the unfolded state, ie, conformational entropy owing to the large ensemble of conformers, and hydration owing to the favorable interaction of the polar atoms with the aqueous environment.

One of the principal driving forces determining the folded structure of a protein is the maintenance of peptide backbone hydrogen bonds on the removal of these bonds from the solvent to the protein interior. This is a force complementary to the hydrophobic effect that forces the nonpolar amino acid side chains away from the solvent into the interior. It is therefore natural to ask what types of structures allow for the maintenance of these hydrogen bonds. In the structure shown it is clear that on opposite edges of the peptide plane there is a hydrogen bond donor (N–H) and a hydrogen bond acceptor (O). Therefore if any set of peptide chains were stretched out side by side, a set of parallel hydrogen bonds between the chains could form. Another type of regular structure that can form is when the peptide chain twists into a long helix such that as one goes up the helix, one edge of the peptide plane always points upward (carboxy terminus) and the other points in the downward direction (amino terminus). The two basic structures of this kind, called the secondary structure, are the β-sheet and α-helix, respectively, shown in Figure 1.

Building on the two secondary structures and using the fact that amino acids can, to a first approximation, be classified as either hydrophilic (polar) or hydrophobic (nonpolar), allows for a simple understanding of the basic architecture of most proteins. Protein folding seeks to maximize the number of backbone hydrogen bonds while minimizing the number of exposed nonpolar amino acid side chains. Making the assumption that by starting with only helices and sheets the first part of the score has been optimized, only the number of solvent-exposed nonpolar residues has to be minimized.

There are basically three kinds of helices: hydrophobic, hydrophilic, and amphipathic. In the last, the two faces have opposite charge distributions. Sheets can be either hydrophobic or amphipathic. These structural units are strung together by a set of small connecting loops. The number of ways that these units

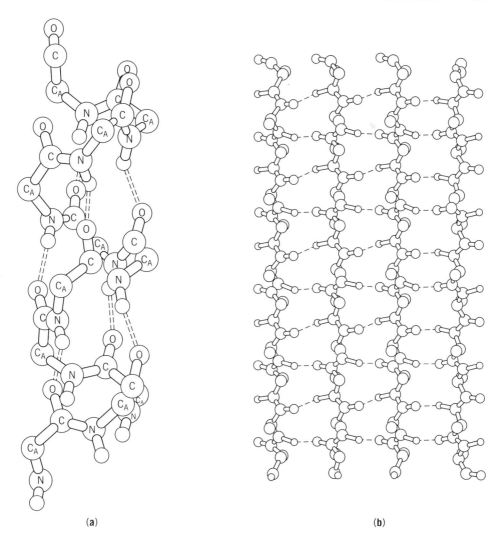

(a)   (b)

Fig. 1. The two principal elements of secondary structure in proteins. (**a**) The α-helix stabilized by hydrogen bonds between the backbone of residue i and $i + 4$. There are 3.6 residues per turn of helix and an axial translation of 150 pm per residue. C_A represents the carbon connected to the amino acid side chain, R. (**b**) The β-sheet showing the hydrogen bonding pattern between neighboring extended β-strands. Successive residues along the chain point alternately up and down the plane of the paper.

can pack against each other, subject to the constraints of charge complementarity and exposure to solvent, gives rise to the various topologies of protein structures, referred to as its tertiary structure.

Depending on the predominance of the type of secondary structure, proteins can fall into any of three classes. Details are available in Reference 6.

All-α-Proteins. The globins are well-known examples of all-α-proteins, the first protein structure to be solved. Globins consist of eight helices packed together around a heme group to form a box-like structure. The helices are so arranged that the ridges formed by the side chains on one helix fit into the grooves between the side chains of another helix. This packing arrangement, first predicted by Crick, has been proved to be true in the general packing of helices in proteins. The simplest motif of this class is the four-helix bundle found in the oxygen-carrying proteins hemerythrin and cytochrome b562. In these structures, hydrophobic side chains from all four helices point to the center of the bundle, forming a hydrophobic core. α-Keratin, the main component of hair, consists of a number of long helices coiled around one another.

Most of the membrane-spanning regions of integral membrane proteins appear to be all-helical. Prototypical of such proteins is bacteriorhodopsin, which consists of seven transmembrane helices arranged in a circle, approximately perpendicular to the membrane. Owing to the hydrophobic nature of the membrane, the helices consist largely of hydrophobic residues except where they bind to the retinal molecule. The loops connecting the helices are more hydrophilic as these are in contact either with the polar head groups of the lipids or the aqueous environment. The complex photosynthetic reaction center from bacteria is made up of four subunits, three of which consist of membrane-spanning helices.

All-β-Proteins. The general topology of all-β-proteins as a structural class of proteins consists of two or more β-sheets packed against each other. The mode of packing depends on the nature of the faces of the sheets. In the immunoglobulin fold, found in recognition molecules of the immune system, two sheets pack against each other to form a β-sandwich having a hydrophobic core and polar surface. Structures of proteins of unrelated functions have revealed an interesting motif described as a four-blade β-propeller (7).

α/β-Proteins. The most frequent domain structures observed in proteins are those having a mixture of helices and sheets. One of the largest of all domain structures is the eight-stranded α/β-barrel structure which consists of a core of eight parallel β-strands surrounded by eight helices, forming a very symmetrical structure. A large number of proteins have the α/β-sandwich structures in which a β-sheet is packed against either a single row of helices or is sandwiched between two such rows.

Larger proteins usually have two or more structural units termed domains, each domain having structures similar to single-domain proteins. The interaction between individual domains is much less extensive than that within a single domain. In many cases each domain is responsible for carrying out a specific function.

In some proteins, such as hemoglobin, separate polypeptide chains must associate for the chains to be functional. This forms a quaternary structure.

Cofactors. Frequently proteins exist in their native state in association with other nonprotein molecules or cofactors, which are crucial to their function. These may be simple metal ions, such as Fe^{n+} in hemerythrin or Ca^{2+} in calmodulin; a heme group, as for the globins; nucleotides, as for dehydrogenases, etc.

Native Conformation. Proteins assume their unique native conformation either during or soon after biosynthesis. This process, known as protein folding, can usually be reproduced *in vitro* under suitable conditions such as pH and temperature. The exact mechanism of folding remains a mystery as of 1996 and a great deal of research is devoted to this field. It is possible that *in vivo* protein folding follows the same sequence as its synthesis, ie, from the amino to the carboxy terminus. On the other hand, there is evidence to suggest that certain segments fold first and serve as seeds around which the rest of the protein folds. The folding pathway of bovine pancreatic trypsin inhibitor (BPTI) has been extensively studied because of this protein's small size and the presence of three disulfide bonds. However, experimental results are ambiguous as to whether folding intermediate states are largely native-like or not. Good reviews on this subject can be found in Reference 8.

A number of proteins are known to pass through a transient intermediate state, the so-called molten globule state. The precise structural features of this state are not known, but appear to be compact, and contain most of the regular structure of the folded protein, yet have a large side-chain disorder (9).

Many proteins frequently require the assistance of other protein molecules called molecular chaperonins, for assuming the fine tertiary structure *in vivo*. In *E. coli*, two such chaperonin molecules bind transiently to newly synthesized polypeptide monomers, preventing them from aggregating prematurely, until the polypeptides attain their folded state (10).

Protein Function

Proteins can be broadly classified into fibrous and globular. Many fibrous proteins serve a structural role (11). α-Keratin has been described. Fibroin, the primary protein in silk, has β-sheets packed one on top of another. Collagen, found in connective tissue, has a triple-helical structure. Other fibrous proteins have a motile function. Skeletal muscle fibers are made up of thick filaments consisting of the protein myosin, and thin filaments consisting of actin, troponin, and tropomyosin. Muscle contraction is achieved when these filaments slide past each other. Microtubules and flagellin are proteins responsible for the motion of cilia and bacterial flagella.

Globular proteins have biological function which they carry out by direct interaction with ligands that may be small atoms or large macromolecules. The protein without the bound ligand is known as the apo form, whereas the bound state is called the holo form. Many proteins are capable of binding more than one ligand, but usually do so on separate domains. The binding sites may be located in the protein interior as in the case of cofactors such as the hemes. Larger ligands most often bind at the surface including in the interdomain region. There is steric and physicochemical complementarity between the interacting surfaces resulting in close packing and favorable polar interactions. The structure of a protein domain generally does not alter significantly on binding. An exception is calmodulin, in which a large hinge movement has been observed between its two lobes on binding to a peptide analogue of myosin kinase (12). Sometimes the binding of a ligand at a particular site of a protein can have an effect on the affinity of binding of another ligand at a second site. The related site may be on a different domain of a multidomain protein or a different subunit of an oligomeric protein. This is termed allostery. The allosteric effect can be explained in terms of a conformational change at the primary site inducing a change at the secondary site. Allosteric control is frequently observed in gene regulation by a feedback mechanism. The binding of a repressor or activator to the DNA sites controlling the synthesis of a protein is modulated by its binding to the synthesized protein itself.

Binding sites comprise only a small fraction of the structure of most proteins. Most of the rest of the protein is responsible for providing the framework of the protein and in the process offering it stability. Consequently, functionally related proteins have the same overall topology, yet differ in the residues that form their binding sites, providing them their substrate specificity. During the folding process, functional residues distant in sequence are brought spatially close together. The chemical methods for determining these residues usually involve the covalent modification of the protein. The reactivity of the various groups to different reagents in the presence and absence of the ligand gives an indication of the binding of the group to its ligand. This assumes that the structure of the protein is unaffected by the modification. A more reliable method is affinity labeling, where a reactive group is incorporated into the ligand and its association with the group is monitored.

More detailed information about ligand binding has come from the three-dimensional (3-D) structures of protein–ligand complexes. In a number of cases where the reaction is too fast to be followed using these methods, the substrate is replaced by an analogue or inhibitor where the mode of binding is presumably similar. The structures of proteins have shown that in a number of instances the same ligand, typified by heme, can be bound to different proteins having no structural similarity.

Oxygen Binding.   The earliest investigations on proteins were carried out on the globins which bind oxygen molecules through the heme group. Its porphyrin ring packs against hydrophobic residues in the interior of the protein. The iron atom of the heme binds to O_2 on one side, and on the other to a histidine (His) residue that is one of only two residues conserved among all globins. This complex is stabilized by the interaction of O_2 with another His. Carbon monoxide, CO, which binds to isolated heme much more strongly than does O_2, has a much lower affinity for the globins owing to steric hindrance of the His residue. Theoretical studies have suggested that fluctuations in the protein structure produce channels for O_2 to diffuse in and out of its interior.

The allosteric effect is seen in hemoglobin which can exist in two quaternary structural states: oxygenated (R) or deoxygenated (T). The binding of one O_2 or some other effector to one of the subunits stabilizes the R form as compared to the T form. Binding of a second and third O_2 stabilizes it even further.

Enzymes.   Enzyme catalysis has been studied since the late 1800s and a great deal is known about this subject (13). Enzymes increase the rate of equilibrium of a chemical reaction by decreasing the barrier between the free energies of the products and reactants. This is believed to be achieved by the tighter binding of the enzyme to the transition state than to the substrate or product. An enzyme can function at very low concentrations by catalyzing the same reaction numerous times. The course of an enzymatic reaction can be studied using steady-state kinetics, where the dependence of the velocity of a reaction on substrate concentration is followed (see KINETIC MEASUREMENTS). An alternative method is relaxation spectrometry, in which the reaction is allowed to come to equilibrium, after which a rapid shift is made to one of the thermodynamic variables. The time course of the relaxation to the new equilibrium state can be followed using a variety of techniques (see BIOCATALYSIS (SUPPLEMENT)).

A mechanism for enzyme action, the lock-and-key theory, was proposed by Emil Fischer in the 1890s, in which it was suggested that the complementary shapes of the enzyme and substrate help each to recognize the other. An alternative theory is that of induced fit, in which the substrate, on approaching the binding site of the enzyme, induces a conformational change making the reaction favorable. This was first observed in hexokinase where phosphorylation of glucose by adenosine triphosphate (ATP) was accomplished by a drastic rotation of the two domains of its structure. The crystal structures of a number of enzymes complexed with ligands have greatly aided the understanding of their mechanisms of action.

Molecules of the Immune System. One of the amazing capabilities of vertebrates is the capacity to recognize and bind foreign molecules called antigens for removal from the system. This is carried out by antibodies or immunoglobulin molecules, remarkable proteins in that having a limited gene pool, ie, about 1000 segments in humans, the body is able to generate millions of such molecules. Each immunoglobulin is a Y-shaped molecule composed of two heavy and two light chains. Each of the chains is further composed of a constant and variable domain. The variable segments have hypervariable regions or complementarity determining regions (CDRs) which provide specificity to the antibody. The large repertoire of the antibody population is produced by the different rearrangements of the various available segments.

Many antibody structures in the presence and absence of antigens are known. All the domains have basically the same structure, termed the immunoglobulin fold, which consists of a β-sandwich held together by a disulfide bridge, having the CDR loops at the top of the sandwich. The variable loops from the different segments come together to form antigen-binding sites of vastly different shapes and sizes (see IMMUNOASSAY; IMMUNOTHERAPEUTIC AGENTS).

The major histocompatibility complex (MHC) molecules are responsible for binding foreign antigens and presenting these to T-cell receptor (TCR) molecules, which trigger a series of events leading to eventual destruction of the foreign invader. MHC molecules accomplish this by possessing domains of immunoglobulin folds that form the antigen-binding site and other domains that are responsible for binding to TCR. The dual function allows a distinction between self-proteins and foreign invaders.

DNA-Binding Proteins. The process of differentiation, by which a single cell multiplies to form different types of cells at different generations, is accomplished by the turning on and off of genes at various stages. The expression and regulation of the genetic information are carried out by proteins which bind to segments of DNA that surround the gene coding for the protein to be synthesized. These can either act as repressors or activators, depending on whether they prevent or help the RNA polymerase from binding to the DNA and initiate its transcription. Both these proteins have the same helix-turn-helix motif in symmetrically related dimers that fit neatly into successive turns of the so-called major groove of the DNA double helix. The DNA undergoes a distortion on binding which is induced by nonspecific protein–DNA interactions involving the backbone atoms. The actual affinity, however, is dependent on the particular sequence of DNA.

Another class of DNA-binding proteins are the polymerases. These have a nonspecific interaction with DNA because the same protein acts on all DNA sequences. DNA polymerase performs the dual function of DNA replication, in which nucleotides are added to a growing strand of DNA, and acts as a nuclease to remove mismatched nucleotides. The domain that performs the nuclease activity has an α/β-structure, a deep cleft that can accommodate double-stranded DNA, and a positively charged surface complementary to the phosphate groups of DNA. The smaller domain contains the exonuclease active site at a smaller cleft on the surface which can accommodate a single nucleotide.

Another common DNA-binding motif is the zinc finger which is found in many gene-regulating proteins. It consists of a helix β-hairpin motif, stabilized by a chelated Zn^{2+} ion. Multiple zinc fingers can occur in tandem, making contact with base pairs along the major groove of DNA.

Many transcription factors bind to DNA as dimers held together by the so-called leucine (Leu) zipper region, consisting of a pair of coiled-coil α-helices, where Leu occurs every two turns of the helix. The neighboring basic region appears to make a transition from random coil to helix on binding to DNA. In the case of another regulatory segment of DNA called the TATA-box, significant unwinding of the DNA occurs on binding to TATA-binding protein (TBP) (Fig. 2) (14).

Signal Transduction Proteins. Biological systems have the ability to respond to changes in external concentration of hormones (qv), growth factors, or other molecules or stimuli. These ligands bind to specific transmembrane receptors initiating a cascade of biochemical processes that produce an intracellular signal. For example, when light falls on the receptor protein rhodopsin, it transmits the signal to multiple copies of a transducing protein termed G-protein, which in turn, interacts with effector proteins that further amplify the signal.

Growth factors function by activating their receptors to phosphorylate other proteins. This results in the initiation of genetic transcription, thereby causing the cell to grow. Mutations to certain normal receptor genes can result in oncogenes whose products can induce uncontrolled cell growth.

Homology

Proteins that carry out the same or similar functions in different organisms generally have very similar structures. Such proteins also are often encoded by similar sequences of amino acids. These similarities are a result of the shared evolutionary history of the organisms and of their genes. Such similar proteins are termed homologous, and are believed to have a common ancestor. There is no direct way of determining whether two proteins are truly homologous. However, when the sequences of any group of proteins are recognizably similar it seems much more likely that they are homologues, having a common ancestor, than that they independently evolved to look the same. There are also cases where the last common ancestor appears to have been so long ago that the sequences are no longer recognizably similar, yet they are believed to be homologous.

In cases where the common function is well understood, such as the binding and transport of oxygen by the globins, identifying homology between proteins is relatively straightforward. In other cases it must be inferred from the degree of

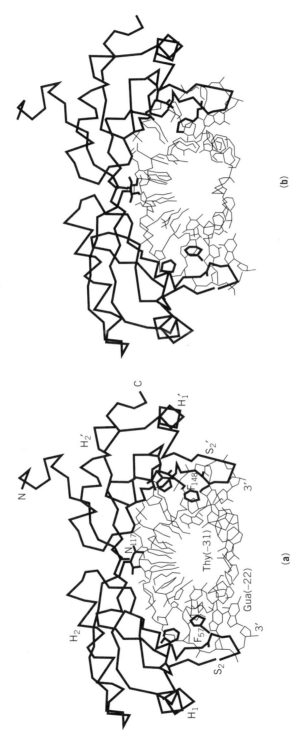

Fig. 2. Two views of the structure of the TATA-box binding protein (TBP)–DNA complex, where TATA is the nucleotide sequence thymine–adenosine–thymine–adenosine. The helices and strands of the protein (—) can be clearly seen. The DNA (—) is partially unwound upon binding to the protein (14).

structural and sequence similarity of two proteins. In the absence of any other information, only the degree of similarity between proteins at the sequence level is available to infer probable homology. Whereas two proteins having 30% or more of their amino acids identical over the whole sequence certainly implies homology, the converse is not necessarily true. For example, in the most distantly related globins, human alpha hemoglobin and bacterial leghemoglobin, there are only two amino acids which have not been modified out of over 100 amino acids, since the time of the last common ancestor. Yet the 3-D structures of these proteins are very similar and the two conserved amino acids are the very two which play the most important functional role.

Comparison of sequences is done by finding an alignment, ie, a position by position correspondence, which brings the maximum number of identical or similar amino acids into correspondence. This can be done mathematically (15). Favorable scores are assigned for the pairing of identical or similar residues in the alignment; penalties are given for positions where insertions or deletions are required to bring other positions into favorable alignment. The scores are based on the physicochemical similarity of the amino acids. In cases of very distantly related proteins such alignments are often ambiguous. In those cases, attempts are made only to identify those regions along the sequence, or even only those positions that form the pattern of conserved amino acids, which are essential to the encoded function and/or structure. Sequence similarity alignments and functional diagnostic patterns form two of the most important tools of modern molecular biology and protein chemistry.

Use of modern sequence similarity alignment methods, for example, has allowed the functional identification of nearly 60% of all genes determined in various international genome projects, such as those in humans, the flat worm, yeast (qv), and the bacterium *H. influenza* (16). Well over half of the human genes share a common ancestor with those of these other organisms. This ability to translate the function of new sequences by reading it from a previously understood one, has provided a biological equivalent of the Rosetta stone. The various diagnostic pattern recognition methods have, in addition, allowed the identification of mutations most likely to be responsible for many genetic diseases or defects.

A comparison of proteins of similar function has shown that the three-dimensional structure of the proteins is more conserved than sequence. It is the structure that defines the shape. The function is thus assured by providing a particular shaped surface which fits the other molecular components with which the protein interacts. However, the core or basic internal structure of two proteins may be quite similar in terms of α-helices and β-sheets, yet the surfaces quite different. This is because the core structure acts as a stable scaffolding onto which loops of varying size and amino acid chemistry can be attached. Thus structural similarity of the cores of proteins does not necessarily imply homology. In fact, only ~125 distinct folds have thus far been observed. These each belong to one of the nine superfolds shown in Figure 3 (18). There appears to be about a 40% chance that a newly discovered sequence adopts a fold similar to one already observed. Thus again, knowledge of existing structures, like sequences, allows inferences about new structures. Structures are cataloged in various protein data banks (17,19).

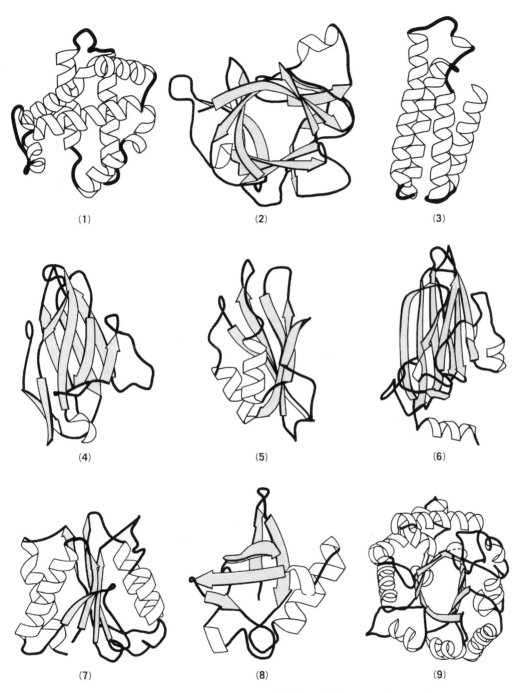

Fig. 3. Representation of the nine principal folds which recur in protein structures, where the codes of the representative proteins taken from the Brookhaven Protein Data Bank (PDB) (17) are given in parentheses (18): (**1**) globin (1thb); (**2**) trefoil (1i1b); (**3**) up–down (256b); (**4**) immunoglobulin folds (2rhe); (**5**) α/β-sandwich (1aps); (**6**) jelly roll (2stv); (**7**) doubly wound (4fxn); (**8**) UB α/β-roll (1ubq); and (**9**) TIM barrel (7tim).

Structure Determination

The most common methods used to study the structure of proteins are as follows:

Technique	Information
electron microscopy	low resolution 3-D structure
x-ray and neutron diffraction	high resolution 3-D structure
electron diffraction	medium resolution 3-D structure
nmr	high resolution 3-D structure
cd/ord	secondary structure
infrared spectroscopy	secondary structure

A detailed account is given in Reference 20. The techniques giving the most detailed 3-D structural information are x-ray and neutron diffraction, electron diffraction and microscopy (qv), and nuclear magnetic resonance spectroscopy (nmr) (see ANALYTICAL METHODS; MAGNETIC SPIN RESONANCE; X-RAY TECHNOLOGY).

Diffraction.   When a beam of x-rays, neutrons, or electrons strikes a specimen, the beam is diffracted by the component atoms. If the atoms are arranged in a regular array as in a crystal, the diffracted rays interfere constructively in certain directions giving rise to a pattern. The rays can be recorded using photodetectors (qv) or on photographic film. The structure of the molecule that gave rise to the pattern can then be determined by calculating the Fourier transform of the diffraction spots. The intensities of the diffraction spots can be measured directly and different methods have been developed to determine the relative phases. X-ray crystallography is the most widely used diffraction method, yielding structural information to atomic detail. The most difficult part of this technique is in the crystallization of the protein, a process that is poorly understood. When the molecules are arranged in a fibrous form as in collagen, muscle, or some viruses, x-ray fiber diffraction is used. Neutron diffraction may be employed to locate the positions of the hydrogen atoms. Structures determined by electron diffraction are limited in resolution, but have the advantage that the phases can be derived directly from the images concurrently obtained by electron microscopy.

Nuclear Magnetic Resonance.   Nmr has developed into a powerful technique for protein structure determination. Nmr spectra are obtained by placing a sample in a magnetic field and applying a radio-frequency pulse. This pulse perturbs the equilibrium nuclear magnetization of atoms having nuclei of nonzero spin, such as ^1H and ^{13}C. The signals emitted as the system returns to equilibrium can be converted to a set of resonances at different frequencies by Fourier transform. The frequencies, called chemical shifts, are indicative of the type of nucleus as well as its chemical environment. Thus nmr can yield structural information about the protein. Both 2-D and 3-D nmr have been developed to yield distances between atoms and thus obtain the atomic positions of a set of possible protein structures. One of the principal difficulties is in resolving the spectral peaks. The superiority of nmr over x-ray crystallography lies in its ability to determine the structure of a protein in the solution state, the natural state of the living cell (see MAGNETIC SPIN RESONANCE).

Structure Prediction

Although the techniques described have resulted in the determination of many protein structures, the number is only a small fraction of the available protein sequences. Theoretical methods aimed at predicting the 3-D structure of a protein from its sequence therefore form a very active area of research. This is important both to understanding proteins and to the practical applications in biotechnology and the pharmaceutical industries.

The most obvious approach for predicting the folded structure of a protein would be to search for its lowest energy conformation. In principle, knowledge of quantum chemistry should allow the necessary calculations to be carried out. However, the sheer size of the problem involving hundreds of thousands of interatomic interactions makes this extremely difficult. Simpler approaches, which fall into two basic categories, are used. The first is based on simplifying the energy calculations and conformational search, which usually involves reducing all the atomic level forces to simple classical mechanical forces. The second approach is not to attempt directly to predict the structure, but instead to use existing knowledge of protein structures to propose models most compatible with a given sequence.

The energy, E, of a protein can be expressed as the sum of different components:

$$E_{\text{tot}} = E_{\text{bl}} + E_{\text{ba}} + E_{\text{ta}} + E_{\text{vdw}} + E_{\text{es}} + E_{\text{hb}}$$

where the first three terms represent the energy associated with the deviation of bond lengths (bl), bond angles (ba), and torsion angles (ta) from equilibrium; E_{vdw} and E_{es} are the van der Waals and electrostatic interactions between nonbonded atoms, respectively; and E_{hb} is the hydrogen bond energy.

Because of the marginal stability of the folded conformation over the unfolded state, results are crucially dependent on the accuracy of these potentials. Much effort has been devoted to the development of force fields (21). The procedures commonly used to minimize the energy functions are described in Reference 22. The values for the various parameters were either experimentally derived or theoretically estimated from values for small molecules obtained in a manner which cannot readily be used for macromolecules owing to effects such as solvation, which are difficult to estimate. Thus, this approach has met with limited success in the prediction of structures. However, it has proved to be useful in obtaining good structures in cases where approximate models are already available, such as by x-ray diffraction, nmr, or molecular modeling (qv). This method has also been successfully used in studies on substrate binding and the effect of amino acid mutations on protein stability (see PROTEIN ENGINEERING).

Attempts have also been made at predicting the secondary structure of proteins from the propensities for residues to occur in the α-helix or the β-sheet (23). However, the assignment of secondary structure for a residue only has an average accuracy of about 60%. A better success rate (70%) is achieved when multiple-aligned sequences having high sequence similarity are available.

The conformation an amino acid adopts depends on the residues in its neighborhood. This was made clear from a study of identical pentapeptides in unrelated proteins which were observed to adopt the same conformation only 20% of the time (24). The effect that residues neighboring in sequence or in space have on the conformation is not well understood.

Even when the secondary structure of a protein is known, there are a large number of ways that this structure can be packed together. Studies dealing with the identification of the topological constraints in the packing of helices and sheets have revealed certain patterns, but as of this writing accurate prediction is not possible.

Comparative Modeling.   Given the limited success of prediction schemes, much effort has been turned to comparative modeling, ie, for a given sequence, identifying an approximately correct fold from among the different possible folds observed in nature. This technique is comparatively easy when homologous proteins having a known structure can be identified, but if no such protein is available it becomes a daunting task.

A review of the methods used in homology modeling is available (25). The various steps are as follows: (*1*) align the new sequence with its homologues of known structure imposing the constraint of no deletion/insertion in helical or strand segments; (*2*) model the backbone atoms of the conserved regions, ie, buried helices and strands and functionally important residues, from the mean positions in the homologues; (*3*) model the turns and loops connecting helices and strands, made difficult owing to conformational flexibility (approaches include database search of all known structures or, from energetic considerations, using the so-called loop closure method (26); (*4*) model the side-chain atoms based on rotamer libraries obtained from observed structures; and (*5*) refine the model by energy minimization.

Nonhomologous Extension Modeling.   In the case of protein sequences that have no clear sequence homologues with known structure, the problem becomes more challenging. One approach is to identify only the tertiary structural class to which a protein sequence belongs rather than its detailed structure. In some cases this is done from an analysis of the difference in properties of amino acids in the various classes (27). More recently a method has been developed which, instead of estimating the likelihood of a single or short segment of amino acids adopting a particular folded structure, estimates it for the entire sequence (28). This holds promise because any relevant structural information about the amino acids can be included in the calculations to increase the accuracy of prediction.

A related approach seeks to determine the compatibility of a given sequence by threading it through a structure. To each residue position in the structure an environment is associated which may be characterized by its secondary structure, level of exposure to solvent, nature of preferred amino acids in its neighborhood, etc. A score is assigned to each amino acid in these positions. All possible alignments of the sequence having the structure are made and the alignment having the optimal total score is taken as the most compatible structure (Fig. 4). Several research groups are engaged in threading studies using different score functions. Only limited success has been achieved.

Simplified models for proteins are being used to predict their structure and the folding process. One is the lattice model where proteins are represented as self-avoiding flexible chains on lattices, and the lattice sites are occupied by the different residues (29). When only hydrophobic interactions are considered and the residues are either hydrophobic or hydrophilic, simulations have shown

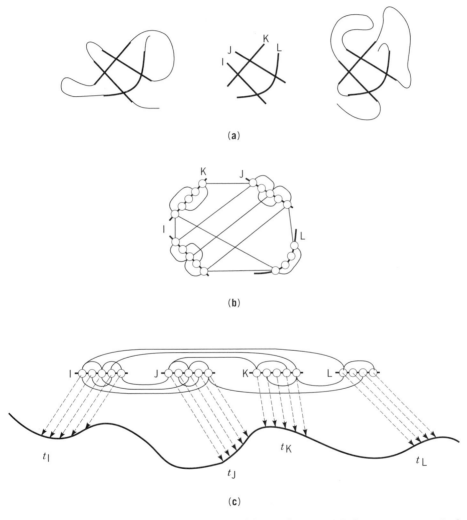

Fig. 4. Schematic of the protein threading problem, where (——) I–L represent particular core segments and (—) corresponds to variable loop segments: (**a**) the variable loop regions from two proteins are removed leaving a single core; (**b**) the interacting residues in the core are shown connected; and (**c**) one possible threading, t, of the sequence through the core where the dashed arrows represent structural positions occupied by the amino acids in the sequence.

that, as in proteins, the structures with optimum energy are compact and few in number. An additional component, hydrogen bonding, has to be invoked to obtain structures similar to the secondary structures observed in nature (30).

Practical Applications

Study of the structure and function of proteins is possible because of the revolution in molecular biology, which has enabled the cloning of genes and the expression of their products in large quantities. Great strides have also been made in the area of peptide synthesis, improvements in experimental techniques for structure determination, and computational speed. Areas of molecular and structural biology are rapidly expanding (see PROTEIN ENGINEERING).

Development of Drugs. The vast majority of drugs available were derived from large-scale screenings. In many cases little is known about the target protein or other macromolecule, or the mode of action. The type of molecule for use as a drug is inferred from the biochemical mechanism of a disease. This information is then used to screen databanks containing thousands of compounds. Modifications to the leading candidates are then made and tested further for efficacy. It can take from 6 to 12 years for a drug to come to the market (31) (see ANTIBIOTICS; ANTIPARASITIC AGENTS).

Rational drug design seeks to decrease this time lag and the resulting cost of development by rejecting unsuitable compounds. The design of drugs from first principles is not possible, but progress has been made. The first step is to obtain the quantitative structure-activity relationships (QSAR) between the biological activity of the protein and its chemical properties, particularly in its active site. The leading candidates are then examined further. The task is made simpler if the structural information of the protein is known, particularly in its complexed state. The increasing computing power and the resulting advance in the development of molecular modeling techniques have made the dream of *de novo* design of novel molecules a distinct possibility (32). Good reviews on molecular modeling (qv) techniques are available (33).

New and Improved Biomaterials. *Protein Computers.* The membrane protein bacteriorhodopsin holds great promise as a memory component in future computers. This protein has the property of adopting different states in response to varying optical wavelengths. Its transition rates are very rapid. Bacteriorhodopsin could be used both in the processor and storage, making a computer smaller, faster, and more economical than semiconductor devices (34).

Biomechanical Machines. The mechanical properties of fibrous polypeptides could be put to use for the commercial production of fibers (qv) that are more elastic and resilient than available synthetics (see SILK). The biochemical properties of proteins could also be harnessed for the conversion of mechanical energy to chemical energy (35).

Enzymes for Extreme Conditions. The possibility of using enzymes from extremophiles, which thrive in oil wells, hot temperatures, freezing conditions, etc, is being explored for the removal of environmental contaminants and survival at extreme temperatures (see WASTES, HAZARDOUS WASTE TREATMENT; BIOREMEDIATION (SUPPLEMENT)).

Conversion of Biomass for Fuels. Many predictions for protein uses in the 1970s have already been fulfilled (36). Biomass is being used for fuel, for example (see FUELS FROM BIOMASS). Many predictions were again made in 1995 (37).

BIBLIOGRAPHY

"Proteins" in *ECT* 1st ed., Vol. 11, pp. 226–248, by H. B. Vickery, The Connecticut Agricultural Station; in *ECT* 2nd ed., Vol. 16, pp. 610–640, by M. V. Tracey, Commonwealth Scientific and Industrial Research Organization, Australia; in *ECT* 3rd ed., Vol. 19, pp. 314–341, by P. L. Pellett, University of Massachusetts.

1. C. B. Anfinsen, *Science* **181**, 223 (1973).
2. R. Scopes, *Protein Purification: Principles and Practice*, 2nd ed., Springer-Verlag, Berlin, 1987.
3. F. Sanger, *Adv. Prot. Chem.* **7**, 1 (1952).
4. A. Bairoch and B. Boeckmann, *Nucleic Acids Res.* **20**, 2019–2022 (1992).
5. P. L. Privalov, *Adv. Prot. Chem.* **33**, 167 (1979).
6. C. Branden and J. Tooze, *Introduction to Protein Structure*, Garland Publishers, Inc., New York, 1991.
7. J. Li and co-workers, *Structure* **3**, 541 (1995).
8. P. L. Privalov, *Physical Basis of the Stability of the Folded Conformations of Proteins: Protein Folding*, Freeman, New York, 1992; K. A. Dill, *Biochemistry* **29**, 7133 (1990).
9. O. B. Ptitsyn, *Adv. Prot. Chem.* **47**, 83 (1995).
10. M. Mayhew and F. U. Hartl, *Science* **271**, 161 (1996).
11. R. E. Dickerson and I. Geis, *The Structure and Action of Proteins*, Benjamin/Cummings, 1969.
12. W. E. Meador, A. R. Means, and F. A. Quiocho, *Science* **257**, 1251 (1992).
13. A. Fersht, *Enzyme Structure and Mechanism*, 2nd ed., W. H. Freeman, New York, 1985.
14. J. L. Kim, D. B. Nikolov, and S. K. Burley, *Nature* **365**, 520 (1993).
15. T. F. Smith and M. S. Waterman, *Adv. Appl. Math.* **2**, 482–489 (1981); S. B. Needleman and C. D. Wunsch, *J. Mol. Biol.* **48**, 443 (1970).
16. *Science* **269**, 468 (1995).
17. F. C. Bernstein and co-workers, *J. Mol. Biol.* **112**, 535–542 (1977).
18. C. A. Orengo, D. T. Jones, and J. M. Thornton, *Nature* **372**, 631 (1994).
19. N. Pattabiraman, K. Namboodiri, A. Lowery, and G. P. Gaber, *Prot. Seq. Data Anal.* **3**, 387 (1990).
20. C. R. Cantor and P. R. Schimmel, *Biophysical Chemistry*, Part II, *Techniques for the Study of Biological Structure and Function*, W. H. Freeman, New York, 1980.
21. U. Buckert and N. L. Allinger, *Molecular Mechanics*, American Chemical Society, Washington, D.C., 1982; T. A. Clark, *A Handbook of Computational Chemistry*, John Wiley & Sons, Inc., New York, 1985.
22. D. H. J. Mackay, in G. D. Fasman, ed., *Prediction of Protein Structure and the Principles of Protein Conformation*, Plenum Press, New York, 1989, p. 317.
23. P. Y. Chou and G. D. Fasman, *Ann. Rev. Biochem.* **47**, 251 (1978).
24. W. Kabsch and C. Sander, *Proc. Natl. Acad. Sci. USA* **81**, 1075 (1984).
25. M. S. Johnson, N. Srinivasan, R. Sowdhamini, and T. L. Blundell, *Crit. Rev. Biochem. Mol. Biol.* **29**, 1 (1994).
26. Q. Zheng, R. Rosenfeld, S. Vadja, and C. Delisi, *J. Comp. Chem.* **14**, 556 (1993).
27. K. Nishikawa, Y. Kubota, and T. Ooi, *J. Biochem.* **94**, 981 (1983); C. Zhang and K. Chou, *Protein Sci.* **1**, 401 (1992); I. Dubchak, S. R. Holbrook, and S-H. Kim, *Proteins* (1993).

28. C. Stultz, J. White, and T. F. Smith, *Protein Sci.* **2**, 305–314 (1993).

29. K. A. Dill, *Protein Sci.* **4**, 561 (1995).

30. H. Ramasamy, R. Nambudripad, and T. F. Smith, *An Extended Genetic Algorithm for 3D Conformational Search.*

31. J. A. Dimasi, N. R. Bryant, and L. Lasagna, *Clin. Pharmacol. Ther.* **50**, 471 (1991).

32. J. E. Brody, *The New York Times*, C1 (Nov. 7, 1995).

33. N. C. Cohen, J. M. Blaney, C. Humblet, P. Gund, and D. C. Barry, *J. Med. Chem.* **33**, 883 (1990); P. M. Dean, *Molecular Foundations of Drug-Receptor Interaction*, Cambridge University Press, Cambridge, U.K., 1987.

34. D. W. Urry, in C. G. Gebelin, ed., *Biotechnological Polymers: Medical, Pharmaceutical and Industrial Applications*, Technomic Publishers, Lancaster, Pa., 1993.

35. R. R. Birge, *Computer* **25**, 56 (1992).

36. P. Handler, *Biology and the Future of Man*, Oxford University Press, New York, 1970.

37. D. E. Koshland, Jr., ed., *Science* **267**, 1609 (1995).

RAMAN NAMBUDRIPAD
Beth Israel Hospital

TEMPLE F. SMITH
Boston University

PROTEINS FROM PETROLEUM.   See FOODS, NONCONVENTIONAL.

PROTOZOAL INFECTIONS, CHEMOTHERAPY.   See ANTIPARA-
SITIC AGENTS, ANTIPROTOZOALS.

PRUSSIAN BLUE.   See IRON COMPOUNDS; PIGMENTS, INORGANIC.

PSYCHOPHARMACOLOGICAL AGENTS

Until the early 1950s only rudimentary pharmacotherapy was available for the treatment of significant psychiatric illnesses. As of the middle 1990s, agents effective in treating prevalent psychiatric disorders, such as mood disorders, schizophrenia, anxiety disorders, insomnia, and substance use disorders, as well as dementias of diverse etiologies, are available. Nevertheless, improved drugs to treat such central nervous system (CNS) disorders continue to represent one of the greatest medical needs in modern medicine (see NEUROREGULATORS). The therapy of cognitive disorders, particularly age-related dementia which has intensified in the face of the demographic trend toward an increasingly aged population, is only one area receiving increased medical attention (see ANTIAGING AGENTS; MEMORY-ENHANCING DRUGS). During the latter part of the twentieth century, significant advances were made in the diagnosis and classification of psychiatric illnesses. This has resulted in a much finer differentiation of disorders based on both epidemiological data and clinical response to available psychopharmacological agents as evidenced by the improved classification systems provided in the *Diagnostic and Statistical Manual of Mental Disorders* (DSM-IV) (1) and the *International Statistical Classification of Diseases and Related Health Problems* (2).

Among the most important psychopharmacological agents initially added to the medical armamentarium were the benzodiazepines, for treating anxiety, insomnia, and for inducing sedation (see HYPNOTICS, SEDATIVES, ANTICONVULSANTS, AND ANXIOLYTICS); the phenothiazines and thioxanthenes, for treating schizophrenia; as well as diverse tricyclics and hydrazide derivatives for treating depression. Continued emphasis has been placed on achieving further increases in symptomatic efficacy but, additionally, weight is given to improving the side-effect profile. Together these trends have resulted in a more favorable therapeutic index for many of the newer psychopharmacological agents. Indeed, effective and safe pharmacotherapy is available for many psychiatric disorders. Future advances are expected to focus increasingly on achieving the optimal therapy for the subcategories of disorders, some of which have no approved pharmacotherapy available.

Despite recognized successes of psychopharmacology, significant challenges remain. The causative factors underlying most neuropsychiatric disorders are, as of 1996, only poorly understood. Missing is the discovery of psychopharmacological agents which directly impact on the etiological or pathophysiological processes underlying psychiatric illnesses. There is considerable structural diversity among the compounds approved for therapeutic use for each of the principal psychiatric disorders. Focus herein is placed on the most important prescription drugs marketed worldwide for treatment of psychiatric illnesses.

Anxiolytics, Sedatives, and Hypnotics

Anxiety disorders and insomnia represent relatively common medical problems within the general population. These problems typically recur over a person's lifetime (3,4). Epidemiological studies in the United States indicate that the lifetime prevalence for significant anxiety disorders is about 15%. Anxiety disorders are serious medical problems affecting not only quality of life, but additionally may indirectly result in considerable morbidity owing to association with depression, cardiovascular disease, suicidal behavior, and substance-related disorders.

Insomnia is a related psychiatric illness having potentially serious consequences. In any given year up to one-third of the general population may experience insomnia and consequently considerable impact on quality of life. Potentially serious psychosocial, health, and socioeconomic consequences may follow. Many sedative–hypnotics additionally have a firmly established position within the field of anesthesiology as premedication, inducing agents, and/or for maintenance in intensive care medicine.

Classification of Anxiety Disorders and Insomnia. Anxiety is an emotional state dominated by the perception or anticipation of danger to physical and/or psychic integrity, well-being, and self-esteem or simply the fear of being unable to adequately cope with problems of daily life. In general, anxiety disorders are more prevalent in women than men. Normally anxiety plays an important adaptive role. When irrational and excessive, however, anxiety is maladaptive and can seriously interfere with daily functioning. According to the DSM-IV classification system (1), anxiety disorders are categorized as the following: (1) panic disorder, involving discrete periods of intense fear or discomfort with or without agoraphobia; (2) agoraphobia, involving the fear of open, public places or being in a crowd, without a history of panic disorder; (3) specific phobia, involving marked, persistent, and excessive fear in the presence or in anticipation of a specific object or situation; (4) social phobia, involving severe and persistent fear of humiliating oneself in a social or performance situation; (5) obsessive–compulsive disorder, involving recurrent and persistent intrusion of unwanted thoughts, impulses, or images (obsessions) often accompanied by repetitive behaviors (compulsions) performed in order to reduce anxiety and distress; (6) post-traumatic stress disorder, resulting from experience of intense fear, helplessness, or horror in connection with events that involved actual or threatened serious injury leading to persistent reexperiencing of the traumatic event; (7) acute stress disorder, involving detachment and derealization following exposure to a traumatic event; and (8) generalized anxiety disorder, involving recurring excessive anxiety and worry for at least six months which is difficult to control.

In view of the prominent role played by anxiety as a symptom in all of these disorders, it has been hypothesized that they may, in fact, share a common etiology. Stress and overreaction to stress appears to be involved in the pathophysiology of anxiety. Anxiety disorders often include such clinical features as motor tension, autonomic hyperactivity, hypervigilance, and apprehension. Both physiological and pathological anxiety can be life-threatening when occurring in the face of preexisting organic disorders and may create or perpetuate various physiological dysfunctions.

Insomnia complaints are common in the general population and can be dichotomized into problems of delayed sleep onset and those related to sleep maintenance. Increasing attention is being focused on the adverse daytime effects of insomnia. Sleep disturbances become more common with increased age and are more prevalent in women. Sleep complaints arise from very diverse etiologies which prominently include concomitant primary psychiatric disorders, substance-related disorders, and environmental factors. These must be adequately taken into account when making the decision to symptomatically treat the sleep disorder.

Pharmacological Profiles of Anxiolytics and Sedative–Hypnotics. Historically, chemotherapy of anxiety and sleep disorders relied on a wide variety of natural products such as opiates, alcohol, cannabis, and kawa pyrones. Use of various bromides and chloral derivatives in these medical indications enjoyed considerable popularity early in the twentieth century. Upon the discovery of barbiturates, numerous synthetic compounds rapidly became available for the treatment of anxiety and insomnia. As of this writing barbiturates are in use primarily as injectable general anesthetics (qv) and as antiepileptics. These agents have been largely replaced as treatment for anxiety and sleep disorders.

In the 1950s, meprobamate, a propandiole, and methaqualone, a quinazoline, achieved limited clinical importance. Piperidindiones including glutethimide and methyprylon were also used.

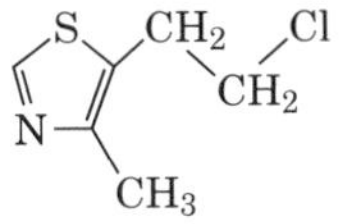

(1)

The short-acting clomethiazole [533-45-9] (**1**), sometimes used as therapy for sleep disorders in older patients, shares with barbiturates a risk of overdose and dependence. Antihistamines, such as hydroxyzine [68-88-2] (**2**), are also sometimes used as mild sedatives (see HISTAMINES AND HISTAMINE ANTAGONISTS). Antidepressants and antipsychotics which have sedative effects are used to treat insomnia when the sleep disorder is a symptom of some underlying psychiatric disorder.

Beginning in the 1960s, benzodiazepine anxiolytics and hypnotics rapidly became the standard prescription drug treatment. In the 1980s, buspirone [36505-84-7] (**3**), which acts as a partial agonist at the serotonin [50-67-9] (5-hydroxytryptamine, 5-HT) type 1A receptor, was approved as treatment for generalized anxiety. More recently, selective serotonin reuptake inhibitors (SSRIs) have been approved for therapy of panic disorder and obsessive–compulsive behavior.

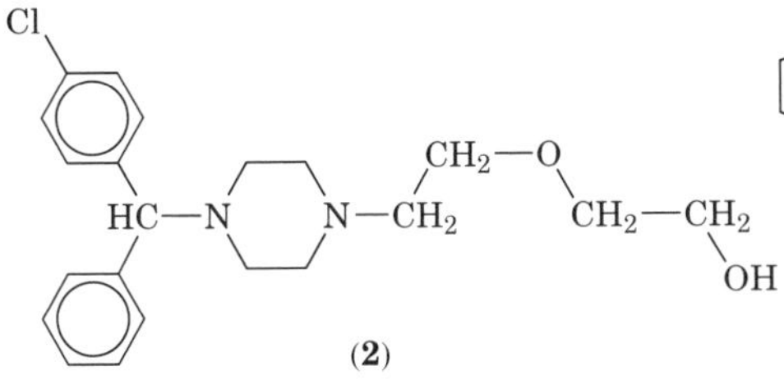

(2)

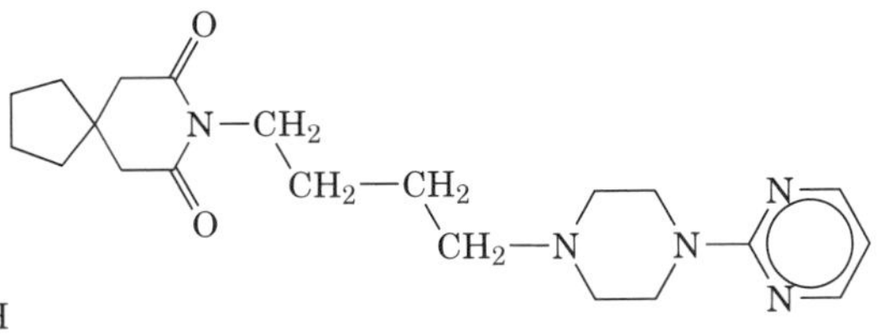

(3)

On the basis of available published clinical results, β-adrenergic blockers, particularly propranolol [525-66-6], have sometimes been used off-label to treat anxiety, especially in those patients in which cardiovascular symptoms predominate (5) (see CARDIOVASCULAR AGENTS). Clinical results also suggest the possible value of the monoamine–oxidase inhibitor phenelzine [51-71-8] in ameliorating panic disorder with agoraphobia (6). As of the middle 1990s, benzodiazepines continue to dominate the therapy of certain anxiety disorders, eg, generalized anxiety disorder and panic disorder, as well as insomnia.

Barbiturates and benzodiazepines have been demonstrated to act by modulating neurotransmission within the γ-aminobutyric acid [55-12-2] (GABA) system, which is the primary inhibitory neurotransmitter system within the central nervous system. In addition, barbiturates act presynaptically to decrease calcium entry, as well as nonsynaptically to decrease voltage-dependent sodium and potassium conductances. The weak sedative effect of the antihistamine hydroxyzine (**2**) is most likely attributable to its antagonism at the histamine-1 receptor. The mechanism of action of meprobamate does not appear to involve the enhancement of GABAergic transmission, whereas methaqualone may do so. Little is known about the mechanism of action of glutethimide and methyprylon.

Barbiturates.   Barbiturates, which by common practice include both oxybarbiturates and thiobarbiturates, are effective as sedative–hypnotics and as anticonvulsants. The clinical effects are the result of allosteric modulation of sites on the $GABA_A$ receptor complex. This increases not only the apparent potency of GABA, but also the maximum effect of GABA. The effects of barbiturates involve the direct alteration of the chloride channel, in contrast to the modulatory effect of benzodiazepines, within the $GABA_A$ receptor complex. Barbiturates enhance GABAergic transmission and at high doses even mimic the effect of GABA itself on chloride conductance by acting via receptors at the post-synaptic membrane, thereby modifying the coupling process between $GABA_A$ receptors and chloride channels or altering the kinetics of the chloride channel. Barbiturates are still occasionally used as alternatives to benzodiazepines in the therapy of anxiety and sleep disorders despite the much narrower therapeutic index of the former drugs and higher liability for abuse and dependence. Barbiturates interact with ethyl alcohol [*64-17-5*] and other central nervous system depressants.

(**4**)

(**5**) R = CH_2CH_3
(**6**) R = CH_2CHCH_2

Barbiturates marketed as psychopharmacological agents are listed in Table 1. The ultrashort-acting barbiturates such as sodium salts of methohexital (**4**), thiopental (**5**), and thiamylal (**6**), are typically used as intravenous anesthetics, often in combination with inhalation agents. The short- to intermediate-acting barbiturates such as secobarbital (**7**), pentobarbital (**8**), amobarbital (**9**), allobarbital (**10**), and aprobarbital (**11**) are prescribed as hypnotics. The long-acting barbiturates phenobarbital (**12**), metharbital (**13**), and methylphenobarbital (**14**) are most commonly used as antiepileptics. Barbiturates, for example butalbital (**15**), are often components in combination preparations together with paracetamol [*103-90-2*] (acetaminophen), acetylsalicylic acid [*50-78-2*] (aspirin), caffeine [*58-08-2*], and/or codeine [*6059-47-8*]; sometimes only

Table 1. Marketed Barbiturates

Agent	Nomen-clature[a]	BAN	CAS Registry Number	Selected trade name(s)[b]	Empirical formula	Refs.
allobarbital	INN, USAN		[52-43-7]	Diadol	$C_{10}H_{12}N_2O_3$	7
amobarbital	INN, JAN, USAN	amylo-barbitone	[57-43-2]	Amytal (sodium)	$C_{11}H_{18}N_2O_3$	8
amobarbital sodium	JAN, USAN		[64-43-7]		$C_{11}H_{17}N_2$-NaO_3	8
aprobarbital	INN		[77-02-1]	Alurate	$C_{10}H_{14}N_2O_3$	9
barbital	INN, JAN	barbitone	[57-44-3]	Veronal (sodium)	$C_8H_{12}N_2O_3$	10
barbital sodium	INN	barbitone sodium	[144-02-5]		$C_8H_{11}N_2$-NaO_3	10
benzo-barbital	INN		[744-80-9]	Benzonal	$C_{19}H_{16}N_2O_4$	11
brallo-barbital	INN		[561-86-4]	Vesperone	$C_{10}H_{11}BrN_2O_3$	12
butalbital	INN, USAN		[77-26-9]	Sandoptal	$C_{11}H_{16}N_2O_3$	13
butallylonal sodium			[3486-86-0]	Tempidorm	$C_{11}H_{14}BrN_2$-NaO_3	14
buthalital sodium	INN		[510-90-7]	Baytinal	$C_{11}H_{15}N_2$-NaO_2S	15
butobarbital		butobar-bitone	[77-28-1]	Neonal, Soneryl	$C_{10}H_{16}N_2O_3$	16
carbubarb	INN		[960-05-4]	Nogexan	$C_{11}H_{17}N_3O_5$	
crotarbital			[1952-67-6]	Kalypnon	$C_{10}H_{14}N_2O_3$	17
cyclobarbital	INN	cyclobar-bitone	[52-31-3]	Cyclodorm, Phanodorm	$C_{12}H_{16}N_2O_3$	18
cyclopento-barbital sodium			[302-34-1]	Cyclopal	$C_{12}H_{13}N_2$-NaO_3	11
eterobarb	INN, USAN	eterobarb	[27511-99-5]	Antilon	$C_{16}H_{20}N_2O_5$	19
febarbamate	INN		[13246-02-1]	Solium, Tymium	$C_{20}H_{27}N_3O_6$	11
heptabarb	INN	heptabar-bitone	[509-86-4]	Medomin	$C_{13}H_{18}N_2O_3$	20
heptobarbital			[76-94-8]	Rutonal(e)	$C_{11}H_{10}N_2O_3$	
hexethal sodium			[144-00-3]	Ortal (sodium)	$C_{12}H_{19}N_2$-NaO_3	21
hexobarbital	INN, JAN	hexobar-bitone	[56-29-1]	Evipal (sodium)	$C_{12}H_{16}N_2O_3$	22
metharbital	INN, JAN, USAN	methar-bitone	[50-11-3]	Gemonil	$C_9H_{14}N_2O_3$	23
methitural	INN		[467-43-6]	Neraval, Thiogenal	$C_{12}H_{20}N_2$-O_2S_2	24
methohexital	INN, USAN	methohexi-tone	[151-83-7]	Brevital (sodium)	$C_{14}H_{18}N_2O_3$	25
methohexital sodium	USAN		[309-36-4]		$C_{14}H_{17}N_2NaO_3$	25
methylpheno-barbital[c]	INN	methylpheno-barbitone	[115-38-8]	Mebaral, Prominal	$C_{13}H_{14}N_2O_3$	26

Table 1. (Continued)

Agent	Nomenclature[a]	BAN	CAS Registry Number	Selected trade name(s)[b]	Empirical formula	Refs.
narcobarbital sodium			[3329-16-6]	Eunaron, Narcotal	$C_{11}H_{14}BrN_2NaO_3$	27
nealbarbital	INN	nealbarbitone	[561-83-1]	Cenobal, Nevental	$C_{12}H_{18}N_2O_3$	28
pentobarbital	INN, USAN	pentobarbitone	[76-74-4]		$C_{11}H_{18}N_2O_3$	
pentobarbital calcium	JAN		[24876-35-5]		$C_{22}H_{34}CaN_4O_6$	29
pentobarbital sodium	USAN		[57-33-0]	Nembutal (sodium)	$C_{11}H_{17}N_2NaO_3$	29
phenallymal			[115-43-5]	Alphenate, Sanudorm	$C_{13}H_{12}N_2O_3$	30
phenobarbital	INN, JAN, USAN	phenobarbitone	[50-06-6]		$C_{12}H_{12}N_2O_3$	31
phenobarbital sodium	INN, JAN, USAN		[57-30-7]	Luminal (sodium)	$C_{12}H_{11}N_2NaO_3$	31
phetharbital	INN		[357-67-5]	Fedibaretta, Pyrictal	$C_{14}H_{16}N_2O_3$	
probarbital sodium	INN		[143-82-8]	Ipral (sodium)	$C_9H_{13}N_2NaO_3$	32
propallylonal			[545-93-7]	Noctal	$C_{10}H_{13}BrN_2O_3$	33
propylbarbital	INN		[2217-08-5]	Propal, Propanal	$C_{10}H_{16}N_2O_3$	
proxibarbal	INN		[2537-29-3]	Axeen, Ipronal	$C_{10}H_{14}N_2O_4$	34
secbutabarbital sodium[d]	INN		[143-81-7]	Barbased, Butisol Sodium	$C_{10}H_{15}N_2NaO_3$	35
secbutabarbital[e]		secbutobarbitone	[125-40-6]		$C_{10}H_{16}N_2O_3$	
secobarbital	INN, USAN		[76-73-3]		$C_{12}H_{18}N_2O_3$	36
secobarbital sodium	JAN, USAN	quinalbarbitone sodium	[309-43-3]	Seconal (sodium)	$C_{12}H_{17}N_2NaO_3$	
talbutal	INN, USAN		[115-44-6]	Lutawin	$C_{11}H_{16}N_2O_3$	37
tetrabarbital	INN		[76-23-3]	Butysal, Butysedal	$C_{12}H_{20}N_2O_3$	
thialbarbital	INN	thialbarbitone	[467-36-7]	Intranarcon, Thialpenton	$C_{13}H_{16}N_2O_2S$	11
thialbarbital sodium			[3546-29-0]		$C_{13}H_{15}N_2NaO_2S$	
thiamylal sodium	JAN, USAN		[337-47-3]	Surital	$C_{12}H_{17}N_2NaO_2S$	38
thiobutabarbital sodium			[947-08-0]	Brevinarcon	$C_{10}H_{15}N_2NaO_2S$	39
thiopental sodium	INN, JAN, USAN	thiopentone sodium	[71-73-8]	Pentothal (sodium)	$C_{11}H_{17}N_2NaO_2S$	40

Table 1. (*Continued*)

Agent	Nomen-clature[a]	BAN	CAS Registry Number	Selected trade name(s)[b]	Empirical formula	Refs.
vinbarbital	INN	vinbarbitone	[125-42-8]	Devinal (sodium)	$C_{11}H_{16}N_2O_3$	41
vinylbital	INN	vinylbitone	[2430-49-1]	Bykonox, Speda	$C_{11}H_{16}N_2O_3$	42

[a]International nonproprietary name (INN), Japanese accepted name (JAN), United States adopted name (USAN), and British approved name (BAN). [b]Only selected salts are included. Trade names shown can represent the base or any of the salts included in "Nomenclature". [c]USAN and JAN = mephobarbital. [d]USAN = butabarbital sodium. [e]USAN = butabarbital.

combination preparations are available. Among the most common adverse effects of barbiturates are ataxia, confusion, drowsiness, hangover, skin rash, and nausea. Many adverse effects of high doses of barbiturates are exaggerations of direct actions observed at therapeutic doses.

Structure number	R	R′	R″
(7)	H	CH_2CHCH_2	$CHCH_3CH_2CH_2CH_3$
(8)	H	CH_2CH_3	$CHCH_3CH_2CH_2CH_3$
(9)	H	CH_2CH_3	$CH_2CH_2CH(CH_3)_2$
(10)	H	CH_2CHCH_2	CH_2CHCH_2
(11)	H	$CH(CH_3)_2$	CH_2CHCH_2
(12)	H	CH_2CH_3	C_6H_5
(13)	CH_3	CH_2CH_3	CH_2CH_3
(14)	CH_3	CH_2CH_3	C_6H_5
(15)	H	CH_2CHCH_2	$CH_2CH(CH_3)_2$

Benzodiazepines.   The marketed benzodiazepine anxiolytics and sedative–hypnotics (see Table 2) act via agonism at benzodiazepine receptors (BZRs) within the central nervous system to yield a wide spectrum of therapeutic actions including anxiolytic, hypnotic, muscle relaxant, and anticonvulsant effects (43). The BZR is a distinct binding site on the $GABA_A$-receptor complex. The BZR can exert either positive or negative modulation on the $GABA_A$ receptor depending on the ligand. Ligands having positive allosteric modulatory activity are classified as BZR agonists. These increase the affinity of GABA to its binding site. Compounds having structures different from those of benzodiazepines can also act via the BZR, for example, the full agonists zopiclone (**16**) and zolpidem (**17**). Figure 1 shows the synthesis of zolpidem, the first FDA approved nonbenzodiazepine hypnotic which acts via the BZR. For marketed benzodiazepines, see Table 2.

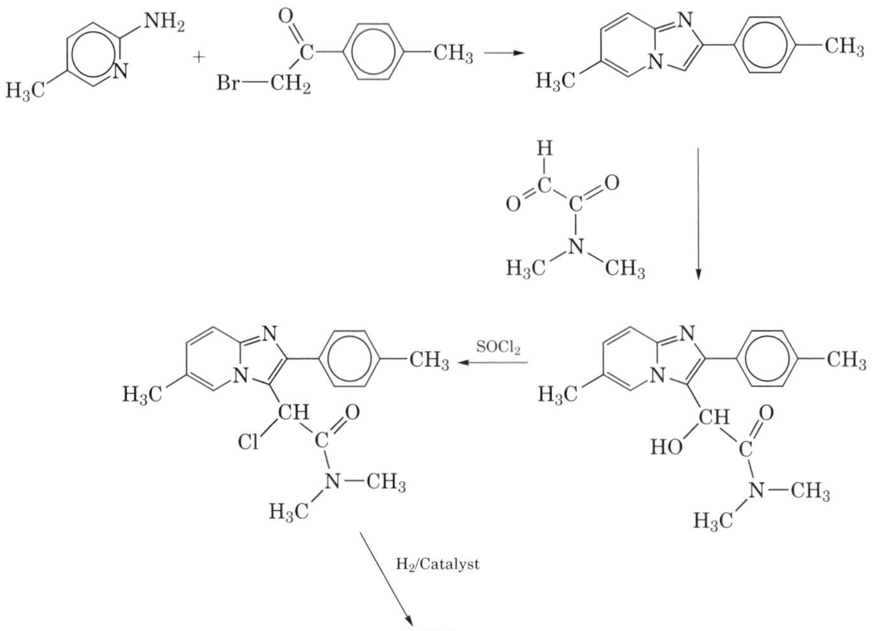

Fig. 1. Synthesis of zolpidem (**17**). 2-Amino-5-methylpyridine and 4-methyl-phenacyl-bromide give the imidazo[1,2-a]pyridine derivative which is then added to N,N-dimethyl-2-oxo-acetamide. The reaction of the adduct and SOCl₂ leads to a chloroacetamide derivative which is then reduced by catalytic hydrogenation to (**17**). This product is transformed into the hemitartrate salt of zolpidem (44–46).

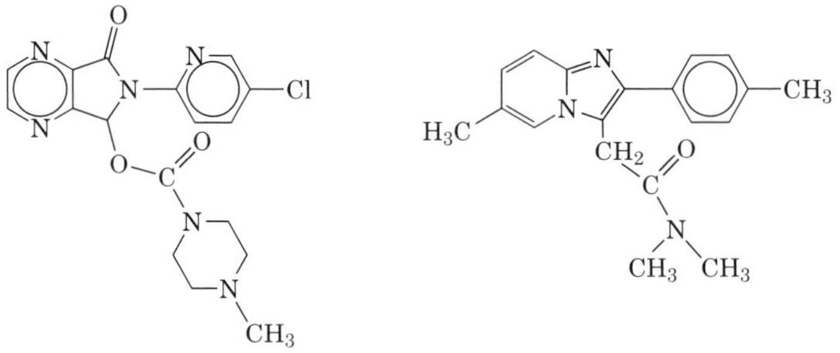

(**16**)　　　　　　　　　　　　(**17**)

Another class of ligands called inverse agonists is found among β-carbolines and benzodiazepine derivatives. These act to depress the function of the GABAₐ receptor complex. None of these compounds, however, is marketed. Other BZR ligands bind to the allosteric site without inducing any appreciable alteration of the GABAₐ receptor gating function. However, these compounds selectively and competitively block the action of both BZR agonists and inverse agonists. Accordingly, they are named BZR antagonists. The imidazobenzodiazepine flumazenil [78755-81-4] (**18**) is the only representative in clinical use. Flumazenil, a specific BZR antagonist having exceptionally good tolerance and effectiveness as an

Table 2. Benzodiazepines Marketed as Psychopharmacological Agents

Agent	Nomen-clature[a]	CAS Registry Number	Selected trade name(s)[b]	Empirical formula	Refs.
alprazolam	INN, BAN JAN, USAN	[28981-97-7]	Xanax	$C_{17}H_{13}ClN_4$	47
bentazepam	INN, USAN	[29462-18-8]	Thiadipone, Tiadipona	$C_{17}H_{16}N_2OS$	48
bromazepam	INN, BAN, JAN, USAN	[1812-30-2]	Lexomil, Lexotan, Lexotanil	$C_{14}H_{10}BrN_3S$	49
brotizolam	INN, BAN, JAN, USAN	[57801-81-7]	Lendorm	$C_{15}H_{10}BrClN_4O$	50
camazepam	INN	[36104-80-0]	Albego, Panevril	$C_{19}H_{18}ClN_3O_3$	51
chlordiazepoxide	INN, BAN, JAN, USAN	[58-25-3]	Librium	$C_{16}H_{14}ClN_3O$	52
chlordiazepoxide hydrochloride	BAN, JAN, USAN	[438-41-5]	Librium	$C_{16}H_{15}Cl_2N_3O$	52
cinolazepam	INN	[75696-02-5]	Gerodorm	$C_{18}H_{13}ClFN_3O_2$	53
clobazam	INN, BAN, USAN	[22316-47-8]	Frisium, Urbanyl	$C_{16}H_{13}ClN_2O_2$	54
clonazepam	INN, BAN, JAN, USAN	[1622-61-3]	Rivotril	$C_{15}H_{10}ClN_3O_3$	55
clorazepate dipotassium	INN, JAN USAN	[57109-90-7]	Tranxen(e), Tranxilium	$C_{16}H_{11}ClK_2N_2O_4$	56
clotiazepam	INN, JAN	[33671-46-4]	Clozan, Trecalmo	$C_{16}H_{15}ClN_2OS$	57
cloxazolam	INN, JAN	[24166-13-0]	Betavel, Cloxam	$C_{17}H_{14}Cl_2N_2O_2$	58
delorazepam	INN	[2894-67-9]	Briatum	$C_{15}H_{10}Cl_2N_2O$	59
diazepam	INN, BAN, JAN,USAN	[439-14-5]	Valium, Valrelease	$C_{16}H_{13}ClN_2O$	60
doxefazepam	INN	[40762-15-0]	Doxans	$C_{17}H_{14}ClFN_2O_3$	61
estazolam	INN, JAN USAN	[29975-16-4]	Nuctalon, Prosom	$C_{16}H_{11}ClFN_4$	62
ethyl loflazepate	INN, JAN	[29177-84-2]	Meilax, Victan	$C_{18}H_{14}ClFN_2O_3$	63
etizolam	INN, JAN	[40054-69-1]	Depas, Pasaden	$C_{17}H_{15}ClN_4S$	64
fludiazepam	INN, JAN	[3900-31-0]	Erispan	$C_{16}H_{12}ClFN_2O$	65
flumazenil	INN, BAN, USAN	[78755-81-4]	Anexate, Romazicon	$C_{15}H_{14}FN_3O_3$	66
flunitrazepam	INN, BAN, JAN, USAN	[1622-62-4]	Narcozep, Rohypnol	$C_{16}H_{12}FN_3O_3$	67
flurazepam	INN, BAN, JAN	[17617-23-1]		$C_{21}H_{23}ClFN_3O$	68
flurazepam hydrochloride	USAN	[1172-18-5]	Dalmadorm, Dalmane,	$C_{21}H_{25}Cl_3FN_3O$	68
flurazepam monohydro-chloride		[36105-20-1]	Dalmate, Dormodor[c]	$C_{21}H_{24}Cl_2FN_3O$	68
flutazolam	INN, JAN	[27060-91-9]	Coreminal	$C_{19}H_{18}ClFN_2O_3$	69
flutemazepam	INN	[52391-89-6]	Somnal	$C_{16}H_{12}ClFN_2O_2$	

Table 2. (Continued)

Agent	Nomen-clature[a]	CAS Registry Number	Selected trade name(s)[b]	Empirical formula	Refs.
flutoprazepam	INN, JAN	[25967-29-7]	Restar, Restas	$C_{19}H_{16}ClFN_2O$	70
halazepam	INN, BAN, USAN	[23092-17-3]	Paxipam	$C_{17}H_{12}ClF_3N_2O$	71
haloxazolam	INN, JAN	[59128-97-1]	Somelin	$C_{17}H_{14}BrFN_2O_2$	72
ketazolam	INN, BAN, USAN	[27223-35-4]	Solatran, Unakalm	$C_{20}H_{17}ClN_2O_3$	73
loprazolam	INN, BAN	[61197-73-7]	Dormonoct, Somnovit	$C_{23}H_{21}ClN_6O_3$	
loprazolam mesylate		[61197-93-1]		$C_{24}H_{25}ClN_6O_6S$	74
lorazepam	INN, BAN, JAN, USAN	[846-49-1]	Ativan	$C_{15}H_{10}Cl_2N_2O_2$	75
lorazepam pivalate[d]		[57773-81-6]		$C_{29}H_{19}Cl_2N_2O_4$	75
lormetazepam	INN, BAN, JAN, USAN	[848-75-9]	Noctamid(e)	$C_{16}H_{12}Cl_2N_2O_2$	76
medazepam	INN, BAN, JAN	[2898-12-6]	Nobrium	$C_{16}H_{15}ClN_2$	77
medazepam hydrochloride	USAN	[2898-11-5]		$C_{16}H_{16}Cl_2N_2$	77
metaclazepam	INN	[65517-27-3]		$C_{18}H_{18}BrClN_2O$	
metaclazepam hydrochloride		[61802-93-5]	Talis	$C_{18}H_{19}BrCl_2N_2O$	78
mexazolam	INN, JAN	[31868-18-5]	Melex	$C_{18}H_{16}Cl_2N_2O_2$	79
midazolam	INN, BAN, JAN	[59467-70-8]	Dormicum, Hypnovel, Versed[e]	$C_{18}H_{13}ClFN_3$	80
midazolam hydrochloride	USAN	[59467-96-8]		$C_{18}H_{14}Cl_2FN_3$	80
midazolam maleate	USAN	[59467-94-6]		$C_{22}H_{17}ClFN_3O_4$	80
nimetazepam	INN, JAN	[2011-67-8]	Erimin	$C_{16}H_{13}N_3O_3$	81
nitrazepam	INN, BAN, JAN, USAN	[146-22-5]	Mogadon	$C_{15}H_{11}N_3O_3$	82
nordazepam	INN	[1088-11-5]	Nordaz, Vegesan	$C_{15}H_{11}ClN_2O$	83
oxazepam	INN, BAN, JAN, USAN	[604-75-1]	Adumbran, Serax	$C_{15}H_{11}ClN_2O_2$	84
oxazolam	INN, JAN	[24143-17-7]	Convertal, Quiadon	$C_{18}H_{17}ClN_2O_2$	85
phenazepam		[51753-57-2]		$C_{15}H_{10}BrClN_2O$	86
pinazepam	INN	[52463-83-9]	Domar, Duna	$C_{18}H_{13}ClN_2O$	87
prazepam	INN, BAN, JAN, USAN	[2955-38-6]	Centrax, Demetrin	$C_{19}H_{17}ClN_2O$	88
quazepam	INN, BAN, USAN	[36735-22-5]	Dormalin	$C_{17}H_{11}ClF_4N_2S$	89
temazepam	INN, BAN, USAN	[846-50-4]	Restoril	$C_{16}H_{13}ClN_2O_2$	90
tetrazepam	INN	[10379-14-3]	Musaril, Myolastan	$C_{16}H_{17}ClN_2O$	91

Table 2. (*Continued*)

Agent	Nomen-clature[a]	CAS Registry Number	Selected trade name(s)[b]	Empirical formula	Refs.
tofisopam	INN, JAN	[22345-47-7]	Grandaxin	$C_{22}H_{26}N_2O_4$	92
triazolam	INN, BAN, JAN, USAN	[28911-01-5]	Halcion	$C_{17}H_{12}ClN_4$	93

[a]International nonproprietary name (INN), British approved name (BAN), Japanese accepted name (JAN), and United States adopted name (USAN). [b]Only selected salts are included. Trade names shown can represent the base or any of the salts included in "Nomenclature". [c]Trade names for listed flurazepam agents. [d]Also called pivazepam. [e]Trade names for listed midazolam agents.

antidote of mono-overdose with BZR agonists, is also used for shortening post-operational unconsciousness following anesthesia involving a BZR agonist. The synthesis of flumazenil is shown in Figure 2.

Benzodiazepines, ie, the full BZR agonists, are prescribed for anxiety, insomnia, sedation, myorelaxation, and as anticonvulsants (97). Those benzodiazepines most commonly prescribed for the treatment of anxiety disorders are lorazepam (**19**), alprazolam (**20**), diazepam (**21**), bromazepam (**22**), chlorazepate (**23**), and oxazepam (**24**). These drugs together represent about 70% of total

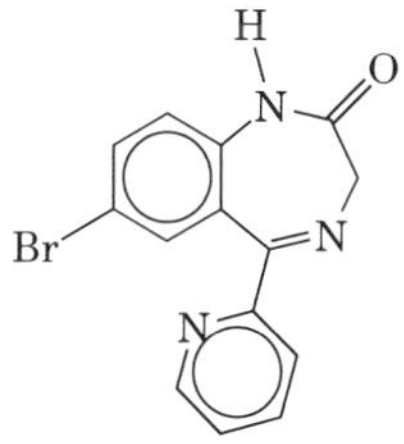

(**22**) 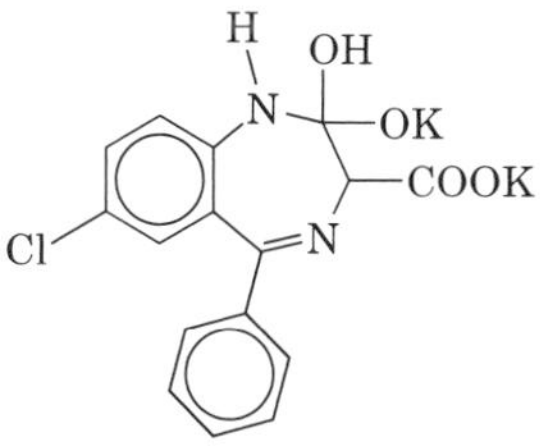   (**23**)

worldwide tranquilizer unit sales. Those benzodiazepines used most frequently in treating insomnia are triazolam (**25**), temazepam (**26**), flunitrazepam (**27**), lormetazepam (**28**), estazolam (**29**), and flurazepam (**30**), which together represent approximately one-third of total worldwide unit sales in the diverse market

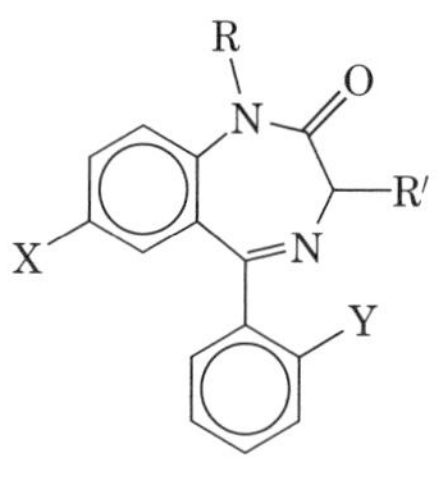

Structure number	R	R'	X	Y
(**19**)	H	OH	Cl	Cl
(**21**)	CH_3	H	Cl	H
(**24**)	H	OH	Cl	H
(**26**)	CH_3	OH	Cl	H
(**27**)	CH_3	H	NO_2	F
(**28**)	CH_3	H	Cl	Cl
(**30**)	$CH_2CH_2N(CH_2CH_3)_2$	H	Cl	F

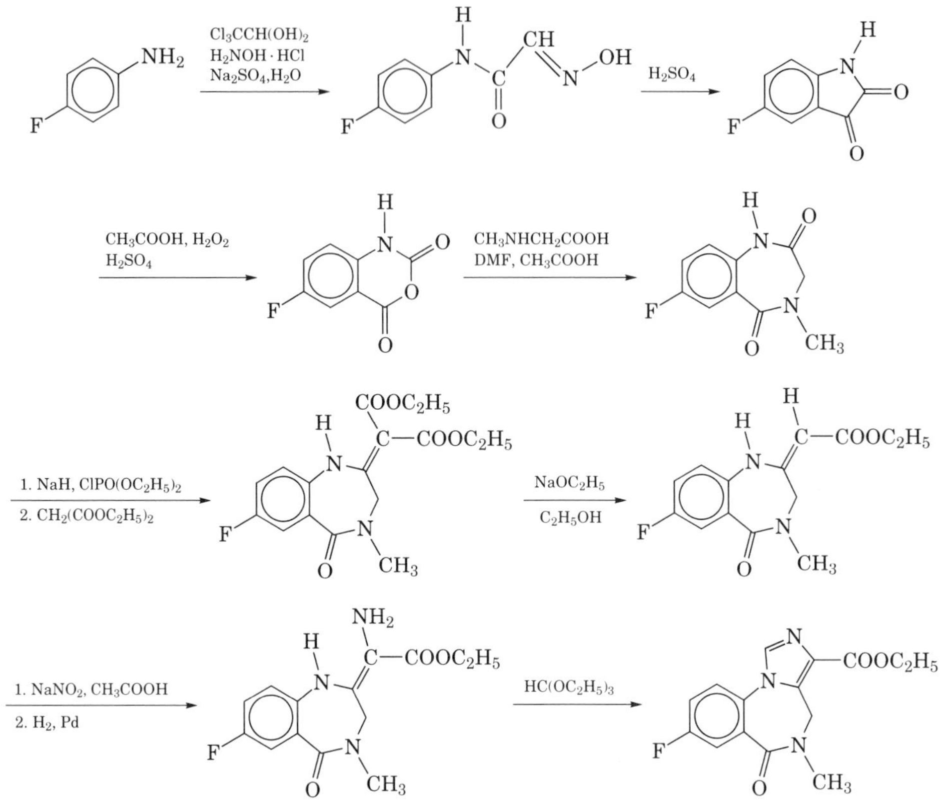

Fig. 2. Synthesis of flumazenil (**18**). The isonitrosoacetanilide is synthesized from 4-fluoroaniline. Cyclization using sulfuric acid is followed by oxidization using peracetic acid to the isatoic anhydride. Reaction of sarcosine in DMF and acetic acid leads to the benzodiazepine-2,5-dione. Deprotonation, phosphorylation, and subsequent reaction with diethyl malonate leads to the diester. After selective hydrolysis and decarboxylation the resulting monoester is nitrosated and catalytically hydrogenated to the aminoester. Introduction of the final carbon atom is accomplished by reaction of triethyl orthoformate to yield (**18**) (94–96).

of sedative–hypnotics. Zopiclone (**16**) and zolpidem (**17**) are approved as hypnotics. Combination preparations are available for some benzodiazepines, eg, diazepam, clorazepate, chlordiazepoxide, and oxazepam. In view of the broad

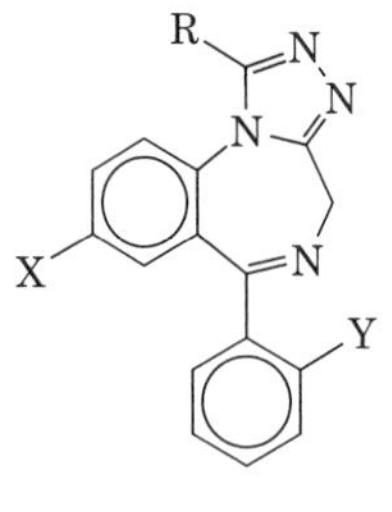

Structure number	R	X	Y
(**20**)	CH$_3$	Cl	H
(**25**)	CH$_3$	Cl	Cl
(**29**)	H	Cl	H

therapeutic index of benzodiazepines, adverse effects can be largely avoided by optimizing dosage for an individual patient. Adverse effects of BZR agonists are primarily those related to central nervous system depression and include ataxia, myorelaxation, and drowsiness. Abuse and dependence problems can also arise. Interactions occur with ethanol and other central nervous system depressants.

Buspirone. Buspirone (**3**) hydrochloride has been approved for the symptomatic management of generalized anxiety disorder (Table 3). This drug is of special interest because it does not exert its therapeutic actions via modulation of the GABA$_A$ receptor complex. This compound is structurally unrelated to the benzodiazepines, barbiturates, or the other anxiolytics and sedative–hypnotics discussed. The anxiolytic effect of buspirone may result from partial agonism at the 5-HT$_{1A}$ receptor, although its pharmacological profile exhibits some similar-

Table 3. Miscellaneous Anxiolytics and Sedative–Hypnotics

Agent	Nomen- clature[a]	CAS Registry Number	Selected trade name(s)[b]	Empirical formula	Refs.
buspirone	INN, BAN	[36505-84-7]	Bespar, Buspar	C$_{21}$H$_{31}$N$_5$O$_2$	98
buspirone hydrochloride	USAN	[33386-08-2]		C$_{21}$H$_{32}$ClN$_5$O$_2$	98
chloral hydrate	BAN, JAN, USAN	[302-17-0]	Chloradorm, Notec	C$_2$H$_3$Cl$_3$O$_2$	99
clomethiazole	INN[c]	[533-45-9]	Distraneurine, Heminevrin	C$_6$H$_8$ClNS	100
etomidate	INN, BAN, USAN	[33125-97-2]	Amidate	C$_{14}$H$_{16}$N$_2$O$_2$	101
glutethimide	INN, BAN, USAN	[77-21-4]	Doriden	C$_{13}$H$_{15}$NO$_2$	102
hydroxyzine hydrochloride	JAN, USAN	[2192-20-3]	Atarax	C$_{21}$H$_{29}$Cl$_3$N$_2$O$_2$	103
ketamine	INN, BAN	[6740-88-1]		C$_{13}$H$_{16}$ClNO	104
ketamine hydrochloride	JAN, USAN	[1867-66-9]	Ketalar	C$_{13}$H$_{17}$Cl$_2$NO	104
meprobamate	INN, BAN, JAN, USAN	[57-53-4]	Equanil, Miltown	C$_9$H$_{18}$N$_2$O$_4$	105
methaqualone	INN, BAN, USAN	[72-44-6]	Optimil, Torinal	C$_{16}$H$_{14}$N$_2$O	106
methyprylon	INN, USAN[d]	[125-64-4]	Noludar, Nolurate	C$_{10}$H$_{17}$NO$_2$	107
propofol	INN, BAN, USAN	[2078-54-8]	Diprivan	C$_{12}$H$_{18}$O	108
zolpidem	INN, BAN	[82626-48-0]	Ambien,	C$_{19}$H$_{21}$N$_3$O	109
zolpidem tartrate	USAN	[99294-93-6]	Lorex, Stilnox	C$_{42}$H$_{48}$N$_6$O$_8$	109
zopiclone	INN, BAN, JAN	[43200-80-2]	Imovane, Zimovane	C$_{17}$H$_{17}$ClN$_6$O$_3$	110

[a]International nonproprietary name (INN), British approved name (BAN), Japanese accepted name (JAN), and United States adopted name (USAN). [b]Only selected salts are included. Trade names shown can represent the base or any of the salts included in "Nomenclature". [c]BAN = chlormethiazole. [d]BAN = methylprylone.

ity to that of dopaminergic receptor antagonists. Antipsychotics are sometimes used to treat anxiety disorders.

The pharmacological profile of buspirone in both animals and humans differs substantially from that of the benzodiazepine anxiolytics. Buspirone lacks anticonvulsant, myorelaxant, and hypnotic effects. It also produces less sedation resulting in less psychomotor impairment in conjunction with ethanol consumption. Adverse effects include headache, dizziness, nervousness, light-headedness, excitement, and nausea. A lagtime of a week or more before anxiolysis is achieved makes buspirone inappropriate for use when rapid action is required. Additionally, some evidence indicates that buspirone may be less effective than BZR agonists in treating severe anxiety or in patients having prior treatment experience with BZR agonists. Buspirone represents about 2% of total worldwide tranquilizer unit sales.

Selective Serotonin Reuptake Inhibitors. In view of the mechanism of action of selective serotonin reuptake inhibitors (SSRIs), it appears that the resulting increased availability of the neurotransmitter serotonin within the synaptic cleft is responsible for the pharmacological effects of this drug class. However, in view of the delayed onset therapeutic action of SSRIs it has been hypothesized that their therapeutic effects may, in fact, be primarily dependent on alterations in receptor density/sensitivity, changes in second messenger pathways, and/or modifications at the level of gene expression. Although originally developed and predominantly used as antidepressants, SSRIs have been increasingly used in treating panic disorder, eg, fluoxetine (**31**), or obsessive–compulsive disorder, eg, fluvoxamine (**32**).

(**31**)   (**32**)   (**33**)

SSRIs are well tolerated. Adverse effects for compounds in this class include nervousness, tremor, dizziness, headache, insomnia, sexual dysfunction, nausea, and diarrhea. In addition, the tricyclic antidepressant clomipramine (**33**), which is a potent nonselective serotonin reuptake inhibitor, is approved for treatment of obsessive–compulsive disorder.

Use of Sedative–Hypnotics in General Anesthesia. In addition to use in the therapy of anxiety and insomnia, sedative–hypnotics are widely used, when given intravenously in appropriate dosages, to induce general anesthesia. Local or regional anesthesia is induced differently by agents of the cocaine class. The optimal anesthetic agent would combine unconsciousness, amnesia, analgesia, loss of both sensory and autonomic reflexes, and muscle relaxation having minimal cardiovascular and respiratory disturbance. No single anesthetic agent combines all these desired features, thus anesthesiologists often employ a cocktail of drugs.

Pharmacological Profiles of Sedatives Used in Anesthesiology. Anesthetic management of a surgical patient includes premedication, induction, maintenance, emergence, and recovery. Light premedication provides sedation, decreased anxiety, amnesia, and decreased parasympathetic outflow. Oral or parenteral benzodiazepines are sedative–hypnotics used preoperatively for premedication when strong analgesia is not required. The rapid onset of action of intravenous anesthetics makes them ideally suited for induction of anesthesia. Agents having a short half-life such as midazolam (**34**) and propofol (**35**) are often used as maintenance anesthetics. Anesthetics can also be administered in combination with intravenous narcotics to obtain appropriate levels of analgesia.

(**34**)   (**35**)

Combinations of barbiturates and benzodiazepine tranquilizers or even antihistaminergics having sedative properties are sometimes used. Furthermore, infusion of anesthetics can be used to provide long-term anesthesia for intensive care medicine. The antagonist flumazenil (**18**) is available to reverse the effects of anesthetics of the benzodiazepine class.

Barbiturates. The ultrashort-acting barbiturates (see Table 1) methohexital (**4**), thiopental (**5**), and thiamylal (**6**) are used for induction and maintenance of anesthesia. Their mechanism of pharmacological action at the $GABA_A$ receptor has been described. Barbiturates exhibit a relatively low therapeutic ratio and provide no protection against painful stimuli. They are often given in combination with other anesthetics. Thiopental is widely used owing to the rapidity with which patients pass through the stages of anesthesia. Apnea is usually the first sign of overdose. Cardiovascular depression may occur after intravenous bolus administration of barbiturates.

Benzodiazepines. Benzodiazepine derivatives (see Table 2) have gained popularity as intravenous anesthetics because of their limited effects on the cardiovascular and respiratory systems. Intravenous benzodiazepines can be used alone for minor procedures causing discomfort but not pain (eg, endoscopy), or in combination with local anesthetics. Midazolam is the most widely utilized benzodiazepine in intravenous anesthesia. Midazolam (**34**) is a BZR full agonist. It has a short plasma elimination half-life, is water soluble, and has marked amnestic and anxiolytic properties. Diazepam (**21**), lorazepam (**19**), and flunitrazepam (**27**) are less widely used than midazolam intravenously owing to poor water solubility, slower onset, occasional pain during injection, and slower recovery.

Other Sedative Agents.   Propofol (**35**) (see Table 3) is a rapidly acting phenol derivative having a short recovery time which readily penetrates the brain. It is very lipid soluble and almost insoluble in water so it is formulated in a fat emulsion. Propofol lacks appreciable amnestic, anxiolytic, or analgesic effects. There is sometimes pain upon injection of the emulsion formulation and anesthesia induction using propofol can reduce blood pressure. Propofol has been hypothesized to act via allosteric modulation of brain $GABA_A$ and glycine receptors (111), as well as on Na^+ and K^+ channels in nerve membranes and directly on membrane lipids (112). Ketamine (**36**) induces dissociative anesthesia, ie, profound analgesia in the absence of sleep, and no loss of protective reflexes and little cardiovascular depression. Unpleasant dreams and post-operative hallucinations are serious disadvantages of ketamine, but these can be reduced by concomitant use of a benzodiazepine or an opiate. Anesthesia using ketamine is rapid and of short duration. Etomidate (**37**) induces short-lasting anesthesia without analgesia so that supplementary analgesic premedication is usually needed. Its use is confined to induction and brief procedures owing to adverse inhibition of adrenocortical steroidogenesis upon prolonged infusion.

(**36**)   (**37**)

Improved anxiolytics, hypnotics, and sedatives are expected to continue to be introduced. Future improvements in symptomatic therapy are expected to occur in increasingly modest increments. The therapeutic efficacy of benzodiazepines in generalized anxiety disorder, as well as stress-induced and situational anxiety, is well established and potential benefits in panic and phobic disorders are judged increasingly positively (113). Benzodiazepine receptor agonists, both benzodiazepines and nonbenzodiazepines, continue to be the predominant prescription drugs used to treat insomnia. Proper dosage and treatment duration can considerably reduce the undesirable effects of drugs acting as agonists at BZRs, ie, sedation, ethanol potentiation, abuse liability, and physical dependence. The potential advantages offered by the BZR partial agonist approach could represent an important advance by maintaining full therapeutic efficacy while further minimizing undesirable effects. Clinical investigations carried out in anxiety and sleep disorders using such BZR partial agonists as divaplon [*90808-12-1*], saripidem [*103844-86-6*], and abecarnil [*111841-85-1*] are expected to provide the basis for such an evaluation.

The possibility of the existence of subtypes of $GABA_A$ receptors differentially sensitive to BZR ligands and exhibiting differential involvement in neuronal populations responsible for anxiety, sleep, or sedation versus undesired drug effects remains a promising avenue which could yield an improved therapeutic index. The clinical introduction of buspirone (**3**) has emphasized the potential value of mechanisms other than those based on $GABA_A$ receptor function as a way of achieving anxiolytic activity. As of 1996, a number of related compounds acting through 5-HT$_{1A}$ receptor partial agonism and antagonism were in various stages of clinical development for this indication. Furthermore, exploratory approaches focusing on cholecystokinin receptor antagonism, 5-HT$_2$ receptor antagonism, 5-HT$_3$ receptor antagonism, neuropeptide Y$_1$ receptor agonism, neuronal nicotinic acetylcholine receptor subtype agonism, melatonin receptor agonism, N-methyl-D-aspartate receptor antagonism, as well as neurosteroids (eg, epalons) as the basis of novel treatment strategies for anxiety disorders and/or insomnia were actively being followed preclinically and even in clinical investigations (114). Finally, increasing attention is being given to the positive results from clinical investigations of various marketed antidepressants, eg, tricyclics and monoamine–oxidase inhibitors, in panic disorder and phobias (115).

Antidepressants

Depression is a common psychiatric disorder. The lifetime risk of developing a depressive episode is estimated to be as high as 8–12% for men and 20–26% for women (116). Depression, one of the most widespread of all life-threatening disorders, is almost always a factor in the more than 30,000 suicides that occur annually in the United States alone (117).

    Classification of Depressive Disorders. The most common mood disorders are major or unipolar depression and manic–depressive illness or bipolar depression. There are other affective disorders, such as dysthymia, which are generally treated with available antidepressants. Major depression is characterized by symptoms such as depressed mood, diminished interest or pleasure in nearly all activities, decreased appetite with weight loss, fatigue, loss of energy, feelings of worthlessness, diminished ability to concentrate, and recurrent thoughts of death or suicidal ideation (118). In contrast, manic episodes are characterized by expansive mood, grandiosity, inflated self-esteem, pressured speech, flight of ideas, and poverty of sleep. The symptomatology of depression is often complicated by the presence of other psychiatric disorders such as anxiety and psychosis.

    Pharmacological Profiles of Antidepressants. Depression is believed to result from a decreased neurotransmission, ie, a lack of sufficient noradrenaline [51-41-2], serotonin [50-67-9], and/or dopamine [51-61-6] concentration at critical synapses in the brain, particularly within the limbic system. This led to the catecholamine and indoleamine hypotheses of depression. The various classes of antidepressant treatments, by blocking transmitter uptake by the neuron, or by slowing down their degradation, can increase neurotransmitter concentration to a normal level and consequently ameliorate depression (119). In the past, amphetamine [300-62-9] (amfetamine is INN) was sometimes used to treat depression, but tolerance develops to its mood-elevating effects and it has high addictive liability, potential to induce agitation, and can worsen somatic and neurovegetative symptoms of depression. Amphetamine acts via stimulation of dopamine, and to a lesser extent noradrenaline, release from neurons and concomitantly via the blockade of reuptake and metabolism by monoamine–oxidase of these two catecholamines. There are no compelling clinical data which demonstrate efficacy of psychostimulants such as amphetamine, methylphenidate, or pemoline in treating depression (120). It has also been speculated that other neurotransmitters and neuromodulators, such as neuropeptides and prostaglandins (qv), may also be involved in the etiology of depressive disorders.

Antidepressant therapy is usually associated with a delay of up to about two to three weeks before the onset of a clear beneficial effect. Therefore, it seems unlikely that the acute biochemical effects of antidepressants are responsible for the actual therapeutic response to these agents. It seems more probable that the action of antidepressants results from an adaptive process to long-term exposure to the drugs, ie, down- or up-regulation of receptor density, receptor sensitivity, second messenger pathways, and/or gene expression.

Treatment of Major Depression. Drugs commonly used for the treatment of depressive disorders can be classified heuristically into two main categories: first-generation antidepressants with the tricyclic antidepressants (TCAs) and the irreversible, nonselective monoamine–oxidase (MAO) inhibitors, and second-generation antidepressants with the atypical antidepressants, the reversible inhibitors of monoamine–oxidase A (RIMAs), and the selective serotonin reuptake inhibitors (SSRIs). Table 4 lists the available antidepressants.

First-Generation Antidepressants. Tricyclic Antidepressants. Imipramine (**38**) was introduced in the late 1950s as one of the first pharmacotherapies for depression. At that time, chlorpromazine [50-53-3] was the first effective antipsychotic treatment to be discovered. Researchers looked for similar chemical structures and imipramine was found to be effective in the symptomatic treatment of depression. Over the years, other congeners, such as desipramine (**39**), amitriptyline (**40**), and dothiepin (**41**), were synthesized and shown to be clinically efficacious antidepressant drugs (121). These substances, known under the general rubric of tricyclic antidepressants, share a basic chemical structure comprising

Structure number	X	Y	R
(**38**)	CH_2	N	$CH_2(CH_2)_2N(CH_3)_2$
(**39**)	CH_2	N	$CH_2(CH_2)_2NHCH_3$
(**40**)	CH_2	CH	$CH(CH_2)_2N(CH_3)_2$
(**41**)	S	CH	$CH(CH_2)_2N(CH_3)_2$

a three-ring core. TCAs remain the mainstay of treatment for depression which represent more than half of total worldwide antidepressant unit sales, despite drawbacks such as anticholinergic side effects, ie, blurred vision, dry mouth, constipation, or confusion; cardiovascular side effects, ie, postural hypotension, dizziness, hypertension, or antiarrhythmic effect; sedation; and weight gain.

Monoamine–Oxidase Inhibitors. In the mid-1950s, tuberculosis patients with depression being treated with iproniazid (**42**) were occasionally reported to become euphoric. This observation led to the discovery of irreversible monoamine–oxidase (MAO) inhibiting properties. Hydrazine and nonhydrazine-related MAO inhibitors were subsequently shown to be antidepressants (122).

Table 4. Marketed Antidepressants

Agent	Nomen- clature[a]	CAS Registry Number	Selected trade name(s)[b]	Empirical formula	Refs.
amfebutamone	INN[c]	[34911-55-2]		$C_{13}H_{18}ClNO$	
bupropion hydrochloride	USAN	[31677-93-7]	Wellbutrin	$C_{13}H_{19}Cl_2NO$	123
amineptine	INN	[57574-09-1]	Survector, Directim, Maneon	$C_{22}H_{27}NO_2$	124
amitriptyline	INN, BAN	[50-48-6]	Laroxyl, Loxaryl	$C_{20}H_{23}N$	
amitriptyline hydrochloride	JAN, USAN	[549-18-8]		$C_{20}H_{24}ClN$	125
amoxapine	INN, BAN, JAN, USAN	[14028-44-5]	Asendin, Demolox, Moxadil	$C_{17}H_{16}ClN_3O$	126
butriptyline	INN, BAN	[35941-65-2]	Evadene, Evadyne	$C_{21}H_{27}N$	127
butriptyline hydrochloride	USAN	[5585-73-9]		$C_{21}H_{28}ClN$	127
citalopram	INN, BAN	[59729-33-8]	Cipramil, Seropram	$C_{20}H_{21}FN_2O$	128
clomipramine	INN, BAN	[303-49-1]	Anafranil, Hydiphen	$C_{19}H_{23}ClN_2$	129
clomipramine hydrochloride	USAN, JAN	[17321-77-6]		$C_{19}H_{24}Cl_2N_2$	129
demexiptiline	INN	[24701-51-7]	Deparon, Tinoran	$C_{18}H_{18}N_2O$	130
desipramine	INN, BAN	[50-47-5]	Norpramin, Pertofran(e)	$C_{18}H_{22}N_2$	131
desipramine hydrochloride	USAN, JAN	[58-28-6]		$C_{18}H_{23}ClN_2$	131
dosulepin	INN[d]	[113-53-1]	Prothiaden, Tihilor	$C_{19}H_{21}NS$	132
dothiepin hydrochloride	USAN	[897-15-4]		$C_{19}H_{22}ClNS$	132
doxepin	INN, BAN	[1668-19-5]	Adapin, Quitaxon, Sinequan	$C_{19}H_{21}NO$	133
doxepin hydrochloride	USAN	[1229-29-4]		$C_{19}H_{22}ClNO$	133
fluoxetine	INN, BAN, USAN	[54910-89-3]	Prozac	$C_{17}H_{18}F_3NO$	134
fluoxetine hydrochloride	USAN	[59333-67-4]	Prozac	$C_{17}H_{19}ClF_3NO$	134
fluvoxamine	INN, BAN	[54739-18-3]	Fevarin, Floxyfral	$C_{15}H_{21}F_3N_2O_2$	135
imipramine	INN, BAN	[50-49-7]	Tofranil	$C_{19}H_{24}N_2$	
imipramine hydrochloride	JAN, USAN	[113-52-0]		$C_{19}H_{25}ClN_2$	136
indalpine	INN, BAN	[63758-79-2]	Upstene	$C_{15}H_{20}N_2$	137
iprindole	INN, BAN, USAN	[5560-72-5]	Prondol, Tertran	$C_{19}H_{28}N_2$	138
iproclozide	INN, BAN	[3544-35-2]	Iproclozide, Sursum	$C_{11}H_{15}ClN_2O_2$	139
iproniazid	INN, BAN	[54-92-2]	Ipronid, Marsilid	$C_9H_{13}N_3O$	140
isocarboxazid	INN, BAN, USAN	[59-63-2]	Marplan	$C_{12}H_{13}N_3O_2$	141

Table 4. (*Continued*)

Agent	Nomen-clature[a]	CAS Registry Number	Selected trade name(s)[b]	Empirical formula	Refs.
lofepramine	INN, BAN	[23047-25-8]	Gamanil,	$C_{26}H_{27}ClN_2O$	142
lofepramine hydrochloride	JAN, USAN	[26786-32-3]	Tymelyt	$C_{26}H_{28}Cl_2N_2O$	142
maprotiline	INN, BAN, USAN	[10262-69-8]	Ludiomil	$C_{20}H_{23}N$	143
maprotiline hydrochloride	JAN, USAN	[10347-81-6]	Ludiomil	$C_{20}H_{24}ClN$	143
metapramine	INN	[21730-16-5]	Prodastene, Timaxel	$C_{16}H_{18}N_2$	144
mianserin	INN, BAN	[24219-97-4]	Bolvidon, Tolvon	$C_{18}H_{20}N_2$	
mianserin hydrochloride	USAN, JAN	[21535-47-7]		$C_{18}H_{21}ClN_2$	145
milnacipran	INN	[92623-85-3]	Dalcipran	$C_{15}H_{22}N_2O$	146
minaprine	INN, BAN, USAN	[25905-77-5]	Cantor, Caprim, Isopulsan	$C_{17}H_{22}N_4O$	147
minaprine dihydro-chloride	USAN	[25953-17-7]		$C_{17}H_{24}Cl_2N_4O$	147
mirtazapine	INN, BAN, USAN	[61337-67-5]	Remergon	$C_{17}H_{19}N_3$	148
moclobemide	INN, BAN USAN	[71320-77-9]	Aurorix, Manerix, Moclamine	$C_{13}H_{17}ClN_2O_2$	149
nefazodone	INN, BAN	[83366-66-9]	Dutonin,	$C_{25}H_{32}ClN_5O_2$	150
nefazodone hydrochloride	USAN	[82752-99-6]	Serzone	$C_{25}H_{33}Cl_2N_5O_2$	150
nialamide	INN, BAN	[51-12-7]	Niamid, Nuredal, Surgex	$C_{16}H_{18}N_4O_2$	151
nortriptyline	INN, BAN	[72-69-5]	Allegron,	$C_{19}H_{21}N$	
nortriptyline hydrochloride	JAN, USAN	[894-71-3]	Aventyl, Pamelor	$C_{19}H_{22}ClN$	152
noxiptiline	INN[e]	[3362-45-6]	Agedal, Elronon, Sipcar	$C_{19}H_{22}N_2O$	153
opipramol dihydro-chloride	INN, BAN	[315-72-0]	Ensidon,	$C_{23}H_{29}N_3O$	154
opipramol dihydro-chloride	USAN	[909-39-7]	Oprimol	$C_{23}H_{31}Cl_2N_3O$	154
paroxetine	INN, BAN, USAN	[61869-08-7]	Paxil, Seroxat, Tagonis	$C_{19}H_{20}FNO_3$	155
phenelzine	INN, BAN	[51-71-8]	Kalgan,	$C_8H_{12}N_2$	156
phenelzine sulfate	USAN	[156-51-4]	Nardil	$C_8H_{14}N_2O_4S$	156
pirlindole	INN	[60762-57-4]	Lifril, Pyrazidol	$C_{15}H_{18}N_2$	157

Table 4. (Continued)

Agent	Nomen-clature[a]	CAS Registry Number	Selected trade name(s)[b]	Empirical formula	Refs.
propizepine	INN	[10321-12-7]	Depressin, Vagran	$C_{17}H_{20}N_4O$	158
protriptyline	INN, BAN	[438-60-8]	Concordin,	$C_{19}H_{21}N$	
protriptyline hydrochloride	USAN	[1225-55-4]	Triptil, Vivactil	$C_{19}H_{22}ClN$	159
quinupramine	INN	[31721-17-2]	Kevopril, Kinupril	$C_{21}H_{24}N_2$	160
sertraline	INN, BAN	[79617-96-2]	Lustral, Zoloft	$C_{17}H_{17}Cl_2N$	
sertraline hydrochloride	USAN	[79559-97-0]		$C_{17}H_{18}Cl_3N$	161
setiptiline	INN	[57262-94-9]	Tecipul	$C_{19}H_{19}N$	162
tianeptine	INN	[66981-73-5]	Stablon	$C_{21}H_{25}ClN_2O_4S$	163
toloxatone	INN	[29218-27-7]	Humoryl	$C_{11}H_{13}NO_3$	164
tranylcypromine	INN, BAN	[155-09-9]	Parnate, Parnitene	$C_9H_{11}N$	156
trazodone	INN, BAN	[19794-93-5]	Desyrel, Molipaxin	$C_{19}H_{22}ClN_5O$	
trazodone hydrochloride	JAN, USAN	[25332-39-2]		$C_{19}H_{23}Cl_2N_5O$	165
trimipramine	INN, BAN, USAN	[739-71-9]	Stangyl,	$C_{20}H_{26}N_2$	166
trimipramine maleate	JAN, USAN	[521-78-8]	Surmontil	$C_{24}H_{30}N_2O_4$	166
venlafaxine	INN, BAN	[93413-69-5]		$C_{17}H_{27}NO_2$	
venlafaxine hydrochloride	USAN	[99300-78-4]	Effexor	$C_{17}H_{28}ClNO_2$	167
viloxazine	INN, BAN	[46817-91-8]	Emovil,	$C_{13}H_{19}NO_3$	
viloxazine hydrochloride	USAN	[35604-67-2]	Vialan, Vivalan	$C_{13}H_{20}ClNO_3$	168

[a]International nonproprietary name (INN), British approved name (BAN), Japanese accepted name (JAN), and United States adopted name (USAN). [b]Only selected salts are included. Trade names shown can represent the base or any of the salts included in "Nomenclature". [c]BAN = bupropion. [d]BAN = dothiepin. [e]BAN = noxiptylene.

Three other clinically effective irreversible MAO inhibitors have been approved for treatment of major depression: phenelzine (**43**), isocarboxazid (**44**), and tranyl-cypromine (**45**).

(**42**)   (**43**)

(**44**)   (**45**)

Chronic use of these irreversible MAO inhibitors has been associated with life-threatening toxicity, ie, hepatotoxicity and hypertensive crisis. Interactions with tyramine contained in food and other drugs have severely limited use of irreversible MAO inhibitors. These MAO inhibitors are also nonselective, inhibiting both MAO-A and MAO-B isoenzymes. Furthermore, they interfere with the hepatic metabolism of many drugs.

Second-Generation Antidepressants.   The frequency of adverse effects associated with first-generation antidepressants and the lack of patient compliance arising from such adverse effects led to the development of a number of second-generation antidepressants.

Atypical Antidepressants.   Structurally diverse drugs such as the tetracyclic mianserin (**46**) and various bicyclic and tricyclic compounds such as trazodone (**47**), venlafaxine (**48**), nefazodone (**49**), and amfebutamone (**50**) are atypical antidepressants. The exact mechanism of action is unclear but probably

(**46**)

(**47**)   (**48**)

(**49**)   (**50**)

involves actions at serotonin, noradrenaline, and/or dopamine synapses (169). Such drugs exhibit reduced toxicity, improved patient compliance, and lower toxicity in overdose than first-generation antidepressants.

Amfebutamone (**50**) (bupropion) is an antidepressant drug that is structurally and mechanistically different from the agents previously approved for treating this disorder. It exhibits a greater effect on the neuronal reuptake of dopamine than on other biogenic amines even though this property does not seem to, by itself, account for its antidepressant effects. In clinical studies, it was found to be as effective as the standard drugs used in the treatment of major depression, but has fewer side effects. It is reported to be particularly useful in patients resistant to other agents, as well as in patients with atypical depression. It is not sedating or cardiotoxic and does not exhibit anticholinergic effects or produce weight gain. Potential adverse effects include seizures and psychosis.

Nefazodone (**49**) is a phenylpiperazine derivative exhibiting a pharmacological profile that is distinct from that of first-generation agents, as well as the more selectively acting second-generation agents, ie, serotonin or noradrenaline reuptake inhibitors. Nefazodone acts both as a serotonin receptor type 2 antagonist and as a serotonin reuptake inhibitor. It appears to be free of cardiotoxicity and is well tolerated even at high doses. It lacks the typical anticholinergic side effects of the tricyclic antidepressants, as well as serotonergic and noradrenergic mediated side effects.

Venlafaxine (**48**) is a structurally novel phenylethylamine derivative that strongly inhibits both noradrenaline and serotonin reuptake. It lacks anticholinergic, antihistaminergic, and antiadrenergic side effects. As compared to placebo, most common adverse events are nausea, somnolence, dizziness, dry mouth, and sweating. Venlafaxine-treated patients also experienced more headaches and nausea, but less dry mouth, dizziness, and tremor than patients treated with comparator antidepressants.

Selective Serotonin Reuptake Inhibitors. Serotonin plays a pivotal role in the physiological regulation of mood. The selective serotonin reuptake inhibitors (SSRIs) resulted from an effort by the pharmaceutical companies to develop drugs exhibiting a higher degree of selectivity for the central serotonin transporter (170). Owing to serotonergic uptake blocking properties, a net enhancement of serotonergic function results from the administration of SSRIs. The drug fluoxetine (**31**), the first SSRI approved by the U.S. Food and Drug Administration for the treatment of major depression, was rapidly followed by several other SSRIs. The synthesis of fluoxetine (Prozac), marketed as a racemate and the commercially most important antidepressant, is shown in Figure 3.

SSRIs are widely used for treatment of depression, as well as, for example, panic disorders and obsessive–compulsive disorder. These drugs are well recognized as clinically effective antidepressants having an improved side-effect profile as compared to the TCAs and irreversible MAO inhibitors. Indeed, these drugs lack the anticholinergic, cardiovascular, and sedative effects characteristic of TCAs. Their main adverse effects include nervousness/anxiety, nausea, diarrhea or constipation, insomnia, tremor, dizziness, headache, and sexual dysfunction. The most commonly prescribed SSRIs for depression are fluoxetine (**31**), fluvoxamine (**32**), sertraline (**52**), citalopram (**53**), and paroxetine (**54**). SSRIs together represent about one-fifth of total worldwide antidepressant unit sales.

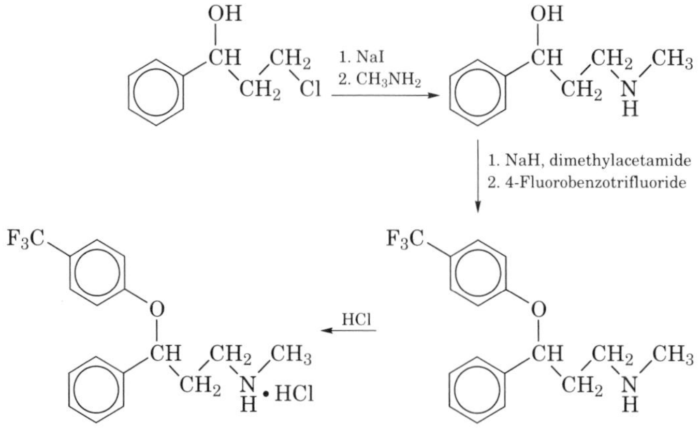

Fig. 3. Synthesis of fluoxetine (**31**). 3-Chloro-1-phenyl-1-propanol reacts with sodium iodide to afford the corresponding iodo derivative, followed by reaction with methylamine, to form 3-(methylamino)-1-phenyl-1-propanol. To the alkoxide of this product, generated using sodium hydride, 4-fluorobenzotrifluoride is added to yield after work-up the free base of the racemic fluoxetine (**31**), thence transformed to the hydrochloride (**51**) (171–174).

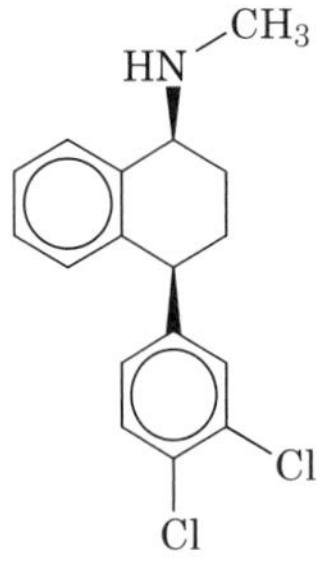

Reversible Inhibitors of Monoamine Oxidase. Selective MAO-A inhibitors, which are reversible (so-called RIMAs), have also been developed, therefore substantially reducing the potential for food and drug interactions. Indeed, the tyramine-potentiating effects of these drugs is much reduced compared with the irreversible MAO inhibitors. The RIMAs represent effective and safer alternatives to the older MAO inhibitors. The only marketed RIMAs are toloxatone [*29218-27-7*] and moclobemide (**55**). The latter is used widely as an effective, well-tolerated antidepressant.

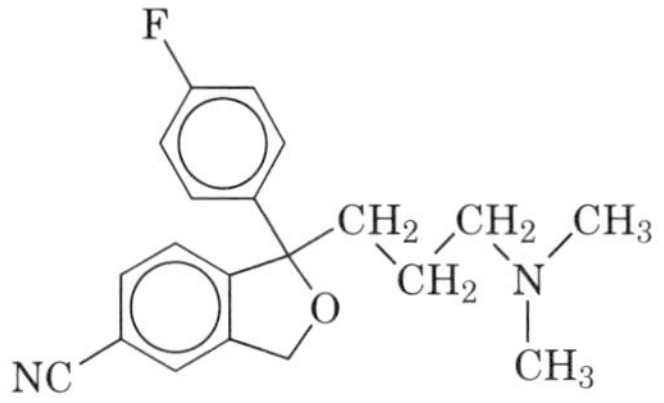

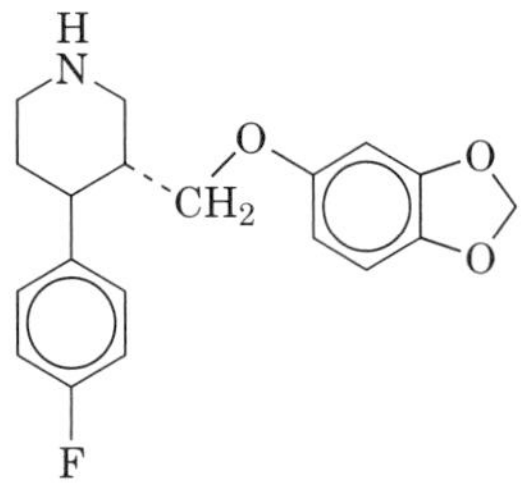

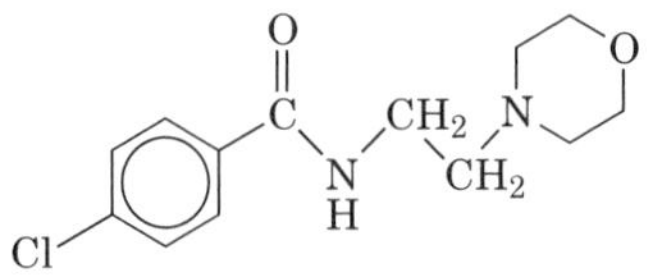

(**55**)

The second-generation antidepressants, particularly RIMAs and SSRIs, are much less toxic in overdose than the older TCAs and irreversible MAO inhibitors. However, similar to first-generation antidepressants, the therapeutic effect only becomes manifest after several weeks. Up to one-third of depressed patients are nonresponders. Ideally, an antidepressant would combine a more rapid onset of action with greater clinical efficacy and a higher responder rate, as well as even better tolerability.

Treatment of Manic–Depressive Illness. Since the 1960s, lithium carbonate [10377-37-4] and other lithium salts have represented the standard treatment of mild-to-moderate manic–depressive disorders (175). It is effective in about 60–80% of all acute manic episodes within one to three weeks of administration. Lithium ions can reduce the frequency of manic or depressive episodes in bipolar patients providing a mood-stabilizing effect. Patients are maintained on low, stabilizing doses of lithium salts indefinitely as a prophylaxis. However, the therapeutic index is low, thus requiring monitoring of serum concentration. Adverse effects include tremor, diarrhea, problems with eyes (adaptation to darkness), hypothyroidism, and cardiac problems (bradycardia–tachycardia syndrome).

Lithium ions are hypothesized to act by reducing the coupling between receptors and their G-proteins. Because several neurotransmitter receptors share common G-protein-regulated second-messenger signaling systems, lithium ions could simultaneously correct the alterations at synapses associated with depression and mania by a single action on the function of specific G-proteins. Lithium ions might also act via their interruption of the phosphatidylinositide cycle, leading to a depletion of membrane inositol and phosphoinositide-derived second-messenger products, thereby reducing signaling through those receptor systems dependent on the formation of these products. The antiepileptic valproic acid [99-66-1] and its salts have also been reported to be therapeutically useful in treating mania (176), possibly via enhancement of GABA metabolism in the brain. Valproate semisodium [76584-70-8] (divalproex sodium) has been approved in the United States for treatment of manic episodes associated with bipolar depression.

Other agents are also used for the treatment of manic–depressive disorders based on preliminary clinical results (177). The antiepileptic carbamazepine [298-46-4] has been reported in some clinical studies to be therapeutically beneficial in mild-to-moderate manic depression. Carbamazepine treatment is used especially in bipolar patients intolerant to lithium or nonresponders. A majority of lithium-resistant, rapidly cycling manic–depressive patients were reported in one study to improve on carbamazepine (178). Carbamazepine blocks noradrenaline reuptake and inhibits noradrenaline exocytosis. The main adverse events are those found commonly with antiepileptics, ie, vigilance problems, nystagmus, ataxia, and anemia, in addition to nausea, diarrhea, or constipation. Carbamazepine can be used in combination with lithium. Several clinical studies report that the calcium channel blocker verapamil [52-53-9], registered for angina pectoris and supraventricular arrhythmias, may also be effective in the treatment of acute mania. Its use as a mood stabilizer may be unrelated to its calcium-blocking properties. Verapamil also decreases the activity of several neurotransmitters. Severe manic depression is often treated with antipsychotics or benzodiazepine anxiolytics.

Future Outlook for Antidepressants. Third-generation antidepressants are expected to combine superior efficacy and improved safety, but are unlikely to reduce the onset of therapeutic action in depressed patients (179). Many drugs in clinical development as antidepressive agents focus on established properties such as inhibition of serotonin, dopamine, and/or noradrenaline reuptake, agonistic or antagonistic action at various serotonin receptor subtypes, presynaptic α_2-adrenoceptor antagonism, or specific monoamine–oxidase type A inhibition. Examples include buspirone (**3**) (only approved for treating anxiety disorders) and ipsapirone [95847-70-4], acting as 5-HT$_{1A}$ receptor partial agonists. Antagonism at 5-HT$_{1B/1D}$ receptors is another approach being pursued.

Modulation of second-messenger pathways is also an attractive target upon which to base novel antidepressants. Rolipram [61413-54-5], an antidepressant in the preregistration phase, enhances the effects of noradrenaline though selective inhibition of central phosphodiesterase, an enzyme which degrades cyclic adenosine monophosphate (cAMP). Modulation of the phosphatidyl inositol second-messenger system coupled to, for example, 5-HT$_{2A}$, 5-HT$_{2B}$, or 5-HT$_{2C}$ receptors might also lead to novel antidepressants, as well as to alternatives to lithium for treatment of mania. Novel compounds such as inhibitors of S-adenosyl-methionine or central catechol-O-methyltransferase also warrant attention.

Neuropeptides also represent innovative targets for drug design in the antidepressant field. A well-known finding in biological psychiatry is hyperactivity of the hypothalamo–pituitary–adrenal axis in patients with endogenous depression. Thyrotropin-releasing hormone (TRH), corticotropin-releasing hormone (CRH), and adrenocorticotropic hormone (ACTH) are hypothesized to be involved in the etiology of depression (see HORMONES). Therefore, modulators of these neuropeptide receptors may represent interesting opportunities for novel antidepressants. Another tripeptide, melanocyte-stimulating hormone-release inhibiting factor-1 (MIF-1), has been claimed to be rapidly effective in the treatment of depressive illness. A number of other compounds such as calcium antagonists, N-methyl-D-aspartate receptor antagonists, vasopressin and melatonin receptor ligands, angiotensin-converting enzyme inhibitors, and adenosine receptor antagonists have been reported to exhibit antidepressant-like properties in animal models and thus provide drug targets worthy of study.

Antipsychotics

Schizophrenia is perhaps the most debilitating psychiatric illness in modern medicine, affecting about 1% of the general population. Many of those affected require institutionalization (180). Unfortunately, the compounds available to treat this disorder are not fully effective in treating the spectrum of symptoms in all patients. Adverse effects are also a problem (181). In addition, available antipsychotic (neuroleptic) drugs (Table 5) can at most only provide symptomatic relief.

Classification of Psychoses. Schizophrenia is characterized in the DSM-IV classification system (1) by a number of symptoms of which typically two or more should be present for most of the time during a period of one month: delusions, hallucinations, disorganized speech, grossly disorganized or catatonic

Table 5. Marketed Antipsychotics

Agents	Nomen-clature[a]	CAS Registry Number	Selected trade name(s)[b]	Empirical formula	Refs.
acetophenazine	INN	[2751-68-0]		$C_{23}H_{29}N_3O_2S$	
acetophenazine maleate	USAN	[5714-00-1]	Tindal	$C_{31}H_{37}N_3O_{10}S$	182
amisulpride	INN	[71675-85-9]	Solian	$C_{17}H_{27}N_3O_4S$	183
benperidol	INN, BAN, USAN	[2062-84-2]	Glianimon	$C_{22}H_{24}FN_3O_2$	184
bromperidol	INN, BAN, JAN, USAN	[10457-90-6]	Impromen	$C_{21}H_{23}BrFNO_2$	185
bromperidol decanoate	BAN, USAN	[75067-66-2]	Impromen	$C_{31}H_{41}BrFNO_3$	185
chlorpromazine	INN, BAN, USAN	[50-53-3]	Largactil,	$C_{17}H_{19}ClN_2S$	186
chlorpromazine hydrochloride	INN, BAN, JAN, USAN	[69-09-0]	Thorazine	$C_{17}H_{20}Cl_2N_2S$	186
chlorprothixene	INN, BAN, JAN, USAN	[113-59-7]		$C_{18}H_{18}ClNS$	
chlorprothixene acetate		[58889-16-0]	Taractan, Truxal	$C_{20}H_{22}ClNO_2S$	182
chlorprothixene hydrochloride	JAN	[6469-93-8]		$C_{18}H_{19}Cl_2NS$	
clopenthixol	INN, BAN, USAN	[982-24-1]	Sordinol	$C_{22}H_{25}ClN_2OS$	187
clotiapine	INN, JAN[c]	[2058-52-8]	Entumin, Etumina	$C_{18}H_{18}ClN_3S$	188
clozapine	INN, BAN, USAN	[5786-21-0]	Leponex, Clozaril	$C_{18}H_{19}ClN_4$	189
cyamemazine	INN	[3546-03-0]	Tercian	$C_{19}H_{21}N_3S$	190
dixyrazine		[2470-73-7]	Esucos	$C_{24}H_{33}N_3O_2S$	191
droperidol	INN, BAN, JAN, USAN	[548-73-2]	Inapsine	$C_{22}H_{22}FN_3O_2$	192
flupentixol	INN[d]	[2709-56-0]	Fluanxol	$C_{23}H_{25}F_3N_2OS$	193
flupentixol decanoate		[30909-51-4]	Fluanxol	$C_{33}H_{43}F_3N_2O_2S$	193
fluphenazine	INN, BAN	[69-23-8]		$C_{22}H_{26}F_3N_3OS$	
fluphenazine enanthate	JAN, USAN	[2746-81-8]		$C_{29}H_{38}F_3N_3O_2S$	
fluphenazine decanoate	USAN	[5002-47-1]	Prolixin	$C_{32}H_{44}F_3N_3O_2S$	194
fluphenazine hydrochloride	BAN, JAN, USAN	[146-56-5]		$C_{22}H_{28}Cl_2F_3N_3OS$	
fluspirilene	INN, BAN, USAN	[1841-19-6]	Imap	$C_{29}H_{31}F_2N_3O$	195
haloperidol	INN, BAN, JAN, USAN	[52-86-8]	Haldol	$C_{21}H_{23}ClFNO_2$	196
haloperidol decanoate	BAN, JAN, USAN	[74050-97-8]	Haldol	$C_{31}H_{41}ClFNO_3$	196
levomepromazine	INN[e]	[60-99-1]		$C_{19}H_{24}N_2OS$	
levomepromazine hydrochloride	JAN	[4185-80-2]	Nozinan	$C_{19}H_{25}ClN_2OS$	197
levomepromazine maleate	JAN	[7104-38-3]	Nozinan	$C_{23}H_{28}N_2O_5S$	197

Table 5. (Continued)

Agents	Nomen-clature[a]	CAS Registry Number	Selected trade name(s)[b]	Empirical formula	Refs.
loxapine	INN, BAN, USAN	[1977-10-2]		$C_{18}H_{18}ClN_3O$	
loxapine succinate	USAN	[27833-64-3]	Loxapac	$C_{22}H_{24}ClN_3O_5$	198
loxapine hydrochloride		[54810-23-0]		$C_{18}H_{19}Cl_2N_3O$	
melperone	INN, BAN	[3575-80-2]		$C_{16}H_{22}FNO$	
melperone hydrochloride		[1622-79-3]	Eunerpan	$C_{16}H_{23}ClFNO$	199
mesoridazine	INN, BAN, USAN	[5588-33-0]	Serentil	$C_{21}H_{26}N_2OS_2$	200
mesoridazine besylate	USAN	[32672-69-8]	Serentil	$C_{27}H_{32}N_2O_4S_3$	200
molindone	INN, BAN	[7416-34-4]		$C_{16}H_{24}N_2O_2$	
molindone hydrochloride	USAN	[15622-65-8]	Moban	$C_{16}H_{25}ClN_2O_2$	201
mosapramine	INN[f]	[89419-40-9]	Cremin	$C_{28}H_{35}ClN_4O$	202
nemonapride	INN, JAN	[93664-94-9]	Emirace	$C_{21}H_{26}ClN_3O_2$	203
oxypertine	INN, BAN, JAN, USAN	[153-87-7]	Forit	$C_{23}H_{29}N_3O_2$	204
perazine	JAN	[84-97-9]	Taxilan	$C_{20}H_{25}N_3S$	205
periciazine	INN[g]	[2622-26-6]	Neuleptil	$C_{21}H_{23}N_3OS$	206
perphenazine	INN, BAN, JAN, USAN	[58-39-9]	Trilafon	$C_{21}H_{26}ClN_3OS$	207
pimozide	INN, BAN, JAN, USAN	[2062-78-4]	Orap	$C_{28}H_{29}F_2N_3O$	208
pipamperone	INN, BAN, USAN[h]	[1893-33-0]	Dipiperon	$C_{21}H_{30}FN_3O_2$	209
pipamperone hydrochloride	JAN	[2448-68-2]	Dipiperon	$C_{21}H_{32}Cl_2FN_3O_2$	209
prochlorperazine	INN, BAN, JAN, USAN	[58-38-8]	Compazine	$C_{20}H_{24}ClN_3S$	210
promazine	INN, BAN	[58-40-2]	Talofen,	$C_{17}H_{20}N_2S$	211
promazine hydrochloride	USAN	[53-60-1]	Sparine	$C_{17}H_{21}ClN_2S$	211
prothipendyl	INN, BAN	[303-69-5]	Dominal	$C_{16}H_{19}N_3S$	190
risperidone	INN, BAN, USAN	[106266-06-2]	Risperdal	$C_{23}H_{27}FN_4O_2$	212
sulpiride	INN, BAN, JAN, USAN	[15676-16-1]	Dogmatil	$C_{15}H_{23}N_3O_4S$	213
sultopride	INN	[53583-79-2]	Barnetil	$C_{17}H_{26}N_2O_4S$	214
thiopropazate	INN, BAN	[84-06-0]	Dartal	$C_{23}H_{28}ClN_3O_2S$	204
thioridazine	INN, BAN, USAN	[50-52-2]	Melleril	$C_{21}H_{26}N_2S_2$	215
thioridazine hydrochloride	JAN, USAN	[130-61-0]	Melleril	$C_{21}H_{27}ClN_2S_2$	215
tiapride	INN, BAN	[51012-32-9]	Gramalil,	$C_{15}H_{24}N_2O_4S$	216
tiapride hydrochloride	BAN	[51012-33-0]	Tiapridal	$C_{15}H_{25}ClN_2O_4S$	216

Table 5. (*Continued*)

Agents	Nomen-clature[a]	CAS Registry Number	Selected trade name(s)[b]	Empirical formula	Refs.
timiperone	INN, JAN	[57648-21-2]	Tolopelon	$C_{22}H_{24}FN_3OS$	217
tiotixene	INN, JAN[i]	[5591-45-7]	Navane	$C_{23}H_{29}N_3O_2S_2$	218
trifluoperazine	INN, BAN	[117-89-5]		$C_{21}H_{24}F_3N_3S$	
trifluoperazine hydrochloride	INN, BAN, USAN	[440-17-5]	Stelazine	$C_{21}H_{26}Cl_2F_3N_3S$	219
triflupromazine	INN, USAN[j]	[146-54-3]	Psyquil	$C_{18}H_{19}F_3N_2S$	220
zotepine	INN, JAN	[26615-21-4]	Lodopin	$C_{18}H_{18}ClNOS$	221
zuclopenthixol	INN, BAN[k]	[53772-83-1]	Cisordinol, Sedanxol	$C_{22}H_{25}ClN_2OS$	222

[a]International nonproprietary name (INN), British approved name (BAN), Japanese accepted name (JAN), and United States adopted name (USAN). [b]Only selected salts are included. Trade names shown can represent the base or any of the salts included in nomenclature column of the table. [c]BAN, USAN = chlothiapine. [d]BAN = flupenthixal. [e]BAN, USAN = methotrimeprazine. [f]JAN = mosapramine hydrochloride. [g]BAN = pericyazine; JAN = propericiazine. [h]Also called fluoropipamide. [i]BAN, USAN = thiothixene. [j]BAN = fluopromazine. [k](Z)-isomer of clopenthiol.

behavior, and/or negative symptoms such as affective flattening, alogia, and avolition. Disorders of thought processes in schizophrenics can be seen in their writing and speech, taking the form of illogical thought patterns. Schizophrenics often run together a number of logically unconnected topics and use words in a meaningless way or indeed use meaningless words. Disorders of thought, however, can also take the form of elaborate fantasies, often on profound topics and matters of principle. There may be a flattening of emotion and patients may remain impassive when confronted with situations that would normally elicit a marked response. Patients often report loss of will power and sometimes feel that they are being controlled by forces outside themselves. There are varying losses of coordination of body movements which can range from awkwardness to chaotic overactivity to catatonia. Delusions and auditory hallucinations are common.

As for most psychiatric disorders, speculations concerning the pathology of the disease come from the limited understanding of the pharmacology of drugs effective in symptomatically treating the disorder. However, because the most effective antipsychotic drugs interact with numerous molecular targets, their mechanism of action in treating this disorder remains unclear.

Pharmacological Profiles of Antipsychotics. Most compounds used for antipsychotic therapy can be assigned to one of four structurally distinct groups. These are phenothiazines, eg, chlorpromazine (**56**); thioxanthenes, eg, chlorpro-thixene (**57**); diphenylbutylpiperidines, eg, pimozide (**58**); and butyrophenones, eg, haloperidol (**59**). Those antipsychotics most widely used clinically are included in Table 5. These compounds represent more than 97% of total units sold world-wide, but only about half of the antipsychotics clinically available. In view of the dopaminergic blocking action of compounds from these classes, some are used predominantly or even solely as antiemetics, including alizapride [59338-93-1], clebopride [55905-53-8], domperidone [57808-66-9], metoclopramide [364-62-5], oxypendyl [5585-93-3], and promethazine [60-87-7].

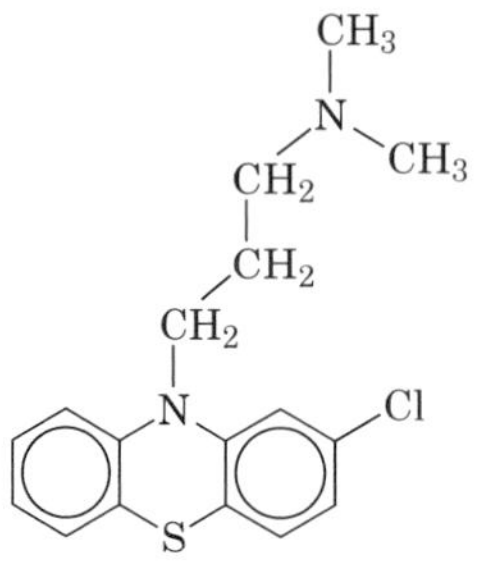

(56)

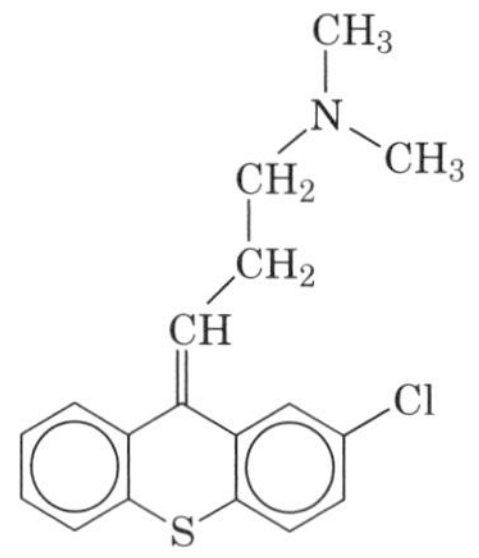

(57)

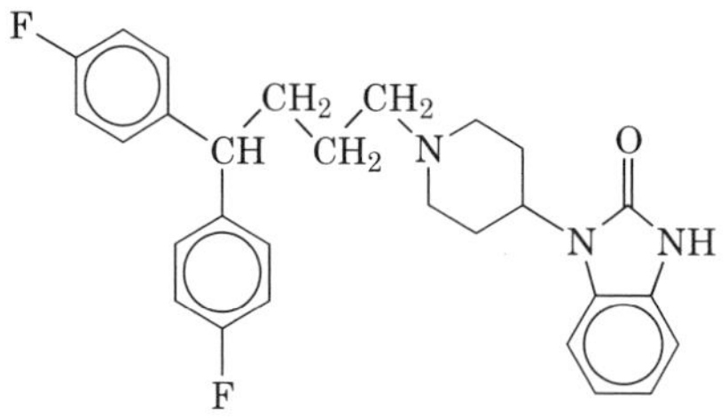

(58)

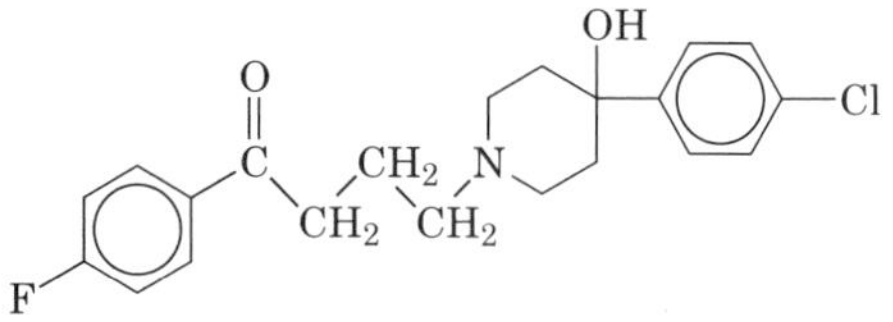

(59)

There is a good correlation between the affinity of antipsychotics for the dopamine D_2 receptor and their clinically effective dose used in the treatment of schizophrenia, suggesting the special importance of this neurotransmitter system with respect to schizophrenia (223). There are two principal dopaminergic pathways in the brain which project forward from cell bodies in the midbrain: the nigrostriatal dopaminergic pathway and the mesocorticolimbic dopaminergic pathway. It has been hypothesized that antipsychotic activity results from the blockade of dopaminergic transmission in the mesolimbic pathway, whereas blockade of dopaminergic transmission in the nigrostriatal pathway gives rise to extrapyramidal side effects such as tardive dyskinesia, akathisia, and Parkinsonian-like symptoms often associated with antipsychotic treatment (224). Therefore, antidopaminergic properties may account for both therapeutic actions and adverse effects of these drugs. However, these effects may be separable. The target of preclinical research is an antipsychotic drug that lacks extrapyramidal side effects. One such atypical antipsychotic is clozapine (**60**) which is effective in treating most of the symptoms of schizophrenia and has minimal or no concomitant extrapyramidal side effects.

Correlation between clinical effectiveness and receptor affinities, however, can be seen with other receptors in addition to the dopamine D_2 receptor. These include other dopaminergic receptors, as well as noradrenergic and serotonergic receptors. For example, most antipsychotics also have high affinity for α_1-adrenoceptors and 5-HT_{2A} receptors (225). Some antipsychotics have been shown to be selective for the α_{1B}-adrenoceptor versus the α_{1A}-adrenoceptor, for example, spiperone [749-02-0] (226) and risperidone (**61**) (227).

(60)   (61)

More interesting is the mechanism of action of atypical antipsychotics, particularly clozapine (**60**). It is also probably the least selective antipsychotic used in the clinic, having high affinity for numerous receptors including 5-HT$_{2A}$, 5-HT$_{2C}$, 5-HT$_6$, and 5-HT$_7$; dopamine D_1, D_2, D_3, and D_4 receptors; H_1; all five subtypes of muscarinic cholinergic receptors; and α_1-adrenoceptors. One or more of these receptors might contribute to the atypical therapeutic profile of clozapine. With the possible exception of the dopamine D_4 receptor, the high affinity of clozapine for these other sites is not shared by other atypical antipsychotics; for example, risperidone (**61**) has low affinity for the 5-HT$_6$ receptor and melperone [3575-80-2] has low affinity for the 5-HT$_7$ receptor (228). In the case of the dopamine D_4 receptor, there is also a good correlation between the affinities of most antipsychotics for the dopamine D_4 receptor and their clinically active dose. Although this may suggest that the dopamine D_4 receptor is an important receptor with respect to antipsychotic effects, at the same time it argues against the likelihood that high affinity for this receptor is solely responsible for the atypical profile of clozapine. Careful evaluation of the affinity of clozapine (**60**) for the dopamine D_2 receptor reveals that, in fact, the compound has less than one logarithmic unit of binding selectivity for the dopamine D_4 receptor versus the dopamine D_2 receptor (229,230).

It has been suggested that high affinity for certain 5-HT receptors is a possible method for reducing extrapyramidal side effects observed with antipsychotics. Liability for extrapyramidal side effects has been hypothesized to be related to the ratio of the affinity between dopamine D_1/D_2 receptor binding sites and 5-HT$_{2A}$ receptor binding sites (231). Risperidone (**61**) has high affinity for dopamine D_2 and 5-HT$_{2A}$ receptors and in clinical trials has been reported to have only a weak tendency to induce extrapyramidal side effects; however, clear demonstration of its advantage in treating all the various symptoms of schizophrenia awaits further clinical results.

Future Outlook for Antipsychotics. Preclinical work is focused on the development of atypical antipsychotics having a favorable clinical profile similar to that of clozapine (**60**). As of the mid-1990s, a wide variety of potential drug targets are being pursued (181). Compounds exhibiting high affinity binding to several receptors such as 5-HT$_{2A}$, dopamine D_1 and dopamine D_2 receptors, ie, risperidone-like compounds, as well as compounds that have high affinity and selectivity for only a specific receptor, eg, dopamine D_4, 5-HT$_{1A}$, 5-HT$_{2C}$, or 5-HT$_6$ receptors, are under consideration. In addition, the so-called σ-receptor provides an interesting drug target. It is located throughout the central nervous system as well as in peripheral tissues. Its biochemical and physiological functions are as yet poorly understood. The identification of three binding site subtypes has added to the complexity of this research area. However, because many antipsychotic agents bind to σ-sites, it has been postulated that selective σ-receptor ligands might prove useful in treating schizophrenia. It has furthermore been hypothesized that neurotensin analogues, opioid peptides, and cholecystokinin receptor agonists might exhibit antipsychotic effects.

Drugs for Treating Substance Use Disorders

It has been estimated that about one out of every seven adults in the United States abuses or is dependent on alcohol with an additional one out of every 20 persons abusing or dependent on other drugs (232). Certainly, the abuse and dependence on ethyl alcohol [64-17-5], nicotine [54-11-5], cocaine [50-36-2], and heroin [561-27-3] (diacetylmorphine), in particular, have produced devastating socioeconomic consequences. The similarity of psychosocial factors leading to abuse of these drugs, despite their diverse pharmacological mechanisms, suggests the predominant role played by the former in both the development and maintenance of substance abuse disorders, and consequently has resulted in an emphasis on social and behavioral treatment approaches (233). Aside from the clear therapeutic value of pharmacological treatment for both severe intoxication and acute withdrawal phenomena, protracted substance abuse and dependence may also prove amenable to therapy with pharmacologic agents.

Classification of Substance-Related Disorders. The DSM-IV classification system (1) divides substance-related disorders into two categories: (*1*) substance use disorders, ie, abuse and dependence; and (*2*) substance-induced disorders, intoxication, withdrawal, delirium, persisting dementia, persisting amnestic disorder, psychotic disorder, mood disorder, anxiety disorder, sexual dysfunction, and sleep disorder. The different classes of substances addressed herein are alcohol, amphetamines, caffeine, cannabis, cocaine, hallucinogens, inhalants, nicotine, opioids, phencyclidine, sedatives, hypnotics or anxiolytics, polysubstance, and others. On the basis of their significant socioeconomic impact, alcohol, nicotine, cocaine, and opioids have been selected for discussion herein.

Pharmacological Profiles of Agents Used to Treat Substance Use Disorders. The subjective effects of drugs leading to their abuse are the result of modulation of brain neurotransmitter systems (234). Alcohol, as well as a number of other central nervous system (CNS) depressants such as benzodiazepines and barbiturates, enhances central GABAergic activity which normally serves as the main inhibitory neurotransmitter system in the brain. Nicotine exerts its effects on brain function via nicotinic acetylcholine receptors. Cocaine and other CNS stimulants facilitate the presynaptic release of catecholamines and block their reuptake by neurons. Depletion of neuronal stores of dopamine and noradrenaline has been hypothesized to underlie the depression accompanying withdrawal. Opiates act at brain receptors, eg, μ-opioid receptors are hypothesized to be most important for the subjective effects of opioids, reduce noradrenergic tone, and their withdrawal results in rebound hyperactivity of this system in both the brain and periphery.

One approach taken to the treatment of chronic alcoholism relies on a form of aversion therapy involving prophylactic administration of inhibitors of aldehyde dehydrogenase, such as disulfiram [97-77-8] (**62**) and calcium carbimide [156-62-7], $CaCN_2$. Any consumption of alcohol then results in an accumulation of acetaldehyde yielding flushing hypotension, tachycardia, nausea, and vomiting. However, it has been reported that the efficacy of disulfiram in reducing relapse is only modest, primarily owing to noncompliance (235). More recently, the μ-opiate antagonist naltrexone [16590-41-3] (**63**) was approved in the United States for the treatment of alcohol abuse. Clinical results suggest that relapse rate in alcoholics may also be reduced by SSRIs, dopaminergic agents, lithium salts, β-adrenergic blockers, carbamazepine, and hydroxyzine.

(**62**)   (**63**)

An approved pharmacologic approach to stopping tobacco smoking, ie, nicotine dependence, involves the substitution of nicotine-containing chewing gum or a dermal patch with subsequent gradual reduction of the nicotine dose until dependence has been eliminated. The treatment of chronic CNS stimulant abuse and dependence is based primarily on the control of craving by support and minimizing contact with situations apt to trigger craving. Possible adjunctive pharmacologic treatments which hold promise include tricyclic antidepressants and carbamazepine [298-46-4].

A common strategy for treating chronic opiate addiction involves the substitution of methadone which can either be provided as maintenance therapy or tapered until abstinence is achieved. Naltrexone and buprenorphine [52485-79-7] have also been used in this manner. The α_2-adrenergic agonist clonidine [4205-90-7] provides some relief from the symptoms of opiate withdrawal, probably the result of its mimicking the inhibitory effect of opiates on the activity of *locus coeruleus* neurons.

Future Outlook for Pharmacologic Treatment of Abuse and Dependence. The importance of the psychosocial dimension in predisposing individuals toward substance use disorders and subsequently maintaining the disorder cannot be overestimated. Additionally, genetic influences have been found to exert an important influence on liability for drug abuse. A high comorbidity of psychiatric illnesses with substance use disorders further complicates therapeutic interventions in such patients (236).

The roles of pharmacologic treatments have been mostly restricted to countering overdose and ameliorating symptoms of acute withdrawal, ie, substance-induced disorders. Effective reduction of craving for abused substances has proved difficult to achieve except through substitution strategy which may reduce some problems but fails to eliminate dependence. However, in view of the immensity of the social and medical problems, the search for pharmacological treatment remains a worthy goal. Elucidation of the neurobiological bases of reward/aversion processes, the lowest common denominator for abuse and dependence, may yield anticraving drugs and drugs effective in reducing dependence (237).

Cognition Enhancers

Cognitive impairment can arise through diverse pathophysiological mechanisms including, most prominently, primary degenerative dementias such as Alzheimer's disease, Parkinson's disease, and Pick's disease as well as multi-infarct dementia, cerebral vascular accidents, CNS trauma, tumors, and other causes. Until recently agents for symptomatic therapy have not received general recognition as efficacious and, although preventative measures for dementias with cardiovascular etiologies could sometimes be applied, as of the mid-1990s no drugs were approved for prevention or slowing of the progression of dementias of primary degenerative origin.

In view of the devastating socioeconomic consequences of senile dementia of Alzheimer's type (SDAT), this has received considerable attention (238), although cognitive disorders can occur throughout the life span owing to diverse etiologies. SDAT is the most common dementing disease, affecting one of every six individuals over 65 years of age. A common feature is the dysfunction or death of selective populations of neurons. For example, basal forebrain cholinergic neurons are vulnerable and their degeneration results in loss of hippocampal and cortical cholinergic function.

The acetylcholinesterase inhibitor tacrine (**64**) was approved for the treatment of mild-to-moderate SDAT in the United States in 1993 followed by several other countries. The acetylcholinesterase inhibitor galanthamine (**65**), which has long been in clinical use in Austria for the treatment of indications such as facial neuralgia and residual poliomyelitis paralysis, has also been approved for use in

(**64**)   (**65**)

Alzheimer's disease in that country (239). There are a variety of other drugs marketed in one or more countries for this or related indications (240). Codergocrine mesylate, a mixture of three dihydrogenated ergot alkaloids, has been marketed worldwide since the 1950s for treating cognitive impairment or dementia. Other drugs such as meclophenoxate, bifemelane, vincamine, vinpocetine, cyclandelate, naftidrofuryl, pyritinol, idebenone, and indeloxazine are used in some countries. Piracetam (**66**) and a number of other structurally related pyrrolidi-nones, eg, aniracetam (**67**), oxiracetam (**68**), and pramiracetam (**69**) constitute the nootropic class of cognition enhancers used in treating dementia. Table 6 lists selected drugs in use as cognition enhancers (see MEMORY-ENHANCING DRUGS).

(**66**)   (**67**)   (**68**)   (**69**)

Pharmacological Profiles of Cognition Enhancers.   The common denomi-nator of most of the cognition enhancers marketed or in clinical development is the ability to reverse learning and memory deficits in animal models. However, the putative neurobiological mechanisms of action of these compounds cover a wide range of mechanisms. The so-called cholinergic hypothesis has evolved with the accumulation of evidence that cognitive deficits associated with aging or with degenerative disorders result, at least in part, from impaired brain cholinergic function. A variety of drug development approaches focusing on cholinomimetic mechanisms have been clinically investigated.

Putative mechanism of action	Representative substances
cholinesterase inhibitors	physostigmine [57-47-6]
	E2020 [120011-70-3]
	eptastigmine [121652-76-4]
cholinomimetics	arecoline [63-75-2]
	bethanechol [590-63-6]
	AF102B [107233-08-9]
	SDZ 212-713 [123441-03-2]
acetylcholine precursors	choline [62-49-7]
	citicoline [987-78-0]
	lecithin [8002-43-5]
acetylcholine releasers	aniracetam [72432-10-1]
	acetyl-L-carnitine [5080-50-2]
	nefiracetam [77191-36-7]
	DUP996 [105431-72-9]
nicotine receptor agonists	nicotine [54-11-5]
neuropeptides	antigalanin agents
	somatostatin [51110-01-0]
	thyrotropin-releasing
	hormone [24305-27-9]

However, clinical results with compounds enhancing cholinergic function have not been overly convincing (272). In the case of tacrine, however, the beneficial

Table 6. Marketed Drugs Used as Cognition Enhancers

Agent	Nomen-clature[a]	Other name(s) or salts	CAS Registry Number	Selected trade name(s)[b]	Empirical formula	Refs.
amfetamine	INN	amphetamine (BAN); amphetamine sulfate (USAN)	[300-62-9]	Benzadrine	$C_9H_{13}N$	241
aniracetam	INN, USAN		[72432-10-1]	Draganon	$C_{12}H_{13}NO_3$	242
bifemelane	INN	bifemelane hydrochloride (JAN)	[90293-01-9]	Alnert, Celeport	$C_{18}H_{23}NO$	243
cinnarizine	INN, BAN, JAN, USAN		[298-57-7]	Aplactan	$C_{26}H_{28}N_2$	244
clonidine	INN, BAN, USAN	clonidine hydrochloride (BAN, JAN, USAN)	[4205-90-7]	Catapres-TTS	$C_9H_9Cl_2N_3$	245
cyclandelate	INN, BAN, JAN		[456-59-7]	Capilan	$C_{17}H_{24}O_3$	246
ergoloid mesylate	USAN	codergocrine mesylate (BAN); dihydro-ergotoxine (JAN)	[8067-24-1]	Hydergine	c	247
galantamine	INN[d]		[357-70-0]		$C_{17}H_{21}NO_3$	
galantamine hydrobro-mide			[1953-04-4]	Nivalin	$C_{17}H_{22}ClNO_3$	248
guanfacine	INN, BAN	guanfacine hydrochloride (JAN, USAN)	[29110-47-2]	Tenex	$C_9H_9Cl_2N_3O$	249
idebenone	INN, JAN		[58186-27-9]	Aban, Avan	$C_{19}H_{30}O_5$	250
indeloxazine	INN	indeloxazine hydrochloride (JAN, USAN)	[60929-23-9]	Alzene, Elen, Noin	$C_{14}H_{17}NO_2$	251
isoxsuprine	INN, BAN, JAN	isoxsupine hydrochloride (USAN)	[395-28-8]	Vasodilan	$C_{18}H_{23}NO_3$	252
levodopa	INN, BAN, JAN, USAN		[59-92-7]	Larodopa	$C_9H_{11}NO_4$	253
meclofen-oxate	INN, BAN	meclofenoxate hydrochloride (JAN)	[51-68-3]	Brenal, Lucidril	$C_{12}H_{16}ClNO_3$	254

Table 6. (Continued)

Agent	Nomen-clature[a]	Other name(s) or salts	CAS Registry Number	Selected trade name(s)[b]	Empirical formula	Refs.
methyl-phenidate	INN, BAN, JAN	methylphenidate hydrochlorde (USAN)	[113-45-1]	Ritalin	$C_{14}H_{19}NO_2$	255
naftidro-furyl	INN, BAN	nafronyl oxalate (USAN)	[31329-57-4]	Praxilene	$C_{24}H_{33}NO_3$	256
naloxone	INN, BAN, JAN	naloxone hydrochloride (USAN)	[465-65-6]	Narcan	$C_{19}H_{21}NO_4$	257
naltrexone hydro-chloride		naltrexone (INN, BAN, USAN)	[16676-29-2]	Revia, Trexan	$C_{20}H_{24}ClNO_4$	258
nicergoline	INN, BAN, JAN, USAN		[27848-84-6]	Sermion, Varson	$C_{24}H_{26}$-BrN_3O_3	259
nimodipine	INN, BAN, USAN		[66085-59-4]	Nimotop, Periplum	$C_{21}H_{26}N_2O_7$	260
oxiracetam	INN, BAN		[62613-82-5]	Neuromet	$C_6H_{10}N_2O_3$	261
papaverine hydro-chloride	JAN, USAN	papaverine (BAN)	[61-25-6]	Pavabid	$C_{20}H_{22}ClNO_4$	256
pemoline	INN, BAN, JAN, USAN		[2152-34-3]	Cylert	$C_9H_8N_2O_2$	262
pentoxi-fylline	INN, JAN, USAN	oxpentifylline (BAN)	[6493-05-6]	Trental	$C_{13}H_{18}N_4O_3$	263
piracetam	INN, BAN, USAN		[7491-74-9]	Nootropil, Diemil	$C_6H_{10}N_2O_2$	264
pramirace-tam	INN	pramiracetam hydrochloride/ sulfate (USANs)	[68497-62-1]	Remen, Neupramir, Pramistar	$C_{14}H_{27}N_3O_2$	265
procaine	INN, BAN	procaine hydrochloride (JAN, USAN)	[59-46-1]	Gerovital, Novocain	$C_{13}H_{20}N_2O_2$	266
propento-fylline[e]	INN, BAN, JAN		[55242-55-2]	Hextol	$C_{15}H_{22}N_4O_3$	267
pyritinol	INN, BAN	pyrithioxine, pyrithioxine hydrochloride (JANs)	[1098-97-1]	Encephabol	$C_{16}H_{20}N_2$-O_4S_2	268
tacrine	INN, BAN	tacrine hydro-chloride (USAN)	[321-64-2]	Cognex, Tha	$C_{13}H_{14}N_2$	269

Table 6. (*Continued*)

Agent	Nomen- clature[a]	Other name(s) or salts	CAS Registry Number	Selected trade name(s)[b]	Empirical formula	Refs.
vincamine	INN, BAN		[1617-90-9]	Cetal, Devincan	$C_{21}H_{26}N_2O_3$	270
vinpocetine	INN, JAN, USAN		[42971-09-5]	Calan, Cavinton, Ceractin	$C_{22}H_{26}N_2O_2$	271

[a]International nonproprietary name (INN), British approved name (BAN), Japanese accepted name (JAN), United States adopted name (USAN). [b]Only selected salts are included. Trade names shown can represent the base or any of the salts included in "Nomenclature". [c]Mixture of dihydroergocornine mesylate, $C_{32}H_{45}N_5O_8S$; dihydroergocristine mesylate, $C_{36}H_{45}N_5O_8S$; and dihydro-β-ergocryptine mesylate, $C_{33}H_{47}N_5O_8S$. [d]Also named galanthamine. [e]Also named propentophylline.

therapeutic index was sufficient to justify regulatory approval in several countries. Psychostimulants such as pemoline, amphetamine, procaine, and methylphenidate have failed to show cognitive enhancing effects in patients with dementia, except possibly as indirect consequences of mood elevation.

Although of possible value in treating cerebrovascular disorders, vasodilators have not proved effective in the treatment of SDAT. Neither drugs having a primary vasodilatory effect such as papaverine, cinnarizine, cyclandelate, and isoxsuprine, nor those having less clearly defined vascular effects such as meclofenoxate, vincamine, pyritinol, and pentoxifylline were found to be consistently effective. Whereas there is little support for a role of vascular abnormalities in the pathophysiology of SDAT, there are promising clinical results on nicergoline, nimodipine, and propentofylline in treating dementia, whether by cerebrovascular or neuronal mechanisms. Although monoaminergic deficits characterize SDAT, no consistent beneficial effects were demonstrated for α_2-adrenoceptor agonists (eg, clonidine or guanfacine), SSRIs, or levodopa. Promising results have been obtained using the MAO type B inhibitor L-deprenyl, but these require further confirmation. Clinical trials of opiate antagonists (naloxone, naltrexone), neuropeptide analogues such as the somatostatin analogue seglitide [81377-02-8] and the ACTH4-9 analogue ORG 2766 [50913-82-1], as well as with prolyl-endopeptidase inhibitors and angiotensin-converting enzyme inhibitors have so far failed to convincingly demonstrate symptomatic amelioration of dementia. Despite intensive preclinical and clinical efforts, the discovery of highly efficacious drugs for the symptomatic therapy of SDAT has proved elusive. The multifaceted character of this neurodegenerative disease which affects multiple neurochemical systems distributed across different brain regions requires more study (273). Interestingly, attention deficit/hyperactivity disorder (1) has been effectively treated in some patients with amphetamine, clonidine, pemoline, and methylphenidate (274).

It has been hypothesized that treatment with neurotrophic factors might be effective in slowing the progression of degenerative diseases of the nervous system (275). Because nerve growth factor (NGF) is found in the target areas for cholinergic neuronal pathways which degenerate in SDAT and clearly prevents the degeneration of cholinergic neurons in animal models, it has been suggested that increasing the concentration of endogenous NGF or an NGF mimetic could provide beneficial clinical effects in SDAT. Another member of the neurotrophin family, brain-derived neurotrophic factor (BDNF), is widely expressed in populations of neurons in the adult mammalian brain, particularly in the hippocampus and the neocortex, and like NGF has been demonstrated to exert trophic effects on cholinergic neurons of the basal forebrain. Because neurodegeneration in Alzheimer's disease extends beyond cholinergic neurons in the basal forebrain, the potentially more widespread effects of a drug which increases the concentration of BDNF or acts as a BDNF mimetic could prove advantageous.

Alzheimer's disease is characterized by excessive deposition of β-A4 protein resulting from proteolysis of the integral membrane amyloid precursor protein (APP) and point mutations in the APP-gene are linked to some familial forms of Alzheimer's disease (276,277). Drug discovery efforts focus on interference with the presumed pathogenic process of β-A4 deposition (leading to the formation of senile plaques) via antagonism of the toxic effects of β-A4, prevention of inflammatory processes associated with the plaques, or direct inhibition of proteases generating β-A4 from APPs. However, cognitive impairment in Alzheimer's disease, in fact, correlates more closely to the occurrence of neurofibrillary tangles composed of paired helical filaments than to β-amyloid deposits. Methods to prevent the underlying abnormal phosphorylation of tau proteins are under investigation (278). Elucidation of the strong association demonstrated between the apoE4 protein variant and occurrence of late-onset Alzheimer's disease may provide additional insights into the molecular mechanisms underlying the pathogenesis of the disease and potentially also new drug targets (279).

Economic Aspects

Prescription sales of psychopharmacological agents represent approximately one-tenth of the total world pharmaceutical market. Antidepressants, anxiolytics, antipsychotics, and sedative–hypnotics are mature market segments characterized by effective and safe drugs, as well as the very large presence of generic drugs. In contrast, the available drugs for treating drug abuse and dependence and those for treating cognition impairment still leave room for considerable improvement. As a whole, prescription sales of these categories of psychopharmacological agents have increased considerably in the 1990s (Table 7).

Table 7. Worldwide Prescription Drug Sales, $ $\times$ 10^9

Therapeutic class	IMS code	1984[a]	1994[a]	2000[b]
anxiolytics	N5C	1.1	2.1	2.2
sedatives–hypnotics	N5B and N1A2	0.6	1.8	2.6
antidepressants	N6A and N6C	0.7	4.0	5.8
antipsychotics	N5A	0.6	1.6	3.2
anticraving agents[c]	N7B	0.05	0.4	0.6
cognition enhancers	N6D	0.3	1.2	1.8
Total		*3.35*	*11.1*	*16.2*

[a]Data provided courtesy of IMS Global Services. [b]Estimated values by extrapolation. [c]Predominantly antismoking agents.

BIBLIOGRAPHY

"Psychopharmacological Agents" in *ECT* 1st ed., Suppl. Vol., pp. 720–743, by M. Gordon and G. E. Ullyot, Smith, Kline & French Laboratories; in *ECT* 2nd ed., Vol. 16, pp. 640–679, by M. Gordon and G. E. Ullyot, Smith, Kline & French Laboratories; in *ECT* 3rd ed., Vol. 19, pp. 342–379, by L. H. Sternbach and W. D. Horst, Hoffmann-La Roche & Co., Inc.

1. American Psychiatric Association, *Diagnostic and Statistical Manual of Mental Disorders*, 4th ed., American Psychiatric Association, Washington, D.C., 1994.
2. World Health Organization, *International Statistical Classification of Diseases and Related Health Problems (ICD-10)*, 10th Rev., World Health Organization, Geneva, Switzerland, 1992.
3. M. R. Rosekind, *J. Clin. Psychiatry* **53**, 4 (1992).
4. A. F. Schatzbert, *J. Clin. Psychiatry* **52**, 5 (1991).
5. M. H. Lader, *J. Clin. Psychiatry* **49**, 213 (1988).
6. D. V. Sheehan, J. Ballenger, and G. Jacobsen, *Arch. Gen. Psychiatry* **37**, 51 (1980).
7. D. Getova and V. Georgiev, *Acta Physiol. Pharmacol. Bulg.* **13**, 43 (1987).
8. I. Haider and I. Oswald, *Br. J. Psychiatry* **118**, 519 (1971).
9. T. D. Yih and J. M. van Rossum, *J. Pharmacol. Exp. Ther.* **203**, 184 (1977).
10. G. Wahlström, *Acta Pharmacol. Toxicol.* **35**, 131 (1974).
11. WHO Expert Committee on Drug Dependence, *World Health Organization Technical Report Series 741*, WHO, Geneva, Switzerland, 1987, pp. 1–64.
12. T. D. Yih and J. M. van Rossum, *Xenobiotica* **6**, 355 (1976).
13. S. M. Chierichetti, G. Moise, M. Galeone, G. Fiorella, and R. Lazzari, *Int. J. Clin. Pharmacol. Ther. Toxicol.* **23**, 510 (1985).
14. W. Holtermann and W. Lochner, *Arzneim. Forsch.* **22**, 1376 (1972).
15. D. W. Barron, J. W. Dundee, W. R. Gilmore, and P. J. Howard, *Br. J. Anaesth.* **38**, 802 (1966).
16. D. D. Breimer, *Eur. J. Clin. Pharmacokinet.* **10**, 263 (1976).
17. T. Walther, F. P. Meyer, K. Puchta, and H. Walther, *Int. J. Clin. Pharmacol. Ther. Toxicol.* **21**, 306 (1983).
18. D. D. Breimer and M. A. Winten, *Eur. J. Clin. Pharmacol.* **9**, 443 (1976).
19. K. D. Wolter, *Epilepsy Res. Suppl.* **3**, 99 (1991).
20. J. Dingemanse, P. H. Hutson, M. W. Langemeijer, G. Curzon, and M. Danhof, *Neuropharmacology* **27**, 467 (1988).
21. S. Toon and M. Rowland, *J. Pharmacol. Exp. Ther.* **225**, 752 (1983).
22. M. van der Graaff, N. P. E. Vermeulen, P. Heij, J. K. Boeijinga, and D. D. Breimer, *Biopharm. Drug Dispos.* **7**, 265 (1986).
23. J. A. Vida, M. L. Hooker, C. M. Samour, and J. F. Reinhard, *J. Med. Chem.* **16**, 1378 (1973).
24. F. R. Preuss and H.-W. Kopsch, *Arzneim. Forsch.* **16**, 858 (1966).
25. Y. Le Normand and co-workers, *Br. J. Clin. Pharmacol.* **26**, 589 (1988).
26. W. D. Hooper, H. E. Kunze, and M. J. Eadie, *Ther. Drug Monit.* **3**, 39 (1981).
27. T. D. Yih and J. M. van Rossum, *J. Pharmacol. Exp. Ther.* **203**, 185 (1977).
28. J. M. Hinton, *Br. J. Pharmacol.* **20**, 319 (1963).
29. J. W. Winer, R. H. Rosenwasser, and F. Jimenez, *Neurosurgery* **29**, 739 (1991).
30. D. J. Harvey, L. Glazener, D. B. Johnson, C. M. Butler, and M. G. Horning, *Drug Metab. Dispos.* **5**, 527 (1977).
31. M. J. Painter, in R. Levy, R. Mattson, B. Meldrum, J. K. Penry, and F. E. Dreifuss, eds., *Antiepileptic Drugs*, 3rd ed., Raven Press, New York, 1989, pp. 329–340.
32. M. Gamski, *Ann. Acad. Med. Gedanesis* **5**, 107 (1975).
33. C. Köppel, J. Tenczer, and K.-H. Beyer, *Arzneim. Forsch.* **35**, 1334 (1985).

34. F. G. Sulman, Y. Pfeifer, and E. Superstine, *Headache* **20**, 269 (1980).

35. O. Strubelt, *Arch. Int. Pharmacodyn. Ther.* **246**, 264 (1980).

36. A. Kales, P. Houri, E. O. Bixler, and P. Silberforb, *Clin. Pharmacol. Ther.* **20**, 541 (1976).

37. American Medical Association, *New and Nonofficial Drugs*, Lippincott, Philadelphia, Pa., 1961, p. 409.

38. E. M. Wertz and co-workers, *Am. J. Vet. Res.* **49**, 1079 (1988).

39. T. Koskela and G. Wahlström, *Acta Pharmacol. Toxicol.* **64**, 308 (1989).

40. M. M. Ghoneim, *Middle East J. Anaesthesiol.* **5**, 351 (1979).

41. M. Cathelin and G. Hosxe, *Thérapeutique* **45**, 39 (1969).

42. D. D. Breimer and A. G. de Boer, *Arzneim. Forsch.* **26**, 448 (1976).

43. W. E. Haefely, in H. Möhler and M. Da Prada, eds., *The Challenge of Neuropharmacology*, Editiones Roche, Basel, Switzerland, 1994, pp. 15–39.

44. U.S. Pat. 4,382,938 (May 10, 1983), J. P. Kaplan and P. George (to Synthélabo).

45. Eur. Pat. 50,563 (Apr. 28, 1982), J. P. Kaplan and P. George (to Synthélabo).

46. P. George and co-workers, *Farmaco* **46**(Suppl. 1), 277 (1991).

47. G. W. Dawson, S. G. Jue, and R. N. Brogden, *Drugs* **27**, 132 (1984).

48. M. P. Fernandez-Tomé, J. A. Fuentes, R. Madronero, and J. del Rio, *Arzneim. Forsch.* **25**, 926 (1975).

49. L. M. Sonne and P. Holm, *Int. Pharmacopsychiatry* **10**, 125 (1975).

50. M. S. Langley and S. P. Clissold, *Drugs* **35**, 104 (1988).

51. R. Ferrini, G. Miragoli, and B. Taccardi, *Arzneim. Forsch.* **24**, 2029 (1974).

52. L. O. Randall, W. Schallek, G. A. Heise, E. F. Keith, and R. E. Bagdon, *J. Pharmacol. Exp. Ther.* **129**, 163 (1960).

53. B. Saletu, G. Kindshofer, P. Anderer, and J. Grunberger, *Int. J. Clin. Pharmacol. Res.* **7**, 407 (1987).

54. R. N. Brogden, R. C. Heel, T. M. Speight, and G. S. Avery, *Drugs* **20**, 161 (1980).

55. J. E. Blum, W. Haefely, M. Jalfre, P. Polc, and K. Schärer, *Arzneim. Forsch.* **23**, 377 (1973).

56. K. D. Charalampous, W. Tooley, and C. Yates, *J. Clin. Pharmacol.* **13**, 114 (1973).

57. M. Nakanishi, T. Tsumagawa, S. Shuto, T. Kenjo, and T. Fukuda, *Arzneim. Forsch.* **22**, 1905 (1972).

58. T. Kamioka, H. Takagai, S. Kobayashi, and Y. Suzuki, *Arzneim. Forsch.* **22**, 884 (1972).

59. U. Traversa, L. De Angelis, and R. Vertua, *J. Pharm. Pharmacol.* **29**, 504 (1977).

60. J. W. Dundee and S. R. Keilty, *Int. Anesthesiol. Clin.* **7**, 91 (1969).

61. S. R. Bareggi and co-workers, *Int. J. Clin. Pharmacol. Res.* **6**, 309 (1986).

62. T. Roth, ed., *Am. J. Med.* **88**(3a), 1S–48S (1990).

63. J. P. Chambon and co-workers, *Arzneim. Forsch.* **35**, 1572 (1985).

64. T. Tsumagari and co-workers, *Arzneim. Forsch.* **28**, 1158 (1978).

65. M. Otsuka, T. Tsuchiya, and S. Kitagawa, *Arzneim. Forsch.* **25**, 755 (1975).

66. E. Geller and D. Thomson, eds., *Eur. J. Anaesthesiol.* (Suppl. 2), 1–332 (1988).

67. M. A. K. Mattila and H. M. Larni, *Drugs* **20**, 353 (1980).

68. D. J. Greenblatt, R. I. Shader, and J. Koch-Weser, *Clin. Pharmacol. Ther.* **17**, 1 (1975).

69. T. Mitsushima and S. Ueki, *Folia Pharmacologica Japonica* **74**, 959 (1978).

70. T. Sukamoto, K. Aikawa, K. Ito, and T. Nose, *Folia Pharmacologica Japonica* **76**, 447 (1980).

71. W. E. Fann, W. M. Pitts, and J. C. Wheless, *Pharmacotherapy* **2**, 72 (1982).

72. T. Kamioka, I. Nakayama, T. Hara, and H. Takagi, *Arzneim. Forsch.* **28**, 838 (1978).

73. D. M. Gallant, R. Guerrero-Figueroa, and W. C. Swanson, *Curr. Ther. Res. Clin. Exp.* **15**, 123 (1973).

74. B. G. Clark, S. G. Jue, G. W. Dawson, and A. Ward, *Drugs* **31**, 500 (1986).
75. B. Ameer and D. J. Greenblatt, *Drugs* **21**, 162 (1981).
76. K. Ohata, T. Murata, S. Kohno, and H. Sakamoto, *Pharmacometrics* **29**, 913 (1985).
77. L. O. Randall, C. L. Scheckel, and W. Pool, *Arch. Int. Pharmacodyn. Ther.* **185**, 135 (1970).
78. G. Buschmann, U. G. Kuhl, and O. Rohte, *Arzneim. Forsch.* **35**, 1643 (1985).
79. T. Kamioka, I. Nakayama, S. Akiyama, and H. Takagi, *Psychopharmacology* **52**, 17 (1977).
80. L. Pieri and co-workers, *Arzneim. Forsch.* **31**, 2180 (1981).
81. M. Otsuka, T. Tsuchiya, and S. Kitagawa, *Arzneim. Forsch.* **23**, 645 (1973).
82. L. Kangas and D. D. Breimer, *Clin. Pharmacokinet.* **6**, 346 (1981).
83. A. N. Nicholson, B. M. Stone, C. H. Clarke, and H. M. Ferres, *Br. J. Clin. Pharmacol.* **3**, 429 (1976).
84. F. J. Ayd, ed., *Int. Clin. Psychopharmacol.* **5**, 1 (1990).
85. Y. Sakai, T. Deguchi, N. Iwata, M. Mori, and Y. Nishijima, *Nippon Yakurigaku Zasshi* **66**, 706 (1970).
86. T. S. Kalinina, T. L. Garibova, and T. A. Voronina, *Behav. Pharmacol.* **5**(Suppl. 1), 112 (1994).
87. J. M. Janbroers, *Clin. Ther.* **6**, 434 (1984).
88. R. C. Robichaud, J. A. Gylys, K. L. Sledge, and I. W. Hillyard, *Arch. Int. Pharmacodyn. Ther.* **185**, 213 (1970).
89. S. I. Ankier and K. L. Goa, *Drugs* **35**, 42 (1988).
90. F. Fraschini and B. Stankov, *Pharmacol. Res.* **27**, 97 (1993).
91. P. E. Keane, J. Simiand, M. Morre, and K. Bizière, *J. Pharmacol. Exp. Ther.* **245**, 962 (1988).
92. H. L. Goldberg and R. J. Finnerty, *Am. J. Psychiatry* **136**, 196 (1979).
93. G. E. Pakes, R. N. Brogden, R. C. Heel, T. M. Speight, and G. S. Avery, *Drugs* **22**, 81 (1981).
94. U.S. Pat. 4,316,839 (Feb. 23, 1982), M. Gerecke and co-workers (to Hoffmann-La Roche Ltd.).
95. W. Hunkeler and co-workers, *Nature* **290**, 514 (1981).
96. W. Hunkeler, *Chimia* **47**, 141 (1993).
97. L. E. Hollister, B. Müller-Oerlinghausen, K. Rickels, and R. I. Shader, *J. Clin. Psychopharmacol.* **13**(Suppl. 1) (1993).
98. K. L. Goa and A. Ward, *Drugs* **32**, 114 (1986).
99. J. Silverman and W. W. Muir, *Lab. Anim. Sci.* **43**, 210 (1993).
100. B. R. Ballinger, *Br. Med. J.* **300**, 456 (1990).
101. P. A. J. Janssen, C. J. E. Niemegeers, K. H. L. Schellekens, and F. M. Lenaerts, *Arzneim. Forsch.* **21**, 1234 (1971).
102. F. Gross, J. Tripod, and R. Meier, *Schweiz. Med. Wochenschr.* **85**, 305 (1955).
103. M. Ferreri, E. G. Hantouche, and M. Billardon, *Encéphale* **20**, 785 (1994).
104. E. F. Domino, P. Chodoff, and G. Corssen, *Clin. Pharmacol. Ther.* **6**, 279 (1966).
105. W. Haefely, R. Schaffner, P. Polc, and L. Pieri, in F. Hoffmeister and G. Stille, eds., *Psychotropic Agents, Part II: Anxiolytics, Gerontopsychopharmacological Agents, and Psychomotor Stimulants*, Springer-Verlag, Berlin, 1981, pp. 263–283.
106. S. S. Brown and S. Goenechea, *Clin. Pharmacol. Ther.* **14**, 314 (1973).
107. W. T. Brown, *Can. Med. Assoc. J.* **102**, 510 (1970).
108. M. S. Langley and R. C. Heel, *Drugs* **35**, 334 (1988).
109. H. D. Langtry and P. Benfield, *Drugs* **40**, 291 (1990).
110. L. Julou, J. C. Blanchard, and J. F. Dreyfus, *Pharmacol. Biochem. Behav.* **23**, 653 (1985).
111. T. G. Hales and J. J. Lambert, *Br. J. Pharmacol.* **104**, 619 (1991).

112. F. Veintemilla, F. Elinder, and P. Arhem, *Eur. J. Pharmacol.* **218**, 59 (1992).
113. L. E. Hollister, B. Müller-Oerlinghausen, K. Rickels, and R. I. Shader, *J. Clin. Psychopharmacol.* **13**(Suppl. 1), 1S (1993).
114. M. Mosconi, C. Chiamulera, and G. Recchia, *Int. J. Clin. Pharmacol. Res.* **13**, 331 (1993).
115. C. S. Dommisse and P. E. Hayes, *Clin. Pharm.* **6**, 196 (1987).
116. J. H. Boyd and M. M. Weissman, *Arch. Gen. Psychiatry* **38**, 1039 (1981).
117. C. Holden, *Science* **254**, 1450 (1991).
118. M. Roth, *Pharmacopsychiatry* **25**, 18 (1992).
119. E. Richelson, *J. Clin. Psychiatry* **55**(Suppl. A), 34 (1994).
120. R. J. Chiarello and J. O. Cole, *Arch. Gen. Psychiatry* **44**, 286 (1987).
121. A. J. Azzaro and H. E. Ward, in C. R. Gray and R. E. Stitzed, eds., *Modern Pharmacology*, Little, Brown & Co., Boston, Mass., 1994, pp. 397–411.
122. D. L. Murphy, C. S. Aulakh, N. A. Garrick, and T. Sunderland, in H. Y. Meltzer, ed., *Psychopharmacology: The Third Generation of Progress*, Raven Press, New York, 1987, pp. 545–552.
123. S. G. Bryant, B. G. Guernsey, and N. B. Ingrim. *Clin. Pharmacol.* **2**, 525 (1983).
124. S. Garattini and T. Mennini, *Clin. Neuropharmacol.* **12**, S13 (1989).
125. A. Coppen, K. Ghose, and A. Jorgensen, *Prog. Neuropsychopharmacol.* **3**, 191 (1979).
126. S. G. Jue, G. W. Dawson, and R. N. Brogden, *Drugs* **24**, 1 (1982).
127. F. Herr, K. Voith, and J. Jaramillo, *J. Med.* **2**, 258 (1971).
128. R. J. Milne and K. L. Goa, *Drugs* **41**, 450 (1991).
129. D. McTavish and P. Benfield, *Drugs* **39**, 136 (1990).
130. A. Martin, J. M. Masson, P. Jusseaume, M. Belonde, and C. Voisinet, *Ann. Med. Psychol.* **139**, 1023 (1981).
131. D. S. Janowsky and B. Byerley, *J. Clin. Psychiatry* **45**, 3 (1984).
132. S. G. Lancaster and J. P. Gonzalez, *Drugs* **38**, 123 (1989).
133. R. M. Pinder, R. N. Brogden, T. M. Speight, and G. S. Avery, *Drugs* **13**, 161 (1977).
134. P. E. Stokes, *Clin. Ther.* **15**, 216 (1993).
135. M. I. Wilde, G. L. Plosker, and P. Benfield, *Drugs* **46**, 895 (1993).
136. S. C. Rogers and P. M. Clay, *Br. J. Psychiatry* **127**, 599 (1975).
137. J. Rigal, *Ann. Med. Psychol.* **143**, 664 (1985).
138. M. I. Gluckman and T. Baum, *Psychopharmacologia* **15**, 169 (1969).
139. Y. Pelicier, J. C. Scotto, and Y. Girard, *Sem. Ther.* **41**, 21 (1965).
140. G. Mathe and co-workers, *Biomed. Pharmacother.* **41**, 13 (1987).
141. J. R. Davidson, E. L. Giller, S. Zisook, and J. E. Overall, *Arch. Gen. Psychiatry* **45**, 120 (1988).
142. S. G. Lancaster and J. P. Gonzalez, *Drugs* **37**, 124 (1989).
143. R. M. Pinder, R. N. Brogden, T. M. Speight, and G. Avery, *Drugs* **13**, 321 (1977).
144. P. Dick, *Encéphale* **4**, 41 (1978).
145. R. N. Brogden, R. C. Heel, T. M. Speight, and G. S. Avery, *Drugs* **16**, 273 (1976).
146. A. Stenger, J. P. Couzinier, and M. Briley, *Psychopharmacology* **91**, 147 (1987).
147. K. Bizière, J. P. Kan, J. Souilhac, J. P. Muyard, and R. Roncucci, *Arzneim. Forsch.* **32**, 824 (1982).
148. S. L. Dickinson, *Drug News Perspect.* **4**, 197 (1991).
149. A. Filtor, D. Faulds, and K. L. Goa, *Drugs* **43**, 561 (1992).
150. A. S. Eison, M. S. Eison, J. R. Torrente, R. N. Wright, and F. D. Yocca, *Psychopharmacol. Bull.* **26**, 311 (1990).
151. J. C. Rowe, *Proc. Soc. Exp. Biol. Med.* **101**, 832 (1959).
152. C. Nordin, L. Bertilsson, and B. Siwers, *Clin. Pharmacol. Ther.* **42**, 10 (1987).
153. F. Hoffmeister, F. Wuttke, and G. Kroneberg, *Arzneim. Forsch.* **19**, 846 (1969).
154. P. C. Waldmeier, *J. Pharm. Pharmacol.* **34**, 391 (1982).

155. K. L. Dechant and S. P. Clissold, *Drugs* **41**, 225 (1991).
156. G. B. Baker, R. T. Coutts, K. F. McKenna, and R. L. Sherry-McKenna, *J. Psychiat. Neurosci.* **17**, 206 (1992).
157. M. D. Mashkovsky and N. I. Andrejeva, *Arzneim. Forsch.* **31**, 75 (1981).
158. J. M. Lwoff, C. Larousse, P. Simon, and J. R. Boissier, *Therapie* **26**, 451 (1971).
159. S. C. Risch, L. Y. Huey, and D. S. Janowsky, *J. Clin. Psychiatry* **40**, 58 (1979).
160. A. Des Lauriers, J. F. Chevalier, and G. Garreau, *Ann. Med. Psychol.* **140**, 262 (1982).
161. D. Murdoch and D. McTavish, *Drugs* **44**, 604 (1992).
162. K. Yamada and T. Furukawa, *Nippon Yakurigaku Zasshi* **97**, 31 (1991).
163. M. I. Wilde and P. Benfield, *Drugs* **49**, 411 (1995).
164. A. M. Cesura and A. Pletscher, *Prog. Drug Res.* **38**, 171 (1992).
165. M. Haria, A. Fitton, and D. McTavish, *Drugs Aging* **4**, 331 (1994).
166. M. Gastpar, *Drugs* **38**, 43 (1989).
167. S. A. Montgomery, *J. Clin. Psychiatry* **54**, 119 (1993).
168. R. M. Pinder, R. N. Brogden, T. M. Speight, and G. S. Avery, *Drugs* **13**, 401 (1977).
169. D. J. Heal and W. R. Buckett, *Int. J. Geriatr. Psychiatry* **6**, 431 (1991).
170. K. Rickels and E. Schweizer, *J. Clin. Psychiatry* **51**, 9 (1990).
171. U.S. Pat. 4,018,895 (Apr. 19, 1977), B. B. Molloy and K. K. Schmiegel (to Eli Lilly & Co.).
172. U.S. Pat. 4,626,549 (Dec. 2, 1986), B. B. Molloy and K. K. Schmiegel (to Eli Lilly & Co.).
173. D. W. Robertson, J. H. Krushinski, R. W. Fuller, and J. D. Leander, *J. Med. Chem.* **31**, 1412 (1988).
174. D. S. Risley and R. J. Bopp, *Anal. Profiles Drug Subst.* **19**, 193 (1990).
175. J. W. Jefferson and J. H. Greist, *Drugs* **6**, 448 (1994).
176. H. G. Pope, Jr., S. L. McElroy, P. E. Keck, Jr., and J. I. Hudson, *Arch. Gen. Psychiatry* **48**, 62 (1991).
177. R. M. Julien, in R. M. Julien, ed., *A Primer of Drug Action*, W. H. Freeman & Co., New York, 1995, pp. 216–232.
178. R. M. Post, T. W. Uhde, P. P. Roy-Birne, and R. Joffe, *Psychiat. Res.* **21**, 71 (1987).
179. D. Leysen and R. M. Pinder, *Annu. Rep. Med. Chem.* **29**, 1 (1994).
180. L. Ereshefsky, T. K. Tran-Johnson, and M. D. Watanabe, *Clin. Pharm.* **9**, 682 (1990).
181. G. P. Reynolds and C. Czudek, *Adv. Pharmacol.* **32**, 461 (1995).
182. J. Schmutz and C. W. Picard, in F. Hoffmeister and G. Stille, eds., *Psychotropic Agents, Part I: Antipsychotics and Antidepressants*, Springer-Verlag, Berlin, 1980, pp. 3–26.
183. A. Delcker, M. L. Schoon, B. Oczkowski, and H. J. Gaertner, *Pharmacopsychiatry* **23**, 125 (1990).
184. P. A. J. Janssen, C. J. E. Niemegeers, and K. H. L. Schellekens, *Arzneim. Forsch.* **15**, 104, 1196 (1965).
185. P. Benfield, A. Ward, B. G. Clark, and S. G. Jue, *Drugs* **35**, 670 (1988).
186. P. A. Dixon, E. Oforah, and R. Makanjuola, *Br. J. Clin. Pharmacol.* **14**, 273 (1982).
187. B. E. Leonard, in T. R. E. Barnes, ed., *Antipsychotic Drugs and their Side-Effects*, Academic Press, London, 1993, pp. 3–63.
188. N. C. Moore and S. Gershon, *Clin. Neuropharmacol.* **12**, 167 (1989).
189. A. Fitton and R. C. Heel, *Drugs* **40**, 722 (1990).
190. E. F. Domino, in D. H. Efron, ed., *Psychopharmacology: A Review of Progress, 1957–1967*, U.S. Government Printing Office, Washington, D.C., 1968, pp. 1045–1056.
191. T. V. Mikhailova, A. I. Terekhina, and A. P. Gilev, *Farmakol. Toksikol.* **32**, 28 (1969).
192. P. A. J. Janssen, C. J. E. Niemegeers, K. H. L. Schellekens, F. J. Verbruggen, and J. M. Van Nueten, *Arzneimittel-Forschung* **13**, 205 (1963).

193. T. J. Crow and E. C. Johnstone, *Br. J. Pharmacol.* **59**, 466 (1977).
194. G. E. Hogarty and co-workers, *Arch. Gen. Psychiatry* **45**, 797 (1988).
195. N. Russell, J. Landmark, H. Merskey, and T. Turpin, *Can. J. Psychiatry* **27**, 593 (1982).
196. R. Beresford and A. Ward, *Drugs* **33**, 31 (1987).
197. S. Courvoisier, R. Ducrot, J. Fournel, and L. Julou, *C. R. Soc. Biol. (Paris)* **151**, 1378 (1957).
198. M. P. Bishop, G. M. Simpson, C. W. Dunnett, and H. Kiltie, *Psychopharmacology* **51**, 107 (1977).
199. L. Bjerkenstedt, C. Härnryd, V. Grimm, G. Gullberg, and G. Sedvall, *Arch. Psychiat. Nervenkr.* **226**, 157 (1978).
200. S. Gershon, G. Sakalis, and P. A. Bowers, *J. Clin. Psychiatry* **42**, 463 (1981).
201. J. L. Claghorn, *J. Clin. Psychiatry* **46**, 30 (1985).
202. A. Miyamoto, K. Kitawaki, K. Nagao, H. Koida, and K. Nagao, *Med. Consult. New Rem.* **28**, 183 (1991).
203. S. Usuda, N. O'uchi, K. Koshiza, F. Wanibuchi, and T. Konishi, *Jpn. Pharmacol. Ther.* **18**, 135 (1990).
204. P. A. J. Janssen and F. H. L. Awouters, in P. L. Munson, ed., *Principles of Pharmacology: Basic Concepts and Clinical Applications*, Chapman & Hall, New York, 1995, pp. 289–308.
205. H. Hadass, H. Hippius, W. Mauruschat, B. Müller-Oerlinghausen, and L. Rosenberg, *Pharmakopsychiatrie* **7**, 17 (1974).
206. A. B. Eppel and R. Mishra, *Can. J. Psychiatry* **29**, 508 (1984).
207. L. B. Hansen, N. E. Larsen, and N. Gulmann, *Psychopharmacology* **78**, 112 (1982).
208. R. McCreadie, M. Mackie, D. Morrison, and J. Kidd, *Br. J. Psychiatry* **140**, 280 (1982).
209. P. A. Janssen and F. H. Awouters, *Arzneim. Forsch.* **44**, 269 (1994).
210. B. R. S. Nakra, A. J. Bond, and M. H. Lader, *J. Clin. Pharmacol.* **15**, 449 (1975).
211. G. Sgaragli and co-workers, *Drug. Metab. Dispos.* **14**, 263 (1986).
212. A. Claus and co-workers, *Acta Psychiatr. Scand.* **85**, 295 (1992).
213. A. J. Wagstaff, A. Fitton, and P. Benfield, *CNS Drugs* **2**, 313 (1994).
214. L. Fouks, *Actual. Psychiatr.* **11**, 54 (1981).
215. G. M. Realmuto, W. D. Erickson, A. M. Yellin, J. H. Hopwood, and L. M. Greenberg, *Am. J. Psychiatry* **141**, 440 (1984).
216. E. D. Peselow and M. Stanley, *Adv. Biochem. Psychopharmacol.* **35**, 163 (1982).
217. T. Kariya and co-workers, *J. Int. Med. Res.* **11**, 66 (1983).
218. R. Bergling, T. Mjorndal, L. Oreland, U. Rapp, and S. Wold, *J. Clin. Pharmacol.* **15**, 178 (1975).
219. A. DiMascio, L. L. Havens, and G. L. Klerman, *J. Nerv. Ment. Dis.* **136**, 168 (1963).
220. J. J. Piala, J. P. High, G. L. Hassert, J. C. Burke, and B. N. Carver, *J. Pharmacol. Exp. Ther.* **127**, 55 (1959).
221. Y. Higashi, Y. Momotani, E. Suzuki, and T. Kaku, *Pharmacopsychiatry* **20**, 8 (1987).
222. F. J. Bereen, F. B. Harte, J. Maguire, and A. N. Singh, *Pharmatherapeutica* **5**, 62 (1987).
223. P. Seeman, T. Lee, M. Chau Wong, and K. Wong, *Nature* **261**, 717 (1976).
224. F. J. White and R. Y. Wang, *Science* **221**, 1054 (1983).
225. B. M. Cohen and J. F. Lipinski, *Life Sci.* **39**, 2571 (1986).
226. A. D. Michel, D. N. Loury, and R. L. Whiting, *Br. J. Pharmacol.* **98**, 883 (1989).
227. A. J. Sleight, W. Koek, and D. C. H. Bigg, *Eur. J. Pharmacol.* **238**, 407 (1993).
228. B. L. Roth and co-workers, *J. Pharmacol. Exp. Ther.* **268**, 1403 (1994).
229. M.-B. Assie, A. J. Sleight, and W. Koek, *Eur. J. Pharmacol.* **237**, 183 (1993).
230. A. Malmberg, D. M. Jackson, A. Eriksson, and N. Mohell, *Mol. Pharmacol.* **43**, 749 (1993).

231. H. Y. Meltzer, S. Matsubara, and J. Lee, *J. Pharmacol. Exp. Ther.* **251**, 238 (1989).
232. L. N. Robins and co-workers, *Arch. Gen. Psychiatry* **41**, 949 (1984).
233. P. R. Martin, D. M. Lovinger, and G. R. Breese, in P. L. Munson, ed., *Principles of Pharmacology: Basic Concepts & Clinical Applications*, Chapman & Hall, New York, 1995, pp. 417–452.
234. Group for the Advancement of Psychiatry Committee on Alcoholism and the Addictions, *Am. J. Psychiatry* **148**, 1291 (1991).
235. R. K. Fuller and co-workers, *JAMA* **256**, 1449 (1986).
236. D. A. Regier and co-workers, *JAMA* **264**, 2511 (1990).
237. E. J. Nestler, B. T. Hope, and K. L. Widnell, *Neuron* **11**, 995 (1993).
238. W. G. Erwin, *Clin. Pharm.* **3**, 497 (1984).
239. *SCRIP* **2028**, 27 (1995).
240. W. H. Moos, R. E. Davis, R. D. Schwarz, and E. R. Gamzu, *Med. Res. Rev.* **8**, 353 (1988).
241. H. Coper and W. M. Herrmann, *Pharmacopsychiatry* **21**, 211 (1988).
242. L. Amaducci and C. G. Gottfries, eds., *Drug Invest.* **5**(Suppl. 1) (1993).
243. S. Yamagami and co-workers, *Drugs Exp. Clin. Res.* **17**, 217 (1991).
244. A. Bernard and J. M. Goffaer, *Clin. Trials J.* **5**, 945 (1968).
245. W. McEntee and R. Mair, *Ann. Neurol.* **27**, 466 (1980).
246. G. Davies and co-workers, *Age Ageing* **6**, 156 (1977).
247. J. R. Hughes, J. G. Williams, and R. D. Currier, *J. Am. Geriatr. Soc.* **24**, 490 (1976).
248. T. Thomsen, U. Bickel, J. P. Fischer, and H. Kewitz, *Dementia* **1**, 46–51 (1990).
249. J. Schlegel, E. Mohr, J. Williams, U. Mann, M. Gearing, and T. N. Chase, *Clin. Neuropharmacol.* **12**, 124 (1989).
250. A. Nagaoka, Y. Nagai, N. Yamazaki, M. Miyamoto, and Y. Kiyota, *Drug Dev. Res.* **14**, 373 (1988).
251. M. Yamamoto and M. Shimizu, *Neuropharmacology* **26**, 761 (1987).
252. P. Cook and I. James, *N. Engl. J. Med.* **305**, 1508, 1560 (1981).
253. V. Kristensen, M. Olsen, and A. Theilgaard, *Acta Psychiatr. Scand.* **55**, 41 (1977).
254. J. R. Wittenborn, *J. Ner. Ment. Dis.* **169**, 139 (1981).
255. T. Crook, S. Ferris, G. Sathananthan, A. Raskin, and S. Gershon, *Psychopharmacology* **52**, 251 (1977).
256. J. A. Yesavage, L. Hollister, and E. Buriane, *Arch. Gen. Psychiatry* **36**, 220 (1979).
257. P. N. Tariot, T. Sunderland, H. Weingartner, D. L. Murphy, M. R. Cohen, and R. M. Cohen, *Arch. Gen. Psychiatry* **43**, 727 (1986).
258. B. T. Hyham, P. J. Eslinger, and A. R. Damasio, *J. Neurol. Neurosurg. Psychiatry* **48**, 1169 (1985).
259. B. Saletu and co-workers, *Psychopharmacology* **117**, 385 (1995).
260. T. A. Ban and co-workers, *Prog. Neuro-Psychopharmacol. Biol. Psychiat.* **14**, 525 (1990).
261. A. Giotti, ed., *Clin. Neuropharmacol.* **9**(Suppl. 3) (1986).
262. C. Eisdorfer, J. F. Conner, and F. L. Wilkie, *J. Gerontol.* **23**, 283 (1968).
263. G. Feine-Haake, *Pharmatherapeutica* **3**, 4651 (1983).
264. M. W. Vernon and E. M. Sorkin, *Drugs Aging* **1**, 17 (1991).
265. R. J. Branconnier, J. O. Cole, E. C. Dessain, K. Spera, S. Ghazvinian, and D. DeVit, *Psychopharmacol. Bull.* **19**, 726 (1983).
266. P. Goodnick and S. Gershon, *J. Clin. Psychiatry* **45**, 196 (1984).
267. H.-J. Möller, I. Mauer, and B. Saletu, *Pharmacopsychiatry* **27**, 159 (1994).
268. K. J. Martin, *J. Int. Med. Res.* **11**, 55 (1983).
269. K. L. Davis and co-workers, *N. Eng. J. Med.* **327**, 1253 (1992).
270. S. Hagstadius, L. Gustafson, and J. Risberg, *Psychopharmacology* **83**, 321 (1984).
271. D. M. Coleston and I. Hindmarch, *Drug Dev. Res.* **14**, 191 (1988).

272. B. E. Leonard, *Med. Sci. Res.* **18**, 663 (1990).
273. B. W. Volger, *Clin. Pharm.* **10**, 447 (1991).
274. M. R. Jacobs and K. O'B. Fehr, *Addiction Research Foundation's Drugs and Drug Abuse*, 2nd ed., Addiction Research Foundation, Toronto, Canada, 1987, pp. 3–640.
275. P. J. Isackson and K. D. Murray, *Drug News Perspect.* **7**, 585 (1994).
276. M.-C. Chartier-Harlin and co-workers, *Nature* **353**, 844 (1991).
277. S. Gandy and P. Greengard, *Int. Rev. Neurobiol.* **36**, 29 (1994).
278. K. S. Kosik, *Trends Neurosci.* **14**, 218 (1991).
279. E. H. Corder and co-workers, *Science* **261**, 921 (1993).

J. R. Martin
T. Godel
W. Hunkeler
F. Jenck
J.-L. Moreau
A. J. Sleight
U. Widmer
F. Hoffmann-La Roche, Ltd.

PULP

Pulp is the raw material for the production of paper (qv), paperboard, fiber-board, and other similar manufactured products (see WOOD-BASED COMPOSITES AND LAMINATES). In purified form, it is a source of cellulose (qv) for rayon (see FIBERS, REGENERATED CELLULOSICS), cellulose esters (qv), and other cellulose-derived products. Pulp is obtained from plant fiber and is therefore a renewable resource (see CHEMURGY). Fibrous plants have been used as a source for writing materials, eg, papyrus, since the earliest Babylonian and Egyptian civilizations. The origin of papermaking, which is the formation of a cohesive sheet from the rebonding of separated fibers, has been attributed to Ts'ai-Lun in China in 105 AD, who used bamboo, mulberry bark, and rags as the ingredients. Around the same time in Europe, parchment, ie, writing material made from the skins of goats, sheep, or other animals, was used. During the Middle Ages, the use of rags and rope supplanted animal skins. The use of wood (qv) as a source of papermaking fiber was not commercially applied until the mid-1800s (see PAPERMAKING MATERIALS AND ADDITIVES). The principal wood-pulping processes in use as of the mid-1990s are stone groundwood, soda, SO_2 or acid sulfite, and the sulfate or kraft processes, developed in 1844, 1853, 1866, and 1870, respectively. Since their development, the basic processes have been modified numerous times and the technology has been highly refined. However, the scientific base for this technology is considerably slower in development, largely because of the physical and chemical heterogeneity of wood and the complexity of its component polymers and their interactions.

For most industries, the environmental and energy concerns of the latter half of the twentieth century have led to significant changes in the operation of pulp and paper mills as well as to much research effort toward developing the cleanest and most energy-efficient methods of production. In the 1970s and 1980s, the practical result was the development of end-of-pipe methods, eg, scrubbers, precipitators, primary and secondary treatments for waste water, and holding ponds, to minimize the discharge of pollutants. Significant advances have included the development of thermomechanical and pressurized groundwood processes for improved pulp strength and quality, increased use of high yield pulps, whole-tree utilization, extended delignification, and a dramatic increase in the use of recycled fiber (see RECYCLING, PAPER). The chemical pulping industry has increased the use of pulping additives such as anthraquinone and minimized the discharge of malodorous sulfur compounds during pulping. Significant advances in bleaching technology include oxygen and ozone delignification, increased use of hydrogen peroxide, enzymatic bleaching, and reductions in the use of chlorine compounds (see BLEACHING AGENTS, PULP AND PAPER). Notable in this area is the development of elemental chlorine-free (ECF) and totally chlorine-free (TCF) bleaching technology and significant reductions in water use. The goal is effluent-free processing.

Wood is the original source of 99% of the pulp fiber produced in the United States. Although virtually any wood can be pulped by some process, there are certain species commonly used for pulp because of desirability of fiber, ease of pulping, availability, or less competition with other wood products. The common pulpwoods in the United States are listed in Table 1.

Table 1. Pulpwood Species by Main U.S. Pulp-Producing Regions

Region	Softwoods		Hardwoods	
	Dominant	Secondary	Dominant	Secondary
Northeast	spruce and fir	hemlock, tamarack, and white pine	oak, hickory, and maple	aspen and poplar
South	yellow pines		oaks and gums	
Northwest	Douglas fir and hemlock	true firs and spruce	red alder	
Lake states	jack pine and red pine	white pine and tamarack	red oak, aspen, and maple	birch

Wood

In terms of abundance and suitability for pulping, there are two chief botanical classifications of trees: the softwoods or evergreens, which are gymnosperms, and the hardwoods or broad-leaved deciduous trees, which are dicotyledon angiosperms. The chemistry and anatomy of wood vary somewhat with the species of tree, but there are gross similarities between the two classifications. The softwoods, which are preferred for most pulp products because of their longer fibers, generally contain a higher (26–32% on an extractives-free basis) percentage of lignin (qv) and a lower (14–17%) percentage of hemicellulose (qv) than the hardwoods. The latter contain 17–26% lignin and 18–27% hemicellulose.

Anatomy and Morphology. A cross section of pine is shown in Figure **1a** as a representation of the anatomy of softwoods. The main cell type is the axially aligned tracheid (TR). Although in botanical terminology tracheids are not considered true fibers, these are the papermaking fibers from softwoods and

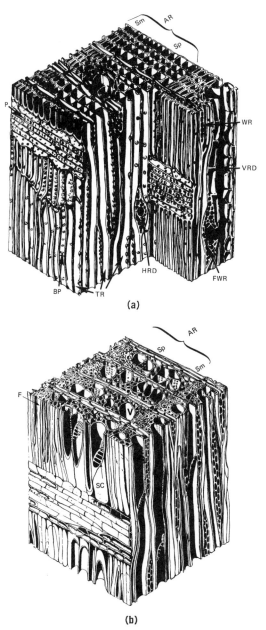

(a)

(b)

Fig. 1. Schematic section, where AR = annual ring, BP = bordered pits, F = wood fiber, FWR = fusiform wood ray, HRD = horizontal resin ducts, P = primary wall, SC = scalariform plate, Sm = summerwood, Sp = springwood, TR = tracheid, VRD = vertical resin ducts, and WR = wood ray, of (**a**) pine, a softwood, and (**b**) yellow poplar, a hardwood.

are referred to herein as fibers, according to the common practice of the industry. Other cell types in softwoods are the ray cells, ie, the fusiform wood ray (FWR) and wood ray (WR) cells, and the longitudinal and epithelial parenchyma, ie, the cells surrounding the horizontal and vertical resin ducts (HRD and VRD, respectively).

As a tree grows, the cells are produced in concentric lamella in the cambium layer, which is between the bark and the wood. In the spring, when moisture is plentiful and the tree is growing rapidly, the tracheid cell wall is thin (3–4 μm) and the hollow center or lumen is relatively large (26–43 μm). This portion is called springwood (Sp) or earlywood. During the summer or later in the growing season, the cell wall thickness increases to 8–12 μm and the outside diameter decreases from 29–47 μm in short-leaf pine. These cells form summerwood (Sm) or latewood. The sequential combination of seasonal cell types leads to the characteristic annual ring (AR) of trees, which is more or less distinct in softwoods, depending on the species.

In the living tree, nutrients flow through the cells of the sapwood. The pattern of liquid conductance is important for the penetration of chemicals in the initial stage of chemical pulping. In softwoods, liquid is transferred from rays to tracheids and between tracheids through tiny voids or pits (P) in the cell wall. Usually, pits in adjacent cells are aligned so that a passage between the lumens of the two cells is formed, which is blocked only by a thin pit membrane of intercellular substance. In bordered pits (BP), this membrane contains a thickened circular portion called a torus, which functions as a check valve to seal the passage against a return flow of liquid.

Hardwoods have a more varied and complex arrangement of cells than softwoods. Figure 1**b** is a cross section of the structure of yellow poplar, a typical hardwood. The main structural element of hardwoods is the wood fiber (F), which is significantly shorter than the softwood tracheid (1–2 mm vs 3–6 mm) and generally thinner, ca 20 μm in diameter. The true fibers are uniform throughout the annual ring. Hardwoods also contain a sizable proportion of short, large-diameter cells or vessels (V) through which sap is transported and which are categorized on the basis of pore (vessel) size. Ring-porous hardwoods, ie, oaks, hickory, chestnut, elm, black locust, and ash, have vessels that are larger in springwood than in summerwood. In diffuse-porous hardwoods, ie, yellow poplar, aspen, maple, gums, and basswood, the pores are fairly uniform in size and evenly distributed throughout the annual ring. Vessels have open ends or a connecting grate-like tissue called a scalariform plate (SC). Hardwood fibers have simple pits, which are smaller, and do not contain the torus system of the softwood bordered pits. This is related to the fact that the vessels perform the primary liquid-transport function in hardwoods. Thus, the cross-fiber liquid flow is greatly restricted in hardwoods in contrast with softwoods. In general, there is less differentiation in springwood and summerwood fibers than in those of softwoods. The AR is shown by one or two layers of terminal parenchyma cells that form at the end of the growing season.

Upon maturation of both softwoods and hardwoods, the parenchyma cells at the core die. This portion of the wood is called heartwood and often contains polyphenols, flavones, and other colored compounds that do not occur in the contrasting sapwood. A clear, visual distinction usually exists between heartwood and sapwood, depending on the species. Heartwood compounds, eg, dihydro-quercetin (taxifolin, 2-(3,4-dihydroxyphenyl)-2,3-dihydro-3,5,7-trihydroxy-4*H*-1-benzopyran-4-one) in Douglas fir, may cause problems in pulping or bleaching. These compounds cause difficulty in sulfite delignification and produce chromophoric groups that are resistant to bleaching.

Other distinct classes of wood in a tree include the portion formed in the first 10–12 years of a tree's growth, ie, juvenile wood, and the reaction wood formed when a tree's growth is distorted by external forces. Juvenile fibers from softwoods are slightly shorter and the cell walls thinner than mature wood fibers. Reaction wood is of two types because the two classes of trees react differently to externally applied stresses. Tension wood forms in hardwoods and compression wood forms in softwoods. Compression wood forms on the side of the tree subjected to compression, eg, the underside of a leaning trunk or branch. Tension wood forms on the upper or tension side. Whereas in compression wood, the tracheid cell wall is thickened until the lumen essentially disappears, in tension wood, true fiber lumens are filled with a gel layer of hemicellulose.

The chemical compositions of the reaction woods are also different. Lignin, which contributes significantly to the compressive strength of wood, is present to a greater extent in compression wood. Cellulose and hemicelluloses, which are largely responsible for tensile strength, are present in greater quantity in tension wood. Normally, these types of wood are a minor portion of the total amount being pulped, and their influence on the average pulp properties is insignificant. However, in certain tree stands, such as short-rotation coppice or high elevation regions, the juvenile wood or the reaction wood becomes a significant portion of the total, and allowance must be made for the different pulping and fiber characteristics. In some cases, specific parts of the tree are deliberately used to provide for desired pulp properties in the final product.

Chemical Composition. The chemical components found in wood follow the general scheme shown in Figure 2 and fall into two categories: low molecular weight substances and macromolecular material. In wood from temperate zones the portion of the high polymeric compounds building up the cell wall accounts for 95 to 99% of the wood material. For tropical wood this value may decrease to an average of 90%. Of this wood, 65 to 75% consists of polysaccharides, mainly cellulose and hemicelluloses. The exact composition depends on the species and age of the tree but considerable variation also exists within a single tree or a single stand of trees of the same species. The basic structural element of the cell

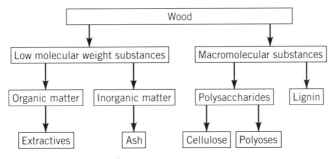

Fig. 2. General scheme of the chemical wood components (1).

wall is cellulose. Lignin and hemicelluloses are distributed throughout the cell wall in an incompletely understood manner. The intercellular substance, which is primarily lignin, must be softened or dissolved to free individual fibers.

Wood also contains 3–10% of extracellular, low molecular weight constituents, many of which can be extracted from the wood using neutral solvents and therefore are commonly called extractives. These include the food reserves, the fats and their esters in parenchyma cells, the terpenes and resin acids in epithelial cells and resin ducts, and phenolic materials in the heartwood. Resin materials occur in the vessels of some hardwood heartwood. Tannins usually occur in bark but may also be in the wood of some species, such as redwood and oak.

Many of these chemicals are recovered as by-products of the pulping operation, eg, tall oil (qv) and turpentine (see TERPENOIDS). Some cause problems in pulping or bleaching, eg, the heartwood phenols react with lignin during acid sulfite pulping more rapidly than the pulping chemicals do and inhibit the lignin solubilization reaction. Condensed lignins and phenolics are dark or form colored salts with metal ions. These compounds are not easily bleached and consume excess bleaching chemical. In alkaline pulping, the fatty acids and their glycerides form sodium soaps, thereby causing foam problems that reduce washing and evaporator efficiency (see SOAP). Western red cedar contains a group of tropolones, eg, thujic acid (5,5-dimethyl-1,3,6-cycloheptatriene-1-carboxylic acid), which necessitates the use of special corrosion-resistant alloys in the digester. High rosin-containing species, eg, pine, cause problems in the production of mechanical pulps such as groundwood and thermomechanical pulps owing to the tacky nature of the rosin. Such materials cause problems during paper production and form impurities in paper products.

All trees contain inorganic minerals as essential nutrients. Generally, minerals amount to less than 0.5% of the weight of the wood. A larger amount occurs in the bark. The principal constituents are calcium, magnesium, and phosphorus. Heavy metals and many other elements are present in smaller amounts. The metals, present as carbonates, silicates, oxalates, and phosphates, usually occur in the intercellular region. They may also be associated with the carboxylic acids of lignin and carbohydrates in the cell wall (see MINERAL NUTRIENTS).

Other Fiber Sources. Wood is the primary source of fiber for pulp, but other resources are also used in areas where wood supply is sparse. Plant sources of fiber other than wood include grasses, canes, and straws. Pulp fibers can be extracted from almost any vacular plant found in nature. However, a high yield of fibers is a necessity if the plant is to have economic importance. Pulp is produced from such nonwood sources (2) as straws and grasses, eg, rice, esparto, wheat, and sabai; canes and reeds, eg, primarily bagasses or sugar cane; several varieties of bamboo; bast fibers, eg, jute, flax, kenaf, linen, ramie, and cannabis; leaf fibers, eg, abaca or manila hemp and sisal; and seed fibers, eg, cotton (qv) and cotton linters. Cotton and cotton linters were the very first fibers used for commercial papermaking. As of this writing (ca 1996), those are still used in the manufacture of high quality paper. The introduction of synthetic fibers such as rayon and nylon and the use of certain other noncellulosic fibers such as polyethylene and polypropylene have further widened the range of raw materials and characteristics of the final paper products.

The use of alternative fibers depends on the ability of the fibers to bond to one another with sufficient strength to form a cohesive sheet. However, practical considerations determine whether pulp from a particular plant source is commercially feasible. These include the characteristics of the fiber, such as strength and optical properties, supply, yield of desirable fibers, waste generated, and the ability to store the fibers without degeneration.

Secondary Fiber. Increasing costs of raw fiber, legislative mandates for recycling (qv), and availability of inexpensive waste papers have contributed to the increased use of recycled fibers. The use of secondary fiber in the United States has grown from 22% of pulp from all sources in 1978 to close to 31% (24 × 10^6 metric tons) in 1993 (3). Recycled fibers are sometimes used in special writing papers, but the principal use is for the manufacture of linerboard, newsprint, tissue, cereal boxes, towels, and molded paper products such as paper plates and egg cartons (see PACKAGING, CONTAINERS FOR INDUSTRIAL MATERIALS).

Effective utilization of secondary fibers involves processing to sort the waste and removal of contaminants such as ink, adhesives, coatings, toner, and plastic films. Because fiber strength and substance significantly decreases with each step in recycling, efforts are being directed toward optimizing sheet properties and protecting the integrity of the fiber.

Pulp Fibers

Pulp fibers are classified according to the method of manufacture as mechanical, chemical, chemimechanical, and semichemical. Factors that determine the type used in a given grade of paper are mainly economic, such as product requirements and properties. Mechanical pulps are produced by mechanical difibrillization of wood and are characterized by low fiber strength and small particle size (Fig. 3). Virtually all of the wood's chemical components remain in the pulp. Because of the small particle size, mechanical pulps exhibit high opacity and good printing characteristics, but tend to discolor with age.

Chemical pulps are produced in a digester where the wood is cooked in pressurized vessels using heat and chemicals to break the intercellular structure of the wood and extractives. The objective is to remove the lignin from the fibers without degrading the carbohydrate content of the wood.

A photomicrograph of Douglas fir kraft pulp is shown in Figure 4**a**. Even though most of the cell wall lignin has been removed from the tracheids, the physical structure remains virtually unaffected. The long, narrow shape and tapered ends are characteristic of the structure in wood with the bordered pits visible on the cell wall. When made into paper, the fiber is collapsed into a ribbon-like structure, as shown in Figure 4**b**. The cellular elements are bonded together by hydrogen bonding at the crossovers or intersection of the fibers. Upon collapse, the surface area for bonding between fibers is increased and, very probably, new intrafiber bonding is established within the cell wall and between the collapsed lumen surfaces. The extent of collapsibility and potential bonding is largely determined by the wood species and the pulping process used. Mechanical pulps retain the stiffness and rigidity of wood, whereas chemical pulp fibers, from which nearly all of the lignin is removed, are flexible and collapse easily.

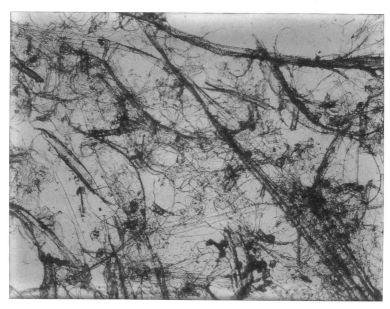

Fig. 3. Photomicrograph of spruce stone groundwood. Magnification = 100×. Courtesy of G. Abbott, University of Maine.

A hardwood kraft pulp of white oak is shown in Figure 5. The fibers are finer and shorter than the softwood tracheid and many more cell types are evident. Noticeable in a porous hardwood pulp are the short and wide vessels. These vessel segments are not long enough to bond across numerous fibers (few crossovers) and, therefore, are more easily lost from a paper sheet. In printing, this leads to picking, ie, the ink-coated vessel segment adheres to the ink printing plate more strongly than it does to the paper, leaving a white, unprinted spot behind. Papers made from hardwood fibers are weaker than those made from softwoods, but the former provide better formation or uniformity, smoothness, bulk, and opacity. Pulps of different types are frequently blended to produce specific properties for a given product.

Structure of the Cell Wall. The interior structure of the cell wall is shown in Figure 6. The interfiber region is the middle lamella (ML). This region, rich in lignin, is amorphous and shows no fibrillar structure when examined under the electron microscope. The cell wall is composed of structurally different layers or lamellae, reflecting the manner in which the cell forms. The newly formed cell contains protoplasm, from which cellulose and the other cell wall polymers are laid down to thicken the cell wall internally. Thus, there is a primary wall (P) and a secondary wall (S). The secondary wall is subdivided into three portions, the S_1, S_2, and S_3 layers, which form sequentially toward the lumen. Viewed from the lumen, the cell wall frequently has a bumpy appearance. This is called the warty layer and is composed of protoplasmic debris. The warty layer and exposed S_3 layer are sometimes referred to as the tertiary wall.

Within the cell wall, the different layers possess distinct characteristic patterns, which can be observed under the electron microscope. Microfibrils, which are linear elements of varying orientation and composed of chains of

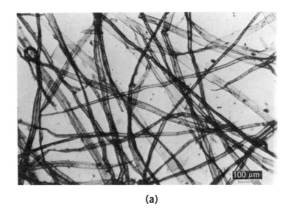

(a)

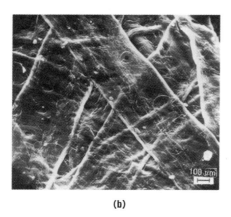

(b)

Fig. 4. (**a**) Photomicrograph of Douglas fir kraft pulp; (**b**) electron micrograph of Douglas fir pulp collapsed into paper.

cellulose, are observed. In the primary wall, which is $0.03-0.10$-μm thick in coniferous tracheids, microfibrils are embedded sparsely in an amorphous area. The orientation of these microfibrils is essentially perpendicular to the axis of the fiber. The S_1 layer contains both left- and right-handed helical windings of microfibrils. In the relatively thick S_2 layer, the microfibrils have a high degree of parallel orientation at a slight angle to the fiber axis. This fibril angle varies with species and pulping treatment and is correlated with the mechanical strength of the fiber. The smaller the fibril angle and the greater the proportion of S_2 layer, eg, in summerwood fibers, the greater the tensile strength. Transitions between the subdivisions of the secondary wall are not sharp, and the fibril angle gradually changes orientation between the layers (4).

The fibrillar structure within the cell wall is presumed to result from the highly oriented macrostructure of cellulose, although there is evidence that hemicellulose can also contribute to the fibrillar structure (5). The three-dimensional lignin polymer shows no tendency to crystallize. It is not known precisely how lignin and hemicelluloses are distributed throughout the cell wall. However, several models have been proposed (1). One representation (Fig. 7) suggests that

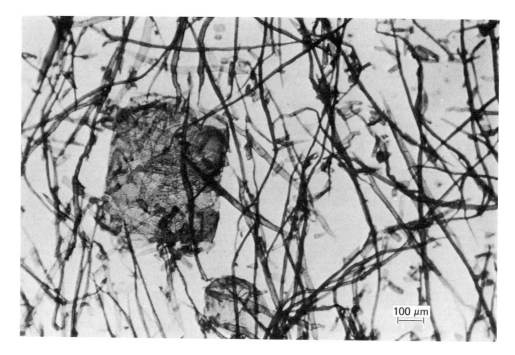

Fig. 5. Photomicrograph of white oak kraft pulp.

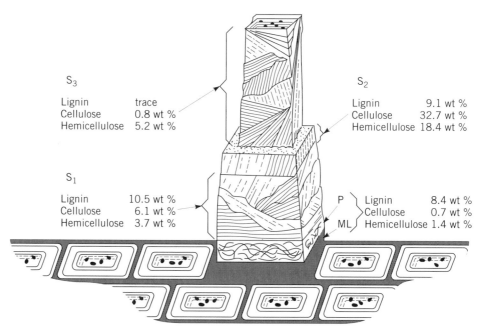

S_3

Lignin trace
Cellulose 0.8 wt %
Hemicellulose 5.2 wt %

S_2

Lignin 9.1 wt %
Cellulose 32.7 wt %
Hemicellulose 18.4 wt %

S_1

Lignin 10.5 wt %
Cellulose 6.1 wt %
Hemicellulose 3.7 wt %

P Lignin 8.4 wt %
Cellulose 0.7 wt %
ML Hemicellulose 1.4 wt %

Fig. 6. Interior structure of the cell wall of Scotch pine, where S = secondary wall, P = primary wall, and ML = middle lamella. Chemical composition of cell wall: lignin, 28.0 wt %; cellulose, 40.3 wt %; and hemicellulose, 28.7 wt %. Extractives, not shown, are 3.0 wt %.

502

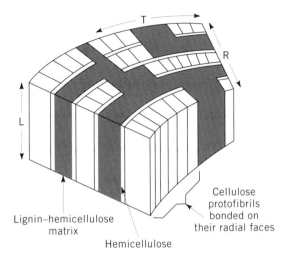

Fig. 7. Diagrammatic representation of the ultrastructure of fiber wall components, where L = longitudinal, R = radial, and T = transverse plane of fiber. Microfibrils are laid down as ribbons having their flat faces parallel to the cell periphery. Hemicelluloses and lignin are deposited between the microfibrils (6,7).

cellulose exists as microfibrils which are laid down in the form of flat ribbons parallel to the middle lamella or cell periphery. Elementary fibrils (protofibrils) are bonded together on their radial face on sets of three or four to comprise a microfibril about 10–12-nm wide and 3–4-nm thick. Immediately surrounding the fiber wall microfibrils is a layer of hemicellulose encrustant in intimate contact with a lignin–hemicellulose matrix.

It has been demonstrated by gel-permeation and absorption techniques that the cell wall is filled with microcapillary pores having openings of up to ~2 nm in diameter in greenwood (8,9). These pore openings essentially close upon drying of the wood. The pore volume increases in the presence of water and certain chemicals, eg, alkalies. During chemical pulping, the dissolved chemicals diffuse into the submicroscopic capillary water and react with the lignin polymer. Polymer fragments, which become solubilized, diffuse out of the cell wall if they are small enough. Frequently, and perhaps unavoidably, lignin removal is accompanied by the removal of hemicelluloses.

Chemical Constituents of Cell Wall. Variation in chemical composition across the cell wall is also shown in Figure 6. The principal constituents of cellulose, hemicellulose, and lignin are present throughout the cell wall but in different proportions. Cellulose is not present in the interfiber middle lamella, which is virtually all lignin. The S_3 layer is essentially all carbohydrates (qv), especially hemicelluloses, having little or no lignin.

Cellulose. Cellulose (qv) is a straight-chain β-1→4-linked D-glucan of high molecular weight, the value of which varies with the source and method of isolation. For native wood cellulose (Fig. 8), the value is ca 8,000–10,000 weight-average degree of polymerization (DP_w). The supramolecular structure of cellulose, ie, the alignment of the cellular chains to form the cell wall fibrils, has long been an enigma. As of this writing, various physical and chemical

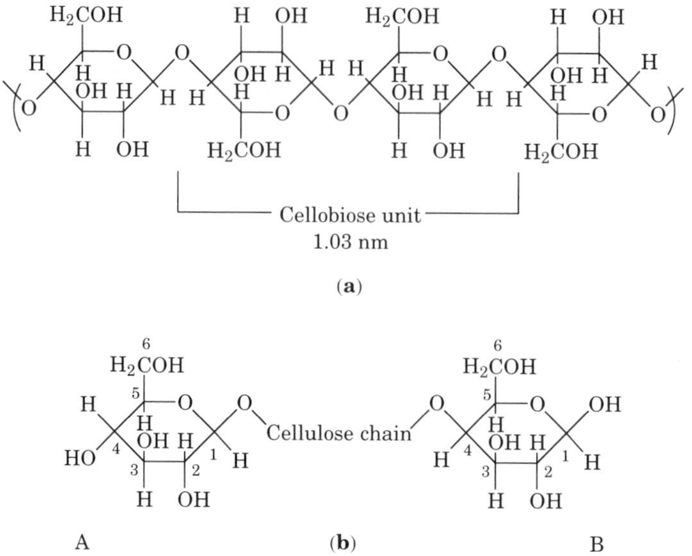

Fig. 8. Formula of cellulose: (**a**) the central part of the molecular chain, and (**b**) end groups of the molecule denoting the carbon numbering scheme where A = the nonreducing, and B = the reducing end (1).

properties have yet been reconciled into a universally acceptable, coherent theory. X-ray diffraction patterns show a distinct crystallinity for various cellulose preparations. There are, however, at least four different polymorphic forms that are identified as cellulose I, II, III, and IV. Different preparations of cellulose or different lignocellulosic samples also have varying degrees of crystallinity. These phenomena are ultimately related to chemical accessibility and physical properties of the pulp fiber.

Hemicelluloses. The hemicelluloses are lower ($DP_w = 100-200$) molecular weight mixed-sugar polysaccharides (see HEMICELLULOSE). Hemicelluloses are of various types and their exact nature varies with tree species and location in the tree. However, there are types common to hardwoods and to softwoods, as listed in Table 2. In some instances, the hemicelluloses are mixtures of closely related polymers. The general structural characteristics of the most common hemicelluloses are summarized in Figure 9. Hemicelluloses are largely responsible for hydration and development of bonding during beating, ie, cutting and fibrillating, of chemical pulps. Pulps containing high percentages of hemicellulose are typically low in tear strength but develop high hydrogen-bonding strength between fibrous elements. This leads to pulps having high internal bond strength.

Minor Constituents. Other hemicellulosic materials occur in extracellular regions of wood. In softwoods, such materials include a $1{\rightarrow}3,1{\rightarrow}6$-linked arabinogalactan. The compound middle lamella, which is composed of the middle lamella and the primary cell wall, contains three hemicelluloses called the pectic-group substances. The group of pectin compounds comprises the galacturanans, the galactans, and the arabinans. Compression wood, which is highly lignified, also contains a high proportion of galactan. Larch is unusual among the softwoods in

Table 2. Composition of the Carbohydrate Fraction of Wood, wt %[a]

Component	Hardwoods	Softwoods
cellulose	40–47	38–42
O-acetyl-4-O-methylglucuronoxylan	20–30	
O-acetylgalactoglucomannan		16–20
glucomannan	2–3	
4-O-methylglucuronoarabinoxylan		8–11
pectic materials	1	1
arabinogalactan	trace	2–3
galactoglucomannan	trace	
starch	trace	trace

[a]Wt % except where noted as trace.

that it contains significant amounts of the water-soluble $1\rightarrow3,1\rightarrow6$-linked arabinogalactan.

Lignin. Lignin, a highly branched alkylaromatic thermoplastic polymer, is incompletely characterized. The structural elements and linkages have been summarized (1,10–12). Experimental difficulties associated with the characterization of lignin as it exists in wood are extensive because of the variety of bond types and possible condensation or degradation during attempted isolation and purification from the polysaccharides with which lignin is intimately associated. Model compound studies have suggested that the basic aromatic skelal unit in lignin is the C_9 phenylpropane unit (10–15). In softwood lignin, the majority of C_9 units have one methoxy group adjacent to the phenolic hydroxyl (PhOH) groups, guaiacyl propane, shown here as a dimer of two C_9 units where R = hydrogen (14).

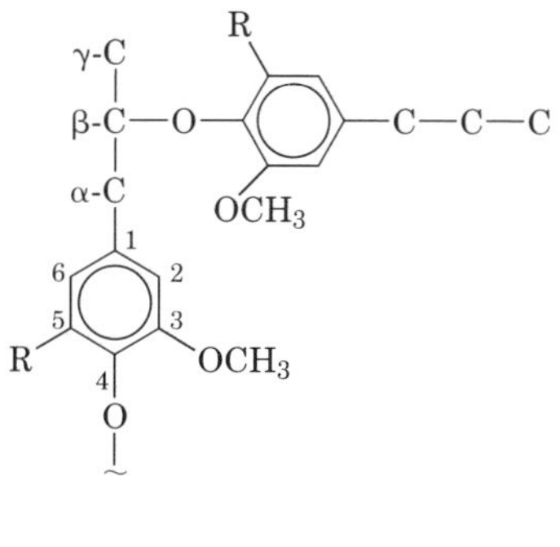

On average, there are about 0.95 methoxy groups per C_9 unit in softwood lignin. Hardwood lignin has a higher methoxyl content, and units of the syringyl type where R = OCH_3 have been isolated from degradative studies. The ratio of OCH_3 to C_9 units in hardwood lignins is approximately 1.3 to 1.7 (see LIGNIN).

The aromatic nature of lignin contrasts with the aliphatic structure of the carbohydrates and permits the selective use of electrophilic substitution reactions, eg, chlorination, sulfonation, or nitration. A portion of the phenolic hydroxyl units, which are estimated to comprise 30 wt % of softwood lignin,

(a)

α-D-CH₃-GlupU
1
↓
2

α-L-Araf
1
↓
3

→ 4-β-D-Xylp-1 → 4-β-D-Xylp-1 → 4-β-D-Xylp-1 → 4-β-D-Xylp-1 → 4-β-D-Xylp-1-

(a)

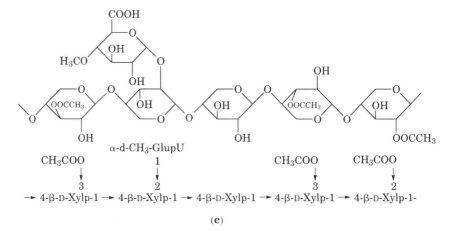

α-D-Galp
1
↓
6

H₃CCOO
↓
2

H₃CCOO
↓
3

→ 4-β-D-Manp-1 → 4-β-D-Glup-1 → 4-β-D-Manp-1 → 4-β-D-Manp-1 → 4-β-D-Glup-1-

(b)

α-d-CH₃-GlupU
1
↓
2

CH₃COO
↓
3

CH₃COO
↓
3

CH₃COO
↓
2

→ 4-β-D-Xylp-1 → 4-β-D-Xylp-1 → 4-β-D-Xylp-1 → 4-β-D-Xylp-1 → 4-β-D-Xylp-1-

(c)

Fig. 9. Partial structural formulas and shorthand notations for principal hemicelluloses found in wood, where the sugar units are noted as β-D-xylopyranose (Xylp), 4-O-methyl-α-D-glucopyranosyluronic acid (GlupU), α-L-arabinofuranose (Araf), β-D-glucopyranose (Glup), β-D-mannopyranose (Manp), and β-D-galactopyranose (Galp) for (a) arabino-4-O-methylglucuronoxylan from softwood, (b) O-acetyl-galactoglucomannan from softwood, and (c) O-acetyl-4-O-methylglucuronoxylan from hardwood (1).

506

are unsubstituted. In alkaline systems the ionized hydroxyl group is highly susceptible to oxidative reactions.

Reductive processes involving intermediate quinonemethides are important in alkaline pulping. Highly colored quinones (qv) and other conjugated chromophores are readily formed during alkaline pulping. These constituents impart the brown color to kraft pulps and necessitate subsequent bleaching operations. The benzylic α-carbon position is reactive and can be substituted by nucleophilic reagents. A common depolymerization site is the β-aryl ether link labeled above, the cleavage of which can be facilitated by neighboring-group participation of substituents on either the α- or γ-carbons.

The molecular weight of lignin in the wood, ie, of protolignin, is unknown. In addition to difficulties of isolation and purification, the polymer exhibits strong solvent, ionic, and associative effects in solution. An unequivocal method of measurement has not been developed. The polymer properties of lignin and its derivatives have been discussed (10,16).

Lignin–Polysaccharide Complexes. The physical association of lignin with carbohydrates is strong, even when purified samples of the separated polymers are remixed. The existence of covalent bonds between lignin and carbohydrates has been difficult to demonstrate unequivocally, although the accumulated evidence is strong that such bonds exist (1). The most likely linkages to lignin are to the side chain at the α- or β-carbons or glycosidic links to phenolic hydroxyls. For the carbohydrates, the evidence is fairly good that uronic acids form esters with lignin and that ether bonding occurs at the C-6 position of hemicellulose hexoses (17). There is also evidence that arabinose and xylose are bonded to softwood lignin. Ultrastructural models for lignin–polysaccharide complexes have been proposed (18,19). Approximately one phenylpropane unit in 30 in softwoods is calculated to be bonded to carbohydrates (20). If no bonding to cellulose is assumed, there are about 3–6 bonds to lignin in each hemicellulose chain.

Wood Procurement and Preparation

Wood procurement and preparation for use in the manufacture of pulp begins with the purchase of roundwood or wood chips from large woodlots or tree plantations, as shown in Figure 10. Wood is processed to produce pulp under specified conditions such as tree species and size, cleanliness, and needs of the product. Attempts are made to utilize as much of the tree as possible. Different species of trees are often mixed to give optimum properties in the pulp. The feed must meet standards of cleanliness regarding bark, foliage, rotten wood, and bits of plastic because these impurities cause dirt specks, structural and surface imperfections, increased chemical consumption, and loss of strength in the finished paper. The standard required in the feed depend both on the pulp grade and pulping process.

Operations of cleaning, debarking, and production of wood chips of uniform dimension all depend on the pulping process to be used. Grinding wood against a stone requires clean roundwood bolts of the same length as the width of the stone. Chemical and semichemical pulping call for a chip about 25 mm (1 in.) in length and about 3–7 mm (0.125–0.25 in.) in thickness, as does refiner pulping in which chips enter between ribbed rotating disks to be defibered. Wood

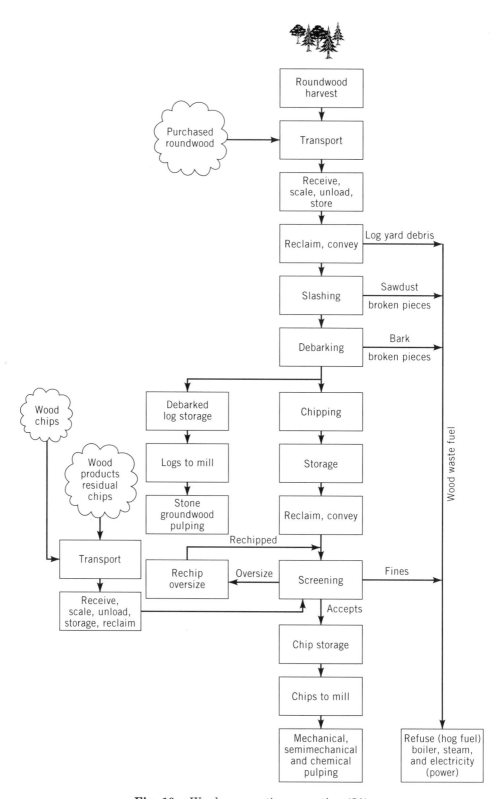

Fig. 10. Wood preparation operation (21).

508

subdivision and dimensioning operations include merchandizing, slashing, debarking, chipping, and screening. The woodyard receives, measures, inventories, stores, and provides steady supplies of wood for pulping. It also utilizes reject wood, eg, bark, fines, and oversize, to recover residual heating value as fuel to produce process steam or electrical power, or sends the reject wood out for final disposal.

Harvesting. Many factors influence the procedures and the choice of equipment used to harvest wood. These include physical considerations such as area of land to be harvested, tree size, terrain, climate, and environmental constraints, as well as commercial considerations such as demand for fresh wood, the need to supply a continuous flow to the mill, and maintaining inventory and a healthy corps of suppliers. Wood processing operations of harvesting, topping and delimbing, debarking, and chipping are typically highly mechanized and combined into as few individual steps as possible. Many of these operations take place on-site in the woods.

In an optimized harvesting situation, such as a company-owned pine plantation on level ground in the southeastern United States, where pulp chipping is the only objective, the operation can take place on-site. Trees are felled by feller-bunchers at a rate of about 100 or more trees per hour, hauled to the roadside by grapple skidders that drag one end of the load, and delimbed at over 75 m³/h. Delimbing of over 200 m³/h is achievable if poor quality can be tolerated. Unbarked trees can be chipped at 24 m/min, yielding 230 to 540 metric tons per day of whole tree chips. When clean chips are needed, trees can be barked at 24 m/min and then chipped separately, yielding 140–180 t/d. An operation of this type uses equipment costing $500,000–750,000 or more and requires sophisticated planning to optimize costs and ensure minimal environmental impact.

Pressures on forest resources and competing uses of fresh wood have resulted in greater recovery of sawlogs from trees that are formerly considered a fiber resource. On the west coast of North America, the larger trees are primarily used for lumber and plywood, leaving only the residues for pulp chips. In order to minimize butt damage, many operations use sawheads rather than shears on the feller-bunchers.

In stands where partial cuttings are made, cut-to-length systems are used in which a delimber and cut-off saw are incorporated into the sawhead. The products are bucked, separated into piles, and then hauled to the roadside on a forwarder, which carries the whole load. The limbs and tops are left on-site for better nutrient recycling.

After being transported to the roadside, stems can be trucked tree-length or bucked in bundles to short lengths of 4, 5, or 8 ft (1.2, 1.5, or 2.4 m) and loaded directly onto a truck. Because the cost of long-distance transportation is high, many producers set up concentration yards where both company and purchased wood can be merchandized, ie, cut into multiple products, and/or chipped. Shipment of chips from the yard to the mill by rail or by vans designed for highway transport is generally more cost-effective and safer than hauling roundwood or stems. A developing technology involves the use of chain-flail debarker–delimbers feeding directly into a chipper that can produce an average of 100 loads of chips a week. Peak production is about 150 loads (18–20 t (40,000–45,000 lbs) per load) from southern pine thinning operations.

It is not always possible to harvest trees year-round. In eastern and central Canada, much of the forested land has shallow, wet soils. The majority of harvesting is therefore done in winter when the ground is frozen. Harvest in other areas may also be seasonal.

Less mechanization is used on smaller tracts of individually owned land, where the terrain is mountainous or otherwise difficult to reach, or where thinning as opposed to clear-cutting is the preferred silvicultural practice. The harvesting operation may be done by the landowner or small contractor using chainsaws and tractors. In some operations where minimal damage to the forest is critical, horses may be used. Sale of pulpwood to the mill operation is usually through dealers.

Sustainable Forest Management.   The forest products industry has placed a high priority on developing sustainable forest practices that optimize yield and fiber properties while preserving soil, wildlife habitat, and species diversity. These practices begin with selection of species that give optimum fiber properties, disease and insect resistance, and high yields. The industry is also focusing research on genetic engineering (qv), land management to protect biodiversity of species and gene pools, and conservation of areas having unique ecosystems. Sustainable forest practices also include harvesting techniques that conserve soil resources and productivity.

Slashing.   Wood is slashed or cut to provide a fit for subsequent conveying, debarking, chipping, and other processing equipment. Slashing is done in the woods or the woodyard and typically uses either circular or chainsaws to cut single large logs or several smaller stems bundled together. The slashing operation is sometimes combined with merchandizing where woodyards feed pulp, lumber, and plywood mills. Larger-diameter and straight logs are sent to lumber and plywood mills; smaller, crooked, decayed, or defective wood is cut out and sent to the pulp mill.

Debarking.   The bark of most pulpwood species is dark, containing extractives and little fiber, often contaminated with dirt or grit. Bark accompanying the wood into the mill causes wear in the equipment and lowers the strength and appearance of the pulp. The acid sulfite process is not very effective in disintegrating and decolorizing bark fragments. Alkaline pulping is more effective in reducing dirt specs caused by bark but requires additional cooking and chemicals, and consequently decreases pulp yield and fiber strength. Mechanical pulping has little effect on bark except to subdivide it, which often multiplies the dirt count. Mechanical pulping processes that use refiners are especially sensitive to grit contained in bark because of the small clearance between refiner plates, where grit causes excessive wear.

Specifications on bark content for debarked wood typically vary between 0.5–2.0% based on the dry weight of wood and bark. To achieve these standards, the efficiency of bark removal must be high for most processes (21). The lowest specifications are for wood that is to be bleached eventually. Meeting bark-removal standards has become more of a problem as the average log diameter has become smaller. Varying amounts of bark can be tolerated under some circumstances, for instance in thin-barked species and such final product as fiberboard or corrugated medium, where a dirtier, lower quality pulp can be accepted. When the demand for fibers is great, whole-tree chipping is sometimes employed to increase the fiber yield. In this process, the entire tree, including bark, limbs, and leaves, is passed through the chipper. However, the higher fiber yield is costly because more extensive screening and cleaning are necessary. The presence of bark necessitates higher chemical consumption in pulping and bleaching and a higher load on the recovery system. Bark has a higher ash content than wood, and entraps more silicates from sand and dirt, particularly if the tree has been dragged on the ground. The silicates cause severe erosion problems in processing and cleaning equipment, mud settling problems in lime cycle, and evaporator scaling (22).

Logs are sometimes debarked in the woods where they decompose, returning humus and nutrients to the soil. But most often the logs are debarked at the mill or sawmills where the waste bark is used as fuel. Debarkers are classified as either rotary drum debarkers or in-line debarkers. Drum barkers are large open-ended cylinders that are rotated on a sloped axis. This enables logs entering at one end to move toward and out the other. The bark, which is broken off by rubbing and pummeling between the logs themselves and between the logs and the cylinders, drops out through slots in·the cylinder. These units have a relatively high capacity. In-line barking units, which remove the bark from one log at a time, may be classified as either hydraulic or mechanical-friction-type debarkers. Hydraulic debarkers remove the bark by jets of water at pressures in excess of 6800 kPa (100 psi). This system is used only for very large logs such as those available on the west coast of North America. For friction debarkers, such as cutterhead and ring (shear) debarkers, special tools are pressed against the log and rotated around it as the log is passed through, processing logs at a linear throughput capacity of 40–60 m/min.

Chipping. Wood is chipped to reduce it to a size that maximizes penetration of processing chemicals without excessive cutting or damage to the fibers. In the case of mechanical pulp produced from refiners, the refiner disks only accept a piece of wood small enough to enter between them. Debarked wood used in chemical or refiner pulping needs to be chipped into small, uniform pieces to allow quick impregnation with chemicals and steam in the digester. Otherwise, the outside of the wood would be overcooked, thus weakening the fiber, and the inside would be undercooked, leaving chunks of unseparated fiber.

Cutting chips too small has significant disadvantages. Each cut causes damage by severing and crushing fibers lying across the cut and produces more shortened fibers per unit of volume, thus reducing fiber strength. Chips that are too small, eg, sawdust, fines, and pin chips, tend to agglomerate and plug chemical pulping equipment, notably some continuous digesters (23). Special nonplugging digester systems are used to pulp fine chips, sawdust, and nonwood annual plant material such as straw and bagasse.

The most common range of dimension for mill-cut chips is between 16–32 mm (0.625–1.25 in.) in length along the grain and 3–10 mm (0.125–0.375 in.) in width. There is considerable variation in both length and width in commercial chips. It is widely recognized that chip thickness is the most important dimension affecting penetration of chemicals and chip screening systems have been modified to remove overthick and underthin chips.

Chippers may be classified as disk or drum chippers, depending on mechanical design (21). In the Norman (Carthage Machine Company) disk chipper design the cutting knives are mounted on the face of a large disk that is rotated on a horizontal axis. The log enters from a chute so arranged that the knives sequentially cut across the log at 37° to the fiber axis. The chips that form pass through slots in the disk. The knife severs the fibers as it enters the wood and, because of its wedge-like shape, places an increasing edge compression on the cut fibers. This pressure is relieved by shear along the grain. Thus, the thickness of the chip is related both to the shearing forces at which the wood yields and to the length of the chip, which is normally specified and set by the arrangement of the knives.

Increased attention to the role of chip thickness in pulping has led to attempts to control that dimension by use of a drum chipper. In the Anglo drum chipper, logs of fixed length are fed from a magazine to a large, rotating blade where the wafers are cut while passing through the drum. The log is presented parallel to the drum axis. The cutting element is a gouge having a flat bottom and straight sides, mounted in rows on the surface of the drum and pointed in the direction of drum rotation.

In the spiral chipper, logs are fed endwise and are sequentially engaged from both sides by a series of winged knives mounted on two cones having their apexes facing each other and their common axis perpendicular to the log. The main knife edge peels, and the wing knife cuts across the fibers, resulting in chips of uniform length and thickness but less compression damage.

Oversized chips are sometimes sliced to control thickness in a slicing rechipper. The uniform thickness results in more even penetration of pulping liquors and therefore less screen rejects in the resultant pulp.

Chip Screening. Because of considerable variation in dimension, chips must be screened before proceeding to the digester where they are cooked under conditions set for average-sized chips. A wide range of chip dimension results in a higher proportion of overcooked and undercooked chips. Additionally, oversized or undersized chips can cause mechanical problems in some types of equipment. Screening removes oversized and undersized chips, narrowing the size distribution and removing undesirable foreign matter, eg, knots, bark particles, and dirt.

In screening, the chips are normally fed first to a flat screen surface having holes of a size that retain the oversized and pass the acceptable and undersized. This material is sent to screen surfaces having holes of a smaller size, retaining the acceptable while passing the undersized. The screen surfaces are inclined and agitated in a gyratory or vibratory motion so that each chip travels across the array of holes and is either discharged or passed through a hole during its dwell time. Each hole has two dimensions, length and width, which can be the same, eg, round, or different, eg, a slot. Each chip has three dimensions, two of which, ie, length and thickness, are usually different, whereas the third dimension, ie, width, varies. A chip can pass through a hole if all three dimensions are smaller than the hole or if two dimensions are smaller and the largest dimension is perpendicular to the hole. Otherwise, the chip is retained (see SEPARATION, SIZE).

In conventional systems, chips are fed to the screen and separated into oversize (slivers), acceptable chips, and dust (pins). Screening is performed based on fiber length, thickness, or both. Disk screens have been developed to give chips of uniform thickness. Commercial screens may be categorized as flat, inclined screens, drum screens, and disk thickness screens. Disk scalping screens, a rugged version of the disk thickness screen, can be used ahead of the screening system to protect the latter by removing metal and wood chunks.

Processing Chip Screen Rejects. Overlong chips are usually reprocessed in a smaller version of the disk-type roundwood chipper. Overthick chips are reprocessed in a chip slicer. The product from these reprocessing operations returns to the main chip flow ahead of the screens (see Fig. 10). The fines are sent to a hog boiler as fuel, or else rescreened. Pin chips are metered back into the main chip flow or sent to a fine-particle pulping system.

Wood Residuals. Chips, sawdust, and other residuals such as planar shavings are used as a primary source of fiber for some pulp mills. Chips are screened and placed in a purchased chip pile. Sawdust and other residuals obtained from wood processing plants must be cooked separately and require special digesters and handling equipment. These materials may also be burned as hog fuel.

Mechanical Pulp

Mechanical pulp is a generic term covering a variety of pulps having a yield range from about 85 to nearly 100%. Mechanical pulps are categorized by size reduction (qv) equipment used in manufacture. Either grinding on a pulpstone (Fig. 11) or in a rotating-disk attrition mill called a refiner (Fig. 12) is employed. The highest yield is obtained for the classical stone groundwood (SGW) pulp and a modification termed pressurized groundwood (PGW), as well as for the refiner mechanical pulp (RMP) and its modification, thermomechanical pulp

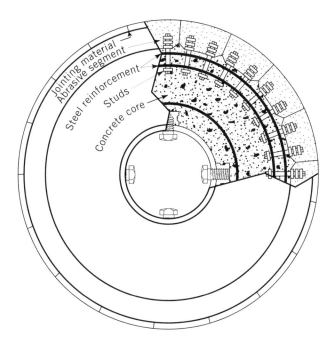

Fig. 11. Cross section showing construction details of a ceramic pulpstone (22).

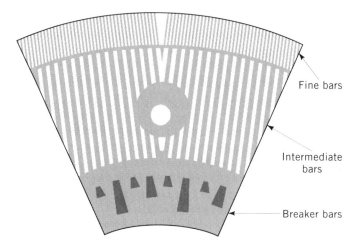

Fig. 12.   Refiner plate design.

(TMP). Chemical pretreatment of wood causes a drop in yield, the magnitude of which depends on the severity of treatment and also on the manufacturing process. Chemithermomechanical pulp (CTMP) is obtained by a comparatively mild chemical treatment followed by pressurized refining. If a stronger chemical treatment is applied at an elevated temperature, the yield may drop to as low as 85%. The resultant chemimechanical pulp (CMP) is obtained when the treated chips are refined under either atmospheric or pressurized conditions.

Basic Principles.   Different manufacturing processes give rise to pulps having different properties. A common feature in mechanical pulps is a woody character caused by the presence of lignin in the fibers. Mechanical treatment disrupts the wood structure into fragments of variable dimensions. A rough classification of the pulp according to particle size gives three main types of material: fines, fibers, and shives.

The particle size distribution of the fibers can be fitted mathematically to statistical probability functions, the upper limit being determined by the fiber length of the wood. Particle size analysis of mechanical pulp, commonly carried out by a screening test, is being done more and more using optical instruments for automatic fiber length counting and testing for shives. Because the lignin-containing fibers are stiff and woody in character, no appreciable internal fibrillation can be achieved by beating or other mechanical means. In contrast to chemical pulp fibers, the ability of mechanical pulp to form close contacts and bonding between fibers is dependent on fibrillation of the outer surface of the fibers. The final properties of mechanical pulp depend on size distribution and the formation of a fine fraction having high specific surface area which increases rapidly with decreasing particle size.

Mechanical pulp has unique properties that make it desirable in numerous paper products. Mechanical pulp is composed mainly of fiber bundles and fragments, some whole individual fibers, and high quantities of fines, which are particles less than ca 76 μm (-200 mesh). Mechanical pulps derive their special properties because of the mixed nature of the particle sizes present and because of the presence of lignin. Desirable properties include a small average particle length and a relatively stiff fiber that prevents close packing. The presence of lignin causes mechanical pulps to regain their shape when compressed and then released, giving a cushioning effect. Mechanical pulps absorb ink readily and uniformly. These properties give papers manufactured from mechanical pulps high bulk (cm^3/g), high opacity, and excellent printing characteristics. Undesirable properties of mechanical pulps are relatively low tensile strength, a harsh feel, and lack of permanence. Mechanical pulps are relatively low in cost because of the high yield.

Drainage Rates.   Mechanical pulps are manufactured to different drainage rate targets given by the Canadian Standard Freeness test, and vary from a very coarse or free (rapid drainage rate) pulp to a very fine or slow (low drainage rate) pulp. Variations in properties are produced by changing production conditions such as chemical pretreatment, grinding, refining, and screening. A high freeness mechanical pulp generally has low strength, high bulk, high drainage rate, and good resiliency; a mechanical pulp of low freeness has considerably more strength, low bulk, and a low drainage rate.

Methods of Production.   In the grinding process, debarked logs are forced against a revolving abrasive stone that pulls the fibers off the log, separating them from the wood matrix. In a refiner, process wood chips are fed between two metal disks, of which at least one disk is rotating, and the wood fibers are separated by the action of grooves and bars located on the surface of the two disks. Freeness and pulp properties depend on the energy and number of impacts received by the chips during their residence in the refiner. In both types of processes, the crude pulp must be upgraded by removing undesirable materials before the accepted fiber can be used.

Bleaching Mechanical Pulps.   Mechanical pulps are bleached to a higher brightness using oxidative or reductive bleaching agents. The bleaching is done in a lignin-conserving manner, called brightening, in which the chromophores, ie, molecules having conjugated double bonds, are modified and little bond cleavage or solubilization of the wood lignin occurs. Oxidative bleaching may use hydrogen peroxide (P) and sodium hypochlorite (H) solutions. In reductive bleaching, sodium hydrosulfite (dithionite, Y) or formamidine sulfinic acid (FAS) may be used. One-stage bleaching is common and usually requires mixing, heating, pumping equipment, and a bleach tower, but no pulp washer. Ancillary equipment for bleach solution preparation is also required. Mechanical pulp brightness can be significantly increased by one stage of bleaching, which traditionally uses hydrosulfite or peroxide. For higher grades of paper, two stages of bleaching are required, which include steps for pulp washing and neutralization. Paper containing mechanical pulps has poor brightness stability even after bleaching, particularly in the presence of uv radiation, and is unsuitable for fine papers. Where light stability is important, the more expensive bleached chemical pulps are normally used (see BLEACHING AGENTS, PULP AND PAPER).

Pulp Characteristics.   Properties of mechanical pulp vary significantly. Pulps that depend on grinding have a higher content of fine material. Refiner pulps, notably refiner mechanical pulps (RMP) and thermomechanical pulps (TMP), generally have a smaller content of fine material and the long fibers tend to be more ribbon-like. Chemithermomechanical (CTMP) and chemimechanical

(CMP) pulps generally have higher quantities of longer fiber and less fines than TMP and RMP. The characteristics of a mechanical pulp also depend on the wood species used, the quality of the wood, the processing conditions, and the amount of mechanical energy applied.

Stone Groundwood Pulping.   A flow sheet for a typical stone groundwood process is shown in Figure 13. The peeled logs sometimes go through a drum washer to remove any remaining bark or dirt. The clean, wet wood is pressed against a wetted, rotating grindstone, and the axis of the wood is parallel to the axis of the stone. After grinding, the pulp is passed on first to coarse (bull)

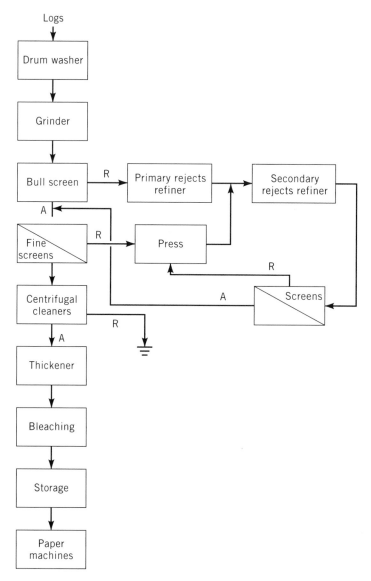

Fig. 13.   Typical groundwood flow sheet, where A = accepts and R = rejects; □ represents a unit operation and ⊠ represents a screen (22).

screens, which remove large splinters, and then to finer screens that remove bundles of fibers or shives. The accepted material is cleaned in centrifugal cleaners to remove short, stubby material called chop, as well as inorganic debris (grit) that comes from the grindstone. The final accepted pulp, which must be thickened for storage, may be bleached or used directly on the paper machine. The rejects from the bull and fine screens are thickened and refined in rejects refiners to produce acceptable pulp and to increase the pulp yield. Multiple stage cleaning and screening systems are common. Rejects from the last cleaning stage are sent to the sewer.

Pulpstones. Improvements have been made in the composition and speed of the grinding wheel, in methods of feeding the wood and pressing it against the stone, in control of power to the stones, and in the size and capacity of the units. The first pulpstones were manufactured from quarried sandstone, but have been replaced by carbide and alumina embedded in a softer ceramic matrix, in which the harder grit particles project from the surface of the wheel (see ABRASIVES). The abrasive segments are made up of three basic manufactured abrasive: silicon carbide, aluminum oxide, or a modified aluminum oxide. Synthetic stones have the mechanical strength to operate at peripheral surface speeds of about 1200–1400 m/min (3900 to 4600 ft/min) under conditions that consume 0.37–3.7 MJ/s (500–5000 hp) per stone.

Grinders. Grinders are characterized according to the two methods of applying force to the pulpwood against the revolving stone: intermittent or continuous. Pocket (Fig. 14**a**) and magazine (Fig. 14**b**) grinders are common intermittent machines. Pocket grinders can contain two, three, or even four pockets, of which two is most prevalent. In the hydraulic magazine grinder (Fig. 14**b**), the most common type of intermittent grinder, logs are fed from above out of a magazine. In pocket and magazine grinders the logs are pressed against the rotating stone by a piston using hydraulic pressure. The logs caught between the foot and the stone wear away. When the foot reaches the end of its travel, it rapidly moves back. In magazine grinders more logs drop down automatically and the cycle is repeated. Two magazines are provided on either side of the stone and pulp from the first magazine is removed by showers ahead of the second magazine. The period of interrupted pressure, ie, when the foot is reset, is about 20 seconds. In the case of pocket grinders the pocket must be refilled when the foot is disengaged from the stone. High magazine chain (Fig. 14**c**) and toothed ring (Fig. 14**d**) grinders have been developed to eliminate this interruption.

Automation. The feeding of logs to the individual grinders has traditionally involved considerable labor. Modern grinders employ an automated feeding line which requires only one or two operators to handle a complete line of grinders. The logs, delivered by a series of belts and pushing devices, are available in bunches of appropriate size at each grinder. These bunches are dropped to the magazine when required.

Theory of Grinding. In the grinding process, the land (compression) and groove (relaxation) areas on the stone pass rapidly over the wood surface, which goes against the surface of the log, as shown in Figure 15. The wood fibers are compressed each time a land area passes over the surface. Each time a groove passes over the wood fiber it releases the surface of the wood. The grit diameter is about 0.2 mm, ie, about 10 times the fiber diameter (Fig. 16**a**). The repeated

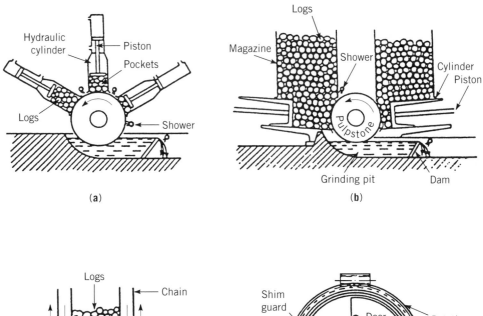

Fig. 14. Types of grinders: (**a**) three-pocket, (**b**) magazine, (**c**) chain, and (**d**) ring (22).

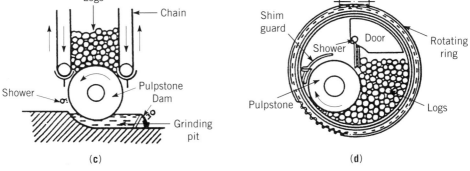

Fig. 15. Stone pattern, where ⬭ represents silicon carbide abrasive, and ⬢, aluminum oxide abrasive (22).

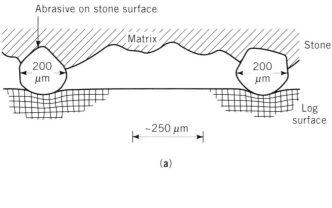

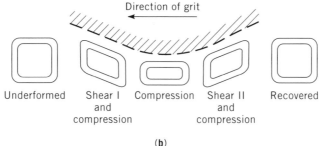

Fig. 16. (**a**) Representation of idealized grinding zone conditions. The wood surface during grinding is deformed between 25–50 μm by a grit of 100-μm radius of curvature under normal grinding pressures. The average distance between conditioned grits is ~625 μm. The grits are separated from the wood by a thin water film (22). (**b**) Stages in the cyclic complex straining of a fiber in its transverse plane, and occurring along part of its length, when a conditioned grit passes over. This straining, comprising reverse shear and compression, leads to fatigue failure in the cell wall (22).

compression–decompression action generates heat and softens the wood matrix before it reaches the grinding surface, where the loosened fibers are carried from the wood (Fig. 16**b**).

Grinder Variables. The quality of pulp depends on wood species, moisture content, and grinder variables such as peripheral stone speed, grit size and number per unit area, and pattern on the stone surface. Process variables that affect pulp quality include grinding pressure; pit consistency, ie, consistency in the space immediately below the grinder (2–6%); and temperature (40–80°C). The combination of moisture and raised temperature tends to soften the lignin.

Temperatures in the immediate grinding zone can be 180–190°C. Movement of water and the removal of pulp control and dissipate the heat, thus preventing charring of the wood. It is advantageous to allow the water temperature in the pit to rise to 70–80°C, which permits a faster grinding rate, lower energy consumption, and higher pulp strength than are possible at lower temperatures. Because groundwood pulps are obtained in yields of about 95%, the cost is relatively low. The principal cost other than wood is energy, which is ca 6,520–11,340 MJ/t (46–80 hp·d/short ton) for normal paper grades (22). Somewhat less energy is used in producing pulps for hardboard and low density wallboards, which contain coarser fiber than paper-grade pulps.

Principal Uses. Groundwood pulp contains a considerable proportion (70–80 wt %) of fiber bundles, broken fibers, and fines in addition to the individual fibers. Principal uses of paper-grade pulps are in newsprint, magazine papers including coated publication grades, board for folding and molded cartons, wallpapers, tissue, and other similar products.

Pressurized Groundwood Pulping. In the late 1970s and early 1980s, it was demonstrated that pressurizing the region surrounding a stone grinder to 100–300 kPa (1–3 atm) markedly improved the strength and fiber integrity of the pulp. Pressurization, accomplished by using air, retains the advantages of low energy consumption and high opacity (24). Figure 17 shows the modification of a Tampella-Great Northern type of pocket grinder to a pressurized system.

Further Modifications. There have been attempts to improve the pulp strength and decrease the energy requirement of the groundwood process by the addition of chemicals, eg, sodium carbonate or sodium sulfite. Although some benefits can be obtained, the additional costs and chemical disposal problems have not been justified. Presteaming of wood is practiced in Europe to add moisture and to soften the lignin, which is especially advantageous for certain hardwoods.

Refiner Mechanical Pulping. The conventional groundwood process requires bolts of roundwood as raw material. In the 1950s, the refiner mechanical pulping (RMP) process was developed. This produced a stronger pulp and utilized various supplies of wood chips, either purchased sawmill chips (sometimes sawdust) or chips produced at the mill site. However, the energy requirement of RMP is higher, and the pulp does not have the opacity of stone groundwood fibers. The process is similar to that for the stone groundwood pulping process. Chips are screened for oversize material in order to avoid nonuniform feeding problems. In some cases, the fines are removed to improve pulp quality. The screened chips are washed to reduce wear on the pulping equipment.

Refiner Plates. The refiner plates have a construction of the type shown in Figure 12. The plates are constructed of hard steel alloys. Alloys of cast white iron and cast stainless steel are used to achieve the desired energy transfer and to resist rounding of the leading edges of the bar, which degrade the quality of the pulp. The plates have three regions of different bar design, depending on the required function: breaker bars, intermediate bars, and fine bars. Bars in different regions are characterized by bar width, pitch between bars, and bar depth. Breaker bars are coarse and widely spaced. Intermediate and fine bars have succeedingly smaller pitch and finer bar width. Sometimes bars are included between bars. The plates are paired face-to-face with a small interval or gap between them.

Refining Action. One disk rotates against a stationary disk, or both disks move in a counterrotating manner. Chips are fed through channels near the shaft in one of the disks and move toward the periphery of the plates while undergoing attrition. The chips are broken down first into matchstick-like fragments by

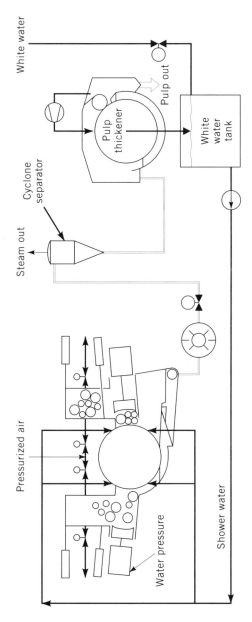

Fig. 17. Modified Tampella–Great Northern grinder (22).

White water

Pulp out

Pulp thickener

White water tank

Cyclone separator

Steam out

Pressurized air

Water pressure

Shower water

521

the action of the breaker bars, then into progressively smaller bundles as the chips move through the intermediate and fine bar sections. They emerge from the periphery as single fibers or fiber fragments, including ribbons and fibrils that were formed by unraveling the spiral fiber walls of individual fibers. This process is termed fibrillation. These thin, flexible elements considerably improve the bonding properties of the mechanical pulps. Although it is possible to make refiner mechanical pulp in a single stage, normally two or three refining units are used in series.

Process Conditions. Chips enter at a moisture content of about 50 wt % and water is added at the eye of the disks to give a consistency or solids content of 20–40 wt %. At this consistency in the refiner, considerable water evaporates as a result of the heat generated, and the steam exits with the pulp, thus assisting the centrifugal movement. Some steam also vents out of the inlet, which helps soften the incoming chips. Untreated chips, however, are brittle, and considerable fiber fracture and debris formation occur. After two stages of refining, the pulp is treated in a manner similar to that for stone groundwood: screening, cleaning, thickening, rejects refining, and sometimes bleaching (see Fig. 13). The main difference is that the attrition step uses two stages of refining rather than the rotating stone arrangement. Also, the chips are washed prior to refining. The second refining step is required to lower the pulp freeness, reduce shives, and improve the physical properties of the pulp.

Thermomechanical Pulping. A modification of RMP is thermomechanical pulping (TMP). If chips are presteamed to 110–150°C, they become malleable and do not fracture readily under the impact of the refiner bars. Thermoplasticization of the wood occurs when the wood is heated to a temperature approximating the glass-transition point of wet lignin. When these chips are fiberized in a refiner at high (30%) consistency, whole individual fibers are released. Separation occurs at the middle lamella, and the same ribbon-like material described for RMP is produced from the S_1 layer (see Fig. 6) of the cell wall. The amount of fibrillation depends on the refining conditions and is critical to the properties of the pulp. This material has a high light-scattering coefficient, although it is lower than that of groundwood, and is highly flexible, giving good bonding and surface smoothness to the paper. The increased proportion of long fibers improves the strength properties of TMP pulps, but the fibers in this fraction are stiff and contribute little to bonding. There is much less fiber fragmentation than in groundwood pulps or in those produced by RMP. Because processed chips sit in wood piles and are subsequently heated (steamed) to higher temperatures than groundwood, the brightness of these chips is lower and the absorption coefficient for visible light is higher than for groundwood pulps. Generally, the higher the temperature and the longer the residence time, the darker is the pulp produced. TMP pulps produced from the same wood species as groundwood often require bleaching, whereas the groundwood pulps usually have sufficient brightness to meet product demands, especially in Canada and the Northeast of United States.

The TMP process is similar to the RMP process except that after chip washing, a steaming vessel is inserted to achieve the thermoplasticization of the lignin in the wood. The production of thermomechanical pulps increased dramatically after the introduction of this method in the early 1970s. Because these pulps can be substituted for conventional groundwood pulps in newsprint blends to give a stronger paper, lower quantities of the more expensive, lower yield chemical pulps are required.

Energy Recovery. Energy usage is high in the TMP process, ca 12,757–15,591 MJ/t (90–110 hp·d for each oven-dried short ton (ODUST)), depending on species and final pulp freeness. This makes energy recovery a significant factor in the pulping processes. Energy may be recovered to produce hot water for mill use and clean steam for use in drying paper by operating refiners at up to 400–500 kPa (4–5 bars) pressure and using interstage cyclones to remove fiber solids from the raw steam generated in the refining stages. This raw steam is then used in conjunction with heat recovery equipment to generate clean steam in a reboiler or hot water in a steam condenser.

Linting and Latency Problems. Two special conditions of thermomechanical and refiner mechanical pulps must be controlled. The first of these is linting or dusting, which can be minimized by increasing refiner energy or by pulp cleaning to remove shives and produce pulp having a higher surface area called the higher fine fraction. The second condition is latency. Pulps that are refined at high consistency can become kinked and curled because the wood is heated above the lignin glass-transition temperature and the pulp is subsequently cooled before the fibers have time to relax. This condition, known as latency, when the fibers retain the deformed state, is relieved by heating or maintaining the pulp at temperatures above 60°C at low (3–4%) consistency and providing sufficient relaxation time in a latency chest.

Chemithermomechanical and Chemimechanical Pulping. The strength properties of thermomechanical pulps (TMP) can be increased by a mild pretreatment with sodium sulfite at pH 9–10 or by using other chemicals. Typically, the chips are impregnated with Na_2SO_3, steamed to 130–170°C, then refined. The yield for chemithermomechanical pulps (CTMP) is typically 90–92%, which is 2–3% lower than in TMP. Chemimechanical pulps result when the cooking is allowed to proceed to give diminished yields in the range of 85–92%.

A range of properties can be obtained by adjusting process variables that affect pulp yield and particle size. In general, CTMP have a greater long-fiber fraction and lower fines fraction than a comparable TMP. The intact fibers are more flexible than TMP fibers, giving better sheet-forming and bonding properties. The mild chemical treatment also allows the production of a long-fibered pulp having fast drainage, good absorption properties, and sufficient softness for tissue-grade pulp. Chemimechanical pulps have even a greater proportion of long fibers but less fines, and, because of lower lignin content, are more conformable in webs. Chemithermomechanical and chemimechanical pulpings are particularly suitable for pulping high density hardwoods.

In the late 1980s and early 1990s, production of high brightness (70–80 ISO) chemithermomechanical pulps began using the alkaline peroxide process (25,26). In this process, called bleached chemithermomechanical pulping (BCTMP), solutions of caustic and hydrogen peroxide are impregnated into hardwood chips in a special device called a pressafiner, a screw press having both a compression and an expansion zone that enhance chemical impregnation. The chips are compressed and allowed to reexpand like a sponge, filling the voids with H_2O_2 and sodium hydroxide, NaOH. The chips are then refined and processed in the normal fashion.

Chemical Pulps

Processing chemical pulps normally involves wood and liquor preparation steps, delignification (pulping), pulp washing, knot removal, screening, cleaning, thickening, and, in the case of bleachable grades, residual lignin removal and brightening to eliminate chromophoric or light absorbing groups. Often, more than 50% of the mass of the wood is dissolved in chemical pulping. This allows recovery of these solids to generate steam and electrical energy for use in the mill. Spent chemicals are recovered for reuse in the process.

In chemical pulps, sufficient amount of lignin is dissolved from the middle lamella to allow the fibers to separate, using little, if any, mechanical action. However, a portion of the cell wall lignin is retained in the fiber, and an attempt to remove this during digestion can result in excessive degradation of the pulp. For this reason, ca 2–3 wt % lignin is normally left in hardwood chemical pulps and 3–6 wt % in softwood chemical pulps. The lignin is subsequently removed by bleaching in separate processing if completely delignified pulps are to be produced.

The concentration of the cooking liquor in contact with the wood influences the rate of delignification. Because of the time required for diffusion of the chemical through the wood structure and the depletion of the reagent concentration as it penetrates the chip, delignification proceeds more slowly at the center of the chip. This is particularly apparent in the case of oversize chips. In order to prevent overcooking of the principal portion of the pulp, digestion is normally halted before the centers of these larger chips are adequately delignified. The resultant pulp thus contains a portion of nondefibered wood fragments, which are separated by screening (rejects) and returned to the digester if low kappa number pulp is to be produced, or are fiberized mechanically if high kappa pulp is desired. Knots, high in lignin and thus difficult to delignify, are removed in a special screen termed a knotter and returned to the digester or reprocessed in the refiner.

Kraft Process. The dominant chemical wood pulping process is the kraft or sulfate process. The alkaline pulping liquor or digesting solution contains about a three-to-one ratio of sodium hydroxide, NaOH, and sodium sulfide, Na_2S. The name kraft, which means strength in German, characterizes the stronger pulp produced when sodium sulfide is included in the pulping liquor, compared to the pulp obtained if sodium hydroxide alone is used, as in the original soda process. The alternative term, ie, the sulfate process, is derived from the use of sodium sulfate, Na_2SO_4, as a makeup chemical in the recovery process. Sodium sulfate is reduced to sodium sulfide in the recovery furnace by reaction with carbon in black liquor.

Chemistry of Delignification. The chemistry of delignification is complex and, despite the extensive literature, not completely understood. A variety of lignin model compounds have been studied and the results compared with the observed behavior of lignin during pulping (1,10–12,16).

   In the presence of alkali, the acidic phenolic units in lignin are ionized. At temperatures above 120°C, quinonemethides form from the phenolic units, as shown in equation 1, where Ar is an aryl group.

$$\text{(structure)} \xrightarrow{\text{OH}^-} \text{(structure)} + \text{OCH}_3^- \tag{1}$$

   In the presence of sulfide or sulfhydryl anions, the quinonemethide is attacked and a benzyl thiol formed. The β-aryl ether linkage to the next phenylpropane unit is broken down as a result of neighboring-group attack by the sulfur, eliminating the aryloxy group which becomes reactive phenolate ion (eq. 2). If sulfide is not present, a principal reaction is the formation of the stable aryl enol ether, ArCH=CHOAr. A smaller amount of this product also forms in the presence of sulfhydryl anion.

$$\text{(structure)} \longrightarrow \text{(structure)} + \text{ArO}^- \tag{2}$$

   Some cleavage takes place even if the phenolic hydroxyl is blocked as an ether link to another phenylpropane unit and quinonemethide formation is prevented. If the α- or γ-carbon hydroxyl is free, alkali-catalyzed neighboring-group attack can take place with epoxide formation and β-aryloxide elimination. In other reactions, blocked phenolic units are degraded if an α-carbonyl group is present.
   Under acid or alkaline catalysis, condensation reactions take place homolytically within the lignin polymer and very likely between lignin and carbohydrates or extraneous components. These reactions are undesirable in delignification, and prevention of such condensation accelerates pulping. Formaldehyde, generated in the formation of the arylenol ether (eq. 3), also causes condensation and cross-linking of phenylpropane units. The highly reactive quinonemethide intermediates are susceptible to attack by any carbanion. In kraft pulping, this type of condensation is minimized by the effective competition for the quinonemethide by the sulfhydryl anion.

$$\text{(3)}$$

Reactions of Polysaccharides.   Although the goal of chemical pulping is to degrade the lignin polymer selectively, reactions with the polysaccharides are unavoidable and not always undesirable. Arabinogalactan is soluble in hot water and is therefore rapidly lost in any chemical pulping operation. In the alkaline kraft process, there is little reaction of carbohydrates with the nucleophilic sulfide of sulfhydryl ions, as in the case of delignification. However, most of the alkali consumption results from reaction with carbohydrates and resinous materials (12). The acetyl groups on glucomannan in softwoods and xylan in hardwoods are easily saponified by alkali, and the uronic acid ester linkages to lignin are apparently cleaved at room temperature. Some hemicelluloses are extracted from wood by aqueous alkaline solutions. In addition, there are types of alkaline-degradative reactions of polysaccharides that occur at elevated temperatures.

The reducing-end units (see Fig. 8) are highly labile in alkaline solutions. After an initial attack by hydroxide ions at the hemiacetal function, C-1, a series of enolizations and rearrangements leads to deoxy acids, ie, saccharinic acids, and fragmentation. Substituents on one or more hydroxyl groups influence the direction, rate, and products of reaction.

Mechanism of the initial reaction, known as alkaline peeling, is shown in equation 4. Enolizations and tautomerizations take place easily because of the contiguous hydroxyl groups. The hydroxyl or substituted hydroxyl on the second, ie, β-carbon, from a carbonyl group is released from the molecule by β-elimination.

$$\text{(4)}$$

Glycosidic substituents are released more rapidly than are free hydroxyls. Thus, the remainder of the chain in a 1→3 or 1→4-linked polysaccharide is eliminated from the reducing-end unit undergoing this reaction. A new reducing end is created on the chain and the reactions may be repeated. This leads to a phenomenon known as peeling, in which polysaccharide chains are progressively shortened by the sequential reaction and loss of the reducing-end units. If there are no interfering structures on the chain, such as C-2 branching or substitution, peeling continues until the chain dissolves and is destroyed or until a hydroxyl group is eliminated in a competing (stopping) reaction and an alkali-stable saccharinic acid forms by a benzylic acid rearrangement (eq. 5).

$$
\begin{array}{ccc}
\underset{\underset{\underset{R}{|}}{\underset{CH_2}{|}}{\overset{\overset{O}{\parallel}}{\underset{C=O}{HC}}}
& \xrightarrow{\ OH^-\ } &
\underset{\underset{\underset{R}{|}}{\underset{CH_2}{|}}}{\underset{C=O}{\overset{HO}{\underset{H-C-O^-}{}}}}
& \longrightarrow &
\underset{\underset{\underset{R}{|}}{\underset{CH_2}{|}}}{\underset{HCOH}{\overset{HOOC}{}}}
\end{array}
\qquad (5)
$$

The 4-*O*-methylglucuronoarabinoxylans (see Fig. 9**a**) are substituted at C-2 in the xylan backbone by the uronic acid group. This inhibits the tautomeric formation of a keto group at that position and, therefore, inhibits peeling. If the xylan unit is not substituted at the C-2, but does carry an arabinose substituent at C-3, the arabinose is eliminated as a xylometasaccharinic acid end unit forms, and peeling is inhibited. Galactoglucomannan (see Fig. 9**b**), however, includes unsubstituted C-2 and C-3 positions and is highly susceptible to the alkaline peeling sequence. This is reflected in the much higher relative loss of galactoglucomannan during kraft pulping, as shown in Table 3. Cellulose is unsubstituted at C-2 and C-3, but its higher molecular weight, crystallinity, and inaccessibility minimize the extent of degradation.

Polysaccharides are also susceptible to alkali-catalyzed glycosidic hydrolysis. Although much slower than acid-catalyzed hydrolysis, the rate of alkaline cleavage becomes significant at 170°C, which is the commonly used temperature in kraft pulping. When a chain is cleaved, a new reducing end group is formed and the peeling units are degraded by peeling with each hydrolytic cleavage (27).

Chemical Penetration and Topochemistry. Chemical penetration and topochemical processes are important in all chemical and semichemical pulping operations because of the solid nature of the wood components and the liquid or gaseous state of the pulping reagents. The initial liquid transport through wood chips is through the bordered pits of softwoods and the vessels of hardwoods. Liquid penetration of sapwood is higher than that of heartwood, especially in the case of some hardwood species where the heartwood vessels are plugged with tyloses. Penetration is also affected by the specific gravity of the wood. In general, penetration into softwoods is faster than into hardwoods, especially in the cross-fiber direction. For the strongly alkaline cooking liquors of the kraft process, diffusion takes place at nearly equal rates in all directions.

Table 3. Yields^a of Carbohydrate Components from Loblolly Pine

		Process	
Component, wt %	Wood	Chlorite holocellulose	Kraft
4-*O*-methylglucuronoarabinoxylan	11	5.8	3.9
galactoglucomannan	17	15.4	4.4
cellulose	41	40.0	36.8

^{a}Wt % based on original wood.

To make pulping as homogeneous as possible, an effort is made to achieve uniform penetration before pulping temperatures are reached. This may be done by presteaming chips at or above atmospheric pressure, applying hydrostatic pressure to chips submerged in pulping liquors, and allowing sufficient time during heating to process temperature.

When the chips are saturated with pulping liquors, the fiber lumens are filled with liquid. However, for fiber separation, the lignin in the intercellular middle lamella must be dissolved. Pulping is accomplished by transport of reagents through the submicroscopic pores of the fiber wall, contact and reaction with lignin in the cell wall and middle lamella, and transport of dissolved fragments back to the bulk solution in the lumen and that surrounding the chip. In the digester, the chips and liquor are kept in constant motion relative to one another to facilitate this transport. The exact topochemical sequence depends on the process (28). In kraft pulping, very little middle lamella lignin is lost until about half of the cell wall lignin has been removed. Then, middle lamella delignification becomes rapid. At the end of the cooking phase, the residual lignin is almost entirely in the cell wall. In the sulfite processes, particularly those involving neutral sulfite, the dissolution of lignin begins in the cell wall, but middle lamella lignin removal begins earlier than in the kraft process. There is much less difference in the rate of delignification in the two regions.

Pulping. Solutions of sodium sulfide and sodium hydroxide are in equilibrium:

$$H_2O + Na_2S \rightleftharpoons NaHS + NaOH \qquad \text{or} \qquad H_2O + S^{2-} \longrightarrow HS^- + OH^-$$

Aqueous sodium sulfide is therefore itself a source of hydroxide ions. This must be considered in adjusting the chemical charge. A system has been developed in the North American industry to put sodium hydroxide and sodium sulfide on an equivalent alkali basis by expressing both as their equivalent weight of sodium oxide, Na_2O. The sum of Na_2S and $NaOH$ calculated as their Na_2O equivalent is called the active alkali ($NaOH + Na_2S$). The percentage of sodium sulfide in the mixture, when both Na_2S and $NaOH$ are expressed as Na_2O, is known as the sulfidity, $Na_2S/(NaOH + Na_2S)$. The chemicals are charged as a percentage of the wood on an oven-dry (OD) basis in terms of active alkali. The actual concentration of pulping chemicals relative to lignin is determined by the liquor-to-wood ratio (L/W) as the total liquid weight, which includes the moisture originally present in the chips, divided by the dry wood weight. Because one-half a formula weight of Na_2S is equivalent to a formula weight of $NaOH$, the effective alkali is the sum of sodium hydroxide and one-half the sodium sulfide, $NaOH + 1/2\ Na_2S$. Both are expressed as the percentage of sodium oxide equivalent based on the wood. In practice, in recycled chemicals, called white liquor, about 15% of the sodium is present as sodium carbonate from incomplete causticizing. Other sodium sulfur compounds are present in small amounts as thiosulfate, sulfate, and sulfite.

Chemical charge, liquor composition, time of heatup, and time at temperature of reaction are all functions of the wood species or species mix being digested and the intended use of the pulp. A typical set of conditions for southern pine chips in the production of bleachable-grade pulp for fine papers is active alkali, 18%; sulfidity, 25%; and liquor-to-wood ratio, 4:1. Time of heatup is 90 min to 170°C; time at temperature of reaction, 90 min at 170°C. Hardwoods require less vigorous conditions primarily because of the lower initial lignin content.

Both batch and continuous digesting systems are in operation. In the 1970s and early 1980s continuous digesters usually were installed in new mills for savings in labor and steam. However, continuous digesters did not completely replace the batch systems, partly because of the high capital cost of replacement, but also because the batch digestion method has advantages that include flexibility to process and product changes, lower maintenance costs, and higher turpentine yields. The advantages of continuous digesters include uninterrupted process flow, higher pulp yields and heat recovery, relative ease of automation, and in-line processing, eg, partial washing and blow-line refining.

Batch Cooking. Traditionally, the primary disadvantage of conventional batch cooking has been relatively higher steam consumption, high labor, and air pollution (qv) effect owing to blowing of the whole cooked batch from high pressure and temperature. During the late 1980s and early 1990s, there was a resurgence in batch cooking because of advances in materials of construction, control technology, and energy conservation through liquor displacement. Notable advances include stainless steel cladding of digester vessels, distributive control, steam packing, cold blowing, and blow heat recovery. These advancements have reduced corrosion as well as labor and steam requirements, made possible the use of larger digesters, and improved pulp quality and uniformity. Steam (qv) requirements in the 1990s are equivalent for batch and continuous pulping systems.

The capacity of batch digester loads can be increased up to 40% by using steam packing, which raises production efficiency and reduces labor costs (29). Cold blowing or low energy cooking (Fig. 18) can lead to steam savings of 40–50% relative to conventional hot blowing systems, but as of this writing (ca 1996) this process is not widely practiced. Pollution control is made easier by eliminating flashing of steam between cooks. When the cook is complete, wash filtrate from the washers is pumped into the bottom of the digester, displacing the hot black liquor which is pumped from the top into a hot black-liquor accumulator. About 75% of the black liquor is displaced before blowing. The blow valve is then opened and the digester emptied. Because the temperature has been decreased before blowing, the internal pressure is low and some external agent such as air is applied to the digester to remove the pulp. Hot black liquor from the accumulator tank is used to preheat white liquor before it is pumped to the evaporators. After blowing, the digester is refilled with chips, packed with a steam packer, and presteamed to remove air and improve impregnation. Hot black liquor from the accumulator and preheated white liquor is subsequently charged to the digester to satisfy the active (effective) alkali and liquor to wood requirements of the cook. The combined liquors and wood in the digester have a starting temperature of 135–145°C. Thus the rise to cooking temperature is reduced approximately 50%, requiring much less steam for heating the cooking liquor.

By transferring most of the black liquor to the accumulator, the odorous gases in the liquor are fed to the evaporator instead of being flashed off with the steam in conventional hot blowing. Pulp viscosity and strength are improved

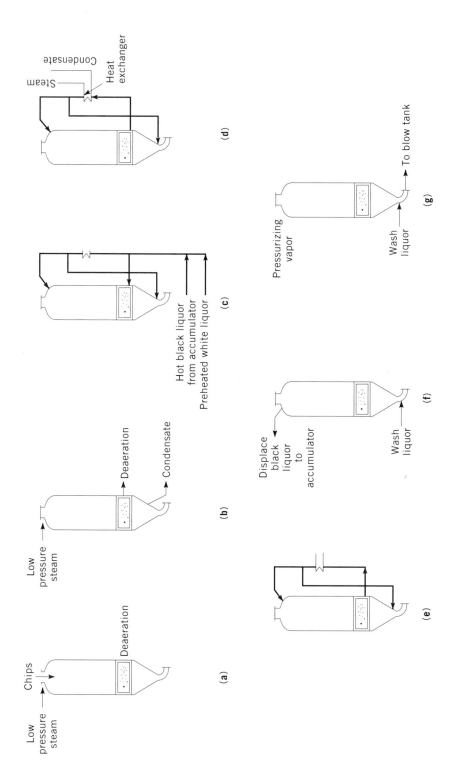

Fig. 18. Low energy cooking sequence: (**a**) chip filling with steam packing, (**b**) steaming, (**c**) charging of cooking liquor, (**d**) heating to cooking temperature, (**e**) cooking period at 170°C, (**f**) filling the digester with wash liquor, and (**g**) cold blowing of digester (29).

530

using a cold blow technique because the forces acting on the pulp fibers are less violent compared to hot blowing. The widespread use of distributive computer control has permitted low energy batch cooking to become competitive with continuous cooking.

Continuous Cooking. A schematic of the widely used Kamyr continuous kraft pulping system is presented in Figure 19, showing a single-vessel hydraulic digester having two stages of diffusion washing and high density brown-stock pulp storage. The chips are continuously steamed first in a chip bin at atmospheric pressure and then at low pressure in a steaming vessel where turpentine and gases are vented to the condenser. The chips are then brought to a digester pressure of ca 1000 kPa (150 psi) via a high pressure feeder and picked up in a stream of recycled liquor to which white liquor, ie, fresh pulping solution, makeup has been added. This stream carries through to the top of the digester, where the recycling feed liquor is extracted from the chips. The chips, having the balance of the liquor, flow continuously down through the digester, first passing through an impregnation zone (Fig. 19, section I) where liquor is impregnated into the chips for a period of 0.5–1.0 h. The temperature is then raised to ca 170°C in a top heating zone (section II) by recirculating the liquor. This step requires ca 1.5 h. Subsequently, the chips are held at about 170°C for 1.5–2.5 h while passing through a cooking zone (section III). At this point, the digestion is essentially complete. The chips next pass continuously through a high heat, countercurrent washing zone (section IV) for a duration that varies from 1 to 4 h. The black liquor or spent pulping solution leaves the digester from the top of the cooking zone (section III) through screens, whereas the digested and partially washed chips leave the bottom of the digester through a blow unit or outlet device. The force of the release is sufficient to fiberize the cooked chips. Cooling the chips prior to fiberization is necessary to ensure adequate pulp strength properties. The pulp can go directly to a blow tank or, more likely, to two stages of pulp washing in a diffusion washer as shown in Figure 19. From the diffusion washer the pulp travels to high density intermediate brown-stock storage. Numerous variants are possible for the basic pulping system shown in Figure 19. A separate chemical impregnation vessel is sometimes added (two-vessel hydraulic system) for increased residence time and better impregnation of chemicals. Sometimes the upper heat exchangers are eliminated and the bulk of the heating of the chips is done by direct steam injection (vapor/liquid-phase digester). Alternatively, vapor/liquid cooking can be accomplished in a two-vessel system in which the second vessel is added for improved impregnation.

Modified Continuous Cooking. Theoretical studies have demonstrated that improved kraft pulping can be obtained by several process modifications: (*1*) low initial alkali concentrations; (*2*) high sulfidity at the outset of the bulk phase of delignification; (*3*) lowered temperature in the cooking zone; and (*4*) reduced concentration of dissolved lignin and sodium salts in the later part of the cook (30). Modifications to the continuous digester were developed to produce pulp under these specified theoretical conditions. Design and the liquor profiles associated with these modifications, termed modified continuous cooking (MCC), are shown schematically in Figure 20.

The white-liquor addition is split among the feed, the transfer circulation, and a new countercurrent cooking zone. This split addition provides a more

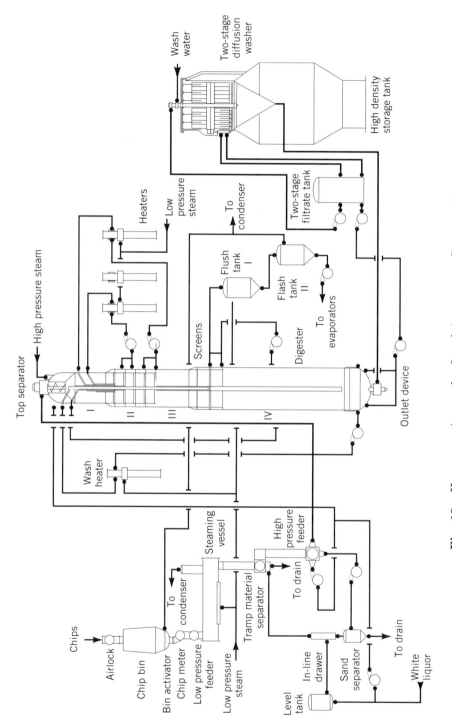

Fig. 19. Kamyr continuous kraft pulping system. See text (29).

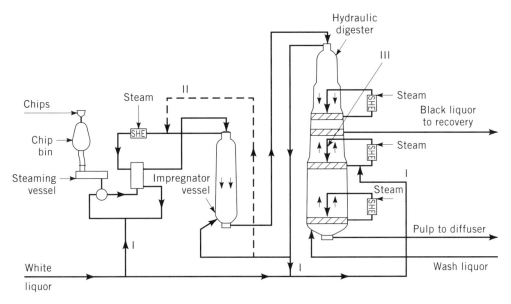

Fig. 20. Modified continuous cooking, where section I shows white liquor additions; II, recirculation; and III, countercurrent zone. SHE = steam heat exchanger (29,31).

uniform alkali concentration over the entire cook (Fig. 21). The countercurrent zone produces a low concentration of lignin and sodium salts in the later part of the cook, thus minimizing lignin condensation (see eq. 3). These modifications provide a higher viscosity or higher intrinsic fiber strength and easier pulp bleaching compared to pulp produced by the conventional continuous process.

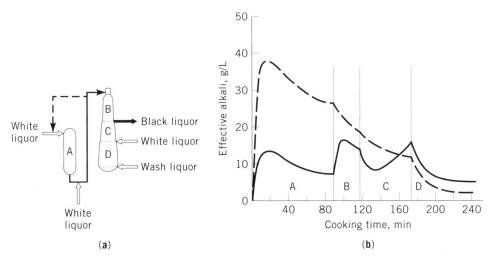

Fig. 21. Concentration of effective alkali in chips during digestion, where A represents the impregnation vessel; B, heating zone; C, cooking zone; and D, washing zone: (**a**) digestion schematic defining digester zones; (**b**) effective alkali in chips vs cooking time for (——) MCC and (— — —) Kamyr unmodified process (29,31).

For MCC cooking, the kappa number leaving the digester is about 25, compared to 30 for conventional cooking of softwood. Most Kamyr pulping systems since 1986 either have this feature incorporated in the design or can allow for conversion.

Isothermal Cooking.   Isothermal cooking (ITC) was introduced in the 1990s as a further modification of the MCC process. In isothermal cooking, the temperature is kept constant throughout the digester, including the high heat countercurrent washing zone, which occurs from the end of the heating zone to the exit of the digester. For ITC cooking, the temperature can be about 10°C lower than that for the MCC process, ie, about 160°C. This reduces the total steam consumption for the digester and also permits the kappa number exiting the digester to be reduced from 25 to about 17 for softwood and still maintain high fiber strength. Cooking to lower kappa number can decrease the yield by 0.1–2.0%.

Delignification Kinetics.   The rate of delignification depends on the concentrations of hydroxide and hydrosulfide ions as well as the lignin content of the wood. However, if the values for the residual lignin remaining in the wood, ie, wt % of original wood, are plotted against time on a semilog scale and assuming that the delignification follows first-order kinetics, the data are described approximately by three straight lines, as shown in Figure 22. During the first and most rapid rate, ie, initial delignification, there is a rapid consumption of alkali by easily soluble carbohydrates and reactive functional groups, eg, esters. The initially removed lignin is probably associated with the solubilized fraction of hemicellulose. In the initial phase, large losses of hemicelluloses occur and the lignin content of the partially cooked wood actually increases. In commer-

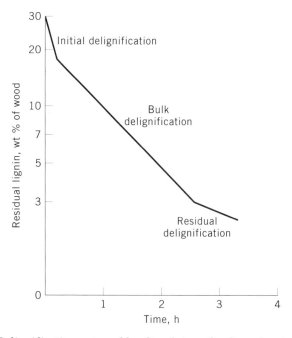

Fig. 22.   Delignification rates of kraft pulping of softwoods at 160°C (32).

cial practice, this initial phase is usually completed during the extended heatup period. The second and slower rate is the bulk delignification, in which the decrease in effective alkali is much slower. Most of the lignin is removed during this time at the final pulping temperature. The amount of lignin removed relative to carbohydrates is greater in this second phase than in initial delignification. Some readsorption of de-esterified polysaccharides does occur. The lignin that remains, ie, residual lignin, is more difficult to remove than either initial or bulk-phase lignin. Pulping is usually terminated as this stage is approached. If desired, the residual lignin is removed in a bleaching operation where more selective, but more expensive, chemicals are employed.

Digester Control. For control purposes at constant sulfidity and alkali charge, the delignification rate is treated as a homogeneous reaction, which is first order with respect to the lignin, L, remaining in the wood, $-dL/dt = kL$. The influence of time, t, and temperature, T (Kelvin), has been incorporated into one term, called the H-factor (33).

The H-factor method utilizes an Arrhenius-type equation, $k = A\exp(-E_a/RT)$, for relating the change in the first-order rate constant, k, to changes in absolute temperature, T. The constant k is assumed to have a value of unity at 100°C, permitting evaluation of the frequency factor, A. Thus the first-order rate constant, k, becomes a relative rate constant, k_r. It is further assumed that the activation energy, E_a, for the delignification reactions has a value of 134 kJ/mol (32 kcal/mol) expressing the dependence on temperature. The relative rate constant can be expressed mathematically as

$$\ln k_r = (43.2 - 16{,}113/T)$$

and relative rates at other temperatures can then be calculated. The H-factor is estimated by evaluating the integral expression,

$$H = \int_0^{t_{\max}} k_r\, dt$$

where $t_{\max}$ is the total cooking time. Allowance can be made for the effect of varying heatup times and maximum temperatures on the cooking process. For example, the kraft cooking of softwoods to a bleachable-grade pulp requires an H-factor of ca 2000 hours. The H-factor does not take into account the effect of chemical concentrations or diffusion effects on pulping rates, and numerous modifications that include additional factors have been proposed (34). The H-factor and modifications are designed to be programmed into computer control systems (see PROCESS CONTROL).

Process Sensors. Developing analytical methods suitable for automation and, preferably, continuous monitoring has been a significant factor since the 1980s. The main control problem in pulping is the variability of the raw material, including the moisture content of the wood, the species mix, and the amount of bark, decay, knots, dirt, and other extraneous material. Continuous monitoring of moisture, weight, and density as well as on-line determination of the effective alkali can be accomplished. This information can be combined with computed

H-factors to give reproducible, automated control of both batch and continuous digesters. Modern control schemes typically monitor time and temperature for estimation of the *H*-factor. Alkali concentration is estimated from alkali analyzers, eg, electromechanical titrators, and conductivity sensors at various levels of time in a batch digester or position in a continuous digester. These measurements coupled with process control (qv) algorithms permit predictions for the lignin concentration and yield from the continuous digester and also the time for blowing batch digesters.

Residual Lignin Sensors. Determination of the lignin remaining in the wood chips or pulp after the cooking operation traditionally required sampling (qv) and analysis of large heterogeneous particles. This was done by sampling, washing, and direct determination of the lignin content by gravimetric analysis or the determination of reducing power of the pulp using a strong oxidizing agent such as potassium permanganate, $KMnO_4$, in empirical tests. The permanganate and kappa number tests give the volume in mL of 0.1 *N* $KMnO_4$ capable of reacting with one gram of pulp under standard test conditions. Empirical relationships exist for relating the kappa number to the Klason lignin content of the pulp:

$$\% \text{ lignin} = 0.147 \times \text{kappa number}$$

Technology has been introduced for on-line estimation of the kappa number based on absorption of ultraviolet light (35). This breakthrough in optical sensor technology permits closed-loop feedback control of digesters from on-line measurement of the kappa number.

Pulp Processing. Following pulping, the acceptable pulp fibers must be separated from a variety of contaminants. Uncooked knots and fiber bundles (shives) are removed by screening, pulping liquor is removed by washing, and denser particles separated by centrifugal cleaning. Different grades of pulp, eg, higher yield, medium yield, and bleachable grades, require different processing.

Washing. Washing removes pulping liquor either by dilution and displacing the dirty liquor with cleaner wash water, termed displacement washing, or by diffusion washing. The latter involves contacting the dirty pulp with relatively clean water for extended periods of time to permit diffusion and desorption of pulping chemicals and organic material from the pores and cell wall of the pulp. Displacement washing usually involves pressing or vacuum filtration (qv) to assist in removing the dirty water from the pulp being washed. Multiple (typically three) stages in a countercurrent arrangement provide the highest washing efficiency and the lowest shower water dilution. The water must be removed in the evaporator section of the recovery process. Operating variables include the pulp itself, consistency, solids loading, shower volume and temperature, air entrainment, and foam.

Washing equipment typically includes cylinder or drum washers, diffusion ring washers, pressure diffusers, pressure washers, and belt washers. Figure 23 illustrates a countercurrent washing system. Either pressure or rotary vacuum filters may be used. The pulp in the blow tank has a consistency between 9 and 11%, depending on whether it came from a continuous or batch digester,

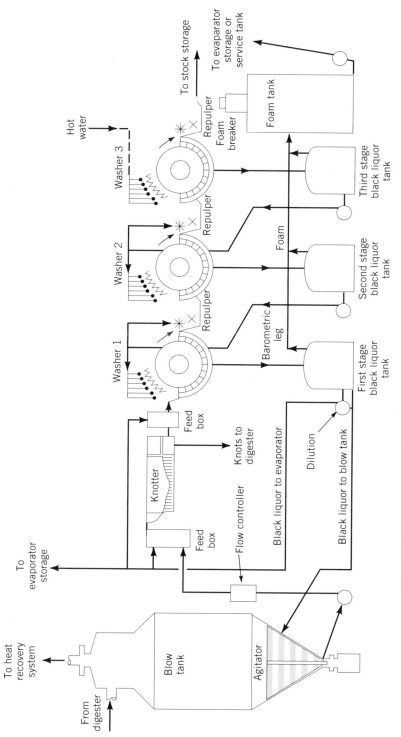

Fig. 23. Flow sequence for countercurrent rotary brown-stock washers (29).

537

respectively. The pulp is diluted in steps to about 1% before entering the first washer vat. It is extracted to about 10% by thickening. Washing liquid from the showers is displaced through the pulp sheet. The pulp is reextracted and then discharged to the next stage. Pulp consistency on leaving any filter stage varies between 9 and 18%, depending on the equipment used. This procedure is repeated for each step or stage in the countercurrent system. Fresh or process water is applied in the last, ie, the third, stage. The thick recovered solids are sent to the evaporator off the first-stage black liquor tank. Excess water in any stage removed from the pulp is recycled to the showers of the preceding stage. The dilution factor, typically 2–4:1 water to pulp on a weight basis, is the net water that is added to the washing system. The higher the dilution factor is, the cleaner the pulp and the lower the chemical (soda) losses, but also the greater the amount of water to be evaporated from the residual liquor in the chemical recovery system. Figure 19 shows the diffusion ring washing used in the Kamyr system. Therein fresh wash water is fed to the second stage of washing where it contacts the pulp for an extended period (1–2 h) and then is sent to the top filtrate tank. From there, the wash water goes to the first stage of diffusion washing for another 1–2 h of contact, back to the filtrate tank, and eventually to the bottom washing zone of the digester via a filtrate tank.

Screening. Shives and other oversize material are removed from good pulp by screening. This makes the pulp more suitable for the intended paper or board products. The biggest oversized particles in the pulp are knots, ie, uncooked wood particles, retained on a screen having 9-mm holes. Knots are removed before washing and fine screening. In high yield pulps, the knots are broken down in refiners and/or fiberizers. In low yield pulps, knots are removed in special coarse screens called knotters.

The main purpose of fine screening is to remove shives that have a length of 1–3 and a width of 0.10–0.15 mm and that are stiffer than fibers. Pressure screens are the most common type used for fine screening. Screen efficiency depends greatly on consistency and rejects rate. Many screening strategies are used but the most common arrangement raises the primary screen rejects level to maximize efficiency and production. The design of the screening basket, ie, hole or slot size, and rotor speed are key variables that determine screening efficiency and thus pulp cleanliness. A three-stage cascade screening is common and involves recycling the accepts to the inlet of the previous stage. Rejects become the feed to the next stage.

Cleaning. The objective of centrifugal cleaning is to remove small unwanted particles still remaining in the pulp following screening. Centrifugal cleaners remove rejects that have different densities from the pulp fibers. The undesirable material consists of sand, grit, bark, small shives, fly ash, rust, scales, and all other constituents that can lower the appearance of the pulp. Normally, only high quality bleachable-grade pulp stocks are cleaned. For lower grades such as linerboard, where appearance is not so important, centrifugal cleaning is usually not applied. Consistency, rejects level, pressure drop, and system design affect performance. Three-stage cascade cleaning systems are common.

Thickening and Storage. After screening and cleaning, the pulp has a low consistency, often between 0.5 and 1%. The pulp slurry must be thickened by removing water prior to further processing. Pulp storage is generally termed low or high density. Low density corresponds to a pulp consistency range of 2.5–4 wt %. At these consistencies, the pulp can be pumped by centrifugal pumps and agitated. High density storage corresponds to a consistency of 10–14 wt %. This latter pulp cannot be blended by agitation, but can be pumped by displacement pumps or by special centrifugal pumps termed medium consistency (MC) pumps. From unbleached storage, the pulp is sent to the paper mill for brown grades or to the bleach plant for bleached grades. There is a variety of equipment used for thickening, depending on the level of consistency required. Gravity thickeners usually can achieve discharge consistencies of 4–8%; vacuum filters, 12–15%; and presses, greater than 20%.

Recovery System. Kraft pulping depends on its associated recovery process for producing the digestion liquor. There are four important functions in kraft recovery: (*1*) the organic substances, ie, lignin and carbohydrate fragment molecules, dissolved during pulping are destroyed, thus eliminating a major potential source of stream pollution; (*2*) large quantities of steam are generated by burning the dissolved organic solids, thus supplying a large portion of the steam used in the pulp mill; (*3*) sodium hydroxide is regenerated for reuse in the pulping process; and (*4*) sulfur compounds present in the black liquor are converted to sodium sulfide, which is reused in the pulping process.

The kraft recovery process is shown schematically in Figure 24. The principal steps in the kraft recovery process include obtaining concentrations of weak black liquor or spent liquor from the pulp washers by evaporation (qv), burning the concentrated liquor in a recovery furnace, dissolving smelt from the furnace to form green liquor, causticizing the green liquor with lime to form white liquor, and burning lime mud to recover lime. The recovery system, together with its heat and material balance, has been described in detail (29). The physical and chemical change in the unit operations are summarized in Figure 25 (36). There are three chemical cycles involving sodium, sulfur, and calcium ions.

Much of the equipment used in the recovery system is identical with or closely related to equipment used in other chemical industries. This includes multiple-effect evaporators, and forced-circulation concentrators, causticizing equipment, and lime kiln. The function and nature of equipment essentially unique to the kraft recovery system are discussed herein.

Black Liquor Processing. After washing, the black liquor is subjected to multiple-effect evaporation. Older mills used direct contact of the concentrated liquor with the recovery-furnace flue gases. This was done as the last stage in the evaporator sequence and was coupled with black liquor oxidation. This technology is rapidly being replaced. Under conditions of direct-contact evaporation, residual sodium hydrosulfide partially hydrolyzes to hydrogen sulfide and sodium hydroxide, whereas sodium methanethiolate, CH_3SNa, partially hydrolyzes to methyl mercaptan and sodium hydroxide. In order to prevent the escape of the volatile hydrogen sulfide and methyl mercaptan with the flue gas, the black liquor is oxidized using air or oxygen, giving sodium thiosulfate (eq. 6) and dimethyl disulfide (eq. 7).

$$2\ NaHS + 2\ O_2 \longrightarrow Na_2S_2O_3 + H_2O \tag{6}$$

$$2\ CH_3SH + 1/2\ O_2 \longrightarrow CH_3SSCH_3 + H_2O \tag{7}$$

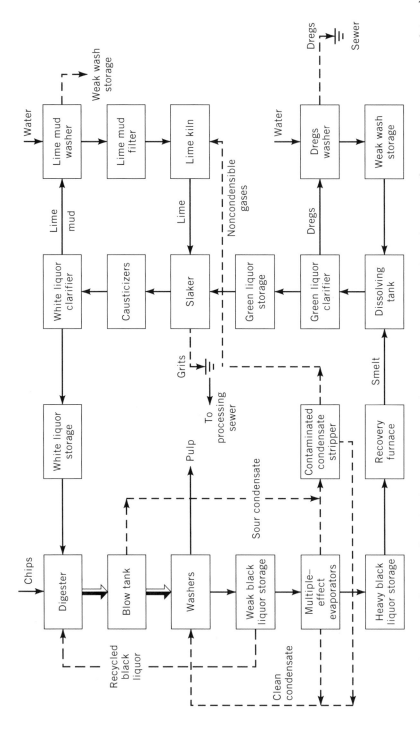

Fig. 24. Kraft process design, where (——) represents a primary process stream; (-----), a secondary process stream; and (⟹), black liquor and pulp (36).

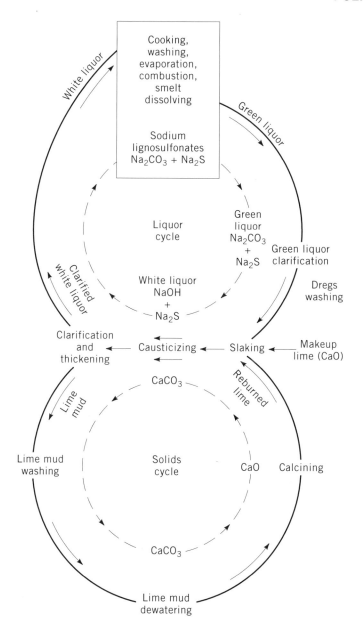

Fig. 25. Causticizing process cycle (36).

Black liquor oxidation is normally carried out by passing a stream of liquor and a stream of air or oxygen to a tower fitted with perforated plates or with suitable packing to allow repeated, intimate contact of liquor and gas. Because of the presence of fatty acid sodium salts, ie, soaps, foaming may occur, particularly when unseasoned pine is used. A number of designs have been used to rectify this problem (see also DEFOAMERS). Less foaming is obtained at a higher solids concentration. This can be achieved by recycling a portion of the evaporated liquor through the oxidation tower. Black liquor oxidation must achieve 99% efficiency to meet emission standards where direct contact evaporation is employed. This is difficult to accomplish. The pH of the black liquor must also be kept high to minimize the hydrolysis to volatile compounds.

As the solids concentration of black liquor is increased as a result of evaporation, the liquor becomes increasingly viscous. For this reason, the evaporation is conducted in two separate operations. Multiple-effect evaporation is used to bring the liquor to a concentration of ca 45–50 wt % solids; forced-circulation evaporation, or additional effects, is used to bring the liquor to ca 65–70 wt % solids, ie, the concentration at which the liquor is fired in the recovery furnace (see EVAPORATION). In addition to problems of high viscosity which inhibits pumping, black liquor evaporators are sometimes troubled with scaling formation in which insoluble deposits coat the heat-transfer surfaces, thus minimizing throughput.

Evaporation of the black liquor by direct contact with the hot recovery-furnace flue gases was a common procedure and involved either cascade evaporators or venturi scrubbers. However, this method proved to be a principal source of air pollution and, although various schemes were proposed to minimize the emissions problems, the use of direct-contact evaporators for concentrating black liquor have generally been eliminated in favor of the use of a concentrator (see AIR POLLUTION CONTROL METHODS). Concentrators, sometimes referred to as finishers, are steam-heated evaporators designed to produce liquor suitable for firing directly. Contact between the black liquor and the flue gas is avoided and concentrators, therefore, require fresh stream. Many different types of concentrators have been used since their introduction in the late 1960s. Long-tube vertical evaporators (LTV) were originally used, but these are being replaced by plate-type falling-film evaporators and crystallizer–evaporators in newer installations.

Recovery Furnace. The function of the recovery furnace (Fig. 26) is to burn the organic matter, recover chemicals in the form of sodium carbonate and sodium sulfide, and generate steam through utilization of the heat liberated. The steam is generated at ca 1360 kPa (200 psi) for direct use as process steam or as superheated steam at up to 8160 kPa (1200 psi) for driving a turbine for electric power generation. The turbine exhaust steam is used as process steam.

In the recovery furnace and after addition of makeup sodium sulfate, the evaporated liquor is heated to ca 120°C and is fired through one or more special oscillating nozzles or spray oscillators in the front of the furnace. These spray oscillators throw the droplets to the far and side walls. The remaining water quickly evaporates. The dried liquor increases to a certain depth on the wall and then drops off from its own weight in chunks and falls toward the hearth, which is covered with char. Primary air is admitted through a series of ports. Thus the depth of the charred mass is controlled by burning. Reducing conditions are maintained in this zone so that the sodium- and sulfur-containing compounds are largely reduced to sodium sulfide (eq. 8).

$$Na_2SO_4 + 2\ C \longrightarrow Na_2S + 2\ CO_2 \tag{8}$$

The inorganic residue melts at ca 990–1020°C, depending on composition, and runs continuously from the furnace through smelt spouts at the lower end

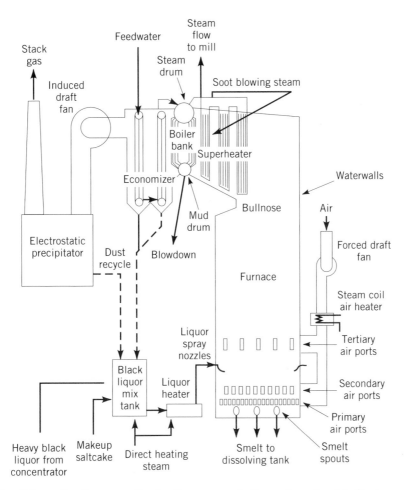

Fig. 26. Schematic diagram of a kraft recovery boiler, where (——) indicates primary and (———) secondary process streams (29).

of a sloping hearth. Secondary air is admitted at a higher level to limit the height of the bed and also to effect gas-phase reactions, eg, conversion of CO to CO_2 and H_2S to SO_2. Tertiary air is added at an even higher point and at relatively high velocity to ensure mixing of the gases and to complete oxidation. Whereas a small excess of oxygen is required to ensure complete combustion of the gases, a large excess is avoided because it favors conversion of SO_2 to SO_3. Furnace temperature is ca 1250°C. Approximately 10% of the sodium salts is volatilized in the form of Na_2O, which quickly reacts with CO_2 to form Na_2CO_3. The latter reacts in the upper portion of the furnace with SO_3 and SO_2 to form Na_2SO_3 and Na_2SO_4, respectively. The walls of the furnace are water-cooled by an arrangement of fin tubes connected to the water circulation of the boiler. The unit is designed so that the combustion gases are cooled to a temperature below the melting point of the inorganic constituents during the upward passage of the gases through the furnace. When the salts condense on the closely spaced boiler tubes at the top and back of the furnace, they are present as a dust-like solid that can be blown off, rather than as a slag that can block the gas passage. Automatically timed soot blowers, which impinge jets of steam or compressed air on the tubes, traverse the boiler banks, thereby cleaning them. The released ash falls either toward the hearth or to a special hopper, where it is collected and added to the black liquor for recycle to the furnace.

Proper control of the reactions in the furnace is important. Reducing conditions are necessary at the base of the furnace if the smelt is to contain sodium sulfide rather than the inert sodium sulfate. Slightly oxidizing conditions are needed above the bed, which causes the conversion of volatile sulfur compounds to sulfur dioxide. The latter compound not only reacts with the fume but is absorbed in the direct-contact evaporator, if one is used. Strongly oxidizing conditions in the upper portion of the furnace are undesirable because sulfur trioxide can form in large amounts. Sulfur trioxide reacts with the sodium sulfate fume to form sodium pyrosulfate, which melts at 400°C and forms a glass-like, corrosive covering on the boiler tubes.

$$Na_2SO_4 + SO_3 \longrightarrow Na_2S_2O_7 \tag{9}$$

Accidental introduction of water into the smelt can result in serious explosions. Furnace explosions occur by mistake when weak rather than strong black liquor is fired in the furnace and reaches the pool of smelt on the hearth. Explosions also occur when one or more of the tubes in the wall rupture, allowing water to flow into the unit. This can occur if the furnace is relighted after a blackout results from an interruption of fuel or air and if the furnace has not been purged of the explosive gas mixture that may be present. If the primary explosion ruptures a tube, a large secondary steam explosion results.

Preparation of White Liquor. The smelt is continuously run from the furnace to the dissolving tank where the liquor is strongly agitated to dissipate the heat. The temperature in the tank is readily controlled by dispersion of the hot smelt with a steam jet before the smelt is added to the water. This prevents the explosive mixture that results from addition of water to a pool of hot smelt. Weak white liquor obtained from washing the lime mud in the causticizing plant is used for dissolving the smelt.

The green liquor, ie, the solution obtained on dissolving the smelt, contains an insoluble residue called dregs, which give a dark green appearance. Dregs contain a small amount of carbon plus a number of inorganic constituents, including iron sulfide, manganese dioxide, calcium carbonate, and magnesium aluminum silicate. These constituents originate in the wood and the process water or result from corrosion of equipment. The liquor is separated from the dregs by continuous decantation. The separated dregs are mixed with water and the decanted solution is used for washing the lime mud at a subsequent stage.

The clarified green liquor is used for slaking the lime. For one to two hours, the mixture is passed through a series of tanks equipped with agitators. The causticizing reaction is as follows:

$$CO_3^{2-} \text{ (aq)} + Ca(OH)_2 \text{ (s)} \rightleftharpoons 2\,OH^- \text{(aq)} + CaCO_3 \text{ (s)} \tag{10}$$

The white liquor is separated from the calcium carbonate by decantation in a clarifier and is then available for a new cooking cycle. The underflow from the clarifier, which contains the calcium carbonate and is referred to as lime mud, is diluted with water and passed to a second clarifier known as the lime mud washer. The clarified weak white liquor (weak wash) goes to storage and then enters the dissolving tank. The lime mud residue from the lime mud washer is passed to a rotary filter and subsequently to the lime kiln where calcium carbonate is converted back to calcium oxide, thus completing the lime cycle.

$$CaCO_3 \xrightarrow{\triangle} CaO + CO_2 \tag{11}$$

Modifications to the Recovery Cycle.   The recovery system is a principal capital cost in a kraft mill. Consequently, any recovery process that is less expensive to build can improve pulping economics. There have been numerous attempts to improve the kraft recovery process. Two examples are the direct alkaline recovery scheme (DARS) and the autocausticizing scheme using sodium borates (37). Both schemes eliminate the lime loop of the conventional kraft mill. As of 1996, neither is commercially used.

Uses.   The kraft process provides wide adaptability to the pulping of virtually any wood species and yields pulps that are suitable for a broad range of products. Packaging products have stringent strength requirements, and the use of kraft pulps in linerboard has enabled the replacement of wooden cases by corrugated cartons. Similarly, bags and multiwalled sacks of kraft pulp have replaced those made of cotton or jute. After partial or complete bleaching, kraft pulps are often blended with less expensive but weaker pulps, eg, groundwood, to add enough strength to meet product requirements. For example, in newsprint, kraft pulps must have sufficient strength to be run on modern, high speed printing presses. The brown color of kraft pulps is undesirable for applications such as writing and printing papers, but kraft pulps can be bleached to remove the residual lignin and other chromophores. The bleached white product does not darken on aging, although a small amount of brightness reversion takes place.

By-Products.   There are three stages within the pulping operation at which wood-derived chemicals can be recovered as by-products. Turpentine is obtained from the relief of gases after an initial steaming of chips in the digester. Better yields of turpentine are obtained from batch digesters than from continuous systems. Pines and firs give the best yields. Turpentine is composed principally of unsaturated bicyclic hydrocarbons, of which ca 90% are α- and β-pinenes and 5–12% other terpenes.

In the initial black liquor concentration, saponified fatty and resin acid salts separate as tall oil soaps (see TALL OIL). These soaps can be skimmed from the aqueous spent liquor, acidified, and refined to give a crude tall oil composed of resin acids, chiefly abietic and neoabietic; fatty acids, chiefly oleic and linoleic; and an unsaponifiable fraction made of phytosterols, alcohols, and hydrocarbons. Tall oil is fractionated primarily into fatty acids (see CARBOXYLIC ACIDS).

The final source of by-products is the spent liquor. Possible utilization of kraft black liquors as a source of chemicals has been evaluated, but the mixture is so complex and the market value of the individual components too low to warrant the extensive separation and purification procedures required. The organic constituents are mainly the various fragments and degradation products of lignin and of the carbohydrates, eg, saccharinic acids and their low molecular weight analogues such as lactic acid. The dissolved alkali lignin, which is still polymeric, can be separated from black liquor by acidification and can be further purified or chemically modified to give useful materials. As of this writing, the production from only a few mills is sufficient to meet world demand.

Another product from kraft black liquor is dimethyl sulfide. This chemical, which is added to natural gas to give it odor, is also oxidized to produce the versatile solvent, dimethyl sulfoxide (see SULFOXIDES).

Modifications. The reasons for improving any pulping process generally fall into two primary categories: the desire to improve pulp properties and the need to improve process economics. The desire for better pulp properties is not a significant issue for the kraft process because the pulp is known for its high level of strength. Economic considerations, however, are increasingly important as international markets and competition grow. The capital requirements for an economically competitive pulp mill are high and trends suggest that 1000 t/d mills are needed to remain competitive in expanding markets. Capital needs for a new mill complex exceeded $1 billion in 1985. Hence there is great interest in modifications that can reduce process complexity and operation expense.

Specifically related to pulp properties and environmental modification, two areas of special interest in the chemistry of the kraft process have been studied for many years. These are carbohydrate yield and low or nonsulfur pulping, the latter to lessen the environmental burden by reducing the formation of malodorous organic sulfur compounds. Low carbohydrate yield, which leads to higher wood cost and increases the price of the pulp, results largely from the alkaline peeling reactions. Because peeling is initiated at the hemiacetal or reducing end of the polysaccharide chains, modification of this functional group inhibits peeling and thereby increases the yield of carbohydrate material in alkaline pulping. A number of methods to protect carbohydrates have proven effective in the laboratory, including glycosidation, reduction, reductive thiolation, and oxidation. To be effective as a process modification, however, the method must be compatible with kraft pulping and chemical recovery systems. It must also be sufficiently inexpensive so that the cost is more than offset by the increased yield, energy requirements, and other factors.

Polysulfide Process. One modification to the kraft process being applied commercially is the polysulfide process (38). Under alkaline conditions and relatively low temperature (100–120°C), polysulfides oxidize the active end group of the polysaccharide polymer to an alkali-stable aldonic acid. This reaction, known for many years (39), was not produced on a commercial scale until the development of an efficient method for *in situ* generation of the polysulfide in kraft white liquor.

When elemental sulfur is added to a solution of sodium sulfide and sodium hydroxide, the sulfur dissolves and forms a mixture of complexes according to the following general formula:

$$Na_2S + x\,S^0 \xrightarrow{OH^-} Na_2S_{x+1} \qquad (12)$$

where x is 1–4, depending on equilibrium conditions and the amount of sulfur added. The compound Na_2S_{x+1} is an oxidizing agent which under the conditions of kraft pulping converts the hemiacetal function to a relatively alkali-stable aldonic acid. The reaction intermediate involves oxidation at C-2 (see Fig. 8). Thus, the epimeric acids are obtained. The increase in yield in the polysulfide process is proportional to the amount of added sulfur up to ca 10% based on wood. Because of the sensitivity of the glucomannans to the peeling sequence, the greatest increase comes from their stabilization in softwoods (40).

The additional sulfur for polysulfide pulping can upset the sodium–sulfur balance in the kraft recovery cycle and increase sulfur emission problems. In the MOXY (Mead Corp.) process, polysulfide is formed from kraft white liquor by catalytic oxidation of sodium sulfide in the white liquor using air.

$$2\,Na_2S + O_2 + 2\,H_2O \longrightarrow 2\,S^0 + 4\,NaOH \tag{13}$$

The sulfur then reacts to form the polysulfide according to equation 12. The key is the use of a catalyst to promote the formation of elemental sulfur. Commercial systems are based on the use of air with an activated carbon catalyst (41). The need for additional sulfur is eliminated, but the sulfur level is limited to Na_2S_2. Under these conditions, yield gains of 2–2.5% based on wood are achieved, but sulfur emissions are lower because of the reduced amount of Na_2S (38).

Anthraquinone. In the early 1970s, the addition of sodium anthraquinone-2-sulfonate to sodium hydroxide solutions was reported (42). This accelerates pulping to a rate similar to that obtained by using sodium sulfide, enhancing yield without any loss of strength. The effect is partially additive, because improved yields and pulp properties are obtained from the kraft process when small amounts of the quinone salt are added to kraft pulping liquors. The sulfonate salt is soluble in the alkaline solutions. A patent has been issued for the use of unmodified anthraquinone (AQ) in alkaline pulping (43). The underivatized AQ, although less soluble in alkaline solutions, is much more effective in accelerating pulping. Anthraquinone (qv) is being used as a kraft pulping additive in commercial production.

Mechanism of Anthraquinone Acceleration. The mechanism for the dual function of AQ has been the subject of much research (29). Anthraquinone is an effective pulping accelerator in very small quantities and functions as a catalyst in the process. It is generally accepted that AQ functions in a complex redox sequence.

Anthraquinone has limited solubility in alkaline solutions, but anthrahydroquinone (AHQ) (9,10-anthracenediol) is highly soluble in aqueous alkali. AQ cycles between its insoluble oxidized form and its soluble reduced form. The first step in the sequence is the reaction of AQ and the reducing end of a carbohydrate. This stabilizes the carbohydrate against alkaline peeling and produces the reduced form, AHQ, which is soluble in alkali. AHQ reacts with the quinone-methide segment of the lignin polymer, increasing the rate of delignification, in a manner analogous to the sulfhydryl anion in kraft pulping. At the same time, AHQ is converted back to AQ, which can then participate again in the redox cycle. The chemistry of these reactions has been described in many technical papers (44–51).

Economics vary, depending on location and country. In Japan, for example, the quinone of choice is the disodium salt of 1,4-dihydro-9,10-dihydroxyanthracene (DDA). Unlike AQ, DDA is directly soluble in white liquor and thus readily penetrates the chip. AQ works well with both hardwoods and softwoods. Addition rates as low as 0.025% by weight give satisfactory increases (ca 1–2%) in pulping rate and yield, especially for hardwoods. In the United States in 1987, the FDA put a limitation to the amount of AQ that can be added to most pulp to 0.11%. This typically is the amount added to softwoods; lesser amounts are employed with hardwoods.

Because AQ promotes delignification, it can be used for general debottlenecking around the kraft pulp mill and whenever a reduction in any of several process parameters is desired. AQ can be added to white liquor to compensate for a reduction in effective alkali, sulfidity, or H-factor. For example, if high sulfidity is causing an air quality problem, AQ can be added to reduce the sulfidity while maintaining a similar production rate. As of this writing, technology for AQ recovery that can lower costs does not exist.

Extended Delignification Low Lignin Pulping.   When chlorinated compounds are used in bleaching kraft pulp, it may be advantageous to remove as much lignin as possible in the pulping stage, ie, to combine the pulping and bleaching processes. This strategy has several advantages, including savings in chemicals in the subsequent bleaching process, increased energy savings, and environmental benefits. More black liquor solids are recovered and used to generate steam and electrical power. Lower lignin content going to the bleach plant results in reduced quantities of organic material leaving the mill in the effluents. However, extended delignification causes a reduction in yield. As a result, some mills choose not to pulp below 24 kappa number. Also, shifts to oxygen delignification and bleach plant closure may cause mills to use another balance between pulping and bleaching.

The normal limit for lignin removal in the digester is set both by pulp strength requirements, usually indicated by pulp viscosity, and by yield requirements. As noted earlier, direct control of hydroxide and hydrosulfide concentrations can maximize lignin removal while maintaining pulp strength. In commercial practice this means control of various liquor streams during the process. Liquor addition and recycle in a Kamyr digester using the MCC process can be used to cook to lower final lignin content without significant loss of strength. This is termed extended cooking (29,52). Extended delignification technology is practiced using rapid displacement heating (RDH) (Beloit Fiber Systems) and other competitive systems for modern batch digesters which are competitive with the MCC process. Kappa number values as low as 22 for softwood and 14 for hardwood can be achieved using extended delignification.

Isothermal cooking (ITC), an extension of MCC technology (53,54), can significantly reduce the kappa number going to the bleach plant. Brown-stock kappa numbers of 13–17 for softwood and 8–10 for hardwood on a commercial scale have been reported (53,54). Combining extended delignification of wood in the digester, either in batch cooking using the RDH process or by MCC and ITC technology, with oxygen delignification of the resulting brown stock can produce softwood pulps of 8–10 going to the bleach plant. Hardwood pulps are even lower. The objective of extended delignification is to have little or no lignin in the pulp leaving the digester.

Modified Soda Pulping.   Wood pulping using caustic soda solutions was the first chemical pulping process. The beneficial effect of including sodium sulfide in the liquors was discovered soon after. The soda process is used to advantage in pulping some hardwood species and nonwood plants, but wood pulping using solely sodium hydroxide was never widely used. The volatile malodorous sulfur compounds hydrogen sulfide, H_2S; methyl mercaptan; dimethyl sulfide, CH_3SCH_3; and dimethyl disulfide are produced as undesirable by-products of kraft pulping. Odorous gas streams are burned either in the lime kiln or in the recovery boiler. Research efforts aimed at processes for eliminating the use of sulfur have continued. Because of the environmental concerns about odors, two possible variants of the soda process have been developed: the soda–oxygen and the soda–anthraquinone processes.

Soda–Oxygen Process.   Oxygen in aqueous alkaline solutions is an effective, readily available, and innocuous delignifying agent. However, it is a gas of low solubility and degrades pulp polysaccharides. The diffusion of oxygen into the chips is the rate-determining step, and diffusion is exacerbated by the rapid surface reaction with lignin, which depletes the oxygen supply. This problem can be minimized by increasing the oxygen pressure, maximizing gas contact, chipping the wood into thin wafers (1–2-mm thick), or decreasing the rate of the delignification reaction by lowering temperatures (125–140°C) and pH (pH 7–9). The last steps increase the reaction time to eight hours or more.

A high yield chemical pulp, eg, 52–53% bleached yield from softwoods, can be obtained, but strength properties are inferior to those obtained from the kraft process. If a protector, eg, potassium iodide, is added, an additional 2–3% yield is obtained, as is an improvement in all strength properties. The gas penetration problem can be minimized if fiberization is accomplished before treatment with oxygen. Oxygen treatment of virtually all types of semichemical and mechanical pulps has been explored (55). Caustic, sodium bicarbonate, and sodium carbonate have been used as the source of base (56,57). In all cases, the replacement of the kraft by these other processes has not been justified over the alternative of pollution abatement procedures.

Oxygen bleaching following pulp delignification in a digester has been widely practiced since the late 1970s (58). The use of oxygen allows the greater retention of polluting materials, including aromatics within the mill.

Soda–Anthraquinone.   A few mills worldwide use soda pulping of hardwoods. In such cases, the addition of anthraquinone is immediately justifiable in terms of increased yield and upgraded pulp quality. The conversion of existing kraft mills is not as simple because AQ contributes no alkalinity to the process as sulfide does, and most kraft causticizing systems would have to be expanded by about 33%. This conversion is probably not justifiable in terms of the yield gain. The greatest benefit from AQ is for new mills in which expenditures for air pollution abatement devices can be reduced.

Sulfite Pulping.   In the original sulfite pulping process, dating back to a 1867 patent (59), wood was pulped using an aqueous solution of SO_2 and lime. Because calcium sulfite has limited solubility above pH 2, an excess of SO_2

gas was maintained in the digester to keep the pH below this level. Thus the process was contrasted with the kraft or soda processes as being an acid process. Because bases other than calcium can be used with SO_2 solutions, sulfite pulping therefore refers to a variety of processes in which the full pH range is utilized for all or part of the pulping. Magnesium sulfite has decreasing solubility above pH 5, but sodium and ammonium sulfites are soluble at pH 1–14.

SO_2–Water System.   When SO_2 is dissolved in water, a series of equilibria are established as follows:

$$SO_2 + H_2O \rightleftharpoons SO_2 \cdot H_2O \rightleftharpoons H_2SO_3 \rightleftharpoons H^+ + HSO_3^- \rightleftharpoons 2\,H^+ + SO_3^{2-} \qquad (14)$$

When a base is added to sulfurous acid, ie, the $SO_2 \cdot H_2O$ system, first bisulfite, and then monosulfite, is formed. Using sodium hydroxide as an example,

$$H_2O + SO_2 \rightleftharpoons H_2SO_3 \qquad (15)$$

$$H_2SO_3 + NaOH \rightleftharpoons NaHSO_3 + H_2O \qquad (16)$$

$$NaHSO_3 + NaOH \rightleftharpoons Na_2SO_3 + H_2O \qquad (17)$$

At room temperature, the bisulfite pH inflection point occurs at pH 4.5 and the monosulfite at pH 9. Analogous equations can be written for magnesium, calcium, and ammonia. The starting raw materials, in addition to sulfur, are sodium hydroxide, magnesium oxide, calcium carbonate, or ammonia, depending on the base used. The four commercial bases used in the sulfite process are compared in Table 4.

Table 4. Comparison of Bases for Sulfite Pulping[a]

Property	Calcium	Magnesium	Sodium	Ammonium
SO_2 absorption system	complex	relatively simple	simple	simple
pH range for digestion	<2	<2	0–14	0–14
rate of pulping	intermediate	intermediate	slow	fast
level of screenings	moderate	moderate	low	low
scaling tendency	high	moderate	low	low
ease of liquor incineration	difficult	simple	complex	simple
recovery				
base	no	yes	yes	no
SO_2	no	yes	yes	yes

[a]Ref. 60.

Delignification Chemistry.   The chemical mechanism of sulfite delignification is not fully understood. However, the chemistry of model compounds has been studied extensively, and attempts have been made to correlate the results with observations on the rates and conditions of delignification (61). The initial reaction is sulfonation of the aliphatic side chain, which occurs almost exclusively at the α-carbon by a nucleophilic substitution. The substitution displaces either a hydroxy or alkoxy group:

$$\text{HCOR} \quad + HSO_3^- \longrightarrow \quad \text{HCSO}_3^- \quad + ROH \qquad (18)$$

Sulfonation can take place at the γ-carbon, if that carbon is activated by an α-keto group or by extended conjugation through a double bond between the α- and β-carbons. The reaction does not occur through a quinonemethide under acidic conditions. However, quinonemethide intermediates become important at neutral or alkaline pH. The addition of the ionic sulfonic acid group imparts water solubility to fairly large lignin fragments, but there must be some depolymerization to permit the physical extraction from the cell wall.

The depolymerization mechanism is less well understood. The sulfonated lignin polymer is subject to both acid- and alkali-catalyzed hydrolysis as well as sulfitolysis, which occurs at 180°C, but little depolymerization takes place. It appears that lower molecular weight lignosulfonates are dissolved first, while the higher molecular weight fraction is being slowly degraded (62). Acid-catalyzed condensations of the benzylic carbonium ion compete with sulfite and bisulfite delignification.

Acid-catalyzed glycosidic hydrolysis of polysaccharides is much more rapid than alkaline hydrolysis. Sulfite pulping at low pH must be carried out at a relatively low (135°C) temperature to avoid excessive polysaccharide depolymerization. The carbohydrate yield is higher in sulfite pulping, eg, 46% for a bleached softwood pulp compared with ca 44% for bleached kraft, because there is no peeling. The reducing-end functions of polysaccharide chains are oxidized to aldonic acid end units by bisulfite. However, a low hemicellulose pulp suitable for producing cellulose derivaties, eg, rayon (see FIBERS, CELLULOSE ESTERS) or cellulose acetate (see CELLULOSE ESTERS), can be made by a two-stage cooking phase such as the Sivola process. The Sivola process is acidic in the first stage, which reduces the average degree of polymerization (DP) of the hemicelluloses, but alkaline in the second stage, which dissolves and removes the hemicellulose. A more important stabilization mechanism in sulfite pulping is deacetylation of the softwood galactoglucomannan, in which the dissolved hemicellulose tends to crystallize or be redeposited on the cellulose microfibrils if the acetate substituents are removed. This may be promoted if a mildly alkaline first stage is followed by a more acidic stage, as in the Stora process.

Sulfite Process. An outline of the sulfite process can be found in Figure 27. Sulfur is burned to SO_2, which is cooled and sent to an absorption tower where base is added to form the storage acid. The liquor is transferred to a low pressure recovery tower and then to a high pressure accumulator where heat, water, and SO_2 from previous runs are absorbed to form the actual cooking liquor. The digester has been filled with chips, presteamed to remove air, and filled with cooking liquor. By direct steam injection or indirect heating, the digester is heated to temperature. A portion of the digester acid, known as side-relief,

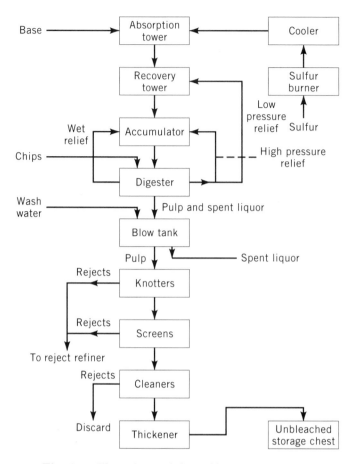

Fig. 27. Flow sheet of the sulfite process (60).

is withdrawn and returned to the accumulator. Subsequently, the pressure is controlled by relieving gas from the top of the digester, which is also returned to the accumulator. Near the end of the cook, blowdown is achieved by relieving gas at a rapid rate through the top of the digester to the accumulator. When the pressure is sufficiently reduced, a bottom blow valve is opened and the residual gas is used to discharge the contents of the digester to the blow pit. From this step on the pulp is treated in a normal manner and stored in the unbleached storage chest.

Depending on the ratio of SO_2 to base, sulfur in the cooking liquor can be present in several forms. There may be dissolved SO_2; sulfurous acid, H_2SO_3; bisulfite, HSO_3^-; and monosulfite, SO_3^{2-}, ions. The bisulfite may be present alone or in combination with two other forms. Conventionally, the liquor composition is described in terms of percent total, free, and combined sulfur dioxide. Total SO_2 is the sum of all forms. Free SO_2 represents the dissolved SO_2, H_2SO_3, plus one-half of the SO_2 present as bisulfite, because bisulfite decomposes to form H_2SO_3 and the sulfite salt of the base.

$$2\ NaHSO_3 \longrightarrow Na_2SO_3 + H_2SO_3 \tag{19}$$

Combined SO_2 represents the SO_2 present as monosulfite plus, for the same reason, one-half of the SO_2 present as bisulfite. All are expressed as percent concentration of SO_2 in the liquor, ie, g of $SO_2/100$ mL of liquor, in which the specific gravity of the liquor is taken to be 1.0.

Sulfite pulps have properties that are desirable for tissues and top quality, fine papers. Because sulfite pulping is not as versatile as kraft pulping, various options have been developed, and the choice of a specific process is dependent on individual mill situations. The unbleached pulp has high (60+) brightness compared to kraft pulp and is easily bleached. However, it is limited to select wood species. The heartwood of pine, Douglas fir, and cedars are not easily pulped. Additionally, pulps produced from hardwood have limited economic value because of low strength.

Chemical Recovery. There are advantages and disadvantages to each of the base systems employed in sulfite pulping (see Table 4). Each has its own potential recovery systems except the calcium system, which is obsolete. Calcium-based liquors can be burned, but scaling problems are severe, and conversion of the calcium sulfate to CaO is not economical.

In all sulfite pulping systems, any excess SO_2 in the digester can be collected easily from the blow gas by adsorption in fresh liquors. Ammonia-based liquors can be incinerated under oxidizing conditions to recover SO_2, but the ammonia is lost as nitrogen gas and water. Magnesium recovery is relatively simple because of the ready conversion of magnesium salts to MgO during combustion. Two types of systems are in use: a furnace and boiler, eg, the Babcock and Wilcox recovery unit; and a fluidized bed, eg, the Copeland process. In the MgO recovery process (Fig. 28), the first-stage washer filtrate, having 13–15% solids, enters the weak red liquor storage tank. The liquor is concentrated in a multiple-effect evaporator system and transferred to a strong liquor tank at the concentration required for boiler operation. The as-fired solids may vary between 60–65%. The liquor is fired at about 110°C (230°F) and can be preheated by a direct steam heater. The last stage of evaporation can be a direct-contact evaporator or a concentrator.

Chemical recovery in sodium-based sulfite pulping is more complicated, and a large number of processes have been proposed. The most common process involves liquor incineration under reducing conditions to give a smelt, which is dissolved to produce a kraft-type green liquor. Sulfide is stripped from the liquor as H_2S after the pH is lowered by CO_2. The H_2S is oxidized to sulfur in a separate stream by reaction with SO_2, and the sulfur is subsequently burned to reform SO_2. Alternatively, in a pyrolysis process such as SCA-Billerud, the H_2S gas is burned directly to SO_2. A rather novel approach is the Sonoco process, in which alumina is added to the spent liquors which are then burned in a kiln to form sodium aluminate. In anther method, used particularly in neutral sulfite semichemical processes, fluidized-bed combustion is employed to give a mixture of sodium carbonate and sodium sulfate, which can be sold to kraft mills as makeup chemical.

Many other recovery alternatives have been proposed that include ion exchange (qv), pyrolysis, and wet combustion. However, these have not gained

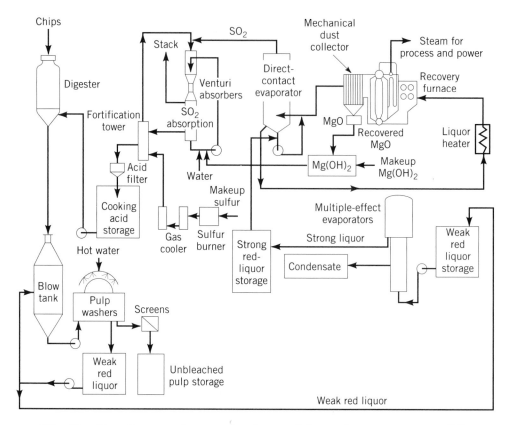

Fig. 28. Flow diagram of magnesium-base pulping and recovery process (36).

general acceptance. A limited number of calcium-based mills are able to utilize their spent pulping liquors to produce by-products such as lignosulfates for oil-well drilling muds, vanillin, yeast, and ethyl alcohol (see PETROLEUM; VANILLIN).

Modifications and Outlook. As a result of the discovery of the catalytic effect of anthraquinone in kraft and soda pulping, addition of AQ to the various sulfite pulping methods has also been investigated. In the neutral to strongly alkaline range, a catalytic effect is observed. The most significant observation is the marked increase in carbohydrate yield obtainable from AQ addition to neutral or slightly alkaline sulfite liquors. This produces a pulp having strength properties similar to kraft except for decreased tear, which is typical of high hemicellulose pulps. Such a process eliminates the air pollution problems of the kraft process and permits direct-contact evaporation of the liquors. However, somewhat stronger cooking conditions, eg, a higher *H*-factor, are required. Thus, the potential exists for sulfite pulp to regain some of the competitive market that has been lost to kraft pulp.

Uses. The market for sulfite pulps is but a small fraction of that for kraft pulp. Full chemical sulfite pulps are used in tissue and sanitary paper products because of fiber softness gained through this cooking process. Low hemicellulose pulps, ie, pulps that have high cellulose content, are marketed as dissolving pulp for cellulosic derivatives. High hemicellulose sulfite pulps are used to make glassine and greaseproof papers because of their ease of refining. Some sulfite chemical pulp is blended in newsprint, especially in the northeastern United States and Canada, and, to an increased extent, in printing and fine papers. Sulfite pulps are easily bleached to a higher brightness than kraft. Therefore they are used in fine paper and board grades that require higher brightness, eg, high quality stationery and business cards. The brightness of sulfite pulps is also stable, making this pulp a good choice for products such as photographic paper and decorative laminates.

Miscellaneous Acid Pulping Processes. Delignification may be accomplished by electrophilic substitution of the aromatic nuclei by nitric acid (qv). The pulp produced is similar in quality to acid sulfite pulp, and the spent nitrated liquors have good fertilizer value (see FERTILIZERS). This process is not used in the United States but may find application in favorable situations. Oxidative delignification by chlorine dioxide is conducted at acidic pHs. It is selective and gives bleached pulps having high hemicellulose contents. The process was evaluated briefly as holopulping in the United States and Japan (63) but is not practiced commercially because of high cost.

Solvent Pulping. The chemical processes considered so far all use water as the principal solvent in the process and are the main methods of commercial pulping. But environmental concerns and the high capital costs of constructing new kraft facilities have driven research into new pulping technologies that are flexible, cost-effective, and still capable of producing pulp of high quality. An alternative method that shows promise is solvent pulping, which uses nonaqueous solvents to remove the lignin from wood. This type of pulping does not necessitate chemical reactions to make the lignin hydrophilic.

The principal solvents that have been used are alcohols such as ethanol, methanol, and propanol, and organic acids such as formic or acetic acid, but other solvents include esters, ethers, phenols, cresols, and some amines. Even solvents such as CO_2 and NH_3 in the supercritical fluid state have been tried as solvents.

Three processes that have been developed are the ASAM (alkali–sulfite–anthraquinine–methanol), the Organocell (sodium hydroxide, methanol, and anthraquinone), and the Alcell (water and ethanol) processes. The ASAM process has been tested at the pilot scale (5 t/d); the Alcell and Organocell processes have been tested at the precommercial scale (15 t/d). In general, these processes appear to give greater yields, probably owing to carbohydrate conservation, lower lignin content, and strength qualities that approach those of kraft pulps. Advantages cited by the developers include higher initial brightness, easier bleaching, and environmental benefits from sulfur-free emissions (64). Softwoods, because of the higher lignin content, are generally difficult to pulp by this method. However, the ASAM and Organocell processes are successful on softwoods as well as on hardwoods.

Because of an advantage in pulp bleachability, solvent pulping lends itself well to oxygen, ozone, and enzyme bleaching, reducing or greatly diminishing adsorbable organic halide (AOX) discharges (65).

Traditionally, recovery of the pulping solvent and finding uses for the recovered lignin have been stumbling blocks to commercialization of solvent pulping. Loss of solvent leads to poor process economics, which is the preliminary obstacle to commercialization. Recovery of lignin by-products has also been suggested as an alternative to burning but few promising by-products have been developed. Another drawback of solvent pulping is the low degree of hydration of the fiber by the organic solvents used to remove the lignin. This has generally made the strength properties lower in quality than those obtained from the kraft process.

Semichemical Pulping

The distinctions between semichemical and high yield chemical processes are small and are more a matter of gradation between the mechanical and full chemical procedures. A semichemical process is essentially a chemical delignification process in which the chemical reactions are stopped at a point where mechanical treatment is necessary to separate fibers from the partially cooked chips. Any chemical pulping process can be used to produce semichemical pulp. Reagents most commonly used are sodium sulfite, sodium hydroxide, sodium carbonate, and kraft green and white liquors. The usual procedure in semichemical pulping is to treat the wood chips using a pulping reagent in batch or continuous digesters. Pulping time, pulping temperature, and/or chemical-to-wood ratio are adjusted to obtain the desired pulp yield. Semichemical pulps, although less flexible, resemble chemical pulps more than mechanical pulps because they are not as dependent on rupture of the fiber wall for bonding. Typically, yield is 60–85% and lignin content is 15–20%. Lignin is concentrated on the fiber surface.

Because of shortages in fiber supply and consequent escalation of wood costs, high yield processes are used whenever possible. The high residual lignin is, however, responsible for the stiffness of the fibers and sensitivity to light aging. Semichemical pulping is particularly suited to the more abundant hardwood species. Softwoods require much more chemical treatment and refining.

Processes Other Than the Neutral Sulfite Process. The traditional method for producing semichemical pulps has been the neutral sulfite semichemical (NSSC) process. However, many other processes exist for producing the same products. The main difference is the composition of the pulping liquor. A few of the more important technologies are as follows. (1) In the acid sulfite process, the cooking liquor consists of a solution of bisulfite plus free sulfur dioxide. Delignification is the predominate reaction as described under sulfite pulping. (2) In the bisulfite process, bisulfite pulping liquor contains no free sulfur dioxide. The mechanism of the delignifying reactions is similar to that in acid sulfite pulping. High (53–70%) yield is produced in Canadian and a few U.S. newsprint mills from a mixture of spruce and balsam fir using magnesium or sodium as a base. (3) In the alkaline sulfite process, the cooking liquor contains sodium sulfite and any or all of the following: sodium carbonate, sodium hydroxide, and sodium sulfide. The proportions are mixed to that the pH falls between 9 and 12. Because the reaction rate is lower than that for kraft pulping, the cooking times are longer. Pulp having high strength properties suitable for linerboard can be produced using this technology (4). Finally, in the kraft–semichemical process, semichemical pulps are readily produced by the kraft process. The pulps are darker, lower in strength, and contain more lignin and less hemicellulose than conventional NSSC pulps. These high yield pulps are often produced in 65–80% yield range and have kappa number values from 55 to 110 mL 0.1 N KMnO$_4$. They are generally produced from coniferous woods and the pulp is often used for linerboard and corrugating media. White liquor from an associated kraft mill is the usual chemical reagent used at about one half the normal amount of active alkali. Other cooking conditions such as time and temperature are usually the same as in normal alkaline pulping.

NSSC Pulping. The characteristics of semichemical pulps are especially suited to the production of corrugated medium, which is the raw material for the fluted center ply of corrugated boxes. Neutral sulfite semichemical pulping was developed specifically as a semichemical process for corrugation medium and lends itself to small mills having minimal capital investment. For many years, this was the only semichemical pulping process.

The chemistry of NSSC pulping is similar to that of sulfite chemical pulping. It involves selective sulfonation and partial hydrolysis of the lignin. The selectivity is high, and corrosion is minimized if the pH is maintained at ca 7. The sodium sulfite, Na$_2$SO$_3$, solution is usually buffered with sodium bicarbonate, Na$_2$CO$_3$, on a weight ratio of 5 to 1. Additional buffer forms from the wood as pulping progresses, because the carbohydrate acetyl groups hydrolyze, wood acids dissolve, and lignosulfonates form.

As for other chemical pulping, continuous and batch digesters are used. In continuous pulping, the temperature is higher (190°C) and the time is shorter (10–12 min) than in batch systems (170°C for ca 20–60 min). At the higher temperatures, complete penetration of pulping chemicals into the chips is especially important. A screw feeder is used to compress the chips in the presence of pulping liquors. When the pressure is released, the liquid is rapidly absorbed by the wood. After being cooked, the chips are fiberized or screw-pressed up to 60% consistency before being fiberized in two or more refining stages before washing.

Demands for pollution abatement forced the development of chemical recovery methods for all sulfite pulping processes, including NSSC. One response to environmental concerns was the development of other semichemical pulping methods. Semichemical pulping can be performed using kraft green liquor. This procedure is especially advantageous if the semichemical pulp mill has access to an existing kraft recovery system. Neutral sulfite semichemical liquors can also be fed into a kraft recovery system as makeup for the kraft chemicals that are consumed, but the balance is unfavorable and the relative amount of NSSC liquors that can be utilized is small.

Semichemical pulping can be accomplished by a sulfur-free system of sodium hydroxide and sodium carbonate. The sodium carbonate is recovered by simple incineration, and sodium hydroxide is added as makeup. Advantages in recovery operation are obtained if potassium hydroxide is added occasionally to maintain ca 20 mol % potassium carbonate (66). Lastly, semichemical pulping can be accomplished by using kraft white liquor of low strength.

Bleaching

The objective of bleaching is to whiten the pulp without reducing the intrinsic strength of the fibers comprising the pulp, or as a means to change other pulp properties. Pulps vary considerably in their color after pulping, depending on the wood species, method of processing, and extraneous components. For many paper types, particularly printing grades, bleaching of the raw pulp is required. The brightness standard is measured as the reflectance of light in the blue range (457 nm) from a thick pad of pulp compared with magnesium oxide as 100% white (R_∞). The reflectance, R_∞, of an infinitely thick (>170 g/m^2) pad of pulp is given by the Kubelka-Munk theory (67):

$$R_\infty = 1 + \left(\frac{K}{S}\right) - \left(\left(\frac{K}{S}\right)^2 + 2\left(\frac{K}{S}\right)\right)^{1/2} \tag{20}$$

where K is the light absorption coefficient and S is the light scattering coefficient at the reference wavelength (457 nm). Both K and S are in units of m^2/g. The fundamental objective of pulp bleaching technology is to lower the absorption coefficient, K, while maintaining the pulp scattering coefficient, S, essentially constant. The greater the scattering coefficient of the pulp, the greater the pulp opacity. These objectives must be accomplished economically and in a sound environmental manner (see BLEACHING AGENTS, PAPER AND PULP).

   Brightness Scales.   There are numerous brightness scales used in commercial operations (29). The three most common are General Electric (GE), electronic reflectance photometer (Elrepho), and the International Standards Organization (ISO). In the United States, the TAPPI or GE scale is widely used on paper grades, whereas ISO is commonly used on pulp. The GE scale is gradually being replaced by the ISO scale. The TAPPI T-452 method specifies the optical geometry to be 45° for the illuminating light and 90° (perpendicular) for the reflected light or normal viewing. Reflectance is compared to a standard magnesium oxide powder and given as % GE brightness units. In Canada, the Canadian Pulp and Paper Association (CPPA) method is used, which specifies that the sample should be diffusely illuminated using a highly reflecting, integrating sphere. Reflected light is measured at 90° to the sample. Reflectance is compared to absolute reflectance from an imaginary, perfectly reflecting, perfectly diffusing surface. Brightness is given in terms of ISO brightness units based on ISO 2469. When MgO powder is used as the standard (98–99% of absolute reflectance), units are Elrepho brightness or MgO brightness. In European countries, the Scandinavian Pulp, Paper, and Board Testing Committee (SCAN) standard method is widely used (Scan method C11:75). This is similar to the CPPA method and brightness is given in ISO brightness units. The Elrepho scale gives slightly higher (about 0.5–1%) values than the GE or TAPPI scale for chemical pulps. Values obtained from the ISO scale are lower than values obtained from the TAPPI and Elrepho scales. An approximate relationship between ISO brightness and SCAN (Elrepho) brightness is given in the literature (60).

$$\text{ISO brightness} = 0.974 \times \text{SCAN brightness} + 0.1 \tag{21}$$

   Bleaching Methods.   There are basically two types of bleaching operations or methods: those that chemically modify the chromophoric groups by oxidation or reduction but remove very little lignin or other substances from the fiber, and those that complete the delignification and remove pitch and some carbohydrate material. In the special case of dissolving-pulp production, bleaching is a final purification of cellulose, and most of the residual hemicellulose is removed (see also BLEACHING AGENTS). Both techniques result in the absorption coefficient for the pulp, K, being reduced and the reflectance factor, R_∞, raised.

Mechanical Pulps. Mechanical pulps are bleached in a lignin-conserving manner, known as brightening. The chromophores are treated with oxidizing agents such as hydrogen peroxide (P) and hypochlorite (H) solutions, and reducing agents such as sodium hydrosulfite (dithionite, Y) and formamidine sulfinic acid (FAS). Even though the brightness of mechanical pulps can be significantly increased, the primary impediment to greater utilization of these pulps for fine papers remains poor brightness stability. Again, where light stability is important, the more expensive bleached chemical pulps are normally used.

Chemical Pulps. Removal of lignin is the goal of bleaching chemical pulp. It may be viewed as an extension of the pulping process. Pulping is discontinued before lignin removal is complete because the selectivity of pulping, ie, the rate of lignin removal relative to the rate of carbohydrate degradation, decreases at low lignin contents. Bleaching is inherently more selective than pulping and can be safely used to remove the last traces of lignin from the pulp.

A fundamental difference between pulping and bleaching concerns the fate of the lignin and other organic materials removed from the wood or pulp. The organic materials removed during pulping are burned in the recovery boiler to generate energy and are therefore an asset to the pulp mill. Material removed during bleaching is difficult to recycle to the recovery system and is usually treated before discharge. This occurs because bleaching has traditionally been accomplished by elemental chlorine and chlorine-containing compounds that contaminate the effluent with chloride ion. If these effluents are recycled, the concentration of chloride ion in the liquor cycle can build up to high levels and result in corrosion and safety problems in the recovery boiler.

If all of the lignin, pitch, carbohydrate degradation products, and other chromophores and uv-absorbing materials are removed, a very white (over 90% ISO), highly color-stable pulp can be obtained. Because the lignin is a modest reducing agent under normal bleaching conditions, all of the reagents used for bleaching full chemical pulps, with the exception of caustic, are oxidizing agents. These oxidizing agents function to lower the molecular weight of the lignin macromolecule and render it hydrophilic so that it has a propensity to dissolve in the bleaching liquor. Because the carbohydrates are also susceptible to oxidation, bleaching is accomplished under the mildest conditions possible and the final depolymerization and removal are performed gently in several stages. It is not clear whether residual lignin is structurally different from the bulk lignin or whether the observed decrease in delignification rate during pulping is caused by other factors. In either case, the remaining lignin is intimately associated with the cell wall polysaccharides. Use of caustic in bleaching stages functions to hydrolyze chlorolignin and solubilize lignin fragments.

Bleach Plant Control Parameters. Pulp properties are measured at various stages of the bleaching process and serve as the basis of control actions. The most important are residual lignin content; cellulose degree of polymerization, ie, chain length or molecular weight; optical properties, ie, brightness and sometimes opacity; and cleanliness. Lignin content is measured indirectly as the amount of an oxidizing agent, potassium permanganate, that can be consumed by the pulp. Either of two methods, the kappa number or K (permanganate) numbers are used. Cellulose degree of polymerization, DP, is measured indirectly as the viscosity of a solution of the pulp in a cellulose solvent. DP correlates with pulp strength if it is below some threshold value that varies from one type of pulp to another. Cleanliness is measured by counting or visually estimating the surface area of particles that appear as imperfections, ie, bark, knots, shives, pitch, and dirt, in a pulp sheet. Strength properties are measured on the finished pulp. Properties are determined as a function of pulp freeness, TAPPI standard T-227, by the use of the beating curve. Important properties such as tensile, burst, fold, and internal tear resistance are measured by following TAPPI standard T-220.

Bleaching Chemicals and Nomenclature. Table 5 lists the principal chemicals used in bleaching. Bleaching agents are used in a variety of combinations. Each compound represents a stage. Because of the variety of staging combinations that have been developed, a set of commonly accepted abbreviations are adopted (68). For example, the five-stage sequence *CEDED* refers to an initial chlorination of the lignin under acidic conditions (*C*), followed by alkaline hydrolysis and extraction (*E*) of the chlorinated lignin, mild oxidation using chlorine dioxide (*D*), a second alkaline extraction (*E*), and, finally, brightening using chlorine dioxide (*D*). The *CEDED* sequence was the conventional bleach plant during the 1960s and 1970s. Numerous other sequences have been developed and used since then.

Delignification and Brightening Stages. Bleach plants for chemical pulp are usually divided into two parts according to the function. The early stages in any bleach plant are delignification stages, for example, the *CE* stages in the *CEDED* sequence. In the delignification stages, large amounts of lignin are

Table 5. Chemicals Used in Bleaching[a]

Name	CAS Registry Number	Formula	Symbol	Form used	Lignin function[b]
chlorine	[7782-50-5]	Cl_2	C	gas	chlorinates
chlorine dioxide	[10049-04-4]	ClO_2	D	7–10 g/L (aq)	brightens and solubilizes
oxygen	[7782-44-7]	O_2	O	gas[c]	solubilizes
hydrogen peroxide	[7722-84-1]	H_2O_2	P	2–5 wt % (aq)	brightens
hypochlorous acid	[7790-92-3]	HOCl	M	(aq)	brightens and solubilizes
sodium hypochlorite	[7681-52-9]	NaOCl	H	40–50 g/L (aq)	brightens and solubilizes
ozone	[10028-15-6]	O_3	Z	gas	brightens and solubilizes
sodium hydroxide	[1310-73-2]	NaOH	E	5–10 wt % (aq)	solubilizes[d]

[a]Ref. 29.
[b]All oxidize lignin unless noted.
[c]Used with NaOH.
[d]Does not oxidize; hydrolyzes chlorolignin.

removed but the brightness is not appreciably raised. The later stages in the bleach plant are termed brightening stages and are intended to remove chromophoric groups to further lower the absorption coefficient, K. Little lignin is removed in these stages but large increases in brightness are achieved. In the brightening stages, chromophoric groups on the lignin that are intimately associated, ie, bound, to the hemicelluloses are destroyed. In the *CEDED* sequence, the brightening stages are the *DED* portion of the sequence. This is illustrated in Figure 29. If large amounts of an oxidizing agent are applied in one stage in an attempt to achieve complete delignification and high brightness gain in one stage, a plateau in the response is obtained. This has necessitated multiple-stage bleaching. Partial sequences for a variety of bleach plants are as follows (69).

Delignification	Brightening
CE	*H*
CE	*HEH*
CE	*DED*
CE$_\mathrm{O}$	*HEDP*
C$_\mathrm{D}$*E*	*D*
D$_\mathrm{C}$*E*$_\mathrm{O}$	*DD*
DE$_\mathrm{OP}$	*DD*
OCE	*HDED*
OCE$_\mathrm{OP}$	*DED*
OC$_\mathrm{D}$	*DEPD*
ODE$_\mathrm{O}$	*D*
OZE$_\mathrm{OP}$	*P*

Important Variables in a Bleaching Stage. For any stage the important bleaching variables are time, temperature, concentration, pH of the chemical reactions, and the degree of mixing. Concentration is determined by pulp consistency and amount of bleaching chemicals applied as a percentage of the pulp. In the delignification stages, the rate of lignin removal can be expressed as the rate of decrease in the kappa number per unit time, t, following the laws of mass action. For example, for chlorine dioxide bleaching, the rate of kappa number decrease is given empirically by equation 22 (29):

$$-\left(\frac{d\,\mathrm{kappa}}{dt}\right) = k[\mathrm{ClO_2}]^{0.5}[\mathrm{Cl^-}]^{0.3}[\mathrm{H^+}]^{0.2}\mathrm{kappa}^5 \qquad (22)$$

where the terms in brackets are given in concentrations and k is the reaction rate coefficient that depends on temperature. In delignification reactions there is a rapid initial delignification that falls off and then proceeds at a much slower rate. The kinetics of the brightening stage is similar in form to the expressions for the delignification stages except that the rate is expressed as the disappearance of chromophoric groups, defined as the absorption coefficient at 457 nm, K_{457}.

Equivalent Weight of Bleaching Agents. When applied to pulp bleaching, equivalent weight relates to the mass of chemical that can accept one mole of electrons supplied by the lignin, ie, that weight which theoretically has the potential to accomplish a specified amount of lignin oxidation. For example,

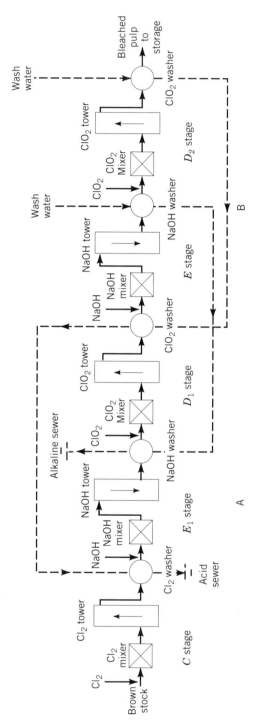

Fig. 29. *CEDED* pulp bleaching sequence: A, delignifying, and B, brightening stages.

562

Cl_2, ClO_2, and O_2, perhaps the three most important bleaching agents, have equivalent weight defined by the following equations.

$$Cl_2 + 2\,e^- \longrightarrow 2\,Cl^- \tag{23}$$

$$ClO_2 + 4\,H^+ + 5\,e^- \longrightarrow Cl^- + 2\,H_2O \tag{24}$$

$$O_2 + H_2O + 4\,e^- \longrightarrow 4\,OH^- \tag{25}$$

Bleaching efficiency is the degree to which the chemical realizes this potential by actually removing lignin or increasing brightness. Reactivity reflects the rate and extent of bleaching of which a chemical is capable. Selectivity is the rate of bleaching relative to the rate of carbohydrate degradation. Also important is the ability of the chemical to break down shives and knots as well as to remove dirt, along with the potential for harming the environment.

The equivalent weight and approximate delignifying efficiencies of the principal oxidizing bleaching agents are listed in Table 6. One kg of ClO_2 is theoretically equivalent to 2.63 kg of chlorine, 2.63 being the ratio of the equivalent weight of chlorine (35.5) to chlorine dioxide (13.5). When used to substitute for about half of the chlorine in the chlorination stage, chlorine dioxide is also more efficient, further enhancing its advantage over chlorine. This efficiency improvement is only observed when chlorine dioxide is added and allowed to react with the pulp before the chlorine is added. At very low or very high levels of substitution, however, ClO_2 is comparable in efficiency to chlorine. Oxygen and hydrogen peroxide are somewhat less efficient than chlorine. In the case of oxygen, this disadvantage is compensated by a much lower cost per equivalent.

Table 6. Equivalent Weight and Efficiency of Oxidizing Bleaching Agents[a]

Chemical	Equivalent weight	Equivalent chlorine	Efficiency
Cl_2	35.5	1.00	high
ClO_2	13.5	2.63	high
O_2	8.0	4.44	low
H_2O_2	17.0	2.09	low
NaOCl	37.2	0.93	medium
HOCl	26.2	1.35	medium
O_3	8.0	4.44	medium

[a] Ref. 69.

Approximate indication of reactivity and selectivity are summarized in Table 7. Chlorine and ozone are extremely reactive, whereas oxygen and hydrogen peroxide tend to react more slowly and to be more limited in the amount of lignin removed. Chlorine dioxide and, to a lesser extent, chlorine are highly selective; ozone tends toward poor selectivity. Hypochlorous acid and, to a lesser extent, oxygen require the addition of inhibitors to achieve satisfactory selectivity. Hypochlorite is selective only if the pH is maintained sufficiently high and consumption is limited. Approximate indication of particle bleaching ability and perceived environmental hazard are also listed in Table 7. Chlorine dioxide, chlorine, and hypochlorite offer good particle removal ability, whereas ozone and

Table 7. Characteristics of Bleaching Agents[a]

Oxidizing agent	Reactivity	Selectivity	Particle bleaching ability	Environmental concern
Cl_2	high	high	high	high
ClO_2	medium	high	high	medium
O_2	low	medium	medium	low
H_2O_2	low	high	low	low
NaOCl	medium	medium	high	high
HOCl	medium	medium	medium	medium
O_3	high	low	low	low

[a]Ref. 69.

hydrogen peroxide are poor in this respect. Regarding environmental effects, chlorine and hypochlorite are seen as having detrimental effects. As a result, there is a trend toward replacing both with chlorine dioxide, oxygen, ozone, and hydrogen peroxide.

Chlorine Stage.   Chlorine oxidizes lignin (see Table 5) without dissolving or weakening the pulp and the oxidation products are easily removed by dissolving in sodium hydroxide. Chlorination involves substitution of the aromatic ring and side chains and oxidative depolymerization through ether cleavages and ring openings. Chlorine also reacts by both ionic and free-radical mechanisms. However, chlorine reacts predominantly by substitution rather than oxidation, and therefore tends to be less degradative to cellulose than, for example, a reagent such as sodium hypochlorite which functions primarily through oxidation. Chlorine was typically added according to the kappa factor, which is the ratio of chlorine addition to the kappa number of the incoming pulp.

$$\% \ Cl_2 \text{ on pulp} = \text{kappa factor} \times \text{pulp kappa number} \qquad (26)$$

Typically a kappa factor of 0.18–0.22 is used, depending on whether hardwood or softwood is bleached. Another notable feature of the chlorination stage is that the reactions between chlorine and lignin are extremely fast. Thus it is extremely important to achieve good mixing of chlorine with the pulp. Otherwise, the pulp in areas of high chlorine concentration would be overchlorinated. Also, because chlorine diffusion is slow compared to the rate of chemical reaction, pulp strength suffers on account of carbohydrate degradation resulting from an excess of chlorine. Often 5 to 10% of the chlorine demand is added as chlorine dioxide in the chlorination stage (C_D) to function partly as a free-radical scavenger to protect carbohydrates and to increase delignification.

Chlorination can be carried out at 25°C or below. However, the reaction is exothermic and, as mills have used filtrate recycle, operating temperatures have unavoidably risen. Retention times are 30–60 minutes but decrease as temperature increases. In most mills the retention time cannot be changed because the tower is upflow in design. The normal pulp consistency is 3–4%, but the trend is toward higher (ca 10%) consistency or gas-phase chlorination. Target pH in the chlorination stage (also C_D) is about 1.8.

Extraction Stage. The washing step after the chlorination stage removes the chlorine. In alkaline extraction, about 1.5–3.5% NaOH based on pulp is used. This typically amounts to about 55% of the Cl_2 demand used in the chlorination stage (% NaOH $\cong 0.55 \times$ % Cl_2). The target pH exiting the reaction tower is about 10.5. Because sulfonated lignins are more water soluble, the extraction is carried out at 20–40°C for sulfite pulps and about 50–60°C for kraft pulp. Most extraction stages are carried out at medium (about 12%) consistency. Residual chlorine is also removed by NaOH, thus arresting the bleaching process. The NaOH, however, dissolves not only some of the resins in the pulp but also some of the hemicelluloses in the pulp. This reduces pulp tensile strength but improves tear strength not because of fiber degradation but because the hemicelluloses bond chemically to give added tensile strength in the final paper product. To overcome this loss in hemicellulose polymers, starch is added on the wet end of the paper machine as a dry strength additive.

Chlorine Dioxide Stages. Another commonly used bleaching agent, chlorine dioxide gas, is essentially nondegrading to cellulose and is capable of producing highly bright pulps. Reaction of pulp and ClO_2 is conducted at 60–80°C for 2–4 h at a target pH of 3.5–4 and a consistency of 12–15%. Typical reaction time is three hours in the second chlorine dioxide stage (D_2) and one to two hours in the first chlorine dioxide stage (D_1). Target pH value in the first stage is 3.5; in the second, about 4.0. For sulfite and bisulfite pulps, because the initial brightness is high, a final brightness of about 85 ISO can be attained using the *CED* sequence. For kraft pulp, a second caustic extraction (E_2), followed by a second chlorine dioxide stage (D_2), is typically employed to bring the kraft pulp brightness to 90 ISO brightness using the *CEDED* sequence.

Formation of chlorinated organic compounds is minimal in chlorine dioxide stages. However, the gas is toxic, highly corrosive, and explodes at high concentration. For safety, ClO_2 is generated at the point of use in low concentration. It is made by reaction of sodium chlorate with a suitable reducing agent in a wide variety of processes. The essential reaction in the generation of ClO_2 is as follows:

$$ClO_3^- + 2\,H^+ + e^- \longrightarrow ClO_2 + H_2O \qquad (27)$$

The electron for the basic molecular reaction is supplied by various reducing agents, depending on the process used. SO_2 is used in the Mathieson process, methanol in the Solvay and R8 processes, and chlorine ion, Cl^-, in the R2, R3, and R7 processes. In the R7 and integrated processes, the source of the chloride ion is a mixture of NaCl and HCl. Typical by-products are elemental chlorine, which is scrubbed in a tail gas stream, and crystalline salt cake, which is often used in the kraft process. Chlorine dioxide yield is typically 87–97%, depending on the process. The R3 process, illustrated in Figure 30, is typical of ClO_2 generation technology. The sodium chlorate is made from NaCl, water, and energy. The use of proper vents and corrosion-resistant materials of construction, eg, titanium, is required.

Hypochlorite Stage. The sequence *CEH* came into use during the 1930s and was dominant with sulfite pulping in the 1950s. During the 1960s the sequence *CEHEH* was the mainstay of the kraft pulping industry, giving brightness values of 82–85 (GE). Calcium rather than sodium hypochlorite was used

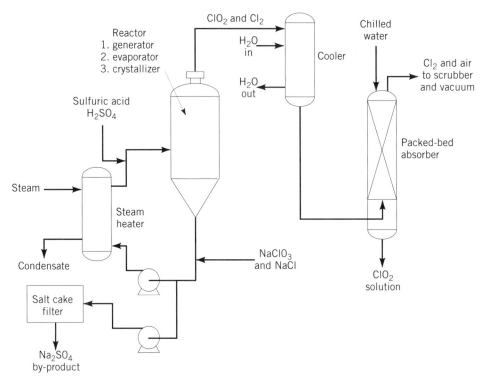

Fig. 30. The R3 process of producing ClO_2 (29).

because of lower cost. In the 1990s, sodium hypochlorite is sometimes used following the first extraction stage or is included in an oxidative extraction stage (*E/H*). In hypochlorite bleaching, sodium hypochlorite has all but replaced calcium hypochlorite as a bleaching agent because of scaling. Hypochlorites attack lignin with vigor, but also attack cellulose. For this reason, during the 1970s, most mills replaced hypochlorite with chlorine dioxide, which attacks lignin selectively. Because of carbohydrate attack, pH is carefully controlled in hypochlorite bleaching and is normally maintained at about 10.5 to minimize degradation to the carbohydrates. Conditions typically used in hypochlorite bleaching of chemical pulps are 1–2% based on pulp, 50°C, 12% consistency, 30-min retention time, and end pH greater than 10. Hypochlorite bleaching is being phased out for most grades of pulp except that used in dissolving grades where the hypochlorite stage is employed to control the molecular weight of the cellulose.

Peroxide Stage.   Peroxides in the form of both hydrogen and sodium peroxide have been used to bleach mechanical pulps since about 1940. As a bleaching agent for chemical pulps, both kraft and sulfite, peroxides were first introduced in 1950 as a means of obtaining increased brightness and better brightness stability in market pulps. For instance, 90+ (GE) brightness could be obtained from sulfite pulp using the *CEHP* sequence. For kraft pulp, peroxide stages were present in such sequences as *CEHDP*, *CEHEDP*, and *CEDE_PD* to obtain high brightness pulps and also to stabilize brightness.

An extremely versatile reagent, hydrogen peroxide, which can be safely shipped and stored in aqueous solutions of up to 70% concentration, is relatively nonvolatile and releases only water and oxygen on decomposition. Sodium peroxide is almost as convenient to ship and handle, having sodium hydroxide and oxygen as the decomposition products. Advantages of using peroxides include improved brightness stability, especially when used in a final stage; little or no degradation of the cellulose; and low capital investment.

The perhydroxyl ion, OOH^-, derived from the dissociation of hydrogen peroxide in alkaline solution, is generally considered the active agent in the bleaching of wood pulp using hydrogen peroxide. This ionization may be expressed by the following reversible reaction:

$$H_2O_2 \rightleftharpoons H^+ + OOH^- \tag{28}$$

where the dissociation constant is 2×10^{-12} at room temperature and increases with temperature (67). The formation of OOH^- ions is promoted by neutralization of the H^+ ion with an alkali. The reaction can be reversed by the addition of acid. Limited dissociation occurs at pH values of 9 and below. This indicates alkaline conditions for peroxide bleaching. However, an upper pH limit exists as a result of competing, nonreversible reactions:

$$H_2O_2 \longrightarrow H_2O + 1/2\ O_2 \tag{29}$$

$$2\ OOH^- \longrightarrow O_2 + 2\ OH^- \tag{30}$$

This decomposition of hydrogen peroxide into water and oxygen, favored at pH values above 11.5, is catalyzed by the presence of heavy-metal ions, particularly copper, iron, nickel, and manganese, and certain enzymes (70). Experience has indicated that the desirable first reaction (eq. 28) can be obtained relative to the undesirable second (eq. 29) and third (eq. 30) reaction when the pH of the peroxide solution is maintained between 10 and 11 during bleaching. Thus the target end pH as the peroxide leaves the tower is often 9.5–10.5. To accomplish this buffers are sometimes added to the bleach liquor. Also, because tramp metal ions decompose the peroxide, chelating agents (qv) are often added to tie up metal ions. A popular sequestering agent such as sodium tripolyphosphate (STPP) and chelating agents such as ethylenediaminetetraacetic acid (EDTA), diethylenetriaminepentaacetic acid (DTPA), and N-(hydroxyethyl)ethylenediaminetriacetic acid (HEDTA) have been used to complex metal ions. The bleaching action of peroxide has been attributed to its reactions with the chromophoric groups in the lignin matrix (71–73).

Bleaching conditions for a typical peroxide stage are 1–3% H_2O_2 based on pulp; 60–80°C temperature; medium (10–16%) consistency; 30–180-min time; 0.05% $MgSO_4$ as a carbohydrate protector; 1–2% sodium silicate, Na_2SiO_3; and 0.5% DTPA. Brightness gain depends on the brightness of the pulp going into the stage and the type of pulp being bleached. In totally chlorine-free bleaching, either an acid wash (A stage) or a chelant wash (Q stage) is used to control heavy metals preceding a P stage.

High Levels of ClO₂ Substitution.   Chlorination produces an effluent containing a mixture of chlorinated organic compounds, much of which is ultimately discharged to a receiving water body. One measure of the amount of such compounds is adsorbable organic halides (AOX), which effectively estimates organically bound chlorine. About 10% of the elemental chlorine consumed by the pulp ends up as effluent AOX. Control of these products for environmental reasons became a significant issue early in the 1990s.

Substitution of chlorine dioxide for chlorine in the first (chlorination) bleaching stage (C_D) is a strategy that has been widely adopted in North America. The objective of this strategy is to minimize discharge of chlorinated organic compounds. Chlorine dioxide may be substituted for part or all of the chlorine in the chlorination stage. Typically, 50% or higher substitution is practiced commercially in a D_C stage. Many mills are using 100% chlorine dioxide substitution in the first stage of bleaching. When the D stage replaces the traditional C stage in the bleaching sequence, it becomes elemental chlorine-free (ECF).

The basis for the use of high levels of ClO_2 substitution to effect AOX reduction is shown in equations 23 and 24. Five electrons are accepted by each atom of chlorine, Cl(IV), when present in the form of ClO_2 (eq. 24) versus only one electron per atom of chlorine, Cl(0), when present in the form of Cl_2 (see eq. 23), thus accounting for equivalent weights (Table 6), ie, one kg of ClO_2 may be substituted for (is equivalent to) each 2.63 kg of Cl_2, assuming equal bleaching efficiency. This strategy results in significantly lower levels of organic chlorides, as measured by the AOX discharged from bleach plants (Fig. 31). High levels of ClO_2 in the chlorination stage have become common commercial practice.

Short-Sequence Bleaching.   A wide variety of sequences may be used for bleaching. In the mid-1980s, short-sequence bleaching was introduced in an effort to reduce capital, steam use, chemical consumption, and energy cost. For hardwood, $D_C E_O D$ is an example of short-sequence bleaching. For softwood, the

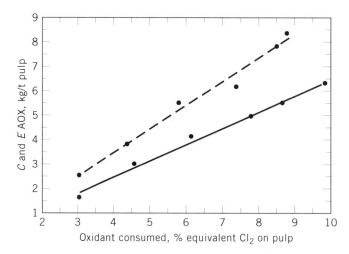

Fig. 31. AOX formation in chlorination (*C*) and extraction (*E*) stages for 30 kappa softwood kraft pulp chlorinated at 4% consistency in mixer, using (– – –) 100% Cl_2 and (——) 50% Cl_2/50% ClO_2 (74).

analogous sequence is $D_C E_O DD$, or, in the limit of zero chlorine, $DE_O DD$. Such sequences are replacing the traditional $CEDED$ sequence popular in the 1960s and 1970s. In addition to the $D_C E_O D$ and $D_C E_O DD$ sequences typical of modern mills in the United States, there are the $ODE_{OP} D$ and $OD_C E_O D$ sequences. In these sequences, an oxygen delignification stage replaces the first two stages of bleaching. In the third stage of the sequence, oxygen is added to the extraction stage, followed by the fourth-stage addition of chlorine dioxide. Often hydrogen peroxide (P) is added to the E_O state to create the E_{OP} stage with further enhanced oxidizing potential.

TCF and ECF Bleaching. Dramatic changes in bleaching technologies have resulted from the discovery in 1986 of polychlorinated dioxins and dibenzofurans as unintended by-products of pulp mill bleach plants. Two compounds of concern are 2,3,7,8-tetrachlorodibenzodioxin (2,3,7,8-TCDD) and 2,3,7,8-tetrachlorodibenzofuran (2,3,7,8-TCDF). Subsequently, many mills have abandoned the use of elemental chlorine altogether. Others have eliminated chlorine dioxide as well. In the late 1980s and the early 1990s, two grades of pulp that minimize discharge of adsorbable organic halides began to be produced. These are designated as elemental chlorine-free (ECF) and totally chlorine-free (TCF) pulps. The $ODE_O D$ is an example of ECF bleaching. Changes in processes and technologies include ozone delignification, improved oxygen delignification, peroxide bleaching, enzyme bleaching, and extended cooking. Some of these changes have resulted in cost savings from reductions in power use, space requirements, as well as chemical and water use.

Typical ECF sequences are $DEDED$, $DE_{OP} DP$, $ODE_{OP} P$, and $ODE_O D$. Totally chlorine-free pulps of high brightness, but inferior strength and higher cost, can be produced by using a combination of oxygen and ozone delignification, hydrogen peroxide, and enzyme bleaching technology as the various stages in the bleaching sequence. Sequences such as $OZEP$, $OE_O ZP$, and $E_{OP} ZQP$ are operating commercially, where Q represents a chelation stage used to control the presence of metal ions. These and other novel chemicals all have contributed to significant industry efforts to minimize chlorinated discharges to the environment. As of 1996, bleaching technology is changing rapidly.

Oxygen Bleaching. Oxygen bleaching, an effective alternative to chlorine bleaching, is becoming increasingly common in the pulp industry, where there is a move away from elemental chlorine. Oxygen bleaching eliminates about half of the total chlorine used (eq. 25) by removing about half of the residual lignin in the unbleached pulp before the pulp enters the chlorination stage. Another significant feature of oxygen bleaching is that it produces an effluent that is uncontaminated by chloride ion and is therefore recyclable to the recovery system. If the oxygen stage is followed by good washing and the filtrate on the brown-stock washers is used, substantial reductions in biochemical oxygen demand (BOD), chemical oxygen demand (COD), and color can be achieved as well as reductions in the chlorinated organic compounds. The primary reason that oxygen can normally be used to remove only about half of the lignin in the unbleached pulp is that oxygen is less selective than chlorine. The selectivity of oxygen is lower in the presence of free transition-metal ions such as iron, manganese, and copper, which catalyze cellulose degradation reactions. Such catalysis can be minimized by adding magnesium salts, eg, $MgSO_4$ and $MgCO_3$, and by strict control of transition metals (54). Oxygen bleaching reactions are conducted in a solution of sodium hydroxide in both medium (10–12%) and high (25–30%) consistency systems. One of the main advantages of oxygen bleaching is the compatibility of the spent liquors with the kraft recovery system, although existing recovery boilers are often overloaded and cannot take more solids without loss of production.

High Consistency Oxygen. In the 1970s and early 1980s, nearly all of the oxygen bleaching systems were high (25–30%) consistency systems. Because the solubility of oxygen in aqueous solutions is low and decreases with increasing temperature, special attention was given to maintaining adequate supplies of oxygen in solution. Raising the consistency of the pulp to 20–27% provided the necessary surface area and reduced water film thickness for interfacial transfer of oxygen, a gas of low solubility in water. High consistency oxygen delignification is normally conducted at a temperature of about 110°C and a pressure of about 0.6 MPa (90 psig). Approximately 0.15% oxygen and a similar amount of caustic is needed for each unit of kappa number reduction. Magnesium ion is added to about the 0.1% level based on oven-dry pulp. Reaction time can vary between 30–90 min, depending on the reactivity of the pulp; 60 min is typical. The pulp must be pressed, fluffed, and then bleached in a gas-phase oxygen reactor. A high consistency oxygen delignification system is illustrated in Figure 32. A significant cost associated with this approach is that of the upstream dewatering equipment.

Medium Consistency Oxygen. The cost of dewatering, autooxidation, and combustion, especially of pulp impurities such as pitch, has led to the development of processes that operate at medium (10–12%) consistency. Although all of the oxygen needed for reaction in medium consistency oxygen bleaching systems cannot be dissolved into the liquid phase at 12% consistency, new mixers make it possible to disperse sufficient oxygen in the pulp suspension as small bubbles. Subsequent dissolution and reaction take place as the liquid-phase oxygen is depleted of oxygen. In medium consistency systems the pulp is thoroughly washed to minimize carryover of black liquor solids that compete with lignin for reaction with oxygen. In medium consistency systems (Fig. 33) the reaction is conducted in the presence of 2–3 wt % NaOH. Oxygen consumption is about 110 kg equivalent chlorine per metric ton of pulp, about 2.5% oxygen based on oven-dry pulp. Reaction conditions are typically one hour at 85–110°C to reduce the lignin content of softwood kraft pulp from 6 to ca 3 wt %. Because pulp saccharides are susceptible to oxygen degradation, magnesium salts are usually added as protectors. Catalytic metal concentration can be minimized by incorporating a preliminary acid wash or A stage but this is not being practiced commercially.

Compared to the high consistency systems, the medium consistency systems consume more alkali and oxygen, give a somewhat lower degree of delignification, but have a lower capital cost. Greater delignification can be achieved by using two-stage medium consistency systems, which is commercially practiced (75).

Oxidative Extraction Stages. Addition of oxygen to a conventional extraction stage (E_O) found widespread application in the 1980s. In a typical E_O stage, about 5 kg of oxygen per metric ton of pulp is added to a medium consistency mixer at the base of an upflow tower or an upflow preretention tube ahead of a

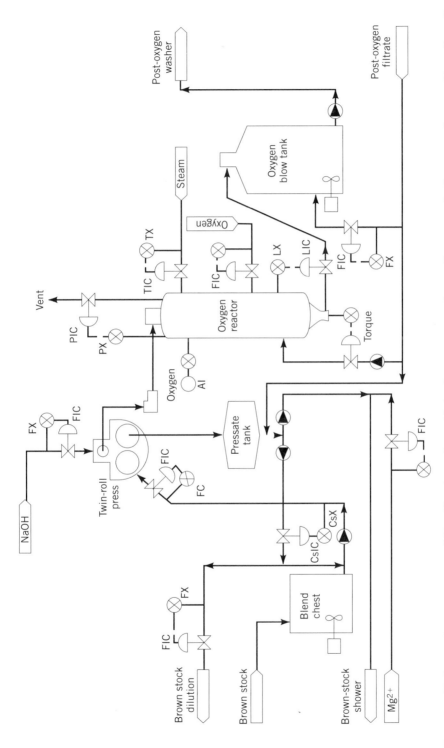

Fig. 32. High consistency oxygen delignification where AI is the oxygen analyzer, C = control, Cs = consistency, F = flow, I = indicator, L = level, P = pressure, T = temperature, and X = sensor (58).

571

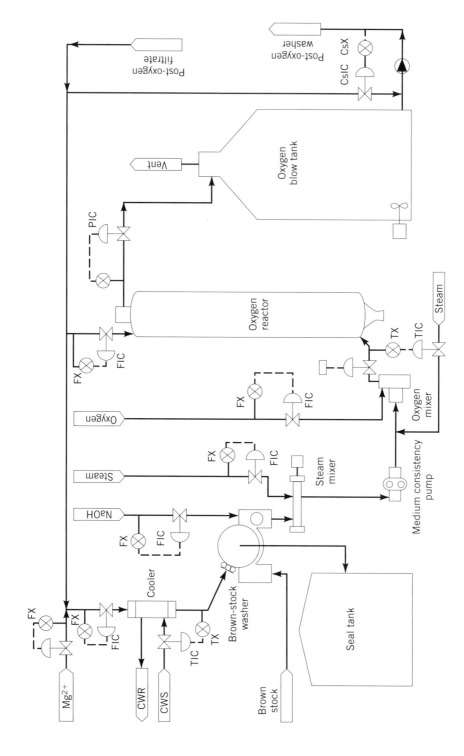

Fig. 33. Medium consistency (10–16% solid by wt) oxygen delignification where CW = cooling water, R = return (outlet), and S = source (inlet). See Fig. 32 for other definitions (58).

downflow tower. This results in a savings of perhaps 3.5 kg of chlorine dioxide in subsequent stages or, alternatively, in a higher fully bleached pulp brightness. Such result can be enhanced by adding hydrogen peroxide together with the oxygen, ie, the E_{OP} stage.

In mid-1990s, oxidative reinforcement of the caustic extraction stage has been employed as a way to further reduce the amount of chlorine applied in the chlorination stage. The effect is enhanced by increasing the temperature in the extraction stage, adding peroxide, pressurizing the upflow tube, increasing the reaction time, or by a combination of these measures. These measures blur the distinction between a conventional oxygen predelignification stage and a reinforced extraction stage. The distinction becomes even more unclear if a pressurized peroxide stage (*PO*) is used, which utilizes caustic, oxygen, and hydrogen peroxide at elevated temperature (105–110°C) and 500 kPa (5 bars) pressure (54). Excellent pulp brightening has been reported.

Ozone Bleaching. Ozone delignification has become important in nonchlorine bleaching sequences. Like oxygen, ozone is a strong oxidizing agent that effectively strips color and breaks down residual lignin. But unlike oxygen, it has the advantage of reacting at atmospheric pressure and room temperature. Typically, ozone is applied at medium (10–12%) consistency at dosage rates of 0.3 to 0.7% based on pulp. High consistency ozonation is also practiced commercially. Some problems involve ozone's tendency to depolymerize cellulose. The drive to develop ozone treatments continues as the industry moves toward elemental chlorine-free and totally chlorine-free technologies. Pulp mills may combine ozone treatment and extended delignification together with oxygen treatment. There were nine commercial installations as of June 1994 that employed ozone in bleaching, having capacity of about 6600 t/d. All but two installations were in Sweden and Finland. A common sequence for kraft pulp is *OZEP* and *ZEP* for sulfite pulp, but the majority of the installed sequences are the *OPZP*-type sequence.

Enzymatic Bleaching. Emerging technologies in biobleaching showing promise for application to the bleach plant have been used commercially. Enzymes such as xylanase that selectively degrade hemicellulose can produce pulps that are easily bleached to high brightness while significantly reducing the amount of chlorine compounds necessary for bleaching. The use of enzyme pretreatment reduces emissions of chlorinated organic substances and can result in cost savings to mills (76). However, the reduction in emission of AOX and the cost savings are mill-specific and may not occur in some situations.

The most commonly used enzyme, xylanase, is produced by batch fermentations (qv) of *Streptomyces lividans*, a high xylanase-producing, cellulase-free microorganism. Xylanase does not appear to have any direct delignifying effect. The mechanisms of action are not completely understood. Most likely, xylosidic bonds are catalytically hydrolyzed, allowing for improved removal of lignin during alkali extraction, washing, or bleaching stages downstream. This improved delignification may possibly result from an interruption of lignin–carbohydrate bonds. Other postulated effects include the removal of reprecipitated xylan on the surface of fibers, which allows for easier access to lignin, and the enlargement of intrafiber pores, which permits the release of lignin fragments. The action of enzymes such as xylanase is highly selective (77).

Derived from bacteria strains or from fungi, each xylanase has a set of specific optimal conditions for activity. Factors that affect enzyme treatment are temperature, pH, enzyme dosage, dispersion, consistency, and reaction time. Conditions are comparatively easy to accommodate, highly flexible, and easily adapted to existing equipment and process conditions. Typical temperatures for enzyme treatment range from 40–55°C; pH, from 4–6. The enzymatic process is easy to apply as pulp comes out of the brown-stock washer and enters the high density storage chest. This compatibility to available processes eliminates the need for expensive capital improvements and retrofitting (78).

Enzymatic reaction time is about two hours. Efficient mixing using screw conveyors or static mixers improves effectiveness. Factors that influence the effectiveness of biobleaching include wood source, bleach sequence, and enzyme type. Benefits, primarily in enhancement of pulp bleaching, are well known. This allows for reductions of chlorine bleaching chemicals and, consequently, reduction in AOX emissions. Other reported advantages are higher brightness ceilings, increased viscosity, excellent specificity of action, yields and strength characteristics comparable to conventionally bleached pulps, and improved mill flexibility (78).

Further research directed toward optimizing bleach sequences, as well as development of biotechnologies to produce other enzymes that can directly delignify pulps with high specificity of action, can be expected.

Bleach Plant Equipment.   Bleach plants are large, complex chemical operations that involve multiple stages (see Fig. 29). Each stage within the bleach plant requires pumping of the pulp, a mixer for introduction of the chemical and mixing of the reagents with the pulp, a tower or reactor where the chemical reaction takes place, and a washer for removing residual chemicals and dissolved lignin fragments. A good review of bleach plant equipment is available (29).

Towers.   Towers are required to provide retention times from 30 minutes, as in the chlorination and hypochlorite stages, to as much as five hours in some chlorine dioxide stages. Upflow and downflow towers are common. Upflow towers are often used for medium consistency stages. These are particularly advantageous when a hydrostatic head must be maintained on a gaseous solution, as in chlorine dioxide stages.

Pumps.   Conventional centrifugal pumps (qv) are used extensively for low consistency stock. However, the upper limit for conventional centrifugal pumps is approximately 5% consistency. Positive displacement pumps are used for pulp of high consistency. These pumps, termed high density or thick stock, operate on a screw and gear principle. The IMPCO Cloverotor pump can convey pulp at a consistency as high as 30%. When medium (8–14%) consistency pulp suspensions are subjected to sufficient shear force, these are fluidized and behave like water with regard to pressure drop (79). This led to the development of the Kamyr medium consistency centrifugal pulps, which are widely used for discharges for towers and pumping (79,80).

Mixers.   Mixers are widely used in bleach plants to add steam; solutions of bleaching chemicals such as chlorine dioxide, sodium hypochlorite, and sodium hydroxide; and bleaching gases such as oxygen and chlorine. Steam mixers are used to add steam directly to medium consistency pulp. Steam mixers, consisting of one or two rotating shafts having pegs mounted on the shafts, are used to break up the pulp and allow steam to intermingle with it. Static mixers have been used for low consistency application of chlorine and chlorine dioxide. Chlorine dioxide was mixed using pump-through mixers for application to medium consistency bleaching. The mixers in these applications are being replaced by a variety of high intensity mixers that have been developed to give intimate contact between the bleaching agent and the pulp.

Washers. The pulp is usually washed after each stage of bleaching to remove dissolved lignin and spent bleaching chemicals. Carryover of chemicals and black liquor solids interferes with bleaching in successive stages and can lead to increased chemical consumption. For example, carryover of hypochloric acid from the chlorination stage can lead to increased sodium hydroxide consumption in the extraction stage. Rotary drum filters are the most common equipment used for this purpose and resemble the rotary drum washers used in brownstock washing.

Displacement Bleach Plants. Multiple-stage bleaching can be accomplished in a single upflow tower in what is termed displacement bleaching. In this technology, commercialized by Kamyr Inc., bleaching solution is allowed to pass downward through the pulp, displacing the solution previously in the pulp. To achieve multiple-stage bleaching, a succession of bleaching agents is passed through the bed of pulp which proceeds upward through a series of diffusion washers all contained within a single tower. Displacement bleaching for the *CEDED* sequence can be conducted in a single upflow tower, and washing takes place only after the last stage. The advantage of displacement bleaching is reduced capital cost and lower steam use. Disadvantages of displacement bleaching are nonuniform bleaching and higher chemical consumption compared to using a series of towers and washers (81).

Pollution Control in the Bleachery. The quantity of water necessary for bleaching, and consequently the volume of effluents, has been decreased significantly by various schemes for recycle of liquors, eg, pulp washing using dilute spent liquors and countercurrent flow. Effort is underway to close bleach plants and further reduce water consumption.

Environmental Management and Pollution Prevention

Environmental protection in the pulp industry manages the amount and type of discharges to soil, water, and air. Management begins in the field when the fiber source is planted to provide high yield trees that can be harvested with minimum impact to soil resources and surrounding areas. Management of clear cutting and protection of critical habitat are considerations and techniques of silviculture. Biogenetic engineering researchers are working to develop fiber resources that grow quickly and produce high quality fibers. In addition, the rapid development of processes to use secondary fiber effectively has placed the industry at the forefront of global efforts to recycle waste paper, conserve forest resources, and protect watersheds.

Most industrialized nations regulate emissions to the environment, principally solids, toxic substances, and other chemicals measured as BOD and COD. At the facility and in the woodyard, nonpoint sources of runoff are contained and treated. As the fiber source passes through consecutive stages of processing, each operation is designed to minimize waste, energy use, and chemical discharges by means of efficient use of water and raw materials, recycling, substituting chemicals such as oxygen for chlorine, scrubbing, and other process controls.

Discharges to receiving water bodies must meet water quality standards to protect drinking water and aquatic life. Effluents contain lignin derivatives from pulping and bleaching operations and carry nutrients, suspended solids, color bodies, and factors that affect pH, dissolved oxygen, foaming, and other chemicals bearing upon water quality in the receiving water bodies. After dioxin was discovered, the industry made significant efforts to reduce and in some cases to eliminate the use of elemental chlorine by substituting chlorine dioxide, oxygen, ozone, hydrogen peroxide, or alternative bleaching agents in various sequences. Methods in common practice include extended delignification in the digester and manipulating the extent of delignification at succeeding stages, enzyme treatment, better brown-stock washing, and optimal use of countercurrent washing. These process changes have significantly decreased the discharge of organic compounds measured as BOD or COD.

Treatment of Liquid Effluents.   Before pulp mill effluents can be released to the environment they must be treated. Primary treatment involves the use of settling ponds or tanks in which suspended solids settle out of the liquid effluent. Solids can be composted and spread on land, converted to other useful products, or incinerated. Secondary effluent treatment includes oxidation and aeration in shallow basins having wide areas or in smaller areas using mechanical agitators and spargers to oxygenate fluids before release. Biological filter systems can be used to remove organic compounds and heavy metals, and often the process can be accelerated by adding nutrients and by using oxygen rather than air. In some areas natural wetland systems have been designed to achieve this. Other means of treatment include lime coagulation and the elevation of pH to precipitate organic color bodies as calcium lignates. Once precipitated, the sludge is dewatered and incinerated to destroy organics.

Air Pollutants.   Air pollution regulations are primarily concerned with minimizing particulates, odor, ground-level ozone precursors, acid rain, and smog. Priority pollutants include chemicals such as acetone, methanol, sulfuric acid, sulfur dioxide, hydrogen sulfide and other reduced sulfur compounds, and nitrogen dioxide emissions, which are monitored closely. Volatile organic compounds such as alcohols, terpenes, and phenols can be controlled by incineration. Sour condensates are collected, steam stripped, and reduced sulfur compounds destroyed by incineration in the lime kiln (see Fig. 24). Similarly, sulfur-bearing off-gases are collected and destroyed by incineration. Monitoring is often performed on-line using continuous monitoring devices for stack effluents as well as ambient air. Pollution control equipment includes catalytic and thermal systems to destroy pollutants, electrostatic precipitators to remove particulate matter, cyclonic separators, venturi scrubbers, and fabric filters. Pollutants in molecular form are removed primarily by adsorption (qv), ie, dry scrubbing, and packed bed and venturi-type scrubbers, ie, wet scrubbing, from the process stream for reuse or for destruction (see AIR POLLUTION CONTROL METHODS).

In the United States, the Clean Air Act of 1990 requires plants to reduce emissions of 189 toxic and carcinogenic substances such as chlorine, chloroform, and 2,3,7,8-TCDD (dioxin) by 90% over the 1990s. The U.S. Environmental

Protection Agency is working to develop standards based on maximum achievable control technologies and the industry has invested billions of dollars in capital investments to retrofit or rebuild plant equipment to meet these measures.

Closed-Mill Concept. The closed-mill concept, or water circuit closure, has been studied by the pulp and paper industry for many years. In some parts of the paper manufacturing process, up to 98% of the water is recycled within the process, eg, the wet end of the paper machine. However, in the pulp mill, especially kraft mills, effluents are produced owing to the need to purge from the system various metals that come in with the wood, as well as organic by-products from the pulping process, additives, and especially chloride ions that originate in the bleach plant.

One notable attempt in Canada to close a kraft mill by recycling bleach plant effluents through the kraft chemical recovery process failed on account of the buildup of corrosive materials (2). A primary barrier is limited tolerance for recycling of chlorides in the conventional kraft system. However, advances in low kappa cooking, ie, extended delignification, and TCF bleaching have stimulated interest in minimizing the environmental impact of kraft mills (2). Theoretically, it should be possible to recycle TCF bleach plant effluents to the recovery boiler. It is also possible to deal with chloride buildup in non-TCF pulping. Removal of chloride from electrostatic precipitator dust can be achieved by several processes, eg, Champion's BFR process. However, recovery of bleach plant filtrates causes an additional dry solids load to the recovery boiler. It is also known that increasing system closure can increase considerably the risk of corrosion from tramp ions, especially from potassium and chloride ions that occur naturally in the wood. These questions and the increased use of hydrogen peroxide, oxygen, and ozone as bleaching agents have stimulated a reevaluation of the metals present in bleach plant filtrates and of the methods for their control and removal. Techniques such as sequestering, crystallization (qv), ion exchange (qv), and evaporation are being considered. Miller Western's Meadow Lake BCTMP mill is an example of successful closure of the water cycle in a mechanical pulp mill.

Economic Aspects

Pulp production and per capita consumption of paper and board for 1992 is shown in Table 8. The United States, Canada, Sweden, Finland, and Norway make up the North American and Scandinavian (NORSCAN) countries and produced about 63% (22.8 million tons) of the world output. Market share is growing for producers in Latin and South America, Western Europe, Asia, and Africa. These areas provide low cost pulp from state-of-the-art mills. Mills in the third world countries often enjoy the benefits of plentiful, fast-growing tree species, such as eucalyptus and tropical pines, and lower operating and labor costs (3).

The United States is the largest importer of pulp, in 1993 importing 5.5 million tons, principally bleached and semibleached kraft pulp. It is also the largest producer of pulp, paper, and allied products. Table 9 shows U.S. pulp production for 1983–1992.

Market prices are extremely volatile and follow cycles of advance and decline related to fluctuations in global inventories and world economic forces.

Table 8. Principal Pulp Producing and Consuming Countries in 1992, t × 10³ᵃ

Country	Pulp production	Paper and paperboard consumption
United States	59,282	78,757
Canada	22,841	5,317
China	11,985	19,464
Japan	11,200	28,318
Sweden	9,589	1,742
CIS[b]	6,800	5,955
Brazil	5,368	3,962
France	2,609	9,092
South Africa	2,320	1,554
Germany	2,240	15,649

[a]Ref. 82.
[b]Commonwealth of Independent States.

Table 9. U.S. Production of Wood Pulp by Type, t × 10³ᵃ

Pulp type	1983	1986	1989	1992
AFPA paper and board grades				
sulfite				
bleached	1,306	1,275	1,373	1,427
unbleached	310	164	190	
sulfate				
bleached and semibleached	21,107	24,158	26,949	29,703
unbleached	19,590	20,481	21,669	22,228
groundwood				
stone and refiner	3,225	3,010	3,058	2,917
thermomechanical	1,842	2,466	2,971	3,584
semichemical	3,812	4,191	4,363	4,101
α-Cellulose and dissolving pulps				
	1,261	1,257	1,425	1,383
Total AFPA grades	*52,453*	*57,002*	*61,173*	*65,343*
Other grades				
defibrated or exploded	620	230	175	25
groundwood E	400	230	175	175
Total other grades	*1,050*	*800*	*600*	*600*
Total	*53,503*	*57,802*	*62,598*	*65,943*

[a]Ref. 83.

For example, for northern bleached softwood kraft, which is considered a global benchmark pulp grade, market price went from $840/t in 1990 down to $440/t in 1993 as a result of oversupply and worldwide recession (3). By 1995 virtually all of the price losses had been recovered.

Environmental concerns and regulations have had a significant impact on global markets for pulp. In Germany, consumer demand has shifted toward totally chlorine-free pulps in order to eliminate pollution from harmful dioxin compounds. For example, Sweden used 150,000 metric tons of chlorine in 1987; in 1993, none. The growth in markets for ECF and TCF pulps is indicated in the market growth from 4.3 million tons in 1990 to 27.9 million metric tons in 1994. However, TCF pulp remains a small factor in overall market pulp. As of 1996, the significance of TCF pulp is almost entirely in Europe (76). Canada has brought dioxins to immeasurable levels using TCF production.

Increasing consumption of recovered paper is expected to remain a significant factor in patterns of fiber production and demand, and is likely to increase as governments legislate increased recovery of fiber from municipal solid waste. Abundant sources of secondary fiber offer world papermakers a less expensive alternative to virgin fiber.

Improvements in profitability of mills result from continuing technological advances providing higher yields, conservation of water, optimization of chemical processes, and more efficient use of energy.

Energy.   The pulp and paper industry is a significant industrial consumer of energy. Much of the energy used, however, is self-generated from combustion of residues such as spent pulping liquors, bark, hogged wood, and other nonprocessed wood. For example, in the United States in 1994, pulp and paper industry was ranked the third highest, behind metals and chemicals, in energy consumption, amounting to 8% of overall manufacturing costs. Total energy usage of U.S. industry in 1992 was ca 275×10^{15} J (2600×10^{12} Btu). Of this, 56.2% was supplied from self-generated and residue fuels (82). This energy is produced on-site in company-owned electric plants or in cogeneration arrangements where steam is provided to the mill and electricity to local utilities. Some companies are large woodlands owners and can utilize wood as process fuel. The industry in general has increased the use of biomass by 17% in the form of bark and hogged wood. Black liquor solids provide about 40% of the fuel used for steam and power generation.

Energy use varies by grade as does the amount of residue available for energy use. Thermomechanical pulp (TMP) for newsprint and chemimechanical pulps use a relatively higher amount of energy that must be purchased from outside sources. Unbleached kraft pulp uses less energy because half of the energy requirement is supplied by burning black liquor solids or other residues. Recycled pulps consume less energy but do not supply usable by-products such as black liquor (84).

Significant changes are likely to occur in the energy self-sufficiency of the industry, resulting in an increased dependency on purchased power. Whereas new installations and retrofits of existing technologies can result in greater energy efficiency, demands from process changes, oxygen/ozone generation requirements, and increased environmental regulations may create even greater demand (84). Industry efforts in the area of energy use are expected to focus on increased energy conservation, greater utilization of biomass and other renewable energy sources, and process improvements such as better drying efficiency. Emerging technologies are most likely to have a great impact on energy use and generation. Black liquor and biomass gasification designs are likely to play a significant role in the future. Many expect these systems to operate at considerably greater efficiency than conventional boilers. In this type of gasification system, black liquor and other types of biomass are fed directly into the gasifier, where cleaned steam exits to drive gas turbines and produce process steam and more power further downline (84).

BIBLIOGRAPHY

"Pulp" in *ECT* 1st ed., Vol. 11, pp. 250–277, by R. S. Harch, Hudson Pulp and Paper Corp.; in *ECT* 2nd ed., Vol. 16, pp. 680–727, by G. H. Tomlinson II, Domtar Ltd.; in *ECT* 3rd ed., Vol. 19, pp. 379–419, by J. Minor, U.S. Forest Products Laboratory.

1. D. Fengel and G. Wegener, *Wood-Chemistry, Ultrastructure, Reactions*, Walter De Gruyter Inc., Berlin, 1984, pp. 132–181.
2. G. A. Smook, *Handbook for Pulp and Paper Technologists*, 2nd ed., TAPPI Press, Atlanta, Ga., 1992.
3. USDOC, International Trade Administration, *U.S. Industrial Outlook*, Paper and Allied Products, Washington, D.C., 1994, Chapt. 10.
4. C. E. Dunning, *TAPPI J.* **52**, 1326 (1969).
5. T. F. Bobbitt, J. H. Nordin, M. Roux, J. F. Revol, and R. H. Marchessault, *J. Bacteriol.* **132**(2), 691 (1977).
6. A. J. Kerr and D. A. I. Goring, *Cell. Chem. Technol.* **9**, 563–573 (1975).
7. M. J. Kocurek and C. F. B. Stevens, "Properties of Fibrous Raw Materials and Their Preparation," *Pulp and Paper Manufacture*, 3rd ed., Vol. 1, TAPPI Press, Atlanta, Ga., 1983.
8. J. E. Stone and A. M. Scallan, *J. Polym. Sci. Part C* **11**, 13 (1965).
9. H. Tarkow, W. C. Feist, and C. F. Southerland, *For. Prod. J.* **16**(10), 61 (1966).
10. E. Sjöström, *Wood Chemistry—Fundamentals and Applications*, Academic Press, Inc., New York, 1981, Chapt. 4, pp. 68–82.
11. K. V. Sarkanen and C. H. Ludwig, *Lignins—Occurrence, Formation, Structure and Reactions*, John Wiley & Sons, Inc., 1971, New York, pp. 165–235.
12. S. A. Rydholm, *Pulping Processes*, Interscience Publishers, New York, 1965, pp. 166–218.
13. J. Gierer, *Svensk Papperstidn.* **73**, 571 (1970).
14. J. Yan, *Macromolecules* **14**, 1438–1445 (1981).
15. E. Adler, *Wood Sci. Technol.* **11**, 169 (1977).
16. D. A. I. Goring, in Ref. 11, pp. 695–768.
17. J. L. Minor, *Wood Chem. Technol.* **2**(1), 1–16 (1982).
18. D. Fengel, *Holzforschung* **30**, 1–6, 73–78, 143–148 (1976).
19. B. Kosiková, L. Zakunta, and D. Joniak, *Holzforschung* **32**, 15–18 (1978).
20. J. R. Obst, *TAPPI J.* **65**(4), 109 (1982).
21. W. S. Fuller, in K. J. Kocurek and C. F. B. Stevens, eds., *Wood Procurement Recovery and Measurement in Pulp & Paper Manufacture*, 3rd ed., Vol. 1, TAPPI Press, Atlanta, Ga., 1983, p. 91.
22. R. A. Leask and M. J. Kocurek, *Pulp and Paper Manufacture*, Vol. 2, *Mechanical Pulping*, 3rd ed., The Joint Committee of the Paper Industry, TAPPI/CPPA, Atlanta, Ga., 1987.
23. N. Hartler and Y. Stade, in J. V. Hatton, ed., *Chip Quality Monograph*, TAPPI Press, Atlanta, Ga., 1979, Chapt. 13.
24. J. C. W. Evans, *Pulp Pap.* **54**(6), 76 (1980).
25. W. L. Bohn, *Proceedings of the 1990 Annual Meeting*, TAPPI Press, Atlanta, Ga., 1990, pp. 47–52.
26. S. Heimburger and co-workers, *Proceedings of the 1993 TAPPI Pulping Conference*, TAPPI Press, Atlanta, Ga., 1993, pp. 205–213.
27. O. Franzon and O. Samuelson, *Svensk Papperstidn* **60**(23), 872 (1957).
28. A. R. Procter, W. Q. Yean, and D. A. I. Goring, *Pulp Pap. Can.* **68**, T445 (1967).
29. T. M. Grace, E. W. Malcolm, and M. J. Kocurek, *Pulp and Paper Manufacture*, Vol. 5, *Alkaline Pulping*, 3rd ed., TAPPI/CPPA, Atlanta, Ga., 1989, p. 121.

30. P. Sandstrom, H. Lundberg, and A. Teder, *Development of Modified Kraft Process with a Mathematical Model for Continuous Digesters*, SCAN Forest Report No. 441, Swedish Forest Products Lab, Stockholm, Sweden, 1985; A. Teder and P. Sandstrom, *Tappi J.* **68**(1), 94–95 (Jan. 1985).

31. B. Johansson and co-workers, *Pulp Pap.* **58**(11), 124–127 (Nov. 1994).

32. P. J. Kleppe, *Tappi J.* **53**(1), 35–47 (1970).

33. K. E. Vroom, *Pulp Pap. Can.* **88**(3), 228 (1957).

34. C. H. Wells, *TAPPI J.* **73**(4), 181 (1990).

35. M. J. Ducey, *Pulp Pap.* **63**(2), 77 (Feb. 1, 1989).

36. S. C. Stultz and J. B. Kitto, *Steam*, 40th ed., Babcock and Wilcox Co., Barberton, Ohio, 1992.

37. J. E. Westling and W. T. McKean, "The Direct Alkaline Recovery System (DARS), an Alternative to Kraft Recovery," *1982 AIChE Fall Meeting*, Los Angeles, Calif., Nov. 14–19, 1982; J. E. Janson, *Pa. Puu.* (8), 495 (1979).

38. R. P. Green and Z. C. Prusas, *Pulp Pap. Can.* **76**(9), T272 (1975).

39. E. Hagglund, *Svensk. Papperstid.* **49**, 204 (1946); P. J. Kleppe and K. Kringstad, *Norsk Skogind.* **17**(11), 428 (1963).

40. N. Sanyer and J. F. Laundrie, *TAPPI J.* **47**(10), 640 (1964).

41. G. C. Smith, R. P. Green, and S. E. Knowles, *Paper Trade J.* **159**(13), 38 (1975).

42. B. Bach and G. Fiehn, *Zell. Pap.* **21**(1), 3 (1972).

43. H. H. Holton, *Pulp Pap. Can.* **78**(10), T218 (1977); U.S. Pat. 4,012,280 (Mar. 15, 1977) (to Canadian Industrial Ltd.).

44. L. L. Landucci, *TAPPI J.* **63**(7), 95 (1980).

45. J. Gierer, O. Lindeberg, and I. Noren, *Holzforschung* **33**, 213 (1979).

46. L. Lowendahl and O. Samuelson, *TAPPI J.* **61**(2), 19 (1978).

47. F. L. A. Arbin, L. R. Schroeder, N. S. Thompson, and E. W. Malcolm, *TAPPI J.* **63**(4), 152 (1980).

48. T. J. Fullerton and S. P. Ahern, *J. Chem. Soc. Chem. Commun.*, 457 (1979).

49. I. Gourang, R. Cassidy, and C. W. Dence, *TAPPI J.* **62**(7), 43 (1979).

50. D. Dimmel, *J. Wood Chem. Technol.* **5**, 1 (1985).

51. T. J. Fullerton, *J. Wood Chem. Technol.* **4**(1), 61 (1984).

52. B. F. Greenwood, *Proceedings of the 1988 AIChE Forest Products Division*, Washington, D.C., Nov. 27–Dec. 2, 1988, pp. 9–44.

53. B. Dillner, *Papermaker*, 35–36 (Feb. 1993).

54. S. Andtbacka and P. Tibbling, *Pap. Puu.* **76**(9), 580–584 (1994).

55. R. Marton, A. Brown, and S. Granzow, "Oxygen Pulping of Thermomechanical Fiber," *1974 Non-Sulfur Pulping Symposium*, Oct. 16–18, Madison, Wis., TAPPI Press, Atlanta, Ga.

56. T. S. Nagano, S. Miyao, and K. Takeda, "Hopes Oxygen Pulping Process," in Ref. 55.

57. J. L. Minor and N. Sanyer, "Oxygen Pulping of Wood Chips with Sodium Carbonate," in Ref. 55; J. P. Hanson and S. I. Kokolich, "Semichemical Pulping With Na_2CO_3 and NaOH Combinations," in Ref. 55.

58. *1987 International Oxygen Delignification Conference, June 7–12, San Diego, Calif.*, TAPPI Press, Atlanta, Ga., 1987, pp. 1–13.

59. Brit. Pat. 385 (1867), B. C. Tilghman.

60. J. P. Casey, ed., *Pulp and Paper Chemistry and Chemical Technology*, 3rd ed., Vol. 1, John Wiley and Sons, Inc., New York, 1980, p. 301.

61. G. Gellerstedt, *Svensk Papperstidn.* **79**(16), 537 (1976), and references therein.

62. W. Q. Yean and D. A. I. Goring, *Svensk Papperstidn.* **71**(20), 739 (1968).

63. R. P. Whitney, N. S. Thompson, G. A. Nicholls, and S. T. Han, *Pulp Pap.* **43**(8), 68 (1969).

64. P. Stockburger, *TAPPI J.* **76**(6), 71–74 (June 1993).

65. G. W. Kutney, *Pulp Pap.*, 85–86 (Jan. 1995).

66. P. E. Shick, paper presented at *TAPPI Conference on Alkaline Pulping and Secondary Fibers*, Washington, D.C., Nov. 7–10, 1977.

67. W. H. Rapson, *Tappi Mono. Ser.* (27) (1963).

68. R. C. Van Lee, *TAPPI J.* **70**(11), 185 (1987).

69. T. J. McDonough, *Kraft Pulp Bleaching Technology: A Brief Overview of Basic Principles and Current Trends*, Institute of Paper Chemistry, Atlanta, Ga., 1992.

70. R. P. Singh, *The Bleaching of Pulp*, 3rd ed., TAPPI Press, Atlanta, Ga., 1979, p. 212.

71. W. H. Rapson, *TAPPI J.* **39**(5), 289 (1956).

72. R. H. Reeves and I. A. Pearl, *TAPPI J.* **48**(2), 121 (1965).

73. T. D. Spittler and C. W. Dence, *Svensk Papperstidn.* **80**(9), 275 (1977).

74. P. F. Earl and D. W. Reeve, *TAPPI Proceedings, 1989, Environmental Conference*, TAPPI Press, Atlanta, Ga., 1989.

75. S. Martikainen, O. Pikka, and Nummenaho, *Pap. Puu., Paper Timber* **76**(9), 564–562 (1994).

76. T. J. McDonough, *TAPPI J.* **78**(3) (Mar. 1995).

77. C. Daneault, C. Leduc, and J. L. Valade, *TAPPI J.* **77**(6), 125–129 (June 1994).

78. D. J. Senior, J. Hamilton, R. L. Bernier, and J. P. Manoir, "Reduction in Chlorine Use During Bleaching of Kraft Pulps Following Xylanase Treatment," in *Bleaching: A TAPPI Press Anthology*, TAPPI Press, Atlanta, Ga., 1994.

79. J. Gullichsen and E. Harkonen, *TAPPI J.* **64**(6), 69–72 (1981).

80. J. Gullichsen, E. Harkonen, and T. Niskanen, *TAPPI J.* **64**(9), 113–116 (1981).

81. W. H. Rapson and C. B. Anderson, *TAPPI J.* **49**(8), 329–334 (Aug. 1966).

82. C. Brookshaw and V. Pelletier, eds., *Pulp and Paper 1994 North American Factbook*, 1994.

83. *1993 Statistics of Paper, Paperboard, and Woodpulp*, American Forest and Paper Association, Washington, D.C.

84. R. L. Erickson and D. R. Raymond, *Papermaker*, 28–30 (Mar. 1995).

General References

References 1, 2, 9, 10, 11, 12, 21–23, 29, 56, and 58 are excellent general references.

A. J. Panshin and C. de Zeeuw, *Textbook of Wood Technology*, 3rd ed., McGraw-Hill Book Co., Inc., New York, 1970.

J. V. Hatton, ed., *Pulp and Paper Technology Series*, No. 5, *Chip Quality Monograph*, Joint Textbook Committee of the Paper Industry, TAPPI Press, Atlanta, Ga., 1979.

J. d' A. Clark, *Pulp Technology and Treatment for Paper*, Miller Freeman Publications, San Francisco, Calif., 1978.

D. Fengel and G. Wegener, *Wood Chemistry and Ultrastructure Reactions*, Walter de Gruyter, New York, 1983.

R. A. Leask and M. J. Kocurek, eds., *Pulp and Paper Manufacture*, 3rd ed., Vol. 2, *Mechanical Pulping*, Joint Textbook Committee of the Paper Industry, TAPPI Press, Atlanta, Ga., 1987.

G. Hough, ed., *Chemical Recovery in the Alkaline Pulping Processes*, TAPPI Press, Atlanta, Ga., 1985.

D. Kiiskibai, *Pap. Puu Paper Timber*, **76**(9), 574–579 (1994).

G. Hough, ed., *Chemical Recovery in the Alkaline Pulping Process*, TAPPI Press, Atlanta, Ga., 1985, p. 197.

JOSEPH M. GENCO
University of Maine

PUMPS

Pumps are used in a wide range of industrial and residential applications. Pumping equipment is extremely diverse, varying in type, size, and materials of construction. There have been significant developments in the area of pumping equipment since the early 1980s. There are materials for corrosive applications (1,2); modern sealing techniques (3,4); improved dry-running capabilities (5,6) of sealless pumps, which are magnetically driven or canned motor types; and applications of magnetic bearings in multistage high energy pumps (7,8). The passage of the Clean Air Act of 1990 by the U.S. Congress, a heightened attention to a safe workplace environment, and users' demand for better equipment reliability have all led to improved mean time between failures (MTBF) and scheduled maintenance (MTBSM).

Classification

One general source of pump terminology, definitions, rules, and standards is the Hydraulic Institute (HI) Standards (9), approved by the American National Standards Institute (ANSI) as national standards. A classification of pumps by type, as defined by the HI, is shown in Figure 1.

Pumps are divided into two fundamental types based on the manner in which pumps transmit energy to the pumped media: kinetic or positive displacement. In the first type, a centrifugal force of the rotating element, called an impeller, impels kinetic energy to the fluid, moving the fluid from pump suction to the discharge. The second type uses the reciprocating action of one or several pistons, or a squeezing action of meshing gears, lobes, or other moving body, to displace the media from one area into another, ie, moving the material from suction to discharge. Sometimes the terms inlet, for suction, and exit or outlet, for discharge, are used. The pumped medium is usually liquid. However, many designs can handle solids in suspension, entrained or dissolved gas, paper pulp, mud, slurries, tars, and other exotic substances, which, at least by appearance, do not resemble a liquid. Nevertheless, an overall liquid behavior must be exhibited by the medium in order to be pumped. In other words, the medium must have negligible resistance to tensile stresses.

The Hydraulic Institute classifies pumps by type, not by application. The user, however, must ultimately deal with specific applications. Often, based on personal experience, preference for a particular type of pump develops. This preference is passed on in the particular industry. For example, boiler feed pumps are usually of a multistage diffusor barrel type, especially for the medium and high energy (over 750 kW (1000 hp)) applications, although volute pumps in single- or multistage configurations, having radially or axially split casings, have also been applied successfully. Examples of pump types and applications and the reasons behind applicational preferences follow.

Operating Conditions

Before a pump selection can be made, the duty conditions must be specified. These include type of fluid, density or specific gravity, temperature, viscosity, flow, inlet and outlet pressures, and presence of solids or corrosive/erosive

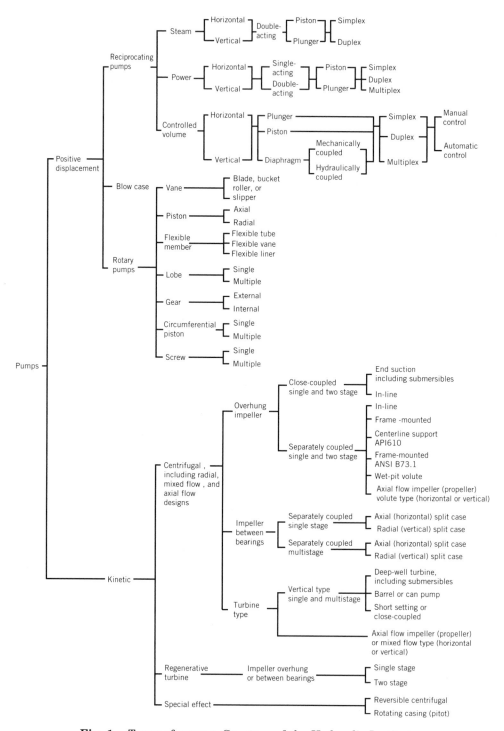

Fig. 1. Types of pumps. Courtesy of the Hydraulic Institute.

material in the liquid. For a typical process installation, an accurate estimate of the pumping system is required, starting with the source of the liquid (tank, vessel, pipeline, basin, etc) through the planned system layout to the terminal point (see PIPELINES; PIPING SYSTEMS). Piping sizes must be determined, based on suitable flowing velocities for the fluid or for the fluid mixture in the case of slurries. Pressure drops through the entire piping system must be estimated, including all valves, fittings, process equipment such as heat exchangers, heaters, and boilers, and any losses through orifices or control valves. Difference in pressures between the suction and discharge location, together with changes in elevation between these two, completes the various elements of the system and makes it possible to estimate or design a complete pumping system.

Most process plant engineers utilize some form of preprinted pump calculation worksheet or a computer to aid in the collection of information. This helps to ensure that, for a given system, all possible pumping situations, alternative routes, alternative pressures or temperatures, varying flowing rates, varying fluid properties, etc, are considered. Good examples of pumping systems evaluation are available (10). The most severe or limiting case is then chosen as the pump-rated condition. At the same time, however, care must be taken not to demand the pump to operate at such a variety of conditions that it is forced to perform either below the minimum allowable flow or too far out in the flow. For best reliability, extra pumps should be included whenever needed to limit the operation of each pump to within the allowable flow limits.

Capacity. Pumps deliver a certain capacity, Q, sometimes referred to as flow, which can be measured directly by venturi, orifice plate (11), or magnetic meters (12) (see FLOW MEASUREMENT). The indirect way to determine capacity is often used. Whereas this method is less accurate than applying a flow meter, it often is the only method available in the field. The total head is measured and the capacity found from the pump head–capacity (H–Q) curve (Fig. 2). More

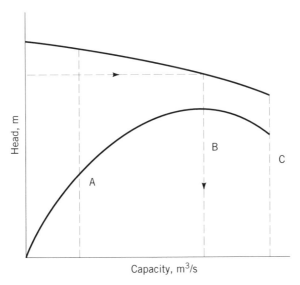

Fig. 2. Head–capacity curve, where A represents the minimum allowable capacity; B, the best efficiency point (BEP); and C, the maximum allowable capacity.

recently, sonic flow meters (13) have been used, which can be installed on the piping without the need for pipe disassembly. These meters are simple to use, but require relatively clean single-phase liquid for reliable measurements.

Head.   The true meaning of the total developed pump head, H, is the amount of energy received by the unit of mass per unit of time (14). This concept is traceable to compressors and fans, where engineers operate with enthalpy, a close relation to the concept of total energy. However, because of the almost incompressible nature of liquids, a simplification is possible to reduce enthalpy to a simpler form, a Bernoulli equation, as shown in equations 1–3, where g is the gravitational constant, SG is specific gravity, γ is the density equivalent, H_s is suction head, H_d is discharge head, and H is the pump head, ie, the difference between H_d and H_s.

$$H_s = \frac{p_{s,g}}{\gamma} + \frac{V_s^2}{2\,g} + Z_s = \frac{p_{s,g} \times 2.31}{SG} + \frac{V_s^2}{2\,g} + Z_s \tag{1}$$

$$H_d = \frac{p_{d,g}}{\gamma} + \frac{V_d^2}{2\,g} + Z_d = \frac{p_{d,g} \times 2.31}{SG} + \frac{V_d^2}{2\,g} + Z_d \tag{2}$$

$$H = H_d - H_s = \frac{p_{d,g} - p_{s,g}}{\gamma} + \frac{V_d^2}{2\,g} - \frac{V_s^2}{2\,g} + (Z_d - Z_s) \tag{3}$$

As can be seen from these equations and Figure 3, the head ingredients are static (pressure, p), dynamic (velocity, V), and elevation, Z, components. A good explanation of this subject is available (15).

Power.   There are two main ways to measure the power delivered by the driver to the pump. The first method is to install a torque meter between the pump and the driver. A torque meter is a rotating bar having a strain gauge to measure shear deformation of a torqued shaft. Discussion of the principle of torque meter operation is available (16). The benefit of this method is direct and accurate measurements. The power delivered to the pump from the driver is calculated from torque, T, and speed (rpm) in units of brake horsepower, ie, BHP (eq. 4a) when T is in lbs·ft, and kW (eq. 4b) when T is N·m.

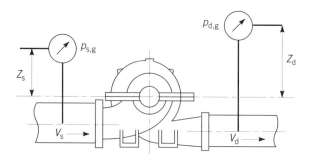

Fig. 3.   Definition of the variables used to determine pump head where (———) represents the positions for measurement. See text (eqs. 1–3).

$$\text{BHP} = \frac{T \times \text{rpm}}{5252} \tag{4a}$$

$$\text{kW} = \frac{T \times \text{rpm}}{9545} \tag{4b}$$

The disadvantages of this method are the need for a torque meter, longer total length of the pumping unit, and greater susceptibility to misalignment and vibrations. This method is used only at a manufacturer's test facilities or research laboratories. It is not used in the field.

In the second method, the pump and the motor are coupled directly, and either power (in kilowatts) or the current, I, and voltage, U, are measured at the motor terminals. To determine the power actually transmitted into a pump, the motor power factor (PF) and efficiency ($\text{Eff}_{\text{motor}}$) must be known. These values are usually taken from the motor manufacturer's calibration curves (17).

For a three-phase electric motor, the horsepower is calculated as

$$\text{kW} = \frac{I \times U \times (3)^{1/2} \times \text{PF} \times \text{Eff}_{\text{motor}}}{1000} \tag{5}$$

$$\text{BHP} = \frac{\text{kW}}{0.746} \tag{6}$$

If the pump manufacturer uses motors for pump power measurement, these motors are calibrated to determine the horsepower from the electric power reading and calibration curves. Such test motors are recalibrated periodically, ensuring the same degree of accuracy as shown by the torque meters.

Efficiency. A portion of the power delivered by the driver to the pump is spent to overcome hydraulic losses, ie, fluid friction, separation, and mixing; volumetric losses, ie, leakage across wear rings; and mechanical losses, ie, bearings friction, internal rubbing, and mechanical seal or packing friction losses. The disk friction (18) of the impeller rotating inside the casing is hydraulic in nature, but is customarily grouped together with mechanical power losses because these latter losses are caused by the fluid trapped between the impeller shrouds and casing walls. The difference between the driver power and the pump losses equals the power delivered to the fluid, ie, the pumped medium.

$$\text{kW}_{\text{fluid}} = \text{kW} - \text{kW}_{\text{losses}} \text{ (metric)} \tag{7a}$$

$$\text{HP}_{\text{fluid}} = \text{BHP} - \text{HP}_{\text{losses}} \text{ (U.S.)} \tag{7b}$$

The pump overall efficiency is then defined as

$$\text{Eff} = \frac{\text{kW}_{\text{fluid}}}{\text{kW}} = \frac{Q \times H \times \text{SG}}{\text{kW} \times 0.102} \tag{8a}$$

when Q is in m³/s and H in m; and

$$\text{Eff} = \frac{\text{HP}_{\text{fluid}}}{\text{BHP}} = \frac{Q \times H \times \text{SG}}{\text{BHP} \times 3960} \tag{8b}$$

when Q is in gal/min and H in ft. More details on overall pump efficiency, as well as the various components, ie, hydraulic, volumetric, and mechanical, can be found in the literature (14).

Specific Speed. A review of the dimensionless analysis (qv) as related to pumps can be found in Reference 14. One of these nondimensional quantities is called the specific speed. The universal dimensionless specific speed, Ω_s, is defined as in equation 9:

$$\Omega_s = \frac{\Omega(Q)^{0.5}}{(gH)^{0.75}} \tag{9}$$

Ω, Q, g, and H may be in any consistent set of units. In equation 9, Ω, the rotational speed, is in rad/s; Q, in m^3/s; H, in m; and g, the gravitational acceleration, is 9.81 m/s^2. The specific speed is calculated at the best efficiency point for the maximum impeller diameter. In the United States, it is customary to drop the gravitation constant and represent specific speed, NS, in the following form:

$$NS = \frac{N(Q)^{0.5}}{H^{0.75}} \tag{10}$$

where N is the rotational speed in rpm, Q is in gal/min, and H in ft.

In Europe and elsewhere, specific speed is calculated as

$$nS = \frac{N(Q)^{0.5}}{H^{0.75}} \tag{11}$$

where n, the rotational speed, is in rpm; Q and H are in m^3/s and m, respectively. The various types of specific speed are all interconvertible:

$$NS = \Omega_s \times 2733 \qquad nS = \Omega_s \times 53.0 \qquad NS = nS \times 51.6 \tag{12}$$

For double suction pumps, using the HI convention, Q is taken as the total pump flow, although some publications use half-flow, ie, flow per impeller eye. For multistage pumps, the developed head must be taken per stage for the NS calculation. By definition (eq. 9), high head, low flow pumps have low specific speed; low head, high flow pumps, such as turbine and propeller pumps, have high specific speed.

Pump efficiency is related to specific speed, flow, and surface roughness (19). Numerous attempts have been made to relate efficiency of any pump to these parameters in a single formula or chart. In practice, however, too many variations exist in design, casting quality, and other factors to allow for a single correlation. Nevertheless, approximate charts have been developed for average achieved efficiencies of commercial pumps, as shown in Figure 4 (20). Although this chart is limited to single-stage pumps, it is a useful tool for making quick efficiency estimates and comparisons.

Suction and Suction Specific Speed. Just as head, capacity, power, and efficiency describe a discharge performance of a pump, a net positive suction head (NPSH) characterizes the suction performance (14). In order to be pumped, the fluid must be in liquid state. No or very little vaporization should occur.

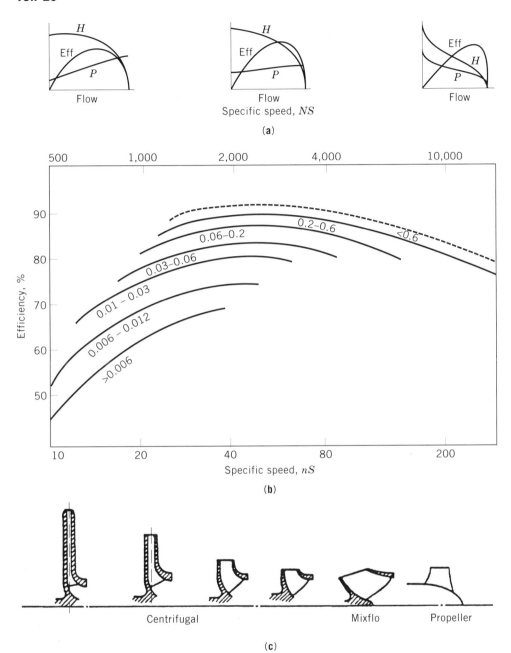

Fig. 4. Chart for efficiency estimates and curve shapes, where (**a**) represents curve shapes showing the relationship between efficiency (Eff), head (*H*), and power (*P*) as a function of flow; (**b**) specific speed, where the numbers represent flow in m^3/s; and (**c**) impeller profiles.

If the liquid pressure drops below a substance's vapor pressure in a pump, vaporization starts and eventually either the pump ceases to operate, or its operation is accompanied by loss of head, excessive noise, and vibrations (21). The vaporization of the liquid should not be confused with the presence of entrained or dissolved gas, often air, that may come out of the solution at lowered pressure. This gas can also cause the loss of performance, although no cavitation damage takes place (22,23). Vaporization of liquid usually happens at the pump suction, and more often at the impeller entrance near the leading edges of the impeller blades, where high local velocities owing to curvature effects cause pressure drop so that vapor bubbles begin to form. As these bubbles move downstream into a higher pressure region, they collapse, exerting a high pressure within a microscopic area on the impeller material. Such hammering by collapsing bubbles against the pump internal surfaces occurs at frequencies ranging from 600 Hz to 25 kHz, and causes cavities, ie, damaging loss of material (24).

It is difficult to determine exactly the areas of localized pressure reductions inside the pump, although much research has been focused on this field. It is easy, however, to measure the total fluid pressure (static plus dynamic) at some convenient point, such as pump inlet flange, and adjust it in reference to the pump centerline location. By testing, it is possible to determine the point when the pump loses performance appreciably, such as 3% head drop, and to define the NPSH at that point, which is referred to as a required NPSH (NPSHR). The available NPSH (NPSHA) indicates how much suction head over vapor pressure head is available, referenced to an accepted datum, as defined by the Hydraulic Institute:

$$\text{NPSHA} = H_{\text{static}} + H_{\text{dynamic}} + Z_{\text{s}} - H_{\text{vapor}} = H_{\text{s}} + \frac{V_{\text{s}}^2}{2\,g} + Z_{\text{s}} - H_{\text{vapor}} \quad (13)$$

where Z_{s} is the elevation of gauge centerline above or below the centerline of the impeller (see Fig. 3).

Such a definition of NPSHA (eq. 13) is convenient. In many applications, the pumpage is taken from the vessel, where static pressure is known, and the dynamic (velocity) head is zero. If the vessel is open to the atmosphere, then the static pressure is simply equal to atmospheric pressure plus elevation. Hydraulic losses between the vessel surface and pump inlet must be calculated, in order to determine total head at the pump suction (near the flange), and adjusted to the datum (centerline) reference plane. NPSHA must always be greater than NPSHR, and a safety margin should be added. Generally, signs of cavitation appear on the suction side (visible side, looking at impeller eye) of the impeller vanes. However, sometimes cavitation appears on the pressure side, in areas of casing inlet, especially for double-suction pumps, and even at the volute near cutwaters and diffusor tips (25). The vane–suction-side cavitation is usually attributed to positive incidence angles (blade inlet angle being higher than the incoming flow angle) during operation at low flow. The damage on the pressure side of vanes is the result of the impeller eye being too big or the pump operating at flow that is too low for the given impeller eye, which makes the impeller sensitive to backflow at the shroud (26). Backflow causes redistribution of the flow field at the inlet of the impeller in such a way that the effective flow angles near root streamlines (closer to the shaft) become too high, switching flow separation from the suction to pressure side and causing vortexing and damage.

Similar to the concept of the specific speed, a suction specific speed, S, is defined as

$$S = \frac{N(Q)^{0.5}}{\text{NPSHR}^{0.75}} \tag{14}$$

A U.S. definition for S is accepted worldwide. For conversion to metric,

$$nSS = \frac{S}{51.6} \tag{15}$$

According to HI convention, for double-suction pumps, Q is half of the total pump flow, ie, taken per impeller eye. The value of S is calculated at the best efficiency point (BEP) at maximum impeller diameter.

Typically, suction specific speed of centrifugal pumps is within the 6,000–14,000 range (U.S. definition). Special inducer designs (27), installed at the inlet, can raise the suction specific speed to 20,000 or even higher to satisfy the requirements of low suction pressures. Pumps designed for any suction specific speed run fully satisfactorily if operated close to the BEP (see Fig. 2). Unfortunately, the higher the suction specific speed, the less tolerant the pumps are to off-peak operation, ie, lower than BEP capacities (28). In practice, pumps seldom operate at the BEP, but rather at a wide range of capacities. This practical consideration has prompted pump users to specify the upper limit of S not to be exceeded. This limit is typically between 9,500 and 11,000. Special impeller designs can desensitize the pump to flow separation. This requires a comprehensive flow analysis, which involves computation flow dynamics (CFD) and research laboratory testing.

Affinity Laws. Pump performance is affected by the rotating speed. When speed increases, the flow increases linearly, and the head increases as a square of the speed (14).

$$\frac{Q_2}{Q_1} = \frac{N_2}{N_1} \qquad \frac{H_2}{H_1} = \left(\frac{N_2}{N_1}\right)^2 \qquad \frac{\text{BHP}_2}{\text{BHP}_1} = \left(\frac{N_2}{N_1}\right)^3 \tag{16}$$

A similar relationship guides the change in performance when the impeller outside diameter (OD) is cut:

$$\frac{Q_2}{Q_1} = \frac{\text{OD}_2}{\text{OD}_1} \qquad \frac{H_2}{H_1} = \left(\frac{\text{OD}_2}{\text{OD}_1}\right)^2 \qquad \frac{\text{BHP}_2}{\text{BHP}_1} = \left(\frac{\text{OD}_2}{\text{OD}_1}\right)^3 \tag{17}$$

In the case of speed changes, the pump efficiency is not affected except for a minor change owing to Reynolds number change, but the diameter cut may reduce the efficiency appreciably on account of increased gap and losses between the impeller OD and a collector (casing or diffusor).

Types of Pumps

Kinetic. Kinetic pumps, which act by impelling a fluid from one location to another, can be centrifugal, regenerative turbine, or special effect (see Fig. 1).

Overhung Impeller. Close-Coupled Single-Stage Horizontal End Suction.
A closed-coupled pump has an impeller mounted directly on the shaft of the
driver, thus eliminating the need for pump-bearing housing. The driver bearings take all pump loads. These pumps are used for relatively light-duty services.
They are often applied for sanitary and corrosive pumping requirements because
of a clean-in-place (CIP) capability, ie, the pump can be flushed (cleaned) without
much disassembly. This type of pump can handle liquids and semisolids having
entrained vapors, pump viscous products, and sustain good vacuum characteristics under varying conditions. Designs are simple, and maintenance is easy.
These centrifugal pumps use only four basic parts: pump hosing or casing, impeller, shaft, and a seal. The shaft, connected to the power source, eg, a motor,
rotates the impeller inside the casing.

Close- and Separately Coupled Single-Stage Vertical In-Line. Close-
coupled pump designs have been largely replaced by separately coupled pumps.
The latter eliminate hydraulic loads on the motor bearings and are also less
prone to hazard if pumpage splashes on the motor, potentially causing fire in
the event of a seal failure. The pump and the motor are separated by the bearing frame, providing better rotor rigidity and resulting in lower deflections and
improved life of seals and bearings. These centrifugal pumps find applications
in petrochemical and refinery plants.

Frame-Mounted. For medium and severe applications, when nozzle loads
and thermal transients tend to impose high stresses and internal deflections
inside the pump, frame-mounted designs are used. Typically, these designs use
conventional impellers, but recessed impeller designs are also available (Fig. 5).

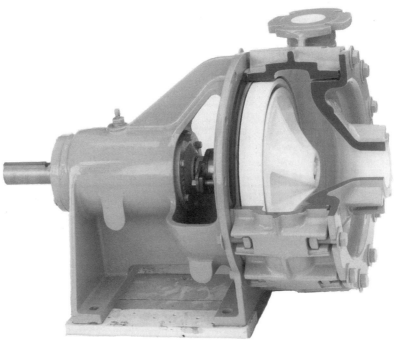

Fig. 5. Frame-mounted pump having recessed impeller. Courtesy of Wemco Pump Co.

Recessed impeller design allows for solid handing. This design has somewhat lower efficiency in comparison to a conventional impeller design. In the latter, the impeller takes up all available space in the casing, giving much tighter clearances between the impeller and walls. Tighter clearance reduces leakage and vortexing, thus resulting in better efficiency.

Standard Chemical Pump. In 1961, the American National Standards Institute (ANSI) introduced a chemical pump standard (29), known as ANSI B73.1, that defined common pump envelope dimensions, connections for the auxiliary piping and gauges, seal chamber dimensions, parts runout limits, and baseplate dimensions. This definition was to ensure the user of the availability of interchangeable pumps produced by different manufacturers, as well as to provide plant designers with standard equipment. A typical ANSI chemical pump, known as of the mid-1990s as ASME B73.1M-1991, is shown in Figure 6.

Centerline-Mounted Refinery Pump. For refinery applications, temperatures, pipe loads, and product flammability are the prime factors in selecting a pump. The pump-mounting feet are located close to the centerline. As a result, the pump thermal growth is uniform on both sides of the centerline, leading to minimal distortions. The design is governed by the American Petroleum Institute (API) standard 610, "Centrifugal Pumps for General Refinery Service" (30).

Wet-Pit Volute, Submersibles, and Dry-Pit Designs. For sump pumping applications, three basic approaches are taken. In the first, called a wet-pit design (Fig. 7**a**), the pump is lowered to the pumping depth, but the motor is

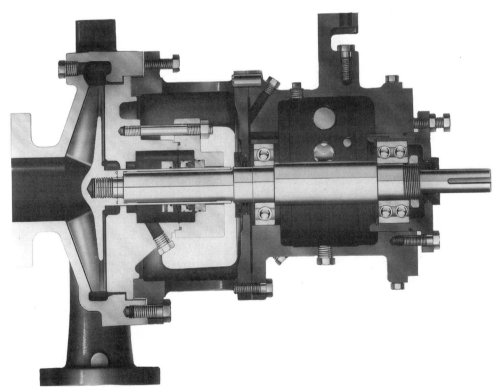

Fig. 6. ANSI chemical pump design. Courtesy of Goulds Pumps, Inc.

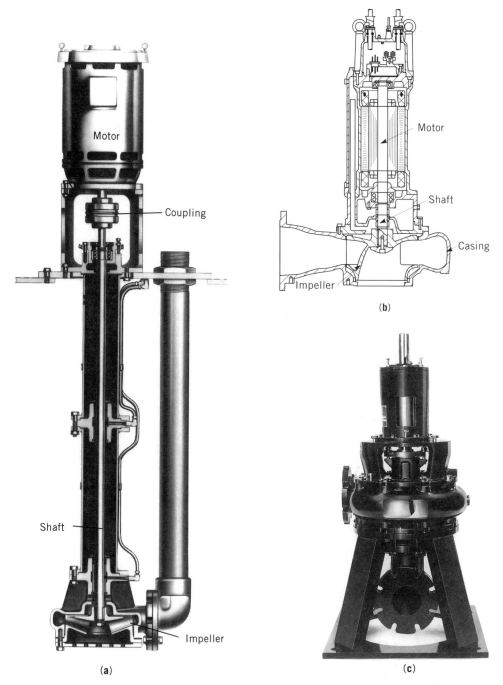

Fig. 7. (**a**) Wet-pit pump design. Courtesy of Goulds Pumps, Inc. (**b**) Submersible pump. Courtesy of Homa Pump Technology. (**c**) Dry-pit pump. Courtesy of Patterson Pump Co.

mounted above the highest anticipated liquid level and coupled to the pump impeller via a long shaft or a series of connected shaft sections. The advantage of this arrangement is the ability to utilize a standard vertical motor. The disadvantages are the long shaft and greater susceptibility to high vibrations, unless special care is taken to determine lateral critical speeds (31).

For greater depths or as an alternative to the wet-pit design, a submersible pump (Fig. 7**b**) having a special water-tight motor can be used. Significant improvements have occurred in the designs of these motors so that the submersible pump is a reliable option to a wet-pit pump.

Another option is the dry-pit design (Fig. 7**c**). This pump is installed in a dry pit and connected to a well via a pipe. Because the dry pit is usually dug out wider than the wet pit, +enough room is available for pump maintenance, troubleshooting, and repair without pulling the pump to the surface for servicing.

Axial-Flow Propeller-Type. The axial-flow propeller-type pumps are used to handle large volumes of pumped liquid at low head. Flow rates of ≥ 12 m^3/s ($\geq 200,000$ gal/min) can be accommodated.

Impeller Between Bearings. These pumps are grouped into single- or multistage designs, and each type is available in either axially or radially split configurations. Sometimes axially split designs are referred to as horizontally split. The term axial, however, is broader. The axial split is not always in the horizontal plane, but can be at an angle to it.

Single-Stage. Axially split pumps are designed for convenient maintenance, inspection, replacement of the impeller or wear parts, and easy access to pump internals. The suction and discharge piping is connected to the lower half of the casing, and does not need to be disturbed for rotor removal. Owing to the nature of the axially split configuration and the large area of the gasket required to seal the two halves, these pumps are limited to relatively moderate developed heads (150 m (500 ft)) and temperatures under 175°C (350°F). They are used as cooling-water pumps at process plants and paper mills, for paper stock transfer, as pipeline pumps of desalinization plants, and for a variety of other applications. Sizes vary from small (1.2 m^3/min (300 gal/min)) to large (>300 m^3/min (80,000 gal/min)).

For the high pressure applications as well as more critical hot-temperature applications such as oil refineries service, radially split casing designs are required. These designs, governed by API 610 specification, allow better sealing at the vertically split joint by using a fully confined gasket. Because these pumps see higher pressures than the axially split pumps, they are generally of a more robust design, such as thicker walls, centerline mounting, ribbing, and stiffer baseplates. A comparison between these two designs is shown in Figure 8.

Multistage Designs. There are traditional differences in the design philosophy between the axially and the radially split multistage pumps. Historically, the axially split design was preferred in the United States, whereas a radially split (segmental) design was preferred in Europe. There are exceptions, of course, and both designs have been applied successfully worldwide.

An axially split pump typically has a volute, and offers the advantage of easy maintenance because of quick upper-half casing and rotor removal that is similar to a single-stage version. The piping, attached to the nozzles located at the lower half of the casing, remains undisturbed.

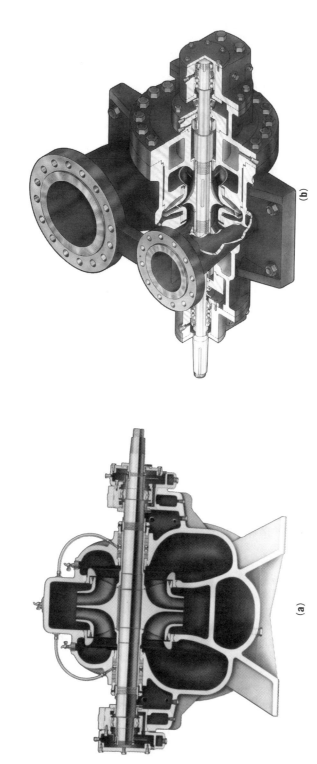

Fig. 8. Single-stage between-bearings double-suction pumps: (**a**) axially split design; and (**b**) radially split design. Courtesy of Goulds Pumps, Inc.

(a)

(b)

Radially split construction is typically a diffusor design, offering the advantage of reduced radial thrust, which is better from the standpoint of lowering shaft deflections, resulting in longer life of seals and bearings. The reason for radial thrust reduction is an inherent geometrical symmetry of the diffusor (having multiple vanes), which results in uniform pressure distribution at the impeller periphery. The volute design has an uneven pressure distribution, which results in radial load. The double volute has two tongues, called cutwaters, which, as compared to a single volute, results in more uniform pressure distribution around the impeller periphery. This distribution is not as uniform as in a diffusor design. The disassembly of the diffusor pump is more involved because a complete pump must be disconnected from the piping for service. Both designs are limited to temperatures around 175°C (350°F), as determined by gasket sealing and safety considerations. For more critical and higher temperature applications such as boiler feed services and refinery applications of volatile or hot (>175°C) temperatures, a segmental ring design is enclosed into a cast or forged barrel, which becomes a pressure-containing vessel having a fully confined gasket. This gasket seals the ends via suction and discharge heads.

Verticals- or Turbine-Type. Historically, all diffusor-type pumps were called turbine pumps (32) because hydraulic turbines have nozzle vanes and resemble the diffusors. Although the word turbine, as applied to vertical pumps, is obsolete, this definition is still widely used in the industry as of the 1990s. Applications for the turbine pumps include agricultural irrigation, storm water drainage, hot-well and booster pumps in power plants, refinery applications for low NPSH requirements, and more. Designs vary in complexity, cost, and robustness, depending on applications.

Deep-Well Turbine and Canned Version. By stacking up the stages, heads up to 1100 m (3500 ft) can be achieved. The deep-well pump is submerged into the well, and liquid comes into contact with the pump body and the well walls. Alternatively, a pump pit may be of dry design, and a can, ie, an enclosure around the pump, having connections to the source of liquid is provided (Fig. 9).

Axial-Flow Propeller-Type. For high capacity, single-stage propeller-type vertical pumps are used. One application of this design is as an intake of cooling water from a river or a holding basin via a pipeline.

Specialty Pumps. There are a multitude of other pump designs in the family of kinetic pumps. Many of these designs have been around for years; others continue to emerge. Some of these special-purpose pumps are described herein.

Vacuum Pumps. Vacuum pumps are more related to compressors than to pumps. What keeps these pumps in the hydraulic family is that they utilize a liquid ring concept to evacuate gas, usually air, from the vessel or pipe. These can be of a single-, two-, or multistage design. A typical two-stage design can function to 735 mm (29 in.) Hg of vacuum. The pump size determines how fast the desired vacuum can be reached in a given size vessel. Explanation of the operating principle can be found in Reference 33.

Self-Priming Pumps. Self-priming pumps are designed to operate in two regimes. The first, utilizing a liquid ring principle, is as a vacuum pump for a short period of time, during which the gas, eg, air, is being evacuated. This is called the priming cycle. When the pump is fully primed, a second regime begins automatically. The pump operates just as a conventional centrifugal

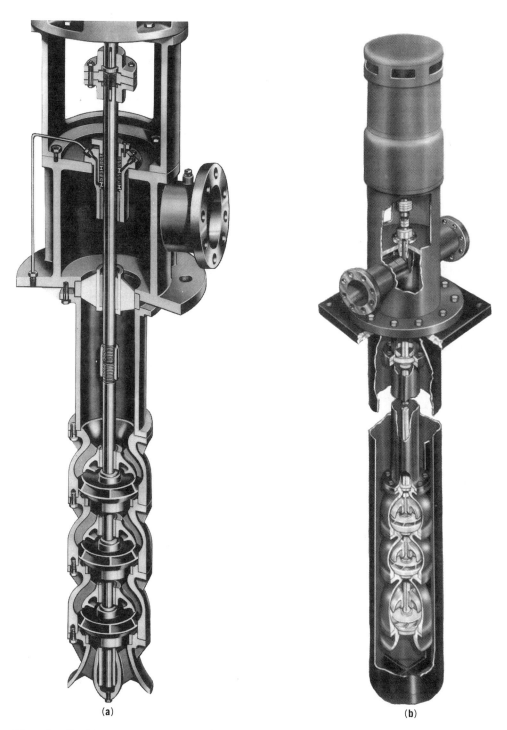

(a) (b)

Fig. 9. Turbine-type pumps. (**a**) Deep-well wet. Courtesy of Goulds Pumps, Inc. (**b**) Canned version. Courtesy of Johnston Pump.

pump. Typically, such pumps can self-prime up to a 6–8-m (20–25-ft) lift, and take 30–90 s of priming time. The pump must be filled with liquid manually once, for initial prime, to ensure that there is liquid in the casing, and to create a liquid ring, which is required for air removal during the priming cycle.

Disk Pumps. When pumping shear-sensitive or highly viscous fluids, it is desirable to reduce internal turbulence caused by the vanes. The disk pump design relies on the centrifugal frictional effect of a vaneless disk. Whereas the efficiency of this pump is lower than that of similar centrifugal pumps having vanes, it is often the only solution to certain pumping applications.

Jet Pumps. A jet pump is a clever way of using a high pressure motive fluid and a venturi to aspire a low pressure fluid to an intermediate pressure level. The high pressure fluid can be liquid, vapor, or gas, which enters the pump suction and exchanges its potential energy to a kinetic energy of the fluid. There are other unique types of kinetic pumps, serving different applicational niches of industrial and residential needs. Among these are reversible centrifugal, rotating casing (pitot), regenerative turbines, and inertia pumps (32).

Positive Displacement Pumps. Positive displacement pumps follow HI convention (see Fig. 1). As a rule, these pumps work against significantly higher pressures and lower flows than do kinetic, particularly centrifugal, pumps. Positive displacement pumps also operate at lower rotational speeds. There are many types of positive displacement pumps, for which designs are constantly being developed. Some of these are discussed herein.

Reciprocating Pumps. A classic example of a positive displacement pump is that of the reciprocating pump (Fig. 10), in which the principle of operation is probably the easiest to understand. These pumps are the most familiar to a nonspecialist. The energy of the reciprocating piston of this pump generates pressure, causing liquid to move from the chamber in which the fluid is acted upon by the piston, to the discharge piping. On the return stroke, the space vacated by the retrieving piston is filled by the liquid, which is moved by the suction pressure. Required valving and timing mechanisms supply necessary opening and closing of ports to provide overall transfer of liquid from suction to discharge.

By the nature of the design, the biggest problem with the application of reciprocating pumps is high pressure pulsations. To reduce pulsations, multiplex (multipiston) pumps are used. The frequency of pressure pulses is decreased in multiplex pumps owing to an increase in the number of pistons properly spaced in phase. Designs of up to 16 cylinders are available. In addition, special pressure-dampening devices are commonly used near the suction and discharge flanges, as well as at certain critical locations along the piping, where the pressure pulsations would otherwise be unacceptably high (34). Pistons are generally used to 7 MPa (1000 psi), and plungers from 7 to 207 MPa (1,000–30,000 psi).

Rotary Vane Pumps. As pump designers tried to combine the high through-put capabilities of kinetic, mostly centrifugal, pumps and the high pressure capabilities of positive displacement pumps, various designs spanning these two extremes emerged. The rotary vane pump features an eccentric rotor having multiple plates (vanes). The rotation of the rotor translates into a reciprocating movement of pistons. This results in pumping action.

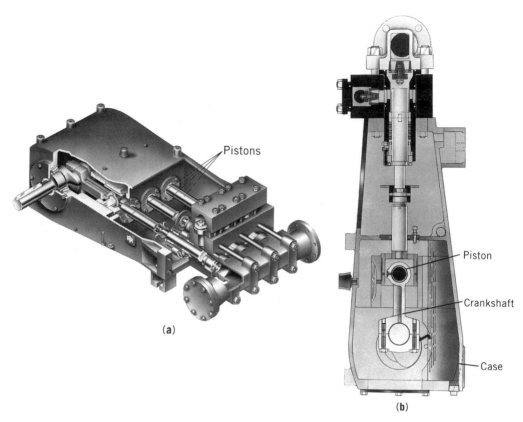

(a)

(b)

Fig. 10. (**a**) Reciprocating pump and (**b**) cross-sectional view. Courtesy of Ingersoll-Dresser Pumps.

Gear Pumps. Liquid enters the inlet port, fills the space between gear teeth, and is trapped in these spaces by the gear teeth and pump housing. It is carried around to where the gear teeth mesh, which forces the liquid out through the discharge opening. The rotor speeds are greater than those for reciprocating pumps, leading to a smaller pump size for the same throughput. Pressure pulsations are also significantly reduced.

Rotary Lobe Pumps. Rotary lobe pumps are similar to gear pumps in principle. These pumps have an added advantage of noncontacting metal parts by use of external gears, which reduces the wear, but adds complexity. Low wear and improved shear characteristics make these pumps applicable to the food industry, where cleanliness and absence of contamination are required. These pumps produce a strong pulsing flow that must be addressed.

Screw Pumps. Constructed by using external timing gears and bearings, two-screw pumps are suited for capacities between 0.1–35 m^3/min (20–9000 gal/min), high suction lift capabilities, and differential pressures of up to 10 MPa (1500 psig). Available in a wide range of materials, two-screw pumps are utilized in a variety of applications in the petrochemical, refining, offshore, and production industries.

Progressive Cavity Pumps. Capable of handling viscosities up to 1,000,000 mPa·s(=cP), progressive cavity pumps (Fig. 11) are suitable for transferring fluids containing high percentages of solids. Additionally, progressive cavity pumps impart extremely low shear into the fluid being handed as a result of their gentle, pulsation-free transfer of the media through the pumping elements.

Fig. 11. Progressive cavity pump. Courtesy of Roper Pump Co.

Flexible Tube. The simplicity of design and the absence of seals and valves make the flexible tube or peristaltic pump a good choice for low capacity and low pressure applications in the pharmaceutical industry or wherever shear-sensitive or moderately abrasive fluids are pumped. Because of the continuous flexing of the tube, the tube material of construction presents a challenge regarding life cycle. For the same reason, pressures are kept relatively low.

Flexible Impeller Pumps. Flexible impeller pumps are designed for general-use service. They can be utilized for pumping in either direction, have low pulsations, and are easy to service and maintain. These pumps are typically limited to low capacities and pressures.

Diaphragm Pumps. Diaphragm pumps (Fig. 12) translate the reciprocating action of a piston into displacement of hydraulic oil (see HYDRAULIC FLUIDS). This causes controlled flexure of the diaphragm within the pumping chamber and the equivalent displacement of process fluid. The diaphragm is hydraulically balanced and statically sealed at its outer diameter. As is the case with other reciprocating pumps, one-way check valves are used. These pumps can operate at discharge and suction pressures up to 35 MPa (5000 psi) using polytetrafluoroethylene (PTFE) diaphragms, or even higher pressures using metal diaphragms. The pumps are available in a wide range of sizes. Flow can range from a few mL/h up to 170 m³/min (45,000 gal/min). Such low capacities result in a formal way of designating capacity specification in the pumping industry, ie, using capacity per hour, rather than minutes. Common uses are for metering and for controlled-volume process services in the chemical process industries.

Condensate Recovery Pumps. The condensate recovery pump is a unique design that utilizes a motive steam or an air supply to move condensate. It can be

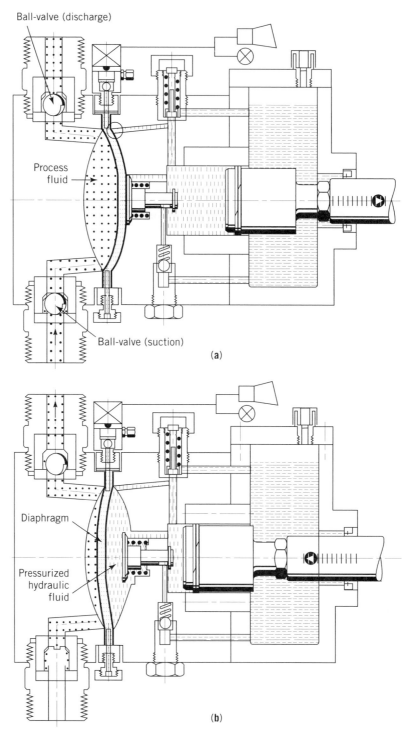

Fig. 12. Diaphragm pump: (**a**) suction cycle, and (**b**) discharge cycle. Courtesy of American Lewa Inc.

applied to capacities up to 23,000 kg/h (50,000 lb/h), providing valuable recovery of condensate or other liquids, which otherwise would be wasted.

Application Examples

Selection of pump for a given application is not a trivial task. Often more than one pump type can accomplish the required job. Thus a final choice on a pump type is often a result of personal experience and usage history. As a rule of thumb, the choice of a kinetic, such as centrifugal, or a positive displacement pump is made on the basis of the specific speed. Whereas specific speed is applicable primarily for centrifugal but not positive displacement pumps, the NS value can be used as a guide. Generally, for calculated values of specific speed, eg, $nS > 10$ ($NS > 500$), kinetic-type pumps are usually selected. For $nS < 10$ ($NS < 500$), positive displacement pumps are typically applied.

From the definition of specific speed (eqs. 9 and 10), it follows that reciprocating pumps operate at high pressures and low flow rates. Conversely, centrifugal pumps are applied at lower pressures and higher flow rates. Many rotary pumps are selected for viscous liquids having pressures equal to or less than, and capacities lower than, centrifugal pumps. However, these limits are relative and a gray area exists as some pump types cross boundaries into the domain of other types.

Plastic Pumps for Corrosive Liquids. The main limitation is lower pressure and temperature capabilities, although developments in new materials are likely to improve these limitations. If plain carbon steel or iron pump metallurgy is applied for corrosive pumpage, the damage to shafts, impellers, casings, and other wetted parts can be quick and extensive, seriously affecting performance, efficiency, and reliability (see CORROSION AND CORROSIVE CONTROL). Although corrosive damage can be reduced upon the use of special alloys and/or metals, pump costs generally rise appreciably. For some applications, even a trace of metallic contamination cannot be tolerated. For these reasons, as well as for extremely broad chemical resistance, a variety of polymer pump designs have been developed.

Nonmetallic centrifugal pumps utilize polymers for all components coming into contact with the pumping media. Even the steel (qv) shaft is sleeved with plastic to isolate the shaft from the fluid. Mechanical seals are reversed so that the metal face does not come into contact with the corrosive liquid. A wide variety of nonmetallics have become available for such applications (35). These pumps are furnished with casing, flanges, and an impeller molded from solid homogeneous polymers, eg, polypropylene. Most polymers are limited to chemical applications of $\leq120°C$ ($\leq250°F$).

Pumps for Leakage Prevention. When leakage to atmosphere must be contained to essentially zero, magnetically coupled or canned motor sealless pumps can be used. These pumps do not have rotating seals between the pumped liquid and the atmosphere.

In magnetic drive designs, the torque is transmitted from the motor and drive magnet assembly to a driven magnetic assembly having a driven shaft and impeller. The drive and driven assemblies are separated by the containment shell. The joint between the shell and the adapter or casing is sealed by an

O-ring or gasket, which is stationary and provides reliable sealing of leakage, as compared to rotating seals between the shaft and the stationary parts. An example of such a pump is shown in Figure 13.

Pumps for Corrosive or Toxic Applications.   For some applications, pumpage is not only severely corrosive but also toxic, so that the pumpage must be contained with zero leakage (see WASTE TREATMENT, HAZARDOUS WASTES). One possibility is to employ a magnetically driven pump having a nonmetallic, usually Teflon, liner at the liquid end.

Another possibility is a canned motor pump, which is essentially a pump and an induction electric motor built together into a single self-contained unit. The pump impeller is mounted on one end of the rotor shaft that extends from the motor section into the pump casing. The rotor is submerged in the fluid being pumped. The stator is isolated from the pumped fluid by a corrosion-resistant nonmagnetic alloy liner. Bearings are submerged and continually lubricated by the process fluid.

Canned motor pumps have only one moving part, a combined rotor–impeller assembly driven by the magnetic field of the induction motor. A small portion of the pumped fluid is allowed to recirculate through the rotor cavity to cool the motor and lubricate the bearings. On some designs this recirculation fluid can be filtered to extend the life of the bearings. Canned motor pumps have fewer bearings as compared to other pump types and do not require any special alignment. A close-coupled centrifugal pump can be compact, and operates with less noise because there are no external moving components. This type of pump

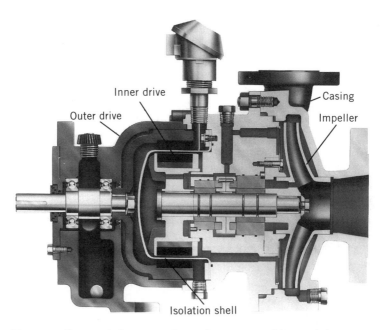

Fig. 13.   Magnetically coupled pump, where the outer and inner drives are magnets. A pumped liquid provides lubrication to bearings of the inner (driven) rotor. Courtesy of Goulds Pumps, Inc.

finds many applications in hazardous and toxic pumpage because the design can offer secondary containment of the process fluid. High pressure designs are also available, which utilize backup sleeves to support the stator liner. Developments in diagnostic systems permit monitoring of the internal clearances of these pumps so that simple parts replacement can be carried out long before a costly failure occurs.

Pumps for Slurries. In mining and minerals applications (see MINERAL RECOVERY AND PROCESSING), abrasive wear by hard particles such as rocks, coal (qv), and other slurries can cause quick damage to a pump's liquid end unless special abrasion-resistant materials are utilized. The wet-end parts, such as the casing, can either be made of hard metals or use solid liners. For certain applications, rubber-lined casings have been successful. Rubber has a unique quality to resist wear because of its ability to comply and resist the shearing wear of particles. Rubber linings often provide good choice for applications where both abrasion and corrosion can occur, such as desulfurization pumps for liquid ash transport at power plants. Because the flow of slurries is multiphased, ie, composed of a liquid having solids and entrained gases, the nature of such a pumping regime is unstable. This results in fluctuating radial and axial loads, which have detrimental effect on pump bearings and seals. To withstand these loads, a heavy-duty support frame having heavy-duty bearings, often roller rather than ball bearings, are used. These designs provide rugged, dependable operation for most demanding applications.

Trash or Contractor Pumps. For general applications where trash and debris are present in the pumped liquid, medium-size submersible centrifugals or diaphragm self-priming pumps can be utilized. These pumps may be equipped with inlet screens to prevent grit and sand from entering the internal parts of the pump.

Metering Pumps. For small flow rates, such as dosing chemical additives where precise control is required, progressive cavity self-contained pumping units are used. These can often handle shear-sensitive fluids or liquids containing abrasive particles. These pumps are not as widely publicized or generally as well known in the literature as other pump types.

Pumps for Nonclogging Applications. When pumping stringy fibrous materials, it is desirable to keep the internal passages as wide as possible. All sharp edges must be blended, and impeller vanes made thicker at the inlets for contouring purposes. Additionally, special designs are available, where the entire rotating assembly can be removed and the clog cleared.

In certain applications, such as canning in food processing (qv), the pumped solids, ie, fruits, vegetables, etc, must survive the pumping action. At the other extreme, when preservation of pumped solid particles is undesirable, special chopper pumps are used. These pumps break up the incoming solids into smaller pieces, thus enabling the main impeller to pump without plugging. Chopper pumps are also useful in the chemical industry for handling waste sump applications, eg, where solids tend to plug standard nonclogging pumps. These chemical industry applications include detergent cakes, hand cleaners with pumice, latex skins, lead oxide slurries, paint sludges, plastics, and shredded filter pads. Progressing cavity pumps are also popular in these applications.

Couplings and Seals

Various coupling designs are available to transmit torque from the driver, eg, electric motor, to a pump. In order to contain the pumped fluid inside the pump and prevent the pumpage from leaking, several types of sealing methods are used. A few options are described herein.

Couplings. *Gear-Type Couplings.* Gear-type couplings (Fig. 14**a**) are usually used for heavy-duty applications having high torques. Gear mesh allows a small parallel misalignment between shafts, usually less than 0.2 mm (0.008 in.). Gear requires lubrication (see LUBRICATION AND LUBRICANTS), usually of a grease type.

Disk Coupling. Diaphragms are used to transmit torque between the inside and outside diameters of the flexible element in disk coupling (Fig. 14**b**). Up to 0.1-mm (0.004-in.) parallel misalignment can usually be tolerated. Whereas no oil lubrication is required as compared to gear couplings, disks are limited to moderate torques.

Grid-Type Couplings. Another type of flexible couplings, similar to disk couplings, is the grid type (Fig. 14**c**). These are applied for moderate loads and allow parallel misalignment up to 0.1 mm (0.005 in.). Lubrication is also required.

Special Types. Many other coupling types are available. These vary in degree of complexity, maintenance time, torque capabilities, and price. A good comparison study on coupling is available (36). An elastomer-type coupling is one of the most popular types used, particularly in the chemical industry.

Among relatively new designs are magnetic couplings, sometimes referred to as magnetic drives. When these drives are applied to magnetically driven pumps, a containment shell isolates the inner assembly (driven) from the outer (drive). The revolving magnetic field, which actually performs a coupling action, induces eddy currents in the shell if the shell is metallic. The eddy currents result in shell heating, ie, power loss, and lower efficiency. The heat is rejected into pumped fluid, raising the fluid temperature and the vapor pressure. If vapor pressure of the fluid becomes higher than the liquid pressure, vaporization (flashing) occurs, and no more heat is removed from the shell. Shell overheating and distortion occur and may lead to failure. To prevent eddy currents, polymer, ceramic, or similar shells are made. Whereas eddy currents do not occur in nonmetallic shells, these shells are limited in pressure and are susceptible to thermal shocks. This limits application to the low temperature pumpage, and excludes these shells from more demanding applications such as in refineries.

A unique solution combining the strength of metals with electrical nonconductivity of plastics became available in the early 1990s. In this design, a shell is made of two layers. The inner layer is made of a set of rings, compressed together, and insulated by gaskets. These gaskets contain eddy currents within each small ring, resulting in significant overall reduction in losses. The outer ring has axial slots, which similarly break up large eddy currents into several small ones. The total power loss is small because it is proportional to the square of the eddy current (Ohm's law). The inner shell takes up the radial stresses, and the outer shell takes up the longitudinal stresses. The liquid is contained inside the shell and is sealed statically by gaskets between the rings.

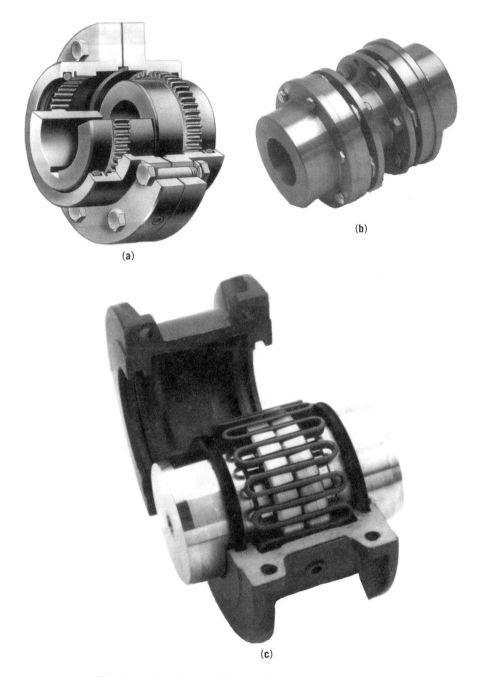

Fig. 14. Couplings: (**a**) gear; (**b**) disk; and (**c**) grid.

A common requirement for all coupling types, although to a different degree, is the need for good alignment. Alignment minimizes coupling strain and prevents shaft bending. The latter can cause seal leakage and reduced bearing life owing to the high loads caused by misalignment. A straight-edge coupling alignment is sometimes used, especially for low horsepower pumps. However, studies in pumping equipment reliability have demonstrated (37) that significant improvements in bearings and seal life can be obtained by proper maintenance and installation procedures. Better coupling alignment of pumps can be achieved by using precision alignment tools and even laser alignment (38) for critical high energy pumps. Good baseplate installation methods make alignment much easier. Improper installation, on the other hand, leads to uneven mounting of the equipment, ie, pump and driver. This can result in high casing stresses, excessive bearing loads, seal distortions, and a reduction in reliability and equipment life.

Sealing. *Packings.* The oldest method of sealing is through the use of packings. It is also the simplest and least expensive. A packing material is inserted into a stuffing box, which may include a lantern ring to provide injected liquid to flush the box from particulates or to cool it. To prevent the shaft from rubbing wear by packing material, hardened sleeves are often used over the shaft (39). Some packing designs include graphite or Teflon particles for better self-lubricating properties to minimize wear damage. Packings need a certain amount of leakage for lubrication, although the latest developments in packings design have reduced this leakage significantly. Packings are typically limited to non-corrosive, nontoxic applications such as paper mills and waterworks. However, some packing materials, eg, Teflon, have been used in corrosive environments.

Mechanical Seals. Mechanical seals provide good sealing for a wide variety of applications. Leakage rates are considerably less than for packing materials, but the mechanical seals are more expensive. Applied in refineries as standard, mechanical seal emissions can be as low as 50 ppm, much lower than the allowable minimum established by the Clean Air Act (40). The sealed faces are kept together by the force of springs or bellows, or, for unbalanced seals, hydraulically. Mechanical seals can be installed in single, double (Fig. 15), or tandem configuration. Good descriptions of the principles of operation and application guidelines are available (41,42). If the single seal fails, the pumpage leaks out to the atmosphere. Double and tandem designs have a barrier fluid between the inner and outer seal. In the event of inner seal failure, the outer seal takes over

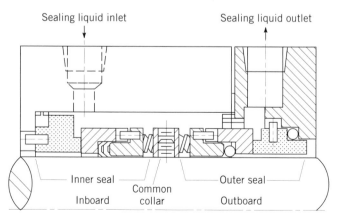

Fig. 15. Double-seal arrangement for mechanical seals. Courtesy of Durametallic Corp.

the sealing, the increased seal cavity pressure can be detected, and the pump can be shut down.

Dynamic Sealing. In pulp (qv) and paper (qv) industry applications, it is often desirable to prevent paper fibers from reaching the sealing chamber. Fibers can cause plugging, overheating, and failure. A device known as an expeller is often successfully utilized in such applications. Based on the liquid ring principle (33), the dynamic seal or expeller keeps a liquid ring within its passage. The radial height of this ring is automatically determined in such a way that the centrifugal force caused by the rotating liquid mass is balanced by the differential pressure across it, ie, the pressure difference between the expeller outer diameter, which is equal to the pump impeller back hub pressure, and the sealing chamber pressure. If the impeller has balancing holes and the sealing chamber is connected to the atmosphere via packing, a differential pressure between pump suction pressure and the atmosphere exists across the expeller. This determines a certain liquid height within the expeller. Thus, if the suction pressure increases, the amount of liquid in the expeller correspondingly increases. At some level of suction pressure the liquid fills the expeller completely, and the expeller reaches the limit of its sealing capability. Upon further increase in suction pressure, paper fibers, if contained in pumpage, can reach inside the seal chamber. Therefore, the application of the dynamic seal is limited by suction pressure.

Application Guidelines

Pumps are designed to give trouble-free operation for a long period of time. The ANSI B73.1M pumps are designed for a bearing life of no less than two years (29), and API 610 pumps for a minimum of five years (30). However, in real applications, a typical mean time between failures (MTBF) is often found to be significantly less, and sometimes it is as short as a few weeks. Whereas in some installations the seals last from three to four years, in others these are replaced monthly. The reason for such wide variations in pump component life is often not poor pump design but equipment misapplication.

Inlet Piping Configurations. An incorrect piping configuration can create undesirable flow distortions that, especially for double-suction pumps, means turbulence. Air pockets trapped in the pipe can cause uneven flow, vibrations, and pump damage.

Low Flow Operation. The optimum operation of a pump is near the best efficiency point. Some manufacturers' curves indicate the minimum allowable continuous stable flow (MCSF) limits for every pump (43). In the 1980s, the processing industry experienced a reduction in flow requirement as a result of business downturn and installation capacity downsizing. The pumping equipment, however, was generally not replaced by smaller pumps, but was forced to operate at reduced flow rates, often below allowable MCSF. This has resulted in increased failure rates and reduced pump component life.

There are two main reasons why a pump should not operate below its MCSF: (*1*) the radial force (radial thrust) is increased as a pump operates at reduced flow (44,45). Depending on the specific speed of a pump, this radial force can be as much as 10 times greater near the shut off, as compared to that near the BEP; and (*2*) the low flow operation results in increased turbulence and internal flow separation from impeller blades. As a result, highly unstable axial and radical fluctuating forces take place.

The relationship between force and the life of ball bearings is cubic (46), ie, doubling the force decreases bearing life eight times. In addition, the increased shaft deflections cause distortion at the faces of mechanical seals, causing leakage and failure. To reduce the radial thrust, some manufacturers have developed special casing designs refered to as circular volutes. These designs result in significantly reduced loads owing to a more even pressure distribution around the impeller periphery, as compared to conventional expanding volutes. The radial thrust reduction can be as high as 85%, depending on the design.

Impeller–Diffusor Gaps. A tight gap between the impeller outer diameter and the diffusor inner diameter can result in high pressure pulsations, vibrations, noise, and vane breakage (47–51). By modifying gaps A and B, as shown in Figure 16 (52), it is possible to reduce significantly these adverse effects for problematic pumps in the field. This is especially true for high energy equipment, such as boiler feed pumps.

Cavitation. The subject of cavitation in pumps is of great importance. When the liquid static pressure is reduced below its vapor pressure, vaporization takes place. This may happen because (1) the main stream fluid velocity is too high, so that static pressure becomes lower than vapor pressure; (2) localized velocity increases and static pressure drops on account of vane curvature effect, especially near the inlets; (3) pressure drops across the valve or is reduced by

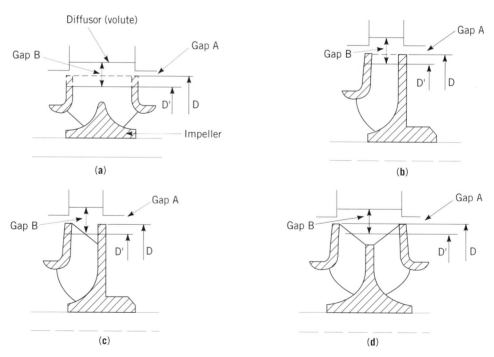

Fig. 16. Gaps A (between impeller shrouds and diffusor wall) and B (between impeller and diffusor vane tips) corrective option where D = diameter and D′ is diameter reduction: (**a**) overall diameter reduction; (**b**) vane diameter reduction; (**c**) angle-cut, single-suction impeller; and (**d**) angle-cut, double-suction impeller. Courtesy of E. Nelson and *Pumps and Systems Magazine.*

friction in front of the pump; or (*4*) temperature increases, giving a corresponding vapor pressure increase.

Pump manufacturers publish data indicating a minimum required net positive suction head (NPSHR) to ensure that suction pressure is adequate and that pumps do not cavitate. A user must therefore calculate the available net positive suction head (NPSHA) by going through the system losses analysis. The NPSHA should always be greater than the NPSHR plus a safety margin. A recommended margin added to NPSHA is 1.5 m (5 ft) or 35% over NPSHR, whichever is greater (53). There are several good reasons for this. The NPSHR published in manufacturers' curves is based on pump tests. These tests, in accordance with HI Standards (9), defines NPSHR as equal to the value at which the pump actually loses 3% of its developed head. However, the incipient cavitation, ie, when first cavitation bubbles just begin to form, starts significantly before a 3% drop in head occurs. The value of the incipient NPSH is difficult to determine in commercial testing. Sophisticated research methods have been applied (54), but despite extensive research there is no single accepted method to predict incipient cavitation for different pump types or different specific speeds. Figure 17 illustrates the phenomenon of incipient cavitation as it progresses toward fully cavitating pump and eventual loss of performance at 3% head drop (55).

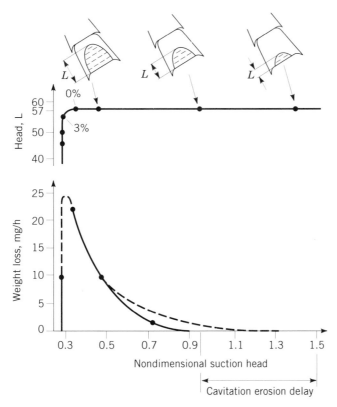

Fig. 17. Cavitation phenomenon in pumps showing cavitation bubble distribution and rate of weight loss as a function of cavitation coefficient at constant speed where (—) represents actual and (———) theoretical measurements and *L* is bubble length growth. Courtesy of S. Florancic and *Pumps and Systems Magazine*.

A further complication arises from the fact that the point of maximum damage to the pump by cavitational pitting happens somewhere between the incipient cavitation and the point of complete performance loss (56). Whereas it is difficult to predict incipient cavitation, it is even more difficult to determine the point of maximum damage. Many attempts have been made, but no method has been accepted. It has been reported that the damage is related to fluid velocity to the fifth or sixth power (57).

The vapor pressure may be dependent on the amount of the dissolved, not the entrained, air in the liquid. This issue is important to applications of cooling-water double-suction pumps (58,59). Because of the unknowns, a safety margin is always recommended for use to minimize the effects of cavitation.

For existing installations where cavitation is known to be a problem but the cost of modification to the system is prohibitive, an effective way to resist cavitation is by upgrading pump metallurgy. For example, upgrading from cast-iron construction to Type-316 stainless steel can improve the impeller life by as much as 10 times. CA15 or CD4MCu alloys can improve this even further (32). Special designs of impellers and volutes can be made to reduce internal velocities and prevent or minimize cavitation damage. These designs usually require a computer analysis of three-dimensional effects of flow inside the impeller and casing (56). Sophisticated computational fluid dynamics (qv) methods (CFD) having complete viscous flow solutions are employed (60).

Reliability. There has been a significant rise in interest among pump users in the 1990s to improve equipment reliability and increase mean time between failures. Quantifiable solutions to such problems are being sought (61). Statistical databases (qv) have grown, improved by continuous contributions of both pump manufacturers and users. Users have also learned to compile and interpret these data. Moreover, sophisticated instrumentation has become available. Examples are vibration analysis and pump diagnostics.

Pump vibration measurements and analysis of the amplitude–frequency spectrum have led to the determination of possible equipment ailments. Examples of frequency relationships to particular pump vibration problems are available (62). For example, a one-time (1×) frequency corresponding to a running speed usually indicates the presence of rotor unbalance, a 2× peak may mean a bent shaft, and a ∼0.4× may be a symptom of an oil whip at the sleeve bearing.

Economic Aspects

When selecting and purchasing pumping equipment, there are one-time capital investment as well as operating and maintenance costs to consider. The cost of pumps and auxiliaries, such as driver, coupling, baseplate, and sealing equipment, is capital investment, a one-time cost. Replacement of worn or broken parts, energy to operate, and service costs are maintenance and operating costs. A lower first time cost, ie, a less expensive pump, may not necessarily indicate a lower total cost. Pump prices vary from one design to another and from one manufacturer to the next, but as a rule of thumb, an estimate can be made. Figure 18 gives an estimate in terms of prices for three types of pumps.

As an approximation, 50% of the list price should be added to account for auxiliaries such as motor, seal, baseplate, and coupling. Furthermore, the material upgrade from ductile iron construction to stainless steel roughly doubles

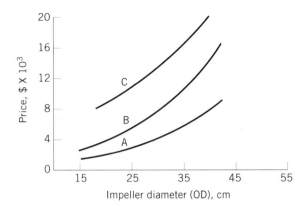

Fig. 18. Cost estimate curves for three types of pumps: A, chemical end suction ANSI iron option (iron); B, split-case between-bearings type (iron); and C, API end-suction type-316 stainless steel.

the price. Pump manufacturers use price book multipliers, eg, discounts, to arrive at a selling price. These vary, but as an estimate, a 30% discount is reasonable. U.S. pump manufacturers include the following.

Company	Location
Tri-Clover Inc.	Kenosha, Wis.
Goulds Pumps, Inc.	Seneca Falls, N.Y.
WEMCO Pump Co.	Sacramento, Calif.
Patterson Pump Co.	Toccoa, Ga.
Homa Pump Technology	Stamford, Conn.
KSB, Inc.	Richmond, Va.
Ingersoll-Dresser Pumps	Phillipsburg, N.J.
Johnston Pump	Brookshire, Tex.
Gorman-Rupp Co.	Mansfield, Ohio
Gast Manufacturing Corp.	Benton Harbor, Mich.
Roper Pump Co.	Commerce, Ga.
Fristam Pumps Inc.	Middleton, Wis.
Bornemann Pumps Inc.	Monroe, N.C.
Barnant Co.	Barrington, Ill.
ITT Jabsco	Costa Mesa, Calif.
American Lewa, Inc.	Holliston, Mass.
Spirax Sarco Inc.	Allentown, Pa.
Vanton Pump	Hillside, N.J.
Chempump	Warrington, Pa.
Vaughan Co., Inc.	Montesano, Wash.
A-Line Mfg. Co. Inc.	La Porte, Tex.
Durametallic Corp.	Kalamazoo, Mich.

Excellent sources of information are also available (9,63).

Cost Example. If a single-stage overhung pump having a 200-mm (8-in.) impeller outside diameter (OD) and 316-ss metallurgy is of interest, then, from Figure 18, an iron construction pump cost is approximately $2000 in 1995 dollars. Doubling that cost for the stainless steel gives \$4000. Adding 50% for motor, coupling, baseplate, and the seal makes the price \$6000. Finally, applying a 70% multiplier (30% discount), a final selling price in the range of \$4200 can be expected.

NOMENCLATURE

Symbol	Definition	Units
Eff	pump efficiency	
$\text{Eff}_{\text{motor}}$	motor efficiency	
g	gravitation constant	9.81 m/s^2 (32.174 ft/s^2)
H	total developed pump head	m
H_d	total discharge head	m
H_s	total suction head	m
H_{dynamic}	velocity head $\left(\frac{V^2}{2g}\right)$	m
$H_{\text{loss,A–B}}$	hydraulic losses from A to B	m
H_{static}	static head	m
H_{vapor}	vapor head	m
HP_{fluid}	net power delivered to fluid being discharged	kW (hp)
HP_{loss}	power losses inside the pump	kW (hp)
N	rotational speed	rpm
NPSH	net positive suction head	m
NPSHA	net positive suction head available	m
NPSHR	net positive suction head required	m
NS	specific speed in U.S. notation	
nS	specific speed in metric notation	
nSS	suction specific speed in metric notation	
OD	outside diameter of impeller	mm
γ	density conversion (SG/2.31)	psi
Ω	rotational speed	rad/s
Ω_s	universal dimensionless specific speed	
$p_{d,g}$	discharge static gauge pressure	N/m^2
$p_{s,g}$	suction static gauge pressure	N/m^2
PF	power factor of electric motor	
Q	capacity	m^3/s
S	suction specific speed in U.S. notation	
SG	specific gravity	
T	torque	N·m
V_d	discharge velocity	m/s
V_s	suction velocity	m/s
Z_d	discharge gauge elevation (+) or submergence (−) from pump centerline	m
Z_s	suction gauge elevation (+) or submergence (−) from pump centerline	m

BIBLIOGRAPHY

"Pressure Technique (Pumps)" in *ECT* 1st ed., Vol. 11, pp. 123–126, by L. H. Garnar, Worthington Corp.; "Pressure Technique (Compressors)," pp. 115–123, by E. L. Case and

J. Charls, Jr., Worthington Corp.; "Pumps and Compressors (Pumps)" in *ECT* 2nd ed., Vol. 16, pp. 728–741, by R. Kobberger, Worthington Corp.; "Pumps and Compressors (Compressors)," pp. 741–762, by J. J. Julian and J. F. Hendricks, Worthington Corp.; in *ECT* 3rd ed., Suppl. Vol. pp. 753–785, by R. Neerken, The Ralph M. Parsons Co.

1. R. D. Beck, *Plastic Product Design*, 2nd ed., Van Nostrand Reinhold Co., Inc., New York, 1980.
2. *Mater. Eng.*, (Dec. 1992).
3. *The Sealing Technology Guidebook*, 9th ed., Durametallic Corp., Kalamazoo, Mich., 1991.
4. *Engineered Sealing Products*, John Crane Catalog No. 80, John Crane, Morton Grove, Ill., 1991.
5. J. W. Veness, C. J. Steer, and S. Rose, *Proceedings of 9th International Pump Users Symposium*, Houston, Tex., 1992.
6. J. Lorenc and L. Nelik, *Proceedings of 11th International Pump Users Symposium*, Houston, Tex., 1994.
7. L. Nelik and co-workers, *Proceedings of NASA Magnetic Suspension Technology Conference*, Hampton, Va., 1991.
8. L. Nelik and co-workers, *Proceedings of Power Plant Pumps Symposium*, Tampa, Fla., 1991.
9. *Hydraulic Institute Standards for Centrifugal, Rotary & Reciprocating Pumps*, Hydraulic Institute, Parsippany, N. J., 1994.
10. J. R. Krebs, *Pumps and Systems*, (Feb. 1994).
11. H. S. Bean, ed., *Fluid Meters* 6th ed., ASME, New York, 1971.
12. *Rosemount Product Catalog*, Rosemount, Eden Praire, Minn., 1993.
13. *Controlitron Product Catalog*, Controlitron, Hauppauge, N.Y. 1993.
14. A. J. Stepanoff, *Centrifugal and Axial Flow Pumps*, 2nd ed., John Wiley & Sons, Inc., New York, 1948.
15. H. P. Bloch, *Process Plant Machinery*, Butterworth & Co., Publishers Ltd., Kent, U.K., 1989.
16. *Himmelstein Product Catalog*, Himmelstein, Hoffman Estates, Ill., 1990.
17. A. E. Knowlton, *Standard Book for Electrical Engineers*, 7th ed., McGraw-Hill Book Co., Inc., New York, 1941.
18. H. Schlichting, *Boundary-Layer Theory*, 7th ed., McGraw-Hill Book Co., Inc., New York, 1979.
19. H. H. Anderson, *Centrifugal Pumps*, 3rd ed., Trade & Technical Press Ltd., London, 1980.
20. J. T. McGuire, *Pumps and Systems*, (Nov. 1993).
21. W. O'Keefe, *Power* (Feb. 1992).
22. C. Cappellino, D. Roll, and G. Wilson, *Proceedings of 9th International Pump Users Symposium*, Houston, Tex., 1992.
23. D. Florjancic, *Influence of Gas and Air Admission on the Behavior of Single- and Multi-Stage Pumps*, Sulzer Research, No. 1970.
24. P. de Haller, *Escher-Wyss News*, 77, (May/June 1993).
25. C. J. Mansell, *I Mech E* (June 10, 1974).
26. R. Palgrave and P. Cooper, *Proceedings of 3rd International Pump Users Symposium*, Houston, Tex., 1986.
27. J. H. Doolin, *Hydrocarbon Process.* (Jan. 1984).
28. W. H. Fraser, *Proceedings of ASME Meeting*, 1981.
29. "Specification for Horizontal End Suction Centrifugal Pumps for Chemical Process," ASME B73.1M, Washington, D.C., 1991.
30. *API Standard 610, Centrifugal Pumps for General Refinery Service*, 7th ed., 1989, 8th ed., 1995, Washington, D.C.
31. F. Ehrich and D. Childs, *Mechanical Eng.* (May 1984).

32. I. J. Karassik and co-workers, eds., *Pump Handbook*, McGraw-Hill Book Co., Inc., New York, 1976.
33. W. H. Faragallah, *Liquid Ring Vacuum Pumps and Compressors*, Gulf Publishing, 1985.
34. R. L. Smith, *Mechanical Eng.* (Jan. 1991).
35. *Modern Plastics Encyclopedia 1992*, Hightstown, N.J., 1991.
36. J. Lorenc, *Proceedings of 8th International Pump Users Symposium*, Houston, Tex., 1991.
37. L. Nelik, *Proceedings of 2nd International Conference on Reliability*, Houston, Tex., 1993.
38. "Optalign Reference Manual," *Pumps and Systems* (1992).
39. L. Nelik and B. Buchanan, *Proceedings of 2nd International Conference on Coatings* 1993.
40. Technical data, International Technical Services, Rochester, N.Y., Dec. 1992.
41. N. Barnes, R. Flitney, and B. Nau, *Proceedings of 9th International Pump Users Symposium*, Houston, Tex., 1992.
42. J. Heatley and M. Giffrow, *Proceedings of 7th International Pump Users Symposium*, Houston, Tex., 1990.
43. D. J. Vlaming, *A Method for Estimating the Net Positive Suction Head Required by Centrifugal Pumps*, ASME 81-WA/FE-32, Washington, D.C., 1981.
44. A. Agostinelli, D. Nobles, and C. Mockridge, *ASME Transactions* (Apr. 1960).
45. H. Iversen, R. Rolling, and J. Carlson, *ASME Transactions* (Apr. 1960).
46. *SKF General Catalog No. 4000US (Bearings)*, SKF, King of Prussia, Pa., 1991.
47. U. Bolleter and A. Frei, *Proceedings of Seminar on Vibrations in Centrifugal Pumps*, London, U.K., 1990.
48. E. Makay, *Power* (July 1987).
49. E. Makay, *Machine Design* (May 13, 1971).
50. E. Makay and D. Nass, *Power* (Sept. 1982).
51. E. Makay and J. Barrett, *Proceedings of 1st International Pump Users Symposium*, Houston, Tex., 1984.
52. W. E. Nelson, *Pumps and Systems* (Jan. 1993).
53. R. Hart, *Best Practice: Centrifugal Pumps. NPSH Definitions and Specifications*, Du Pont Internal Specification, Wilmington, Del., May 18, 1992.
54. D. Florjancic, "Net Positive Suction Head for Feed Pumps", *Sulzer Report*, 1984.
55. S. Florjancic and A. Clother, *Pumps and Systems* (Sept. 1993).
56. L. Nelik, *Pumps and Systems* (Mar. 1995).
57. P. Cooper and F. Antunes, *Proceedings of EPRI Symposium on Power Plant Feed Pumps*, Cherry Hill, N.J., 1982.
58. W. R. Penny, *Inert Gas in Liquid Mars Pump Performance* (July 1978).
59. C. C. Chen, *Chem. Eng.* (Oct. 1993).
60. T. Dahl and L. Nelik, *Computer Aided Design of a Centrifugal Pump Impeller*, ASME 85-WA/FE-10, Washington, D.C., 1985.
61. M. Smith, *World Pumps* (Nov. 1993).
62. E. Makay, *Power* (1980).
63. *Pump and Systems Magazine*, Fort Collins, Colo.

General References

W. Micheletti and J. Davis, *Pumps and Systems* (Feb. 1994).
R. Blong and B. Manion, *Pumps and Systems* (Dec. 1993).

LEV NELIK
Roper Pumps Company

PURGATIVES. See GASTROINTESTINAL AGENTS.

PURINES. See ANTIBIOTICS, NUCLEOSIDES AND NUCLEOTIDES; CHEMO-THERAPEUTICS, ANTICANCER; NUCLEIC ACIDS.

PYRAZOLES, PYRAZOLINES, AND PYRAZOLONES

The compounds of this article, ie, five-membered heterocycles containing two adjacent nitrogen atoms, can best be discussed according to the number of double bonds present. Pyrazoles contain two double bonds within the nucleus, imparting an aromatic character to these molecules. They are stable compounds and can display the isomeric forms, (**1**) and (**2**), when properly substituted. Pyrazoles are scarce in nature when compared to the imidazoles (**3**), which are widespread and have a central role in many biological processes.

(**1**)   (**2**)   (**3**)

Pyrazolones, also containing two double bonds, are predominantly in the keto form (**4**), although they can also exist in the enol form (**5**).

(**4**)   (**5**)

Pyrazolines have only one double bond within the nucleus and, depending on the position of the double bond, can exist in three separate forms: 1-pyrazoline [2721-43-9] (**6**), 2-pyrazoline [109-98-8] (**7**), and 3-pyrazoline [6569-23-9] (**8**).

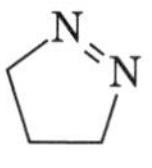

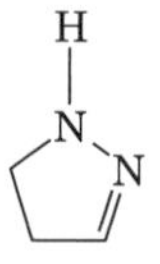

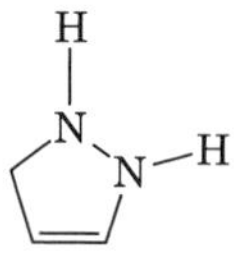

(6)   (7)   (8)

Neither pyrazolidines (**9**), which have no double bonds, nor pyrazoline diones (**10**), with two double bonds, and pyrazolidine triones (**11**), which have three double bonds, are covered in this article.

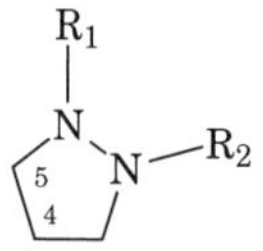

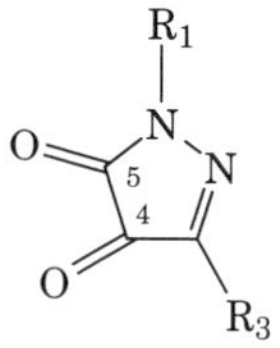

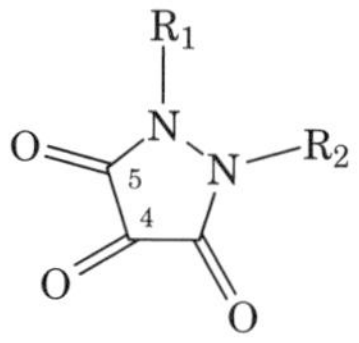

(9)   (10)   (11)

Despite their scarcity in nature, the title compounds have found use in many applications, including pharmaceuticals, agricultural chemicals, and dyes.

Extensive reviews have been published, covering the literature to about 1967 (1–3). Pyrazoles and the benzopyrazoles have been well reviewed in References 4 and 5. More up-to-date reviews, though much narrower in scope, have been published on pyrazole oxides (6), dihydropyrazoles as insecticides (7), the anticancer drugs anthrapyrazoles (8,9), and pyrazole sulfonylureas as herbicides (10).

Theoretical Methods

A number of theoretical studies on the reactivity of pyrazoles have been published (11). However, due to the difficulties involving these calculations, the studies often only approximate the actual reactions occurring in the laboratory. A summary of the calculated electron densities of the pyrazole molecule is available (11). Using the Dewar modified intermediate neglect of differential overlap (MINDO) method, optimized geometries have been obtained that approximate those observed from microwave spectroscopy and neutron diffraction experiments. Localization of the π-system has been established by Hückel molecular orbital (HMO) calculations. In accordance with experimental results, total electronic density values suggest electrophilic substitution at position 4. Pyrazole has a binding energy of 0.309 atomic units. It is less basic than imidazole due to greater destabilization of bonding after protonation as found by the MINDO method. Electron densities and bond orders of N-substituted pyrazoles have been calculated by the HMO method (12). The complete neglect of differential overlap (CNDO/2) procedure has been used to calculate π- and σ-electron densities on protonated and neutral N-methylpyrazole (13). Also, the energy required to localize π-electrons at the position where substitution takes place has been determined (12). Calculations for the electrophiles NO_2^+ and Br^+ include the electrostatic energy required to move the electrophile to the position where substitution takes place.

Structural Elucidation

Among the modern procedures utilized to establish the chemical structure of a molecule, nuclear magnetic resonance (nmr) is the most widely used technique. Mass spectrometry is distinguished by its ability to determine molecular formulas on minute amounts, but provides no information on stereochemistry. The third most important technique is x-ray diffraction crystallography, used to establish the relative and absolute configuration of any molecule that forms suitable crystals. Other physical techniques, although useful, provide less information on structural problems.

Nuclear Magnetic Resonance Spectroscopy. The literature up to 1984 has been reviewed (14).

^{1}H-Nmr. This topic is best discussed in terms of the formulas (**12**) and (**13**). In symmetrical pyrazoles with no substituents on N_1, ie, (**12**), where $R_1 = H$ and $R_3 = R_5$, the peaks attributed to the R_3 and R_5 positions are the same on account of tautomerism; the peak due to R_4 is generally at a higher field. In N-substituted nonsymmetrical pyrazoles, ie, (**12**) and (**13**), where $R_3 \neq R_5$, nmr has been widely used to distinguish between the two possible isomeric structures.

(**12**)   (**13**)

The following generalizations have been useful in determining the structures of isomeric pyrazoles. In compounds of type (**12**), where $R_3 \neq R_5$ and $R_1 \neq H$, the coupling constant, $J_{4,5}$, is larger than $J_{3,4}$ and increases with an increase in the electron-withdrawing property of the substituent on N_1. The peak due to hydrogen at C_3 is normally broad because of the nuclear quadrupole relaxation effect of nitrogen. When R_4 is not hydrogen, the signal for the proton at C_3 is smaller than that for the proton at C_5; when R_4 is hydrogen, the signal at C_3 shows poorer resolution than that at C_5. The absorptions due to substituents at C_5 display solvent shifts that are greater than the solvent shifts displayed by substituents at C_3 (14). Table 1 summarizes the chemical shifts of some representative pyrazoles.

The main application of nmr in the field of pyrazolines is to determine the stereochemistry of the substituents and the conformation of the ring. For pyrazolones, nmr is useful in establishing the structure of the various tautomeric forms. Table 2 summarizes the chemical shifts of a few representative derivatives.

^{13}C-Nmr. The availability of superconducting magnets combined with spectrometers operating at up to 600 MHz has led to great advancements in

Table 1. ^{1}H-Nmr Chemical Shifts for Selected Pyrazoles (12)^a

Compound	R_1	R_3	R_4	R_5	$\delta(^1H)^b$, ppm R_1	R_3	R_4	R_5
pyrazole	H	H	H	H	12.64	7.61	6.31	7.61
3-methyl	H	CH$_3$	H	H		2.32	6.06	7.48
4-methyl	H	H	CH$_3$	H		7.36	2.09	7.36
3,5-dimethyl	H	CH$_3$	H	CH$_3$		2.21	5.76	2.21
3-phenyl	H	C$_6$H$_5$	H	H		7.77	6.53	7.52
4-fluoro	H	H	F	H	12.40	7.49		7.49
4-ethoxy	H	H	OC$_2$H$_5$	H	10.96	7.25	3.92; 1.36	7.25
3-cyano	H	CN	H	H	12.37		6.80	7.81
1-methyl	CH$_3$	H	H	H	3.88	7.49	6.22	7.35
1-phenyl	C$_6$H$_5$	H	H	H	7.2–7.6	7.72	6.46	7.87
1-acetyl	CH$_3$CO	H	H	H	2.70	7.71	6.44	8.25
1–amino	NH$_2$	H	H	H	6.30	7.30	6.06	7.30
1,3-dimethyl	CH$_3$	CH$_3$	H	H	3.80	2.23	5.95	7.22
1,5-dimethyl	CH$_3$	H	H	CH$_3$	3.73	7.36	5.98	2.22
1,3,4,5-tetra-methyl	CH$_3$	CH$_3$	CH$_3$	CH$_3$	3.68	2.14	1.89	2.12
1,3-dimethyl-5-phenyl	CH$_3$	CH$_3$	H	C$_6$H$_5$	3.80	2.30	6.09	7.43
1,5-dimethyl-3-phenyl	CH$_3$	C$_6$H$_5$	H	CH$_3$	3.80	7.78	6.30	2.30

^{a}Ref. 15.
^{b}Solvent is CDCl$_3$.

Table 2. ^{1}H-Nmr Chemical Shifts for Selected Pyrazolones and Pyrazolines^a

Compound	$\delta(^1H)^b$, ppm R_1	R_3	R_4	R_5
Pyrazolone				
3-methyl-1-phenyl-2-pyrazolin-5-one	7.3–7.8	2.12	3.37	
5-ethoxy-3-methyl-1-phenylpyrazole	7.0–7.9	2.26	5.50	4.14; 1.41
2,3-dimethyl-1-phenyl-3-pyrazolin-5-one	7.30	2.20	5.37	
Pyrazoline				
unsubstituted-Δ^2	5.33	6.88	2.65	3.31
1,3-dimethyl-Δ^2	2.74	1.95	2.96	3.58
1-formyl-3-methyl-Δ^2	8.81	2.05	2.88	3.90
1,2,3-trimethyl-Δ^3	2.55	1.71	4.60	3.65
1,2,3,5,5-pentamethyl-Δ^3	2.47	1.80	4.53	1.17

^{a}Ref. 16.
^{b}Solvent is CDCl$_3$.

^{13}C-nmr spectroscopy. Pyrazole [288-13-1] itself shows one or two peaks for C$_3$ and C$_5$ depending on solvent and temperature, illustrating the similarity of the tautomers present in solution. When a substituent at N$_1$ is present, the chemical shifts at C$_3$ and C$_5$ move to higher (ppm) values (deshielded, lower fields) as the electron-withdrawing property of the N$_1$ substituent increases. Compared to ^{1}H-nmr, ^{13}C-nmr is much more precise in distinguishing between two isomers, eg,

3-CH$_3$ vs 5-CH$_3$ or N-CH$_3$. Table 3 presents the chemical shifts for a number of representative pyrazoles.

Table 3. ^{13}C-Nmr Chemical Shifts for Pyrazoles (1)^a

Compound	Solvent	$\delta(^{13}$C), ppm from TMS^b			
		C-1	C-3	C-4	C-5
pyrazole	CH$_2$Cl$_2$		134.6	105.8	134.6
	HMPT ($-17°$C)		138.1	103.9	127.6
	DMSO-d_6 (29°C)		138.6	104.5	128.2
3(5)-methylpyrazole	CH$_2$Cl$_2$		144.4 (CCH$_3$)	105.3	135.8
3(5)-nitropyrazole	acetone-d_6			102.4	132.6
3(5)-aminopyrazole	DMSO-d_6		154.0 (CNH$_2$)	91.5	132.0
1-methylpyrazole	CDCl$_3$	38.4	138.7	105.1	129.3
1-benzylpyrazole	CDCl$_3$	55.8 (CH$_2$)	139.4	105.7	129.1
1-phenylpyrazole	CDCl$_3$		140.7	107.3	126.2
1-acetylpyrazole	CDCl$_3$	21.3^c; 169.2^d	143.6	109.3	127.8
1-nitropyrazole	DMSO-d_6		141.6	109.8	126.8
1,3,5-trimethylpyrazole	CH$_2$Cl$_2$	36.3	147.5	105.4	139.8
3-amino-1-methylpyrazole	DMSO-d_6	38.0	155.7	92.2	131.5
5-bromo-1-phenylpyrazole	CDCl$_3$		141.1	110.2	112.5

^{a}Ref. 17. ^{b}TMS = tetramethylsilane. ^{c}CH$_3$. ^{d}CO.

Newer Developments.   The nmr spectra of four pyrazole derivatives using 98% sulfuric acid as the solvent has been recorded (18); proton chemical shifts, ^{1}H–^{1}H coupling constants, ^{13}C chemical shifts, and ^{1}H–^{13}C coupling constants of the pyrazolium derivatives were determined. Ten pyrazoles bearing a formyl or hydrazone group at C$_4$ and a benzyl, p-methoxybenzyl, or phenyl group attached to N$_1$, have been prepared. The ^{13}C signals of these derivatives were compared to pyrazoles unsubstituted at C$_4$ (19), and it was concluded that a formyl group produces a large effect on C$_4$ and lesser effect on C$_5$ and C$_3$ (19).

Nuclear Overhauser enhancement (NOE) spectroscopy has been used to measure the through-space interaction between protons at C$_5$ and the protons associated with the substituents at N$_1$ (20). The method is also useful for distinguishing between isomers with different groups at C$_3$ and C$_5$. Reference 21 contains the chemical shifts and coupling constants of a considerable number of pyrazoles with substituents at N$_1$ and C$_4$. NOE difference spectroscopy (^{1}H) has been employed to differentiate between the two regioisomers [153076-45-0] (**14**) and [153076-46-1] (**15**) (22). ^{15}N-nmr spectroscopy also has some utility in the field of pyrazoles and derivatives.

(**14**)   (**15**)

Mass Spectrometry. As of 1996, ms characteristics of pyrazoles and derivatives had not been described in depth. The fate of unsubstituted pyrazole (**23**) in the mass spectrometer operated in the electron ionization mode may be depicted as follows:

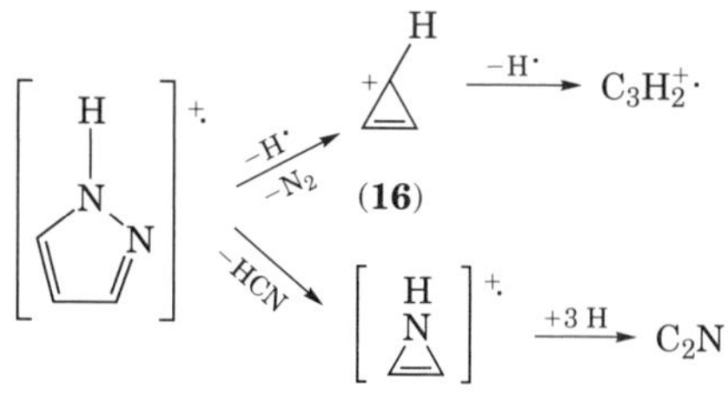

Loss of nitrogen occurs to give the cyclopropene (**16**), which loses a proton to yield the charged species $C_3H_2^+$. Loss of HCN leads to the unsaturated aziridine (**17**), which further decays to the C_2N^+ fragment. In the presence of a methyl group at N_1, ring enlargement can occur to give (**18**), which degrades to the charged species $C_4H_5^+$, or degradation proceeds to a mixture of N-methylaziridine (**19**) and the charged species $C_3H_4N^+$.

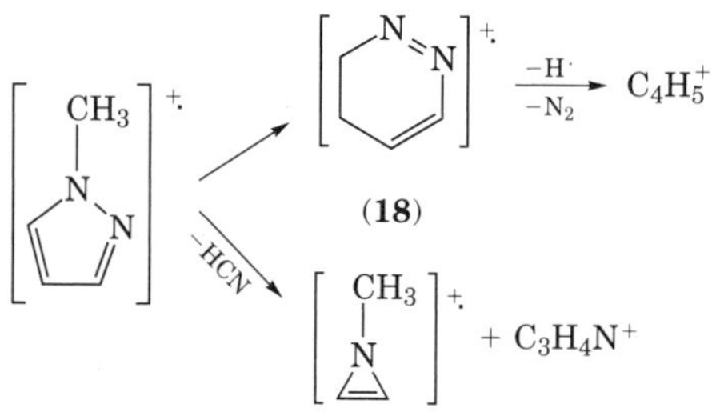

X-Ray Diffraction. Because of the rapid advancement of computer technology (qv), this technique has become almost routine and the structures of moderately complex molecules can be established sometimes in as little as 24 hours. An example illustrating the method is offered by Reference 24. The reaction of the acrylate (**20**) with phenyldiazo derivatives results in the formation of pyrazoline (**21**). The stereochemistry of the substituents and the conformation of the ring can only be established by single crystal x-ray diffraction.

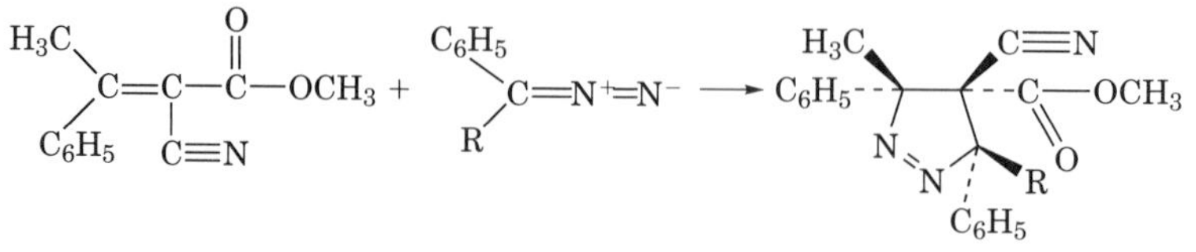

Miscellaneous Techniques. The use of ultraviolet (uv) and infrared (ir) spectroscopy has diminished drastically as newer and more powerful procedures have been introduced. However, uv is still useful in studying the tautomeric structures and ionization constants of pyrazoles. Studies of the effect of substituents on the absorption and fluorescence properties of di- and triphenyl-substituted pyrazolines showed that electron-withdrawing groups attached to phenyl groups such as $-CN$, $-SO_2CH_3$, and $-OCOCH_3$ cause a profound quenching of fluorescence (25). Infrared spectroscopy was used in theoretical studies on pyrazoles and deuterated pyrazoles, and also to facilitate structure determination. Infrared was also used to characterize in solution the various tautomeric forms of pyrazolones (double bond within or outside the ring). The three isomeric pyrazolines could be distinguished by their characteristic $N{=}N$ and $C{=}N$ vibrations (26). Other methods useful in structural elucidation are microwave spectroscopy and dipole moment measurements, but these are seldom used in the field of pyrazoles.

Physical Properties

Pyrazoles in general are stable compounds, as demonstrated by pyrazole itself, which distills at 186°C at atmospheric pressure. The boiling point (bp) increases with an increase in the number of alkyl substituents on carbon. N-Methylation decreases both the bp and the melting point (mp) as a result of the elimination of hydrogen bonding. Thus 3-methylpyrazole [1453-58-3] has a bp of 205°C and the isomeric N-methylpyrazole [930-36-9] has a bp of 127°C. Pyrazoles with substituents at C_3 (C_5) are tautomeric mixtures and form azeotropes. The solubility of pyrazole in H_2O is about 1 g/mL, but it is much less soluble in organic solvents. Pyrazole is a weak base ($pK_a = 2.5$) and can be protonated by strong acids; strong bases yield metal salts. The pyrazolines resemble the pyrazoles in their physical properties. They are liquids with a high bp or low mp. Pyrazolines are basic and the ease of protonation is dependent on the position of the double bond. Most pyrazolones are solids and the mp usually decreases in the presence of substituents at N_1. Simple low molecular weight pyrazolones are soluble in hot water and the higher mol wt materials are soluble in most organic solvents. Hydrogen bonding has strong influence on the predominant tautomeric form. 3-Pyrazolones are more basic than the isomeric 5-pyrazolones.

Chemical Reactivity

Pyrazoles. The chemical reactivity of the pyrazole molecule can be explained by the effect of individual atoms. The N-atom at position 2 with two electrons is basic and therefore reacts with electrophiles. The N-atom at position 1 is unreactive, but loses its proton in the presence of base. The combined two N-atoms reduce the charge density at C_3 and C_5, making C_4 available for electrophilic attack. Deprotonation at C_3 can occur in the presence of strong base, leading to ring opening. Protonation of pyrazoles leads to pyrazolium cations that are less likely to undergo electrophilic attack at C_4, but attack at C_3 is facilitated. The pyrazole anion is much less reactive toward nucleophiles, but the reactivity to electrophiles is increased. Some of the more general chemical properties of the pyrazole molecule are as follows. Chlorination of pyrazole yields 4-chloropyrazole [*15878-00-4*] (**22**) and bromination can produce mono-, di-, or tribromo pyrazoles [*17635-44-8*] (**23**). 3-Methylpyrazole on treatment with chlorine in acetic acid yields the pentachloropyrazole derivative (**24**) [*80294-32-2*].

(**22**)   (**23**)   (**24**)

1,4-Dinitropyrazole [*35852-77-8*] (**25**) can be reduced, or can react with amines to give 5-substituted pyrazoles. Friedel-Craft reactions introduce acyl groups at C_4. Halogens at C_3 or C_5 can be displaced only if there is an activating group at C_4; bromine at C_4 can be replaced by hydrogen or a carboxyl group after treatment with butyllithium. In the presence of base, alkyl and acyl halides react at N_1 to yield alkyl and acyl derivatives, respectively. Treatment with alkyl chloroformate leads to (**26**) and reaction with isocyanates yields (**27**).

(**25**)   (**26**)   (**27**)

The pyrazole ring is resistant to oxidation and reduction. Only ozonolysis, electrolytic oxidations, or strong base can cause ring fission. On photolysis, pyrazoles undergo an unusual rearrangement to yield imidazoles via cleavage of the N_1–N_2 bond, followed by cyclization of the radical intermediate to azirine (27).

1-Nitropyrazoles undergo a thermal intramolecular rearrangement leading to the migration of the nitro group from nitrogen (N_1) to carbon (C_3 and C_4) (**28**). Thermolysis of 1-nitro-3,4,5-tribromopyrazole [*104599-40-8*] (**28**) in boiling benzene gives a mixture consisting of di- and tribromopyrazole derivatives, (**29**) and (**30**), with the nitro group having migrated from N_1 to C_4.

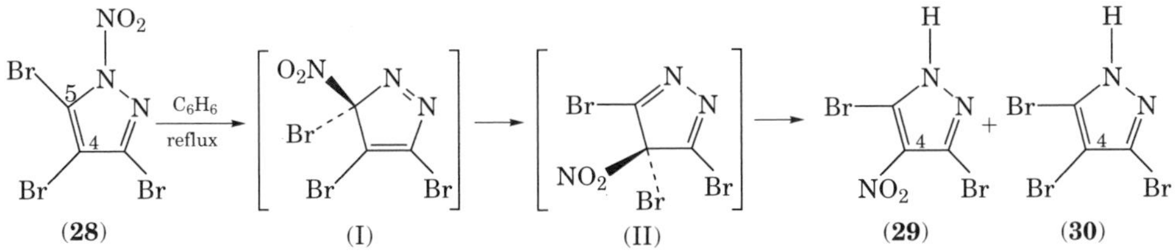

This reaction probably proceeds via the bromo nitro intermediate (I), which under certain conditions rearranges to intermediate (II) by a second migration of the nitro group from C_5 to C_4. Loss of Br or NO_2 can lead to a neutral pyrazolyl radical, followed by phenylation on nitrogen to yield 1-phenylpyrazole derivatives such as 1-phenyl-3,5-dibromo-4-nitropyrazole [104599-39-5] or 1-phenyl-3,4,5-tribromopyrazole. On changing solvents from benzene to acetonitrile, besides the previously identified compounds 3,5-dibromo-4-nitropyrazole [104599-36-2] (29) and 3,4,5-tribromopyrazole [17635-44-8] (30), 3,4-dibromo-5-nitropyrazole [104599-37-3] is also isolated, further indicating the presence of (I) as an intermediate. Reaction of 4-chloro-3,5-dimethyl-1-nitropyrazole [153813-97-9] with three equivalents of 1,8-diazabicyclo[5.4.0]undec-7-ene (DBU) or 1,5-diazabicyclo[4.3.0]non-5-ene (DBN) in acetonitrile leads to two unexpected compounds, (31) and (32), respectively (29). Their formation takes place via the nucleophilic attack of DBU or DBN on the diazafulvene intermediate, which in turn arises by the elimination of HNO_2 from the starting pyrazole (Fig. 1).

Oxidation of N_1-substituted pyrazoles to 2-substituted pyrazole-1-oxides using various peracids (30) facilitates the introduction of halogen at C_3, followed by selective nitration at C_4. The halogen atom at C_3 or C_5 is easily removed by sodium sulfite and acts as a protecting group. Formaldehyde was used to direct

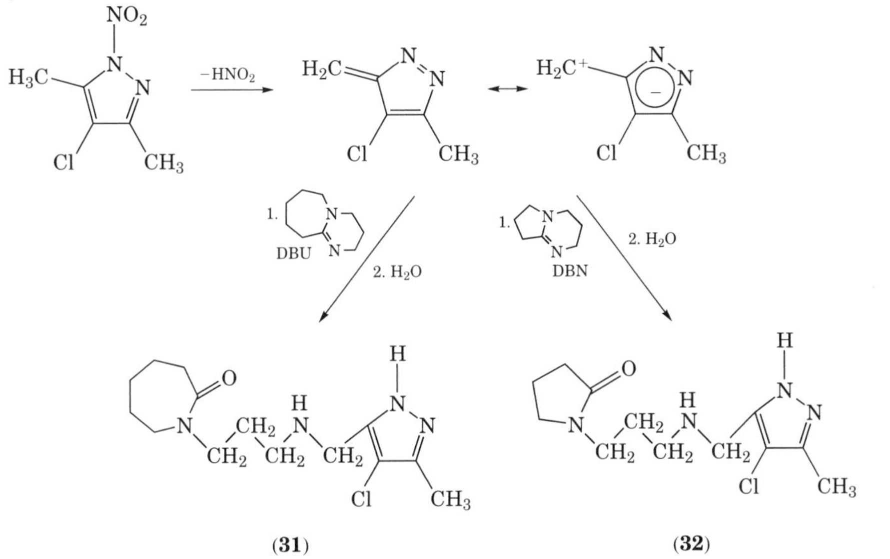

Fig. 1.   Intermediary of diazafulvene in reactions of 1-nitropyrazoles.

the selective introduction of electrophiles at C_5 in a simple one-pot procedure (31). The SO_2NH_2 moiety has been introduced selectively at C_5 by treating ethyl 1-methylpyrazole-4-carboxylate with lithium diisopropylamide (LDA), followed by SO_2, N-chlorosuccinimide, and 28% aqueous ammonium hydroxide (32).

Pyrazolines. The chemical properties of pyrazolines are governed by their relative instability. They readily undergo ring cleavage, and are easily reduced and oxidized. Loss of nitrogen occurs in pyrazolines lacking a substituent at N_1 to give a mixture of olefins and cyclopropanes, the latter being predominant. This elimination occurs near the mp and can be catalyzed by uv light, aluminum oxide, and many other substances. Mild reduction of pyrazolines leads to pyrazolidines. Sodium–alcohol, tin–HCl, or Raney nickel cause ring cleavage, yielding diamines or aminonitrile derivatives. Pyrazolines are easily oxidized to pyrazoles by many reagents, such as bromine, permanganate, and lead tetraacetate. Besides pyrazole formation, rearrangements or side-chain oxidations may also occur. Oxidation with peracids produce N-oxides. Pyrazolines lacking a substituent at N_1 undergo reactions typical of secondary amines, such as acylation, benzoylation, nitrosation, carbamate, and urea formation. An interesting reaction of pyrazoline has been described (33). Reaction of (33) with excess diphenylketene gives a 60% yield of the rearranged (34); the structure of (34) has been determined by x-ray crystallography.

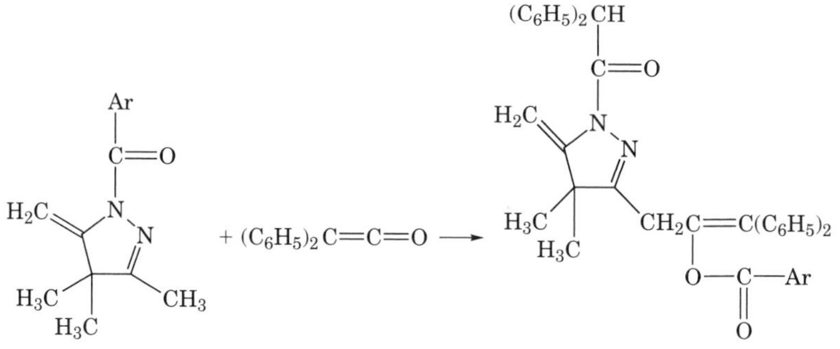

(33)   (34)

With dichloroketene, the reaction takes a different path leading to the hydroxypyrazoline (35). A mechanism to explain these results has been proposed (33).

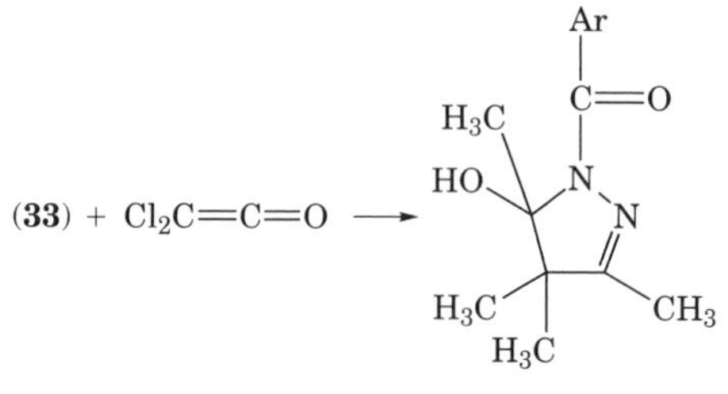

(33) + $Cl_2C{=}C{=}O$ →

(35)

Pyrazolones. The oxo derivatives of pyrazolines, known as pyrazolones, are best classified as follows: 5-pyrazolone, also called 2-pyrazolin-5-one [137-44-0] (**36**); 4-pyrazolone, also called 2-pyrazolin-4-one [27662-65-3] (**37**); and 3-pyrazolone, also called 3-pyrazolin-5-one [137-45-1] (**38**). Within each class of pyrazolones many tautomeric forms are possible; for simplicity only one form is shown.

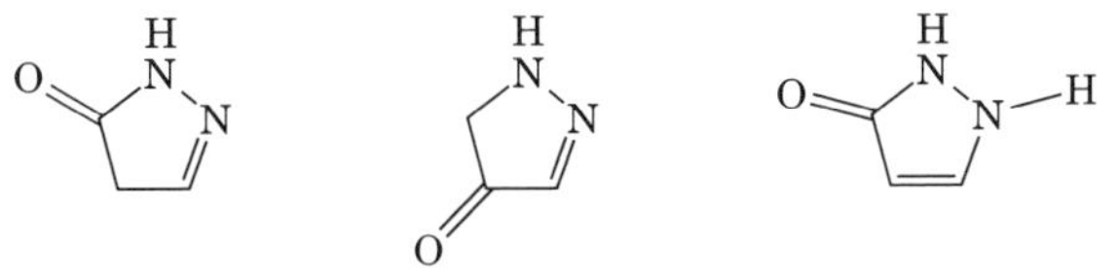

(**36**)   (**37**)   (**38**)

Substitution at N_1 decreases the possible number of tautomers: for 3-pyrazolones, two tautomeric forms are possible, (**39**) and (**40**), which in nonpolar solvents are both present in about the same ratio. 5-Pyrazolones exhibit similar behavior.

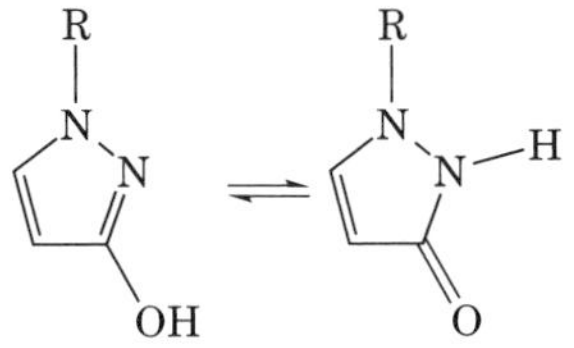

(**39**)   (**40**)

In 4-pyrazolones, the enol form predominates, although the keto form has also been observed.

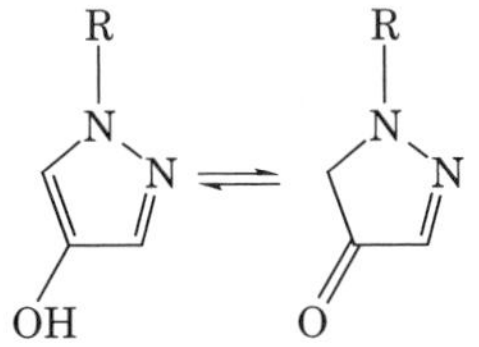

enol   keto

The tautomeric character of the pyrazolones is also illustrated by the mixture of products isolated after certain reactions. Thus alkylation normally takes place at C_4, but on occasion it is accompanied by alkylation on O and N. Similar problems can arise during acylation and carbamoylation reactions, which also favor C_4. Pyrazolones react with aldehydes and ketones at C_4 to form a carbon–carbon double bond, eg (**41**). Coupling takes place when pyrazolones react with diazonium salts to produce azo compounds, eg (**42**).

(41)   (42)

Compounds of type (**42**) are widely used in the dye industry (see AZO DYES). The Mannich reaction also takes place at C_4, as does halogenation and nitration. The important analgesic aminoantipyrine [*83-07-8*] (**43**) on photolysis in methanol undergoes ring fission to yield (**44**) (27).

(43)   (44)

Synthesis

Although there are many publications devoted to the synthesis of pyrazoles and related compounds, most can be classified into one of four principal categories, with the first class being by far the most important: (*1*) from the reaction of hydrazine or its derivatives with β-bifunctional compounds, or compounds that give rise to such functionality; (*2*) by 1,3-dipolar cycloaddition, usually involving diazo compounds; (*3*) by ring-opening of more complex systems already containing the pyrazole nucleus; and (*4*) by chemical, thermal, or photochemical rearrangement of other monocyclic heterocycles. Examples from each class follow.

From Hydrazines and β-Bifunctional Compounds.   One of the oldest examples in this class is the reaction of a β-diketone with a substituted hydrazine to give a pyrazole (eq. 1).

(1)

If $R_A \neq R_B$, a mixture of the two isomeric pyrazoles is obtained. An excellent method to prepare pyrazole [288-13-1] consists in treating 1,1,3,3-tetramethoxypropane (masked malondialdehyde) with hydrazine (eq. 2).

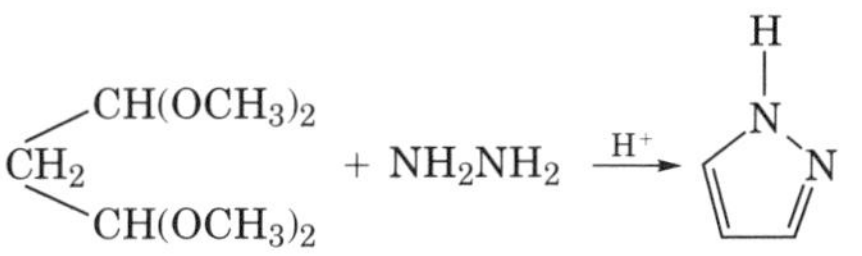

$$
\begin{array}{c}
\text{CH(OCH}_3)_2 \\
| \\
\text{CH}_2 \\
| \\
\text{CH(OCH}_3)_2
\end{array}
+ \text{NH}_2\text{NH}_2 \xrightarrow{\text{H}^+}
\text{(pyrazole ring)}
\tag{2}
$$

Reaction of an acid chloride with trimethylsilylacetylene produces an α,β-ethynyl ketone, which on treatment with substituted hydrazines yields a mixture of 1,5- and 1,3-substituted pyrazoles (34). The ratio is dependent on the reaction conditions (eq. 3).

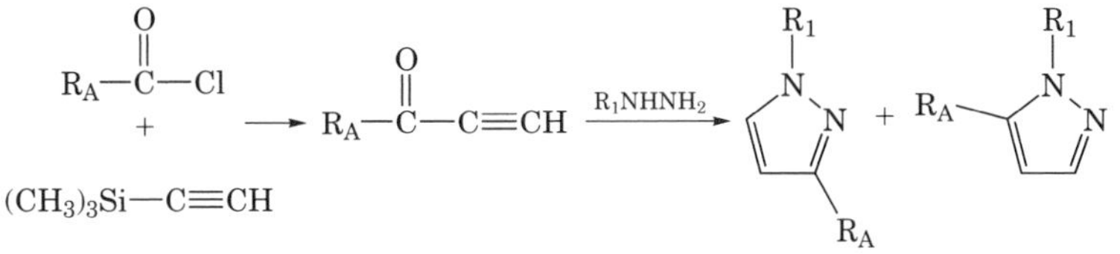

$$
\begin{array}{c}
\text{O} \\
\parallel \\
R_A-\text{C}-\text{Cl} \\
+ \\
\text{(CH}_3)_3\text{Si}-\text{C}\equiv\text{CH}
\end{array}
\longrightarrow
R_A-\overset{\text{O}}{\underset{}{\text{C}}}-\text{C}\equiv\text{CH}
\xrightarrow{R_1\text{NHNH}_2}
\text{(pyrazoles)}
\tag{3}
$$

Unsaturated ketones react with phenylhydrazines to form hydrazones, which under acidic conditions cyclize to pyrazolines (35). Oxidation, instead of acid treatment, of the hydrazone with thianthrene radical cation (TH·⁺) perchlorate yields pyrazoles; this oxidative cyclization does not proceed via the pyrazoline (eq. 4).

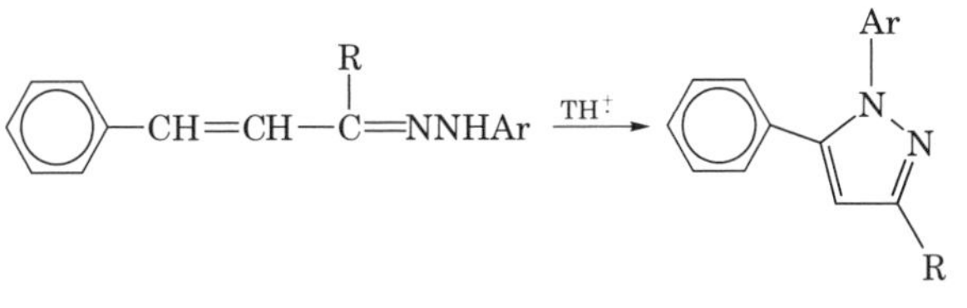

$$
\text{C}_6\text{H}_5-\text{CH}=\text{CH}-\overset{R}{\underset{}{\text{C}}}=\text{NNHAr}
\xrightarrow{\text{TH}^{\ddagger}}
\text{(pyrazole)}
\tag{4}
$$

Reaction of a β-ketoester with gaseous ammonia (1) gives an enamine, which on treatment with methylhydrazine (2) yields an 85:15 mixture of 3-hydroxy- and 5-hydroxy-1-substituted pyrazoles (36) (eq. 5). Previously β-ketoesters furnished mainly the 5-hydroxy isomer.

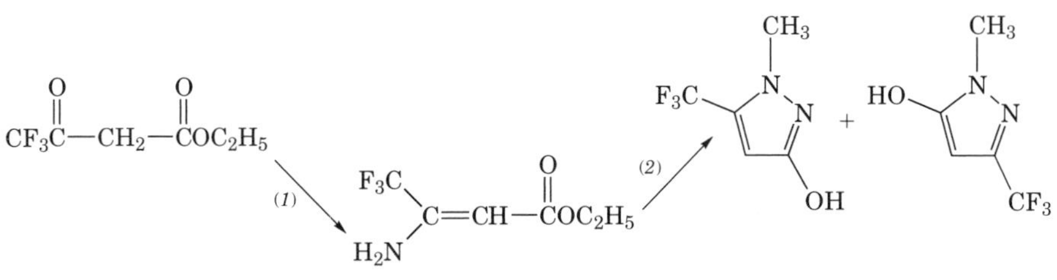

$$
\begin{array}{c}
\text{O} \quad\quad \text{O} \\
\parallel \quad\quad \parallel \\
\text{CF}_3\text{C}-\text{CH}_2-\text{COC}_2\text{H}_5 \\
\downarrow (1) \\
\begin{array}{c}
\text{F}_3\text{C} \\
\diagdown \\
\text{C}=\text{CH}-\overset{\text{O}}{\underset{}{\text{COC}_2\text{H}_5}} \\
\diagup \\
\text{H}_2\text{N}
\end{array}
\xrightarrow{(2)}
\text{(3-hydroxy)} + \text{(5-hydroxy)}
\end{array}
\tag{5}
$$

Treatment of a hydroxyiminoimine with one equivalent of hydrazine hydrate gave a nitrosopyrazole, which on addition of excess hydrazine hydrate yielded a 4-aminopyrazole derivative (eq. 6) (37).

$$\tag{6}$$

The need for pyrazoles substituted with the trifluoroacetyl group led to the reaction of ethoxyvinyl ether with trifluoroacetic anhydride, yielding 4-ethoxy-1,1,1-trifluoro-3-buten-2-one (38); this further reacted with aldehyde *tert*-butylhydrazones, and after cyclization at room temperature under mildly acidic conditions the pyrazoles were obtained in satisfactory yields (eq. 7). Further treatment with H_2SO_4 removed the *tert*-butyl group, thus providing an opportunity for further derivatization at N_1.

$$\tag{7}$$

Reaction of 4-amino-1-azabutadienes (**45**) with various hydrazine salts at 60°C leads to the expected pyrazoles (**46**) without isolation of the hydrazone intermediate (eq. 8) (39).

$$\tag{8}$$

On reaction with *N*-methyl-*N*-phenylhydrazine, however, the hydrazone (**47**) can be isolated, which on further treatment with anhydrous trifluoroacetic acid gives an *N*-alkenylpyrazole (**48**) (eq. 9).

$$(47) \qquad\qquad (48) \tag{9}$$

Instead of a β-dicarbonyl compound, triphenylphosphorane derivatives have been used to synthesize 5-alkoxy-substituted pyrazoles (40). Thus, reaction of ethyl chloroglyoxalate phenylhydrazone (**49**) with carbomethoxymethylene triphenylphosphorane (**50**) furnishes the 5-methoxypyrazole (**51**) in an 84% yield (eq. 10).

$$(49) \qquad\qquad (50) \qquad\qquad (51) \tag{10}$$

Structure proof is based on spectroscopic means and an independent synthesis of (**51**). Reaction of phenylhydrazine with dimethyl acetylenedicarboxylate gives a hydrazone, which is then cyclized to 1-phenyl-3-methoxycarbonyl-5-methoxypyrazole [*113246-37-0*] (**52**) (eq. 11). This is followed by transesterifi-

$$(52) \tag{11}$$

cation with ethanol, yielding (**51**). A second synthesis utilizing phosphorus has also been developed (41). 1,4-Addition of triphenylphosphine to conjugated azoalkenes gives rise to 3-hydrazono-2-triphenylphosphoranylidenebutanoates (**53**). The latter, on treatment with acid chlorides or anhydrides, leads to the intermediate (**54**), which cyclizes spontaneously to the phosphonium betaine (**55**). Loss of triphenylphosphine oxide produces the desired pyrazoles (**56**) in good to excellent yields.

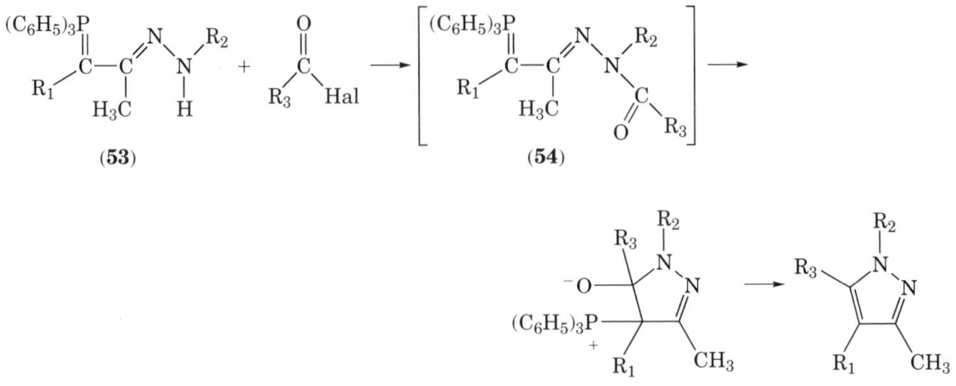

(53)   (54)

(55)   (56)

From Diazo Compounds via 1,3-Dipolar Cycloaddition. This method has been utilized widely in heterocyclic chemistry. Pyrazoline (**57**) has been synthesized by reaction of ethyl diazoacetate (**58**) with α,β-unsaturated ester in the presence of pyridine (eq. 12) (42).

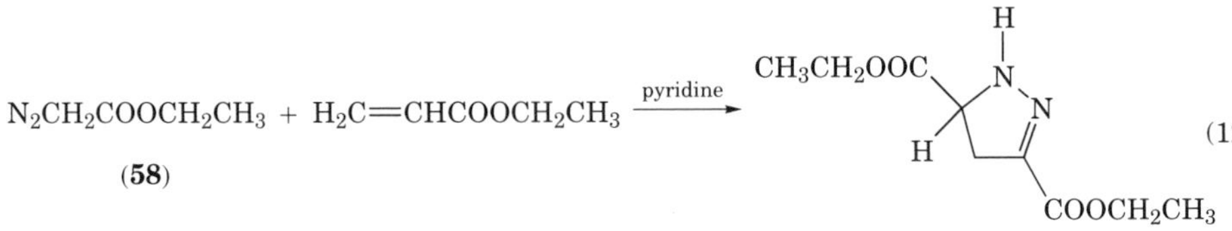

$$N_2CH_2COOCH_2CH_3 + H_2C{=}CHCOOCH_2CH_3 \xrightarrow{\text{pyridine}}$$

(**58**)

(12)

(**57**)

Reaction of (**58**) with unsaturated nitrile (**59**) produces 5-cyanopyrazoline (**60**), which on treatment with sodium ethoxide eliminates hydrogen cyanide to provide the pyrazole (**61**) in high yield (eq. 13).

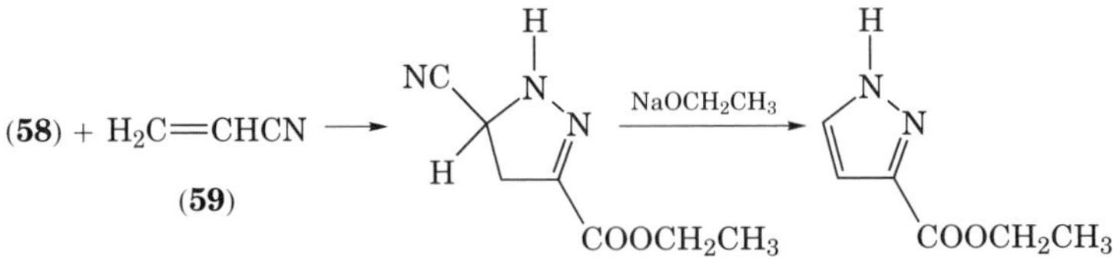

$$(\mathbf{58}) + H_2C{=}CHCN \longrightarrow$$

(**59**)

(13)

(**60**)   (**61**)

Reaction of aminoacetonitrile hydrochloride with sodium nitrite provides diazoacetonitrile (**62**). The product undergoes a 1,3-dipolar cycloaddition with diethyl fumarate to yield a pyrazoline intermediate, which without isolation reacts with ammonia in water to furnish the pyrazole [*119741-54-7*] (**63**) (eq. 14) (43).

$$(14)$$

Dehydration with $POCl_3$, followed by N-alkylation with isobutylene, gives the cyano ester intermediate, which on treatment with methylamine yields (**64**), a compound having good herbicidal activity (eq. 15).

$$(15)$$

From Multiring Systems Containing Pyrazoles. The pyrazolopyrimidine (**65**) on treatment with diazomethane forms the cyclopropane (**66**), which undergoes a ring-opening reaction with potassium hydroxide to yield the pyrazole (**67**) (eq. 16) (44).

$$(16)$$

From Other Heterocycles by Rearrangement. Although there are numerous examples of pyrazole syntheses from other monocyclic heterocycles by chemical, thermal, or photochemical means, such examples are only of limited practical value on account of high cost. Reaction of diaziridinone (**68**) with the sodium salt of malondinitrile yields the di-t-butylaminopyrazolinone (**69**) (eq. 17) (45).

$$(17)$$

The pyrazolone-3-carboxylic acid (**71**) has been isolated by reaction of oxazolone (**70**) with hydrazonyl chloride (eq. 18) (46).

$$(18)$$

Heating methylhydrazine with an *S*-methylpyrimidine derivative (**72**) in an HCl methanolic solution produces in a yield of over 50% the 4-acetylpyrazole (**73**) (eq. 19) (47).

$$(19)$$

Another example is a claim of possible industrial application for preparing 1-cyclohexyl-3,5-dimethylpyrazole [*79580-49-7*] (**75**) and similar compounds from 1,2,6-thiadiazine-1,1-dioxide (**74**) by extrusion of SO_2 (eq. 20) (48). This process has the added advantage of not requiring hydrazine derivatives as reactants.

$$(20)$$

Health Factors

Pyrazole is considered a toxic material because in rats it causes hepatomegaly, anemia, and atrophy of the testis. It also inhibits the enzyme alcohol dehydrogenase, leading to severe hepatotoxic effects and liver necrosis when administered in combination with alcohol. Chronic administration of bromopyrazoles to rats, even in the absence of ethanol, causes significant increases in liver size. Pyrazolones with a free NH group are easily nitrosated and give rise to nitrosamines, which cause tumors in the liver of test animals. The analgesics antipyrine [60-80-0] (**76**) and aminopyrine [58-15-1] (**77**), if admixed with nitrites, are mutagenic when tested *in vitro*; however, when tested in the absence of nitrites, negative results are obtained (49).

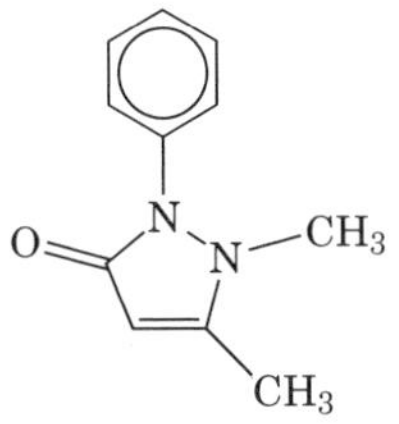

(**76**) 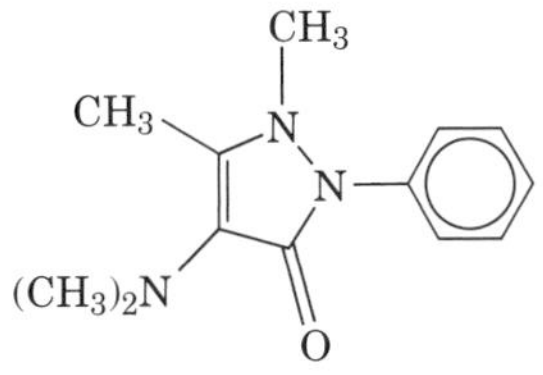   (**77**)

Pyrazolone-type drugs, such as phenylbutazone and sulfinpyrazone, are metabolized in the liver by microsomal enzymes, forming glucuronide metabolites that are easily excreted because of enhanced water solubility.

Applications

Pyrazoles, pyrazolines, and pyrazolones have all found wide use in many fields. Their greatest utility resides in pharmaceuticals, agrochemicals, dyes (textile and photography), and to a lesser extent in plastics. Figure 2 summarizes some of the pharmaceuticals that incorporate the pyrazole nucleus. Their main uses are as antipyretic, antiinflammatory, and analgesic agents. To a lesser extent, they have shown efficacy as antibacterial/antimicrobial, antipsychotic, antiemetic, and diuretic agents (50). The analgesic aminopyrine (**77**), the antipyretic dipyrone [5907-38-0] (**78**), and the antiinflammatory phenylbutazone [50-33-9] (**79**), though once widely prescribed, are rarely used in the 1990s on account of their tendency to cause agranulocytosis.

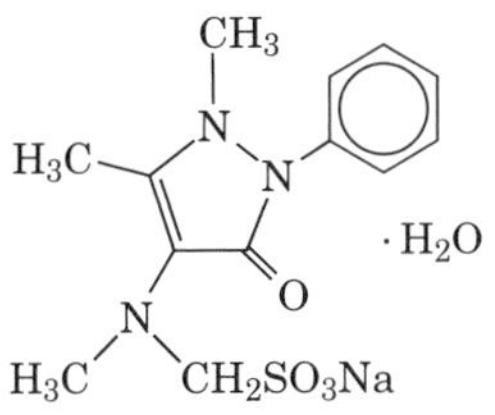

(**78**) 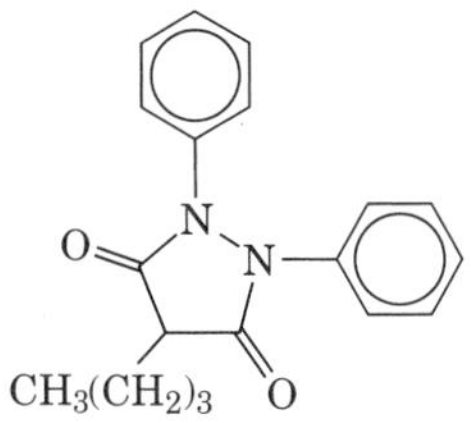   (**79**)

4-Methylpyrazole has been investigated as a possible treatment for alcoholism. The structure–activity relationship (SAR) associated with a series of

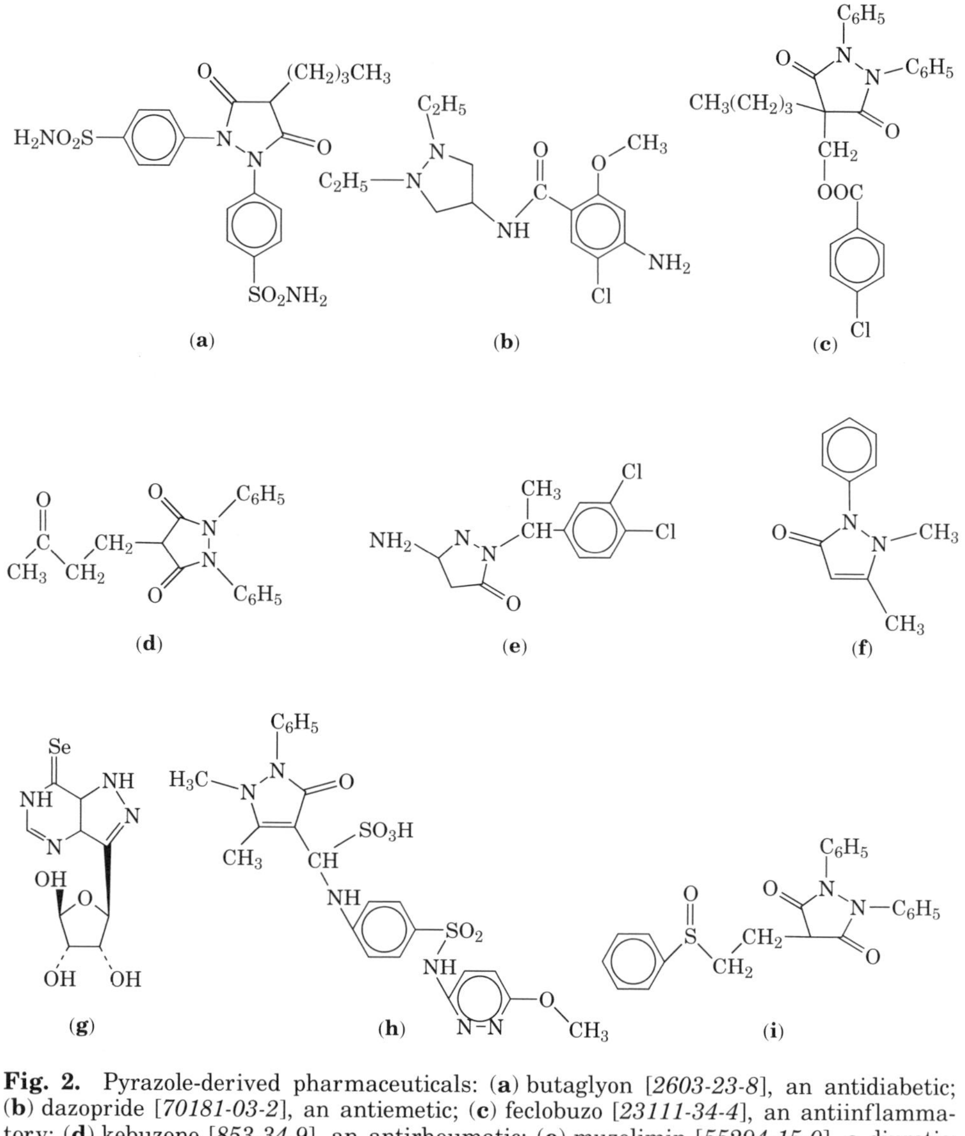

Fig. 2. Pyrazole-derived pharmaceuticals: (**a**) butaglyon [2603-23-8], an antidiabetic; (**b**) dazopride [70181-03-2], an antiemetic; (**c**) feclobuzo [23111-34-4], an antiinflammatory; (**d**) kebuzone [853-34-9], an antirheumatic; (**e**) muzolimin [55294-15-0], a diuretic; (**f**) phenazobz [20610-63-3], an antiasthmatic; (**g**) selenofob [39102-63-1], an antibiotic; (**h**) sulfamazo [65761-24-2], an antiseptic; and (**i**) sulfinpyrazone [57-96-5], an antigout preparation (50).

pyrazoles has been examined in a 1992 study (51). These compounds were designed as nonprostanoid prostacyclin mimetics to inhibit human platelet aggregation. In this study, 3,4,5-triphenylpyrazole was linked to a number of alkanoic acids, esters, and amides. From the many compounds synthesized, triphenyl-1*H*-pyrazole-1-nonanoic acid (**80**) was found to be the most efficacious candidate (IC$_{50}$ = 0.4 μM).

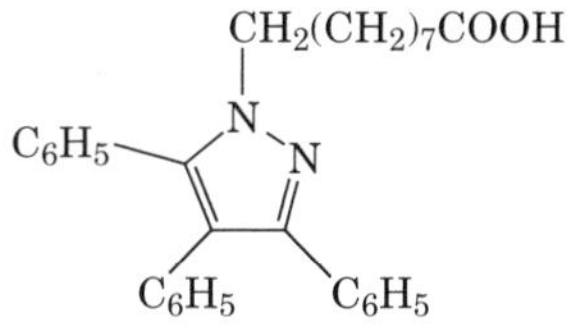

$$CH_2(CH_2)_7COOH$$

(**80**)

At Merck Research Laboratories, the imidazole ring in losartan (**81**, R = *n*-butyl), a novel clinical candidate against hypertension, was replaced with a pyrazole ring (52). Some of the best compounds are represented by formula (**82**), where R = *n*-butyl and R′ = 2,6-dichlorophenyl, 2-chlorophenyl, or 2-trifluoromethylphenyl.

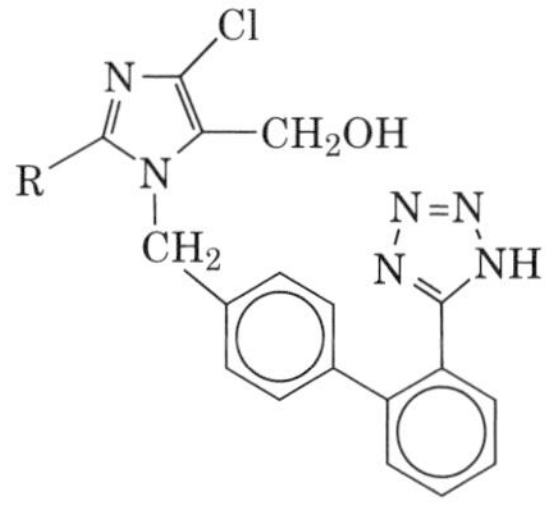

(**81**) 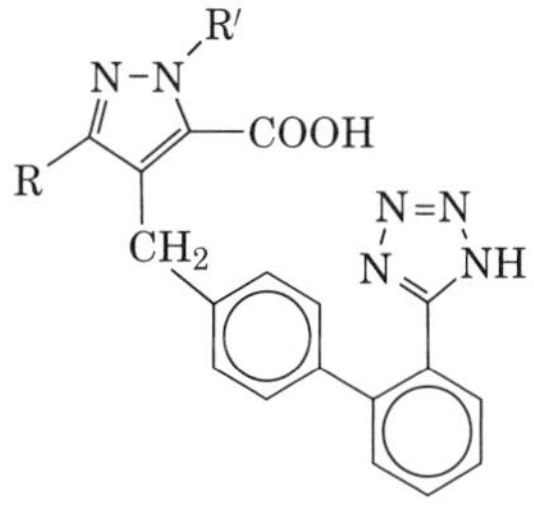   (**82**)

Compounds containing the pyrazole nucleus have also found utility in agriculture. The organophosphate and carbamoyl functionalities, which impart insecticidal activity through linkage to many organic molecules, have been attached to the pyrazole nucleus to yield compounds of type (**83**), where R = S or O, and type (**84**). These compounds act by interfering with acetylcholinesterase in the cholinergic synapses (7,53).

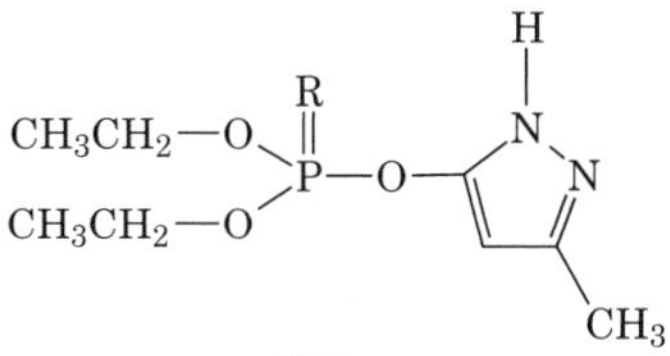

(**83**) 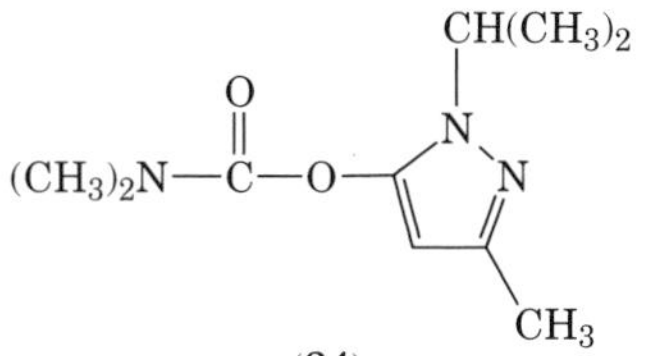   (**84**)

Another class of insecticides is being developed in the 1990s (54); the 3,4-diphenyl-substituted pyrazoline (**85**) is a representative member. The 5-cyanopyrazole [*98477-07-7*] (**64**) is being developed as a preemergent corn and post-emergent cereal herbicide (55). 3-Substituted pyrazole-5-sulfonylureas are being developed as herbicides against broad-leaf weeds and sedges in corn (10). The 3-phenylpyrazole (**86**) is a representative of a great many amidic pyrazoles having excellent preemergent herbicidal activity.

(85)   (86)

Dyes based on the pyrazolone nucleus were discovered in 1884 and are still in use in the 1990s. The majority of these dyes have an azo linkage attached at C_4, eg, (**87**) and (**88**). Table 4 summarizes the production and sales figures for four important dyes.

Table 4. U.S. Production and Sales of Selected Pyrazolone Derivatives[a]

Product	CAS Registry Number	Year	Production, t	Sales Quantity, t	Sales Unit value, $/kg
acid yellow 17[b]	[6359-98-4]	1983	100	84	5.08
		1985	97	105	5.81
		1986	37	83	6.61
		1987	90	75	7.09
acid yellow 23[c]	[1934-21-0]	1983	70	74	4.62
		1985	46	79	4.26
		1986	113	94	4.25
		1987	139	118	4.33
pigment orange 13[d]	[3520-72-7]	1983	112	139	8.21
		1985	128	118	9.42
		1986	120	112	8.52
		1987	171	160	9.93
		1991	58	51	23.74
pigment red 38[e]	[6358-87-8]	1983	128	143	10.94
		1985	150	134	11.19
		1986	87	129	5.40
		1987	177	151	11.25
		1991	72	72	25.51

[a]Ref. 56.
[b]Structure (**87**), where R_1 = Cl and R_2 = CH_3.
[c]Structure (**87**), where R_1 = H and R_2 = COONa.
[d]Structure (**88**), where R = CH_3.
[e]Structure (**88**), where R = $COOCH_2CH_3$.

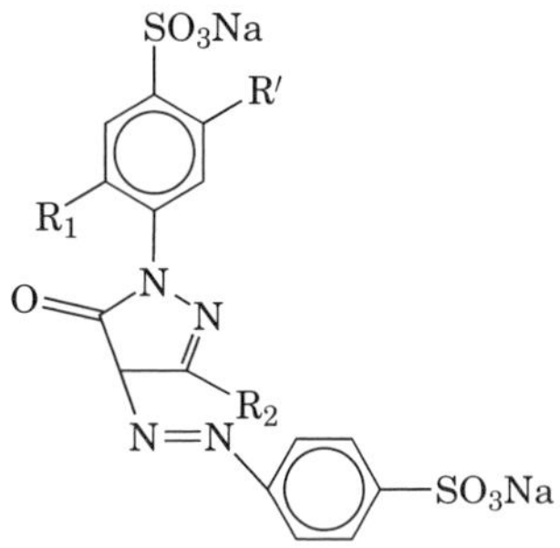

(87)

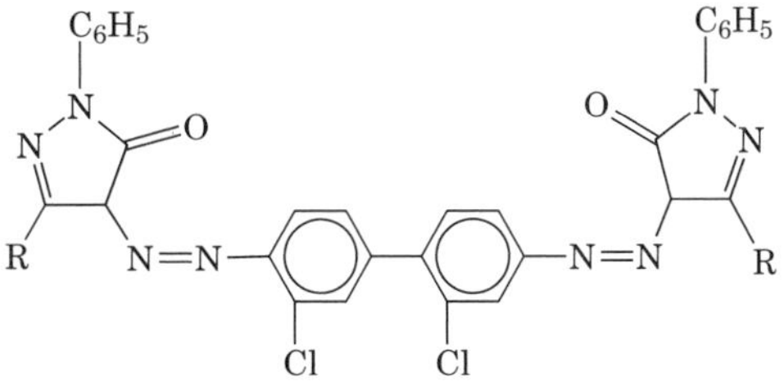

(88)

BIBLIOGRAPHY

"Pyrazoles, Pyrazolines, Pyrazolones" in *ECT* 2nd ed., Vol. 16, pp. 763–779, by P. F. Wiley, The Upjohn Co.; in *ECT* 3rd ed., Vol. 19, pp. 436–453, by G. Kornis, The Upjohn Co.

1. T. L. Jacobs, in R. C. Elderfield, ed., *Heterocyclic Compounds*, Vol. 5, John Wiley & Sons, Inc., New York, 1957, p. 45.
2. A. N. Kost and I. I. Grandberg, in A. R. Katritzky and A. J. Boulton, eds., *Advances in Heterocyclic Chemistry*, Vol. 6, Academic Press, Inc., New York, 1966, pp. 347–429.
3. L. C. Behr, R. Fusco, and C. H. Jarboe, in R. H. Wiley, ed., "Pyrazoles, Pyrazolines, Pyrazolidines, Indazoles and Condensed Rings," Vol. 22 of A. Weissberger, ed., *The Chemistry of Heterocyclic Compounds*, Wiley-Interscience, New York, 1967.
4. J. Elguero, in A. R. Katritzky and C. W. Rees, eds., *Comprehensive Heterocyclic Chemistry*, Vol. 4, Pergamon Press, Oxford, U.K., 1984, pp. 167–303.
5. A. R. Katritzky, *Handbook of Heterocyclic Chemistry*, Pergamon Press, Oxford, U.K., 1985.
6. A. Kotali and P. G. Tsoungas, *Heterocycles* **29**, 1615–1648 (1989).
7. R. M. Jacobson, *Spec. Publ. R. Soc. Chem.* **79**, 206–211 (1990).
8. I. R. Judson, *Anti-Cancer Drugs* **2**, 223–231 (1991).
9. K. Reszka, J. A. Hartley, and J. W. Lown, *Biophys Chem.* **35**, 313–323 (1990).
10. S. Yamamoto and co-workers, *ACS Symp. Ser.* **504**, 34–42 (1992).
11. Ref. 4, p. 173.
12. I. L. Finar, *J. Chem. Soc.* (B), 725–732 (1968).
13. R. E. Burton and I. L. Finar, *J. Chem. Soc.* (B), 1692–1695 (1970).
14. Ref. 4, pp. 182–197.
15. Ref. 4, p. 184.
16. Ref. 4, p. 187.
17. Ref. 4, p. 191.
18. J. Elguero, M. L. Jimeno, and G. I. Yranzo, *Magn. Reson. Chem.* **28**, 807–811 (1990).
19. A. Echevarria, J. Elguero, and W. Meutermans, *J. Heterocycl. Chem.* **30**, 957–960 (1993).
20. W. Holzer, *Tetrahedron* **47**, 1393–1398 (1991).
21. G. Heinisch and W. Holzer, *Heterocycles*, **27**, 2443–2457 (1988).
22. W. T. Ashton and G. A. Doss, *J. Heterocycl. Chem.* **30**, 307–311 (1993).
23. Ref. 4, p. 203.
24. W. Nagai and co-workers, *J. Heterocycl. Chem.* **31**, 225–231 (1994).
25. D. E. Rivett, J. Rosevear, and J. F. K. Wilshire, *Aust. J. Chem.* **36**, 1649–1658 (1983).
26. Ref. 4, pp. 197–202.

27. Ref. 4, p. 220.

28. J. P. H. Juffermans and C. L. Habraken, *J. Org. Chem.* **51**, 4656–4660 (1986).

29. H. Lammers, P. Cohen-Fernandes, and C. L. Habraken, *Tetrahedron* **50**, 865–870 (1994).

30. M. Begtrup, P. Larsen, and P. Vedso, *Acta. Chem. Scand.* **46**, 972–980 (1992); M. Sprecher, rev., *Chem. Tracts, Org. Chem.* **6**, 145–146 (1993).

31. A. R. Katritzky, P. Lue, and K. Akutagawa, *Tetrahedron* **45**, 4253–4262 (1989).

32. S. Yamamoto and co-workers, *J. Heterocycl. Chem.* **28**, 1849–1852 (1991).

33. S. Mitkidou and co-workers, *J. Org. Chem.* **55**, 4732–4735 (1990).

34. R. D. Miller and O. Reiser, *J. Heterocycl. Chem.* **30**, 755–763 (1993).

35. A. C. Kovelesky and H. J. Shine, *J. Org. Chem.* **53**, 1973–1979 (1988).

36. B. J. Gaede and L. L. McDermott, *J. Heterocycl. Chem.* **30**, 49–54 (1993).

37. R. T. Chakrasali, C. S. Rao, H. Ila, and H. Junjappa, *J. Heterocycl. Chem.* **30**, 129–134 (1993).

38. Y. Kamitori and co-workers, *J. Heterocycl. Chem.* **30**, 389–391 (1993).

39. J. Barluenga and co-workers, *J. Chem. Res. (S)*, 124–125 (1985).

40. A. Padwa and J. G. MacDonald, *J. Heterocycl. Chem.* **24**, 1225–1227 (1987).

41. O. A. Attanasi and co-workers, *J. Chem. Res. (S)*, 192–193 (1994).

42. M. P. Doyle, M. R. Colsman, and R. L. Dorow, *J. Heterocycl. Chem.* **20**, 943–946 (1983).

43. E. V. P. Tao, J. R. Rizzo, and J. Aikins, *J. Heterocycl. Chem.* **27**, 1655–1656 (1990).

44. Ref. 4, p. 285.

45. M. Komatsu and co-workers, *J. Org. Chem.* **57**, 7359–7361 (1992).

46. F. Clerici and co-workers, *Heterocycles* **27**, 1411–1419 (1988).

47. G. Menichi and co-workers, *J. Heterocycl. Chem.* **23**, 275–279 (1986).

48. J. F. Klein and co-workers, *Can. J. Chem.* **71**, 410–412 (1993).

49. H. G. Berscheid and co-workers, in F. T. Coulston and J. F. Dunne, eds., *The Potential Carcinogenicity of Nitrosatable Drugs WHO Symposium, Geneva, Switzerland, June 1978*, Ablex Publishing Corp., Norwood, N.J., 1980, p. 121.

50. *The DERWENT Standard Drug File of Known Drugs*, Derwent, London, 1994.

51. N. A. Meanwell and co-workers, *J. Med. Chem.* **35**, 389–397 (1992).

52. W. T. Ashton and co-workers, *J. Med. Chem.* **36**, 3595–3605 (1993).

53. Ref. 4, p. 297.

54. A. Bosum-Dybus and H. Neh, *Liebigs Ann. Chem.*, 823–825 (1991).

55. M. P. Lynch and co-workers, *ACS Symp. Ser.* **443**, 144–157 (1991).

56. *Synthetic Organic Chemicals, United States Production and Sales: 1983, USITC Public. 1588; 1985 USITC Public. 1892; 1986, USITC Public. 2009; 1987 USITC Public. 2118; 1991 USITC Public. 2607*, U.S. International Trade Commission, Washington, D.C., Feb. 1993.

GABE I. KORNIS
Pharmacia & Upjohn Inc.

PYRIDINE AND PYRIDINE DERIVATIVES

Since the early twentieth century, pyridine derivatives have been commercially important, but most prominently so during World War II and thereafter. Many pyridines of commercial interest find application in market areas where bioactivity is important, as in medicinal drugs and in agricultural products such as herbicides, insecticides, fungicides, and plant growth regulators. However, pyridines also have significant market applications outside the realm of bioactive ingredients. For instance, polymers made from pyridine-containing monomers are generally sold on the basis of their unique physical properties and function, rather than for any bioactivity. Pyridines can be classified as specialty chemicals because of a relatively lower sales volume than commodity chemicals. They are most often sold in the marketplace as chemical intermediates used to manufacture final consumer products.

Pyridine compounds are defined by the presence of a six-membered heterocyclic ring consisting of five carbon atoms and one nitrogen atom. The carbon valencies not taken up in forming the ring are satisfied by hydrogen atoms. The arrangement of atoms is similar to benzene except that one of the carbon–hydrogen ring sets has been replaced by a nitrogen atom. The parent compound is pyridine itself (**1**). Substituents are indicated either by the numbering shown, 1 through 6, or by the Greek letters, α, β, or γ. The Greek symbols refer to the position of the substituent relative to the ring nitrogen atom, and are usually used for naming monosubstituted pyridines. The ortho, meta, and para nomenclature commonly used for disubstituted benzenes is not used in naming pyridine compounds.

(**1**)

Important commercial alkylpyridine compounds are α-picoline (**2**), β-picoline (**3**), γ-picoline (**4**), 2,6-lutidine (**5**), 3,5-lutidine (**6**), 5-ethyl-2-methylpyridine (**7**), and 2,4,6-collidine (**8**). In general, the alkylpyridines serve as precursors of many other substituted pyridines used in commerce. These further substituted pyridine compounds derived from alkylpyridines are in turn often used as intermediates in the manufacture of commercially useful final products.

(**2**)   (**3**)   (**4**)

(5)   (6)   (7)   (8)

As is the case with most specialty organic compounds, pyridine sales are generally not publicized, and industrial processes for their manufacture are either retained as trade secrets or patented (see PATENTS AND TRADE SECRETS). Up to about 1950, most pyridines were isolated from coal-tar fractions; however, after 1950 synthetic manufacture began to take an ever-increasing share of products sold. By 1988, over 95% of all pyridines were produced by synthetic methods.

Pyridine was first synthesized in 1876 (1) from acetylene and hydrogen cyanide. However, α-picoline (**2**) was the first pyridine compound reported to be isolated in pure form (2). Interestingly, it was the market need for (**2**) that motivated the development of synthetic processes for pyridines during the 1940s, in preference to their isolation from coal-tar sources. The basis for most commercial pyridine syntheses in use can be found in the early work of Chichibabin (3). There are few selective commercial processes for pyridine (**1**) and its derivatives, and almost all manufacturing processes produce (**1**) along with a series of alkylated pyridines in admixture. The chemistry of pyridines is significantly different from that of benzenoids. Pyridines undergo some types of reaction that only highly electron-deficient benzenoids undergo, and do not undergo some facile reactions of benzenoids, such as Friedel-Crafts alkylation and C-acylation, for example.

Physical Properties

The physical properties of pyridines are the consequence of a stable, cyclic, 6-π-electron, π-deficient, aromatic structure containing a ring nitrogen atom. The ring nitrogen is more electronegative than the ring carbons, making the two-, four-, and six-ring carbons more electropositive than otherwise would be expected from a knowledge of benzenoid chemistries. The aromatic π-electron system does not require the participation of the lone pair of electrons on the nitrogen atom; hence the terms weakly basic and π-deficient used to describe pyridine compounds. The ring nitrogen of most pyridines undergoes reactions typical of weak, tertiary organic amines such as protonation, alkylation (qv), and acylation.

Liquid pyridine and alkylpyridines are considered to be dipolar, aprotic solvents, similar to dimethylformamide or dimethyl sulfoxide. Most pyridines form a significant azeotrope with water, allowing separation of mixtures of pyridines by steam distillation that could not be separated by simple distillation alone. The same azeotropic effect with water also allows rapid drying of wet pyridines by distillation of a small forecut of water azeotrope.

Pyridine. Many physical properties of pyridine are unlike those of benzene, its homocyclic counterpart. For instance, pyridine has a boiling point 35.2°C higher than benzene (115.3 vs 80.1°C), and unlike benzene, it is miscible with water in all proportions at ambient temperatures. The much higher dipole moment of pyridine relative to benzene is responsible, in significant part, for the higher boiling point and water solubility. Benzene and pyridine are aromatic compounds having resonance energies of similar magnitude, and both are miscible with most other organic solvents. Pyridine is a weak organic base ($pK_a = 5.22$), being both an electron-pair donor and a proton acceptor, whereas benzene has little tendency to donate electron pairs or accept protons. Pyridine is less basic than aliphatic, tertiary amines. Table 1 lists some physical properties of pyridine, and Table 2 compares physical properties of pyridine to some alkyl- and alkenylpyridine bases.

Other Pyridine Bases. The nucleophilicity and basicity of pyridines can be reduced by large, sterically bulky groups around the nitrogen atom, such as *tert*-butyl in the 2- and 6-positions. Sterically undemanding groups like methyl tend to increase basicity relative to parent pyridine, as expected. Electron-withdrawing substituents can also reduce pyridine basicity and nucleophilicity. 2,6-Dichloropyridine [2402-78-0] is sterically hindered and also contains strong electron-withdrawing substituents. As such, it cannot be titrated by acid, even by using extremely acidic media such as perchloric acid in acetic acid solvent.

Generally, hydrophobic substituents on the pyridine ring reduce water solubility, polar ones capable of hydrogen bonding as acceptor or donor, increase it.

Table 3 gives the corresponding physical properties of some commercially important substituted pyridines having halogen, carboxylic acid, ester, carboxamide, nitrile, carbinol, aminomethyl, amino, thiol, and hydroxyl substituents.

Quantitative Structure–Property Relationships. A useful way to predict physical property data has become available, based only on a knowledge of molecular structure, that seems to work well for pyridine compounds. Such a prediction can be used to estimate real physical properties of pyridines without

Table 1. Physical Properties of Pyridine[a]

Physical property	Value
enthalpy of fusion at −41.6°C, kJ/mol[b]	8.2785
enthalpy of vaporization, kJ/mol[b]	
at 25°C	40.2
115.26°C	35.11
critical temperature, °C	346.8
critical pressure, MPa[c]	5.63
enthalpy of formation, gas at 25°C, kJ/mol[b]	140.37
Gibbs free energy of formation, gas at 25°C, kJ/mol[b]	190.48
heat capacity, gas at 25°C, J/(K·mol)	78.23
ignition temperature, °C	550
explosion limit, %	1.7–10.6
surface tension, liquid at 25°C, mN/m(=dyn/cm)	36.6
viscosity, liquid at 25°C, mPa(=cP)	0.878
dielectric constant, liquid at 25°C, ϵ	13.5
thermal conductivity, liquid at 25°C, W/(K·m)	0.165

[a]Ref. 4.
[b]To convert kJ to cal, divide by 4.184.
[c]To convert MPa to atm, multiply by 9.87.

Table 2. Physical Properties of Pyridine, Alkyl-, and Alkenylpyridine Derivatives[a]

Compound	CAS Registry Number	Structure number	Mol wt	Freezing point, °C	Boiling point, °C	Density, d_4^{20}	Index of refraction, n_D^{20}	Water solubility[b] at 20°C, g/100 mL	pK_a at 20°C, in water[c]	Water azeotrope Bp[d], °C	Water azeotrope Water, %
pyridine	[110-86-1]	(1)	79.10	−41.6	115.3	0.9830	1.5102	e	5.22	93.6	41.3
2-picoline[f,g]	[109-06-8]	(2)	93.13	−66.7	129.5	0.9462	1.5010	e	5.96	93.5	48
3-picoline[g]	[108-99-6]	(3)	93.13	−18.3	143.9	0.957	1.5043	e	5.63	96.7	63
4-picoline[f,g]	[108-89-4]	(4)	93.13	3.7	144.9	0.9558	1.5058	e	5.98	97.4	63.5
2,3-lutidine	[83-61-9]	(58)	107.16	−15.5	161.5	0.9491	1.5085	13.3	6.57		
2,4-lutidine	[108-47-4]		107.16	−64	158.7	0.9325	1.5000	e	6.63	71.5	67.4
2,5-lutidine	[589-93-5]	(5)	107.16	−15.7	157	0.9331	1.5005	10.0	6.40		
2,6-lutidine	[108-48-5]		107.16	−6.1	143.7	0.923	1.4977	e	6.72	93.3	51.5
3,4-lutidine	[583-58-4]		107.16	−10.6	179.1	0.9534	1.5081	5.2	6.46		
3,5-lutidine	[591-22-0]	(6)	107.16	−6.6	172.7	0.944	1.5049	3.3	6.15		
2,4,6-collidine	[108-75-8]	(8)	121.18	−44.5	170.4	0.913	1.4981	3.6	7.43		
5-ethyl-2-methyl-pyridine	[104-90-5]	(7)	121.18	−70.9	178.3	0.9208	1.4974	1.2		98.4	72
2-vinylpyridine	[100-69-6]	(23)	105.14		110[h]	0.9746	1.5509	2.75	4.98	97.0	62.0
4-vinylpyridine	[100-43-6]		105.14		121[h]	0.988	1.5525	2.91	5.62	98.0	76.6

[a] Ref. 5, unless otherwise noted.
[b] Ref. 6.
[c] Ref. 7.
[d] Ref. 8.
[e] Miscible.
[f] Ignition temperature = 535°C for (2) and 500°C for (4).
[g] Explosion limit = 1.4–8.6% for (2) and 1.3–8.7% for (3) and (4).
[h] At 20 kPa (150 mm Hg).

644

Table 3. Physical Properties of Substituted Pyridines[a]

Compound	CAS Registry Number	Structure number	Mol wt	Freezing point, °C	Boiling point, °C	Density, d_4^{20}	Index of refraction, n_D^{20}	Water solubility[b] at 20°C, g/100 mL	pK_a[c] at 20°C in water
2-chloropyridine	[109-09-1]	(17)	113.55	−46.5	168–170	1.205[d]	1.532	2.5	0.5
2,6-dichloropyridine	[2402-78-0]		147.99	87–89	211			[e]	[f]
2-bromopyridine	[109-04-6]		158.00	192–194	194.8	1.627	1.5714	2.1	0.9
picolinic acid[g]	[98-98-6]		123.11	134–136	[h]	1.48		96	5.39
nicotinic acid[i]	[59-67-6]	(27)	123.11	236–239	[h]	1.473		1.7	4.81
isonicotinic acid[j]	[55-55-1]		123.11	314–315	[h]			0.5	4.86
methyl] picolinate	[2459-07-6]		137.14	18.7	92[k]	1.166	1.519		2.21
ethyl nicotinate	[614-18-6]		151.17	8–10	223–224	1.107	1.5040	5.0	4.48
picolinamide	[1452-77-3]		122.13	110 dec	143[l]	1.39		18	2.10
nicotinamide	[98-92-0]	(26)	122.13	128–131		1.401		120	3.33
2-cyanopyridine	[100-70-9]	(14)	104.11	26–28	212–215	1.081	1.5290		−0.26
3-cyanopyridine	[100-54-9]	(25)	104.11	50–52	206	1.159	1.5252[m]	13.5	1.36
4-cyanopyridine	[100-48-1]	(15)	104.11	79	195	1.03[n]		4.0	1.90
3-pyridylcarbinol	[100-55-0]	(30)	109.13	−6.5	266	1.124	1.5455	[o]	4.90
4-pyridylcarbinol	[586-95-8]		109.13	57–59	146[p]			[o]	5.33
2-picolylamine	[3731-51-9]		108.14	81	203	1.053	1.5431	[o]	8.8
4-picolylamine	[3731-53-1]		108.14	−7.6	229	1.066	1.5493	[o]	4.39
2-mercaptopyridine	[2637-34-5]		111.17	128–130					−1.07
4-mercaptopyridine	[4556-23-4]		111.17	184–186					1.43
2-pyridone	[142-08-5]	(16)	95.10	107	294			>100	1.25
3-hydroxypyridine	[109-00-2]	(40)	95.10	126				3.6	4.9
4-pyridone	[626-64-2]	(38)	95.10	150	181[q]			>100	3.23
2-aminopyridine	[504-29-0]	(35)	94.12	58	211	1.065		>100	6.9
3-aminopyridine	[462-08-8]	(33)	94.12	62.2	248	1.24		>100	5.98
4-aminopyridine	[504-24-5]	(34)	94.12	159	273			8.3	9.2
4-dimethylamino-pyridine	[1122-58-3]	(24)	122.17	112	162[r]	1.250		7.6	9.5

[a]Measured in the laboratories of Reilly Industries, Inc., unless otherwise noted. [b]Ref. 6. [c]Ref. 7. [d]Ref. 9. [e]Insoluble. [f]Cannot be protonated. [g]2-Carboxypyridine. [h]Sublimes. [i]3-Carboxypyridine. [j]4-Carboxypyridine. [k]At 4 Pa (0.03 mm Hg). [l]At 2.7 kPa (20 mm Hg). [m]At 50°C. [n]At 80°C. [o]Miscible. [p]At 1.3 kPa (10 mm Hg). [q]At 0.133 kPa (1 mm Hg). [r]At 6.7 kPa (50 mm Hg).

having to synthesize and purify the substance, and then measure the physical property.

The relationship between the structure of a molecule and its physical properties can be understood by finding a quantitative structure–property relationship (QSPR) (10). A basis set of similar compounds is used to derive an equation that relates the physical property, eg, melting point or boiling point, to structure. Each physical property requires its own unique QSPR equation. The compounds in the basis set used for QSPRs with pyridines have sometimes been quite widely divergent in respect to structural similarity or lack of it, yet the technique still seems to work well. The terms of the equation are composed of a coefficient and an independent variable called a descriptor. The descriptors can offer insight into the physical basis for changes in the physical property with changes in structure.

The same strategy can be used to relate chemical reactivity, catalytic ability, and bioactivity (10) of pyridine compounds with their structure. Although such a prediction is still in its formative stage, early results have been encouraging.

Chemical Properties

Chemical reactivity of pyridines is a function of ring aromaticity, presence of a basic ring nitrogen atom, π-deficient character of the ring, large permanent dipole moment, easy polarizability of the π-electrons, activation of functional groups attached to the ring, and presence of electron-deficient carbon atom centers at the α- and γ-positions. Depending on the conditions of the chemical transformation, one or more of these factors can give rise to the observed chemistry. The chemistry of pyridines can be divided into two categories: reactions at the ring-atomic centers, and reactions at substituents attached to the ring-atomic centers.

REACTIONS AT RING ATOMS

Ring-atomic centers can undergo attack by electrophiles, easily at the ring nitrogen and less easily at ring carbons. Nucleophilic attack is also possible at ring carbons or hydrogens.

Electrophilic Attack at Nitrogen. The lone pair on pyridine (1) ($pK_a = 5.22$) reacts with electrophiles under mild conditions, with protonic acids to give simple salts, with Lewis acids to form coordination compounds such as (9) [2903-67-5], and with transition metals to form complex ions, such as (10) [24444-58-4]. The complex ion pyridinium chlorochromate [67369-53-3] (11) is a mild oxidizing agent suitable for the conversion of alcohols to carbonyl compounds.

(9)   (10)   (11)

Reactive halogen compounds, alkyl halides, and activated alkenes give quaternary pyridinium salts, such as (**12**). Oxidation with peracids gives pyridine *N*-oxides, such as pyridine *N*-oxide itself [*694-59-7*] (**13**), which are useful for further synthetic transformations (11).

(**12**)   (**13**)

Electrophilic Attack at Carbon. Electrophilic attack at a C-atom in pyridines is particularly difficult unless one or more strong electron-donating substituents are attached to the ring. Knowledge of this fact has resulted in the widespread use of pyridine as a solvent for reactions involving electrophilic species. Pyridine undergoes nitration in low yield (15%) to give 3-nitropyridine [*2530-26-9*] (12) whereas pyridine *N*-oxide is nitrated at position 4 to give 4-nitropyridine *N*-oxide [*1124-33-0*] in high yield, and 2- and 4-aminopyridines may be dinitrated (13). The chlorination of pyridine and picolines (**2**), (**3**), and (**4**), on the other hand, is of great commercial importance (14).

Nucleophilic Attack at Carbon or Hydrogen. Only the strongest of nucleophiles (eg, $^-NH_2$) can replace a hydrogen in pyridine. However, N-oxides and quaternary salts rapidly undergo addition, followed by subsequent transformations (12).

The Feely-Beavers procedure (eq. 1) provides a method for the introduction of a cyano group (9), principally at the 2-position, to give compounds such as (**14**).

(**14**)   (**15**)   (1)

A modification of the Reissert-Henze reaction employing benzoyl chloride and $(CH_3)_3SiCN$ gives good yields of 2-cyano pyridine (**14**) from pyridine *N*-oxide (**13**) (15).

Methylpyridinium quaternary salts, such as (**12**), undergo oxidation in alkaline solution in the presence of potassium ferricyanide to give 2-pyridones, eg, *N*-methyl-2-pyridone [*694-85-9*] (16). Frequently nucleophilic attack at position 2 by excess hydroxide leads to ring opening; this and synthetically useful recyclizations have been reviewed (17).

Treatment of pyridine *N*-oxide (**13**) with acetic anhydride leads chiefly to 2-pyridone (**16**) formation (eq. 2) (11).

$$ \text{(13)} \qquad\qquad \text{(16)} \tag{2} $$

Pyridine and its methyl derivatives, (**2**), (**3**), and (**4**), undergo amination with sodium amide at the 2-position eg, 2-amino-3-methylpyridine [*1603-40-3*] and 2-amino-5-methylpyridine [*1603-41-4*] from (**3**) (eq. 3). This Chichibabin reaction is most important for introduction of a 2-amino substituent, which may be replaced readily by many other groups (18).

$$ \text{(3)} \tag{3} $$

The N-oxides readily undergo nucleophilic addition followed by elimination, which forms the basis of several useful syntheses of 2-substituted pyridines. Chlorination of (**13**) with $POCl_3$ to give 2-chloropyridine (**17**) is a good example (eq. 4); some chlorination may occur also at C-4 (11).

$$ \text{(13)} \qquad\qquad\qquad \text{(17)} \tag{4} $$

Pyridine undergoes 2- and 4-alkylation with Grignard reagents, depending on whether free metal is present (19). Free metal gives mixtures or exclusive 4-alkylation. Substituent-directed metallation (eq. 5) has become an important approach to the synthesis of disubstituted pyridines (12). For example, 2-fluoropyridine [*372-48-5*] reacts with butyllithium and acetaldehyde to give a 93% yield of alcohol [*79527-61-1*].

$$ \xrightarrow[\text{2. } CH_3CHO]{\text{1. } C_4H_9Li} \tag{5} $$

Treatment of pyridine (**1**) with sodium metal in ethanol gives piperidine [*1910-42-5*] (**18**); however, dimerization to 4,4′-bipyridine [*553-26-4*] (**19**) is favored in aprotic solvents (12).

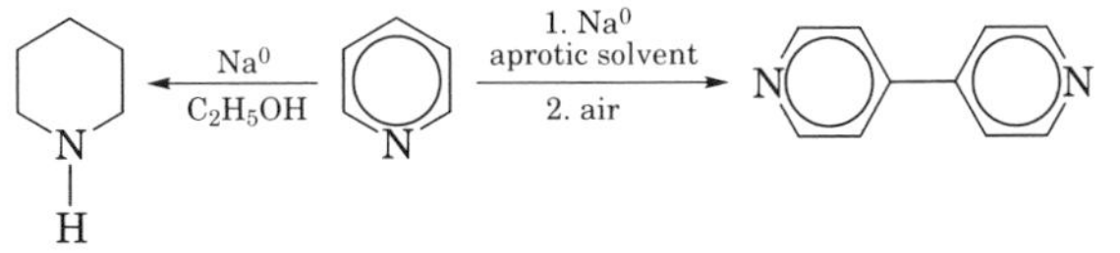

(**18**)   (**1**)   (**19**)

The 4,4′-bipyridine (**19**) formed is a precursor of the important herbicide Paraquat [*1910-42-5*] (**20**) (20).

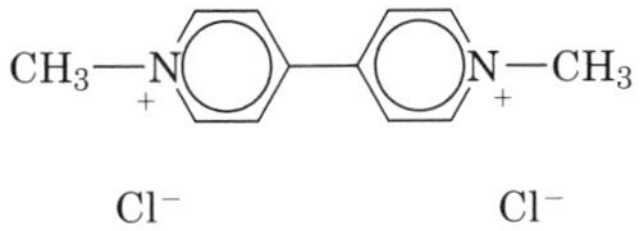

(**20**)

Free-Radical Attack at Carbon. Homolytic substitution of pyridines has not been as thoroughly studied as heterolytic processes have been, owing to low conversions and poor selectivity (21). However, some pyridinium salts have been found to undergo potentially useful regiospecific reactions (22). Protonated pyridines readily undergo acylation by acyl radicals, generated by abstracting a hydrogen atom from an aldehyde, eg, 2-acetyl-4-cyanopyridine [*37398-49-5*] from protonated 4-cyanopyridine (eq. 6).

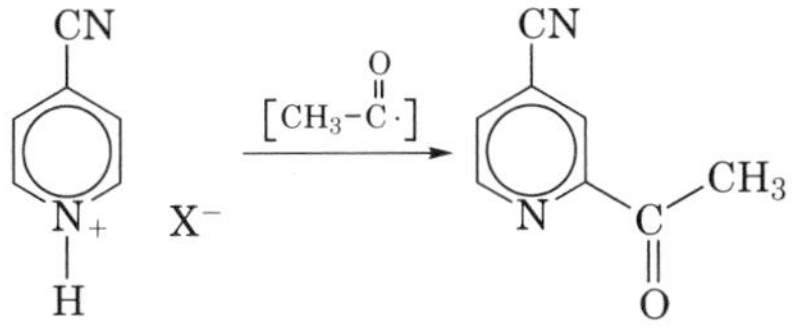

(6)

4-Cyanopyridine (**15**) reacts with ketones not bearing an α-hydrogen in the presence of sodium metal to afford a tertiary alcohol (a precursor of azacyclonol) in high yield, eg, the reaction of (**15**) and benzophenone yields the tertiary benzyl alcohol [*1620-30-0*] (23):

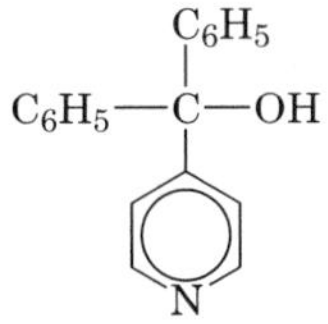

REACTIONS OF SUBSTITUTED PYRIDINES

Carbon Substituents. Alkyl groups at positions 2 and 4 of a pyridine ring are more reactive than either those at the 3-position of a pyridine ring or those attached to a benzene ring. Carbanions can be formed readily at alkyl carbons attached at the 2- and 4-positions. This increased chemical reactivity has been used to form 2- and 4-(phenylpropyl)pyridines, eg, 4-(3-phenylpropyl)-pyridine [2057-49-0] (**21**) (24).

(**4**)   (**21**)

An industrially important example is the condensation of α- (**2**) or γ-picoline (**4**) with aqueous formaldehyde to form the corresponding ethanolpyridines, 2-ethanolpyridine [104-74-2] (**22**) and 4-ethanolpyridine [5344-27-4], respectively, followed by dehydration of the alcohols to give 2- (**23**) or 4-vinylpyridine.

(**2**)   (**22**)   (**23**)

2-Vinylpyridine (**23**) came into prominence around 1950 as a component of latex. Butadiene and styrene monomers were used with (**23**) to make a terpolymer that bonded fabric cords to the rubber matrix of automobile tires (25). More recently, the ability of (**23**) to act as a Michael acceptor has been exploited in a synthesis of 4-dimethylaminopyridine (DMAP) (**24**) (26). The sequence consists of a Michael addition of (**23**) to 4-cyanopyridine (**15**), replacement of the 4-cyano substituent by dimethylamine (taking advantage of the activation of the cyano group by quaternization of the pyridine ring), and base-catalyzed dequaternization (retro Michael addition). 4-Dimethylaminopyridine is one of the most effective acylation catalysts known (27).

(**24**)

Cyanopyridines are usually manufactured from the corresponding picoline by catalytic, vapor-phase ammoxidation (eq. 7) in a fixed- or fluid-bed reactor (28). 3-Cyanopyridine (**25**) is the most important nitrile, as it undergoes partial or complete hydrolysis under basic conditions to give niacinamide [*98-92-0*] (**26**) or niacin (nicotinic acid) (**27**), respectively (29).

$$\text{(3)} \xrightarrow[\text{V}_2\text{O}_5]{\text{O}_2,\text{NH}_3} \text{(25)} \tag{7}$$

$$\text{(25)} \xrightarrow{\text{OH}^-} \text{(26)} \xrightarrow{\text{OH}^-} \text{(27)}$$

Compound (**27**) may also be obtained directly by oxidation of β-picoline (**3**) or by exhaustive oxidation of 5-ethyl-2-methylpyridine (**7**), followed by decarboxylation of the initially formed pyridine-2,5-dicarboxylic acid [*100-26-5*] (**28**) (eq. 8) (30).

$$\text{(7)} \longrightarrow \text{(28)} \xrightarrow{-\text{CO}_2} \text{(27)} \tag{8}$$

Hydrogenation of 3-cyanopyridine (**25**) in the presence of ammonia gives 3-picolylamine [*3731-52-0*] (**29**); however, hydrogenation in the presence of hydrogen chloride affords the corresponding 3-carbinol (**30**) (31).

$$\text{(30)} \xleftarrow[\text{HCl (aq)}]{\text{H}_2} \text{(25)} \xrightarrow[\text{Ni(R)}]{\text{H}_2,\text{NH}_3} \text{(29)}$$

Treatment of cyanopyridines such as (**25**) with a Grignard reagent yields a ketone (32). A carboxylic ester is obtained by reaction of the nitrile (**25**) with sodium alkoxide, followed by hydrolysis (33).

$$\xleftarrow[\text{2. H}_2\text{O}]{\text{1. NaOR}'} \text{(25)} \xrightarrow{\text{RMgBr}}$$

Isonicotinic hydrazide [*54-83-3*] (isoniazid) (**31**) is still an important tuberculostat (34). It may be obtained by reaction of isonicotinic esters such as ethyl isonicotinate [*1570-45-2*], or the 4-nitrile (**15**), with hydrazine.

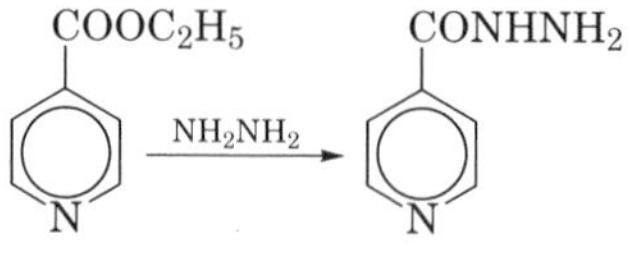

(31)

Nicotinamide [98-92-0] (**26**) and isonicotinamide [1453-82-3] (**32**) undergo Hofmann rearrangements to form 3- (**33**) and 4-aminopyridine (**34**), respectively (**35**). This provides an important route for the manufacture of these amines.

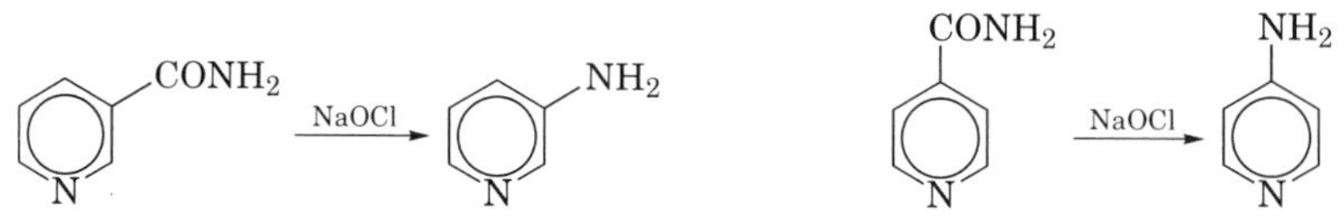

Nitrogen Substituents. 4-Aminopyridine (**34**) and 2-aminopyridine (**35**) react with cold nitric acid to give the corresponding nitramines, which are insoluble in the media. On heating, these nitramines rearrange intermolecularly to nitroaminopyridines having the nitro group mainly adjacent to the amino group (13). From (**34**) the products are the intermediate 4-nitraminopyridine [26482-55-3] and 4-amino-3-nitropyridine [1681-37-4] (eq. 9).

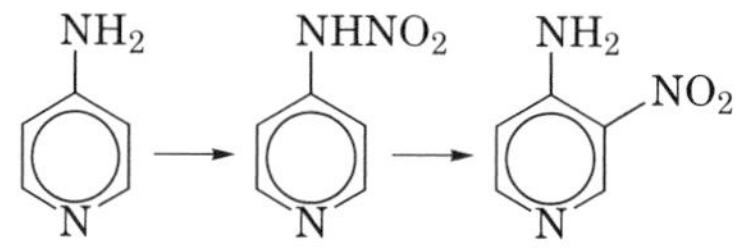

(9)

Reaction of 2-aminopyridine (**35**) with *N*-acetylsulfanilyl chloride, followed by hydrolysis, gives sulfapyridine [144-83-2] (**36**), an antibacterial (36).

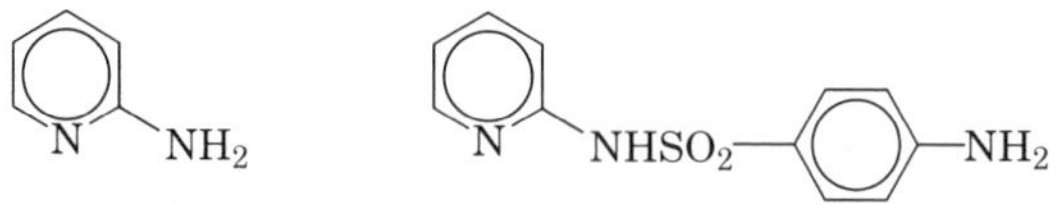

The antibacterial agent nalidixic acid [389-08-2] (**37**) is formed by reaction of 2-amino-6-methylpyridine [1824-81-3] with an alkoxymethylenemalonic ester to form the 1,8-naphthyridine carboxylic ester followed by alkylation and ester hydrolysis (37).

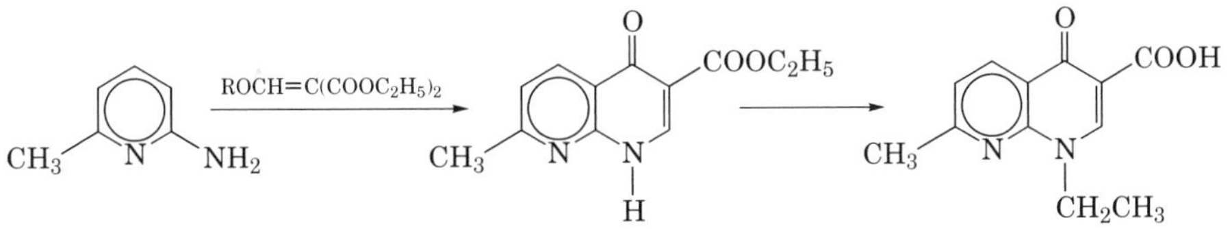

(**37**)

Oxygen Substituents. The presence of oxygen or sulfur attached to the ring can affect the chemistry of those compounds through tautomerism. This phenomenon in the pyridine series has been well studied and reviewed (38). An example of 2-pyridone–2-pyridinol tautomerism was shown in equation 2, compound (**16**).

The compounds 2- (**16**) and 4-pyridone (**38**) undergo chlorination with phosphorus oxychloride; however, 3-pyridinol (**39**) is not chlorinated similarly. The product from (**38**) is 4-chloropyridine [*626-61-9*]. The 2- (**16**) and 4-oxo (**38**) isomers behave like the keto form of the keto–enol tautomers, whereas the 3-oxo (**39**) isomer is largely phenolic-like, and fails to be chlorinated (38).

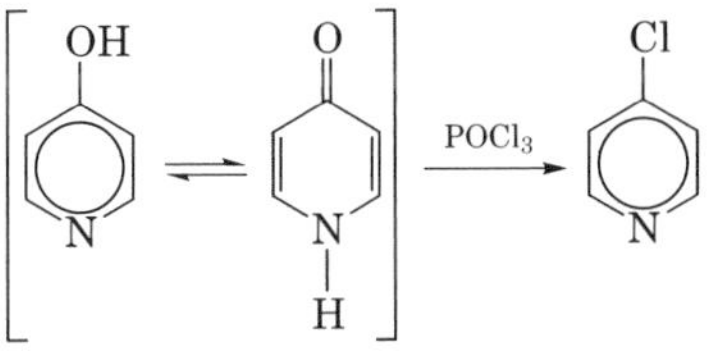

(**38**) 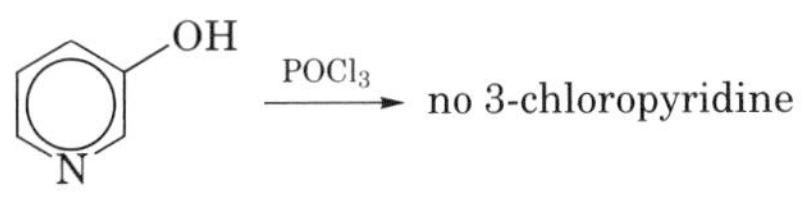   (**39**)

Exclusive O-alkylation of (**16**) (eq. 10) may be achieved using ethyl iodide and silver carbonate (39); the product is 2-ethoxypyridine [*14529-53-4*]. Heating 2- or 4-alkoxypyridines, with or without acid catalyst, induces intermolecular migration of the alkyl group on oxygen to the ring nitrogen to form *N*-alkyl-2- or *N*-alkyl-4-pyridones (39).

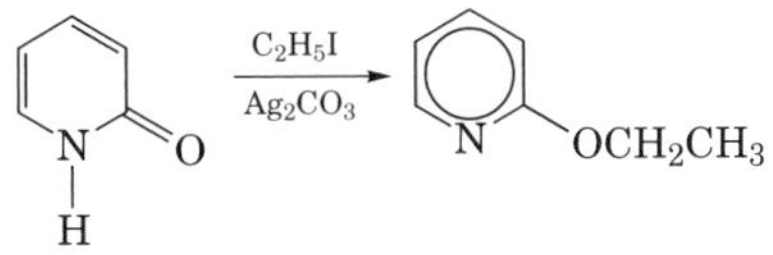

(10)

(**16**)

The N-oxide function has proved useful for the activation of the pyridine ring, directed toward both nucleophilic and electrophilic attack (see AMINE OXIDES). However, pyridine N-oxides have not been used widely in industrial practice, because reactions involving them almost invariably produce at least some isomeric by-products, adding to the cost of purification of the desired isomer. Frequently, attack takes place first at the O-substituent, with subsequent rearrangement into the ring. For example, 3-picoline N-oxide [1003-73-2] (**40**) reacts with acetic anhydride to give a mixture of pyridone products in equal amounts, 5-methyl-2-pyridone [1003-68-5] and 3-methyl-2-pyridone [1003-56-1] (11).

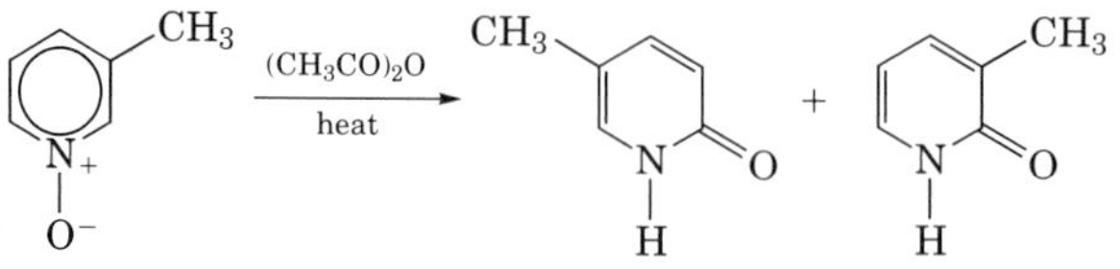

(**40**)

Alkylation at the oxygen atom of N-oxides is also a facile process, and the entire N-substituent may be removed with base (eq. 11).

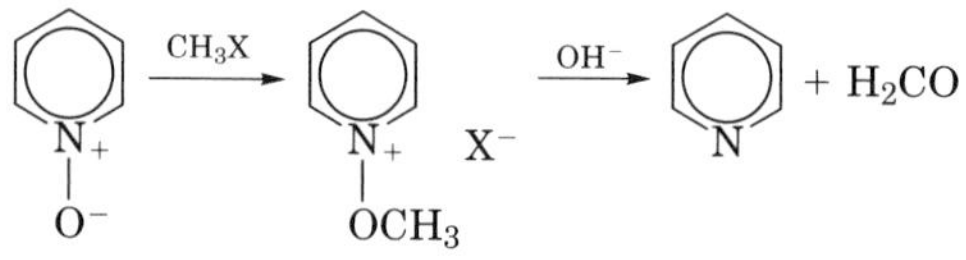

(13)   (1)   (11)

Treatment of N-oxides with phosphorus trichloride provides a good method for deoxygenation (eq. 12) to obtain the free base.

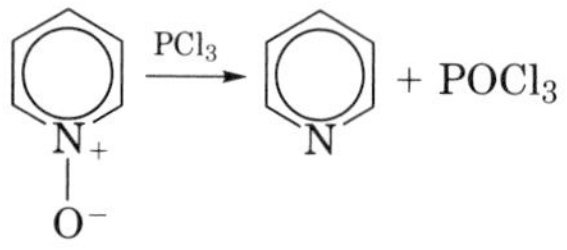

(12)

Sulfur Substituents. Acetylation and alkylation of pyridinethiones usually take place on sulfur (39). An exception to this is 4-pyridinethione [19829-29-9] which is acetylated on nitrogen. Displacement of thioethers can be achieved with hydroxide or amines (eq. 13) (40). Thioether functional groups can also be removed by reduction (39).

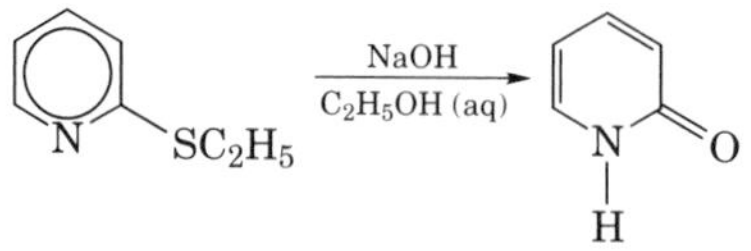

(13)

(**16**)

Oxidation of a pyridinethione gives the corresponding sulfonic acid, eg, 6-carboxy-2-pyridinesulfonic acid [18616-02-9] from 6-carboxy-2-pyridinethione [14716-87-1] (eq. 14) (41).

$$(14)$$

Tetrachloropyridine-4-thiol [*10357-06-1*] (**41**) reacts with chlorine in carbon tetrachloride to give a sulfenyl chloride (**42**), which is fairly stable. The sulfenyl chloride may be converted into a number of derivatives (39).

Halogen Substituents.   Halogen functional groups are readily replaced by nucleophiles, eg, hydroxide ion, especially when they are attached at the α- or γ-position of the pyridine ring. This reaction has been exploited in the synthesis of the insecticide chlorpyrifos [*2921-88-2*] (**43**) (42), and the insecticide triclopyr [*55335-06-3*] (**44**) (14,43). 2,3,5,6-Tetrachloropyridine [*2402-79-1*] reacts with caustic to form the hydroxylated material [*6515-38-4*], which then can be used to form (**44**) and (**43**).

2-Chloropyridine *N*-oxide [*20295-64-1*] reacts with sodium hydrosulfide to give pyrithione [*1121-31-9*] (**45**), the zinc salt of which is used as an antifungal agent, most prominently in shampoos (44).

Amines or ammonia replace activated halogens on the ring, but competing pyridyne [7129-66-0] (**46**) formation is observed for attack at 3- and 4-halo substituents, eg, in 3-bromopyridine [626-55-1] (39). The most acidic hydrogen in 3-halopyridines (except 3-fluoropyridine) has been shown to be the one in the 4-position. Hence, the 3,4-pyridyne is usually postulated to be an intermediate instead of a 2,3-pyridyne. Product distribution (40% (**33**) and 20% (**34**)) tends to support the 3,4-pyridyne also.

(**46**)   (**34**)   (**33**)

The 4-chloro group can be removed selectively when pentachloropyridine [2176-62-7] (**47**) is treated with zinc to form symmetrical 2,3,5,6-tetrachloropyridine (45).

Organometallics. Pentachloropyridine (**47**) forms Grignard reagent (**48**) by the entrainment method (eq. 15) (46). Addition of CO_2 produces 4-carboxy-2,3,5,6-tetrachloropyridine [19340-26-2].

(**47**)   (**48**)   (15)

Metallation of 2-picoline N-oxide [931-19-1] with butyllithium followed by addition of the electrophilic cyclohexanone leads to two carbinol products (47) (4% (**49**) [34277-46-8] and 20% (**50**) [34277-59-3]).

(**49**)   (**50**)

Substituent-directed metallations are being used for the synthesis of disubstituted pyridines. A 2-substituent directs to the 3-position, and a 3-substituent usually directs to the 4-position; however, in the presence of $N,N,N'N'$-tetramethylethylenediamine (TMEDA), 2-metallation may be achieved (12).

Synthesis

Pyridine ring syntheses (48) can be classified into essentially two categories: ring synthesis from nonheterocyclic compounds, and synthesis from other ring systems. The synthesis of pyridine derivatives by transformations on the pyridine ring atoms and side-chain atoms have been considered in the previous section.

Ring Synthesis From Nonheterocyclic Compounds. These methods may be further classified based on the number of bonds formed during the pyridine ring formation. Synthesis of α-picoline (**2**) from 5-oxohexanenitrile is a one-bond formation reaction (eq. 16) (49). The nitrile is obtained by reaction between acetone and acrylonitrile (50). If both reaction steps are considered together, the synthesis must be considered a two-bond forming one, ie, formation of (**2**) from acetone and acrylonitrile in a single step comes under the category of two-bond formation reaction.

$$\text{(16)}$$

The formation of quinoline [*91-22-5*] (**51**) from aniline and acrolein involves formation of two bonds during the ring synthesis.

Synthesis of 4,6-disubstituted-2-picolines and their corresponding nicotin-amides has been developed using β-aminocrotonitrile (**52**) and α,β-unsaturated compounds, where $R^1 = R^2 = $ aryl (51).

The formation of pyridine derivatives from α,β-unsaturated aldehydes and ammonia involves formation of three bonds during the ring synthesis. For example, with an α,β-unsaturated aldehyde, both 2,5-substituted as well as 3,4-substituted pyridines can be obtained, depending on whether a 1,2- (eq. 17) or

1,4-addition (eq. 18) occurs with ammonia. Reactions are performed in the vapor phase with catalysts.

$$\text{(17)}$$

$$\text{(18)}$$

Acrolein and ammonia give β-picoline (**3**, R = H) (eq. 17). Acrolein, ammonia, and acetaldehyde give pyridine (**1**) (eq. 19). Acrolein, ammonia, and propionaldehyde give (**3**) (eq. 20) (52–56). Reactions are performed in the vapor phase with proprietary catalysts.

$$\text{(19)}$$

$$\text{(20)}$$

The vapor-phase synthesis of pyridines and picolines from formaldehyde, acetaldehyde, and ammonia falls in the category of four-bond formation reactions (Fig. 1). Reactions are performed in the vapor phase with proprietary catalysts.

Synthesis From Other Ring Systems.   These syntheses are further classified based on the number of atoms in the starting ring. Ring expansion of dichlorocyclopropane carbaldimine (**53**), where R = H and R′ = aryl, on pyrolysis gives 2-arylpyridines. Thermal rearrangement to substituted pyridines occurs in the presence of tungsten(VI) oxide. In most instances the nonchlorinated product is the primary product obtained (63).

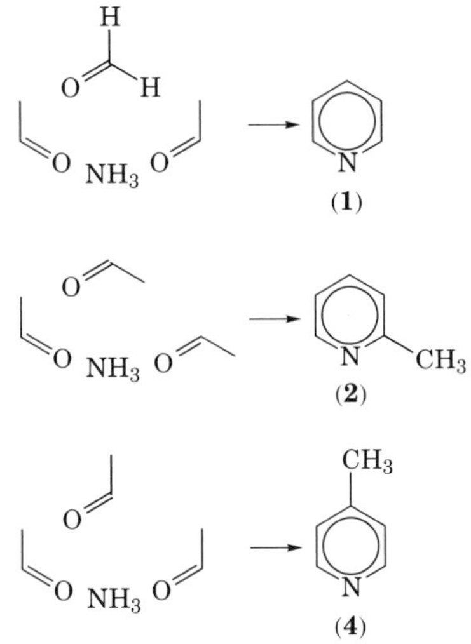

Fig. 1. Four-bond reactions: formaldehyde, acetaldehyde, and ammonia mainly give pyridine (**1**), and acetaldehyde and ammonia give α- (**2**) and γ-picoline (**4**) (57–62).

Azacyclobutenes have been used to generate 1-azabutadienes, which are intramolecularly as well as intermolecularly cyclized to give tetrahydropyridines, eg, hexahydroquinolizin-4-one [87842-80-6] (64,65). In the following, FVP = flash vacuum pyrolysis.

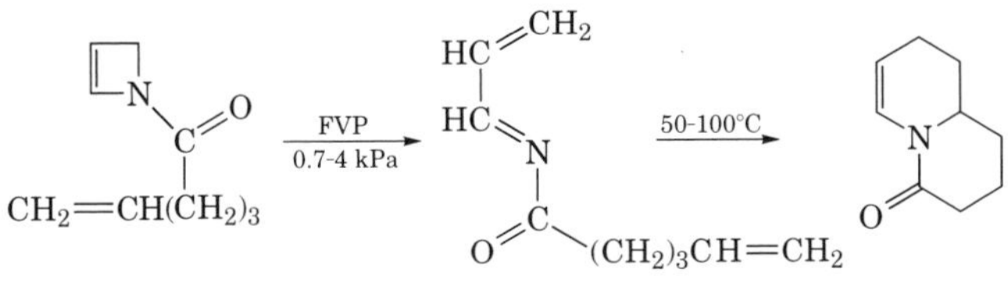

Ring expansion of five-membered ring heterocyclic compounds has been accomplished to form pyridine derivatives. Reaction of tetrahydrofurfuryl alcohol with ammonia gives pyridine (**1**) under dehydrogenating conditions, and gives piperidine (**18**) under reductive conditions.

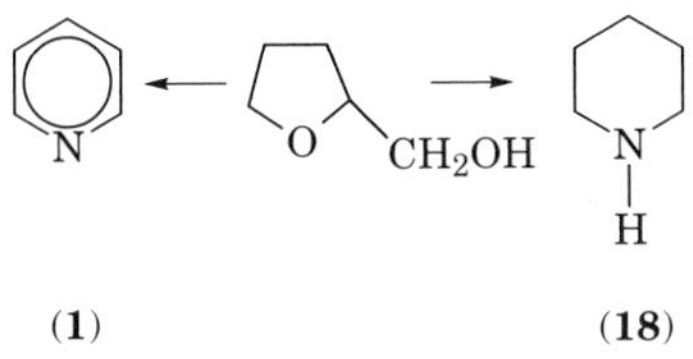

Furfurylamine reacts with hydrogen peroxide and acid to give 3-hydroxypyridine (**39**).

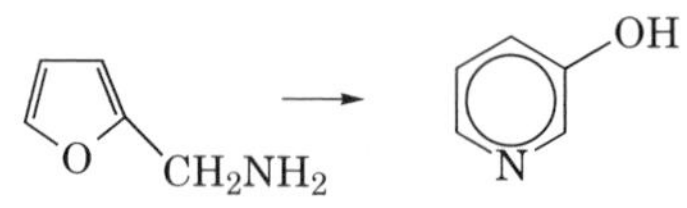

(**39**)

2-Alkyl-3-pyridinols (**54**) are reported to be formed from acyl furans and ammonia under pressure (66–68).

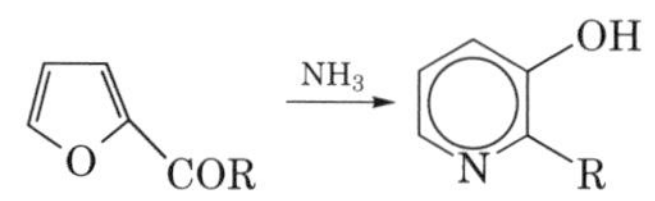

(**54**)

Oxazoles react with dienophiles to give pyridines after dehydration or other aromatization reactions (69,70). A commercially important example is the reaction of a 5-alkoxy-4-methyloxazole with 1,4-butenediol to yield pyridoxine (**55**), which is vitamin B$_6$.

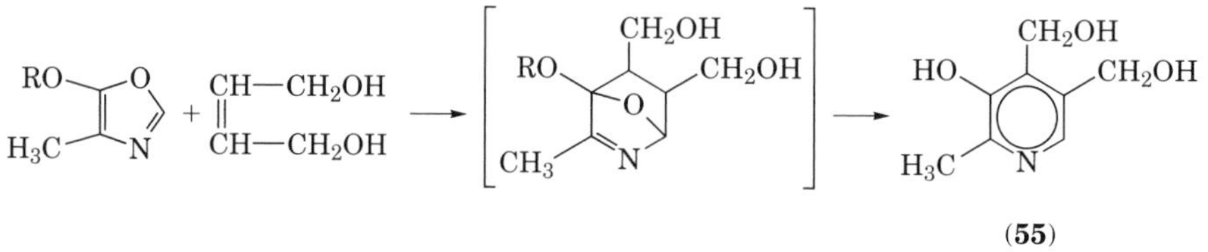

(**55**)

Pyrroles may be ring-expanded to pyridines in reactions having a greater academic than practical interest. Treatment of pyrrole with chloroform and sodium ethoxide (in effect, with dichlorocarbene, :CCl$_2$) gives a low yield of 3-chloropyridine [626-60-8]. A much better yield (33%) is obtained if chloroform and pyrrole are heated together in the vapor phase at 550°C; some 2-chloropyridine (**17**) is also formed (71).

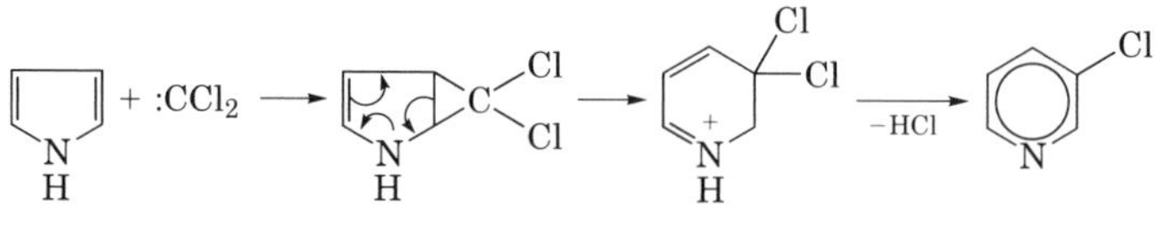

A large number of pyridine derivatives have been obtained by the dehydrogenation of suitable piperidines.

Pyrans and related compounds react with ammonia to give pyridines. A commercially useful example is the reaction of dehydroacetic acid (derived from diketene) with ammonia to give 2,6-dimethyl-4-pyridinone [7516-31-6] via 2,6-dimethyl-4-pyridinone-3-carboxylic acid [52403-25-5]. Chlorination of the pyridone gives clopidol [2971-90-6] (**56**), a coccidiostat (72,73).

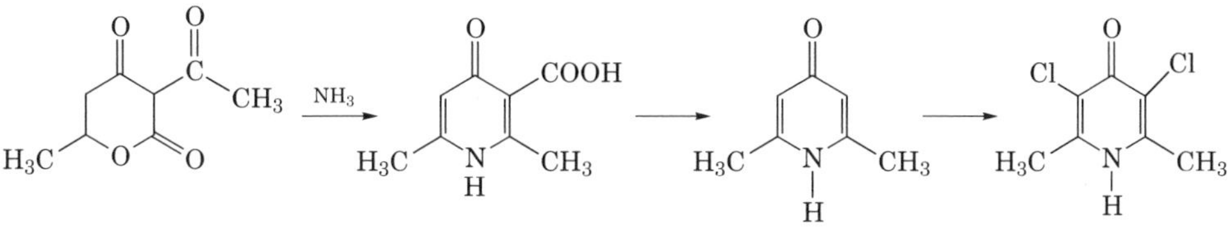

(56)

Manufacture and Processing

There are no natural sources of pyridine compounds that are either a single pyridine isomer or just one compound. For instance, coal tar contains a mixture of bases, mostly alkylpyridines, in low concentrations. Few commercial synthetic methods produce a single pyridine compound, either; most produce a mixture of alkylpyridines, usually with some pyridine (**1**). Those that produce mono- or disubstituted pyridines as principal components also usually make a mixture of isomeric compounds along with the desired material.

Manufacturing methods must suit the scale of manufacture. By the late 1980s, some 6,000 t of pyridine (**1**) were consumed in the United States, and nearly 20,000 t worldwide.

Historical. Pyridines were first isolated by destructive distillation of animal bones in the mid-nineteenth century (2). A more plentiful source was found in coal tar, the condensate from coking ovens, which served the steel industry. Coal tar contains roughly 0.01% pyridine bases by weight. Although present in minute quantities, any basic organics can be easily extracted as an acid-soluble fraction in water and separated from the acid-insoluble tar. The acidic, aqueous phase can then be neutralized with base to liberate the pyridines, and distilled into separate compounds. Only a small percentage of worldwide production of pyridine bases can be accounted for by isolation from coal tar. Almost all pyridine bases are made by synthesis.

Most processes in use in the 1990s make pyridines by condensation of ammonia with aldehydes or ketones either in the vapor phase or in the liquid phase. These processes are based on the pioneering work of Chichibabin (3). Commercial practice of that process was not realized until some 50 years after its discovery (74,75).

Commercial Manufacture of Pyridine. There are two vapor-phase processes used in the industry for the synthesis of pyridines. The first process (eq. 21) utilizes formaldehyde and acetaldehyde as a co-feed with ammonia, and the principal products are pyridine (**1**) and 3-picoline (**3**). The second process produces only alkylated pyridines as products.

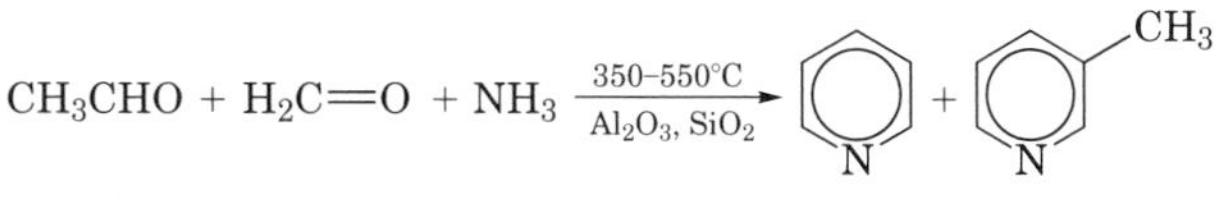

$$CH_3CHO + H_2C{=}O + NH_3 \xrightarrow[\text{Al}_2\text{O}_3,\ \text{SiO}_2]{350-550^\circ C}$$

(1)   (3)

(21)

Acrolein ($CH_2{=}CHCHO$) can be substituted for formaldehyde and acetaldehyde in the above reaction to give similar results, but the proportion of (**3**) is higher than when acetaldehyde and formaldehyde are fed separately. Acrolein may be formed as one of the first steps to pyridine (**1**) and β-picoline (**3**) formation. There are many variations on the vapor-phase synthesis of pyridine itself. These variations are the subject of many patents in the field.

Commercial Manufacture of Specific Pyridine Bases. Condensation of paraldehyde with ammonia at 230°C and autogenous pressure (eq. 22) is used to manufacture 5-ethyl-2-methylpyridine (**7**). This is one of the few liquid-phase processes used in the industry to make relatively simple alkylpyridines, and one of the few processes known to make a single alkylpyridine product selectively.

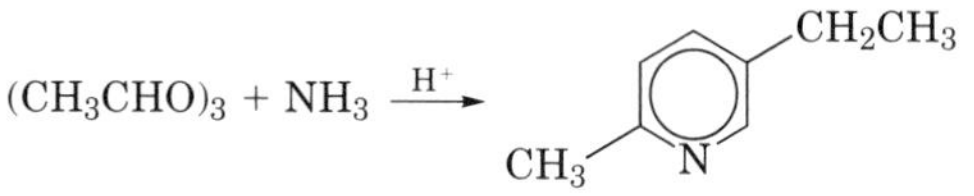
(CH3CHO)3 + NH3
H+
CH2CH3
CH3
N

(**7**)

(22)

The vapor-phase analogue of this liquid-phase reaction (eq. 22) is used to make α- (**2**) and γ-picoline (**4**) (eq. 23). The gas-phase products are different from the liquid-phase products, because acetaldehyde is used in place of its trimer, paraldehyde, and a multivalent metal oxide catalyst is used.

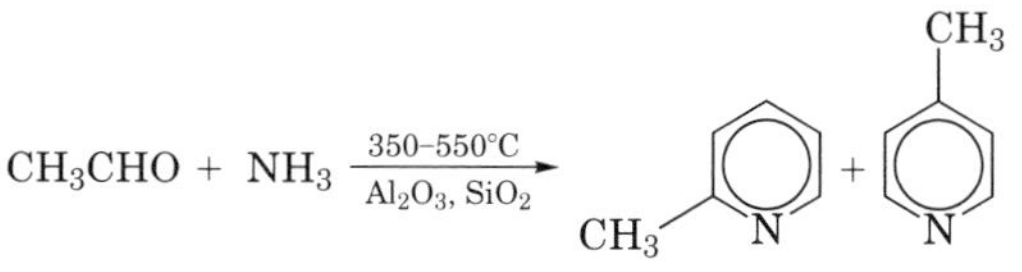
CH3CHO + NH3
350–550°C
Al2O3, SiO2
CH3
N
+
CH3
N

(**2**) (**4**)

(23)

Replacing acetaldehyde with acetone and using a co-feed of formaldehyde and ammonia give mainly 2,6-lutidine (**5**). However, leaving out the formaldehyde results in production of 2,4,6-collidine (**8**) as the primary product.

Another of the few selective syntheses of alkylpyridines is one for α-picoline (**2**) (76). This is a two-step process (eq. 24) where acrylonitrile is used to monocyanoethylate acetone in the liquid phase at 180°C and at autogenous pressure, 2 MPa (300 psig). The monoadduct, 5-cyano-2-pentanone, is then passed over a palladium-containing catalyst to reduce, cyclize, and dehydrogenate, in sequence.

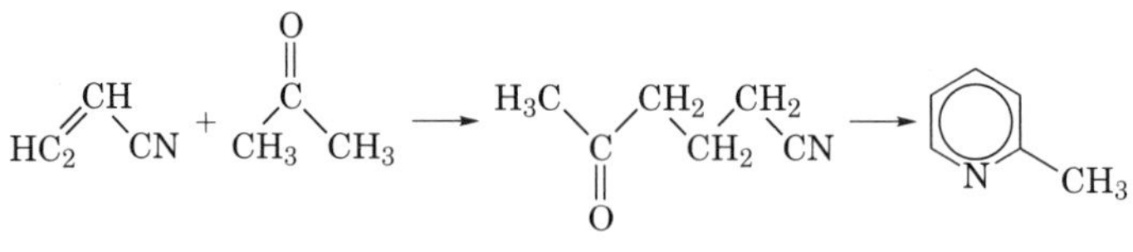
CH
HC2
CN
+
O
C
CH3
CH3
H3C
CH2
CH2
C
O
CH2
CN
N
CH3

(24)

The same methodology can be used to prepare 2,3-lutidine (**58**) by using methyl ethyl ketone in place of acetone.

(58)

By-Products. Almost all commercial manufacture of pyridine compounds involves the concomitant manufacture of various side products. Liquid- and vapor-phase synthesis of pyridines from ammonia and aldehydes or ketones produces pyridine or an alkylated pyridine as a primary product, as well as isomeric alkylpyridines and higher substituted alkylpyridines, along with their isomers. Furthermore, self-condensation of aldehydes and ketones can produce substituted benzenes. Condensation of ammonia with the aldehydes can produce certain alkyl or unsaturated nitrile side products. Lastly, self-condensation of the aldehydes and ketones, perhaps with reduction, can lead to alkanes and alkenes.

Raw Material and Energy Aspects to Pyridine Manufacture. The majority of pyridine and pyridine derivatives are based on raw materials like aldehydes or ketones. These are petroleum-derived starting materials and their manufacture entails cracking and distillation of alkanes and alkenes, and oxidation of alkanes, alkenes, or alcohols. Ammonia is usually the source of the nitrogen atom in pyridine compounds. Gas-phase synthesis of pyridines requires high temperatures (350–550°C) and is therefore somewhat energy intensive.

Production and Shipment

Worldwide production of pyridine bases in the late 1980s was estimated at thousands of tons a year. Production was concentrated mainly in the United States, Western Europe, and Japan. Production statistics are not complete for any of the principal producing areas and trade statistics are also incomplete.

The relative production volumes of pyridine compounds can be ranked in the following order: pyridine (**1**) > β-picoline (**3**) > α-picoline (**2**) > niacin (**27**) or niacinamide (**26**) > 2-vinylpyridine (**23**) > piperidine (**18**). U.S. and Japanese production was consumed internally as well as being exported, mainly to Europe. European production is mostly consumed internally. Growth in production of total pyridine bases is expected to be small through the year 2000.

Shipment Methods and Packaging. Pyridine (**1**) and pyridine compounds can be shipped in bulk containers such as tank cars, rail cars, and super-sacks, or in smaller containers like fiber or steel drums. The appropriate U.S. Department of Transportation (DOT) requirements for labeling are given in Table 4. Certain temperature-sensitive pyridines, such as 2-vinylpyridine (**23**) and 4-vinylpyridine are shipped cold (<-10°C) to inhibit polymerization. Piperidine (**18**) and certain piperidine salts are regulated within the United States by the Drug Enforcement Agency (DEA) (77). Pyridines subject to facile oxidation, like those containing aldehyde and carbinol functionality, can be shipped under an inert atmosphere.

Table 4. U.S. DOT Labeling Regulations

Compound	Structure number	Flash point[a], °C	DOT label[b]
pyridine	(1)	19	red
2-methylpyridine	(2)	27	red
3-methylpyridine	(3)	38	red
4-methylpyridine	(4)	39	red
2-vinylpyridine	(23)	50	white red
4-vinylpyridine	(23)	56	white red
piperidine	(18)	12	white red
2-aminopyridine	(35)	104[c]	poison B

[a]Tagliabue closed cup (TCC) unless otherwise noted.
[b]Red is flammable; white, corrosive.
[c]Cleveland open cup (COC).

Economic Aspects

Although the volume of commercial pyridine compounds is relatively large, economic aspects resemble those of specialty markets more than those of commodities. Commercial transactions occur with little publicity, trade secrets are carefully guarded, and patents proliferate, thus obscuring the industrial processes used for their manufacture.

Pyridine bases are produced in three principal areas of the world: the United States, Western Europe, and Japan. In the United States, there are two principal producers of synthetic pyridine compounds: the privately owned Reilly Industries, Inc. (Indianapolis, Indiana) and Nepera Chemical Company, Inc. (Harriman, New York; a division of Cambrex Chemicals). Reilly Industries is the world's largest producer. Coal-tar-derived pyridine bases are made in only small amounts. In Europe, synthetic pyridine compounds are manufactured by Reilly Chemicals, SA (Belgium), Lonza AG (Switzerland), and DSM (the Netherlands). Small amounts of natural pyridine bases from coal tar can be obtained from a number of sources, including Raschig and Rütgerswerke AG (Germany). In Japan, synthetic pyridines are produced by Koei Chemical Company, Ltd., and by Daicel, Ltd.; natural pyridine bases can be obtained from Nippon Steel.

Prices of pyridine compounds reflect two sources of manufacturing cost. One source is the cost of petroleum-based raw materials used in the manufacture of the pyridine base itself. This factor varies according to the current pricing of petroleum feedstocks in the country of manufacture, and can vary from country to country simultaneously. Long term, the variation in cost tracks the variation in market price of ethylene. The second source is related to the chemistry performed on a pyridine-based raw material to arrive at the desired compound. This cost can be variable, depending on the chemical technology used and the cost of reagents and solvents for the transformation. As a group of compounds, pyridine bases are considered specialty chemical products with prices generally no lower than about $5/kg and going up to $100–150/kg, with an expected inverse relationship between price and volume sold. Because pyridine compounds are specialty products and manufacture is concentrated among a few suppliers, competition between producers is significant. Pricing information on specific products is generally considered proprietary information.

Specifications, Standards, and Quality Control

Most pyridine compounds are sold on the basis of a >98 wt % criterion as analyzed by gas chromatography (gc), freezing point, titration, or hplc analysis. Because many pyridines are sold for specific applications by a single customer, or by a small group of customers, specification for those products is set by agreement. They are not generally published. However, specifications for pyridine products sold to a large group of customers, such as pyridine (**1**) itself; the picolines, (**2**), (**3**), or (**4**); and niacin (**27**) or niacinamide (**26**), are publicly known. The standards for ACS reagent-grade pyridine are shown in Table 5.

Pyridine is also sold as a 1° grade, which means that the boiling point range of 98% of the sample will fall in a 1°C range which includes the normal boiling point of (**1**) (115.3 ± 0.1°C). Niacin (**27**) and niacinamide (**26**), equivalent forms of vitamin B_3, are generally sold under a *U.S. Pharmacopeia* (USP) specification (78). They are also sold as a feed-grade supplement (see VITAMINS).

Table 5. Specifications for ACS Reagent-Grade Pyridine

Physical properties	Specification
assay, wt % by gc	≥99.97
boiling point range, °C	2.0 including 115.3 ± 0.1
solubility in water	no turbidity noted in a 10 wt % soln[a]
residue after evaporation, ppm	≤20
water content, wt %	≤0.1
chloride content, ppm	≤10
sulfate content, ppm	≤10
ammonia content, ppm	≤20
copper content, ppm	≤~5
reducing substances	5 mL of pyridine does not entirely discharge the color of 0.5 mL of 0.1 N $KMnO_4$ solution[a]

[a]Within 30 minutes.

Analytical and Test Methods, and Storage

Most common analytical methods for analysis of the major component or minor components of organic products are used for pyridines. These include gas chromatography, titration, freezing point, nmr, ir, hplc, and gc/ms.

As a class, pyridine compounds tend to darken with storage. The color change is related to the conditions of storage; it is more rapid at higher temperatures and becomes more intense with increasing storage time. Pyridines with certain functional groups tend to be unstable on long storage. Aldehyde and carbinol groups tend to oxidize on exposure to air, and vinyl groups tend to polymerize. Hence, these compounds are stored at low temperature (<−10°C for the vinylpyridines), or under an inert atmosphere (aldehydes and carbinols). Storing vinylpyridines under inert atmosphere is not recommended, because the shelf life can be shortened by doing so.

Health and Safety Factors

Pyridine Acute Toxicology. Pyridine causes gastrointestinal upset and central nervous system (CNS) depression at high levels of exposure. The odor of pyridine can be detected at extremely low concentrations (12 ppb). The LD_{50} (oral, rats) is 891 mg/kg, the LC_{50} (inhalation, rats) is 4000/4 (ppm/h), and the TLV is 15 mg/m^3 (79,80).

Pyridine Chronic Toxicology. All mutagenicity tests have been negative and (**1**) is not considered a carcinogen or potential carcinogen. There have been no reports of adverse health effects on long-term exposure to (**1**) at low concentrations.

Acute Toxicology of Pyridine Derivatives. Table 6 shows the known acute health and safety factors for pyridine derivatives. In general, many pyridines are reasonably safe to handle and do not represent a serious hazard. However, some types of aminopyridines are poisons. Quaternary salts of pyridines can also be toxic. Special care should be exercised when handling bis-quaternary salts, such as (**20**) of 4,4'-bipyridine (**19**), as the fatal effect cannot be reversed after ingestion or exposure. Chloropyridines, especially polychloropyridines, can potentially be mutagenic, teratogenic, and carcinogenic.

Safety Aspects in Handling and Exposure. Pyridine compounds are ubiquitous in the natural environment, and are often found in foods as minor flavor and fragrance components. Some synthetic pyridines are used as food additives (qv) (81). A high proportion of pyridine compounds show some type of bioactivity, albeit mostly minor, such as herbicidal, insecticidal, or medicinal activity. Therefore, all the normal precautions should be exercised when handling pyridines that would be used when handling other organic products that are potentially

Table 6. Acute Toxicity of Pyridine Derivatives

Compound	Structure number	Oral LD_{50} rats, mg/kg	Inhalation LC_{50}, rats, ppm/h	TLV, mg/m^3
2-methylpyridine	(**2**)	790	4000/4	[a]
3-methylpyridine	(**3**)	400–800	8700/2	[a]
4-methylpyridine	(**4**)	1290	1000/4	[a]
2,6-dimethylpyridine	(**5**)	400	7500/1[b]	[a]
5-ethyl-2-methylpyridine	(**7**)	1540	1800/3.7[b]	2
2-vinylpyridine	(**23**)	100–200	5500/1.5[b]	0.05
4-vinylpyridine		100–200	2000/2[b]	[a]
nicotinic acid	(**27**)	7000		[a]
nicotinamide	(**26**)	7000		[a]
2-aminopyridine	(**35**)	200	5/5[c]	0.5
2-chloropyridine	(**17**)	140	100/4[d]	
3-cyanopyridine	(**25**)	1100		
N,N'-dimethyl-4,4'- bipyridinium chloride	(**20**)	57		
cetylpyridinium chloride	(**67**)	200		

[a] Not established.
[b] LC_{100}.
[c] TC_{LO}.
[d] LC_{LO}.

bioactive. Care should be taken to avoid skin or eye contact, ingestion, or inhalation of vapors and dusts. Protective garments and respirators should be worn when handling these materials. Specific and more complete recommendations are available from the manufacturers of each pyridine base.

Particular attention should be paid to aminopyridines, especially unsubstituted ones, which tend to exert severe neurotoxic effects on exposure. Most of them are generally classified as poisons. Exceptions to this rule are known. A notable one is 4-dimethylaminopyridine (DMAP) (**24**), which is widely used in industry as a superior acylation catalyst (27). Quaternary salts of pyridines are usually toxic, and in particular paraquat (**20**) exposure can have fatal consequences. Some chloropyridines, especially polychlorinated ones, should be handled with extra care because of their potential mutagenic effects. Vinylpyridines are corrosive to the skin, and can act as a sensitizer for some susceptible individuals. Niacin (**27**), niacinamide (**26**), and some pyridinecarbaldehydes can cause skin flushing.

Pyridine and alkylpyridines are excellent solvents for many materials, a property that must be taken into account when selecting O-rings, gaskets, and other sealants that are in contact with liquids. Generally, only polytetrafluoroethylene, graphite, and asbestos-based gasket and O-ring materials are acceptable. Most rubbers are rapidly swollen or degraded by liquid alkylpyridines.

Uses

Pyridine and Picolines.   These have been widely used as solvents in organic chemistry and, with increasing frequency, in industrial practice. Pyridine itself is a good solvent that is rather unreactive. The basic nature of pyridine and the picolines makes them ideal acid scavengers. Typically, pyridine is the solvent of choice for acylations (82). Furthermore, for dehydrochlorination reactions and extraction of antibiotics, pyridine is an excellent solvent. Large amounts of pyridine are used as the starting material for agrochemicals and pharmaceuticals (qv). For example, pyridine is a precursor for herbicides such as diquat [2764-72-9] (**59**) (20) and paraquat (**20**) (78), insecticides such as chlorpyrifos (**43**), and antifungal agents such as the zinc salt of pyrithione (**45**) (45).

(**59**)

The primary use of α-picoline (**2**) is as a precursor of 2-vinylpyridine (**23**). It is also used in a variety of agrochemicals and pharmaceuticals, such as nitrapyrin [1929-82-4] (**60**) to prevent loss of ammonia from fertilizers; picloram [1918-02-1] (**61**), a herbicide; and amprolium [121-25-5] (**62**), a coccidiostat.

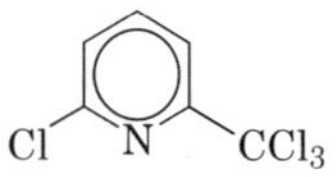

(60)

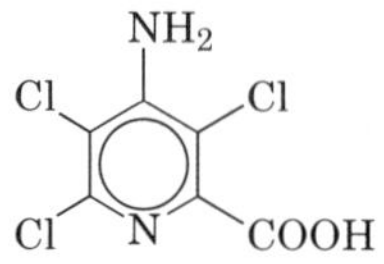

(61)

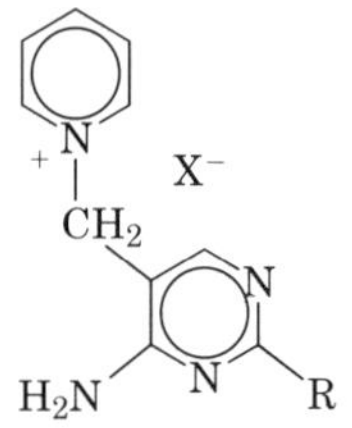

(62)

The predominant use of β-picoline (**3**) is as a starting material for agrochemicals and pharmaceuticals. For example, it is used to make insecticides such as chlorpyrifos (**43**), food additives such as niacin (**27**) and its amide (**26**), and herbicides such as fluazifop-butyl [69806-50-4] (**63**).

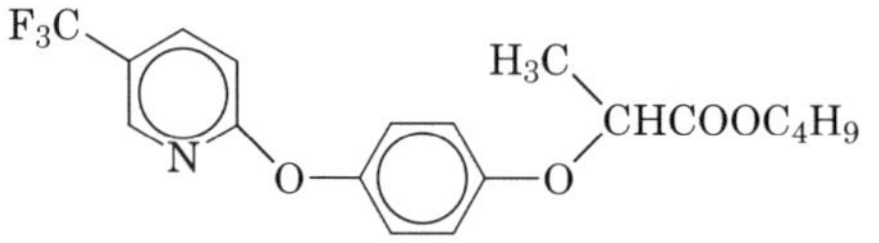

(63)

The main use of γ-picoline (**4**) is in the production of the antituberculosis agent, isoniazid (**31**). Compound (**4**) is also used to make 4-vinylpyridine, and subsequently polymers.

5-Ethyl-2-methylpyridine (**7**) is used as starting material for niacin (**27**). 2,6-Dimethylpyridine (**5**) is used for the antiarteriosclerotic pyridinol carbamate [1882-26-4] (**64**). 2,3,5-Collidine [695-98-7] (**65**) is used in the manufacture of the antiulcer drug, omeprazole [73590-58-6] (**66**) (83).

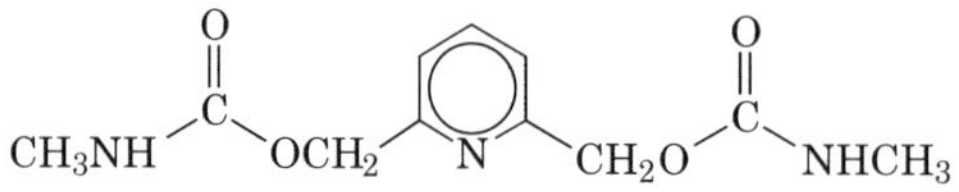

(64)

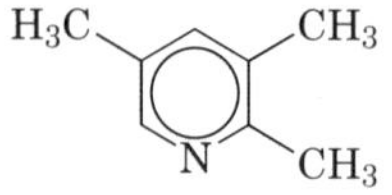

(65)

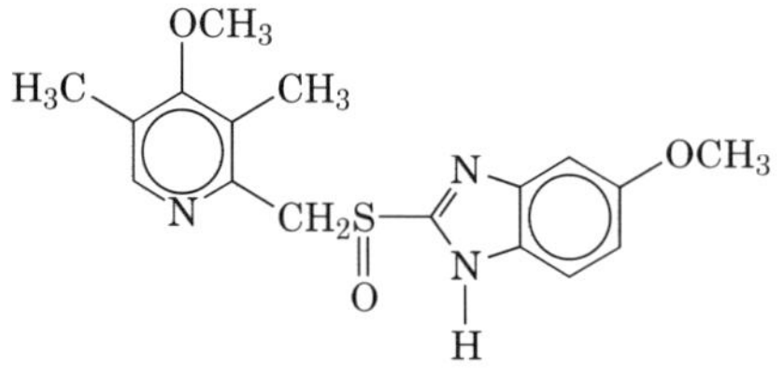

(66)

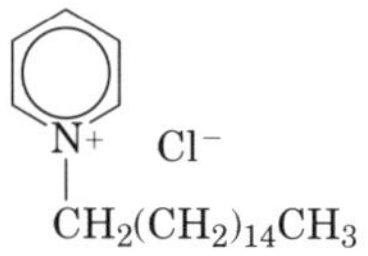

(67)

Quaternary Salts. Herbicides paraquat (**20**) and diquat (**59**) are the quaternary salts of 4,4′-bipyridine (**19**) and 2,2′-bipyridine with methyl chloride and

1,2-dibromoethane, respectively. Higher alkylpyridinium salts are used in the textile industry as dye ancillaries and spin bath additives. The higher alkylpyridinium salt, hexadecylpyridinium chloride [*123-03-5*] (**67**) (cetylpyridinium chloride) is a topical antiseptic. Amprolium (**62**), a quaternary salt of α-picoline (**2**), is a coccidiostat. Bisaryl salts of butylpyridinium bromide (or its lower 1-alkyl homologues) with aluminum chloride have been used as battery electrolytes (84), in aluminum electroplating baths (85), as Friedel-Crafts catalysts (86), and for the formylation of toluene by carbon monoxide (87) (see QUATERNARY AMMONIUM COMPOUNDS).

Pyridine *N*-Oxides. Analgesic and antiinflammatory drugs niflumic acid [*4394-00-7*] (**68**) and pranoprofen [*52549-17-4*] (**69**) are manufactured from nicotinic acid *N*-oxide [*2398-81-4*]. The antiulcer drug omeprazole (**66**) is produced from 2,3,5-trimethylpyridine *N*-oxide [*74409-42-0*]. Zinc pyrithione, the zinc salt of pyrithione (**45**), is a fungicide derived from 2-chloropyridine *N*-oxide (45).

(68)   (69)

Aminopyridines. Aminopyridines are key intermediates for the synthesis of important pharmaceutical and agricultural products (88,89). 2-Aminopyridine (**35**) is used in the production of sulfasalazine [*599-79-1*] (**70**), an antibacterial used for veterinary purposes (90). Compound (**35**) is also used in the synthesis of a number of antihistamines, including methapyrilene hydrochloride [*135-23-9*] (**71**) and pyrilamine maleate [*59-33-6*] (**72**) (91). Picoxicam [*36322-90-4*] (**73**), a nonsteroidal antiinflammatory agent, is based on (**35**) (92) (see Fig. 2).

The synthesis of aldipem [*82626-01-5*] (**74**), an anxiolytic, is based on 2-amino-5-chloropyridine [*1072-98-6*] (93). 2,6-Diaminopyridine [*141-86-6*] is used in the preparation of the urinary tract analgesic phenazopyridine hydrochloride [*136-40-3*] (**75**) (94). Nalidixic acid [*389-08-2*] (**76**), a quinolone carboxylic acid, is derived from 2-amino-6-methylpyridine [*1824-81-3*]. 2-Amino-3-methylpyridine [*1603-40-3*] is is used in the synthesis of pemirolast potassium (**77**), an antiallergic and antiasthmatic medication. 2-Amino-4-methylpyridine [*695-34-1*] is used in the preparation of piketoprofen [*60576-13-8*] (**78**), a nonsteroidal antiinflammatory drug (95). Zopidem [*82626-48-0*] (**79**) is a hypnotic medication based on 2-amino-5-methylpyridine [*1603-41-4*] (96). The synthesis of pirenzepine [*28797-61-7*] (**80**), an antiulcerative drug, proceeds via 2-chloro-3-aminopyridine [*6298-19-4*], an intermediate derived from chlorination of 3-aminopyridine (**33**) (97). 3-Amino-2-chloro-4-methylpyridine [*133627-45-9*] and 2-chloronicotinic acid [*2942-59-8*] are used in the preparation of nevirapine [*129618-40-2*] (**81**), a drug for treatment of AIDS (98). Cephapirin sodium [*24356-60-3*] (**82**) is an antibacterial medicine derived from 4-aminopyridine (**34**) (99). Compound (**34**) is also used in the preparation of the antihypertensive pinacidil [*60560-33-0*] (**83**).

4-Dimethylaminopyridine [*1122-58-3*] (DMAP) (**24**) has emerged as the preferred catalyst for a variety of synthetic transformations under mild con-

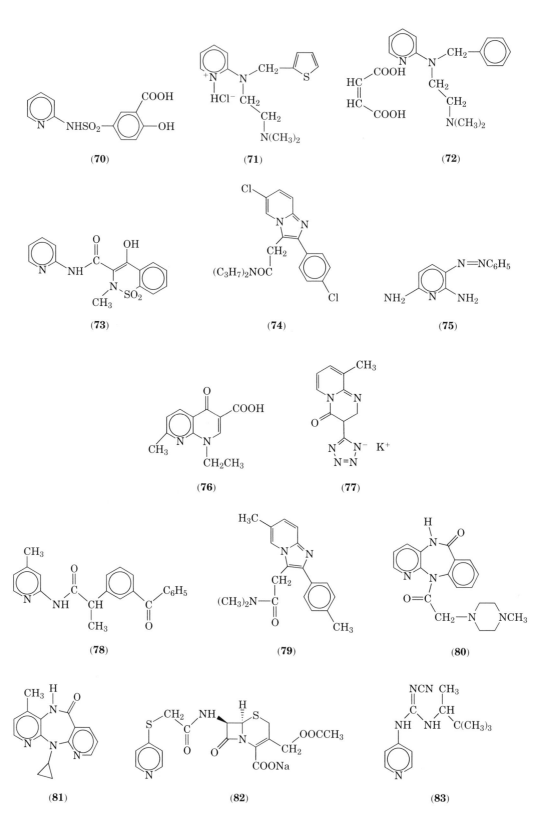

Fig. 2. Pharmaceutical products from aminopyridines.

ditions, particularly acylations, alkylations, silylations, esterifications, polymerizations, and rearrangements (100). POLYDMAP resin [*1122-58-3*], a polymeric version of DMAP, is available, and is as effective as DMAP as a catalyst for acylation reactions. Furthermore, it can be recycled without regeneration more than 20 times with very little loss in activity. POLYDMAP is a trademark of Reilly Industries, Inc.

Chloropyridines.   Figure 3 shows structures arising from chloropyridines, important intermediates in pharmaceuticals and agrochemicals. A significant use of 2-chloropyridine (**17**) is in the production of pyrithione (**49**). The zinc salt is a widely used antifungal agent (44). Phenylacetonitriles and (**17**) have been used in making pheniramine [*86-21-5*] (**84**) and dipyramide [*3737-09-5*] (**85**), both antihistamines. 2,6-Dichloropyridine is used for 8-azaquinoline antibiotics such as enoxalin [*74011-58-8*] (**86**). 2-Chloro-6-trichloromethylpyridine [*1929-82-4*] is the agrochemical nitrapyrin (**60**), which prevents the loss of ammonia from fertilizers. 2,5-Dichloro-6-trichloromethylpyridine [*1817-13-6*] is converted to the herbicide clopyralid [*1702-17-6*] (**87**). 2,3,4,5-Tetrachloro-6-trichloromethylpyridine [*1134-04-9*] is a precursor of the herbicide picloram (**61**). 2,3,5,6-Tetrachloropyridine is used for producing the insecticide chlorpyrifos (**43**) and the herbicide triclopyr (**44**). 2-Chloro-5-trifluoromethylpyridine [*52334-81-3*] is used for the herbicide fluazifop-butyl (**63**). Significant use of 2-chloronicotinic acid is for production of herbicides, such as diflufenican [*83164-33-4*] (**88**). 2-Chloro-5-methylpyridine [*18368-64-4*] is used in the manufacture of the insecticide imidacloprid [*105827-78-9*] (**89**).

Pyridinecarbonitriles, -carboxamides, and -carboxylic Acids.   3-Cyanopyridine (**25**) is used for the production of niacin (**27**), or vitamin B$_3$. 4-Cyanopyridine (**15**) is used for making the antitubercular drug isoniazid (**31**) (101).

3- and 4-Pyridinecarboxamides, (**26**) and (**32**), respectively, are used in making the corresponding aminopyridines, (**33**) and (**34**), respectively, by Hofmann

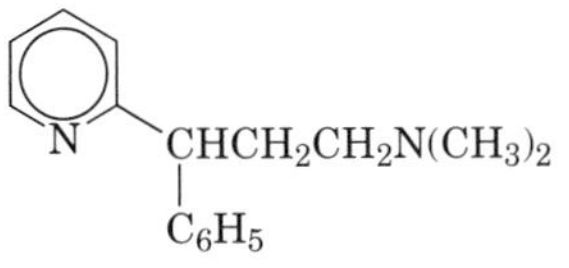

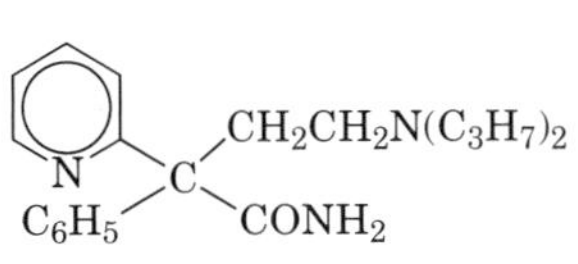

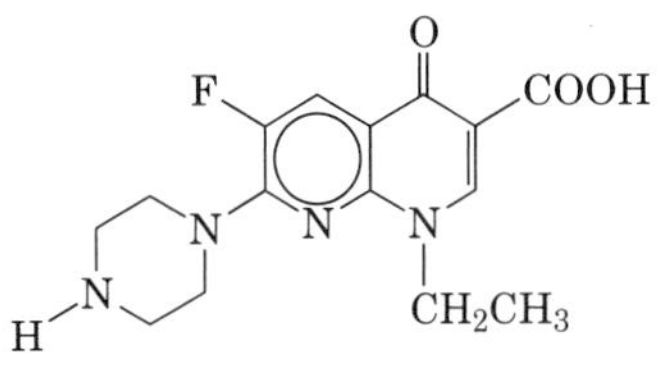

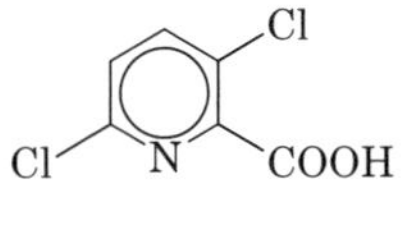

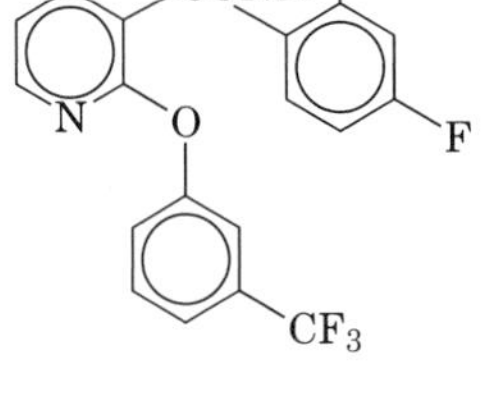

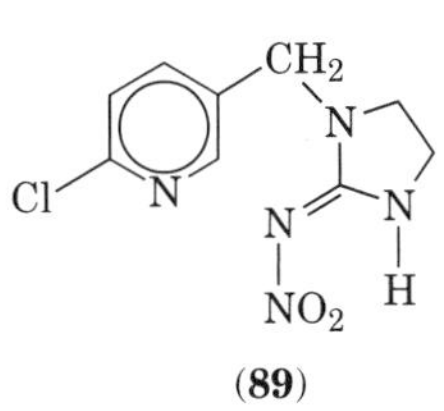

Fig. 3.   Pharmaceuticals and agrochemicals from chloropyridines.

degradation reactions. Niacin (**27**) can be used to make the vasodilator nicorandil [*65141-46-0*] (**90**). Inabenfide [*82211-24-3*] (**91**), a plant growth regulator, is an amide of isonicotinic acid. Terfenadine [*50679-08-8*] (**92**), an antihistamine, and

(**90**)   (**91**)   (**92**)

nialamide [*51-12-1*] (**93**), an antidepressant, are also made from isonicotinic acid (4-carboxypyridine). Picolinic acid (2-carboxypyridine) is an intermediate for local anesthetics such as mepivacaine [*96-88-8*] (**94**; R = CH$_3$) and bupivacaine [*2180-92-9*] (**94**; R = n-C$_4$H$_9$) hydrochlorides. Quinolinic acid [*89-00-9*] (**95**) goes into the herbicide imazapyr [*81334-34-1*] (**96**).

(**93**)   (**94**)   (**95**)   (**96**)

Hydroxy-, Hydroxyalkyl-, and Aminoalkylpyridines. A full discussion of the tautomerism occurring in heterocycles with oxygen and sulfur substituents has been published (38). Equation 2 shows the tautomerism expected in 2-pyridone (**16**) and 4-pyridone (**38**).

2-Pyridone (**16**) is used in the synthesis of the tranquilizer amphenidone [*134-37-2*] (**97**). 3-Pyridiniol (**39**) serves as a precursor to the cholinesterase inhibitors, pyridostigmine bromide [*101-26-8*] (**98**) and distigmine bromide [*15876-67-2*] (**99**). 4-Pyridone (**38**) is a precursor for propyliodone [*587-61-1*] (**100**), an x-ray contrast agent; it is also used as an intermediate for pericyazine [*2622-26-6*] (**101**), a psychotropic agent.

(**97**)   (**98**)   (**99**)

(**100**)   (**101**)

2-Pyridylcarbinol [*586-98-1*] (**102**) is used for the preparation of ibupro-phenpiconol [*64622-45-3*] (**103**), an antiinflammatory drug. 3-Pyridylcarbinol (**30**) is the active component of the vasodilator nicotinyl tartrate [*100-55-0*] (**104**). The bis(carbamate) (**64**) of pyridine-2,6-dimethanol [*6231-18-1*] is an antiarter-iosclerotic drug. The urea derivative (**105**) [*38641-94-0*] of 3-picolylamine (**29**) is a rodenticide.

Ethanolpyridines are used mainly for vinylpyridine manufacture. However, 2-ethanolpyridine (**22**) is used to make the psychotropic agent thioridazine [*50-52-2*] (**106**), and the antispasmodic agent triquizium bromide [*71731-58-3*] (**107**).

(**102**)   (**103**)   (**104**)

(**105**)   (**106**)   (**107**)

Pyridinecarbaldehydes.   Pralidoxime iodide [*51-15-0*] (**108**), a derivative of 2-pyridinecarbaldehyde (**109**), is used as an antidote against organophospho-rous nerve agents (acetylcholinesterase inhibitors). Compound (**109**) is also used in making the laxative bisacodyl [*603-50-9*] (**110**).

(108)   (109)   (110)

Bipyridines. Since the 1960s, the most important commercial agrochemical based on pyridine has been the herbicide paraquat (**20**) which is made from 4,4′-bipyridine (**19**). The isomeric herbicide diquat (**59**) is made by an analogous route, but utilizing 2,2′-bipyridine [366-18-7] as a precursor.

Vinylpyridines. 2-Vinylpyridine (**23**) is a large-volume specialty chemical used primarily for manufacturing terpolymers, eg, styrene, butadiene, and vinylpyridine, used to coat fabric cords of tires and fabric-reinforced rubber products, such as belts (25). Vinylpyridines undergo Michael addition easily, and two of those products using (**23**) as a starting material, (**111**) [114-91-0] and (**112**) [5638-76-6], find use as a veterinary anthelmintic and a vasodilator, respectively (102). 4-Vinylpyridine is used to make 2-(4-pyridyl)ethanesulfonic acid [53054-76-5] (**113**) which serves as a coagulation accelerator for the gelatin layer of photographic plates (103).

(111)   (112)   (113)   (114)

Homopolymers of vinylpyridines find many uses. These homopolymers are soluble in many organic solvents and retain the basic functionality of the pyridine ring. 2-Methyl-5-vinylpyridine [140-76-1] (**114**) was once manufactured and polymerized to form coatings for medicinal tablets. Cross-linked vinylpyridine polymers are generally insoluble in all solvents and can be utilized as catalysts for acylation reactions, as basic tertiary amine catalysts, or as an acid scavenger in reactions where acid is a troublesome by-product. These insoluble catalysts and acid scavengers are especially easy to regenerate to the free-base form for reuse.

Piperidines. A significant use of piperidine (**18**) has been in the manufacture of vulcanization accelerators, eg, thiuram disulfide [120-54-7] (**115**) (see RUBBER CHEMICALS). Mepiquat dichloride [24307-26-4], the dimethyl quaternary salt of (**18**), is used as a plant growth regulator for cotton (qv). Piperidine is used to make vasodilators such as dipyridamole [58-32-2] (**116**) and minoxidil [38304-91-5] (**117**), and diuretics such as etozoline [73-09-6] (**118**).

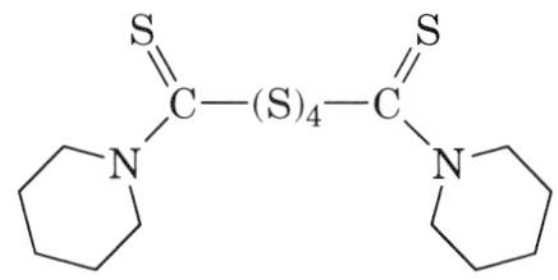

(115)

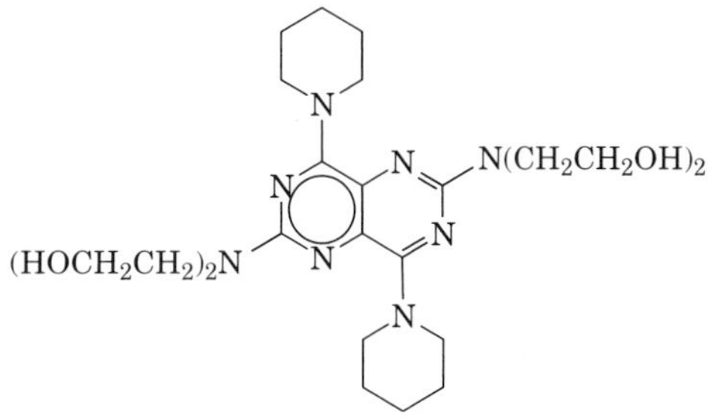

(116)

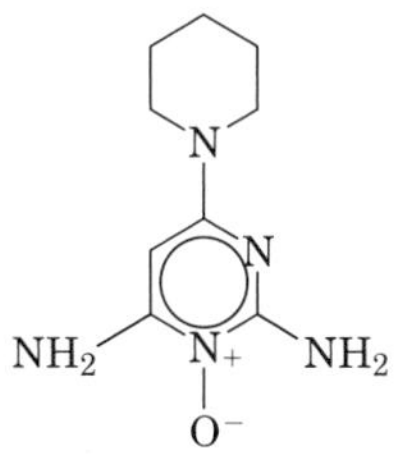

(117)

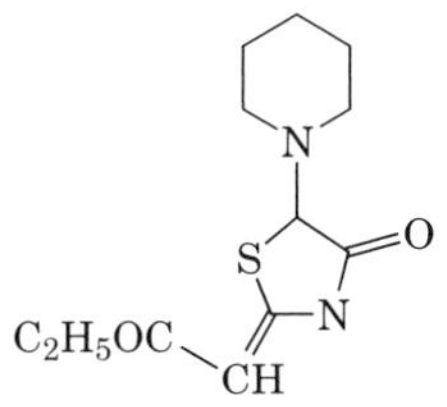

(118)

Both piperidine (**18**) and *N*-formylpiperidine [*2591-86-8*] are used as solvents. *N*-Formylpiperidine is a dipolar, aprotic solvent that has considerably better hydrocarbon solubility than other dipolar, aprotic solvents having formamide or acetamide functionality.

2-Methylpiperidine [*109-05-7*] is employed for making fungicides such as piperalin [*3478-94-2*] (**119**). 2,6-Dimethylpiperidine [*766-17-6*] is used for the antiarrhythmic pirmenol [*61477-94-9*] (**120**). 4-Benzyl piperidine is used to produce ifenprodil tartrate [*23210-56-2*] (**121**), a cerebral vasodilator.

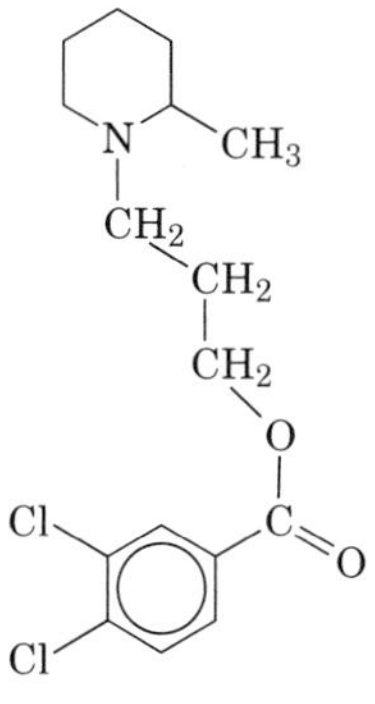

(119)

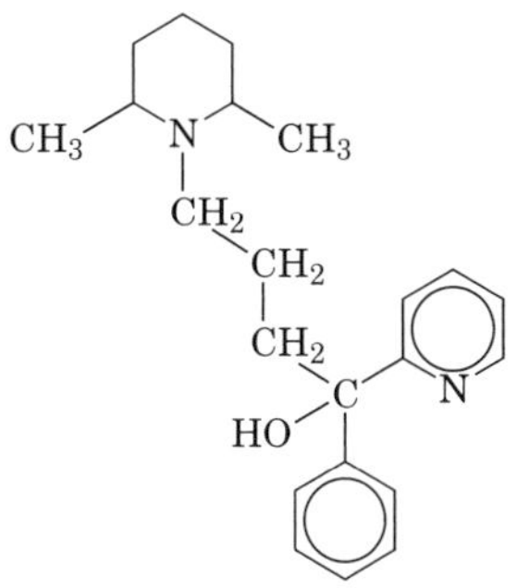

(120)

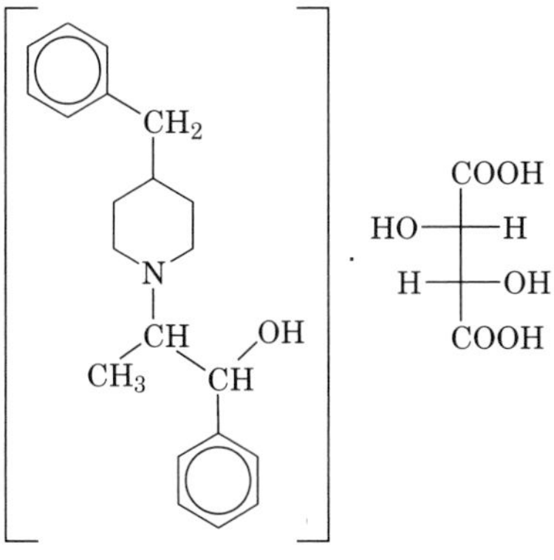

(121)

BIBLIOGRAPHY

"Pyridine and Pyridine Bases" in *ECT* 1st ed., Vol. 11, pp. 278–293, by H. S. Mosher, Stanford University; "Pyridine and Pyridine Derivatives" in *ECT* 2nd ed., Vol. 16, pp. 780–806, by R. A. Abramovitch, University of Alabama; in *ECT* 3rd ed., Vol. 19, pp. 454–483, by G. L. Goe, Reilly Tar & Chemical Corp.

1. W. Ramsay, *Ber. Dtsch. Chem. Ges.* **10**, 736 (1877).
2. T. Anderson, *J. Liebigs Ann. Chem.* **60**, 86 (1846).
3. A. E. Chichibabin, *Zh. Russ. Fiz. Khim. O-va.* **37**, 1229 (1905).
4. A. P. Kudchadker and S. A. Kudchadker, *Pyridine and Phenylpyridines*, API Publication 710, American Petroleum Institute, Washington, D.C., 1979.
5. Technical data, Reilly Industries, Indianapolis, Ind.
6. R. J. L. Andon and J. D. Cox, *J. Chem. Soc.*, 4601 (1952).
7. R. J. L. Andon, J. D. Cox, and E. F. G. Herington, *Trans. Faraday Soc.* **50**, 918 (1954).
8. L. H. Horsley, *Advances in Chemistry Series 116*, American Chemical Society, Washington, D.C., 1973.
9. W. E. Feely and E. M. Beavers, *J. Am. Chem. Soc.* **81**, 4004 (1959).
10. R. Murugan and co-workers, *CHEMTECH*, 17 (June, 1994); C. Hansch and A. Leo, "Exploring QSAR Fundamentals and Application in Chemistry and Biology," *ACS Professional Reference Book*, American Chemical Society, Washington, D.C., 1995.
11. R. A. Abramovitch and E. M. Smith, in R. A. Abramovitch, ed., *Pyridine and Its Derivatives*, Suppl. Part 2, Vol. 14, John Wiley & Sons, Inc., New York, 1974, p. 1.
12. E. F. V. Scriven, in A. R. Katritzky and C. W. Rees, eds., *Comprehensive Heterocyclic Chemistry*, Vol. 2, Pergamon Press, Oxford, U.K., 1984, p. 165.
13. L. W. Deady, M. R. Grimmett, and C. H. Potts, *Tetrahedron* **35**, 2895 (1979), and references cited therein.
14. H. Suschitzky, ed., *Polychloroaromatic Compounds*, Plenum Publishing Corp., New York, 1974.
15. W. K. Fife and E. F. V. Scriven, *Heterocycl.* **22**, 2375 (1984).
16. R. A. Abramovitch and A. R. Vinutha, *J. Chem. Soc.* (B), 131 (1971).
17. A. N. Kost, S. P. Gromov, and R. S. Sagitullin, *Tetrahedron* **37**, 3423 (1981).
18. C. K. McGill and A. Rappa, *Adv. Heterocycl. Chem.* **44**, 1 (1988).
19. D. Bryce-Smith, P. J. Morris, and B. J. Wakefield, *J. Chem. Soc., Perkins Trans. I*, 1977 (1976).
20. L. A. Summers, *The Bipyridinium Herbicides*, Academic Press, Inc., New York, 1980.
21. R. A. Abramovitch and J. G. Saha, *Adv. Heterocycl. Chem.* **6**, 229 (1966).
22. F. Minisci and O. Porta, *Adv. Heterocycl. Chem.* **16**, 123 (1974).
23. G. L. Goe, G. F. Hillstrom, R. Murugan, E. F. V. Scriven, and A. R. Sherman, *Heterocycl.* **37**, 1489 (1994).
24. H. Pines and W. Stalic, *Base-Catalyzed Reactions of Hydrocarbons and Related Compounds*, Academic Press, Inc., New York, 1977.
25. D. B. Wootton, *Dev. Adhes.* **1**, 181 (1977).
26. U.S. Pat. 4,158,093 (June 12, 1979), T. D. Bailey and C. K. McGill (to Reilly Tar & Chemical Corp.).
27. E. F. V. Scriven, *Chem. Soc. Rev.* **12**, 129 (1983).
28. H. Beschke, H. Friedrich, H. Schaefer, and G. Schreyer, *Chem. Ztg.* **101**, 384 (1977) and references therein.
29. C. B. Rosas and G. B. Smith, *Chem. Eng. Sci.* **35**, 330 (1980).
30. Ger. Pats. 2,046,556 and GB 1,276,776 (Apr. 22, 1971), A. Stocker, O. Marti, T. Pfammatter, G. Schreiner, and S. Brander (to Lonza Ltd.).
31. Ger. Pat. 717,172 (Oct. 20, 1954), G. O. Chase (to Roche Products, Ltd.).
32. R. L. Frank and C. Weatherbee, *J. Am. Chem. Soc.* **70**, 3482 (1948).

33. Jpn. Pat. 89 175,968 (July 12, 1989), M. Nozawa (to Koei Chemical Industry Co., Ltd.).

34. G. B. Kauffmann, *J. Chem. Ed.* **55**, 448 (1978).

35. C. F. H. Allen and C. N. Wolf, *Organic Synthesis, Collected Volume IV*, John Wiley & Sons, Inc., New York, 1963, p. 45.

36. C. S. Giam, in Ref. 11, Suppl. Part 3, p. 41.

37. Belg. Pat 612,258 (July 3, 1962), G. Y. Lesher and M. D. Gruett (to Sterling Drug, Inc.).

38. G. L. Goe, C. A. Huss, J. G. Keay, and E. F. V. Scriven, *Chem. Ind.*, 694 (1987); J. Elguero, C. Marzin, A. R. Katritzky, and P. Linda "The Tautomerism of Heterocycles," in A. R. Katritzky and A. J. Boulton, eds., *Advances in Heterocyclic Chemistry*, Suppl. 1, Academic Press, Inc., New York, 1976.

39. M. G. Reinecke, *Tetrahedron* **38**, 427 (1982); R. Murugan, *Metallation, Conformational Analysis, Hydrogen Exchange and Rearrangement in Amides*, Doctoral thesis, University of Florida, Gainesville, 1987; A. Weissberger and E. C. Taylor, eds., *The Chemistry of Heterocyclic Compounds*, in Ref. 11, pp. 448–449.

40. A. V. Voropaeva and N. G. Garbar, *Khim. Geterotsikl. Soedin.*, 677 (1969).

41. J. Delarge, *Farmaco Ed. Sci.* **22**, 1069 (1967).

42. Fr. Pat. 1,360,901 (May 15, 1964), R. H. Rigterink (to Dow Chemical Co.).

43. U.S. Pat. 3,862,952 (Jan. 28, 1975), L. D. Markley (to Dow Chemical Co.).

44. E. Shaw, J. Bernstein, K. Loser, and W. A. Lott, *J. Am. Chem. Soc.* **72**, 4362 (1950); U.S. Pat. 2,745,826 (May 15, 1956), S. Semenoff and M. A. Dollicer (to Olin Mathieson Chemical Corp).

45. U.S. Pat. 3,694,332 (Sept. 26, 1972), V.D. Parker (to Dow Chemical Corp.).

46. L. F. Mikhailova and V. A. Barkash, *J. Gen. Chem. USSR* **37**, 2792 (1967).

47. R. A. Abramovitch, E. M. Smith, E. E. Knaus, and M. Saha, *J. Org. Chem.* **37**, 1690 (1972).

48. G. Jones, in Ref. 12, 395.

49. Eur. Pat. Appl. 9,289 (Apr. 2, 1980), E. J. M. Verheijen and C. H. Van Geersheuvels (to Stamicarbon BV).

50. Neth. Pat. 7,013,453 (Sept. 11, 1970), J. J. M. Deumens and S. H. Groen (to Stamicarbon NV).

51. K. Shibata, I. Katsuyama, H. Izoe, M. Matsui, and H. Muramatsu, *J. Heterocycl. Chem.*, 277 (1993).

52. H. Beschke and H. Friedrich, *Chem. Ztg.* **101**, 377 (1977).

53. Ger. Pat. 1,917,037 (Nov. 20, 1968), G. Swift (to ICI).

54. Jpn. Pat. 7,039,545 (Dec. 12, 1970), Y. Watanabe and S. Takenaka (to Nippon Kayaku).

55. Ger. Pat. 2,054,773 (May 19, 1971), Y. Minato and T. Niwa (to Koei Chemical Co.).

56. Jpn. Pat. 81 26,546 (Mar. 14, 1981), (To Daicel Chemical).

57. Belg. Pat. 845,405 (1975), L. Letartre (to ICI); Ger. Pat. 2,637,363 (1975), L. Letartre (to ICI).

58. Ger. Pat. 2,203,384 (1972), G. Grigoleit, R. Oberkobusch, G. Collin, and K. Matern (to Ruetgerswerke).

59. Eur. Pat. Appl. 131,887 (Jan. 23, 1985), D. Feitler, W. Schimming, and H. Wetstein (to Nepera Chemical Co.).

60. Eur. Pat. 232,182 (1987), S. Shimizu and co-workers (to Koei Chemical Co.).

61. Brit. Pat. 1,216,866 (Dec. 23, 1970), N. Y. Minato and O. S. Yasuda (to Koei Chemical Co.).

62. Jpn. Pat. 7,139,873 (1969), Y. Watanabe, S. Takenada, and K. Koyasu (to Nippon Kayaku).

63. S. Kagabu, S. Naruse, Y. Tagami, and Y. Watanabe, *J. Org. Chem.* **54**, 4275 (1989).

64. M. E. Jung and Y. M. Choi, *J. Org. Chem.* **56**, 6729 (1991).
65. N. J. Sisti, F. W. Fowler, and D. S. Grierson, *Syn. Lett.*, 816 (1991).
66. B. R. Baker and F. J. McEvoy *J. Org. Chem.* **20**, 118 (1955).
67. H. Leditschke, *Chem. Ber.* **85**, 202 (1952).
68. H. Leditschke, *Chem. Ber.* **86**, 123 (1953).
69. N. S. Boodman, J. O. Hawthorne, P. X. Masciantonio, and A. W. Simon in Ref. 11, Vol. 1, p. 183.
70. M. Ya. Karpeiskii and V. L. Florentev, *Usp. Khim.* **38**, 1244 (1969).
71. H. L. Rice and R. E. Londergan, *J. Am. Chem. Soc.* **77**, 4678 (1955).
72. Neth. Pat. Appl. 6,410,223 (Mar. 26, 1965), G. T. Stevenson (to Dow Chemical Co.).
73. W. M. Reid and L. R. McDougald, *Feedstuffs* **53**, 27 (Jan. 12, 1981).
74. F. Brody and P. R. Ruby, in E. Klingsberg, ed., *Pyridine and Its Derivatives*, Interscience Publishers, Inc., New York, 1960.
75. I. Ya. Lazdin'sn and A. A. Avots, *Khim. Geterotsikl. Soedin.*, 1011 (1979).
76. U.S. Pat. 3,780,082 (Dec. 18, 1973), J. J. M. Deumens and S. H. Groen (to Stamicarbon, NV).
77. *Fed. Reg.* **43**, 57922 (Dec. 11, 1978).
78. *The United States Pharmacopeia XX*, The United States Pharmacopeial Convention, Inc., Rockville, Md., 1980, pp. 548–549.
79. *Patty's Industrial Hygiene and Toxicology*, 3rd ed., Vol. 2A, John Wiley & Sons, Inc., New York, 1981, pp. 2699–2690, 2719–2745, 2751–2761.
80. R. J. Lewis, ed., *Registry of Toxic Effects of Chemical Substances*, U.S. Department of Health, Education, and Welfare, U.S. Government Printing Office, Washington, D.C., 1978.
81. *Scientific Literature Review of Pyridine and Related Substances in Flavor Usage*, PB-296005, Food and Drug Administration, U.S. Department of Commerce, Washington, D.C., 1979.
82. A. O. Fitton and J. Hill, *Selected Derivatives of Organic Compounds*, Chapman Hall, London, 1970.
83. Eur. Pat. Appl. 5,129 (Oct. 31, 1979), U. K. Junggren and S. E. Sjostrand (to Aktiebolag Hassle).
84. U.S. Pat. 4,122,245 (Oct. 24, 1979), J. C. Nardi, C. L. Hussey, and L. A. King (to U.S. Dept. of the Air Force).
85. Jpn. Pat. 62 70,593 (1985), Y. Kato, S. Takahashi, and N. Kong (to Nisshin Steel).
86. J. A. Boon, J. A. Levisky, J. L. Pflug, and J. S. Wilkes, *J. Org. Chem.* **51**, 480 (1986).
87. U.S. Pat. 4,554,383 (Nov. 19, 1985), J. F. Knifton (to Texaco).
88. P. Arnall and G. R. Dace, *Mfg. Chem. Aerosol News*, 21–26 (Feb. 1970).
89. D. J. Berry and E. F. V. Scriven, *Spec. Chem.*, 30–31 (Jan.–Feb. 1992).
90. O. H. Siegmund, ed., *The Merck Veterinary Manual*, 5th ed., Merck & Co., Inc., Rahway, N.J., 1979, p. 481.
91. C. P. Huttrer, C. Djerassi, W. L. Beears, R. L. Mayer, and C. R. Scholz, *J. Am. Chem. Soc.* **68**, 1999 (1946).
92. Ger. Offen. 1,943,265 (Aug. 13, 1970), J. G. Lombardino (to S. Chas. Pfizer & Co., Inc.).
93. M. Dimsdale, J. C. Friedmann, P. L. Morselli, and B. Zivkovic, *Drugs of the Future* **13**, 106 (1988).
94. R. N. Shreve, M. W. Swaney, and E. H. Riechers, *J. Am. Chem. Soc.* **65**, 2241 (1943).
95. Brit. Pat. 1,436,502 (May 19, 1976), R. G. W. Spickett, N. A. Vega, and S. J. Prieto (to A. Gallardo, SA).
96. Eur. Pat. Appl. 50,563 (Apr. 28, 1982), J. P. Kaplan and P. George (to Synthelabo, SA).
97. Fr. Pat. 1,505,795 (Dec. 15, 1967), K. Thomae.

98. V. J. Merluzzi and co-workers, *Science*, **250**, 1411 (1990).
99. L. B. Crast, Jr., R. G. Graham, and L. C. Cheney, *J. Med. Chem.* **16**, 1413 (1973).
100. G. Hoefle, W. Steglich, and H. Vorbruggen, *Angew. Chem., Intern'l Ed. Eng.* **17**, 569 (1978); B. Neises and W. Steglich, *Org. Synth.* **63**, 183 (1985).
101. T. P. Sycheva, T. N. Pavlova, and M. N. Shchukina, *Khim. Farm. Zh.* **6**, 6 (1972).
102. A. W. J. Broome and N. Greenhalgh, *Nature* **189**, 59 (1961).
103. Ger. Offen. 2,439,551 (Feb. 26, 1976), W. Himmelman, J. Sobel, and W. Sauerteig (to Agfa-Gavert, AG).

General References

H. S. Mosher, in R. C. Elderfield, ed., *Heterocyclic Compounds*, Vol. 1, John Wiley & Sons, Inc., New York, 1950, p. 397 ff.
E. Klingsberg, ed., *Pyridine and Its Derivatives*, Interscience Publishers, Inc., New York, 1960.
R. A. Abramovitch, ed., *Pyridine and Its Derivatives, Supplement*, John Wiley & Sons, Inc., New York, 1974.
M. H. Palmer, in S. Coffey, ed., *Rodd's Chemistry of Carbon Compounds*, 2nd ed., Vol. IV, Part F, Elsevier Scientific Publishing Co., Amsterdam, the Netherlands, 1976, pp. 1–26.
D. M. Smith, *ibid.*, pp. 27–226.
H. Beschke and co-workers, in *Ulmanns Encyklopaedie der Technischen Chemie*, 4th ed., Vol. 19, Verlag Chemie GmbH, Weinheim, Germany, 1980, pp. 591–617.
S. Shimuzu and co-workers, in *Ulmanns Encyklopaedie der Technischen Chemie*, Vol. A22, VCH Publishers, Inc., Weinheim, Germany, 1993, pp. 399–430.

ERIC F. V. SCRIVEN
JOSEPH E. TOOMEY, JR.
RAMIAH MURUGAN
Reilly Industries, Inc.

PYRIDOXINE, PYRIDOXAL, AND PYRIDOXAMINE. See
VITAMINS.

PYRITE. See IRON; PIGMENTS, INORGANIC; SULFUR; SULFURIC ACID AND SULFUR TRIOXIDE.

PYROCATECHOL. See HYDROQUINONE, RESORCINOL, AND CATECHOL.

PYROGALLOL. See (POLYHYDROXY)BENZENES.

PYROMETALLURGY. See METALLURGY, EXTRACTIVE.

PYROMETRIC CONES. See CERAMICS.

PYROMETRY. See TEMPERATURE MEASUREMENT.

PYROTECHNICS

Pyrotechnics is the field of technology that combines science and art to chemically generate heat, and from that heat create light, color, audible effects, and gas pressure for entertainment, emergency signaling, and military applications. The civilian side of pyrotechnics includes fireworks, highway flares (fusees), air bag inflators, and special effects devices for the entertainment industry. Military and aerospace pyrotechnics include a wide range of devices for illumination, signaling, obscuration, and gas generation. These devices are characterized by rugged construction and greater resistance to adverse environmental conditions with associated higher cost, reliability, and safety than most civilian pyrotechnic devices. In some ways, unlike the related technologies of explosives and propellants, pyrotechnics remains a practical discipline with an incomplete theoretical foundation, and the rigorous application of the various scientific disciplines is taking place gradually (see EXPLOSIVES AND PROPELLANTS).

Principles

The usefulness of pyrotechnic reactions derives from their being exothermic (heat-releasing), self-sustaining, and self-contained. Many chemical reactions require a sustained input of energy or they occur by interaction with other substances provided from external sources; combustion of a fuel such as gasoline or wood with atmospheric oxygen is a typical example. Most pyrotechnic reactions occur independently of any external oxidizer, although in some instances the pyrotechnic effect is enhanced by interaction with the environment. Such characteristics of pyrotechnics are shared by explosives and propellants, which also involve exothermic and self-propagating reactions and, in some cases, the distinction is derived solely from the application of the device and not from the chemical nature of the reaction. Most pyrotechnic devices contain no moving parts and are small and lightweight. Compared with their mechanical analogues, pyrotechnic devices tend to be inexpensive and often are highly reliable. On the other hand, pyrotechnic devices function only once and do not normally lend themselves to reuse.

Pyrotechnics is based on the established principles of thermochemistry and the more general science of thermodynamics. There has been little work done on the kinetics of pyrotechnic reactions, largely due to the numerous chemical and nonchemical factors that affect the burn rate of a pyrotechnic mixture. Information on the fundamentals of pyrotechnics have been published in Russian (1) and English (2–6). Thermochemical data that are useful in determining the energy outputs anticipated from pyrotechnic mixtures are contained in general chemical handbooks and more specialized publications (7–9).

A pyrotechnic composition contains one or more oxidizers in combination with one or more fuels. Oxidizers used in pyrotechnics, such as potassium nitrate, KNO_3, are solids at room temperature and release oxygen when heated to elevated temperatures. The oxygen then combines with the fuel, and heat is generated by the resulting chemical reaction. Chemicals that release fluorine or chlorine on heating, such as polytetrafluoroethylene (Teflon) are also capable of serving as oxidizers, particularly with a metallic fuel. If the released heat is efficiently captured by adjacent pyrotechnic composition, further reaction occurs between oxidizer and fuel, and a self-propagating reaction ensues through the remaining mixture. The onset temperature for a rapid, self-propagating reaction between oxidizer and fuel is termed the ignition temperature of the composition. When a portion of a pyrotechnic mixture reaches this temperature through an external input of energy, such as a burning fuse, hot wire, friction, or impact, the composition undergoes a rapid chemical reaction that is self-sustaining through the remaining composition.

The most commonly used oxidizers in pyrotechnics are listed in Table 1. They are normally ionic solids such as nitrate or perchlorate salts; oxides, chlorates, and chromates are also used. The selection of oxidizer is determined by the desired heat output, reaction rate, and the physical state of the anticipated reaction products. A slower reaction occurs if the oxidizer releases its oxygen only at high temperatures, and if the oxidizer requires a net heat input (is endothermic) in its decomposition. Iron(III) oxide, Fe_2O_3, is an example of a "slow" oxidizer, since it melts at 1565°C and has a heat of decomposition of $+824$ kJ/mol (197 kcal/mol) (7). Potassium chlorate, $KClO_3$, in contrast, melts at approximately 356°C with the evolution of oxygen, and releases approximately -45 kJ/mol (-10.8 kcal/mol) upon decomposition into potassium chloride, KCl, and oxygen. Potassium chlorate ignites and readily reacts with a wide range of fuels, whereas iron(III) oxide can only sustain a self-propagating reaction with the most energetic fuels. Ammonium perchlorate, NH_4ClO_4, is an example of an oxidizer that liberates extensive gaseous reaction products upon its decomposition. It has found wide use as an oxidizer in the solid propellant field, since the thrust generated by a propellant is related to the amount of gas generated by the burning propellant. The solid propellant used in the booster rockets on the U.S. Space Shuttle is 70% ammonium perchlorate by weight.

Pyrotechnic fuels are selected for their heat of combination with oxygen, melting or decomposition temperature, and the physical state of their reaction products. A highly exothermic fuel that combines with oxygen at low temperature tends to be the most reactive in a pyrotechnic system. Pyrotechnic fuels include readily oxidized metals and other elements as well as organic compounds consisting of carbon and hydrogen, sometimes in combination with other elements such as nitrogen and oxygen. Organic compounds produce carbon dioxide and water as two of their primary oxidation products; these are gases above 100°C, and hence gas pressure is generated when they undergo oxidation in a pyrotechnic mixture. This may or may not be desired in the reaction products. Common

Table 1. Representative Oxygen Donors Used in Pyrotechnics

Compound	Formula	CAS Registry Number
ammonium perchlorate	NH_4ClO_4	[7790-98-9]
barium nitrate	$Ba(NO_3)_2$	[13465-94-6]
iron(III) oxide	Fe_2O_3	[1309-37-1]
potassium chlorate	$KClO_3$	[3811-04-9]
potassium perchlorate	$KClO_4$	[7778-74-7]
sodium nitrate	$NaNO_3$	[7631-99-4]
strontium nitrate	$Sr(NO_3)_2$	[10042-76-9]

pyrotechnic fuels include (*1*) metal powders, eg, aluminum, magnesium, titantium, magnesium–aluminum alloy (magnalium), and zirconium; (*2*) elemental fuels, such as carbon (charcoal), sulfur, boron, silicon, and phosphorus; and (*3*) carbon–hydrogen compounds (organic compounds), ie, starch, plastics (poly(vinyl chloride) (PVC)), epoxy, polyesters, and tree gums.

A fuel must be reasonably stable to air and moisture to be usable in pyrotechnic manufacturing. Metals such as sodium or lithium, although attractive from their low atomic masses and high heats of combustion, do not possess the required stability to be safely used in the manufacturing process and would not produce devices with acceptable long-term storage stability unless they were hermetically sealed to prevent atmospheric oxygen and moisture from slowly reacting with the fuel over time. More stable metals such as magnesium, aluminum, titanium, and zirconium do find extensive use in pyrotechnic applications. When designing pyrotechnic mixtures it must be remembered that magnesium reacts in an acidic environment, whereas aluminum is more affected in an alkaline environment. Mixtures that contain magnesium are sometimes stabilized with a few percent of an organic binder that coats and protects the magnesium particles. Similar compositions containing aluminum are more stable, except that these mixtures can undergo a base-accelerated decomposition if they get wet. Boric acid is therefore sometimes added as a stabilizer to mixtures containing a nitrate oxidizer and aluminum powder. Illuminating flare and tracer compositions contain magnesium with barium, strontium, or sodium nitrates in an organic binder, whereas more energetic photoflash compositions are typically prepared with aluminum powder and the more energetic oxidizer potassium perchlorate and used as loose powders rather than consolidated into a tube.

Other typical pyrotechnic fuels include charcoal, sulfur, boron, silicon, and synthetic polymers such as poly(vinyl alcohol) and poly(vinyl chloride). Extensive use has been made of natural products such as starches and gums, and the use of these materials continues to be substantial in the fireworks industry. Military pyrotechnics have moved away from the use of natural products due to the inherent variability in these materials depending on climatic conditions during the growth of the plants from which the compounds are derived.

Pyrotechnic mixtures may also contain additional components that are added to modify the burn rate, enhance the pyrotechnic effect, or serve as a binder to maintain the homogeneity of the blended mixture and provide mechanical strength when the composition is pressed or consolidated into a tube or other container. These additional components may also function as oxidizers or fuels in the composition, and it can be anticipated that the heat output, burn rate, and ignition sensitivity may all be affected by the addition of another component to a pyrotechnic composition. An example of an additional component is the use of a catalyst, such as iron oxide, to enhance the decomposition rate of ammonium perchlorate. Diatomaceous earth or coarse sawdust may be used to slow up the burn rate of a composition, or magnesium carbonate (an acid neutralizer) may be added to help stabilize mixtures that contain an acid-sensitive component such as potassium chlorate. Binders include such materials as dextrin (partially hydrolyzed starch), various gums, and assorted polymers such as poly(vinyl alcohol), epoxies, and polyesters. Polybutadiene rubber binders are widely used as fuels and binders in the solid propellant industry. The production of colored flames is enhanced by the presence of chlorine atoms in the pyrotechnic flame, so chlorine donors such as poly(vinyl chloride) or chlorinated rubber are often added to color-producing compositions, where they also serve as fuels.

Cost is always a factor in selecting a component for a pyrotechnic composition, as is the water-attracting ability of a substance (hygroscopicity). In general, substances that attract water from the atmosphere are avoided in pyrotechnic formulations for a variety of reasons, thereby supporting the centuries-old adage, "Keep your powder dry."

The process of designing a pyrotechnic mixture begins with the selection of oxidizer and fuel, and proceeds to incorporate additional components to achieve the exact pyrotechnic effect and burn rate desired in the end item. It is at this point that pyrotechnics takes on the dual nature of an art and science, and experience is often the only thing that can be relied upon for the solution of a difficult problem.

The classic example of a pyrotechnic composition is black powder, a blend of potassium nitrate, sulfur, and charcoal in a 75:10:15 ratio by weight (10,11). This composition has been used as a propellant for cannons and muskets as well as in fireworks for centuries, and is still used in the 1990s in significant amounts as a propellant, fuse powder, and bursting charge by the fireworks industry. It must be well mixed in the proper proportions to achieve a reactive powder, and even the specific charcoal selected to manufacture the black powder affects its burn behavior. Its batch-to-batch variability, depending on the mixing time, mixing energy, charcoal surface area, and residual water content, has been responsible in large part for the somewhat questionable reputation that the field of pyrotechnics has acquired over the years.

Another matter of concern in pyrotechnic formulations is the possibility of exchange reactions occurring between components. Addition of ammonium salts to compositions containing nitrate oxidizers can produce ammonium nitrate, a very hygroscopic material. The composition then becomes quite prone to pick up water and its performance deteriorates. The addition of an ammonium salt to a chlorate-based formulation can lead to the formation of the highly unstable ammonium chlorate, and spontaneous ignition may occur at room temperature. In designing a pyrotechnic composition, the pyrotechnic chemist must be careful to select components that do not undergo such adverse exchange reactions, which are often aggravated by the presence of residual moisture in the blended mixture.

Theoretical Considerations

The heat output of a pyrotechnic reaction is the difference between the sum of all heats of formation of the products of the reaction and the sum of the heats of formation of the reactants. If the starting and final conditions of pressure and temperature are close to those of the initial surroundings, it is permissible to use the tabulated standard heats of formation. If the reaction products are primarily solids or liquids, a moderate exothermic output is a reliable indication of a tendency to self-propagate. As long as the pyrotechnic process is represented by a primary chemical reaction, the preceding analytical methods are adequate. If, however, chemical processes take place at extreme temperatures and pressures, a large number of simultaneous reactions may exist. As the number of possible reactions increases, so does the mathematical difficulty; most calculations of chemical equilibria are performed with computers, which are programmed for the determination of the minimum free energy attainable at any given composition and temperature (12).

The ignition temperature of a given pyrotechnic mixture is largely determined by the decomposition temperature of the oxidizer and the melting point or decomposition temperature of the fuel. Ignition, or the onset of a self-propagating reaction, occurs when the oxygen from the oxidizer reacts with the fuel and produces enough heat output to rapidly activate the adjacent particles to their ignition temperature. The combination of an easily decomposed oxidizer, such as potassium chlorate, with a fuel that melts or decomposes at low temperature, such as a sulfur or sugar, produces mixtures with ignition temperatures below 200°C. A high melting oxidizer, such as iron oxide, combined with a fuel such as aluminum that does not melt until 600°C or more, results in significantly higher ignition temperatures.

The addition of sulfur or an organic compound to pyrotechnic mixtures normally improves the ignitability by providing a readily ignitable fuel. A retardation of the burn rate may also occur, however, so a minimum percentage of sensitizer should be used. The sensitivity of the mixture to spark, friction, and impact may also be affected by the addition of a second fuel. If sulfur is used, particular care must be exercised in specifying only ground and sieved crystalline material (flour of sulfur) and not sublimed sulfur (flowers of sulfur), as commonly employed in agriculture. The latter may be contaminated with sulfurous and sulfuric acids, which can make some pyrotechnic mixtures unsafe.

Once ignition is achieved, the reaction continues through the remaining composition. This propagation rate, usually referred to as the burn rate, is determined by a complex combination of chemical and physical factors. The burn rate, to a significant extent, is determined by the selection of oxidizer and fuel and the weight ratio of the oxidizer to fuel. The stoichiometric mixture, or that weight ratio corresponding to complete reaction between oxidizer and fuel, generates the maximum heat output, in kJ/g, but it is not always the fastest burning composition. Other factors affecting the burn rate include the particle sizes of the oxidizer and fuel, with finer particles usually yielding the fastest burning mixtures; the extent of homogeneity of the blended pyrotechnic composition, with burn rate increasing as the intimacy of the mixing increases; and the thermal conductivity of the mixture, ie, compositions with high thermal conductivity tend to burn faster. Metals are particularly good thermal conductors, so raising the percentage of metal fuel in a mixture, even well beyond the stoichiometric point, usually speeds up the burn rate of a composition. External temperature, ambient pressure, the extent of residual moisture in the mixture, the nature of the container holding the pyrotechnic composition, the porosity of the composition (determined by the consolidation pressure used to load the material into a tube or other container), and the surface area of the burning composition all affect the rate of burning of a pyrotechnic composition. Because of the large number of factors that affect burn rate, attempts to develop a burn rate equation for pyrotechnic compositions have not met with great success.

Many pyrotechnic reactions develop a gas-phase reaction component as they burn. A flame plume is present above the burning reaction surface as the reaction continues in the vapor phase, and the heat return from this vapor-phase reaction is another critical component of the burn rate of a pyrotechnic mixture. External pressure plays a significant role in determining the burn rate when a vapor-phase reaction component is present. High pressure tends to hold the hot gases near the burning surface, increasing the burn rate. This is the principle behind confining propellants in a chamber as they burn, leading to a high pressure and high burn rate for the propellant. In the open, propellants often are quite calm in burning performance. Not all pyrotechnic reactions, however, depend on a gas phase. Certain reactions, such as between iron powder and barium peroxide, are solid-phase reactions (3).

Light output is achieved in a pyrotechnic mixture by a combination of high flame temperature and the presence in the flame of solid or liquid particles that are heated to incandescence. A flame temperature approaching 3000°C is required for the emission of bright white light. Because the intensity of light emission from an incandescent object is proportional to temperature to the fourth power, a slight increase in flame temperature can result in a significantly brighter flame (13). Particles of magnesium, aluminum, titanium, or zirconium metal in a pyrotechnic composition cause high flame temperatures to be achieved because they are active fuels, and their solid oxidation products, such as MgO or Al_2O_3, incandesce and provide bright light output. Sparks occur if large fuel particles such as aluminum or charcoal are present in a pyrotechnic mixture. These fuel particles continue to burn in air and incandesce as they exit the pyrotechnic flame, leaving a spark-like trail.

Color and Sound Production

Emission of a specific color can be achieved by the presence of a specific atomic or molecular species in the vapor state in the pyrotechnic flame. The atom or molecule undergoes electronic excitation due to the elevated flame temperature, and the atom or molecule subsequently returns to its ground electronic state with the emission of a photon of light of specific wavelength. If this specific emission happens to fall in the visible region of the electromagnetic spectrum, color is perceived by onlookers. The production of color by pyrotechnic means requires the pyrotechnic composition to generate the color-emitting species as a reaction product.

Red flame color, for example, is observed if a species in the flame emits light that falls in the 650–780 nm region; visible light covers the region from ~400–780 nm. The element strontium is usually employed to produce a red pyrotechnic flame, and the primary red-emitting species is $SrCl$, produced in the vapor state at temperatures in the 1500°C range by a combination of Sr and Cl atoms. Strontium nitrate or strontium carbonate are the usual sources of Sr. Other color emitters include $BaCl$ (green, from $Ba(NO_3)_2$), $CuCl$ (blue), and atomic sodium, Na (yellow-orange) from sodium oxalate or sodium aluminum fluoride, Na_3AlF_6 (cryolite) (5,14). Copper carbonate, copper oxide, or copper oxychloride are the usual sources of blue color. A violet hue can be achieved by a combination of $SrCl$ (red) and $CuCl$ (blue) in the pyrotechnic flame. Aluminum, magnesium, or titanium particles produce white sparks. Charcoal particles are used to give gold sparks.

Another unique pyrotechnic effect, a shrill whistle, can be achieved if certain pyrotechnic mixtures are pressed into narrow-diameter tubes and ignited. The escaping gas pulsing out of the tube creates the whistling phenomenon. The mixtures usually contain potassium perchlorate and a salt of an organic acid, such as sodium salicylate or potassium benzoate, all pressed into a narrow cardboard tube (5). The whistling effect is right on the verge of an explosion, and these compositions are quite dangerous to prepare and load into tubes.

Civilian Pyrotechnics

Fireworks are used around the world to celebrate special occasions, and the appeal of lighting up the sky with bright colors, flashes of light, and loud booms appears to be universal. Fireworks have been used for many centuries, and their origin traces back to the discovery of black powder in China or India over 1000 years ago. Black powder was used to make firecrackers and sky rockets in China, and the technology made its way to Europe where a flourishing fireworks industry then developed. The United States adopted fireworks as the traditional means of celebrating Independence Day on the 4th of July, and the tradition has continued for over 200 years. In the United States, fireworks consumption in 1994 was approximately 45,000 t with a retail sales value of approximately $300,000,000. A majority of these fireworks were imported from China, the world's principal fireworks producer. The U.S. fireworks consumption is approximately 67% consumer fireworks and 33% public display fireworks. Many other countries also have a traditional fireworks day, such as Bastille Day (July 14) in France, Guy Fawkes Day (November 5) in Great Britain, Halloween (October 31) in some parts of Canada, New Year's Eve in Germany, and Chinese New Year throughout the world in Chinese communities.

Black powder remains a key component of fireworks and research into the energetics of black powder continues to be performed (10,11). It serves as the composition for fireworks fuses; as the propelling charge in sky rockets, Roman candles, and aerial shells; and as the bursting charge to explode aerial shells high in the air. Modern public displays are largely based on the use of mortar-fired aerial shells which explode into beautiful patterns of color high in the sky. The assembly of various types of fireworks has been described (4,6).

Types of Fireworks. Fireworks are conveniently classified as consumer or display.

Consumer Fireworks. An assortment of small fireworks devices are permitted for use by private citizens in many areas in the United States and elsewhere throughout the world. These devices consist of items such as wire sparklers, fountains, Roman candles, sky rockets, mines, and small aerial shells.

Wire sparklers are wires coated with pyrotechnic composition which are hand-held and produce a gentle spray of gold sparks from iron filings. Fountains are cardboard tubes filled with chemical mixtures that produce a spray of color and sparks extending 2–5 m into the air. Roman candles are cylindrical tubes which repeatedly fire colored stars distances of 5–20 m into the air. These items typically contain 5–12 stars.

Sky rockets are tubes with a stick attached for guidance and stability and which contain a pressed black powder propellant. They rise high into the air when ignited and a burst of color or a report, an audible bang, is normally produced in the air. Alternatively, rockets may contain a different propellant that fires the device into the air with a simultaneous whistling effect. Mines are aerial items that use a black powder propelling charge to fire a burst of colored stars, whistles, or firecrackers into the air from a cardboard tube. A barrage of color and noise results. Small aerial shells are plastic or cardboard spheres that are similar in effect to the type used by professional operators at large displays. These devices are launched from a mortar tube by a black powder propelling charge. A time delay fuse burns as the sphere or canister tumbles into the air, and a burst of color is produced high in the air when a black powder bursting charge explodes, breaking the sphere open and igniting the stars that are mixed with the black powder.

Display Fireworks. Larger versions of the devices sold as consumer fireworks are used for public fireworks displays. The principal item used in fireworks displays is the aerial shell, a sphere or cylinder typically 8–20 cm (3–8 in.) in diameter that is launched several hundred meters into the air (~100–300 m), from a mortar tube, by a propelling charge of black powder. A time fuse burns as the device climbs into the air, and a bursting charge explodes the device high in the air, lighting a shower of stars to produce a spectacular visual effect (15). Alternatively, the shell may contain an explosive charge of salute powder, which produces a loud explosion in the air. Displays also make use of large Roman candles and fountains, which can produce a waterfall-type effect if suspended upside down. Lances are small, cigarette-sized tubes that burn with specific flame colors. Pyrotechnicians attach hundreds of lances to wooden frames to create fire pictures, called set pieces. The pictures can resemble objects like the Liberty Bell or a famous person's face, or spell out messages such as "Good Night."

Theatrical Pyrotechnics or Special Effects. Many spectacular visual and audible effects are produced for stage presentations of both music and drama, and many motion pictures and television shows incorporate pyrotechnic and explosive special effects to liven up the presentation. These spectacular effects are a combination of pyrotechnics, explosives, combustion, and electronics. After the effects are filmed or videotaped, they are often enhanced by slow-motion replay and by the addition of more exciting noise. A real explosion is over in milliseconds, and hence there is a need for electronic enhancement to create a more spectacular effect on the screen.

Bullet effects or bullet hits are electrically initiated devices that simulate bullet impact through the use of various amounts of sensitive explosive composition. These explosive devices can be buried in the ground, attached to walls, or even attached with protective padding to human bodies. The explosions are triggered by firing an electric igniter, and a bullet appears to strike the intended target. Many pyrotechnic special effects resemble fireworks in much of their chemistry and intended effect. However, because the audience is often much closer to special effect devices than they would be to a fireworks display, greater control needs to be taken in manufacturing special effects to ensure that the height of the flame and the maximum distance that sparks fly are closely controlled. These effects must be set up and fired by trained professionals with extensive experience.

Model Rockets and Missiles. Model rockets are another type of pyrotechnic device that have been quite popular since the appearance of the Russian satellite Sputnik in the 1950s. These items are most often sold as kits with modular, preassembled, solid propellant engines. The engines have traditionally been constructed of nonmetallic casings, clay nozzles, and a compressed black powder charge, which is ignited by an electric match. The actual burn time of the propellant is generally less than a second, and the rocket can be propelled quite a distance into the air ($\sim$200 m). Parachutes are often deployed at the height of flight, and the rocket body is recovered and reused. The concept of prepackaging model rocket motors provided a significant element of safety to the hobby of model rocketry, because it eliminated the need for individuals to blend and load their own propellant mixtures, a potentially dangerous operation.

Model rocket engines have appeared in the 1990s that use solid rocket propellant composition similar to the higher technology propellants used in military rockets and missiles. These propellants use ammonium perchlorate in combination with a rubber-like binder. They produce a higher percentage of gaseous products than does black powder, and the use of these new engines has helped advance the hobby of model rocketry into a mature science (16). Many rocket enthusiasts are adults who began firing black powder rockets as children and have continued to pursue rocketry as their avocation into adulthood.

Other Civilian Devices. Pyrotechnic devices also serve an assortment of civilian uses, primarily for signaling. Highway flares, often known as fusees, are made chiefly of strontium nitrate mixed with sawdust, wax, sulfur, and potassium perchlorate and contained in a waterproof cardboard tube. A small quantity of a safety-match composition also is incorporated and, upon ignition by friction, the device burns for up to 30 minutes with a distinctive red flame. Other hand-held and aerial devices are used for marine distress signals, and some type of distress signal must be carried on boats in intercoastal waterways under U.S. Coast Guard regulations. Noisemaking pyrotechnic devices are also used to a limited extent in agriculture to discourage birds and other pests from frequenting fruit orchards and freshly planted fields.

The air bag industry has become one of the principal users of pyrotechnic compositions in the world. Most of the current air bag systems are based on the thermal decomposition of sodium azide, NaN_3, to rapidly generate a large volume of nitrogen gas, N_2. Air bag systems must function immediately (within 50 ms) upon impact, and must quickly deploy a pulse of reasonably cool, nontoxic, unreactive gas to inflate the protective cushion for the driver or passenger. These formulations incorporate an oxidizer such as iron oxide to convert the atomic sodium that initially forms into sodium oxide, Na_2O. Equation 1 represents the reaction.

$$6\ NaN_3 + Fe_2O_3 \longrightarrow 3\ Na_2O + 2\ Fe + 9\ N_2 \tag{1}$$

The actual formulations used by the various companies engaged in the highly competitive air bag business are proprietary, and contain numerous additives to modify the actual performance of the air bag compositions. Many more changes are likely in this new industry.

Regulations. The manufacture, transportation, storage, and use of pyrotechnics are all very highly regulated fields, and the volume and scope of the regulations covering the industry, both military and civilian, has increased enormously since the 1980s. The transportation of all commercial pyrotechnic articles, for both consumer and military use, is under the control of the U.S. Department of Transportation (DOT). All pyrotechnic devices must be officially approved for transportation by DOT, and this approval process involves testing to determine the appropriate classification of the device, as well as its suitability for transportation from a sensitivity point of view. Pyrotechnic devices are classed as "Explosives" for transportation purposes (17).

The U.S. Treasury Department's Bureau of Alcohol, Tobacco and Firearms (BATF) regulates the licensing (qv) and storage of all civilian explosives, including pyrotechnic devices. Consumer fireworks are exempt from most BATF regulations, because they are subject to regulations of the U.S. Consumer Product Safety Commission (CPSC). The CPSC rules cover the labeling, construction, performance, and explosive content of fireworks intended for sale to consumers (18). States and local governments may enact stricter regulations governing consumer fireworks than those adopted by CPSC. The U.S. Occupational Safety and Health Administration (OSHA) regulates the manufacturing of pyrotechnics from an employee safety point of view. Additional U.S. Department of Defense regulations cover the manufacturing of military pyrotechnics.

Military Pyrotechnics

Pyrotechnics have been used for military purposes for many centuries as propellants, explosive charges, time fuses, and for illumination. There are still many uses of pyrotechnic devices in military applications, where they provide portability, storage stability, simplicity of operation, safety, and the reliability required for military scenarios. The devices must be capable of surviving rough handling, weather extremes, and extended storage, yet reliably perform when called on to function.

Pyrotechnic devices are used for light generation in flares, tracers, and flash cartridges; for smoke generation for signaling and obscuration; for heat production for time delay components, incendiary applications, and ignition; and for gas-generation applications. The propellants used in many military devices are also essentially pyrotechnic compositions in nature, containing an oxidizer and fuel designed to burn at a high rate with the generation of a large quantity of hot gas to produce significant thrust. Military pyrotechnics are required to meet rigorous standards of performance, reliability, and storage lifetime, and the materials of construction are generally substantially more sturdy than those used for civilian pyrotechnics.

Many aspects of the performance of military pyrotechnics are measured and analyzed by modern instrumental techniques such as spectrophotometers for light intensity studies, replacing the qualitative, visual evaluations that were formerly used for acceptance of these devices. Military pyrotechnics are also experiencing a shift away from a concentration on the visible region of the electromagnetic spectrum, and are moving into the generation and obscuration of the infrared and microwave/millimeter regions, as modern techniques such as heat-seeking missiles, thermal imaging systems, and night vision equipment have dramatically altered the battlefield scenario. Pyrotechnic devices, such as screening smoke units, have had to adjust with these changes in technology to remain viable and effective. There will undoubtedly be more changes in pyrotechnics as modern warfare becomes more and more instrumental rather than visual in nature.

Military illuminating flares have been based for many years on the energetic reaction between sodium nitrate and magnesium metal. One of the primary reactions is equation 2. This high candlepower composition is blended with an

$$2\ NaNO_3 + 5\ Mg \longrightarrow Na_2O + 5\ MgO + N_2 \qquad (2)$$

organic binder to produce a mixture with good mechanical strength, and the mixture is then consolidated into a tube to produce a long-burning effect rather than a flash of light. A typical composition contains 53–58 wt % magnesium powder, 36–40 wt % sodium nitrate, and 4–8 wt % binder, such as a polyester. Thermodynamic data for flare reactions are given in Reference 13. The intense light emission is a combination of atomic emission from sodium atoms and incandescence from solid particles, such as magnesium oxide, present in the flame. The mixtures are usually prepared to be quite fuel (magnesium) rich in formulation. The excess magnesium, for which there is no sodium nitrate with which to react, vaporizes in the high temperature flame (>3000°C) and vigorously reacts with atmospheric oxygen at the edge of the flame, yielding additional energy and candlepower. The relatively low boiling point (~1100°C) of magnesium metal allows this process to occur efficiently.

Military flare technology has concentrated in recent years on the development and production of decoy flares to protect aircraft against heat-seeking missiles. These flares, which emit a considerable amount of infrared radiation, use a composition based on magnesium metal and polytetrafluoroethylene (PTFE), which is perhaps better known by one of its trade names, Teflon. Here, Teflon serves as the oxidizer, releasing fluorine atoms that energetically combine with magnesium to form magnesium fluoride. These flares ignite and undergo the reaction shown as equation 3. This reaction is quite exothermic and generates hot carbon particles which emit considerable infrared radiant energy. The flame signature resembles that of a jet engine.

$$2n\ Mg + (C_2F_4)_n \longrightarrow 2n\ MgF_2 + 2n\ C \qquad (3)$$

The intensity of a flare is largely determined by its flame temperature, which depends on the stability of the reaction products. The higher the decomposition temperature of the oxide, the higher the flame temperature. Hydrogen, carbon, boron, silicon, and phosphorus form oxides that dissociate at comparatively low temperatures. A flare temperature greater than 3000 K is required to generate gray-body radiation, which is optimum for the spectral sensitivity of the human eye. Metallic fuels such as magnesium, aluminum, zirconium, and titanium are necessary to achieve these high temperatures.

Military signal flares are designed to burn with clearly distinguishable colors, and the chemistry of these compositions is quite similar to that used for color effects in civilian pyrotechnics (14). They normally contain magnesium metal as a fuel to achieve high candlepower output. Standard colors are red, green, and yellow (Table 2). Photoflash bombs are explosive devices that are

Table 2. Color Flare Compositions[a]

Ingredient	Wt %
Red	
magnesium	24.4
potassium perchlorate	20.5
strontium nitrate	34.7
poly(vinyl chloride)	11.4
asphaltum	9.0
Green	
magnesium	21.0
potassium perchlorate	32.5
barium nitrate	22.5
poly(vinyl chloride)	12.0
copper powder	7.0
binder	5.0
Yellow	
magnesium	30.3
potassium perchlorate	21.0
barium nitrate	20.0
sodium oxalate	19.8
asphaltum	3.9
binder	5.0

[a]Ref. 14.

used in night photography and that, for time intervals as short as 0.1 s, provide intense light output. Flash charges are loosely packed powdered mixtures of aluminum powder with potassium perchlorate and barium nitrate, and initiation is by an explosive charge. Flash compositions are among the most hazardous to manufacture and are processed by remote control in heavy-walled bays. With advances in the technology of low light photography, the need for these devices is diminishing. These flash mixtures are also useful for simulating explosions when packaged into smaller units as projectile ground-burst simulators for training purposes.

Smoke-Generating Devices. Smoke generators are used by the military for daytime obscuration and signaling. For field use where portable stable systems are required, pyrotechnic devices are often employed. The primary composition since the 1940s has been HC smoke, which generates a cloud of zinc chloride, $ZnCl_2$, smoke by a series of reactions between hexachloroethane, $C_2Cl_6(HC)$, zinc oxide, and aluminum (3) (eq. 4–6). The zinc regenerated in

$$3\,ZnO + 2\,Al \longrightarrow 3\,Zn + Al_2O_3 \tag{4}$$

$$3\,Zn + C_2Cl_6 \longrightarrow 3\,ZnCl_2 + 2\,C \tag{5}$$

$$C + ZnO \longrightarrow Zn + CO \tag{6}$$

equation 6 then reacts with C_2Cl_6, etc. This mixture is used in smoke grenades and large smoke pots, and it produces a dense, gray-white cloud that is largely zinc chloride. Concerns about the toxicity of the smoke from HC mixtures has led to efforts to find a replacement, but it has proven to be a difficult challenge to find a composition that can equal the obscuring efficiency of the HC mixture.

Signaling smokes may be white or colored. White signal smoke has traditionally been derived from red phosphorus or an HC mixture. New smokes based on the sublimation of terephthalic acid (TA) have been developed by the U.S. military to replace HC smoke for training purposes. Both red phosphorus and HC smokes generate high temperatures and significant flames during their operation, and neither of these are desirable in a training situation; the TA smokes are far better because they generate little if any external flame and function at much lower reaction temperatures.

Grenades and smoke marker devices that produce highly visible plumes of brightly colored smoke are used by the military for signaling and marking purposes. The production of a distinct color is critical in these devices, and the color is produced by the sublimation of an organic dye which, because of its low sublimation temperature, evaporates and recondenses in air as a brilliantly colored smoke cloud. The mixtures must be cool-burning to minimize decomposition of the dye, and are composed of potassium chlorate with sulfur or sugar (qv) as the fuel. They also contain sodium bicarbonate or magnesium carbonate to act as an acid neutralizer, generate carbon dioxide gas to assist in expelling the dye from the combustion zone, and slow the burn rate (5). Some of the organic dyes formerly used in these devices have been replaced due to carcinogenicity concerns (see DYES, NATURAL). Standard military smokes are red, violet, green, and yellow. Orange smoke is widely used in the civilian pyrotechnics industry for marine distress signaling.

Tracer Munitions.   Tracer bullets guide the direction of the fire, aid in range estimation, mark target impact, and act as incendiaries. Tracers can, through preselected tracer colors, also serve for nighttime identification of the combatants. Red strontium-containing tracers are more visible under adverse atmospheric conditions, therefore these are preferred although green tracers based on barium salts also are used. Daylight visibility can be enhanced by increasing the fraction of magnesium in the tracer composition. With the increased use of night vision systems for pilots and ground artillery personnel, interest in tracers with reduced light output (dim tracers) has increased; such devices are readily observable with night vision equipment but almost invisible to others. Conversely, the light output would not overwhelm night vision systems the way standard, high intensity tracer output can.

Incendiary Devices.   Incendiary devices are used to initiate destructive fires in a variety of targets. Small-arms incendiaries are used primarily for starting fires in aircraft fuels. Whereas they are highly effective against subsonic aircraft, such as helicopters, the problem of defeating supersonic aircraft by incendiary action alone is more difficult owing to the high flash point of jet fuels. Most small-arms incendiary compositions consist of a metallic fuel such as a magnesium–aluminum alloy, and an oxidizer such as barium nitrate. The use of pyrophoric fragment generators such as zirconium, titanium, or depleted uranium as incendiary components of projectiles has been the subject of intensive study (19). The reason some metals ignite and burn progressively when fractured is uncertain. Generally, these metals have higher densities than do their oxides, so that the oxides flake from the metal during combustion, thereby exposing fresh metallic surface to oxygen attack.

Special Gas Generators. A special category of pyrotechnic system with military and aerospace applications is the cartridge-actuated device (CAD) which is also called the propellant-actuated device (PAD) (20). The CAD produces mechanical movement for closing electrical switches, opening or closing valves, cutting cables, or the performance of other mechanical work. They usually contain a small quantity of a gas-generating mixture which is ignited electrically. Many of these devices are used in aerospace systems.

Ignition

Pyrotechnic devices are initiated by some type of external energy input. This can range from a match lighting a piece of black powder-containing fuse to a battery sending a surge of current through a circuit, creating a hot spot on a narrow, high resistance section of wire that is coated with a thermally sensitive material (see BATTERIES). Alternatively, impact can be used to ignite a primer, as in a shotgun shell, or the friction generated by rubbing two surfaces together can create a hot spot; a highway flare uses this technique. Lasers (qv), high intensity beams of light, have been investigated more and more as another potentially reliable way to ignite pyrotechnic devices. When very rapid ignition is required, such as propellant in a gun, the method of ignition is usually not pyrotechnic but relies instead on a sensitive explosive mixture. The choice of the ignition method depends on the permissible ignition delay. Igniters containing primary explosives function in microseconds, whereas purely pyrotechnic igniters require milliseconds.

Within a given pyrotechnic device there may be several components such as a propelling charge, a delay column, and the main pyrotechnic effect (perhaps a flare). Each of these components must sequentially undergo ignition by the output from the preceding component in the pyrotechnic train. To ensure reliable ignition transfer, a family of pyrotechnic compositions known as ignition compositions, primes, or first fires has been developed. A small quantity of ignition composition is coated, pressed, or otherwise held in contact with the material to be ignited. The ignition mixture is readily ignited by the energy input from the fuse, delay column, or propellant and then transfers its energy to the next composition, assuring a reliable device. Ignition compositions tend to be relatively slow burning rather than explosive, and tend to generate hot particles as reaction products. These particles spray the next pyrotechnic composition with high temperature particulate matter and produce smooth, reliable ignition of more-difficult-to-light compositions.

Igniter compositions are formulated for specific desired response times, ignition sensitivities, and gas output (21). These mixtures are often rather sensitive to ignition, and require caution in their manufacturing. Examples of igniter materials are mixtures of aluminum, boron, magnesium, silicon, titanium, or zirconium with ammonium or potassium perchlorate, barium or potassium nitrate, barium or lead chromate, and cupric or lead oxides. Representative igniter compositions are shown in Table 3. A traditional igniter is black powder, which is

Table 3. Representative Ignition Compositions

Composition	Ingredients	Wt %
MTV	magnesium	54.0
	Teflon	30.0
	Viton A[a]	16.0
A1A	zirconium	65.0
	iron(III) oxide	25.0
	diatomaceous earth	10.0
B/KNO$_3$	potassium nitrate	70.7
	boron	23.7
	polyester[a]	5.6
black powder	potassium nitrate	75.0
	charcoal	15.0
	sulfur	10.0

[a]Binder.

an efficient flame carrier because of the high fraction of hot solid particles in its combustion products.

Delay Elements. Compositions and characteristics of some delay compositions are listed in Table 4. Many of these contain lead, barium, and chromium compounds, and there is a need for new delay compositions that are more environmentally compatible yet have the excellent reliability and performance properties of the current compositions. In some pyrotechnic devices, a precise time delay is required between the operating stages of the device. An example is the several-second delay designed into hand grenades to provide a delay between the release of the safety pin and the functioning of the grenade. Delay elements are self-contained pyrotechnic devices consisting of an initiator, a pressed column of pyrotechnic composition, and an output charge. Delays are either obturated or vented. An obturated element contains all of the reaction products and, therefore,

Table 4. Standard Delay Compositions[a]

Component	Composition	Wt %
lead tetroxide	Pb$_3$O$_4$	85
silicon	Si	15
binder	[b]	additional 2%
barium chromate	BaCrO$_4$	90
boron	B	10
barium chromate	BaCrO$_4$	40
potassium perchlorate	KClO$_4$	10
tungsten	W	50
lead chromate	PbCrO$_4$	37
barium chromate	BaCrO$_4$	30
manganese	Mn	33

[a]Ref. 5.
[b]Nitrocellulose.

it is not affected by ambient pressure. Obturation is an aid in protecting other components of the hardware from the effects of unwanted smoke and debris, but it also tends to increase the burn rate of the composition. Vented delays are used if large amounts of gaseous products must be disposed of and if the device is used where the effect of changes in ambient pressure can be ignored. Delay elements based on black powder are necessarily vented and their burn rates are affected by the ambient pressure.

Safety Concerns

Safety concerns permeate all aspects of pyrotechnics. Although pyrotechnic materials have less of an explosive output than the energetic materials used for military and civilian blasting applications, the sensitivity of pyrotechnics to ignition from flame or spark is considerably greater than for many of the high explosives. Hence, great caution is needed in the manufacture of pyrotechnic mixtures to avoid possible ignition sources.

Wherever possible, operations involving the mixing and processing of pyrotechnic compositions should be performed remotely and monitored by video camera. Where this is not possible, the protection of personnel should include shielding for the eyes, face, and hands as well as the use of antistatic clothing. Rubber and plastic gloves may present a hazard when used for handling flammable materials because they may melt and stick to damaged skin. Suede leather gloves are preferable as they are wettable and washable (2).

Because pyrotechnic compositions contain their own oxygen, suffocation methods of fire fighting are not effective. Instead, the temperature of the burning material must be brought down below the ignition point of the material. A water deluge is perhaps the best way to accomplish this, as evaporating water is a very effective heat remover. However, this method must be used with great care if burning metal fuels such as magnesium are involved in the fire. An insufficient amount of water actually contributes to the violence of the fire by reacting with the burning metal to generate hydrogen gas, a violently reactive material. Chlorinated hydrocarbons should also be used only with great caution to fight pyrotechnic fires, because metal fuels can also react with those liquids.

Many pyrotechnic mixtures potentially can be ignited by electrostatic discharge during the manufacturing process. Loose pyrotechnic composition is considerably more sensitive to static than is consolidated material. The risk of electrostatic discharge can be minimized by grounding and bonding materials, containers, and personnel. Pyrotechnic processing is generally performed in a humidity range from 45–60%. These values allow a conductive moisture layer to form on work surfaces, which can effectively dissipate static charge accumulations without being too humid such that the component chemicals pick up excessive amounts of water from the atmosphere.

As an additional safety precaution, water must never be used in the blending of mixtures containing zinc or magnesium, nor should it be used with titanium or zirconium powder unless water is present in excess. Otherwise an exothermic reaction between the metal and water might occur. The danger of a dust explosion is present when fine particle size fuels are processed. Of particular concern is the handling of zirconium powder (<10 μm dia), which may be pyrophoric in air. Specific standard operating procedures must be established for each pyrotechnic operation, and personnel must be instructed in the importance of strictly following these procedures. In designing a pyrotechnic facility, it is advisable to assume that an ignition can occur at each stage of the manufacturing process at some point during the lifetime of the plant. The facility should be designed to minimize the consequences to both personnel and the physical plant if an ignition does occur. Separation of plant areas and the minimization of in-process material in each area are two key points of plant safety (qv) (3). All finished devices and all composition awaiting further stages of the manufacturing process must be held in storage magazines where no other manufacturing activities are performed. It is highly unlikely that an incident will begin in a storage magazine, since no work is being performed on the stored materials. Hence, separating the stored materials from work-in-progress helps to minimize the consequences of any event that might occur in manufacturing.

BIBLIOGRAPHY

"Pyrotechnics" in *ECT* 1st ed., Vol. 11, "Military Pyrotechnics," pp. 322–332, by D. Hart, P. Arsenal; "Commercial Pyrotechnics," pp. 332–338, V. C. Allison, Aerial Products, Inc.; in *ECT* 2nd ed., Vol. 16, pp. 824–840, by H. Ellern, Pyrotechnic Consultant (formerly UMC Industries, Inc.); in *ECT* 3rd ed., Vol. 19, pp. 484–499, by A. P. Hardt, Lockheed Missiles and Space Co., Inc.

1. A. A. Shidlovski, *Principles of Pyrotechnics*, Publ. No. AD AOOl 859, Mashinostroyeniye Press, Moscow, Russia, 1973; trans. 1974.
2. H. Ellern, *Military and Civilian Pyrotechnics*, The Chemical Publishing Co., New York, 1968.
3. J. H. McLain, *Pyrotechnics*, The Franklin Institute Press, Philadelphia, Pa., 1980.
4. T. Shimizu, *Fireworks, the Art, Science, and Technique*, Maruzen Co., Ltd., Tokyo, Japan, 1981.
5. J. Conkling, *The Chemistry of Pyrotechnics*, Marcel Dekker, Inc., New York, 1985.
6. R. Lancaster, *Fireworks: Principles and Practice*, 2nd ed., Chemical Publishing Co., New York, 1992.
7. *Handbook of Chemistry and Physics*, 74th ed., CRC Publishing Co., Boca Raton, Fla., 1993.
8. M. W. Chase, Jr., and co-workers,, *JANAF Thermochemical Tables*, 3rd ed., NSRDS-NBS 37, Catalogue No. C 13.48.37, American Chemical Society, Washington, D.C., 1996.
9. D. D. Wagman and co-workers, *Selected Values of Chemical Thermodynamic Properties*, Bureau of Standards Technical Notes 270-3 and 270-4, U.S. Department of Commerce, Springfield, Va., 1968, 1969.
10. M. E. Brown and R. A. Rugunanan, *Propell. Explos. Pyrotech.* **14**, 69 (1989).
11. G. Hussain and G. J. Rees, *Propell. Explos. Pyrotech.* **15**, 43 (1990).
12. F. J. Zeleznik and S. Gordon, *Ind. Eng. Chem.* **60**(6), 25 (June 1968); S. Gordon and B. J. McBride, *Computer Program for the Calculation of Complex Chemical Equilibrium Compositions, Rocket Performance, Incident and Reflected Shocks and Chapman-Jouguet Detonations*, NASA SP-273, Lewis Research Center, NTIS N-71-37775, U.S. Department of Commerce, Springfield, Va., 1971.
13. *Engineering Design Handbook, Military Pyrotechnics Series*, Part One, *Theory and Application*, AMCP 706-185, U.S. Army Materiel Command, Washington, D.C., Apr. 1967, pp. 6–38.

14. H. A. Webster III, *Propell. Explos. Pyrotech.* **10**, 1 (1985).
15. J. A. Conkling, *Chem. Eng. News*, 24 (June 29, 1981).
16. G. H. Stine, *Handbook of Model Rocketry*, 6th ed., John Wiley & Sons, Inc., New York, 1994.
17. *Code of Federal Regulations*, Title 49, Parts 171–173, U.S. Dept. of Transportation, Washington, D.C., 1995.
18. *Code of Federal Regulations*, Title 16, Parts 1500 and 1507, U.S. Consumer Products Safety Commission, Washington, D.C., 1995.
19. W. W. Hillstrom, *Formation of Pyrophoric Fragments*, BRL MR 2306, AD 765447, Ballistics Research Laboratory, Aberdeen, Md., 1973.
20. S. M. Kaye, ed., *Encyclopedia of Explosives and Related Items*, Vol. 7, ADA 019502, Vol. 8, ADA 057762, Vol. 9, ADA 097595, and Vol. 10, U.S. Army Research and Development Command, Dover, N.J., 1960–1982.
21. W. E. Robertson, *Igniter Materials Handbook*, TR 690, AD 51001, Naval Ammunition Depot, Crane, Ind., Sept. 1969.

JOHN A. CONKLING
American Pyrotechnic Association

PYRROLE AND PYRROLE DERIVATIVES

Pyrrole [*109-97-7*], a five-membered, heterocyclic system, is a fundamental structural subunit of many of the most important biological molecules, eg, heme, the chlorophylls, the bile pigments, some naturally occurring antibiotics, many alkaloids, and some enzymes. Pyrrole was first obtained in 1834 from the destructive distillation of coal or bone (1). It was characterized in 1858, and its composition was determined in 1870 (2,3). Early interest in the chemistry of pyrrole began with the discovery that indole (benzopyrrole) is the fundamental nucleus of indigo. Ring positions in pyrrole (**1**) are designated by number or Greek letter.

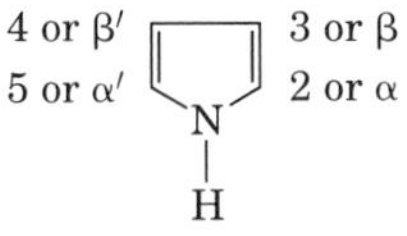

4 or β′   3 or β
5 or α′   N   2 or α
       |
      H

(**1**)

Physical Properties

Pyrrole is a colorless, slightly hygroscopic liquid which, if fresh, emits an odor like that of chloroform. However, it darkens on exposure to air and eventually produces a dark brown resin. It can be preserved by excluding air from the storage container, preferably by displacement with ammonia to prevent acid-catalyzed polymerization. A review of the physical and theoretical aspects of pyrrole is found in Reference 4. Some physical properties of pyrrole are listed in Table 1.

Pyrrole has a planar, pentagonal (C_{2v}) structure and is aromatic in that it has a sextet of electrons. It is isoelectronic with the cyclopentadienyl anion. The π-electrons are delocalized throughout the ring system, thus pyrrole is best characterized as a resonance hybrid, with contributing structures (**1–5**). These structures explain its lack of basicity (which is less than that of pyridine), its

(**1**)   (**2**)   (**3**)   (**4**)   (**5**)

unexpectedly high acidity, and its pronounced aromatic character. The resonance energy which has been estimated at about 100 kJ/mol (23.9 kcal/mol) is intermediate between that of furan and thiophene, or about two-thirds that of benzene (5).

The contributions from the canonical forms have been calculated, and the contributions from the equivalent polar structures (**2**) and (**3**) dominate those of (**4**) and (**5**) (6). Thus, electrophilic substitution is predicted to occur in the α-position, and this is proven experimentally in most cases. Nitrosation and selenocyanation occur at the β-position (7,8).

Many of the physical characteristics of pyrrole indicate at least partial association. In particular, the boiling point is 98°C higher than that of furan. It has been postulated that various associated dimeric and higher structures occur because of hydrogen bonding (9,10).

Pyrrole is soluble in alcohol, benzene, and diethyl ether, but is only sparingly soluble in water and in aqueous alkalies. It dissolves with decomposition in dilute acids. Pyrroles with substituents in the β-position are usually less soluble

Table 1. Physical Properties of Pyrrole

Property	Value
melting point, °C	−23.4
boiling point, °C	129.8
critical temperature, °C	366
density, d_4^{20}, g/mL	0.970
refractive index, n_D^{20}	1.5085
dielectric constant at 20°C, ϵ	8.00
flash point, closed cup, °C	39

in polar solvents than the corresponding α-substituted pyrroles. Pyrroles that have no substituent on nitrogen readily lose a proton to form the resonance-stabilized pyrrolyl anion, and alkali metals react with it in liquid ammonia to form salts. However, pyrrole (pK_a = ca 17.5) is a weaker acid than methanol (11). The acidity of the pyrrole hydrogen is greatly increased by electron-withdrawing groups, eg, the pK_a of 2,5-dinitropyrrole [32602-96-3] is 3.6 (12,13).

The dipole moment varies according to the solvent; it is ca 5.14×10^{-30} C·m (ca 1.55 D) when pure and ca 6.0×10^{-30} C·m (ca 1.8 D) in a nonpolar solvent, such as benzene or cyclohexane (14,15). In solvents to which it can hydrogen bond, the dipole moment may be much higher. The dipole is directed toward the ring from a positive nitrogen atom, whereas the saturated nonaromatic analogue pyrrolidine [123-75-1] has a dipole moment of 5.24×10^{-30} C·m (1.57 D) and is oppositely directed. Pyrrole and its alkyl derivatives are π-electron rich and form colored charge-transfer complexes with acceptor molecules, eg, iodine and tetracyanoethylene (16).

Infrared spectra of pyrrole and its derivatives have been described in detail (13). The N–H absorption of nonassociated pyrroles varies predictably with the substituents and is related to the acidity of the pyrrole. It has been used as evidence for intermolecular association, which results from hydrogen bonding, between pyrrole units. Nuclear magnetic resonance studies have provided evidence for ring currents (17). The concentration and temperature dependence of the hydrogen chemical shifts have been used to estimate the enthalpy of self-association as −6.7 kJ/mol in pyrrole (18). The effect of substituents upon the various chemical shifts has also been reported (19). Pyrrole gives an intense molecular ion in its mass spectrum. Smaller fragments result primarily from ring-size reduction.

Syntheses of Pyrroles

Knorr Synthesis. Condensation of an α-aminoketone with a carbonyl compound was first reported by Knorr (20). This reaction and its modifications are among the most important and widely used methods for the synthesis of pyrroles.

Because the α-aminoketone is subject to self-condensation, the condensation with a β-dicarbonyl derivative (**6**) is usually carried out by generating the α-aminoketone *in situ* through reduction of an oximino derivative (**7**); zinc in glacial acetic acid is used as the reductant. For example, Knorr's pyrrole [2436-79-5] (**8**) forms from (**6**) and (**7**).

Modifications include the use of β-ketoaldehydes as acetals, eg (**9**), which leads to loss of the formyl group (**21**); the product in this example is 5-ethoxycarbonyl-2-methylpyrrole [*3284-51-3*].

The Knorr synthesis is not particularly sensitive to the nature of R and R‴, ie, they may be alkyl, acyl, aryl, or carbalkoxy without significantly affecting the yield. Similarly, good yields are obtained if R′ and R″ are acyl or carbalkoxy, but poor yields are obtained if they are alkyl or aryl.

Hantzsch and Feist Syntheses.   The Hantzsch synthesis of pyrroles involves condensation of an α-haloketone (**10**) with a β-keto ester (**6**) in the presence of ammonia or an amine (**22**).

The Feist synthesis is similar to the Hantzsch method and involves condensation of acyloins, eg (**11**), with aminocrotonic esters, eg (**12**), in the presence of zinc chloride (**23**).

(11)   (12)

Paal-Knorr Synthesis. The condensation of a 1,4-diketone, for example, with ammonia or a primary amine generally gives good yields of pyrroles; many syntheses have been reported (24). The lack of availability of the appropriate 1,4-diketone sometimes limits the usefulness of the reaction.

Other Methods. Newer methods for forming pyrrole and related hetero-cyclic rings include the formation of substituted pyrrole 2-carboxylate esters by condensation of β-dicarbonyl compounds with glycinate esters (25).

A Wittig-type reaction has been used to obtain N-protected 3-hydroxy-pyrroles, which exist as the pyrrolenone (**13**) tautomers (26).

Acetylenic compounds have often been used as precursors to certain pyrroles. Thus, 2-butyne-1,4-diol reacts with aniline in the presence of alumina to produce N-phenylpyrrole [635-90-5] (27).

$$HOCH_2C{\equiv}CCH_2OH + C_6H_5NH_2 \xrightarrow[300°C]{alumina}$$

Acetylenedicarboxylic esters also react with phenylhydroxylamines to give pyrroles (28), eg, N-phenylpyrrole-2,3,4,5-tetracarboxylic acid, tetramethyl ester [*37802-39-4*] (**14**).

$$2\ CH_3OCC{\equiv}CCOCH_3 + C_6H_5NHOH \longrightarrow$$

(**14**)

α-Aminoketones and acetylenic carbonyl compounds cyclize and dehydrate to give pyrroles in high yields by a Michael-type addition (29), eg, dimethyl 5-methyl-4-phenylpyrrole-2,3-dicarboxylate [*53252-73-6*] (**15**) in the following.

(**15**)

Pyrrolines and Pyrrolidines

The pyrrolines or dihydropyrroles can exist in three isomeric forms: 1-pyrroline (3,4-dihydro-2H-pyrrole [*5724-81-2*]) (**16**) is an unstable material that resinifies upon exposure to air; 2-pyrroline (2,3-dihydro-1H-pyrrole [*638-31-3*]) (**17**) is even more unstable; only 3-pyrroline (2,5-dihydro-1H-pyrrole [*109-96-6*]) (**18**) is reasonably stable. 3-Pyrroline boils at 91°C and has a density of 0.9097 g/cm^3 and a refractive index of 1.4664.

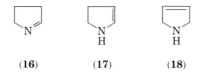

(**16**)   (**17**)   (**18**)

Pyrrolidine [*123-75-1*] (tetrahydropyrrole) (**19**) is a water-soluble strong base with the usual properties of a secondary amine. An important synthesis of pyrrolidines is the reaction of reduced furans with excess amine or ammonia over an alumina catalyst in the vapor phase at 400°C. However, if labile substituents are present on the tetrahydrofurans, pyrroles may form (30).

(19)

Pyrrolidines also can be obtained by reaction of 1,4-dihydroxyalkanes with amines in the presence of dehydrating agents at elevated temperatures or by reaction of primary amines with 1,4-dihaloalkanes. The dry distillation of 1,4-butanediamine dihydrochloride also generates pyrrolidine. Pyrroles can also be catalytically hydrogenated to pyrrolidines.

Reactions of Pyrroles

In keeping with its aromatic character, pyrrole is relatively difficult to hydrogenate, it does not ordinarily serve as a diene for Diels-Alder reactions, and does not undergo typical olefin reactions. Electrophilic substitutions are the most characteristic reactions, and pyrrole has often been compared to phenol or aniline in its reactivity. Acids strong enough to form salts with pyrrole destroy the aromaticity and cause polymerization.

N-Alkylpyrroles may be obtained by the Knorr synthesis or by the reaction of the pyrrolyl metallates, ie, Na, K, and Tl, with alkyl halides such as iodomethane, eg, 1-methylpyrrole [96-54-8]. Alkylation of pyrroles at the other

$$\text{pyrrole} + CH_3I \xrightarrow[\text{DMSO}]{\text{KOH}} \text{N-methylpyrrole} + HI$$

ring positions can be carried out under mild conditions with allylic or benzylic halides or under more stringent conditions (100–150°C) with CH_3I. However, unless most of the other ring positions are blocked, polyalkylation and polymerization tend to occur. N-Alkylation of pyrroles is favored by polar solvents and weakly coordinating cations (Na^+, K^+). More strongly coordinating cations (Li^+, Mg^{2+}) lead to more C-alkylation.

N-Acylation is readily carried out by reaction of the alkali metal salts with the appropriate acid chloride. C-Acylation of pyrroles carrying negative substituents occurs in the presence of Friedel-Crafts catalysts. Pyrrole and alkylpyrroles can be acylated noncatalytically with an acid chloride or an acid anhydride. The formation of trichloromethyl 2-pyrryl ketone [35302-72-8] (**20**, R′ = CCl₃) is a particularly useful procedure because the ketonic product can be readily converted to the corresponding pyrrolecarboxylic acid or ester by treatment with aqueous base or alcoholic base, respectively (31).

The most generally useful method for acylation or formylation of pyrroles is the Vilsmeier-Haack reaction (32,33). The pyrrole is treated with the phosphoryl complex of N,N-dialkylamide and the intermediate imine salt is hydrolyzed.

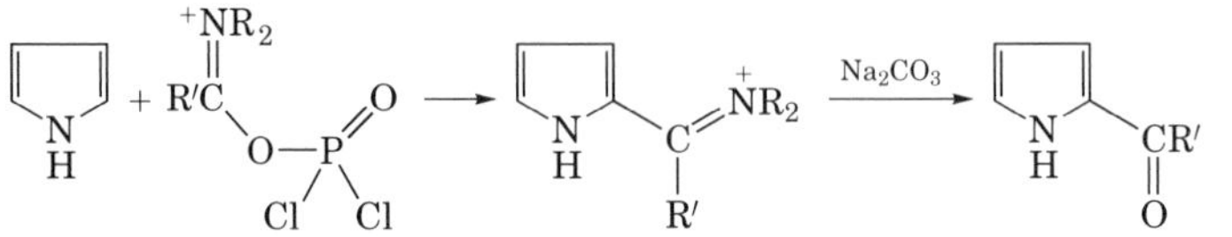

(20)

Nitration of pyrroles by the usual methods leads to extensive degradation. However, nitration can be achieved with an equimolar nitric acid–acetic anhydride mixture at low temperatures. In the case of pyrrole, the reaction leads predominantly to substitution at the β-position (34), ie, in the following: 51% 3-nitropyrrole [5930-94-9] (**21**) and 13% 2-nitropyrrole [5919-26-6] (**22**).

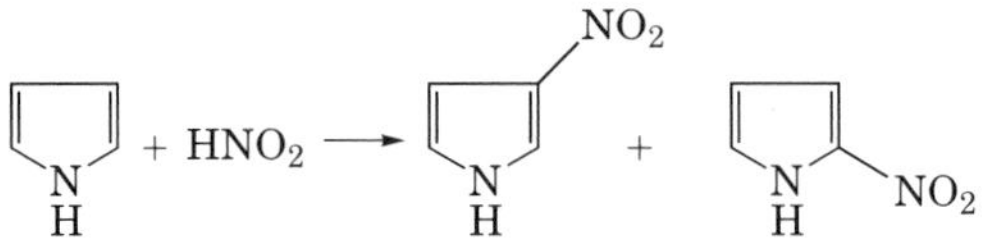

(**21**)   (**22**)

Halogenation reactions usually involve pyrroles with electronegative substituents. Mixtures are usually obtained and polysubstitution products, ie, tetrahalopyrroles, predominate. The monohalopyrroles are difficult to prepare and are not very stable in air or light.

The following rules pertain to electrophilic substitution in pyrroles (35): (1) an electron-withdrawing substituent in the α-position directs substitution to the β'- and α'-positions, (2) an electron-releasing substituent in the α-position directs substitution to the neighboring β-position or to the α'-position, (3) an electron-withdrawing substituent in the β-position leads to substitution in the α'-position, and (4) an electron-releasing substituent in the β-position tends to direct substitution into the neighboring α-position.

Pyrrole can be reduced catalytically to pyrrolidine over a variety of metal catalysts, ie, Pt, Pd, Rh, and Ni. Of these, rhodium on alumina is one of the most active. Less active reducing agents have been used to produce the intermediate 3-pyrroline (36). The 2-pyrrolines are ordinarily obtained by ring-closure reactions. Nonaromatic pyrrolines can be reduced easily with H_2 to pyrrolidines.

Pyrrole oxidizes in air to red or black pigments of uncertain composition. More useful is the preparation of 2-oxo-Δ^3-pyrrolines, which is best carried out by oxidation of the appropriate pyrrole with H_2O_2 in pyridine (37), eg, 3,5-dimethyl-

4-ethyl-3-pyrrolin-2-one [*4030-24-4*] from 2,4-dimethyl-3-ethylpyrrole [*517-22-6*].
Perbenzoic acid oxidizes *N*-methylpyrrole to *N*-methylsuccinimide (38).

Ring openings of pyrrole commonly occur at the carbon–nitrogen bond.
Treatment of pyrrole or 2,5-dimethylpyrrole [*625-84-3*] (**23**, R = CH$_3$) with hydroxylamine leads to ring opening and formation of dioximes (**31**) (39).

Reaction of pyrrole with carbenes yields enlarged ring systems as well as
2-formylpyrrole [*1003-29-8*] (40).

Analytical and Test Methods.   In addition to the modern spectroscopic
methods of detection and identification of pyrroles, there are several chemical tests. The classical Runge test with HCl yields pyrrole red, an amorphous
polymer mixture. In addition, all pyrroles with a free α- or β-position or with
groups, eg, ester, that can be converted to such pyrroles under acid conditions undergo the Ehrlich reaction with *p*-(dimethylamino)benzaldehyde to give purple
products.

Both pyrrole and indole react with selenium dioxide in the presence of nitric
acid to give a deep violet solution. Very small quantities (ca 4×10^{-5} g) of pyrrole
can be detected by this method.

Functional Derivatives

Hydroxypyrroles. Pyrroles with nitrogen-substituted side chains containing hydroxyl groups are best prepared by the Paal-Knorr cyclization. Pyrroles with hydroxyl groups on carbon side chains can be made by reduction of the appropriate carbonyl compound with hydrides, by Grignard synthesis, or by insertion of ethylene oxide or formaldehyde. For example, pyrrole plus formaldehyde gives 2-hydroxymethylpyrrole [27472-36-2] (**24**). The hydroxymethylpyrroles do not act as normal primary alcohols because of resonance stabilization of carbonium ions formed by loss of water.

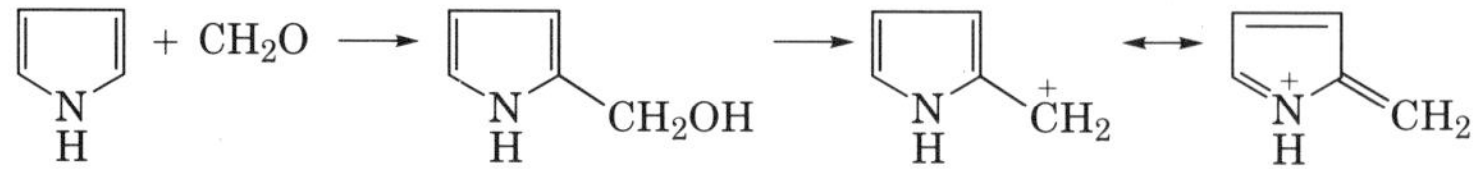

(**1**)   (**24**)

The α-hydroxypyrroles, which exist primarily in the tautomeric pyrrolin-2-one form, can be synthesized either by oxidation of pyrroles that are unsubstituted in the α-position or by ring synthesis. β-Hydroxypyrroles also exist primarily in the keto form but do not display the ordinary reactions of ketones because of the contributions of the polar form (**25**). They can be readily O-alkylated and -acylated (41).

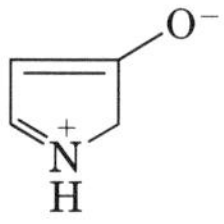

(**25**)

Aldehydes and Ketones. Pyrrole aldehydes and ketones are somewhat less reactive than the corresponding benzenoid derivatives. The aldehydes do not undergo Cannizzaro or Perkin reactions but condense with a variety of compounds that contain active methylene groups. They also react with pyrroles under acidic conditions to form dipyrrylmethenes (**26**). The aldehydes can be reduced to the methyl or carbinol structures. The ketones undergo normal carbonyl reactions.

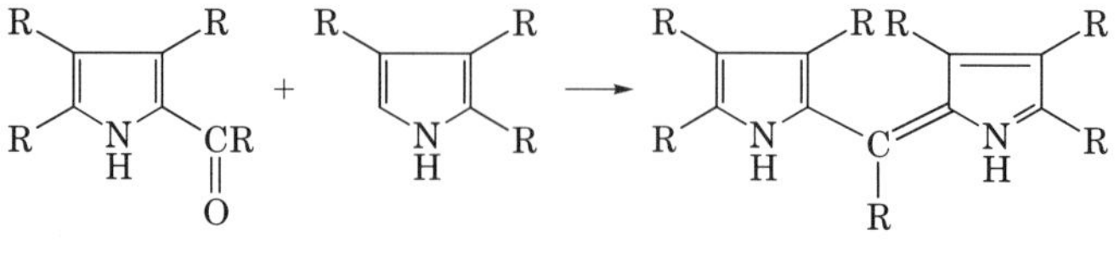

(**26**)

Pyrrole Carboxylic Acids and Esters. The acids are considerably less stable than benzoic acid and often decarboxylate readily on heating. However, electron-withdrawing substituents tend to stabilize them toward decarboxylation. The pyrrole esters are important synthetically because they stabilize the ring and may also act as protecting groups. Thus, the esters can be utilized synthetically and then hydrolyzed to the acid, which can be decarboxylated by heating. Often β-esters are hydrolyzed more easily than the α-esters.

Vinyl Pyrroles. Relatively new synthetic routes based on a one-pot reaction between ketoximes and acetylene in an alkali metal hydroxide–dimethyl sulfoxide (DMSO) system have made vinyl pyrroles accessible. It requires no pyrrole precursors and uses cheap and readily available ketones (42).

The 1-vinylpyrroles are highly reactive and are sensitive to oxygen. Conjugation of the vinyl group with the aromatic ring leads to a greater susceptibility to electrophilic attack. N-Vinylpyrroles have been shown to react additively with alcohols, diols, and hydrosilanes.

N-Vinyl polymers may be used for the preparation of semiconductors (qv). Derivatives of others have biological activity, eg, a derivative of 2-phenyl-1-vinylpyrrole, 2-phenyl-1-(propargyloxyethyl)pyrrole, stimulates motor activity and increases excitation, etc (42) (see VINYL POLYMERS, N-VINYL AMIDES).

2-Vinylpyrroles and 3-vinylpyrroles can be prepared by the base-catalyzed condensation of the corresponding formyl pyrroles with activated methylene groups, CH_2XY, where X,Y = COR, COOH, CN, etc.

Condensed Pyrroles. Pyrroles can be condensed to compounds containing two, three, or four pyrrole nuclei. These are important in synthetic routes to the tetrapyrrolic porphyrins, corroles, and bile pigments and to the tripyrrolic prodigiosins. The pyrrole nuclei are joined by either a one-carbon fragment or direct pyrrole–pyrrole bond.

Bipyrroles. Although four different types of bipyrroles, ie, 1,1′-, 2,2′-, 2,3′-, and 3,3′-, are known, the most important is 2,2′-bipyrrole [*10087-64-6*], which can be made by the Vilsmeier condensation of 2-pyrrolidinone [*616-45-5*] (**27**) (43).

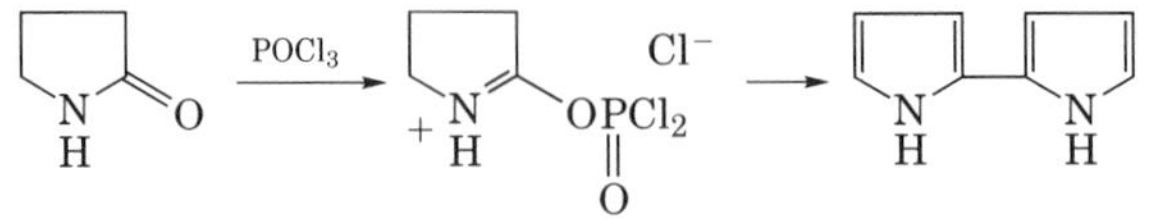

(27)

Other syntheses utilize α-bromo or α-iodo compounds, which condense on heating with copper (44).

Dipyrrylmethanes. The most important dipyrrylmethanes are the 2,2'-derivatives. The parent compound is not very stable, but electron-withdrawing substituents increase its stability considerably. Symmetrical dipyrrylmethanes, eg, the diethyl ester of 5,5'-methylenebis(4-ethyl-3-methyl)pyrrole-2-carboxylic acid [6305-93-7] (28), can be synthesized by acid-catalyzed self-condensation of α-halomethyl-, α-acetoxymethyl-, or α-methoxymethylpyrroles (45). Unsym-

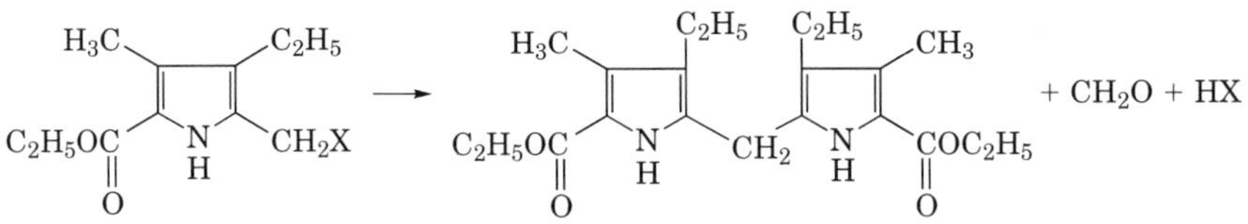

(28)

metrical dipyrrylmethanes are obtained through condensation of α-bromomethyl- or α-hydroxymethylpyrroles with a pyrrole that has an open α- or β-position.

Dipyrrylmethenes. Oxidation of the dipyrrylmethanes, eg, 5-carboxy-3,4'-diethyl-4,3',5'-trimethyl-2,2'-dipyrrylmethane [26030-65-9] (29), by bromine or sulfuryl chloride yields dipyrrylmethenes, eg (30) [80294-31-1].

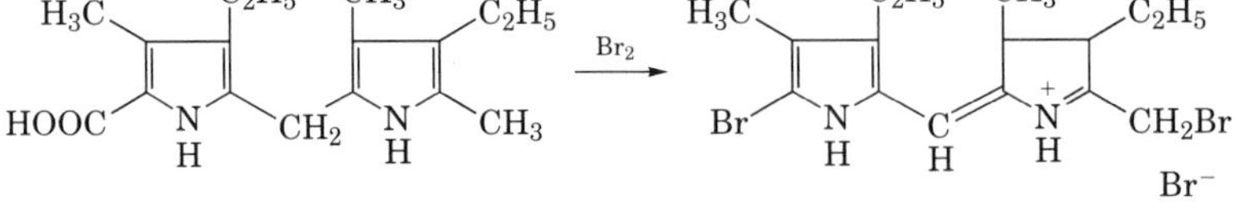

(29)   (30)

Polypyrroles. Highly stable, flexible films of polypyrrole are obtained by electrolytic oxidation of the appropriate pyrrole monomers (46). The films are not affected by air and can be heated to 250°C with little effect. It is believed that the pyrrole units remain intact and that linking is by the α-carbons. Copolymerization of pyrrole with N-methylpyrrole yields compositions of varying electrical conductivity, depending on the monomer ratio. Conductivities as high as $10^4/(\Omega \cdot m)$ have been reported (47) (see ELECTRICALLY CONDUCTIVE POLYMERS).

Because of its physical properties, polypyrrole has been cited as a unique building block for intelligent polymeric materials, ie, it has characteristics which make it capable of sensing, information processing, and response actuation (48).

Natural Products. The prodigiosins are antibacterial and antifungal orange-red pigments based on the basic pyrryl–dipyrrylmethene unit [22187-69-5] (**31**).

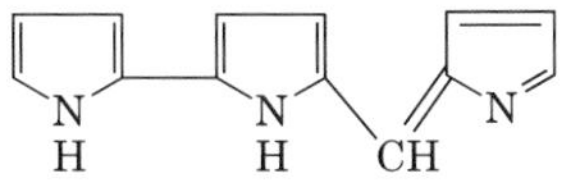

(**31**)

Bile Pigments. The oxidative degradation of heme yields open-chain tetrapyrrole as a waste product in humans and other higher animals. The yellow color of the skin in jaundice victims is caused by the presence of bilirubin [635-65-4] (**32**, R = $(CH_2)_2COOH$).

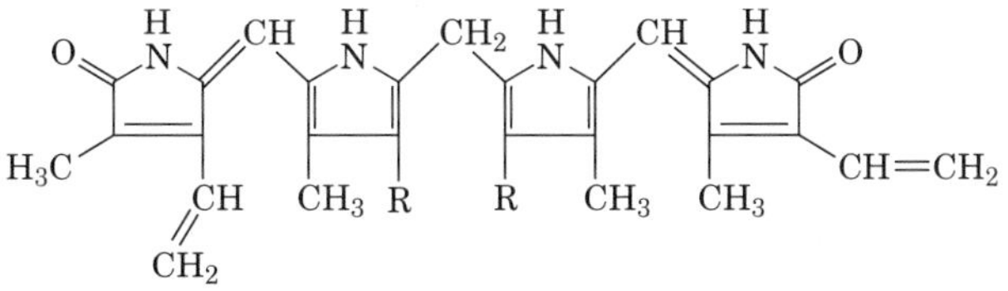

(**32**)

Phthalocyanines. The pyrrole ring system is also the fundamental structural unit of the important group of blue and blue-green pigments known as the phthalocyanines (see PHTHALOCYANINE COMPOUNDS).

Pyrrolidinones and Derivatives

2-Pyrrolidinone (2-pyrrolidone, butyrolactam or 2-Pyrol) (**27**) was first reported in 1889 as a product of the dehydration of 4-aminobutanoic acid (49). The synthesis used for commercial manufacture, ie, condensation of butyrolactone with ammonia at high temperatures, was first described in 1936 (50). Other synthetic routes include carbon monoxide insertion into allylamine (51,52), hydrolytic hydrogenation of succinonitrile (53,54), and hydrogenation of ammoniacal solutions of maleic or succinic acids (55–57). Properties of 2-pyrrolidinone are listed in Table 2. 2-Pyrrolidinone is completely miscible with water, lower alcohols, lower ketones, ether, ethyl acetate, chloroform, and benzene. It is soluble to ca 1 wt % in aliphatic hydrocarbons.

Table 2. Properties of 2-Pyrrolidinone

Property	Value
melting point, °C	25.6
boiling point, °C	
at 0.133 kPa[a]	103
1.33 kPa[a]	122
13.2 kPa[a]	181
101.3 kPa[a]	245
density (l), g/cm^3	
d_4^{25}	1.107
d_4^{50}	1.087
refractive index	1.4860
viscosity, at 25°C, mPa·s(=cP)	13.3
flash point, open cup, °C	129.41

[a]To convert kPa to mm Hg, multiply by 7.5.

Reactions of 2-Pyrrolidinone. Pyrrolidinone undergoes the reactions of a typical lactam, eg, ring opening, attack on the carbonyl group, and replacement of hydrogens alpha to the carbonyl group. Many of the reactions involve the amide. 2-Pyrrolidinone can be polymerized with anionic catalyst systems to polypyrrolidinone (nylon-4) (**33**), which is a high molecular weight linear polymer of potential interest as a textile fiber, film former, and molding compound (58,59) (see POLYAMIDES).

$$n \quad \underset{(\mathbf{27})}{\boxed{\quad}} \longrightarrow \underset{(\mathbf{33})}{-(\mathrm{NHCH_2CH_2CH_2C})_n-}$$

Strong acids or bases catalyze the hydrolysis of 2-pyrrolidinone to 4-aminobutanoic acid [γ-aminobutyric acid [56-12-2] (GABA)]. GABA is involved in the functioning of the brain and nervous system and is of considerable interest as a potential dietary supplement (60).

2-Pyrrolidinone forms alkali metal salts by direct reaction with alkali metals or their alkoxides or with their hydroxides under conditions in which the water of reaction is removed. The potassium salt prepared *in situ* serves as the catalyst for the vinylation of 2-pyrrolidinone in the commercial production of N-vinylpyrrolidinone. The mercury salt has also been described, as have the N-bromo and N-chloro derivatives (61,62).

$$\boxed{\quad} + \mathrm{KOH} \longrightarrow \boxed{\quad} + \mathrm{H_2O}$$

2-Pyrrolidinone can be alkylated by reaction with an alkyl halide or sulfate and an alkaline acid acceptor (63,64). This reaction can be advantageously

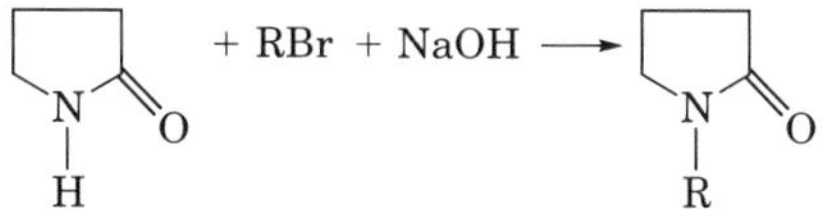

carried out with a phase-transfer catalyst (65). Alkylation can also be accomplished with alcohols and either copper chromite or heterogenous acid catalysts (66,67).

Treatment of 2-pyrrolidinone with an acid anhydride or acyl halide results in N-acylation.

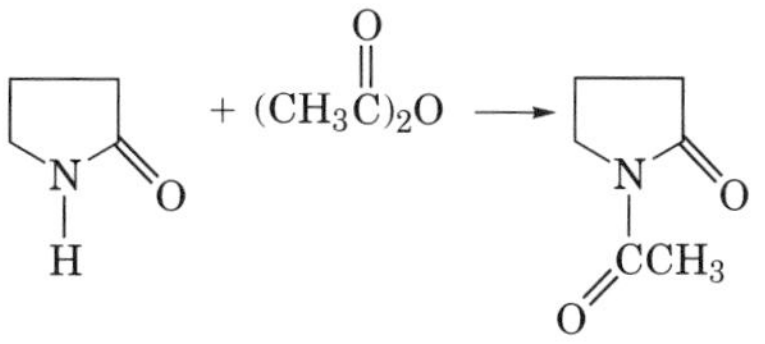

The amide nitrogen readily adds across the carbonyl group of an aldehyde yielding N-hydroxyalkyl-substituted pyrrolidinones (68), eg, *N*-methylol-2-pyrrolidinone [*15438-71-8*] (**34**). In the presence of secondary amines or alcohols, the hydroxyl groups are replaced (69), eg, if diethylamine is present the product is *N*-diethylaminomethyl-2-pyrrolidinone [*66297-50-5*] (**35**).

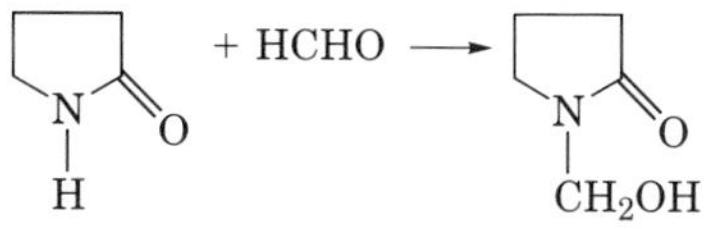

(27) + HCHO + (C₂H₅)₂NH →

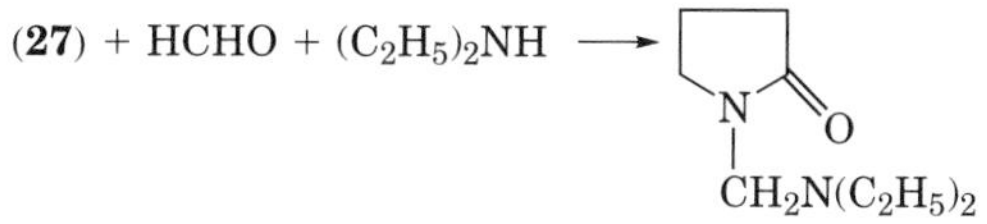

(35)

Treatment of *N*-hydroxymethylpyrrolidinone (**34**) with a chlorinating agent, eg, thionyl chloride, produces chloromethyl pyrrolidinone (**36**), a powerful alkylating agent useful for introducing the methylpyrrolidinonyl group into surface-active materials via quarternization of long-chain surface-active amines (70).

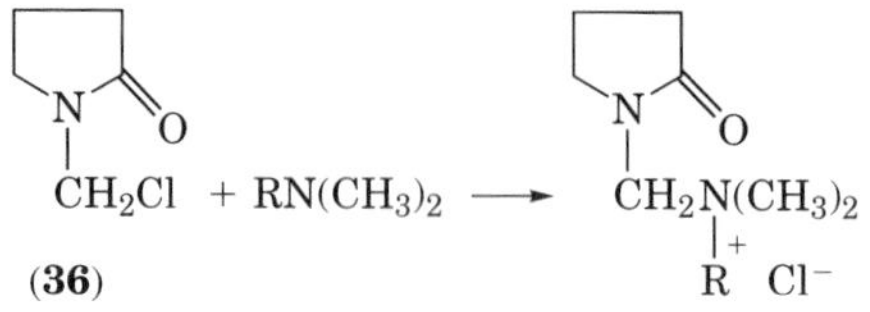

(36)

2-Pyrrolidinone is readily N-alkylated by styrene to give *N*-(2-phenyethyl)-2-pyrrolidinone [*10135-23-6*] (**37**). Additional styrene alkylates the 3-position (71) yielding 1,3-bis(2-phenylethyl)-2-pyrrolidinone [*60548-73-7*] (**38**).

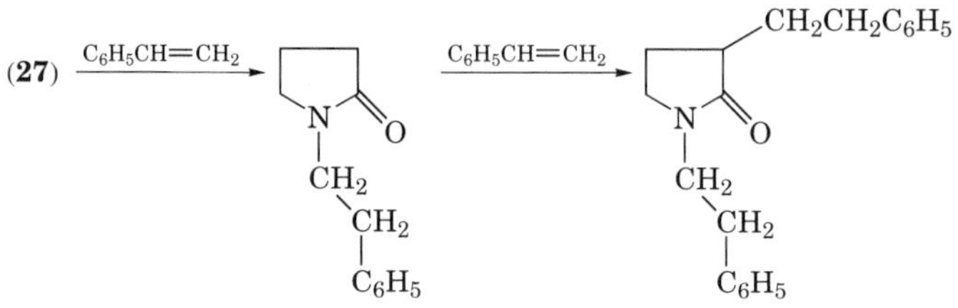

(37)   **(38)**

High temperature hydrogenation with a cobalt catalyst gives pyrrolidine, (**27**) + $H_2 \rightarrow$ (**19**) (72). Under dehydrating conditions, 2-pyrrolidinone condenses with itself to form 1-($\Delta 1'$-pyrrolin-2-yl)-2-pyrrolidinone [*7060-52-8*] (**39**) (73).

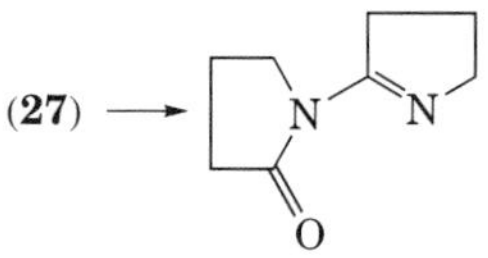

(39)

2-Pyrrolidinone can also condense with primary or secondary amines.

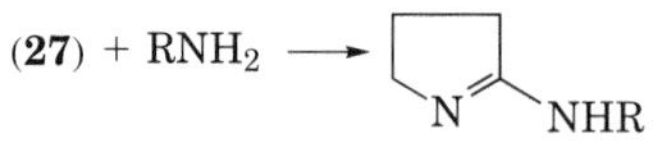

Under suitable conditions, O-alkylation rather than N-alkylation takes place, eg, to form 2-methoxy-1-pyrroline [*5264-35-7*] (**40**) (74–76).

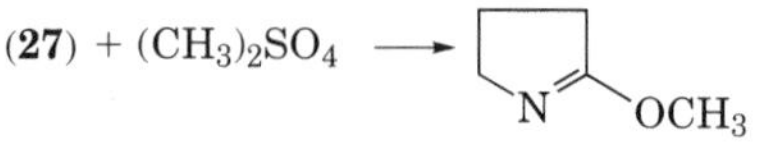

(40)

Manufacture. There are two main 2-pyrrolidinone producers. International Specialty Products (ISP) (GAF Corporation) has manufacturing facilities in Calvert City, Kentucky, and Texas City, Texas, and BASF manufactures it at Ludwigshafen, Germany. Both producers consume most of their production in the manufacture of 1-vinyl-2-pyrrolidinone.

Butyrolactone (**41**) and a moderate excess of ammonia are passed through a reactor at ca 250°C and 8–9 MPa (80–90 atm). Yields of 90–95% have been reported (77). The reaction proceeds in two steps, but the intermediate 4-hydroxybutyramide (**42**) is not ordinarily isolated. Improved yields are obtained

$$\text{(41)} + NH_3 \rightleftharpoons HO(CH_2)_3CONH_2 \longrightarrow \text{(27)} + H_2O$$

(**41**)   (**42**)   (**27**)

if the reaction is carried out in the gas phase on a magnesium silicate catalyst (250–290°C, 0.4–1.4 MPa), owing to supression of the undesirable by-product 4-(N-2-pyrrolidonyl)butyramide (78).

Shipment and Storage. 2-Pyrrolidinone is available in steel drums and in aluminum or stainless-steel tank cars and tank trailers. Because of its high freezing point, bulk shipments are in tanks with heating coils. Heating with hot water rather than steam avoids product discoloration. Steel (qv), stainless steel, and aluminum are satisfactory materials for storage containers. Because 2-pyrrolidinone is hygroscopic, it must be protected from atmospheric moisture.

Specifications and Analytical Methods. The purity of 2-pyrrolidinone is determined by gas chromatography and is specified as 98.5 wt % minimum. Maximum moisture content is specified as 0.5 wt %. Typical purities are much higher than specification.

Health and Safety Factors. Results of acute oral toxicity studies of 2-pyrrolidinone on white rats and guinea pigs show the LD_{50} to be 6.5 mL/kg. Skin patch tests on 200 human subjects indicate that 2-pyrrolidinone is a skin irritant, but there is no indication of sensitizing action. It is a mild eye irritant (79).

Uses. Because of the labile hydrogen on the nitrogen, 2-pyrrolidinone is not as good a solvent as 1-methylpyrrolidinone. Nevertheless, moderate amounts are sold as solvents and as plasticizers (qv) and coalescing agents for polymer emulsion coatings. There is also continuing interest in 2-pyrrolidinone as a monomer for polypyrrolidinone and as a source of 4-aminobutanoic acid. Significant quantities of 2-pyrrolidinone react with ethyl chloroacetate in the preparation of Piracetam (1-acetamido-2-pyrrolidinone [7491-74-9]) (**43**), which is useful for treatment of motion sickness, epilepsy, etc (80). The main use of 2-pyrrolidinone is, however, as an intermediate for the manufacture of 1-vinyl-2-pyrrolidinone.

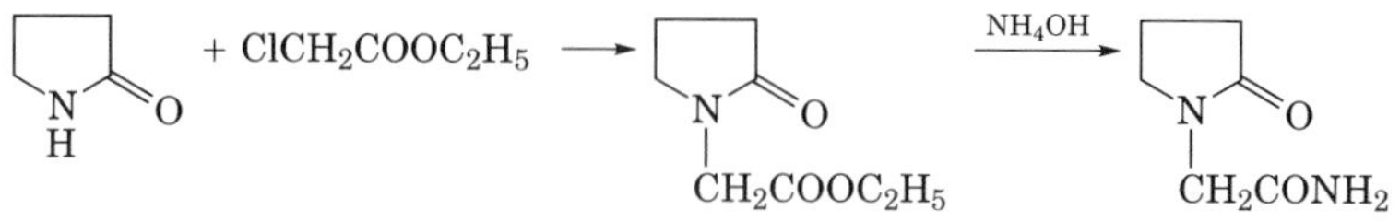

(43)

1-Methyl-2-Pyrrolidinone. *N*-Methyl-2-pyrrolidinone [*872-50-4*] (**44**) (NMP or methyl-2-pyrrolidone, M-Pyrol) was first reported in 1907 as prepared by alkylation of 2-pyrrolidinone with methyl iodide (81). The present commercial route, ie, condensation of butyrolactone with methylamine, was first described in 1936 (50).

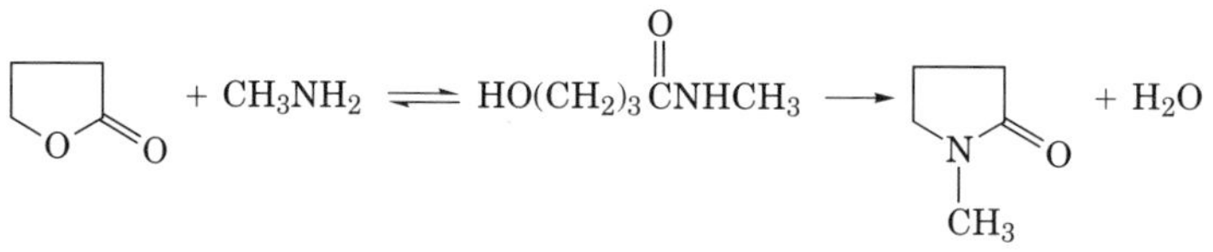

(44)

Other preparative routes include hydrogenation of succinonitrile in the presence of methylamine and hydrogenation of solutions of maleic or succinic acid and methylamine (82,83). Properties are listed in Table 3. 1-Methyl-2-pyrrolidinone is completely miscible with water, lower alcohols, lower ketones, ether, ethyl acetate, chloroform, and benzene. It is moderately soluble in aliphatic hydrocarbons and dissolves many organic and inorganic compounds.

Reactions. Although usually a stable and unreactive solvent, 1-methyl-2-pyrrolidinone can undergo a number of characteristic chemical reactions. In particular, these involve ring opening, attack on the carbonyl group, or replacement of hydrogens alpha to the carbonyl group. Although it is very resistant to hydrolysis under neutral conditions, with strong acids or bases 1-methyl-2-pyrrolidinone can be hydrolyzed to 4-methylaminobutyric acid. Borohydride reduction under suitable conditions yields 1-methylpyrrolidine [*120-94-5*] (**45**) (84).

Table 3. Properties of 1-Methyl-2-Pyrrolidinone

Property	Value
freezing point, °C	−24.4
boiling point, °C	
at 0.133 kPa[a]	41
1.33 kPa[a]	79
13.3 kPa[a]	136
101.3 kPa[a]	202
density, g/cm^3, d_4^{25}	1.028
refractive index, n_D^{25}	1.65
flash point, open cup, °C	95
solubility parameter, δ	11

[a]To convert kPa to mm Hg, multiply by 7.5.

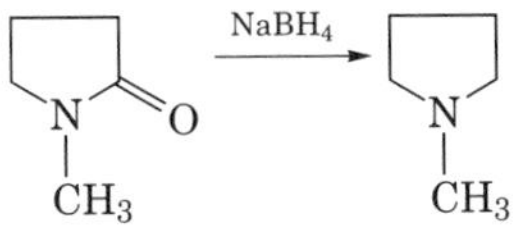

(**44**) (**45**)

1-Methyl-2-pyrrolidinone reacts with chlorinating agents, eg, $COCl_2$, $SOCl_2$, $POCl_3$, and PCl_5, etc, to form the salt (**85**), which then reacts with a variety of compounds (86–88). For example, reaction of (**46**) with primary

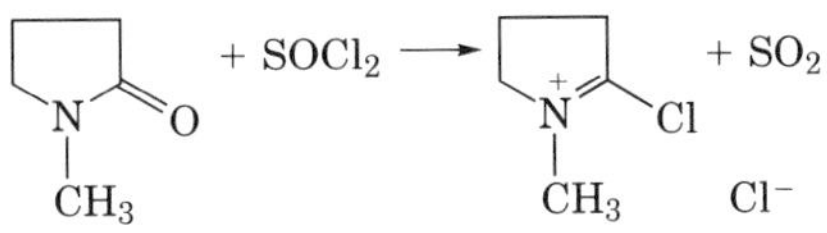

(**44**) (**46**)

amine (RNH_2) yields imines (**47**), and reaction with alkoxides ($RONa$), the corresponding ketals (**48**). Hydroxylamine plus (**46**) gives N-methyl-2-pyrroli-

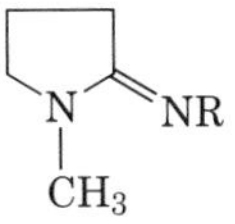

(**47**) 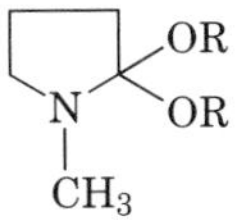 (**48**) 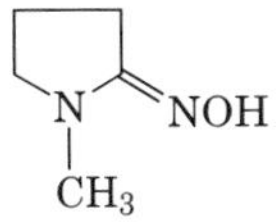 (**49**)

done oxime [*35197-40-1*] (**49**), and the reaction of (**46**) with H_2S replaces the keto oxygen with sulfur (**50**). Methyl-2-thiopyrrolidinone [*10441-57-3*] (**50**) can also be prepared by reaction of 1-methyl-2-pyrrolidinone with sulfur or carbon disulfide at high temperatures and pressures (89,90).

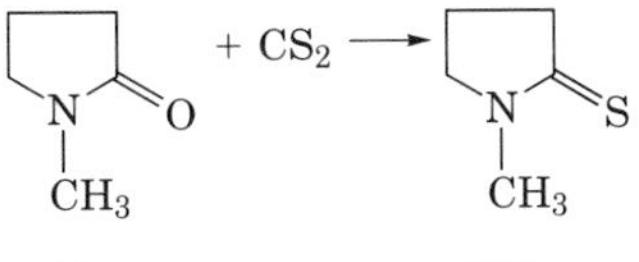

(**44**) (**50**)

Manufacture. 1-Methyl-2-pyrrolidinone (**44**) is manufactured at the same sites and by essentially the same process as described for 2-pyrrolidinone (**27**). In general, capacity for 2-pyrrolidinone and N-methylpyrrolidinone (NMP) involves the same equipment with only moderate differences in auxiliaries.

In addition to ISP and BASF, ARCO has begun producing 1-methyl-2-pyrrolidinone at a plant in Texas. NMP is also made in Japan (Mitsubishi) and in Russia. Annual U.S. production (1991) has been estimated by OSHA at 36,000–39,000 t.

Shipment, Storage, and Price.   1-Methyl-2-pyrrolidinone is available in tank cars or tank trailers as well as in drums. Shipping containers are normally of unlined steel. Rubber hose is unsuitable for handling; standard steel pipe or braided steel hose is acceptable. Ordinarily 1020 carbon steel (0550) is satisfactory as a storage material. Stainless-steel 304 and 316, nickel, and aluminum are also suitable. Methylpyrrolidinone is hygroscopic and must be protected from atmospheric moisture. In September 1994, NMP was listed at $3.89/kg.

Specifications and Analytical Methods.   The purity of 1-methyl-2-pyrrolidinone is determined by gas chromatography and is specified as 99.5 wt % minimum. Maximum moisture content is specified as 0.05 wt % by ir spectroscopy.

Health and Safety Factors.   1-Methyl-2-pyrrolidinone is less toxic than many other dipolar aprotic solvents. The LD_{50} for white rats is 4.2 mL/kg. Although it does not appear to be a sensitizing agent, prolonged contact with skin should be avoided. It is a moderate eye irritant.

Uses.   1-Methyl-2-pyrrolidinone is a dipolar aprotic solvent. It has a high dielectric constant and is a weak proton acceptor. All of its commercial uses involve its strong and frequently selective solvency. It has replaced other solvents of poorer stability, higher vapor pressures, greater flammabilities, and greater toxicities.

The largest use of NMP is in extraction of aromatics from lube oils. In this application, it has been replacing phenol and, to some extent, furfural. Other petrochemical uses involve separation and recovery of aromatics from mixed feedstocks; recovery and purification of acetylenes, olefins, and diolefins; removal of sulfur compounds from natural and refinery gases; and dehydration of natural gas.

Large amounts of NMP are consumed in the polymer industry as a medium for polymerization and as a solvent for finished polymers. Polymers that are soluble in NMP are poly(vinyl acetate), poly(vinyl fluoride), polystyrene, nylon and aromatic polyamides and polyimides (qv), polyesters (qv), acrylics, polycarbonates (qv), cellulose derivatives, and synthetic elastomers. 1-Methyl-2-pyrrolidinone is also useful for cleaning and stripping of magnetic wire coatings and electronic parts as well as in agricultural applications for preparing emulsifiable concentrates. Its low toxicity has allowed it to displace chlorinated solvents in many of these applications as well as in paint and finish removers (qv), where it is gaining increasing popularity (91).

Higher 1-Alkyl-2-Pyrrolidinones.   The hydrophilicity of the pyrrolidinone ring system when combined with the hydrophobicity of a longer chain alkyl group leads to compounds with surfactant properties which enhance the water solubility of hydrophobic materials. Two of these, 1-octyl-2-pyrrolidinone and 1-dodecyl-2-pyrrolidinone, have been made available commercially. These surface-active pyrrolidinones (Surfadones) are finding applications in several areas. They are excellent wetting agents and interact with anionic surfactants to exhibit synergistic effects on both static and dynamic surface-tension reductions (92). 1-Octyl-2-pyrrolidinone is used in hard-surface cleaners, in fountain solutions (graphic arts), and for pigment dispersions (qv) (93,94). It is also used for preparing emulsifiable concentrates in agricultural applications (95) and it inhibits crystallization. 1-Dodecyl-2-pyrrolidinone is used in some shampoos.

A number of other N-substituted 2-pyrrolidinones have been offered commercially or promoted as developmental products. These materials offer different and sometimes unique solvency properties. All are prepared by reaction of butyrolactone with suitable primary amines. Principal examples are listed in Table 4.

Several N-substituted pyrrolidinones eg, ethyl, hydroxyethyl and cyclohexyl, are used primarily in specialized solvent applications where their particular physical properties are advantageous. For example, mixtures of 1-cyclohexyl-2-pyrrolidinone and water exhibit two phases at temperatures above 50°C; below that temperature they are miscible in all proportions. This phenomenon can be used to facilitate some extractive separations. Mixtures of 1-alkyl-pyrrolidinones that are derived from coconut and tallow amines can be used at lower cost in certain applications where they may be used instead of the pure 1-dodecyl-2-pyrrolidinone and 1-octadecyl-2-pyrrolidinone.

1-Vinyl-2-Pyrrolidinone. 1-Vinyl-2-pyrrolidinone (VP) (1-ethenyl-2-pyrrolidinone, *N*-vinyl-2-pyrrolidone, and V-Pyrol) is manufactured by ISP in the United States and by BASF in Germany by vinylation of 2-pyrrolidinone with acetylene. It forms the basis for a significant specialty polymer and copolymer industry and consumes the primary portion of all 2-pyrrolidinone manufactured (see VINYL POLYMERS, *N*-VINYL MONOMERS AND POLYMERS).

Table 4. N-Substituted 2-Pyrrolidinones

N-substituent	CAS Registry Number	Bp, °C	Mp, °C
ethyl	[2687-91-4]	212	−77
vinyl	[88-12-0]	215	13
hydroxyethyl	[3445-11-2]	309	26
isopropyl	[3772-26-7]	219	−28
butyl	[3470-98-2]	244	−106
hexyl	[4838-65-7]	276	−52
cyclohexyl	[6837-24-7]	292	15
octyl	[2687-94-7]	307	−26
dodecyl	[2687-96-9]	361	10

BIBLIOGRAPHY

"Pyrrole and Pyrrole Derivatives" in *ECT* 1st ed., Vol. 11, pp. 339–353, by W. R. Vaughan, University of Michigan; in *ECT* 2nd ed., Vol. 16, pp. 841–858, by E. V. Hort and R. F. Smith, GAF Corp.; in *ECT* 3rd ed., Vol. 19, pp. 499–520, by E. V. Hort and L. R. Anderson.

1. R. Runge, *Ann. Physik*, **31**, 67 (1834).

2. T. Anderson, *Ann.* **105**, 349 (1858).

3. A. Baeyer and H. Emmerling, *Ber.* **3**, 517 (1870).

4. D. J. Chadwick, in R. A. Jones, ed., *Heterocyclic Compounds*, Vol. 48, Part I, John Wiley and Sons, Inc., New York, 1990, pp. 1–103; *The Beilstein Handbook of Organic Chemistry*, Vols. 21–22, Beilstein Institute, Frankfurt/Main, Germany, 1978.

5. A. Jones, in A. R. Katrizky and A. J. Boulton, eds., *Advances in Heterocyclic Chemistry*, Vol. 11, Academic Press, Inc., New York, 1970, p. 386.

6. B. Bak, *Acta Chem. Scan.* **9**, 1355 (1955).

7. H. Fischer and H. Orth, *Die Chemie des Pyrrols*, Vol. 1, Akad, Verlagsges, Leipzig, Germany, 1934.

8. L-B. Agenas and B. Lindgren, *Arkiv. Kemi.* **28**, 145 (1968).

9. A. Marinangelli, *Ann. Chim. Rome* **44**, 211 (1954).

10. M. L. Josien, M. Paty, and P. Pineau, *J. Chem. Phys.* **24**, 126 (1956).

11. G. Yagil, *Tetrahedron* **23**, 2855 (1967).

12. A. Gossauer, *Die Chemie der Pyrrole*, Springer-Verlag, Berlin, 1974, p. 130.

13. Ref. 12, pp. 60–77.

14. L. Janelli and P. G. Orsini, *Gazz. Chim. Ital.* **89**, 1467 (1959).

15. H. Kofod, L. E. Sutton, and J. Jackson, *J. Chem. Soc.*, 1467 (1952).

16. A. Albert, *Heterocyclic Chemistry*, University of London, 1959.

17. J. A. Elvidge and L. M. Jackman, *J. Chem. Soc.*, 859 (1961); J. Elvidge, *Chem. Commun.*, 160 (1965).

18. H. H. Perkampus, U. Kruger, and W. Kruger, *Zeit. fur Naturforschl*, **24B**, 1365 (1965).

19. R. A. Jones, T. Mc. L. Spotswood, and P. Chenychit, *Tetrahedron* **23**, 4469 (1967).

20. L. Knorr and H. Lange, *Ber.* **35**, 2998 (1902); H. Fisher, *Org. Syn.* **15**, 17 (1935); **17**, 96 (1937).

21. H. Fisher and E. Fink, *Z. Physiol. Chem.* **280**, 123 (1944); **283**, 152 (1948).

22. A. Hantzsch, *Ber.* **23**, 1474 (1890).

23. M. Feist and E. Stenger, *Ber.* **35**, 1558 (1902).

24. H. S. Broadbent, W. S. Burnham, R. K. Olsen, and R. M. Shelley, *J. Heterocyc. Chem.* **5**, 757 (1968).

25. G. H. Walizei and E. Breitmaier, *Synthesis*, 337 (1989).

26. W. Flitsch and M. Hobenhorst, *Liebigs Ann. Chem.*, 397 (1990).

27. Yu. K. Yurev, I. K. Karobitsyma, R. D. Ben-Yarik, L. A. Savina, and Pl. A. Akeshin, *Bestn. Mosk. Univ.* **6**(2); *Ser. Fiz. Mat. Estestv. Nauk* (1), 37 (1951).

28. E. H. Huntress, T. E. Lesslie, and W. M. Hearon, *J. Am. Chem. Soc.* **78**, 419 (1956).

29. J. B. Hendrickson and R. Rees, *J. Am. Chem. Soc.* **83**, 1250 (1961).

30. Yu. K. Yurev and I. S. Levi, *Vestn. Mosk. Univ. Ser, Mat. Mekh. Astron. Fiz. Khim.* **11**(2), 153 (1956).

31. S. Clementi and G. Marino, *Tetrahedron* **25**, 4599 (1969).

32. P. Rothemund, *J. Am. Chem. Soc.* **58**, 625 (1936).

33. W. C. Anthony, *J. Org. Chem.* **25**, 2049 (1960); P. A. Burbridge, G. L. Collier, A. H. Jackson, and G. W. Kenner, *J. Chem. Soc. Part B*, 930 (1967).

34. A. R. Cooksey, K. J. Morgan, and D. P. Morrey, *Tetrahedron* **26**, 5101 (1970).

35. A. Treibs and G. Fritz, *Ann.* **611**, 162 (1958).

36. A. H. Corwin, in R. C. Elderfield, ed., *Heterocyclic Compounds*, Vol. 1, John Wiley and Sons, Inc., New York, 1950, pp. 277–342.

37. J. H. Atkinson, R. S. Atkinson, and A. W. Johnson, *J. Chem. Soc.*, 5999 (1964).

38. I. Nabih and E. Helmy, *J. Pharm. Soc.* **56**, 649 (1967).

39. S. P. Findlay, *J. Org. Chem.* **21**, 644 (1956).

40. C. W. Rees and C. E. Smith, *Adv. Heterocyclic Chem.* **3**, 57 (1964).

41. R. Chong and P. W. Clezy, *Aust. J. Chem.* **20**, 935 (1967).

42. B. A. Trofimov, in R. A. Jones, ed., *Heterocyclic Compounds*, Vol. 48, Part II, John Wiley and Sons, Inc., New York, 1990, pp. 131–298.

43. H. Rapaport and G. Castagnoli, Jr., *J. Am. Chem. Soc.* **84**, 2178 (1962); H. Rapaport and J. Bordmer, *J. Org. Chem.* **29**, 2727 (1964).

44. D. Dolphin, R. Grigg, A. W. Johnson, and J. Leng, *J. Chem. Soc.*, 1460 (1965).

45. A. W. Johnson, I. T. Kay, E. Markham, R. Price, and K. Be. Shaw, *J. Chem. Soc.*, 3416 (1959); J. Ellis, A. H. Jackson, A. C. Vain, and G. W. Kenner, *J. Chem. Soc.*, 1935 (1964).
46. A. F. Diaz, K. K. Kanazawa, and G. P. Gardini, *J. Chem. Soc., Chem. Commun.*, 635 (1979).
47. K. K. Kanazawa and co-workers, *J. Chem. Soc. Chem. Commun.*, 854 (1979).
48. P. R. Teasdale and G. G. Wallace, *Chim. Oggi.* **10**(10), 19 (1992).
49. S. Gabriel, *Ber.* **22**, 3335 (1889).
50. E. Spath and J. Lintner, *Ber.* **69**, 2727 (1936).
51. U.S. Pat. 4,110,340 (Aug. 29, 1978), J. F. Knifton (to Texaco Inc.).
52. U.S. Pat. 4,111,952 (Sept. 5, 1978), J. F. Knifton (to Texaco Inc.).
53. U.S. Pat. 4,036,836 (July 19, 1977), J. L. Green (to Standard Oil Co., Ohio).
54. U.S. Pat. 4,181,662 (Jan. 1, 1980), W. A. Sweeney (to Chevron Research Co.).
55. U.S. Pat. 3,812,149 (May 21, 1974), E. J. Hollstein (to Sun Research and Development Co.).
56. U.S. Pat. 3,812,149 (May 21, 1974), E. J. Hollstein (to Sun Research and Development Co.).
57. U.S. Pat. 3,884,936 (May 20, 1975), E. J. Hollstein (to Sun Research and Development Co.).
58. U.S. Pat. 2,638,463 (May 12, 1953), W. O. Ney, Jr., W. R. Nummy, and C. E. Barnes (to GAF Corp.).
59. U.S. Pat. 2,739,959 (Mar. 27, 1956), W. O. Ney, Jr., and M. Crowther (to Arnold, Hoffman and Co.).
60. Span. Pat. 398,322 (Sept. 16, 1974), S. A. Hebron.
61. J. Tafel and M. Stern, *Ber.* **33**, 2228 (1900).
62. U.S. Pat. 3,850,920 (Nov. 26, 1974), W. E. Walles (to The Dow Chemical Co.).
63. T. Tafel and O. Wassmuth, *Ber.* **40**, 2835 (1907).
64. W. Gaffield, L. K. Keefer, and P. P. Roller, *Org. Prep. Proceed. Int.* **9**(2), 49 (1977).
65. J. Palecek and J. Kuthan, *Z. Chem.* **17**(7), 260 (1977).
66. Jpn. Kokai 74 117,459 (Nov. 9, 1974), T. Ayusawa and S. Fukami (to Mitsubishi Petroleum Co.).
67. Jpn. Kokai 76 16,657 (Aug. 1, 1974), S. Dnomoto and Y. Takahashi (to Asahi Chemical Industry Co.).
68. U.S. Pat. 3,073,843 (Jan. 15, 1963), S. R. Buc (to GAF Corp.).
69. Span. Pat. 447,346 (Oct. 16, 1977), A. Surroca and J. Jose (to Farma-Lepori SA).
70. PCT Int. Appl. WO 90 15,797 (Dec. 27, 1990); U.S. Appl. 370,226 (1991), R. K. Chaudhuri, D. J. Tracy, and R. F. Login.
71. A. T. Malkhasyan, G. S. Sukiasyan, S. G. Matinyan, and G. T. Martirosyan, *Arm. Khim. Zh.* **29**(5), 458 (1976).
72. T. Kamiyama, M. Inoue, and S. Enomoto, *Yuki Gosei Kagaku Kyokaishi* **36**(1), 65 (1978).
73. G. Dannhardt, *Arch. Pharm. Weinheim (Germany)* **311**(4), 294 (1978).
74. V. G. Granik, A. M. Zhedkova, N. S. Kuryatov, V. P. Pakhomov, and R. G. Glushkov, *Khim. Geterotsikl. Soedin*, (11), 1132 (1973).
75. U.S. Pat. 3,816,454 (June 11, 1974), Y-H. Wu and W. G. Lobeck (to Mead Johnson and Co.).
76. U.S. Pat. 3,706,766 (Dec. 19, 1972), F. M. Hershenson (to G. D. Searle and Co.).
77. J. W. Copenhaver and M. H. Bigelow, *Acetylene and Carbon Monoxide Chemistry*, Van Nostrand Reinhold, New York, 1949, pp. 163–164.
78. U.S. Pat. 4,824,867 (Apr. 25, 1989), K-C. Liu and P. D. Taylor (to GAF Corp.).
79. *2-Pyrrole*, Technical bulletin 7543-120, Rev. 1, GAF Corp., New York.
80. Brit. Pat. 1,039,113 (Aug. 7, 1966), H. Morren (to UCB Corp.).

81. J. Tafel and O. Wassmuth, *Ber.* **40**, 2839 (1907).

82. U.S. Pat. 4,152,331 (May 5, 1979), P. J. N. Jeijer and L. H. Geurtz (to Stamicarbon BV).

83. U.S. Pat. 3,448,118 (June 3, 1964), G. Chickery, P. Benite, and P. Perras (to Rhône-Poulenc SA).

84. A. Basha, J. Orlando, and S. M. Weinreb, *Syn. Commun.* **7**(8), 549 (1977).

85. H. Eilingsfeld, M. Seefelder, and H. Weidinger, *Angew. Chem.* **78**, 836 (1960).

86. H. Eilingsfeld, M. Seefelder, and H. Weidinger, *Chem. Ber.* **96**, 2671 (1963).

87. U.S. Pat. 3,306,911 (Feb. 28, 1967), R. C. Doss (to Phillips Petroleum).

88. T. Jen, H. Van Hoeven, W. Groves, R. A. McLean, and B. Loev, *J. Med. Chem.* **18**(1), 90 (1975).

89. U.S. Pat. 3,306,911 (Feb. 28, 1967), R. C. Doss (to Phillips Petroleum).

90. J. C. Meslin and G. Duguay, *Bull. Soc. Chim. Fr.* **7–8**(2), 1200 (1976).

91. U.S. Pat. 5,064,557 (Nov. 1991), F. Fusiak (to ISP Corp.).

92. M. J. Rosen, *Langmuir* **7**, 885 (1991).

93. J. C. Hornby and D. Jon, *Soaps / Cosmetics / Chemical Specialties*, Sept. 1992.

94. P. Petter, *Manu. Chem.*, 18–21 (Apr. 1991).

95. K. S. Narayanan and R. K. Chaudhuri, in D. G. Chasin and L. E. Bode, eds., *Emulsifiable Concentrate Formulations for Multiple Active Ingredients, Pesticide Formulations and Application Systems*, Vol. 11 ASTM STP 1112, American Society of Testing and Materials, Philadelphia, Pa., 1990.

General References

R. A. Jones, ed., *Pyrroles in Heterocyclic Compounds*, Vol. 48, Pts. I and II, John Wiley and Sons, Inc., New York, 1990.

R. J. Sundberg, *Prog. Heterocycl. Chem.* **3**, 90–108 (1991).

A. H. Jackson, *Compr. Org. Chem.* **4**, 275 (1979).

M-Pyrol, N-Methyl-2-Pyrrolidone Handbook, GAF Corp., New York, 1972.

LOWELL RAY ANDERSON
KOU-CHANG LIU
ISP Corporation

QUALITY ASSURANCE

The objective of chemical manufacturing is to provide products that perform to expectation. A manufacturing unit is responsible for producing a product. The quality assurance (QA) and quality control (QC) units are designated to assure the product not only meets its stated specification but also performs up to customer expectation. The activities typically performed by the QC laboratory ensure through testing that a product conforms to specification. The QA unit operates in support of the lab activities to assure the correctness of the results and the consistency of the product. The relationship between QA and QC is shown in Figure 1. A review of several quality improvement techniques applicable to the manufacture of chemical products is given herein. Several quality system standards used by chemical manufacturing organizations are also discussed.

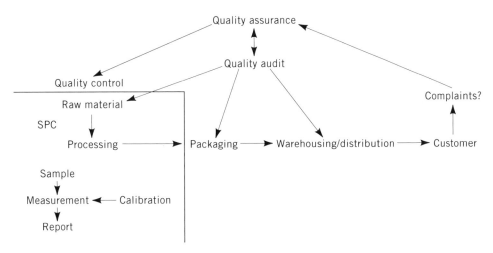

Fig. 1. Quality control and quality assurance oversight activities where SPC = statistical process control.

The twentieth century has often been called the Century of Productivity; the twenty-first century may well be the Century of Quality (1). A discussion of how the chemical industry is organized to develop and manufacture quality products is available (2).

Quality Improvement

The chemical and allied products industry has been the largest U.S. exporting sector since the mid-1980s (3). Quality improvement has evolved as companies, comprising the chemical industry, struggle to maintain this position. The United States has lost the competitive edge in many other technological areas, even when found to be the low cost producer. This loss has been attributed in part to quality of goods and services.

Early Activities. Historically, quality was entrusted to the artisan, who was solely responsible for the products made. Thus financial success often rested on product quality. The industrial revolution replaced this system with one in which product quality was the result of the combined efforts of a group of factory workers. Quality was ensured through the combination of worker skills and production supervisor monitoring.

The adoption of the Taylor system in the late nineteenth century changed the lines of responsibility for product quality (4). This management philosophy was based on using incentives, such as pay based on output, to motivate worker productivity. As the workforce became better educated and labor unions gained strength, it became difficult to motivate workers doing simple, repetitive tasks (5).

The Taylor system, successful in the United States until the end of World War II, resulted in a dramatic increase in productivity. The transfer of responsibility for product quality from production to the QC laboratory allowed production to emphasize productivity, relying on quality control to keep nonconforming products from reaching the customer. In the years following World War II, quality became secondary to productivity (6).

In this period, leading executives in Japan were convinced by W. Edwards Deming, who lectured on statistical methods, and Joseph M. Juran, who taught the principles of managing for quality, that economic recovery could best be achieved through the manufacture of quality products (7,8). During the 1970s, Japanese quality improvement became apparent as manufacturing cost dropped dramatically. Quality and reliability of the manufactured products increased. This change was demonstrated in such products as radios, televisions, and videocassette recorders. It was Japanese automotive quality, however, that caught the undivided attention of U.S. industry (9). As a consequence, the U.S. chemical industry began to respond with improvement initiatives of its own.

Quality in Japan. Japanese economic prowess has been attributed variously to such quality improvement activities as quality circles, statistical process control (SPC), just-in-time delivery (JIT), and zero defects (ZD). However, the real key to success lies in the application of numerous quality improvement tools as part of a management philosophy called Kaizen, which means continuous improvement (10).

During the 1970s and 1980s, U.S. companies tried to adopt Japanese improvement techniques, but not the philosophy. Thus quality circles, ie, problem-solving groups of production workers, were initiated (11,12). When this approach failed to achieve anticipated results, it was replaced by other techniques, such as SPC (13), JIT (14), and ZD. This use of contrasting approaches has been summarized (15).

Quality Techniques.　*Statistical Process Control.*　A properly running production process is characterized by the random variation of the process parameters for a series of lots or measurements. The SPC approach is a statistical technique used to monitor variation in a process. If the variation is not random, action is taken to locate and eliminate the cause of the lack of randomness, returning the process or measurement to a state of statistical control, ie, of exhibiting only random variation.

This technique, widely used in U.S. plants during World War II, helped to ensure reliability and performance of military supplies. Once the war ended, SPC lost favor. However, in the face of rising Japanese product quality, SPC was reintroduced. In the chemical industry the use of SPC continues to grow in popularity as a key element of an ongoing continuous improvement.

Just-in-Time.　Just-in-time closely followed the reintroduction of SPC into the American workplace. The efficacy of JIT was limited in the United States to an emphasis on keeping component inventories low to reduce the cost of inventory (16). More importantly, however, JIT should be used to focus management attention on quality problems.

Prior to JIT, defective parts could be discarded and a replacement taken from inventory. Under JIT there is insufficient inventory to replace defective parts, often leading to a shutdown of the assembly process. The Japanese use this impact to heighten the awareness of nonconforming material and to ensure that the causes of such defective parts are eliminated. U.S. industry, however, often sees the disadvantageous tradeoff between interrupting production to improve quality vs the economic savings from reduced inventory.

Zero Defects.　Whereas zero defects was often interpreted to be a quality goal, its full meaning is to encourage continuous quality improvement (17). When ZD was treated only as a slogan, it failed to have a lasting impact.

As of 1996 the term ZD is no longer heard. One U.S. manufacturing company, however, successfully changed its culture using this approach. In 1987 Motorola adopted "Six Sigma Quality" as the impetus toward quality improvement (18). All work processes were to be continuously improved and the 1992 goal was to reduce defects to this level, which corresponds to 3.4 defects per million. Top management fixed its sights on this aggressive goal, provided employees with the resources to accomplish it, and then closely monitored progress. Their quest was successful, leading them to world class quality, the winning of the Malcolm Baldrige National Quality Award, and, most importantly, resulted in the economic revival of the company.

Quality Function Deployment.　Sometimes referred to as the House of Quality, quality function deployment (QFD) is a technique for translating the voice of the customer into design requirements (19). This is a systematic approach identifying customer expectations and relating the expectations to product properties. The usage of QFD in the chemical industry appears to be growing. QFD results in chemical specifications optimized to assure the material is suitable for its intended use and performs up to customer expectations.

Total Quality Management.   Total quality management (TQM) is the term which encompasses all of the continuous improvement activities with the goal of world class quality. This corporate culture sets up the conditions for a climate favorable to companywide improvement.

Companies following a TQM philosophy place heavy emphasis on employee training for quality improvement. Top management monitors improvement through well-designed metrics. These organizations have a comprehensive employee recognition and reward program to celebrate success. Successful implementation of TQM results in high customer satisfaction, employee involvement and motivation, and financial success.

In the chemical industry, TQM continues to evolve from a corporate-run program to a decentralized one (20). Responsibility for implementation has been turned over to local sites so that often a small corporate staff can provide resources and guidance.

Quality in the Chemical Industry.   The chemical industry uses quality as a strategic tool for financial success. One measure of quality is the degree of product variation from lot to lot. In the chemical industry, it is often difficult to provide product specifications comprehensive enough to ensure product performance in all applications (21). Therefore, the manufacture of product having a minimum of lot-to-lot variability allows the customer to use the product without modifying their formulation or process to accommodate such variation.

Quality Control

Within the chemical industry, quality control (QC) is the systematic monitoring of product conformance to specification through testing (22). The reliability of laboratory testing is essential to effective operation. Therefore proper sampling (qv) and accurate and precise measurement are important. This relatively narrow focus differs from quality assurance where the role involves monitoring all activities that impact product and service quality, including quality control (see Fig. 1).

Quality control inspection is a post-, or at best concurrent, manufacture activity. The QC laboratory reports on in-process and finished product quality based on testing. Thus QC confirms whether a material has been manufactured in conformance. Sometimes QC also assists the production unit in salvaging process material or reworking off-standard finished product.

The purpose of the QC laboratory is to monitor product quality through sampling and analysis. This can sometimes result in an adversarial relationship with the production unit. Thus, the lab must balance competing risks. One, called α-risk, is the risk of rejecting product which in fact meets specification. The competing risk, known as β-risk, is that of erroneously releasing off-standard material that might lead to customer complaints. When QC reports to local plant management these risks are generally well balanced. When, however, QC reports directly to production, there is often a tendency to accept more β-risk for the sake of higher output. Proper operation of the QC laboratory, as it relates to the accuracy of test results, is extremely important to the harmony of the plant, the profitability of the company, and the satisfaction of the customer.

Sampling Plan. The first step to assuring accuracy, ie, the conformity of the measured value to the true or expected value (23), is to obtain a representative sample. Sampling may be the responsibility of either production or of QC. A written sampling plan approved by QC should be followed. It is the responsibility of the QC unit to ensure that samples represent all the material under evaluation.

A thorough sampling plan should describe when the sample is to be taken and how many samples are required. It also specifies from what location within the equipment, such as the manway or discharge valve, to take the sample. The plan should also indicate what sampling equipment and sample container should be used, as well as the type of tests to be performed and the acceptance criteria.

How many samples are taken can be of importance. One sample often suffices where it is known that the material in question is homogeneous for the parameter(s) to be tested, such as for pure gases or bulk solvents. If this is not the case, then statistical sampling should be considered. Samples should be taken from various points within the material, if the material stratifies.

Samples can be analyzed individually or may be combined into a homogeneous composite sample and then analyzed. In either case, only a portion of the sample is customarily used for a given test. Therefore, this material must be representative of the entire lot. Where this may not be the case, as for particle size measurement of powders, it may be important to extract representative test quantities from the bulk sample. For powders this is best done using a spinning riffler; for liquids, it should be carried out only after thorough mixing (see MIXING AND BLENDING; POWDERS, HANDLING; SIZE MEASUREMENT OF PARTICLES).

Calibration. Calibration of lab instruments is important to the accuracy of test results. Calibration, the use of an accepted standard to adjust an instrument or measurement standard so as to improve the accuracy of the instrument or measurement, is an essential requirement of both the U.S. Food and Drug Administration (FDA) Good Manufacturing Practice (GMP) (24) and the ISO 9000 standards (25).

A calibration schedule appropriate for each instrument must be established. The frequency should ideally be based on experience. Until sufficient experience is gained, the schedule is based on customary practices or manufacturer recommendations. The calibration frequency can then be adjusted. If the instrument is often out of calibration, a more frequent schedule should be used; if in calibration, the interval can be extended.

For conformance to GMP or ISO 9000 requirements, instrument calibration must be documented to include identification of the instrument and the calibration standard. The standard must be traceable to a recognized standard such as from the National Institute of Science and Technology (NIST). Calibration documentation also includes the method of calibration and a record of the calibration results.

Replicate Analyses. Confidence in the test result is improved by reducing the measurement variability. This variability in repeat analyses is known as precision. One method to improve the precision of the measurement is to perform complete replicate analyses of the same sample beginning with the sample preparation (26). This is appropriate when the sample is known to be representative of the material sampled. When this is not the case, multiple samples should be taken for analysis.

The average value of the replicates is reported along with the standard deviation, which reflects the variability in the measurement. Large standard deviations relative to the average measurement indicate the need for an action plan to improve measurement precision. This can be accomplished through more replicate measurements or the elimination of the source of variation, such as the imprecision of an instrument or poor temperature control during the measurement.

Statistical Control.   Statistical quality control (SQC) is the application of statistical techniques to analytical data. Statistical process control (SPC) is the real-time application of statistics to process or equipment performance. Applied to QC lab instrumentation or methods, SPC can demonstrate the stability and precision of the measurement technique. The SQC of lot data can be used to show the stability of the production process. Without such evidence of statistical control, the quality of the lab data is unknown and can result in production challenging adverse test results. Also, without control, measurement bias cannot be determined and the results derived from different labs cannot be compared (27).

Statistical control of an analysis or instrument is best demonstrated by SQC of a standard sample analysis. The preferred approach to demonstrate statistical control is to use a reference sample of the subject material that has been carefully analyzed or, alternatively, to use a purchased reference standard. Either material must be stored so that it remains unchanged, eg, sealed in ampuls or septum capped bottles. Periodically a sample can then be reanalyzed by the technique used for routine analysis. These results are plotted in a control chart. Any change in the stability of the test in question results in a lack of randomness of the measurement and becomes apparent as a trend or shift. A change in measurement precision is indicated by an increase in the distance measurements are from the average line (28).

Statistical quality control charts of variables are plots of measurement data, preferably the average result of replicate analyses, vs time (Fig. 2). Time is often represented by the sequence of batches or analyses. The average of all the data points and the upper and lower control limits are drawn on the chart. The control limits are closely approximated by the sum of the grand average: plus for the upper control limit, or minus for the lower control limit, three times the standard deviation.

There are several rules applied to control charts to spot a lack of randomness. The most obvious is a point outside the control limits. A trend such as a run, where at least seven consecutive data points are either above or below the average line, or a trend of seven consecutive points either increasing or decreasing in value, also indicates an out of control situation (29). A lack of randomness is also apparent from a pattern in which there is a repeated sequence of points cycling between rising then falling, or when points tend to cluster around the center line or near the control limits.

The value of control charts is to provide early warning to lab personnel of changes affecting the test results. The drift of an instrument, such as results

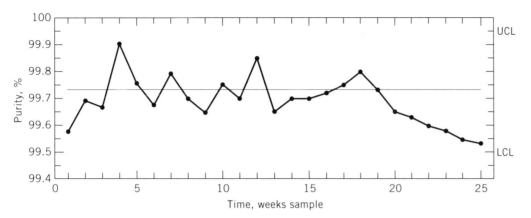

Fig. 2. An SPC control chart of the purity analysis of a reference standard where (—) represents the average value and UCL and LCL are the upper and lower control limits, respectively.

from deterioration of a chromatographic column, or the need for cleaning or servicing an instrument, as in the weakening of the light source in a liquid chromatography detector, can affect results. Control charts can also point to variation in the test results owing to inconsistencies in analyst technique arising from insufficient training. The performance of lab equipment should always be monitored using SPC.

Laboratory Information Management System. The QC lab must analyze raw material, in-process, and finished product samples; adhere to calibration schedules; record data; and perform statistical analyses. These activities lend themselves to the application of software packages such as a laboratory information management system (LIMS) (qv). An inexpensive LIMS is within the reach of even small laboratories.

The LIMS software is essentially a database for tracking, reporting, and archiving lab data as well as scheduling and guiding lab activities. Graphical and statistical treatment of data for improved process control (qv) as well as preparation of certificates of analysis (COA) for the customer are some of the other features of a comprehensive LIMS package (30).

Samples logged into the system are assigned a unique number which is often physically attached to the sample using a bar-code label. The testing protocol is contained in the LIMS and is based on the point in the process from which the sample was taken. This information and the identification of the type of sample enable the LIMS to schedule the testing of the material.

Sample test data are either manually entered into the system or captured from analytical instruments connected to the LIMS. The system performs any necessary calculations and compares the result to the appropriate specification stored in the database. If the comparison indicates the material is in conformance, the system can automatically provide an approval. Otherwise, the LIMS can alert lab supervision to the nonconforming sample analysis.

When the data are already in the computer, tracking lab performance using statistical techniques can be done with little effort. By having the data archived, historical trends can be charted and past process capability compared to current capability. This can be useful in responding to challenges to test results (30). The availability of production data makes periodic comparison of process capability to specification limits easy.

Whereas issues of technician productivity, sample status, and scheduling of analyses are economically important, these take second place to the issue of measurement quality in the laboratory. LIMS software enables activities such as calibration and proper maintenance to be scheduled automatically. By recording the results of these activities in the LIMS, the data to demonstrate proper performance are readily made available. Additionally, the schedule can be adjusted to reflect the actual needs of the individual instruments.

Another benefit of using LIMS is the ease of ensuring that tests are performed according to the most current procedure. Methods can be incorporated into the system and printed either as a recipe or worksheet for the analyst, thus helping to assure consistency between analysts.

Finally, the laboratory expends significant effort communicating results to both internal and external customers. Production, quality assurance, and purchasing all have various information needs ranging from the simple pass/fail decisions to statistical summaries of the data and supplier product quality. Customers expect to receive lot analyses in the form of a COA and often also want their own product-specific information on the document as well. This information can automatically be applied to the COA if entered into the LIMS. Often, a quality-conscious customer wants information about the product in the form of process capability or control charts. Using LIMS, these charts can be provided on demand.

Quality Assurance

The responsibilities of the quality assurance unit generally fall into two categories: support for or improvement of the existing quality system (31). The support activities of QA often include employee training, quality system documentation, method validation and method transfer, audit, and customer complaints.

Whereas QC is responsible for monitoring production, the responsibility of QA encompasses the entire product cycle from development to customer satisfaction (see Fig. 1). This role varies from auditing raw material suppliers to evaluating in-process sampling, equipment calibrations, and statistical process control. QA also evaluates the packaging environment and labeling procedures, as well as warehousing and shipping operations. This quality oversight activity plays a large role in preventing nonconforming product from ever reaching the customer.

Quality assurance must remain independent of manufacturing so that problems can be reported freely to upper management without fear of retribution. QA should have oversight responsibility for QC. A reporting structure helps to ensure the independence of both quality units and conforms to both Good Manufacturing Practice (GMP) and ISO 9000 requirements.

Employee Training.   Employee training related to quality includes many topics. For companies manufacturing pharmaceuticals or chemicals used for pharmaceutical manufacturing, GMP compliance training for employees is mandatory (32). If a company is pursuing ISO 9000 registration, training also covers the ISO standard (33). In many organizations employees are trained in improvement techniques such as statistical process control, tools of quality, and the requirements of the quality system being utilized. The extent of QA involvement ranges from assisting in the development of the curriculum to providing trainers and facilitating implementation. The result is a better-educated workforce that can perform more productively while achieving higher quality output.

Quality System Documentation.   Quality system documentation has two aspects. One is the development of a quality manual, which is a description of how the company's quality system operates (34,35). This manual should also address how the company plans to comply with such requirements as fall under the GMP or ISO 9000 standards.

The quality manual should be organized to facilitate referral to the quality system standard. It should be brief and refer to other documentation for more detail. The manual should be under document control, ie, each page is uniquely identified as to date or revision number and its preparer. It is common practice to offer customers a copy of the manual upon request, ie, the manual should not be proprietary.

The second aspect of quality documentation is to detail how the work processes referred to in the manual are performed. The QA unit is often the organization responsible for issuing a set of procedures designed to assure conformance to the appropriate standards or to company policy. The procedures, often called standard operation procedures (SOP) or quality operating procedures (QOP), should include such topics as customer complaints, audit protocols, stability testing, preparation of COAs, test method validation, etc.

Well-written procedures should begin with a purpose and scope. The procedure section usually follows, describing the work process, and it should be clearly and concisely written. The document should include copies of all forms and a process flow diagram for illustration.

Test Methods.   In addition to that provided by proper sampling and replication of analysis, a test method also has a significant impact on the accuracy and precision of the results. Preferred methods are those which are accepted in the chemical industry such as those from the American Society of Testing Materials (ASTM), Association of Official Analytical Chemists (AOAC), or from compendia such as the *United States Pharmacopoeia* (USP) or the *Food Chemical Codex* (FCC) (36). The use of such methods eliminates the need for method validation.

Method validation is the demonstration, accompanied by a high degree of assurance, that the test method consistently performs as expected. Validation of test methods is especially important when the intended method differs from a compendial or other referenced method and yet is expected to provide equivalent or perhaps better test results. When the reference method is to be used unchanged, no validation is necessary. Thus a primary benefit of using compendial methods as written is not needing to do a method validation.

An analytical method validation study should include demonstration of the accuracy, precision, specificity, limits of detection and quantitation, linearity, range, and interferences. Additionally, peak resolution, peak tailing, and analyte recovery are important, especially in the case of chromatographic methods (37,38).

The method limit of quantitation and limit of detection must be determined as well as the limit of linearity. The limit of quantitation is defined as the level at which the measurement is quantitatively meaningful; the limit of detection is the level at which the measurement is larger than the uncertainty; and the limit of linearity is the upper level of the measurement reliability (39). These limits are determined by plotting concentration vs response.

Interferences in the method can reduce selectivity and thus reliability of the measurement. Therefore it is important to evaluate the method for interferences and to utilize techniques to reduce their impact as well as to make them known to the analyst (40).

The analytical research and development (R&D) unit is often responsible for the preparation and validation of test methods. The R&D lab is not faced with the same pressures for rapid analysis as the QC unit, where pending results often hold up production. In addition, R&D often assigns personnel to specific instruments or techniques, whereas QC generally requires technicians to perform varied analyses. This leads to an expertise on the part of analytical chemists and technicians which is difficult to duplicate in QC. Therefore the R&D test method should be rugged enough to withstand the different environment of the QC lab and still provide valid results.

Method Transfer. Method transfer involves the implementation of a method developed at another laboratory. Typically the method is prepared in an analytical R&D department and then transferred to quality control at the plant. Method transfer demonstrates that the test method, as run at the plant, provides results equivalent to that reported in R&D. A validated method containing documentation eases the transfer process by providing the recipient lab with detailed method instructions, accuracy and precision, limits of detection, quantitation, and linearity.

Preferably the transferring lab provides a sample which has already been analyzed, with the certainty of the results being known (41). This can be either a reference sample or a sample spiked to simulate the analyte. An alternative approach is to compare the test results with those made using a technique of known accuracy. Measurements of the sample are made at the extremes of the method as well as the midpoint. The cause of any observed bias, the statistical difference between the known sample value and the measured value, should be determined and eliminated (42). When properly transferred, the method allows for statistical comparison of the results between the labs to confirm the success of the transfer.

The Tools of Quality. Quality assurance also plays an important role in problem solving and process improvement. To do so, QA personnel must be knowledgeable in the many so-called tools of quality (TOQ) and their application, so as to guide the efforts of process improvement. Many QA organizations are involved in training employees in these techniques to facilitate quality improvement.

Seven of the tools of quality have been summarized (43). The first tool is a flow chart, used to help understand the organizational flow of a procedure or process. A flow chart should be constructed with the full participation of the people who do the work. Its principal benefit is to enable teams, such as problem-solving or productivity improvement teams, to reach a common vision of the work flow. Its use enables the improvement effort to begin with this common understanding. Figure 3 contains an example for manufacture of a polymeric material.

The next tool is a cause and effect diagram (44), illustrated by Figure 4. This diagram is used by teams to relate an effect to its potential causes. Diagram construction often begins with the four main branches shown. These

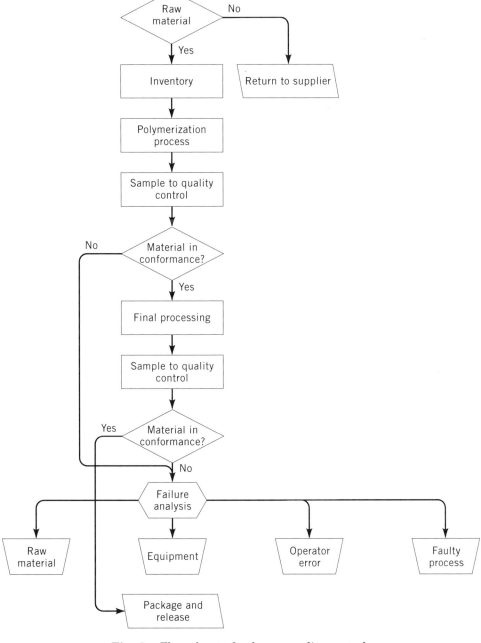

Fig. 3. Flow chart of polymer quality control.

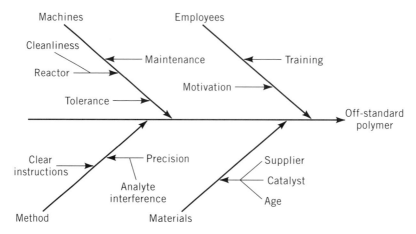

Fig. 4.   Cause and effect diagram of off-standard polymer production.

diagrams resemble the skeleton of a fish and thus are sometimes called fishbone diagrams.All possible primary causes and their associated subcauses and their next-level associated causes should be shown. When the lowest level causes are identified, the team can use the chart for ranking the most likely origin of an effect. Once this is accomplished, other TOQ can be used for the collection of data to confirm the cause leading to the effect in question.

The third tool is the control chart (see Fig. 2)(45). This type of chart is used to display how a measurement, such as purity, or a control parameter, such as temperature, varies with time. The time parameter is often expressed in batch or process order. The technique is used to identify a situation where a variable measurement or series of consecutive measurements differs more than normal variation would predict. This situation is described as being out of statistical control. Correction might be to adjust the equipment to regain control or to investigate the cause of being out of control. The latter might lead to identifying the cause, for example, as a change in raw material, the chemical operator, or the manufacturing process.

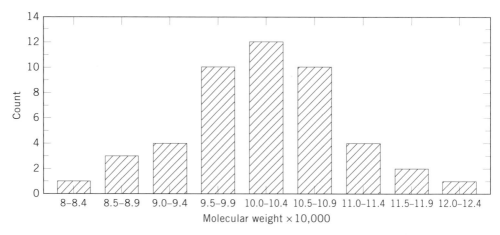

Fig. 5.   Histogram of polymer molecular weight distribution.

The preparation and use of histograms is the fourth TOQ (46). These charts, such as that shown in Figure 5, illustrate the variability of data as a frequency distribution. Thus, the range of variability is broken down into approximately 10 intervals. The total number of measurements in each interval is presented as the height of a bar graph. The resulting pattern shows the spread of the variable and whether it is normally distributed. If the graph is bell-shaped, ie, symmetrical with its peak in the center of the data range, then the data has a normal distribution. If the curve has a different appearance, its shape can often suggest possible causes.

The check sheet shown below, which is tool number five, is a simple technique for recording data (47). A check sheet can present the data as a histogram when results are tabulated as a frequency distribution, or a run chart when the data are plotted vs time. The advantage of this approach to data collection is the ability to rapidly accumulate and analyze data for trends. A check sheet for causes of off-standard polymer production might be as follows:

Attribute	Occurrence
color	XXXX
moisture	XXXXXXXXX
pH	XX
viscosity	XXXXXXX
particle size	X
monomer	XXX
solvent	X

The pareto chart, tool number six, is a special type of histogram (48) where the frequency data is grouped in order of decreasing occurrence or other measures of importance rather than in sequential or numerical order. The chart, an example of which is shown in Figure 6, illustrates the causes in decreasing order of importance. It enables the improvement effort to be focused where it can have the most impact and is an effective management communication tool.

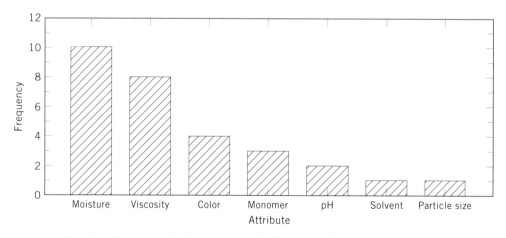

Fig. 6. Pareto analysis of causes of off-standard polymer production.

Pareto charts often illustrate the principle that 80% of the effect is the result of 20% of the causes (49). Thus these charts are valuable in prioritizing improvement activities. Identifying and correcting the 20% often results in an 80% improvement in the measured effect.

The seventh tool is the scatter or correlation diagram also known as an XY plot (50). This plot of one variable vs another is most useful in confirming interrelationships. Thus, scatter diagrams can verify the relationships shown in the cause and effect diagram.

Proper application of one or more of the tools of quality should lead to the elimination of the causes effecting off-standard results and thus to improvement of the process under investigation. It is often the QA professional who is responsible for the introduction and facilitation of the use of these techniques in the chemical corporation.

Quality Audit. Another important responsibility of quality assurance is the audit function. Using the quality audit as a tool, QA can monitor the operation of the manufacturing facility; a toll, ie, contract, manufacturer; or raw material supplier to assure that written procedures are in place and that there is documentation to indicate the procedures are being followed. Properly executed audits allow QA to spot potential weakness in the quality system that could allow errors to occur. Once identified, these weaknesses can then be corrected before they result in nonconformance.

An audit must be exercised with great care lest it become a policing function. To optimize the effectiveness of the audit, proper techniques include the following (51). (1) Initiation: the auditee should be given ample notice of the impending visit; (2) plan: the audit standard, such as GMP or ISO 9000, the scope, the schedule, etc, should be clearly identified; (3) implementation: factual information to document all observations should be carefully collected while objectivity is maintained and casting of blame avoided; (4) wrap-up: the auditee should be informed of any positive, as well as negative, findings immediately and the audit results presented to senior management representatives at an exit meeting; (5) reporting: a written audit report should be available in a timely fashion, typically within two to four weeks; and (6) conclusion: an appropriate period of time, typically four to six weeks, should be allowed for receipt of the auditee corrective action plan.

QA should audit at least the most important suppliers. This type of review often results in the exchange of ideas for improvement to quality systems of both the supplier and the customer. Sometimes such an audit also identifies a supplier having a serious deficiency.

Customer Complaints. A failure in a company's quality system often shows up in the form of a customer complaint. These reports of nonconformances, whether for product or service deficiencies, are typically received by the sales or customer service organization and then channeled to QA. Quality assurance often tracks the progress and coordinates the complaint investigation. Once completed, it is QA that reviews the report of the investigation and corrective action for thoroughness and efficacy. Quality assurance then either passes the information along to the sales or customer service organization or contacts the customer directly.

Quality Systems

Besides internal quality audits, there are audits conducted by external authorities for conformance to established quality systems. The two chief standards affecting the chemical industry are the U.S. Food and Drug Administration Current Good Manufacturing Practice (GMP) regulation and the International Organization for Standardization ISO 9000 series. A quality system performance-related standard is the Malcolm Baldrige National Quality Award (MBNQA).

Good Manufacturing Practice.   The GMPs were issued by the U.S. FDA in 1978 to provide minimum quality standards in the production of pharmaceuticals (qv) for the finished dosage form as well as their ingredients. The standard has been updated periodically.

The GMP details certain requirements for the quality system, such as the independence and responsibilities of the quality control unit. It requires such activities as internal audit to monitor GMP conformance, employee training, complaint investigation, failure analysis, and verification of proper manufacture and test by QC prior to release of the batch.

It is also a guideline to ensure product quality through the suitability of the manufacturing equipment, air and water quality, sanitation, insect and rodent control, and housekeeping. The FDA periodically sends inspectors to audit chemical companies who manufacture bulk pharmaceutical chemicals or inactive ingredients called excipients to ensure conformance. Whereas GMP conformance ensures that the product meets pharmaceutical quality standards, it does not ensure conformance to customer-service-related requirements.

ISO 9000.   The ISO 9000 series of standards was first issued in 1987 and then updated in 1994. The ISO 9000 standard describes the selection criteria for the four standards, ISO 9001–9004. ISO 9001 is composed of 20 items covering manufactured products from design and development in R&D to commercial production and after-sale service or technical support. ISO 9002 has 19 of the items in 9001 describing the requirements of a quality system for the manufacture of an item, such as chemicals, to a specification; only the requirement relating to product development is omitted. Chemical companies seldom certify to ISO 9003, which describes the quality system for a laboratory involved in final inspection and test. Finally, ISO 9004 presents guidelines for total quality management.

Conformance to one of the first three ISO standards is verified by third-party audit for which the auditee pays the auditor a fee. Continued conformance is assured through periodic reaudit by the third party. The ISO 9000 standard is not a product quality standard. Conformance to ISO 9001 or 9002 ensures that the customer receives product in conformance to mutually agreed-upon specifications. Other agreed-upon requirements are also met. The benefit to the certified company is the confidence it provides to its customers that the company has met the minimum requirements for this internationally recognized quality system.

Conformance to ISO 9000 by U.S. companies was led by the chemical industry as a result of the importance of international trade to chemical companies. An ISO 9000 certification was once thought to be necessary for access to the European market (52). At the beginning of 1994, almost 20% of U.S. certifications went to chemical companies. The emphasis on registration continues at a fast pace, as certification is considered an important supplier selection criterion by U.S. chemical companies.

Malcolm Baldrige National Quality Award. The most stringent and comprehensive quality system criterion is described in the Malcolm Baldrige National Quality Award. The award was created by an act of U.S. Congress in 1987 and has been given annually since 1988. The award program is managed by the National Institute of Science and Technology (NIST) and administered by the American Society for Quality Control (ASQC). The award is a competition for the annual honors given to the highest scoring companies who exceed a minimum score. A maximum of two awards each are given in the categories of large manufacturing and small manufacturing company. There is also a separate category for service companies. Selection is based on a written application indicating how the company meets the award requirements. A site visit by a MBNQA audit team seeks to verify the applicant's claim in fulfilling the award criteria.

The winning quality system has three principal features: (1) integration with business strategy; (2) active organizational learning processes tying together all corporate requirements and responsibilities, ie, customer, employee, supplier, productivity, etc; and (3) multidimensional results that contribute to overall business improvement and competitiveness (53). Winners of the award in the large company category have included chemical and allied industry companies and their customers, including Eastman Chemical, Motorola, Xerox, and Millikan Company. Companies having exemplary quality systems have shown benefits such as "better employee relations, higher productivity, greater customer satisfaction, increased market share, and improved quality" (54).

Selection of a Quality System Standard. Conformance to GMP is mandatory for most companies supplying the pharmaceutical industry; conformance to ISO 9000 is an important step toward achieving a world-class quality system and a Malcolm Baldrige Award. There are many differences between the ISO 9000 standards and criteria for the Malcolm Baldrige Award. The most important of these is their purpose. The ISO standard is meant to demonstrate conformance to customer requirements. The Malcolm Baldrige criteria are intended to show that a company is capable of "delivery of ever-improving value to customers; and systematic improvement of company operational performance" (55). This leads to a difference in executive commitment. Winning the Baldrige Award requires significant top management involvement. Conformance to ISO 9000 can often be delegated to a lower level of management.

Economic Aspects

Quality professionals use the term quality cost when discussing waste in a company. Quality cost includes any form of waste associated either with an activity that is unnecessary or an effort that must be corrected. There is internal waste in the form of rework, scrap, failure investigation, downgrading, etc. External waste includes customer complaints, including returned product, price concessions, and complaint investigation. Appraisal costs are those used to ascertain conformance to specification during the manufacturing cycle. Chemical companies have come to recognize the costs associated with wasteful efforts and are attempting to identify and eliminate waste. The savings accrued have a dramatic impact on profitability (see WASTE REDUCTION).

A company focusing on quality control and quality assurance is often attempting to redefine the corporate culture to reflect quality improvement and customer satisfaction. It has been estimated that U.S. companies spend from 20–40% of their sales dollars on various forms of waste (56,57). Quality-related expenses for quality plant laboratories and corporate staff associated with control and improvement activities run about 1–3% of sales (58). Thus, the control and improvement efforts represent the potential for a 10:1 return on investment.

BIBLIOGRAPHY

1. J. M. Juran, *A History of Managing for Quality: The Evolution, Trends, and Future Directions of Managing for Quality*, ASQC Quality Press, Milwaukee, Wis., 1995, p. 597.
2. J. M. Juran and F. M. Gryna, *Juran's Quality Control Handbook*, 4th ed., McGraw-Hill Book Co., Inc., New York, 1988, pp. 28.4–28.38.
3. M. Good, *ACS Keynote Address*, Chicago, Ill., Aug. 1995.
4. Ref. 1, p. 555.
5. Ref. 2, pp. 10.3–10.4.
6. Ref. 1, p. 558.
7. M. Walton, *The Deming Management Method*, Perigee Books, New York, 1986, pp. 10–21.
8. Ref. 1, pp. 532–541 and 579–581.
9. *Ibid.*, pp. 579–583.
10. M. Imai, *Kaizen, The Key to Japanese Economic Success*, McGraw-Hill Book Co., Inc., New York, 1986, pp. 1–5.
11. T. J. Peters and R. H. Waterman, Jr., *In Search of Excellence*, Warner Books, New York, 1982, pp. 241–242.
12. P. B. Crosby, *Quality Without Tears*, McGraw-Hill Book Co., Inc., New York, 1984, p. 10.
13. Ref. 1, pp. 585–586.
14. Ref. 7, p. 75.
15. A. D. Stratton, *Quality Progress*, ASQC, Milwaukee, Wis., Apr. 1990, p. 44.
16. *BusinessWeek*, 176D (May 14, 1984).
17. P. B. Crosby, *Quality is Free*, McGraw-Hill Book Co., Inc., New York, 1979.
18. M. S. Gill, *Business Monthly*, 42–46 (Jan. 1990).
19. W. Lawton, *Chem. Proc. Ind. Div. News*, 2–6 (Spring 1992).
20. K. Sissell, *Chem. Week*, 31–42 (Sept. 20, 1995).
21. *Quality Assurance for the Chemical and Process Industries*, ASQC Chemical Interest Committee, Milwaukee, Wis., 1987, p. 2.
22. Ref. 2, pp. 7.13–7.16.
23. J. K. Taylor, *Quality Assurance of Chemical Measurements*, Lewis Publishers Inc., Chelsea, Mass., 1987, p. 7.
24. "Current Good Manufacturing Practice in Manufacturing, Processing, Packaging, or Holding of Drugs," *Code of Federal Regulations*, Title 21, Part 210, U.S. FDA, Washington, D.C., 1993.
25. International Organization for Standardization, *Quality Systems—Model for Quality Assurance in Production and Installation, ISO 9000:1994*, ANSI/ASQC Q9000, American National Standards Institute, New York, 1994.

26. Ref. 21, pp. 14–16.
27. Ref. 23, Chapt. 3.
28. G. Kateman and F. W. Pijpers, *Quality Control in Analytical Chemistry*, John Wiley & Sons, Inc., New York, 1981, pp. 107–111.
29. K. Ishikawa, *Guide to Quality Control*, Unipub, New York, 1984, pp. 76–77.
30. S. W. Kanick, T. J. Conti, T. R. Spickler, and G. W. Stine, *PI Qual.*, 28–30 (May/June 1993).
31. Ref. 2, pp. 7.16–7.21.
32. "Current Good Manufacturing Practice for Finished Pharmaceuticals," *Code of Federal Regulations*, Title 21, Part 211.25, U.S. FDA, Washington, D.C., 1988.
33. Ref. 25, *Q91 / 1994*, p. 4.18.
34. Ref. 21, pp. 3–4.
35. Ref. 2, pp. 6.40–6.45.
36. Ref. 21, p. 22.
37. *United States Pharmacopoeia*, rev. 23, Washington, D.C., pp. 1982–1983.
38. G. Wernimont, in W. Spendley, ed., *Use of Statistics to Develop and Evaluate Analytical Methods*, Association of Official Analytical Chemists, Arlington, Va., 1985.
39. Ref. 23, pp. 79–82.
40. *Ibid.*, p. 84.
41. *Ibid.*, pp. 193–195.
42. *Ibid.*, pp. 83–84.
43. J. T. Burr, *The Tools of Quality Part I: Going with the Flow(chart)*, in Ref. 15, June 1990, pp. 64–67.
44. J. S. Sarazen, *The Tools of Quality Part II: Cause-and-Effect Diagrams*, in Ref. 15, July 1990, pp. 59–62.
45. P. D. Shainin, *The Tools of Quality Part III: Control Charts*, in Ref. 15, Aug. 1990, pp. 79–82.
46. The Juran Institute, Inc., *The Tools of Quality Part IV: Histograms*, in Ref. 15, Sept. 1990, pp. 75–78.
47. The Juran Institute, Inc., *The Tools of Quality Part V: Check Sheets*, in Ref. 15, Oct. 1990, pp. 51–56.
48. J. T. Burr, *The Tools of Quality Part VI: Pareto Charts*, in Ref. 15, Nov. 1990, pp. 59–61.
49. J. M. Juran, *Indus. Qual. Cont.*, 5 (Oct. 1950).
50. J. T. Burr, *The Tools of Quality Part VII: Scatter Diagrams*, in Ref. 15, Dec. 1990, pp. 87–89.
51. Ref. 2, pp. 9.7–9.14.
52. A. M. Thayer, *Chem. Eng. News*, 10–26 (Apr. 25, 1994).
53. C. W. Reimann and H. S. Hertz, *ASTM Standard. News*, 44 (Nov. 1993).
54. *Management Practices—U.S. Companies Improve Performance Through Quality Efforts*, United States General Accounting Office Report to the Honorable D. Ritter, House of Representatives, Washington, D.C., May 1991, p. 2.
55. Ref. 53, p. 43.
56. Ref. 2, p. 4.2.
57. P. B. Crosby, in Ref. 15, Apr. 1983, p. 39.
58. H. Tilton, *Chem. Mark. Rep.*, 3 (Oct. 31, 1994).

IRWIN SILVERSTEIN
International Specialty Products

QUALITY CONTROL. See QUALITY ASSURANCE.

QUARTZ. See SILICA.

QUATERNARY AMMONIUM COMPOUNDS

There are a vast number of quaternary ammonium compounds or quaternaries (1). Many are naturally occurring and have been found to be crucial in biochemical reactions necessary for sustaining life. A wide range of quaternaries are also produced synthetically and are commercially available. Over 204,000 metric tons of quaternary ammonium compounds are produced annually in the United States (2). These have many diverse applications. Most are eventually formulated and make their way to the marketplace to be sold in consumer products. Applications range from cosmetics (qv) to hair preparations (qv) to clothes softeners, sanitizers for eating utensils, and asphalt emulsions.

Most quaternary ammonium compounds have the general formula $R_4N^+X^-$ and are a type of cationic organic nitrogen compound. The nitrogen atom, covalently bonded to four organic groups, bears a positive charge that is balanced by a negative counterion. Heterocyclics, in which the nitrogen is bonded to two carbon atoms by single bonds and to one carbon by a double bond, are also considered quaternary ammonium compounds. The R group may either be equivalent or correspond to two to four distinctly different moieties. These groups may be any type of hydrocarbon: saturated, unsaturated, aromatic, aliphatic, branched chain, or normal chain. They may also contain additional functionality and heteroatoms. Examples include methylpyridinium iodide [930-73-4] (**1**); benzyldimethyloctadecylammonium chloride [122-19-0] (**2**); and di(hydrogenated tallow)alkyldimethylammonium chloride [61789-80-8] (**3**), where R = C_{14}–C_{18}.

(**1**)   (**2**)   (**3**)

Nomenclature

Quaternary ammonium compounds are usually named as the substituted ammonium salt. The anion is listed last (3). Substituent names can be either common (stearyl) or IUPAC (octadecyl). If the long chain in the compound is from a natural mixture, the chain is named after that mixture, eg, tallowalkyl. Prefixes such as di- and tri- are used if an alkyl group is repeated. Complex compounds usually have the substituents listed in alphabetical order. Some common quaternary ammonium compounds and their applications in patent literature are listed in Table 1.

Properties

Physical Properties.   Most quaternary compounds are solid materials that have indefinite melting points and decompose on heating. Physical properties are determined by the chemical structure of the quaternary ammonium compound as well as any additives such as solvents. The simplest quaternary ammonium compound, tetramethylammonium chloride [75-57-0], is very soluble in water (163) and insoluble in nonpolar solvents. As the molecular weight of the quaternary compound increases, solubility in polar solvents decreases and solubility in nonpolar solvents increases (164–166). For example, trimethyloctadecylammonium chloride [112-03-8] is soluble in water up to 27%, whereas dimethyldioctadecylammonium chloride [107-64-2] has virtually no solubility in water. Appropriately formulated, however, this latter compound can be dispersed in water at relatively high (~15%) levels.

The ability to form aqueous dispersions is a property that gives many quaternary compounds useful applications. Placement of polar groups, eg, hydroxy or ethyl ether, in the quaternary structure can increase solubility in polar solvents.

Higher order aliphatic quaternary compounds, where one of the alkyl groups contains ~10 carbon atoms, exhibit surface-active properties (167). These compounds compose a subclass of a more general class of compounds known as cationic surfactants (qv). These have physical properties such as substantivity and aggregation in polar media (168) that give rise to many practical applications. In some cases the ammonium compounds are referred to as inverse soaps because the charge on the organic portion of the molecule is cationic rather than anionic.

Chemical Properties.   Reactions of quaternaries can be categorized into three types (169): Hoffman eliminations, displacements, and rearrangements. Thermal decomposition of a quaternary ammonium hydroxide to an alkene, tertiary amine, and water is known as the Hoffman elimination (eq. 1a) (170). This reaction has not been used extensively to prepare olefins. Some cyclic olefins, however, are best prepared this way (171). Exhaustive methylation, followed by elimination, is known as the Hoffman degradation and is important in the structural determination of unknown amines, especially for alkaloids (qv) (172).

$$CH_3CH_2CH_2\overset{+}{N}(CH_3)_3 + {}^-OH \underset{\textit{displacement}}{\overset{\textit{elimination}}{\rightleftarrows}} \begin{array}{l} CH_3CH_2{=}CH_2 + N(CH_3)_3 + H_2O \quad\quad (1a) \\[2ex] CH_3CH_2CH_2N(CH_3)_2 + CH_3OH \quad\quad (1b) \end{array}$$

Displacement of a tertiary amine from a quaternary (eq. 1b) involves the attack of a nucleophile on the α-carbon of a quaternary and usually competes

Table 1. Selected Quaternary Ammonium Compounds and Their Applications

Compound	CAS Registry Number	Application and function	Comments	Reference
		Agricultural chemicals		
cetyltrimethylammonium chloride	[112-02-7]	phase-transfer catalyst	used in production of furfuryl alcohol	4
dodecyltrimethylammonium bromide	[1119-94-4]	phase-transfer catalyst	used in production of furfuryl alcohol	4
dicocoalkyldimethylammonium chloride	[61789-77-3]	surfactant, flocculating agent	for herbicidal compositions	5–8
soyaalkyltrimethylammonium chloride	[61790-41-8]	surfactant, flocculating agent	for herbicidal compositions	5–8
2-chloroethyltrimethylammonium chloride	[999-81-5]	surfactant	used in formulations for plant growth regulation	9–11
trimethyloctadecylammonium chloride	[112-03-8]	surfactant	used in formulations for plant growth regulation	9–11
dimethyldioctadecylammonium chloride	[107-64-2]	surfactant, flocculating agent	for sugar liquor purification	12
dihexadecyldimethylammonium chlorides		surfactant, flocculating agents	for sugar liquor purification	12
short-chain alkyl quaternaries		fungicides, mold inhibitors	for protection of seeds and agricultural products	13,14
imidazoline-type quaternaries		fungicides, mold inhibitors	for protection of seeds and agricultural products	13,14
		Chemical industry		
benzylcetyldimethylammonium chloride	[122-18-9]	defoaming agent	used in the precipitation of $NaCO_3$ precursor crystals such as $NaHCO_3$	15
trimethyloctadecylammonium chloride	[112-03-8]	surfactant	for decolorization and purification of halogenated silane	16
C_{10}–C_{18}-ethoxylated quaternaries		surfactants	for coating polystyrene beads in the manufacture of lightweight concrete	17
dibenzyldimethylammonium chloride	[100-94-7]	directing agent	used in the preparation of crystalline silicate catalysts	18–20

741

Table 1. (Continued)

Compound	CAS Registry Number	Application and function	Comments	Reference
DL-3-cyano-2-hydroxy-propyltrimethylammonium hydroxide		optically active counterion	for production of optically active L-carnitine (vitamin B_T) for pharmaceuticals	21
many compounds claimed[a]		antistatic	for antistatic materials containing polyethylene glycol and poly-(vinyl chloride)	22
		Coatings and paints		
dodecyltrimethylammonium chloride	[112-00-5]	leveling agent	enhances the leveling activity of fluorochemical coatings	23
ethoxylated quaternaries		surfactants	stabilizes epoxy–polyamide coatings and imparts sag resistance	24,25
dialkyldiethoxylated quaternaries		foaming agents	for textile coating formulations	26–28
trialkyl (C_8–C_{10}) methyl-ammonium chlorides		surfactants	solubilizes hexavalent chromium compounds in paints	29
cocoalkylmethyldiethoxylated quaternaries		cations	for the preparation of organo-modified clays for paint thickeners	30
		Cosmetics		
dodecyltrimethylammonium chloride	[112-00-5]	combing aids, conditioners, creme rinses, antistatic	for hair conditioning formulations, creme rinses, and shampoos	31–39
quaternary ammonium-substituted sterol derivatives		surfactants	in cosmetic formulations	40
dimethyl and trimethyl alkyl quaternaries		emulsifiers	produces water-in-oil emulsions	41
quaternary ammonium derivatives of lanolinic acid		surfactants	produces high solids, clear costme-tic formulations	42
		Detergents		
dicocoalkyldimethylammonium chloride	[61787-77-3]	surfactant	for liquid crystal detergent composition for cold-temperature cleaning	43

742

Compound	CAS Registry Number	Function	Use	Reference
dialkyldiethoxylatedammonium chlorides		surfactants	for formulations removing road film from transportation vehicles	44
didodecyldimethylammonium chloride	[3401-74-9]	surfactant	removes iron sulfate from metal surfaces and acid cleaning compositions	45–47
octadecyltrimethylammonium chloride	[112-03-8]	surfactant	removes iron sulfate from metal surfaces and acid cleaning compositions	45–47
ditallowalkyldimethylammonium chloride	[68783-78-8]	surfactant	for improved perfume deposition on hard surfaces	48
alkylbenzyldimethyl quaternaries		biocides	for liquid disinfectant laundry detergent composition	49
many compounds claimed[b]		processing aids	incorporates positively charged nitrogen atom into spray-dried nonionic detergents to reduce autoxidation	50
many compounds claimed[c]		surfactants	for hard-surface cleaning formulations	51,52
Electronics				
dodecyltrimethylammonium chloride	[112-00-5]	etching solution, leveling agent	eliminates nonuniformities in etch rates as a result of variations in mass transport conditions	53
hexadecyltrimethylammonium chloride	[112-02-7]	etching solution, leveling agent	eliminates nonuniformities in etch rates as a result of variations in mass transport conditions	53
octadecyltrimethylammonium chloride	[112-03-8]	etching solution, leveling agent	eliminates nonuniformities in etch rates as a result of variations in mass transport conditions	53
hydroxyethyltrimethylammonium hydroxide	[123-41-1]	remover solution	for manufacture of semiconductor devices	54
Mining and flotation				
hydroxy-propoxylated quaternaries		surfactants	purifies calcium carbonate ore by removing silicate impurities through reverse flotation	55,56

Table 1. (Continued)

Compound	CAS Registry Number	Application and function	Comments	Reference
dialkyldimethyl quaternaries		surfactants	purify calcium carbonate ore by removing silicate impurities through reverse flotation	55,56
didodecyldimethylammonium chloride	[3401-74-9]	flocculent	recovery of precious metals by froth flotation	57–62
Paper manufacture				
di(hydrogenated tallow)-alkyldimethylammonium chloride	[61789-80-8]	softener	gives improved absorbency and softness to paper towels and tissues	63–66
(3-chloro-2-hydroxypropyl)-trimethylammonium chloride	[3327-22-8]	surfactant	stabilizes pulp bleaching formulations	67–69
cetyltrimethylammonium chloride	[112-02-7]	flocculent	precipitates lignins and lignin derivatives from pulp mill waste waters	70
Petroleum				
dicocoalkyldimethylammonium chloride	[61789-77-3]	counterion	used to prepare thiomolybdates for antifriction additives in oils and greases	71–73
tallowalkyltrimethyl quaternaries		suspending agents	stabilizes coal-oil slurries	74–78
imidazoline-type quaternaries		suspending agents	stabilizes coal-oil slurries	74–78
ethoxylated quaternaries		suspending agents	stabilizes coal-oil slurries	74–78
dioctadecyldimethylammonium chloride	[107-64-2]	organo-modified clays	used in polyolefin-based greases	79,80
diethoxylatedtallowmethyl quaternaries		dispersants	stabilizes drilling fluids and minimizes solids disintegration in down-hole fluids	81
tetraethoxylated quaternaries		corrosion inhibitors, demulsifiers	crude-oil refining	82
benzyldodecyldimethylammonium hydroxide	[10328-35-5]	surfactant	used in catalyst preparation for the oxidation of mercaptans in sour petroleum distillates	83–89

Compound	CAS Number	Function	Application	Reference
dibenzyldimethylammonium sulfocyanate		suspending agent	for sludge recovery in storage vessels	90
alkylbenzyldimethylammonium chlorides		complexing agents	for regeneration of used oils	91
trimethyloctadecylammonium hydroxide	[15461-40-2]	catalyst	for removing elemental sulfur from refined petroleum products	92,93
many compounds claimed[a]		biocides	for drilling fluids, enhanced oil recovery, and hydraulic fluid compositions	94,97
Pharmaceuticals and dermopharmacy				
quaternary ammonium salicylates		microbiocides	as bactericidal and keratolytic agents in the treatment of dermatoses	98,99
quaternary ammonium retinoates		microbiocides	as bactericidal and keratolytic agents in the treatment of dermatoses	98,99
benzyloctadecyldimethyl-ammonium chloride	[122-19-0]	germicide	in contraceptive formulations	100
dodecyldimethyl (2-phenoxyethyl)-ammonium bromide	[538-71-6]	antimicrobial	in aqueous ocular solutions for contact lenses	101
phosphate-type quaternaries		cleaners and disinfectants	formulated for preoperative skin scrubs, treatment for poison ivy and other skin disorders	102,103
imidazoline-type quaternaries		cleaners and disinfectants	formulated for preoperative skin scrubs, treatment for poison ivy and other skin disorders	102,103
Photographic industry				
disoyaalkyldimethylammonium ethosulfate		antistatic agent	added to developer rolls to increase productivity	104

Table 1. (Continued)

Compound	CAS Registry Number	Application and function	Comments	Reference
didodecyldimethylammonium chloride	[3401-74-9]	cationic activator	in recovery solution for a silver salt photographic plate in offset printing	105
alkynyl-substituted heterocyclic quaternaries		emulsifiers	for silver halide photographic materials	106–108
Polymer industry				
dialkyldiethoxylated quaternaries		antistatics	for producing expandable styrene polymer particles having antistatic and antilumping properties	109–113
didodecyldimethylammonium molybdates		smoke retardants	as additives for vinyl chloride polymer compositions	114
dialkyldimethylammonium chlorides		organoclays	as thixotropic agents in thermoset resins containing unsaturated polyesters	115
Textile manufacture				
dodecyltrimethyl quaternaries		dye bath additives	for dyeing jute backing of multi-level nylon carpet; dyeing assistants for polyamide fibers	116,117
imidazoline-type quaternaries		dye bath additives	for dyeing jute backing of multi-level nylon carpet; dyeing assistants for polyamide fibers	116,117
ethoxylated quaternaries		antistatics	for woven and nonwoven fibers from natural and synthetic sources	118–120
ditallowalkyldimethylammonium chloride	[68783-78-8]	germicide	as disinfecting cleaning intensifier for dry cleaning	121
dodecyltrimethylammonium chloride	[112-00-5]	emulsifier	as cationic emulsifier for fluoro-carbon finishes	122
Other industries				
dicocoalkyldimethylammonium chloride	[61789-77-3]	surfactant in automotive services	a key ingredient in protective polish formulations	123–125

Compound	CAS Number	Use		Reference
hexadecyltrimethylammonium hydroxide	[505-86-2]	surfactant in analytical chemistry	mobile-phase modifier for chromatographic determination of ions by reverse-phase liquid chromatography	126
di(hydrogenated tallow)alkyldimethylammonium chloride (DHTDMAC)	[61789-80-8]	household fabric softening, antistatic agent	for fabric softener compositions at the industrial and consumer level; rinse additive and impregnated substance composition for tumble dryers	127–144
distearyldimethylammonium methosulfate	[3843-16-1]	positive-charge carrier, charge control agent in electro-photographic imaging	as a charge control agent in electrographic toner and developer compositions	145–150
didodecyldimethylammonium chloride	[3401-74-9]	house biocide, fungicide, algicide, and germicide	cleaning and disinfecting formulations	151–159
cetyltrimethylammonium bromide	[57-09-0]	biocide for wastewater treatment	for disinfectant in wastewater treatment	160
many compounds claimed[e]		organoclays for wastewater	for removal of aromatic contaminants, such as benzene from water and air	161
many compounds claimed[f]		coupling agents for rubber	for incorporation of carbon black into natural rubber compositions	162

[a](3-Dodecylamidopropyl)trimethylammonium methylsulfate [10595-49-0] is an example.
[b]Trimethyltallowalkylammonium chloride [8030-78-2] is an example.
[c]Benzyldimethyl-2-hydroxydodecylammonium isononanoate is an example.
[d]Alkylbenzyldimethylammonium chlorides are an example.
[e]Tetramethylammonium chloride [75-57-0] is an example.
[f]Ditallowalkyldimethylammonium chloride [68783-78-8] is an example.

with the Hoffman elimination (173). The counterion greatly influences the course of this reaction. For example, the reaction of propyltrimethylammonium ion with hydroxide ion yields 19% methanol and 81% propylene, whereas the reaction with phenoxide ion yields 65% methoxybenzene and 15% propylene (174).

The Stevens rearrangement (eq. 2) is a base-promoted 1,2-migration of an alkyl group from a quaternary nitrogen to carbon (175,176). The Sommelet-Hauser rearrangement (eq. 3) is a base-promoted 1,2-migration of a benzyl group to the *ortho*-position of that benzyl group (177,178).

$$C_6H_5COCH_2\overset{+}{N}(CH_3)_2 \xrightarrow{\ ^-OH\ } C_6H_5COCHN(CH_3)_2 + H_2O \qquad (2)$$
$$\underset{CH_2C_6H_5}{|} \qquad\qquad \underset{CH_2C_6H_5}{|}$$

$$C_6H_5CH_2\overset{+}{N}(CH_3)_2 \xrightarrow[\text{NH}_3\,(l)]{\text{NaNH}_2} \qquad (3)$$

Naturally Occurring Quaternaries

Naturally occurring quaternary ammonium compounds have been reviewed (179). Many types of aliphatic, heterocyclic, and aromatic derived quaternary ammonium compounds are produced both in plants and invertebrates. Examples include thiamine (vitamin B_1) (**4**) (see VITAMINS); choline (qv) [62-49-7] (**5**); and acetylcholine (**6**). These have numerous biochemical functions. Several quaternaries are precursors for active metabolites.

(**4**)   (**5**)   (**6**)

Thiamine (**4**) functions as a coenzyme in several enzymatic reactions in which an aldehyde group is transferred from a donor to a receptor molecule. The thiazole ring is the focus of this chemistry. Thiamine also serves as a coenzyme in the pyruvate dehydrogenase and α-ketoglutarate dehydrogenase reactions. These take place in the main pathway of oxidation of carbohydrates (qv) in cells.

Choline functions in fat metabolism and transmethylation reactions. Acetylcholine functions as a neurotransmitter in certain portions of the nervous system. Acetylcholine is released by a stimulated nerve cell into the synapse and binds to the receptor site on the next nerve cell, causing propagation of the nerve impulse.

Biochemically, most quaternary ammonium compounds function as receptor-specific mediators. Because of their hydrophilic nature, small molecule quaternaries cannot penetrate the alkyl region of bilayer membranes and must activate receptors located at the cell surface. Quaternary ammonium compounds also function biochemically as messengers, which are generated at the inner surface of a plasma membrane or in a cytoplasm in response to a signal. They may also be transferred through the membrane by an active transport system.

General types of physiological functions attributed to quaternary ammonium compounds are curare action, muscarinic–nicotinic action, and ganglia blocking action. The active substance of curare is a quaternary that can produce muscular paralysis without affecting the central nervous system or the heart. Muscarinic action is the stimulation of smooth-muscle tissue. Nicotinic action is primary transient stimulation and secondary persistent depression of sympathetic and parasympathetic ganglia.

Analytical Test Methods

There are no universally accepted wet analytical methods for the characterization of quaternary ammonium compounds. The American Oil Chemists' Society (AOCS) has established, however, a number of applicable tests (180). These include sampling, color, moisture, amine value, ash, iodine value, average molecular weight, pH, and flash point.

Determination of the activity of quaternary samples falls into four categories (181,182). First is the partition titration in immiscible solvent systems, usually chloroform and water, using an anionic surfactant such as sodium lauryl sulfate [151-21-3] and an anionic dye indicator (183). This process, essentially a microtitration, is often referred to as the Epton titration (184,185). The end point requires considerable practice to detect. Second is the direct titration of the long-chain cation with sodium tetraphenylboron [143-66-8] using an anionic indicator (186). This is a macro method that is convenient and relatively simple to perform. Third is the titration of the halide anion (Volhardt type) or perchloric acid titration in acetic anhydride (187). The fourth category includes colorimetric methods using anionic dyes or indicators and partition solvent systems. These partition/colorimetric analytical methods, which have found widespread use in environmental analysis, have long been used for determining small amounts of quaternary ammonium compounds. Although they are not specific, these methods do indicate the presence of long-chain cationics that have at least one chain composed of eight or more carbon atoms.

The chain length composition of quaternaries can be determined by gas chromatography (qv) (188). Because of low volatility, quaternaries cannot be chromatographed directly, but only as their breakdown products. Quaternary ammonium salts can be analyzed by injecting them directly onto Apiezon grease columns treated with potassium hydroxide. In most cases, long-chain monoalkyl quaternaries break down to form two sets of peaks: short-chain alkyl halides and the long-chain tertiary amines. Long-chain dialkyl quaternaries elute on the basis of total carbon atoms present in the resulting methyldialkyl tertiary amine.

Mass spectral analysis of quaternary ammonium compounds can be achieved by fast-atom bombardment (fab) ms (189,190). This technique relies on bombarding a solution of the molecule, usually in glycerol [56-81-5] or *m*-nitrobenzyl alcohol [*619-25-0*], with argon and detecting the parent cation plus a proton (MH$^+$). A more recent technique has been reported (191), in which information on the structure of the quaternary compounds is obtained indirectly through cluster-ion formation detected via liquid secondary ion mass spectrometry (lsims) experiments.

Liquid chromatography has been widely applied for analysis of quaternaries. Modified reverse-phase columns can provide chain length information, whereas normal-phase chromatography results in groupings of alkyl distributions. Quaternary ammonium compounds can be separated into their mono-, di-, and trialkyl components on a normal-phase silica or alumina column using a conductivity detector (192). Solvent systems generally include tetrahydrofuran, methanol, and acetic acid. Because the conductivity detector is only sensitive to ionic species, solvents and nonionic components of the sample are not seen by the detector. Alternative columns are amino, modified silica, or cation exchange. Evaporative laser light scattering detectors (elsd) have also been utilized for the nonvolatile, nonultraviolet-absorbing quaternaries (193).

Nuclear magnetic resonance (nmr) spectroscopy is useful for determining quaternary structure. The ^{15}N-nmr can distinguish between quaternary ammonium compounds and amines, whether primary, secondary, or tertiary, as well as provide information about the molecular structure around the nitrogen atom. The ^{13}C-nmr can distinguish among oleic, tallow, and hydrogenated tallow sources (194).

Toxicology and Environmental Fate

Some quaternary ammonium compounds are potent germicides (164,195,196), toxic in small (mg/L range) quantities to a wide range of microorganisms. Bactericidal, algicidal, and fungicidal properties are exhibited. Ten-minute-contact kills of bacteria are typically produced by quaternaries in concentration ranges of 50–333 mg/L (197). Acute toxicity at low (1 mg/L) levels has been reported in invertebrates, snails, and fish (198,199). In plant systems, growth inhibition of green algae and great duckweed occurs at 3–5 mg/L (200). Acute oral toxicity data (albino rats) show most structures to have an LD$_{50}$ in the range of 100–5000 mg/kg (201). Many quaternaries are considered to be moderately to severely irritating to the skin and eyes.

Most uses of quaternary ammonium compounds can be expected to lead to these compounds' eventual release into wastewater treatment systems except for those used in drilling muds. Useful properties of the quaternaries as germicides can make these compounds potentially toxic to sewer treatment systems. It appears, however, that quaternary ammonium compounds are rapidly degraded in the environment and strongly sorbed by a wide variety of materials. Under normal circumstances these compounds are unlikely to pose a significant risk to microorganisms in wastewater treatment systems (202). Microbial populations acclimate readily to low levels of quaternary compounds and biodegrade them. Environmental toxicity and stability of these compounds have been described (202,203). Biodegradation data for some common quaternaries are given in Table 2.

Table 2. Biodegradation Data[a] for Quaternaries[b]

Chloride compound	CAS Registry Number	Concentration of actives, mg/L	Quantity biodegraded in 48 h, %
trimethyltallowalkylammonium	[8030-78-2]	5	95.3
trimethylsoyaalkylammonium	[61790-41-8]	5	94.1
cocoalkyltrimethylammonium	[61789-18-2]	5	82.0
di(hydrogenated tallow)alkyl-dimethylammonium	[61789-80-8]	7.5	79.0
dodecyltrimethylammonium	[112-00-5]	5	98.3
trimethyloctadecylammonium	[112-03-8]	5	98.4
hexadecyltrimethylammonium	[112-02-7]	5	81.9
dicocoalkyldimethylammonium	[61789-77-3]	5	80.3
benzyldi(hydrogenated tallow)-alkylmethylammonium	[61789-73-9]	7.5	79.0

[a]Ref. 204.
[b]Aerated, acclimated cultures from raw city sewage.

There are two reasons why the concentration of quaternaries is believed to remain at a low level in sewage treatment systems. First, quaternaries appear to bind anionic compounds and thus are effectively removed from wastewater by producing stable, lower toxicity compounds (205). Anionic compounds are present in sewer systems at significantly higher concentrations than are cations (202). Second, the nature of how most quaternaries are used ensures that their concentrations in wastewater treatment systems are always relatively low but steady. Consumer products such as fabric softeners, hair conditioners, and disinfectants contain only a small amount of quaternary compounds. This material is then diluted with large volumes of water during use.

The threat of accidental misuse of quaternary ammonium compounds coupled with potential harmful effects to sensitive species of fish and invertebrates has prompted some concern. Industry has responded with an effort to replace the questionable compounds with those of a more environmentally friendly nature. Newer classes of quaternaries, eg, esters (206) and betaine esters (207), have been developed. These materials are more readily biodegraded. The mechanisms of antimicrobial activity and hydrolysis of these compounds have been studied (207). Applications as surface disinfectants, antimicrobials, and *in vitro* microbiocidals have also been reported. Examples of ester-type quaternaries are shown in Figure 1.

Synthesis and Manufacture of Quaternaries

A wide variety of methods are available for the preparation of quaternary ammonium compounds (218–220). Significantly fewer can be used on a commercial scale. A summary of the most commonly used commercial methods is given herein.

Quaternary ammonium compounds are usually prepared by reaction of a tertiary amine and an alkylating agent (eq. 4). The most widely used alkylating agents are listed in Table 3. Some of these alkylating reagents pose significant

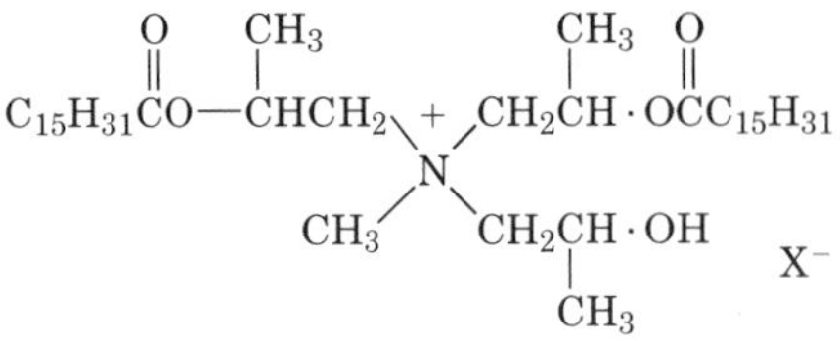

(**7**)

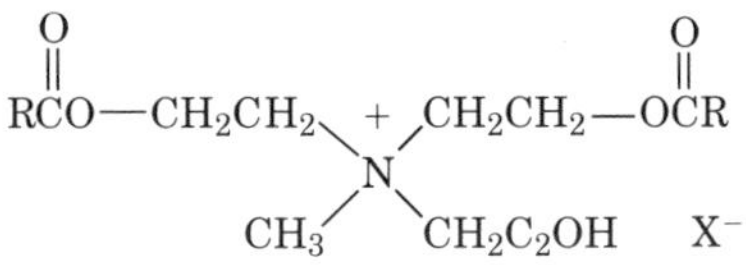

(**8**)

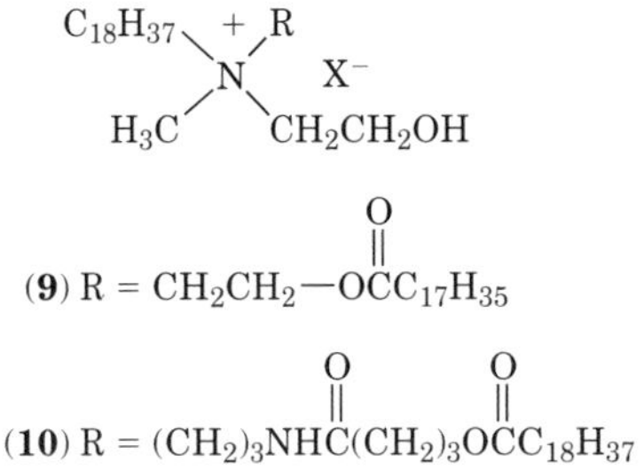

(**9**), (**10**)

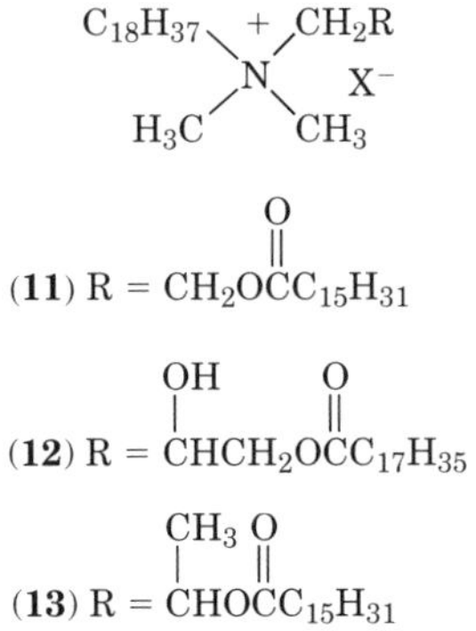

(**11**), (**12**), (**13**)

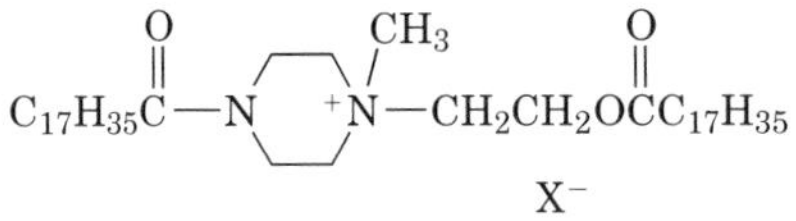

(**14**)

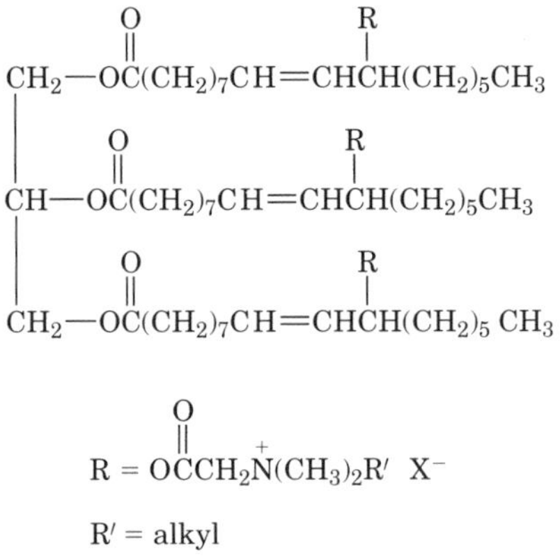

(**15**)

Fig. 1. Quaternary esteramines: (**7**) (208); (**8**) (209,210); (**9**) (211,212); (**10**) (213); (**11**) (214); (**12**) (215); (**13**) (136); (**14**) (216); and (**15**) (217).

Table 3. Typical Alkylating Agents for Preparation of Quaternaries

Alkylating agent	CAS Registry Number	Chemical formula	Quaternary using R_3N	
methyl chloride	[*74-87-3*]	CH_3Cl	$R_3N^+CH_3$	Cl^-
dimethyl sulfate	[*77-78-1*]	$(CH_3)_2SO_4$	$R_3N^+CH_3$	$CH_3SO_4^-$
diethyl sulfate	[*64-67-5*]	$(CH_3CH_2)_2SO_4$	$R_3N^+CH_2CH_3$	$CH_3CH_2SO_4^-$
benzyl chloride	[*100-44-7*]	$C_6H_5CH_2Cl$	$R_3N^+CH_2C_6H_5$	Cl^-

health concerns and require special handling techniques. Alkylation reactions are usually run at moderate (60–100°C) temperatures. When methyl chloride is used, the reactions are often performed under moderate (415–790 kPa (60–115 psi)) pressures.

$$R\!-\!\underset{\displaystyle \overset{\displaystyle R'}{|}}{N}\!-\!R'' + R'''X \longrightarrow R\!-\!\overset{\displaystyle \overset{R'}{|}}{\underset{\displaystyle \underset{R'''}{|}}{\pm N}}\!-\!R'' \quad X^- \tag{4}$$

Equation 4 can be classified as S_N2, ie, substitution nucleophilic bimolecular (221). The rate of the reaction is influenced by several parameters: basicity of the amine, steric effects, reactivity of the alkylating agent, and solvent polarity. The reaction is often carried out in a polar solvent, eg, isopropanol, which may increase the rate of reaction and make handling of the product easier.

Primary and secondary amines are usually converted to tertiary amines using formaldehyde and hydrogen in the presence of a catalyst (eqs. 5 and 6). This process, known as reductive alkylation (222), is attractive commercially. The desired amines are produced in high yields and without significant by-product formation. Quaternization by reaction of an appropriate alkylating reagent then follows.

$$RNH_2 + 2\,CH_2O \xrightarrow[\text{H}_2/\text{catalyst}]{} RN(CH_3)_2 \tag{5}$$

$$R_2NH + CH_2O \xrightarrow[\text{H}_2/\text{catalyst}]{} R_2NCH_3 \tag{6}$$

Dialkyldimethyl and alkyltrimethyl quaternaries can be prepared directly from secondary and primary amines as shown in equations 7 and 8, respectively. This process, known as exhaustive alkylation, is usually not the method of choice on a commercial scale. This technique requires the continuous addition of basic material over the course of the reaction to prevent the formation of amine salts (223,224). Furthermore, products such as inorganic salt and water must be removed from the quaternary. The salt represents a significant disposal problem.

$$R_2NH + 2\,CH_3Cl + NaOH \longrightarrow R_2^+N(CH_3)_2Cl^- + NaCl + H_2O \tag{7}$$

$$RNH_2 + 3\,CH_3Cl + 2\,NaOH \longrightarrow RN(CH_3)_3^+Cl^- + 2\,NaCl + 2\,H_2O \tag{8}$$

Synthesis and Manufacture of Amines. The chemical and business segments of amines (qv) and quaternaries are so closely linked that it is difficult to consider these separately. The majority of commercially produced amines originate from three amine raw materials: natural fats and oils, α-olefins, and fatty alcohols. Most large commercial manufacturers of quaternary ammonium compounds are fully back-integrated to at least one of these three sources of amines. The amines are then used to produce a wide array of commercially available quaternary ammonium compounds. Some individual quaternary ammonium compounds can be produced by more than one synthetic route.

Nitrile Intermediates. Most quaternary ammonium compounds are produced from fatty nitriles (qv), which are in turn made from a natural fat or oil-derived fatty acid and ammonia (qv) (Fig. 2) (see FATS AND FATTY OILS) (225). The nitriles are then reduced to the amines. A variety of reducing agents may be used (226). Catalytic hydrogenation over a metal catalyst is the method most often used on a commercial scale (227). Formation of secondary and tertiary amine side-products can be hindered by the addition of acetic anhydride (228) or excess ammonia (229). In some cases secondary amines are the desired products.

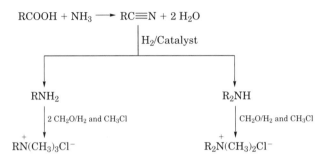

Fig. 2.   Quaternaries from fatty nitriles where R is a fatty alkyl group.

Fats, Oils, or Fatty Acids. The primary products produced directly from fats, oils, or fatty acids without a nitrile intermediate are the quaternized amidoamines, imidazolines, and ethoxylated derivatives (Fig. 3). Reaction of fatty acids or tallow with various polyamines produces the intermediate dialkylamidoamine. By controlling reaction conditions, dehydration can be continued until the imidazoline is produced. Quaternaries are produced from both amidoamines and imidazolines by reaction with methyl chloride or dimethyl sulfate. The ami-

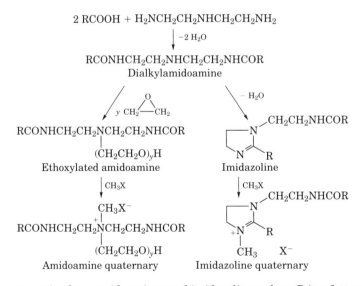

Fig. 3.   Quaternaries from amidoamines and imidazolines where R is a fatty alkyl group.

doamines can also react with ethylene oxide (qv) to produce ethoxylated amidoamines which are then quaternized.

These compounds and their derivatives can be manufactured using relatively simple equipment compared to that required for the fatty nitrile derivatives. Cyclization of amidoamines to imidazolines requires higher reaction temperatures and reduced pressures. Prices of imidazolines are therefore high.

Olefins and Fatty Alcohols. Alkylbenzyldimethylammonium (ABDM) quaternaries are usually prepared from α-olefin or fatty alcohol precursors. Manufacturers that start from the fatty alcohol usually prefer to prepare the intermediate alkyldimethylamine directly by using dimethylamine and a catalyst rather than from fatty alkyl chloride. Small volumes of dialkyldimethyl and alkyltrimethyl quaternaries in the C_8–C_{10} range are also manufactured from these precursors (Fig. 4).

Quaternized Esteramines. Esterquaternary ammonium compounds or esterquats can be formulated into products that have good shelf stability (209). Many examples of this type of molecule have been developed (see Fig. 1).

Quaternized esteramines are usually derived from fat or fatty acid that reacts with an alcoholamine to give an intermediate esteramine. The esteramines are then quaternized. A typical reaction scheme for the preparation of a diester quaternary is shown in equation 9 (210), where R is a fatty alkyl group. Reaction occurs at 75–115°C in the presence of sodium methoxide catalyst. Free fatty acids (230) and glycerides (231) can be used in place of the fatty acid methylester.

$$2\ RCOOCH_3 + N(CH_2CH_2OH)_3 \xrightarrow[-2\ CH_3OH]{} \underset{\underset{CH_2CH_2OH}{|}}{N(CH_2CH_2OOCR)_2} \xrightarrow{(CH_3)_2SO_4} (\mathbf{8}, X = CH_3SO_4^-)\quad (9)$$

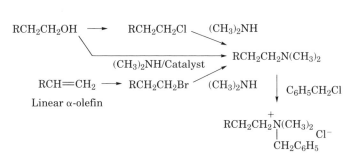

Fig. 4. Quaternaries from α-olefins of fatty alcohols where R is a fatty alkyl group. The product is alkylbenzyldimethyl quaternary.

Economic Aspects

A summary of the U.S. economic perspective, approximate production volumes, average retail prices, and leading producers is given in Table 4. Other commercial products include diallyldimethylammonium chloride [7398-69-8], produced by CPS Chemical Company; di(hydrogenated tallow)alkyldimethylammonium methosulfate [61789-81-9], produced by Akzo Nobel and High Point Chemical Corporation; and tetrabutylammonium bromide [1643-19-2], produced by

Table 4. 1993 U.S. Economic Aspects of Quaternaries[a]

Compound	CAS Registry Number	Production, t	Sales, t	Sales, $ × 10³	Average retail price, $/kg	Manufacturer[b]
benzyldimethyloctadecylammonium chloride	[122-19-0]	826	698	1,696	2.43	Cr, GC, L, P, RP, W
benzyltrimethylammonium chloride	[56-93-9]	1,732	1,132	1,971	1.74	HPC, RSA, Sy, W
benzylcocoalkyldimethylammonium chloride	[61789-71-7]	467	480	1,350	2.81	A, En, Ex, Gr, HP, J, W
assorted benzyldimethylalkylammonium chlorides		8,580	4,348	18,236	4.19	HL, L, P, St, Sy
benzyldimethyltallowalkylammonium chloride	[61789-75-1]	138	140	439	3.13	B, En, W
dicocoalkyldimethylammonium chloride	[61789-77-3]	2,118	1,985	5,090	2.56	A, Ex, J, W
dodecyltrimethylammonium chloride	[112-00-5]		239	702	2.93	A, J, L
hexadecyltrimethylammonium chloride	[112-02-7]	1,420	1,093	2,949	2.70	A, L, W
di(hydrogenated tallow)alkyldimethylammonium chloride	[61789-80-8]	24,009				A, En, W
oxygen-containing quaternaries (excluding amides)		2,431	1,624	4,524	2.79	Et, Ex, HC, K, L, SC
quaternary ammonium compounds with amide linkages		2,702	2,473	4,134	1.67	A, Ex, HPC, HD, I, L
quaternary ammonium compounds (cyclic) not containing oxygen		817	536	1,865	3.48	A, L
trimethyloctadecylammonium chloride	[112-03-8]	27	26	91	3.48	A, Ex, W
trimethyltallowalkylammonium chloride	[8030-78-2]	1,101	931	2,821	3.03	A, En, J, W

[a] Ref. 232.

[b] Manufacturer: A = Akzo Nobel Chemicals Inc.; B = Boehme Filatex., Inc.; CPS = CPS Chemical Co., Inc.; Cr = Croda, Inc.; En = Enenco, Inc.; Et = Ethox Chemicals, Inc.; Ex = Exxon Chemical Americas; GC = Gillette Chemical Co.; Gr = Gresco, Mfg., Inc.; HP = Hart Products Corp.; HC = Henkel Corp.; HPC = High Point Chemical Corp.; HD = Hilton Davis Chemical Corp.; HL = Huntington Laboratories, Inc.; I = Inolex Chemical Co.; J = Jetco Chemicals, Inc.; K = Karlshamms; L = Lonza, Inc.; P = PPG Industries, Inc.; RSA = RSA Corp.; RP = Rhône-Poulenc, Inc.; SC = Sandoz Chemicals Corp.; St = Stepan Co.; Sy = Sybron Chemicals, Inc.; W = Witco Corp.; and Z = Zeeland Chemical, Inc.

Zeeland Chemical. The leading producers of phase-transfer quaternaries are Eastman Kodak Company, Hexcel Corporation, RSA, Chemical Dynamics Corporation, Lindan Chemicals, Henkel Corporation, and Akzo Nobel. From 2,300 to 11,300 metric tons of quaternaries were used as phase-transfer catalysts during 1991 (233). The principal producers of perfluorinated quaternaries are Ciba, 3M Specialty Chemicals, and E. I. du Pont de Nemours & Company, Inc.

The leading U.S. manufacturers in terms of volumes of fatty nitrogen derivatives are Akzo Nobel and Witco. The combined annual production of these two companies accounts for about 80% of the fatty nitrogen products manufactured in the United States. The remaining production is divided among smaller-volume producers. Nearly all fatty nitrile plants require considerable technological expertise and capital investment to operate. The largest-volume product types in the 1990s are di(hydrogenated tallow)alkyldimethylammonium (DHT) quaternaries, which are sold as the chloride or methyl sulfate salts (234). The principal U.S. producers of imidazoline quaternaries are Akzo Nobel, Witco, Croda, and Lonza.

The single largest market for quaternary ammonium compounds is as fabric softeners. In 1993 this market accounted for over 50,000 metric tons of quaternaries in the United States (235). Consumption of these products is increasing at an annual rate of about 2–3%. The hair care market consumed over 9000 metric tons of quaternary ammonium compounds in 1992 (236). The annual consumption for organoclays is estimated at 12,700 metric tons (237). Esterquats have begun to gain market share in Western Europe and growth is expected to continue.

Uses

Uses of quaternary ammonium compounds range from surfactants to germicides and encompass a number of diverse industries (see Table 1).

Fabric Softening. The use of quaternary surfactants as fabric softeners and static control agents can be broken down into three main household product types: rinse cycle softeners; tumble dryer sheets; and detergents containing softeners, also known as softergents. Rinse cycle softeners are aqueous dispersions of quaternary ammonium compounds designed to be added to the wash during the last rinse cycle (131,135,137,143). Original products contained from 3–8% quaternary ammonium compound, typically di(hydrogenated tallow)alkyldimethylammonium chloride [61789-80-8] (DHTDMAC). During the 1980s and early 1990s, rinse cycle softeners went through significant changes as the active concentration of the dispersions was increased to reduce packaging and solid waste. Refills were introduced that had cationic activity of 16–27%. These products are meant to be added to a regular-strength softener bottle (typically 64 ounces (0.9 kg)) and diluted. As of 1995, the latest innovation was the ultra-fabric softener. This product, also formulated to contain from 16–27% cationic activity, is designed to be used without dilution. The actual dosage to the washer was decreased to approximately one fluid ounce (29 mL) from the older standard of 3–4 fluid ounces (87–116 mL). Although DHTDMAC is a widely employed softener, the use of imidazoline and amidoamine quaternaries has increased because these latter are easier to formulate into high active systems. The combination of DHTDMAC, imidazoline, and amidoamine is used to maximize softening performance and facilitate handling and formulation.

In 1991, the European fabric softener market took a sharp turn. Producers in Germany, the Netherlands, and later in Austria and Switzerland voluntarily gave up the use of DHTDMAC (238) because of pressure from local environmental authorities, who gave an environmentally hazardous classification to DHTD-MAC. A number of esterquats were developed as candidates to become successors to DHTDMAC (see Fig. 1). The ester group facilitates biodegradation.

Tumble dryer sheets contain a quaternary ammonium compound formulation applied to a nonwoven sheet typically made of polyester or rayon (130,132) (see NONWOVEN FABRICS). These sheets are added with wet clothes to the tumble dryer. Although these products afford some softening to the clothes, their greatest strength is in preventing static charge buildup on clothes during the drying cycle and during wear. A nonionic surfactant, such as an ethoxylated alcohol or fatty acid, is typically used in combination with the quaternary ammonium compound. The nonionics are known as release agents (qv) or distribution agents. More efficient transfer of the quaternary from the substrate to the drying fabric can be obtained.

Detergents containing softeners are also produced (134) (see DETERGENCY). These softergents are made from complex formulas in order to accomplish both detergency and softening during the wash cycle. These formulations typically contain quaternary ammonium compounds mixed with other materials such as clays (qv) for softening, in conjunction with the typical nonionic and anionic cleaning surfactant. The consumer benefits of quaternaries are fabric softening, antistatic properties, ease of ironing, and reduction in energy required for drying.

Hair Care.   Quaternary ammonium compounds are the active ingredients in hair conditioners (31–39,239). Quaternaries are highly substantive to human hair because the hair fiber has anionic binding sites at normal pH ranges. The use of quaternaries as hair conditioners can be broken down into creme rinses and shampoo conditioners.

Creme rinses are applied to the hair after washing. Frequently used quaternaries in creme rinses are dodecyltrimethylammonium chloride [112-00-5], dimethyloctadecyl(pentaethoxy)ammonium chloride, benzyldimethyloctadecylammonium chloride [122-19-0], and dimethyldioctadecylammonium chloride [107-64-2] (31–34,36).

Conditioning shampoos are formulations that contain anionic surfactants for cleaning hair and cationic surfactants for conditioning (39,236,239). The quaternary ammonium compounds most often used are either trihexadecyl-methylammonium chloride [71060-72-5], ethoxylated quaternaries, or one of the polymeric quaternaries. The polymeric quaternaries have either a natural or a synthetic backbone and numerous quaternary side functions. The polymer may offer an advantage by showing a high degree of affinity to the human hair surface and providing better compatibility with the other ingredients of conditioner shampoos (240).

Regardless of how the conditioner is applied or what the structure of the quaternary is, benefits provided to conditioned hair include the reduction of combing forces, increased luster, and improved antistatic properties.

Germicides.   The third largest market for quaternaries is sanitation (241). Generally, quaternaries offer several advantages over other classes of sanitizing chemicals, such as phenols, organohalides, and organomercurials, in that quaternaries are less irritating, low in odor, and have relatively long activity (242). The first use of quaternaries in the food industry occurred in the dairy industry for the sanitization of processing equipment. Quaternaries find use as disinfectants and sanitizers in hospitals, building maintenance, and food processing (qv) (151–159); in secondary oil recovery for drilling fluids (94–97); and in cooling water applications (see DISINFECTANTS AND ANTISEPTICS; PETROLEUM). Quaternaries have also received extensive attention for use as a general medicinal antiseptics (102) and in the pharmaceutical area as skin disinfectants and surgical antiseptics (103). In addition, quaternaries have been used in the treatment of eczema and other dermatological disorders as well as in contraceptive formulations (see CONTRACEPTIVES) (104) and ocular solutions for contact lenses (qv) (105).

Alkylbenzyldimethyl quaternaries (ABDM) are used as disinfectants (49) and preservatives. The most effective alkyl chain length for these compounds is between 10 and 18 carbon atoms. Alkyltrimethyl types, alkyldimethylbenzyl types, and didodecyldimethylammonium chloride [3401-74-9] exhibit excellent germicidal activity (151–159). Dialkyldimethyl types are effective against anaerobic bacteria such as those found in oil wells (94–97). One of the most effective and widely used biocides is didecyldimethylammonium chloride [7173-57-5].

Organoclays.   Another large market for quaternary ammonium salts is the manufacture of organoclays, ie, organo-modified clays (237). Clay particles are silicate minerals that have charged surfaces and which attract cations or anions electrostatically. Organoclays are produced by ion-exchange (qv) reaction between the quaternary ammonium salt and the surface of the clay particles (79,80,115,161). The quaternary ammonium salt displaces the adsorbed cations, usually sodium or potassium, producing an organo-modified clay. The new modified clay exhibits different behavior from that of the initial clay. Most importantly, it is preferentially wet by organic liquids and not by water.

The main use of these clays is to control, or adjust, viscosity in nonaqueous systems. Organoclays can be dispersed in nonaqueous fluids to modify the viscosity of the fluid so that the fluid exhibits non-Newtonian thixotropic behavior. Important segments of this area are drilling fluids, greases (79,80), lubricants, and oil-based paints. The most used commercial products in this area are dimethyldi(hydrogenated tallow)alkylammonium chloride [61789-80-8], dimethyl(hydrogenated tallow)alkylbenzylammonium chloride [61789-72-8], and methyldi(hydrogenated tallow)alkylbenzylammonium chloride [68391-01-5].

Miscellaneous Uses.   *Phase-Transfer Catalysts.*   Many quaternaries have been used as phase-transfer catalysts. A phase-transfer catalyst (PTC) increases the rate of reaction between reactants in different solvent phases. Usually, water is one phase and a water-immiscible organic solvent is the other. An extensive amount has been published on the subject of phase-transfer catalysts (233). Both the industrial applications in commercial manufacturing processes (243) and their synthesis (244) have been reviewed. Common quaternaries employed as phase-transfer agents include benzyltriethylammonium chloride [56-37-1], tetrabutylammonium bromide [1643-19-2], tributylmethylammonium chloride [56375-79-2], and hexadecylpyridinium chloride [123-03-5].

Polyamine-Based Quaternaries. Another important class of quaternaries are the polyamine-based or polyquats. Generally, polyamine-based quaternaries have been used in the same applications as their monomeric counterparts (245). Discussions, including the use of polymeric quaternaries in laundry formulations (246) and in the petroleum industry as damage control agents (247), have been published.

Perfluorinated Quaternaries. Perfluorinated quaternaries are another important, but smaller, class of quaternary ammonium compounds. In general, these are similar to their hydrocarbon counterparts but have at least one of the hydrocarbon chains replaced with a perfluoroalkyl group. These compounds are generally much more expensive than hydrocarbon-based quaternaries, so they must offer a significant performance advantage if they are to be used. Production volumes of perfluorinated quaternary ammonium compounds are significantly smaller than those of other classes. Many of these quaternaries have proprietary chemical structures. They are used in water-based coating applications to promote leveling, spreading, wetting, and flow control.

BIBLIOGRAPHY

"Quaternary Ammonium Compounds" in *ECT* 2nd ed., Vol. 19, pp. 859–865, by R. A. Reck, Armour Industrial Chemical Co.; in *ECT* 3rd ed., Vol. 19, pp. 521–531, by R. A. Reck, Armak Co.

1. E. Jungermann, ed., *Cationic Surfactants*, Marcel Dekker, Inc., New York, 1969, Chapts. 2–5.
2. *Chemical Economics Handbook*, SRI International, Menlo Park, Calif., 1990, p. 583.8200G; Supplemental data, May 1994, pp. 107–108.
3. J. Fletcher, O. Dermer, and R. Fox, *Nomenclature of Organic Compounds*, ACS, Washington, D.C., 1974, pp. 189–195.
4. U.S. Pat. 5,008,410 (Apr. 16, 1991), K. Tashiro and K. Tanaka (to Sumitomo Chemical Co., Ltd.).
5. Ger. Pat. 2,506,834 (Aug. 21, 1975), R. Ford and T. Tadros (to Imperial Chemical Industries Ltd.).
6. U.S. Pat. 5,078,781 (Jan. 7, 1992), C. Finch (to Imperial Chemical Industries Ltd.).
7. Ger. Pat. 2,304,204 (Aug. 2, 1973), N. Drewe, R. Parker, and T. Tadros (to Imperial Chemical Industries Ltd.).
8. Ger. Pat. 2,352,334 (May 9, 1974), F. Hauxwell, H. Murton, and R. Brain (to Imperial Chemical Industries Ltd.).
9. U.S. Pat. 2,844,466 (July 22, 1958), A. Rogers and co-workers (to Armour and Co.).
10. U.S. Pat. 3,506,433 (Apr. 14, 1970), W. Abramitis and R. A. Reck (to Armour and Co.).
11. U.S. Pat. 4,765,823 (Aug. 23, 1988), K. Lurssen (to Bayer AG).
12. U.S. Pat. 3,698,951 (Oct. 17, 1972), M. Bennett (to Tate and Lyle Ltd.).
13. U.S. Pat. 5,017,612 (May 21, 1987), J. Nayfa (to J. Nayfa).
14. U.S. Pat. 2,740,744 (Apr. 3, 1956), W. Aoramitis and co-workers (to Armour and Co.).
15. U.S. Pat. 3,725,014 (Apr. 3, 1973), R. Poncha and co-workers (to Allied Chemical Corp.).

16. U.S. Pat. 5,210,250 (Aug. 27, 1992), I. Watanuki and co-workers (to Shin-Etsu Chemical Co.).

17. U.S. Pat. 4,398,958 (Aug. 16, 1983), R. Hodson and co-workers (to Cempol Sales Ltd.).

18. U.S. Pat. 4,636,373 (Aug. 30, 1987), M. Rubin (to Mobil Oil Corp.).

19. U.S. Pat. 4,640,829 (Aug. 30, 1987), M. Rubin (to Mobil Oil Corp.).

20. U.S. Pat. 4,642,226 (Mar. 29, 1985), R. Calvert and co-workers (to Mobil Oil Corp.).

21. U.S. Pat. 5,191,085 (Mar. 2, 1993), H. Jakob and co-workers (to Degussa AG).

22. U.S. Pat. 4,661,547 (Apr. 28, 1987), M. Harada, K. Tsukanoto, and E. Tomohiro (to Nippon Rubber Co., Ltd.).

23. U.S. Pat. 4,064,067 (Dec. 20, 1977), A. Lore (to E.I. du Pont de Nemours & Co., Inc.).

24. U.S. Pat. 4,771,088 (Jan. 15, 1987), A. Pekarkik (to Glidden Co.).

25. U.S. Pat. 4,840,980 (Feb. 29, 1988), A. Pekarkik (to Glidden Co.).

26. U.S. Pat. 4,208,485 (June 17, 1980), R. Nahta (to GAF Corp.).

27. U.S. Pat. 4,230,746 (Oct. 28, 1980), R. Nahta (to GAF Corp.).

28. U.S. Pat. 4,198,316 (Apr. 15, 1980), R. Nahta (to GAF Corp.).

29. U.S. Pat. 4,226,624 (Oct. 7, 1980), J. Ohr (to U.S. Navy).

30. U.S. Pat. 4,677,158 (Nov. 12, 1985), S. Tso and co-workers (United Catalysts Inc.).

31. U.S. Pat. 5,019,376 (May 28, 1991), H. Vick (to S. C. Johnson & Son, Inc.).

32. U.S. Pat. 4,818,523 (Apr. 4, 1989), J. Clarke and co-workers (to Colgate-Palmolive Co.).

33. U.S. Pat. 4,886,660 (Dec. 12, 1989), A. Patel and co-workers (to Colgate-Palmolive Co.).

34. U.S. Pat. 4,144,326 (Mar. 13, 1979), O. Luedicke and F. Gichia (to American Cyanamide Co.).

35. U.S. Pat. 5,034,219 (July 23, 1991), V. Deshpande and J. Walts (to Sterling Drug Inc.).

36. U.S. Pat. 4,719,104 (Jan. 12, 1988), C. Patel (to Helene Curtis, Inc.).

37. U.S. Pat. 3,980,091 (Sept. 14, 1976), G. Dasher and co-workers (to Alberto Culver Co.).

38. U.S. Pat. 4,919,846 (Apr. 24, 1990), Y. Nakama and co-workers (to Shiseido Co., Ltd.).

39. U.S. Pat. 3,577,528 (May 4, 1971), E. McDonough (to Zotos International).

40. U.S. Pat. 5,004,737 (Apr. 2, 1991), Y. Kim and B. Ha (to Pacific Chemical Co. Ltd.).

41. U.S. Pat. 5,015,469 (May 14, 1991), T. Yoneyama and co-workers (to Shiseido Co., Ltd.).

42. U.S. Pat. 4,069,347 (Jan. 17, 1978), J. McCarthy and co-workers (to Emery Industries, Inc.).

43. U.S. Pat. 5,035,826 (Sept. 22, 1991), P. Dorbut, M. Mondin, and G. Broze (to Colgate-Palmolive Co.).

44. U.S. Pat. 4,284,435 (Aug. 18, 1981), D. Fox (to S. C. Johnson & Son, Inc.).

45. U.S. Pat. 3,969,281 (July 13, 1976), T. Sharp (to T. Sharp.).

46. U.S. Pat. 4,011,097 (Mar. 8, 1977), T. Sharp (to T. Sharp.).

47. U.S. Pat. 4,541,945 (Sept. 17, 1985), J. Anderson and S. Seiglle (to Amchem Products).

48. U.S. Pat. 4,636,330 (Jan. 13, 1987), J. Melville (to Lever Brothers Co.).

49. U.S. Pat. 4,576,729 (Mar. 18, 1986), L. Paszek and B. Gebbia (to Sterling Drug, Inc.).

50. U.S. Pat. 4,126,586 (Nov. 21, 1978), M. Curtis, R. Davies, and J. Galvin (to Lever Brothers Co.).

51. U.S. Pat. 4,814,108 (Mar. 21, 1989), J. Geke and H. Rutzen (to Henkel Kommanditgesellschaft auf Aktien).

52. U.S. Pat. 4,678,605 (July 7, 1987), J. Geke and S. Seiglle (to Henkel Kommanditgesellshaft auf Aktien).

53. U.S. Pat. 5,127,991 (July 7, 1992), S. Lai and C. Smith (to AT&T Bell Laboratories).

54. U.S. Pat. 5,185,235 (Mar. 30, 1990), H. Sato, K. Tazawa, and T. Aoyama (to Tokyo Ohka Kogyo Co., Ltd.).
55. U.S. Pat. 4,892,649 (Jan. 9, 1990), J. Mehaffey and T. Newman (to Akzo America Inc.).
56. U.S. Pat. 4,995,965 (Feb. 26, 1991), J. Mehaffey and T. Newman (to Akzo American Inc.).
57. U.S. Pat. 3,976,565 (Aug. 24, 1976), V. Petrovich (to V. Petrovich).
58. U.S. Pat. 4,098,686 (July 4, 1978), V. Petrovich (to V. Petrovich).
59. U.S. Pat. 4,225,428 (Sept. 30, 1980), V. Petrovich (to V. Petrovich).
60. U.S. Pat. 3,979,207 (Sept. 7, 1976), J. MacGregor (to Mattey Rustenburg Refiners).
61. U.S. Pat. 4,306,081 (Dec. 15, 1981), G. Rich (to Albee Laboratories, Inc.).
62. U.S. Pat. 4,289,530 (Sept. 15, 1981), G. Rich (to Albee Laboratories, Inc.).
63. U.S. Pat. 4,351,699 (Sept. 28, 1982), T. Osborn (to Procter & Gamble Co.).
64. U.S. Pat. 4,441,962 (Apr. 10, 1984), T. Osborn (to Procter and Gamble Co.).
65. U.S. Pat. 5,240,562 (Aug. 31, 1993), D. Phan and P. Trokhan (to Procter & Gamble Co.).
66. U.S. Pat. 3,916,058 (Oct. 28, 1975), P. Vossos (to Nalco Chemical Co.).
67. U.S. Pat. 4,119,486 (Oct. 10, 1978), R. Eckert (to Westvaco Corp.).
68. U.S. Pat. 5,013,404 (May 7, 1991), S. Christiansen, T. Littleton, and R. Patton (to Dow Chemical Co.).
69. U.S. Pat. 5,145,558 (Sept. 8, 1992), S. Christiansen, T. Littleton, and R. Patton (to Dow Chemical Co.).
70. U.S. Pat. 4,134,786 (Jan. 16, 1979), J. Fletcher and M. Humphrey (to J. Fletcher and M. Humphrey).
71. U.S. Pat. 4,343,746 (Aug. 10, 1982), J. Anglin, Y. Ryu, and G. Singerman (to Gulf Research & Development Co.).
72. U.S. Pat. 4,400,282 (Aug. 23, 1983), J. Anglin, Y. Ryu, and G. Singerman (to Gulf Research & Development Co.).
73. U.S. Pat. 4,343,747 (Aug. 10, 1982), J. Anglin, Y. Ryu, and G. Singerman (to Gulf Research & Development Co.).
74. U.S. Pat. 4,364,742 (Dec. 21, 1982), K. Knitter and J. Villa (to Diamond Shamrock Corp.).
75. U.S. Pat. 4,364,741 (Dec. 21, 1982), J. Villa (to Diamond Shamrock Corp.).
76. U.S. Pat. 4,398,918 (Aug. 16, 1983), T. Newman (to Akzona Inc.).
77. U.S. Pat. 4,478,602 (Oct. 23, 1984), E. Kelley, W. Herzberg, and J. Sinka (to Diamond Shamrock Chemicals Co.).
78. U.S. Pat. 4,575,381 (Mar. 11, 1986), R. Corbeels and S. Vasconcellos (to Texaco Inc.).
79. U.S. Pat. 4,317,737 (Mar. 2, 1982), A. Oswald, G. Harting, and H. Barnum (to Exxon Research and Engineering Co.).
80. U.S. Pat. 4,365,030 (Dec. 21, 1982), A. Oswald and H. Barnum (to Exxon Research & Engineering Co.).
81. U.S. Pat. 4,828,724 (May 9, 1989), C. Davidson (to Shell Oil Co.).
82. U.S. Pat. 3,974,220 (Aug. 10, 1976), L. Heiss and M. Hille (Hoechst AG).
83. U.S. Pat. 4,206,079 (June 3, 1980), R. Frame (to UOP Inc.).
84. U.S. Pat. 4,260,479 (Apr. 7, 1981), R. Frame (to UOP Inc.).
85. U.S. Pat. 4,290,913 (Sept. 22, 1981), R. Frame (to UOP Inc.).
86. U.S. Pat. 4,295,993 (Oct. 20, 1981), D. Carlson (to UOP Inc.).
87. U.S. Pat. 4,354,926 (Oct. 19, 1982), D. Carlson (to UOP Inc.).
88. U.S. Pat. 4,337,141 (June 19, 1982), R. Frame (to UOP Inc.).
89. U.S. Pat. 4,124,493 (Nov. 7, 1978), R. Frame (to UOP Inc.).
90. U.S. Pat. 4,474,622 (Oct. 2, 1984), M. A. Forster (Establisments Somalor-Ferrari Somafer SA).

91. U.S. Pat. 4,376,040 (Mar. 8, 1993), G. Sader (to G. Sader).

92. U.S. Pat. 5,200,062 (Apr. 6, 1993), M.-A. Poirier and J. Gilbert (to Exxon Research and Engineering Co.).

93. U.S. Pat. 4,514,286 (Apr. 30, 1985), S. Wang, G. Roof, and B. Porlier (to Nalco Chemical Co.).

94. U.S. Pat. 4,427,435 (Jan. 24, 1984), J. Lorenz and R. Grade (Ciba-Geigy Corp.).

95. U.S. Pat. 4,470,918 (Apr. 19, 1984), B. Mosier (to Global Marine, Inc.).

96. U.S. Pat. 4,560,761 (Dec. 24, 1985), E. Fields and M. Winzenburg (to Standard Oil Co.).

97. U.S. Pat. 4,526,986 (July 2, 1985), E. Fields and M. Winzenburg (to Standard Oil Co.).

98. U.S. Pat. 4,857,525 (Aug. 15, 1989), M. Philippe and co-workers (to L'Oreal).

99. U.S. Pat. 5,001,156 (Mar. 19, 1991), M. Philippe and co-workers (to L'Oreal).

100. U.S. Pat. 4,321,277 (Mar. 23, 1982), V. Saurino (to Research Lab Products, Inc.).

101. U.S. Pat. 5,165,918 (Jan. 5, 1990), B. Heyl, L. Winterton, and F. P. Tsao (Ciba-Geigy Corp.).

102. U.S. Pat. 5,244,666 (Sept. 14, 1993), J. Murley (to Consolidated Chemical, Inc.).

103. U.S. Pat. 4,941,989 (July 17, 1990), D. Kramer and P. Snow (to Ridgely Products Co., Inc.).

104. U.S. Pat. 5,132,050 (July 21, 1992), R. Baker and J. Thompson (to Lexmark International, Inc.).

105. U.S. Pat. 4,965,168 (Oct. 23, 1990), H. Yoshida, T. Kamada, and O. Kainuma (to Nikken Chemical Laboratory Co., Ltd.).

106. U.S. Pat. 5,035,993 (July 30, 1991), S. Hirano, A. Murai, and S. Suzuki (Fuji Photo Film Co., Ltd.).

107. U.S. Pat. 4,828,973 (May 9, 1989), S. Hirano, A. Murai, and S. Suzuki (Fuji Photo Film Co., Ltd.).

108. U.S. Pat. 4,471,044 (Sept. 11, 1984), R. Parton, W. Gaugh, and K. Wiegers (to Eastman Kodak Co.).

109. U.S. Pat. 4,628,068 (Dec. 9, 1986), H. Kesling and J. Harris (to Atlantic Richfield Co.).

110. U.S. Pat. 4,603,149 (July 29, 1986), H. Kesling and J. Harris (to Atlantic Richfield Co.).

111. U.S. Pat. 4,599,366 (July 8, 1986), H. Kesling and J. Harris (to Atlantic Richfield Co.).

112. U.S. Pat. 4,622,345 (Nov. 11, 1986), H. Kesling and J. Harris (to Atlantic Richfield Co.).

113. U.S. Pat. 5,110,835 (May 5, 1992), M. Walter, K.-H. Wassmer, and M. Lorenz (to BASF AG).

114. U.S. Pat. 4,410,462 (Oct. 18, 1983), W. Kroenke (to B. F. Goodrich Co.).

115. U.S. Pat. 3,974,125 (Aug. 10, 1976), A. Oswald and H. Barnum (to Exxon Research and Engineering Co.).

116. U.S. Pat. 3,981,679 (Sep. 21, 1976), N. Christie and J. Karnilaw (to Diamond Shamrock Corp.).

117. U.S. Pat. 4,441,884 (Apr. 10, 1984), H.-P. Baumann and U. Mosimann (to Sandoz Ltd.).

118. U.S. Pat. 4,104,175 (Aug. 1, 1978), M. Martinsson and K. Hellsten (to Modokemi Aktiebolag).

119. U.S. Pat. 3,972,855 (Aug. 3, 1976), M. Martinsson and K. Hellsten (to Modokemi Aktiebolag).

120. U.S. Pat. 4,104,443 (Aug. 1, 1978), B. Latha, C. Stevens and B. Dennis (to J. P. Stevens & Co., Inc.).

121. U.S. Pat. 4,406,809 (Sept. 27, 1983), K. Hasenclever (to Chemische Fabrik Kreussler & Co., GmbH).
122. U.S. Pat. 4,416,787 (Nov. 22, 1983), R. Marshall, W. Archie, and K. Dardoufas (to Allied Corp.).
123. U.S. Pat. 3,756,835 (Sept. 4, 1973), R. Betty and H. Nemeth (to Akzona Inc.).
124. U.S. Pat. 3,497,365 (Feb. 24, 1970), J. Roselle and W. Wagner (to Armour Industrial Chemicals, Co.).
125. U.S. Pat. 3,551,168 (Dec. 29, 1970), J. Roselle and W. Wagner (to Armour Industrial Chemicals, Co.).
126. U.S. Pat. 5,167,827 (Jan. 8, 1992), B. Glatz (to Hewlett-Packard Co.).
127. U.S. Pat. 3,625,891 (Dec. 7, 1991), M. Walden, W. Springs, and A. Mariahazey-Westchester (to Armour Industrial Chemical Co.).
128. U.S. Pat. 3,505,221 (Apr. 7, 1970), M. Walden, Northbrook, and A. Mariahazey-Westchester (to Armour Industrial Chemical Co.).
129. U.S. Pat. 3,573,091 (Mar. 30, 1971), M. Walden, Northbrook, and A. Mariahazey-Westchester (to Armour and Co.).
130. U.S. Pat. 4,096,071 (June 20, 1978), A. Murphy (to Procter & Gamble Co.).
131. U.S. Pat. 4,203,852 (May 20, 1980), J. Johnson and W. Chirash (to Colgate-Palmolive Co.).
132. U.S. Pat. 4,255,484 (Mar. 10, 1981), F. Stevens (to A. E. Staley Manufacturing Co.).
133. U.S. Pat. 4,772,404 (Sept. 20, 1988), D. Fox, M. Sullivan, and A. Cuomo (to Lever Brothers Co.).
134. U.S. Pat. 4,806,260 (Feb. 21, 1989), G. Broze and D. Bastin (to Colgate-Palmolive Co.).
135. U.S. Pat. 4,844,822 (July 6, 1987), P. Fox and B. Felthouse (to Dial Corp.).
136. U.S. Pat. 4,808,321 (Feb. 28, 1989), D. Walley (to Procter and Gamble Co.).
137. U.S. Pat. 4,895,667 (Jan. 23, 1990), P. Fox and B. Felthouse (to Dial Corp.).
138. U.S. Pat. 4,970,008 (Nov. 13, 1990), T. Kandathil (to T. Kandathil).
139. U.S. Pat. 4,948,520 (Aug. 14, 1990), H. Sasaki (to Lion Corp.).
140. U.S. Pat. 4,986,922 (Jan. 22, 1991), S. Snow and L. Madore (to Dow Corning Corp.).
141. U.S. Pat. 5,132,425 (July 21, 1992), K. Sotoya, U. Nishimoto, and H. Abe (to Kao Corp.).
142. U.S. Pat. 5,151,223 (Sept. 29, 1992), H. Maaser (to Colgate-Palmolive Co.).
143. U.S. Pat. 5,259,964 (Nov. 9, 1993), N. Chavez and I. Oliveros (to Colgate-Palmolive Co.).
144. U.S. Pat. 5,221,794 (June 22, 1993), J. Ackerman, M. Miller, and D. Whittlinger (to Sherex Chemical Co., Inc.).
145. U.S. Pat. 4,256,824 (Mar. 17, 1981), C. Lu (to Xerox Corp.).
146. U.S. Pat. 4,323,634 (Apr. 6, 1982), T. Jadwin (to Eastman Kodak Co.).
147. U.S. Pat. 4,604,338 (Aug. 5, 1986), R. Gruber, R. Yourd, and R. Koch (to Xerox Corp.).
148. U.S. Pat. 4,560,635 (Dec. 24, 1985), T. Hoffend and A. Barbetta (to Xerox Corp.).
149. U.S. Pat. 3,985,663 (Oct. 12, 1976), C. Lu and R. Parent (to Xerox Corp.).
150. U.S. Pat. 4,059,444 (Nov. 22, 1977), C. Lu and R. Parent (to Xerox Corp.).
151. U.S. Pat. 3,970,755 (July 20, 1976), E. Gazzard and M. Singer (to Imperial Chemical Industries Ltd.).
152. U.S. Pat. 4,272,395 (June 9, 1981), R. Wright (to Lever Brothers Co.).
153. U.S. Pat. 4,444,790 (Apr. 24, 1984), H. Green, A. Petrocci, and Z. Dudzinski (to Millmaster Onyx Group, Inc.).
154. U.S. Pat. 4,868,217 (Sept. 19, 1989), S. Araki and co-workers (to Eisai Co., Ltd. Kao Corp.).
155. U.S. Pat. 4,800,235 (Jan. 24, 1989), T. La Marre and C. Martin (to Nalco Chemical Co.).

156. U.S. Pat. 4,847,089 (July 11, 1989), D. Kramer and P. Snow (to D. N. Kramer).
157. U.S. Pat. 4,941,989 (July 17, 1990), D. Kramer and P. Snow (to Ridgely Products Co., Inc.).
158. U.S. Pat. 4,983,635 (Jan. 8, 1991), H. Martin (to M. Howard).
159. U.S. Pat. 5,049,383 (Sept. 17, 1991), H.-U. Huth, H. Braun and F. Konig (to Hoechst AG).
160. U.S. Pat. 4,615,807 (Oct. 7, 1986), F. Haines and D. Forte (to United States Environmental Resources, Corp.).
161. U.S. Pat. 5,286,109 (Dec. 7, 1993), S. Boyd (to S. Boyd).
162. U.S. Pat. 4,602,052 (July 22, 1986), K. Weber and D. Oberlin (to Amoco Corp.).
163. *CRC Handbook of Chemistry and Physics*, 61st ed., CRC Press, Inc., Boca Raton, Fla., 1981, p. C-586.
164. R. Shelton and co-workers, *J. Am. Chem. Soc.* **66**, 753 (1946).
165. R. Reck, H. Harwood, and A. Ralston, *J. Org. Chem.* **12**, 517 (1947).
166. A. Ralston and co-workers, *J. Org. Chem.* **13**, 186 (1948).
167. Ref. 1, Chapts. 7 and 8.
168. Ref. 1, Chapts. 9, 10, and 11.
169. E. White and D. Woodcock, in Patai, ed., *The Chemistry of the Amino Group*, Wiley-Interscience, New York, 1968, pp. 409–416.
170. A. Hofmann, *Ann. Chem.* **78**, 253 (1851); H. Hofman, *Ann. Chem.* **79**, 11 (1851).
171. J. March, *Advanced Organic Chemistry*, 3rd ed., John Wiley & Sons, Inc., New York, 1985, pp. 906–908.
172. A. Cope and E. Trumbull, *Org. Reactions*, **11**, 317 (1960).
173. J. Baumgarten, *J. Chem. Ed.* **45**, 122 (1968).
174. W. Hawart and C. Ingold, *J. Chem. Soc.*, 997 (1927).
175. T. Stevens and co-workers, *J. Chem. Soc.*, 3193 (1928); G. Wittig, R. Mangold, and G. Felletschin, *Ann. Chem.* **560**, 117 (1948).
176. T. Stevens, *J. Chem. Soc.* (2), 2107 (1930); R. Johnson and T. Stevens, *J. Chem. Soc.* (4), 4487 (1955).
177. M. Sommelet, *Compt. Rend.* **205**, 56 (1937); S. Kantor and C. Hauser, *J. Am. Chem. Soc.* **73**, 4122 (1951); C. Hauser and D. Van Eenam, *J. Am. Chem. Soc.* **79**, 5513 (1957); W. Beard and C. Hauser, *J. Am. Chem. Soc.* **25**, 334 (1961).
178. H. Zimmerman, in de Mayo, ed., *Molecular Rearrangements*, Vol. 1, John Wiley & Sons, Inc., New York, 1963, p. 387.
179. U. Anthoni and co-workers, *Comp. Biochem. Physiol.* **99B**, 1–18 (1991).
180. *Official Methods and Recommended Practices of American Oil Chemists' Society*, 4th ed., American Oil Chemists Society, Champaign, Ill., 1990, pp. S 4c-64 and Section T; L. Metcalfe, *J. Am. Oil Chem. Soc.* **61**, 363 (1984).
181. Ref. 1, Chapt. 13.
182. M. Rosen and H. Goldsmith, *Systematic Analysis of Surface Active Agents*, Interscience, New York, 1960.
183. D. Herring, *Laboratory Practice*, **II**, 113 (1962).
184. S. Epton, *Nature* **160**, 795 (1947).
185. S. Epton, *Trans. Faraday Soc.* **44**, 226 (1948).
186. Ref. 182, p. 455.
187. Ref. 182, p. 445.
188. Ref. 1, p. 430.
189. M. Bambagiottialberti and co-workers, *Rapid Commun. Mass Spectrom.* **8**, 439 (1994).
190. A. Tyler, L. Romo, and R. Cody, *Int. J. Mass. Spectrom. Ion Process.* **122**, 25 (1992).
191. D. Fisher and co-workers, *Rapid Commun. Mass Spectrom.* **8**, 65 (1994).
192. V. Wee and J. Kennedy, *Anal. Chem.* **54**, 1631 (1982).

193. G. Szajer, in Perkins, *Analyses of Fats, Oils and Lipoproteins*, American Oil Chemists Society, Champaign, Ill., 1991, Chapt. 18; T. Schmitt, *ibid.*, Chapt. 19.

194. F. Mozayeni, in J. Cross and E. Singer, eds., *Cationic Surfactants*, Marcel Dekker, Inc., New York, 1994, Chapt. 11.

195. R. Shelton and co-workers, *J. Am. Chem. Soc.* **66**, 755 (1946).

196. *Ibid.*, p. 757.

197. W. Sexton, *Chemical Constitution and Biological Activity*, Van Nostrand, Princeton, N.J., 1963.

198. A. Vallejo-Freire, *Science*, **114**, 470 (1954).

199. K. Biesinger and co-workers, *J. Water Pollut. Control Fed.* **48**, 183 (1976).

200. J. Walker and S. Evans, *Marine Pollut. Bull.* **9**, 136 (1978).

201. Technical data, *Toxicity*, Akzo Nobel Chemicals Inc., Dobbs Ferry, N.Y., Apr. 1995.

202. R. Boethling, *Water Res.*, **18**, 1061 (1984); L. Huber, *J. Am. Oil Chem. Soc.* **61**, 377 (1984).

203. J. Cooper, *Ecotoxicity Environ. Safety*, **16**, 65 (1988).

204. Technical data, Akzo Nobel Chemical Inc., Dobbs Ferry, N.Y., Apr. 1995.

205. U.S. Pat. 4,204,954 (May 27, 1980), J. Jacob (to Chemed Corp.) for a patented procedure to remove quaternaries from wastewater.

206. W. Ruback, *Chem. Today*, 15 (May/June 1994).

207. L. Edebo and co-workers, *Ind. Appl. Surfactants*, **III**, 184 (1992).

208. U.S. Pat. 4,787,491 (Dec. 6, 1988), N. Chang and D. Walley (to Procter & Gamble Co.).

209. U.S. Pat. 4,767,547 (Aug. 30, 1988), T. Straathof and A. Konig (to Procter & Gamble Co.).

210. U.S. Pat. 3,915,867 (Oct. 28, 1975), H. Kang, and R. Peterson, and E. Knaggs (to Stepan Chemical Co.).

211. U.S. Pat. 4,339,391 (July 13, 1982), E. Hoffmann and co-workers (to Hoechst AG).

212. U.S. Pat. 4,963,274 (Oct. 16, 1990), W. Ruback and co-workers (to Hüls AG).

213. R. Lagerman and co-workers, *J. Am. Oil Chem. Soc.* **71**, 97 (1994).

214. U.S. Pat. 5,066,414 (Nov. 19, 1991), N. Chang (to Procter & Gamble Co.).

215. U.S. Pat. 4,840,738 (June 20, 1981), F. Hardy and D. Walley (to Procter & Gamble Co.).

216. U.S. Pat. 5,128,473 (July 7, 1992), F. Friedli and M. Watts (to Sherex Chemical Co.).

217. U.S. Pat. 4,857,310 (Aug. 15, 1989), A. Baydar (to Gillette Co.).

218. B. Challis and A. Butler, in Ref. 169, pp. 290–300.

219. M. Gibson, in Ref. 169, pp. 44–55.

220. Ref. 171, p. 364.

221. C. Ingold, *Structure and Mechanism in Organic Chemistry*, Bell, London, 1953, Chapt. 7; A. Streitwieser, *Chem. Rev.* **56**, 571 (1956).

222. W. Emerson, *Organic Reactions*, Vol. IV, John Wiley & Sons, Inc., New York, 1948, Chapt. 3.

223. H. Sommer and L. Jackson, *J. Org. Chem.* **35**, 1558 (1970).

224. H. Sommer, H. Lipp, and L. Jackson, *J. Org. Chem.* **36**, 824 (1971).

225. S. Billenstein and G. Blaschke, *J. Am. Oil Chem. Soc.* **61**, 353 (1984).

226. R. Schroter, in Houben-Weyl, ed., *Methoden der Organischen Chemie*, Vol. XIII, 4th ed., George Thieme Verlag, Stuttgart, Germany, 1957, Chapt. 4; M. Rabinovitz, in Z. V. C. Rappoport, *The Chemistry of the Cyano Group*, Wiley-Interscience, New York, 1970, pp. 307–340; C. De Bellefon and P. Fouilloux, *Catal. Rev.-Sci. Eng.* **36**, 459 (1994).

227. P. Rylander, *Catalytic Hydrogenation Over Platinum Metals*, Academic Press, Inc., New York, 1967, pp. 203–226.

228. F. Gould, G. Johnson, and A. Ferris, *J. Org. Chem.* **25**, 1658 (1960).

229. M. Freifelder, *J. Am. Chem. Soc.* **82**, 2386 (1960).

230. U.S. Pat. 4,830,771 (May 16, 1989), W. Ruback and J. Schut (to Hüls AG).
231. Can. Pat. 1,312,619 (Jan. 12, 1993), M. Hofinger and co-workers (to Hoechst AG).
232. *Synthetic Organic Chemicals*, U.S. International Trade Commission, Washington, D.C., 1994.
233. C. Starks, *Ind. Appl. Surfactants II*, **77**, 165 (1990); C. Starks, ed., *Phase-Transfer Catalysis: New Chemistry, Catalysts and Applications*, American Chemical Society, Washington, D.C., 1987; E. Dehmlov, *Phase-Transfer Catalysis*, Verlag Chemie, Deerfield Beach, Fla., 1983; M. Halpern, *Phase-Transfer Catalysis* in *Ullman's Encyclopedia of Industrial Chemistry*, Vol. A19, VCH V6, New York, 1991; M. Halpern, *Phase-Transfer Catalysis Commun.* **1**, 1 (1995).
234. *Specialty Surfactants Worldwide* in *Specialty Chemicals*, SRI International, Menlo Park, Calif., 1989, pp. 81–94.
235. Ref. 2, p. 657.5000yz; W. Evans, *Chem. Ind.* **27**, 893 (1969); R. Puchta, *J. Am. Oil Chem. Soc.* **61**, 367 (1984).
236. *Cosmetic Chemicals*, in *Specialty Chemicals*, SRI International, Menlo Park, Calif., 1992, p. 73.
237. Technical data, Akzo Nobel Chemicals Inc., Dobbs Ferry, N.Y., Apr. 1995; for a review, see W. Mardis, *J. Am. Oil Chem. Soc.* **61**, 382 (1984).
238. H. Berenbold, *Inform*, **5**, 82 (1994).
239. E. Spiess, *Perfumerie Kusmetik*, **72**, 370 (1991); M. Jurczyk, D. Berger, and D. Damaso, *Cosmetics Toiletries* **106**, 63 (1991).
240. C. Reich and C. Robbins, *J. Soc. Cosmet. Chem.* **44**, 263 (1993).
241. P. D'Arcy and E. Taylor, *J. Pharm. Pharmacol.* **14**, 193 (1962); P. Schaeufele, *J. Am. Oil Chem. Soc.* **61**, 387 (1984).
242. Ref. 1, Chapt. 14.
243. H. Freedman, *Pure Appl. Chem.* **58**, 857 (1986).
244. J. D'Souza and N. Sridhar, *J. Sci. Ind. Res.* **42**, 564 (1983).
245. J. Salamone and W. Rice, in J. I. Kroschwitz, ed., *Encyclopedia of Polymer Science and Engineering*, 2nd ed., John Wiley & Sons, Inc., New York, 1988.
246. R. McConnell, *Soap, Cosmetics, Chem. Special.*, 37 (Apr. 1989).
247. J. K. Borchardt, *Proceedings of the Symposium on Advances in Oil Field Chemistry*, Toronto, Canada, June 5–11, 1988.

MAURICE DERY
Akzo Nobel Chemicals Inc.

QUENCHING OILS. See FEEDSTOCKS, PETROCHEMICALS AND FEEDSTOCKS.

QUINHYDRONE. See QUINONES.

QUININE. See ALKALOIDS.

QUINOLINE DYES. See QUINOLINES AND ISOQUINOLINES.

QUINOLINES AND ISOQUINOLINES

Replacing one carbon atom of naphthalene with an azomethene linkage creates the isomeric heterocycles 1- and 2-azanaphthalene. Better known by their trivial names quinoline [91-22-5] (1) and isoquinoline [119-65-3] (2), these compounds have been the subject of extensive investigation since their extraction from coal tar in the nineteenth century. The variety of studies cover fields as diverse as molecular orbital theory and corrosion prevention. There is also a vast patent literature. The best assurance of continuing interest is the frequency with which quinoline and isoquinoline structures occur in alkaloids (qv) and pharmaceuticals (qv), for example, quinine [130-95-0] and morphine [57-27-2] (see ALKALOIDS).

(1)   (2)

   Alternative names for (1) and (2), benzo[b]pyridine and benzo[c]pyridine, respectively, are little used. These synonyms, like their modern counterparts, suggest an important generalization concerning the chemical behavior of these compounds and their many derivatives. Like naphthalene (qv) and pyridine (qv), (1) and (2) are clearly aromatic, but less intensely than benzene. Resonance energy per π-electron for benzene (0.065 β), naphthalene (0.055 β), and pyridine (0.058 β) may be compared with quinoline (0.052 β) and isoquinoline (0.051 β) (1).

   The close chemical relationship among these structural entities, as well as the uniqueness of (1) and (2), have been evident from the time of the earliest structural studies. Permanganate oxidation of (1) (2) produces 2,3-pyridinedicarboxylic acid (quinolinic acid [89-00-9]) (3), whereas similar treatment of (2) (3) yields a mixture of 3,4-pyridinedicarboxylic acid (cinchomeronic acid [490-11-9]) (4) and phthalic acid.

(1)   (3)

(2)   (4)

The continuing importance of these compounds has led to a number of general and specific reviews. Five volumes of Weissberger's *The Chemistry of Heterocyclic Compounds* are essential reading for studies through 1977 and, in many instances, to 1990 (4).

Comparative Properties

Physical Properties. Both (1) and (2) are weak bases, showing pK_a 4.94 and 5.40, respectively. Their facile formation of crystalline salts with either inorganic or organic acids and complexes with Lewis acids is in each case of considerable interest. Selected physical data for quinoline and isoquinoline are given in Table 1. Reference 4 greatly expands the range of data treated and adds to them substantially.

Chemical Properties.   The presence of both a carbocyclic and a heterocyclic ring facilitates a broad range of chemical reactions for (1) and (2). Quaternary alkylation on nitrogen takes place readily, but unlike pyridine both quinoline and isoquinoline show addition by subsequent reaction with nucleophiles. Nucleophilic substitution is promoted by the heterocyclic nitrogen. Electrophilic substitution takes place much more easily than in pyridine, and the substituents are generally located in the carbocyclic ring.

Table 1. Physical Properties of Quinoline and Isoquinoline[a]

	Value	
Property	Quinoline	Isoquinoline
mp, °C	−15.6	26.5
bp, °C	238	243
ΔH_{vap}, kJ/mol[b]	46.4	49.0
n_D^{20}	1.6268	1.6148
d^{20}, g/cm^3	1.0929	1.0986
K_a	1.25×10^{-5}	3.80×10^{-6}
viscosity at 30°C, mPa·s(=cP)	2.997	3.2528
T_c	509	530

[a]Ref. 4.
[b]To convert J to cal, divide by 4.186.

Quinoline

Reactions. Quinoline exhibits the reactivity of benzene and pyridine rings, as well as its own unique reactions.

Nitration. As an aromatic system, (1) shows important synthetic and mechanistic nitro group chemistry (see NITRATION). The experimental conditions employed usually determine the product structure. At 0°C, mixed acid attacks the protonated quinoline to yield a 1:1 mixture of 5-nitroquinoline [607-34-1] and 8-nitroquinoline [607-35-2] (5). Under less acidic conditions, as with an acetic–nitric acid mixture or with dinitrogen tetroxide, 3-nitroquinoline [17576-53-3] is the primary product (6). In the absence of nitrous fumes, the 3-nitro product is not formed. This result is rationalized as involving initial 1,2-addition of the reagent. If 1,2,3,4-tetrahydroquinoline [635-46-1] is the starting material, mixed acid attacks the aromatic ring and subsequent dehydrogenation produces 7-nitroquinoline [613-51-4] (7). Excellent yields of 3- or 7-nitroquinoline can be obtained selectively under mild conditions. Tetranitratotitanium(IV) at ambient temperature produces the 3-nitro isomer; tetranitratozirconium(IV) gives 90% of the 7-nitro product (8).

In many instances, beginning a synthesis with quinoline N-oxide [1613-37-2] facilitates the preparation of difficult compounds. Quinoline is converted to the N-oxide using hydrogen peroxide in acetic acid, and later reduced to the substituted quinoline. Warm mixed acid gives 4-nitroquinoline 1-oxide [56-57-5] in overall 65% yield (9). Depending on the reducing agent employed, either 4-nitroquinoline [3741-15-9] or 4-aminoquinoline [578-68-7] may be obtained (see AMINES BY REDUCTION).

Sulfonation. The main sulfonation product of quinoline at 220°C is 8-quinolinesulfonic acid [85-48-3]; at 300°C it rearranges to 6-quinolinesulfonic acid [65433-95-6] (10). Optimum conditions for sulfonation, 2 h at 140°C and a 1:4 quinoline/40% (wt) oleum ratio, produces 80% yield. The yield drops to 64% at 130°C with a 1:3 reactant ratio (11). Somewhat higher, but variable, yields of 8-quinolinesulfonic acid hydrochloride [85-48-3] have been reported with chlorosulfonic acid (12).

1,2-Addition. Unlike pyridine, quinoline undergoes facile addition to the nitrogen-containing ring. Allylmagnesium chloride reacts with quinoline in deoxygenated tetrahydrofuran to produce 80% 2-allyl-1,2-dihydroquinoline [55570-23-5] (13). Depending on experimental conditions, either 2-propylquinoline [1613-32-7] or 2-propenylquinoline [57078-89-4] may be obtained. Similar results are observed with vinyl Grignard reagents (14) and with alkyllithium reagents (15). Phosphorylated 1,2-addition products have been obtained by acylating the ring nitrogen and treating the intermediate with trimethyl phosphite and sodium iodide (16).

Treatment of quinoline with cyanogen bromide, the von Braun reaction (17), in methanol with sodium bicarbonate produces a high yield of 1-cyano-2-methoxy-1,2-dihydroquinoline [880-95-5] (**5**) (18). Compound (**5**) is quantitatively converted to 3-bromoquinoline [5332-24-1], through the intermediate (**6**) [66438-70-8]. These conversions are accomplished by sequential treatment with bromine in methanol, sodium carbonate, or concentrated hydrochloric acid in methanol. Similar conditions provide high yields of 3-bromomethylquinolines.

Amination. 2-Aminoquinoline [580-22-3] is obtained from quinoline in 80% yield by treatment with barium amide in liquid ammonia (19). This product, as well as 3-aminoquinoline [580-17-6] and 4-aminoquinoline [578-68-7], may be obtained through nucleophilic substitution of the corresponding chloroquinolines with ammonia.

Halogenation. One review provides detailed discussion of direct and indirect methods for both mono- and polyhalogenation (20). As with nitration, halogenation under acidic conditions favors reaction in the benzenoid ring, whereas reaction at the 3-position takes place in the neutral molecule. Radical reactions in the pyridine ring can be important under more vigorous conditions.

A quinoline–bromine adduct in hot carbon tetrachloride containing pyridine gives a 90% yield of 3-bromoquinoline (21); 3-chloroquinoline [612-59-9] is prepared by an analogous route, but in poorer yield. A quinoline–aluminum chloride complex heated with bromine gives a 78% yield of 5-bromoquinoline [165-18-3] (22). Equal quantities of 5- and 8-bromoquinoline [16567-18-3] are formed when quinoline is treated with one equivalent of bromine in concentrated sulfuric acid containing silver sulfate (23).

Quinolinium chloride in nitrobenzene reacts with excess bromine to give an 81% yield of 3-bromoquinoline (24). It seems likely that the 1,2-dibromo complex is actually being brominated.

Direct iodination or fluorination leads to ill-defined products and fragmentation, respectively. Sandmeyer chemistry and nucleophilic substitution of chloroquinolines have proved to be useful alternative routes.

Oxidation. The synthesis of quinolinic acid and its subsequent decarboxylation to nicotinic acid [59-67-6] (**7**) has been accomplished directly in 79% yield using a nitric–sulfuric acid mixture above 220°C (25). A wide variety of oxidants have been used in the preparation of quinoline *N*-oxide. This substrate has proved to be useful in the preparation of 2-chloroquinoline [612-62-4] and 4-chloroquinoline [611-35-8] using sulfuryl chloride (26). The oxidized nitrogen is readily reduced with DMSO (27) (see AMINE OXIDES).

　　　　(**1**)　　　　　　　　(**3**)　　　　　　　　(**7**)

Quaternary Salts. The ring nitrogen of quinoline reacts with a wide variety of alkylating and acylating agents to produce useful intermediates like *N*-benzoylquinolinium chloride [4903-36-0] (**8**). The quinoline 1,2-adducts, eg, *N*-benzyl-2-cyano-1,2-dihydroquinoline [13721-17-0] (**9**), or Reissert compounds (28), formed with potassium cyanide can produce 2-carboxyquinoline [93-10-7] (**10**) or 2-cyanoquinoline [11436-43-7] (**11**).

　　(**1**)　　　　　　　　　　　(**8**)　　　　　　　(**9**)　　　　　(**10**)

(**11**)

Excellent yields of the former product are also obtained with quinoline *N*-oxide. Improved yields of Reissert compounds are found under phase-transfer conditions (29). The regiochemistry of the method changes dramatically with *N*-alkylquinolinium salts, eg, *N*-methylquinolinium iodide [3947-76-0] (**12**), which form 4-cyanoquinoline [23395-72-4] (**13**) (30), through the intermediary in this example of *N*-methyl-4-cyano-1,4-dihydroquinoline [828-69-3] and *N*-methyl-4-cyanoquinolinium iodide [64275-22-5].

(**12**)   (**13**)

Alkylation and Arylation.   The direct introduction of carbon–carbon bonds in quinoline rings takes place in low yield and with little selectivity. The most promising report involves carboxylic acids with ammonium persulfate and silver nitrate (31).

Reduction.   Quinoline may be reduced rather selectively, depending on the reaction conditions. Raney nickel at 70–100°C and 6–7 MPa (60–70 atm) results in a 70% yield of 1,2,3,4-tetrahydroquinoline (32). Temperatures of 210–270°C produce only a slightly lower yield of decahydroquinoline [2051-28-7]. Catalytic reduction with platinum oxide in strongly acidic solution at ambient temperature and moderate pressure also gives a 70% yield of 5,6,7,8-tetrahydroquinoline [10500-57-9] (33). Further reduction of this material with sodium–ethanol produces 90% of *trans*-decahydroquinoline [767-92-0] (34). Reductions of the quinoline heterocyclic ring accompanied by alkylation have been reported (35). Yields vary widely; sodium borohydride–acetic acid gives 17% of 1,2,3,4-tetrahydro-1-(trifluoromethyl)quinoline [57928-03-7] and 79% of 1,2,3,4-tetrahydro-1-isopropylquinoline [21863-25-2]. This latter compound is obtained in the presence of acetone; the use of cyanoborohydride reduces the pyridine ring without alkylation.

Manufacture From Coal Tar.   Commercially, quinoline is isolated from coal-tar distillates (36). Tar acids are removed by caustic extraction, and the oil is distilled to produce the methylnaphthalene fraction (230–280°C). Washing with dilute sulfuric acid produces sulfate salts, from which the tar bases are liberated by treatment with caustic followed by distillation. Commercial quinoline is at least 90% pure with a distillation range of 2°C from 235–238°C, and a specific gravity of 1.095 at 15.5°C. Chromatographic composition of this product is typically 92% quinoline and 5% isoquinoline by weight. Studies of the composition of crude tar bases in the distillation range of 230–265°C show the presence of at least trace amounts of all monomethylquinolines, 2,8-dimethylquinoline [1763-17-8], and some homologues of isoquinoline (37).

Many forms of chromatography have been used to separate mixtures of quinoline and isoquinoline homologues. For example, alumina saturated with cobalt chloride, reversed-phase liquid chromatography, and capillary gas chromatography (gc) with deactivated glass columns have all been employed (38,39).

A 90% yield of isoquinoline (>95% pure) was reported by treating a crude fraction with hydrochloric acid followed by addition of an alcoholic solution of cupric chloride in a mole ratio of 1:2 $CuCl_2$/isoquinoline (40). A slightly lower yield of 2-methylquinoline [91-63-4] (97.5% pure) was obtained from bituminous coal using 30% aqueous urea to form a clathrate (41).

Syntheses of Quinolines. The large number of alkaloids and medicinal compounds which contain the quinoline ring has created a long and active search for synthetic routes. Several classical routes were developed in the nineteenth century and, with many modifications, are still used.

Skraup Synthesis. This general method, used for many quinolines, consists of heating a primary aniline with glycerol, concentrated sulfuric acid, and an oxidizing agent (42). Often the nitrobenzene corresponding to the aniline employed is used as the oxidant, and iron(II) sulfate is added to moderate the often violently exothermic process. It is probable that the glycerol is dehydrated to acrolein, which undergoes conjugate addition to the aniline. The formation of quinoline has been reported in the 80–90% range. The use of compounds related to acrolein, such as crotonaldehyde and methyl vinyl ketone, allow substituents to be placed in the heterocyclic ring. With ortho- and para-substituted anilines, a single product is usually found; meta-derivatives produce mixtures. The nature of the substituent in the starting material is important. The presence of electron-withdrawing groups favors the production of 5-substituted quinolines, whereas with electron-donating groups the 7-substituted product dominates.

Döbner-von Miller Synthesis. A much less violent synthetic pathway, the Döbner-von Miller, uses hydrochloric acid or zinc chloride as the catalyst (43). As in the modified Skraup, α,β-unsaturated aldehydes and ketones make the dehydration of glycerol unnecessary, and allow a wider variety of substitution patterns. No added oxidant is required. With excess aniline the reaction proceeds as follows:

or as shown below:

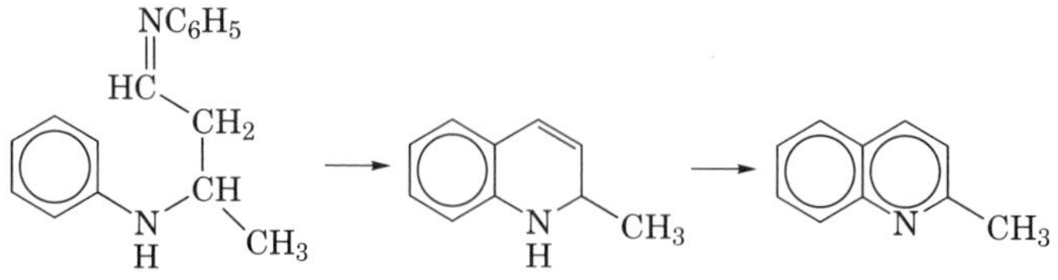

The mechanism of both syntheses has been studied in detail, and well summarized (44,45). Interesting questions remain; for example, in neither of these sequences is it certain whether the carbonyl compound or its Schiff base is undergoing Michael addition.

A number of improvements have been made in these syntheses. For example, the use of ethanolic ferric chloride and zinc chloride produces a good yield of 2-isopropylquinoline [17507-24-3] from isovaleraldehyde (46). The purification of 2-methylquinoline is facilitated through precipitation. A crude quinaldine–hydrochloride and zinc chloride complex is prepared and then treated with aqueous base (47).

Combes Synthesis. When aniline reacts with a 1,3-diketone under acidic conditions a 2,4-disubstituted quinoline results, eg, 2,4-dimethylquinoline [1198-37-4] from 2,4-pentadione. A similar result has been reported using a mixture of an aldehyde and a ketone (48).

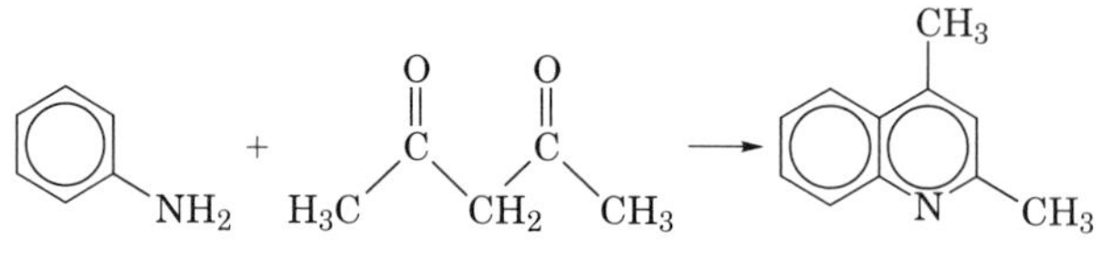

Conrad-Limpach-Knorr Synthesis. When a β-keto ester is the carbonyl component of these pathways, two products are possible, and the regiochemistry can be optimized. Aniline reacts with ethyl acetoacetate below 100°C to form 3-anilinocrotonate (**14**), which is converted to 4-hydroxy-2-methylquinoline [607-67-0] by placing it in a preheated environment at 250°C. If the initial reaction takes place at 160°C, acetoacetanilide (**15**) forms and can be cyclized with concentrated sulfuric acid to 2-hydroxy-4-methylquinoline [607-66-9] (49). This example of kinetic vs thermodynamic control has been employed in the synthesis of many quinoline derivatives. They are useful as intermediates for the synthesis of chemotherapeutic agents (see CHEMOTHERAPEUTICS, ANTICANCER).

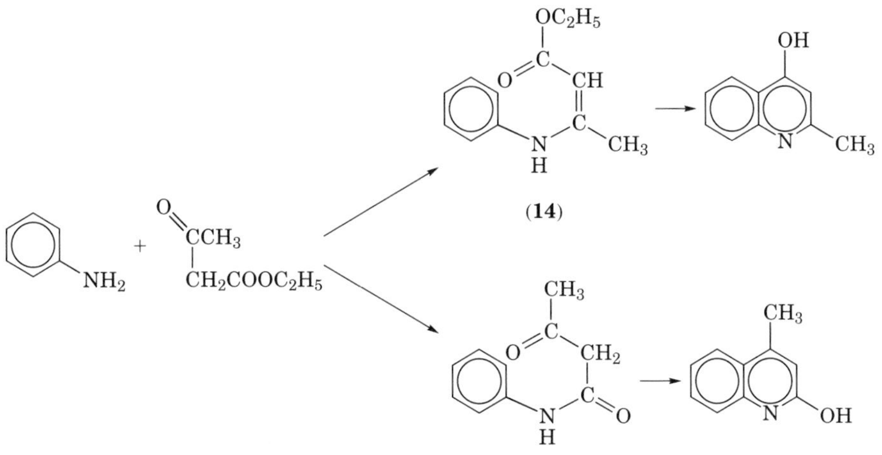

(14)

(15)

Pfitzinger Reaction. Quinoline-4-carboxylic acids are easily prepared by the condensation of isatin [*91-56-5*] (**16**) with carbonyl compounds (50). The products may be decarboxylated to the corresponding quinolines. The reaction of isatin with cyclic ketones has been reported, eg, the addition of cyclohexanone gives the tricyclic intermediate (**17**) [*38186-54-8*], which upon oxidation produces quinoline-2,3,4-tricarboxylic acid [*16880-83-4*] (51).

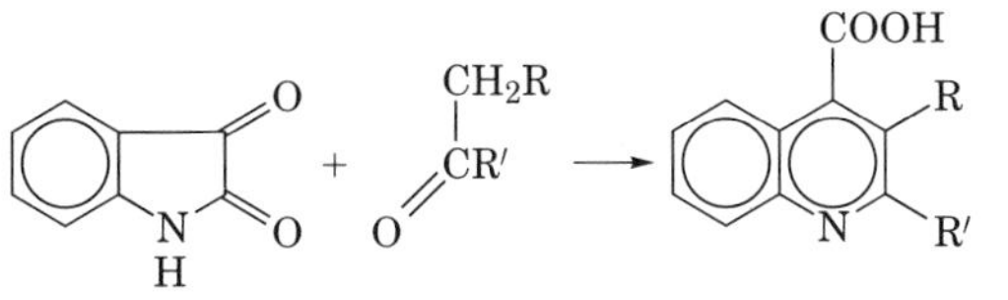

(6)

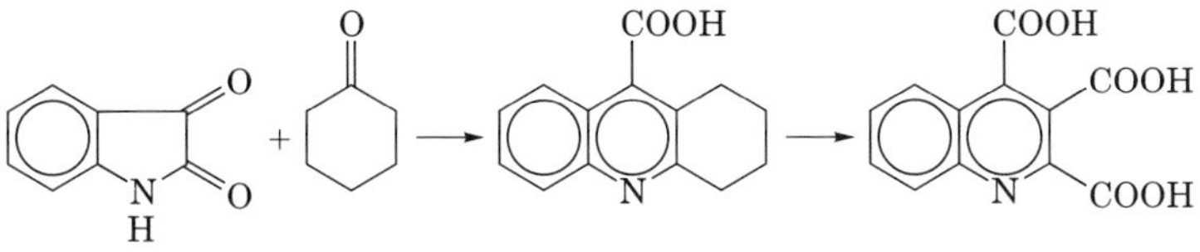

(17)

Frieländer Synthesis. The methods cited thus far all suffer from the mixtures which usually result with meta-substituted anilines. The use of an ortho-disubstituted benzene for the subsequent construction of the quinoline avoids the problem. In the Frieländer synthesis (52) a starting material like 2-aminobenzaldehyde reacts with an α-methyleneketone in the presence of base. The difficulty of preparing the required anilines is a limitation in this approach, but 2-nitrocarbonyl compounds and the subsequent reduction of the nitro group present a useful modification (53).

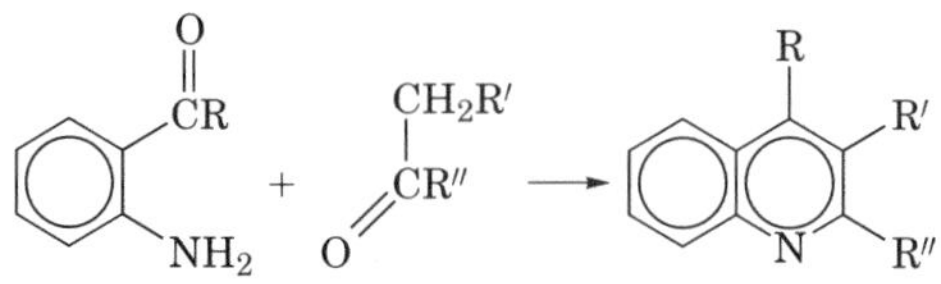

New Synthetic Approaches.　There have been a number of efforts to prepare quinolines by routes quite different from the traditional methods. In one, the cyclization of 3-amino-3-phenyl-2-alkenimines (**18**) using alkali metals leads to modest yields of various 4-arylaminoquinolines (54). Because this structure is found in many natural products and few syntheses of it exist, the method merits further investigation.

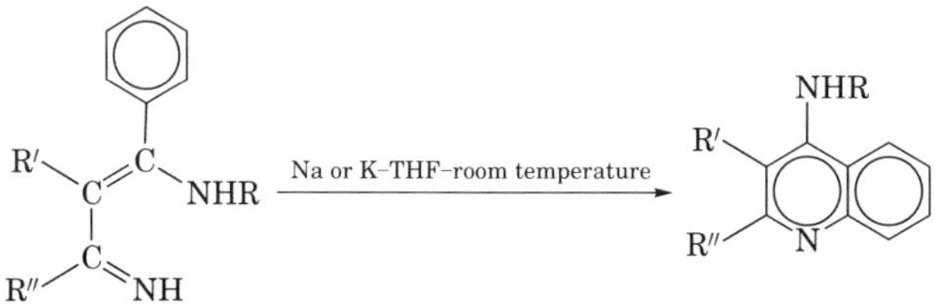

(**18**)

Studies of the synthesis of quinolines using transition-metal catalysts and nonacidic conditions (55) have determined that ruthenium(III) chloride is the most effective of a wide range of catalysts. The reaction between nitrobenzene and 1-propanol or 1-butanol gives 65 and 70% yields of 2-ethyl-3-methylquinoline [27356-52-1] and 3-ethyl-2-propylquinoline, respectively.

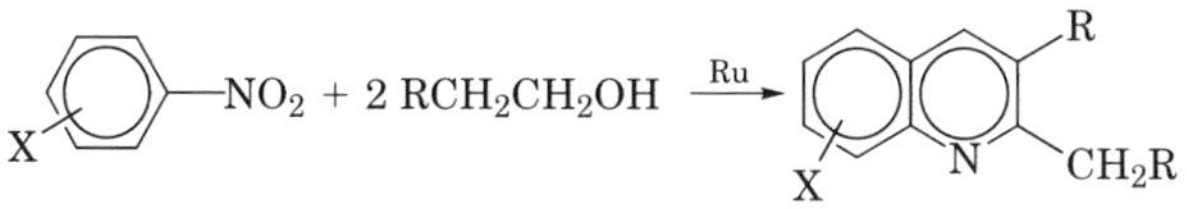

The reaction of *N*-benzylideneaniline (**19**) with alkynes leads to quinolines substituted in the heterocyclic ring (56). Except for benzylidenes bearing nitro substituents, the reaction occurs in good yield and under mild conditions. The method appears capable of elaboration.

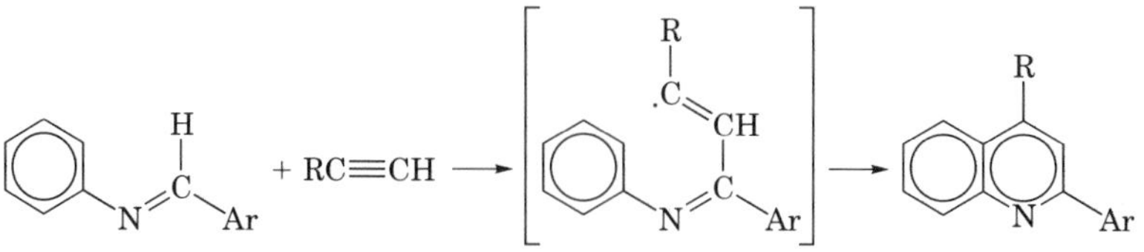

(**19**)

An intramolecular Diels-Alder cyclization produces excellent yields of 2-aminoquinoline-3-carboxylate esters (57). Equally fine yields of the required carbodiimides have been reported, making this an attractive route to an unusual substitution type.

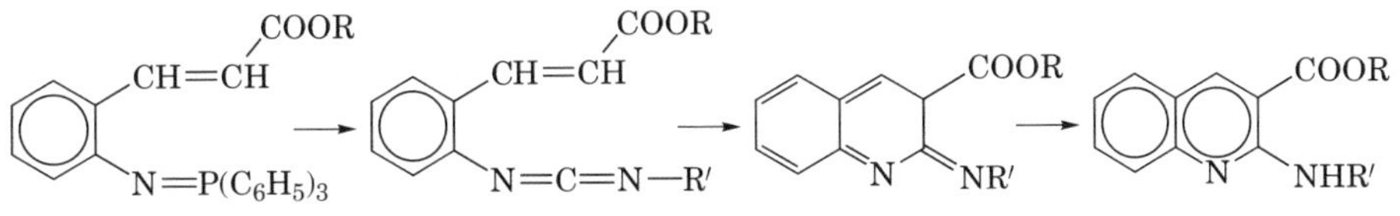

Good yields of 2,4-diaminoquinolines are obtained through either Lewis acid- or base-induced cyclization of 2-amidinobenzonitriles (**20**) (58). The method avoids both the harsh conditions and lack of regiospecificity characteristic of earlier preparations.

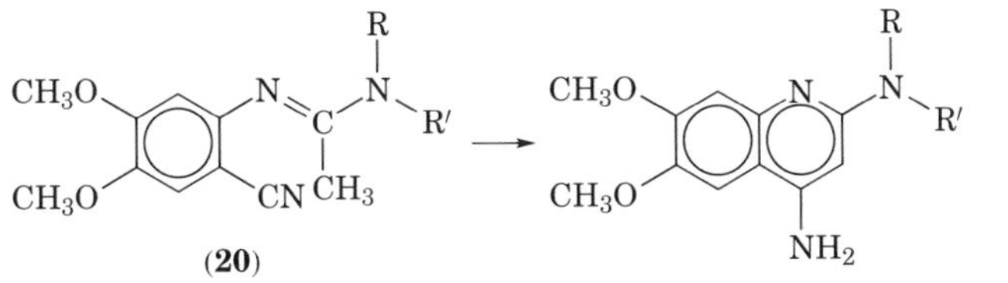

(**20**)

The Lewis acid-catalyzed cyclization of 3-amino-2-alkenimines (**21**) leads to a wide variety of alkyl- and aryl-substituted quinolines (59). The high regiospecificity and the excellent yields obtained make this process promising.

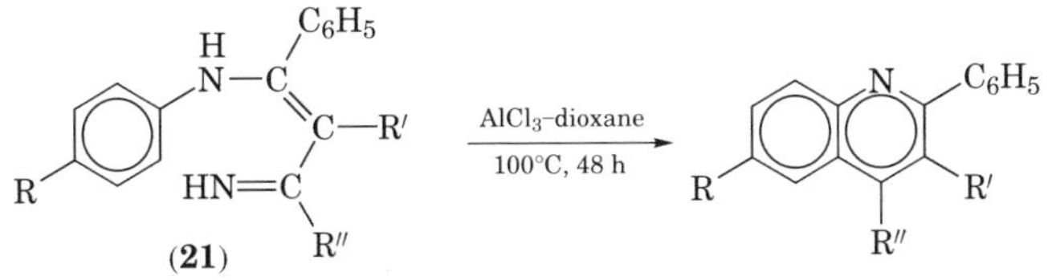

(**21**)

Finally, the importance of quinolinium salts to dye chemistry accounts for the long, productive history of their synthesis. The reaction of *N*-methylformanilide with ketones, aldehydes, ketone enamines, or enol acetates in phosphoryl chloride leads to high yields of *N*-methylquinolinium salts (60).

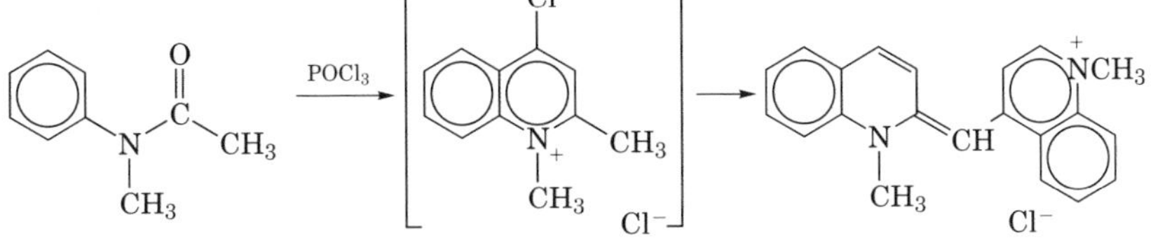

Economic Aspects. There is little evidence of large-scale demand for either quinoline or isoquinoline in 1996. The U.S. Tariff Commission reports

Table 2. Commercially Available Quinolines and Isoquinolines, 1995

Compound	Price, $		
	Lancaster[a]	Aldrich[b]	ACROS[c]
quinoline	10.80/100 g	48.10/500 g	91.20/500 g
2-methylquinoline	36.90/100 g	147.10/500 g	88.00/250 mL
3-methylquinoline	117.50/10 g	133.15/100 g	
4-methylquinoline			138.60/100 g
2-chloroquinoline	92.90/25 g	192.45/25 g	
2-hydroxyquinoline		76.85/5 g	92.55/10 g
4-hydroxyquinoline		84.05/10 g	83.30/10 g
2-quinolinecarboxylic acid	63.70/10 g	69.15/10 g	69.80/10 g
isoquinoline	40.50/100 g	94.00/500 g	124.20/500 g
1-isoquinolinecarboxylic acid	44.80/5 g		133.40/25 g

[a]Ref. 61.
[b]Ref. 62.
[c]Ref. 63.

no longer show separate production or sales data for any quinoline derivative. A number of these compounds are available as fine chemicals; representative examples are found in Table 2. The principal supplier of quinoline and quinoline still residue is Koppers Chemical.

Toxicology. Quinoline is a poison when it enters the body by any of the normal routes, ie, ingestion, or subcutaneous or intraperitoneal injection. Even contact with the skin produces a moderate toxic reaction, and can result in severe irritation. These is evidence that quinoline is mutagenic, and long exposure can produce lung problems. Quinoline vapors cause irritation of the nose, throat, and bronchial tubes; they also cause headaches, nosebleed, and difficult breathing. Higher exposure may cause a possibly fatal buildup of fluids in the lungs. Toxicity data are typically LD_{50} (rat), 330 mg/kg, and dermal LD_{50} (rat), 540 mg/kg.

Quinoline derivatives are also dangerous; for example, 8-quinolinol is especially toxic intraperitoneally, with an LD_{50} (mouse) of 48 mg/kg. This compound is known to cause neoplasma of various parts of the body when ingested, implanted, or administered intravenously (64–67).

Uses. *Antioxidants.* The 1,2-dihydroquinolines have been used in a variety of ways as antioxidants (qv). For example, 1,2-dihydro-2,2,4-trimethylquinoline along with its 6-decyl [81045-48-9] and 6-ethoxy [91-53-2] derivatives have been used as antiozonants (qv) and stabilizers (68). A polymer [26780-96-1] of 1,2-dihydro-2,2,4-trimethylquinoline is used in resins, copolymers, lubricant oils, and synthetic fibers (69). These same compounds react with aldehydes and the products are useful as food antioxidants (70). A cross-linked polyethylene prepared with peroxides and other monomers in the presence of 1,2-dihydro-6-ethoxyquinoline produces polymers with a chemically bonded antioxidant (71).

Corrosion Inhibitors. Steel-reinforcing wire and rods embedded in concrete containing quinoline or quinoline chromate are less susceptible to corrosion (72) (see CORROSION AND CORROSION CONTROL). Treating the surface of metals with 8-hydroxyquinoline [148-24-3] makes them resistant to tarnishing and corrosion (73). Ethylene glycol-type antifreeze may contain quinoline, 2-chloro-, 4-amino-, 8-nitro-, or 8-hydroxyquinoline to prevent corrosion (74).

Agricultural Chemicals. A herbicide possessing activity comparable to 2,4-D is found in compounds like quinolyl esters of N-substituted dithiocarbamic acids (75). These products result when the salt formed from carbon disulfide and primary or secondary aliphatic amines reacts with 2-vinylquinoline [772-03-2]. When applied properly, 7-chloroquinoline [612-61-3] appears to be an environmentally safe weed control with its ability to prevent seed germination (76). The corresponding N-oxide 7-chloroquinoline N-oxide [22614-94-4] has been used to destroy broadleaved weeds. A wide variety of compounds containing the quinoline system are herbicides (qv) (77). Derivatives and salts of 8-quinolinecarboxylic acid [86-59-9] as well as quinolyl carbamates are each useful insecticides; the latter do not appear to affect desirable insects or fish when used moderately (78). The copper salt of 8-hydroxyquinoline is an effective fungicide (see FUNGICIDES, AGRICULTURAL). Very low concentrations of 1-naphthols with hydroxyquinolines repel termites (79). The yield and quality of cotton crops has been improved by coating the seed with quinoline N-oxide before sowing (80) (see GROWTH REGULATORS, PLANT).

Polymers. Quinoline and its derivatives may be added to or incorporated in polymers to introduce ion-exchange properties (see ION EXCHANGE). For example, phenol–formaldehyde polymers have been treated with quinoline, quinaldine, or lepidine (81) (see PHENOLIC RESINS). Resins with variable basic exchange capacities have been prepared by treating Amberlites with 2-methylquinoline (82).

Platinum-group metals (qv) form complexes with chelating polymers with various 8-mercaptoquinoline [491-33-8] derivatives (83) (see CHELATING AGENTS). Hydroxy-substituted quinolines have been incorporated in phenol–formaldehyde resins (84). Stannic chloride catalyzes the condensation of bis(chloromethyl)benzene with quinoline (85).

Methylquinolines react with chloromethylphenyl groups of cross-linked polymers to form anion-exchange resins (86).

Metallurgy. The extraction and separation of metals and plating baths have involved quinoline and certain derivatives (see ELECTROPLATING; METAL SURFACE TREATMENTS; EXTRACTION).

Aldehydes react with 2- or 4-methylquinoline, and the product improves ductility, brightness, and leveling of nickel deposits in baths containing nickel, chloride, and sulfate (87). Soft-zinc electroplating baths have included 8-hydroxyquinaldine (88). Copper-plating baths containing quinoline or benzoquinoline [123-31-9] have shown promise in the manufacture of printed-circuit boards, and electroforming or rotogravure applications. Desirable qualities include smooth, ductile, and uniformly lustrous deposits as well as high throwing power in aqueous acidic baths (89). A bright tin electroplating bath for steel sheets employs 8-hydroxyquinoline as a brightener (90).

The extraction of metal ions depends on the chelating ability of 8-hydroxyquinoline. Modification of the structure can improve its properties, eg, higher solubility in organic solvents (91). The extraction of nickel, cobalt, copper, and zinc from acid sulfates has been accomplished using 8-hydroxyquinoline in an immiscible solvent (92). In the presence of oximes, halo-substituted 8-hydroxyquinolines have been used to recover copper and zinc from aqueous solutions (93). Dilute solutions of heavy metals such as mercury, cadmium, copper, lead, and zinc can be purified using quinoline-8-carboxylic acid adsorbed on various substrates (94).

Polymers containing 8-hydroxyquinoline appear to be selective adsorbents for tungsten in alkaline brines (95). In the presence of tartrate and citrate, quinaldic acid [93-10-7] allows the separation of zinc from gallium and indium (96). Either of these compounds can selectively separate lead and zinc from oxide ores as complexes (97). It is also possible to separate by extraction micro quantities of rhenium(VII), using quinoline in basic solution (98). The presence of large excesses of tungsten(VI), copper(II), vanadium(V), chromium(VI), and molybdenum does not interfere with this process. Cobalt, copper, and nickel have been separated by extraction with 8-sulfamidoquinolines (99).

Catalysts.   A small amount of quinoline promotes the formation of rigid foams (qv) from diols and unsaturated dicarboxylic acids (100). Acrolein and methacrolein 1,4-addition polymerization is catalyzed by lithium complexes of quinoline (101). Organic bases, including quinoline, promote the dehydrogenation of unbranched alkanes to unbranched alkenes using platinum on sodium mordenite (102). The peracetic acid epoxidation of a wide range of alkenes is catalyzed by 8-hydroxyquinoline (103). Hydroformylation catalysts have been improved using 2-quinolone [59-31-4] (104) (see CATALYSIS).

Analytical Reagents.   The chelating property of quinolines, eg, 8-hydroxy derivatives, make them useful in metal gravimetric applications; however, few are any longer of practical importance. Amino- and sulfur-substituted quinolines have also been employed in metal analyses (105,106).

Medicine.   A wide variety of alkaloids (qv) contain the quinoline ring system; this fact accounts, in large measure, for the extensive synthetic research reported (107). In addition to the naturally occurring compounds, a large number of synthetic quinolines, eg, (**22**) and (**23**), have been prepared and studied for use in medicine. Table 3 presents selected examples.

(**22**)   (**23**)

Quinoline Dyes.   The reaction of 2-methylquinoline with phthalic anhydride produces a 2:1 mixture of 2-(2-quinolinyl)-1,3-indandione [83-08-9] (**24**, R = H) and 2-(6-methyl-2-quinolinyl)-1,3-indandione [6493-58-9] (**24**, R = CH$_3$).

Table 3. Quinoline-Derived Drugs Marketed in the United States[a]

Name	CAS Registry Number	Structure	Application	Brand name (manufacturer)
7-chloro-4-[(4-diethylamino-1-methylbutyl)amino]quinoline	[54-05-7]	CH$_3$ / HNCH(CH$_2$)$_3$N(C$_2$H$_5$)$_2$ (quinoline ring, Cl substituent)	antirheumatic	Aralen (Winthrop)
2-butoxyquinoline-4-carboxylic acid, β-diethylaminoethylamide	[61-12-1]	O=CNHCH$_2$CH$_2$N(C$_2$H$_5$)$_2$; OC$_4$H$_9$ (quinoline ring)	local anesthetic	Nupercaine (Ciba)
5-chloro-7-iodo-8-hydroxyquinoline, X = Cl; Y = I	[130-26-7]	X, Y, OH (quinoline ring)	wound and intestinal antiseptic	Vioform (Ciba)

781

Table 3. (Continued)

Name	CAS Registry Number	Structure	Application	Brand name (manufacturer)
5,7-diiodo-8-hydroxyquinoline, X = Y = I	[83-73-8]		intestinal antiseptic	Yodoxin (Glenwood)
5,7-dichloro-8-hydroxyquinoline, X = Y = Cl	[8067-69-4]		intestinal antiseptic, antidiarrheal	Quinolor (Squibb)
4-(4-[ethyl-(2-hydroxyethyl)amino]-1-methylbutylamino)-7-chloroquinoline	[118-42-3]	(22)	antirheumatic, antimalarial	Plaquenil (Winthrop)
1-ethyl-6,7-methylenedioxy-4-oxo-1,4-dihydroquinoline-3-carboxylic acid	[14698-29-4]	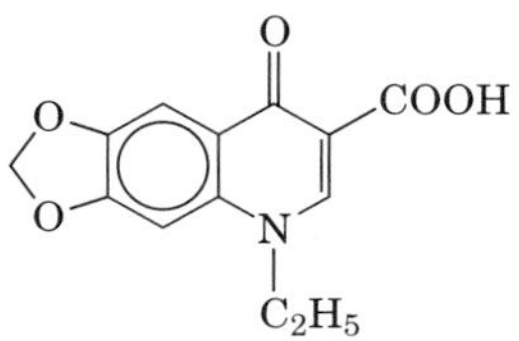 (23)	antibacterial	Utibid (Warner)
2-[2-(2,5-dimethyl-1-phenyl-3-pyrrolyl)vinyl]-6-dimethylamino-1-methylquinolinium pamoate	[3546-41-6]		anthelmintic	Povan (Parke-Davis)

[a]Ref. 108.

782

(24)

This mixture is known as Quinoline Yellow A [8003-22-3] (CI 47000) and is most widely used with polyester fibers (109). Upon sulfonation, the water-soluble Quinoline Yellow S or Acid Yellow 3 [8004-92-0] (CI 47005) is obtained. This dye is used with wool and its aluminum salt as a pigment. Foron Yellow SE-3GL (CI Disperse Yellow 64) is the 3-hydroxy-4-bromo derivative. Several other quinoline dyes are commercially available and find applications as biological stains and analytical reagents (110).

Derivatives. Small amounts of alkyl quinolines are present in the tars resulting from the carbonization and liquefaction of coal (111). Good yields of 4-methylquinoline, 4,6-dimethylquinoline [826-77-7], and 4,8-dimethylquinoline [13362-80-6] are obtained from 4-(diethylamino)-2-butanone and the appropriate aniline. This approach is a promising addition to the traditional syntheses discussed earlier (112). Vinylacetylene reacts with mercuric chloride and either aniline or p-toluidine to yield 4-methyl- and 4,6-dimethylquinoline, respectively (113).

The greater reactivity of 2- and 4-alkylquinolines allows them to condense with benzaldehyde to produce 2-β-styrylquinoline [4945-26-0] and 4-β-styrylquinoline [13362-63-5]. This chemistry is also useful with the pyridine carboxaldehydes to form adequate yields of the corresponding 2-[2-(2-pyridinyl)vinyl]quinoline [13206-41-2], 2-[2-(3-pyridinyl)vinyl]quinoline [1586-51-2], or 2-[2-(4-pyridinyl)vinyl]quinoline [18633-00-6] (114).

The methyl group attached to quinoline undergoes reactions analogous to those of other arenes. The chlorination of 2-methylquinoline in the presence of acetic acid or phosphorus pentoxide produces 88% of pure 2-(trichloromethyl)quinoline [4032-53-5] (115). Methyl groups in the heterocyclic ring are oxidized to carboxylic acids when their palladium chloride complex is treated with hydrogen peroxide (116).

Hydroquinolines. Pyrans formed by reactions of α,β-unsaturated aldehydes with 1-ethoxycyclohexene and treated with hydroxylamine are converted in good yield to 5,6,7,8-tetrahydroquinolines (117). These compounds can be dehydrogenated to the corresponding quinolines. The parent reduced product has been prepared by heating O-allylcyclohexanone oxime (118).

1,2,3,4-Tetrahydroquinoline [635-46-1] shows the properties of a mono-alkylaniline, and its chemistry has been reviewed (119). It forms an N-nitroso derivative, which rearranges readily to 6-nitro-1,2,3,4-tetrahydroquinoline [14026-45-0]. Mild oxidation with potassium permanganate in acetone forms 3,3',4,4'-tetrahydro-1,1'(2H,2'H)biquinoline [34555-59-4], which can be converted to 1',2',3,3',4,4'-hexahydro-1(2H),6'-biquinoline [53899-16-4] by a benzidine-type rearrangement. Indole (qv) is formed by the thermolysis of 1,2,3,4-tetrahyroquinoline in a moist atmosphere at above 650°C (120).

Hydroxyquinolines (*Quinolinols*).   A number of methods have been employed for their preparation. A modified Chichibabin reaction of quinoline in fused KOH–NaOH at 240°C produces 70% of 2-hydroxyquinoline [*59-31-4*] (121). Alternative names based on the facile keto–enol tautomerism of two of these compounds are 2(1*H*)- and 4(1*H*)-quinolinone; none of the other quinolinols show this property. The treatment of 3-hydroxyquinoline [*580-18-7*] with aqueous sodium hydroxide at 300°C gives 88% 2-hydroxyquinoline (122). 7-Chloro-4-hydroxyquinoline [*86-99-7*] can be prepared in 93% yield by ring closure, hydrolysis, and decarboxylation using various acid catalysts (123).

Both 5-hydroxyquinoline [*578-67-6*] and 8-hydroxyquinoline [*148-24-3*] have been prepared in good yields by the acid hydrolysis of the appropriate aminoquinoline at temperatures of 180–235°C (124). The latter compound has been prepared in several different ways, including sulfonation-fusion of quinoline. Hydrolysis of 8-chloroquinoline [*611-33-6*] gives a 93% yield, whereas 80% is obtained in a modified Skraup synthesis with *o*-aminophenol (125,126).

Dihydroxyquinolines are found in nature and may be prepared synthetically. Heating 3,1-benzoxazin-4-ones with strong base or sequential treatment of *N*-acetoacetylanthranilate with base, then acid, produce 2,4-dihydroxyquinolines [*70254-43-2*] (127,128). An enzymatic preparation of 4,5-dihydroxyquinoline has been reported (129).

Direct halogenation of 8-hydroxyquinoline has been used as a route to 5,7-dihalo derivatives. Compounds of this type reported include 5,7-dichloro-8-hydroxyquinoline [*733-76-2*] (130), 5,7-dibromo-8-hydroxyquinoline [*521-74-4*] (131), 5-chloro-8-hydroxyquinoline [*130-16-5*], and 5-chloro-7-iodo-8-hydroxyquinoline [*130-26-7*] (132).

Haloquinolines.   2-Chloroquinoline [*612-62-4*] and 4-chloroquinoline [*611-35-8*] are prepared from the corresponding hydroxyquinolines by reaction with phosphorus oxychloride and phosphorus pentachloride. Reactions of substituted anilines with acrylic acid in the presence of hydroquinone gives the 3-anilinopropionic acid, which is treated with iodine and phosphorus oxychloride to give the corresponding 4-chloroquinoline (133). Chloroanilines have been used and several dichloroquinolines obtained, eg, 4,6-dichloroquinoline [*4203-18-3*], 4,7-dichloroquinoline [*86-98-6*], and 4,8-dichloroquinoline [*21617-12-9*] (134).

Aminoquinolines.   The reduction of nitroquinolines and the displacement of halo derivatives represent the most common methods for the preparation of aminoquinolines. A 72% yield of 8-aminoquinoline [*578-66-5*] has been obtained by treating 8-hydroxyquinoline with ammonium sulfite (135). An interesting rearrangement has been reported in which quinazoline 3-oxide reacts with active methylene compounds to make various 3-substituted 2-aminoquinolines (136).

Aldehydes, Ketones, and Acids.   As with many aromatic compounds, the oxidation of methyl groups is an attractive synthetic route to both aldehydes and carboxylic acids in the quinolines. The hydrolysis of dibromomethyl groups has also been used for aldehydes and the hydrolysis of nitriles for carboxylic acids. Detailed reviews of the synthesis of these compounds have appeared (4).

Biquinolines.   The standard synthetic methods, with bifunctional molecules, eg, aromatic diamines, or using the Ulmann synthesis leads generally to the 2, 2′-bisquinoline structure. Heating the *N*,*O*-di(4-quinolyl)hydroxylamine causes rearrangement to 3, 3′-(4-amino-4′-hydroxy)biquinoline [*64372-81-2*] (137).

Benzoquinolines. Because certain alkaloids are characterized by these more elaborate fused-ring systems, their synthesis has been reviewed in detail (138). Heating 4-methoxy-1-naphthylamine with epichlorohydrin produces a modest yield of 3-hydroxy-6-methoxy-1,2,3,4-tetrahydrobenzo[*h*]quinoline [*38419-41-9*], which can be aromatized, also in small yield, to 6-methoxy-benzo[*h*]quinoline [*38419-43-1*] (139).

Mercaptoquinolines (Quinolinethiols). Mercaptoquinolines are usually prepared using the diazotization of amines, the reduction of sulfides or sulfonyl chlorides, or displacement of active halogen (140–142).

Isoquinoline

The early structural evidence, physical properties, and aromaticity of isoquinoline have been discussed at the beginning of this article. Two additional trivial names are encountered occasionally: 2-benzazine and leucoline. The widespread occurrence of this structure in such important alkaloids as those found in cactus, opium, and curare has created a long-standing interest in its synthesis and properties (4).

Reactions. In general, isoquinoline undergoes electrophilic substitution reactions at the 5-position and nucleophilic reactions at the 1-position. Nitration with mixed acids produces a 9:1 mixture of 5-nitroisoquinoline [*607-32-9*] and 8-nitroisoquinoline [*7473-12-3*]. The ratio changes slightly with temperature (143,144). Sulfonation of isoquinoline gives a mixture with 5-isoquinolinesulfonic acid [*27655-40-9*] as the principal product.

Amination of isoquinoline with sodamide in neutral solvents gives 1-aminoisoquinoline [*1532-84-9*]. This product is converted to 1-hydroxyisoquinoline [*491-30-5*] by diazotization and hydrolysis. Either this compound or 3-methylisoquinoline [*1125-80-0*] can be obtained in excellent yields by the same modified Chichibabin reaction described for 2-hydroxyquinoline (121).

Direct bromination of isoquinoline hydrochloride in a solvent like nitrobenzene gives an 81% yield of 4-bromoisoquinoline [*1532-97-4*] (24). By contrast, bromination of an isoquinoline–aluminum chloride complex with bromine vapor gives a 78% yield of 5-bromoisoquinoline [*34784-04-8*] (22). Continued bromination leads to 5,8-dibromoisoquinoline [*81045-39-8*] and 5,7,8-tribromoisoquinoline [*81045-40-1*].

The oxidation of isoquinoline has also been examined using ruthenium tetroxide. In this instance, the surprising observation that phthalic acid is the only significant product (58%) was made; this fact is both important and difficult to explain (145). Isoquinoline is also oxidized to its N-oxide by peracids. Isoquinoline *N*-oxide [*1532-72-5*] has also been obtained from 2-(2,4-dinitrophenyl)isoquinolinium chloride [*33107-14-1*] by refluxing with hydroxylamine hydrochloride in concentrated hydrochloric acid (146).

The N-oxides of isoquinolines have proved to be excellent intermediates for the preparation of many compounds. Trialkylboranes give 1-alkyl derivatives (147). With cyanogen bromide in ethanol, ethyl *N*-(1- and 4-isoquinolyl)carbamates are formed (148). A complicated but potentially important reaction is the formation of 1-acetonylisoquinoline and 1-cyanoisoquinoline

[*1198-30-7*] when isoquinoline *N*-oxide reacts with methacrylonitrile in the presence of hydroquinone (149). Isoquinoline *N*-oxide undergoes direct acylamination with *N*-benzoylanilinoisoquinoline salts to form 1-*N*-benzoylanilinoisoquinoline [*53112-20-4*] in 55% yield (150). A similar reaction of *N*-sulfinyl-*p*-toluenesulfonamide leads to 1-(tosylamino)isoquinoline [*25770-51-8*] which is readily hydrolyzed to 1-aminoisoquinoline (151).

Isoquinoline can be reduced quantitatively over platinum in acidic media to a mixture of *cis*-decahydroisoquinoline [*2744-08-3*] and *trans*-decahydroisoquinoline [*2744-09-4*] (32). Hydrogenation with platinum oxide in strong acid, but under mild conditions, selectively reduces the benzene ring and leads to a 90% yield of 5,6,7,8-tetrahydroisoquinoline [*36556-06-6*] (32,33). Sodium hydride, in dipolar aprotic solvents like hexamethylphosphoric triamide, reduces isoquinoline in quantitative yield to the sodium adduct [*81045-34-3*] (**25**) (152). The adduct reacts with acid chlorides or anhydrides to give N-acyl derivatives which are converted to 4-substituted 1,2-dihydroisoquinolines. Sodium borohydride and carboxylic acids combine to provide a one-step reduction–alkylation (35). Sodium cyanoborohydride reduces isoquinoline under similar conditions without N-alkylation to give 1,2,3,4-tetrahydroisoquinoline [*91-21-4*]. Reaction of this compound with cyanamide gives an 87% yield of 2-amidino-1,2,3,4-tetrahydroisoquinoline (153).

(**25**)

Isoquinoline also forms Reissert compounds when treated with benzoyl chloride and alkyl cyanide (28), especially under phase-transfer conditions (29). The *N*-phenylsulfonyl Reissert has been converted to 1-cyanoisoquinoline with sodium borohydride under mild conditions (154). When the *N*-benzoyl-1-alkyl derivative is used, reductive fission occurs and the 1-alkylisoquinoline is obtained.

Isoquinoline reacts with aliphatic carboxylic acids photolytically or with a silver catalyst to give excellent yields of alkylation products by decarboxylation (155). This method is useful in the synthesis of 2-benzoylisoquinolines bearing a variety of aromatic substituents in the 1-position (156).

Synthesis of Isoquinoline and Isoquinoline Derivatives. *Bischler-Napieralski Reaction.* This synthetic method involves the cyclodehydration of N-acyl derivatives of *β*-phenethylamines (**26**) to 3,4-dihydroisoquinolines, such as 1-methyl-3,4-dihydroisoquinoline [*2412-58-0*] (**27**) (157). Lewis acids such as phosphorus pentoxide, polyphosphoric acid, or zinc chloride are employed using a dry inert solvent. The 3,4-dihydroisoquinoline can either be dehydrogenated to an isoquinoline or reduced to a tetrahydroisoquinoline.

(26)   (27)

Several modified forms of this synthesis are available. For example, treatment of either isocyanate (**28**) or urethane (**29**) derivatives with phosphoryl chloride followed by stannic chloride has been reported to give the substituted isoquinoline [*80388-01-8*] (158).

(**28**)   (**29**)

The Pictet-Gams method involves the cyclization of β-hydroxy- or β-methoxy-β-phenethylamides, and produces the isoquinoline derivative rather than the reduced form. A further extension of method is based on a methoxy-ethylamine.

Pictet-Spengler Synthesis. An acidic catalyst results in the condensation of β-phenethylamines with carbonyl compounds to give 1,2,3,4-tetrahydroisoquinolines (159).

An enzyme-catalyzed application has been used to prepare the enantiomers of hydroxy-substituted tetrahydroisoquinolines (160). The synthesis of (*S*)-reticuline [*485-19-8*] (**30**) has been reported using similar methodology (161). The substitution of formic acid and paraformaldehyde in this method leads to lower reaction temperatures, freedom from hydrolysis of protective groups, and improved yields (162).

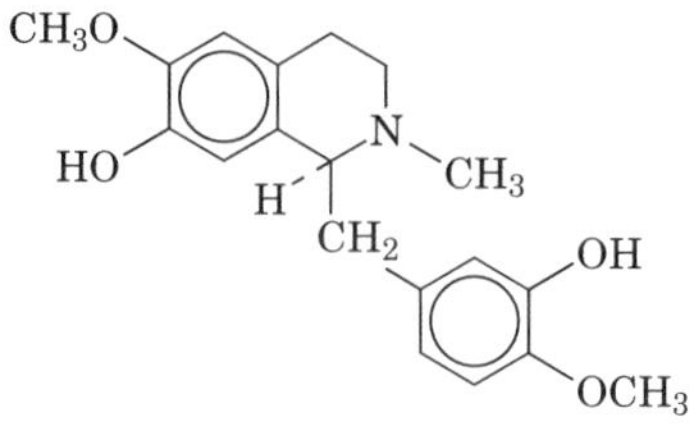

(**30**)

Another limitation in the original method lies in the preparation of 3-alkyl derivatives. This shortcoming has been addressed by two quite different approaches. An investigation of β-arylalkylamines showed that they undergo smooth Pictet-Spengler ring closure (163). Similarly successful studies have been

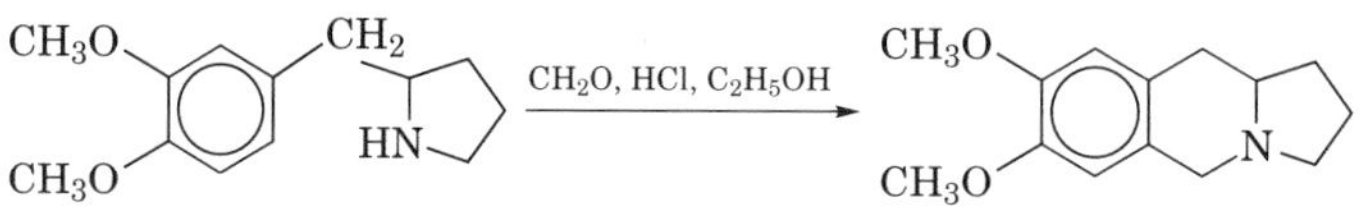

conducted on the flash vacuum pyrolysis of acyloxybenzocyclobutenes (**31**) to 2-formylbenzyl ketones (**32**). The latter react with ammonium acetate to form a variety of 3-alkylisoquinolines in acceptable to excellent yields (164).

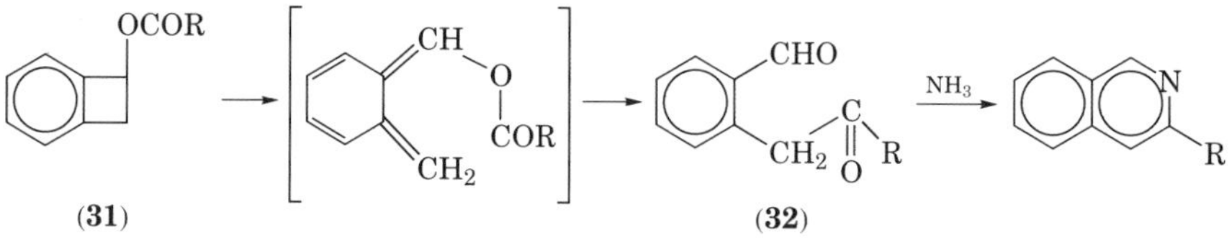

(**31**)   (**32**)

Pomeranz-Fritsch Synthesis. Isoquinolines are available from the cyclization of benzalaminoacetals under acidic conditions (165). The cyclization is preceded by the formation of the Schiff base (**33**). Although the yields are modest, polyphosphoric acid produces product in all cases, and is especially useful for 8-substituted isoquinolines (166).

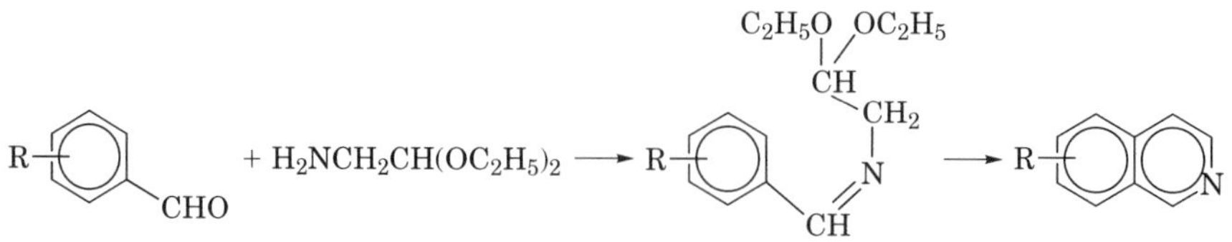

(**33**)

An important modification improves the yields by first reducing the Schiff base to an amine, which is then cyclized to a 4-hydroxy intermediate (167).

This compound is converted to the 1,2,3,4-tetrahydroisoquinoline (**34**). The entire process takes place under mild conditions and in good yield.

A variation involves the reaction of benzylamines with glyoxal hemiacetal (168). Cyclization of the intermediate (**35**) with sulfuric acid produces the same isoquinoline as that obtained from the Schiff base derived from an aromatic aldehyde and aminoacetal. This method has proved especially useful for the synthesis of 1-substituted isoquinolines.

Miscellaneous Synthetic Reactions. A number of *o*-disubstituted ben-zenes have been used to prepare isoquinolines. For example, the Radziszewski method and subsequent dehydration converts *o*-cyanomethylbenzoic acid to homophthalimide [88-97-1] in 90% yield (169). Reaction of the same acid with sodium cyanide at 215°C produces 3-(*o*-carboxylbenzoyl)-4-cyano-1(2*H*)-isoquinoline [81045-38-7] (170). Homophthalimide reacts with acetic anhydride to produce 4-acetylisochroman-1,3-dione, which is converted to 3-methyl-1-(2*H*)-isoquinolone [59816-89-6] or 2,3-dimethyl-1-(2*H*)-isoquinolone [7114-78-5] in good yield (171). These products are also obtained from the cyclization of *trans-β*-styrylisocyanates formed by thermal Curtius reactions of *trans*-cinnamoyl azides (172).

The Ritter reaction with unsaturated carbenium ions under either silver-assisted solvolysis or photolytic conditions leads to excellent yields of isoquino-lines (173). The ease of preparation of the required vinyl bromides makes an attractive route to highly substituted isoquinolines.

Alkyl Isoquinolines. Coal tar contains small amounts of 1-methylisoquinoline [*1721-93-3*], 3-methylisoquinoline [*1125-80-0*], and 1,3-dimethylisoquinoline [*1721-94-4*]. The 1- and 3-methyl groups are more reactive than others in the isoquinoline nucleus and readily oxidize with selenium dioxide to form the corresponding isoquinoline aldehydes (174). These compounds can also be obtained by the hydrolysis of the dihalomethyl group. The 1- and 3-methylisoquinolines condense with benzaldehyde in the presence of zinc chloride or acetic anhydride to produce 1- and 3-styrylisoquinolines. Radicals formed by decarboxylation of carboxylic acids react to produce 1-alkylisoquinolines.

Hydroisoquinolines. In addition to the ring-closure reactions previously cited, a variety of reduction methods are available for the synthesis of these important ring systems. Lithium aluminum hydride or sodium in liquid ammonia convert isoquinoline to 1,2-dihydroisoquinoline (175). Further reduction of this intermediate or reduction of isoquinoline with tin and hydrochloric acid, sodium and alcohol, or catalytically using platinum produces 1,2,3,4-tetrahydroisoquinoline. Other reduction reactions have already been described.

The hydroisoquinolines are more susceptible to ring cleavage than the isoquinolines. Ring cleavage occurs with nitrogen elimination when 3,4-dihydroisoquinolines are heated with alkali and dimethyl sulfate. The resulting *o*-acetylstyrenes form in moderate to excellent yield (176). If there is no 1-substituent, the product is an aldehyde. Ring opening without loss of nitrogen occurs if the 3,4-dihydroisoquinoline is heated with formaldehyde, formic acid, and diethylamine or sodium formate to form the *o*-substituted *N,N*-dimethylphenethylamine (177).

Hydroxyisoquinolines. Hydroxy groups in the 5-, 6-, 7-, and 8-position show phenolic reactions; for example, the Bücherer reaction leads to the corresponding aminoisoquinolines. Other typical reactions include the Mannich condensation, azo-coupling reactions, and nitrosation. Both *O*-methyl and *N*-methyl derivatives are obtained from the methylation of 1-hydroxyisoquinoline, indicating that both tautomeric forms are present. Distillation of various hydroxy compounds, eg, 1- and 4-hydroxyisoquinoline, with zinc dust removes the oxygen. Treatment of 1-isoquinolinol with phosphorus tribromide yields 1-bromoisoquinoline [*1532-71-4*] (178).

Haloisoquinolines. The Sandmeyer reaction is commonly used to prepare chloroisoquinolines from the amino compound. The corresponding hydroxy compounds are also used by treatment with chlorides of phosphorus. The addition of bromine to a slurry of isoquinoline hydrochloride in nitrobenzene gives a 70–80% yield of 4-bromoisoquinoline [*1532-97-4*]. Heating 1-chloroisoquinoline

[*19493-44-8*] with sodium iodide and hydriodic acid gives 1-iodoisoquinoline [*19658-77-6*] (179).

The halogen substituents of isoquinoline undergo typical reactions of similar heterocycles. Heating 1-iodoisoquinoline with sodium nitrite at 100°C gives 1-nitroisoquinoline [*19658-76-5*] (179). Both 1- and 3-bromoisoquinoline react with potassium amide in liquid ammonia to give excellent yields of the corresponding amines (180). The reaction of 5-bromoisoquinoline [*34784-04-8*] with potassium amide in liquid ammonia at −33°C gives 47% 6-aminoisoquinoline [*23687-26-5*] and 21% 5-aminoisoquinoline [*1125-60-6*] (181). This latter amine can be converted to 5-bromoisoquinoline by the Sandmeyer reaction.

Toxicology. Isoquinoline is a poison when ingested or injected intraperitoneally. Even in cases of skin contact it is moderately toxic. As in the case of quinoline, its vapors are irritating to the eyes, nose, and throat. Exposure causes headaches, dizziness, and nausea. Rapid absorption through the skin makes it a dangerous chemical. Its toxicity is oral LD_{50} (rat), 360 mg/kg, and dermal LD_{50} (rabbit), 590 mg/kg (65,66,182,183).

Uses. Isoquinoline and isoquinoline derivatives are useful as corrosion inhibitors, antioxidants, pesticides, and catalysts. They are used in plating baths and miscellaneous applications, such as in photography, polymers, and azo dyes (qv). Numerous derivatives have been prepared and evaluated as pharmaceuticals. Isoquinoline is a main component in quinoline still residue bases, which are sold as corrosion inhibitors and acid inhibitors for pickling iron and steel.

4-Aminoisoquinoline is a component of an ethylene glycol-based corrosion inhibiting antifreeze agent (184) (see ANTIFREEZES AND DEICING FLUIDS). Various compounds related to s-triazolo-[5,1a]isoquinoline [*34784-04-8*] are antioxidants, corrosion inhibitors, and acid acceptors (185). Other effective corrosion inhibitors for steel are 1,2-dihydroisoquinoline-1-phosphonate [*39233-31-3*] and its N-methyl analogue [*39233-30-2*] (186). Compounds related to the catachols, eg, 6,7-dihydroxy-2,3-dimethyl-1,2,3,4-tetrahydroisoquinoline-3-carboxylic acid [*62356-02-9*] have found use as antioxidants for fats and oils (187,188).

A number of isoquinoline derivatives have fungicidal properties, eg, 1,2,3,4-tetrahydroisoquinolines bearing acyl nitrogen substituents like (**36**) [*41910-26-3*] (189). Substituted isoquinolines (**37**) have proved to be effective in controlling undesired vegetation, insects, acarina, and fungi (190).

(**36**)   (**37**)

Isoquinolines are used in catalytic (191), photographic (192), and dye applications (193). A great many alkaloids and synthetic medicinal compounds are isoquinoline derivatives. The principal drugs containing this structure marketed in the United States are listed in Table 4.

Table 4. Isoquinoline-Derived Drugs Marketed in the United States[a]

Name	CAS Registry Number	Structure	Application	Brand name (manufacturer)
2-acetoxy-9,10-dimethoxy-1,3,4,6,7,11b-hexahydro-2H-benzo[a]quinolizin-3-carboxylic acid diethylamide	[63-12-7]		antiemetic, tranquilizer	Emete-Con (Roerig)
1,2,3,4-tetrahydroisoquinoline-2-carboxamidine	[1131-64-2]		muscle relaxant	Declinax (Lilly)
1-(4-ethoxy-3-methoxybenzyl)-6,7-dimethoxy-3-methyl-isoquinoline	[147-27-3]		spasmolytic	Paveril (Lilly)
1-(3,4-diethoxybenzyl-6,7-diethoxyisoquinoline	[486-47-5]		spasmolytic	Ethoquin (Ascher), Tensodin (Knoll)

792

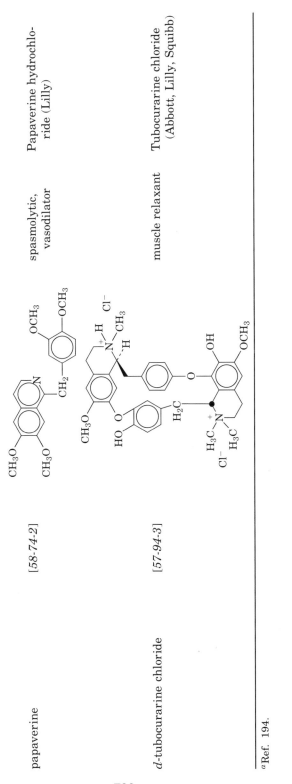

papaverine [58-74-2] spasmolytic, vasodilator Papaverine hydrochloride (Lilly)

d-tubocurarine chloride [57-94-3] muscle relaxant Tubocurarine chloride (Abbott, Lilly, Squibb)

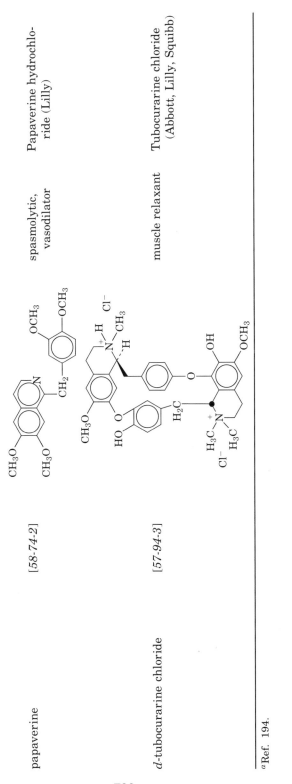[a]Ref. 194.

BIBLIOGRAPHY

"Quinoline and Isoquinoline" in *ECT* 1st ed., Vol. 11, pp. 389–401, by A. G. Renfrew, Mellon Institute; in *ECT* 2nd ed., Vol. 16, pp. 865–885, by M. Kulka, Uniroyal (1966) Ltd.; "Quinolines and Isoquinolines" in *ECT* 3rd ed., Vol. 19, pp. 532–572, by S. N. Holter, Koppers Co., Inc.

1. B. A. Hess, L. J. Schaad, and C. W. Holyoke, *Tetrahedron* **31**, 295–298 (1995).
2. F. F. Runge, *Ann. Phys. Chem. (Poggendorff's Ann.)* **31**, 65–68 (1834).
3. S. Hoogewerff and W. A. van Dorp, *Recl. Trav. Chim. Pays-Bas* **4**, 125–129, 285–293 (1885).
4. G. Jones, "Quinolines," Part I, 1977; Part II, 1982; Part III, 1990; G. Grenthe, "Isoquinolines," Part I, 1981; F. G. Kathawala, G. M. Coppola, and H. F. Schuster, Part II, 1990; and G. M. Coppola and H. F. Schuster, Part III, 1995, in A. Weissberger and E. C. Taylor, eds., *The Chemistry of Heterocyclic Compounds*, Vols. 32 and 38, Wiley-Interscience, New York; D. R. Lide, ed., *Handbook of Chemistry & Physics*, 72nd ed., CRC Press, Boston, Mass., 1992.
5. M. W. Austin and J. H. Ridd, *J. Chem. Soc.*, 4204–4210 (1963).
6. M. J. S. Dewar and P. M. Maitlis, *J. Chem. Soc.*, 2521–2528 (1957).
7. M. Kulka and R. H. F. Manske, *Can. J. Chem.* **30**, 720 (1952).
8. R. G. Coombes and L. W. Russell, *J. Chem. Soc., Perkin Trans. 1*, 1751 (1974).
9. V. N. Konyukhov and M. K. Murshtein, *Izv. Vyssh. Ucheb. Zaved. Khim. Khim. Technol.* **18**, 1267 (1975).
10. D. W. Rangnekar and S. V. Sunthankar, *Indian J. Technol.* **13**, 460 (1975).
11. A. R. Firth and R. D. Smith, *J. Appl. Chem. Biotechnol.* **28**, 857 (1978).
12. O. F. Sidorov, M. K. Murshtein, and L. F. Vasil'chenko, *Izv. Vyssh. Ucheb. Zaved. Khim. Khim. Technol.* **15**, 723 (1972).
13. H. Gilman, J. Eisch, and T. S. Soddy, *J. Am. Chem. Soc.* **81**, 4000 (1959).
14. J. J. Eisch and D. R. Comfort, *J. Org. Chem.* **40**, 2288 (1975).
15. D. I. C. Scopes and J. A. Joule, *J. Chem. Soc. Perkin Trans. 1*, 2810–2811 (1972); C. E. Crawforth, O. Meth-Cohn, and C. A. Russell, *ibid.*, pp. 2807–2810.
16. K. Akiba, Y. Negishi, and N. Inamoto, *Synthesis* 55 (1979).
17. H. A. Hageman, "The Von Braun Cyanogen Bromide Reaction," in R. Adams and co-workers, eds., *Organic Reactions*, Vol. 7, John Wiley & Sons, Inc., New York, 1953.
18. Y. Hamada and M. Sugiura, *Yakugaku Zasshi* **99**, 445 (1979).
19. F. W. Bergstrom, *Chem. Rev.* **35**, 77–277 (1944).
20. M. R. Grimmett, "Halogenation of Heterocycles," in *Advances in Heterocyclic Chemistry*, Vol. 59, Academic Press, Inc., New York, 1994, pp. 286–294.
21. J. J. Eisch, *J. Org. Chem.* **27**, 1318 (1962).
22. M. Gordon and D. E. Pearson, *J. Org. Chem.* **29**, 329 (1964).
23. P. B. D. De la Mare, M. Kiamud-din, and J. H. Ridd, *Chem. Ind. London*, 361 (1958); *J. Chem. Soc.*, 561 (1960).
24. T. J. Kress and S. M. Constantino, *J. Heterocycl. Chem.* **10**, 409 (1973).
25. J. Bialek, *Przem. Chem.* **46**, 526 (1967).
26. G. B. Bochman and D. E. Cooper, *J. Org. Chem.* **9**, 302–309 (1944).
27. M. E. C. Biffin, J. Miller, and D. B. Paul, *Tetrahedron Lett.*, 1015–1018 (1969).
28. W. E. McEwen and R. L. Cobb, *Chem. Rev.* **55**, 511–549 (1955).
29. T. Koizumi and co-workers, *Synthesis*, 497–498 (1977).
30. F. D. Popp, *Adv. Heterocycl. Chem.* **9**, 1 (1968).
31. Ital. Pat. 906,418 (Feb. 1, 1972), F. Minisci and co-workers (to Montedison SpA).
32. R. D. Elderfield, *Heterocyclic Compounds*, Vol. 4, John Wiley & Sons, Inc., New York, 1952, p. 282.
33. J. Z. Ginos, *J. Org. Chem.* **40**, 1191 (1975).

34. E. W. Vierhapper and E. L. Eliel, *J. Am. Chem. Soc.* **96**, 2256 (1974).
35. G. W. Gribble and P. W. Heald, *Synthesis*, 650–652 (1975).
36. R. Ruzicka, *Phenole und Basen-Vorkommen und Gewinnung*, Akademie-Verlag, Berlin, Germany, 1958.
37. J. Vymetal, *Khos. Khim.*, 31–35 (1987).
38. M. Dong and co-workers, *J. Chromatogr. Sci.* **15**, 32 (1977).
39. M. Novotny and co-workers, *Anal. Chem.* **52**, 401 (1980).
40. J. Vymetal and A. Kulhankova, *Chem. Prum.* **26**, 193 (1976).
41. *Ibid.*, **29**, 308 (1979).
42. R. H. F. Manske and M. Kulka, "The Skraup Synthesis of Quinolines," in Ref. 17.
43. I. T. Millar and H. D. Springall, *Sidgwick's Organic Chemistry of Nitrogen*, 3rd ed., Clarendon Press, Oxford, U.K., 1966, Chapt. 24.
44. G. R. Newkome and W. W. Paudler, *Contemporary Heterocyclic Chemistry*, John Wiley & Sons, Inc., New York, 1982, pp. 200–212.
45. T. L. Gilchrist, *Heterocyclic Chemistry*, 2nd ed., John Wiley & Sons, Inc., New York, 1992, pp. 152–166.
46. P. A. Claret and A. G. Osborne, *J. Chem. Technol. Biotechnol.* **29**, 175 (1979).
47. C. M. Leir, *J. Org. Chem.* **42**, 911 (1977).
48. A. L. Katritzky and C. W. Rees, in A. J. Boulton and A. McKillop, eds., *Comprehensive Heterocyclic Chemistry*, Vol. 2, Pergammon Press, Oxford, U.K., 1984, p. 474; Beyer, *J. Prakt. Chem.* **33**, 393 (1886).
49. R. H. Reitsema, *Chem. Rev.* **43**, 43 (1968).
50. Ng. Ph. Buu-Hoi and R. Roger, *J. Chem. Soc.*, 106 (1948).
51. F. A. Al-Tai, G. Y. Sarkis, and J. M. Al-Jsmabi, *Bull. Coll. Sci.* **9**, 55 (1966).
52. P. Friedlander, *Chem. Ber.* **15**, 2572 (1882); Ref. 45, p. 155.
53. P. Caluwe, *Tetrahedron* **36**, 2369–2382 (1980).
54. J. Barluenga, E. Aguilar, J. Joglar, B. Olano, and S. Fustero, *J. Org. Chem.*, **54**, 2596–2598 (1989).
55. Y. Watanabe, Y. Tsuji, and J. Shida, *Bull. Chem. Soc. Jpn.*, **57**, 435–438 (1984), and several earlier articles.
56. R. Leardini, G. F. Pedulli, A. Tundo, and G. Zanardi, *J. Chem. Soc., Chem. Commun.*, 1320–1321 (1984).
57. T. Saito, H. Ohmori, E. Furuno, and S. Motoki, *J. Chem. Soc., Chem. Commun.*, 22–24 (1992).
58. S. F. Campbell, J. D. Hardstone, and M. J. Palmer, *Tetrahedron Lett.*, 4813–4816 (1984).
59. J. Barluenga, H. Cuervo, S. Fustero, and V. Gotor, *Synthesis*, 82–84 (1987).
60. O. Meth-Cohn, *Tetrahedron Lett.*, 1901–1904 (1985).
61. *Chemical Catalog, 1995–1996*, Lancaster Synthesis, Inc., Windham, N.H., 1995.
62. *Aldrich Catalog Handbook of Fine Chemicals 1994–1995*, Aldrich Chemical Co., Inc., Milwaukee, Wis., 1994.
63. *ARCOS Organics Handbook of Fine Chemicals, 1995–1996*, Fisher Scientific, Pittsburgh, Pa., 1995.
64. N. I. Sax and R. J. Lewis, Sr., *Dangerous Properties of Industrial Materials*, 8th ed., Vol. III, Van Nostrand, New York, 1992, pp. 2968–2969.
65. *The Sigma-Aldrich Library of Regulatory and Safety Data*, Vol. 2, Aldrich Chemical Co., Milwaukee, Wis., 1993, pp. 2611–2612.
66. R. P. Pohanish and S. A. Green, eds., *Hazardous Substance Resource Guide*, Gale Research, Detroit, Mich., 1993, pp. 252–253.
67. N. I. Sax and R. J. Lewis, Sr., *Hazardous Chemical Desk Reference*, 3rd ed., Van Nostrand, New York, 1993, pp. 759–760.
68. Ger. Offen. 2,832,126 (Feb. 8, 1979), S. Sato and T. Watanabe (to Bridgestone Tire Co., Ltd.).

69. Brit. Pat. 1,113,360 (May 15, 1968) (to Ciba-Geigy Ltd.).
70. Jpn. Kokai Tokkyo Koho 78,46,839 (Dec. 16, 1978) (to Material Vegyi RSZ).
71. Ger. Offen. 2,439,534 (Mar. 4, 1976), H. U. Voigt (to Kabel- und Metallwerke Gute-hoffnungshuette A-G).
72. Brit. Pat. 1,153,178 (May 29, 1969), I. Norvick.
73. Fr. Pat. 1,529,230 (June 14, 1968), I. G. Rose (to National Research Development Corp.).
74. Ger. Offen. 2,149,138 (Apr. 5, 1973), C. Rasp (to Bayer A-G).
75. G. Buchmann and O. Wolniak, *Pharmazie* **21**, 650 (1966).
76. Ger. Offen. 2,322,143 (Dec. 13, 1973), D. Cartwright and co-workers (to Imperial Chemical Industries, Ltd.).
77. Jpn. Kokai 77,72,821 (June 17, 1977), T. Takematsu and co-workers (to Mitsubishi Petrochemicals Co., Ltd.).
78. Ger. Offen. 2,361,438 (June 26, 1975), A. Studeneer and co-workers (to Hoechst A-G).
79. D. L. Lewis and co-workers, *J. Econ. Entomol.* **71**, 818 (1978).
80. A. S. Sadykov and co-workers, *Uzb. Biol. Zh.* **11**, 19 (1967).
81. Czech. Pat. 158,491 (June 15, 1975), J. Ciernik and D. Ambrus.
82. Jpn. Kokai Tokkyo Koho 79,110,292 (Aug. 29, 1979), H. Kanbara (to Tokyo Organic Chemical Industries, Ltd.).
83. G. V. Myasoedova and co-workers, *Org. Reagenty Anal. Khim. Tezisy Dokl. Vses. Konf. 4th*, **1**, 116 (1976).
84. F. Vernon and K. M. Nyo, *Anal. Chim. Acta* **93**, 203 (1977).
85. N. Grassie and I. G. Meldrum, *Eur. Polym. J.* **4**, 571 (1968).
86. Jpn. Kokai Tokkyo Koho 79,139,989 (Oct. 30, 1979), H. Kanbara (to Tokyo Organic Chemical Industries, Ltd.).
87. Fr. Pat. 1,497,474 (Oct. 13, 1967), P. Enthone.
88. Jpn. Kokai 76,126,936 (Nov. 5, 1976), S. Emori (to Seiken Chemical Co., Inc.).
89. U.S. Pat. 4,009,087 (Feb. 22, 1977), O. Kardos and co-workers (to M & T Chemicals, Inc.).
90. Ind. Pat. 140,766 (Dec. 18, 1976), J. Adhya.
91. U. S. Pat. 4,045,441 (Aug. 30, 1977) and Ger. Offen. 2,710,491 (Sept. 15, 1977), H. J. Richards and B. C. Trivedi (to Ashland Oil, Inc.).
92. Fr. Demande 2,309,645 (Nov. 26, 1976), (to International Nickel Co. of Canada, Ltd.).
93. S. Afr. Pat. 75 00,412 (Dec. 8, 1975), P. L. Mattison (to General Mills Chemicals, Inc.).
94. Jpn. Kokai 73 55,558 (Aug. 4, 1973), N. Yokota and co-workers (to Osaka Soda Co., Ltd.).
95. U.S. Appl. 925,672 (Mar. 2, 1979), W. N. Marchant and P. T. Brooks (to U.S. Dept. of the Interior); P. B. Altringer and co-workers, *Min. Eng.* (*Littleton, Colo.*) **31**, 1220 (1979).
96. E. A. Ostroumov and A. V. Kulumbegashvii, *Zh. Anal. Khim.* **31**, 1338 (1976).
97. Fr. Demande 2,401,228 (Mar. 23, 1979) (to Consiglio Nazionale delle Ricerche).
98. D. Dorz and J. Dobrowolski, *Chem. Anal. Warsaw* **24**, 465 (1979).
99. U.S. Pat. 4,100,163 (July 11, 1978), M. J. Virnig (to General Mills Chemicals, Inc.).
100. U.S. Pat. 3,671,471 (June 20, 1972), S. E. Jamison (to Celanese Corp.).
101. Jpn. Pat. 15,621 (Aug. 29, 1967), R. Wakasa, S. Ishida, and Y. Kitahama (to Asahi Chemical Industry Co., Ltd.).
102. U.S. Pat. 4,000,210 (Dec. 28, 1976), E. E. Sensel and A. W. King (to Texaco, Inc.); Neth. Appl. 76 13,473 (June 6, 1978) (to Texaco Development Corp.).
103. Jpn. Kokai Tokkyo Koho 79,07,764 (Apr. 10, 1979), S. Nagato and co-workers (to Daicel, Ltd.).
104. Ger. Offen. 2,901,347 (July 19, 1979), H. Kojima and co-workers (to Daicel, Ltd.).
105. K. Yamamoto and co-workers, *Bunseki Kagako* **16**, 229 (1967).

106. H. F. Schaeffer, *Microchem. J.* **14**, 90 (1969).

107. A. L. Katritzky and C. W. Rees, in A. J. Boulton and A. McKillop, eds., *Comprehensive Heterocyclic Chemistry*, Vol. 2, Pergammon Press, Oxford, U.K., 1984.

108. J. Elks and C. R. Ganellin, eds., *Dictionary of Drugs*, Chapman and Hall, London, 1990, pp. C-284, C-357, C-244, D-364, D-190, H-120, D-131, and V-52.

109. H. Zollinger, *Color Chemistry*, VCH, Weinheim, Germany, 1987, pp. 52–53.

110. F. J. Green, ed., *The Sigma-Aldrich Handbook of Stains, Dyes, and Indicators*, Aldrich Chemical Co., Milwaukee, Wis., 1990, pp. 253, 256, 294, 318, 402, 579, 603, 608, 610.

111. W. W. Pandler and M. Cheplen, *Fuel* **58**, 775 (1979).

112. B. I. Ardashev and co-workers, *Khim. Geterotsikl. Soedin.*, 857 (1968).

113. N. S. Kozlov and G. P. Korotyshova, *Tezisy. Dokl. Bsec. Konf. Khim. Atsetilena, 5th*, 288 (1975).

114. S. Biniecki and W. Modrzejewska, *Acta Pol. Pharm.* **24**, 561 (1967).

115. Jpn. Pat. 73 05,598 (Feb. 17, 1973), M. Fukushima and co-workers, (to Chisso Corp.).

116. S. Paraskewas, *Synthesis*, 819 (1974).

117. Yu. I. Chumakov and N. B. Bulgakova, *Ukr. Khim. Zh.* **36**, 514 (1970).

118. T. Kusumi, Ko Yoneda, and H. Kakisawa, *Synthesis*, 221 (1979).

119. H. Beschke, *Aldrichimica Acta* **11**, 13 (1978).

120. Ger. Offen. 2,822,907 (Nov. 29, 1979), K. Handrick and G. Koelling (to Bergwerksverband GmbH).

121. J. J. Wandewalle and co-workers, *Chem. Ber.*, **108**, 3898 (1975).

122. K. M. Dyumaev and E. P. Popova, *Khim. Geterotsikl Soedin.*, 85 (1973).

123. Hung. Teljes 6251 (May 28, 1973), A. Ujhidy and co-workers, (to E. Gy. T. Gyogyszervegyeszeti Gyar).

124. Ger. Pat. 2,352,976 (June 20, 1974), F. M. Covelli (to Koppers Co., Inc.).

125. Jpn. Pat. 72 37,436 (Sept. 20, 1972), I. Onishi and co-workers (to Yuki Gosei Kogyo Co., Ltd.).

126. Ger. Offen. 2,545,704 (Apr. 22, 1976), J. M. Cognion (to Ugine Kuhlmann).

127. Fr. Demande 2,009,119 (Jan. 30, 1970) (to Badische Anilin- und Soda-Fabrik A-G).

128. Fr. Demande 2,009,125 (Jan. 30, 1970) and Ger. Offen. 1,924,362 (Nov. 19, 1970), H. J. Sturm and H. Goerth (to Badische Anilin- und Soda-Fabrik A-G).

129. T. Noguchi and co-workers, *J. Biochem. Tokyo* **67**, 113 (1970).

130. Czech. Pat. 180,910 (Sept. 15, 1979), J. Simek and M. Ulrich.

131. Ger. Offen. 2,515,476 (Oct. 21, 1976), H. Jenkner and G. Neisen (to Chemische Fabrik Kalk GmbH).

132. USSR Pat. 591,467 (Feb. 5, 1978), A. P. Zuev and co-workers (to S. Ordzhonikidze, All-Union Scientific Research Chemical-Pharmaceutical Institute).

133. Ger. Offen. 2,843,781 (Apr. 12, 1979), J. Bulidon and C. Pavan (to Roussell-UCLAF).

134. Fr. Pat. 1,514,280 (Feb. 23, 1968), R. Joly, J. Warnant, and B. Goffinet (to Roussell-UCLAF).

135. U.S. Pat. 3,312,708 (Apr. 4, 1967), C. J. Lind (to Allied Chemical Corp.).

136. T. Higashino, U. Nagano, and E. Hayashi, *Chem. Pharm. Bull.* **21**, 1943 (1973).

137. H. Sawanishi and Y. Kamiya, *Chem. Pharm. Bull.* **23**, 2949 (1975).

138. Ref. 107, Vol. 1, pp. 57 and 65.

139. S. Kutkevicius and R. Valite, *Khim. Geterotsikl. Soedin*, 1121 (1972).

140. U.S. Pat. 3,509,159 (Apr. 28, 1970), D. Kealey.

141. S. Kubota and co-workers, *Yakugaku Zasshi* **93**, 354 (1973).

142. A. O. Fritton and F. Ridgway, *J. Med. Chem.* **13**, 1008 (1970).

143. Ref. 6; R. A. Robinson, *J. Am. Chem. Soc.* **69**, 1943 (1947).

144. C. G. LeFevre and R. J. W. LeFevre, *J. Chem. Soc.*, 1470 (1935).

145. D. C. Ayres and A. M. M. Hossain, *J. Chem. Soc. Perkin Trans. I*, 707 (1975).

146. Y. Tamura, N. Tsuijimoto, and M. Uchimura, *Chem. Pharm. Bull.* **19**, 143 (1971).
147. T. Koduo, T. Nose, and M. Hamana, *Yakugaku Zasshi* **95**, 521 (1975).
148. M. Hamana and S. Kumadaki, *Yakugaku Zasshi* **95**, 87 (1975).
149. M. Hamana and co-workers, *Chem. Pharm. Bull.* **23**, 346 (1975).
150. R. A. Abramovitch, R. B. Rogers, and G. M. Singer, *J. Org. Chem.* **40**, 41 (1975).
151. T. Onaka, *Itsuu Kenkyusho Nempo*, 29 (1968).
152. M. Natsume and co-workers, *Tetrahedron Lett.*, 2335 (1973).
153. Brit. Pat. 1,149,463 (Apr. 23, 1969) (to Hoffmann-LaRoche Inc.).
154. I. Saito, Y. Kikugawa, and S. Yamada, *Chem. Pharm. Bull.* **22**, 740 (1974).
155. F. Minisci and co-workers, *Tetrahedron* **27**, 3575 (1971).
156. USSR Pat. 434,076 (June 30, 1974), A. K. Sheinkman and G. V. Samoilenko (to Donetsk State University).
157. W. M. Whaley and T. R. Govindachari, in R. Adams and co-workers, eds., *Organic Reactions*, Vol. 6, John Wiley & Sons, Inc., New York, 1951, Chapt. 2.
158. Y. Tsuda and co-workers, *Heterocycles* **5**, 157 (1976).
159. W. M. Whaley and T. R. Govindachari, in Ref. 157, Chapt. 3.
160. S. Teitel and A. Brossi, *Lloydia* **37**, 196 (1974).
161. M. Konda, T. Shioiri, and S. Yamada, *Chem. Pharm. Bull.* **23**, 1063 (1975).
162. S. Ruchirawat, M. Chaisupakitsin, N. Patranuwatana, J. L. Cashaw, and V. E. Davis, *Synthetic Commun.* **14**, 1221–1228 (1984).
163. J. E. Norlander, M. J. Payne, F. G. Njoroge, M. A. Balk, G. D. Laikos, and V. M. Vishwanath, *J. Org. Chem.* **49**, 4107–4111 (1984).
164. P. Schiess, M. Huys-Francotte, and C. Vogel, *Tetrahedron Lett.*, 3959–3962 (1985).
165. W. J. Gensler, in Ref. 157, Chapt. 4.
166. M. J. Bevis, *Tetrahedron* **25**, 1585 (1969).
167. J. M. Bobbitt and co-workers, *J. Org. Chem.* **30**, 2247, 2459 (1965).
168. E. Schlitter and J. Muller, *Helv. Chim. Acta* **31**, 914 (1948).
169. S. Nanya and E. Maekawa, *Nagoya Kogyo Daigaku Gakuho* **20**, 161 (1968).
170. G. Pangon and co-workers, *Bull. Soc. Chim. Fr.*, 1991 (1970).
171. S. A. Foster, J. Leyshorn, and D. G. Saunders, *Indian J. Chem.* **10**, 1060 (1972).
172. G. J. Mikol and J. H. Boyer, *J. Org. Chem.* **37**, 724 (1972).
173. T. Kitamura, S. Kobayashi, and H. Taniguchi, *Chem. Lettr.*, 1351–1354 (1984).
174. A. Roe and C. E. Teague, *J. Am. Chem. Soc.* **73**, 687 (1951).
175. W. Huckel and G. Graner, *Chem. Ber.* **90**, 2017 (1957).
176. W. J. Gensler and co-workers, *J. Am. Chem. Soc.* **78**, 1713 (1956).
177. J. Gardent, *Comp. Rend.* **243**, 1042 (1956).
178. J. L. Butler, F. L. Bayer, and M. Gordon, *Trans. Ky. Acad. Sci.* **38**, 15 (1977).
179. B. Hayashi and co-workers, *Yakugaku Zasshi* **87**, 1342 (1967).
180. G. M. Sanders, M. Vandijk, and H. J. Den Hertog, *Recl. Trav. Chim. Pays-Bas* **93**, 198 (1974).
181. H. Poradowska and E. Huczkowska, *Synthesis*, 733 (1975).
182. Ref. 64, p. 2061.
183. Ref. 67, p. 573.
184. Ger. Offen. 2,149,138 (Apr. 5, 1973), C. Rasp (to Bayer AG).
185. U.S. Pat. 3,758,480 (Sept. 11, 1973), H. K. Reimlinger and J. J. M. Vanderwalle (to Mallincrodt Chemical Works).
186. U.S. Pats. 3,809,694 (May 7, 1974) and 3,830,815 (Aug. 20, 1974), D. Redmore (to Petrolite Corp.).
187. Ger. Offen. 2,342,474 (Mar. 7, 1974) and Jpn. Kokai 74,39,684 (Apr. 13, 1974), S. Okumura and co-workers (to Ajinomoto Co., Inc.).
188. Jpn. Kokai 76,105,982 (Sept. 20, 1976), I. Chibata and co-workers (to Tanabe Seiyaku Co., Ltd.).

189. U.S. Pat. 3,929,830 (Dec. 30, 1975), S. B. Richter (to Velsicol Chemical Corp.).
190. U.S. Pat. 3,930,837 (Jan. 6, 1976), A. Serban (to ICI Australia Ltd.).
191. Ger. Offen. 2,630,086 (Jan. 12, 1978), R. Platz and co-workers (to BASF AG).
192. Ger. Offen. 2,444,422 (Mar. 27, 1975), N. Kurutachi and A. Arai (to Fuji Photo Film Co., Ltd.).
193. Ger. Offen. 2,650,226 (May 11, 1978), D. Rose (to Henkel und Cie GmbH).
194. Ref. 108, pp. B-102, D-33, D-462, E-150, P-19, and C-309.

K. Thomas Finley
State University of New York, Brockport

QUINONES

These useful compounds have played a central role in both theoretical and practical organic chemistry since the 1840s. In 1838 quinic acid [*36413-60-2*] (**1**) was oxidized to 1,4-benzoquinone (**2**) with manganese dioxide (1).

(**1**)　　　　　(**2**)

Synthesis by oxidation remains the first choice for commercial and laboratory preparation of quinones; the starting material (**1**) provided the generic name quinone. This simple, descriptive nomenclature has been abandoned by *Chemical Abstracts*, but remains widely used (2). The systematic name for (**2**) is 2,5-cyclohexadiene-1,4-dione. Several examples of quinone synonyms are given in Table 1. Common names are used in this article. 1,2-Benzoquinone (3,5-cyclohexadiene-1,2-dione) (**3**) is also prepared by oxidation, often with freshly prepared silver oxide (3). Compounds related to (**3**) must be prepared using mild conditions because of their great sensitivity to both electrophiles and nucleophiles (4,5).

Quinonoid compounds have been thoroughly reviewed (4,6). More recent trends in quinone addition and substitution chemistry were reviewed in 1993 (7). The quinone system in natural products has also been covered (8); there has been a fascinating discussion of the *meta*-quinone problem (9).

The close electrochemical relationship of the simple quinones, (**2**) and (**3**), with hydroquinone (1,4-benzenediol) (**4**) and catechol (1,2-benzenediol) (**5**),

Table 1. Quinone Nomenclature

Common name	CAS Registry Number	Synonym	Structure number
1,4-benzoquinone	[106-51-4]	p-benzoquinone	(2)
1,2-benzoquinone	[583-63-1]	o-benzoquinone	(3)
1,4-naphthoquinone	[130-15-4]	α-naphthoquinone	(9)
4,4'-diphenoquinone[a]	[494-72-4]	diphenoquinone	(17)
o-chloranil[b]	[2435-53-2]	tetrachloro-o-quinone	(22)
p-chloranil[c]	[118-75-2]	tetrachloro-p-quinone	(23)
2,3-dichloro-5,6-dicyano-1,4-benzoquinone[d]	[84-58-2]	DDQ	(24)

[a]Bis-2,5-cyclohexadien-1-ylidene-4,4'-dione; 4-(4-oxo-2,5-cyclohexadien-1-ylidene)-2,5-cyclohexadien-1-one is also used.
[b]3,4,5,6-Tetrachloro-3,5-cyclohexadiene-1,2-dione.
[c]2,3,5,6-Tetrachloro-2,5-cyclohexadiene-1,4-dione.
[d]4,5-Dichloro-3,6-dioxo-1,4-cyclohexadiene-1,2-dicarbonitrile.

respectively, has proven useful in ways extending beyond their offering an attractive synthetic route. Photographic developers and dye syntheses often involve (4) or its derivatives (10). Biochemists have found much interest in the interaction of mercaptans and amino acids with various compounds related to (3). The reversible redox couple formed in many such examples and the frequently observed quinonoid chemistry make it difficult to avoid a discussion of the aromatic reduction products of quinones (see HYDROQUINONE, RESORCINOL, AND CATECHOL).

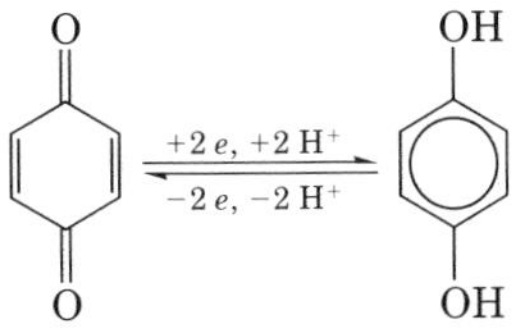

(2)   (4)

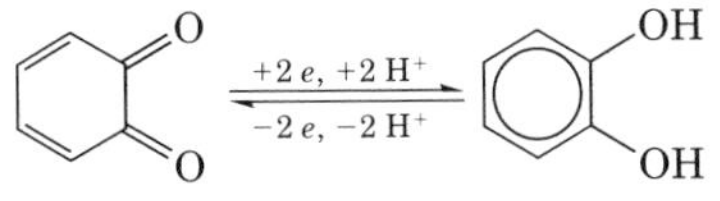

(3)   (5)

An example of the Michael chemistry, typical of all quinones bearing a replaceable hydrogen, is the preparation of a sulfone (6) (in 55% yield), which was ultimately converted to a polystyrene redox polymer (11).

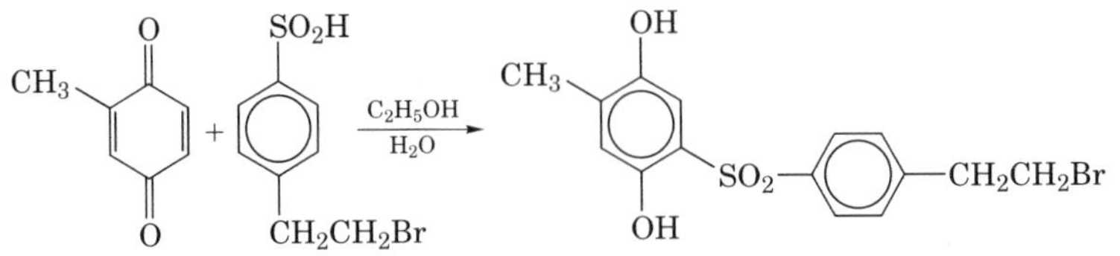

(6)

One of the most exciting discoveries related to quinone/hydroquinone chemistry is their synthesis by biosynthetic routes (12,13). Using bacterial enzymes to convert D-glucose [50-99-7] (**7**) to either 1,2- or 1,4-benzenediol allows the use of renewable raw material to replace traditional petrochemicals. The promise of reduced dependence on caustic solutions and the use of transition-metal catalysts for their synthesis are attractive in spite of the scientific and economic problems still to be solved.

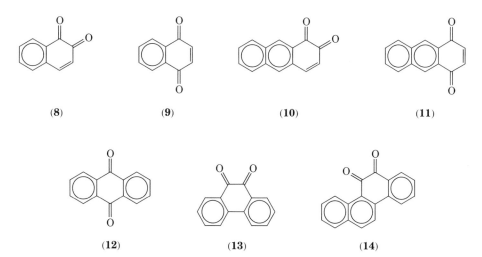

The cross-conjugated system of two α,β-unsaturated carbonyl groups of both 1,2- and 1,4-quinones occurs in many polynuclear hydrocarbons, eg, 1,2-naphthoquinone [524-42-5] (**8**) and 1,4-naphthalenedione [130-15-4] (1,4-naphthoquinone) (**9**) (see Fig. 1). The carbonyl groups may be located in different rings, but occupy positions corresponding to the 1,2- or 1,4-orientation of monocyclic quinones, eg, in naphthalenes such as 2,6-naphthoquinone [613-20-7] (**15**), in biphenyls such as 2,2'-diphenoquinone [59869-78-2] (**16**) and 4,4'-diphenoquinone [494-72-4] (**17**), and in stilbenes such as 4,4'-stilbenequinone [3457-53-2] (stilbene-4,4'-dione) (**18**).

Fig. 1. Quinones of polynuclear hydrocarbons: (**8**) 1,2-naphthoquinone [524-42-5]; (**9**) 1,4-naphthalenedione [130-15-4]; (**10**), 1,2-anthraquinone [655-04-9]; (**11**), 1,4-anthraquinone [635-12-1]; (**12**), 9,10-anthraquinone [84-65-1]; (**13**), 9,10-phenanthraquinone [84-11-7]; and (**14**), 5,6-chrysenequinone [2051-10-7] (5,6-dihydrochrysene-5,6-dione). See text.

(15)   (16)

(17)   (18)

There is a rich literature of nonbenzenoid quinones such as acenaphth-enequinone [*82-86-0*] (1,2-acenaphthylenedione) (**19**) (14,15). More recently the chemistry of the azulene quinones such as 1,5-azuloquinone [*74424-63-8*] (**20**) has emerged as important in its own right (16). A few compounds having no apparent relationship to quinones continue to be misnamed, eg, camphorquinone [*465-29-2*] (2,3-camphandione) (**21**) (17).

(19)   (20)   (21)

Simple quinones have two notable physical properties: odor and color. The 1,4-benzo- and 1,4-naphthoquinones and many of their derivatives have high vapor pressures and pungent, irritating odors. The single-ring compounds are often found as constituents of insects' chemical defense against predators. In general, the 1,2-quinones are vibrant in color, ranging from orange to red, whereas the 1,4-quinones are usually lighter, ie, yellow to orange.

The quinones have excellent redox properties and are thus important oxi-dants in laboratory and biological synthons. The presence of an extensive array of conjugated systems, especially the α,β-unsaturated ketone arrangement, allows the quinones to participate in a variety of reactions. Characteristics of quinone reactions include nucleophilic substitution; electrophilic, radical, and cycloaddi-tion reactions; photochemistry; and normal and unusual carbonyl chemistry.

Physical Properties

Selected physical data for various quinones are given in Table 2; Table 3 gives uv spectral data and redox potentials (Fig. 2). References 19 and 20 greatly expand the range of data treated.

Table 2. Physical Properties of Selected Quinones[a]

Name	CAS Registry Number	Structure number	Color	Melting point, °C	Crystalline form	Solubility	
						Soluble	Insoluble
1,4-benzoquinone	[106-51-4]	(2)	yellow	113, 116	monoclinic prisms	alcohol, ether	water, pentane
1,2-benzoquinone	[583-63-1]	(3)	red	60–70 dec	plates or prisms	ether, benzene	pentane
1,2-naphthoquinone	[524-42-5]	(8)	yellow-red orange	145–147	needles	water, alcohol	ligroin
1,4-naphthoquinone	[130-15-4]	(9)	bright yellow	125, 128.5	needles	alcohol, benzene	water, ligroin
3,4,5,6-tetrachloro-1,2-benzoquinone	[2435-53-2]	(22)	orange-red	133, 122–127			
2,3,5,6-tetrachloro-1,4-benzoquinone	[118-75-2]	(23)	yellow	290, 294	monoclinic prisms	ether	water, ligroin
2,3-dichloro-5,6-dicyano-1,4-benzoquinone	[84-58-2]	(24)	bright yellow	201–202 dec	plates		
2-chloro-1,4-benzoquinone	[695-99-8]	(25)	yellow-red	57	rhombic-hexagonal	water, alcohol	
2,5-dichloro-1,4-benzoquinone	[615-93-0]	(26)	pale yellow	161–162	monoclinic prisms	ether, chloroform	water, alcohol
2,5-dimethyl-1,4-benzoquinone	[137-97-9]	(27)	yellow	125		ether, alcohol	water, alcohol
2-methyl-1,4-benzoquinone	[553-97-9]	(28)	yellow	69	plates or needles	ether, alcohol	water
3-chloro-1,2-naphthoquinone	[18099-99-5]	(29)	red	172 (dec)	needles	alcohol, benzene	water
2,3-dichloro-1,4-naphthoquinone	[177-80-6]	(30)	yellow	193, 195	needles	benzene, chloroform	water, alcohol
2-methyl-1,4-naphthoquinone	[58-27-5]	(31)	yellow	105–107	needles	ether, benzene	water, alcohol

[a]Ref. 18.

Table 3. Spectral (Uv) Data and Redox Potentials for Selected Quinones[a]

Structure number	λ_{max}, nm (ϵ)[b]	$-E_{1/2}$[c], V	
		E_1	E_2
(3)	580 (30), 385 (1585)[d]	0.31	0.90
(22)		0.1	−0.71
(2)	243	0.51	1.14
(25)	251 (7740)	0.34	0.92
(26)	271 (5710)	0.18	0.81
(24)		−0.51	−0.30
(27)	293 (1770), 251 (11650), 220 (4080)	0.67	1.27
(28)	429 (19), 314 (589), 246 (13804)[e]	0.58	1.10
(23)	364 (248), 286 (12600)	−0.01	0.71
(8)	398 (1800), 336 (2280), 248 (20400)	0.56	1.02
(29)	330 (3020), 250 (19953), 246 (20417)[f]	0.71	1.25
(9)	337, 279, 252, 246		
(30)	328, 264, 253, 249, 244[g]	0.77	1.28

[a]See Figure 2; Ref. 18.
[b]Solvent is CH_3OH unless otherwise noted; usually several solvents are given in the works cited.
[c]25°C, SCE, CH_3CN, 0.1 N $(C_2H_5)_4NClO_4$.
[d]Ether. [e]Ethanol. [f]CH_3CN. [g]Cyclohexane.

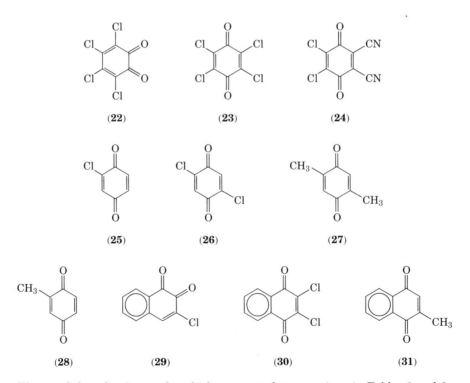

Fig. 2. Selected quinones for which property data are given in Tables 2 and 3.

804

Chemical Properties

Biochemical Reactions. The quinones in biological systems play varied and important roles (21,22). In insects they are used for defense purposes, and the vitamin K family members, eg, vitamin K_1 [*11104-38-4*] (**32**) and vitamin K_2 [*11032-49-8*] (**33**), which are based on 2-methyl-1,4-naphthoquinone, are blood-clotting agents (see VITAMINS, VITAMIN K).

(32)   (33)

Two groups of substituted 1,4-benzoquinones are associated with photosynthetic and respiratory pathways; the plastoquinones, eg, plastoquinone [*4299-57-4*] (**34**), and the ubiquinones, eg, ubiquinone [*1339-63-5*] (**35**), are involved in these processes. Although they are found in all living tissue and are central to life itself, a vast amount remains to be learned about their biological roles.

(34)   (35)

Quinones of various degrees of complexity have antibiotic, antimicrobial, and anticancer activities, eg, aziridinomitosene [*80954-63-8*] (**36**), (−)-2-methyl-1,4-naphthoquinone 2,3-epoxide [*61840-91-3*] (**37**), and doxorubicin [*23214-92-8*] (adriamycin) (**38**) (see ANTIBIOTICS; CHEMOTHERAPEUTICS, ANTICANCER). All of these natural and synthetic materials have stimulated extensive research in synthetic chemistry.

(36)   (37)   (38)

Interest in the use of quinones as oxidants has not been restricted to biosynthetic and biochemical problems. Since the 1960s the literature of quinone oxidation and dehydrogenation chemistry has grown enormously (23,24). The most widely used oxidants have been 1,4-benzoquinone (**2**) (oxidation potential: 711 mV), *p*-chloranil (**23**) (742 mV), and DDQ (**24**) (ca 1000 mV). Low cost, availability, stability, and high oxidation potential characterize these reagents. Although the oxidation potential is of primary importance, it can be outweighed by other considerations. For example, 2,3,5,6-tetracyano-1,4-benzoquinone [*4032-03-5*] has been prepared, but high reactivity and moisture sensitivity have greatly restricted its application. Because the 1,2-quinones tend to undergo Diels-Alder dimerization easily, only *o*-chloranil (**22**) has been used extensively. More complex quinones have been used, eg, 2,5-dibromo-6-isopropyl-3-methyl-1,4-benzoquinone [*29096-93-3*], which has found specific applications. By contrast, anthraquinone (**12**) is important industrially, but rarely used in the synthetic laboratory.

Dehydrogenation.   The oldest and still important synthetic use of quinones is in the removal of hydrogen, especially for aromatization. This method has often been applied to the preparation of polycyclic aromatic compounds, and may be accompanied by unexpected ring expansion (25), eg, in the conversion of 7-difluoromethyl-8,9-dihydro-7(*H*)-cyclopenta[*a*]pyrene [*71511-38-1*] to 7-fluorobenzo[*a*]pyrene [*72297-03-1*] in equation 1.

$$(1)$$

Excellent evidence of the gentle nature of quinones as oxidants in the presence of the thiophene ring, eg (**39**), has been found (26).

Quinones are used extensively in the dehydrogenation of steroidal ketones (27,28). Such reactions are marked by high yield and selectivity. Generally, the results when using nonsteroidal ketones are disappointing. The powerful accelerating effect of electron-donating substituents in dehydrogenation reactions is illustrated in a reaction of DDQ (**24**) with substituted bibenzyls (**40**), where the yield is 10% when R = H, but 85% when R = OCH$_3$ (29,30).

A spectacular example of selective dehydrogenation in the steroid ring system (**42**) has been attributed to stereoelectronic effects (31); the yield is 80%. Several related steroids also show this chemistry. An extensive review containing many additional examples and a mechanistic discussion is available (23).

Oxidation. The use of 1,4-benzoquinone in combination with palladium(II) chloride converts terminal alkenes such as 1-hexene to alkyl methyl ketones in high yield (81%) (32). The quinone appears to reoxidize the palladium.

Although saturated alcohols are sufficiently stable toward quinones to be used as solvents for these oxidation reactions, benzylic (**43**) and allylic alcohols are often readily converted to the corresponding carbonyl compounds (**44**), as shown in equation 2 for benzene systems (33). For the substituents indicated, yields are as follows:

R	R'	Yield, %
H	CH_3	93
OH	H	83
OH	OCH_3	97

$$\text{(43)} \quad + \text{(24)} \xrightarrow{\text{dioxane}} \quad \text{(44)} \quad + \text{(41)} \qquad (2)$$

(43)   (44)

Photochemical Reactions. Increased knowledge of the centrality of quinone chemistry in photosynthesis has stimulated renewed interest in their photochemical behavior. Synthetically interesting work has centered on the 1,4-quinones and the two reaction types most frequently observed, ie [2 + 2] cycloaddition and hydrogen abstraction. Excellent reviews of these reactions, along with mechanistic discussion, are available (34,35).

Cycloadditions. The products of [2 + 2] cycloaddition are usually of the cyclobutane type, eg (**45**) [*80328-12-7*], or spirooxetane type, eg (**46**) [*137378-86-0*].

$$\text{(2) or (23)} + \quad \xrightarrow{h\nu} \quad \text{(45)} \quad + \quad \text{(46)}$$

(45)   (46)

The product distribution appears to depend on the radiation used for quinone excitation, the structure of the quinone, and the quinone–alkene ratio. In the example cited, 1,4-benzoquinone gives only the spirooxetane, whereas chloranil gives both products in amounts related to the ratio of starting materials (36,37).

A delicate stereochemical balance is clear in the high yield photocyclization of one cyclophane configuration (**47**) to a cage compound (**48**), whereas its spatial relative (**49**) leads only to higher molecular weight product (38).

$$\text{(47)} \quad \xrightarrow{h\nu} \quad \text{(48)} \quad \xleftarrow{\;\;/\!\!/\;\;} \quad \text{(49)}$$

(47)   (48)   (49)

Hydrogen Abstraction. These important reactions have been carried out using a variety of substrates. In general, the reactions involve the removal of hydrogen either directly as a hydrogen atom or indirectly by electron transfer followed by proton transfer. The products are derived from ground-state reactions. For example, chloranil probably reacts with cycloheptatrienyl radicals to produce ether (**50**) (39). This chemistry contrasts with the ground-state reaction in which DDQ produces tropylium quinolate in 91% yield (40).

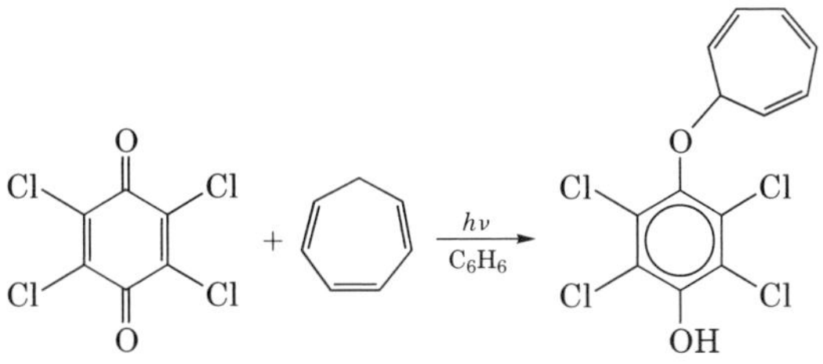

(**50**)

Photocycloaddition. Synthesis of the highly carcinogenic polycyclic hydrocarbons, eg (**51**) [72735-91-2], may be affected by photocycloaddition of 2-bromo-3-methoxy naphthoquinone [26037-61-6] with 1,1-diarylethylenes such as 1,1-bis(p-methoxyphenyl)ethene (41).

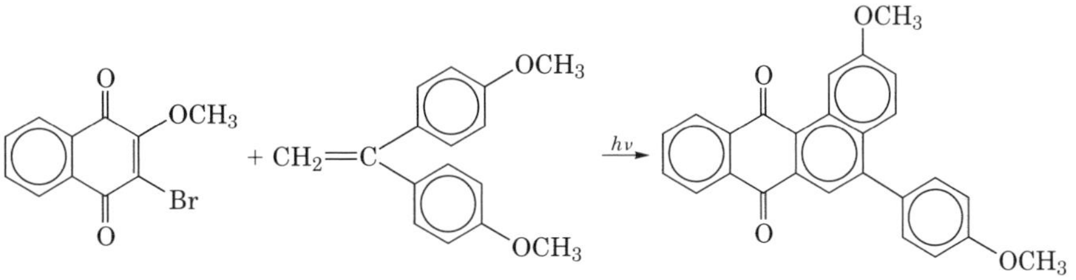

(**51**)

Although the yields are variable and isomeric products are obtained using unsymmetrical ethenes, the reaction conditions are mild and show promise of stereoselectivity.

The role of rose bengal and other sensitizer dyes in the photodimerization of 2-acetyl-1,4-benzoquinone [1125-55-9] involves electron transfer but not singlet oxygen (42) (see DYES, SENSITIZING).

Addition Reactions. The addition of nucleophiles to quinones is often an acid-catalyzed, Michael-type reductive process (7,43,44). The addition of benzenethiol to 1,4-benzoquinone (**2**) was studied by A. Michael for a better understanding of valence in organic chemistry (45). The presence of the reduced product thiophenylhydroquinone (**52**), the cross-oxidation product 2-thiophenyl-1,4-benzoquinone [18232-03-6] (**53**), and multiple-addition products such as 2,5-(bis(thiophenyl)-1,4-benzoquinone [17058-53-6] (**54**) and 2,6-bis(thiophenyl)-1,4-benzoquinone [121194-11-4] (**55**), is typical of many such transformations.

(52)   (53)

(54)   (55)

Recognition of the thio group's key role in biochemistry has led to studies of 1,4-benzoquinone with glutathione, a tripeptide γ-Glu·Cys·Gly (GSH). The cross-oxidation of the initial addition product by excess quinone leads, under physiological conditions, to all three isomeric products (46), ie, the 2,3- and 2,6-isomers as well as the 2,5-disubstituted 1,4-benzoquinone shown.

It appears that only 1-phenyltetrazole-5-thiol (HSPT) (**56**) produces all three isomeric products when the initial hydroquinone product is further oxidized and a second thiol is added (47).

(56)

The complications of subsequent cross-oxidation and further addition depend on both electronic and steric considerations. Thiols and amines generally promote cross-oxidation and tend to be oriented in the 2,5-positions of the resulting quinonoid product. For example, the electron-donating phenyl sulfide linkage makes the hydroquinone product susceptible to oxidation by the quinone starting material as illustrated by the reaction of (**2**) and benzenethiol. The

2,5-bis(phenylthio)-1,4-benzoquinone (**54**) predominates for electronic reasons, but an appreciable amount of the 2,6-isomer (**55**) is also obtained. Electron-withdrawing substituents, eg, arylsulfonyl, effectively prevent further oxidation.

The importance of steric effects in determining the oxidation state of the product can be illustrated by a thioether linkage, eg (**57**). If a methyl group is forced to be adjacent to the sulfur bond, the planarity required for efficient electron donation by unshared electrons is prevented and oxidation is not observed (48). Similar chemistry is observed in the addition of organic nitrogen and oxygen nucleophiles as well as inorganic anions.

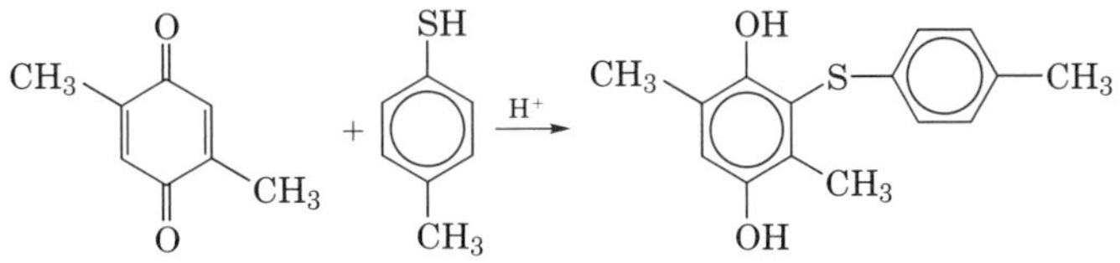

(**57**)

The azide anion, often in hydrazoic acid, adds in good yield and, when strong electron-withdrawing groups are present, shows the normal 2,3-orientation (**58**) (49). The intermediates, useful for further synthetic goals, can, for example, be converted to heterocyclic products, eg (**59**). The yields for various R groups are as follows:

R	Yield of (**58**), %	Yield of (**59**), %
H	66	52
CH$_3$	92	76
OCH$_3$	89	

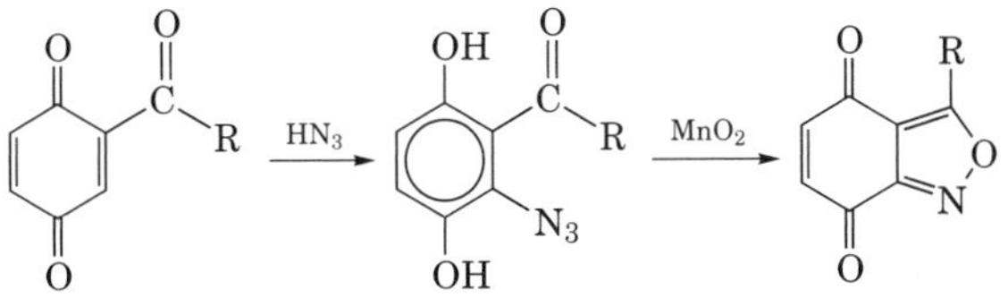

(**58**)   (**59**)

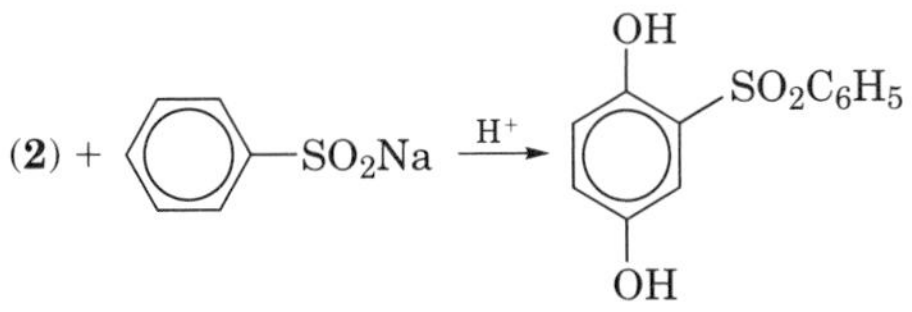

The most extensive mechanistic studies of quinone Michael addition chemistry involve the arylsufinic acids, which yield reduced product (50,51). The sulfones produced in such reactions have been examined electrochemically (48) and kinetically (52). The influence of substitutents in the quinone has also been studied (53).

An especially interesting case of oxygen addition to quinonoid systems involves acidic treatment with acetic anhydride, which produces both addition and esterification (eq. 3). This Thiele-Winter acetoxylation has been used extensively

for synthesis, structure proof, isolation, and purification (54). The kinetics and mechanism of acetoxylation have been described (55). Although the acetylium ion is an electrophile, extensive studies of electronic effects show a definite relationship to nucleophilic addition chemistry (56).

In seeking a synthetic route to an antibiotic antitumor agent, the Thiele-Winter synthon, with 2,3-bis(methoxycarbonyl)-1,4-benzoquinone [77220-15-6] (**60**), was used to introduce a required third oxygen linkage (57). A 67% yield of (**61**) was obtained.

(**60**)   (**61**)

Electrophilic addition to quinones, eg, the reaction of 2-chloro-1,4-benzoquinones with diazonium salts, represents a marked contrast with acetoxylation in product distribution (58). Phenyldiazonium chloride (Ar = C_6H_5) yields 8%

2,3-substitution [80632-59-3], 75% 2,5-substitution [39171-11-4], and 17% 2,6-substitution [80632-60-6]. For p-chlorophenyldiazonium chloride, the pattern is

28% 2,3-substitution [*80632-61-7*], 35% 2,5-substitution [*62120-48-3*], and 37% 2,6-substitution [*80954-64-9*]. For *p*-nitrophenyldiazonium chloride, the pattern is 41% 2,3-substitution [*80632-62-8*], 19% 2,5-substitution [*80632-63-9*], and 40% 2,6-substitution [*80954-64-9*]. Noteworthy is the presence of all three isomeric products, and the product ratio dependence on the aryl substituent. The reaction of 2-methyl-1,4-benzoquinone with a wide variety of diazonium salts produces only the 2,5-isomer (59).

A more recent development in quinone chemistry has been the tandem reaction sequence. In seeking elegant syntheses of complex molecules, careful orchestration of transformations has become essential. The use of the Thiele-Winter reaction in tandem with arylation gives good yields of pharmacologically interesting heterocycles, such as (**62**), from 2,5-dihydroxy-1,4-benzoquinone [*615-94-1*] and pyridines, where R = H or CH_3 (60).

(**62**)

The quinones undergo Michael addition with nitrogen nucleophiles in much the same fashion as sulfur (61). More recent interest in quinones that contain heterocyclic nitrogen has prompted a detailed study of the addition of pyrrolidine to unsymmetrical benzoquinones such as 2-methoxy-3-methyl-1,4-benzoquinone [*2207-57-0*] (**63**) having high regioselectivity (62). The yields are 93% of (**64**) [*77357-35-8*] and 5% of (**65**) [*77357-36-9*].

(**63**)   (**64**)   (**65**)

Information on nucleophilic addition chemistry of quinones and various mechanistic rationalizations have been discussed, and molecular orbital calculations have been proposed as more definitive approaches for explanation and prediction (63).

The synthetic techniques of *in situ* generation of the quinone and utilization of quinone monoacetals avoid the problems of instability, sequential addition, and poor regiospecificity, which often occur in quinone addition reactions. For example, although 1,2-benzoquinones react in much the same fashion as

1,4-benzoquinones, their lower stability makes *in situ* generation advisable. A number of nascent quinone reactions have been reviewed (64).

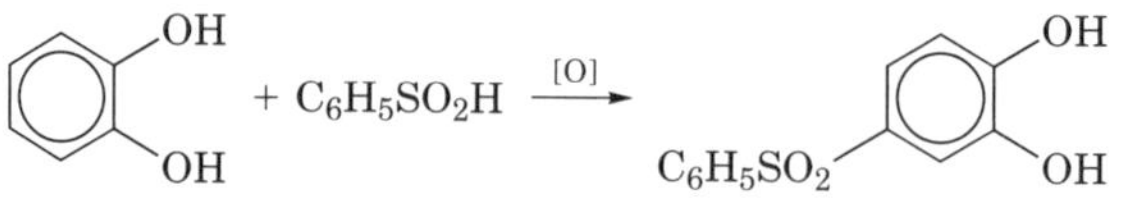

(**5**)

Of special importance is the complex problem of nitrogen addition as related to the drug dopamine, where even model compounds lead to extremely complex chemistry and difficult analytical problems (65).

Quinone monoacetals such as 2-methoxybenzoquinone monoacetal [*64701-03-7*] (**66**) show regiospecific addition of active methylene compounds (66), yielding 83% (**67**) and 63% (**68**) on reactions with ethyl malonate.

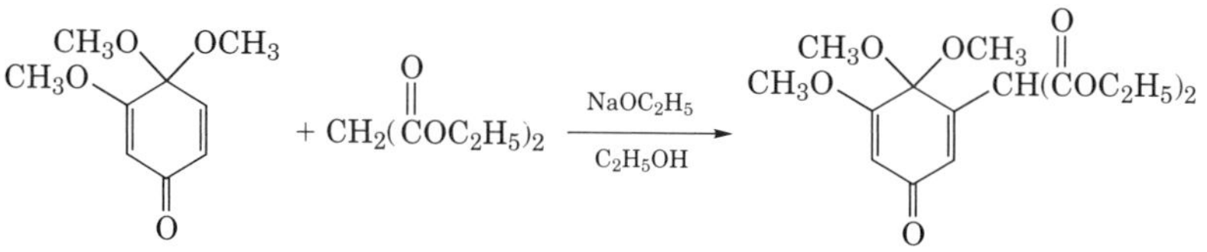

(**66**)   (**67**)

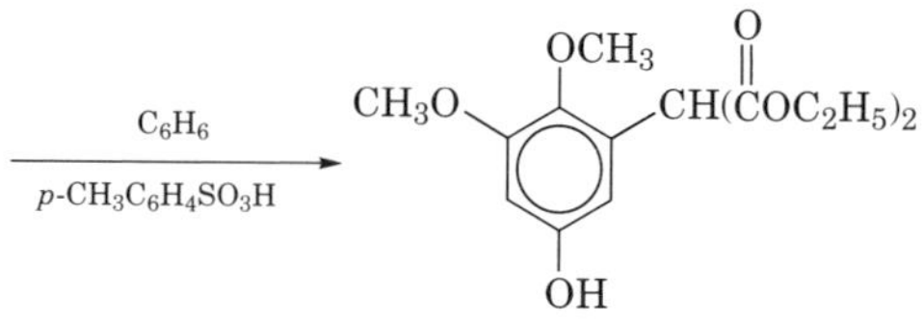

(**68**)

The synthesis of optically active epoxy-1,4-naphthoquinones (**69**) using benzylquininium chloride as the chiral catalyst under phase-transfer conditions has been reported (67). 2-Methyl-1,4-naphthoquinone (R = CH$_3$) (**31**) yields 70% of levorotatory (**37**). 2-Cyclohexyl-1,4-naphthoquinone [*34987-31-0*] yields 85% of (+)-2-cyclohexyl-1,4-naphthoquinone-2,3-epoxide [*73377-78-3*]. 2-Phenyl-1,4-naphthoquinone [*2348-77-8*] yields 92% (−)-2-phenyl-1,4-naphthoquinone-2,3-epoxide [*73377-82-9*].

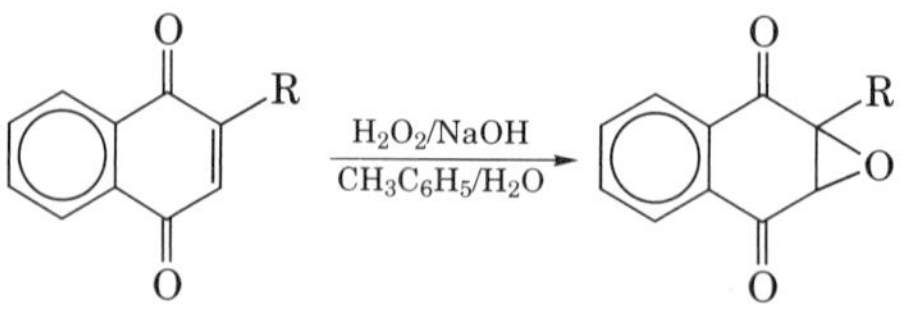

(**69**)

The reaction is characteristic of the usual Michael addition of hydroperoxide anion, yielding enantiomeric excesses up to 45%.

Reactions of quinones with radicals have been explored, and alkylation with diacyl peroxides constitutes an important synthetic tool (68). Although there are limitations, an impressive range of substituents can be introduced in good yield. Examples include alkyl chains ending with functional groups, eg, 50% yield of (**70**) [*80632-67-3*] (69,70).

(**70**)

The importance of quinones with unsaturated side chains in respiratory, photosynthetic, blood-clotting, and oxidative phosphorylation processes has stimulated much research in synthetic methods. The important alkyl- or polyisoprenyltin reagents, eg, (**71**) or (**72**), illustrate significant conversions of 2,3-dimethoxy-5-methyl-1,4-benzoquinone [*605-94-7*] (**73**) to 75% (**74**) [*727-81-1*] and 94% (**75**) [*4370-61-0*] (71–73).

(**71**)   (**74**)

(**72**)   (**75**)

A selective method for preparing unsymmetrical 2-alkynyl-5-alkyl- or aryl-1,4-benzoquinones, eg, 60% yield of (**76**) [*64080-65-5*], involves the sequential introduction of lithium salts, eg, (**77**), to 2,5-dimethoxy-1,4-benzoquinone [*3117-03-1*] (**78**) (74). Because the lithium reagents are generated *in situ* and the intermediates are not isolated, the transformations are showing great promise as pro-drugs (75,76).

The synthesis of natural products containing the quinonoid structure has led to intensive and extensive study of the classic diene synthesis (77). The Diels-Alder cycloaddition of quinonoid dienophiles has been reported for a wide range of dienes (78–80). Reaction of (**2**) with cyclopentadiene yields (**79**) [1200-89-1] and (**80**) [5439-22-5]. The analogous 1,3-cyclohexadiene adducts have been the subject of ^{13}C-nmr and x-ray studies, which indicate the endo–anti–endo stereostructure (81).

Unsymmetrical dienes in this synthesis are often capable of high regioselectivity (eqs. 4 and 5) (82). Reaction of (**81**) with 2-methoxycarbonyl-1,4-benzoquinone [3958-79-0] yields 97% of (**82**) [80328-15-0]. Reaction of (**81**) with 2,3-dicyano-1,4-benzoquinone [4622-04-2] yields 58% of (**83**) [80328-16-1].

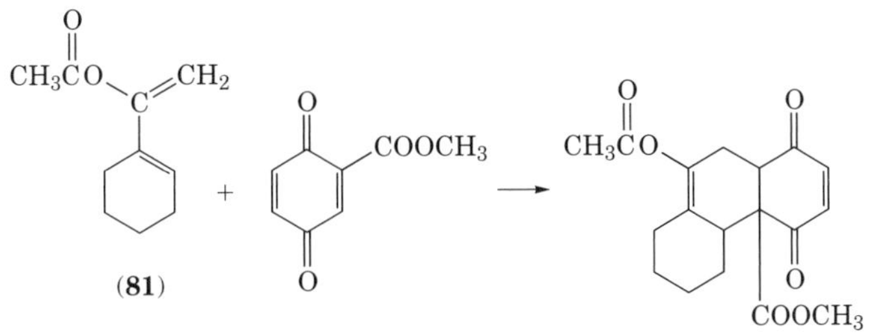

(4)

(81)

(82)

(5)

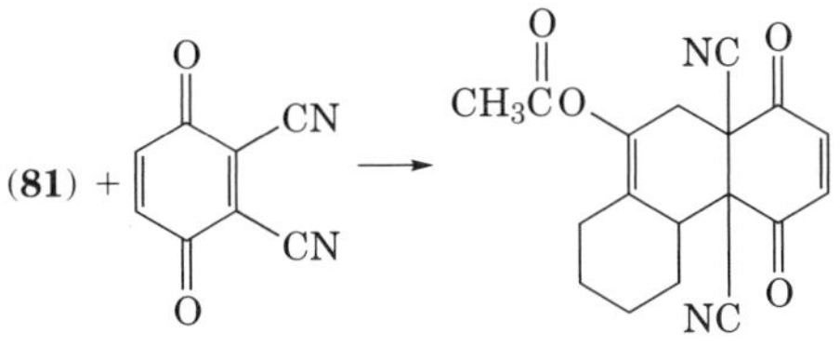

(81) +

(83)

The importance of both electronic and steric effects is clear in cycloadditions as in cross-oxidations. One example is a heterocyclic modification leading to the thermodynamically less stable natural form of juglone derivatives such as ventiloquinones *H* [*124917-64-2*] (**84**) and *I* [*124917-65-3*] (**85**) (83). The yields are 97% (**84**) from 6-chloro-2,3-dimethoxy-1,4-benzoquinone [*30839-34-0*] and 100% (**85**) upon hydrolysis.

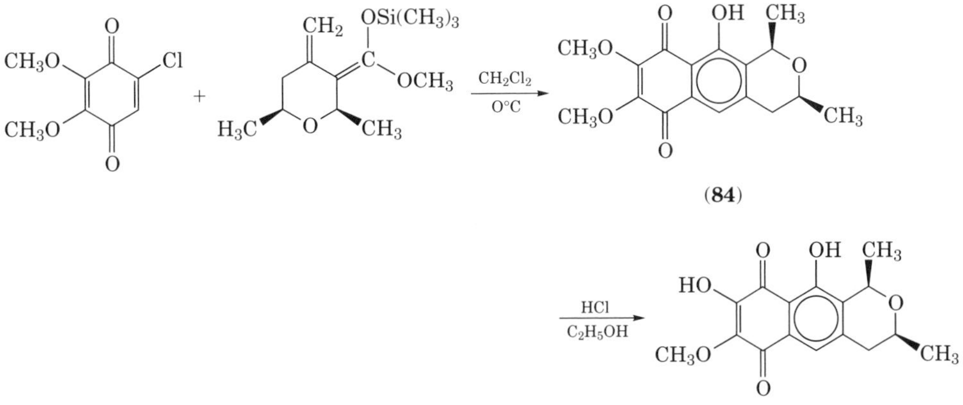

(84)

(85)

Sequential applications of these methods yield naturally occurring anthraquinones, eg, macrosporin [*22225-67-8*] (**86**) in 83% yield from 2-chloro-6-methyl-7-hydroxy-1,4-naphthoquinone [*76665-67-3*] (**87**), which is produced in 78% yield from (**26**) (84).

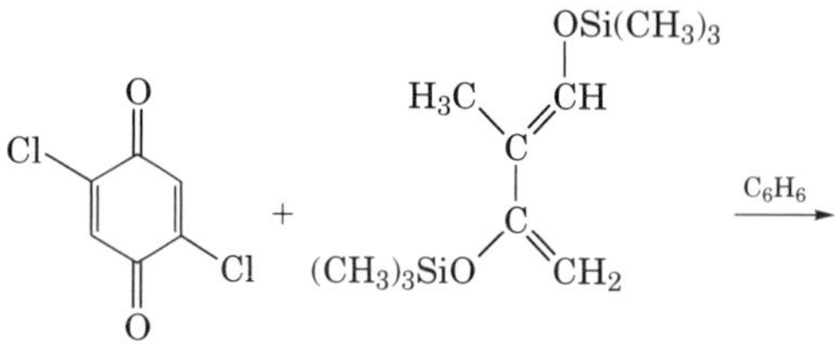

(26)

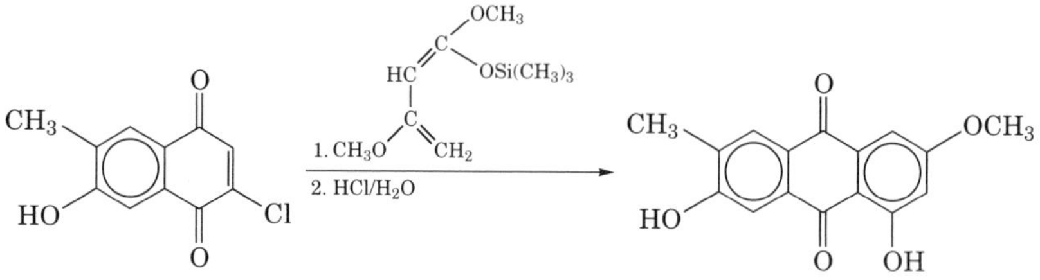

(87)   (86)

The problems associated with predicting regioselectivity in quinone Diels-Alder chemistry have been studied, and a mechanistic model based on frontier molecular orbital theory proposed (85). In certain cases of poor regioselectivity, eg, 2-methoxy-5-methyl-1,4-benzoquinone with alkyl-substituted dienes, the use of Lewis acid catalysts is effective (86).

Especially sensitive quinones can be generated *in situ*; the diene adduct, eg (**88**), can be obtained in excellent yield (87). For R = methoxycarbonyl, carboxaldehyde, and acetyl, the yields are 95, 97, and 100%, respectively.

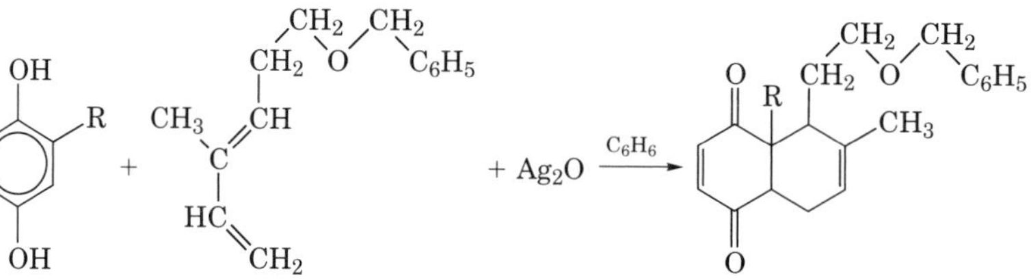

(88)

The potentially important 3,5,5-trialkoxy-1,2,4-trichlorocyclopentadienes, eg (**89**), react with 1,4-benzoquinone (R = R' = H) (**2**) and 1,4-naphthoquinone (R, R' close the ring) (**9**) (88). For (**2**) the yield is 95% of (**90**); for (**9**), 60%.

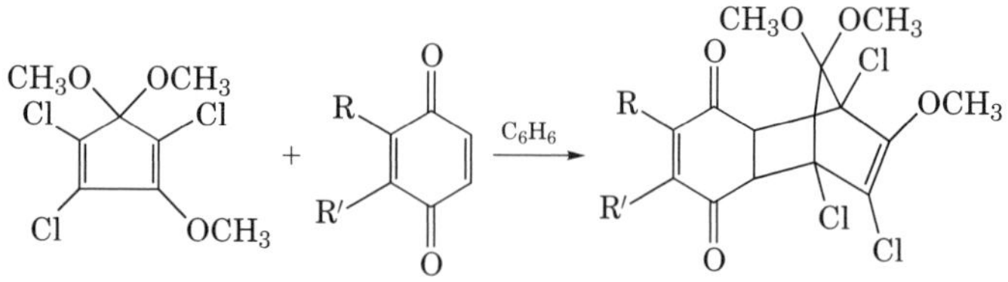

(89)   (90)

Typical diene reactions of 1,2-quinones are also well known (89).

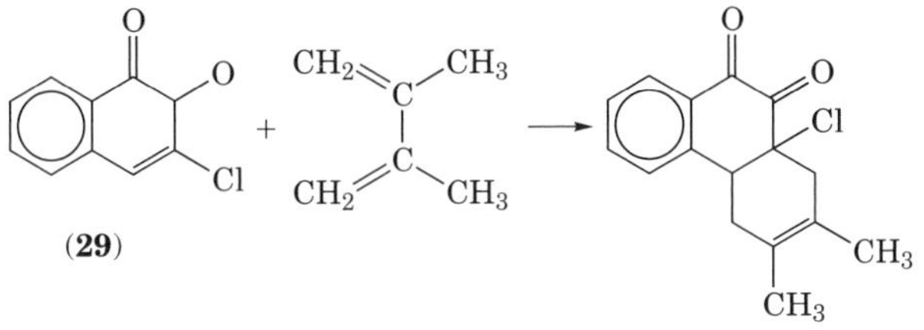

(29)

A wide variety of substituents, eg, CH_3O, Br, CN, CH_3OOC, and NO_2, have been investigated, and the products described (90). The Diels-Alder reactions of 1,2-benzoquinones generated *in situ* have also been reported (91). For R = acetyl, 51% of (91) [*67984-83-2*] has been reported; for R = benzoyl, 65% of (91) [*67984-84-3*].

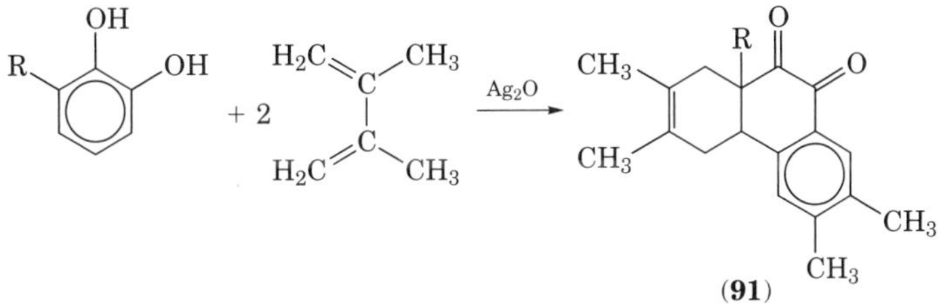

(91)

The formation of heterocycles derived from quinones is an important synthetic technique. The reaction may be intramolecular, eg, the reaction of (92). Either nitrogen products, eg (93) (yields of 85–91% for R = H, CH_3, and C_6H_5) or oxygen products (94) are obtained (92,93). Reactions with enamines have been especially important.

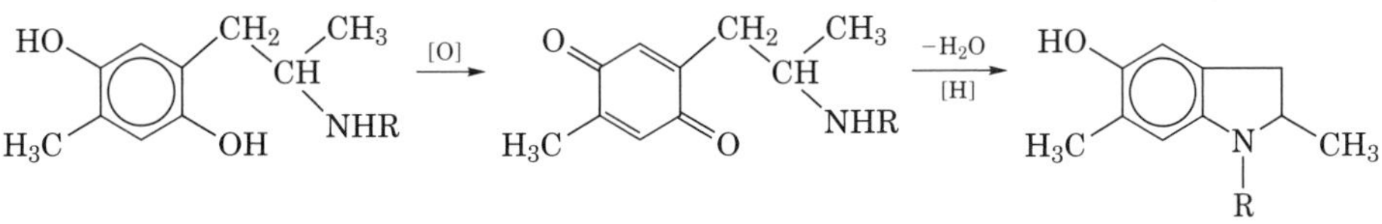

(92)   (93)

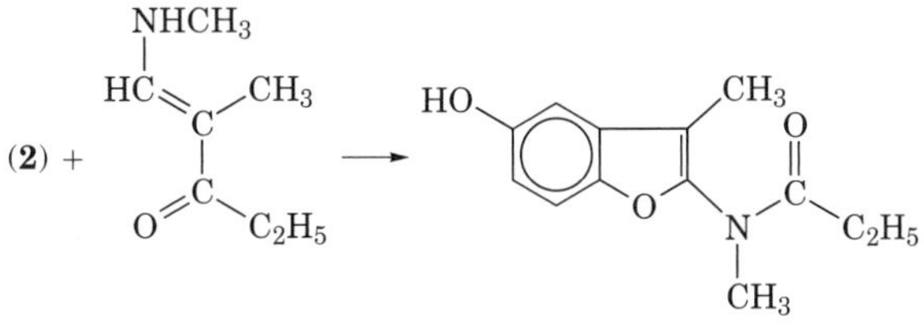

(94)

The addition of thioureas to 1,4-benzoquinones is an excellent preparation of 5-hydroxy-1,3-benzoxathiol-2-ones (94). Monosubstituted 1,4-benzoquinones give good yields (92% in eq. 6) and high regioselectivity.

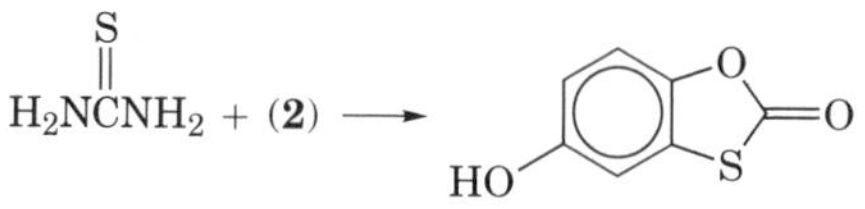

$$H_2NCNH_2 + (2) \longrightarrow \qquad (6)$$

A useful synthesis of benzoxazoles (**95**) was developed in an attempt to oxidize an antibiotic α-amino acid with 3,5-di-*tert*-butyl-1,2-benzoquinone [*3383-21-9*] (**96**) (95). Yields of (**95**) are 80% for R = CH_3 and 32% for R = $CH_2C_6H_5$.

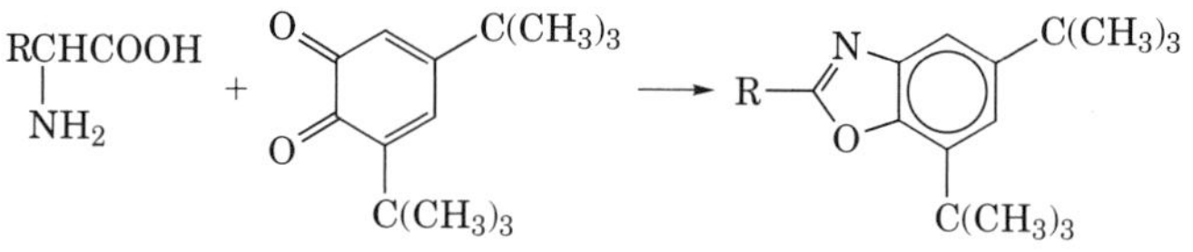

(**96**) (**95**)

A general method for the preparation of 2*H*-isoindol-4,7-diones, eg (**97**) [*72726-02-4*], involves 1,3-dipolar addition of oxazolium-5-oxides (sydnones) (**98**) to 2-methyl-1,4-benzoquinone (**96**). Yield of (**97**) is 37%.

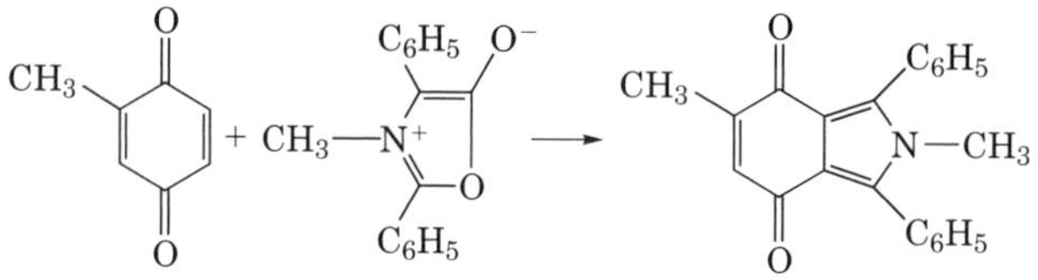

(**28**) (**98**) (**97**)

The formation of oximes of the quinones is strongly influenced by ring substitution. The mono- and dioximes form if at least one position adjacent to the carbonyl group is unsubstituted (97), eg, in the case of (**2**) (R = H) and (**27**) (R = CH_3).

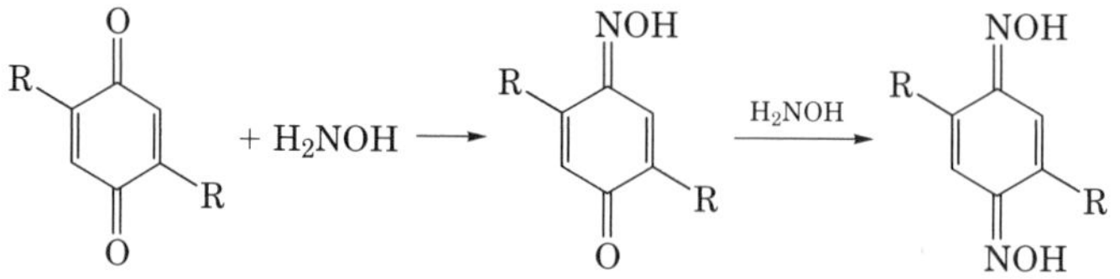

(**2**), (**27**)

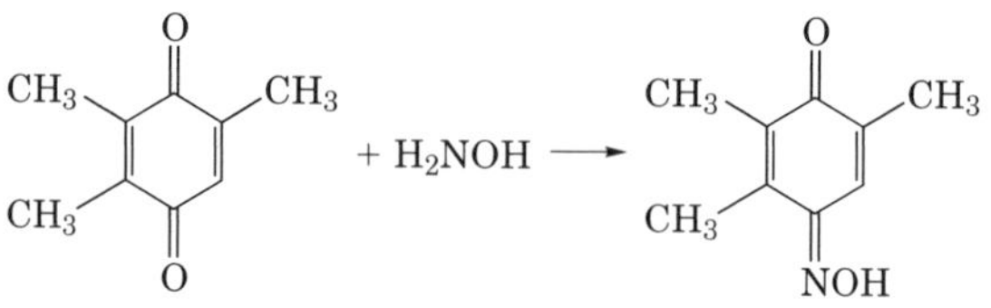

Unlike simple, unhindered carbonyl compounds, the quinones do not yield bisulfite addition products, but undergo ring addition. Another significant carbonyl reaction is the addition of tertiary phosphites under anhydrous conditions (98). The ester product (**99**) is easily hydrolyzed, and the overall sequence produces excellent yields of hydroquinone monoethers.

(**99**)

The valuable addition of cyanotrimethylsilane, $(CH_3)_3SiCN$, to 1,4-benzoquinone can be controlled by using newly developed catalysts (99). Using hydroxyapatite in CH_2Cl_2 at 0°C for five hours gives 74% of (**100**), whereas using Fe^{3+} ion-exchanged montmorillonite in the same solvent at 0°C for 1.5 hours gives 91% of (**101**).

(**100**)   (**101**)

Nucleophilic Substitution Reactions. Many of the transformations realized through Michael additions to quinones can also be achieved using nucleophilic substitution chemistry. In some instances the stereoselectivity can be markedly improved in this fashion (100), eg, in the reaction of benzenethiol with esters ($R^3 = CH_3C{=}O$) and ethers ($R^3 = CH_3$) of 1,4-naphthoquinones. 2-Bromo-5-acetyloxy-1,4-naphthoquinone [77189-69-6], $R' = Br$, yields 75% of 2-thiophenyl-5-acetyloxy-1,4-naphthoquinone [71700-93-1], $R' = SC_6H_5$. 3-Bromo-5-methoxy-1,4-naphthoquinone [69833-10-9], $R^2 = Br$, yields 82% of 3-thiophenyl-5-methoxy-1,4-naphthoquinone [112740-62-2] $R^2 = SC_6H_5$.

The nitrogen substitution chemistry of 2,3-dichloro-1,4-naphthoquinone (**30**) has been widely employed (101). Although the product mixtures are sometimes complex, many examples illustrate the excellent results possible in specific instances; eg, 3-nitro-12H-benzo[b]phenoxazine-6,11-dione [75197-88-5]

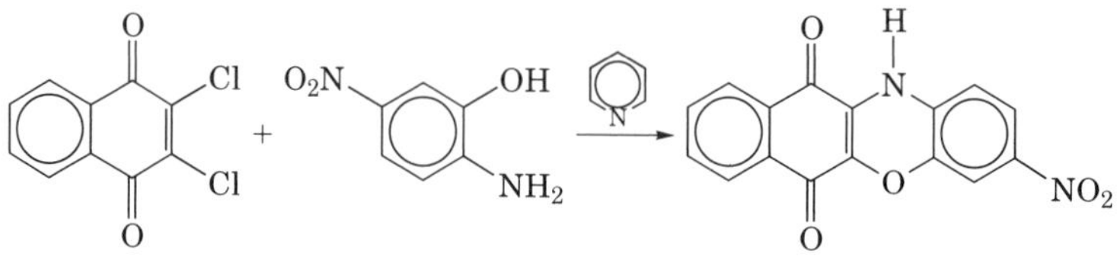

(**30**)

Analogous chemistry with 2,3-dibromo-1,4-naphthoquinone [13243-65-7] (**102**) has been used in the synthesis of mitomycin antibiotics; for example, (**103**) [72866-63-8] has been obtained in 91% yield (102).

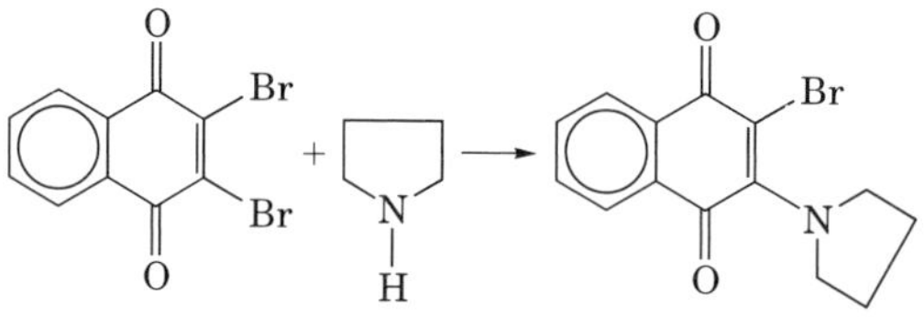

(**102**)   (**103**)

Syntheses

Syntheses of quinones often involve oxidation because this is the only completely general method (103). Thus, in several instances, quinones are the reagents of choice for the preparation of other quinones. Oxidation has been especially useful with catechols and hydroquinones as starting materials (23,24). The preparative

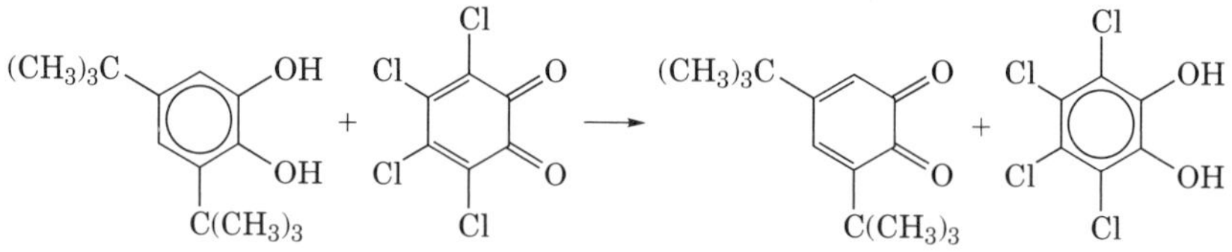

(**96**)

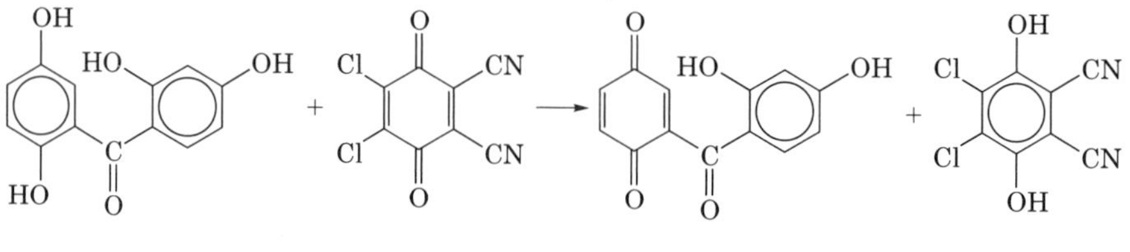

utility of these reactions depends largely on the relative oxidation potentials of the quinones (104,105).

For the preparation of ≤10 g of a quinone, the oxidation of a phenol with Fremy's salt (**104**) in the Teuber reaction is the method of choice (106). A wide

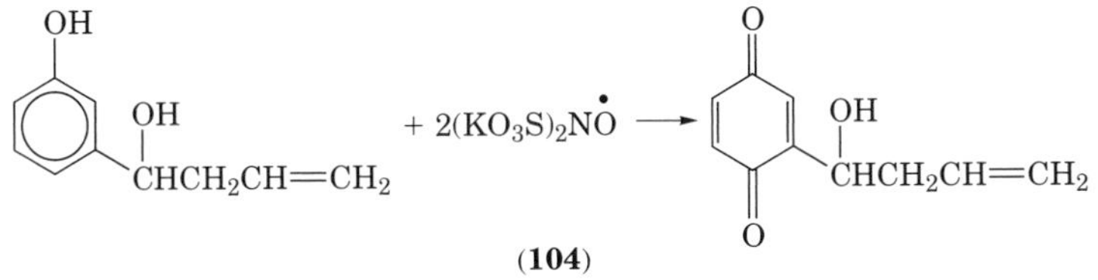

(**104**)

range of phenols has been used, including some having 4-substituents. The latter usually produce 1,2-benzoquinones, but the 4-chloro- and 4-*tert*-butyl groups can be eliminated, resulting in 1,4-benzoquinones. The yield for simple phenols is frequently in excess of 70%, and some complex phenols show highly selective oxidation. With an occasional exception, substituents and side chains are not attacked by Fremy's salt (107,108). The extraordinary delicacy and selectivity possible are illustrated by a conversion, (**105**) [*58785-57-2*] to (**106**) [*58785-59-4*], achieved only with Fremy's salt (109).

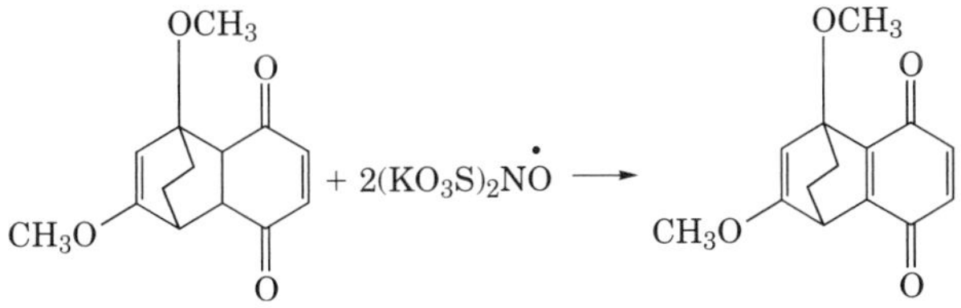

(**105**)   (**106**)

In small-scale syntheses, a wide variety of oxidants have been employed in the preparation of quinones from phenols. Of these reagents, chromic acid, ferric ion, and silver oxide show outstanding usefulness in the oxidation of hydroquinones. Thallium(III) trifluoroacetate converts 4-halo- or 4-*tert*-butylphenols to 1,4-benzoquinones in high yield (110). For example, 2-bromo-3-methyl-5-*t*-butyl-1,4-benzoquinone [*25441-20-3*] (**107**) has been made by this route.

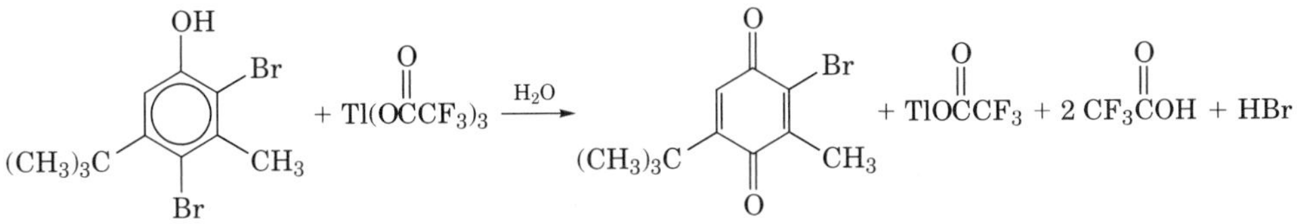

(**107**)

Thallium trinitrate oxidizes naphthols and hydroquinone monoethers, respectively, to quinones and 4,4-dialkoxycyclohexa-2,5-dienones, eg, 4,4-dimethoxy-2-methyl-2,5-cyclohexadienone [*57197-11-2*] (**108**) (111,112). The yield of (**108**) is 89%. Because the monoacetal is easily converted to the quinone, the yield of 5-hydroxy-1,4-naphthoquinone [*481-39-0*] is 64%.

(**108**)

The oxidation of 4-bromophenols to quinones can also be accomplished using periodic acid (113). A detailed study of this reagent with sterically hindered phenols provided insight about the quinonoid product (114). The highest yield of 2,6-di-*t*-butyl-1,4-benzoquinone [*719-22-2*] is for the case of R = OCH_3. The stilbene structure [*2411-18-9*] is obtained in highest yield for R = H.

The anodic oxidation of hydroquinone ethers to quinone ketals yields synthetically useful intermediates that can be hydrolyzed to quinones at the desired stage of a sequence (76). The yields of intermediate diacetal are 83% for chlorine and 75% for bromine.

Derivatives of the natural product juglone [*77189-69-6*], eg (**109**), have been obtained in 90% yield in a single reaction involving halogenation and oxidation by *N*-bromosuccinimide (115).

(**109**)

The dimethyl ethers of hydroquinones and 1,4-naphthalenediols can be oxidized with silver(II) oxide or ceric ammonium nitrate. Aqueous sodium hypochlorite under phase-transfer conditions has also produced efficient conversion of catechols and hydroquinones to 1,2- and 1,4-benzoquinones (116), eg, 4-*t*-butyl-1,2-benzoquinone [*1129-21-1*] in 92% yield.

Manufacture

With the exceptions of 1,4-benzoquinone and 9,10-anthraquinone, quinones are not produced on a large scale, but a few of these are commercially available (see ANTHRAQUINONE). The 1995 prices of selected quinones are listed in Table 4. Most of the compounds are prepared by the methods described herein. The few large-scale preparations involve oxidation of aniline, phenol, or aminonaphthols, eg (**110**), from which (**8**) is obtained in 93% yield.

Table 4. Commercially Available Quinones

Quinone	Structure number	Price in 1995, $	
		ARCOS[a]	Aldrich[b]
1,4-benzoquinone	(**2**)	46.50/kg	46.95/kg
1,2-naphthoquinone	(**8**)	152.60/25 g	36.65/5 g
1,4-naphthoquinone	(**9**)	43.80/500 g	43.35/500 g
9,10-anthraquinone	(**12**)	31.00/kg	36.25/250 g
p-chloranil	(**23**)	76.60/500 g	61.00/100 g
o-chloranil	(**22**)	145.20/100 g	24.50/5 g
2,3-dichloro-5,6-dicyano-1,4-benzoquinone	(**24**)	145.40/100 g	146.90/100 g
2-methyl-1,4-benzoquinone	(**28**)	42.80/100 g	42.00/100 g
2,3-dichloro-1,4-naphtho-quinone	(**30**)	25.20/250 g	11.95/100 g

[a]Ref. 117.
[b]Ref. 17.

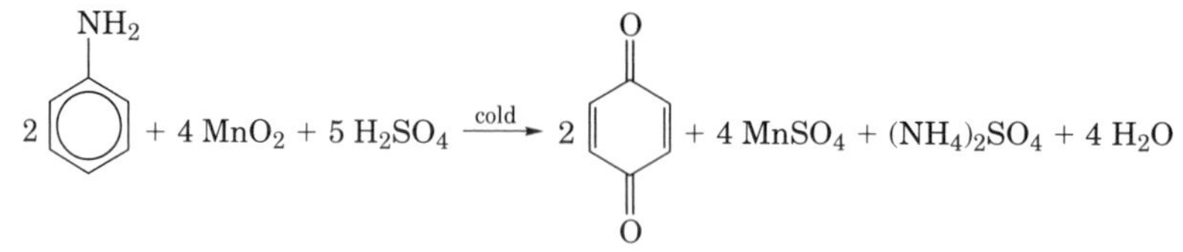

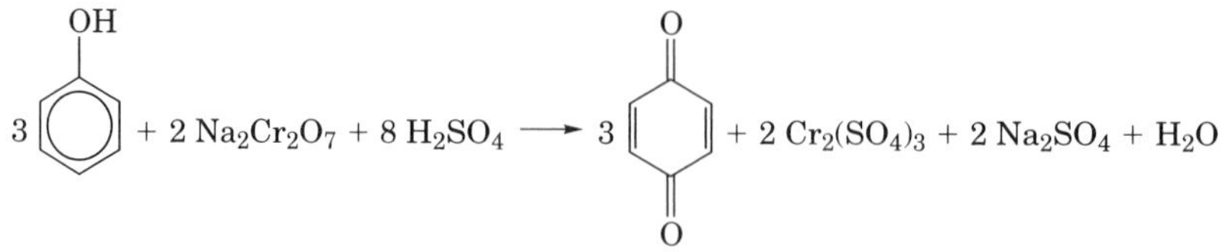

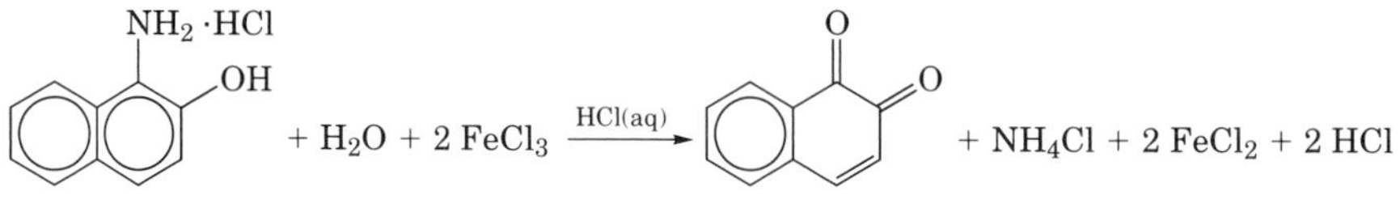

$$2 \; C_6H_4(NH_2) + 4\,MnO_2 + 5\,H_2SO_4 \xrightarrow{\text{cold}} 2 \; C_6H_4O_2 + 4\,MnSO_4 + (NH_4)_2SO_4 + 4\,H_2O$$

$$3 \; C_6H_5OH + 2\,Na_2Cr_2O_7 + 8\,H_2SO_4 \longrightarrow 3 \; C_6H_4O_2 + 2\,Cr_2(SO_4)_3 + 2\,Na_2SO_4 + H_2O$$

$$\text{(110)} + H_2O + 2\,FeCl_3 \xrightarrow{\text{HCl(aq)}} \text{(8)} + NH_4Cl + 2\,FeCl_2 + 2\,HCl$$

(110)   (8)

In the case of 1,4-benzoquinone, the product is steam-distilled, chilled, and obtained in high yield and purity. Direct oxidation of the appropriate unoxygenated hydrocarbon has been described for a large number of ring systems, but is generally utilized only for the polynuclear quinones without side chains. A representative sample of quinone uses is given in Table 5.

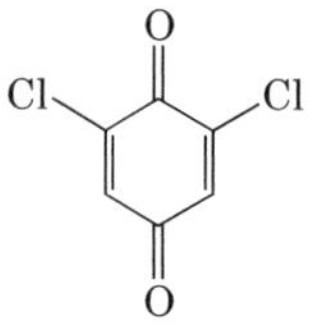

(111) 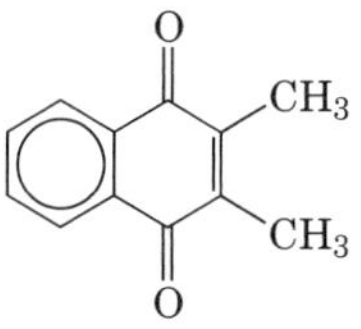   (112)

Table 5. Uses of Selected Quinones

Quinone	Structure number	Use
1,4-benzoquinone	(2)	oxidant, amino acid determination
2-chloro-, 2,5-dichloro-, and 2,6-dichloro-1,4-benzoquinone	(25), (26), (111)[a]	bactericides
2,3-dichloro-5,6-dicyano-1,4-benzoquinone	(24)	oxidation and dehydrogenation agent
2-methyl- and 2,3-dimethyl-1,4-naphthoquinone	(31), (112)[b]	vitamin K substitutes, antihemorrhagic agents

[a] CAS Registry Number = [697-91-6].
[b] CAS Registry Number = [2197-57-1].

Health and Safety Factors

Because of the high vapor pressure of the simple quinones and their penetrating odor, adequate ventilation must be provided in areas where these quinones are handled or stored. Quinone vapor can harm the eyes, and a limit of 0.1 ppm of 1,4-benzoquinone in air has been recommended. Quinone in either solid or solution form can cause severe local damage to the skin and mucous membranes. Swallowing benzoquinones may be fatal; the LD_{50} in rat is 130 mg/kg orally and 0.25 mg/kg intravenously. There is insufficient data concerning quinones and cancer. The higher quinones are less of a problem because of their decreased volatility (118–120).

BIBLIOGRAPHY

"Quinones" in *ECT* 1st ed., Vol. 11, pp. 410–424, by J. R. Thirtle, Eastman Kodak Co.; in *ECT* 2nd ed., Vol. 16, pp. 899–913, by J. R. Thirtle, Eastman Kodak Co.; in *ECT* 3rd ed., Vol. 19, pp. 572–606, by K. T. Finley, State University College at Brockport.

1. A. Woskressenski, *Justus Liebigs Ann, Chem.* **27**, 268 (1838).
2. J. H. Fletcher, O. C. Dermer, and R. B. Fox, *Nomenclature of Organic Compounds*, American Chemical Society, Washington, D.C., 1974.
3. J. Cason, *Org. Reactions* **4**, 305–361 (1948).
4. S. Patai, ed., *The Chemistry of Quinonoid Compounds*, John Wiley & Sons, Inc., New York, 1974.
5. R. H. Thomson, in Ref. 4, pp. 123–124.
6. S. Patai and Z. Rappoport, eds., *The Chemistry of Quinonoid Compounds*, Vol. 2, John Wiley & Sons, Inc., New York, 1988.
7. K. T. Finley, in S. Patai, ed., *Supplement E: The Chemistry of Hydroxyl, Ether and Peroxide Groups*, Vol. 2, John Wiley & Sons, Inc., New York, 1993, pp. 1027–1134.
8. R. H. Thomson, *Naturally Occurring Quinones*, 3rd ed., Chapman and Hall, London, 1987.
9. J. A. Berson, in Ref. 6, pp. 455–536.
10. H. H. Szmant, *Organic Building Blocks of the Chemical Industry*, John Wiley & Sons, Inc., New York, 1989, Chapt. 9.
11. I. H. Spinner, W. D. Raper, and W. Metanomski, *Can. J. Chem.* **41**, 483–494 (1963).
12. K. M. Draths and J. W. Frost, *J. Amer. Chem. Soc.* **113**, 9361–9363 (1991).
13. K. M. Draths, T. L. Ward, and J. W. Frost, *J. Amer. Chem. Soc.* **114**, 9725–9726 (1992).
14. T. A. Turney, in Ref. 4, pp. 857–876.
15. H. N. C. Wong, T.-L. Chan, and T.-Y. Luh, in Ref. 6, pp. 1501–1563.
16. L. T. Scott, in Ref. 6, pp. 1385–1417.
17. *Aldrich Catalog Handbook of Fine Chemicals 1994–1995*, Aldrich Chemical Co., Inc., Milwaukee, Wis., 1994.
18. D. R. Lede, ed., *Handbook of Chemistry and Physics*, 73rd ed., Boca Raton, Fla., 1992–1993; J. G. Grasselli and W. M. Ritchey, eds., *Atlas of Spectral Data and Physical Constants for Organic Compounds*, 2nd ed., CRC Press, Cleveland, Ohio, 1975; L. Meites and P. Zuman, eds., *CRC Handbook Series in Electrochemistry*, Vols. 1–4, CRC Press, Boca Raton, Fla., 1980.
19. S. Berger and A. Rieker, in Ref. 4, pp. 163–229.
20. S. Berger, P. Hertl, and A. Rieker, in Ref. 6, pp. 29–86.
21. R. Bentley and I. M. Campbell, in Ref. 4, pp. 683–736.

22. H. Inouye and E. Leistner, in Ref. 6, pp. 1293–1349.

23. H.-D. Becker, in Ref. 4, pp. 335–423.

24. H.-D. Becker and A. B. Turner, in Ref. 6, pp. 1351–1384.

25. M. S. Newman, V. K. Khanna, and K. Kanakarajan, *J. Am. Chem. Soc.* **101**, 6788–6789 (1979).

26. A. I. Kosak and co-workers, *J. Am. Chem. Soc.* **76**, 4450–4454 (1954).

27. U.S. Pat. 3,365,499 (Jan. 23, 1968), W. H. Clement and C. M. Selwitz (to Gulf Research Development Co.).

28. H.-D. Becker, A. Björk, and E. Adler, *J. Org. Chem.* **45**, 1596–1600 (1980).

29. E. J. Agnello and G. D. Laubach, *J. Am. Chem. Soc.* **82**, 4293–4299 (1960).

30. D. Burn, D. N. Kirk, and V. Petrov, *Proc. Chem. Soc.* 14 (1960).

31. E. A. Braude, A. G. Brook, and R. P. Linstead, *J. Chem. Soc.* 3569–3574 (Oct. 1954).

32. J. W. A. Findlay and A. B. Turner, *Organic Syntheses*, Vol. 49, John Wiley & Sons, Inc., New York, 1973, pp. 428–431.

33. W. Brown and A. B. Turner, *J. Chem. Soc. C*, (11), 2057–2061 (1971).

34. J. M. Bruce, in Ref. 4, pp. 465–538.

35. K. Maruyama and A. Osuka, in Ref. 6, pp. 759–878.

36. D. Bryce-Smith, A. Gilbert, and M. G. Johnson, *J. Chem. Soc. C*, (5), 383–389 (1967).

37. D. Bryce-Smith and A. Gilbert, *Tetrahedron Lett.* (47), 3471–3473 (1964).

38. H. Irngartinger and co-workers, *Angew. Chem. Int. Ed. Engl.* **13**, 674 (1974).

39. G. O. Schenck and co-workers, *Angew. Chem. Int. Ed. Engl.* **1**, 516 (1964).

40. D. H. Reid and co-workers, *Tetrahedron Lett.* (15), 530–535 (1961).

41. M. Maruyama, T. Otsuki, and K. Mitsui, *J. Org. Chem.* **45**, 1424–1428 (1980).

42. K. Maruyama and N. Narita, *J. Org. Chem.* **45**, 1421–1424 (1980).

43. K. T. Finley, in Ref. 4, pp. 877–1144.

44. K. T. Finley, in Ref. 6, pp. 537–717.

45. A. Michael, *J. Prakt. Chem.* **79**, 418 (1909) and **82**, 306 (1910).

46. S. S. Lau and co-workers, *Mol. Pharmacol.* **34**, 829–835 (1988).

47. H. S. Wilgus and co-workers, *J. Org. Chem.* **29**, 594–600 (1964).

48. E. R. Brown, K. T. Finley, and R. L. Reeves, *J. Org. Chem.* **36**, 2849–2853 (1971).

49. R. Cassis and co-workers, *Synth. Commun.* **17**, 1077 (1987).

50. O. Hinsberg, *Ber.* **27**, 3259 (1894) and **28**, 1315–1318 (1895).

51. R. F. Porter and co-workers, *J. Org. Chem.* **29**, 588–594 (1964).

52. Y. Ogata, Y. Sawaki, and M. Isone, *Tetrahedron* **25**, 2715–2721 (1969) and **26**, 731–736 (1970).

53. J. M. Bruce and P. Lloyd-Williams, *J. Chem. Soc. Perkin Trans. 1*, 2877–2784 (1992).

54. J. F. W. McOmie and J. M. Blatchly, *Organic Reactions*, Vol. 19, John Wiley & Sons, Inc., New York, 1972, pp. 199–277.

55. H. A. E. Mackenzie and E. R. S. Winter, *Trans. Faraday Soc.* **44**, 159–171 (1948).

56. H. S. Wilgus and J. W. Gates, Jr., *Can. J. Chem.* **45**, 1975–1980 (1967).

57. K. A. Parker, D. M. Spero, and K. A. Koziski, *J. Org. Chem.* **52**, 183–188 (1987).

58. J. F. Bagli and P. L'Ecuyer, *Can. J. Chem.* **39**, 1037–1348 (1961).

59. P. Brassard and P. L'Ecuyer, *Can. J. Chem.* **37**, 1505–1507 (1959).

60. N. Farfán, E. Ortega, and R. Contreras, *J. Heterocycl. Chem.* **23**, 1609 (1986).

61. M. L. Iskander, H. A. A. Medien, and S. Nashed, *Microchem. J.* **36**, 368 (1987) and **39**, 43 (1989).

62. J. R. Luly and H. Rapoport, *J. Org. Chem.* **46**, 2745–2752 (1981).

63. M. D. Rozeboom, I.-M. Tegmo-Larsson, and K. N. Houk, *J. Org. Chem.* **46**, 2338–2345 (1981).

64. W. H. Wanzlick, in W. Foerst, ed., *Newer Methods of Preparative Organic Chemistry*, Vol. IV, Academic Press, Inc., New York, 1968, pp. 139–154.

65. M. K. Manthey, S. G. Pyne, and R. J. W. Truscott, *Aust. J. Chem.* **42**, 365 (1989).

66. K. A. Parker and S.-K. Kang, *J. Org. Chem.* **45**, 1218–1224 (1980).
67. H. Pluim and H. Wynberg, *J. Org. Chem.* **45**, 2498–2502 (1980).
68. L. F. Fieser and A. E. Oxford, *J. Am. Chem. Soc.* **64**, 2060–2065 (1942).
69. L. F. Fieser and R. B. Turner, *J. Am. Chem. Soc.* **69**, 2338–2341 (1947).
70. D. H. R. Barton, D. Bridow, and S. Z. Zard, *Tetrahedron*, **43**, 5307–5314 (1987).
71. K. Maruyama and Y. Naruta, *J. Org. Chem.* **43**, 3796–3798 (1978).
72. Y. Naruta, *J. Org. Chem.* **45**, 4097–4104 (1980).
73. Y. Naruta, *J. Am. Chem. Soc.* **102**, 3774–3783 (1980).
74. H. W. Moore, Y.-L. L. Sing, and R. S. Sidhu, *J. Org. Chem.* **45**, 5057–5064 (1980).
75. K. F. West and H. Moore, *J. Org. Chem.* **47**, 3591–3593 (1982).
76. H. W. Moore and co-workers, *J. Org. Chem.* **52**, 2537–2549 (1987).
77. K. T. Finley, in Ref. 7, pp. 1063–1091.
78. W. Albrecht, *Justus Liebigs Ann. Chem.* **348**, 31 (1906).
79. O. Diels and K. Alder, *Justus Liebigs Ann. Chem.* **460**, 98 (1928).
80. O. Diels and co-workers, *Chem. Ber.* **62**, 2337 (1929).
81. R. K. Hill and co-workers, *J. Org. Chem.* **45**, 1593–1596 (1980).
82. M. F. Ansell, B. W. Nash, and D. A. Wilson, *J. Chem. Soc.* 3006–3012 (1963).
83. M. Blouin, M.-C. Béland, and P. Brassard, *J. Org. Chem.* **55**, 1466–1471 (1990).
84. C. Brisson and P. Brassard, *J. Org. Chem.* **46**, 1810–1814 (1981).
85. I.-M. Tegmo-Larsson, M. D. Rozeboom, and K. N. Houk, *Tetrahedron Lett.* **22**, 2043–2046 (1981).
86. J. S. Tou and W. Reusch, *J. Org. Chem.* **45**, 5012–5014 (1980).
87. G. A. Kraus and M. J. Taschner, *J. Org. Chem.* **45**, 1174–1175 (1980).
88. U. O'Connor and W. Rosen, *J. Org. Chem.* **45**, 1824–1828 (1980).
89. M. F. Ansell and R. A. Murray, *Chem. Commun.*, (19), 1111–1112 (1969).
90. M. F. Ansell and R. A. Murray, *J. Chem. Soc. C*, (8), 1429–1437 (1971).
91. R. Al-hamdang and B. Ali, *Chem. Commun.* (9), 397 (1978).
92. U. Kuckländer and co-workers, *Chem. Ber.* **120**, 1791–1795 (1987).
93. U. Kuckländer and K. Kuna, *Chem. Ber.* **120**, 1601–1602 (1987).
94. P. T. S. Lau and M. Kestner, *J. Org. Chem.* **33**, 4426–4431 (1968).
95. M. C. Zwan and co-workers, *J. Org. Chem.* **43**, 509–511 (1978).
96. J. A. Myers and co-workers, *J. Org. Chem.* **45**, 1202–1206 (1980).
97. F. Kehrmann, *Chem. Ber.* **21**, 3315 (1888).
98. F. Ramirez and S. Dershowitz, *J. Am. Chem. Soc.* **81**, 587–590 (1959).
99. M. Onaka and co-workers, *Chem. Lett.* 1393 (1989).
100. S. Laugraud and co-workers, *J. Org. Chem.* **53**, 1557–1560 (1988).
101. N. L. Agarwal and W. Schäfer, *J. Org. Chem.* **45**, 2155–2161 and 5139–5143 (1980).
102. S. N. Failing and H. Rapoport, *J. Org. Chem.* **45**, 1260–1270 (1980).
103. Y. Naruta and K. Maruyama, in Ref. 6, p. 242.
104. M. F. Ansell and co-workers, *J. Chem. Soc. C*, 1401–1414 (1971).
105. R. C. Ellis, W. B. Whalley, and K. Ball, *Chem. Commun.* (16), 803–804 (1967).
106. H. Zimmer, D. C. Lankin, and S. W. Horgan, *Chem. Revs.* **71**, 229–246 (1971).
107. O. Dann and H.-J. Zeller, *Chem. Ber.* **93**, 2829–2833 (1960).
108. J. M. Bruce, D. Creed, and K. Dawes, *J. Chem. Soc. C*, (12), 2244–2252 (1971).
109. V. H. Powell, *Tetrahedron Lett.* (40), 3463–3465 (1970).
110. A. McKillop, B. P. Swann, and E. C. Taylor, *Tetrahedron*, **26**, 4031 (1970).
111. A. McKillop and co-workers, *J. Org. Chem.* **41**, 282–287 (1976).
112. D. J. Crouse and co-workers, *J. Org. Chem.* **46**, 1814–1817 (1981).
113. P. T. Perumal and M. V. Bhatt, *Snythesis*, (3), 205–206 (1979).
114. H.-D. Becker and K. Gustafsson, *J. Org. Chem.* **44**, 428–432 (1979).
115. S. W. Heinzman and J. R. Grunwell, *Tetrahedron Lett.* **21**, 4305–4308 (1980).
116. F. Ishii and K. Kishi, *Synthesis*, (9), 706–708 (1980).

117. *ARCOS Chemical Catalog*, Fisher Scientific Co., 1995–1996.

118. R. E. Lenga, ed., *The Sigma-Aldrich Library of Chemical Safety Data*, 2nd ed., Aldrich Chemical Co., Inc., Milwaukee, Wis., 1988.

119. *Toxic and Hazardous Industrial Chemicals Safety Manual*, The International Technical Information Institute, Tokyo, Japan, 1989.

120. N. I. Sax and R. J. Lewis, Sr., eds., *Dangerous Properties of Industrial Materials*, 7th ed., Vol. II, Van Nostrand Reinhold, Co., Inc., New York, 1989.

General References

Refs. 5, 6, 7, and 8.

K. Thomas Finley
State University of New York, Brockport

R-ACID.   See Azo dyes; Naphthalene derivatives.

RADIATION CURING

The interaction of electromagnetic radiation with organic substrates is of widespread interest and has broad commercial applications. The use of electromagnetic radiation to alter the physical and chemical nature of a material is sometimes termed radiation curing technology. In radiation curing, electromagnetic radiation interacts with organic substrates to develop cross-linked or solvent-insoluble network structures. For example, a preformed thermoplastic polymer that interacts directly with certain types of ionizing (high energy) radiation from a given source of energy can develop into cross-linked or network structures having higher melting points, improved heat resistance, and improved chemical resistance than the original thermoplastic polymer starting materials (Fig. 1; Table 1) (1–3).

Similar types of high energy radiation can be used to cross-link (cure) organic coatings, providing that the liquid coating composition contains reactive vinyl-unsaturated components such as allyl, acrylate, methacrylate, styrene, and fumarate reactive groups (4–7). Radiation curing technology involves consideration of at least four main variables: type of radiation source; organic substrate to be irradiated; kinetics and mechanisms of radiation energy–organic substrate interactions; and final chemical, physical, and mechanical properties associated with the formation of three-dimensional (cured) network structures.

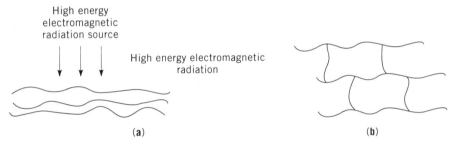

Fig. 1. (a) Interaction of high energy electromagnetic radiation with a preformed thermoplastic polymer to develop (b) cross-linked network polymer structures.

Table 1. Comparison of Physical Properties for Conventional 105°C PVC and Irradiated PVC Wire Compounds[a]

	Property values	
Physical property	Conventional 105°C PVC	Irradiated PVC
tensile strength, MPa[b]	20.7	20.7
elongation, %	250	200
solder iron resistance, time to fail at 349°C, 0.68-kg load, s	<1	>300
insulation resistance, $\Omega/30$ m	1000	1000
cut-through resistance, time to cut through at 105°C, 0.125-mm chisel, s	<5	>600
heat resistance, % retention of elongation after 168 h at 136°C	50	75

[a]Ref. 3.
[b]To convert MPa to psi, multiply by 145.

Radiation and Electromagnetic Radiation Sources

Radiation curing, as applied to the cross-linking of polymers or coating materials, involves the full spectrum of electromagnetic radiation energies to effect chemical reactions. These forms of radiation energy include ionizing radiation, ie, α-, β-particles and γ-rays from radioactive nuclei; x-rays; high energy electrons; and nonionizing radiation such as are associated with uv, visible, ir, microwave, and radio-frequency wavelengths of energy (Table 2) (8–10).

Table 2. Electromagnetic Spectrum[a]

Types of radiation	Wavelengths, nm	Frequency, Hz	Energy, eV
gamma ray	10^{-4}–10^{-2}	10^{19}–10^{22}	10^{5}–10^{8}
electron beam	10^{-3}–10^{-1}	10^{18}–10^{21}	10^{4}–10^{7}
x-ray	10^{-2}–10	10^{16}–10^{19}	10^{2}–10^{5}
ultraviolet	10–400	10^{15}–10^{16}	5–10^{2}
visible	400–750	10^{15}	1–5
infrared	750–10^{5}	10^{12}–10^{14}	10^{-2}–1
microwave	>10^{6}	10^{11}–10^{12}	<10^{-2}
radio frequency	>10^{6}	<10^{11}	<10^{-2}

[a]Refs. 8 and 9.

One of the most common sources of α-, β-, and γ-rays is radioactive nuclei, such as those listed in Table 3. The only significant ionizing radiation having limited commercial polymer-coating curing applications is the γ-ray produced from either ^{60}Co or ^{137}Cs radioactive nuclei. Detailed descriptions of these sources, eg, energies, cost of operation, shielding requirements, comparisons between ^{60}Co and ^{137}Cs efficiencies, and reactor geometries, are available (8,11–13) (see Radioisotopes).

X-rays can be produced through deceleration of high speed electrons through the electric field of an atomic nucleus. Various types of accelerator equipment capable of producing x-rays are listed in Table 4. High voltage electron accelerators have a distinct advantage over γ-ray and certain types of x-ray processing equipment. High energy electrons produced by machine acceleration, in comparison with radioisotopes, can be applied easily to industrial processes for the following reasons: easy on–off switching capability; less shielding is required than with gamma radiation; accelerator beams are directional and less penetrating than γ- or x-rays; and electron radiation provides high dose rates, ie, maximum penetration per unit density of material, and is well suited for on-line, high speed processing applications.

Table 3. Radioactive Nuclei[a]

Isotopes	Half-life	Type of radiation
polonium-210	138 d	α, γ-rays
radium-226	1620 yr	α, γ-rays
cesium-137	30 yr	β, γ-rays
cobalt-60	5.27 yr	β, γ-rays
strontium-90	28 yr	β-rays

[a]Ref. 11.

Table 4. X-Ray Processing Equipment[a]

Accelerator	Energy, MeV
x-ray machine	0.05–0.3
resonant transformer	0.1–3.5
Van de Graaf accelerator	1–5
betatron	10–300
linear electron accelerator	3–630

[a]Refs. 11, 14, and 15.

Commercially available high or low energy electron-processing equipment includes dynamitron, dynacote, insulating-core transformers, linear accelerators, Van de Graaf accelerator, pelletron, laddertron, and planar cathodes (16). Manufacturers of high voltage electron-accelerator equipment are High Voltage Engineering Corporation (Burlington, Massachusetts); Radiation Dynamics, Inc. (Edgewood, New York); Energy Science, Inc. (Woburn, Massachusetts); and Radiation Polymer Company (RPC) (Hayward, California). The two most common high energy electron accelerators in use in the 1990s are the scanned beam configuration (Radiation Dynamics) and the linear or planar cathode system (Energy Sciences and RPC).

Electromagnetic radiation, ie, from photons in the uv and visible ranges, can also produce chemical changes, but these energies do not cause direct ionization of organic substrates. Chemical reaction depends on the ability of the organic substrate or photoactive compounds to absorb light energy and to undergo photophysical processes involving electronically excited states and photochemical processes, which ultimately result in the formation of reactive intermediates (free radical or acid catalysts) that cause the organic substrate cross-linking reactions to occur.

Ultraviolet light sources are based on the mercury vapor arc. The mercury is enclosed in a quartz tube and a potential is applied to electrodes at either end of the tube. The electrodes can be of iron, tungsten, or other metals and the pressure in a mercury vapor lamp may range from less than 0.1 to >1 MPa (<1 to >10 atm). As the mercury pressure and lamp operating temperatures are increased, the radiation becomes more intense and the width of the emission lines increases (17).

Visible light sources can be obtained from high pressure mercury arcs by addition of rare gases or metal halides, which increase the number of emission lines in the 350–600-nm region of the spectrum. Fluorescent lamps, tungsten halide lamps, and visible lasers are also used for light-induced photochemical reactions as applied to the curing of polymers and coating technologies (18).

Infrared radiation ($\lambda = 0.7 - 400~\mu$m) has been used to fabricate plastics and to cure coating systems for a wide variety of commercial applications. In these applications, either gas fuel or electricity generates the infrared radiation (9,19–21). Gas-fired infrared-generating systems include radiant tube burners, surface combination burners, and direct-fired refractory burners. Electrically powered infrared-generating systems are short-wave emitter lamps, radiant metallic rods, and ceramic, quartz, or glass tubes (see INFRARED TECHNOLOGY).

The energy density of microwaves is proportional to the square of the electrical field intensity at a given point of reference. Microwaves can generate thermal energy through resistive losses in a conductor, magnetic losses in magnetic materials, and dielectric losses in materials having high dielectric constants (19). Microwave heating equipment consists of the following five main elements: a power supply, which converts 440 V, three-phase, 60 Hz to 1–20 kV dc; a high frequency generating system (magnetron or klystron tube circuits); a high frequency transmission system (microwave waveguides); a control system; and work application fixtures.

Radio-frequency energies (4–5 MHz (1–100 W)) initiate glow discharge, ie, plasma polymerization reactions for a wide variety of organic starting materials. These types of energetic gas-phase reactions produce thin cross-linked films that have a broad range of useful physical properties (22,23). Typical plasma or glow discharge monomer polymerization or polymer modification reaction equipment consists of the following components: a radio-frequency power source, a standing-wave ratio bridge, a matching network, inductive coupling or capacitor electrodes, and a vacuum system (see MICROWAVE TECHNOLOGY; PLASMA TECHNOLOGY).

Mechanisms of Radiation Energy–Organic Substrate Interaction

High energy interaction with organic substrates produces excited states which undergo secondary reactions, eg, electron capture, charge neutralization, intermolecular and intramolecular energy-transfer processes, ion formation, and molecular dissociation to produce free-radical intermediate species. The resulting chemical reactions are caused by the excited species and the formation of reactive intermediates (11). Ionizing radiation can produce excited molecules and secondary reactions through the following direct interaction processes:

$$A{:}B \xrightarrow{\;\gamma\text{-ray, x-ray, or high energy electrons\;}} (A{:}B)^* \text{ (excited state)}$$

$$(A{:}B)^* \xrightarrow{\hspace{3cm}} (A{\cdot}B)^{+\ddagger} \text{ (radical - ion)} + e^- \text{ (electron ejection)}$$

$$(A{\cdot}B)^{+\ddagger} \xrightarrow{\hspace{3cm}} A^+\text{(ion)} + B{\cdot}\text{ (free radical)}$$

$$(A{:}B)^* + C{:}D \longrightarrow A{:}B + (C{:}D)^* \text{ (energy transfer)}$$

$$(A{\cdot}B)^+ + e^- \longrightarrow (A{:}B)^* \text{ (electron capture)}$$

$$C{:}D + e^- \longrightarrow (C{:}D)^- \text{ (electron capture)}$$

$$(A{\cdot}B)^+ + (C{:}D)^- \longrightarrow (A{:}B)^* + (C{:}D)^* \text{ (charge neutralization)}$$

$$(A{:}B)^* \longrightarrow A{\cdot} + B{\cdot} \text{ (molecular dissociation)}$$

In the case of photochemical reactions, light energy must be absorbed by the system so that excited states of the molecule can form and subsequently produce free-radical intermediates (24,25) (see PHOTOCHEMICAL TECHNOLOGY).

$$A{:}B \xrightarrow[\text{uv or visible light energies}]{h\upsilon \text{ (photons)}} (A{:}B)^* \text{ (excited states)}$$

$$(A{:}B)^* \longrightarrow A{\cdot} + B{\cdot} \text{ (free-radical intermediates)}$$

It is also possible to use special photoactive catalysts that absorb light energy and produce cation or acid-reactive intermediates.

$$\text{photoactive catalyst} \xrightarrow{h\upsilon \text{ (photons)}} [\text{excited states}]^*$$

$$[\text{excited states}]^* \longrightarrow \text{acids or cations}$$

The difference between excited states produced by ionizing radiation and those produced photochemically is that an incident photon does not have sufficient energy to eject an electron completely from the molecule but only displaces it into a new orbital farther from the nucleus. Ionizing radiation produces the same types of excited states as photochemical processes, but ionizing radiation also produces higher excited states, ie, of more intrinsic energy, which cannot be formed directly by absorption of light energies (11,26). Energies emitted from an infrared source are transmitted directly to the surface of an organic substrate where they are absorbed, reflected, and/or transmitted, thereby causing vibrational and rotational molecular processes, which are subsequentially converted into heat (19,27,28).

Microwave curing of organic substrates involves dielectric loss of energy, which results in heat formation. In an oscillating electric field, organic substrates having high dipole moments, ie, high dielectric constants and high tan δ power factors, align, rotate, and realign, and these changes cause internal molecular friction and conversion of the electromagnetic energy into thermal energy. Rapidly oscillating electric fields cause a greater rate of conversion to heat than do lower frequency electromagnetic waves. Because microwave fields vary from 1–10 GHz, the rate of electromagnetic energy conversion to heat energy is significant (19).

Microwave or radio frequencies above 1 MHz that are applied to a gas under low pressure produce high energy electrons, which can interact with organic substrates in the vapor and solid state to produce a wide variety of reactive intermediate species: cations, anions, excited states, radicals, and ion radicals. These intermediates can combine or react with other substrates to form cross-linked polymer surfaces and cross-linked coatings or films (22,23,29).

Curing Polymers with γ-Rays, X-Rays, and High Energy Electrons

Radiation curing of preformed polymers using ionizing radiation processing equipment can result in two types of chemical change that are associated with cross-linking and degradation reaction mechanisms. Cross-linking reaction mechanisms on preformed polymer substrates usually involve removal of hydrogen atoms to form a macroradical intermediate. These macroradical intermediates can then couple to form a single molecule. This coupling results in an increase in the original average molecular weight of the starting polymer.

macroradicals

$$\text{polymer}-CH_2CH_2-\text{polymer} \xrightarrow[\text{(direct absorption)}]{\text{ionizing radiation}} \begin{cases} \text{polymer}-\overset{\cdot}{C}HCH_2-\text{polymer} \\ \text{polymer}-\overset{\cdot}{C}HCH_2-\text{polymer} \end{cases} + H_2$$

$$\downarrow$$

$$\begin{array}{l} \text{polymer}-CHCH_2-\text{polymer} \\ \quad\quad | \; \text{cross-linking site} \\ \text{polymer}-CHCH_2-\text{polymer} \end{array}$$

If irradiation continues, the original polymer substrate is transformed into one gigantic molecule of infinite molecular weight having lower solvent solubility, higher melting point, and improved physical properties over the original material (30). Enhancement of cross-linking can be facilitated through the use of multifunctional vinyl monomers or oligomers, $(CH_2\!=\!CH)_nR$, (meth)acrylates, allyl methacrylate, divinylbenzene, etc, which copolymerize and propagate much more rapidly than in a direct coupling reaction to form greater amounts of gel or cross-linked material at lower dose rates and shorter reaction times (31).

$$\text{polymer} + (CH_2\!=\!CH)_nR \xrightarrow[\text{radiation}]{\text{ionizing}} \text{rapid gel formation}$$

Radiation-induced degradation reactions are in direct opposition to cross-linking or curing processes, in that the average molecular weight of the preformed polymer decreases because of chain scission and without any subsequent

$$\underset{\text{(high molecular weight)}}{\text{polymer}-\text{polymer}} \xrightarrow{\text{ionizing radiation}} \underset{\text{(low molecular weight)}}{\text{polymer}} + \underset{\text{(low molecular weight)}}{\text{polymer}}$$

recombination of its broken ends. In order for efficient radiation curing of a polymer to take place, these degradation processes must be minimized in favor of the desired cross-linking reaction (32,33).

Examples of typical radiation curable polymer systems, experimental conditions, and applications are listed in Table 5.

Curing of Coatings with Electron Beams, γ-Ray, X-Ray, and Planar Cathodes

In conventional gas oven and other heat energy sources associated with the thermal curing of coatings, a mixture of polymers, cross-linking oligomers, catalysts, additives, pigments, and fillers are dissolved or dispersed in organic or water-based solvents to form a coating system. The coating is applied to a substrate and the solvents are thermally removed. The coating cross-links into a three-dimensional network by an energy-rich chemistry, which requires a high degree of thermal energy to convert the polymers into those having useful commercial properties. Much of the energy is absorbed by the substrate before heat reaches the polymers to initiate the curing chemistry (Fig. 2) (4,5,42).

High energy electron-curable coatings generally consist of multifunctional acrylate or methacrylate unsaturated polymers. They differ from conventional

Table 5. Ionizing Radiation Interactions with Polymeric Substrates[a]

Radiation source	Polymer	Results	Reference
^{60}Co	polyethylene	6% gel content at 12 kGy	31,34
	polyethylene plus 0.5 wt % ethylene glycol diacrylate	30% gel content at 12 kGy	34
	polyethylene plus 4 wt % allyl methacrylate	improved heat stability and tensile strength over polyethylene without added cross-linking monomer	31,34
	polypropylene	onset of gel formation requires 500 kGy with large amounts of degradation products	31,35
	polypropylene plus allyl acrylate	70% gel content at 50 kGy	31,35
electron beam	poly(vinyl chloride)	electrical applications	36,37
	cross-linked silicone rubber	cable termination cover and other electrical applications	15,38
	polyester	degradation and cross-linking reactions are correlated with chemical structure of the polyester	32
	polyethylene, neoprene, and silicone rubbers	heat-shrinkable articles for film packaging and electrical connector applications	39,40
electron beam and ^{60}Co	polyethylene	10–800 kGy, 6.4-mm thick slabs; electrical applications	34
	poly(vinylidene fluoride)s, polyimides, ethylene–alkyl acrylate copolymers, nylons, and natural polymers	wide range of applications	8,15
x-rays	polysulfones	photoresist technologies	41

[a] 1 Gy = 100 rad (10 kGy = 1 Mrad).

837

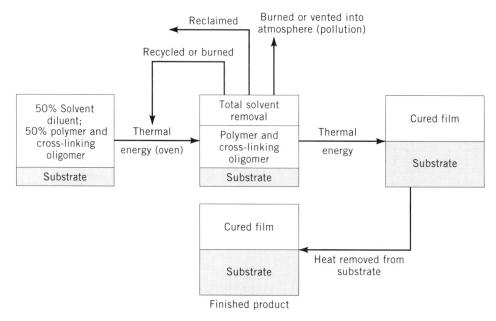

Fig. 2. Conventional thermally cured, organic solvent base coating system.

coatings in that the solvents for the polymers are high boiling, usually non-volatile, and 100% coreactive with themselves and with other organic components in the film. The curing process for these coatings is a free-radical chain reaction. Ionizing radiation from the processing equipment is absorbed directly in the coating and generates the free radicals that initiate the curing process.

$$AB \rightsquigarrow \begin{array}{c} AB^+ \\ \updownarrow e^- \\ AB \cdot \end{array} \longrightarrow A\cdot + B\cdot$$

Coating   Initiating free radicals

Because electron energies of only 100 eV or less are required to break chemical bonds and to ionize or excite components of the coating system, the shower of scattered electrons produced in the coating leads to a intense population of free radicals throughout the coating. These initiate the polymerization reaction and this polymerization process results in a dry, three-dimensional cross-linked coating (Fig. 3). In this process, most of the energy is absorbed into the coating and is not lost to the substrate, as is the case in thermal curing reactions (5). Neither cobalt-60 nor x-ray energy sources are used in radiation curable coating systems; x-rays are used in photolithographic processes (see LITHOGRAPHIC RESISTS (SUPPLEMENT)).

Electron Beam. An electron beam processing unit (Fig. 4) consists mainly of a power supply and an electron beam acceleration tube. The power supply increases and rectifies line current and the accelerator tube generates and

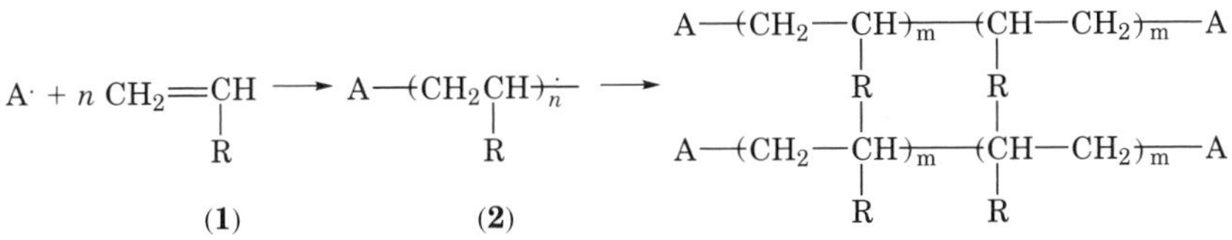

$$A\cdot + n\ CH_2{=}\underset{\underset{R}{|}}{CH} \longrightarrow A{-}(CH_2CH{)}_{\overline{n}}^{\cdot} \longrightarrow$$

$$A{-}(CH_2{-}\underset{\underset{R}{|}}{CH}{)}_{\overline{m}}{-}(\underset{\underset{R}{|}}{CH}{-}CH_2{)}_{\overline{m}}{-}A$$

$$A{-}(CH_2{-}\underset{\underset{R}{|}}{CH}{)}_{\overline{m}}{-}(\underset{\underset{R}{|}}{CH}{-}CH_2{)}_{\overline{m}}{-}A$$

$$(\mathbf{1}) \qquad\qquad (\mathbf{2})$$

Cured coating

Fig. 3. Polymerization initiation and propagation by radiation-generated free radicals. A is the initiating radical produced by irradiating the liquid coating. (**1**) represents the liquid monomer–unsaturated polymer reactive coating system. R is functional. (**2**) is the growing polymer chain (free radical). The cured coating is a solid polymeric network film structure.

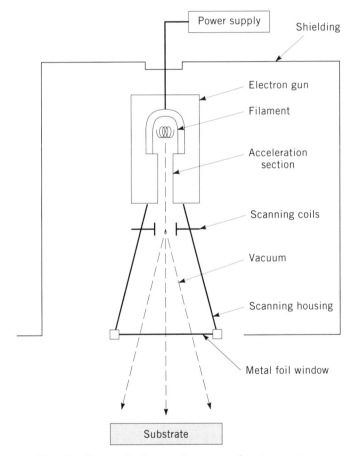

Fig. 4. Scanned electron beam accelerator system.

focuses the beam and controls the electron scanning. The beam is produced when high voltage energizes a tungsten filament, thereby causing electrons to be produced at high rates. These fast electrons are concentrated to form a high energy beam and are accelerated to full velocity inside the electron gun. Electromagnets on the sides of the accelerator tube allow deflection or scanning of the beam as with a television tube. Scanning widths and depths vary from 61–183 cm to 10–15 cm, respectively. The scanner opening is covered with a thin metal foil, usually titanium, that allows passage of electrons but maintains the high vacuum required for high free-path lengths. Characteristic power, current, and dose rates of accelerators are 200–500 kV, 25–200 mA, and 10–100 kGy/s (1–10 Mrad/s), respectively.

Electron processors have several disadvantages. The most severe of these is the large area that must be shielded, because any surface enclosing the electron accelerator scanner acts as a source of x-rays generated by electrons scattered to the wall, and these emissions are along the entire length of the system. Another disadvantage is the large space requirement for housing the equipment.

Advantages of the electron beam processor are its ability to penetrate thick and highly pigmented coatings. It is used to cross-link reactive unsaturated polymers, nonreactive thermoplastic polymers, insulation, and wire-cable coverings (4,9,16) (see INSULATION, ELECTRIC–WIRE AND CABLE COVERINGS).

Electrocurtain. The electrocurtain processor (Energy Sciences) (Fig. 5) is a high (150 kV) voltage electron tube that provides a continuous strip of energetic electrons from a linear filament or cathode, which is on the axis of symmetry of the system. The cylindrical electron gun shapes and processes the electron stream, in a grid-controlled structure. The stream is then accelerated across a

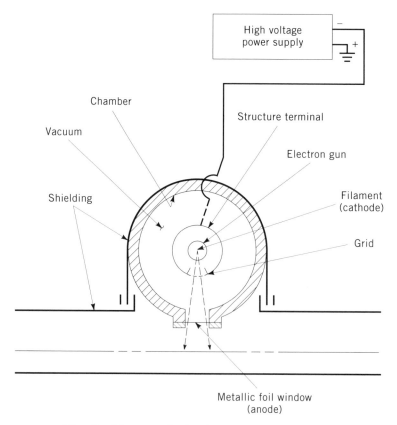

Fig. 5.   Planar cathode electrocurtain processor.

vacuum gap to a metal window where it emerges directly into air and travels onto the product. The energetic electrons from the processor are injected directly in the coating, where they create the initiating free radicals. In the liquid-phase systems, the polymerization process propagates until the activity of the growing chain is terminated. These energetic electrons can penetrate many different types of pigmented coatings and are capable of producing through-cure to the substrate–polymer coating interface. Both the electron beam and the electrocurtain cure pigmented films, but the power of the electrocurtain is substantially less, having a maximum-curing-film thickness range of ca 0.025–0.36 mm.

In the electrocurtain processor, the shielding is clad directly to the tube housing. Housing space is relatively small, because a shielded tube 25 cm in diameter replaces the 3-m high structure required for the scanned electron beam apparatus. The electrocurtain has a more flexible geometry and can be adapted more readily to many different types of curing applications (43).

Multiple-Planar Cathode Processors. The design criterion for this electron accelerator system is a planar array of concentrated cathode control grid elements. The modular cathode construction allows for broad-beam (250-cm wide) processing of materials using powers of 30 kGy (3 Mrad) at 300 m/min. The system (RPC Industries) also includes integrated shielding and high terminal voltages (44).

Coatings Ingredients. Ingredients of liquid high energy electron radiation-curable coatings are analogues of components contained in conventional solvent-based thermal-curing coating systems. In conventional solvent-based coatings systems, a preformed polymer (usually 3,000–25,000 mol wt) is dissolved in an organic solvent (30–80% solids), and a cross-linking oligomer and various flow agents, catalysts, pigments, etc, are added. The coating is applied to a substrate by conventional methods, eg, spray, roll coating, and flow coating, and subsequently is cured in gas or ir thermal ovens. Curing of conventional solvent-based coatings involves solvent removal and thermal initiation of chemical reactions between the polymers and cross-linking oligomers that are involved in developing final film properties through three-dimensional network formation of cross-linking sites (45,46) (see Fig. 2).

Monomers and Cross-Linking Agents or Oligomers. The monomer in high energy electron radiation-curable coatings is the analogue of the solvent in conventional paint. Although, like a solvent, it is a medium for all of the other ingredients and provides the necessary liquid physical properties and rheology, the monomer differs in that it enters the copolymerization and is not lost on cure. Most radiation-curable monomers (ca 100–500 mol wt) contain single unsaturation sites and are high boiling acrylate esters, although in some special coatings styrene can be used as a monomer. Monomers are usually very pure compounds that have well-defined molecular weights and boiling points.

Cross-linking agents or cross-linking oligomers in conventional thermosetting coatings formulations are usually melamine resins, ie, they involve acid, hydroxyl transetherification cross-linking; amine–amide hardeners that include oxirane (epoxy) ring opening; and blocked isocyanate prepolymers. The term oligomer represents materials that do not have well-defined molecular weights but rather consist of a distribution of low molecular weight subunits that are associated with the chemical structure of the parent molecule. The term cross-linking agents/oligomers represents molecules that have multiple (two or more) functional groups that can enter into a chemical reaction. Cross-linking agents or oligomers in radiation-curing systems are similar to single, vinyl monomers except they contain di-, tri-, or multifunctional unsaturation sites (ca 200–1000 mol wt). These multifunctional components cause polymer propagation reactions to grow into three-dimensional network structures of a cured film. Monofunctional (single vinyl or acrylic/methacrylic acid ester unsaturation site) monomers include the following (47,48):

tetrahydrofurfuryl methacrylate or acrylate	lauryl methacrylate or acrylate
cyclohexyl methacrylate or acrylate	octyl acrylate
n-hexyl methacrylate or acrylate	2-phenoxyethyl acrylate or methacrylate
2-ethoxyethyl acrylate or methacrylate	glycidyl methacrylate
isodecyl methacrylate or acrylate	isobornyl methacrylate or acrylate
2-methoxyethyl acrylate	benzyl acrylate
2(2-ethoxyethoxy)ethyl acrylate	isooctyl acrylate
stearyl acrylate or methacrylate	polypropylene glycol monomethacrylate or monoacrylate

Cross-linking agents or oligomers with multifunctional unsaturation sites include the following:

triethylene glycol dimethacrylate or acrylate	ethoxylated bisphenol A dimethacrylate or diacrylate
ethylene glycol dimethacrylate	di/trimethylolpropane tetraacrylate
tetraethylene glycol dimethacrylate or diacrylate	ethoxylate trimethylolpropane triacrylate
polyethylene glycol dimethacrylate	propoxylated trimethylolpropane triacrylate
1,3-butylene glycol diacrylate or dimethacrylate	propoxylated neopentyl glycol diacrylate
1,4-butanediol diacrylate or methacrylate	glyceryl propoxy triacrylate
diethyl glycol diacrylate or dimethacrylate	tris(2-hydroxy ethyl) isocyanurate trimethacrylate
1,6-hexanediol diacrylate or dimethacrylate	pentaerythritol tetraacrylate
neopentyl glycol diacrylate or dimethacrylate	trimethylolpropane trimethacrylate or triacrylate
polyethylene glycol (200) diacrylate	tris(2-hydroxyethyl) isocyanurate triacrylate
triethylene glycol diacrylate	dipentaerythritol pentaacrylate
tripropylene glycol diacrylate	

Polymers. The molecular weights of polymers used in high energy electron radiation-curable coating systems are ca 1,000–25,000 and the polymers usually contain acrylic, methacrylic, or fumaric vinyl unsaturation along or attached to the polymer backbone (4,48). Aromatic or aliphatic diisocyanates react with glycols or alcohol-terminated polyether or polyester to form either isocyanate or hydroxyl functional polyurethane intermediates. The isocyanate functional polyurethane intermediates react with hydroxyl functional polyurethane and with acrylic or methacrylic acids to form reactive polyurethanes.

The chemical class of the polymer backbone varies with coating requirements, and the different chemical classes give the same overall properties as a related thermal cure coating system (Table 6). Examples of high energy electron or ionizing radiation-curable coating systems, experimental conditions, and their application are listed in Table 7 (46–48). Reactive coating formulation ingredients are (*1*) single-functional vinyl monomers such as 2-ethylhexyl acrylate, styrene, *N*-vinylpyrrolidinone, vinyltoluene, and lauryl methacrylate; (*2*) multifunctional vinyl monomers such as 1,6-hexanediol diacrylate, tetraethylene glycol diacrylate, trimethylolpropane triacrylate, and pentaerythritol triacrylate; and (*3*) unsaturated polymers such as maleic–fumaric acid unsaturated polyesters, acrylic copolymers containing pendent vinyl unsaturation, epoxy acrylates, and polyurethane acrylates.

Reactive (unsaturated) epoxy resins (qv) are reaction products of multiple glycidyl ethers of phenolic base polymer substrates with methacrylic, acrylic, or

Table 6. Performance Characteristics of Thermally Curable Polymer Backbone Systems

System	Cost	Chemical resistance	Physical properties	Outdoor durability
epoxies	medium	excellent	very good	very poor
acrylics	low–medium	very good	good	very good
polyesters	low	fair–good	very good	good
polyurethanes				
aliphatic	high	very good	excellent	very good
aromatic	high	very good	excellent	very poor

Table 7. Ionizing Radiation Curing of Coatings

Coating composition	Conditions	Reference
65 wt % unsaturated polyester, 35 wt % vinyl monomer[a]	cured using 300-keV electrons at 200 kGy/min[b]	49,50
acrylic copolymers containing pendent unsaturation[c] and 35–45 wt % of a vinyl monomer[a]	cured using a total dosage of 150-kGy[b] electron beam; used to coat plastic substrates	51
acrylate monomers, ie, acrylate unsaturated epoxy and acrylate unsaturated polyurethanes	electron curtain curing; used in general coating applications	52
monomers, ie, polyfunctional vinyl intermediates	metal, wood, and plastic finishing applications using electron-beam processes	53,54
vinyl monomer	^{60}Co-curing of monomer; vacuum impregnation of wood	55

[a]2-Ethylhexyl acrylate or styrene.
[b]1 Gy = 100 rad (10 kGy = 1 Mrad).
[c]0.5–1.75 mol double bonds per 1000 mol wt.

fumaric acids. Reactive (unsaturated) polyester resins are reaction products of glycols and diacids (aromatic, aliphatic, unsaturated) esterified with acrylic or methacrylic acids (see POLYESTERS, UNSATURATED). Reactive polyether resins are typically poly(ethylene glycol (600) dimethacrylate) or poly(ethylene glycol (400) diacrylate) (see POLYETHERS).

Curing with Ultraviolet, Visible, and Infrared Processing Equipment

Polymers. Upon direct absorption of uv or visible wavelengths of light, polymer substrates undergo chain scission and cross-linking. Cross-linking or curing of preformed polymeric materials, ie, thermoplastics, can be markedly enhanced through the use of special photosensitive molecules that are mixed into the polymer matrix or that chemically attach to the backbone of the polymer chains. These special photosensitive molecules absorb uv or visible light energies much more efficiently than the polymer; they rapidly form excited states that undergo photochemical reactions which in turn form reactive free-radical intermediates that effect polymer dimerization or cross-linking. When compounded into the preformed polymer matrix, these special photosensitive molecules, eg, benzophenone, can undergo light-induced radical abstraction or insertion reactions that result in coupling of the polymer chains and in network formation (48,56,57) (Fig. 6) (see also UV STABILIZERS).

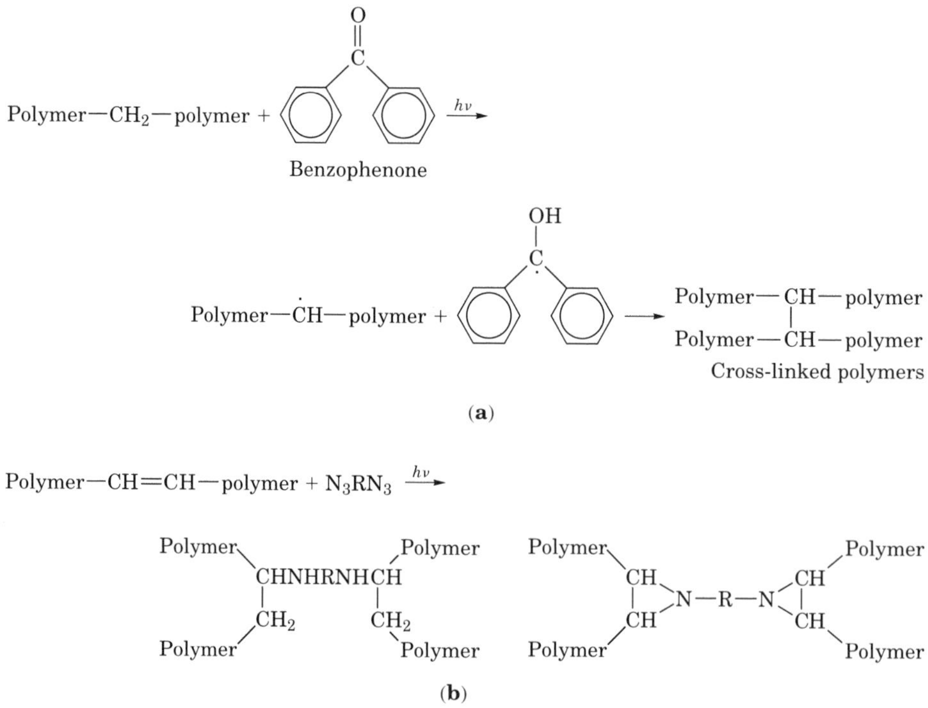

Fig. 6. Coupling of polymer chains via (**a**) photoinduced hydrogen abstraction free-radical reactions and (**b**) nitrene insertion/addition reactions.

Similar types of cross-linking reactions are observed for polymers to which photosensitive molecules are chemically attached to the backbone of the polymer structure (Fig. 7). Radiation curing of polymers using uv and visible light energies is used widely in photoimaging and photoresist technology (Table 8) (58,59).

Infrared processing is involved with thermoforming or heat-bonding of thermoplastic polymeric materials together. These polymer heat-forming or melting processes do not usually cure the polymer but only cause physical changes and maintain the original thermoplastic characteristics of the polymer (19).

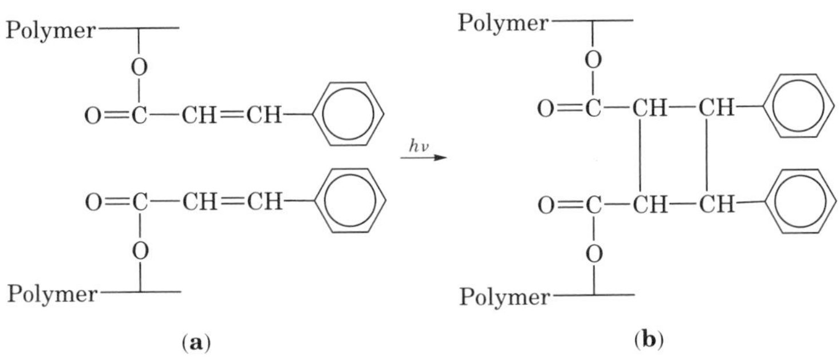

Fig. 7. Photodimerization of light-sensitive functional groups attached to polymer backbone structures: (**a**) polymer containing photosensitive cinnamic ester linkage, and (**b**) cross-linked polymer.

Table 8. Photocurable Polymer Systems For Photoresist Applications

Polymers	Comments	Reference
poly(vinyl cinnamate)s	uv- and visible light-induced photodimerization reactions; used in negative photoresist technologies	60
poly(vinyl cinnamylideneacetate)s	visible light-induced photodimerization reactions; 50% more efficient than poly(vinyl cinnamate)s	61
polychalcones	photodimerization or addition reactions; used in negative photoresist technologies	62
polycoumarins	photodimerization or addition reactions; used in negative photoresist technologies	63
polystilbenes	photodimerization or addition reactions; used in negative photoresist technologies	64
cyclized rubber	cross-linked with bis-azide–nitrene insertion and addition reactions	65
allylic esters	cross-linked with bis-azide–nitrene insertion reactions	66
thiolene polymers	unsaturated polymers (allyl or vinyl functionality) cross-linked with multifunctional mercaptans and photosensitizer molecules	67
phenolic polymers and acid functional acrylic resins	quinone diazide, solubility inhibitors, and photosensitizers for light-induced hydrophobic or hydrophilic reactions associated with positive photoresist technology	68,69

In order to cure, ie, form three-dimensional network structures through chemical changes on polymer systems with ir radiation, it is necessary to design a reactive functionality within the polymer structure so that coupling reactions can take place between the polymer chains as shown in the following reaction:

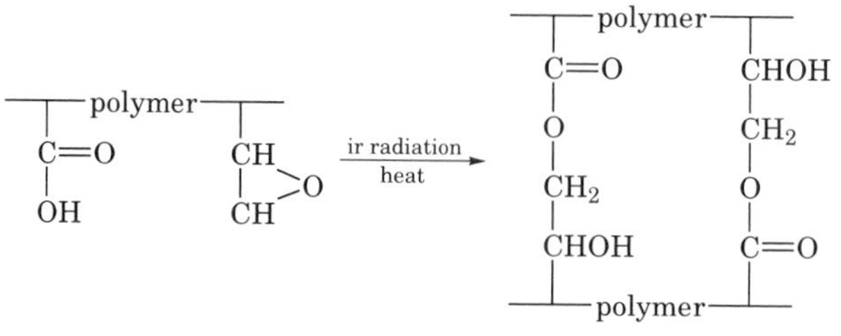

cross-linked polymer

Certain polymeric structures can also be blended with other coreactive polymers or multifunctional reactive oligomers that affect curing reactions when exposed to ir radiation. These coreactive polymers and cross-linking oligomers undergo condensation or addition reactions, which cause the formation of network structures (Table 9) (4,5,47).

Table 9. Thermally Curable Polymer Systems

Polymers	Comments	Reference
aromatic epoxy polymers plus acid functional polymers or oligomers	poor outdoor weatherability; good chemical resistance	70
aromatic epoxy polymers plus polyfunctional amines	excellent corrosion resistance	71
hydroxyl-containing polymers and melamine oligomers	good outdoor weatherability	72
hydroxyl-containing polymers plus blocked isocyanates	good outdoor weatherability	71
carboxyl-containing polymers plus oxazolines	good outdoor weatherability	71

Coatings. There are five characteristics of uv and visible light energy irradiation or photocuring of coatings: (1) the stable light source must be capable of producing uv and visible wavelengths of light, ie, near and far uv (200–400 nm) to visible (400–700 nm), with sufficient power or intensity to be commercially feasible; (2) the photoinitiator (PI) must be capable of absorbing uv and visible-light radiation at appropriate wavelengths of energy as emitted from the light source; (3) in a free-radical mechanism process, active free-radical intermediates must be produced through the action of light absorption by the photochemically active photoinitiator; these free radicals initiate polymerization of unsaturated monomers, oligomers, and polymers, and photoinitiators are not required in high energy electron-curing processes; (4) unsaturated, high boiling acrylate or methacrylate monomers, oligomers, cross-linking agents, and low molecular weight polymers comprise the fluid, low viscosity, light-curable coating system and are similar to the coating materials used in the high energy electron-curing process; and (5) after free-radical initiation of the reactive liquid coating system, the monomers propagate into a fully cured, cross-linked solid coatings or films (5,73,74).

$$\text{photoinitiator} \xrightarrow{h\nu} \text{[free-radical intermediates]}$$

$$\text{[free-radical intermediates]} + \text{unsaturated acrylate monomers,} \longrightarrow \text{cured films} \\ \text{polymers, etc}$$

In a cationic photoinitiated reactive system, the photoactive catalyst releases a cation or acid intermediate, which then interacts with low molecular weight epoxy resins or vinyl ether-substituted monomers, oligomers, cross-linkers, and polymers, thus making up the liquid coating composition. After release of the acid or cationic catalyst, the reactive liquid epoxy or vinyl ether coating propagates into a cured film structure in an analogous manner as the free-radical-initiated/cured coating system.

$$\text{photoactive catalyst} \longrightarrow \text{[acid or cation intermediates]}$$

$$\text{[acid or cation intermediates]} + \text{epoxy resins or unsaturated vinyl} \longrightarrow \text{cured films} \\ \text{ether monomers, polymers, etc}$$

These photoinitiators or photocatalysts are usually added to the reactive coating formulations in concentration ranges from less than 1 to 20 wt % based on the total formulation.

Light Source.   The light source normally used in commercial photocuring reactions is the medium pressure mercury arc lamp having a quartz or Vicor envelope. These lamps may contain electrodes for electrical-to-light energy conversion or may be electrodeless, in which case a radio-frequency wave causes mercury atom excitation and subsequent light emission. The normal power input levels are 40–160 W/cm arc length, resulting in sharp peak outputs having ca 10-nm bandwidths. The main peaks are at 313, 365, 404, 436, 546, and 578 nm, with relative outputs of between 2–80 W/cm over the length of the arc. Many other types of uv and visible light sources can be used for photopolymerization reactions, eg, low pressure mercury arcs, flash lamps, fluorescent or germicidal lamps, and lasers (qv).

A different type of light source based on silent electrical gas discharge phenomena has been developed by Asea Brown Boveri Corporation (75). Ultraviolet light radiation is generated in these types of lamps through the formation of excited molecular complexes (excimers or exciplexes) that do not have stable ground states. These excimers can be created in electrical discharges in a rare-gas (helium, neon, argon, krypton, xenon) environment. Because these exciplexes are unstable, they disintegrate within a nanosecond time frame and undergo a spontaneous emission of light energy. A silent discharge configuration allows the maintenance of a high pressure nonequilibrium discharge in which the plasma can be optimized to sustain a stable excimer formation process. A silent discharge

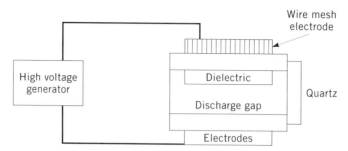

Fig. 8. Schematic of a silent discharge lamp.

configuration has at least one insulator or dielectric barrier, a transparent gas containment compartment, and a space between the electrodes (Fig. 8).

Discharges initiate in the gap region between the electrodes and then spread into surface discharges at the surface of the dielectric layer. The bright surface discharges are on the order of a few mm in diameter and the power ranges from these lamps can be from a few watts to several kilowatts over a variety of narrow wavelengths of light energy. Complete reviews of light sources used in photopolymerization reactions are available (17,18,76).

Light-wavelength output energies from various sources are small compared to electron-beam or electrocurtain processors (2–5 to thousands of eV). The energy associated with 365-nm wavelengths of light is equivalent to 3.4 eV or 343 kJ/mol (82 kcal/mol) and is sufficient to cause selective carbon bond rearrangements and cleavage of aromatic carbonyl–alkyl carbon bonds (aromatic C(O)–alkyl) (77).

Light energy alone is not sufficient to cause direct, efficient monomer initiation reactions. Commercial light-induced curing reactions require the use of a special photosensitive compound in the coatings formulation. These photosensitive compounds or photoinitiators are an integral part of the formulation and the cost of a light-sensitive radiation-curable coating system. The type and amount of photoinitiator also influences the relative rate of cure and the final properties of the cured film or coating.

Photoinitiators. Many theories of photoinitiated polymerization reactions using different light-sensitive catalysts have been reviewed (77,78,79,80). There are two general classes of photoinitiators: those that undergo direct photofragmentation upon exposure to uv or visible light irradiation and produce active free-radical intermediates, and those that undergo electron transfer followed by proton transfer to form a free-radical species. The absorption bands of the photoinitiators should overlap the emission spectra of the various commercial light sources (80).

The alkyl ethers of benzoin undergo direct photofragmentation upon absorption of uv energy at ca 360 nm to produce two free-radical intermediates.

Other similar structures undergo the following photofragmentation rearrangement decomposition processes:

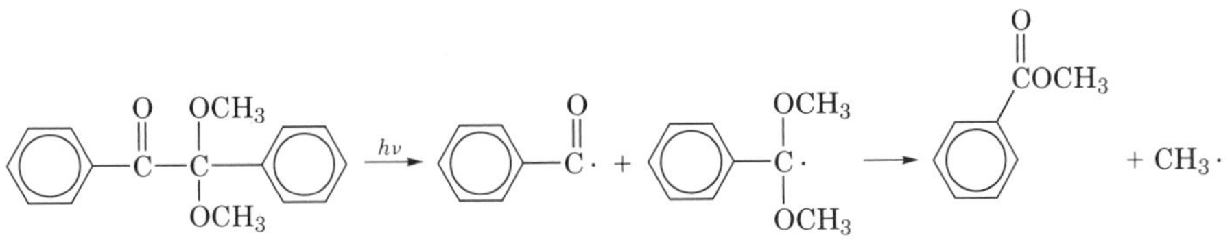

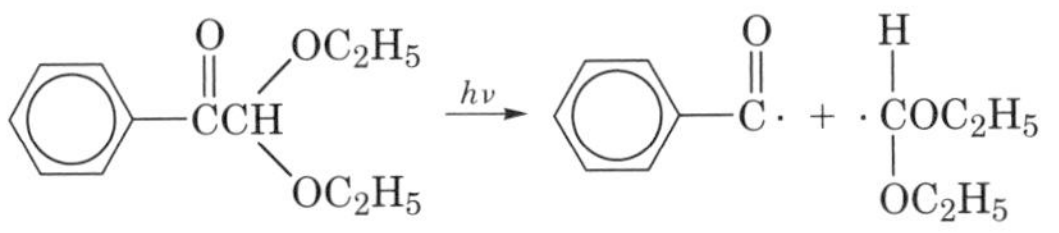

The second type of photoinitiators, ie, those that undergo electron transfer followed by proton transfer to give free-radical species, proceed as follows, where k_{st} is the rate constant for intersystem crossing.

$$(C_6H_5)_2C{=}O \xrightarrow{h\nu} (C_6H_5)_2C{=}O_{S_1} \xrightarrow{k_{st}} (C_6H_5)_2{=}O_{T_1} \xrightarrow{2(CH_3CH_2)_3N}$$

$$(C_2H_5)_3\ddot{N}[(C_6H_5)_2\dot{C}{-}\overset{-}{\ddot{O}}\underset{\substack{|\\ HCHCH_3}}{\overset{+}{\dot{N}}}(C_2H_5)_2] \longrightarrow (C_6H_5)_2\dot{C}OH + CH_3\dot{C}H\ddot{N}(C_2H_5)_2$$

exciplex

Benzophenone in its ground state (S_0) undergoes absorption of uv energy (340–360 nm) and is excited to its singlet excited states (S_1) followed by intersystem crossing ($k_{st} \cong 1$) to its triplet excited state (T_1). From the triplet excited state, benzophenone forms an encounter complex, or exciplex, with the ground-state alkyl amine, which then undergoes electron transfer from the nitrogen to the excited carbonyl followed by proton transfer from the carbon α to the nitrogen atom. This results in reduction of the benzophenone to form the semibenzpinacol radical and radical formation on the carbon α to the nitrogen atom. Both of these free-radical species can initiate or terminate polymerization of acrylic monomers. These free-radical species can also cross-couple or dimerize to form unreactive compounds. Excited-state aromatic ketones can abstract hydrogen atoms directly from the backbone structure of ethers and other organic substrates. These hydrogen radical abstraction reactions can lead to initiation or termination of the polymerization process. Photoinitiators having absorption capabilities in the visible light energy range are based on dyes, quinones, diketones, and heterocyclic chemical structures (80).

Examples of typical photoinitiator systems used to cure reactive coating systems are as follows (80,81). The reactive systems are primarily unsaturated acrylic acid esters of different alcohol and polymer structures.

Photoinitiators (electron transfer)	Photoinitiators (photofragmentation)
benzophenone	alkyl ethers of benzoin
halogenated benzophenones	benzil dimethyl ketal
amino functional benzophenones	α-hydroxyacetophenone
fluorenone derivatives	2-hydroxy-2-methylphenol-1-propanone
anthraquinone derivatives	2,2-diethoxyacetophenone
xanthone derivatives	1-phenyl-1,2-propanedione-2-(O-ethoxy-
thioxanthone derivatives	carbonyl) oxime
camphorquinone	2-methyl-1-(4-methylthio) phenyl)-2-
benzil	morpholino propanone-1
diphenoxy benzophenone	2-benzyl-2-N,N-dimethylamino-1-
	(4-morpholinophenyl) butanone
	α-amino acetophenones
	halogenated acetophenone derivatives
	sulfonyl chlorides of aromatic compounds
	acylphosphine oxides and bis-acyl
	phosphine oxides
	bisimidazoles

Photoactive Catalysts for Acid or Cation Generators

A variety of monomer and polymer structures can polymerize (cure) when exposed to an acid or cation intermediate species.

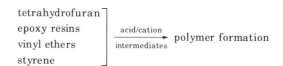

The photoactive catalyst systems commonly used to cure epoxy resins and multifunctional vinyl ether materials include aryldiazonium salts ($ArN{=}N^+X^-$), and triarylsulfonium and diaryliodonium salts ($Ar_3S^+X^-$ and $Ar_2I^+X^-$). Other cation-intermediate-generating catalyst systems are cyclopentadienyl iron(II)–arene hexafluorophosphate complexes, phenylphosphonium benzophenone salts, and pentafluoro phenyl borate anions associated with aryl sulfonium cations (81). An example of how these photoactive catalyst materials initiate the cure of an epoxy functional monomer or polymer is shown in Figure 9. These types of acid-initiated curing reactions for epoxy resins and vinyl ether containing monomers or polymers have been used in coatings and photoresist and printed circuit technologies (78,81) (see EPOXY RESINS).

Formulation Design for Free-Radical Cured Systems

Light-induced (free-radical intermediates), radiation-curable coating systems are similar to those used in high energy, electron-radiation-cured coating materials. The reactive coating ingredients in both the light and high energy electron curing

$$Ar_3S^+X^- \xrightarrow{h\nu} Ar_2S^+_\cdot \longrightarrow Ar + X^-$$

$$Ar_2S^+_\cdot + RH \longrightarrow Ar_2S^+H + R\cdot$$

$$Ar_2S^+H \longrightarrow Ar_2S + H^+$$

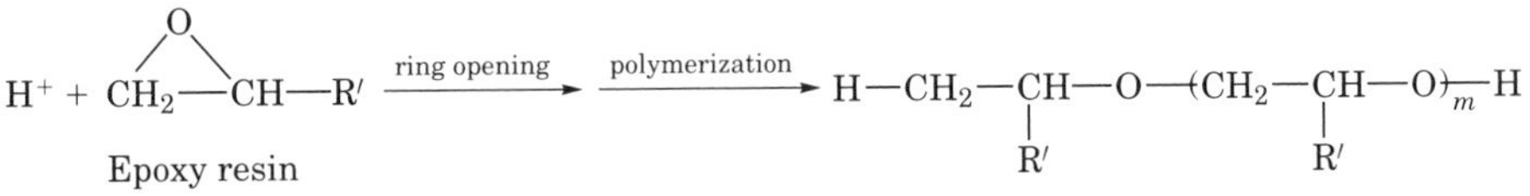

$$H^+ + \underset{\text{Epoxy resin}}{CH_2\!-\!CH\!-\!R'} \xrightarrow[\text{ring opening}]{} \xrightarrow{\text{polymerization}} H\!-\!CH_2\!-\!\underset{R'}{CH}\!-\!O\!-\!(CH_2\!-\!\underset{R'}{CH}\!-\!O)_m\!-\!H$$

Fig. 9. Initiation of epoxy cure. Irradiation of a triaryl sulfonium salt produces a radical cation that reacts with an organic substrate RH to produce a cation capable of releasing a proton. The proton initiates ring-opening polymerization. $X^- = BF_4^-$, PF_6^-, AsF_6^-, and $S_6F_6^-$.

processes utilize combinations of single vinyl unsaturated monomers, multifunctional vinyl-substituted cross-linking oligomers, and a variety of unsaturated polymer structures. The only significant difference between high energy cured coatings and light energy cured coatings is the use of a photoinitiator that absorbs the light energy and initiates the start of the curing process.

Although the primary function of the single vinyl ((meth)acrylate) unsaturated monomer component in the formulation is to lower the viscosity of the coating system, the choice of monomer, however, can strongly influence the adhesion of the cured coating to the substrate, the surface energy of the formulation, the barrier properties of the cured film, and the kinetic cure-rate response capability for the entire system. Cross-linking agents or oligomers (multifunctional vinyl or acrylate unsaturated structures) in the formulation tend to give rapid cure rates (formation of three-dimensional networks) and hard surfaces, but strongly contribute to the shrinkage of the final film structure during and after the curing process takes place. This shrinkage of the film produces internal stresses that lead to a loss of adhesion of the coating to the surface of the substrate (82).

High molecular weight polymeric materials in the coating formulation increase the viscosity of the reactive liquid coating system, which strongly influences the method of applying the coating to a substrate. High molecular weight polymers in a coating formulation impart toughness, flexibility, and adhesion capabilities to a cured film structure.

An idealized formulation guideline showing possible interaction of single vinyl unsaturated monomers, multifunctional unsaturated vinyl cross-linking agents or oligomeric molecules, and higher molecular weight polymeric materials is represented by Figure 10.

The design of a commercial coating formulation relies heavily on the science and technology associated with the design of monomers, cross-linking oligomers, and polymers that have an optimized balance of chemical and physical properties for a specific product or application. Empirical correlations of chemical structures of monomers, oligomers, and polymers with their chemical and physical properties have been developed; advanced experimental design techniques are increasingly used to create complex coating compositions that can contain as many as 30 variables in the final system (82–85).

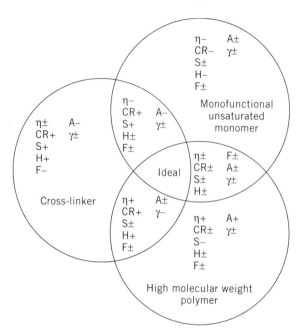

Fig. 10. Generalized formulation design outline for radiation-curable coatings and adhesive systems. The cross-linker is a multifunctional unsaturated cross-linking agent or oligomer. η = viscosity; CR = cure rate; S = shrinkage; H = hardness; F = flexibility; A = adhesion; γ = surface energy; + = increase; and – = decrease.

The majority of the commercial radiation-curable coating systems (high energy electron and light-cured coating systems) are clear or contain silica for lowering the surface gloss properties of the final cured film structure. The addition of colorizing pigments or opacifying and extended pigments (titanium dioxide, metal carbonates, etc) to a high energy electron coating formulation usually has only a minor effect on the overall cure rate and the final film performance capabilities of the total system. This is not the case, however, for pigmented coatings and inks that are cured with ultraviolet light energy. Many pigments commonly used in decorative coating formulations have large light absorption as well as scattering and reflectance properties that interfere with the ability of the photoinitiator to absorb the light energy required to initiate the polymerization reactions which ultimately create the final cured film structure. Thus, in pigmented coating formulations that are cured by ultraviolet light energies, the special photoinitiator systems being selected must have either the ability to absorb energy directly in the presence of a pigment that has the same absorption bands (the photoinitiator must then have a higher extinction coefficient than the pigment at the same maximum absorbance wavelength), or an absorption band that is outside the region where the pigment absorbs the light energy (86).

Examples of typical coating compositions and their applications in high energy, electron-radiation-curable systems (no photoinitiator required) and light-induced (photoinitiator required) curing reactions are available (87). Although the coating formulation for a photoinitiated acid/cationic curing resin system uses the same methodology and experimental design constraints as those described for the radiation curing of coatings based on free-radical intermediate reactions, the polymers and materials used in the cationic/acid curing system consist of single functional vinyl ethers or epoxy monomers, multifunctional vinyl ethers, and epoxy resins and polymers.

Applications and Markets

Since the early 1980s, the conversion of electrical energy into infrared, ultraviolet, and high energy electromagnetic radiation has gained worldwide acceptance as an efficient and economical method for modifying polymeric materials. These radiation-modified polymer systems are associated with many different types of products produced under a wide diversity of manufacturing operations. From a consumer point of view, almost every day a polymeric product that has been manufactured or processed by some form of radiation energy is encountered, for example:

Wood furniture (ir-, electron beam-, and uv-curable coatings)

Beer and beverage can labels (ir- and uv-curable inks)

Metal pipe coatings (uv-curable coatings)

Packaging, eg, paper, foil, and film (uv and electron-beam inks and coatings)

Floor coatings (uv-curable coatings)

Printed publications (ir- and uv-curable inks)

Graphic screen printing ink applications on mirrors (uv-curable inks)

Foamed plastic insulation (electron-beam-irradiated plastics)

Adhesive tapes (electron-beam-cured adhesives)

Rubber tires (electron-beam-irradiated rubber)

Food packaging (electron-beam-irradiated polyolefins)

Wire (electron-beam-irradiated polyolefins)

Telephone cable and optical fibers (uv-curable coatings)

Computers (uv and electron-beam polymer resist materials)

Recording tape (electron-beam-curable coatings)

A summary of applications, markets, and products where radiation-curing technologies have found commercial success may be summarized as follows:

Adhesives (82)   adhesives and encapsulants
  anaerobics   optical fiber coatings and adhesives
  pressure sensitive adhesives (PSA)   wire and cable coatings
  hot melt PSA   dry film solder masks
  flocks   magnetic and optical media

Coatings (48)
  furniture coatings
  "no wax" floor tile coatings
  protective topcoats
  metal decoration
  particle board sealer
  coil coatings
  beverage can coatings
  urethane automotive coatings
  glass coatings
  coatings on plastics
Electronics (9)
  integrated circuits
  photoresists and photolithography
  marking masks

conformal coatings
Inks (88)
  letterpress
  screen
  flexographic
  lithographic
  gravure
Packaging (paper and plastic)
  packaging adhesives
  release coatings
  barrier coatings
Photochemical machining (89)
  micromechanical parts
  optical waveguides
  three-dimensional model production
    (stereolithography)

Infrared Radiation Curing of Coatings. Thermal curing of a conventional coating system requires solvent removal and chemical cross-linking of polymers or oligomers. Commercial processors emit short wavelengths (0.7–2.0 μm), medium wavelengths (2–4 μm), and long wavelengths (4–1000 μm) of infrared radiation. The shortwave ir is near the visible end of the spectrum and, therefore, higher in energy and more penetrating than either the medium or the long wavelength emissions. The shortwave ir radiation is also reflective and can be focused for improved efficiency. A possible difficulty in using shortwave ir processing equipment is the substrate color selectivity characteristics. A dark-colored substrate or coating absorbs more thermal energy than a light-colored substrate; such color selectivity can lead to differential curing processes. Medium wavelengths of ir radiation are not substrate-color-sensitive, but are more difficult to control in terms of reflection or focusing, and somewhat less efficient for certain coating cure requirements than shortwave ir radiation (90). Many conventional coatings systems, such as powder coatings and single- and two-component 100% reactive liquid epoxy or polyurethane coatings, that contain polymer functionalities which cure through thermal processes involved with gas-fired oven technologies (see Tables 6 and 9) can also be cured or processed efficiently using ir radiation energies (91).

Curing with Microwave or Radio-Frequency Processing Equipment

Polymer surfaces can be easily modified with microwave or radio-frequency-energized glow discharge techniques. The polymer surface cross-links or oxidizes, depending on the nature of the plasma atmosphere. Oxidizing (oxygen) and nonoxidizing (helium) plasmas can have a wide variety of effects on polymer surface wettability characteristics (92).

One particularly promising approach to the development of improved coatings for use as metal primers, in fiber treatment, in packaging materials, etc, involves the use of vacuum plasma deposition techniques. Thin (micrometers thick), pinhole-free, polymeric coatings can be developed from a variety of organic moieties, both monomeric and nonmonomeric. Depending on the applications, such coatings can be developed to provide corrosion protection, abrasion and chemical resistance, improved adhesion, barrier properties, etc. With the plasma technique, both capacitively coupled and inductively coupled radio-frequency-energized vacuum chambers can be used for the deposition. In the capacitively coupled chamber, the substrate to be coated is attached to one of the electrodes. Monomer is then introduced into the chamber such that an operating pressure of ca 13 Pa (ca 0.1 mm Hg) is maintained. A glow discharge between the electrodes polymerizes a coating on the substrate. In the inductively coupled reactor, radio-frequency energy is supplied by coils surrounding the chamber. The glow discharge, therefore, fills the entire chamber and coats any item placed in the chamber.

Plasma deposition has been used to prepare acrylic, silicone, and fluorocarbon coatings on a variety of substrates in thicknesses of up to ca 8 μm. Coating variables include applied power, substrate temperature, deposition time, and monomer pressure. The coatings are similar to those produced from the same monomers by conventional polymerization techniques. However, there are structural differences and the resulting coatings are highly cross-linked. All exhibit excellent chemical resistance and good adhesion to the different substrates. Certain coatings have good optical clarity. Coating hardness appears to vary appreciably with monomer type and deposition conditions (9,22,23,93,94).

Plasma processing technologies are used for surface treatments and coatings for plastics, elastomers, glasses, metals, ceramics, etc. Such treatments provide better wear characteristics, thermal stability, color, controlled electrical properties, lubricity, abrasion resistance, barrier properties, adhesion promotion, wettability, blood compatibility, and controlled light transmissivity.

BIBLIOGRAPHY

"Radiochemical Technology" in *ECT* 2nd ed., Vol. 17, pp. 53–64, by W. H. Beamer, The Dow Chemical Co.; "Radiation Curing" in *ECT* 3rd ed., Vol. 19, pp. 607–624, by V. D. McGinniss, Battelle Columbus Laboratory.

1. A. Chapiro, *Radiation Chemistry of Polymeric Systems*, John Wiley & Sons, Inc., New York, 1962.
2. J. E. Wilson, *Radiation Chemistry of Monomers, Polymers, and Plastics*, Marcel Dekker, Inc., New York, 1974.
3. A. Singh and J. Silverman, eds., *Radiation Processing of Polymers*, Hanser Publishers, New York, 1992; L. F. Rossetti, AFP-SME Technical Paper, "Irradiated Wave, Cable and Tubing," Association for Finishing Processes of SME, FC76-510, Society of Manufacturing Engineers, Dearborn, Mich., 1976.
4. V. D. McGinniss, L. J. Nowacki, and S. V. Nablo, *ACS Symposium* **107**, 51–70 (1979).
5. V. D. McGinniss, *National Symposium on Polymers in the Service of Man*, American Chemical Society, Washington, D.C., 1980, pp. 175–180.
6. G. A. Senich and R. E. Florin, *J. Macromol. Sci.* **C24**(2), 240–324 (1984).

7. J. C. Colbert, ed., *Modern Coating Technology: Radiation Curing, Electrostatic, Plasma, and Laser Methods, Chemical Technology*, Noyes Data Corp., Park Ridge, N.J., 1982.

8. D. R. Randell, ed., *Radiation Curing of Polymers*, Royal Society of Chemistry, Burlington House, London, 1987.

9. V. D. McGinniss, "Radiation Curing: State-of-the-Art Assessment," Report prepared for the Electric Power Research Institute, Palo Alto, Calif., EPRI Report No. EM-4570, 1986.

10. James H. O'Donnell, ed., *Irradiation of Polymeric Materials: Processes, Mechanisms and Applications*, American Chemical Society, Washington, D.C., 1993.

11. J. W. T. Spinks and R. J. Woods, An Introduction to Radiation Chemistry, John Wiley & Sons, Inc., New York, 1964.

12. A. Danno, in *Radiation Chemistry and Its Applications*, IAEA Technical Report Series No. 84, International Atomic Energy Agency, Vienna, Austria, Apr. 1967, p. 2341.

13. S. Jefferson and co-workers, in *Advances in Nuclear Science and Technology*, Vol. 4, Academic Press, New York, 1968, pp. 335–338.

14. G. W. Grodstein, *X-Ray Attenuation Coefficients from 10 keV to 100 MeV*, Circ. 583, U.S. National Bureau of Standards, Washington, D.C., 1957.

15. Richard Bradley, *Radiation Technology Handbook*, Marcel Dekker, Inc., New York, 1984.

16. A. F. Readdy, Jr., "Application of Ionizing Radiations in Plastics and Polymer Technology," Plastic Report R41, Plastics Technical Evaluation Center, Picatinny Arsenal, Dover, N.J., 1971.

17. W. E. Elanbass, *Light Sources*, Crane, Russak & Co., Inc., New York, 1972.

18. V. D. McGinniss, in S. P. Pappas, ed., *UV Curing Science and Technology*, Technology Marketing Corp., Stamford, Conn., 1978, pp. 96–132.

19. A. F. Readdy, Jr., *Plastics Fabrication by Ultraviolet, Infrared, Induction, Dielectric and Microwave Radiation Methods*, Plastic Report R43, Plastics Technical Evaluation Center, Picatinny Arsenal, Dover, N.J., 1972.

20. R. W. Pray, *A New Look at Infrared*, SME Publication FC78-543, Association for Finishing Processes of SME, Dearborn, Mich., 1978.

21. J. R. Bush, "Electric Infrared Process Heating: State-of-the-Art Assessment," Report prepared for the Electric Power Research Institute, Palo Alto, Calif., EPRI Report No. EM-4571, 1986.

22. J. R. Hollahan and A. T. Bell, *Techniques and Applications of Plasma Chemistry*, John Wiley & Sons, Inc., New York, 1974.

23. M. Shen and A. T. Bell, eds., *Plasma Polymerization*, American Chemical Society, Washington, D.C., 1979.

24. N. J. Turro, *Modern Molecular Photochemistry*, Benjamin/Cummings Publishing Co., Menlo Park, Calif., 1978.

25. Charles E. Hoyle and James F. Kinstle, eds., *Radiation Curing of Polymeric Materials*, American Chemical Society, Washington, D.C., 1990.

26. R. O. Bolt and J. D. Carroll, *Radiation Effects on Organic Materials*, Academic Press, Inc., New York, 1963.

27. J. F. Kinstle, *Paint Varn. Prod.* **63**(6), 17 (1973).

28. S. P. Pappas, ed., *Radiation Curing Science and Technology*, Plenum Press, New York, 1992.

29. L. F. Thompson and K. G. Mayhan, *J. Appl. Polym. Sci.* **16**, 2291 (1972).

30. G. Alder, *Science*, **141**, 321 (1963).

31. G. Odian and B. S. Bernstein, *Nucleonics*, **21**, 80 (1963).

32. G. F. D'Alelio, R. Haberli, and G. F. Pezdirtz, *Effect of Ionizing Radiation on a Series of Saturated Polyesters*, NASA-SP-58, National Aeronautics and Space Administration, Washington, D.C., 1964.

33. V. D. McGinniss, in H. F. Mark and co-workers, eds., *Encyclopedia of Polymer Science and Engineering*, 2nd ed., Vol. 4, pp. 418–449, John Wiley & Sons, Inc., New York, 1986.
34. B. S. Bernstein and co-workers, *J. Appl. Polym. Sci.* **10**, 143 (1966).
35. B. S. Bernstein and J. Lee, *Ind. Eng. Chem. Prod. Res. Deu.* **6**, 211 (1967).
36. M. Izumi and co-workers, *Sumitomo Electr. Tech. Res.* **9**, 50 (Mar. 1967).
37. V. L. Lanza and R. M. Halperin, *Proceedings of 13th Annual Symposium on Communication Wires and Cables*, Asbury Park, N.J., Dec. 1963.
38. "Self-Adhering Silicone Rubber Tape" in *General Electric Company Bulletin*, Insulating Materials Data PD 1302, General Electric Co., Schenectady, N.Y.
39. V. L. Lanza and P. M. Cook, *Proceedings of 9th Annual Symposium on Communication Wire and Cable*, Asbury Park, N.J., Nov.–Dec. 1960.
40. R. C. Becker, *Plast. World*, **35**, 48 (Feb. 1977).
41. "Photopolymers: Principles, Processes and Materials," *Proceedings of Regional Technical Conference of the Society of Plastic Engineers*, Ellenville, N.Y., 1976.
42. A. Banov, *Paints and Coatings Handbook*, Structures Publishing Co., Farmington, Mich., 1978.
43. S. V. Nablo, SME Technical Paper, FC75-311, Society of Manufacturing Engineers, Dearborn, Mich., 1975.
44. W. J. Ramler, *J. Radiat. Curing*, **1**(3), 129–135 (Aug. 1984).
45. A. J. Chompff and S. Newman, eds., *Polymer Networks*, Plenum Press, New York, 1971.
46. A. R. Shultz, in E. M. Fettes, eds., *Chemical Reactions of Polymers*, John Wiley & Sons, Inc., New York, 1964; R. W. Tess and G. W. Pochlein, ed., *Applied Polymer Science*, 2nd ed., American Chemical Society, Washington, D.C., 1985.
47. J. L. Gardon and J. W. Prane, eds., *Nonpolluting Coatings and Coating Processes*, Plenum Press, New York, 1973.
48. P. K. T. Oldering, ed., *Chemistry and Technology of UV and EB Formulation for Coatings, Inks and Paints*, SITA Technology, Ltd., London, 1991, Vols. I–IV; Sartomer Monomer/Oligomer Photoinitiator Product Line Literature, Sartomer Co., Exton, Pa., 1991; D. H. Solomon, *The Chemistry of Film Formers*, John Wiley & Sons, Inc., New York, 1967.
49. W. Burlant and J. H. Hinsch, *J. Polym. Sci.* **A2**, 2135 (1964).
50. *Ibid.*, **A3**, 3587 (1965).
51. S. S. Labana and E. O. McLaughlin, *J. Elastoplast.* **2**, 3 (1970).
52. S. V. Nablo and E. P. Tripp, *Radiat. Phys. Chem.* **9**, 325 (1977).
53. T. J. Miranda and T. F. Huemmer, *J. Paint Technol.* **41**, 118 (1969).
54. K. H. Morganstern, *Proceedings of SAE Engineering Congress*, Detroit, Mich., 1967.
55. J. H. Frankfort, *Proceedings of Information Meeting on Irradiated Wood-Plastic Materials*, Report ORNL-11C-7, Chicago, Ill., 1965.
56. S. S. Labana, ed., *Ultraviolet Light Induced Reactions in Polymers*, American Chemical Society, Washington, D.C., 1976.
57. *Radiation Curing in Polymer Science and Technology*, Vols. I–IV, J. P. Fouassier and J. F. Rabek, eds., Elsevier Science Publishing Co., Inc., New York, 1993.
58. E. D. Feit, in S. P. Pappas, ed., *UV Curing: Science and Technology*, Technology Marketing Corp., Stamford, Conn., 1978, pp. 229–256.
59. L. F. Thompson and M. J. Bowden, *Introduction to Microlithograph*, 2nd ed., American Chemical Society, Washington, D.C., 1994; H. Ito, S. Tagawa, and K. Horie, eds., *Polymeric Materials for Microelectronic Applications: Science and Technology*, American Chemical Society, Washington, D.C., 1995.
60. K. Nakamura, T. Sakata, and S. Kikuchi, *Bull. Chem. Soc. Jpn.* **41**, 1765 (1968).
61. T. A. Shankoff and A. M. Trozzolo, *Photo. Sci. Eng.* **19**, 143 (1975).

62. M. Tsuda, *J. Polym. Sci.* **A2**, 2907 (1964).
63. G. A. Delzenne, in N. Bikales, ed., *Encyclopedia of Polymer Science and Technology*, Supp. 1, John Wiley & Sons, Inc., New York, 1976.
64. F. A. Stuber and co-workers, *J. Appl. Polym. Sci.* **13**, 2217 (1969).
65. F. C. Deschryver, N. Boens, and G. Smith, *J. Polym. Sci.* **A1**, 1939, 1970.
66. S. Shimizm and G. R. Bird, *J. Electrochem. Soc.* **124**, 1394 (1977).
67. W. S. DeForest, *Photoresists: Materials and Processes*, McGraw-Hill Book Co., Inc., New York, 1975.
68. J. Kosar, *Light-Sensitive Systems*, John Wiley & Sons, Inc., New York, 1965, pp. 194–320.
69. L. F. Thompson, C. G. Wilson, and J. M. J. Frechet, *Materials for Microlithography*, American Chemical Society, Washington, D.C., 1984; M. Biswas and T. Uryu, *J. Macromol. Sci.* **C26**(2), 249–352 (1986).
70. U.S. Pat. 3,888,943 (June 10, 1975), S. S. Labana (to Ford Motor Corp.).
71. S. Gabriel, *J. Oil Col. Chem. Assoc.* **59**, 52 (1976).
72. N. J. H. Gulpen and A. J. VanDeWerff, *J. Paint Technol.* **47**, 81 (1975).
73. V. D. McGinniss and D. M. Dusek, *J. Paint Technol.* **46**, 23 (1974).
74. C. G. Roffey, *Photopolymerization of Surface Coatings*, John Wiley & Sons, Inc., New York, 1982; C. Decker and K. Moussa, *J. Coatings Tech.* **65**(819), 49–57 (1993).
75. U. Kogelschatz, *Appl. Surface Sci.* **54**, 410–423 (1992).
76. L. R. Koller, *Ultraviolet Radiation*, John Wiley & Sons, Inc., New York, 1965.
77. J. Kopecky, *Organic Photochemistry: A Visual Approach*, VCH Publishers, Inc., New York, 1992; S. P. Pappas and V. D. McGinniss, in Ref. 45, Chapt. 1, pp. 1–22.
78. V. D. McGinniss, *Photogr. Sci. Eng.* **23**(3), 124 (1979).
79. V. D. McGinniss, *J. Radiat. Curing*, **2**, 3 (1975).
80. V. D. McGinniss, *Develop. Polym. Photochem.* **3**, 1–52 (1982); H. J. Hageman, *Prog. Org. Coatings*, **13**, 123, 1985.
81. C. Chang and co-workers, in L. J. Calbo, ed., *Handbook of Coatings Additives*, Vol. 2, Marcel Dekker, Inc., New York, 1992; J. V. Crivello and J. H. W. Lam, *J. Polym. Sci.* **17**, 977 (1979); J. L. Dektar and N. P. Hacker, *J. Chem. Soc. Chem. Commun.*, 1591 (1987); C. Priou and co-workers, *Polym. Mat. Sci. Eng.* **72**, 417, 1995.
82. V. D. McGinniss and A. Kah, *Polym. Eng. Sci.* **17**(7), 478–483 (1977); V. D. McGinniss, in L. H. Lee, ed., *Adhesive Chemistry*, Plenum Publishing Corp., 1984, pp. 363–377.
83. V. D. McGinniss, *Org. Coatings Plast. Chem.* **39**, 529–537 (1978); V. D. McGinniss, *Org. Coatings Appl. Polym. Sci.* **46**, 214–223 (1982).
84. D. W. VanKrevelen, *Properties of Polymers*, Elsevier Science Publishing Co., Inc., New York, 1972.
85. G. C. Derringer, *J. Quality Tech.* **12**(4), 214–219 (1980).
86. V. D. McGinniss, in S. S. Labana, ed., *International Symposium of UV Light-Induced Polymer Reactions*, American Chemical Society, Washington, D.C., 1976.
87. C. H. Carder, *Paint Varn. Prod.* **64**(8), 19 (1974).
88. S. G. Wentink and S. D. Koch, ed., *UV Curing in Screen Printing for Printed Circuits and the Graphics Arts*, Technology Marketing Corp., Norwalk, Conn., 1981.
89. P. F. Jacobs, *Rapid Prototyping and Manufacturing: Fundamental of Stereolithography*, Society of Manufacturing Engineers, Dearborn, Mich., 1992.
90. W. C. Hankins, *J. Oil. Col. Chem. Assoc.* **60**, 300 (1977).
91. R. J. Dick and co-workers, *J. Coatings Tech.* **66**(831), 23–38 (1994).
92. J. F. Kinstle, *J. Radiat. Curing*, **2**(2), 15 (1975).
93. V. D. McGinniss, *Proceedings of 20th International Conference in Organic Coatings Science and Technology*, Athens, Greece, 1994, pp. 331–346.
94. V. D. McGinniss, *Paint Coatings Ind.* **10**(7), 34–39 (1994).

General References

Proceedings of Radcure 84 Conference, Atlanta, Ga., published by the Association for Finishing Processes of SME, Dearborn, Mich., 1984.
Proceedings of Radcure 86 Conference, Baltimore, Md., published by Association for Finishing Processes of SME, Dearborn, Mich., 1986.
Proceedings of RadTech 88 North American Conference, New Orleans, La., published by RadTech International, Northbrook, Ill., 1988.
Proceedings of RadTech 90 North American Conference, Vols. 1 and 2, Chicago, Ill., published by RadTech International, Northbrook, Ill., 1990.
Proceedings of RadTech 92 North American Conference, Vols. 1 and 2, Boston, Mass., published by RadTech International North America, Northbrook, Ill., 1992.
Proceedings of RadTech 94 North American Conference, Vols. 1 and 2, Orlando, Fla., published by RadTech International North American, Northbrook, Ill., 1994.
Suppliers of radiation-curable materials: CIBA-GEIGY (photoinitiators), Aceto Corp. (photoinitiators), BASF (photoinitiators), Henkel (photoinitiators), Sartomer (monomers, cross-linking agents/oligomers, polymers), Henkel (cross-linking agents/oligomers, polymers), Cargill (polymers).

VINCENT D. MCGINNISS
Battelle Columbus Laboratory

RADIOACTIVE TRACERS

Radiochemical tracers, compounds labeled with radioisotopes (qv), have become one of the most powerful tools for detection and analysis in research, and to a limited extent in clinical diagnosis (see MEDICAL IMAGING TECHNOLOGY). A molecule or chemical is labeled using a radioisotope either by substituting a radioactive atom for a corresponding stable atom in the compound, such as substituting ^3H for ^1H, ^{14}C for ^{12}C, ^{32}P or ^{33}P for ^{31}P, and ^{35}S for ^{32}S, for example, or by adding a radioactive atom to a molecule, such as iodinating a protein or peptide using ^{125}I. In some cases radioactive labeling results in substituting an atom using a noncorresponding, but chemically similar, radioactive isotope, such as replacing ^{16}O with ^{35}S.

Radiometric detection technology offers high sensitivity and specificity for many applications in scientific research. The radioactive emission of the labeled compound is easily detected and does not suffer from interference from endogenous radioactivity in the sample. Because of this unique property, labeled compounds can be used as tracers to study the localization, movement, or transformation of molecules in complex experimental systems.

The use of radioactive tracers was pioneered by Georg von Hevesy, a Hungarian physical chemist, who received the Nobel Prize in 1943 for his work on radioactive indicators (1). Radioisotopes have become indispensable components of most medical and life science research strategies, and in addition the technology is the basis for numerous industries focused on the production and detection of radioactive tracers. Thousands of radioactive tracers have been synthesized and are commercially available. These are used worldwide in tens of thousands of research laboratories.

Properties

Any radioactive nuclide or isotope of an element can be used as a radioactive tracer, eg, chromium-51 [14392-02-0], cobalt-60 [10198-40-0], tin-110 [15700-33-1], and mercury-203 [13982-78-0], but the preponderance of use has been for carbon-14 [14762-75-5], hydrogen-3 [10028-17-8] (tritium), sulfur-35 [15117-53-0], phosphorus-32, and iodine-125 [14158-31-7]. More recently phosphorus-33 has become available and is used to replace sulfur-35 and phosphorus-32 in many applications. By far the greater number of radioactive tracers produced are based on carbon-14 and hydrogen-3 because carbon and hydrogen exist in a large majority of the known natural and synthetic chemical compounds.

The properties of the commonly produced and used radioactive isotopes are listed in Table 1. The half-lives, and therefore the specific activities, of these nuclides vary over many orders of magnitude. It is this variation, coupled with variation in decay energy, which determines the suitability of a nuclide for the various applications and detection strategies. For example, techniques requiring the highest sensitivity of detection utilize ^{32}P, ^{33}P, ^{35}S, or ^{125}I. In addition, physical properties, along with the chemical form and biological properties, determine the approaches used for safe handling and waste disposal. The radioactive isotopes having short half-lives, such as ^{32}P or ^{33}P, do not pose a serious disposal problem because these can be held for decay and ultimately disposed of as cold

Table 1. Properties of the Principal Radioactive Isotopes[a]

Isotope	CAS Registry Number	Half-life, $t_{1/2}$	Specific activity, Bq/mmol[b]	Energy, MeV[c], %	Decay product
^3H	[10028-17-8]	12.35 yr	1.07×10^{12}	0.019, 100	^3He
^{14}C	[14762-75-5]	5730 yr	2.31×10^9	0.156, 100	^{14}N
^{32}P	[14596-37-3]	14.29 d	3.36×10^{14}	1.71, 100	^{32}S
^{33}P	[15749-66-3]	25.4 d	1.89×10^{14}	0.249, 100	^{33}S
^{35}S	[15117-53-0]	87.44 d	5.51×10^{13}	0.167, 100	^{35}Cl
^{45}Ca	[13966-05-7]	163 d	2.95×10^{13}	0.257, 100	^{45}Sc
^{51}Cr	[14392-02-0]	27.70 d	1.73×10^{14}	0.320[d], 10	^{51}V
^{125}I	[14158-31-7]	60.14 d	7.99×10^{13}	0.03, 20	^{125}Te
				0.02, 13	
				0.027[e], 114	
				0.031, 27	
				0.035[d], 7	

[a] Refs. 2 and 3.
[b] To convert Bq to Ci, divide by 3.70×10^{10}.
[c] Maximum energies for β^- emissions unless otherwise noted.
[d] γ-Emission.
[e] X-ray emission.

waste. On the other hand, the disposal of ^{14}C and ^{3}H low level radioactive waste has become a problem for generators in the United States owing either to the unavailability of operating disposal sites or to the cost of disposal.

Syntheses

Syntheses of radioactive tracers involve all of the classical biochemical and synthetic chemical reactions used in the synthesis of nonradioactive chemicals. There are, however, specialized techniques and considerations required for the safe handling of radioactive chemicals, strategic synthetic considerations in terms of their relatively high cost, and synthesis scale constraints governed by specific activity requirements.

The radioactive isotopes available for use as precursors for radioactive tracer manufacturing include barium [^{14}C]-carbonate [*1882-53-7*], tritium gas, [^{32}P]-phosphoric acid or [^{33}P]-phosphoric acid [*15364-02-0*], [^{35}S]-sulfuric acid [*13770-01-9*], and sodium [^{125}I]-iodide [*24359-64-6*]. It is from these chemical forms that the corresponding radioactive tracer chemicals are synthesized. [^{14}C]-Carbon dioxide, [^{14}C]-benzene, and [^{14}C]-methyl iodide require vacuum-line handling in well-ventilated fume hoods. Tritium gas, [^{3}H]-methyl iodide, sodium borotritide, and [^{125}I]-iodine, which are the most difficult forms of these isotopes to contain, must be handled in specialized closed systems. Sodium [^{35}S]-sulfate and sodium [^{125}I]-iodide must be handled similarly in closed systems to avoid the liberation of volatile [^{35}S]-sulfur oxides and [^{125}I]-iodine. Adequate shielding must be provided when handling [^{32}P]-phosphoric acid to minimize exposure to external radiation.

A multistep synthesis is strategically designed such that the labeled species is introduced as close to the last synthetic step as possible in order to minimize yield losses and cost. Use of indirect reaction sequences frequently maximizes the yield of the radioactive species at the expense of time and labor.

Many applications in tracer technology require products of high specific activity, ie, compounds having a high degree of substitution of specific atoms with radioisotopes. For many labeled compounds nearly 100% labeling can be achieved at one or more locations in a molecule using ^{14}C, ^{3}H, ^{32}P, ^{33}P, or ^{35}S, for example, using carrier-free precursors. The mass quantities of high specific activity compounds are usually low, even for those required for commercial production. Micro or semimicro synthetic methods are thus used in the manufacturing process.

Synthetic chemical approaches to the preparation of carbon-14 labeled materials involve a number of basic building blocks prepared from barium [^{14}C]-carbonate (2). These are carbon [^{14}C]-dioxide; [^{14}C]-acetylene; [U−^{14}C]-benzene, where U = uniformly labeled; [1- and 2-^{14}C]-sodium acetate, [^{14}C]-methyl iodide, [^{14}C]-methanol, sodium [^{14}C]-cyanide, and [^{14}C]-urea. Many complicated radiotracers are synthesized from these materials. Some examples are [1-^{14}C]-8,11,14-eicosatrienoic acid [*3435-80-1*] from [^{14}C]-carbon dioxide, [ring-U−^{14}C]-phenylisothiocyanate [*77590-93-3*] from [^{14}C]-acetylene, [7-^{14}C]-norepinephrine [*18155-53-8*] from [1-^{14}C]-acetic acid, [4-^{14}C]-cholesterol [*1976-77-8*] from [^{14}C]-methyl iodide, [1-^{14}C]-glucose [*4005-41-8*] from sodium [^{14}C]-cyanide, and [2-^{14}C]-uracil [*626-07-3*], [*27017-27-2*] from [^{14}C]-urea. All syntheses of the basic radioactive building blocks have been described (4).

The introduction of tritium into molecules is most commonly achieved by reductive methods, including catalytic reduction by tritium gas, [$^3H_2$], of olefins, catalytic reductive replacement of halogen (Cl, Br, or I) by $^3H_2$, and metal [^3H] hydride reduction of carbonyl compounds, eg, ketones (qv) and some esters, to tritium-labeled alcohols (5). The use of tritium-labeled building blocks, eg, [^3H] methyl iodide and [^3H]-acetic anhydride, is an alternative route to the preparation of high specific activity, tritium-labeled compounds. The use of these techniques for the synthesis of radiolabeled receptor ligands, ie, drugs and drug analogues, has been described in detail in the literature (6,7).

Iodination of organic compounds using iodine-125 gives radiotracers that are in most cases modified forms of the compound being traced. Iodine is found in only a relatively small number of the naturally occurring compounds of interest. The radiotracer and the unlabeled parent substance must be determined to behave identically before acceptance of any derived data is valid. In the case of thyroxine, which is a naturally occurring iodine-containing substance, labeling with ^{125}I is achieved by exchange with [^{125}I]-NaI (see THYROID AND ANTITHYROID PREPARATIONS).

Noniodine-containing substances that are to be iodinated must have a moiety that can be iodinated directly, eg, phenol, imidazole, pyrimidine, etc, and can react with such reagents as [^{125}I]-KI$_3$, [^{125}I]-NaI–lactoperoxidase, [^{125}I]-ClI, etc. An alternative method for iodination is the use of the reactive Bolton-Hunter reagent [60285-92-9], ie, [^{125}I] iodinated p-hydroxyphenylpropionic acid N-hydroxysuccinimide ester. Proteins are most readily labeled with iodine and in most cases their properties are unaffected by iodination. [^{32}P]-Phosphorus and [^{35}S]-sulfur are introduced mostly through the use of biosynthetic techniques acting upon [^{32}P]-phosphate and [^{35}S]-sulfate. Such reagents as [^{32}P]-PCl$_5$, [^{32}P]-POCl$_3$, [^{32}P]-PCl$_3$, [^{32}P]-P$_2$S$_5$, [^{35}S]-H$_2$S, [^{35}S]-Na$_2$SO$_4$, [^{35}S]-Na$_2$SO$_3$, [^{35}S]-KSCN, and [^{35}S]-P$_2$S$_5$ have been prepared and are either available routinely as synthetic precursors or as custom syntheses from the principal radiochemical suppliers.

Biosynthetic techniques utilizing enzymes isolated from plants, animals, or microorganisms have made possible the synthesis of many labeled compounds of biological interest. Enzymes are uniquely suited for such syntheses because the enzymes are biological catalysts that act with high specificity and at low substrate concentrations. The enzymes can be used for the synthesis of many high specific activity tritiated nucleosides, nucleotides, nucleotide sugars, and leukotrienes as well as most ^{32}P- and ^{33}P-labeled nucleotides and ^{35}S-labeled nucleotide analogues. Many of these products have been synthesized at near-theoretical specific activity using multistep enzymatic procedures. Enzymes used in the production of radioactive tracers are isolated from various sources depending on the specific enzyme of interest. Bacteria, such as E. coli, sea urchin sperm, carrots, bull testicles, rat liver, and human blood are some of the varied sources. Some of these enzymes have been cloned in bacteria, making isolation and purification less complex than using natural sources. Some of the enzymes used for labeled compounds production are available from commercial suppliers such as Boehringer Mannheim (Indianapolis, Indiana), Sigma Chemical Company (St. Louis, Missouri), and Pharmacia (Uppsala, Sweden).

Microbiological procedures which exploit the ability of bacteria and photosynthetic algae to incorporate exogenous labeled precursors such as $^{14}CO_2$, $^{35}SO_4^{2-}$, and $^{32}PO_4^{3-}$ can be used to label complex molecules in cells such as proteins (qv) and nucleic acids (qv), which are then processed to give labeled constituents such as uniformly labeled ^{14}C-amino acids, ^{14}C-nucleotides, ^{14}C-lipids, ^{35}S-amino acids, etc (8).

Even higher organisms can be used for the production of labeled compounds. Plants, tobacco, or *Canna indica*, for example, when grown in an exclusive atmosphere of radioactive carbon dioxide, [$^{14}CO_2$], utilize the labeled precursor as the sole source of carbon for photosynthesis. After a suitable period of growth, almost every carbon atom in the plant is radioactive. Thus, plants can serve as an available source of ^{14}C-labeled carbohydrates (9).

Purification

The small synthetic scale used for production of many labeled compounds creates special challenges for product purification. First, because of the need for use of micro or semimicro synthetic procedures, the yield of many labeled products such as high specific activity tritiated compounds is often low. In addition, under such conditions, side reactions can generate the buildup of impurities, many of which have chemical and physical properties similar to the product of interest. Also, losses are often encountered in simply handling the small amounts of materials in a synthetic mixture. As a consequence of these considerations, along with the variety of tracer chemicals of interest, numerous separation techniques are used in purifying labeled compounds.

For products having relatively low specific activity, such as some compounds labeled with ^{14}C and which are synthesized on the scale of several millimoles, classical organic chemical separation methods may be utilized, including extraction, precipitation, and crystallization. For separation of complex mixtures and for products having high specific activity, such as those labeled with tritium, ^{125}I, ^{32}P, etc, chromatographic methods utilizing paper, thin layers, or columns packed with silica and ion-exchange (qv) resins are more useful. For many applications, the method of choice is high pressure liquid chromatography (hplc) using columns packed with resins designed for ion-exchange and normal phase chromatography, or materials designed for reverse-phase separations. Hplc has the advantages of high resolution and speed; however, it requires costly chromatography columns and instrumentation.

Decomposition

Decay products of the principal radionuclides used in tracer technology (see Table 1) are not themselves radioactive. Therefore, the primary decomposition events of isotopes in molecules labeled with only one radionuclide/molecule result in unlabeled impurities at a rate proportional to the half-life of the isotope. For ^{14}C and ^{3}H, impurities arising from the decay process are in relatively small amounts. For the shorter half-life isotopes the relative amounts of these impurities caused by primary decomposition are larger, but usually not problematic because they are not radioactive and do not interfere with the application of the tracer compounds. For multilabeled tritiated compounds the rate of accumulation of labeled impurities owing to tritium decay can be significant. This increases with the number of radioactive atoms per molecule.

More problematic for suppliers and users of tracer chemicals is the decomposition of the labeled compounds caused by radiolysis during storage. This phenomenon is the result of the dissipation of the energy released in the surrounding media by decay of a radionuclide forming reactive species such as free radicals, which can cause chemical bond cleavage in other labeled molecules in the surrounding microenvironment. This mechanism of decomposition is usually referred to as secondary decomposition and is the most significant factor affecting the stability of tracers labeled with weak β-emitting radionuclides such as tritium.

A related mechanism of degradation involves the direct interaction of the radioactive emission with other tracer molecules in the preparation. This phenomenon is likely to occur in high specific activity compounds stored at high radiochemical concentrations in the absence of free-radical scavengers.

Many tracer chemicals are inherently unstable even as the unlabeled forms. Susceptibility of a chemical to hydrolysis, oxidation, photolysis, and microbiological degradation needs to be evaluated when designing suitable storage conditions for the labeled compound. Factors that reduce radiolytic degradation, such as dispersal in solution, are apt to increase chemical degradation or instability.

A great deal of empirical information has been developed on storage conditions and additives which reduce the rate of decomposition of tracer chemicals. In general, low storage temperatures, dispersal in solution to reduce radiochemical concentration, and the addition of free-radical scavengers and antioxidants (qv), such as ethanol, benzyl alcohol, ascorbic acid, and mercaptoethanol among others, have been shown to stabilize many classes of ^{3}H-, ^{14}C-, ^{32}P-, and ^{35}S-labeled compounds (10,11). The choice of stabilizer often needs to be assessed against possible interference with the use of the tracer chemical. For example, high concentrations of ethanol inhibit enzymatic systems used with ^{32}P- and ^{35}S-labeled nucleotides. For these products alternative stabilizers, such as Tricine (N-[2-hydroxy-1,1-bis(hydroxymethyl)ethyl]glycine [5704-04-1]), have been used by commercial suppliers. Some novel proprietary formulations have been developed by manufacturers of tracer chemicals which allow the shipment of some ^{32}P-, ^{33}P-, and ^{35}S-labeled compounds at ambient temperatures.

Detection and Quantitation

The methods for detection and quantitation of radiolabeled tracers are determined by the type of emission, ie, β-, γ-, or x-rays, the tracer affords; the energy of the emission; and the efficiency of the system by which it is measured. Detection of radioactivity can be achieved in all cases using the Geiger counter. However, in the case of the radionuclides that emit low energy betas such as ^{3}H, large amounts of isotopes are required for detection and accurate quantitation of a signal. This is in most cases undesirable and impractical. Thus, more sensitive and reproducible methods of detection and quantitation have been developed.

Liquid scintillation counting is by far the most common method of detection and quantitation of β-emission (12). This technique involves the conversion of the emitted β-radiation into light by a solution of a mixture of fluorescent materials or fluors, called the liquid scintillation cocktail. The sensitive detection of this light is affected by a pair of matched photomultiplier tubes (see PHOTODETECTORS) in the dark chamber. This signal is amplified, measured, and recorded by the liquid scintillation counter. Efficiencies of detection are typically 25–60% for tritium; >90% for ^{14}C, ^{35}S, and ^{32}P; and 60–70% for ^{125}I. A lesser used technique for the detection and quantitation of β-emissions is planchette counting. A film of the sample on a planchette, a flat metal pan, is brought into proximity, but at a fixed distance, to a proportional counter. The emissions are measured and recorded. Typical efficiencies are ^{14}C, ca 30–40%; ^{35}S, ca 30–40%; and ^{32}P, ca 50%.

The detection and quantitation of γ-emission from ^{125}I is accomplished by well counting. A thallium-activated sodium iodide crystal, having a well or drilled hole which contains the sample, converts the emission to light. The light is amplified, counted, and recorded using a photomultiplier tube. The efficiency for ^{125}I is typically 70%.

The β-emission of ^{32}P is energetic enough in its passage through water to emit light (Cherenkov effect). This emission can be measured by a photomultiplier tube with a typical efficiency of ca 40%.

The nonquantitative detection of radioactive emission often is required for special experimental conditions. Autoradiography, which is the exposure of photographic film to radioactive emissions, is a commonly used technique for locating radiotracers on thin-layer chromatographs, electrophoresis gels, tissue mounted on slides, whole-body animal slices, and specialized membranes (13). After exposure to the radiolabeled emitters, dark or black spots or bands appear as the film develops. This technique is especially useful for tritium detection but is also widely used for ^{14}C, ^{35}S, ^{32}P, ^{33}P, and ^{125}I. Instrumentation is also available by which the location of radioactive spots or bands on gels and membranes can be imaged without use of film.

Gas-flow counting is a method for detecting and quantitating radioisotopes on paper chromatography strips and thin-layer plates. Emissions are measured by interaction with an electrified wire in an inert gas atmosphere. All isotopes are detectable; however, tritium is detected at very low ($\sim$1%) efficiency.

Other methods of sensitive detection of radiotracers have been developed more recently. Fourier transform nmr can be used to detect ^3H (nuclear spin 1/2), which has an efficiency of detection $\sim$20% greater than that of ^1H. This technique is useful for ascertaining the position and distribution of tritium in the labeled compound (14). Field-desorption mass spectrometry (fdms) and other mass spectral techniques can be applied to detection of nanogram quantities of radiolabeled tracers, and are well suited for determining the specific activity of these compounds (15).

Determination of the radiochemical purity of labeled compounds is usually carried out using various chromatographic techniques, such as paper or thin-layer chromatography, or hplc (see CHROMATOGRAPHY). These procedures involve separation of the radioactive components of a sample using an appropriate elution solvent, followed by quantitation of the radioactivity in the separated fractions or bands by gas-flow counting, liquid scintillation counting, or autoradiography.

Economic Aspects

Radioactive tracers account for about 20% of the worldwide market for consumables and reagents for life science research. In 1994 the value was estimated at about $300 million. The principal full line manufacturers are Du Pont–NEN Research Products (Boston, Massachusetts) and Amersham International (Amersham, U.K.). These companies share roughly equally about 85% of the radiochemicals worldwide market. In addition to an extensive line of catalog products, these suppliers offer custom labeling and custom synthesis services. The rest of the market is shared by producers of a limited range of products or services, such as ICN Biomedicals (Costa Mesa, California) and American Radiolabeled Chemicals (St. Louis, Missouri).

The growth rate for tracers labeled with short-lived isotopes such as ^{32}P and ^{35}S was about 10–15% per year from 1990 through 1994. This trend reflects the increased use of these radiochemicals for research in molecular biology and genetics; ^{125}I-labeled tracers have also exhibited similar growth rates in this period. On the other hand, the market for ^{14}C- and ^{3}H-labeled chemicals essentially leveled off. The overall growth rate for all tracer chemicals was estimated at ~5%/yr for 1990–1994.

Yearly consumption of isotopes for production of tracer chemicals varies depending on the radionuclide. For example, about 1,100 GBq (30 Ci) of ^{33}P was expected to be utilized worldwide in 1995 compared with ca 7,400 TBq (200,000 Ci) of tritium.

Health and Safety Factors

Allowable external radiation doses are described in Reference 16. Depending on the quantities used and type of operation, the more energetic emissions of ^{32}P (β-ray), ^{125}I (γ- and x-rays), and ^{51}Cr (γ-ray) may require appropriate shielding to minimize personnel exposure. The β-rays of ^{35}S, ^{14}C, ^{33}P, ^{45}Ca, and ^{3}H are of low enough energy to require no shielding. All isotopes, however, present toxicity problems if taken into the body. Both tritium and ^{125}I are particularly problematic because of their volatility as $[^{3}\text{H}]\text{-H}_2\text{O}$ and $[^{125}\text{I}]\text{-I}_2$. However, personal safety precautions are related to the relative quantities of radioactive materials handled. Basic laboratory procedures to be followed protect the user from oral ingestion, skin contact, self-injection, and inhalation. The use of closed systems; well-ventilated work areas, ie, hoods and glove boxes; disposable gloves; disposable lab coats, etc; and neat work habits provide a safe working environment (17). Personnel monitoring may include urine and exhaled breath (CO_2) analysis, thyroid uptake (^{125}I) radioassay, use of dosimeters for detecting radiation from ^{32}P, and contamination surveys using appropriate instruments. The choice of test and frequency of testing vary with quantities and use (18).

Protection of the environment from uncontrolled radioactive release is also a consideration in the use of radiotracers. The quantity and concentration of radionuclides that may be discharged into sewer systems is limited by regulations of the Nuclear Regulatory Commission (NRC). Similarly, airborne emission limits have been established by the NRC for nonrestricted areas. Limits of surface contamination must be established to provide a safe workplace for users (19). The application of the as-low-as-reasonably-achievable (ALARA) principle to the above draws on the creative talents of the user to regard the limits as nonapproachable barriers and not as tolerable maxima for discharge.

U.S. radiation protection guidelines are established by the National Council on Radiation Protection and Measurement (NCRP) and are based on the recommendations of the International Commission on Radiological Protection (ICRP). The National Research Council also sponsors a report from its advisory committee on the biological effects of ionizing radiations (20).

Uses

The detectability of minute quantities of radiolabeled tracers makes possible the determination of microquantities of substances. The most effective use of radiotracers has been in biomedical research. For example, a radiolabeled, non-metabolized tracer for glucose, such as [1-^{3}H]-2-deoxyglucose [77590-94-4], is administered to a test animal to identify areas of brain activity corresponding to particular external stimuli. An external stimulus is given and the animal is sacrificed. The brain is frozen, sectioned, and exposed to x-ray film (autoradiography), and the location of the radioactivity is noted. In this way it is possible to relate changes in glucose metabolism, which reflects brain activity, to a stimulus, and a stimulus−response map of the brain can be constructed. This procedure is referred to as neuroanatomical mapping. In a similar way animal tissues can be processed and incubated with tritium or ^{14}C-labeled drugs or drug analogues (ligands) to localize the site of action or receptors, for drugs, hormones, and neurotransmitters (21).

Labeled drugs and ligands using ^{3}H, ^{14}C, and ^{125}I, are also widely used for screening for potential new drugs. In this procedure the labeled drug is incubated with a receptor preparation *in vitro* and the radioactivity bound to the receptor is quantitated using liquid scintillation counting. By including an unlabeled drug candidate compound in the mixture and measuring the extent of displacement or inhibition of the binding of the radioactive tracer, the potential pharmacologic activity of the candidate compound can be assessed. Thousands of chemicals can be tested using such receptor binding assays in a week employing automated, high throughput screening procedures.

In addition to being used for screening purposes for new drug development, labeled drugs and ligands are widely used by pharmaceutical companies for metabolic, bioavailability, and toxicological studies to support new drug applications for FDA approval.

Radiotracers have also been used extensively for the quantitative microdetermination of blood serum levels of hormones (qv), proteins, neurotransmitters, and other physiologically important compounds. Radioimmunoassay, which involves the competition of a known quantity of radiolabeled tracer, usually ^{125}I or ^{3}H, with the unknown quantity of serum component for binding to a specific antibody that has been raised against the component to be determined, is used in the microdetermination of physiologically active materials in biological samples (see IMMUNOASSAY).

The determination of the presence of reverse transcriptase in virus-infected cells can be done using labeled nucleotide triphosphates. Reverse transcriptase is an enzyme capable of synthesizing DNA from RNA and it is thought to play an important role in virus-mediated cell modification. This enzyme is also a marker enzyme for HIV, the virus implicated in causing acquired immunodeficiency syndrome (AIDS). The procedure utilizes radiolabeled nucleotides with nonlabeled substrates to synthesize tagged DNA. The degree of radioactive incorporation reflects the reverse transcriptase activity.

The availability of high specific activity ^{32}P-, ^{33}P-, and ^{35}S-labeled nucelotides has been crucial in the development and wide application of techniques for labeling deoxyribonucleic acid (DNA) and ribonucleic acid (RNA), the pivotal targets of research in molecular biology and genetics (see GENETIC ENGINEERING). These labeled compounds are substrates for most of the enzymes which synthesize, modify, or degrade nucleic acids (qv). Thus a multitude of approaches have been devised for incorporating the tracers into DNA and RNA fragments. The labeled DNA or RNA is usually used in one of two principal applications: structure or sequence analysis of the labeled fragment, or use as a probe to detect sequences in an unknown sample which are complementary to the labeled fragment. For example, the use of [γ-^{32}P]-adenosine 5'-triphosphate [2964-07-0] as a source of phosphate for end labeling DNA is an integral part of the Maxam-Gilbert sequencing procedure. In this method, the [^{32}P] end-labeled DNA is selectively cleaved by reagents specific for the rupture of selected chemical bonds between the various base sequences. Gel electrophoresis of the reaction mixtures followed by autoradiographic detection of the fragments allows the determination of the sequence of the bases (see ELECTROSEPARATIONS).

In another method for sequencing, the DNA is labeled at internal base positions by enzymes which replicate DNA strands in the presence of specific chain terminators, thus producing an array of labeled fragments which reflect the sequence of the DNA. The rest of the procedure is the same as described for the chemical cleavage method (22).

Similar procedures are in use for labeling DNA and RNA either internally or at terminal positions for use as probes for complementary sequences in samples immobilized on nylon or nitrocellulose membranes (23). In this procedure the membrane-bound DNA or RNA is exposed to a solution of the labeled probe, and then autoradiographed using x-ray film to determine the binding of the probe. Often the immobilized nucleic acid is prepared by blotting specifically processed and separated fragments from gel or agarose electrophoresis slabs to the membrane. Probing such a blot with specific labeled DNA or RNA fragments can be used to identify the location of genes in DNA or to generate patterns which are unique to organisms or even individuals. Thus, this technique forms the basis for DNA fingerprinting, a useful method for typing tissues for organ transplants, for determining maternity or paternity relationships of an offspring with contested parents, or for forensic analysis of blood and other bodily fluids (see FORENSIC CHEMISTRY) (24).

Radioactive tracers also are used in agriculture. A test field containing a food crop is sprayed with either an organic fertilizer, pesticide, or fungicide that is laced with the appropriate radioactive tracer. Run off, leaching, or contamination of the water table can then be determined by measuring radioactivity in local ponds or rivers. Possible incorporation of these chemicals into the plants also is easily determined as is longevity or breakdown in the soil (see FERTILIZERS; FUNGICIDES, AGRICULTURAL; PESTICIDES; SOIL CHEMISTRY OF PESTICIDES).

In the petroleum industry, the size of an underground oil deposit is determined by the injection of radiolabeled substances into a well head. The occurrence of radioactivity in the oil–water mixture, which is pumped out of adjoining wells, gives an indication of the pocket size of the oil deposit (see PETROLEUM).

Concerns over safe handling of radioactive materials and issues around the cost and disposal of low level radioactive waste has stimulated the development of nonradiometric products and technologies with the aim of replacing radioactive tracers in research and medical diagnosis (25). However, for many of the applications described, radioactive tracer technology is expected to continue to be widely used because of its sensitivity and specificity when compared with colorimetric, fluorescent, or chemiluminescent detection methods.

BIBLIOGRAPHY

"Radioactive Drugs and Tracers" in *ECT* 2nd ed., Vol. 17, pp. 1–9, by S. Hopfan, Memorial Hospital for Cancer and Allied Diseases; "Radioactive Tracers" in *ECT* 3rd ed., Vol. 19, pp. 633–639, by R. E. O'Brien, New England Nuclear Corp.

1. G. Hevesy, *Radioactive Indicators*, Interscience Publishers, New York, 1948.
2. *Handbook of Radioactivity Measurements Procedures*, NCRP Report No. 58, 2nd ed., National Council on Radiation Protection and Measurement, Bethesda, Md., 1985.
3. *Radionuclide Transformations, Annals of the ICRP*, ICR Publication 38, Vol. 11–13, Pergamon Press, New York, 1983.
4. A. Murray and D. Williams, *Organic Synthesis With Isotopes*, Interscience Publishers, New York, 1958.
5. E. A. Evans, *Tritium and Its Compounds*, 2nd ed., John Wiley & Sons, Inc., New York, 1974.
6. Y. P. Wan and S. D. Hurt, in P. J. Marangos, I. C. Campbell, and R. M. Cohen, eds., *Brain Receptor Methodologies, Part A*, Academic Press, Inc., Orlando, Fla., 1984, pp. 21–32.
7. C. Filer, S. Hurt, and Y. P. Wan, in M. Williams, R. A. Glennon, and P. B. M. W. M. Timmermans, eds., *Receptor Pharmacology and Function*, Marcel Dekker, Inc., New York, 1989, pp. 105–135.
8. K. C. Tovey, G. H. Spiller, K. G. Oldham, and N. Lucas, *Biochem. J.* **142**, 47 (1974).
9. M. Calvin and J. A. Bassham, *The Photosynthesis of Carbon Compounds*, Benjamin, New York, 1962.
10. *Guide to the Self-Decomposition of Radiochemicals*, Amersham International plc, Amersham, U.K., 1992.
11. *DuPont Guide to Radiochemical Storage and Handling*, DuPont/NEN Research Products, Boston, Mass., 1990.
12. Y. Kobayashi and D. V. Maudsley, *Biological Applications of Liquid Scintillation Counting*, Academic Press, Inc., New York, 1974.
13. A. Rogers, *Techniques of Autoradiography*, Elsevier Scientific Publishing Co., Amsterdam, the Netherlands, 1979.
14. E. A. Evans, D. C. Warrell, J. A. Elvidge, and J. R. Jones, *Handbook of Tritium NMR Spectroscopy and Applications*, John Wiley & Sons, Inc., New York, 1985.

15. A. L. Burlingame, D. S. Millington, D. L. Norwood, and D. H. Russell, *Anal. Chem.* **62**, 268R (1990).

16. "Standards for Protection Against Radiation," *Code of Federal Regulations*, Title 10, Part 20, Washington, D.C., 1994.

17. K. J. Connor and I. S. McLintock, *Radiation Protection Handbook for Laboratory Workers*, HHSC Handbook No. 14, H and M Scientific Consultants, Ltd., Leeds, U.K., 1994.

18. J. Shapiro, *Radiation Protection*, 3rd ed., Harvard University Press, Cambridge, Mass., 1990.

19. *Health Physics Surveys for By-Product Material at NRC Licensed Processing and Manufacturing Plants*, U.S. Nuclear Regulatory Commission Regulatory Guide 8.21, Office of Standards Development, Washington, D.C., 1979.

20. *Health Effects of Exposure to Low Levels of Ionizing Radiation BEIR V*, National Research Council, National Academy Press, Washington, D.C., 1990.

21. M. Herkenham, in Ref. 6, pp. 127–152.

22. J. Sambrook, E. F. Fritsch, and T. Maniatis, *Molecular Cloning*, 2nd ed., Cold Spring Harbor Laboratory Press, Cold Spring Harbor, N.Y., 1989, Chapt. 13.

23. Ref. 22, Chapt. 10 and 11.

24. G. H. Keller and M. M. Manak, *DNA Probes*, 2nd ed., Stockton Press, New York, 1993.

25. E. Party and E. L. Gershey, *Health Phys.* **69**, 1 (1995).

LASZLO BERES
DuPont–NEN Products

RADIOCHEMICAL ANALYSIS AND TRACER APPLICATIONS.
See RADIOACTIVE TRACERS.

RADIOCHEMICAL TECHNOLOGY. See RADIATION CURING.

RADIOGRAPHY. See X-RAY TECHNOLOGY.

RADIOIMMUNOASSAY. See IMMUNOASSAY; MEDICAL DIAGNOSTIC REAGENTS; RADIOACTIVE TRACERS.

RADIOISOTOPES

A radioisotope is an atom the nucleus of which is not stable and which decays to a more stable state by the emission of various radiations. Radioactive isotopes, also called nuclides or radionuclides, are important to many areas of scientific research, as well as in medical and industrial applications (see RADIOACTIVE TRACERS; RADIOPHARMACEUTICALS).

The study of radioactive isotopes has revealed much information about the forces of nature. Prior to the twentieth century, the gravitational force that exists between bodies having mass and the electromagnetic force that causes charged and magnetic bodies to attract or repel each other were known. Studies of radioactive decay have led to the understanding of two more forces: the strong force between protons and neutrons which holds nuclei together; and the weak force, responsible for radioactive decay by the β-decay process. Advances in knowledge continue in several areas, including studies of the rare decay modes of double β-decay and cluster emission.

Radioactive isotopes have found application in many fields of scientific research besides nuclear physics. Naturally occurring isotopes are used to determine the ages of, for example, trees that lived a few thousand years ago and rocks that were formed a few million years ago. Radioisotopes can be inserted into molecules and tracked as they proceed through a biological process, through the measuring of the radiations emitted. Similarly, chemical and physical processes can be studied by the addition of radioisotopes as tracers in forms that mimic the regular constituents of the process. Measurements of the radiation levels can lead to determination of the distribution of the chemical in the process. Radioisotopes have also been injected into underground water systems in order to determine the flow patterns and rates in geological formations.

There are many industrial applications for radioisotopes. A simple example is the determination of the thickness of some item that is being formed, or of the density of a flow stream in a chemical process, by measuring the attenuation of β- or γ-radiation as it passes through the material. Another version of this type of density measurement is the search for cracks or thin spots in welds by measuring the variation in the intensity of the radiation passing through it. Radioisotopes are also employed to analyze fine arts objects (see FINE ART EXAMINATION AND CONSERVATION; NONDESTRUCTIVE EVALUATION).

Radioisotopes have become very important in the practice of modern medicine, for both diagnosis and treatment. Some diagnoses are done by injecting a radionuclide in a biochemical form such that it goes to a particular organ, and the measured radiation then allows the functional level of that organ to be determined. A common treatment is to expose a portion of the body, for example a tumor, to radiation from a radioisotope with the source either internal or external to the body. Another usage involves radioactively labeled antibodies (see IMMUNOASSAY).

Some of the heavy radioisotopes, namely those of uranium and plutonium, are used as the fuel in nuclear reactors (qv) which are used by commercial power companies to produce electricity. These radioisotopes have also been used as the critical components in nuclear weapons.

Knowledge about the radiations from each isotope is important because as the uses of the radioisotopes have increased, it has become necessary to develop sensitive and accurate detection methods designed to determine both the presence of these materials and the amount present. These measurements determine the amount of radiation exposure of the human body or how much of the isotope is present in various places in the environment. For a discussion of detection methods used see References 1 and 2.

History

In 1895 Roentgen observed that penetrating radiation was produced by his Crookes tube, a device in which electrons were accelerated and then stopped in a target. Unlike visible light, this radiation passed through paper (qv), wood (qv), and even thin metal sheets. The name x-rays was applied to it because the nature of the radiation was unknown, and that name has been used for over a hundred years. This new phenomenon caused great excitement among scientists and the general public. Members of the medical profession immediately made use of it to study the broken and malformed bone structures of their patients, as well as to attempt to treat tumors and other medical problems. It soon became clear that x-rays were useful in industry to determine the internal structure of manufactured items. There was great fascination with this radiation, as for instance in the entirely novel ability to observe one's own bone structure (see X-RAY TECHNOLOGY).

Hazards were discovered by experience. It was soon found that x-rays produced skin burns and lesions that healed only slowly; later it was learned that tumors could be produced. Within the medical community, it became clear that guidelines were needed for the use of x-rays, especially for medical workers, who were often present while the x-rays were produced.

In 1896, only a few months after Roentgen announced the observation of x-rays, Becquerel reported the additional observation of penetrating radiation emitted from certain natural materials, a phenomenon that Marie Curie would later name radioactivity. This phenomenon had a much less glamorous development. Over a three-year period, Becquerel published three articles, decided there was little else to learn about it, and went on to the study of other fields. During this period, only a few other articles were published on radioactivity and radioisotopes, but hundreds of articles were published on x-rays.

The uranium ores from which this new radiation was discovered were fluorescent, and x-ray tubes fluoresced; thus, an early hypothesis was that the visible fluorescence and the new penetrating radiation were related and would occur together. Becquerel did a series of careful experiments showing that penetrating radiation also came from some materials that did not fluoresce.

There were two immediate questions to be investigated concerning this radioactivity. The first question, about the nature of the radiation, was mainly a physics problem; the second, from where did the radiation come, was addressed by chemical methods. Ernest Rutherford and associates led the physics studies with the early discovery of two types of rays, called alpha (α) and beta (β). These were easily distinguished by the fact that the α-rays were stopped by paper or a thin foil, whereas the β-rays were more penetrating. In 1900, Paul Villard discovered the presence of an even more penetrating radiation, called gamma (γ)-rays. In the early chemical studies of radioactivity, the α-rays were very useful for measuring the amount of radioactive material present because these produced a strong response in the ionization detectors available at that time.

The more penetrating β-rays were easily studied. In 1899 their direction of deflection in a magnetic field was observed, indicating the negative charge. Then Becquerel was able to deflect β-rays in electric and magnetic fields and thereby determined the charge-to-mass ratio. This ratio showed that the mass was much smaller than that of any atom and corresponded to that of electrons.

In 1903, Rutherford and associates were finally able to deflect the α-rays by electric and magnetic fields, showing that these are positively charged. Measurement of the charge-to-mass ratio indicated that α-rays were of atomic dimensions. In 1908 definitive experiments showed α-rays to be doubly charged helium atoms, ie, helium nuclei.

The nature of γ-rays, basically the same as x-rays, was only slowly determined. Crystal diffraction had been used to show that x-rays are part of the electromagnetic spectrum and to measure their wavelengths. Because γ-rays are generally of a much higher energy, ie, a shorter wavelength, it was necessary to improve the crystal diffraction method before it could be used in 1913 to show that γ-rays are also electromagnetic radiation. The relationship of the x- and γ-rays to the other regions of electromagnetic radiation is illustrated in Figure 1. In some portions of the spectrum it is useful to discuss radiation in terms of its continuous wave properties of frequency and wavelength. For γ- and x-rays, however, it is usually more useful to use the discrete photon properties, such as the energy of a single photon.

It was shown early on that uranium and thorium are radioactive, independent of their chemical and physical form. At the same time, it was shown that some ores had more activity than others for a given amount of uranium. Starting with knowledge of the fact that the mineral pitchblende has a higher specific radioactivity, defined as the activity per unit mass as exemplified by the rate of discharge of ionization detectors per gram of material, than other uranium ores, Marie Curie chemically processed tons of the ore in order to extract portions having different chemical properties. She and her associates were able to extract a fraction that was chemically similar to bismuth, to be named polonium, and another fraction that was chemically like barium, to be called radium. From repeated purification steps, she was able to produce a very pure radium fraction that had 10^6 times the specific radioactivity of uranium, and from optical

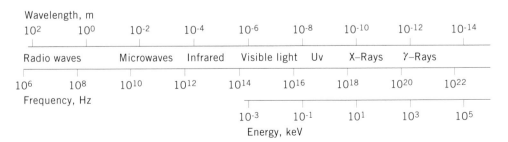

Fig. 1. Electromagnetic spectrum (1).

spectrometry, others were able to show that her samples were essentially pure radium. Measurement of the atomic mass as 225 gave the first proof that radium was not a new form of barium, which has a mass of about 137, but rather a new element having a mass between that of lead (qv) and uranium (see URANIUM AND URANIUM COMPOUNDS).

The same chemical separation research was done on thorium ores, leading to the discovery of a completely different set of radioactivities. Although the chemists made fundamental distinctions among the radioactivities based on chemical properties, it was often simpler to distinguish the radiation by the rate at which the radioactivity decayed. For uranium and thorium the level of radioactivity was independent of time. For most of the radioactivities separated from these elements, however, the activity showed an observable decrease with time and it was found that the rate of decrease was characteristic of each radioactive species. Each species had a unique half-life, ie, the time during which the activity was reduced to half of its initial value.

By this time, the Periodic Table of elements was well developed, although it was considered a function of the atomic mass rather than atomic number. Before the discovery of radioactivity, it had been established that each natural element had a unique mass; thus it was assumed that each element was made up of only one type of atom. Some of the radioactivities found in both the uranium and thorium decays had similar chemical properties, but because these had different half-lives it was assumed that there were different elements. It became clear, however, that if all the different radioactivities from uranium and thorium were separate elements, there would be too many to fit into the Periodic Table.

To resolve this problem two concepts needed to be developed. The first, accepted in 1910, was that in a radioactive decay the original element is transmuted to another element and upon β^--decay the decaying element moves up one unit in the Table, and that upon α-decay, the element moves down two units. The second concept was that for each element there can be atoms having different nuclear properties. These were first called isotopes in 1913. Using these concepts, the various natural decay chains began to fall into place. The uranium and thorium chains are shown in Tables 1 and 2. Although early researchers had reason to believe that only elements above Pb in the Periodic Table were radioactive, the development of far more sensitive measurement methods has shown that natural radioactivity exists throughout the Periodic Table. Natural radioisotopes, excluding the actinides (see ACTINIDES AND TRANSACTINIDES), are listed in Table 3.

Nuclear Physics Properties

The Atom.   Before the studies of Rutherford, there were several models of the atom. It was clear that the atom contained electrons because these could be extracted from the atoms. In order to have the whole atom be electrically neutral, there had to be an equal amount of positive charge. Because the only forces known were gravity and electromagnetism, atomic models had to be carefully contrived to have an assembly of positive and negative charges that would stay together. From the scattering of α-particles as these struck thin metal foils, it was deduced that almost all of the mass of the atom along with the positive

Table 1. Naturally Occurring Radioisotopes in the ^{238}U Chain

Parent[a]	Original name	Half-life	Decay mode[b]	Branching, %	Daughter
^{238}U		4.468×10^9 yr	α	100	
^{234}Th	U X1	24. 10 d	β^-	100	
^{234m}Pa	U X2	1. 17 min	IT	0.13	^{234}Pa
			β^-	99.87	^{234}U
^{234}Pa		6.70 h	β^-	100	
^{234}U	U II	2.445×10^5 yr	α	100	
^{230}Th	ionium	7.578×10^4 yr	α	100	
^{226}Ra	radium	1600 yr	α	100	
^{222}Rn	Ra emanation	3.8235 d	α	100	
^{218}Po	RaA	3.05 min	α	99.980	^{214}Pb
			β^-	0.020	^{218}At
^{218}At		1.6 s	α	99.9	^{214}Bi
			β^-	0.1	^{218}Rn
^{218}Rn		35 ms	α	100	^{214}Po
^{214}Pb	RaB	26.8 min	β^-	100	^{214}Bi
^{214}Bi	RaC	19.9 min	β^-	99.979	^{214}Po
			α	0.021	^{210}Tl
^{214}Po	RaC'	164.3 μs	α	100	^{210}Pb
^{210}Tl	RaC''	1.30 min	β^-	100	^{210}Pb
^{210}Pb	RaD	22.3 yr	β^-	99.9999983	^{210}Bi
			α	0.0000017	^{206}Hg
^{210}Bi	RaE	5.013 d	β^-	99.99987	^{210}Po
			α	0.00013	^{206}Tl
^{210}Po	RaF	138.378 d	α	100	^{206}Pb
^{206}Hg		8.2 min	β^-	100	^{206}Tl
^{206}Tl		4.2 min	β^-	100	
^{206}Pb	RaG	c			

[a]Each radioisotope decays to the next entry in the table, unless otherwise noted in the last column.
[b]IT = internal transition.
[c]Stable.

Table 2. Naturally Occurring Radioisotopes in the ^{232}Th Chain

Parent[a]	Original name	Half-life	Decay mode	Branching, %	Daughter
^{232}Th		1.405×10^{10} yr	α	100	
^{228}Ra	mesothorium I	5.75 yr	β^-	100	
^{228}Ac	mesothorium II	6.13 h	β^-	100	
^{228}Th	radiothorium	1.9131 yr	α	100	
^{224}Ra	thorium X	3.66 d	α	100	
^{220}Rn	thoron	55.6 s	α	100	
^{216}Po	ThA	0.145 s	α	100	
^{212}Pb	ThB	10.64 h	β^-	100	
^{212}Bi	ThC	60.55 min	β^-	64.06	^{212}Po
			α	35.94	^{208}Tl
^{212}Po	ThC'	0.298 s	α	100	
^{208}Tl	ThC''	3.053 min	β^-	100	
^{208}Pb	ThD	b			

[a]Each radioisotope decays to the next one in the table, unless otherwise noted in the last column.
[b]Stable.

Table 3. Naturally Occurring Radioisotopes[a]

Radioisotopes	Abundance, %	Half-life, yr	Decay mode[b]	Q-value[c], keV	Stable daughter
^{40}K	0.012	1.277×10^9	β^-	1311.6	^{40}Ca
			ϵ	1505.0	^{40}Ar
^{87}Rb	27.83	4.80×10^{10}	β^-	273.3	^{87}Sr
^{113}Cd	12.2	9.3×10^{15}	β^-	322	^{113}In
^{115}In	95.7	4.41×10^{14}	β^-	495	^{115}Sn
^{123}Te	0.89	$>10^{13}$	ϵ	52	^{123}Sb
^{138}La	0.089	1.35×10^{11}	β^-	1041	^{138}Ce
			ϵ	1749	^{138}Ba
^{144}Nd	23.8	2.4×10^{15}	α	1910.3	^{140}Ce
^{147}Sm	15.1	1.06×10^{11}	α	2310.5	^{143}Nd
^{148}Sm	11.3	8×10^{15}	α	1986.2	^{144}Nd
^{152}Gd	0.20	1.08×10^{14}	α	2206.2	^{148}Sm
^{174}Hf	0.16	2.0×10^{15}	α	2504	^{170}Yb
^{176}Lu	2.6	3.60×10^{10}	β^-	1186.5	^{176}Hf
^{180}Ta	0.012	$>10^{13}$	β^-	710	^{180}W
			ϵ	865	^{180}Hf
^{186}Os	1.58	2.0×10^{15}	α	2816.5	^{182}W
^{187}Re	62.60	5×10^{10}	β^-	2.64	^{187}Os
^{190}Pt	0.013	6×10^{11}	α	3243	^{186}Os

[a]Excluding radon, radium, actinides, and transactinides.
[b]ϵ = electron capture.
[c]Q-value = decay energy.

charge was contained in a very small volume called the nucleus. This nucleus has a diameter of about 10^{-12} cm, compared to about 10^{-8} cm for that of an atom. It has been found that for all isotopes the diameter of the nucleus is about $2.4 \times 10^{-13} A^{1/3}$·cm, where A is the atomic mass. This corresponds to a density on the order of 10^{14} g/cm^3. The space outside the nucleus, which is almost all of the atom's volume, is filled with a cloud of electrons.

This model had an immediate nuclear problem because the positive charges in the nucleus repel each other. The nucleus should thus blow itself apart. This model clearly required a new force to hold the particles in the nucleus together.

At the same time, the laws of classical electromagnetism suggested the opposite problem for the atom, owing to the attraction between the positively charged nucleus and the negatively charged electrons in the cloud. If these electrons were not in rapid motion around the nucleus, they would fall directly into the nucleus. And, if the electrons were in such motion, classical electromagnetism required that they radiate energy (called bremsstrahlung), lose their energy, and spiral into the nucleus. In either case the atom should collapse into the size of the nucleus. Therefore, some new laws of physics were needed to go along with the new nuclear force.

In 1913 Niels Bohr proposed a system of rules that defined a specific set of discrete orbits for the electrons of an atom with a given atomic number. These rules required the electrons to exist only in these orbits, so that they did not radiate energy continuously as in classical electromagnetism. This model was extended first by Sommerfeld and then by Goudsmit and Uhlenbeck. In 1925

Heisenberg, and in 1926 Schrödinger, proposed a matrix or wave mechanics theory that has developed into quantum mechanics, in which all of these properties are included. In this theory the state of the electron is described by a wave function from which the electron's properties can be deduced.

The structure of the particles inside the nucleus was the next question to be addressed. One step in this direction was the discovery of the neutron in 1932 by Chadwick, and the determination that the nucleus was made up of positively charged protons and uncharged neutrons. The number of protons in the nucleus is known as the atomic number, Z. The number of neutrons is denoted by N, and the atomic mass is thus $A = Z + N$. Another step toward describing the particles inside the nucleus was the introduction of two forces, namely the strong force that holds the protons and neutrons together in spite of the repulsion between the positive charges of the protons, and the weak force that produces the transmutation by β-decay.

The results of the theory of quantum mechanics require that nuclear states have discrete energies. This is in contrast to classical mechanical systems, which can have any of a continuous range of energies. This difference is a critical fact in the applications of radioactivity measurements, where the specific energies of radiations are generally used to identify the origin of the radiation. Quantum mechanics also shows that other quantities have only specific discrete values, and the whole understanding of atomic and nuclear systems depends on these discrete quantities.

Conservation of Energy. Because the naturally occurring radioactive materials continued to emit particles, and thus the associated energy, without any decrease in intensity, the question of the source of this energy arose. Whereas the conservation of energy was a firmly established law of physics, the origin of the energy in the radioactivity was unknown.

The actual source of this energy became clear using the postulate of Einstein that mass can be converted to energy, and vice versa, once knowledge of the nuclear masses involved became available. The specific relationship is $E = mc^2$ were E is the equivalent energy, m is the mass, and c is the velocity of light. Whereas this conservation law was challenged when the β-decay process was studied, it was finally shown that all types of radioactive decay result in a state of lower energy for the emitting atom and nucleus.

Angular Momentum and Parity. Another set of properties of each atomic and nuclear state and the associated radiations are the parity and the angular momentum, or spin. These are quantum mechanical properties the origins of which are not discussed here in detail, but their phenomenological results are important. The parity, which depends on the reflection symmetry of the wave function, is designated by Π and is either $+$ or $-$ (or $+1$ and -1) for any state. The spin or angular momentum is a vector quantity denoted by $\boldsymbol{J}$ for an atomic or nuclear level and by $\boldsymbol{L}$ for a γ-ray or other radiation. The lengths of these vectors, or the maximum component in any direction, are $[J(J + 1)]^{1/2}\hbar$ and $[L(L + 1)]^{1/2}\hbar$, where $\hbar = h/2\pi$ and $h =$ Planck's constant. The usual practice is to simply denote the length of $\boldsymbol{J}$ and $\boldsymbol{L}$ by J and L.

For any nuclear decay, such as the emission of a γ-ray, the angular momentum and parity must be conserved. Therefore, if $\boldsymbol{J}_i, \Pi_i$ and $\boldsymbol{J}_f, \Pi_f$ are the spins and parities of the initial and final levels, and $\boldsymbol{L}$ and Π_γ are the angular momentum and parity carried off by the γ-ray,

$$\boldsymbol{J}_i = \boldsymbol{J}_f + \boldsymbol{L}$$

and

$$\Pi_i = \Pi_f \Pi_\gamma$$

In terms of the magnitudes of the spin vectors, this means that

$$|J_f - L| \le J_i \le J_f + L$$

These rules have very distinct influences on the decays of nuclear states.

Properties of Particles. From the research of the early part of the twentieth century, the existence of several types of particles was firmly established, and the properties were determined. The particles that are involved in the decay of radioisotopes are given in Table 4. An additional type of conservation is that in all atomic and nuclear decays, the number of nucleons, ie, protons and neutrons, is conserved and the number of leptons, ie, electrons and neutrinos, is also conserved.

Another consequence of the quantum theory of the atomic and nuclear systems is that no two protons, or two neutrons, can have exactly the same wave function. The practical application of this rule is that only a specific number of particles can occupy any particular atomic or nuclear level. This prevents all of the electrons of the atom, or protons and neutrons in the nucleus, from deexciting to the single lowest state.

Half-Lives and Decay Constants. Each nuclear state, whether an unstable ground state or an excited level, has a characteristic probability of decay per unit time, λ, which is known as the decay constant. For a level that decays by more than one mode, each mode has a partial decay constant, λ_i, such that $\lambda = \Sigma_i \lambda_i$.

Very early in the study of radioactivity it was determined that different isotopes had different λ values. Because the laws of gravity and electromagnetism were deterministic, an initial concept was that when each radioactive atom was created, its lifetime was determined, but that different atoms were created having different lifetimes. Furthermore, these different lifetimes were created such

Table 4. Properties of Stable Particles Associated With Radioactive Decay

Name	Symbol	Energy, MeV	Charge	Spin
proton	p	938	+1	1/2
neutron[a]	n	940	0	1/2
electron	e^- or β^-	0.511	−1	1/2
positron	e^+ or β^+	0.511	+1	1/2
electron neutrino	ν	ca 0.0	0	1/2
electron antineutrino	$\bar{\nu}$	ca 0.0	0	1/2
photon	γ or x	0.0	0	1

[a]A free neutron has a half-life of 10.4 minutes, but a neutron is stable when bound in a nucleus.

that a collection of nuclei decayed in the observed manner. Later, as the probabilistic properties of quantum mechanics came to be accepted, it was recognized that each nucleus of a given radioactive species had the same probability for decay per unit time and that the randomness of the decays led to the observed decay pattern.

Experimental measurements have shown that the following description of the decay is correct. If at any time, t, there are a large number of nuclei, $N(t)$, in the same state which has the decay constant, λ, then the change in the number of nuclei in this state in a short time, dt, is

$$dN = \lambda N(t)\, dt$$

If N_0 is the number of nuclei in this state at time $t = 0$, the number of nuclei in this state at any later time, t, is

$$N(t) = N_0 e^{-\lambda t}$$

and the associated activity, or the decay rate, is

$$A(t) = |dN/dt| = \lambda N(t) = \lambda N_0 e^{-\lambda t}$$

In a description of nuclear properties, the half-life, $t_{1/2}$, is quoted rather than the decay constant. This quantity is the time it takes for one-half of the original nuclei to decay. That is,

$$N(t_{1/2}) = N_0/2 = N_0 e^{-\lambda t_{1/2}}$$

which has the solution $t_{1/2} = \ln 2/\lambda$. For the N_0 nuclei that exist at $t = 0$, the average or mean life, τ, is

$$\tau = \int_0^\infty N_0 e^{-\lambda t}\, dt = -\frac{1}{\lambda}\left[e^{-\lambda t}\right]_0^\infty = \frac{1}{\lambda}$$

During this decay process the number of daughter nuclei, D, is increasing at the same rate that the parent is decaying. So,

$$dD(t)/dt = +\lambda N(t) = +\lambda N_0 e^{-\lambda t}$$

If at $t = 0$ there are no daughter atoms, then

$$D(t) = N_0(1 - e^{-\lambda t})$$

and for t very large, all of the original radioactive atoms are converted to daughter atoms.

In almost all cases λ is unaffected by any changes in the physical and chemical conditions of the radionuclide. However, there are special conditions that can influence λ. An example is the decay of ^{7}Be that occurs by the capture of an atomic electron by the nucleus. Chemical compounds are formed by interactions between the outer electrons of the atoms in the compound, and different compounds have different electron wave functions for these outer electrons. Because ^{7}Be has only four electrons, the wave functions of the electrons involved in the electron-capture process are influenced by the chemical bonding. The change in the ^{7}Be decay constant for different compounds has been measured, and the maximum observed change is about 0.2%.

For any nuclide that decays only by this electron capture process, if one were to produce an atom in which all of the electrons were removed, the effective λ would become infinite. An interesting example of this involves the decay of ^{54}Mn in interstellar space. For its normal electron cloud, ^{54}Mn decays with a half-life of 312 d and this decay is by electron capture over 99.99% of the time. The remaining decays are less than 0.0000006% by β^+-decay and a possible branch of less than 0.0003% by β^--decay. In interstellar space some ^{54}Mn atoms have all of their electrons stripped off so they can only decay by these particle emissions, and therefore their effective half-life is greater than 3×10^5 yr.

In spite of these special cases, in all applied uses the decay constants and half-lives can be considered to be independent of the physical and chemical environment.

The above decay equations apply to the simple case of a radionuclide that is decaying without being replenished. There are many cases in which a nuclide is both being produced and decaying at the same time. One example is the case where one radioactive nuclide is produced by the decay of another nuclide (see Tables 1 and 2). If there are $N_1(0)$ atoms of nuclide 1, the parent, having decay constant λ_1, decays to nuclide 2, the daughter, having decay constant λ_2, then at $t = 0$ there are $N_2(0)$ such atoms present. Then

$$\frac{dN_1(t)}{dt} = -\lambda_1 N_1(t)$$

and

$$\frac{dN_2(t)}{dt} = +\lambda_1 N_1(t) - \lambda_2 N_2(t)$$

The solutions to these equations give the number of atoms of each nuclide that are present at time t as

$$N_1(t) = N_1(0)e^{-\lambda_1 t}$$

and

$$N_2(t) = N_2(0)e^{-\lambda_2 t} + [N_1(0)\lambda_1/(\lambda_2 - \lambda_1)](e^{-\lambda_1 t} - e^{-\lambda_2 t})$$

The activities of the two nuclides are then

$$A_1(t) = \lambda_1 N_1(0)e^{-\lambda_1 t}$$

and

$$A_2(t) = \lambda_2 N_2(0)e^{-\lambda_2 t} + [N_1(0)\lambda_2\lambda_1/(\lambda_2 - \lambda_1)](e^{-\lambda_1 t} - e^{-\lambda_2 t})$$

If one has a sample which at $t = 0$ contains only the parent, and the parent half-life is much longer than that of the daughter, the daughter activity grows exponentially and then decays as the parent decays. At values of t that are large compared to the daughter half-life, the activities reach a constant ratio and the daughter decays at a rate that is larger than the parent by the ratio $\lambda_2/(\lambda_2 - \lambda_1)$. This situation is called transient equilibrium. If the two half-lives are comparable, the ratio of the activities does not establish a useful equilibrium. If the parent half-life is much shorter than that of the daughter, the initial short-lived sample becomes a long-lived sample of just the daughter nuclei.

If there are more than two members of the decay sequence, as for the natural chains from ^{238}U and ^{232}Th, the activity of each member depends on the initial abundances and the half-lives of all of the previous members.

Atomic Levels and Their Decay. There are many commonalities between the properties of atomic and nuclear levels and between their respective decays. Each level has a quantum mechanical wave function which describes its properties. It is common practice to illustrate the atomic and nuclear information in terms of level diagrams.

The levels in which the individual atomic electrons reside are illustrated in Figure 2**a** for an atom of nickel, Ni, in its ground state. Atomic levels are commonly grouped by energy. The lowest energy, most tightly bound level is labeled K; the next group of three closely spaced levels is labeled L_i; etc. If energy from some external source is transferred to one of the two K-shell electrons, the electron can be excited to an upper state, or be completely removed from the atom, as illustrated in Figure 2**b**. The complete removal of an electron from shell X is referred to as ionization in the X shell. In either case the atom not in the ground state is in an excited state. The atom promptly returns to a lower energy state by having an outer electron transfer to the K shell. The vacancy that this creates in an outer shell is then filled by another electron from a still higher energy level. If the original excitation completely removes the K-shell electron, the atom must attract an electron from the surrounding material if it is to become electrically neutral and return to its ground state. This process is illustrated in Figure 2**c** by the emission of three consecutive x-rays.

The electrons in each atomic level have specific spins or angular momenta and parities. These properties define the characteristics of the level and the x-ray spectrum produced. Figure 3 illustrates a typical x-ray spectrum where the four peaks above channel 1000 are from the filling of a hole in the K shell by electrons from the L subshells, tagged by K_α, and the higher lying shells, tagged K_β. Once a single hole is created in the K shell, the energies and intensities of these four peaks are completely determined by the wave functions of the atomic electrons and not by the mode of excitation of the K-shell electron. The spin and parity of the electrons in the L_2 and L_3 subshells allow these electrons to fill the K-shell hole, but the spin of the electrons in the L_1 subshell make it very unlikely for them to fill a K-shell hole. The K_α region in Figure 3, therefore, has two peaks, not three.

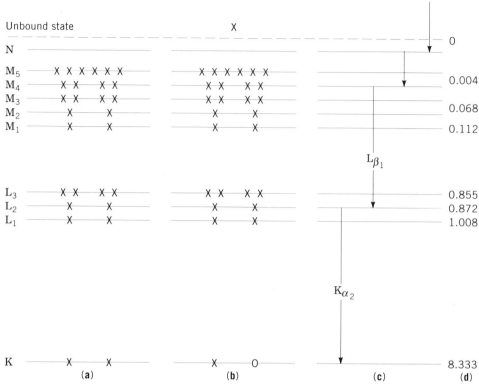

Fig. 2. Energy level diagram where K–N correspond to electron energy levels for an atom, X to electrons in a particular energy level, and 0 to an empty slot in an energy level (1). Above the dashed line is the unbound state. (**a**) An atom of Ni, 28 electrons, in the lowest energy or ground state; (**b**) an ion of Ni where on electron from the K level has been excited to the unbound state; (**c**) the process by which Ni returns to the ground state where each arrow represents a transition for an electron from one level to another; and (**d**) the energies of the levels in keV from which the energy of the emitted x-rays may be calculated.

Modes of Nuclear Decay and Radiation

The decay of radioisotopes involves both the decay modes of the nucleus and the associated radiations that are emitted from the nucleus. In addition, the resulting excitation of the atomic electrons, the deexcitation of the atom, and the radiations associated with these processes all play a role. Some of the atomic processes, such as the emission of K x-rays, are inherently independent of the nuclear processes that cause them. There are others, such as internal conversion, where the nuclear and atomic processes are closely related.

There are four modes of radioactive decay that are common and that are exhibited by the decay of naturally occurring radionuclides. These four are α-decay, β^--decay, electron capture and β^+-decay, and isomeric or γ-decay. In the first three of these, the atom is changed from one chemical element to another; in the fourth, the atom is unchanged. In addition, there are three modes of decay that occur almost exclusively in synthetic radionuclides. These are spontaneous fission, delayed-proton emission, and delayed-neutron emission.

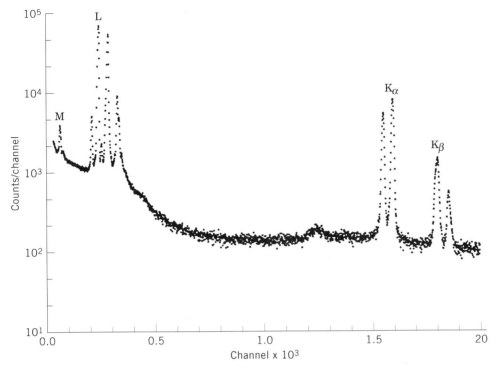

Fig. 3.   Photon spectrum, measured by using a planar Ge detector for x-rays emitted in the radioactive β-decay of ^{207}Bi.

Lastly, there are two exotic, and very long-lived, decay modes. These are cluster emission and double β-decay. In all of these processes, the energy, spin and parity, nucleon number, and lepton number are conserved. Methods of measuring the associated radiations are discussed in Reference 2; specific methods for γ-rays are discussed in Reference 1.

α-Decay.   In α-decay the parent atom of atomic number Z and mass A emits an α-particle, a ^{4}He nucleus having $Z = 2$ and $A = 4$, and becomes an atom having atomic number $Z - 2$ and mass $A - 4$. From the conservation of energy, the energy of the α-particle is

$$E(\alpha) = Q - E_1 - E_R$$

where Q is the energy equivalent of the difference in mass of the parent atom and the sum of the masses of the daughter atom and a ^{4}He atom, E_1 is the energy of the populated level in the daughter nucleus, and E_R is the recoil energy imparted to the atom. The recoil energy is

$$E_R = E(\alpha){\cdot}M_\alpha/(M - M_\alpha)$$

where M_α and M are the masses of an ^{4}He atom and the daughter atom, respectively. The parent often decays by α-emission to several distinct levels in the daughter nucleus. The α-particle can carry off any number of units of angular momentum or spin, so this decay can occur to any daughter states (except between 0^+ and 0^- states). The probability for decay to each daughter level depends on the specific properties of the corresponding wave functions, but the strongest decays usually occur to levels having the same spin as the parent level.

In this decay process, only one particle is emitted and, because energy is conserved, for each level in the daughter nucleus there is a unique α-particle energy. This means that a measurement of the differences in the energies of the α-particles emitted in a radioactive decay gives explicitly the differences in the energies of the levels in the daughter nucleus.

β^--Decay. In this decay, a β^--particle is emitted from the nucleus and the parent atom with Z and A is transmuted into the daughter atom of $Z + 1$ and A. The emitted β^--particle is an ordinary electron having the properties listed in Table 4. This decay mode is equivalent to converting a neutron into a proton. There can be β^--branches to the different levels in the daughter nucleus. If this decay were analogous to α-decay, it would result in a series of mono-energetic electron groups corresponding to the daughter levels. However, early experiments showed that these electrons had a continuous spectrum. This could be interpreted to mean either that the decay involved one or more unidentified particles, or that energy was not conserved.

Pauli proposed that two particles were emitted, and Fermi called the second one a neutrino, ν. The complete process therefore is $n \rightarrow p + e^- + \bar{\nu}$. Owing to the low probability of its interacting with other particles, the neutrino was not observed until 1959. Before the β^--decay takes place there are no free leptons, so the conservation of leptons requires that there be a net of zero leptons afterward. Therefore, the associated neutrino is designated an antineutrino, $\bar{\nu}$; that is, the emitted electron (lepton) and antineutrino (antilepton) cancel and give a net of zero leptons.

The decay energy is shared between the electron and the antineutrino and each particle has an energy distribution that extends from 0 to the total energy available, namely $Q - E_1$. The β-decay theory, developed by Fermi, provides a prediction of the energy distribution of the electrons from various classifications of β-decays, and these predictions are verified by the experiments. The β-decays are classified by the change in nuclear spin and parity as noted in Table 5. The specific probability for β^--decay to each of the levels of the daughter nucleus depends on the individual wave functions. However, the probability decreases rapidly as the available energy decreases, as well as by the factors shown in Table 5 for the various classifications. Thus, the strongest β-branches are those to levels near the ground state of the daughter and those that have the smallest spin changes between the parent and daughter levels.

The masses of the neutrinos have generally been considered to be exactly 0, but modern theory and some more recent experiments suggest the masses may be nonzero, but still on the order of 1 eV. Because the neutrinos have such a small mass and no electrical charge, they interact primarily by the weak interaction. This means that their interaction probability is very small and they typically pass through a mass as large as the earth without interacting. Therefore, they are not useful for any measurements related to radioactive decay.

Table 5. Classification of β-Decays[a]

Classification	ΔJ[b]	$\Delta\Pi$[c]	Relative emission probability
allowed	0,1	no	$\equiv 10^2$
1st forbidden	0,1	yes	10^1
unique 1st forbidden	2	yes	10^{-2}
2nd forbidden	2	no	10^{-5}
unique 2nd forbidden	3	no	10^{-7}
3rd forbidden	3	yes	10^{-11}
unique 3rd forbidden	4	yes	10^{-12}
4th forbidden	4	no	10^{-15}

[a]Values are for decay to a $Z = 60$ daughter nuclide and a decay energy of 500 keV and are estimated from data in Ref. 3.
[b]ΔJ = change in spin or angular momentum.
[c]$\Delta\Pi$ = change in parity. Parity is either + or −; change in parity either occurs or does not occur.

Electron Capture and β^+-Decay. These processes are essentially the inverse of the β^--decay in that the parent atom of Z and A transmutes into one of $Z - 1$ and A. This mode of decay can occur by the capture of an atomic electron by the nucleus, thereby converting a proton into a neutron. The loss of one lepton (the electron) requires the creation of another lepton (a neutrino) that carries off the excess energy, namely $Q - E_1 - E(e^-)$, where the last term is the energy by which the electron was bound to the atom before it was captured. So the process is equivalent to $p + e^- \rightarrow n + \nu$. The experimental signature for this process is the emission of x-rays from the atom as the resulting hole in an atomic shell is filled. If the decay energy to the level is well above the binding energy of the K electrons, the capture occurs from all atomic shells, but it is primarily from the K shell. If the decay energy to a level is below the K-shell binding energy, then capture only occurs from the higher atomic shells. The relative probability of capturing an electron from a specific atomic shell can be calculated theoretically, and some examples for the K and L shells are given in Table 6.

In this process only one particle is emitted, so the energy spectrum of the neutrinos consists of discrete lines and in principle the energies of the levels

Table 6. Ratios for K and L Electron Capture, and for β^+-Emission and Electron Capture[a]

Decay energy, keV	Classification	K/L capture	capture/β^+
500	allowed and 1st forbidden	6.2	no β^+
	unique 1st forbidden	4.9	
1500	allowed and 1st forbidden	6.9	286
	unique 1st forbidden	6.5	1900
2000	allowed	7.0	20.0
3000	allowed and 1st forbidden	7.1	1.96
	unique 1st forbidden	6.9	6.1
4000	allowed and 1st forbidden	7.1	0.56
	unique 1st forbidden	7.0	1.50

[a]Calculations are for $Z = 60$ daughter nuclide. Values are from computer code that calculates values from relations in Ref. 4.

in the daughter nucleus could be determined from this spectrum. However, the detection of neutrinos is very difficult, so this is not a practical possibility.

The same transmutation can also occur by positron, or β^+-emission, in which the process is $p \rightarrow n + \beta^+ + \nu$. As in β^- decay, the β^+ and ν share the available energy of $Q - E_1 - 2m_0c^2$, where m_0c^2 is the energy equivalent of the mass of an electron, or 511.0 keV. The $2\,m_0c^2$ term occurs because the conversion of the proton into a neutron effectively requires the creation of an electron, in addition to the positron that is created. The β^+-particle is the antiparticle of the electron having the properties listed in Table 4. Because the β^+ is an antilepton and the associated neutrino is a lepton, leptons are conserved. As in the case of β^--emission, the energy distribution of the β^+ for decay to any level is continuous from 0 to the available energy and the probability of this decay to each of the daughter levels depends on the available energy, the spins, and the parities. If the decay energy is more than 1022 keV, both electron capture and β^+-decay can occur to a level and the relative probabilities of the two can be calculated from the β-decay theory. Some examples of the theoretical electron-capture/β^+ ratios are included in Table 6.

γ-Decay. In the γ-decay mode, a nucleus in an excited state decays to a lower energy state via the emission of a γ-ray, and the Z and the A are unchanged. Although to first order the γ-ray carries off all of the available energy, a small amount is transferred to the atom, which recoils. This recoil energy is $E_r = 0.5368 \times 10^{-6}E_\gamma^2/A$, where A is the atomic mass and E_γ is the γ-ray energy; both energies are in keV. An isomeric transition or isomeric decay is a γ-ray transition from a nuclear state that has a significantly long half-life, for example longer than 1 μs.

Typically, γ-rays of interest in nuclear transitions have energies from ca 50 to 2500 keV, but some may be as low as 0.003 keV, which is near the visible light region (see Fig. 1), and as high as 12,000 keV. Extraterrestrial photons have been observed up to 10^{18} keV. As for all electromagnetic radiation, a γ-ray has electric and magnetic properties that are indicated by their character or multipolarity, commonly designated as E1, E2, E3, E4, M1, M2, M3, or M4, etc. Here the E and M stand for their electric and magnetic character and the number refers to the multipole order of the γ-ray, where 1 indicates a dipole, 2 a quadrupole, 3 an octupole, and 4 a hexadecapole transition. This number also indicates the units of angular momentum, L, carried off by the γ-ray. The transitions that do not involve a change in the parity are M1, E2, M3, and E4; the others do involve a change in parity. In many cases the character of a γ-ray is essentially of one type, but in other cases it can be made up of a mixture of two, or even three, types. Parity conservation means that these two component mixtures are of the type M1 + E2, E1 + M2, etc.

For specific models of the nucleus, it is possible to compute theoretical wave functions for the states. For a model that assumes that the nucleus is spherical, the general properties of these wave functions have been used to compute theoretical estimates of the half-lives for γ-rays of the various multipolarities. Some values from the Weisskopf estimate of these half-lives are shown in Table 7. These half-lives decrease rapidly with the γ-ray energy, namely, as E_γ^{2L+1} and, as Table 7 shows, increase rapidly with L. This theoretical half-life applies only to the γ-ray decay, so if there are other modes of decay of the associated level,

Table 7. Calculated Partial Lifetimes for γ-Ray Transitions[a,b]

	E_γ lifetimes, s			
Multipolarity	10 keV	100 keV	1,000 keV	10,000 keV
E1	3.1×10^{-10}	3.1×10^{-13}	3.1×10^{-16}	3.1×10^{-19}
E2	1.9×10^{-1}	1.9×10^{-6}	1.9×10^{-11}	1.9×10^{-16}
E3	$2.0 \times 10^{+8}$	$2.0 \times 10^{+1}$	2.0×10^{-6}	2.0×10^{-13}
E4	$3.0 \times 10^{+17}$	$3.0 \times 10^{+8}$	3.0×10^{-1}	3.0×10^{-10}
M1	2.2×10^{-8}	2.2×10^{-11}	2.2×10^{-14}	2.2×10^{-17}
M2	$1.4 \times 10^{+1}$	1.4×10^{-4}	1.4×10^{-9}	1.4×10^{-14}
M3	$1.4 \times 10^{+10}$	$1.4 \times 10^{+3}$	1.4×10^{-4}	1.4×10^{-11}
M4	$2.1 \times 10^{+19}$	$2.1 \times 10^{+10}$	$2.1 \times 10^{+1}$	2.1×10^{-8}

[a]Ref. 1.

[b]Weisskopf estimates for mass 100 and a few selected energies.

such as another γ-ray or internal conversion, the observed half-life of a level is expected to be shorter than this.

Decay Schemes. For nuclear cases it is more useful to show energy levels that represent the state of the whole nucleus, rather than energy levels for individual atomic electrons (see Fig. 2). This different approach is necessary because in the atomic case the forces are known precisely, so that the computed wave functions are quite accurate for each particle. For the nucleus, the forces are much more complex and it is not reasonable to expect to be able to calculate the wave functions accurately for each particle. Thus, the nuclear decay schemes show the experimental levels rather than calculated ones. This is illustrated in Figure 4, which gives the decay scheme for ^{60}Co. Here the lowest level represents the ground state of the whole nucleus and each level above that represents a different excited state of the nucleus.

The ^{60}Co nucleus decays with a half-life of 5.27 years by β^- emission to the levels in ^{60}Ni. These levels then deexcite to the ground state of ^{60}Ni by the emission of one or more γ-rays. The spins and parities of these levels are known from a variety of measurements and require that the two strong γ-rays of 1173 and 1332 keV both have E2 character, although the 1173 γ could contain some admixture of M3. However, from the theoretical lifetime shown in Table 7, the E2 contribution is expected to have a much shorter half-life and therefore also to dominate in this decay. Although the emission probabilities of the strong 1173- and 1332-keV γ-rays are so nearly equal that the difference cannot be determined by a direct measurement, from measurements of other parameters of the decay it can be determined that the 1332 is the stronger. Specifically, measurements of the continuous electron spectrum from the β^--decay have shown that there is a branch of 0.12% to the 1332-keV level. When this, the weak γ-rays, the internal conversion, and the internal-pair formation are all taken into account, the relative emission probabilities of the two strong γ-rays can be determined very accurately, as shown in Table 8.

A decay scheme for ^{131}I is shown in Figure 5. The low lying states of the daughter ^{131}Xe have widely varying spins and parities, from $1/2^+$ to $11/2^-$, resulting in a variety of multipolarities for the γ-rays. The half-life of the 341-keV level is 2.1×10^{-12} s and that of the 164-keV level is 11.9 d. This large range

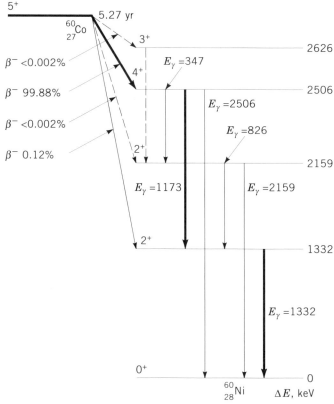

Fig. 4. Decay scheme of ^{60}Co as an example of β^--decay, showing the spins and parities of the levels populated in the daughter nucleus and the energies in keV of these levels, where (→) represents the principal decay mode, (→) an alternative mode, and (--→) is a highly improbable transition.

results from the strong dependence of the γ-ray lifetime on the γ-character. Specifically, a half-life of ps for the M1 + E2 γ-ray of 177 keV and of Ms for the 164-keV M4 γ-ray. ^{131}I is often of interest both in medical diagnostics and as a monitor for release of fission products from a nuclear reactor. In these applications it may be important to take into account that the parent has a half-life of 8 d, while the half-life of the 164-keV level is 11 d. Therefore, the latter γ-ray is not in equilibrium with the parent.

Spontaneous Fission. The spontaneous splitting of a heavy nucleus into two large fragments, usually accompanied by release of a few neutrons, is termed spontaneous fission. A typical example is as follows:

$$^{252}_{97}\text{Cf} \longrightarrow ^{144}_{55}\text{Cs} + ^{106}_{42}\text{Mo} + 2\,^{1}_{0}n$$

This process occurs only for very heavy nuclei where the mass of the original nucleus is significantly larger than that of all the products. This excess mass appears as kinetic energy and in the excitation energy of the products. The large

Table 8. γ-Ray Energies, E_γ, and Emission Probabilities, P_γ, for ^{24}Na and ^{60}Co[a]

Parent nuclide	E_γ, keV	$P_\gamma{}^b$, %	Multipolarity	ICC[c] $\times 10^3$	IPFC[b,d,e] $\times 10^3$	Total transition probability,[b] %
^{24}Na	1997.2	0.0015 (3)	M1, E2	0.017	0.0	0.0015 (3)
	1368.626	99.9936 (4)	E2	0.010	0.045 (4)	99.9991 (4)
	2754.007	99.872 (7)	E2	0.0028	0.68 (4)	99.940 (7)
	2870.4	0.00028 (6)	M1 + E2	0.0026		0.00028 (6)
	3867.5	0.057 (8)	M1 + E2	0.0015		0.057 (8)
	4237.9	0.0009 (2)	E2	0.0016		0.0009 (2)
^{60}Co	347.13	0.0075 (4)	E2	5.54	0.0	0.0075 (4)
	826.10	0.0076 (8)	M1 + E2	0.330	0.0	0.0076 (8)
	1173.228	99.85 (3)	E2	0.168	0.0062	99.87 (3)
	1332.490	99.9826 (7)	E2	0.128	0.034	99.9988 (7)
	2158.57	0.0012 (2)	E2	0.0491		0.0012 (2)
	2505.690	0.0000020 (4)	E4	0.086		0.0000020 (4)

[a]Refs. 5 and 6.
[b]Uncertainty in last digit or digits is shown in parentheses.
[c]ICC = internal conversion coefficient.
[d]IPFC = internal-pair formation coefficient.
[e]Values not given are small compared to the uncertainty in P_γ.

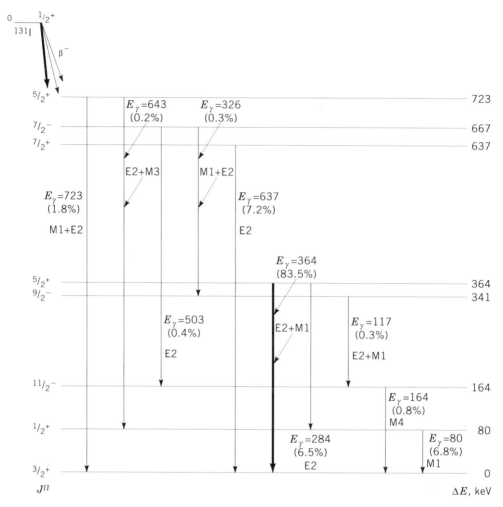

Fig. 5. Decay scheme of ^{131}I, showing the energies, spins, and parities of the levels populated in the daughter nucleus, ^{131}Xe, and the energies in keV, emission probabilities (in %), and multipolarities of the γ-ray transitions. There is a strong dependence of the γ-ray lifetime on the γ-character. The M1 + E2 γ-ray of 177 keV has a half-life of 2.1 ps; the half-life of the 164-keV M4 γ-ray is 1.03×10^6 s.

fragments emit γ-rays until the ground states are reached. The resulting fission-product nuclei are generally radioactive also. For example, in the subsequent β^--decay chains of $^{144}_{55}$Cs, the corresponding half-lives are ^{144}Cs (1 s) $\rightarrow$ ^{144}Ba (11 s) $\rightarrow$ ^{144}La (40 s) $\rightarrow$ ^{144}Ce (284 d) $\rightarrow$ ^{144m}Pr (7 min) $\rightarrow$ ^{144}Pr (17 min) $\rightarrow$ ^{144}Nd (stable), and ^{106}Mo (8 s) $\rightarrow$ ^{106}Tc (36 s) $\rightarrow$ ^{106}Ru (372 d) $\rightarrow$ ^{106}Rh (29 s) $\rightarrow$ ^{106}Pd (stable). This decay mode is observed in several U (Z = 92) and Pu (Z = 94) isotopes and many of the more neutron-rich isotopes of the elements having higher atomic number. Except for U, all of these elements are synthetic.

Delayed Proton and Neutron Decays. By means of a variety of nuclear reactions, as well as the spontaneous fission of synthetic nuclides, large numbers of isotopes of some elements have been produced. For example, whereas the only stable isotope of Cs ($Z = 55$) is ^{133}Cs ($N = 78$), all of the Cs isotopes from ^{114}Cs where $N = 59$ and $t_{1/2} = 0.57$ s, to ^{148}Cs where $N = 93$ and $t_{1/2} = 0.13$ s, have been observed. At the low mass end of this series, the last proton is only loosely bound, and at the high mass end, the last neutron is only loosely bound.

Although protons and neutrons are not emitted from the ground states of these isotopes, there are many cases where particles are emitted from excited states. For example, ^{114}Cs decays by electron capture and β^+-emission to excited levels in ^{114}Xe, and in 7% of these cases protons are emitted from the excited states which thereby populate levels in ^{113}I. In the other 93% of the decays, γ-deexcitation leads to the ^{114}Xe ground state. At the other extreme, ^{147}Cs decays by β^--decay to excited states in ^{147}Ba, and in as much as 43% of these cases neutrons are emitted, which thereby populate levels in ^{146}Ba. In the other 57% of the decays, γ-deexcitation leads to the ^{147}Ba ground state.

Alternatives to γ-Ray Emission. γ-Ray emission results in the deexcitation of an excited nuclear state to a lower state in the same nuclide, ie, no change in Z or A. There are two other processes by which this transition can take place without the emission of a γ-ray of this energy. These are internal conversion and internal pair formation. The internal-conversion process involves the transfer of the energy to an atomic electron.

Internal-Pair Formation. The process of internal-pair formation occurs only rarely. It provides a good demonstration of the equivalence of mass and energy as proposed by Einstein. For γ-transitions of less than 1022 keV, this process cannot occur; but for transitions of over 1022 keV, a small fraction of the decays occur by the creation of an electron–positron pair. Of the available energy, 1022 keV is used to create the two particles, and the remainder appears as kinetic energy of the two particles. The probability of the internal-pair formation (IPF) is given by the ratio of the number of pairs emitted to the number of γ-rays emitted. Table 9 gives the theoretical ratios for a few γ-ray energies, multipolarities, and atomic numbers. These IPF coefficients (IPFC) increase with E_γ, in contrast to internal-conversion coefficients, which decrease with increasing energy. From the small size of the ratios it is clear that this effect is generally not important. However, it is of interest in very precise computations of some γ-ray emission probabilities and in the study of this effect itself. One case where the values are important is the calculation of the γ-ray emission probabilities for the decay of ^{24}Na, as shown

Table 9. Internal-Pair Formation Coefficients[a,b]

E_γ, keV	Coefficient			
	E1, $Z = 20$	E1, $Z = 50$	M1, $Z = 20$	M1, $Z = 50$
1100	0.0000132	0.0000037	0.0000007	0.0000005
1400	0.000190	0.000156	0.0000392	0.0000485
1800	0.000496	0.000458	0.000168	0.000206
2600	0.00102	0.000992	0.000504	0.000587
4000	0.00168	0.00166	0.00104	0.00116
8000	0.00271	0.00270	0.00199	0.00212

[a] Ref. 7.
[b] The ratio of the number of e^- + β^+ pairs to the number of γ-rays emitted.

in Table 8, where these coefficients are larger than the corresponding internal conversion coefficients. In contrast, for ^{60}Co these coefficients are smaller than the internal-conversion coefficients and are therefore much less significant.

Exotic Decays. In addition to the common modes of nuclear decay, two exotic modes have been observed. These decay modes are of theoretical interest because their long half-lives place strict constraints on the details of any theory used to calculate them.

Cluster emission is an exotic decay that has some commonalities with α-decay. In α-decay, two protons and two neutrons that are moving in separate orbits within the nucleus come together and leak out of the nucleus as a single particle. Cluster emission occurs when other groups of nucleons form a single particle and leak out. Several of the observed decays are shown in Table 10. The emitted clusters include ^{14}C, ^{20}O, ^{24}Ne, ^{28}Mg, and ^{32}Si. The rare nature of these decays is indicated by the ratio of their probability to that of α-decay. These ratios, the inverse of the values in the last column, are 10^{-9} to 10^{-16}.

Double β-decay is the other class of exotic decay. The members of this class that are of interest are those in which the parent nuclide is stable against the single β-decay because the corresponding decay energy is negative, or because the single β-decay is very highly hindered by spin-selection rules. Table 11 shows three cases where double β-decay has been observed and others where a significant lower limit has been established for the corresponding half-life.

Table 10. Exotic Radioactive Decay and Emissions of Clusters

			Half-life[a], yr		
Parent	Cluster	Daughter	Cluster decay	α-Decay[b-d]	$t_{1/2\,\text{cluster}}/t_{1/2\,\text{d}}$
^{221}Fr	^{14}C	^{207}Tl	9.2×10^{6}	9.3×10^{-6}	9.9×10^{11}
^{221}Ra	^{14}C	^{207}Pb	4.8×10^{5}	8.9×10^{-7}	5.4×10^{11}
^{222}Ra	^{14}C	^{208}Pb	3.2×10^{3}	1.2×10^{-6}	2.7×10^{9}
^{223}Ra	^{14}C	^{209}Pb	4.4×10^{7}	3.1×10^{-2}	1.4×10^{9}
^{224}Ra	^{14}C	^{210}Pb	2.6×10^{8}	1.0×10^{-2}	2.6×10^{10}
^{225}Ac	^{14}C	^{211}Bi	4.4×10^{9}	2.7×10^{-2}	1.6×10^{11}
^{226}Ra	^{14}C	^{212}Pb	7.0×10^{13}	1.6×10^{3}	4.4×10^{10}
^{228}Th	^{20}O	^{208}Pb	1.7×10^{13}	1.9	8.9×10^{12}
^{230}Th	^{24}Ne	^{206}Hg	1.3×10^{17}	7.5×10^{4}	1.7×10^{12}
^{231}Pa	^{23}F	^{208}Pb	3.3×10^{18}	3.3×10^{4}	9.7×10^{13}
^{231}Pa	^{24}Ne	^{207}Tl	2.5×10^{15}	3.3×10^{4}	9.7×10^{10}
^{232}U	^{24}Ne	^{208}Pb	7.9×10^{12}	6.9×10^{1}	1.1×10^{11}
^{233}U	^{24}Ne/^{25}Ne	^{209}Pb/^{208}Pb	2.2×10^{17}	1.6×10^{5}	1.4×10^{12}
^{234}U	^{28}Mg	^{206}Hg	1.7×10^{18}	2.5×10^{5}	6.8×10^{12}
^{234}U	^{24}Ne/^{26}Ne	^{210}Pb/^{208}Pb	2.7×10^{18}	2.5×10^{5}	1.1×10^{13}
^{235}U	^{24}Ne/^{25}Ne	^{211}Pb/^{210}Pb	8.2×10^{19}	7.0×10^{8}	1.2×10^{11}
^{236}Pu	^{28}Mg	^{208}Pb	1.5×10^{14}	2.9	5.2×10^{13}
^{238}Pu	^{32}Si	^{206}Hg	6.0×10^{17}	8.8×10^{1}	6.8×10^{15}
^{238}Pu	^{28}Mg/^{30}Mg	^{210}Pb/^{208}Pb	4.7×10^{17}	8.8×10^{1}	5.3×10^{15}

[a]Ref. 8, unless otherwise specified.
[b]Ref. 9. [c]Ref. 10.
[d]Ref. 11.

Table 11. Experimental Half-Life Results for Double β-Decay

Parent	Daughter	β-Decay energy, keV	Mode[a]	Half-life,[b] yr	$\beta\beta$-Decay energy[c], keV
^{48}Ca	^{48}Ti	$+289$	$\beta\beta_0$	$>2 \times 10^{21}$	4267
			$\beta\beta_2$	$>3 \times 10^{19}$	
^{76}Ge	^{76}Se	-923	$\beta\beta_0$	$>5 \times 10^{21}$	2045
			$\beta\beta_2$	1.2×10^{21c}	
^{82}Se	^{82}Kr	-89	$\beta\beta_0 + \beta\beta_2$	2×10^{20}	3003
			$\beta\beta_0$	$>3 \times 10^{21}$	
^{130}Te	^{130}Xe	-407	$\beta\beta_0 + \beta\beta_2$	2×10^{21}	2543

[a]The symbol $\beta\beta_2$ signifies the emission of 2 βs and 2 neutrinos, and $\beta\beta_0$ signifies the emission of 2 βs and no neutrinos.
[b]Ref. 12, unless otherwise indicated.
[c]Ref. 13.

The decay energy indicates that for ^{48}Ca the single β-decay is possible, but it would be between states with $J^{II} = 0^+$ and $J^{II} = 6^+$ and this large spin change results in a very long half-life. In the other cases the single β-decay is forbidden by energy conservation.

The observation of double β-decay has been of particular interest because there are different concepts of how this decay can occur. The theoretical half-life depends on what concept is correct. Also, the predictions of the theory influence not only the design of the experiments used to observe the decay, but also the interpretation.

From the discussion on lepton conservation, it follows that the two emitted βs would be accompanied by two antineutrinos. This case is denoted $\beta\beta_2$. However, there is an alternative theoretical mechanism by which the two βs could be emitted without any neutrino, denoted $\beta\beta_0$. The experimental methods that are used to look for the double β-decay mode are often more sensitive to one of these decay modes than the other. The difference in the expected energy distribution of the electrons is clear from the fact that in the first case the total decay energy is divided between four particles, including the two antineutrinos that cannot be observed; in the second, it is only divided between the two electrons. As more exotic modes of decay are measured and even larger limits are placed on some of the half-lives, the constraints on theory become even stronger.

Combined Nuclear and Atomic Processes

There are two processes where nuclear and atomic contributions are interrelated. These are the emission of electrons from the atomic shells as an alternative to the emission of a photon and the emission of bremsstrahlung photons in the β-decay process.

Internal Conversion. As an alternative to the emission of a γ-ray, the available energy of the excited nuclear state can be transferred to an atomic electron and this electron can then be ejected from the atom. The kinetic energy of this electron is $E_\gamma - E_b$ where E_b is the energy by which the electron was bound to the nucleus. Because the atomic electrons exist in a series of discrete levels commonly denoted by K, L_1, L_2, L_3, M, etc (see Fig. 2) for each γ-ray there is a series of discrete internal-conversion lines that can be observed in an appropriate electron spectrometer.

The intensity of a conversion line can be expressed relative to that of the associated γ-ray as the internal-conversion coefficient (ICC), denoted as α. For example, α_K is the ratio of the number of electrons emitted from the K atomic shell to the number of photons emitted. For the other atomic levels, the corresponding conversion coefficients are denoted by α_{L1}, α_{L2} α_M, etc. The total conversion coefficient is $\alpha = \Sigma_i \alpha_i$, where the sum includes all atomic shell or subshells. The conversion coefficients have strong dependences on the γ-ray energy, multipolarity, atomic number, and electron shell, and these range over many orders of magnitude. Therefore, the measurement of internal-conversion coefficients is a sensitive way to determine the multipolarity of the γ-ray.

A sample of theoretical conversion coefficients is given in Table 12, illustrating the great range of values. The most common case where two multipolarities compete is M1 + E2. For example, the α_K for M1 and E2 are nearly equal at $Z = 60$ and $E = 150$ keV. It is, however, relatively easy to determine the fraction of each of these components that is present if the relative intensity of the three L conversion lines can be measured. For the E2 portion the three lines are almost equal, $\alpha_{L_1} = 0.035$, $\alpha_{L_2} = 0.050$, and $\alpha_{L_3} = 0.048$, whereas for the M1 portion they change by factors of 10, namely, $\alpha_{L_1} = 0.050$, $\alpha_{L_2} = 0.0037$, and $\alpha_{L_3} = 0.00072$. The values for the mixed transition are just the weighted sum of these values. For example, for a mixture of 50% of each of the M1 and the E2, the coefficients are $\alpha_{L_1} = 0.042$, $\alpha_{L_2} = 0.027$, and $\alpha_{L_3} = 0.024$.

In addition to the possible multipolarities discussed in the previous sections, internal-conversion electrons can be produced by an E0 transition, in which no spin is carried off by the transition. Because the γ-rays must carry off at least one unit of angular momentum, or spin, there are no γ-rays associated with an E0 transition, and the corresponding internal-conversion coefficients are infinite. The most common E0 transitions are between levels with $J_i = J_f = 0$ where the other multipolarities cannot contribute. However, E0 transitions can also occur mixed with other multipolarities whenever $J_i^{\Pi} = J_f^{\Pi}$. For example, when $J_i = J_f = 2$ with $\Delta \Pi = 0$, transitions of an E0 + M1 + E2 character are sometimes observed.

Internal Bremsstrahlung.   Another type of radiation that originates in the atomic electron cloud as a direct consequence of a nuclear process is a continuous photon spectrum known as bremsstrahlung. This radiation is caused by the sudden change in the electric field on each electron around the atom resulting from the change in the nuclear charge associated with α, β^-, or electron-capture decay. This sudden change produces a continuous bremsstrahlung spectrum of photons that range in energy from 0 up to the decay energy. The probability of this process occurring is quite small, and only a small fraction of the decay energy is carried off by this radiation.

Table 13 gives calculated bremsstrahlung spectra for several radioisotopes. Because the decay energies of ^{54}Mn, ^{56}Cr, ^{177}Ta, and ^{177}Yb are all similar, the spectra can be compared. Electron capture gives a much lower intensity in the low energy bins than the β^--decay. The ^{54}V has a much higher decay energy and, therefore, a higher bremsstrahlung yield.

Table 12. Calculated Internal-Conversion Coefficients for γ-Rays[a,b]

E_γ, keV	Multi-polarity	α_K	α_{L_1}	α_{L_2}	α_{L_3}	α
			Values for Z = 30			
20	E1	6.02	0.485	0.0696	0.125	6.80
	E2	150.0	11.8	22.6	39.1	234.0
	E3	2960.0	185.0	3400.0	5770.0	13600.0
	M1	5.41	0.539	0.0344	0.0132	6.09
	M2	241.0	34.9	3.39	9.19	296.0
	M3	6280.0	1630.0	238.0	2060.0	10840.0
150	E1	0.0158	0.00149	0.00004	0.000065	0.0176
	E2	0.126	0.0117	0.00110	0.00147	0.143
	E3	0.844	0.0766	0.0222	0.0262	0.988
	M1	0.0189	0.00188	0.00006	0.000026	0.0212
	M2	0.139	0.0146	0.00090	0.000613	0.158
	M3	0.969	0.108	0.0109	0.0142	1.12
1000	E1	0.000115	0.000011	0	0	0.000128
	E2	0.000265	0.000026	0	0	0.000296
	E3	0.000553	0.000054	0	0	0.000619
	M1	0.000227	0.000022	0	0	0.000253
	M2	0.000516	0.000051	0	0	0.000576
	M3	0.00108	0.000108	0	0	0.00120
			Values for Z = 60			
20	E1		1.06	0.759	1.18	3.80
	E2		4.65	774.0	1100.0	2420.0
	E3		1530.0	111000.0	156000.0	364000.0
	M1		17.7	1.58	0.325	25.0
	M2		1920.0	126.0	1030.0	4020.0
	M3		77800.0	5510.0	422000.0	684000.0
150	E1	0.0730	0.00781	0.00105	0.00130	0.0859
	E2	0.356	0.0350	0.0501	0.0476	0.527
	E3	1.47	0.139	1.06	0.972	4.16
	M1	0.392	0.0500	0.00367	0.000717	0.461
	M2	2.55	0.420	0.0494	0.0510	3.22
	M3	13.1	2.95	0.519	1.51	19.6
1000	E1	0.000810	0.00010	0	0	0.000940
	E2	0.00192	0.00023	0.000022	0.000012	0.00225
	E3	0.00400	0.00050	0.000105	0.000034	0.00480
	M1	0.00299	0.00037	0.000015	0	0.00348
	M2	0.00732	0.00095	0.000059	0.000012	0.00861
	M3	0.0148	0.00202	0.000191	0.000059	0.0177
			Values for Z = 90			
20	E1			*	3.06	5.24
	E2			*	10600.0	18200.0
	E3			*	1130000.0	2580000.0
	M1			*	2.19	117.0

Table 12. (*Continued*)

E_γ, keV	Multi-polarity	α_K	α_{L_1}	α_{L_2}	α_{L_3}	α
	M2			*	26700.0	55400.0
	M3			*	1.61E+7	2.48E+7
150	E1	0.147	0.0180	0.00740	0.00596	0.189
	E2	0.241	0.0701	0.781	0.463	2.04
	E3	0.376	0.936	15.1	7.24	33.0
	M1	5.06	0.858	0.0915	0.00491	6.32
	M2	21.2	6.76	0.853	1.22	33.3
	M3	43.3	40.4	6.85	40.1	167.0
1000	E1	0.00288	0.00041	0.000044	0.000023	0.00351
	E2	0.00775	0.00124	0.000467	0.000096	0.0102
	E3	0.0170	0.00321	0.00234	0.000292	0.0248
	M1	0.0294	0.00482	0.000440	0.000023	0.0364
	M2	0.0651	0.0119	0.00156	0.000163	0.0832
	M3	0.114	0.0233	0.00425	0.00113	0.152

[a]From a computer file based on calculations published in Ref. 14.
[b]A 0 indicates the value is greater than 0, but less than 0.00001, a blank indicates that the value is 0 because the γ-ray energy is too low to eject the corresponding electrons, and an * means the γ-ray energy is just above the binding energy so conversion occurs, but is too low to allow application of the theoretical tables.

Atomic Decays and Radiations

Of the modes of nuclear decay discussed, two produce excitation of the atomic electron system. In electron capture an electron goes into the nucleus, leaving a hole in the electron shell. In internal conversion, energy is transferred to an electron, which is thereby ejected from the atom. The electron system promptly returns to its ground state by a series of processes that radiate away the excess energy. This energy can be evidenced in a series of electromagnetic photons, called x-rays, or by transfer of energy to additional electrons, which are then ejected. Here it is assumed that the atom is in an environment in which other electrons are available that can be attracted to this positively charged atom so that it becomes neutral and reaches its ground state. In a nonconducting medium or in empty space, the atom can in fact stay charged (see X-RAY TECHNOLOGY).

X-Rays.   If an x-ray is emitted, it has an energy, ΔE, equal to the difference in the binding energies of the two atomic shells, $E_i - E_j$. If the original hole is in the K shell, the x-ray is called a K x-ray; if the hole is in the L shell it is an L x-ray. Because the hole can be filled by an electron from any of the several outer shells, x-ray spectra contain a large number of discrete lines.

The energies of the x-rays depend on the properties of the atom and not on those of the nucleus. Thus the energies of the x-rays do not give any information about the nuclear process that created the original hole. Also, because all K-shell holes are equivalent, the relative emission probabilities of the various K x-rays are determined by the atomic properties. These do not give information about the related nuclear process. An exception is the case when only one γ-ray transition

Table 13. Calculated Internal-Bremsstrahlung Spectra[a]

Decay mode	Energy, keV	Average IB[b] energy, keV	Energy bin[c], keV	Average energy in bin[c], keV	Photon intensity, %
			^{54}Mn		
EC	1377	0.038	10–20	0.000052	0.00037
			20–40	0.000131	0.00044
			40–100	0.00151	0.0021
			100–300	0.021	0.0103
			300–542	0.0149	0.0041
			^{56}Cr		
β^-	1617	0.90	10–20	0.025	0.17
			20–40	0.048	0.166
			40–100	0.129	0.20
			100–300	0.31	0.18
			300–600	0.25	0.060
			600–1300	0.116	0.0156
			1300–1508	0.00026	0.000020
			^{54}V		
β^-	7042	3.5	10–20	0.044	0.31
			20–40	0.087	0.30
			40–100	0.25	0.38
			100–300	0.72	0.40
			300–600	0.81	0.19
			600–1300	1.08	0.126
			1300–2500	0.44	0.027
			2500–3387	0.00054	0.00021
			^{177}Ta		
EC	1158	0.53	10–20	0.00029	0.0023
			20–40	0.0035	0.0102
			40–100	0.28	0.50
			100–300	0.040	0.021
			300–600	0.112	0.025
			600–1158	0.098	0.0133
			^{177}Yb		
β^-	1398	0.56	10–20	0.018	0.126
			20–40	0.034	0.120
			40–100	0.091	0.141
			100–300	0.21	0.119
			300–600	0.143	0.035
			600–1300	0.050	0.0067
			1300–1398	0.0000053	0.00000040

[a]Data are from Ref. 15.
[b]IB = internal bremsstrahlung.
[c]The bin is the range of energy over which the portion of the spectrum is summed or integrated.

produces all of the observed x-rays. In this case the x-ray intensity gives the K-shell internal-conversion coefficient. In some cases careful measurements of the relative emission rates of the L x-rays can give information about the related nuclear process. However, this is only useful if most of the L-shell holes are produced directly by internal conversion in the L subshells rather than by the filling of holes in the K shell.

When the hole in the ith shell is filled by an electron from the jth shell, there is a hole in the latter shell that will in turn be filled by an electron from a higher kth shell. This may result in the emission of a second x-ray, such that one hole in an inner electron shell can result in a cascade of several x-rays having ever-decreasing energies.

For holes in the ith shell, the fraction of the holes that result in x-rays when that hole is filled with an outer electron is called the fluorescent yield, ω_i, for example ω_K and ω_{L_1}. The quantity ω_K has been computed theoretically, but the best values come from a simultaneous evaluation of the measured and theoretical values. The value of ω_K varies smoothly with the atomic number Z, and the fluorescence yields for each L subshell are smaller than the ω_K at the same Z. Table 14 gives values of the K and L_1 shell binding energies, ω_K, ω_{L_2}, and relative emission probabilities of the K_α and K_β x-rays as a function of the atomic number, Z.

Auger Electrons. The fraction of the holes in an atomic shell that do not result in the emission of an x-ray produce Auger electrons. In this process a hole in the ith shell is filled by an electron from the jth shell, and the available energy is transferred to a kth shell electron, which in turn is ejected from the atom with a kinetic energy $= E_i - E_j - E_k$. Usually, the most intense Auger electron lines are those from holes in the K shell and involve two L-subshell electrons. For example, assume an initial hole in the K shell is filled by an electron from the L_1 subshell and that the Auger process results in the ejection of an electron from the L_3 subshell. The kinetic energy of the latter electron is then equal to $E(K) - E(L_1) - E(L_3)$, and the electron is denoted as a KL_1L_3 Auger electron. The probability of producing a KXY Auger electron from a hole in the K shell is simply $1 - \omega_K$.

This process results in two holes in the L-electron shell, which then results in two L x-rays, two more Auger electrons, or one of each. In either case, there are more holes in the M or higher electron shells. Because the energies of the atomic

Table 14. Selected Values of the K and L_1 Electron-Binding Energies, K- and L_2-Shell Fluorescent Yields, and K_α/K_β x-Ray Intensity Ratio[a]

Atomic number, Z	Binding energy, keV		Fluorescent yield		X-ray intensity ratio, K_α/K_β
	K	L_1	ω_K	ω_{L_2}	
1	0.013				
10	0.8701	0.0485	0.018		
20	4.0381	0.4378	0.163	0.00033	9.6
30	9.6586	1.1936	0.474	0.011	8.2
40	17.9976	2.5316	0.730	0.028	5.9
50	20.2001	4.4647	0.862	0.065	4.9
60	43.5689	7.1260	0.921	0.124	4.4
70	61.3323	10.4864	0.951	0.222	4.0
80	83.1023	14.8393	0.965	0.347	3.6
90	109.6500	20.4720	0.971	0.479	3.5
100	141.9260	27.5740	0.976	0.506	3.4

[a]Values are from Ref. 15.

electrons are discrete, the Auger electrons appear as discrete lines, although there are many lines which may be difficult to distinguish. The energies of the Auger electrons and their intensities relative to the corresponding x-rays depend only on atomic parameters. Thus they do not give any information about the nuclear transitions that produced them beyond that already available from the x-ray spectrum.

Secondary Radiations

The previously discussed radiations have their origin in the atom in which the original decay took place. If the radiation reaching a detector is measured, there are other radiations that are observed.

External Bremsstrahlung. When a charged particle is decelerated or accelerated, it produces a continuous photon spectrum called external bremsstrahlung. When α-, β^--, or β^+-particles are emitted, they are scattered in the surrounding material, and in the process they produce bremsstrahlung radiation. This spectrum is quite similar to the internal bremsstrahlung spectrum from β- or electron-capture decay. It is continuous, ranging from 0 to the decay energy. Most of the photons have low energies. The number of photons produced per particle emitted depends on the atomic number of the material in which the particles are scattered and the Q-value of the decay.

Compton Scattered Photons. When a γ-ray interacts in material, there are several processes that may occur. One of these is Compton scattering, in which part of the γ-ray energy is transferred to an atomic electron, and the part remaining is carried off by a secondary photon, as illustrated in Figure 6. The conservation of energy and momentum requires specific relations between the scattering angles for the photon and electron and their energies. Compton-scattered photons are generated in the surrounding material and observed in any γ-ray detector.

X-Rays and Annihilation Radiation. The interaction of γ-rays with matter produces the x-rays that are characteristic of the atoms in the material in which the interactions take place. Such x-rays appear in measured spectrum.

The β^+-particles that are emitted in the β^+-decay mode are slowed down in the material around the source. When these reach very low velocities they interact with an ordinary electron and the pair is annihilated. The corresponding

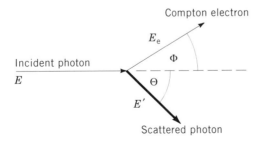

Fig. 6. Schematic illustration of the relationships of the original γ-ray and the scattered radiations for Compton scattering where E is the energy of the incident photon, E_e is the energy of the recoiling electron, and E' is the energy of the scattered photon.

energy of $2 \times m_0 c^2$, or 1022 keV, is normally released in the form of two photons of 511 keV each, emitted in opposite directions.

In addition to Compton scattering, γ-rays having energies above 1022 keV interact with matter by a process called pair production, in which the photon is converted into a positron and an electron. The γ-ray energy in excess of the 1022 keV needed to create the pair is shared between the two new particles as kinetic energy. Each β^+-particle is then slowed down and annihilated by an electron producing two 511-keV photons.

Background Radiation. If the radiation from a radioactive source is measured, the spectrum also includes contributions from the radiations from the surrounding environment. This includes radiations from the radioactivity in the materials in and around the detector, including the structure of the building or nearby earth. There is also cosmic radiation that comes from space and interacts with the earth and atmosphere to produce radiations that may enter the detector, and thus is observed.

Decay Data

Most areas of research and applications involving the use of radioisotopes require a knowledge of what radiations come from each isotope. The particular application determines what type of information is needed. If the quantity of a radionuclide in a particular sample or at a particular location is to be determined and this value is to be determined from the γ-ray spectrum, the half-life of the nuclide and the energies and intensities or emission probabilities of the γ-rays of interest must be known. Usually it is preferable to use the γ-rays for an assay measurement because the α- and β-rays are much more readily absorbed by the source material, and may not reach the sample surface having their original energies. Once these energies are altered they cannot be used to identify the parent radionuclide.

In research environments where the configuration and activity level of a sample can be made to conform to the desires of the experimenter, it is now possible to measure the energies of many γ-rays to 0.01 keV and their emission rates to an uncertainty of about 0.5%. As the measurement conditions vary from the optimum, the uncertainty of the measured value increases. In most cases where the counting rate is high enough to allow collection of sufficient counts in the spectrum, the γ-ray energies can still be determined to about 0.5 keV. If the configuration of the sample is not one for which the detector efficiency has been directly measured, however, the uncertainty in the γ-ray emission rate may increase to 5 or 10%.

For most applications, the decay parameters of the radionuclides of interest are known with sufficient accuracy for their uncertainties not to be of any concern. Typically this is the case if the uncertainty in the half-life is 0.1% or less and the uncertainty in the γ-emission probability is 1% or less.

There are single-volume sources available that contain the half-lives and γ-ray intensities for all radionuclides (15,17–19). These data can also be obtained electronically from the NUDAT (8) or Evaluated Nuclear Structure Data File (ENSDF) (17) databases (see DATABASES).

Table 15 shows data for several radionuclides the γ-rays of which are often used to calibrate the efficiency of γ-ray detectors. For a number of these γ-rays the very high accuracy arises because the γ-ray occurs in essentially 100% of the decays of the nuclide, and only small corrections are needed to deduce the γ-emission probability. In other cases the accuracy is high because a number of careful measurements have been made. The γ-emission rate from a calibration source also depends on the decay rate of the source, and for these nuclides the uncertainty in the source activity is often the larger uncertainty.

The half-lives, γ-ray energies, and γ-ray emission probabilities given in Table 15 are what is needed if the amount of a radioisotope present in a sample is to be measured. However, there are other uses of radionuclides where additional data concerning the decay are needed. If a radionuclide is to be injected or implanted *in vivo*, it is necessary to have data on all of the radiations produced to be able to assess the impact on the cell structure. Table 16 gives samples of the data that can be useful in this latter case. Such information can be obtained from some of the references above. There are also computer codes that can use the decay data from the ENSDF database to produce this type of information for any radionuclide, eg, RADLIST (21).

Applications

Dating. The idea of using the radiation from an object to determine its age goes back to the early years after the discovery of radioactivity. In approximately 1905 the ages of certain ores were estimated from the radioactivity. Although these estimates were too low by a large factor, they were good enough to challenge the concept that the earth had been created only a few thousand years ago.

There are two methods of using radioactive decays to determine the age of an object. A method applicable to formerly living organisms is based on the assumption that the specific activity, eg, disintegrations per gram of sample of a particular isotope, is known for the time the organism was alive. The time since death is then determined by comparing this to the measured specific activity. The other method, applicable to nonliving objects, determines age from the amounts of both the radioactive parent and the usually stable daughter isotope present in a sample. In this latter case all of the daughter atoms must be assumed to come from the parent, or an independent method of determining what fraction of the daughter atoms are from the decay of the parent must be available. Table 17 lists half-lives of several radioisotopes that have been used for dating.

The use of ^{14}C to determine the lapse of time since a living plant or animal died was developed early on. ^{14}C, produced in the atmosphere by the ^{14}N + $n \rightarrow {}^{14}$C + p reaction, becomes distributed in the air and ground along with the stable isotopes of carbon, ^{12}C (98.9%) and ^{13}C (1.1%). The ^{14}C is then taken up by living bodies in food along with the stable isotopes of carbon. When the organism dies, however, it ceases taking up ^{14}C, so the amount present decreases with the characteristic half-life. In the dating of old trees, it can be assumed that the specific activity of ^{14}C is the same in trees, or at least in the specific types of trees, in the 1990s as it was a few thousand years ago. Because

Table 15. Decay Data for Radionuclides Useful for the Energy and Efficiency Calibration of γ-Ray Detectors

Radionuclide	$t_{1/2}{}^{a-c}$	$E_\gamma{}^{c,d}$, keV	$P_\gamma{}^{c,e}$, %
^{22}Na	950.6 (3) d	1274.537 (7)	99.940 (14)
^{24}Na	14.970 (11) h	1368.625 (5)	99.9936 (4)
		2754.008 (11)	99.872 (7)
^{46}Sc	83.79 (3) d	889.271 (2)	99.984 (1)
		1120.537 (3)	99.97 (2)
^{51}Cr	27.702 (3) d	320.0824 (4)	9.87 (5)
^{54}Mn	312.16 (5)	834.838 (5)	99.975 (11)
^{57}Co	271.79 (9) d^f	14.4127 (4)	9.16 (15)^f
		122.06065 (12)	85.60 (17)
		136.4736 (3)	10.68 (8)
^{60}Co	5.2713 (11) yr	1173.228 (3)	99.85 (3)
		1332.492 (4)	99.9826 (7)
^{65}Zn	244.15 (9) d	1115.539 (2)	50.60 (22)
^{85}Sr	64.849 (4) d^f	514.0048 (22)	98.4 (4)^f
^{88}Y	106.630 (25) d^f	898.036 (4)	94.0 (3)^f
		1836.052 (13)	99.36 (3)
^{109}Cd	462.6 (7) d^f	88.0336 (10)	3.63 (2)^f
^{113}Sn	115.1 (3) d	391.698 (3)	63.56 (19)
^{137}Cs	30.037 (28) yr	661.657 (3)	85.03 (19)
^{139}Ce	137.641 (20) d	165.856 (6)	79.92 (4)
^{152}Eu	13.51 (3) yr^f	121.7817 (3)	28.37 (13)^f
		244.6975 (8)	7.53 (4)
		344.2785 (12)	26.57 (11)
		411.1163 (11)	2.238 (10)
		~443.96^g	3.125 (14)
		778.9040 (18)	12.97 (6)
		867.373 (3)	4.214 (25)
		~964.05^g	14.63 (6)
		1085.828 (6)	10.13 (5)
		1089.737 (5)	1.731 (9)
		1112.069 (3)	13.54 (6)
		1212.948 (11)	1.412 (8)
		1299.142 (8)	1.626 (11)
		1408.006 (3)	20.85 (9)
^{198}Au	2.6943 (8) d^f	411.80205 (17)	95.57 (47)^f
^{203}Hg	46.595 (13) d^f	279.194 (3)	81.48 (8)^f
^{228}Tl and daughters	1.9116 (16) yr^f	238.632 (2)	43.5 (4)^f
		583.187 (2)	30.6 (2)
		727.330 (9)	6.69 (9)
		860.564 (5)	4.50 (4)
		1620.735 (10)	1.49 (5)
		2614.511 (10)	35.86 (6)
^{241}Am	432.2 (7) yr^f	26.3446 (2)	2.4 (1)^f
		159.5409 (2)	36.0 (4)

[a]Ref. 6, unless otherwise noted.
[b]For conversion from days to years, one year is 365.2422 days.
[c]Uncertainty in last digit or digits is shown in parentheses.
[d]Ref. 5.
[e]Ref. 20, unless otherwise noted.
[f]Ref. 17.
[g]In a Ge detector spectrum, this peak is a multiplet.

Table 16. Decay Data for Radioisotopes of Interest in Applications

Radio-isotope	$t_{1/2}$	Primary γ-ray E_γ, keV	$P_\gamma{}^{a,b}$, %	Decay energy, keV	Particle type[c]	Average energy[d], keV	Average γ-ray energy[e], keV
^{75}Se	119.62 (24) d[f]	10.52 K^g	49.0 (16)	863.9	e^-	14.2	392
		96.7340	3.41 (4)[f]		IB	0.03	
		121.1155	17.1 (1)[f]				
		136.0001	58.8 (3)[f]				
		264.6576	59.0 (2)[f]				
		279.5422	25.0 (1)[f]				
		400.6572	11.5 (1)[f]				
^{99m}Tc	6.01 (1) h[h] (^{99}Mo 2.7 d)	140.5110	89.06 (24)	142	e^-	14.2	123.3
^{131}I	8.02070 (11) d	284.305	6.14 (6)	970.8	β^-	182	382
		364.489	81.7 (8)		e^-	10.0	
		636.989	7.17 (11)		IB	0.113	
^{201}Tl	72.912 (1) h	70.1 $K_\alpha{}^g$	73.7 (15)	482	e^-	48	92
		80.3 $K_\beta{}^g$	20.4 (7)		IB	0.47	
		135.34	2.565 (24)				
		167.43	10.00				

[a]Ref. 12, unless otherwise noted.
[b]Uncertainty in last digit or digits is shown in parenthesis.
[c]The particle notation is β^- for electrons from β^--decay, e^- for internal-conversion electrons, and IB for photons from internal bremsstrahlung.
[d]Ref. 15.
[e]Ref. 5 for ^{75}Se and ^{99m}Tc, and Ref. 12 for ^{131}I and ^{201}Tl.
[f]Ref. 20.
[g]X-ray.
[h]Value has been reported to vary up to 0.3% owing to chemical environment.

Table 17. Radionuclides Useful for Determining the Age of Materials

Parent	Half-life, yr	Stable daughter	Natural abundance of daughter, %	Intermediate nuclides
^{14}C	5.73×10^3	^{14}N	0.0	no
^{235}U	7.1×10^8	^{207}Pb	20.8	yes
^{238}U	4.5×10^9	^{206}Pb	26.3	yes
^{40}K	1.274×10^{10}	^{40}Ar	99.6	no
^{232}Th	1.39×10^{10}	^{208}Pb	51.6	yes
^{87}Rb	5.0×10^{10}	^{87}Sr	7.0	no

^{14}C decays by pure β-decay, which has a decay energy of only 155 keV, care must be taken in the measurement of the activity. The material must be processed to extract the carbon and prepare a thin solid or gaseous sample that allows the β-particles to reach the detector. The time since the sample was last in equilibrium with the ^{14}C in the atmosphere is simply the time it has taken for its specific activity to decrease from the value for currently living trees of the same type, to the measured value. The range of time where this method is most useful is around one half-life, ie, ca 6000 yr.

Since this method was perfected, several other dating methods that do not involve radioactivity have been developed. The one applied to trees is the counting of the annual growth rings. In comparing results from ^{14}C dating and ring counting, discrepancies were identified. From these differences it was concluded that the rate of ^{14}C production in the atmosphere was not the same in the past as it is in the late twentieth century, but the ring counting data have been used to determine the historical atmospheric concentrations of ^{14}C. Although this knowledge has made the use of ^{14}C dating more complex, requiring corrections to the simple calculated values, the method is still a useful one.

The other radioisotopes in Table 17 have much longer half-lives and can be used to determine the ages of materials that are on the order of a million years old. Therefore, these are especially useful in dating rocks. In these cases the amount of the parent present can be determined either by measuring its radiation or by counting the number of atoms. The daughters are stable, so their abundance must be ascertained by counting the atoms, or by some chemical assay method. In all of these cases there is more than one element of the same mass, thus a simple mass spectrometer measurement cannot be used until the daughter atoms are chemically purified (see MASS SPECTROMETRY).

In these cases what is usually measured is not the time of the original formation of the rocks, but the time at which the parent and daughter elements were last separated. That is, if the rocks were remelted at some point in their history in a manner that removed the daughter elements, this would be the age measured. For the ^{40}K measurements this is especially important. The daughter is a gas and thus could escape at any time when it was not sealed in. The decay sequences of ^{232}Th, ^{235}U, and ^{238}U all have gaseous members in their decay chains (see Tables 1 and 2), but the final members are solids. Therefore, in these cases it is only necessary that the gaseous members be retained for periods of several of their half-lives for the measured ages to be unaffected by loss of the gases. In the ^{238}U chain, the ^{222}Rn has a half-life of 3.8 d, so that diffusion out of the region of the ^{238}U may occur, but in the ^{232}Th chain the ^{220}Rn has a half-life of only 55 s, so that diffusion would be less likely.

All of these daughter isotopes occur naturally; thus it is always necessary to show that in the rocks of interest the daughter element present originated only from the decay of the parent, or the fraction of the daughter isotope that is from the decay of the parent must be separately determined. For example, the presence of ^{87}Sr in the absence of the more abundant ^{88}Sr (82.7%) would clearly indicate that its origin is from the decay of ^{87}Rb. For the Pb isotopes, a more detailed set of measurements might be needed. In this case the normal Pb abundances can be altered by the presence of any or all of the indicated Th and U isotopes, so not only the abundances of the Pb isotopes, but also those of the potential parent isotopes must be determined.

Medical Diagnoses.   There are many radioisotopes that are used for medical diagnosis (see RADIOACTIVE TRACERS; RADIOPHARMACEUTICALS). An example is the monitoring of blood flow to various regions of the heart using ^{201}Tl, which has a half-life of 3.0 d. One procedure involves having the patient walk on a treadmill at a pace which forces the body and the heart to work quite hard. Then the thallium compound is injected into the blood stream and is attracted to the active heart muscle. Using a multidetector γ-ray measurement system that can be rotated around the reclining patient, the approximate origin of the γ-rays from within the heart is measured. A computer analysis of these data produce computer-assisted tomography (CAT) scan pictures which can be displayed to show the distribution of the radioactivity in different regions of the heart. If a coronary artery is blocked, the radioactive blood does not reach the corresponding part of the heart as quickly as it reaches other portions. Although the spatial resolution of the measurement and analysis system is not fine enough to give detailed information about blockages along each artery, it does identify the portions of the heart that are not receiving normal blood flow. The decay data for ^{201}Tl (see Table 16) can be used in the calculation of the radiation dose received by the various organs of the patient. Because ^{201}Tl decays by electron capture, a large fraction of the decay energy of 482 keV is emitted in the form of neutrinos, which do not contribute to the patient dose.

The radioisotopes ^{99m}Tc and ^{131}I (see Table 16) are often used for medical purposes. ^{99m}Tc has a half-life of only 6 h, which would normally make it difficult to transport from a production facility to the medical facility. However, one can supply the longer-lived 2.7-d ^{99}Mo in a chemical form that allows one to separate out, generate or milk, the daughter ^{99m}Tc when the latter is needed.

Another medical use of radioisotopes, such ^{60}Co, is to irradiate certain tissues within the body. An intense source of ^{60}Co in a heavily shielded facility provides a highly collimated beam of γ-rays that impinge on a tumor in order to kill its cells.

BIBLIOGRAPHY

"Isotopes" in *ECT* 1st ed., Vol. 8, pp. 89–104, by D. H. Templeton, University of California; "Radioisotopes" in *ECT* 1st ed., Second Supplement Volume, pp. 681–708, by A. Beerbower, Esso Research and Engineering Co.; in *ECT* 2nd ed., Vol. 17, pp. 64–130, by A. Beerbower and A. E. von Rosenberg, Esso Research and Engineering, Co.; in *ECT* 3rd ed., Vol. 19, pp. 682–785, by M. J. Martin, Oak Ridge National Laboratory.

1. K. Debertin and R. G. Helmer, *Gamma- and X-ray Spectrometry With Semiconductor Detectors*, North-Holland, Amsterdam, the Netherlands, 1988.
2. G. F. Knoll, *Radiation Detection and Measurement*, 2nd ed., John Wiley & Sons, Inc., New York, 1989.
3. S. Raman and N. B. Gove, *Phys. Rev. C* **7**, 1995 (1973).
4. N. B. Gove and M. J. Martin, *Nucl. Data Tbls. A* **10**, 206 (1971).
5. R. G. Helmer and C. Van der Leun, technical data, Idaho National Engineering Laboratory, Idaho Falls, Idaho, 1995.
6. R. G. Helmer, technical data, *International Decay Data Evaluation Project*, 1995.
7. P. Schlütter and G. Soff, *At. Data Nucl. Data Tbls.* **24**, 509 (1979).
8. B. Buck, A. C. Merchant, S. M. Perez, and P. Tripe, *J. Phys. G, Nucl. Particle Phys.* **20**, 351 (1994).
9. E. Hourany and co-workers, *Phys. Rev. C* **52**, 267 (1995).
10. P. B. Price and co-workers, *Phys. Rev. C* **46**, 1939 (1992).
11. *Nuclear Data Sheets*, Academic Press, San Diego, Calif., *Evaluated Nuclear Structure Data File* (*ENSDF*), a computer database of nuclear structure data evaluated by an international network of evaluators, is maintained at the National Nuclear Data Center, Brookhaven National Laboratory. *NUDAT* is a computer database of decay data extracted from the *ENSDF*.

12. E. D. Commins and P. H. Bucksbaum, *Weak Interactions of Leptons and Quarks*, Cambridge University Press, Cambridge, U.K., 1983, p. 389.

13. A. Balysh and co-workers, *Phys. Lett. B* **322**, 176 (1994).

14. F. Rösel, H. M. Fries, K. Alder, and H. C. Pauli, *At. Data Nucl. Data Tbls.* **21**, 92 (1978); *ibid.*, **21**, 292 (1978).

15. E. Browne and R. B. Firestone, in V. S. Shirley, ed., *Table of Radioactive Isotopes*, John Wiley & Sons, Inc., New York, 1986.

16. Ref. 15, Appendix C.

17. *ENSDF*, in Ref. 11.

18. U. Reus and W. Westmeier, *At. Data Nucl. Data Tbls.* **29**, 1 and 205 (1983).

19. G. Erdtmann and W. Soyka, *The Gamma Rays of the Radionuclides, Tables for Applied Gamma Ray Spectrometry*, Verlag Chemie, Weinheim, Austria, 1979.

20. *X-ray and Gamma-Ray Standards for Detector Calibration* Report IAEA-TECDOC-619, International Atomic Energy Agency, Vienna, Austria, 1991.

21. T. W. Burrows, *The Program RADLIST*, Brookhaven National Laboratory report BNL-NCS-52142, Brookhaven National Laboratory, Upton, N.Y., Feb. 1988.

RICHARD G. HELMER
Idaho National Engineering Laboratory

RADIOPAQUES

Medical examination of soft tissues or organs by nonsurgical means often requires the introduction of a special agent which makes the detection system responsive to detail in the tissue of interest. Diagnostic imaging agents include those used in magnetic resonance, ultrasound, radionuclide imaging, and x-ray technology (qv) (see MEDICAL IMAGING TECHNOLOGY). Radiopaques for x-ray imaging, more commonly called x-ray contrast media or radiographic contrast agents, are examples of such diagnostic agents. These chemicals absorb x-rays strongly. When they accumulate in the target area, they create a contrast in the x-ray image thereby permitting visual examination of the target organ. Hence the classification as contrast agents or contrast media.

Absorption of x-radiation is an atomic phenomenon related to the atomic number of the absorbing atom (1) such that the heavier elements are, in general, more efficient at absorbing x-rays. Except for barium sulfate [7727-43-7], all other radiopaque agents in use as of the mid-1990s are organic derivatives of iodine. The iodine atoms function as the x-ray absorbers, and the organic moiety can be manipulated to provide desirable characteristics of a contrast agent and to decrease toxicity and physiological side-effects. Table 1 contains a summary of the more important radiographic procedures and contrast agents for x-ray visualization of various tissues and organs.

Table 1. Radiographic Procedures and Corresponding Radiopaques

Procedure	Organ/region	Radiopaque agents
angiography	blood vessels	sodium or meglumine salt of diatrizoic acid, iothalamic acid, metrizoic acid, and ioxaglic acid; iopamidol, iohexol, ioversol, iopromide, iomeprol, iopentol, ioxilan, iobitridol
arteriography	arteries	
aortography	aorta	
ventriculography	ventricles of the heart	
venography (phlebography)	veins	
urography	urinary tract	
computed tomography	body, head	
myelography	subarachnoid space of the spinal cord	meglumine salt of iothalamic acid[a] and iocarmic acid[a]; metrizamide,[a] iohexol, iopamidol, iotrol, iodixanol
cholecystography	gallbladder	iopanoic acid, iocetamic acid, sodium or calcium iopodate, sodium tyropanoate, meglumine iodipamide, ioglycamide, iodoxamate
cholangiography	bile ducts	
gastrointestinal radiography	alimentary tract	barium sulfate, sodium or meglumine diatrizoate, iohexol
arthrography	joints	meglumine diatrizoate, meglumine iothalamate, sodium and meglumine salt of ioxaglic acid, iohexol
hysterosalpingography	uterus and fallopian tubes	ethiodol, meglumine diatrizoate-meglumine iodipamide mixture, sodium and meglumine salts of iothalamic, diatrizoic and ioxaglic acids, iohexol

[a]Not commonly used as of the mid-1990s.

Angiography and Urography

Angiographic contrast media (CM) are administered intravascularly for the radiographic visualization of blood vessels to evaluate vascular abnormalities. Uses include cerebral, coronary, pulmonary, renal, visceral, and peripheral arteriography, aortography, ventriculography, and venography. Because of the very high concentrations of contrast media required for angiography, such materials must have high water solubility. More dilute formulations of these same agents are used for urography. In urographic procedures, the CM are injected intravenously and their excretion via the kidneys is visualized radiographically as an evaluation of renal function (excretory urography). Alternatively, the CM are instilled via catheters directly into the lower urinary tract (retrograde pyelography). Excretory urography is the more common procedure. Following intravascular administration, the angiographic–urographic CM are distributed in the extracellular space and subsequently excreted unchanged, principally in the urine.

The historical development of angiographic–urographic CM (2) proceeded from inorganic forms of strontium bromide through sodium iodide and thorium dioxide to organic compounds. The inorganic bromide and iodide compounds produced painful and life-threatening adverse reactions. The radioactive nature of thorium resulted in severe but delayed effects, eg, liver cancer. Attempts to eliminate the adverse effects resulting from iodide ions, yet keep the heavy atom x-ray opacity of iodine, led to the consideration of iodinated organic compounds. Because the chemotherapeutic drug Selectan [60154-05-4] (1) was iodinated and known to be excreted in the urine, it was tested as an intravenous urographic agent and proved to give adequate pictures (3). An improved derivative of Selectan called Uroselectan [80462-95-9] (2) followed (3). Then iodopyracet [300-37-8] (3) and sodium iodomethamate [519-26-6] (4), each containing two iodine atoms per molecule for higher radiographic efficacy, were widely used in the 1930s and 1940s (4–6). Sodium methiodal [126-31-8], ICH_2SO_3Na, was also in clinical use for a time (7).

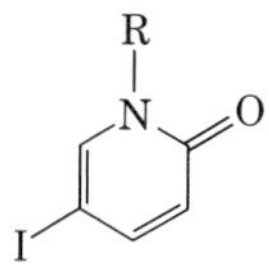

(1) R = CH_3
(2) R = $CH_2COO^-Na^+$

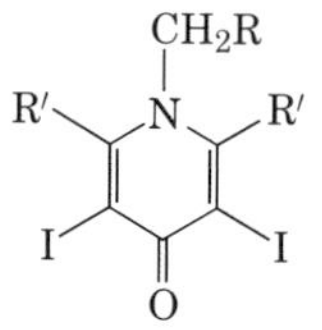

(3) R = COOH · $NH(CH_2CH_2OH)_2$
    R′ = H
(4) R = H
    R′ = COO^-Na^+

High Osmolality Contrast Media. An important advance in radiopaques came with the synthesis of aminotriiodobenzoic acid and its acetylated derivative, acetrizoic acid [85-36-9] (5) (8,9). Aqueous solutions of sodium acetrizoate possessed the thermal stability so that they could be autoclaved (10) with minimal decomposition. The higher iodine content, ie, 3 atoms/molecule, increased the contrast efficiency, and the clinical safety of acetrizoate was improved over that of the earlier urographic agents.

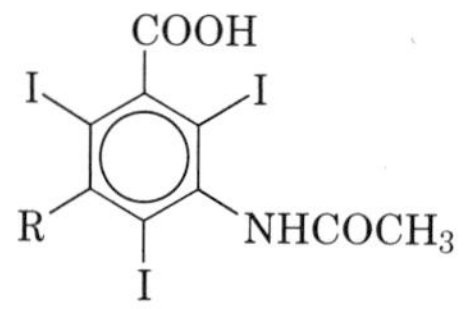

(5) R = H
(6) R = $NHCOCH_3$
(7) R = $CONHCH_3$
(8) R = $N(CH_3)COCH_3$

Further improvements in the late 1950s and early 1960s led to the development of three derivatives of acetrizoate that comprise a group of important angiographic and urographic ionic contrast media: diatrizoic acid [117-96-4] (6), iothalamic acid [2276-90-6] (7), and metrizoic acid [1949-45-7] (8). These compounds, in which the hydrogen on the ring is replaced by more hydrophilic moieties, were found to be less toxic than acetrizoate, because modification made them less amenable to protein binding. The preparations of these compounds are straightforward and cost efficient. Starting with a derivative of nitrobenzoic acid, reduction of the nitro group to an amine is followed by iodination and acylation. The R group can be chemically altered in the reaction sequence, depending on the substitution requirements (10,11).

$$R\text{-}C_6H_3(COOH)(NO_2) \xrightarrow{H_2/Pd} R\text{-}C_6H_3(COOH)(NH_2) \xrightarrow{ICl} R\text{-}C_6(COOH)(I)_3(NH_2) \xrightarrow{(CH_3CO)_2O} R\text{-}C_6(COOH)(I)_3(NHCOCH_3)$$

These ionic contrast media are synthesized and purified as free acids. As such, they are highly insoluble in water, which facilitates isolation and purification. In commercial formulations, these compounds are prepared as sodium or meglumine salts in order to provide solubility. Sequestering agents such as EDTA are added to bond to trace inorganic impurities that can cause catalytic deiodination (12). Buffers may also be included to ensure physiological pH conditions (12). Table 2 contains the important physical and biological (intravascular LD_{50}, median lethal dose) properties of the ionic agents.

Osmolality, a measure of the number of particles in a solution, is approximately proportional to the sum of the concentrations of all molecular and ionic particles present. As shown in Table 2, the osmolality of ionic monomers is about 1500 mOsm/kg at concentrations of 280–300 mg/mL on a mg of iodine basis. These ionic compounds dissociate into sodium or meglumine cations and the benzoate anions. Each ion contributes to the overall osmolality of the diagnostic solution in a ratio of three iodine atoms delivered as two particles, the cation and the anion. In the range of concentrations required for good x-ray visualization, the high osmolality of these ionic agents relative to plasma, which has an osmolality of approximately 300 mOsm/kg H_2O (17,18), and surrounding tissues causes leaching of water across semipermeable membranes, resulting in undesirable physiological effects (19,20). Hence, this class of agents is known as high osmolality contrast media (HOCM), or ratio-1.5 CM, ie, three iodines per two particles. Clinical trials indicate that many adverse effects owing to ionic agents can be attributed to high osmolality (18–29). Such effects include vasodilation, hemodilution, crenation of red blood cells, and disruption of endothelial integrity. The pain and heat sensation generated upon intravascular injection of the ionic agents has been correlated with the vasodilation and vascular endothelial damage induced by the high osmolality of the contrast solutions (25,30,31).

Low Osmolality Contrast Media. An ideal intravascular CM possesses several properties: high opacity to x-rays, high water solubility, chemical stability, low viscosity, low osmolality, and high biological safety. Low cost and patentability are also important for commercial agents. The newer nonionic and

Table 2. Properties of Ionic Monomeric Radiopaques for Angiography and Urography

Name		Company name	Iodine concentration, 60% wt/vol, mg/mL	Osmolality, mOsm/kg H_2O	LD_{50}, iv mice, g/kg[a]	Ref.
Generic	Trade					
acetrizoate sodium	Urokon	Mallinckrodt Chemical Works			5.5	6
diatrizoate meglumine	Angiovist, Reno-M-60, Hypaque-M 60	Berlex Labs, Squibb Diagnostics[b] Sterling-Winthrop Pharmaceuticals[c]	282	1400, 1500	8.4[d]	13
diatrizoate meglumine (52%)/ sodium (8%)	MD-60, Renografin-60	Mallinckrodt Medical, Squibb Diagnostics[c]	292	1420,[e] 1539[f]		13
iothalamate meglumine	Conray	Mallinckrodt Medical	282	1400	8.0[g,h] 11.5[g,i]	16
metrizoate meglumine			278	1660	9.1[h,j]	

[a] Grams of iodine per kilogram of body weight.
[b] Bracco Diagnostics, as of 1996.
[c] Nycomed Imaging, as of 1996.
[d] Value is for diatrizoate sodium (6).
[e] Value for Renografin-60.
[f] Value for MD-60.
[g] Values are for iothalamate sodium.
[h] Ref. 14.
[i] Ref. 15.
[j] Value is for metrizoate sodium (Isopaque, Sterling-Winthrop Pharmaceuticals) (14).

low osmolar agents represent an advanced class of compounds in the development of x-ray contrast media.

Development of nonionic compounds to eliminate ionicity, reduce osmolality, and hence minimize adverse effects, such as painful reactions associated with the injection of ionic agents, was proposed in 1968 (22,32). These triiodobenzene derivatives contain no ionizable carboxyl moiety, and their high water solubility and biological safety are achieved by employment of highly polar, hydrophylic groups. Because these compounds do not dissociate in solution, approximately half the osmolality of the ionic agents results at equivalent concentrations. Therefore, these nonionic agents are referred to as low osmolality contrast media (LOCM) or ratio-3 CM, ie, three iodine atoms per particle.

Introduced in 1975, metrizamide [*31112-62-6*] (**9**) was the first clinically successful nonionic agent (33). It possesses lower subarachnoid and acute intravascular toxicity than the ionic agents and was used for myelographic applications. However, this compound is unstable under autoclaving conditions in aqueous solution and therefore requires dispensing as a lyophilized powder which must be reconstituted prior to use. Intensive research in the 1970s and 1980s produced safer and more stable agents, eg, iopamidol [*60166-93-0*] (**10**) (26,34–37), iohexol [*66108-95-0*] (**11**) (28,37–40), and ioversol [*87771-40-2*] (**12**) (29,41). These three nonionic LOCM have become the primary products utilized for angiographic–urographic procedures.

(**9**)

(**10**) R = CONHCH(CH$_2$OH)$_2$
R′ = NHCOCH(OH)CH$_3$ (L-form)
(**11**) R = CONHCH$_2$CH(OH)CH$_2$OH
R′ = N(COCH$_3$)CH$_2$CH(OH)CH$_2$OH
(**12**) R = CONHCH$_2$CH(OH)CH$_2$OH
R′ = N(COCH$_2$OH)CH$_2$CH$_2$OH

Tables 3 and 4 summarize the important properties of the LOCM in use or under development as of this writing (ca 1996). As shown in Table 4, these agents provide lower osmolality, approximately 550–700 mOsm/kg at 300 mg/mL on a mg of iodine basis, and reduced toxicity relative to the ionic monomeric agents (see Table 2). Clinical studies (17–20,26,28,29,64–67) have demonstrated that the nonionic agents offer a significant margin of safety, have fewer side effects, and provide a much-improved level of comfort to the patients, compared to the ionic species. The increased hydrophilicity of these LOCM also contributes to a reduction in adverse physiological effects by limiting binding to proteins and other biomolecules (17,43,58,68–70). The partition coefficient data in Table 4 reflect the correlation between hydrophilicity and intracisternal neurotoxicity. The more hydrophilic agents, ie, those having the lower octanol/water partition coefficients, are generally less neurotoxic.

Contrast materials of low osmolality can be classified into three chemical types: (*1*) nonionic monomers, such as iopamidol, iohexol, ioversol, iopromide [*73334-07-3*] (**13**) (71), iomeprol [*78649-41-9*] (**14**) (42), iopentol [*89797-00-2*] (**15**) (51), ioxilan [*107793-72-6*] (**16**) (53,60,63), and iobitridol [*136949-58-1*] (**17**) (72); (*2*) monoionic dimers, such as ioxaglic acid [*59017-64-0*] (**18**) (27,73); and (*3*) nonionic dimers, iotrolan [*79770-24-4*] and iodixanol [*92339-11-2*], both of which are being investigated for intravascular and intrathecal (myelographic) uses.

Table 3. Properties of Low Osmolality Contrast Media (LOCM) for Angiography and Urography[a]

Name		Company name	Molecular weight	Iodine content, %	Solubility in water, % wt/vol	Viscosity,[b] 300 mg/mL, cps	
Generic	Trade					20°C	37°C
metrizamide	Amipaque	Sterling-Winthrop Pharmaceuticals[c]	789.10	48.2	80 (35)	11.7 (42)	6.2 (43)
iopamidol	Isovue	Squibb Diagnostics[d]	777.09	49.0	89 (35)	8.8 (44)	4.7 (44)
iohexol	Omnipaque	Winthrop Pharmaceuticals[c]	821.14	46.4	>120 (45)	11.8 (46)	6.3 (46), 5.6 (47)
ioversol	Optiray	Mallinckrodt Medical	807.12	47.2	>125 (48)	8.2[e] (49)	5.5 (49)
iopromide	Ultravist	Berlex Labs	791.12	48.1		8.7 (50)	4.8 (43)
iomeprol	Iomeron	Bracco	777.09	49.0	>100 (42)	7.5 (42)	4.2 (42)
iopentol	Imagopaque	Nycomed Imaging	835.17	45.6	100 (51)	13.2 (52)	6.5 (52)
ioxilan	Oxilan	Cook Imaging	791.12	48.1			4.9 (53)
iobitridol	Xenetix	Guerbet Laboratory	835.17	45.6			6.0 (47)
ioxaglic acid, meglumine, and sodium salts	Hexabrix	Mallinckrodt Medical and Guerbet Laboratory	1268.90	60.0		15.7[f] (54)	7.5[f] (54)

[a]References cited are given in parentheses.
[b]Readings are on a weight of iodine per volume of solution basis; cps = cycles per second of viscometer.
[c]Nycomed Imaging, as of 1996.
[d]Bracco Diagnostics, as of 1996.
[e]Viscosity reading at 25°C.
[f]Viscosity readings at 320 mg/mL on a weight of iodine basis.

Table 4. Osmolality, Toxicity, and Partition Coefficients of LOCM for Angiography and Urography[a]

| Name | | | Concentration,[b] mg/mL | Osmolality, mOsm/kg H$_2$O | LD$_{50}$[b] | | | Partition coefficient, octanol/water, $\times 10^4$ |
Generic	Trade	Company name			Intravascular, mice g/kg	Intracisternal,[c] rats mg/kg		
metrizamide	Amipaque	Sterling-Winthrop Pharmaceuticals[d]	300	490 (43)	15 (33,55) 12.1 (57)	150 (48,56)		190 (56) 390[e] (58)
iopamidol	Isovue	Squibb Diagnostics[f]	250 300 370	520 (44) 620 (44) 800 (44)	21.8 (35,59) 17 (48) 17–18.5 (53) 16.4 (57)	800 (48,56)		19 (56) 25 (52) 38[e] (58)
iohexol	Omnipaque	Winthrop Pharmaceuticals[d]	240 300 350	520 (46) 670 (46) 690 (47) 840 (46)	23.4 (40) 15 (48) 17 (47) 17.9 (53) 18.5 (59)	977 (48,56)		8 (56) 10 (52) 19[e] (58)
ioversol	Optiray	Mallinckrodt Medical	240 300 320 350	500 (49) 650 (49) 700 (49) 790 (49)	20 (41) 16 (48) 19.6 (59)	>1200 (48,56)		4 (56) 10[e] (58)
iopromide	Ultravist	Berlex Labs	300 350	610 (60) 760 (60)	11.5–13.0 (53) 16.5 (57) 18.5 (59)	122 (61)		89[e] (58)
iomeprol	Iomeron	Bracco	250 300 350	450 (42) 540 (42) 630 (42)	19.9 (59)			

913

Table 4. (Continued)

Generic	Trade	Company name	Concentration,[b] mg/mL	Osmolality, mOsm/kg H_2O	LD$_{50}$[b] Intravascular, mice g/kg	LD$_{50}$[b] Intracisternal,[c] rats mg/kg	Partition coefficient, octanol/water $\times 10^4$
iopentol	Imagopaque	Nycomed Imaging	300	640 (52)	22 (51)		70 (52)
			350	660 (51)	19.5 (62)		
ioxilan	Oxilan	Cook Imaging	300	810 (52)	15.7 (59)		
			350	560 (63)	18.8 (53)		
				690 (63)			
iobitridol	Xenetix	Guerbet Laboratory	300	695 (47)	16.8 (47)		
ioxaglic acid, meglumine, and sodium salts	Hexabrix	Mallinckrodt Medical and Guerbet Laboratory	320	600 (54)	13.5 (32)		10[e] (58)
					10.2 (59)		

[a]References cited are given in parentheses.
[b]Values given are on a weight of iodine basis.
[c]Describes administration of CM into the cisterna magna to assess the neurotoxicity of the CM.
[d]Nycomed Imaging, as of 1996.
[e]These partition coefficient data were derived from the log P values reported therein.
[f]Bracco Diagnostics, as of 1996.

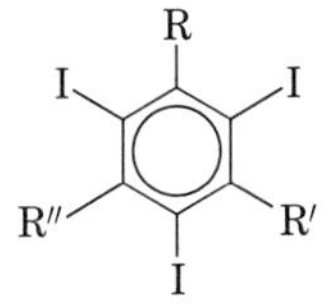

(**13**) R = CON(CH$_3$)CH$_2$CH(OH)CH$_2$OH

R′ = CONHCH$_2$CH(OH)CH$_2$OH

R″ = NHCOCH$_2$OCH$_3$

(**14**) R = R′= CONHCH$_2$CH(OH)CH$_2$OH

R″ = N(CH$_3$)COCH$_2$OH

(**15**) R = R′= CONHCH$_2$CH(OH)CH$_2$OH

R″ = N(COCH$_3$)CH$_2$CH(OH)CH$_2$OCH$_3$

(**16**) R = CONHCH$_2$CH(OH)CH$_2$OH

R′ = CONHCH$_2$CH$_2$OH

R″ = N(COCH$_3$)CH$_2$CH(OH)CH$_2$OH

(**17**) R = R′= CON(CH$_3$)CH$_2$CH(OH)CH$_2$OH

R″ = NHCOCH(CH$_2$OH)$_2$

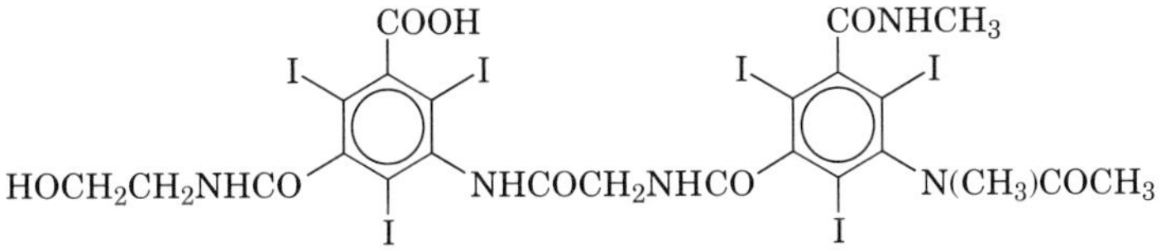

(**18**)

All nonionic contrast media used as of the mid-1990s are amide derivatives of triiodinated 5-aminoisophthalic acid. Therefore, 5-nitroisophthalic acid (**19**) serves as the starting material and, after reduction and iodination, the intermediates are chemically manipulated to provide the desired final products. The general preparation of triiodoisophthalamide radiopaque agents (74) is illustrated by the syntheses of iopamidol, iohexol, ioversol, and ioxaglic acid. In the synthesis of iopamidol (**10**) (26,35), reduction of 5-nitroisophthalic acid yields the corresponding amino diacid, which is then iodinated to give 5-amino-2,4,6-triiodoisophthalic acid (**20**). Activation of the aromatic nucleus by the amino group is required for the iodination. Treatment of the latter with thionyl chloride results in the dichloride, which is subjected to *N*-acylation with L-2-acetoxypropionyl chloride (*S*-configuration) to afford the diacid chloride (**21**). Amidation of (**21**) with 2-amino-1,3-propanediol (serinol) followed by hydrolysis of the ester group using aqueous sodium hydroxide gives iopamidol (**10**).

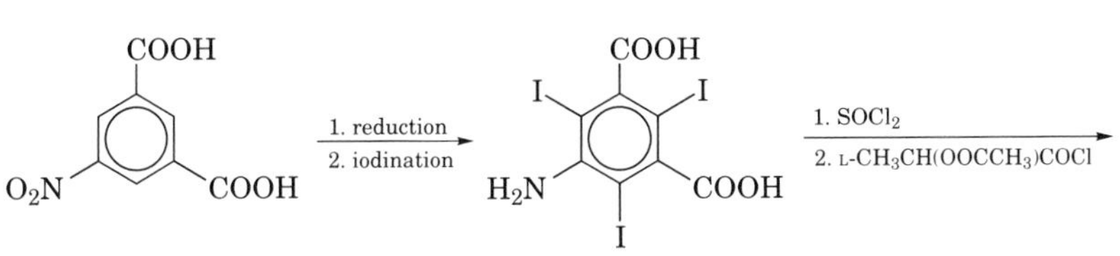

(**19**)   1. reduction   2. iodination   (**20**)   1. SOCl$_2$ 2. L-CH$_3$CH(OOCCH$_3$)COCl

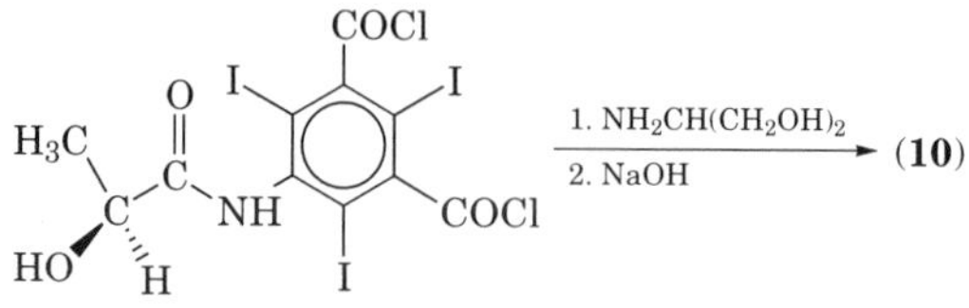

$$\text{L-form } (\textbf{21})$$

$$\xrightarrow[\text{2. NaOH}]{\text{1. NH}_2\text{CH(CH}_2\text{OH)}_2} (\textbf{10})$$

Iohexol (**11**) is prepared (39,40) starting with aminolysis of the nitro-diester (**22**), which is prepared by esterification of the nitro-diacid (**19**). Reduction of the nitro group followed by iodination of the resulting amino-diamide gives the key triiodinated intermediate (**23**). Peracylation of (**23**) using acetic anhydride and catalytic amounts of sulfuric acid followed by saponification of the O-acyl groups with aqueous sodium hydroxide produces the acetamido compound (**24**). N-Alkylation of compound (**24**) using 3-chloro-1,2-propanediol in the presence of sodium methoxide yields iohexol (**11**).

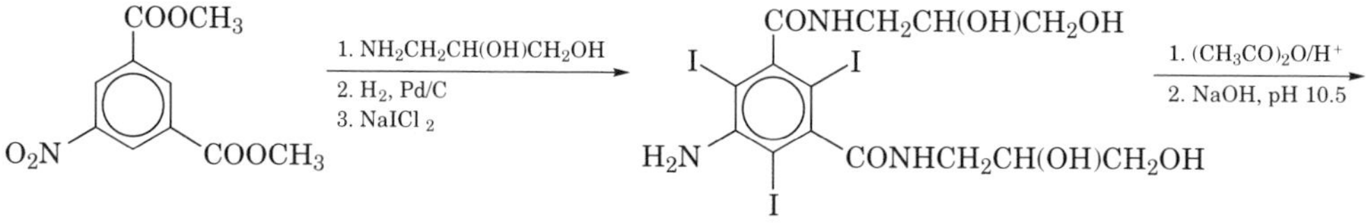

(**22**)

$$\xrightarrow[\substack{\text{2. H}_2\text{, Pd/C} \\ \text{3. NaICl}_2}]{\text{1. NH}_2\text{CH}_2\text{CH(OH)CH}_2\text{OH}}$$

(**23**)

$$\xrightarrow[\text{2. NaOH, pH 10.5}]{\text{1. (CH}_3\text{CO)}_2\text{O/H}^+}$$

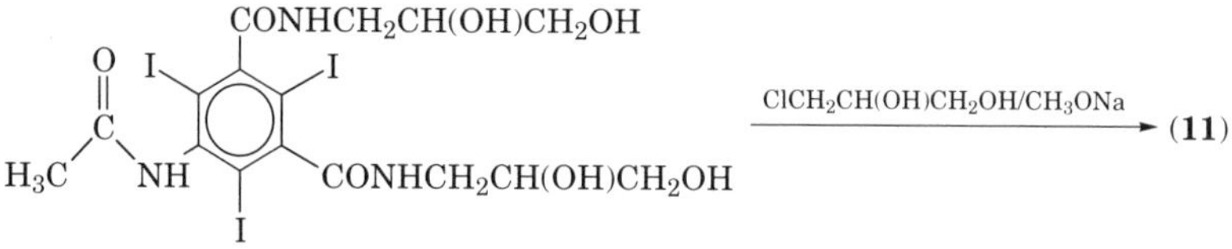

(**24**)

$$\xrightarrow{\text{ClCH}_2\text{CH(OH)CH}_2\text{OH/CH}_3\text{ONa}} (\textbf{11})$$

In the preparation of ioversol (**12**) (41), the key intermediate (**23**) is prepared from the diacid (**20**) by the action of thionyl chloride followed by 3-amino-1,2-propanediol. The alcohol groups of (**23**) are protected as the acetates (**25**), which is then N-acylated with acetoxyacetyl chloride and deprotected in aqueous methanol with sodium hydroxide to yield (**26**). N-alkylation of (**26**) produces ioversol (**12**).

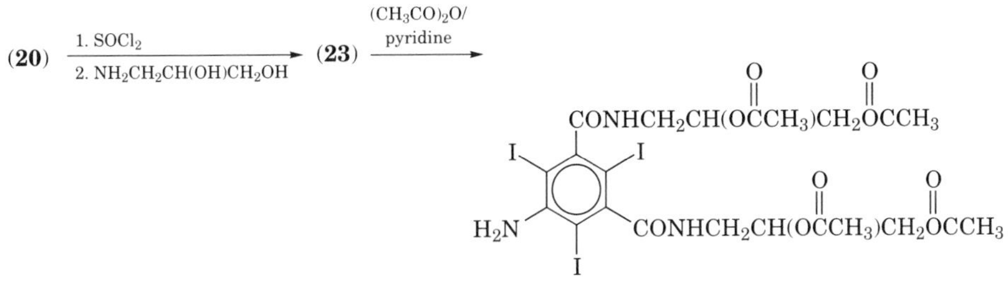

$$(20) \xrightarrow[\text{2. NH}_2\text{CH}_2\text{CH(OH)CH}_2\text{OH}]{\text{1. SOCl}_2} (23) \xrightarrow[\text{pyridine}]{(CH_3CO)_2O/}$$

CONHCH$_2$CH(OCCH$_3$)CH$_2$OCCH$_3$ (with two C=O)

CONHCH$_2$CH(OCCH$_3$)CH$_2$OCCH$_3$ (with two C=O)

H$_2$N

I (iodine substituents)

(25)

$$\xrightarrow[\text{2. NaOH}]{\text{1. CH}_3\text{COOCH}_2\text{COCl}}$$

CONHCH$_2$CH(OH)CH$_2$OH

CONHCH$_2$CH(OH)CH$_2$OH

HOCH$_2$ NH

C=O

I (iodine substituents)

$$\xrightarrow[]{\text{NaOH/ClCH}_2\text{CH}_2\text{OH}} (12)$$

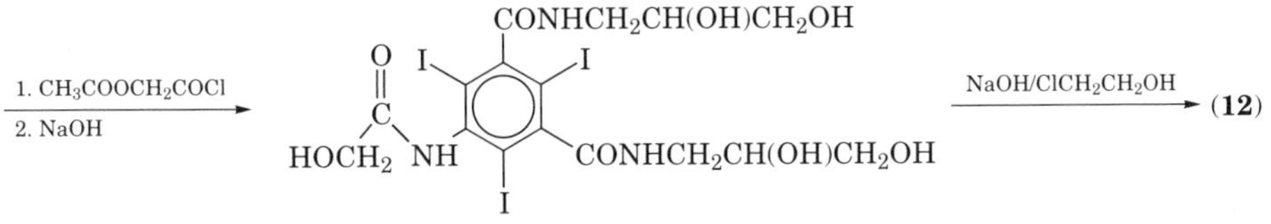

(26)

The final step in the preparation of both iohexol and ioversol involves some very interesting chemistry, ie, the N-alkylation of an acylamino-triiodoisophthalamide. Derivatizing the aromatic core, particularly using hydrophilic groups, confers desired properties such as water solubility for angiographic agents.

The steric bulk of the three iodine atoms in the 2,4,6-triiodobenzene system and the amide nature of the 1,3,5-substituents yield rotational isomers of the 5-*N*-acyl-substituted 2,4,6-triiodoisophthalamides. Rotational motion in the bonds connecting the side chains and the aromatic ring is restricted. These compounds also exhibit stereoisomerism when chiral carbon atoms are present on side chains. (*R*,*S*)-3-Amino-1,2-propanediol is incorporated in the synthesis of iohexol (**11**) and ioversol (**12**) and an (*S*)-2-hydroxypropanoyl group is used in the synthesis of iopamidol (**10**). Consequently, the resulting products contain a mixture of stereoisomers, ie, *d,l*-pair and meso-isomers, or an optical isomer.

Ioxaglic acid (**18**), a monoionic dimer, is also widely used in angiographic–urographic applications. Because it contains six iodine atoms in two dissociated particles, (**18**) is classified as a ratio-3 LOCM. The key steps of its synthesis are as follows (73):

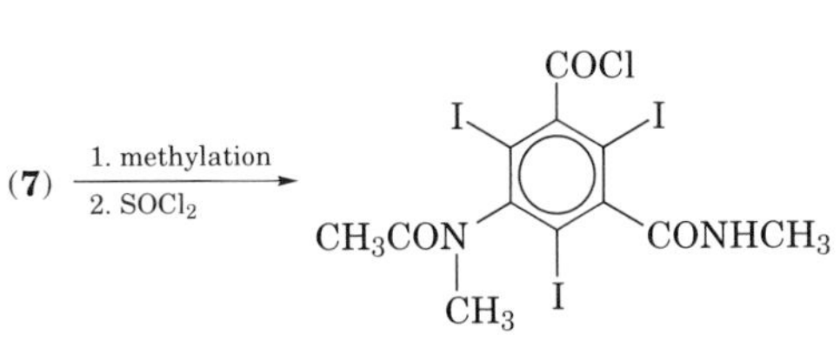

$$(7) \xrightarrow[\text{2. SOCl}_2]{\text{1. methylation}}$$

COCl

CH$_3$CON

CH$_3$ I

CONHCH$_3$

I (iodine substituents)

(27)

$$(27) + (28) \longrightarrow (18)$$

Economic Considerations.   The accumulation of clinical and animal data over many years has resulted in a universal consensus that LOCM are safer and cause less patient discomfort than HOCM (17–20,26–29,64–67,70,75). Although all research efforts are directed toward LOCM, these remain substantially more expensive than the HOCM (17). The LOCM are fundamentally more difficult to synthesize and purify and, therefore, more expensive to manufacture. The expenses associated with the research and development of new drugs are substantial. Licensing fees and royalties paid to the companies owning the LOCM patents inflict additional cost. These factors make LOCM 10–20 times more expensive than HOCM (65,75). Until cost can be reduced, the main benefit of LOCM may be to provide an additional margin of safety to those patients at higher risk (19). However, the LOCM reduce the risk of all levels of adverse reactions (66), not just those that are life-threatening. Thus, managing the adverse reactions of HOCM can become more costly than simply curtailing adverse reactions to CM by routine use of LOCM (75).

Myelography

The administration of a contrast agent into the subarachnoid space permits delineation of the spinal cord and is used for diagnosis of diseases of the nervous system and spinal canal (76). As early as 1919, air or other gases were used to provide negative contrast. Because gases are inadequately miscible with cerebrospinal fluid (CSF), large amounts of CSF had to be removed for the gas to occupy the space and effectively define the region. The occurrence of severe side effects and visualization restrictions contraindicates gas myelography.

Development of positive contrast material for myelography began with water-insoluble radiographic agents such as bismuth salts, colloidal silver, thorium dioxide, iodinated poppyseed oil, and the oil-based iophendylate [1320-11-2] (Pantopaque) (29). Because of high density and viscosity, iophendylate forms a cohesive oily mass that does not mix with CSF and can be repositioned by the action of gravity. Incomplete penetration of iophendylate around nerve roots and other narrow crevices may, however, produce inadequate definition.

After the procedure, the iophendylate is removed as much as possible by aspiration. Any remaining material is absorbed very slowly, tending to remain fixed in position, and can produce chronic tissue irritation.

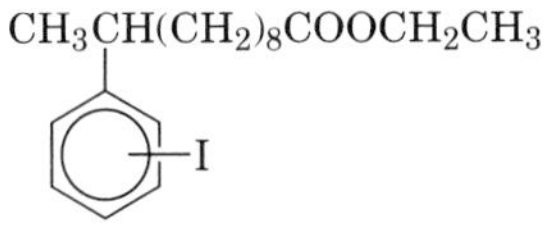

$$CH_3CH(CH_2)_8COOCH_2CH_3$$

(**29**)

Water-soluble contrast media (CM) are preferred because of effective mixing with CSF, plus the radiopaque is absorbed and effectively excreted in the urine, and does not have to be physically removed from the subarachnoid space after the procedure. Sodium methiodal, the first water-soluble agent used for myelography, produced neurotoxicity problems when exposed to the cells of the spinal cord and brain, thus limiting utility to the lumbar region and requiring the application of spinal or general anesthesia.

Soon after iothalamic acid (**7**) was introduced as a urographic–angiographic agent, it was recognized that its meglumine salt produced fewer neurotoxic effects than sodium methiodal, and iothalamate meglumine replaced sodium methiodal in myelographic procedures. The meglumine salt of iocarmic acid [*54605-45-7*] (**30**) also demonstrated decreased neurotoxicity (77). These two ionic agents were used extensively throughout the 1970s.

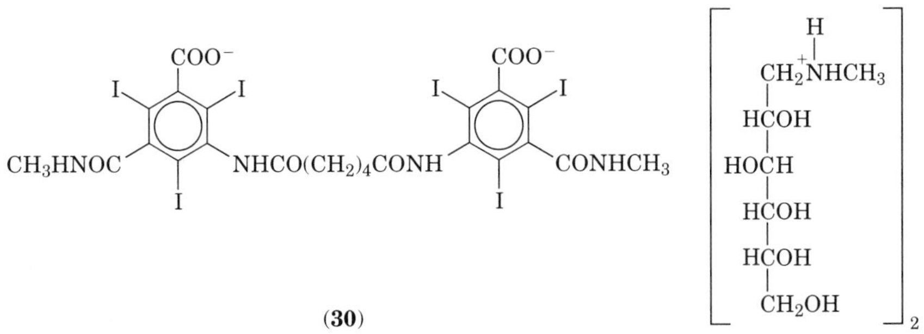

(**30**)

Iothalamate meglumine and iocarmate meglumine, both used clinically, are accompanied by significant adverse effects, such as muscle spasms and convulsions. These effects are related to the ionic nature of these agents and their hyperosmolality, which may disrupt the electrolyte balance in the CSF and the central nervous system. The nonionic CM offer a class of agents characterized by the absence of ionic charge and reduced osmolality. Metrizamide (**9**) substantially increases patient safety and decreases both acute and chronic adverse reactions, but high cost and instability in solution are drawbacks to use. The second-generation water-soluble nonionic CM, iopamidol (**10**) and iohexol (**11**), further decrease patient risk and offer lower cost and autoclaving stability. These last two agents are the approved and most widely used myelographic CM as of this writing (ca 1996).

An improved intrathecal radiographic agent should be nonionic, hydrophilic, and isoosmolar with CSF at the concentrations needed for radiography (78,79). The development of nonionic demeric media is focused on further reducing osmolality, concomitantly lowering the neurotoxicity. A new series of candidates for myelographic agents is available. Iotrolan (**31**) (79,80) and iodixanol (**32**) (81), available as of this writing in some European countries, are nonionic dimers that appear to possess higher neural tolerance and fewer side-effects in myelographic applications owing to the isotonic nature of these agents. Table 5 summarizes the important properties of these dimers. These compounds have six iodine atoms per molecule (ratio-6 CM) and exhibit the lowest osmolality of all water-soluble CM. On the other hand, they possess the highest viscosity, owing to large molecular sizes.

(**31**) R = CONHCH(CH$_2$OH)CH(OH)CH$_2$OH

(**32**) R = CONHCH$_2$CH(OH)CH$_2$OH

Cholecystography and Cholangiography

Radiographic studies of the gallbladder and bile duct with radiopaques are called cholecystography and cholangiography, respectively (85,86). Cholecystographic agents are administered orally for the evaluation of gallbladder abnormalities, such as gallstones. Cholangiographic agents are administered intravenously to produce opacification of the cystic and common bile ducts (85). Because of different biochemical transformations, oral cholecystographic agents and intravenous cholangiographic agents have distinct chemical requirements and pharmacokinetic properties.

Oral Agents.   The orally administered media are absorbed through the gastrointestinal tract and, after entering the portal venous circulation, are taken up by the liver. Following hepatocyte uptake, the cholecystographic agents are metabolized (conjugated) to form their glucuronide derivatives and excreted through bile into the gallbladder. Therefore, cholecystographic agents require both hydrophilic and lipophilic properties. Hydrophilicity provides adequate water-solubility, necessary for initial dissolution of the CM in the bulk-water phase of the small intestine. Lipophilicity permits the diffusion of CM through

Table 5. Properties of Nonionic Dimeric Contrast Media[a]

Name				Iodine content, %	Concentration,[b] mg/mL	Osmolality, mOsm/kg H$_2$O	Viscosity, cps 20°C/37°C	LD$_{50}$,[b] iv rats, g/kg
Generic	Trade	Company name	Mol wt					
Iotrolan	Isovist	Schering AG	1626	46.8	240	270 (82)	6.8/3.9 (82)	26[c] (83)
					300	320 (82)	16.4/8.1 (82)	12.7 (83)
Iodixanol	Visipaque	Nycomed Imaging	1550	49.1	300	200 (81)	18.9/8.7 (81)	>21 (84)
					350	220 (81)		

[a]References cited are given in parentheses.
[b]Values given are on a weight of iodine basis.
[c]In mice.

921

the lipid membrane of the gastrointestinal mucosa and influences the protein binding necessary for blood transport, liver uptake, and biliary excretion. Suitable oral cholecystographic CM are monomers of amino-triiodobenzene derivatives having alkanoic acid substituents (Table 6). Iopanoic acid [*96-83-3*] (**33**), developed in the early 1950s, represents the first of a series of these oral biliary contrast media (88). Other commercially available agents include ipodoic acid [*5587-89-3*] (**34**) (89), iocetamic acid [*16034-77-8*] (**35**) (90) and tyropanoic acid [*27293-82-9*] (**36**) (91). These agents each contain an aliphatic carboxylic acid group capable of ionizing to form a water-soluble salt for increased solubility and intestinal absorption. Iopanoic acid is the least water-soluble agent of the group and has the highest albumin binding (92). The agents developed subsequently each display increased water solubility. In contrast to angiographic–urographic agents that contain substituents at the 1, 3, and 5 positions in the aromatic ring, position 5 is unsubstituted in cholecystographic agents. This structural feature plays an important role in imparting lipophilic binding with serum albumin and hence facilitating hepatobiliary vs renal excretion of the cholecystographic media.

Intravenous Agents. Intravenous administration of biliary contrast agents circumvents the relatively slow absorption of the intestinal system (85) and allows for rapid and efficient heptocyte uptake and biliary excretion. Structurally, the intravenous biliary agents are dimers of triiodobenzene derivatives and differ only in the composition of the methylenic linkage. The relatively strong dibasic acid is ionized at physiological pH and, as a meglumine salt, is suitably soluble in water for intravenous administration. As for the oral agents, a high affinity for biliary excretion is based on the polar as well as lipophilic nature of the compounds, provided by the carboxylate groups and the unsubstituted 5 and 5′ positions on the dimers. Although it is not commercially readily available, the meglumine salt of iodipamide [*606-17-7*] (**37**) (Cholografin) (93,94) is the only cholangiographic agent used in the United States. Ioglycamide [*2618-25-9*] (**38**) (Biligram) (93,94) and iodoxamate [*51764-33-1*] (**39**) (Cholovue) (93,95) were previously used in these applications.

(**37**) R = (CH$_2$)$_4$
(**38**) R = CH$_2$OCH$_2$
(**39**) R = (CH$_2$CH$_2$O)$_4$CH$_2$CH$_2$

Although still valuable for selected studies, cholecystography and cholangiography have been largely replaced by other diagnostic modalities. Methologies include ultrasound, computed tomography, magnetic resonance, and radionuclide techniques (see MAGNETIC SPIN RESONANCE; MEDICAL IMAGING TECHNOLOGY; RADIOACTIVE TRACERS; ULTRASONICS).

Table 6. Chemical Structures and Solubilities of Cholecystographic Agents

| Name | | | Structure | | | Aqueous |
Generic	Trade	Company name	number	R	R′	solubility,[a] pH 7.4 and 37°C, mmol/L
iopanoic acid	Telepaque	Sterling-Winthrop Pharmaceuticals[b]	(33)	$CH_2CH(CH_2CH_3)COOH$	NH_2	0.61
ipodoic acid	Oragrafin	Squibb Diagnostics[c]	(34)	CH_2CH_2COOH	$N{=}CHN(CH_3)_2$	1.87[d] 3.75[e]
iocetamic acid	Cholebrine	Mallinckrodt Medical	(35)	$N(COCH_3)CH_2CH(CH_3)COOH$	NH_2	8.61
tyropanoic acid	Bilopaque	Sterling-Winthrop Pharmaceuticals[b]	(36)	$CH_2CH(CH_2CH_3)COOH$	$NHCOCH_2CH_2CH_3$	26.48[e]

[a]Refs. 85 and 87.
[b]Nycomed Imaging, as of 1996.
[c]Bracco Diagnostics, as of 1996.
[d]Calcium salt.
[e]Sodium salt.

Gastrointestinal Radiography

In the early development of radiopaques, barium sulfate [7727-43-7] was introduced for use in imaging the gastrointestinal (GI) tract (2,96). This compound has remained the agent of choice for gastrointestinal radiography (97). Barium sulfate forms a colloidal suspension and is administered orally when the regions of interest reside in the upper GI tract, and rectally when the lower GI tract is the focus. Being chemically inert and practically insoluble in water, barium sulfate demonstrates negligible absorption from the digestive system and produces few physiological side effects. It is excreted in the feces unchanged.

Two types of imaging techniques are routinely used. Single-contrast imaging is performed using a large volume of low density barium sulfate preparation to fill the entire lumen of the GI segment, to produce full-column opacification. Double-contrast imaging utilizes a smaller amount of a high density, low viscosity barium preparation to coat the mucosal surface. Air or carbon dioxide, through the oral administration of commercially available sodium bicarbonate preparations for upper GI procedures, is then administered to distend the region and provide a negative contrast. In this way, surface detail of the GI tract is finely delineated.

Because the regions of the alimentary tract vary widely in pH and chemical composition, many different commercial formulations of barium sulfate are available. The final preparations of varying viscosity, density, and formulation stability levels are controlled by the different size, shape, uniformity and concentration of barium sulfate particles and the presence of additives. The most important additives are suspending and dispersing agents used to maintain the suspension stability. Commercial preparations of barium sulfate include bulk and unit-dose powders and suspensions and principal manufacturers are E-Z-EM (Westbury, New York), Lafayette-Pharmacol, Inc. (Lafayette, Indiana), and Picker International, Inc. (Cleveland, Ohio).

Extravasation of barium sulfate into the peritoneal cavity through a perforated GI tract can produce serious adverse reactions. When a perforation is suspected, the use of a water-soluble iodinated contrast medium is indicated. In this case, oral or rectal administration of sodium or meglumine-sodium salts of diatrizoic acid (**6**) and oral use of iohexol (**11**) are the preferred procedures.

Computed Tomography

In computed tomography (CT) (98) the usual x-ray film image is replaced by sets of digitized matrices which represent the x-ray attenuation through the body. Multiple x-ray projections are utilized. After the data are computer-analyzed, cross-sectional views of the target organ(s) can be generated. The advantage of CT over the more conventional x-ray imaging technique is the greater contrast sensitivity to attenuation changes. However, because film is a continuous medium whereas the CT images are derived from digital picture elements (pixels), resolution of very small structures generated from a finite number of pixels can be limited using CT, as compared to conventional film-screen radiography.

The CT procedure can be performed with or without the use of intravenous contrast media. Contrast-enhanced CT involves the administration of a radiopaque to increase the degree of contrast between anatomical structures and to improve the differentiation between pathological and physiological phenomena. In general, because of the increased sensitivity of CT compared to film methods, lower concentrations of CM are indicated. The same water-soluble CM used in angiography and urography are successfully utilized to enhance contrast in CT (see Table 1). Because the CM reaches the various vessels and organs, eg, the brain, liver, and kidneys, at varying intervals, timing between CM administration and the collection of data is crucial. The CM gradually approaches equilibrium with body fluids and the resultant nonspecific opacification decreases the contrast sensitivity and curtails the clinically useful imaging time period.

In other applications of CT, orally administered barium sulfate or a water-soluble iodinated CM is used to opacify the GI tract. Xenon, atomic number 54, exhibits similar x-ray absorption properties to those of iodine. It rapidly diffuses across the blood brain barrier after inhalation to saturate different tissues of brain as a function of its lipid solubility. In preliminary investigations (99), xenon gas inhalation prior to brain CT has provided useful information for evaluations of local cerebral blood flow and cerebral tissue abnormalities. Xenon exhibits an anesthetic effect at high concentrations but otherwise is free of physiological effects because of its nonreactive nature.

Arthrography

The radiological visualization of joint cavities using contrast media is termed arthrography (100). Single-contrast arthrographic techniques utilize direct injection of a water-soluble contrast agent that readily mixes with the synovial fluid, producing opacification of the joint surfaces and cavity. The CM is then rapidly absorbed and excreted in the urine. Double-contrast arthrography involves the removal of the joint fluid and injection of a water-soluble positive contrast agent to coat the surfaces of the joint, followed by the introduction of air or carbon dioxide as a negative contrast medium to fill the cavity. Double-contrast techniques can result in a finer delineation of surface contours than the single-contrast method, especially when combined with computed tomography. The water-soluble CM used in arthrography include meglumine diatrizoate (**6**), meglumine iothalamate (**7**), sodium and meglumine ioxaglate (**18**), and iohexol (**11**). Adverse physiological effects, such as pain and swelling, are related to the hyperosmolality and chemical toxicity of the radiographic agent. The use of nonionic, ratio-3 LOCM minimizes these adverse effects (101).

Hysterosalpingography

Hysterosalpingography describes the radiological examination of the uterus and fallopian tubes for the purpose of detecting structural abnormalities and for the evaluation of fallopian tube patency. The CM for intrauterine administration include the oily agent Ethiodol, which consists of a mixture of ethyl esters of the iodinated fatty acids of poppy seed oil. The two main iodinated components of Ethiodol are diiodoethylstearate (**40**) and monoiodoethylstearate (**41**).

Iodipamide meglumine mixed with diatrizoate meglumine (Sinografin) is specifically indicated for the hysterosalpingographic procedure to provide adequate viscosity and proper retention of the agent. Other water-soluble agents include iohexol and meglumine-sodium salts of iothalamate, diatrizoate, and ioxaglate (see Table 1). Use of water-soluble agents eliminates the risk of adverse effects, such as granulomas and pulmonary embolism, resulting from prolonged retention of an oily agent (102).

$$CH_3(CH_2)_4\overset{\overset{\displaystyle I}{|}}{C}H(CH_2)_3\overset{\overset{\displaystyle I}{|}}{C}H(CH_2)_7COOCH_2CH_3 \qquad CH_3(CH_2)_7\overset{\overset{\displaystyle I}{|}}{C}H(CH_2)_8COOCH_2CH_3$$

(**40**)   (**41**)

Contrast Media Under Development

The continuing search for new radiopaque agents possessing desirable properties promises to benefit many aspects of diagnostic medicine. Permutations in the side-chain substituents affect the physicochemical and biological properties of the triiodobenzene CM. The presence of a primary carboxamide substituent, $-CONH_2$, enhances water solubility because of its polarity, whereas its small size does not interfere with intermolecular hydrophobic interactions. Hydrophobic bonding between molecules may produce aggregation of the CM in solution, thereby lowering the osmolality (103,104). Some heterocyclic substituents may confer an anticoagulant property which is desirable in a nonionic CM for angiographic applications (105,106). Water-insoluble particulate CM, derived from esters of metrizoic acid and iodipamide, have potential for contrast-enhanced CT scanning of the liver (107,108). In animal experiments, contrast-carrying liposomes using several CM have been demonstrated to be taken up by the reticuloendothelial system, thus functioning as imaging agents for the liver and spleen (109–115). Iodinated polymers having increased molecular weights that keep the material trapped in the vascular space longer are being investigated as blood-pool agents (116–119). Perfluorooctylbromide [423-55-2] (perflubron), $C_8F_{17}Br$, wherein bromine is the radiopaque element (98,120), has also been studied for blood-pool applications as well as for lymph node imaging using CT (121,122).

Other imaging techniques such as magnetic resonance and ultrasound have opened up avenues of tremendous potential for contrast medium enhancement (123). Ultrasound contrast media developments have centered around encapsulated air micro-bubbles. Magnetic resonance contrast agents involve metal–ligand complexes and have evolved from ionic to nonionic species, much as radiopaques have.

Radiopaqeus, including LOCM, HOCM, and barium sulfate, accounted for $\$3.32 \times 10^9$ in world revenue and $\$1.22 \times 10^9$ in U.S. revenue (124).

BIBLIOGRAPHY

"Radiopaques" in *ECT* 2nd ed., Vol. 17, pp. 130–142, by James Ackerman, Sterling-Winthrop Research Institute; in *ECT* 3rd ed., Vol. 19, pp. 786–801, by J. Ackerman, Sterling-Winthrop Research Institute.

1. *International Tables for X-ray Crystallography*, Vol. III, Kynoch Press, Birmingham, U.K., 1974, pp. 161 ff; distributed by Kluwer Academic Publishers, Dordrecht, the Netherlands.
2. W. H. Strain, in P. K. Knoefel, ed., *International Encyclopedia of Pharmacology and Therapeutics*, Sect. 76, Vol. 1, Pergamon Press, Oxford, U.K., 1971, Chapt. 1, pp. 1–22, for a review of the historical development of radiopaques.
3. M. Swick, *J. Am. Med. Assoc.* **95**, 1403 (1930).
4. H. L. Abrams, *Abrams Angiography: Vascular and Interventional Radiology*, 3rd ed., Vol. 1, Little, Brown and Co., Boston, Mass., 1983, pp. 3–14.
5. R. G. Grainger, *Br. J. Radiol.* **55**, 1 (1982).
6. J. O. Hoppe, A. A. Larsen, and F. Coulston, *J. Pharmacol. Exptl. Therap.* **166**, 394 (1956).
7. A. von Lichtenberg, *Br. J. Urol.* **3**, 119 (1931).
8. V. H. Wallingford, *J. Am. Pharmacol. Assoc.* **42**, 721 (1953).
9. V. H. Wallingford, H. G. Decker, and M. Kruty, *J. Am. Chem. Soc.* **74**, 4365 (1952).
10. G. B. Hoey and K. R. Smith, in M. Sovak, ed., *Radiocontrast Agents*, Springer-Verlag, New York, 1984, pp. 23–125.
11. G. B. Hoey, P. E. Wiegert, and R. D. Rands, in Ref. 2, Chapt. 2, pp. 23–131.
12. D. P. Swanson, P. C. Shetty, D. J. Kastan, and N. Rollins, in D. P. Swanson, H. M. Chilton, and J. H. Thrall, eds., *Pharmaceuticals in Medical Imaging*, Macmillan Co., New York, 1990, Chapt. 1, pp. 1–77.
13. H. W. Fischer, *Radiol.* **159**, 561 (1986).
14. J. O. Hoppe, L. P. Duprey, W. A. Borisenok, and J. G. Bird, *Angiography* **18**, 257 (1967).
15. G. B. Hoey, R. D. Rands, G. B. DeLaMater, D. W. Chapman, and P. E. Wiegert, *J. Med. Chem.* **6**, 24 (1963).
16. Technical data, Mallinckrodt Medical Inc., St. Louis, Mo., 1989.
17. B. L. McClennan, *Am. J. Roentgenol* **155**, 225 (1990).
18. R. F. Spataro, *Radiol. Clin. North Am.* **22**, 365 (1984).
19. B. L. McClennan, *Radiology* **162**, 1 (1987).
20. P. Dawson, *Invest. Radiology* **19**, S293 (1984).
21. H. W. Fischer, R. F. Spataro, and P. M. Rosenberg, *Arch. Intern. Med.* **146**, 1717 (1986).
22. T. Almén, *J. Theor. Biol.* **24**, 216 (1969).
23. T. Almén, *Invest. Radiol.* **15**(suppl), S283 (1980).
24. E. C. Lasser, J. H. Lang, A. E. Hamblin, S. G. Lyon, and M. Howard, *Invest. Radiol.* **15**(suppl), S2 (1980).
25. T. S. Padayachee, J. F. Reidy, D. H. King, M. Reeves, and R. G. Gosling, *Clin. Radiol.* **34**, 79 (1983).
26. *Invest. Radiol.* **19**(suppl), S164–S285 (1984).
27. *Ibid.*, pp. S289–S392.
28. *Invest. Radiol.* **20**(suppl), S2–S121 (1985).
29. *Invest. Radiol.* **24**(suppl), S1–S76 (1989).
30. J. C. Holder and G. V. Dalrymple, *Invest. Radiol.* **16**, 508 (1981).
31. I. J. Gordon and J. L. Wescott, *Radiology* **124**, 43 (1977).
32. T. Almén, *Invest. Radiology* **20**(suppl), S2 (1985).
33. Ger. Offen 2,031,724 (Jan. 7, 1971), T. Almén, J. Haavaldsen, and V. Nodal (to Nyegaard & Co.).
34. Ger. Offen 2,547,789 (Jan. 24, 1976), E. Felder and D. E. Pitrè (to Savac AG).
35. U.S. Pat. 4,001,323 (Jan. 4, 1977), E. Felder, R. S. Vitale, and D. E. Pitrè (to Savac AG).
36. E. Felder, M. Grandi, D. Pitrè, and G. Vittadini, in K. Florey, ed., *Analytical Profiles of Drug Substances*, Vol. 17, Academic Press, Inc., New York, 1988, pp. 115–154.

37. T. K. Kawada, *Drug Intell. Clin. Phar.* **19**, 525 (1985).
38. Ger. Offen 2,726,196 (June 10, 1977), V. Nordal and H. Holtermann.
39. Brit. Pat. 1,548,594 (June 11, 1976), V. Nordal and H. Holtermann.
40. U.S. Pat. 4,250,113 (Feb. 10, 1981), V. Nordal and H. Holtermann.
41. U.S. Pat. 4,396,598 (Aug. 2, 1983), Y. Lin.
42. U.S. Pat. 4,352,788 (Oct. 5, 1982), E. Felder, R. S. Vitale, and D. Pitrè (to Bracco Industria Chimica).
43. M. Sovak, in Z. Parvez, ed., *Contrast Media: Biologic Effects and Clinical Applications*, Vol. 1, CRC Press, Boca Raton, Fla., 1987, pp. 27–45.
44. Technical data, Squibb Diagnostics, Princeton, N.J., 1993.
45. J. Haavaldsen, V. Nordal, and M. Kelly, *Acta Pharm. Suec.* **20**, 219 (1983).
46. Technical data, Sanofi-Winthrop, New York, 1993.
47. B. Bonnemain, personal communication, Guerbet Laboratory, Aulnay-sous-Bois, France, Nov. 1994.
48. W. H. Ralston, M. S. Robbins, J. Coveney, and M. Blair, *Invest. Radiol.* **24**(suppl), S2 (1989).
49. Technical data, Mallinckrodt Medical Inc., St. Louis, Mo., 1994.
50. S. P. H. Eggleton and E. Puchert, *X-Ray Contrast Media Made Clear*, Schering AG, Germany, 1992.
51. Eur. Pat. Appl. 105,752 A1 (Apr. 18, 1984), K. Wille (to Nyegaard & Co.).
52. K. Skinnemoen, *Acta Radiol.* **370**(suppl), 33 (1987).
53. U.S. Pat. 4,954,348 (Sept. 4, 1990), M. Sovak and R. Ranganathan.
54. Technical data, Mallinckrodt Medical Inc., St. Louis, Mo., 1988.
55. U.S. Pats. 3,701,771 (Oct. 31, 1972) and 4,021,481 (May 3, 1977), T. Almén, J. Haavaldsen, and V. Nordal.
56. M. D. Adams, W. H. Ralston, J. W. Miller, and J. A. Ferrendelli, *Proceedings of the Second International Symposium on Contrast Media*, Osaka, Japan, Nov. 9–10, 1990; *Excerpta Medica*, 17–23 (1991); M. D. Adams, W. H. Ralston, J. W. Miller, and J. A. Ferrendelli, *Invest. Radiol.* **25**(suppl), S86 (1990).
57. W. Mützel and U. Speck, *Fortschr. Geb. Röntgenstr. Nuklearmed. Ergänzungsbd.* **118**, 11 (1983).
58. B. Bonnemain and co-workers, *Invest. Radiol.* **25**(suppl), S104 (1990).
59. A. Morisetti, P. Tirone, F. Luzzani, and C. de Haën, *Eur. J. Radiol.* **18**(suppl), S21 (1994).
60. M. Sovak, *Invest. Radiol.* **23**(suppl), S79 (1988).
61. J. H. Wible, Jr., S. J. Barco, D. E. Scherrer, J. K. Wojdyla, and M. D. Adams, *Eur. J. Radiol.* **19**, 206–211 (1995).
62. A. A. Michelet, S. Salvesen, and T. Renaa, *Acta Radiol.* **370**(suppl), 41 (1987).
63. R. W. Katzberg and co-workers, *Invest. Radiol.* **25**, 46 (1990).
64. R. F. Spataro, *Urol. Radiol.* **10**, 2 (1988).
65. B. F. King, G. W. Hartman, B. Williamson, A. J. LeRoy, and R. R. Hattery, *Mayo Clin. Proc.* **64**, 976 (1989).
66. H. Katayama and co-workers, *Radiology* **175**, 621 (1990).
67. T. Almén, *Invest Radiol.* **29**(suppl), S37 (1994).
68. P. K. Knoefel, in Ref. 2, Chapt. 3, pp. 131–145.
69. H. Levitan and S. I. Rapoport, *Acta Radiol. Diagn.* **17**, 81 (1976).
70. T. Almén, *Am. J. Cardiol.* **66**, 2F (1990).
71. W. Krause, *Invest. Radiol.* **29**(suppl), S21–S32 (1994), and other articles in this supplement, pp. S68–S117.
72. Eur. Pat. Appl. 437,144 A1 (Dec. 26, 1990), M. Schaefer, M. Dugat-Zrihen, M. Guillemot, D. Doucet, and D. Meyer (to Guerbet SA).
73. U.S. Pat. 4,065,554 (Dec. 27, 1977), G. Tilly, M. J. C. Hardouin, and J. Lautrou (to Guerbet Laboratories).

74. P. Blaszkiewicz, *Invest. Radiol.* **29**(suppl), S51 (1994).

75. D. R. Dakins, *Diag. Imag.*, 43 (Feb. 1991).

76. D. P. Swanson and R. S. Boulos, in Ref. 12, Chapt. 4, pp. 125–154, and references therein.

77. G. B. Hoey and co-workers, *J. Med. Chem.* **9**, 964 (1966).

78. M. Sovak, *Invest. Radiol.* **19**(suppl), S134 (1984).

79. M. Sovak, R. Ranganathan, and B. Hammer, *Invest. Radiol.* **19**(suppl), S139 (1984).

80. U.S. Pat. 4,341,756 (July 27, 1982), M. Sovak and R. Ranganathan (to University of California).

81. Eur. Pat. Appl. 108,638 A1 (May 16, 1984), P-E. Hanson, H. Holtermann, and K. Wille (to Nyegaard & Co.); Eur. Pat. 108,638 B1 (July 7, 1986).

82. *Isovist™ technical brochure*, Schering Health Care, West Sussex, U.K.

83. M. Sovak, R. Ranganathan, and U. Speck, *Radiology* **142**, 115 (1982).

84. J. O. Nossen, T. Aakhus, K. J. Berg, N. P. Jorgensen, and E. Andrew, *Invest. Radiol.* **25**(suppl), S113 (1990).

85. D. P. Swanson and S. M. Simms, in Ref. 12, Chapt. 6, pp. 184–220.

86. V. G. Urich and U. Speck, *Progr. Pharm. Clin. Pharm.* **8**, 307 (1991).

87. J. O. Janes, J. M. Dietschy, R. N. Berk, P. M. Loeb, and J. L. Barnhart, *Gastroenterology* **76**, 970 (1979).

88. T. R. Lewis and S. Archer, *J. Am. Chem. Soc.* **71**, 3753 (1949).

89. H. Priewe and A. Poljak, *Chem. Ber.* **93**, 2347 (1960).

90. J. A. Korver, *Rec. Trav. Chim. Pays-Bas* **87**, 308 (1968).

91. J. O. Hoppe, J. H. Ackerman, A. A. Larsen, and J. Moss, *J. Med. Chem.* **13**, 997 (1970).

92. J. H. Lang and E. C. Lasser, *Invest. Radiol.* **2**, 396 (1967).

93. J. L. Barnhart, in Ref. 10, pp. 367–418.

94. H. Priewe, R. Rutkowski, K. Pirner, and K. Junkmann, *Chem. Ber.* **87**, 651 (1954).

95. E. Felder, D. Pitrè, L. Fumagalli, and E. Lorenzotti, *Farmaco Ed. Sci.* **28**, 912 (1973).

96. C. Bachem and H. Gunther, *Z. Rontgenk. Rad. Forschr.* **12**, 369 (1910).

97. D. P. Swanson and R. D. Halpert, in Ref. 12, Chapt. 5, pp. 155–183.

98. D. P. Swanson and M. B. Alpern, in Ref. 12, Chapt. 3, pp. 99–124.

99. J. S. Meyer, L. A. Hayman, M. Yamamoto, F. Sakai, and S. Nakajima, *Am. J. Roentgenol.* **135**, 239 (1980).

100. D. P. Swanson and B. I. Ellis, in Ref. 12, Chapt. 7, pp. 221–252.

101. F. M. Hall, D. I. Rosenthal, R. P. Goldberg, and G. Wyshak, *Am. J. Roentgenol.* **136**, 59 (1981).

102. H. Y. Yune, in R. E. Miller and J. Skucas, eds., *Radiographic Contrast Agents*, University Park Press, Baltimore, Md., 1977, pp. 307–321.

103. M. Sovak, R. C. Terry, J. G. Douglass, and L. Schweitzer, *Invest. Radiol.* **26**(suppl), S159 (1991).

104. Eur. Pat. Appl. 406,992 A2 (June 29, 1990), M. Sovak (to Schering Aktiengesellschaft).

105. Eur. Pat. Appl. 431,838 A1 (Nov. 29, 1990), R. S. Ranganathan, E. R. Marinelli, T. Arunachalam, and R. Pillai (to Squibb & Sons).

106. Eur. Pat. Appl. 516,050 A2 (May 26, 1992), R. S. Ranganathan, P. W. Wedeking, M. F. Tweedle, and R. Pillai (to Squibb & Sons).

107. M. S. Sands, M. R. Violante, and G. Gadeholt, *Invest. Radiol.* **22**, 408 (1987).

108. P. Leander and co-workers, *Invest. Radiol.* **28**, 513 (1993).

109. S. E. Seltzer, *Invest. Radiol.* **23**(suppl), S122 (1988).

110. D. Revel and co-workers, *Invest. Radiol.* **25**(suppl), S95 (1990).

111. T. Gjoen and co-workers, *Invest. Radiol.* **25**(suppl), S98 (1990).

112. R. Passariello and co-workers, *Invest. Radiol.* **25**(suppl), S92 (1990).

113. A. S. Janoff and co-workers, *Invest. Radiol.* **26**(suppl), S167 (1991).
114. S. E. Seltzer and co-workers, *Invest. Radiol.* **26**(suppl), S169 (1991).
115. W. Krause, A. Sachse, S. Wagner, U. Kollenkirchen, and G. Rössling, *Invest. Radiol.* **26**(suppl), S172 (1991).
116. Int. Pat. Appl. WO 88/06162 (Aug. 25, 1988), D. Paris, J-M. Nigretto, B. Bonnemain, D. Meyer, and D. Doucet.
117. D. Doucet, D. Meyer, C. Chambon, and B. Bonnemain, *Invest. Radiol.* **26**(suppl), S53 (1991).
118. D. Revel and co-workers, *Invest. Radiol.* **26**(suppl), S57 (1991).
119. J. Lautrou and co-workers, *Invest. Radiol.* **25**(suppl), S109 (1990).
120. D. M. Long, E. C. Lasser, C. M. Sharts, F. K. Multer, and M. Nielsen, *Invest. Radiol.* **15**, 242 (1980).
121. R. F. Mattrey, *Invest. Radiol.* **26**(suppl), S55 (1991).
122. G. Hanna and co-workers, *Invest. Radiol.* **29**(suppl), S33 (1994).
123. D. D. Shaw, *Invest. Radiol.* **28**(suppl), S138 (1993).
124. Frost and Sullivan, *World Contrast Media Market, New Dynamics in an Evolving Market*, Mallinkrodt Medical, St. Louis, Mo., 1994.

General References

J. Ackerman, *Diagnostic Agents*, in A. Burger, ed., *Medicinal Chemistry*, 3rd ed., Part II, John Wiley & Sons, Inc., New York, 1970, Chapt. 67, pp. 1686–1699.
P. K. Knoefel, ed., *International Encyclopedia of Pharmacology and Therapeutics*, Sect. 76, Pergamon Press, Oxford, U.K., 1971.
Z. Parvez, ed., *Contrast Media: Biologic Effects and Clinical Applications*, 3 vols., CRC Press, Boca Raton, Fla., 1987.
M. Sovak, ed., *Radiocontrast Agents*, Springer-Verlag, New York, 1984.
M. Sovak, in M. Sovak, ed., *Handbook of Experimental Pharmacology*, Vol. 73, Springer-Verlag, New York, 1984.
D. P. Swanson, H. M. Chilton, and J. H. Thrall, eds., *Pharmaceuticals in Medical Imaging*, Macmillan Co., New York, 1990.

YOULIN LIN
Mallinckrodt Medical, Inc.

RADIOPHARMACEUTICALS

Radiopharmaceuticals form the chemical basis of the medical specialty of nuclear medicine, a group of techniques used primarily for diagnosis, but also to a lesser degree in the treatment of disease. *In vivo* diagnostic information is obtained by intravenous injection of compounds tagged with rapidly decaying radioactive isotopes. The biological distribution of these compounds is then determined using a gamma-camera. This distribution usually takes a form that is organ- and lesion-specific, although this is not always the case. From the distribution of radioactivity and its behavior over time, it is possible to obtain information about the presence, progression, and state of disease. Localized changes in the shape or concentration of radioactivity in a given organ or structure reflect alterations in either organ anatomy or local function (see RADIOACTIVE TRACERS; RADIOISOTOPES).

Unlike other medical imaging modalities such as x-rays and x-ray computed tomography (see X-RAY TECHNOLOGY), ultrasound (see ULTRASONICS), or magnetic resonance imaging (mri) (see MAGNETIC SPIN RESONANCE), nuclear medicine studies provide information about the functional state of tissues rather than primarily anatomy or shape of regions of differing fixed properties (see MEDICAL IMAGING TECHNOLOGY). X-rays and computed tomography yield images that are largely reflective of tissue density, a property that tends to change significantly only with injury, deformity, or advanced disease. Ultrasound imaging, based on reflections arising from the density gradients present in the body, tends to reveal the borders of regions of differing density. Mri tends to measure a gross parameter, water content of tissues, and thus also tends to be reflective of the anatomy of the tissue rather than its function. Although these modalities can yield enhanced information by the use of injected contrast agents, these agents serve primarily to visualize anatomic structures that could not be detected without them. They are not generally used to obtain functional information. This limitation results primarily from the relative insensitivity of these modalities to the small concentrations of contrast, and the even smaller variation in that concentration, that are needed to reveal functional information.

One of the great advantages of techniques using radiopharmaceuticals is extreme sensitivity. Compared to other *in vivo* diagnostic methodologies, nuclear medicine techniques can provide critical information upon introduction of many orders of magnitude less compound into the body. This is possible because the external measurement of radioactivity is accomplished by detection of individual photons, each of which represents the disintegration of a single atom. In order to obtain a sufficient signal to noise ratio, mri imaging must excite approximately one nucleus in 10^6. Thus for each atom producing a readable signal, approximately one million others must be present that produce no signal. Total injected quantities of contrast for mri and computed tomography are in the milligram to tens of grams range.

It is possible to achieve an excellent measure of the quantity of radioactivity upon disintegration of several hundred thousand atoms, ie, 10^{-18} moles. In practice, many effects, including limited sensitivity when imaging, a decay time that is longer than the measurement intervals, and absorption and scatter of photons in transit, result in the need for higher concentrations. Nonetheless, total injected quantities of tracer in the picomole, ie, nanogram range for many compounds, are common.

Nuclear medicine has found applications in measurements of cardiac mechanical function and localized tissue perfusion; in renal, hepatobiliary, and pulmonary function; and in investigations of localized perfusion in the brain and peripheral vasculature. Nuclear medicine has seen considerable use in finding and following solid tumors and their metastases, particularly those that localize in bone, as well as in identifying regions of abnormal bone metabolism. Therapy using radiopharmaceuticals can reduce pain in bony metastases, ablate hyperactive thyroid tissue in Grave's disease, and cure thyroid cancer. Newer agents have provided the capability to identify a broader range of tumors, and agents on the horizon promise increasing access to this information. Agents under development promise to aid in the discovery and localization of intravascular thrombosis, both arterial and venous, and in localization of the sites of focal infection and of tissues that are hypoxic but remain viable.

A number of excellent general texts are available that describe nuclear medicine and radiopharmaceuticals and the various technologies that underlie both (1–6).

Efficacy

In common with other pharmaceuticals, the two primary requirements for success are safety and efficacy. Safety is determined both by chemical toxicity and radiation dose delivered to the patient.

It is truly possible to imagine the characteristics of an ideal radiopharmaceutical only in the context of a specific disease and organ system to which it might be applied. Apart from the physical factors related to the radioisotope used, the only general characteristic that is important in defining the efficacy of these materials is the macroscopic distribution in the body, or biodistribution. This time-dependent distribution at the organ level is a function of many parameters which may be divided into four categories: factors related to delivery of the radiopharmaceutical to a particular tissue; factors related to the extraction of the compound from circulation; factors related to retention of the compound by that tissue; and factors determined by clearance. The factors in the last set are rarely independent of the others.

Delivery factors include tissue perfusion and competing extraction by nontarget tissues, including unwanted binding to blood components. Factors related to extraction are highly dependent on the method of extraction; although in general, greater lipophilicity is related to more efficient tissue extraction from circulation. Some compounds simply diffuse across cell membranes, either passively or as in the case of ^{99m}Tc-sestamibi [109581-73-9], a moderately lipophilic monovalent cation, driven by transmembrane potentials. Others, such as the thallous ion, a potassium analogue, are actively incorporated, usually through a pathway intended for a chemically similar substrate. Still others are not extracted to any significant degree except by excretion pathways, when the target is an accessible intravascular one.

Retention, too, is highly tissue-specific. Sometimes, the extraction mechanism is also the retention mechanism, as for ^{99m}Tc-sestamibi, which is retained in mitochondria as long as transmembrane potentials remain intact. Others are separate. ^{18}F-2-Fluorodeoxyglucose enters the cell by the same pathway as glucose, but is trapped because it is not a substrate for hexokinase, preventing further intracellular metabolism.

Excretion factors are often related to lipophilicity. More lipophilic compounds tend to be excreted by the liver into the bile, resulting in elimination ultimately in the feces. As this is a relatively slow process, much of the radioactivity having a shorter half-life decays before being eliminated. Polar compounds are more likely to be excreted by the kidneys.

The biodistribution of the ideal radiopharmaceutical would show extremely rapid distribution via the circulatory system to the organs of interest and little distribution to others. It would be rapidly extracted by the organ or tissue of interest differentially in a way that reflects the disease process of interest. Ideally, abnormalities should be defined by substantial increases in the concentration of the agent because lesions defined by a decrease from surrounding concentrations, ie, cold-spot imaging, are more difficult to image.

The compound should be rapidly cleared from the circulation, but not so quickly as to preclude complete extraction in the tissue of interest, to keep background radioactivity to a minimum. The excretion pathway should be preferentially via the renal system unless this obscures structures of interest. Renal excretion is faster and elimination of excreted radioactivity from the body by frequent voiding can reduce the overall body burden more quickly than can fecal excretion via the biliary tree and gut.

Kinetics of the biodistribution must be compatible with the practical aspects of hospital routine and imaging capability. In the case of a diagnostic agent, maximal lesion contrast, maximal radioactivity concentrations in tissue of interest, and minimal background radioactivity during the time imaging takes place are desired. This is primarily limited by practical considerations from several hours to as much as a day following administration, for the beginning of imaging, and about 40 minutes imaging time. One-hour imaging studies can be done, but such procedures tend to be stressful for the patient. Imaging times greater than an hour are generally not practical as few people can remain still for that long, even when healthy.

For therapeutic compounds, the considerations are similar, although timing requirements are quantitatively different. A diagnostic agent must retain the optimal or at least acceptable biodistribution only during the imaging time. For a therapeutic, the agent must exhibit a high target-to-nontarget ratio during the entire time following administration until the radioactive decay of the radioisotope reduces the radiation delivery rate to negligible values.

Nuclear Medicine Studies

Nuclear medicine studies may reveal information that is primarily anatomic in nature, or indicate the function of an organ on a regional basis (Table 1). These studies may be intended to identify new disease, confirm or deny suspected disease, or follow the progress of treatment or the course of disease. The diseases may be relatively benign or extremely serious and can range from widespread medical problems such as ischemic heart disease to rarities such as Legge-Perthe's disease and malignant pheochromocytoma (7).

Nuclear Medicine Images. Two commonly performed nuclear medicine studies are indicative of the way in which radioactive diagnostic agents may be used to elucidate the cause of a specific complaint. In Figure 1, a bone scan is used to identify the cause of acute lower back pain. Plain radiographs showed demineralization of the bones, but no specific abnormalities could be identified to account for the pain. The increased uptake in the sacral bone is in the pattern of the letter H, corresponding to the shape of this bone. This pattern, called the

Table 1. Medicine Diagnostic Studies and Associated Radiopharmaceuticals

Study	Disease target(s)	Organ	Radiopharmaceutical
myocardial perfusion	coronary artery disease, myocardial infarction	heart	^{201}Tl–Cl ^{82}Rb–Cl ^{99m}Tc-sestamibi ^{99m}Tc-teboroxime ^{99m}Tc-tetrafosmin
bone	bone metastases, osteomyelitis, stress fractures	skeleton	^{99m}Tc-medronate ^{99m}Tc-oxidronate
gallium study	primary tumors and metastases, infection, inflammation	any	^{67}Ga-citrate
white cell study	infection, inflammation	any	^{111}In-WBC[a] ^{99m}Tc-WBC[b]
renal study	impaired renal function	kidney	^{111}In-DTPA ^{131}I-iodohippurate ^{99m}Tc-DTPA ^{99m}Tc-gluceptate ^{99m}Tc-mertiatide
brain perfusion	stroke, other perfusion abnormalities	brain	^{123}I-iofetamine ^{99m}Tc-exametazime ^{99m}Tc-bicisate
hepatobiliary study	acute cholecystitis	gall bladder	^{99m}Tc-disofenin ^{99m}Tc-mebrofenin
lung perfusion–ventilation (V/Q)	pulmonary embolism	lung	^{99m}Tc-albumin aggregated (perfusion) and ^{133}Xe (ventilation)
tumor localization	certain neuroendocrine tumors	any	^{111}In-pentetreotide
tumor localization	nonliver metastases of colorectal or ovarian cancer	any	^{111}In-satumomab pendetide
liver–spleen	metastatic disease of the liver	liver, spleen	^{99m}Tc-albumin colloid
tumor localization	pheochromocytoma neuroblastoma	any	^{131}I-iobenguane sulfate
thyroid study	thyroid carcinoma, hyperthyroidism	thyroid	^{131}I–NaI, ^{123}I–NaI
cardiac ejection fraction/wall motion	coronary artery disease, myocardial infarction, cardiomyopathy, other diseases affecting muscle function	heart	^{99m}Tc-red blood cells[c] ^{99m}Tc-sestamibi

[a] ^{111}In-Oxine.
[b] ^{99m}Tc-Exametazime.
[c] Using ^{99m}Tc–TcO$_4^-$ and stannous pyrophosphate.

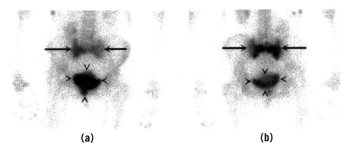

(a)   (b)

Fig. 1. Bone scan of a 75-year-old woman presented with acute onset of low back pain employing either ^{99m}Tc-medronate [25681-89-4] or ^{99m}Tc-oxidronate [14255-61-9]. The bone scan of (**a**) the anterior and (**b**) the posterior pelvis shows increased uptake in the region of the sacral bone (arrows). The bladder (arrowheads) is a normal route of tracer excretion and is also prominently identified in the image.

H sign, is typical of sacral insufficiency fracture, a fracture which is the result of severe bone loss. In Figure 2, a lung or V/Q scan is used to rule out or verify the suspicion of pulmonary embolism. The lung scan uses ^{99m}Tc macroaggregated albumin for the perfusion phase and either ^{133}Xe gas, ^{81}Rb gas, or ^{99m}Tc aerosol for the ventilation stage. This ventilation/perfusion mismatch is a result of a pulmonary embolism, ie, blood clots typically from the veins in the legs traveling through the body and lodging in the lungs.

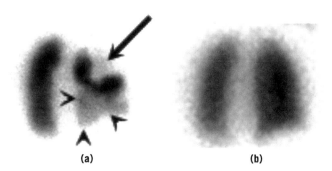

(a)   (b)

Fig. 2. Right posterior oblique lung or V/Q scan of a 65-year-old man presented with acute onset of shortness of breath and pleuritic chest pain (pain with breathing). (**a**) The perfusion lung scan showing defects in both the right upper lobe (arrow) and in the right lower lobe (arrowheads). (**b**) The ventilation scan in the same projection showing normal ventilation.

Isotopes for Nuclear Medicine

Radioactive isotopes are characterized by a number of parameters in addition to those attributable to chemistry. These are radioactive half-life, mode of decay, and type and quantity of radioactive emissions. The half-life, defined as the time required for one-half of a given quantity of radioactivity to decay, can range from milliseconds to billions of years. Except for the most extreme conditions under very unusual circumstances, half-life is independent of temperature, pressure, and chemical environment.

Radioactivity is equal to the rate of decay of a given radioisotope. This quantity is proportional to the number of radioactive atoms present, so that for a single isotope,

$$A = N \cdot \lambda$$

where A is the radioactivity in Becquerels and one Becquerel (1 Bq) is equal to one disintegration per second (the Curie is equal to 3.7×10^{10} Bq); N is the number of radioactive atoms; and λ is the probability of decay per second. The probability of decay is constant, thus

$$\frac{dN}{N} = \lambda \, dt$$

and integrating with respect to time,

$$N = N_0 e^{-\lambda t}$$

where N_0 is the initial number of radioactive atoms present at the beginning of the time interval, t. When N/N_0 is 0.5, the resulting value of t is the half-life, $t_{1/2}$, the time during which a radioactive material decays to one-half its original amount. Radioactive materials are often listed by half-life rather than by decay constant.

Substituting the following result,

$$N = N_0 e - ((t \cdot \ln 2)/t_{1/2})$$

from which it may be observed that both radioactivity and the quantity of radioactive atoms present decrease exponentially and are proportional to one another.

Diagnostic Radioisotopes.   In order to be useful as an *in vivo* diagnostic agent, a radioisotope needs to possess a number of characteristics. (*1*) It must be practical to produce in sufficient quantities so that the cost of a patient dose is within acceptable limits. (*2*) It must be available at a concentration, ie, specific activity, sufficient to allow labeling reactions to proceed and to permit practical volumes of injectate for the required amount of radioactivity. (*3*) Its half-life must be sufficiently short so that the patient receives a radiation dose that is within acceptable limits. The half-life must be long enough, however, so that there is not unacceptable loss of isotope from decay during transportation between the site of production and the patient. It must also be long enough that, for a given chemical form and biological system of interest, there is time for the radioactivity to reach the required distribution in the organs of interest. (*4*) It must yield an acceptable radiation dose in patients for the injected chemical form and amount needed for a particular patient procedure. *In vivo* studies always require consideration of radiation dose. Although most *in vivo* studies involve creating images of the radioactivity distribution in the patient using a gamma camera, there are some, such as the Shilling test for circulating levels of cyanocobalamin, that do not. (*5*) It must be possible to manufacture the radioisotope in a form that is compatible with labeling methodologies and be of sufficient radiophysical, ie, isotopic, and radiochemical purity. (*6*) It must emit radiation that can be detected readily in a way that the required information can be extracted.

Types of Diagnostic Isotopes. Isotopes used in nuclear medicine may be characterized by the source used to produce the radioactive isotope, by whether the isotopes are produced at a central location and shipped or at the clinic, or by the type of emission and thus the equipment used to detect them. The first of these, the sources, are summarized in Table 2. Some isotopes may be produced by more than one method.

Isotopes arising from reactors are produced by the addition of neutrons. These lie below the line of stability, thus relatively few diagnostically useful isotopes are produced by reactor methods. These isotopes tend to decay by β^--emission, a mode that results in a generally higher radiation dose per disin-

Table 2. Diagnostic Radioisotopes by Means of Production

Radioisotope	Principal decay modes[a]	Half-life
Accelerator		
commercially available		
^{201}Tl	EC	72 h
^{111}In	EC	67 h
^{133}Xe	EC	5.27 d
^{67}Ga	EC	77.9 h
on-site production		
^{11}C	β^+	20.4 min
^{13}N	β^+	10.05 min
^{15}O	β^+	124 s
^{18}F	β^+	112 min
Generator		
commercially available		
^{99m}Tc	IT	6.02 h
^{81m}Kr	IT	13 s
^{68}Ga	β^+	68 min
^{82}Rb	β^+	1.25 min
not commercially available		
^{191m}Ir	IT	5 s
^{178}Ta	EC	9.35 min
^{77m}Se	IT	17.5 s
^{113m}In	IT	104 min
Reactor		
commercially available		
^{133}Xe	EC	5.27 d
^{125}I	EC	60 d
^{131}I	β^-	8.08 d
not commercially available		
^{135m}Ba	IT	28.7 h
^{195m}Pt	IT	3.5 d
	β^+	112 min

[a]IT = isomeric transition, EC = electron capture, β^+ = positron emission, and β^- = beta decay.

tegration than other modes. Accelerator-produced radionuclides are formed by the addition of protons. These tend to lie above the line of stability, resulting in a preference for decay by positron, β^+, emission and electron capture. The remainder are produced by generators. The parent isotopes of those produced by generator are often reactor products.

Accelerator-Produced Isotopes. Particle accelerators cause nuclear reactions by bombarding target materials, which are often enriched in a particular stable isotope, with rapidly moving protons, deuterons, tritons, or electrons. Proton reactions are most commonly used for production purposes.

For accelerator-produced isotopes, the cost of the isotope is directly related to the time and total energy necessary to produce a unit of radioactivity. A portion of the capital cost of building the cyclotron is also added. Owing to the cost and siting difficulties inherent in a large production-scale cyclotron, there are relatively few cyclotrons. The need to ship the resulting products long distances tends to dictate that these products have relatively long half-lives. The most commonly used accelerator-produced isotopes are ^{201}Tl [*15064-65-0*] as thallous chloride for myocardial perfusion, ^{133}Xe [*14932-42-5*] for pulmonary ventilation studies, ^{67}Ga as gallium citrate for imaging of inflammation and certain tumors, and ^{111}In for white cell labeling, studies using antibodies, and other receptor antagonists. In addition, ^{123}I is used in a form bound to an amphetamine derivative, ^{123}I-iofetamine [*75917-92-9*], for imaging localized brain perfusion.

Alternatively, smaller on-site cyclotrons can provide immediate access to substantial quantities of short half-life positron-emitting isotopes such as ^{11}C, ^{13}N, ^{15}O, and ^{18}F. Less expensive than large, centralized machines, these smaller cyclotrons remain in the neighborhood of \$1,500,000. In addition, the annihilation radiation produced indirectly by these isotopes is best imaged using a positron camera. This device has both advantages and disadvantages. Although single-photon emission computed tomography (SPECT) imaging of positron emitters is widely available commercially, this method suffers from poor sensitivity and resolution. Better resolution has been achieved using coincidence methods and dual-headed conventional cameras, although sensitivity remains low and injected activity is restricted owing to count rate limitations.

These cyclotron-produced positron-emitting isotopes of biologically common elements can be used to create direct analogues of molecules of biological interest, obviating the need for complex labeling chemistry involving large nonphysiological moieties. Although this approach is attractive, direct substitution into physiological molecules often results in a tracer that is rapidly metabolized. This creates an unstable biological distribution that can be difficult to image, thus derivatives are often used. The most prominent of these is ^{18}F-2-fluorodeoxyglucose which serves as a marker of glycolytic activity and can create images of viable but stunned myocardial tissue and a variety of tumors.

The combination of a PET camera and small cyclotron facility provides ready access to the ability to image the *in vivo* distribution of an enormous variety of molecules. The importance of this technique to research is substantial.

Generator-Produced Isotopes. Although the isotopes from which these are made are derived by one of the other methods described, generators may be considered separately as a production method. Such systems can produce large amounts of radioactivity at high concentrations inexpensively. To create a generator system, a parent–daughter pair of isotopes must be identified in which the daughter has the desirable properties of a diagnostic radiopharmaceutical and the parent has a similar or longer half-life and is sufficiently different chemically that separation is simple. The most commonly used radionuclide, ^{99m}Tc, is produced in a generator system from the parent ^{99}Mo. When configured as a generator, the ^{99}Mo, half-life 66 h, is immobilized and the ^{99m}Tc, half-life 6.02 h, eluted periodically. The long half-life, 200,000 yr, of the daughter ^{99}Tc ensures that only minute amounts of radioactivity remain following injections of ^{99m}Tc-labeled compounds.

$$^{99}\text{Mo} \longrightarrow {}^{99m}\text{Tc} \longrightarrow {}^{99}\text{Tc}$$

A chromatographic generator is a shielded column loaded with a parent radionuclide. The column material is chosen so that the daughter radionuclide can be eluted while the parent remains on the column. The most prevalent example is the ^{99}Mo/^{99m}Tc generator, in which the parent ^{99}Mo-molybdate is adsorbed on an alumina column and the column periodically eluted with saline. The daughter ^{99m}Tc-pertechnetate is selectively eluted resulting in a stock solution of ^{99m}Tc-pertechnetate in saline. Such generators are made to produce radioactivity levels as high as 100 GBq or more (>3 Curies) in 5 mL. The generator delivers the isotope as TcO_4^- in saline at neutral pH. Although there are clinical applications of ^{99m}Tc as TcO_4^-, it is primarily used in combination with kits for the preparation of complexes for specific functions.

The 140 KeV photons emitted by ^{99m}Tc are accompanied by few conversion and Auger electrons and no beta-particles which increase the radiation dose without adding to the imaged information, and no gamma- or x-rays of other energies. These other gamma- or x-rays, if substantially lower in energy, also add to the radiation dose while being too heavily absorbed to contribute significantly to the image. If significantly higher in energy, they interact poorly with the detector and force compromises in collimator design that reduce detector resolution and efficiency.

^{99m}Tc has a number of other properties that make it the most nearly ideal isotope for diagnostic imaging. The energy is high enough that a substantial fraction of emitted photons escapes the patient without interaction, but low enough that the detector material can absorb virtually all that strike it with little scatter. The half-life is long enough for most preparation procedures that might be considered in a clinical setting and also long enough for convenient storage throughout the day and for equilibration under virtually any physiological process. ^{99m}Tc is easily produced and inexpensive and its variety of oxidation states provide ample opportunity for labeling technologies for many different types of ligands. It has few drawbacks, but among them is the fact that it is large and the even larger chelate complexes necessary to bind it to molecules with biological activity can make difficult the task of doing so without compromising their function. This means that it is rarely if ever possible to create a simple technetium analogue of a molecule of known function.

Other generator systems are possible and many have been constructed, but none thus far has yielded the combination of low cost, high utility, concentration, and activity of the ^{99}Mo–^{99m}Tc system.

Reactor-Produced Isotopes.   Relatively few radioisotopes for clinical use are produced as the direct result of reactor-based nuclear processes, although ^{99}Mo, the parent of ^{99m}Tc is produced by this method. These processes are either irradiation of source material using thermal, epithermal, or fast neutrons, or chemical separation of isotopes of interest from spent fuel. The former process results in nuclear reactions in which the target atom absorbs a neutron and emits either a gamma-ray, one or more protons, or another neutron and is then transmuted into a new element or a metastable excited form of the original. In the second, the desired material may be the result of neutron irradiation of the fuel or a fission product produced as a result of the fission of the ^{235}U or ^{237}Pu (see NUCLEAR REACTORS).

The isotopes produced by reactors are not common as biological markers. The absorption of neutrons generally results in radioisotopes (qv) that lie below the line of stability on the chart of the nuclides. These tend to decay by β^{-}-emission, an undesirable mode for a diagnostic.

The only commonly used radioisotope in this class is ^{131}I, used in small ($\sim$18.5 MBq (500 μCi) injected dose) quantities as a diagnostic for the evaluation of thyroid function. The compound is administered as NaI and these procedures are only possible owing to the favorable biological distribution of iodide. Up to 25% of the entire injected dose of iodide is accumulated in the thyroid with a very slow washout; the rest is rapidly excreted in the urine. No other compound exhibits so high a ratio of concentration in a target tissue to that of other tissues.

Therapeutic Radioisotopes.   Isotopes used for therapy must possess many of the same characteristics as those used for diagnosis (Table 3). Because the purpose is to deliver radiation dose to a tissue, radiation dosimetry considerations are different from those of the diagnostic isotope. However, the criteria given for diagnostic radioisotopes remain true with the exception of number six and with a redefinition of acceptable dose. Acceptable dose for a therapeutic radioisotope means a radiation dose sufficient to cause the required level of tissue damage in the tissue of interest and as little damage as possible in all others. It is not necessary to be able to detect the distribution *in vivo*, although this can be used, when available, to confirm the concentration and thereby the dose administered to the lesion. Some therapeutic isotopes, notably ^{89}Sr, emit only beta-particles yielding little externally detected radiation. Conversely, externally detectable radiation makes the patient a potential hazard to others until the radioactivity level has decreased significantly. This may require hospitalization, increasing costs.

Table 3. Isotopes Used for Internal Therapy

Isotope	CAS Registry Number	Therapeutic application
^{131}I	[10043-66-0]	thyroid carcinoma, hyperthyroidism, antibody therapies
^{32}P	[14596-37-3]	polycythemia vera
^{89}Sr	[14158-27-1]	pain palliation in metastatic carcinoma
^{186}Re	[14998-63-1]	pain palliation in metastatic carcinoma
^{153}Sm	[15766-00-4]	pain palliation in metastatic carcinoma
^{165}Dy	[13967-64-1]	synovectomy

Labeling Chemistry of Radiopharmaceuticals

There are three general types of radiopharmaceuticals: elemental radionuclides or simple compounds, radionuclide complexes, and radiolabeled biologically active molecules. Among the first type are radionuclides in their elemental form such as ^{81m}Kr and ^{127}Xe or ^{133}Xe, and simple aqueous radionuclide solutions such as ^{125}I or ^{131}I-iodide, ^{201}Tl-thallous chloride, ^{82}Rb-rubidium(I) chloride [*14391-63-0*], ^{89}Sr-strontium(II) chloride, and ^{99m}Tc-pertechnetate. These radiopharmaceuticals are either used as obtained from the manufacturer in a unit dose, ie, one dose for one patient, or dispensed at the hospital from a stock solution that is obtained as needed from a chromatographic generator provided by the manufacturer.

The second type of radiopharmaceuticals are radionuclide complexes. These are almost exclusively metal radionuclides complexed to a ligand or ligand system. The use of a ligand or ligand system allows the physical and chemical properties of the radionuclide to be modified. These properties include the stability, charge, oxidation state, and lipophilicity. Many of the uncomplexed metal radionuclides are not stable in biological media, forming a variety of species with differing biological properties, which confounds the ability to target the radiopharmaceutical to a particular organ or disease state. Metal complexes can be made that remain a single discreet species *in vivo* or that undergo a specific chemical transformation that is central to their utility. A variety of ligands have been used to form radionuclide complexes, ranging from monodentate to octadentate, and both homoleptic and binary ligand systems have been used. Examples of the resulting radionuclide complexes include cationic, neutral, and anionic complexes and cover the range from very hydrophilic to very lipophilic. The vast majority of these examples are coordination complexes of ^{99m}Tc. A variety of ligand systems, ligand denticities, and technetium oxidation states are employed. Some are shown in Figure 3. Those complexes shown in Figure 3**a**–**c** have been approved for routine clinical use in humans (see Table 1). The complexes in Figure 3**d** and 3**e** are under active clinical development as of this writing (ca 1996).

Certain neutral technetium complexes can be used to image cerebral perfusion (Fig. 4). Those in Figure 4**a** and 4**b** have been approved for clinical use. Two other complexes (Fig. 4**c** and 4**d**) were tested in early clinical trials, but were not developed further. An effective cerebral perfusion agent must first cross the blood brain barrier and then be retained for the period necessary for image acquisition. ^{99m}Tc-bicisate is retained owing to a stereospecific hydrolysis in brain tissue of one of the ester groups to form the anionic complex TcO(ECD)$^-$, which does not cross the barrier. This mechanism of retention is termed metabolic trapping.

Several hydrophilic, anionic technetium complexes can be used to perform imaging studies of the kidneys. ^{99m}Tc-Mertiatide (Fig. 5**a**) is rapidly excreted by active tubular secretion, the rate of which is a measure of kidney function. ^{99m}Tc-succimer (Fig. 5**b**), on the other hand, accumulates in kidney tissue thus providing an image of kidney morphology.

A number of other technetium-99m complexes are utilized in nuclear medicine procedures for which the structures of the complexes have not been

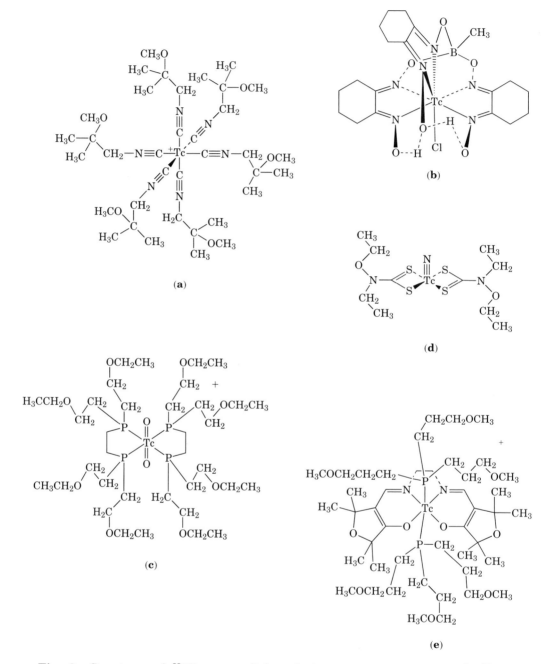

Fig. 3. Structures of ^{99m}Tc myocardial perfusion agents: (**a**) Tc(I)(MIBI)$_6^+$ (^{99m}Tc-sestamibi), where MIBI = 2-methoxyisobutylisonitrile; (**b**) Tc(III)Cl(CDO)(CDOH)$_2$BMe (^{99m}Tc-teboroxime [10471-22-5]), where CDO = 1,2-cyclohexanedione dioxime and BMe = methylboronic acid adduct; (**c**) Tc(V)O$_2$(1,2-bis(bis(2-ethoxyethyl)phosphino)-ethane)$_2^+$ (^{99m}Tc-tetrofosmin [127455-27-0]); (**d**) Tc(V)N(EtO(Et)dtc)$_2$, where EtO(Et)dtc = N,N-ethoxyethyldithiocarbamate; and (**e**) Tc(III)(1,2-bis((dihydro-2,2,5,5-tetramethyl-3(2H)-furanonato)methylene)amino)ethane)(TMPP)$_2^+$ (^{99m}Tc-furifosmin), where TMPP = tris(3-methoxypropyl)phosphine.

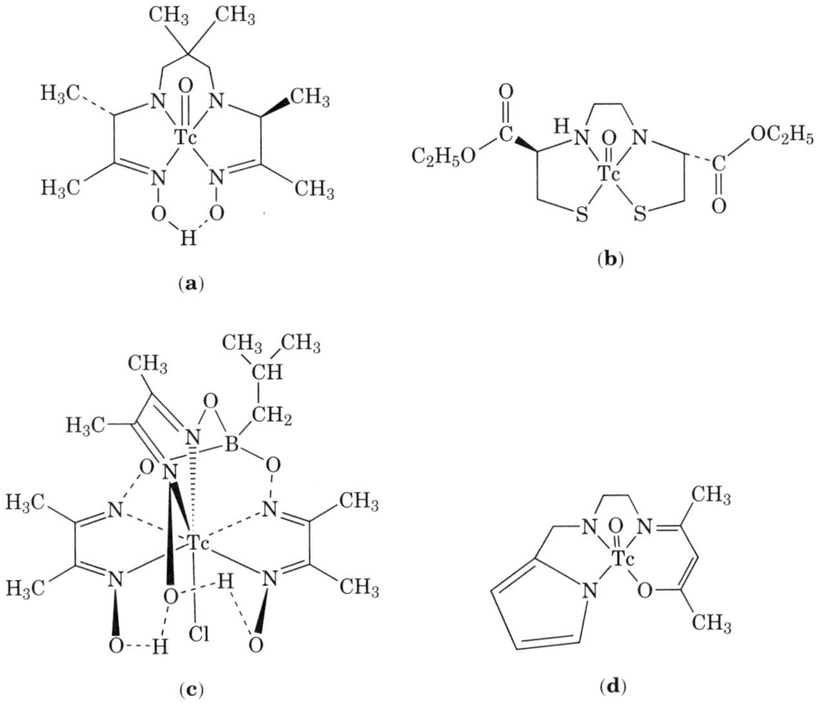

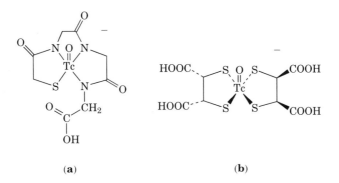

Fig. 4. Structures of ^{99m}Tc cerebral perfusion agents: (**a**) Tc(V)O(HMPAO) (^{99m}Tc-exametazine [*105613-48-7*]), where HMPAO = 3,6,6,9-tetramethyl-4,8-diazaunde-cane-2,10-dione dioxime; (**b**) Tc(V)O(ECD) (^{99m}Tc-bicisate [*121281-41-2*]), where ECD = *N,N'*-1,2-ethylenediylbis(L-cysteine) diethylester; (**c**) Tc(III)Cl(DMG)(DMGH)$_2$BiBu, where DMG = dimethylglyoxime and BiBu = isobutylboronic acid adduct; and (**d**) Tc(V)O(*N*-(2(1*H*-pyrolylmethyl))*N'*-(4-pentene-3-one-2)ethane-1,2-diamine).

Fig. 5. Structures of ^{99m}Tc renal imaging agents: (**a**) Tc(V)O(MAG$_3$)$^-$ (^{99m}Tc-mertiatide [*66516-09-4*]), where MAG$_3$ = *N*-2-mercaptoacetyltriglycine; and (**b**) one isomer of Tc(O)(dmsa)$_2^-$ (^{99m}Tc-succimer), where dmsa = dimercaptosuccinic acid.

unambiguously established. Structural determination requires synthesis and characterization of the chemically identical complexes using ^{99}Tc, a long-lived β-particle emitting isotope available in gram quantities. However, it is difficult logistically to maintain the reaction stoichiometries used in the synthesis of the technetium-99m complexes, frequently molar ligand excess of 1000–10,000 (*vide infra*), when performing the syntheses using macroscopic quantities of ^{99}Tc. This can lead to differences in the chemistry observed in the different concentration regimes. Without the characterization of the ^{99}Tc analogues, the structures of the ^{99m}Tc complexes can only be inferred. Examples of such complexes having inferred structures include the Tc-iminodiacetic acid complexes (^{99m}Tc-mebrofenin [78266-06-5], ^{99m}Tc-disofenin [65717-97-7], and ^{99m}Tc-lidofenin). These are hepatobiliary imaging agents having different substituents on the iminodiacetic acid ligands. The Tc-diphosphonate complexes (^{99m}Tc-medronate, ^{99m}Tc-oxidronate), bone imaging agents having different substituents on the diphosphonate ligands, Tc-DTPA (^{99m}Tc-penetate [12775-34-7], DTPA = diethylenetriaminepentaacetic acid), Tc-glucoheptonate (^{99m}Tc-gluceptate [87-74-1]), and ^{99m}Tc-pyrophosphate, also fall into this category.

Several complexes of other radionuclides are also widely used in nuclear medicine. In(III)(oxine)$_3$ (^{111}In-oxyquinoline) is used to label white blood cells for imaging sites of infection or inflammation, In(III)(DTPA)$^{2-}$ (^{111}In-penetate) is used for radiographic cisternography, and a Ga(III)-citrate complex is used for imaging tumors. The structures of the two indium complexes have been determined by synthesis and characterization of the nonradioactive indium analogues. The structure of the Ga complex is not known. Two complexes of high energy β-particle emitting radionuclides are in clinical trials, ^{186}Re-HEDP (HEDP = hydroxyethylidinediphosphonate) and ^{153}Sm-(EDTMP) (EDTMP = ethylenediamine-N,N,N',N'-tetrakis(methylenephosphonic acid)) for use in bone pain palliation. Neither structure has been unambiguously determined.

The procedures used to synthesize metal radionuclide complexes vary depending on the chemical reactivity of the radionuclide in its available form. Some radionuclides are available as metal halide solutions and can be formed by reaction with the ligand(s) directly. An example of the direct complexation approach is the synthesis of In(III)(oxine)$_3$. In(III)-chloride and 8-hydroxyquinoline react by adjusting the pH to 7. Owing to the dilute concentration of the In–chloride solution, the hydroxyquinoline is added in large excess to achieve a reasonable reaction rate. In contrast, pertechnetate must be reduced to form complexes. Typically this is achieved by addition of the ligand(s) and a reducing agent, such as tin(II) chloride. Both the ligand and the reducing agent are added in large molar excess. If the rate of pertechnetate reduction is significantly more rapid than the rate of complexation, reduced technetium species accumulate in the solution. These reduced species are not stable in aqueous solution, undergoing hydrolysis and disproportionation to form pertechnetate and TcO$_2$ (Tc-colloid), insoluble, colloidal, Tc(IV)-oxide. The formation of this by-product can be prevented by including in the reaction mixture another ligand that forms complexes more rapidly. The reduced technetium species are then complexed by the additional ligand, termed a transfer ligand (preventing hydrolysis and disproportionation), and subsequently converted to the final ligand complex. For example,

$$^{99m}\text{Tc-pertechnetate} + \text{tartrate} \xrightarrow{\text{Sn(II)}} {}^{99m}\text{Tc-tartrate} \xrightarrow{\text{MAG}_3} {}^{99m}\text{TcO(MAG}_3)^-$$

where MAG$_3$ is N-2-mercaptoacetyltriglycene.

Complexes of radionuclides of high specific activity are synthesized in very dilute aqueous solution. For example, solutions of technetium complexes typically range from 10^{-7} to 10^{-6} M. Most techniques, such as nmr, ir, etc, do not have the necessary sensitivity to analyze for such dilute species. Therefore, radionuclide complexes are analyzed almost exclusively by chromatography (qv), using the radioactive emission as the detectable label. High performance liquid chromatography (hplc) and thin-layer chromatography (tlc) are used extensively. Hplc chromatograms are obtained by monitoring the column effluent with a radioactivity detector, such as a sodium iodide or Geiger-Mueller probe. For tlc, the plate is either scanned using a probe or divided into sections such that the amount of radioactivity in each section can be counted in an ionization chamber. The limit of detection for ^{99m}Tc by these methods is generally ~37 kBq (1 μCi) or ~10^{-15} mol.

The third type of radiopharmaceuticals are radiolabeled biologically active molecules. Proteins, antibodies, peptides, and enzyme substrates are modified using a gamma-emitting radionuclide to image biological function. The method of labeling is chosen to have the least impact on the biological activity. Typical radionuclides include ^{11}C, ^{13}N, ^{15}O; the radiohalogens, such as ^{18}F, ^{123}I, ^{125}I, and ^{131}I; and the metal radionuclides, ^{99m}Tc and ^{111}In.

Biologically active molecules may be labeled either directly or indirectly. For direct labeling, that is, incorporating the radiolabel directly into the compounds, the labeling may be isotopic or nonisotopic. For isotopic labeling, one group already present in the molecule is substituted with or exchanged for the radioisotope. For nonisotopic labeling, that is, incorporating the radiolabel into the compounds through a chelator or other functional group which has been incorporated into the compounds, the radioisotope is added to the molecules without substituting with or exchanging for an already existing group.

Generally, labeled compounds are prepared by procedures which introduce the radionuclide at a late stage of the synthesis. This allows for maximum radiochemical yields, and reduces the handling time of radioactive material. When dealing with short half-life isotopes, a primary consideration is the time required to conduct synthetic procedures and purification methods.

Various procedures may be employed in preparing the radiolabeled compounds where the radiolabel is a halogen. Some common synthetic methodologies for isotopic halogen labeling of aromatic compounds or aromatic groups in peptides or proteins are iododediazonization, iododeboration, iododestannylation, iododesilation, iododethallation, and halogen-exchange reactions. The most common synthetic methodology for nonisotopic halogen labeling of aromatic compounds is iododeprotonation or electrophilic aromatic substitution reaction. These methods and additional procedures are described in the literature (8,9).

Alternatively, radiohalogen-labeled compounds may be prepared by way of isotopic labeling from the unlabeled bromo or iodo derivatives by various two-step reaction sequences. Examples include the use of trialkylsilyl synthons as described in References 10–13, and the use of boronic acid synthons as described in References 14 and 15.

Metal radionuclide labeling of biologically active molecules can be performed either directly or indirectly. For example, for proteins or antibodies that contain disulfide linkages, the addition of a reducing agent such as stannous ion,

2-mercaptoethanol, or dithiotheitol generates free thiol groups on the molecules. These free thiol groups can be directly labeled with ^{99m}Tc by reaction with pertechnetate and an additional reducing agent or with a ^{99m}Tc transfer ligand complex. The advantage of direct labeling is that it is easy to carry out. There are two main potential disadvantages. Reduction of the native disulfide linkages to form the free thiols that bind the technetium may adversely affect the biological properties of protein, antibody, or polypeptide. Also, particularly for larger molecules, a number of free thiol-containing sites may be formed that when reacted with the technetium result in binding with a varying degree of stability. Whether these disadvantages become manifest depends on the biologically active molecule chosen to be labeled.

There are two approaches for indirect labeling of biologically active molecules with metal radionuclides. Both approaches require that the molecules be chemically modified to attach the chelator without significantly affecting their biological activity. The first involves attaching a metal chelator to the molecule and then reacting the resulting conjugate with the radionuclide. The procedures for the reactions are similar to those used to synthesize the corresponding metal chelate complexes. An example of this approach is the synthesis of ^{111}In-labeled octreotide (^{111}In-pentetreotide [*13906-04-1*]) for imaging neuroendrocrine tumors.

The second indirect approach involves forming a metal radionuclide complex, chemically modifying the complex to generate a reactive site on the complex, and then reaction of the modified complex with the biologically active molecule. This approach is termed the preformed chelate approach or retrolabeling. It requires several more steps than the other approaches, but the molecule is not present during the synthesis of the metal radionuclide complex, which is an advantage if the molecule cannot survive the reaction conditions used for forming the complex (Fig. 6).

Radiolabeled biologically active molecules are analyzed by chromatographic techniques similar to those used for metal radionuclide complexes. Hplc and tlc methods for large molecules generally utilize different chromatographic media to minimize adsorption of the analyte. Common hplc methods include size-exclusion chromatography, gel-permeation chromatography, ion-exchange chromatography, and affinity chromatography.

An important consideration for all radiopharmaceuticals and especially radiolabeled biologically active molecules is specific activity. There are two types

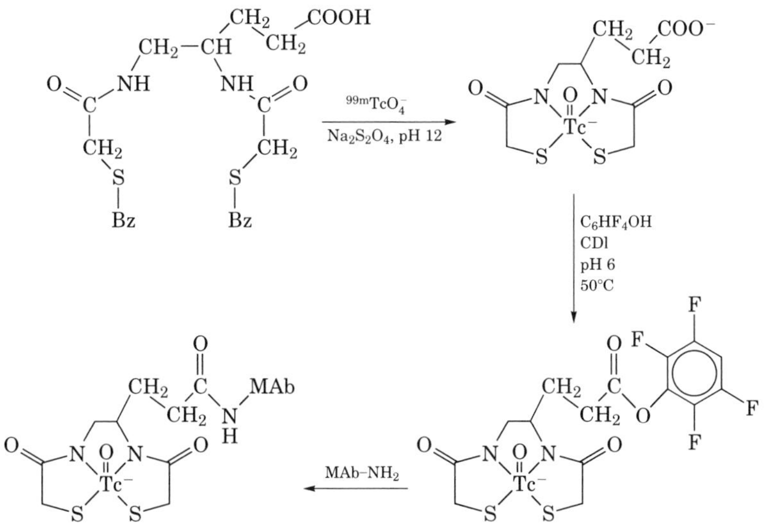

Fig. 6. General reaction scheme for the preformed chelate approach where Bz = benzoyl and MAb = monoclonal antibody.

of specific activity: radionuclidic and biological. Radionuclidic specific activity refers to the ratio of the number of atoms of a particular radioisotope to the total number of atoms of the element. For ^{99m}Tc, the radionuclidic specific activity is the number of ^{99m}Tc atoms to the total number of ^{99m}Tc and ^{99}Tc atoms. Because all isotopes of an element are chemically identical, a low specific activity may lead to a low yield in the synthesis of a radiopharmaceutical if a significant proportion of the reagents is consumed by the undesired isotopes.

Biological specific activity refers to the ratio of the number of radiolabeled biologically active molecules to the total number of biologically active molecules. If the biological activity is dependent on the interaction of the molecule with a receptor or binding site, the excess unlabeled molecules may compete with the labeled molecules, limiting the number of detectable molecules that can be localized at the receptor or binding site. Also, the excess unlabeled molecules may cause an undesired biological response, limiting the number of molecules that can be present in the radiopharmaceutical. Therefore, both a high radionuclidic and biological specific activity is preferred.

Instrumentation for Imaging and Counting

Detection of the emissions from radiopharmaceuticals is dependent on the type of emission. In practice, only those isotopes emitting either γ-ray or x-ray photons, either directly from the decay reaction or indirectly from annihilation of an emitted positron, are useful for *in vivo* imaging, although β-emissions may be of use for *in vitro* purposes. These high energy photons are ionizing radiation and interact with matter via only two mechanisms. For those photons emitted by commonly available radionuclides, the predominant mechanisms are the photo-electric effect and Compton scattering. The photoelectric effect is an interaction between the photon and inner shell electrons in which the original photon is completely absorbed leaving the atom in an excited state, followed by deexcitation by emission of low energy electrons or fluorescence photons. Compton scattering is an interaction with loosely bound valance electrons resulting in acceleration of the electron and in general a deflection and energy loss for the photon. The direction of the scattered photon is related to that of the electron and to the energies of both, but is nearly isotropic at low energies, with respect to the trajectory of the original photon.

Measurement of Injected Patient Dose.   The patient's injected dose, as distinguished from the radiation dose, must be within 10% of the prescribed dose for a given procedure by law in the United States. This is measured using a large ionization chamber of cylindrical geometry. An ionization chamber is a space filled with air or other gas in which a constant electric field is present. Gamma-rays or other ionizing particles passing through the gas produce electron–ion pairs. The electric field is of sufficient magnitude that these cannot recombine but rather migrate through the gas until they are collected by the electrodes producing the field. Thus, the more electron–ion pairs produced in the gas, the more current flows in the external circuit. This current is a measure of the ionization and therefore of the radioactivity present. If the chamber is arranged so that the ionization produced is relatively independent of the position and shape of the source (thus the cylindrical well) the device can be calibrated for each isotope.

The modern ionization chamber, called a dose calibrator in this application, is capable of linear measurements of radioactivity having a precision in the range of several percent coefficient of variation over a range of 370 kBq (10 μCi) to at least 370 GBq (10 Ci). This extraordinary range is the chief advantage of this instrument. It may only be used when the sample is known to have only a single isotope. It has no capacity to distinguish radiation from different isotopes.

Counting Systems.   Counting of small samples of γ-emitting radioisotopes is commonly accomplished in biology and medicine using a scintillation detector. This is a block of crystalline sodium iodide with a small proportion of thallium, generally cylindrical, and either 5 cm in diameter and 5 cm high (2 $\times$ 2 in.) or 7.6 cm in diameter and 7.6 cm high (3 $\times$ 3 in.). It is hermetically sealed in an aluminum enclosure having one glass wall on a cylinder end. Often the other end has a well or a blind hole centered on the face and a depth of about two-thirds the thickness of the crystal and a diameter that is convenient for a sample tube, usually 10 mm. The glass panel of the crystal enclosure is coupled to a photomultiplier tube that records the bursts of light produced each time a γ-ray interacts to produce ionization in the crystal.

Samples in tubes are placed in the well of the crystal and the number of light bursts observed over a given time recorded. This value is proportional to the number of disintegrations that occurred during that time and therefore to the amount of radioactivity in the sample. Mixtures of isotopes emitting γ-rays of different energies can be assayed in the same tube. This is made possible by the proportionality between the intensity of each light burst with the energy deposited by the photon that created it. If the photon was completely absorbed in the detector, either in a single event or several successive ones, then the recorded light burst yields information about the γ-ray energy and thereby the isotope that produced it.

This type of device is commonly limited to a maximum radioactivity of approximately $3.7–7.4 \times 10^4$ Bq (1–2 μCi). Its lower limit is primarily determined by the background count rate and the patience of the investigator. It can be used to separate the contributions of as many as five to six isotopes in a single sample. These counting systems are normally highly automated, incorporating mechanical sample changers with a capacity of several hundred sample tubes and computers controlling the process and recording the result.

Single-Photon Imaging Systems. There are many materials and schemes for detecting photons, but for nuclear medicine one instrument, the γ-camera, is used almost exclusively. This device consists of a disk or rectangle of crystalline sodium iodide with a small proportion of thallium, generally 10-mm thick and in rectangles of up to 40×60 cm, hermetically sealed in an aluminum enclosure with one glass wall on a large surface. γ-Rays absorbed in the crystal result in the emission of a burst of light photons over a period of several hundred nanoseconds. These are detected by an array of photomultiplier tubes and the centroid of the light distribution, an estimate of the γ-ray interaction point, may be determined by hardware interpolation among this array. For each γ-ray detected, a point on an image plane is incremented in intensity.

In order to produce an image, photons must escape the patient without interaction. Because scattering yields a nearly isotropic distribution of scattered photon directions, any photon undergoing scatter within the patient no longer carries information about its point of origin and must be rejected by an imaging apparatus. Because the light intensity produced for each γ-ray is proportional to its energy, that energy may be determined for each event. This permits the system to reject γ-rays having energy lower than expected, an indication of scatter prior to detection.

γ-Rays cannot be refracted, only scattered or absorbed, thus it is not possible to create a lens in the conventional sense. The collimator, a tall lead honeycomb between the crystal and the patient but very close to the crystal, serves this purpose. Rather than refracting the light to form an image on a focal plane, as does a light camera, the collimator simply discards all rays having a direction not perpendicular to (close to) the crystal surface. This results in a photon flux at each point on the crystal that is proportional to the integral of the radioactivity concentration along a line perpendicular to the crystal surface, with appropriate modification for attenuation in the body.

Collimators for image formation are extremely inefficient compared to conventional lenses. This is a natural result of the physics of the interactions between γ-ray photons and matter, in this case lead or tungsten. Whereas a lens for image formation using light can refract the light and concentrate it, a γ-ray collimator can only discard those photons not traveling approximately perpendicular to the crystal surface. As a result, the vast majority of photons are excluded from the imaging process.

Although performance varies with the isotopes for which they are intended, and with the balance in the design between resolution and efficiency, the overall sensitivity of a γ-camera collimator is on the order of 5000 counts/(MBq·min) (several hundred counts/(μCi·min)). In terms of photons detected per photon emitted, this is equivalent to about 2×10^{-4}. In other words, about two photons out of 10,000 emitted arrives at the crystal. This necessitates exposure times that range from several minutes to the better part of an hour. Fortunately, the large number of photons available from a modest injected radioactive dose more than offsets the poor detector sensitivity. The camera's ability to resolve small objects, however, is ultimately limited by the collimator inefficiency.

The resolution of the γ-camera is determined by both the collimator and the crystal/photomultiplier tube combination. In general, the resolution is significantly poorer than that of other modalities such as magnetic resonance imaging (mri) and x-ray computerized tomography (ct). For single-photon imaging, a system resolution of the order of 6–7 mm full-width at half-maximum (FWHM), the width of the spot that results in an image from a very small point of radioactivity, at the surface of the collimator and approximately 1 cm FWHM, 10 cm away is common when a higher resolution collimator is used.

The image is created either on film by allowing each detected photon to brighten a corresponding point on an oscilloscope screen viewed by high speed film, either conventional medical transparency or rapid development film, or by computer. The computer method is becoming nearly universal and consists of incrementing an appropriate location in a two-dimensional array in memory upon detection of each photon. This is then available for display, processing, calculation, and printing. The resulting digital image is created by transferring the contents of this array to an array of picture elements (pixels) on the computer display, in which increasing values are translated to colors of increasing shades of gray.

Several types of studies may be performed using modern equipment. Static studies are simply views of the body from various perspectives, arranged so as to yield the maximum isolation of a given structure of interest. Dynamic studies are composed of a series of images of a single organ or structure, repeated over time so that replay of them in sequence yields a motion picture-like change. Gated studies are similar to dynamic studies, except that they are used for the imaging of structures, primarily the heart, that have a rhythmic movement. Data are acquired into a series of images for a single period, then data for subsequent periods are added to these images in a synchronized fashion.

Single-photon emission computed tomography (SPECT) studies are acquired by rotating the γ-camera around the patient's long axis. These data are then used to reconstruct the radioactivity distribution in three dimensions. This may be displayed as slices of radioactivity concentration or rendered so as to present the appearance of a solid volume.

As a modality, SPECT imaging represents a growing segment of nuclear medicine studies. Although it does not require a specific agent, many agents yield significantly better information when SPECT acquisitions are used. SPECT eliminates the uncertainty that occurs in planar images when organs at different depths cannot be distinguished. Because overlapping organs no longer pose a problem, contrast and visualization are both improved. Although it is not yet possible to obtain truly quantitative estimates of radioactivity concentration using SPECT because of scatter and attenuation, compensation methods are improving and the modality should compete with positron imaging in the area.

Several specialized devices exist, among both commercialized systems and those used for research, that yield improved SPECT images for specific organs. For many of these, the brain is the target and systems using rings of scintillating detectors or arrays of individually collimated scanning scintillation detectors can create images with significantly better resolution or, in some cases, dynamic SPECT images.

Positron Imaging. Creating images of distributions of positron emitters requires a somewhat different type of apparatus. Positron cameras use many of the same technologies as do cameras for other isotopes, but there is a broader array of methods and physical arrangements. All of these systems take advantage of the physical characteristics of positrons.

Positrons are emitted having energies that can range from several tens of keV to several MeV. As for all charged particles, they travel a distance that is proportional to their starting energy, leaving a trail of ionization as they slow down. This distance can range from less than one micrometer to several millimeters in tissue. As the positron energy decreases, its probability of undergoing an annihilation reaction with an electron increases. This reaction results in the emission of two 511 keV γ-rays traveling essentially in opposite directions, yielding conservation of both momentum and energy. The rest energies of the electron and the positron are both 511 keV.

Detectors are arranged on opposite sides of the patient and circuitry detects coincidence events in which photons are detected simultaneously within the time resolution of the system. These events mark annihilations that have occurred somewhere along the line connecting the detectors. If there are many detectors at different angles, this data may then be used, as is SPECT, to reconstruct the three-dimensional distribution of the radioisotope. Because virtually all positron imaging is tomographic, that is, a three-dimensional reconstruction or literally slice writing, positron imaging is normally called positron emission tomography (PET).

The camera actually images the annihilation events, not the radioactive decay events directly. Thus imaging of high energy positron emitters can have a limiting resolution owing to the range of the positron.

Although energy resolution is rarely employed in positron camera systems, scatter is not normally a problem. This is because of the very short time window within which two photons must arrive in order to be counted. At low decay rates, the incidence of accidental events is very low, rising only slightly for those that occur as the result of scatter. Some systems employ time-of-flight measurements of the time difference between the arrival of the two photons to obtain additional information about the location of an annihilation along the line. This has been used to improve resolution and statistical accuracy. Resolution is in the range of 3–4 mm and is less dependent on position than is SPECT (16).

Because few scatter events are recorded, attenuation compensation is relatively easier for PET using an external positron emitting source. As a result, the technology for quantitative determinations of radioactivity distributions is significantly more advanced in PET imaging. Technology development for SPECT, however, is improving this parameter.

PET imaging systems are somewhat more complex, and therefore more expensive than are SPECT systems, and the price factor is generally between two and three. The primary cost premium associated with these systems, however, is the need for a cyclotron and its attendant staff combined with the relative complexity of radiopharmaceutical preparation for short half-life isotopes. As of

1996, there are considerable hurdles blocking widespread regulatory approval and full reimbursement of PET studies.

Other Instrumentation.   *Health Physics Instruments.*   Minimization of radiological hazards to personnel and patients in a nuclear medicine environment requires appropriate instrumentation. It is in many ways easier to protect people from nuclear agents than chemical ones because the great sensitivity of measuring apparatus permits the identification and localization of extremely small quantities quite readily. The general armamentarium of radiation monitoring includes a survey meter, generally employing a gas-media detector, such as a Geiger-Müller tube, or a small scintillation detector if additional sensitivity is required. Also needed is a system for assay of wipe tests. This consists of wiping a defined area of the bench, floor, and countertop and assaying the wipes using a survey meter or specialized system for the detection of removable radioactive contamination. An air-ionization chamber survey meter, called a cutie pie from qT-π for charge, time, and four-pi geometry, is necessary if high radiation fields are likely to be present. A thyroid counter is necessary if radioiodine is used in the facility. Finally, area monitors of appropriate type are necessary wherever radioactivity is used. These must include the capability for detecting all radionuclides used in the facility, including heavy gases such as ^{133}Xe.

Thyroid Uptake Systems.   Studies involving absolute thyroid uptake can be performed without imaging using small amounts of ^{131}I or ^{123}I and a simple scintillation probe. This is calibrated using a phantom, ie, a model of a portion of the human body, loaded with the isotope being used. This instrument is also useful for assaying thyroid exposure to radioiodine among personnel.

Safety

For virtually all radiopharmaceuticals, the primary safety consideration is that of radiation dosimetry. Chemical toxicity, although it must be considered, generally is a function of the nonradioactive components of the injectate. These are often unreacted precursors of the intended radioactive product, present in excess to facilitate the final labeling reaction, or intended product labeled with the daughter of the original radioactive label.

Radiation Dosimetry.   Radioactive materials cause damage to tissue by the deposition of energy via their radioactive emissions. Thus, when they are internally deposited, all emissions are important. When external, only those emissions that are capable of penetrating the outer layer of skin pose an exposure threat. The biological effects of radiation exposure and dose are generally credited to the formation of free radicals in tissue as a result of the ionization produced (17).

By definition (18), radiation dose represents the energy deposited per mass of tissue and has the units of Gray, where 1 Gy = 10 μJ/g (1 rad = 0.01 Gy). Exposure is a material-independent measure of the radiation incident on a volume as defined by the electrical charge it produces in air. Exposure is measured in charge per mass, ie, C/kg (one Röentgen (1 R = esu/g dry air at STP) = 2.58 × 10^{-4} C/kg). Effective dose is a measure of radiation effect that includes biological factors, weighted on an organ-by-organ basis, and factors related to the microscopic distribution of the energy deposition and the effects of dose rate. Its unit is the Sievert (same dimensions as the Gray) or the rem (same dimensions as the rad).

Radiation dose for radiopharmaceuticals in a given patient can only be estimated. There are significant variations in dose from internally deposited radionuclides arising from the shape and location of organs and of the circulation within them. Thus, over the years, computer modeling methods have been developed to obtain estimates of dose, primarily under the auspices of the Medical Internal Radiation Dose (MIRD) Committee of the Society of Nuclear Medicine. These methods employ models of human beings of various sizes and both sexes, data on the biological distribution of a particular radioactive compound, physical data on the radionuclide, and the results of Monte Carlo modeling of different types of emissions from each organ of interest. Monte Carlo codes for calculating the transport and absorption of radiation from radioactivity were developed primarily for weapons and nuclear reactor design; however, they have proven to be readily adaptable to medical applications. These codes are extremely well validated and most have undergone many years of testing. In addition, studies have been conducted to validate the results generated specifically for the human models as well.

Nonetheless, these methods only estimate organ-averaged radiation dose. Any process which results in high concentrations of radioactivity in organs outside the MIRD tables or in very small volumes within an organ can result in significant error. In addition, the kinetic behavior of materials in the body can have a dramatic effect on radiation dose and models of material transport are constantly refined. Thus radiation dosimetry remains an area of significant research activity.

The actual hazard from radiation exposure experienced by the patient in a nuclear medicine study is relatively small. The dose for a commonly performed study, the bone scan (see Fig. 1), ranges with tissue from approximately 650 μGy (65 mrad) for breast to 46 mGy (4.6 rad) for the bone surfaces. The International Committee on Radiation Protection has compiled standard values for radiation dose, making these available in book form (19,20). Calculations of effective dose have been derived (21). These yield a single number that is weighted by organ radiosensitivity and radiation type. Although disputed as absolute measures for nuclear medicine risk because the weights are based on long-term occupational exposure, effective dose provides a convenient way of comparing risk among different types of nuclear medicine studies.

Chemical Toxicity. Radiopharmaceuticals are subject to the same requirements for safety as are other pharmaceuticals, and are tested for chemical toxicity in much the same manner. It is generally understood, however, that patients are likely to receive relatively few doses of any given radiopharmaceutical so that the effects of long-term chronic exposure to the compound rarely need be assessed. Safety margins, that is, the ratio of the administered dose to the lowest dose that produces an observable effect, are usually on the order of 100 or more.

Formulation and Packaging

The diversity of radionuclide half-life and chemical nature of commonly used radiopharmaceuticals demands a variety of formulation matrices, packaging containers, and storage conditions. The containers, ingredients, and processes used in these products must meet the stringent requirements for parenteral pharmaceuticals, as well as provide safe conditions for storage, handling, and disposal of the radioactive material.

Route of administration is a primary factor in formulation design. Orally administered compounds are commonly given in either solution or gel capsules. ^{99m}Tc sulfur colloid bound to egg albumin is given as a solid meal. Parenteral products are given in solution with either water or a dilute alcohol as the solvent. The solution is ideally sterile, isotonic, and nonpyrogenic. For multidose preparations, it must also include a bactericidal additive. Alternatively, the injection can be given as colloidal suspensions (^{99m}Tc sulfur colloid or ^{99m}Tc albumin colloid) or as microaggregates (^{99m}Tc albumin) or microspheres. Pulmonary function is tested using radioactive gases such as ^{95}Kr and ^{133}Xe which are supplied in glass vials, metal cylinders, or aerosols (qv).

The strategy for formulation may require, in addition to the optimal concentration of radionuclide and/or active ingredient, a selection of appropriate ionic strength, buffer, stabilizers, reactants, reductants, and bulking agents. The following parameters of the final product are then assayed: total radioactivity, specific radioactivity, carrier concentration, pH, quality control for final and intermediate components, and radionuclidic purity. The solution is dispensed into containers, sealed, packaged, and stored under appropriate conditions. The liquid contents of the nonradioactive kits are either frozen or lyophilized to achieve greater stability, and an inert gas is often added. The final contents must be sterile, nonpyrogenic, isotonic, and aqueous for parenteral use.

The container size is selected for either unit or multiple dose use. The type and extent of the shielding of the radionuclide container depends on the path length and energy of the emitted particles, as well as the amount of the individual product. The β-emitters are typically stored in thick-walled glass containers; γ-emitters require an additional layer or layers of lead shielding. These external layers are designed to provide protection for shipping and storage as well as convenient access to the container during use.

The labeling on the containers and packages for use and storage are regulated by the Nuclear Regulatory Commission. The Federal Department of Transportation regulates the conditions of shipment, and requires a Transportation Index on the label that gives a rapid indication of the rheological hazard of the package. There are additional regulations at state and local levels, as well as those by the Federal Drug Administration and the U.S. Public Health Service. The container labeling includes the name of the preparation, the amount of radioactivity, the half-life, expiration date, calibration date, and the statement "Caution—Radioactive Material", as well as a statement that correction must be made for radioactive decay.

Direct. Some radionuclides are packaged in solution for direct sampling (qv) via a septum and injection into the patient. Gallium-67 is a marker of inflammation, infection, and various tumor types. Its half-life is 78.3 h and it is supplied as the gallium citrate salt. Indium-111 chloride is supplied for the labeling of white blood cells. The ^{111}In chloride is mixed with oxine (9-hydroxyquinoline) to form a lipophilic, cationic ^{111}In oxine complex, which enters the white blood cell. The complex dissociates within the cell, and the cationic In^{3+} ion is trapped within the cell, owing to its charge.

Technetium-99m pertechnetate is produced from a ^{99}Mo/^{99m}Tc TcO$_4$ generator. The product, ^{99m}Na TcO$_4$, may be supplied directly or by elution from an on-site generator. Upon iv administration, it accumulates in the stomach, small and large bowel, salivary glands, thyroid gland, choroid plexus, sweat glands, and kidney. This distribution is used to assess a variety of conditions including thyroid transport, bowel function, and blood brain barrier integrity. An additional important use of ^{99m}Tc pertechnetate is in the radiolabeling of a variety of substances to form ^{99m}Tc radiopharmaceuticals.

Thallium-201 is a potassium analogue as the thallous ion. This radionuclide is used for the diagnosis and localization of myocardial perfusion and parathyroid function. It is supplied as the thallous chloride salt in an aqueous injectable solution and has a half-life of 73.1 h. Sodium iodide I-131, used for the treatment of hyperthyroidism and selected cases of thyroid cancer, is supplied in capsules for oral administration containing 37–1850 MBq (1–50 mCi) of I-131. I-131 MIBG (iobenguane sulfate [77679-27-7]) is used for localization of primary or metastatic pheochromocytomas and neuroblastomas. It is supplied in a sterile, nonpyrogenic solution. I-123 iofetamine is a brain imaging agent and is useful for evaluating nonlacunar stroke within 96 h of focal neurological deficit. It is supplied as a single vial injectable solution, and is stored at 5–30°C. I-131 iodohippurate is used as a diagnostic aid in determining renal function, renal blood flow, and urinary obstruction. The appearance, concentration, and excretion of the tracer in the kidney can be followed, and an index of renal vascular competence and renal evacuation may also be estimated.

Strontium-89 chloride is a calcium analogue that rapidly clears from the blood and is taken up into bone mineral, particularly in areas of active osteogenesis, as well as primary bone tumors and metastases. It is used for relief of bone pain in patients having painful skeleton bone metastases. It is supplied in an injectable solution.

Kits. Kits for the preparation of radiopharmaceuticals are a convenient solution to synthesis of products containing short-lived radionuclides (eg, ^{111}In, ^{123}I, ^{99m}Tc) bound to a nonradioactive moiety. The labeling step is performed either at a commercial radiopharmacy, or within the institutional nuclear medicine laboratory. The kits are usually stored as a frozen solution or lyophilized product. The material of interest is then metered out into kit dosages. The kit vials are thawed or reconstituted and mixed with the appropriate radionuclide.

Many kits contain the indicated biologically active ingredient in a lyophilized form with stannous chloride. A ^{99m}Tc-labeled radiopharmaceutical, which can be used for six hours, is formed when mixed with ^{99m}Tc pertechnetate. Preparation of the agent is at room temperature, unless otherwise stated.

Technetium-99m. Available ^{99m}Tc kits are listed below.

Technetium-99m sestamibi is used in myocardial perfusion imaging for the evaluation of ischemic heart disease. It is prepared from a lyophilized kit containing tetrakis(2-methoxy isobutyl isonitrile) copper(I) tetrafluoroborate stored under nitrogen. Upon reconstitution with up to 5.6 GBq (150 mCi) of ^{99m}Tc pertechnetate, the product is formed by boiling for 10 minutes.

Technetium-99m oxidronate is a bone imaging agent used to demonstrate areas of altered osteogenesis. It is rapidly cleared from the blood and taken up in areas of bone that are undergoing osteogenesis. The kit is a vial containing a lyophilized powder where sodium oxidronate is the active ingredient.

Technetium-99m teboroxime is a myocardial imaging agent and is excreted primarily by the hepatobiliary system. It is rapidly taken up by the myocardium and mostly washes out within 30 minutes. Imaging protocols are performed immediately after injection. The product is a lyophilized mixture of boronic acid, dioxine, and other excipients, and the agent is formed with a heating step.

Technetium-99m pentetate (N,N-bis[2-[bis(carboxymethyl)amino]ethyl]-glycinato(5-)) distributes into the extracellular space and is excreted by glomerular filtration through the kidney. It is indicated for kidney imaging, brain imaging, assessment of renal perfusion, and estimation of glomerular filtration rate. The active ingredient is pentetate calcium trisodium and the kit contents are stored under nitrogen. The product is formed by addition of up to 7.4 GBq (200 mCi) of ^{99m}Tc pertechnetate.

Technetium-99m gluceptate is used in brain and kidney imaging. Sodium gluceptate is the active ingredient. The product is formed by the addition of up to 7.4 GBq (200 mCi) of ^{99m}Tc pertechnetate.

Technetium-99m disofenin is used for hepatobiliary imaging. Disofenin (2,6-diisopropylphenylcarbamoylmethyliminodiacetic acid) is the active ingredient. Product formation is accomplished by addition of up to 3.7 GBq (100 mCi) of ^{99m}Tc pertechnetate.

Technetium-99m albumin colloid is cleared by the reticuloendothelial (RE) cells and is used for visualization of the RE system of the liver, spleen, and bone marrow. The product is formed by the addition of up to 2.8 GBq (75 mCi) of ^{99m}Tc pertechnetate.

Technetium-99m medronate (^{99m}Tc methylene diphosphonate) is used as a bone imaging agent to delineate areas of altered osteogenesis. The product is formed by the addition of up to 7.4 GBq (200 mCi) of ^{99m}Tc pertechnetate.

Technetium-99m albumin aggregated (^{99m}Tc-macroaggregated albumin) is trapped in the pulmonary alveolar bed and is used as a lung imaging agent. The aggregated particles (10–90 μm dia) are formed by denaturation of human albumin in a heating and aggregation process. The ^{99m}Tc complex is formed by the addition of up to 1850 MBq (50 mCi) of ^{99m}Tc pertechnetate.

Technetium-99m pyrophosphate is used for bone imaging. The compound appears to have an affinity for the hydroxyapatite crystals within bone, and is formed by addition of up to 7.4 GBq (200 mCi) pertechnetate.

Technetium-99m mertiatide (N-[N-[N-[(benzoylthio)acetyl]glycyl]glycine) is a renal imaging agent. It is excreted by the kidneys via active tubular secretion and glomerular filtration. The kit vial is reconstituted by using 740–3700 MBq (20–100 mCi) of ^{99m}Tc pertechnetate and boiling for 10 minutes.

Technetium-99m mebrofenin is an iminodiacetic acid derivative used as a hepatobiliary agent. The kit is supplied as a single vial containing lyophilized mebrofenin. The reconstituted kit has 18-hour usage, owing to the preservative, propylparaben.

Technetium-99m exametazime [(RR,SS)-4,8-diaza-3,6,6,9-tetramethyl-undecane-2,10-dione bisoxime] is used as an adjunct in the detection of altered regional cerebral perfusion in stroke. The kit for the preparation of the radiopharmaceutical is supplied as a single dose vial.

Technetium-99m tetrafosmin (^{99m}Tc-(V)O$_2$ (1,2-bis(bis(2-ethoxyethyl)phosphino)ethane) (see Fig. 3**d**)) is a myocardial perfusion agent. It is used as an adjunct in the diagnosis and localization of myocardial ischemia and/or infarction.

Technetium-99m bicisate is a brain imaging agent that is used for localization of stroke. The lyophilized kit contains ethyl cysteine dimer as the active ingredient.

Indium-111. Kits for labeling using other radionuclides include two indium-111 compounds. Indium-111 pentetreotide is used for the scintigraphic localization of primary and metastatic neuroendocrine tumors bearing somatostatin receptors. For octreotide DTPA, the active agent is supplied in a lyophilized kit with gentisic acid, citrate buffer, and inositol.

Indium-111 satumomab pendetide [148805-91-8] is used for determining location and extent of extrahepatic malignant disease in patients having known colorectal or ovarian cancer. The compound is a conjugate produced from the murine monocolonal antibody, CYT-099 (MAb B72.3), an antibody directed to a high molecular weight tumor-associated glycoprotein (TAG 72). This is a two-vial kit, stored as refrigerated solutions. The active ingredient is in one vial and sodium acetate in the other.

On-Site. Positron-emitting radionuclides are short-lived (see Table 2) and are produced either by a generator (^{82}Rb, ^{62}Cu, ^{68}Ga) or within a cyclotron at a given institution for immediate formulation and use (^{15}O, ^{13}N, ^{11}C, ^{18}F). Because carbon, nitrogen, and oxygen can all be produced as positron emitters, the technique is amenable to following biochemical processes directly. Commonly used positron-emitting radiopharmaceuticals include the following:

Radiopharmaceutical	Application
$[^{15}O]$–O_2	cerebral O_2 extraction and metabolism
$[^{15}O]$–CO	cerebral and myocardial blood volume
$[^{15}O]$–H_2O	cerebral and myocardial blood flow
$[^{13}N]$–NH_3	myocardial blood flow
$[^{11}C]$–butanol	cerebral blood flow
$[^{18}F]$–2-fluorodeoxyglucose	cerebral and myocardial metabolism; tumor localization
$[^{82}Rb]$–Rb^+	myocardial blood flow
$[^{68}Ga]$–gallium citrate	plasma volume transferrin

Future Directions

Radiopharmaceuticals. Research in radiopharmaceuticals is ongoing. Progress in radiopharmaceuticals is derived not only from the identification of new compounds but, also in new uses for existing, approved agents. An example of the latter is the use of ^{99m}Tc-exametazime for the labeling of white blood cells in the investigation of localized inflammatory disease and occult focal infection. The initial use for which this compound was approved was the investigation of perfusion abnormalities in the brain. For some time, more than half of the sales of ^{99m}Tc-exametazime was for use in white cell labeling at a time when this compound had not been approved for this purpose. Physicians, however, have the latitude to use any approved product as they see fit, consistent with training and ethical obligations. Thus, although ^{99m}Tc-exametazime for labeling white blood cells is an indication that was only approved by the U.S. Food and Drug Administration in 1995, its use for this purpose has been widespread in the United States for some years.

Another example is the use of ^{99m}Tc-sestamibi, approved for use in the evaluation of coronary artery disease and myocardial infarction, in patients with breast cancer. Use in breast cancer is under investigation by a number of physicians. The data are not yet sufficient to determine the efficacy of this agent in this setting. Its safety, of course, has already been demonstrated as part of its initial evaluation for heart disease.

A number of promising new approaches to *in vivo* diagnosis have appeared. Intravascular thrombosis in the femoral veins of the leg, resulting in detached emboli traveling to the heart and ultimately the lungs, can result in pulmonary embolism (PE). Blockage of the circulation in the lungs results in symptoms ranging from coughing and pain to sudden death. Diagnosis of the underlying deep vein thrombosis (DVT) in the legs is often difficult. Even when the patient presents at the hospital with symptoms suggestive of PE, the procedure for making a definitive diagnosis can be complex. A ventilation/perfusion study, involving two separate acquisitions with two separate isotopic agents, is usually required, then possibly a pulmonary angiogram, and then an ultrasound procedure to confirm the presence of clotting in the leg veins. A single agent that identifies thrombosis directly by binding to it would provide a better answer, fewer steps, and thus result in lower costs and more rapid initiation of appropriate therapy.

A number of promising approaches have emerged in the early to mid-1980s that may ultimately provide an agent. In general these approaches have focused on compounds that bind to the platelet glycoprotein (GP) IIb/IIIa receptor. This receptor on the platelet surface is responsible for its binding to fibrin. It is expressed in large numbers in activated platelets but is present only in small numbers on circulating platelets. Some of these compounds are derived from several of a series of small peptides known to block this receptor by binding to it. The peptides incorporate the recognition sequence of amino acid residues, arginine–glycine–aspartic acid, in either a linear or cyclic configuration. Initial work in this area was presented by two companies in 1994 (22,23). Another approach uses the snake venom-derived compound bitistatin, a member of the integrin family of polypeptides.

Another type of agent that is undergoing investigation is one that localizes in regions of tissue hypoxia. This is anticipated to find use in the evaluation of tissues that are chronically ischemic but potentially viable should blood flow be restored completely. This occurs in coronary artery disease, and in particular after reperfusion with tissue plasminogen activator (tPA) or streptokinase following myocardial infarction. It also may have applications in the brain during the period following a stroke, and in localizing tumors with hypoxic regions.

General trends in radiopharmaceutical research emphasize the use of small peptides. These molecules, of which the agents mentioned for thrombosis localization are an example, exhibit rapid and specific binding, and rapid blood clearance, two important parameters for a successful radiopharmaceutical. Peptides are readily labeled with ^{99m}Tc and lend themselves to formulation as lyophilized kits that can be rapidly and reliably reconstituted. Possible targets for these molecules are quite varied, ranging from atherosclerotic plaque to β-amyloid (for Alzheimer's disease), to a variety of somatic receptors the populations of which are increased or decreased in disease.

Much research has been directed toward the development of targeted antibodies. The results have been less fruitful than was hoped. Radiopharmaceutical imaging requires not only a sufficiently high concentration of agent in the lesion or abnormalities, but also a sufficiently low concentration in the surrounding normal tissues to permit a contrast that proves discernible in the final image. Naturally, for imaging in which the abnormality reveals itself as a decrease in radioactivity, the reverse must be true. Thus the excretion pathways, through which the vast majority of the radioactivity passes, are as influential in the final result as the behavior in the lesion. In the case of antibodies, imaging is impeded by slow clearance of the antibody from the circulation, despite rapid binding to the target. As a result, the contrast is rarely sufficient on the first day after administration to obtain an image. One solution to this has been to use ^{111}In as the radiolabel. This isotope has a 67-h physical half-life, a time better matched to the clearance of the drug. The slow clearance of the antibody requires the patient to be injected one day and return a day or two later for imaging. This is considerably less than ideal from a practical standpoint. The half-life and physical emissions of the isotope results in a generally higher radiation dose per MBq injected than is usually the case for shorter half-life isotopes such as ^{99m}Tc. This restricts injected doses to between 37 and 110 MBq (1–3 mCi). The higher energy of the γ-rays from ^{111}In combined with the lower photon emission rate resulting from the lower injected radioactivity (owing to the longer half-life) yields a poorer image than is possible with ^{99m}Tc.

Instrumentation. The future of radiopharmaceuticals is highly dependent on imaging instrumentation. Instrumental methods can evolve rapidly. Performance and characteristics of these instruments are important in the choice of disease target, lesion size, location, and contrast.

As of 1995, single-photon imaging instrumentation could produce excellent three-dimensional, ie, SPECT, images having a resolution between 7 and 15 mm, depending on location and methodology, using an injected dose of roughly 740 MBq (20 mCi) of ^{99m}Tc. Straying from these parameters results in poorer performance. It is possible to decrease injected doses to the neighborhood of 37–110 MBq (1–3 mCi) and photon energies to approximately 80 KeV, but at the cost of poorer resolution and a more photon deficient, ie, statistically more grainy image. Higher energies also yield poorer resolution when energies are above ~200 KeV. Higher injected doses are not feasible owing to the increase in radiation dose.

One active area of instrumentation research is the absolute quantification of radioactivity, which principally involves identifying methods for correcting properly for attenuation and scatter. This should permit more accurate estimates of radiation dose for therapeutic radioisotopes. Another important area related to instrumentation is the trend toward the fusion of images from a variety of imaging modalities, such as magnetic resonance imaging or x-ray computed tomography, with PET and SPECT data. Registration and simultaneous display of images permits a more accurate definition of the location of a given lesion, particularly in a complex geometry such as that of the brain. It has the potential to increase a physician's confidence in the information provided by the nuclear medicine data and allow more accurate localization of lesions for surgical treatment.

A number of devices suggest the possibility of improvement in the basic limitations of resolution and sensitivity for single-photon instrumentation. One device (24) employs an array of pinholes in a hemispherical shield that lies inside a hemispherical solid-state detector array. Simulations and initial experience using early models have suggested that the device could achieve a resolution in the brain of less than 3 or 4 mm and possibly as low as 1 mm.

Another entirely different class of devices may extend the range of radioisotopic imaging in a different way. These rely on the scatter of primary photons in the detector to extract information about the trajectory of each incoming γ-ray. Back-projection of these trajectories into the conic surface representing all possible paths can yield a reconstruction of the linear integrals of the radioactivity along each path. Regrouping and conventional reconstruction can then yield three-dimensional concentration information. These devices, known as double-Compton imaging systems or Compton scatter imaging devices were, as of early 1996, in an extremely early stage of development. A wide variety of detector types and geometries were under investigation. In addition, the mathematical problem of accurate reconstruction had not yet been solved. If successful, these devices promise the ability to image isotopes having γ-ray energies greater than 1 MeV without a collimator, suggesting considerable sensitivity as well.

Economic Aspects

Radiopharmaceutical manufacture in the United States was dominated as of the mid-1990s by three primary suppliers: Du Pont Pharma, Mallinckrodt, and Amersham International. The last was through its Medi-Physics subsidiary. Bracco Diagnostics is also a supplier, as is CIS-US. Cytogen has a product in this group as well. In addition, Golden Pharmaceuticals supplies Indium-131 for clinical use. These companies plus a number of smaller firms are engaged in research and development activities involving radiopharmaceutical products. In Europe, the suppliers include the same companies as operate in the United States with the addition of Medgenix in an alliance with Du Pont Pharma. Here, CIS, based in France, plays a larger role as do a number of smaller companies, many of them nationally based. Nordion, a Canadian firm, is a supplier of the ^{99}Mo used in generators and other raw materials as well as some clinical products both in the United States and Europe. In Japan, the primary suppliers are Nihon-Medi-Physics and Daiichi. Many of these companies also supply products to other parts of the world. The degree varies considerably from one country to another.

Table 4 is a breakdown of radiopharmaceutical studies by imaging areas from the late 1980s through the early 1990s. The cost of a given radiopharmaceutical, and therefore its relative contribution to the total market size, varies considerably. Newer agents, such as those for tumor localization, ^{99m}Tc myocardial perfusion, and regional brain indications, involve newer and more expensive agents.

Table 4. Nuclear Medicine Studies Performed in the United States, 1988–1993[a]

Organ system	Total U.S. studies $\times 10^3$						
	1988	1989	1990	1991	1992	1993	1994[b]
bone	2,397	2,611	3,794	3,794	3,844	3,939	3,624
brain	73	101	130	109	104	121	136
cardiovascular	2,233	2,529	2,878	3,109	3,720	3,874	4,106
liver	466	440	514	457	381	306	275
renal/hepatobiliary	687	782	850	827	882	853	853
respiratory	1,233	1,184	1,105	1,052	1,015	976	996
tumor localization	0	0	0	0	21	88	264
other	1,089	1,190	667	752	802	710	710
Total	*8,179*	*8,838*	*9,938*	*10,100*	*10,769*	*10,867*	*10,964*

[a]Courtesy of TMG Marketing.
[b]Values are estimated.

Radiopharmaceuticals may be sold either directly or through nuclear pharmacies. These entities, some of which are owned by manufacturers, provide radiopharmaceuticals in unit dose form. In the United States, both Mallinckrodt and Amersham own nuclear pharmacies in many cities. In addition the market is also served by Syncor, an independent nuclear pharmacy that has a nonexclusive strategic alliance with Du Pont Pharma.

For radiopharmaceuticals, such as ^{201}Tl–Cl, the role of the nuclear pharmacy is simply to draw a dose in an appropriate syringe, providing quality control and primary injected dose readings. For other products the service involves preparing radiopharmaceuticals from kits, maintaining ^{99}Mo/^{99m}Tc generators, and associated quality control. When radiopharmaceuticals and kits are sold directly, the hospital department, typically the nuclear medicine department, is responsible for all quality control. In either case, the hospital and the attending physician are responsible for verifying that the dose is correct and is administered to the patient for whom it is intended.

The total U.S. market for radiopharmaceuticals in 1993 was $406 million. Radiopharmaceuticals sold directly to U.S. hospital nuclear medicine departments accounted for approximately 25% of the total sales, or about $98 million. Another $308 million were sold via nuclear pharmacies. Of direct sales among the primary U.S. manufacturers, one had approximately 49% of the market, another approximately 34%, and the third about 9%.

The total U.S. market grew from about $332 million in 1991 to $372 million in 1992 to $406 million in 1993. Uncertainties in health care financing and restructuring in 1994 may have led to stabilization or a slight decline.

The unique information available from nuclear medicine studies suggests continued steady growth in sales. However, products introduced since the late 1980s account for only 8% of total nuclear medicine studies (25), suggesting that work in product development must be more targeted toward providing information that is critical in patient management and/or must concentrate on situations affecting a large percentage of the populace.

BIBLIOGRAPHY

"Radioactive Drugs and Tracers" in *ECT* 2nd ed., Vol. 17, pp. 1–8, by S. Hopfan, Memorial Hospital for Cancer and Allied Diseases; "Radioactive Drugs" in *ECT* 3rd ed., Vol. 19, pp. 625–632, by A. M. Green and I. Gruverman, New England Nuclear Corp.

1. P. J. Early and D. B. Sodee, *Principles and Practice of Nuclear Medicine*, 2nd ed., Mosby, St. Louis, Mo., 1995.
2. N. Alazraki and F. S. Mishkin, *Fundamentals of Nuclear Medicine*, The Society of Nuclear Medicine, New York, 1984.
3. M. P. Iturralde, *Dictionary and Handbook of Nuclear Medicine and Clinical Imaging*, CRC Press, Boca Raton, Fla., 1990.
4. G. Saha, *Fundamentals of Nuclear Pharmacy*, 2nd ed., Springer-Verlag, New York, 1984.
5. P. H. Cox, ed., *Radiopharmacy and Radiopharmacology Yearbook*, Gordon and Breach, New York, 1985.
6. A. R. Fritzberg, ed., *Radiopharmaceuticals: Progress and Clinical Perspectives*, CRC Press, Boca Raton, Fla., 1986.
7. F. Tenenbaum and co-workers, *J. Nucl. Med.* **36**, 1 (Jan. 1995).
8. E. B. Merkushev, *Synthesis*, 923 (1988).
9. R. H. Seevers and co-workers, *Chem. Rev.* **82**, 575 (1982).
10. S. R. Wilson and co-workers, *J. Org. Chem.* **51**, 4833 (1986).
11. D. S. Wilbur and co-workers, *J. Label. Compound. Radiopharm.* **19**, 1171 (1982).
12. S. Chumpradit and co-workers, *J. Med. Chem.* **34**, 877 (1991).
13. S. Chumpradit and co-workers, *J. Med. Chem.* **32**, 1431 (1989).
14. G. W. Kabalka and co-workers, *J. Label. Compound. Radiopharm.* **19**, 795 (1982).
15. K. H. Koch and co-workers, *Chem. Ber.* **124**, 2091 (1991).
16. Z. H. Cho, J. P. Jones, and M. Singh, *Foundations of Medical Imaging*, John Wiley & Sons, Inc., New York, 1993, pp. 201–232.
17. A. B. Brill, *Low-Level Radiation Effects: A Fact Book, The Subcommittee on Risks of Low-Level Ionizing Radiation*, The Society of Nuclear Medicine, New York, 1982.
18. G. Shani, *Radiation Dosimetry Instrumentation and Methods*, CRC Press, Boca Raton, Fla., 1991, pp. 1–2.
19. *1988 Radiation Dose to Patients from Radiopharmaceuticals*, ICRP Publication 53, Pergamon Press, Oxford, U.K., 1988.
20. *Radiation Dose to Patients from Radiopharmaceuticals*, ICRP Publication 62, 1991 Addendum 1 to Publication 53, Pergamon Press, Oxford, U.K., 1991.
21. *1990 Recommendations of the International Commission on Radiological Protection*, ICRP Publication 60, Pergamon Press, Oxford, U.K., 1990.
22. T. D. Harris and co-workers, *Society of Nuclear Medicine Annual Meeting*, Orlando, Fla., Abstract 1005B, 1994, p. 245P.
23. P. Muto and co-workers, *Society of Nuclear Medicine Annual Meeting*, Orlando, Fla., Abstract 315, 1994, p. 80P.
24. R. K. Rowe and co-workers, *J. Nucl. Med.* **34**(3), 474–480 (1993).
25. K. Kasses, *J. Nucl. Med.* **36**, 5 (May 1995).

JOEL L. LAZEWATSKY
PAUL D. CRANE
D. SCOTT EDWARDS
Du Pont Merck Pharmaceutical Company

RADIOPROTECTIVE AGENTS

There has been substantial progress in the area of radioprotective agents since the early 1980s, especially in terms of the biological and mechanistic evaluation of the large number of compounds synthesized prior to that time. This information has also begun to direct the development of a new generation of derivative compounds. Advances have been made in understanding the fundamental mechanisms of chemical radioprotector action, an area long dominated by thiol compounds (see THIOLS). Nonthiol protectors, including protease inhibitors, vitamins (qv), metalloelements, and calcium antagonists, are playing a larger role as of the mid-1990s. There has been an explosion of interest in biological, as opposed to chemical, modifiers of radiation injury. Some of these biologics act best when given prior to irradiation; many can modulate radiation injury when given after irradiation, presumably by affecting the recovery and repopulation of critical tissue elements. Biologics thus afford an opportunity for therapeutic intervention following accidental radiation exposure (1) as well as in radiation therapy (XRT). Demonstrations of a potential therapeutic advantage from combining chemical radioprotectors, which decrease the extent of initial damage, and nonthiol biologics, which accelerate tissue recovery, provides an attractive approach to radioprotection (2).

Thiols

Some important radioprotective thiols are listed in Table 1. The most effective compounds have a sulfhydryl, $-SH$, group at one terminus and a strong basic function, usually an amino group, at the other. The general structure of these aminothiols is $H_2N(CH_2)_xNH(CH_2)_ySH$, where x is optimally 3 and y is optimally 2 or 3. Because of lower toxicity, clinical interest has focused on pro-drugs in

Table 1. Radioprotective Thiols and Phosphorothioates

Compound[a]	CAS Registry Number	Structure
Thiols		
dithiothreitol (DTT)	[27565-41-9]	$HSCH_2CH(OH)CH(OH)CH_2SH$
2-mercaptoethanol (WR-15504)	[60-24-2]	$HOCH_2CH_2SH$
cysteamine (MEA, WR-347)	[156-57-0]	$H_2NCH_2CH_2SH$
2-((aminopropyl)amino)ethanethiol (WR-1065)	[31098-42-7]	$H_2N(CH_2)_3NHCH_2CH_2SH$
WR-255591	[117062-90-5]	$CH_3NH(CH_2)_3NHCH_2CH_2SH$
WR-151326	[120119-18-8]	$CH_3NH(CH_2)_3NH(CH_2)_3SH$
Phosphorothioates		
WR-638	[3724-89-8]	$H_2NCH_2CH_2SPO_3H_2$
WR-2721	[20537-88-6]	$H_2N(CH_2)_3NHCH_2CH_2SPO_3H_2$
WR-3689	[20751-90-0]	$CH_3NH(CH_2)_3NHCH_2CH_2SPO_3H_2$
WR-151327	[82147-31-7]	$CH_3NH(CH_2)_3NH(CH_2)_3SPO_3H_2$

[a]WR = Walter Reed Army Institute of Research (4).

which the −SH group has been derivatized. Phosphorothioates such as WR-2721, WR-3689, and WR-151327 are the most practicable (3). These pro-drugs are dephosphorylated by enzymes such as alkaline phosphatase, generating the corresponding active free thiols WR-1065, WR-255591, and WR-151326. Initially developed for military applications, the emphasis for the thiols (qv) has shifted to their potential use in XRT to protect against damage to normal tissues, based on reports that concentrations of WR-2721 that are relatively well tolerated by mice can protect normal tissues more than tumors (4).

General Mechanisms of Radioprotection by Thiols.   Thiols protect mammalian cells from radiation effects primarily by reducing the severity of the initial damage inflicted to genomic deoxyribonucleic acid (DNA). The hydroxyl radical, $OH^{\bullet}$, is a primary cause of damage to cellular DNA. Thiols, RSH; thiolate anions, RS^-; and disulfides, RSSR; react rapidly with $OH^{\bullet}$ and may thus protect cells by scavenging $OH^{\bullet}$ before this radical can react with DNA. Scavenging of secondary radicals may also contribute to protection. Thiols may also enhance cell survival by chemically repairing DNA radicals caused both by $OH^{\bullet}$ and by the direct ionization of DNA. Either an H atom, $H^{\bullet}$, or an electron can be transferred from the thiol to a DNA radical (see eq. 1). These reactions are embodied in the fixation-repair model, the cornerstone of which is the hypothesis that, in the absence of oxygen, DNA radicals, which are presumed to be responsible for radiation lethality, are either inherently irreparable (eq. 2); can be chemically repaired, usually by reaction with a thiol (eq. 1); or can be fixed by a competing reaction with oxygen (eq. 3). Direct evidence for such a competition between oxygen and thiols for DNA radicals in mammalian cells has come from fast-kinetic studies (5).

$$\begin{array}{l} \longrightarrow + \; RSH/RS^- \; \longrightarrow \; DNA/DNA^- + RS^{\bullet} \hfill (1) \\[2mm] \boxed{DNA^{\bullet}} \; \longrightarrow \; DNA \; lesion \hfill (2) \\[2mm] \longrightarrow + \; O_2 \; \longrightarrow \; DNAO_2^{\bullet} \; \longrightarrow \; DNAO_2H \; \longrightarrow \; DNA \; lesion \hfill (3) \end{array}$$

A variety of additional factors are considered in derivatives of this model, such as fixation by species other than oxygen, repair by species other than thiols, and subsets of DNA radicals that do not react with oxygen or thiols (6). In addition to increasing the cells' capacity for $OH^{\bullet}$ scavenging and chemical repair, thiols can undergo autooxidation, resulting in the depletion of oxygen and a decrease in O_2-fixation reactions. Although oxygen depletion is unlikely to represent a general mechanism for radioprotection *in vitro*, it may be a significant factor in tissues and tumors *in vivo* where oxygen levels are much lower.

The outcome of rapid radiation chemical processes in mammalian cells is to cause a variety of longer-lived physical alterations in the DNA. Of these, double-strand breaks (DSBs) appear to be most frequently involved in cell killing if not correctly repaired. In general, thiols protect against DSB induction in proportion to their effect on cell killing (7), although there are exceptions (8).

Modulation of the Killing of Mammalian Cells by Thiols.   Important aspects of the effects of exogenous thiols on clonogenic cell survival following exposure to low linear energy transfer (LET) radiations include the following.

(*1*) Thiols must be added before or within a very short time after irradiation to protect against cell killing. This is apparent from conventional cell survival data (9) but is even better illustrated by kinetic studies showing that 2-mercaptoethanol (see Table 1) protects oxic V79 cells when added just before but not 7 milliseconds after irradiation (5).

(*2*) Subcellar distribution may be an important aspect of the biological activity of thiols. Cationic thiols such as MEA and WR-1065 protect cultured cells from radiation injury at much lower concentrations than the uncharged thiols such as 2-mercaptoethanol and DTT. This behavior cannot be explained by differences in the efficiency of OH$^{\bullet}$ scavenging or chemical repair (10). The variable term may be the thiol concentration close to the DNA, where cationic thiols could be sufficiently concentrated to scavenge OH$^{\bullet}$ effectively and/or to participate in chemical repair. Uncharged thiols should be neither condensed nor excluded, whereas anionic thiols such as glutathione [70-18-8] (GSH) should be excluded from the vicinity of the DNA. A variety of experimental data are consistent with this counterion condensation model (10,11), although other studies suggest that GSH is inherently limited compared with MEA in its ability to protect against DSB induction (12).

(*3*) Protection by thiols depends on the concentration of oxygen and of the thiol. Thiols generally protect cells at intermediate oxygen concentrations, [O_2], to a greater extent than fully oxic or anoxic cells (6,13). A bell-shaped relationship has also been observed between thiol concentration and oxygen enhancement ratio (OER) (6,14). Discrepancies among earlier studies with respect to the effect of thiols on the OER may thus be a consequence of the different thiol concentrations used.

(*4*) Alteration in the levels of endogenous thiols is not required for protection by exogenous thiols. A number of thiols elevate intracellular GSH (15,16), although this is not always the case (17). Elevation of endogenous thiol levels is generally attributed to the exogenous thiol reducing cystine in the growth medium to cysteine, followed by transport of the cysteine into the cell and stimulation of *de novo* GSH synthesis. In most of these studies the addition of buthionine sulfoximine, an inhibitor of γ-glutamyl cysteine synthase and thus of GSH synthesis, abolished the elevation of GSH but had little effect on protection by the exogenous thiol (15,16). GSH elevation is thus unlikely to be a mechanism of radioprotection.

(*5*) Different types of cells can be protected by thiols to different degrees. WR-1065 protects normal human fibroblasts, but not fibrosarcoma tumor cells, against the DNA-damaging and lethal effects of x-irradiation *in vitro* (18). Cell-type differences in chromatin organization and DNA-drug associations may be responsible. WR-1065 also protects the DSB-repair-deficient xrs5 Chinese hamster ovary (CHO) mutant much less efficiently than the parent line against killing by x-rays and neutrons (19). In contrast, WR-1065 has an equally protective effect on two human glioblastoma lines (20), one of which is DSB-repair-deficient because of a defective DNA-dependent protein kinase activity. Two human squamous cell carcinoma lines, which have different DSB-repair capability, are also protected equally by MEA (8).

(*6*) Thiols can protect against the lethal effects of high LET radiations. Pretreating cells with thiols such as WR-255591 protects against injury by high

LET radiations such as neutrons (21), but invariably the magnitude of protection is lower than with x- or γ-rays. It is generally presumed that the greater severity of the DNA lesions induced by high LET radiations makes these less amenable to chemical modification.

(7) Thiols appear to be effective modulators of apoptosis and oxidative injury to the cell membrane. Apoptotic cell death may be important in the radiation responsiveness of some tumors and normal tissues. The identification of agents that can decrease apoptosis in normal cells while not affecting or enhancing the response of tumor cells has clear therapeutic benefits (22). Alterations in thiol redox status regulate various aspects of cell function, including signal transduction, transcriptional activity, and the apoptotic pathway (23). WR-1065 protects against internucleosomal DNA fragmentation and loss of viability when added to thymocytes that had been exposed to γ-radiation. Pretreatment with WR-1065 does not protect against DNA fragmentation. Clearly the post-irradiation effect is unrelated to the ability of the thiol to protect against DNA damage. Rather, WR-1065 may either inactivate the endonuclease responsible for DNA degradation, alter chromatin structure so that the internucleosomal region is not accessible to the endonuclease, or regulate cation mobilization (24). Pretreatment with WR-1065 exerted an equal protective effect on the M059J and M059K human glioblastoma cell lines. The former cells undergo significant apoptosis following irradiation, which is a minor mode of death in the latter. Thus in this case preirradiation WR-1065 appears to modulate apoptotic death (20).

Although DNA damage may trigger apoptosis (25), non-DNA targets such as the cell membrane may also be involved (26). Some studies have demonstrated effects of thiols on non-DNA targets. WR-1065 and GSH (both at 5 mM) have efficiently inhibited lipid peroxidation in mitochondrial membranes (27). WR-1065, N-acetylcysteine [616-91-1] (NAC), and DTT have protected against ^{90}Sr β-ray-induced perturbation in synchronously contracting chick embryo cardiac myocytes, which is suggested to reflect protection at the cell membrane/cytosol level (28).

Modulation of Cellular Recovery and Stress Responses. Thiols have frequently been suggested to alter cellular radiosensitivity on a longer time scale than can be explained by free-radical events. This may be because thiols can also produce profound changes in cell metabolism, including effects on cell progression, DNA synthesis, and protein synthesis (29,30). These combined effects are often referred to as biochemical shock and, in some cases, may contribute to protection by allowing additional time for the repair of DNA damage. However, the lack of protection against cell killing by thiols added immediately after irradiation (5) is inconsistent with such a scenario. Biochemical shock may therefore be more important for antimutagenic or anticarcinogenic activity which, in contrast to effects on survival, are manifested when the thiol is added after irradiation (9,31). The post-irradiation modulation of transformation and mutation, but not of cell killing, may relate to the hypothesis that the initial response of cells is to initiate constitutive DNA repair processes required for survival, regardless of the mutagenic/carcinogenic consequences (32). The subsequent slow phase of repair, which requires protein induction, may correct potentially carcinogenic lesions that are missed or misrepaired during the initial error-prone phase. Biochemical shock may thus allow additional time for such processes to function. Cell division is an important step in the fixation of mutational and transformational events.

It is becoming apparent that alterations in redox status may provide a more general control of cell metabolism, analogous to that mediated by phosphorylation. Thiols may thus affect cellular radiosensitivity by modulating the activity of a variety of proteins involved in the signaling pathways that regulate cell progression, DNA repair, apoptosis, etc (23). Radiation itself activates a cascade of signaling events that alter the expression and/or activity of specific genes and proteins involved in cell cycle progression and apoptosis, and of growth factors that stimulate the proliferation of surviving cells. This can lead to enhanced survival or cell death, depending on which pathway a cell follows, which in turn depends on the cell type and its environment, as well as on the extent of DNA damage (25). Protein kinase activation and subsequent phosphorylation reactions appear to play an important role in these responses. Thiols can attenuate the radiation-induced activation of these pathways by a number of mechanisms. By scavenging $OH^{\bullet}$, thiols prevent the oxidation of membrane phospholipids and associated generations of arachidonate and diacylglycerol, which activate protein kinase C (PKC), and inositol triphosphate, which causes Ca^{2+} mobilization and activation of PKC. The activation of c-jun, a cellular proto-oncogene transcription factor, by x-rays has been suggested to be mediated by reactive oxygen intermediates based on the inhibition of this effect by NAC (33). Similarly, WR-1065 blocks the activation of PKC in CHO cells exposed to 20 Gy (1 Gy = 100 rad = cGy) of x-rays (34). Signal transduction pathways can also be activated by DNA damage, and are thus sensitive to the effects of thiols on DNA damage.

Modulation of Mutation.   Because mutations in cells that survive radiation exposure may be expressed as undesirable late effects in XRT, the potential use of thiols to protect against mutation and transformation has generated considerable interest. The phosphorothioates WR-2721 and WR-151327 inhibit the induction of tumors in mice exposed to both low and high LET radiations (35,36). WR-2721 given to mice 30 min before irradiation also protects against hypoxanthine phosphoribosyltransferase (HPRT) mutations induced by both γ-rays ($\sim$2.4-fold) and neutrons ($\sim$1.4-fold) (37). The free thiols WR-1065 and WR-151326 also display antimutagenic (9) and antitransforming (38) activity *in vitro*. Studies have shown several important aspects. (*1*) Timing. Adding aminothiols to V79 cells up to 3 h after irradiation has no effect on survival following either γ-rays or neutrons, but markedly reduces the induction of HPRT mutations for both radiation types (9,31). WR-2721 also protects against neutron-induced HPRT mutations *in vivo* when it is given 3 h after irradiation (37). These data suggest a role in the treatment of individuals who have already been exposed to radiation. (*2*) Concentration. WR-1065 protects against transformation and HPRT mutations at much lower concentrations than are necessary to protect against cell killing (16). (*3*) LET dependency. Thiols are equally effective against neutron and γ-ray-induced HPRT mutations *in vitro* (31). Protection by WR-151326 and WR-1065 against neutron-induced transformation *in vitro* has also been reported (38). (*4*) Form of the drug responsible. Although WR-1065 per se, rather than its disulfide WR-33278 (N,N''-(dithiodi-2,1-ethanediyl)bis-1,3-propanediamine), appears to be responsible for protecting V79 cells against killing by γ-rays, this distinction is not so clear-cut for CHO cells (16). The polyamine spermine and WR-33278 are structurally similar agents capable of binding to DNA. When introduced into CHO cells by electroporation at 0.01–0.001 mM either 30 min prior to or 3 h following exposure to 7.5 Gy (750 rad), both agents protect against HPRT mutations (39). (5) Alterations in the distribution of mutations by thiols. A different distribution of HPRT mutations was observed in cells irradiated in the presence or absence of MEA. Thiols preferentially protect against those events that generate large-scale molecular changes as opposed to point mutations (40).

Preclinical and Clinical Studies using Phosphorothioates. Because radiation does not discriminate between normal and tumor cells, XRT may cause adverse normal tissue reactions that can limit the intensity of treatment and also be life-threatening. An improved therapeutic index would be possible if the differential response between the tumor and irradiated normal tissues could be increased. Promising agents in this context include the phosphorothioates WR-2721, WR-151327, and WR-3689, which protect a variety of dose-limiting normal tissues, both early and late responding (3,4,13,41–43). An enormous literature has accumulated relating to the effects of phosphorothioates on animals and humans. On the basis of these studies, several generalizations can be made.

(1) Phosphorothioates generally protect normal tissues more than tumors. Tumor protection reported in some animal studies can partly be explained by physiological effects of the particular drugs, which are specific to rodents (4). WR-2721 does not appear to protect human and most animal tumors, apparently because of the low availability of the drug to tumor cells (4). Many tumors appear to have a reduced capillary density (44), which may mean that these tumors have altered levels of alkaline phosphatase, the enzyme that converts WR-2721 to WR-1065. A reduced ability of thiols to protect the hypoxic cells characteristic of many tumors may also contribute to their selectivity for normal tissues. The observation that WR-1065 protects cultured normal human fibroblasts, but not fibrosarcoma tumor cells, suggests that additional factors may contribute to the selectivity of radioprotection by WR-2721 *in vivo* (18).

(2) Thiols protect different normal tissues to different degrees. Differences in the extent of protection of normal tissues by WR-2721 have been attributed to differential drug accumulation. A potential mediator of this selectivity is alkaline phosphatase. However, there is no simple relationship between protection and tissue drug concentration or alkaline phosphatase levels (42). Another potential factor is tissue oxygenation, or more specifically the oxygen concentration, pO_2, of the critical target cells for radiation injury. Protection by WR-2721 appears to be maximal when the pO_2 of target cells is at or just above the transition region from hypoxic to oxic sensitivity, ie, the K-value, which may be different in different tissues (13,42,43). The likelihood that many normal tissues exist at intermediate oxygenation may in fact contribute to the therapeutic ratio. The importance of considering the target cells in this regard is illustrated by the murine bone marrow (BM), which is regarded as well oxygenated but is efficiently protected by WR-2721. However, the BM stem cells may exist in a more hypoxic environment than the committed lineage-specific progenitor cells and the proliferating cells (45).

(3) Fractionated and low dose-rate XRT schedules may be less amenable to modification. There has long been a concern that radioprotective agents are less effective when fractionated irradiation is employed. The loss of protection

of normal tissues by WR-2721 found in some studies is probably a result of the reduced drug dosage used with each fraction to avoid cumulative drug toxicity (43). The same appears to be true in clinical fractionated XRT, where the daily dose of WR-2721 would have to be reduced by ~50% to minimize toxicity (4,44). WR-2721 efficiently protects the BM after acute doses and could thus be effective in radioimmunotherapy, where myelosuppression is the dose-limiting toxicity, for protecting against the early phase of injury associated with the initial systemic distribution of the radioisotope. Although initial results showed promise (42), subsequent studies, in which WR-2721 was given to mice prior to an $LD_{90/30}$ dose of radiolabeled antibody, followed by further injections of the drug every 4 h up to 72 h, failed to show significant benefit (46). A use for WR-2721 has also been suggested in single high dose intraoperative XRT, in which dogs given WR-2721 in their duodenal lumen prior to irradiation showed significantly reduced normal tissue injury (47).

(*4*) Clinical studies of WR-2721 using XRT for certain types of cancer show promise, but toxicity remains a concern. Although systemic WR-2721 is generally well tolerated as a single iv infusion, the achievable concentration of WR-2721 is still limited by toxic side effects. These include nausea, vomiting, sneezing, a warm or flushed feeling, mild somnolence, a metallic taste in the mouth, transient hypocalcemia, and allergic reactions (4). Emesis may be moderately severe, but can be reduced by antiemetics such as ondansetron and dexamethasone. The potentially dose-limiting toxicity is hypotension, which is rapidly reversed by discontinuing the drug.

WR-2721 shows promise for patients with inoperable, unresectable, or recurrent rectal adenocarcinoma who receive pelvic irradiation (48). In these studies there is no evidence of tumor protection. Some acute reactions are less frequent in the WR-2721-treated group. There are no moderate or severe late effects in the WR-2721-treated group, compared to 14% in the XRT-only group. Patients receiving WR-2721 15 min before each fraction show no significant hypotension or hematologic toxicity up to 18 months. Mild to moderate emesis occurred in 80% of the courses. These encouraging results are being extended worldwide, and trials are planned for patients having head and neck or lung cancer, with and without concomitant antiemetics (5,48). WR-2721 also shows promise for protecting the BM of patients receiving total body irradiation (TBI) for lymphoma or chronic lymphocytic leukemia (CLL) (49,50).

(*5*) Although WR-2721 is generally superior to other phosphorothioates in protecting normal tissue in mice, other drugs also show some potential. No phosphorothioate is markedly superior to WR-2721. However, WR-3689 is equivalent in most respects and is less toxic (41). Whereas both WR-2721 and WR-3689 are behaviorally toxic at radioprotective doses in rats, the combining of these with caffeine appears promising for ameliorating these effects (2,51).

Other phosphorothioates such as WR-44923 (41), WR-638, and WR-77913 (52) provide radioprotection comparable to WR-2721 in the small intestine of mice. Similarly, WR-151327 effectively protects the intestine and BM and modulates late normal tissue reactions in mice (53,54). WR-151327 is active orally and is undergoing preclinical testing (4,55). Clinical trials are anticipated.

(*6*) Alternative methods of delivering radioprotective agents may provide an improved therapeutic index. Topical or regional application can circumvent problems related to systemic toxicity and tumor protection, eg, during irradiation of the chest wall after breast tumor surgery. Unfortunately, WR-2721 is ineffective topically for skin protection, presumably because of inadequate penetration. Potential enhancers of transdermal drug delivery (qv) have been screened in rodents, although even the best WR-2721 levels attained may be insufficient for protection (56). WR-2721 applied topically before irradiation protects the mucosa of the large bowel of rats by ~1.8-fold (57). A subsequent phase-I/II evaluation of WR-2721 as a protector of the rectosigmoid mucosa, when given topically in enema form daily before fractionated XRT to the pelvis, showed no unfavorable side effects or toxicity in patients given WR-2721 daily during a five-week course of external beam XRT. However, WR-2721 did not protect the rectal mucosa from radiation damage, possibly because of the low drug doses used (58).

Another approach to targeting phosphorothioates can be found in the demonstration that intraventricular injection of WR-2721, WR-3689, or WR-77913 into the brains of rats protects against hind-limb paralysis caused by irradiation of the cervical spinal cord (59). This strategy could be useful during XRT for noncentral nervous system (non-CNS) tumors where the spinal cord is in the radiation field. Thiols can also be targeted at the subcellular level. A series of compounds in which an aminothiol is bound to the DNA-binding chromophores quinoline and acridine have been synthesized and shown to protect against guanine oxidation more efficiently than the parent aminothiol (60).

(7) Phosphorothioates attenuate the effects of neutrons *in vivo*, but are generally less protective than for x-rays. WR-2721 and WR-151327 protect against neutron lethality in mice and against damage to various normal tissues and tumors (53,54,61). Whereas WR-2721 is generally superior for protecting against low LET radiations, other drugs may be equal or better protectors for high LET radiations. WR-151327 is more effective than WR-2721 in protecting against gastrointestinal (GI) injury caused by neutrons (53), whereas the phosphorothioate WR-2529 exhibited the optimal protection in the $LD_{50/30}$ assay and, in view of its lower toxicity, is a significantly better protector than WR-2721 (61).

Other Sulfur-Containing Compounds. *Glutathione and Cysteine Analogues.* Endogenous thiols such as GSH and cysteine [7048-04-6] are present in all mammalian cells and perform a variety of metabolic functions, including the protection of cellular components from the damaging effects of radiation and other oxidative stresses (62). Radioprotection *in vitro* cannot generally be achieved by adding cysteine or GSH per se, as these agents can be quite cytotoxic and often have irregular uptake into cells. However, GSH given to rats 15 min prior to irradiation does protect the parotid glands from radiation injury (63). Similarly, a randomized pilot trial indicates that administration of GSH (1200 mg iv) before XRT reduces the occurrence of diarrhea resulting from damage to the GI mucosa of patients undergoing surgery for endometrial cancer and receiving adjuvant XRT to the pelvis (64).

An alternative method for elevating intracellular GSH levels is to use agents that promote GSH synthesis, such as 2-oxothiazolidine-4-carboxylate (OTZ), a cysteine delivery system, or glutathione ethyl ester [118421-50-4] or methyl ester, which is readily taken up by a variety of cell types and converted to GSH (62). When OTZ has been used to elevate intracellular GSH by two- to three-fold, conflicting results have been obtained for oxic cells where either modest or no protection has been found (65). However, GSH should be more effective at lower O_2 tensions. Some protection of V79 cells by OTZ at low or intermediate oxygenation, but none at high $[O_2]$, was observed. The conversion of OTZ to cysteine requires the activity of 5-oxoprolinase. Whereas abundant in normal tissues, this enzyme is deficient in many experimental tumors, which may give rise to a therapeutic advantage (62).

Another alternative is NAC, which enters cells, is rapidly hydrolyzed to cysteine, and can then enter the GSH-synthetic pathway. Radioprotective effects of NAC have been reported in a variety of *in vitro* systems (28,66) and in mice (67). Treating cultures of nonadherent low density human BM cells with NAC (2 mg/mL) before and after irradiation protects granulocyte/macrophage colony-forming units (CFU-GM) by ~1.56-fold (68). A radioprotective effect has also been obtained using NAC applied topically on normal human skin (69). In contrast, NAC given to patients with inoperable nonsmall-cell lung cancer who were receiving hyperfractionated XRT (60 Gy in 48 fractions over 32 days) had little effect on either normal tissue reactions or tumor responses (70). Ribose-cysteine (RibCys), a cysteine pro-drug that stimulates GSH biosynthesis, shows some ability to protect against radiation-related deaths and anastomotic leaks in a swine model of surgical injury to the rectosigmoid (71).

2-Mercaptopropionylglycine. Preirradiation administration of 2-mercapto-propionylglycine [*1953-02-2*] (MPG) protects against alterations in the esophageal epithelium of mice receiving a single dose of 7.5 Gy of γ-rays TBI (72) and ameliorates the depletion of various types of ovarian follicles (73). Combining nontoxic doses of MPG and WR-2721 can lead to improved protection of tissues such as BM and GI tract (74).

2-Aminoethyl Isothiuronium Bromide Hydrobromide and 5-Hydroxy-L-Tryptophan. Synergistic protection by the thiol compound aminoethylisothi-uronium bromide hydrobromide (AET) and 5-hydroxy-L-tryptophan (5-HTP) at doses of each agent that were individually ineffective and nontoxic has been described (75,76). Treating mice with AET (20 mg/kg) plus 5-HTP (100 mg/kg ip) 30 min before TBI modified the decline in sperm counts, significantly decreased the level of sperm abnormalities, and protected against sterility associated with oligospermia. This same combination of agents markedly protected normal hemopoietic mouse tissues, showing a ~1.76-fold increase in $LD_{50/30}$ as well as enhanced recovery of 10-day spleen colony-forming units (CFU-S), but did not protect a transplanted mammary carcinoma or a significantly modified injury to rat splenic tissue following 8 Gy TBI. This combination notably protected against micronucleus induction in the femoral BM cells of mice exposed to 8 or 12 Gy of γ-rays TBI. Both the magnitude and duration of radioprotection in the 30-day survival assay were further improved by the administration of hydroxylamine, a decarboxylase inhibitor, prior to the combination of AET/5-HTP (76).

Other Thiols. A variety of sulfur-containing organic compounds have been synthesized and screened for radioprotective activity. Compounds having a high ability to protect mice against lethal doses of TBI γ-radiation have been identified among a series of quinolinium and pyridinium bis(methylthio) and methylthio amino derivatives (77). Several 1,2-dithiol-3-thione and dithioester compounds are also highly effective protectors of EMT6 cells *in vitro*, and in some cases are differentially cytotoxic toward hypoxic cells (78). Lipoic acid [*1077-28-7*]

(6,8-dimercaptooctanoic acid), a lipophilic endogenous disulfide that can be re-duced to the dithiol dihydrolipoic acid [7516-48-5], protects against free-radical-mediated injury both *in vivo* and *in vitro*. Lipoic acid given before, but not after, irradiation markedly protects murine hemopoietic tissues in the $LD_{50/30}$ (~1.26-fold), endogenous CFU-S (~1.5-fold), and exogenous CFU-S (~1.34-fold) assays (79). Dihydrolipoic acid, but not lipoic acid, increases the survival of x-irradiated V79 cells, thus suggesting that dihydrolipoic acid is the active cellular radiopro-tective agent.

Other Classes of Protectors

Tempol. Nitroxides are low molecular weight, stable free radicals that protect against various biological manifestations of oxidative stress, including mutagenic effects. Tempol [2226-96-2], a water-soluble piperidine nitroxide derivative having nonspecific radical-scavenging and superoxide dismutase (SOD) activity, protects cultured aerobic, but not hypoxic, cells against radiation-induced killing (80). Protection does not depend on intracellular thiols and does not involve O_2-depletion. Tempol reacts with peroxyl radicals and can also oxidize DNA-bound metal ions, thereby interfering with $OH^\bullet$ generation. Ana-logues such as 3-aminomethyl-Proxyl and Tempamine show superior protective activity to Tempol in V79 cells. These analogues have amino groups that may confer an affinity for DNA comparable to that of the aminothiols, a sugges-tion supported by equilibrium dialysis studies and by the ineffectiveness of the negatively charged derivative 3-carboxy-Proxyl (80).

Tempol protects against alopecia in guinea pigs, possibly by directly pro-tecting the hair follicle stem cells. Tempol administered 5–10 min before TBI protected mice by ~1.3-fold based on $LD_{50/30}$ values, but did not compromise the ability of radiation to control the RIF fibrosarcoma, presumably because of enhanced reduction of the drug in tumor cells (80). A potential application for nitroxides as selective normal tissue radioprotectors, especially in view of the modest radiosensitizing effect in hypoxic cells, has been suggested. Nonclassi-cal radioprotectors such as Tempol, Zn aspartate, and Ca^{2+} antagonists, which are effective protectors against TBI at low concentrations, may act by inhibiting apoptosis (81).

The potential for improved radioprotection by combining Tempol with growth factors such as stem cell factor (SCF), which protects by quite differ-ent mechanisms, has been examined in mice (82). Both SCF alone, given 20 and 4 h before and 4 h after TBI, and Tempol alone, given 10 min before TBI, in-creased 30-day survival, but protection was greater than additive when the two agents were combined.

Captopril. The antihypertensive agent captopril [62571-86-2] ((2S)-1-(3-mercapto-2-methylpropionyl)-L-proline) is an inhibitor of angiotensin-converting enzyme (ACE) (see CARDIOVASCULAR AGENTS; ENZYME INHIBITORS). ACE converts angiotensin I to angiotensin II, a potent vasoconstrictor. Captopril is well tol-erated and approved by the U.S. FDA for chronic use in humans. ACE in-hibitors protect against radiation-induced injury to lung and skin as well as early hemodynamic changes after local renal irradiation (83–85). Post-irradiation captopril also ameliorates radiation nephropathy in rats receiving 15–27 Gy

(1500–2700 rad) in 12 fractions, especially in animals having less extensive renal injury. Captopril and other ACE inhibitors may also be useful for the treatment and prophylaxis of nephropathy, which is a significant cause of late morbidity in BM transplantation patients receiving total body irradiation (TBI) (83).

Captopril protects against pulmonary endothelial dysfunction and fibrosis and reduces the severity of the histopathological reaction in the irradiated rat lung. When added to the feed after the final irradiation, it protected against various manifestations of early lung reaction in rats receiving fractionated hemithorax irradiation (60 or 80 Gy (6000 or 8000 rad) in 10 daily fractions) (84). Captopril prevented or attenuated hypertensive reactions possibly in part by limiting edema in the irradiated lung. However, this drug is not able to modify all aspects of acute radiation lung damage, as it had no effect on the early increase in permeability surface area and delayed the onset of hypoperfusion only slightly.

Whereas daily doses of captopril following irradiation ameliorates several indexes of cardiac damage measured three and six months after 20 Gy (2000 rad), it fails to prevent the progressive functional deterioration of the heart, thus indicating that captopril is ineffective for the treatment of late cardiac complications (85). Captopril administered continuously in the drinking water from day seven before irradiation protects against acute (3- and 5-day) damage to the jejunal mucosa of mice exposed to 9 or 15 Gy (900 or 1500 rad) TBI (86).

Assigning a specific mechanism to the various therapeutic effects of captopril is difficult because the drug has diverse biological effects. Both the antihypertensive activity via ACE inhibition and the nonspecific thiol function may contribute. The thiol group is unlikely to contribute in cases where captopril is administered only after irradiation. Inhibition of ACE may be important mechanistically because enalapril [75847-73-3], a nonthiol ACE inhibitor, is an effective radioprotector in some situations. The ability of captopril to control hypertension may be a factor in kidney response and may play a minor role in protection of lung, but is unlikely to contribute to protection of skin. Comparing antihypertensives that are or are not ACE inhibitors indicates that, whereas hydrochlorothiazide [58-93-5], a diuretic, and enalapril, an ACE inhibitor, are equally effective in preventing hypertension, only the latter prevents proteinuria.

Captopril does appear to be capable of acting as a traditional thiol under some conditions. Pretreatment using captopril inhibited the induction of single-strand break (SSB)s and DSBs in plasmid pBR322 DNA and protected against the loss of clonogenicity of cultured human SCL-1 keratinocytes irradiated with neutrons, apparently by an $OH^{\bullet}$ scavenging mechanism (66).

Metallothioneins. The metallothioneins, a group of low ($<10,000$) molecular weight proteins containing $\sim30\%$ cysteine residues, are efficient $OH^{\bullet}/O_2^{\bullet-}$ scavengers. Transcription of metallothionein genes can be induced by a variety of agents, including heavy metals such as cadmium, Cd; glucocorticoids (GC); interferon (IFN); Ha-*ras*; and mediators of the inflammatory stress response such as tumor necrosis factor (TNF) and interleukin 1 (IL-1). In some but not all studies induction of metallothioneins prior to irradiation *in vitro* is associated with a radioresistant phenotype (87). Cadmium-treated mice also show a significantly increased $LD_{50/30}$ for x-rays (88). Bismuth nitrate modifies both acute and late effects of x-rays (30-day lethality, number of endogenous CFU-S, and induction of thymic lymphomas) in mice by inducing metallothionein in the target tissues, but does not protect against GI lethality (89). In contrast to these reports of radioresistance following gene induction, no change in radiosensitivity has been observed in several studies where mammalian cells are transfected with metallothionein genes (90).

Protease Inhibitors. Protease inhibitors such as antipain [149116-08-5], a tripeptide analogue derived from actinomycetes, and the soybean-derived Bowman-Birk inhibitor [37330-34-0] family exhibit antitransforming and anticarcinogenic activity *in vivo* and *in vitro* at nontoxic doses but generally do not modify cell killing (91). The Bowman-Birk inhibitor suppressed x-ray-induced transformation *in vitro* at nanomolar concentrations. Many protease inhibitors suppress radiation-induced transformation in a variety of *in vitro* systems. The most effective of these is chymostatin, a quite specific and potent inhibitor of chymotrypsin. Protease inhibitors are also antiteratogenic, decreasing the induction of exencephaly in irradiated mice (92).

Although the mechanism by which protease inhibitors block carcinogenesis is unknown, these inhibitors appear to be able to reverse the initiating event, presumably by stopping an ongoing process begun by the carcinogen exposure. At the subcellular level, protease inhibitors suppress c-myc transcripts in irradiated C3H10T^{1}/$_2$ cells and alter radiation-induced gene amplification in a manner that correlates with their ability to suppress transformation (91). Protease inhibitors can suppress transformation even when applied after carcinogen exposure both *in vitro* and *in vivo*, provided that the cells are able to proliferate at the time of exposure to the inhibitor (91).

Antioxidant Vitamins. The natural antioxidant vitamins A, C, and E (retinoic acid, ascorbic acid, and α-tocopherol, respectively), as well as the vitamin A dietary precursor β-carotene, exhibit a range of radioprotective effects (see VITAMINS). These protect against lethality, mutation, and transformation in a variety of cultured cell types and laboratory animals. Effects have been reported for both pre- and post-irradiation treatments (3,93). Although vitamins protect much less efficiently than WR-2721, the vitamins have low toxicity and may thus be useful in some situations. Protection against 30-day lethality has been observed in mice given vitamin A or β-carotene within two days after TBI (93). Post-irradiation, but not preirradiation, treatment with vitamin E also enhances the 30-day survival of mice exposed to 8 Gy (800 rad) of γ-rays, which suggests an enhancement of recovery of BM function. Although such preirradiation activity is likely to involve radical scavenging, these mechanisms are unlikely to explain the ability of these agents to protect when administered post-irradiation. Vitamin C, however, does suppress radical formation in irradiated hamster cells (94).

Supplemental vitamin A, begun prior to or directly after local irradiation (30 Gy (3000 rad) to the hind limb), decreases radiation-induced toxicity. Supplemental vitamin A or β-carotene also diminishes systemic toxicity owing to local x-radiation in tumor-bearing mice without diminishing the antitumor effect of the radiation. In addition to the effects on lethality ($LD_{50/30}$) and survival time, these vitamins also protect rodents against gastric and intestinal bleeding, adrenal gland hypertrophy, thymic involution, lymphocytopenia, weight loss, and carcinogenesis. Vitamin A and β-carotene can stimulate immune reactions, including those directed against tumors. In some instances this results in an enhanced antitumor effect of local irradiation (93).

A number of studies have shown that vitamins moderate the induction of chromosomal aberrations by radiation. Vitamins C and E given orally to mice either 2 h before, immediately after, or 2 h after 1 Gy (100 rad) of γ-ray TBI significantly reduce the frequencies of micronuclei and chromosomal aberrations in BM cells. Vitamin E is the more effective (95). Administration of vitamins C and E within 5 min of irradiation is as effective as pretreatment. Protection by vitamin C has also been shown in humans. Whereas chronic treatment of rats using vitamin C (100 or 300 mg/(kg/d)) for six months prior to TBI protects against chromosomal aberrations, vitamin E is not radioprotective in this setting (96).

There are numerous reports of the effects of antioxidant vitamins on transformation. Vitamin C suppresses x-ray-induced transformation when C3H10T^{1}/$_2$ cells are treated daily for one week following irradiation (97), suppresses transformation by γ-rays or neutrons, and prevents the promotion of radiation-induced transformation by 12-O-tetradecanoylphorbol 13-acetate (TPA), but has no effect on cell survival (98). In these studies, the continuous presence of vitamin C for a critical period appears to be necessary for suppression of transformation. Vitamin C may act on the promotion stage of transformation which, for TPA, may involve activation of PKC or production of $O_2^{\bullet}$ (98).

Some radioprotective agents can ameliorate the effects of radionuclides, which irradiate tissues chronically. Many radionuclides emit low energy Auger electrons that may have severe biological effects. Vitamins A and C, as well as traditional thiol radioprotectors such as MEA, offer significant protection against the effects of tissue-incorporated radionuclides on murine spermatogenesis (99). The radioprotective ability of vitamin C appears to depend on the dose rate and the radiation type.

One vitamin E analogue, TROLOX, inhibits radiation-induced apoptosis in murine thymocytes (26). Chicks given vitamin E prior to exposure to a sublethal dose (2.25 Gy (225 rad)) of γ-radiation demonstrate a more rapid recovery from damage to the thymus (100).

The potential for combining vitamins and aminothiols shows some promise. Vitamin E (100 IU/kg) injected subcutaneously (sub-q) either 1 h before or within 15 min after irradiation significantly increases the 30-day survival of mice. Improved protection has been obtained when WR-3689 (150–225 mg/kg) is also given ip 30 min before irradiation (101). Preirradiation treatment using vitamin E and cysteine in combination also offers better radioprotection than the individual agents against alterations in various hematological parameters (102). An additional potential benefit is that vitamin A and β-carotene may actually inhibit some of the side effects, such as ulceration and bleeding, associated with the use of aminothiols (93).

Metalloelements. Radioprotection by the essential metals Cu, Fe, Mn, and Zn and their intracellular complexes has been reviewed (103). Metalloelement-dependent enzymes include SOD, catalase, metallothioneins, lipoxygenases and cyclo-oxygenases involved in arachadonic acid metabolism, alkaline phosphatase, DNA polymerase and gyrase involved in DNA synthesis, and Zn-finger proteins involved in the regulation of transcription. These may modulate recovery from radiation-induced injury, and the depletion of their activity may partially account for the biological effects of ionizing radiation, as well as explaining the radioprotective activity of exogenous metal compounds. Cytokine-mediated redistribution of metalloelements may also be an important factor in radiation response. The SOD-mimetic tetra(3,5-diisopropylsalicylato)dicopper(II), $Cu(II)_2(DIPS)_4$, protects mice by ~ 1.2-fold at nontoxic doses and protects against radiation carcinogenesis. Protection by $Cu(II)_2(DIPS)_4$ may be related to the induction of hypoxia, stimulation of lympho/hemopoiesis, and/or Cu-dependent effects on the transcription and translation of stress response genes, rather than to its SOD activity. Various other essential metal compounds such as $ZnCl_2$ and $Zn(II)(DIPS)_2$ exhibit some degree of radioprotective activity in the mouse $LD_{50/30}$ assay (103,104). The involvement of zinc compounds in transcription and translation may be important for cellular/tissue recovery processes.

Metalloelement complexes may be useful for the post-irradiation treatment of radiation injury, based on the observation that several of these compounds accelerate recovery of, among other things, lympho/hemopoiesis. Preirradiation $Mn_3(O)(DIPS)_6$ increases the survival of γ-irradiated mice (103). Treatment of mice that have been exposed to an $LD_{50/30}$ dose of γ-rays plus $Mn_3(O)(DIPS)_6$ either 1 or 3 h after irradiation also increases survival, which supports the hypothesis that this compound is an effective radiorecovery agent (105). Again, this increase in survival may result from the resynthesis of radiation-depleted Mn-dependent enzymes that facilitate the recovery of immunocompetence and tissue repair, as reported for $Cu(II)_2(DIPS)_4$.

Significant data support the approach of combining metalloelement complexes such as $MnCl_2$, zinc aspartate, and $ZnCl_2$, with aminothiols such as MEA and WR-2721 to minimize acute toxicity and improve radioprotective efficacy by minimizing tumor protection (2,103). Nontoxic doses of metalloelements appear to decrease the toxicity of aminothiols (103). Combining WR-2721 and $ZnCl_2$ results in enhanced normal-tissue radioprotection (2). Intraperitoneal injection of mice with $ZnCl_2$ and WR-2721 also decreases the extent of tumor protection observed using WR-2721 alone, which suggests an additional favorable therapeutic gain factor for this combination (103). Although normal tissue specificity has been shown for zinc aspartate with and without WR-2721 (106), the level of zinc aspartate that humans can tolerate is not known.

A variety of other metals and their complexes have been studied for radioprotective activity. Among these are copper glycinate, strontium chloride, $ZnNa_3$-diethylenetriaminepentaacetate (ZnDTPA), and selenium, which has been studied because of its relationship to endogenous antioxidant mechanisms, especially GSH peroxidase and vitamin E.

Calcium Antagonists. The potential use of Ca^{2+} antagonists as radioprotective agents has been suggested based on the importance of maintaining Ca^{2+} homeostasis for cell viability following a variety of cytotoxic insults (107). Alterations of cytosolic Ca^{2+} levels can result in changes in the activity of Ca^{2+}-dependent degradative enzymes, which may contribute to cell death, and of the PKC-mediated signal transduction pathway. Although the role of perturbations in Ca^{2+} homeostasis in cellular radiation response remains poorly defined, the realization that modulation of intracellular Ca^{2+} can prevent apoptosis in some cell types has stimulated even greater interest in Ca^{2+} antagonists (22,25).

A variety of drugs that inhibit cellular Ca^{2+} uptake exhibit radioprotective activity. The Ca^{2+} channel blocker diltiazem [33286-22-5] protects mice against radiation lethality when administered sub-q or ip, and even orally, 10 or 30 min preirradiation (108). The Ca^{2+} channel blockers nifedipine [21829-25-4] and nimodipine [66085-59-4] are also effective. Protection by diltiazem may result from inhibition of the cellular influx of Ca^{2+} subsequent to membrane injury or to free-radical scavenging. Diltiazem is both a thiol and a Ca^{2+} antagonist. Diltiazem, however, has some effectiveness when administered after irradiation, a feature that distinguishes it from other thiols. Flunarizine [52468-60-7], a Ca^{2+} antagonist that also enhances local blood flow, is not radioprotective in these studies, which indicates that radioprotection is not a general property of Ca^{2+} antagonists, and that some specificity exists (108).

As for all radioprotective agents, the clinical usefulness of these drugs depends on an ability to protect normal tissue better than tumor cells at nontoxic levels. Diltiazem, nifedipine, nimodipine, and nitrendipine do not appear to alter the radioresponsiveness of human tumors grown in immunosuppressed mice (108). Although providing only modest normal tissue potection, Ca^{2+} antagonists have mild side effects and may thus be useful in combination with other agents. Synergistic effects have been observed when diltiazem is combined with zinc aspartate, dimethyl sulfoxide (DMSO), and nifedipine. In fact, zinc aspartate combined with diltiazem protects to a similar extent to combinations of zinc aspartate with WR-2721 (108).

Adenosine Analogues. Exogenous adenine nucleotides are moderately radioprotective when given to animals shortly before irradiation. Protection appears to be mediated by extracellular adenosine receptors that are coupled to the inhibition/activation of adenylate cyclase, which in turn regulates intracellular cyclic adenosine monophosphate (AMP), which is itself radioprotective *in vitro* (109). The combination of AMP, a soluble adenosine pro-drug, and dipyridamole, an inhibitor of adenosine uptake, leads to an elevation of extracellular adenosine and activation of cell surface adenosine receptors. The combination enhances the proliferation of hemopoietic cells, including CFU-GM, in nonirradiated mice. In mice evaluated 24 h after 3 Gy (300 rad), a greater number of DNA-synthesizing hemopoietic cells have been observed when dipyridamole-AMP is given 60–90 min prior to irradiation. This drug combination also protects against the increase in free polynucleotide levels in the thymus and spleen of mice receiving 1 Gy (100 rad) TBI (110). Dipyridamole-AMP given either 15 or 60 min preirradiation is radioprotective in mice based on $LD_{50/30}$ values (~1.11-fold), endogenous CFU-S survival, and post-irradiation BM CFU-GM recovery (109). Protection is also apparent for fractionated irradiation upon repeated drug administration. In addition to the stimulation of hemopoietic recovery, protection appears to involve induction of hypoxia resulting from effects on the cardiovascular system. However, noradrenaline given along with dipyridamole-AMP eliminates hypoxia induction but preserves the radioprotective action of dipyridamole-AMP in terms of hemopoietic recovery and partially with respect to survival enhancement (111).

Administration of dipyridamole-AMP to mice 5–25 min after 1 Gy (100 rad) of TBI γ-irradiation is also protective, as indicated by plasma thymidine levels and the amount of saline soluble polynucleotides in the thymus (112). Adding dipyridamole-AMP to *in vitro* irradiated suspensions of thymocytes enhances the rejoining of DNA strand breaks (112). These post-irradiation effects are presumably mediated by the activation of extracellular adenosine receptors.

Exogenous adenosine triphosphate (ATP) protects various tissues of mice subsequently exposed to a lethal TBI dose of neutrons (113). ATP (700 mg/kg ip) increases the 30-day survival from 40 to 85%, protects against damage to the testes, and increases the seven-day survival from 26 to 86%. The increase in activity of acid phosphatase after neutron irradiation, an indicator of lytic processes, is ameliorated by ATP in both the testes and small intestine. ATP also differentially ameliorates the increase in cholinesterase activity observed in rhabdomyosarcoma tumors and the small intestine of neutron-irradiated mice. ATP-MgCl$_2$ (60 μmol/kg iv) given to pigs receiving preoperative fractionated external-beam pelvic XRT also protects against various manifestations of colorectal injury (114).

Priming with Other Antitumor Agents.   Priming animals with low doses of some chemotherapeutic agents enhances resistance to subsequent exposures to radiation (see CHEMOTHERAPEUTIC AGENTS, ANTICANCER). Priming mice with vincristine [57-22-7] increases the radioresistance of the GI epithelium and of the BM, primarily by accelerating post-irradiation recovery of the hemopoietic stem- and progenitor cells, rather than altering their intrinsic radiosensitivity (115). Giving vincristine 24 h prior to irradiation provides optimal radioprotection of 12-day CFU-S, possibly by initiating a recruitment of stem cells into the cycle prior to irradiation. At the time of irradiation, cells may be in a more radioresistant phase (S) of the cell cycle, or have a decreased tendency to apoptose or increased radical-scavenging capacity. Many biologics that prime against radiation are also potent immunostimulants that activate macrophages and cells of the immune system to release cytokines, such as IL-1, that can confer radioprotection and enhance hemopoietic stem cell recovery. The similarity between priming with vincristine and these immunostimulants suggests that radioprotection by vincristine may be cytokine-mediated. A number of antitumor agents increase cytokine production by macrophages (116). Because cytokines such as IL-1 and TNF can induce MnSOD (115), it is possible that this could mediate priming by vincristine. Although treatment of nontumor-bearing mice with vincristine increases their MnSOD activity on a time scale similar to that for radioprotection, no such increase has been found in tumor-bearing mice, which suggests that increased antioxidant enzyme activity is not responsible for this vincristine effect (115).

An additive acceleration of post-irradiation regeneration of BM function (white blood cells, 12-day CFU-S, burst-forming unit-erythroid (BFU-E), colony-forming unit-fibroblast (CFU-F), colony-forming unit-granulocyte/erythroid/-macrophage/megakaryocyte (CFU-GEMM), and CFU-GM), has been observed in the BM of mice treated with a combination of vincristine and lithium prior to 4.5 Gy (4500 rad) TBI (117). Whereas vincristine does not appear to increase stromal production of hemopoietic growth factors, lithium does enhance the production of some of these factors, which may contribute to its activity.

Methylxanthines. Pentoxifylline [*6493-05-6*] (Trental) a synthetic dimethylxanthine derivative, is a hemorrheologic agent that can prevent or ameliorate late radiation injury to soft tissues and lung in animals and humans, and may be useful in the management of fibrotic sequelae, particularly if administered prophylactically or in the inflammatory early stages of fibrosis (118–120). Pentoxifylline administered concurrently with and following irradiation has reduced high grade late soft tissue injury in a murine experimental model (118). Pentoxifylline given just prior to irradiation and continued for 40 weeks has also decreased late lung injury in rats (120). In humans, pentoxifylline is used as an interventional therapy for persistent soft tissue ulceration or necrosis. The drug treatment is initiated ~30 weeks after irradiation (119). Pentoxifylline may influence late radiation injury via several mechanisms, including protection of endothelial cells soon after radiation exposure by enhancing blood flow in injured microvasculature and down-regulating TNF (121). In contrast to its effects on late radiation injury, pentoxifylline had little or no effect on acute skin or lung reactions in animals (118,120,122). Pentoxifylline, given at various times after a single dose of 20–24 Gy (2000–2400 rad) x-rays, also fails to moderate the development of late rectal ulcers in rats (123). Pentoxifylline generally increases tumor radiosensitivity (124).

Pentoxifylline is structurally related to other methylxanthine derivatives such as caffeine [*58-02-2*] (1,3,7-trimethylxanthine), theobromine [*83-67-0*] (3,7-dimethylxanthine), and theophylline [*58-55-9*] (3,7-dihydro-1,3-dimethyl-1H-purine-2,6-dione or 1,3-dimethylxanthine), which also show radioprotective activity in some instances, suggesting that methylxanthines as a drug class may radioprotect through a common mechanism (see ALKALOIDS). In a retrospective analysis of cervical and endometrial cancer patients receiving primary or adjuvant XRT, no association between caffeine consumption and incidence of acute radiation effects has been found. However, there was a decreased incidence of severe late radiation injury in cervical cancer patients who consumed higher levels of caffeine at the time of their XRT (121). The observed lack of correlation between caffeine consumption and acute radiation effects is consistent with laboratory investigations using pentoxifylline.

Superoxide Dismutase. Superoxide dismutase (SOD) exhibits radioprotective activity in a variety of systems, including protection against 30-day lethality in mice and some late radiation effects in humans when given after irradiation (125). The mechanistic basis for these effects is controversial. The preirradiation activity of SOD, and its activity in cultured cells, has generally been attributed to radical-scavenging effects, whereas its activity when given to animals or patients after irradiation is probably related to its antiinflammatory and/or immunostimulatory properties.

Several studies have reported radioprotective effects following transfection of SOD genes into mammalian cells (126). MnSOD protects against events occurring as early as 24 h after irradiation, which suggests that it may prevent apoptosis. Whereas the presence of SOD during irradiation increases the resistance of murine leukemia BCL1 cells *in vitro*, injecting BCL1-bearing mice with SOD (100 mg/kg iv) 30 min prior to irradiation does not protect against the development of leukemia. Thus, SOD may be useful in normal tissue protection in patients receiving TBI as part of the conditioning prior to allo-geneic BM transplantation (127).

Liposomally encapsulated SOD has better pharmacological properties than the free enzyme and has been tested as an antiinflammatory agent. A suggested clinical benefit of liposomal SOD in two patients with severe fibrosis and necro-sis caused by exposure to high dose XRT has been confirmed in a clinical trial of systemic CuZnSOD for the treatment of patients with preexisting severe symp-tomatic radiation-induced fibrosis of the skin and underlying tissue (128). A three-week course of liposomal SOD caused some regression of fibrosis in all subjects. Although several drugs, including antiinflammatory agents or vascu-lar modifiers such as captopril, may be useful for the management of fibrotic sequelae, particularly if administered prophylactically or in the inflammatory early stages of fibrosis, SOD is the first agent having a demonstrated ability to reduce preexisting fibrosis. A preliminary account of a second clinical study using topical liposomal SOD also suggested an improved control of fibrosis (129). Little is known of the effects of liposomal SOD on tumors.

Chinese Herbal Medicines. Many traditional Chinese medicines have been screened for radioprotective activity in experimental animals. In one study of more than a thousand Chinese herbs, a number of agents increased the sur-vival rate of dogs exposed to a lethal dose of γ-rays by 30–40%, and some symp-toms of radiation injury were ameliorated. These effects are potentially related to stimulation of the hemopoietic and immune systems (130). Extracts of five Chinese drug plants, as well as aspirin, effectively protected mice exposed to 7.5–8.0 Gy (750–800 rad) of γ-radiation, and increased survival rates by 8–50% (131). Several Chinese traditional medicines, administered ip before or after ir-radiation, protected against lipid peroxidation in a variety of mouse tissues, in-cluding BM, liver, and spleen, as well as in mouse liver microsomal suspensions irradiated *in vitro* (132). Jen-Sheng-Yang-Yung-Tang, when administered ip at 1 mg/g after γ-irradiation, appeared to facilitate the recovery of cellular im-munocompetence in mice. When injected ip at 1 mg/(g·d) for seven days after irradiation, this herb accelerated the recovery of hemocyte counts and 10-day exogenous CFU-S in 4 Gy (400 rad) irradiated mice (133). Both the aqueous (2 mg/mL in the drinking water) and alkaloid fractions (5.4 mg/d po) of Panax ginseng extract protected the jejunal crypts of γ-irradiated mice (both as pre- and post-irradiation treatments). The incidence of radiation-induced micronuclei in splenic lymphocytes was also reduced by pretreatment with both fractions (134). Thus, Panax ginseng may act as a scavenger and may also promote the repair or regeneration of damaged cells following γ-rays.

Antibiotics. Although not strictly speaking radioprotective agents, antibi-otics (qv) are an important component of the treatment of radiation injuries. By preventing or delaying the onset of systemic infection owing to endogenous and exogenous organisms, these agents can allow greater recovery of tissues such as BM and intestine (135). Indeed, mortality following exposure to radiation in the Chernobyl nuclear accident was mainly attributed to bacteriemia under im-munocompromised conditions (136). Antibiotics such as penicillin and synthetic antibacterials such as quinolone (see ANTIBACTERIAL AGENTS, SYNTHETIC), alone and in combination, can reduce bacterial translocation from the intestine, treat the subsequent sepsis, and reduce mortality (137,138). Combined glycopeptide, ie, vancomycin or teicoplanin, and antimicrobial, ie, the quinolone L-of loxacin, therapy has also been evaluated in mice exposed to mixed-field neutrons plus γ-rays (138).

Other Radioprotective Chemicals. The bis-methylthio- and methylthio-amino-derivatives of 1-methylquinolinium iodide and 1-methylpyridinium-2-dithioacetic acid provide reasonable protection to mice at much lower doses than the aminothiols, which suggests a different mechanism of action (139). One of these compounds, the 2-(methylthio)-2-piperidino derivative of the 1-methyl-2-vinyl quinolinium iodide (VQ), interacts with supercoiled plasmic DNA primarily by intercalation. Minor substitutions on the aromatic quinolinium ring system markedly influence this interaction. Like WR-1065, VQ is positively charged at physiological pH, and the DNA-binding affinities of VQ and WR-1065 appear to be similar.

Cimetidine [51481-61-9], a histamine H2 receptor antagonist, given at 15 mg/kg ip 2 h prior to irradiation of mice using 0.25–1 Gy (25–100 rad) of γ-rays, protects ~1.5-fold against the induction of micronuclei in erythrocytes (140). Cimetidine also protects ~1.5-fold against lymphoid tissue injuries in mice receiving 1–8 Gy TBI. Protection may involve radical scavenging, although cimetidine may also augment the proliferative and cytotoxic response of lymphocytes and can prevent the interaction of histamine with leukocytes and endothelial cells and block inflammatory reactions (see HISTAMINES AND HISTAMINE ANTAGONISTS).

Diethyldithiocarbamate [20624-25-3] (DDC) is both an inhibitor of SOD and a thiol, and exerts both radiosensitizing and radioprotective properties in mice, depending on factors such as the time of its administration relative to irradiation. For neutrons, DDC shows only protective effects (141). DDC (1 mg/g ip) given 30 min before 15 Gy (1500 rad) also protects mouse jejunal crypt cells and reduces the frequency of micronuclei in splenic lymphocytes (134).

A 24-h pretreatment of V79 cells using the DNA methylation-disrupting agent 5-azacytidine decreases the cells' 5-methylcytosine content by 50% and protects them by ~1.8-fold from killing by γ-radiation, possibly by activating repair enzymes (142). Deproteinized calf blood serum (ActoHorm), given to rats at various times after 20–24 Gy (2000–2400 rad) of x-rays, stimulates regeneration of the mucosal epithelium and may be effective in promoting the healing of the GI mucosa (123). A slight radioprotective effect of inosine given ip to mice shortly before γ-irradiation can be enhanced by magnesium aspartate, apparently owing to the additive vasodilatory activity of the two agents (143).

Glutamine has been widely examined as a potential agent for enhancing intestinal repair following radiation injury, although its value in this regard remains to be clearly established (144). Glutamine does exert radioprotective effects in cultured CHO cells.

Radioprotective effects of the bibenzimidazoles Hoechst 33342 [23491-52-3] and 33258 [23491-45-4] in cultured cells have been reported. These drugs bind in the minor groove of DNA and protect against SSBs in pBR 322 plasmid DNA, providing a possible basis for the observed protection (145). The pattern of protection suggests both a general reduction in SSBs throughout the entire plasmid, probably owing in part to OH$^\bullet$ scavenging by unbound ligand, as well as a binding site-specific component that may involve H$^\bullet$ donation from the ligand to 4'-sugar radicals (145). DNA binding represents a focal feature for the design of more efficient radioprotectors, and substitution of these bibenzimidazoles, eg, halogenation, markedly alters their protective activity.

Several antiulcer drugs have been evaluated for their ability to protect against acute skin reactions, such as erythema and moist desquamation, which are significant problems during XRT of superficially located tumors. Atropine [51-55-8] (dl-hyoscamine), a naturally occurring alkaloid, is the prototypal antimuscarinic agent. It can decrease gut motility and, when given with irradiation, reduce the extent of deformation of villous shape and protect against crypt cell depletion (146). Sucralfate, an antiinflammatory agent that activates cell proliferation, when applied as a cream to breast cancer patients receiving postoperative electron beam XRT to their chest wall, significantly decreases acute skin reaction and enhances skin recovery (147). Intraperitoneal injection of irsogladine maleate, an antiulcer drug, immediately after irradiation and then daily for three days protects intestinal stem cells by ~1.16-fold, which suggests that the drug may be useful for alleviating GI injuries in radiation accident victims (148).

Many agents that alter blood flow to tissues have been examined as possible radioprotective agents. BW12C, ie, 5-(2-formyl-3-hydroxyphenoxyl) pentanoic acid, a substituted benzaldehyde that stabilizes oxyhemoglobin and reduces oxygen delivery to tissues, is of interest as a possible potentiator of bioreductive agents and/or hyperthermia. Owing to these changes in oxygen availability, BW12C can act as a protector against radiation-induced injury to normal tissues. However, variable results have been reported with respect to BM12C's ability to protect normal tissues and tumors (149).

Cytokines and Related Factors as Radioprotective Agents

Although thiols dominated the field of radioprotection from the 1950s through the 1980s, radioprotective, or rather radiomodulating, activity of biologics, and especially cytokines, has received considerable attention in the 1990s (3,150). The term cytokines herein means proteins that modify cellular responses through ligand–receptor interactions, including growth factors and interleukins, and discussion is extended to immunomodulating agents, which function mainly through the generation of cytokines. Eicosanoids are included because these can be generated during cytokine responses and mediate such diverse processes as vasoregulation and inflammation. These are potent protectors of jejunal and hemopoietic stem cells when given prior to irradiation, and are therefore implicated as potential mediators of radioprotection by cytokines, although this does not seem to be the case for IL-1 (151). Some cytokines are most effective as radioprotectors when given between one and several days prior to irradiation; others work best when administered after irradiation. Biologics generally protect tissues by 1.3-fold or less, whereas WR-2721 can protect by as much as 2.7-fold. However, clinically, protection as low as 1.1-fold could be beneficial. The ability to increase the tumor dose by 10% while maintaining the same level of complications consequently translates into a significant increase in tumor control rates (135). Certain cytokines, such as G-CSF, SCF, and GM-CSF, are well tolerated at effective doses (44). Others, such as TNF-α and IL-1, are more toxic, especially if administered systemically, as is consistent with their roles in inflammation and septic shock.

Elucidation of the mechanisms of cytokine action can be complicated. Cytokine action may be indirect and mediated by the products of responding cells, such as other cytokines, eicosanoids, and other biologically active molecules. The specific cellular response therefore feeds into a network, involving soluble factors, cell adhesion molecules, and extracellular matrix components, that determines cell maturation, proliferation, and even death, within tissues. Disturbing this delicately balanced network can therefore cause wide ranging effects (152). Tissue response to a cytokine is determined by the receptor profiles of the cells, the microenvironmental signals that are present and induced, and the programmed agendas of the cells in the tissue. Many cytokines cause a cellular influx which forms part of the response, and radiation itself induces imbalances in the system by directly generating positive cellular responses, including cytokine expression, and causing cell death. It is axiomatic that cytokine responses are generally highly predictable and reproducible within any one tissue, and can readily be investigated. Anticytokine antibodies can be used to determine the role of individual cytokines in radioprotection. Cytokine and cytokine receptor knockout mice have been invaluable in defining those situations in which individual cytokines are critically involved and in determining the extent of redundancy.

Any one cytokine can have more than one function (pleiotropy). A few examples from the lympho/hemopoietic system clearly demonstrate the selective and specific action of these factors in context. For example, stem cell factor (SCF) or c-kit ligand, IL-1, and IL-6 are involved in the proliferation, differentiation, and functional activation of the earliest pluripotent stem cells. IL-3 (multi-CSF) stimulates colony formation by more committed myeloid progenitors. IL-7 is needed for the clonal development of T and B lymphocyte precursors; erythropoietin (EPO) for red cells; granulocyte–macrophage colony-stimulating factor (GM-CSF) for macrophages, granulocytes, and dendritic cells; and granulocyte colony-stimulating factor (G-CSF) for granulocytes. The CSFs therefore differ in their lineage specificity and the maturity of their primary targets. This, in turn, reflects the cellular receptor profile and the other accessory factors that are needed for development of each cell type (151).

G-CSF and GM-CSF. Because of availability in recombinant form and enormous clinical potential in both XRT and chemotherapy, hemopoietic growth factors have moved rapidly from the laboratory to the clinic, in spite of their expense (44). Clinical use of cytokines is largely limited to accelerating lympho/hemopoietic regeneration using exogenous CSFs, in particular G-CSF and GM-CSF (135). Such agents have the potential to protect selectively normal hemopoietic and other tissues, allowing higher doses of radiation or chemotherapy to be delivered to the tumor. They also have roles in the management of disease.

Injection of these factors rescues mice (153), dogs (154), and monkeys (155) from radiation injury. Whether GM-CSF and G-CSF can protect mice against supralethal doses of irradiation is controversial and appears to vary with the strain, although there is little evidence that these cytokines, on their own, increase stem cell survival in addition to accelerating recovery of progenitor cells.

GM-CSF is optimally protective when given before irradiation (150), and G-CSF is even more potent than GM-CSF when given to mice prior to an $LD_{100/30}$ TBI dose (156). The most dramatic increases in survival of mice and neutrophil recovery after irradiation are found when multiple doses of G-CSF are administered (157). In animals given 4 Gy (400 rad) TBI, continuous treatment using G-CSF over 21 days initiated shortly after TBI causes complete and sustained recovery (154). Several other studies have reported that G-CSF enhances recovery from neutropenia in mice caused by single high dose TBI (158) and fractionated TBI (159). Daily treatment of rats using sub-q G-CSF for 21 days following fractionated upper hemibody irradiation prevents neutropenia induced by 15 doses of 3 Gy/d (300 rad/d) (160). Administration of G-CSF to tumor-bearing rats also maintains the white blood cell counts above critical levels, enabling completion of the irradiation schedule and better tumor response. Thus G-CSF may be useful with conventional XRT schedules.

Similar results have been reported in sublethally and lethally irradiated dogs, where G-CSF reduced the severity and duration of neutropenia and the duration of thrombocytopenia (161). G-CSF increases the survival of lethally irradiated animals by inducing earlier recovery of neutrophils and platelets. GM-CSF also decreases the severity and duration of neutropenia in dogs exposed to 2.4 Gy (2400 rad) TBI, but does not influence monocyte or lymphocyte recovery (162), indicating its expected selective action.

If G-CSF and GM-CSF can prevent neutropenia caused by XRT, their use may allow more intensive XRT to be given to cancer patients. In cancer patients, the selectivity factor has further relevance because cytokines can stimulate tumor growth. *In vitro* experiments show that some human myeloid malignancies and solid tumors produce myeloid growth factors, proliferate in response to growth factors, or express growth factor receptors. However, studies using G-CSF and GM-CSF have shown no protective effect against a variety of tumors (163). Another potential concern is that G-CSF and GM-CSF, in stimulating the proliferation of BM progenitors, could sensitize the progenitors to the effects of cytotoxic agents. Some clinical data support this concern (44) and further investigation is needed for the optimal sequence administration of these agents relative to the cytotoxic therapy. An additional concern is that G-CSF and GM-CSF could act to promote carcinogenesis or tumor progression if given close to the time of irradiation.

Clinically, GM-CSF or G-CSF have been used to accelerate recovery after chemotherapy and total body or extended field irradiation, situations that cause neutropenia and decreased platelets, and possibly lead to fatal septic infection or diffuse hemorrhage, respectively. G-CSF and GM-CSF reproducibly decrease the period of granulocytopenia, the number of infectious episodes, and the length of hospitalization in such patients (152), although it is not clear that dose escalation of the cytotoxic agent and increased cure rate can be reliably achieved. One aspect of the effects of G-CSF and GM-CSF is that these agents can activate mature cells to function more efficiently. This may, however, also lead to the production of cytokines, such as TNF-α, that have some toxic side effects. In general, both cytokines are reasonably well tolerated. The side effect profile of G-CSF is more favorable than that of GM-CSF. Medullary bone pain is the only common toxicity.

The role of cytokine therapy in the management of radiation accident victims has been summarized (152). In Goiânia in Brazil in 1987, eight radiation accident victims were treated with GM-CSF one month after radiation exposure. Marked increases in granulocyte production were induced in five persons, although this did not prevent death.

Interleukin-1 α **and** β. IL-1 has radioprotective activity toward BM and other tissues (151,164). IL-1 is produced in response to endotoxin, other cytokines, and microbial and viral agents, primarily by monocytes and macrophages. Other nucleated cells can also produce it. IL-1 appears to play an important role in the regulation of normal hemopoiesis directly by stimulating the most primitive stem cells and indirectly by stimulating other hemopoietic factors, including G-CSF, GM-CSF, M-CSF, and IL-6.

Although IL-1 protects a number of normal tissues against radiation injury, its effects on BM are the best characterized. IL-1 provides varying degrees of protection, depending on the timing; ~20 h prior to irradiation is optimal (164). Synergy has been found with other cytokines, most notably TNF-α, IL-6, and SCF. Multiple daily injections of IL-1 α preceding irradiation are more effective than single doses in promoting both BM progenitor cell survival and granulocyte recovery (165). Protection may involve a number of mechanisms. One is the stimulation of BM progenitor cells such that more of these are in the radioresistant S-phase of the cell cycle (151). Another is the induction by IL-1 of a number of radioprotective substances such as prostaglandins (PGs), metallothionein, scavenging acute-phase proteins, GSH, and SOD, as well as other hemopoietic growth factors (151). The protective effects of preirradiation IL-1 in murine BM cells and human cell lines correlate closely with the induction of MnSOD (126), and growth factors induced from accessory cells that constitute the hemopoietic microenvironment can enhance repopulation of the immune and hemopoietic systems after irradiation. Possible effects of IL-1 on radiation-induced apoptosis in tissues have yet to be reported, but cytokines can clearly affect the tendency of cells to undergo this form of death. The ability of IL-1 to accelerate the reconstitution of murine BM following lethal doses of radiation, which may be through stimulating production of CSFs, can also impact survival. Although IL-1 can protect mice from acute lethality and from CFU-GM damage caused by TBI, it has no significant effect on immediate CFU-GM survival or on the level of radiation-induced DNA strand breaks in BM cells (166).

A critical step in radioprotection involves the IL-1 receptors. Monoclonal antibodies to the type 1 IL-1 receptor block IL-1-induced radioprotection (167). Although this receptor is not present on BM cells, it is present on fibroblasts, which suggests that the effects of IL-1 on stem cells may be largely indirect and mediated by stromal cell activation (168). Anti-IL-1 receptor (type 1) also sensitizes normal mice to the effects of TBI, which suggests that endogenous IL-1 has an intrinsic radioprotective role. IL-6 induction by IL-1, but not CSF levels, is inhibited, which supports the concept that G-CSF and GM-CSF are insufficient by themselves at radioprotecting stem cells and indicates a contributory role for IL-6. Anti-IL-6 antibody blocks IL-1 and TNF-induced radioprotection and also decreases the intrinsic radioresistance of mice, as does anti-TNF-α (169).

Anti-SCF antibody similarly abrogates lipopolysaccharide- and IL-1-induced radioprotection (170) and sensitized mice to radiation. Such effects are not obtained using anti-IL-3, anti-IL-4, or anti-GM-CSF antibodies. SCF, IL-1, IL-6, and TNF-α have acknowledged interactive roles in the normal development of BM stem cells, and their radioprotective activities seem consistent with these roles, as is the ability of CSFs to promote further hemopoietic development at higher hierarchical levels.

Clinical use of IL-1, IL-6, and TNF-α, is limited by associated systemic toxicity. SCF seems better tolerated. It may be possible to develop derivatives of other cytokines that are less toxic. The synthetic nonapeptide VQGEESNDK (position 163–171 of human IL-1 β) (see AMINO ACIDS; PROTEIN ENGINEERING) increases the survival of mice when injected 20 h before or immediately after exposure to 8.5 Gy (850 rad) TBI (171). Although the nonapeptide is less effective than IL-1 β, it does not exhibit the IL-1-like inflammatory side effects of the whole molecule.

IL-1 can radioprotect murine and human BM progenitors *in vitro* as well as *in vivo* (126). Also, BM cells from donor mice treated with IL-1 *in vitro* prior to irradiation show an increased ability to rescue irradiated recipient mice following BM transplantation, which suggests that IL-1 exerts its protective effects directly on the hemopoietic cells and protects both short-term and long-term repopulating stem cells (126). In fact, most hemopoietic growth factors protect their respective target cells from the effects of irradiation *in vitro*, a phenomenon that may reflect the tendency of the cells to apoptose in the absence of growth factor.

The effects of IL-1 in accelerating recovery of BM hemopoiesis in mice have been characterized (172). Injection of IL-1 20 h prior to sublethal irradiation promotes an earlier CFU-S/CFU-GM recovery in the BM and spleen, and markedly affects BM cellularity and mobilization of progenitor cells (172). Differences have been found between strains and administration protocols, especially with respect to BM CFU-GM numbers.

Synergy has been reported between IL-1, IL-3, and SCF, in enhancing the survival of lethally irradiated mice engrafted with 2×10^6 BM cells and immediately given cytokines once daily for five days (173). SCF alone does not enhance survival, and IL-1 or IL-3 has limited effect. Pretreating mice with thymopentin, a synthetic pentapeptide derivative of thymopoietin, enhances the protective effect of IL-1 α, as indicated by 30-day survival (174).

Extrapolation of experimental observations using cytokines to the clinical setting, ie, to patients with altered physiological conditions, may not be possible (175). Whereas IL-1 administered 24 h prior to a lethal dose of TBI increases 30-day survival of normal mice, it does not protect tumor-bearing animals known to have altered hemopoiesis. The accelerated repopulation of eight-day CFU-S, CFU-GEMM, and CFU-M, following sublethal TBI seen in control mice, has not been seen in tumor-bearing animals. The failure of IL-1 to protect tumor-bearing animals is not a result of elevated plasma prostaglandin (PG) levels caused by the tumor. On the positive side, doses of IL-1 that protect BM and oral mucosa do not protect the RIF-1 and SCCVII mouse tumors (176) which indicates a degree of selectivity in effects.

In addition to BM, IL-1 protects murine intestinal crypt cells (177,178), oral mucosa (176), and lung (179) against radiation injury. Optimal protection of colony-forming duodenal crypt cells was observed when IL-1 was given

~13–25 h before irradiation, although protection was observed up to 20 h after irradiation, presumably involving accelerated recovery of crypt cells (177). The mechanism by which IL-1 protects GI and hemopoietic tissues appears to be different. Higher doses of IL-1 are required for the protection of the BM than for that of the intestine, perhaps reflecting different IL-1 receptor densities. Protection of the murine intestine by IL-1 is dependent on the mouse strain, the IL-1 dose, and the method used to assess protection (178).

Because the GI syndrome after TBI includes a hemopoietic component, the observed protection of the jejunum may partly be a result of the IL-1 protection of the BM. IL-1 given before abdominal irradiation also increases survival, which suggests that protection against GI syndrome by IL-1 is at least partially independent of its effect on BM (178). IL-1 also protects mouse duodenal crypt cells against fractionated irradiation (176). There is one report that IL-1 given ip prior to abdominal irradiation increases the incidence of peritoneal adhesion formation (180), which indicates that the effects of cytokines may be site- and time-dependent.

IL-1 increases the 24-h thymidine labeling index, with or without localized irradiation, in several normal mouse tissues, including the lip and tongue mucosal basal cell layers, crypt cells of the duodenum, alveolar cells of the lung, hepatocytes, and basal skin cells, but not in the RIF-1 tumor (176).

Stem Cell Factor. Stem cell factor (SCF), also known as c-kit ligand, is the ligand for a tyrosine kinase-associated receptor encoded by the c-kit protooncogene, and stimulates primitive multipotential hemopoietic stem cells. It does not act as a colony-stimulating factor (CSF) for these cells but rather primes cells to respond to other cytokines, such as IL-1. Mutations in white spotting and steel mice that affect hemopoiesis have been influential in characterizing SCF. These strains display an increased sensitivity to lethal doses of irradiation, which is thought to be the result of a lack of the apoptosis-suppressing effects of SCF, in addition to a lack of proliferative effects (181). SCF is radioprotective in mice on its own (182) and in concert with other cytokines.

Mice treated with SCF show improved long-term survival, more rapid hemopoietic recovery after irradiation, and a much reduced incidence of septicemia. The optimum SCF schedule, which involves both pre- and post-irradiation treatments, protects by ~1.3-fold. Administration of IL-1 and SCF to mice 18 h before lethal TBI results in synergistic radioprotection in terms of both survival and recovery of c-kit$^+$ BM cells (183). Anti-SCF antibody inhibits IL-1-induced radioprotection, which indicates that endogenous SCF is necessary for radioprotection by IL-1. In turn, radioprotection by SCF is reduced by anti-IL-1-receptor antibody, which indicates that endogenous IL-1 contributes to radioprotection by SCF. SCF, unlike IL-1, does not induce hemopoietic CSFs, IL-6, or MnSOD, which suggests that SCF and IL-1 may protect BM cells by different pathways. The c-kit mRNA and SCF binding to BM cells is elevated within 2–4 h of IL-1 administration. Thus, the synergy between SCF and IL-1 may depend on IL-1 and SCF-induced increases in numbers of c-kit$^+$ stem and progenitor cells that survive lethal irradiation.

SCF increases absolute colony number and surviving fraction of CFU-E, CFU-G, and CFU-GM in irradiated human BM. An increase in the fraction of CD34$^+$ cells in the radioresistant S-phase has been noted, which suggests a possible mechanism (184). A cautionary note has been sounded about attempting to predict interactions between SCF and CSFs in hemopoietically deprived individuals (185). Although SCF synergizes with GM-CSF or GM-CSF and IL-3 to increase CFU-GM *in vitro*, no such effect has been found *in vivo*.

SCF protects mice from GI death after irradiation and increases the number of surviving mucosal crypt stem cells (186) in a manner similar to IL-1.

Other Lympho/Hemopoietic Cytokines. IL-3 shares a common signaling receptor chain with GM-CSF, but stimulates the proliferation, differentiation, and function of a less mature, multipotential, myeloid progenitor cell population. The broader activity of IL-3 suggests that it may be of greater benefit than the more lineage-restricted G-CSF and GM-CSF in accelerating BM recovery after injury. Administration of IL-3 to sublethally irradiated mice induces cell recovery in the thymus (187), and in primates, IL-3 effectively decreases the period of thrombocytopenia after drug or radiation-induced BM aplasia, a feature that G-CSF or GM-CSF do not possess, although IL-1 and IL-6 do. Because of IL-3's broader activity, a combination of IL-3 and later-acting CSFs is expected to be efficacious at enhancing post-irradiation recovery of platelets and neutrophils. The sequence of administration is critical for combined effects of IL-3 and GM-CSF in primates receiving 4.5 Gy (450 rad) (188).

Administration of IL-12 before lethal γ-irradiation of mice protects against hemopoietic death and increases the number of BM cells at six days after irradiation, but sensitizes for GI injury (189). The protective effects are abrogated by anti-IL-1 receptor or anti-SCF antibodies but not by anti-IFN-γ antibodies. The sensitizing effect of IL-12 may be due to its ability to prime for TNF and IL-6, and can be abolished by anti-IFN antibody.

TNF-α also protects mice against the lethal effects of radiation (164). TNF-α given before sublethal irradiation reduces the decline of neutrophils and total blood counts and accelerates the recovery of peripheral blood cells (190). TNF-α also alters the radiosensitivity of murine GI progenitors (191).

Other Cytokines. Basic FGF (b-FGF or FGF-2) belongs to a class of heparin-binding growth factors associated with growth and differentiation of a number of cell types, including endothelial cells and fibroblasts, and a number of *in vivo* processes, including angiogenesis and wound healing. Its presence is required for endothelial cell culture *in vitro* and protects against the ability of radiation to induce apoptosis (192). Basic FGF has no effect on the repair of SSBs or DSBs but is required for the shoulder on the survival curve, in essence allowing repair of potentially lethal damage. Protection appears to be mediated by PKC activation. Radiation induces b-FGF production by these cells, allowing a survival-associated positive autocrine loop to develop.

Basic FGF protects the endothelial cell lining of the pulmonary microvasculature after iv injection into mice receiving lethal doses of whole-lung irradiation (192). b-FGF given immediately before and within the first two hours after irradiation prevents the clinical syndrome of radiation pneumonitis, presumably by modulating apoptosis in the endothelial cells (192). In contrast, b-FGF fails to protect against classical radiation pneumonitis in two strains of mice having different susceptibilities to lung injury (193). Furthermore, there is little evidence of radiation-induced apoptosis in the lungs of either strain. The reasons for this discrepancy are not clear, but the lung-irradiation conditions, including the doses and the fields, are different in the two studies (193). Minor, but possibly significant, differences exist between the mouse strains (192,194).

Basic FGF can also stimulate murine hemopoietic progenitors *in vitro*. It is synergistic with hemopoietic growth factors such as GM-CSF, EPO, and Meg-CSF and has radioprotective activity *in vivo*, increasing the number of day-9 and day-12 CFU-S from lethally irradiated animals (195). Furthermore, b-FGF combined with GM-CSF protects against the killing of murine and human CFU-GM exposed to radiation *in vitro* (195).

Another member of the FGF family, FGF-4, protects against radiation-induced cell killing and enhanced the G_2 arrest when overexpressed in a human adrenal cortical carcinoma cell line (196). This effect is again manifested as the appearance of a shoulder on the survival curve, although neither the control nor the transfected cells undergo radiation-induced apoptosis. No differences in the yield or repair of either SSBs or DSBs have been observed.

Whereas epidermal growth factor (EGF) enhances the radiosensitivity of human squamous cell carcinoma cells *in vitro* (197), addition of EGF to hormone-deprived MCF-7 breast cancer cells prior to irradiation results in increased radioresistance (198). An anti-EGF-receptor monoclonal antibody blocks the ability of EGF to enhance growth and radioresistance. Tumor cells, the growth of which is stimulated by EGF, appear to be protected; those where growth is inhibited are sensitized (198).

Other Agents.   Many agents that nonspecifically enhance immunological and hemopoietic responses can also function as radioprotectants. Immunomodulatory substances such as endotoxin, glucan, muramyl dipeptide, bacillus of Calmette and Guerin (BCG), and OK-432 given to mice prior to, and in many cases after, irradiation result in an increased survival beyond 30 days. Such substances are believed to protect by activating macrophages and inducing the production of endogenous cytokines such as IL-1, TNF-α, G-CSF, GM-CSF, and IFN in irradiated animals that subsequently act synergistically to stimulate the proliferation of the target cells. Although it is likely that there are differences in the pattern of cytokines that are expressed following administration of these agents, and in their mechanisms of action, in general these are poorly described. In some cases radioprotection may be attributed in part to the mobilization of stem cells from BM to spleen, as opposed to stimulating proliferation *in situ* (164). Others may modulate oxidative processes, and some exhibit both prooxidant and antioxidant properties.

OK-432.   A lyophilized preparation of an attenuated strain of *Streptococcus haemolyticus* available commercially as Picinabil, OK-432, is a cytokine inducer having strong immunomodulatory activity. OK-432 protects *in vivo* against BM suppression in medulloblastoma patients receiving whole-axis irradiation and in mice given TBI (199). Protection of mice has been observed when the drug is given ip from 48 h before to 24 h after irradiation. Treating irradiated mice using OK-432 accelerates recovery of neutrophil count as well as improving host defense systems, which may lead to the prevention of bacteriemia. OK-432 also protects 10-Gy-irradiated murine BM CFU-GM *in vitro* when the drug is added from 24 h before to 3 h after irradiation (199). It was concluded that OK-432 stimulates BM cells to produce GM-CSF *in vitro* by a direct contact effect. Immediate post-irradiation OK-432 also increases the 30-day survival of mice (200).

Because the use of antibiotics (qv) to prevent bacteriemia can be suboptimal in immunocompromised animals, the activity of a concomitant immunomodulator, ie, OK-432, and the antibiotic aztreonam has been examined. Post-irradiation OK-432 plus aztreonam greatly increases the 30-day survival of lethally irradiated mice compared to OK-432 alone (200), presumably reflecting a cooperative action of the bactericidal effect of aztreonam and improved host defense by OK-432. Thus, combining an immunomodulator with a broad-spectrum antibiotic may be a useful strategy for treating patients having radiation injuries.

Ammonium Trichloro(dioxoethylene-O,O')tellurate. AS101, ammonium trichloro(dioxoethylene-O,O')tellurate, is a minimally toxic synthetic organotellurium compound that stimulates the production of a variety of cytokines and is radioprotective when injected into mice prior to sublethal and lethal irradiation (201). The compound appears to act partly by inducing hemopoietic progenitors into the radioresistant S-phase, and partly by stimulating CFU-S to proliferate and self-renew. AS101 stimulates DNA repair replication in spleen and BM cells of mice *in vivo* and in spleen cells *in vitro* (201). It protects cells from DSB induction and enhances their ability to rejoin DSBs. AS101 also enhances the ability of irradiated splenocytes, both *in vitro* and *in vivo*, to repair DNA damage, possibly by increasing DNA polymerase activity.

Tetrachlorodecaoxide. Tetrachlorodecaoxide (TCDO) administered iv from days 4–11 after TBI using γ-rays protects rats against the acute lethal effects of TBI, mainly by accelerating the regeneration of the hemopoietic system and by preventing severe hemorrhage in the lungs and the GI tract. Hair loss is also reduced (202). TCDO also protects against colon damage even when administered days or weeks after irradiation, probably by stimulating the regeneration of colon epithelial cells. TCDO (1 mL/kg) given iv daily within a week after 3 Gy (300 rad) TBI promotes recovery of BM and spleen cells of sublethally irradiated mice (203). TCDO greatly affects spleen weight and cellularity, as well as endogenous CFU-S. These mice are protected from lethality caused by TBI by 1.12–1.18-fold and from endogenous CFU-S depletion by 1.4–1.5-fold. TCDO may enhance regeneration of hemopoietic stem cells by stimulating the immune system with the associated release of cytokines such as IL-1 and TNF. It may also protect partly by suppressing PG production (203). TCDO given to mice as single or multiple iv injections one day before or after irradiation actually radiosensitizes a murine fibrosarcoma tumor, whereas TCDO given on days 17–21 after irradiation protects against leg contracture, which suggests a therapeutic advantage.

Not only is TCDO a potent therapeutic agent in acute radiation syndrome, but treatment using TCDO from days 4–11 after TBI increases the survival rate in rats for up to one year, protects against the development of late GI ulcers, and also reduces the development of γ-ray-induced leukemias and malignant epithelial tumors, but not sarcomas (202). The anticarcinogenic effect of TCDO may be related to the inhibition of PGs, which promote carcinogenesis, or to immunostimulation, which may result in a more effective elimination of malignant cells.

Endotoxin and Muramyl Dipeptide Derivatives. Bacterial cell wall constituents such as the lipopolysaccharide endotoxin and muramyl dipeptide, which stimulate host defense systems, show radioprotective activity in animals (204). Although endotoxin is most effective when given ~24 h before irradiation, it provides some protection when administered shortly before and even after radiation exposure. Endotoxin's radioprotective activity is probably related to its lipid component, and some of its properties may result from PG and leukotriene induction (204).

Muramyl tripeptide phosphatidylethanolamine (MTP-PE), a synthetic analogue of muramyl dipeptide and an effective systemic macrophage activator, induces a variety of cytokines such as IL-1, IL-6, and TNF, as well as PGE_2 (205). Preirradiation treatment of mice using MTP-PE encapsulated in liposomes, which can intensify radioprotective ability, stimulates the monocyte/macrophage system and accelerates the recovery of hemopoietic cells. The recovery phase is associated with BM granulocyte hyperplasia, accelerated erythropoiesis in the spleen, and recovery of granulocyte counts in the peripheral blood. The drug has no immediate effect on the number of day-8 exogenous BM CFU-S, but stimulates the regeneration of the surviving endogenous CFU-S and BM CFU-GM (205). Optimum survival of both endogenous CFU-S and mice (~1.17-fold protection) has been observed with 200 μg MTP-PE given ip 24 h before irradiation. Some protection has been observed when the drug is injected 8 h after irradiation. Combining MTP-PE (24 h) and indomethacin (INDO) (24 and 3 h) prior to irradiation exerts an additional radioprotective effect (205).

Glucan and Derivatives. Glucan (β, 1-3 polyglucose), an immunomodulator and hemopoietic stimulant isolated from *S. cerevisiae*, enhances resistance to lethal irradiation in animals. This effect is mediated primarily by an enhanced host resistance to life-threatening opportunistic infections and accelerated hemopoietic regeneration. Glucan protects a variety of indexes for damage to the murine hemopoietic system (206). Glucan and several glucan derivatives protect against the induction of micronuclei in polychromatic erythrocytes of the mouse BM. Significant protection by iv glucan is observed when the drug is given 1 h after irradiation (207).

Carboxymethylglucan, a soluble glucan derivative (208), enhances hemopoietic recovery in sublethally irradiated mice and increases the survival of lethally irradiated animals when given 24 h prior to γ-irradiation. Post-irradiation treatment using carboxymethylglucan also improves survival when used in combination with preirradiation cystamine. Carboxymethylglucan in combination with diclofenac, an inhibitor of PG production, when given to mice 1 day before γ-irradiation demonstrates at least additive radioprotective effects on hemopoietic recovery and mouse survival. These effects may be a consequence of increased hemopoietic cell proliferation as a result of the concomitant inhibition of PG production and the release of growth factors.

Significant protection of mice by several polysaccharides other than glucan isolated from *S. cerevisiae* has been described (209). A 2.16-fold protection in the $LD_{50/30}$ assay is observed for one modifier, MNZ, when given 15 min prior to irradiation. Glucan protects 2.25-fold in this same protocol. Many of these polysaccharides may act through activation of the complement system, rather than directly on cells.

Broncho-Vaxom. The bacterial extract Broncho-Vaxom (BV) protects murine hemopoietic tissue by 1.18-fold in the $LD_{50/30}$ assay. Maximum protection is achieved when the drug is given 24 h before irradiation (210). Pretreating mice with BV before sublethal irradiation increases the number of endogenous

CFU-S. The optimal CFU-S survival again is observed when BV is given 24 h before irradiation. BV does not affect day-9 BM CFU-S survival immediately after irradiation. However, 5–12 days after irradiation, the number of day-9 CFU-S is ~twofold higher in BV-treated mice. During this period BV-treated mice also show increased BM cellularity and accelerated BM CFU-GM regeneration (210). Combining BV (24 h) and INDO (24 and 3 h) prior to irradiation causes an additional radioprotective effect (210). BV given 24 h before irradiation also accelerates peripheral blood recovery, but does not affect thymus recovery.

BV, like many immunostimulatory radioprotective agents, activates macrophages (206). It has been suggested that most immunomodulators that affect macrophages affect radioprotection not only by direct macrophage activation and enhancing the nonspecific host defense mechanisms, but also by inducing the release of cytokines that directly or indirectly enhance additional hemopoietic and immunologic activities (206). Therefore, a possible mechanism of radioprotection is BV-induced secretion of IL-1 and PGs. *In vitro* studies have shown an increased IL-1 and PG production, reaching the maximum within 24 h of BV administration. If a similar situation occurs *in vivo*, then the peak period of IL-1 and PG production coincides with the maximal protective effect.

Lactobacillus Casei Preparations. Lactobacillus casei LC9018, prepared from heat-killed Lactobacillus casei YIT9018, exerts radioprotective effects in a variety of systems. Given to mice immediately after TBI, LC9018 causes a sustained increase in serum colony-stimulating activity followed by an enhanced repopulation of CFU-GM in the femoral BM and spleen (211). The numbers of blood leukocytes, erythrocytes, and platelets are increased earlier in the treated mice than in the controls, and survival is increased significantly. A single sub-q injection of LC9018, given before or after irradiation, increases the survival of mice given 8.5 Gy TBI. Similar protection is observed when LC9018 is administered between 2 d before to 9 h after irradiation, although the preirradiation treatment is slightly better. Increases in the weight of the spleen and in the number of endogenous 8- and 12-day CFU-S suggest that the radioprotective effect is based on enhanced recovery of hemopoietic tissues.

LC9018 proves to be an effective agent for adjuvant immunotherapy when combined with XRT in a randomized, controlled, comparative study of 228 patients with stage IIIB cervical cancer (212). Not only does LC9018 enhance tumor regression by XRT, but the combination therapy also prolongs survival and the relapse-free interval compared to XRT alone. Side effects of combined LC9018 include fever and skin lesions at the injection site, but no severe symptoms are noted. Radiation-induced leukopenia is also less severe in the LC9018-combined group than in the XRT-alone group.

Other Immunostimulatory Agents. Other immunomodulatory substances having demonstrated radioprotective activity include Shigoka (*Acanthopanax senticosus* Harms), Bestatin, Ivastimul (an extract from chlorococcal algae), trehalose dimycolate (TDM) and a synthetic analogue (S-TDM), dextran sulfate (a synthetic heparinoid polyanion), the bacterial immunomodular Nocardia delipidated cell mitogen (NDCM), the lipopeptide lauroyl-L-Ala-γ-D-Glu-L,L-A2pm (LtriP), AM5 (a protein-associated polysaccharide present in the biologic AM3), thrombopoietin, Thymex L (a thymic preparation), adenochrome monoaminoguanidine methanesulfonate (AMM) (an activator of CFU-GM), and serum thymic factor (FTS). A clinical protocol combining vitamin A, IFN, and XRT for advanced cervical cancer produces severe proctitis that necessitates dose reductions. However, in mice treated with vitamin A (100 μg/d) and IFN (3×10^4 units/d) for five days before TBI, there is a modest (1.1–1.4-fold) protection of the bowel, assessed by maintenance of body weight and colon crypt cell survival (213).

Arachidonic Acid Metabolites. *Prostaglandins.* Various bioactive lipids, especially the prostaglandins (qv) (PGs) and leukotrienes (LTs), the principal eicosanoid products of the arachidonic acid cascade, are radioprotective. Indeed, protection of hemopoietic tissue by eicosanoids can approach that achieved using WR-2721 (see Table 1) (214,215). Most attention has focused on the PGs, which are synthesized via the cyclooxygenase pathway and are among the most promising of the newer radioprotective agents for clinical use during cancer treatment (215). The mechanisms by which eicosanoids protect against radiation damage are obscured by their wide range of both pathological and normal physiological effects on tissues and animals. Protection by PGs is seen only when administered prior to irradiation. WR-2721 protects in mg quantities, whereas PGs protect in μg quantities. Thus the PGs are probably not direct protectors; rather, they induce secondary changes within cells that lead to protection (215). Many eicosanoids are vasoactive and some are proinflammatory, although these effects may or may not be observed at radioprotective concentrations. Receptor-mediated mechanisms are clearly important, and many cell types in various tissues express eicosanoid receptors. Eicosanoid receptor expression may be regulated by cellular differentiation, which may in turn influence the location of protection within a tissue. For the E-series PGs, the GC receptor appears to be important (215). Loss of PG-induced protection in cultured cells appears to correlate with the loss of these PG receptors. Membrane effects may be involved in PG-induced protection (215). Whether or not hypoxia plays a significant role in eicosanoid radioprotection is a contentious issue (216,217). PGs may also increase intracellular cyclic adenosine monophosphate (cAMP) levels (218), although elevated cAMP alone does not lead to PG-induced radioprotection of bovine aortic endothelial cells (219). Some eicosanoids may act as free-radical scavengers.

PGE$_2$ and Derivatives. PGE$_2$ and several related PGs protect against radiation injury in the rodent intestine with respect to both crypt clonogen survival and LD$_{50/6}$ (215). Protection of hemopoietic tissue has also been reported using the exogenous CFU-S assay (215), and 40 μg of 16,16-dimethyl PGE$_2$ (DMPG) given 30 min preirradiation increases LD$_{50/30}$ by ~1.7-fold (220). Both systemic and topical DMPG given 1 h prior to irradiation protect mice from γ-ray-induced alopecia after single doses. Similar changes are observed using fractionated XRT and repeated drug administration prior to each fraction (215). Topical DMPG and WR-1065 protect by 1.25–2.0-fold against alopecia resulting from fractionated XRT (221). Post-irradiation hair regrowth is increased by both WR-2721 and eicosanoids. It is unclear whether increased regrowth is solely the result of initial radiation protection of hair follicles, other cell populations and/or vasculature, or also occurs through accelerated repair processes. Although PGs are radioprotective when given either systemically or topically, the former is slightly more effective. Topical PGs or WR-1065 may, however, be clinically more useful in protecting against scalp alopecia, and perhaps radiation dermatitis/mucositis of the oral cavity, rectum, and bladder, when these tissues lie within the treatment field. Studies using topical viprostol, another PGE_2 analogue, also suggest that the use of selected vehicles may lead to negligible blood concentrations of isotope-labeled compound (222).

A 2-h incubation with another PGE_2 analogue, nocloprost (9β-chloro-DMPG) protects normal human fibroblasts but has no effect on the survival of colon adenocarcinoma cells exposed to 10 Gy (1000 rad) (218). Nocloprost protects against radiation-induced DSBs in normal cells but not in tumor cells. Moreover, incubation using nocloprost for 2 h after irradiation enhances the rate of DSB rejoining in fibroblasts but not in adenocarcinoma cells. These data possibly reflect a different distribution of PG receptors on the plasma membrane of the two cell types.

PGE_1 and Derivatives.   Synthetic PGE_1 analogues such as misoprostol actually protect better than the naturally occurring compounds (215). Misoprostol is an effective radioprotector of clonogenic cells in the rodent intestine, BM, and hair follicle (215). Given sub-q 2 h prior to irradiation, it protects against the killing of Syrian hamster embryo (SHE) cells exposed *in utero* to x-rays by ~1.5-fold, and protects ~20-fold against oncogenic transformation, compared to 4-6-fold for WR-1065 (216). Although the mechanisms by which misoprostol modulates transformation are unknown, arachadonic acid regulates *ras* proto-oncogene function (223). Some PGs stimulate *ras*-GAP (guanosine triphosphatase (GTPase) activating protein) and thereby switch off the gene product, p21 *ras*.

A clinical trial to evaluate misoprostol as a protector of normal tissue during a course of XRT in cancer patients suggests a reduction in acute normal tissue injury (215). A randomized, prospective, double-blind study indicates that topical misoprostol, administered as an oral rinse ~15–20 min before irradiation using conventional 2-Gy (200 rad) fractions, five days a week over 6–7 weeks, significantly protects the oral mucosa from radiomucositis, a frequently observed normal tissue complication during XRT for head and neck cancer (215).

Topical misoprostol directly administered into the operatively exteriorized intestinal lumen of rats 30 min prior to 11 Gy (1100 rad) of x-rays reduces the severity of acute radiation effects, as evidenced by the increased numbers of surviving crypt cells and mucosal height five days after irradiation (224). Topical misoprostol does not reduce blood flow. The attractive possibility of combining misoprostol with antioxidants has been reported (225). Two novel antioxidants, the 21-amino steroid U-74500A and the vitamin E-like compound U-78518F, as well as some nonsteroid antiinflammatory drugs (NSAIDs) and methylprednisolone, are radioprotective for the rat intestine in these same assays when given by the lumenal route. Misoprostol alone, WR-2721 alone, or the combination of the two agents, also increases the survival of intestinal clonogenic cells and animal survival following exposure to neutrons (225). PGE_1 analogues also protect the arterial wall of rabbits irradiated with single or repeated doses up to 10 Gy (1000 rad) (226). Administration of either PGE_1 or its active metabolite 13,14-DH-PGE_1 (5 μg/kg either 6 h pre- or post-irradiation) reduces radiation-induced mitotic activity and extracellular matrix and glycosaminoglycan formation.

Leukotrienes. Leukotrienes, products of the lipoxygenase pathway, are generally less radioprotective than the PGs, with the exception of LTC_4, which is among the most potent of the naturally occurring eicosanoids (214). LTC_4 radioprotects V79 hamster cells *in vitro* and mouse CFU-S and intestinal crypt cells *in vivo* (215,227). Protection factors of 1.65 for exogenous CFU-S and 2.01 for CFU-GM are obtained with mice given 400 μg/kg LTC_4 (214). In addition, 200 μg/kg LTC_4 given 5 min before irradiation protects 1.45 fold in the $LD_{50/30}$ assay (228). In all cases, protection is observed only when the LTC_4 is administered prior to radiation exposure. The mechanism of LTC_4 radioprotection is unknown, although *in vitro* data suggest that binding to a specific LTC_4 receptor is involved. Differences in the time course for protecting CFU-S versus $LD_{50/30}$ imply different mechanisms of action. LTC_4 may enhance the survival of mice partially via physiological changes such as induction of hypoxia (217). However, LTC_4 mediates a number of physiological changes, including inflammatory responses and cardiovascular changes, that may also be important.

Other Bioactive Lipids. Linoleic acid [*60-33-3*] (*cis*-9,*cis*-12-octadecadienoic acid), an essential fatty acid, can act as a radioprotective agent of BM while being toxic to certain tumor cells. Both irradiation (0.5 Gy (50 rad)) and exogenous linoleate generate increased levels of the oxygenated product of linoleate, 13-hydroxyoctadecadienoate (13-HODE), by macrophages (229), most likely through a 15-lipoxygenase-mediated pathway. This pathway seems to serve not only as a means of dealing with free radicals that are generated in cells but also as a mediator of oxidative stress responses.

Nontoxic doses of γ-linolenic acid [*506-26-3*] (6,9,12-octadecatrienoic acid) protect against radiation injury to pig skin (230). Pigs receiving 3 mL/d orally of an active oil containing 9% γ-linolenic acid for four weeks before and for 16 weeks after localized β-irradiation of the skin show less severe (1.13–1.24-fold) acute reactions, ie, erythema and moist desquamation, than pigs receiving linoleic acid or pigs that are only treated prior to irradiation. A similar reduction in the severity of acute skin injury is seen in pigs treated for 10 weeks after irradiation. Late skin damage, ie, late erythema or dermal necrosis, is also decreased by 1.14–1.51-fold.

Mice fed a diet containing the hexaisoprenoid cholesterol precursor squalene [*111-02-4*] (2,6,10,15,19,23-hexamethyl-2,6,10,14,18,22-tetracosahexaene) for 14 days prior to and 30 days after 6–8 Gy (600–800 rad) TBI also show some cellular and systemic radioprotection, including prolonged survival (231).

Eicosanoid Blocking Agents. A number of studies have documented the radioprotective effects of eicosanoid blocking agents such as the NSAIDs (215). Prophylactic administration of INDO, an inhibitor of PG synthesis, delays or reduces radiation mucositis owing to head and neck and thoracic XRT and experimental radiation esophagitis (232,233). Whereas INDO radiosensitizes murine solid tumors that have a high level of eicosanoid production, apparently by stimulating antitumor immune reactions, it has little effect on tumors having low eicosanoid production (233). In contrast, pretreatment of mice using INDO has no effect on the radiosensitivity of hair follicles, jejunum, and soft tissues of the extremities, whereas it protects hemopoietic tissue in both the $LD_{50/30}$ and endogenous CFU-S assays and lung (233). It also protects against carcinogenesis (233). Protection of hemopoietic tissue is not a result of an effect of INDO on the radiosensitivity of, or on the number of, BM stem cells, although the number of stem cells in the spleen is increased. Rather, removal of immunosuppressive PGs by INDO may augment immune functions and release IL-1 and other cytokines that can stimulate the proliferation of hemopoietic stem cells in the spleen. Enhanced hemopoietic recovery is also reported in mice given INDO after irradiation (234). INDO decreases indexes of early radiation effects on pulmonary endothelial cell function *in vivo* 7–8 h after exposure of rabbits to 30 Gy (3000 rad) to the chest (235).

In contrast to these protective effects, acute high dose (25 mg/kg) INDO injected ip 30 min prior to γ-irradiation decreases the LD_{50} of mice from ~6.5 Gy (650 rad) to ~4.5 Gy (450 rad) (232).

The ability of various NSAIDs to protect against mortality owing to radiation pneumonitis in mice receiving 19 Gy (1900 rad) to the thorax (bilaterally) has been examined (236). Treatments are continuous from 10 weeks after irradiation. Each inhibitor has the expected effect on arachidonate levels in the lungs. The 5-lipoxygenase inhibitor diethylcarbamazine and the LTD_4/LTE_4 receptor antagonist LY171883 markedly reduce mortality. The cyclooxygenase inhibitors piroxicam and ibuprofen are marginally protective, whereas INDO accelerates mortality, and aspirin reduces mortality.

Nordihydroguaiaretic acid, a lipoxygenase inhibitor, also protects against radiation-induced effects on hemopoiesis in mice (237). INDO and the lipoxygenase inhibitor esculetin show different effects on hemopoiesis (238). Whereas INDO augments the *in vitro* proliferation of BM CFU-GM from irradiated mice, esculetin is inhibitive. Similarly, post-irradiation INDO stimulates CFU-S and CFU-GM in lethally irradiated mice, whereas esculetin inhibits CFU-GM. When given ip 1 h before 5 Gy (500 rad), INDO enhances the post-irradiation recovery of CFU-S, CFU-GM, peripheral blood granulocytes, and nucleated BM cells, whereas esculetin has no effect or even inhibits recovery. These effects are not the result of changes in the intrinsic radiosensitivity of the CFU-S or CFU-GM. These results suggest that the balance between PG and leukotriene production may be important in determining the regulatory role of arachidonic acid metabolites in hemopoiesis in irradiated organisms. Such effects may be mediated through the control of cytokine production.

Steroids and Glucocorticoids. Glucocorticoid (GC) steroids, potent antiinflammatory and immunosuppressive agents, inhibit the hydrolysis of membrane phospholipids by phospholipase A2, which is the initial step in the generation of both lipoxygenase and cyclooxygenase products of arachidonic acid. GCs are used clinically to treat radiation pneumonitis, although the response rate is variable 239). Steroids can prevent death from radiation pneumonitis in animals, but their withdrawal prior to the end of the usual period of pneumonitis results in accelerated mortality (240). Conflicting observations on the ability of GCs to prevent radiation injuries may be partly related to how these drugs are administered after irradiation and to drug dosage (241).

Continuous low dose post-irradiation administration of the synthetic GC dexamethasone (DEX) in the drinking water is beneficial in treating radiation-induced lung and kidney injuries, deleterious for spinal cord injury, and of little value against hepatic injury (241,242). DEX delays the development of anemia and uremia, the impairment of glomerular filtration, and mortality resulting from renal failure after partial-body or local-kidney irradiation (241). Protection against lethal nephropathy is unexpected considering earlier reports that radiation nephropathy is enhanced by post-irradiation GCs. Chronic low dosage DEX therapy probably delays mortality after kidney irradiation in part by suppressing acute and chronic inflammation reactions and by stimulating physiological responses that compensate for the physiological derangements which develop after radiation exposure of this organ, and may be an effective therapeutic approach to delaying the development of lethal radiation nephropathy. In a more detailed study, various DEX treatments in the post-irradiation drinking water have increased survival times and delayed the development of kidney dysfunction. The most effective treatment was 94 μg/L DEX for 88 days (242). GCs such as DEX can down-regulate inflammatory eicosanoids and cytokines such as TNF-α, and thereby suppress the radiation-induced inflammatory response that perpetuates radiation damage in some tissues. Knowledge of the dose of GC and timing needed for suppression of radiation-induced cytokine responses is becoming available (243) and this may help to explain some of the discordant observations. Clinically, prolonged treatment using high doses of DEX may offset some beneficial effects of this steroid.

In an attempt to better define the mechanisms of delay of the development of lethal radiation nephropathy by DEX, the effect of post-irradiation DEX on the kidneys of rats receiving 20 Gy intraoperative bilateral local kidney irradiation has been examined (244). DEX significantly protects against loss of body weight, renal mass, renal function, and the development of anemia, but has no effect on edema or fibrosis. DEX-treated irradiated kidneys have significantly less damage to the glomeruli and tubules. The results suggest that DEX prevents the destruction of the nephron.

GCs such as methylprednisolone can delay radiation-induced lung injury in rodents, even when given well after lethal lung irradiation (236,240). DEX also protects against lung injury (30–90 day lethality after partial-body irradiation) (241), possibly by preventing the increase in capillary permeability and consequently the leakage of protein and fluid into the pleural cavity. This is because steroids do protect against radiation-induced vascular injury. In a rat model, ip administration of methylprednisolone three times weekly during weeks 3–8 after irradiation suppresses pneumonitis, delays the rise in tissue mast cell number, does not affect fibrosis at 20 weeks, supporting the dissociation of the pneumonitis and fibrosis phases (245). In contrast, WR-2721 protects better against late fibrosis than against pneumonitis (246). This dissociation may arise because the two different responses are mediated by different target cell populations (239). The observation that pneumonitis can be suppressed almost completely when GCs are administered after the development of significant interstitial edema may be important clinically, and combining steroids and WR-2721 may alleviate the entire course of the radiation-induced lung syndrome (245).

As regards GC treatment of central nervous system (CNS) radiation injury, short-term high dose DEX therapy of spinal-cord-irradiated animals delays the progression of injury to complete paraplegia and results in a transitory improvement in motor function of paralyzed animals (247). In contrast, chronic low dose DEX shortens the latent period between spinal cord irradiation and paralysis without affecting the tolerance dose (241), which suggests that DEX promotes the development of CNS radiation injury without changing the sensitivity of the target cells. The difference between these two studies may result from the equivalent daily GC doses used, which were much (~1000 times) larger in the former study. High doses may be needed to completely suppress inflammatory responses. On a more cautious note, administration of prednisolone acetate, a synthetic GC, after 2.84-Gy (284 rad) TBI results in an increased incidence of myeloid leukemias in mice. However, corticosterone, a GC secreted by cells, shows no such activity (248).

Miscellaneous Radioprotective Agents. Steroid hormones have been extensively studied for their ability to protect against infertility in males receiving XRT (see HORMONES). A variety of results have been reported (249). Pretreatment using a GnRH antagonist protects spermatogonial stem cell function from single doses of x-rays. Pretreatment using testosterone also protects spermatogonial stem cells from four daily x-ray fractions, whereas sub-q medroxyprogesterone and testosterone pretreatments protect against a single 3-Gy (300-rad) dose of x-rays. Similarly, 1.5–2.2 fold protection of spermatogonial stem cells has been reported in rats implanted with encapsulated testosterone and estradiol six weeks prior to irradiation. Such protection perhaps involves alterations in oxygen levels, GSH levels, and DNA repair activity in the stem cells. In contrast, no protection against damage to spermatogenesis by pretreatment using either testosterone or estrogen is observed after single γ- or x-ray doses.

The potential for normal brain tissue injury is one of the limiting factors in the use of XRT for brain tumors. Pentobarbital is a cerebral radioprotectant in rodent and primate models after single doses, but is associated with significant risks. Of alternative barbiturates, thiopental given to rats receiving 70-Gy (7000-rad) whole-brain irradiation in a single fraction enhances the 30-day survival similarly to pentobarbital, whereas ethohexital and phenobarbital show no radioprotective activity (250).

Combinations of Biologics and Thiols or Other Agents

Significant clinical promise lies in the concept of combining low (nontoxic) doses of radioprotective drugs having different, but complementary, mechanisms of action to achieve better protection than using either agent alone. In particular, combining an agent such as IL-1 that has the potential to stimulate stem cell proliferation before or after irradiation with an agent such as WR-2721 that protects the individual stem cells is appealing. This is illustrated in the diagram in Figure 1. The critical level below which the integrity of the tissue, and thus the viability of the animal, may be compromised is also shown. The balance between the gradual removal of the aging mature cells and their replacement from the regenerating stem cell pool determines whether the tissue remains sufficiently intact to maintain its integrity. An important general observation is that a combined effect can be seen only following radiation doses that do not eradicate most of the stem cells. Several biologics have been shown to exhibit additive or synergistic effects in combination with aminothiols, and may thus be potentially useful for reducing the risks associated with myelosuppression induced by XRT.

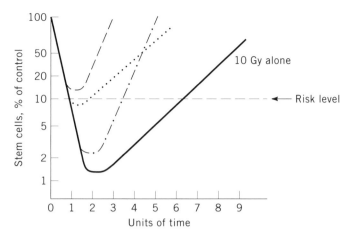

Fig. 1. Hypothetical kinetics of depletion of bone marrow (BM) stem cells following exposure to a lethal total-body irradiation (TBI) of 10 Gy (1000 rad) of x-rays (—) in the absence of radioprotective agents; (· · ·) when treated with WR-2721 30 min prior to irradiation; (— · —) when treated with IL-1; and (– – –) when treated with WR-2721 and IL-1.

Glucan Plus WR-2721/Aminothiols. Combinations of WR-2721 and glucan, given prior to irradiation, are additive in protecting hemopoietic tissue (206). Glucan can be administered up to ~1 h after irradiation. Whereas WR-2721 and glucan alone protect ~1.37-fold and ~1.08-fold, respectively, the combination of WR-2721 and glucan results in ~1.52-fold protection. Repopulation of hemopoietic cells in mice treated with both agents appears to occur faster. Combining selenium, Se; glucan; and WR-2721, which protect by different mechanisms, also gives an additive or synergistic radioprotection of endogenous CFU-S and accelerates BM and splenic CFU-GM regeneration (2). At least additive protection is also observed with the combination of preirradiation po cystamine and post-irradiation ip glucan in mice receiving sublethal and lethal TBI, as evidenced by enhanced hemopoietic recovery and survival (251).

G-CSF Plus WR-2721. Because glucan is a potent inducer of several hemopoietic cytokines, it has been suggested that specific hemopoietic cytokines may also stimulate hemopoiesis and synergize with WR-2721 (252). Based on $LD_{50/30}$ values, combining WR-2721 (4 mg/mouse ip 30-min preirradiation) and G-CSF (2.5 μg/(mouse·d) sc from days 1–16 post-irradiation) protects by 1.64 fold, which is greater than the additive effects of G-CSF (1.06-fold) and WR-2721 (1.44-fold) alone (252). BM and splenic CFU-S and CFU-GM recoveries are also accelerated in mice treated with WR-2721 plus G-CSF, and these animals exhibit the fastest recovery of mature functional blood elements (252). Thus, G-CSF does accelerate hemopoietic reconstitution from WR-2721-protected stem and progenitor cells, further increasing survival. WR-2721 and G-CSF also show additive effects in dogs receiving TBI, which has led to its evaluation in leukemia patients (253).

Other Immunomodulating Agents Plus WR-2721/Thiols. Combining WR-2721 (200 mg/kg ip 30-min preirradiation) and BV (25 mg/kg ip 24-h preir-radiation) demonstrates at least additive radioprotection in the $LD_{50/30}$ assay

(254). Protection is optimal when WR-2721 is given 30 min before irradiation and BV 24 h before or 4–8 h after irradiation. The combined treatments are also more effective in accelerating BM CFU-GM recovery. Endotoxin produces an additive effect when combined with the aminothiol AET (204). IL-1 enhances survival ($LD_{50/30}$) when given in combination with WR-2721 at radiation doses of 15–16 Gy (1500–1600 rad) causing hemopoietic and GI injury (2). Whereas IL-1 alone is not protective when given 30 min before irradiation, simultaneous administration of IL-1 and WR-2721 30 min before irradiation gives greater than additive protection. Synergistic protection in the mouse $LD_{50/30}$ assay is reported for combinations of WR-2721 and 5-hydroxytryptamine, as well as for WR-2721 plus DMPG, whereas combinations of DMPG with cysteine, glucan, GSH, or 5-hydroxytryptamine are only additive (255).

Prostaglandins Plus WR-2721.   Results on the efficacy of combining PGs and aminothiols are variable. Although combining WR-2721 and DMPG has no further effect on the radiosensitivity of mouse jejunal crypt cells beyond that observed using WR-2721 alone, six-day survival is increased in a greater than additive fashion (215). This drug combination is slightly less than additive in the 30-day survival assay (2). Misoprostol effectively protects intestinal clonogenic cells in combination with WR-2721 and increases the survival of intestinal clonogenic cells and mice exposed to neutrons (225). In contrast, WR-2721 followed by DMPG prior to neutrons does not improve the survival of mice over those receiving only WR-2721 (256).

Indomethacin Plus Thiols.   Preirradiation INDO, which stimulates hemopoietic stem cell proliferation, and WR-2721 combines to give a greater-than-additive protection of endogenous CFU-S and increases survival (257). WR-2721 does not compromise the ability of INDO to potentiate local tumor control in mice bearing FSa tumors, nor does INDO compromise the ability of WR-2721 to protect against hair loss or leg contracture. Clearly, the combination of INDO and WR-2721 may be valuable if hemopoietic toxicity is the limiting factor in XRT. Combined preirradiation INDO and cystamine also synergistically enhances the recovery of hemopoiesis in sublethally irradiated mice (258). These effects, however, do not translate into an increased survival, apparently because of the GI toxicity of the drug combination. For less toxic combinations, such as diclofenac and WR-2721, additive protection of survival has been observed (258).

Bryostatin and Hemopoietic Growth Factors.   Bryostatin 1 is a macrocyclic lactone PKC activator that, in combination with hemopoietic growth factors, exerts a variety of effects on radiosensitivity both *in vitro* and *in vivo* (259). A concomitant 24-h preirradiation treatment using GM-CSF-bryostatin 1 or other PKC activators enhances *in vitro* radioprotection of day-14 CFU-GM. Treating cells using GM-CSF-bryostatin 1 immediately after irradiation also results in some radioprotection. GM-CSF-bryostatin 1 pretreatment also protects enriched ($CD34^+$) progenitors. In contrast to the results with GM-CSF, preincubation of cells using bryostatin 1 and the GM-CSF/IL-3 fusion protein PIXY 321 does not lead to increased radioprotection of total day-14 CFU-GM. However, the combination treatment selectively augments the radioprotective capacity of the hybrid cytokine toward noneosinophilic elements. Bryostatin 1 is itself radioprotective in lethally irradiated mice and can augment the radioprotective capacity of GM-CSF, although marked strain differences have been observed.

BIBLIOGRAPHY

"Radioprotective Agents" in *ECT* 3rd ed., Vol. 19, pp. 801–832, by D. L. Klayman, Walter Reed Army Institute of Research, and E. S. Copeland, National Institutes of Health.

1. R. P. Gale and A. Butturini, *Exp. Haematol.* **18**, 958 (1990).
2. J. F. Weiss and co-workers, *Int. J. Radiat. Biol.* **57**, 709 (1990).
3. T. R. Sweeney, *A Survey of Compounds from the Antiradiation Drug Development Program of the U.S. Army Medical Research and Development Command*, Walter Reed Army Institute of Research, Washington, D.C., 1979.
4. L. M. Schuchter and J. H. Glick, *Biol. Ther. Cancer Updates* **3**, 1 (1993).
5. R. J. Hodgkiss and co-workers, *Int. J. Radiat. Biol.* **52**, 735 (1987).
6. D. Murray, in E. A. Bump and K. Malker, eds., *Radioprotectors: Chemical Biological and Clinical Perspective*, CRC Press, Inc., Boca Raton, Fla., 1996.
7. D. Murray and co-workers, *Int. J. Radiat. Biol.* **65**, 419 (1994).
8. M. F. M. A. Smeets and co-workers, *Radiat. Res.* **140**, 153 (1994).
9. D. J.. Grdina and co-workers, *Carcinogenesis*, **6**, 929 (1985).
10. R. C. Fahey, *Pharmacol. Ther.* **39**, 101 (1988).
11. J. A. Aguilera and co-workers, *Radiat. Res.* **130**, 194 (1992).
12. E. A. Bump and co-workers, *Radiat. Res.* **132**, 94 (1992).
13. J. Denekamp, A. Rojas, and G. Stevens, *Pharmacol. Ther.* **39**, 59 (1988).
14. K. D. Held, *Radiation Research*, Vol. 2, Taylor and Francis, London, 1987, pp. 689–694.
15. A. Russo and J. B. Mitchell, *Int. J. Radiat. Oncol. Biol. Phys.* **10**, 1243 (1984).
16. D. J. Grdina and co-workers, *Carcinogenesis*, **16**, 767 (1995).
17. M. F. Dennis and co-workers, *Int. J. Radiat. Biol.* **56**, 877 (1989).
18. X. Zhang, P. P. Lai, and Y. C. Taylor, *Int. J. Radiat. Oncol. Biol. Phys.* **24**, 713 (1992).
19. P. J. Meechan and co-workers, *Radiat. Res.* **140**, 437 (1994).
20. D. Murray and co-workers, *Abstracts of the 43rd Annual Meeting of the Radiation Research Society*, San Jose, Calif., Abstract P25-394, 1995, p. 203.
21. S. C. vanAnkeren, L. Milas, and D. Murray, *Int. J. Radiat. Oncol. Biol. Phys.* **16**, 1205 (1989).
22. R. E. Meyn and co-workers, *Radiat. Res.* **136**, 327 (1993).
23. N. Sato and co-workers, *J. Immunol.* **154**, 3194 (1995).
24. N. Ramakrishnan and G. N. Catravas, *J. Immunol.* **148**, 1817 (1992).
25. I. Szumiel, *Int. J. Radiat. Biol.* **66**, 329 (1994).
26. N. Ramakrishnan, D. E. McClain, and G. N. Catravas, *Int. J. Radiat. Biol.* **63**, 693 (1993).
27. L. Tretter and co-workers, *Int. J. Radiat. Biol.* **57**, 467 (1990).
28. M. L. Griem and co-workers, *Abstracts of the 43rd Annual Meeting of the Radiation Research Society*, San Jose, Calif., Abstract P07-99, 1995, p. 129.
29. J. S. Murley, D. J. Grdina, and P. J. Meechan, *Radiat. Res.* **126**, 223 (1991).
30. J. C. Livesey and D. J. Reed, *Advances in Radiation Biology*, Vol. 13, Academic Press, Inc., San Diego, 1987, pp. 285–353.
31. D. J. Grdina and co-workers, *Radiat. Res.* **117**, 251 (1989).
32. D. A. Boothman, E. N. Hughes, and A. B. Pardee, *Mutation and the Environment*, Part E, 319 (1990).
33. R. Datta and co-workers, *Biochem.* **31**, 8300 (1992).
34. D. J. Grdina and J. S. Murley, *Abstracts of the 42nd Annual Meeting of the Radiation Research Society*, Nashville, Tenn., 1994, p. 54.
35. N. R. Hunter, R. Guttenberger, and L. Milas, *Int. J. Radiat. Oncol. Biol. Phys.* **22**, 795 (1992).
36. B. A. Carnes and J. Grdina, *Int. J. Radiat. Biol.* **61**, 567 (1992).

37. D. J. Grdina and co-workers, *Carcinogenesis*, **13**, 811 (1992).
38. E. K. Balcer-Kubiczek and co-workers, *Int. J. Radiat. Biol.* **63**, 37 (1993).
39. N. Shigematsu, J. L. Schwartz, and D. J. Grdina, *Mutagenesis*, **9**, 355 (1994).
40. C. M. Denault, T. R. Skopek, and H. L. Liber, *Radiat. Res.* **136**, 271 (1993).
41. D. Q. Brown and co-workers, *Pharmacol. Ther.* **39**, 157 (1988).
42. J. S. Rasey and co-workers, *Pharmacol. Ther.* **39**, 33 (1988).
43. F. A. Stewart, *Int. J. Radiat. Oncol. Biol. Phys.* **16**, 1195 (1989).
44. L. M. Schuchter, W. E. Luginbuhl, and N. J. Meropol, *Seminars Oncol.* **19**, 742 (1992).
45. M. G. Cipolleschi, P. D. Sbarba, and M. Olivetto, *Blood*, **82**, 2031 (1993).
46. C. C. Badger and co-workers, *Radiat. Res.* **128**, 320 (1991).
47. F. E. Halberg and co-workers, *Int. J. Radiat. Oncol. Biol.* **12**, 211 (1986).
48. M. M. Kligerman and co-workers, *Int. J. Radiat. Oncol. Biol. Phys.* **22**, 799 (1992).
49. L. R. Coia, D. Q. Brown, and J. Hardiman, *NCI Monogr.* **6**, 235 (1988).
50. L. Coia and co-workers, *Int. J. Radiat. Oncol. Biol. Phys.* **22**, 791 (1992).
51. D. L. Palazzolo and K. S. Kumar, *Proc. Soc. Exp. Biol. Med.* **203**, 304 (1993).
52. O. A. Mendiondo, A. M. Connor, and P. Grigsby, *Acta Radiol. Oncol.* **21**, 319 (1982).
53. C. P. Sigdestad and co-workers, *Radiat. Res.* **106**, 224 (1986).
54. S. Matsushita and co-workers, *Int. J. Radiat. Oncol. Biol. Phys.* **30**, 867 (1994).
55. D. Green, D. Bensely, and P. Schein, *Cancer Res.* **54**, 738 (1994).
56. A. Lamperti and co-workers, *Radiat. Res.* **124**, 194 (1990).
57. H. G. France, R. L. Jirtle, and C. M. Mansbach, *Gastroenterol.* **91**, 644 (1986).
58. G. S. Montana and co-workers, *Cancer*, **69**, 2826 (1992).
59. A. M. Spence and co-workers, *Pharmacol. Ther.* **39**, 88 (1988).
60. A. Laayoun and co-workers, *Int. J. Radiat. Biol.* **66**, 259 (1994).
61. C. P. Sigdestad, A. M. Connor, and C. S. Sims, *Int. J. Radiat. Oncol. Biol. Phys.* **22**, 807 (1992).
62. A. Meister, *Cancer Res. (Suppl.)* **54**,   (1994).
63. R. Arima and R. Shiba, *Int. J. Radiat. Biol.* **61**, 695 (1992).
64. D. De Maria, A. M. Falchi, and P. Venturino, *Tumori.* **78**, 374 (1992).
65. A. Russo and co-workers, *Radiat. Res.* **103**, 232 (1985).
66. M. Spotheim-Maurizot and co-workers, *Radiat. Environ. Biophys.* **32**, 337 (1993).
67. L. E. Blank, J. Haveman, and M. Zanwijk, *Radiother. Oncol.* **10**, 67 (1987).
68. C. Selig, W. Nothdurft, and T. M. Fliedner, *J. Cancer. Res. Clin. Oncol.* **119**, 346 (1993).
69. J. Kim and co-workers, *Semin. Oncol.* **10**, 86 (1983).
70. P. Maasilta and co-workers, *Radiother. Oncol.* **25**, 192 (1992).
71. J. K. Rowe and co-workers, *Dis. Colon Rectum*, **36**, 681 (1993).
72. S. Kumar and co-workers, *Strahlenther. Onkol.* **168**, 610 (1992).
73. S. Mathur, K. Nandchahal, and H. C. Bhartiya, *Acta Oncol.* **30**, 981 (1991).
74. P. G. S. Prasanna and P. Uma Devi, *Radiat. Res.* **133**, 111 (1993).
75. M. L. Gupta and A. Ghose, *J. Radiat. Res. (Tokyo)* **34**, 295 (1993).
76. S. K. Basu, M. N. Srinivasan, and K. Chuttani, *Ind. J. Exp. Biol.* **31**, 837 (1993).
77. W. O. Foye, *Ann. Pharmacother.* **26**, 1144 (1992).
78. B. A. Teicher and co-workers, *Br. J. Cancer*, **62**, 17 (1990).
79. N. Ramakrishnan, W. W. Wolfe, and G. N. Catravas, *Radiat. Res.* **130**, 360 (1992).
80. S. M. Hahn and co-workers, *Cancer Res.*, **54**, 2006S (1994).
81. R. E. Langley and co-workers, *Radiat. Res.* **136**, 320 (1993).
82. J. Liebmann and co-workers, *Radiat. Res.* **137**, 400 (1994).
83. J. E. Moulder and co-workers, *Radiat. Res.* **136**, 404 (1993).
84. W. F. Ward and co-workers, *Radiat. Res.* **135**, 81 (1993).
85. R. Yarom and co-workers, *Radiat. Res.* **133**, 187 (1993).
86. S.-C. Yoon and co-workers, *Int. J. Radiat. Oncol. Biol. Phys.* **30**, 873 (1994).

87. M. J. Renan and P. I. Dowman, *Radiat. Res.* **120**, 442 (1989).
88. J. Matsubara, Y. Tajima, and M. Karasawa, *Radiat. Res.* **111**, 267 (1987).
89. O. Kagimoto and co-workers, *J. Radiat. Res.* **32**, 417 (1991).
90. J. Koropatnick and J. Pearson, *Mol. Pharmacol.* **44**, 44 (1993).
91. A. R. Kennedy, *Cancer Res.* **54**, 1999S (1994).
92. E. Von Hofe, R. Brent, and A. R. Kennedy, *Radiat. Res.* **123**, 108 (1990).
93. E. Seifter and co-workers, *Pharmacol. Ther.* **39**, 357 (1988).
94. T. Yoshimura and co-workers, *Radiat. Res.* **136**, 361 (1993).
95. L. Sarma and P. C. Kesavan, *Int. J. Radiat. Biol.* **63**, 759 (1993).
96. S. M. el-Nahas, F. E. Mattar, and A. A. Mohamed, *Mutat. Res.* **301**, 143 (1993).
97. M. Yasukawa, T. Terashima, and M. Seki, *Radiat. Res.* **120**, 456 (1989).
98. H. Tauchi and S. Sawada, *Int. J. Radiat. Biol.* **63**, 369 (1993).
99. R. S. Harapanhalli and co-workers, *Radiat. Res.* **139**, 115 (1994).
100. K. Rana, A. Sood, and N. Malhotra, *Ind. J. Exp. Biol.* **31**, 847 (1993).
101. V. Srinivasan and J. F. Weiss, *Int. J. Radiat. Oncol. Biol. Phys.* **23**, 841 (1992).
102. A. A. Shaheen and S. M. Hassan, *Strahlenther. Onkol.* **167**, 498 (1991).
103. J. R. J. Sorenson, *Radiat. Res.* **132**, 19 (1992).
104. J. R. J. Sorenson and co-workers, *Eur. J. Med. Chem.* **28**, 221 (1993).
105. T. D. Henderson and co-workers, *Radiat. Res.* **136**, 126 (1993).
106. G. L. Floersheim and A. Bieri, *Br. J. Radiol.* **63**, 468 (1990).
107. F. Battaini and co-workers, *Pharmacol. Ther.* **39**, 385 (1988).
108. G. L. Floersheim, *Radiat. Res.* **133**, 80 (1993) and **135**, 439 (1993).
109. M. Pospisil and co-workers, *Radiat. Res.* **134**, 323 (1993).
110. B. Hosek and co-workers, *Radiat. Environ. Biophys.* **31**, 289 (1992).
111. M. Pospisil and co-workers, *Physiol. Res.* **42**, 333 (1993).
112. J. Bohacek, B. Hosek, and M. Pospisil, *Life Sci.* **53**, 1317 (1993).
113. D. Szeinfeld and N. de Villiers, *Strahlenther. Onkol.* **169**, 311 (1993).
114. A. J. Senagore and co-workers, *Surgery*, **112**, 933 (1992).
115. R. M. Johnke and co-workers, *Int. J. Radiat. Oncol. Biol. Phys.* **20**, 369 (1991).
116. A. Mantovani and A. Vecchi, *Progr. Drug Res.* **35**, 487 (1990).
117. R. M. Johnke and R. S. Abernathy, *Int. J. Cell. Cloning*, **9**, 78 (1991).
118. M. Dion, D. Husey, and J. Osborne, *Int. J. Radiat. Oncol. Biol. Phys.* **17**, 101 (1989).
119. M. Dion and co-workers, *Int. J. Radiat. Oncol. Biol. Phys.* **19**, 401 (1990).
120. W.-J. Koh and co-workers, *Int. J. Radiat. Oncol. Biol. Phys.* **31**, 71 (1995).
121. K. J. Stelzer and co-workers, *Int. J. Radiat. Oncol. Biol. Phys.* **30**, 411 (1994).
122. W. F. Ward and co-workers, *Radiat. Res.* **129**, 107 (1992).
123. S. Tamou and K. R. Trott, *Strahlenther. Onkol.* **170**, 415 (1994).
124. C. W. Song and co-workers, *Radiat. Res.* **130**, 205 (1992).
125. A. Petkau, *Br. J. Cancer*, **55** (Suppl. VIII), 87 (1987).
126. J. R. Zucali, *Leuk. Lymphoma*, **13**, 27 (1994).
127. L. Weiss and co-workers, *Leuk. Lymphoma*, **10**, 477 (1993).
128. S. Delanian and co-workers, *Radiother. Oncol.* **32**, 12 (1994).
129. B. Perdereau and co-workers, *Bull. Cancer*, **78**, 526 (1991).
130. B. J. Wang, *Mem. Inst. Oswaldo Cruz*, **86** (Suppl. 2), 165 (1991).
131. H. F. Wang and co-workers, *J. Ethnopharmacol.* **34**, 215 (1991).
132. C. M. Wang, S. Ohta, and M. Shinoda, *Chem. Pharm. Bull.* (*Tokyo*) **40**, 493 (1992).
133. H. Y. Hsu, D. M. Hau, and C. C. Lin, *Am. J. Chin. Med.* **21**, 269 (1993).
134. S. H. Kim and co-workers, *In Vivo*, **7**, 467 (1993).
135. J. H. Hendry, *Semin. Oncol.* **4**, 123 (1994).
136. R. P. Gale, *J. Amer. Med. Assoc.* **285**, 625 (1987).
137. I. Brook and T. B. Elliot, *Radiat. Res.* **128**, 100 (1991).
138. I. Brook, S. P. Tom, and G. D. Ledney, *Int. J. Radiat. Biol.* **64**, 771 (1993).

139. C. E. Swenberg, S. Birke, and N. E. Geacintov, *Radiat. Res.* **127**, 138 (1991).
140. H. Mozdarani and A. Gharbali, *Int. J. Radiat. Biol.* **64**, 189 (1993).
141. C. Kent and G. Blekkenhorst, *Free Rad. Res. Commun.* **12–13**, 595 (1991).
142. J. F. Kalinich, G. N. Catravas, and S. L. Snyder, *Int. J. Radiat. Biol.* **59**, 1217 (1991).
143. M. Pospisil and co-workers, *Physiol. Res.* **40**, 445 (1991).
144. A. H. McArdle, *Gut*, **35**, S60 (1994).
145. R. F. Martin and L. Denison, *Int. J. Radiat. Oncol. Biol. Phys.* **23**, 579 (1992).
146. K. E. Carr and co-workers, *J. Submicrosc. Cytol. Pathol.* **23**, 569 (1991).
147. A. Maiche, O. P. Isokangas, and P. Grohn, *Acta Oncol.* **33**, 201 (1994).
148. A. Kurishita and co-workers, *Strahlenther. Onkol.* **168**, 728 (1992).
149. G. Stevens and co-workers, *Australas. Radiol.* **38**, 199 (1994).
150. T. E. Goffman and co-workers, *Cancer Res.* **50**, 7735 (1990).
151. R. Neta, *Pharmacol. Ther.* **39**, 261 (1988).
152. A. Wilson, *Int. J. Radiat. Biol.* **63**, 529 (1993).
153. F. M. Uckun and co-workers, *Blood*, **75**, 638 (1990).
154. F. G. Schuening and co-workers, *Blood*, **74**, 1308 (1989).
155. R. L. Monroy and co-workers, *Exp. Hematol.* **16**, 344 (1988).
156. K. G. Waddick and co-workers, *Blood*, **77**, 2364 (1991).
157. J. Laver and co-workers, *Treatment of Radiation Injuries*, Plenum Press, New York, 1990.
158. M. Y. Patchen and co-workers, *Int. J. Cell Clon.* **8**, 107 (1990).
159. M. Fushiki and co-workers, *Int. J. Radiat. Oncol. Biol. Phys.* **18**, 353 (1990).
160. K. Kabaya and co-workers, *Radiat. Res.* **139**, 92 (1994).
161. T. J. MacVittie and co-workers, *Int. J. Radiat. Biol.* **57**, 723 (1990).
162. W. Nothdurft and co-workers, *Int. J. Radiat. Biol.* **61**, 519 (1992).
163. G. J. Lieschki and A. W. Burgess, *N. Engl. J. Med.* **327**, 99 (1992).
164. R. Neta, *Biother.* **1**, 41 (1988) and *Progr. Clin. Biol. Res.* **352**, 471 (1990).
165. S. R. Dalmau, C. S. Freitas, and D. G. Tabak, *Bone Marrow Transpl.* **12**, 551 (1993).
166. S.-G. Wu, A. Tuboi, and T. Miyamoto, *Int. J. Radiat. Biol.* **56**, 485 (1989).
167. R. Neta and co-workers, *Blood*, **76**, 57 (1990).
168. C. M. Dubois and co-workers, *Exp. Hematol.* **21**, 303 (1993).
169. R. Neta and co-workers, *J. Exp. Med.* **175**, 689 (1992).
170. R. Neta and co-workers, *Blood*, **81**, 324 (1993).
171. D. Frasca and co-workers, *Radiat. Res.* **128**, 43 (1991).
172. G. N. Schwartz and co-workers, *Radiat. Res.* **121**, 220 (1990).
173. D. D. Wu, B. Bosch, and A. Keating, *Exp. Hematol.* **22**, 202 (1994).
174. N. Barbera and co-workers, *Pharmacol. Toxicol.* **72**, 256 (1993).
175. C. J. Kovacs and co-workers, *J. Leukoc. Biol.* **51**, 53 (1992).
176. M. S. Zaghloul, M. J. Dorie, and R. F. Kallman, *Int. J. Radiat. Oncol. Biol. Phys.* **26**, 417 (1993) and **29**, 805 (1994).
177. S.-G. Wu and T. Miyamoto, *Radiat. Res.* **123**, 112 (1990).
178. D. B. Roberts, E. L. Travis, and S. L. Tucker, *Radiat. Res.* **135**, 56 (1993).
179. M. J. Dorie, G. Bedarida, and R. F. Kallman, *Radiat. Res.* **128**, 316 (1991).
180. W. H. McBride and co-workers, *Cancer Res.* **49**, 169 (1989).
181. N. S. Yee, I. Paek, and P. Besmer, *J. Exp. Med.* **179**, 1777 (1994).
182. K. M. Zsebo and co-workers, *Proc. Natl. Acad. Sci. (U.S.)* **89**, 9464 (1992).
183. R. Neta and co-workers, *Immunol.* **153**, 1536 (1994).
184. B. R. Leigh, S. L. Hancock, and S. J. Knox, *Cancer Res.* **53**, 3857 (1993).
185. M. L. Patchen and co-workers, *Biother.* **7**, 13 (1993).
186. B. R. Leigh and co-workers, *Radiat. Res.* **142**, 12 (1995).
187. G. Doria and co-workers, *Stem Cells (Dayt.)* **11** (Suppl. 2), 93 (1993).
188. A. M. Farese and co-workers, *Blood*, **82**, 3012 (1993).

189. R. Neta and co-workers, *J. Immunol.* **153**, 4230 (1994).
190. L. Slordal, D. J. Warren, and M. A. S. Moore, *J. Immunol.* **142**, 833 (1989).
191. S. L. Hancock, *Abstracts of the 40th Annual Meeting of the Radiation Research Society*, Salt Lake City, Utah, Abstract P-06-5, 1992, p. 19.
192. Z. Fuks and co-workers, *Cancer Res.* **54**, 2582 (1994).
193. P. G. Tee and E. L. Travis, *Cancer Res.* **55**, 298 (1994).
194. W. H. McBride and co-workers, *Cancer Res.* **50**, 2949 (1990).
195. V. S. Gallicchio and co-workers, *Int. J. Cell Cloning*, **9**, 220 (1991).
196. M. Jung and co-workers, *Cancer Res.* **54**, 5194 (1994).
197. T. T. Kwok and R. M. Sutherland, *Br. J. Cancer*, **64**, 251 (1991).
198. R. Wollman and co-workers, *Int. J. Radiat. Oncol. Biol. Phys.* **30**, 91 (1994).
199. M. Nose and co-workers, *Int. J. Radiat. Oncol. Biol. Phys.* **29**, 631 (1994).
200. A. Kurishita, T. Ono, and A. Uchida, *Int. J. Radiat. Biol.* **63**, 413 (1993).
201. Y. Kalechman and co-workers, *Radiat. Res.* **136**, 197 (1993).
202. S. R. Kempf, R. E. Port, and S. Ivankovic, *Radiat. Res.* **139**, 226 (1994).
203. K. A. Mason, S. Murphy, and L. Milas, *Radiat. Res.* **136**, 229 (1993).
204. E. J. Ainsworth, *Pharmacol. Ther.* **39**, 223 (1988).
205. P. Fedorocko, *Int. J. Radiat. Biol.* **65**, 465 (1994).
206. M. L. Patchen and co-workers, *Pharmacol. Ther.* **39**, 247 (1988).
207. D. Chorvatovicova, *Strahlenther. Onkol.* **167**, 612 (1991).
208. M. Pospisil and co-workers, *Exp. Hematol.* **20**, 891 (1992).
209. J. R. Maisin, *Frontiers of Radiation Biology*, VCH, Weinheim and Balaban, Rehovot, 1989, pp. 165–173.
210. P. Fedorocko and co-workers, *Immunopharmacol.* **28**, 163 (1994).
211. K. Tsuneoka and co-workers, *J. Radiat. Res.* **35**, 147 (1994).
212. T. Okawa and co-workers, *Cancer*, **72**, 1949 (1993).
213. K. A. Mason and P. J. Tofilon, *Cancer Chemother. Pharmacol.* **33**, 435 (1994).
214. T. L. Walden Jr., M. L. Patchen, and T. J. MacVittie, *Radiat. Res.* **113**, 388 (1988).
215. W. R. Hanson, *Prostaglandin Inhibitors in Tumor Immunology and Immunotherapy*, CRC Press, Inc., Boca Raton, 1994, pp. 171–186.
216. R. C. Miller, P. LaNasa, and W. R. Hanson, *Radiat. Res.* **139**, 109 (1994).
217. T. L. Walden Jr., *Radiat. Res.* **132**, 359 (1992).
218. N. Zaffaroni and co-workers, *Radiat. Res.* **135**, 88 (1993).
219. D. B. Rubin and co-workers, *Radiat. Res.* **125**, 41 (1991).
220. T. L. Walden Jr., M. Patchen, and S. L. Snyder, *Radiat. Res.* **109**, 440 (1987).
221. F. D. Malkinson, L. Geng, and W. R. Hanson, *J. Invest. Dermatol.* **101**, 135S (1993).
222. G. Nicolau and co-workers, *Skin Pharmacol.* **2**, 22 (1989).
223. J.-W. Han, F. McCormick, and I. G. Macara, *Science*, **252**, 576 (1991).
224. J. P. Delaney, M. E. Bonsack, and I. Felemovicius, *Am. J. Surg.* **166**, 492 (1993) and *Radiat. Res.* **137**, 405 (1994).
225. W. R. Hanson and D. J. Grdina, *Radiat. Res.* **128**, S12 (1991).
226. I. Neumann and co-workers, *Agents Actions Suppl.* **37**, 58 (1992).
227. T. L. Walden and J. F. Kalinich, *Pharmacol. Ther.* **39**, 379 (1988).
228. T. L. Walden, *Ann. N.Y. Acad. Sci.* **524**, 431 (1988).
229. K. S. Iwamoto and W. H. McBride, *Radiat. Res.* **139**, 103 (1994).
230. J. W. Hopewell and co-workers, *Br. J. Cancer*, **68**, 1 (1993).
231. H. M. Storm and co-workers, *Lipids*, **28**, 555 (1993).
232. G. L. Floersheim, *Radiat. Res.* **139**, 240 (1994).
233. L. Milas and co-workers, *Adv. Space Res.* **12**, 265 (1992).
234. M. Pospisil and co-workers, *Strahlenther. Onkol.* **165**, 627 (1989).
235. S. E. Orfanos and co-workers, *Toxicol. Appl. Pharmacol.* **124**, 112 (1994).
236. N. J. Gross, N. O. Holloway, and K. R. Narine, *Radiat. Res.* **127**, 317 (1991).

237. A. Kozubik and co-workers, *Exp. Hematol.* **21**, 138 (1993).
238. A. Kozubik and co-workers, *Int. J. Radiat. Biol.* **65**, 369 (1994).
239. N. J. Gross, *Ann. Intern. Med.* **86**, 81 (1977).
240. N. J. Gross, K. R. Narine, and R. Wade, *Radiat. Res.* **113**, 112 (1988).
241. J. P. Geraci and co-workers, *Radiat. Res.* **129**, 61 (1992).
242. J. P. Geraci and co-workers, *Radiat. Res.* **131**, 186 (1992).
243. J.-H. Hong and co-workers, *Int. J. Radiat. Oncol. Biol. Phys.* **33**, 619 (1995).
244. J. P. Geraci, M. S. Mariano, and K. L. Jackson, *Radiat. Res.* **134**, 86 (1993).
245. H. E. Ward and co-workers, *Radiat. Res.* **146**, 22 (1993).
246. E. L. Travis and co-workers, *Int. J. Radiat. Oncol. Biol. Phys.* **10**, 243 (1984).
247. J. Y. Delattre and co-workers, *Brain*, **111**, 1319 (1988).
248. M. Seki and co-workers, *Radiat. Res.* **127**, 146 (1991).
249. B. Kurdoglu and co-workers, *Radiat. Res.* **139**, 97 (1994).
250. J. J. Olson and co-workers, *Neurosurgery*, **30**, 720 (1992).
251. M. Pospisil and co-workers, *Folia. Biol. (Praha)* **37**, 117 (1991).
252. M. L. Patchen, T. J. MacVittie, and L. M. Souza, *Int. J. Radiat. Oncol. Biol. Phys.* **22**, 773 (1992).
253. T. J. MacVittie, *Proc. AACR*, **33**, 505 (1992).
254. P. Fedorocko, P. Brezani, and N. O. Mackova, *Int. J. Immunopharmacol.* **16**, 177 (1994).
255. J. R. Maisin, C. Albert, and A. Henry, *Radiat. Res.* **135**, 332 (1993).
256. L. K. Steel and co-workers, *Radiat. Res.* **115**, 605 (1988).
257. P. C. Besa, N. R. Hunter, and L. Milas, *Radiat. Res.* **135**, 93 (1993).
258. A. Kozubik, M. Pospisil, and J. Netikova, *Strahlenther. Onkol.* **167**, 186 (1991).
259. S. Grant and co-workers, *Blood*, **83**, 663 (1994).

DAVID MURRAY
University of Alberta,
Cross Cancer Institute

WILLIAM H. MCBRIDE
UCLA Medical Center

RADIUM. See RADIOACTIVITY, NATURAL.

RADON. See HELIUM GROUP, GASES.

RARE-EARTH ELEMENTS. See LANTHANIDES.

RASCHIG RINGS. See ABSORPTION; DISTILLATION.

RAYON. See FIBERS, REGENERATED CELLULOSICS.

REACTOR TECHNOLOGY

Reactor technology comprises the underlying principles of chemical reaction engineering (CRE) and the practices used in their application. The focuses of reactor technology are reactor configurations, operating conditions, external operating environments, developmental history, industrial application, and evolutionary change. Reactor designs evolve from the pursuit of new products and uses, higher conversion, more favorable reaction selectivity, reduced fixed and operating costs, intrinsically safe operation, and environmentally acceptable processing.

Early in the development of chemical reaction engineering, reactants and products were treated as existing in single homogeneous phases or several discrete phases. The technology has evolved into viewing reactants and products as residing in interdependent environments, a most important factor for multiphase reactors which are the most common types encountered.

Many, but not all, reactor configurations are discussed. Process design, catalyst manufacture, thermodynamics, design of experiments (qv), and process economics, as well as separations, the technologies of which often are applicable to reactor technology, are discussed elsewhere in the *Encyclopedia* (see CATALYSIS; SEPARATION; THERMODYNAMICS).

Besides stoichiometry and kinetics, reactor technology includes requirements for introducing and removing reactants and products, efficiently supplying and withdrawing heat, accommodating phase changes and material transfers, assuring efficient contacting of reactants, and providing for catalyst replenishment or regeneration. Consideration must be given to physical properties of feed and products (vapor, liquid, solid, or combinations), characteristics of chemical reactions (reactant concentrations, paths and rates, operating conditions, and heat addition or removal), the nature of any catalyst used (activity, life, and physical form), and requirements for contacting reactants and removing products (flow characteristics, transport phenomena, mixing requirements, and separating mechanisms).

All the factors are interdependent and must be considered together. Requirements for contacting reactants and removing products are a central focus in applying reactor technology; other factors usually are set by the original selection of the reacting system, intended levels of reactant conversion and product selectivity, and economic and environmental considerations. These issues should be taken into account when determining reaction kinetics from laboratory and bench-scale data, designing and operating pilot units, scaling up to large units, and ultimately in designing, operating, and improving industrial plant performance.

Reactor Types and Characteristics

Specific reactor characteristics depend on the particular use of the reactor as a laboratory, pilot plant, or industrial unit. All reactors have in common selected characteristics of four basic reactor types: the well-stirred batch reactor, the semibatch reactor, the continuous-flow stirred-tank reactor, and the tubular reactor (Fig. 1). A reactor may be represented by or modeled after one or a

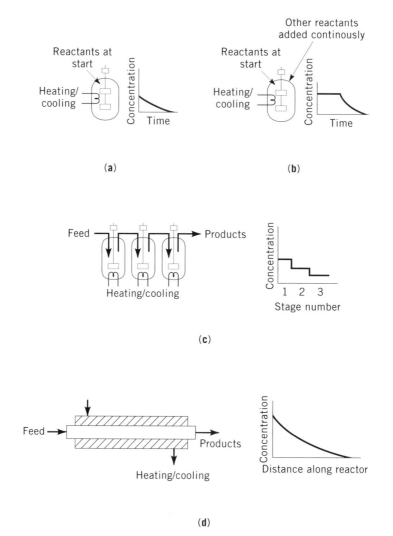

Fig. 1. Reactor types: (**a**) batch, (**b**) semibatch, (**c**) continuous-flow stirred-tank, and (**d**) tubular.

combination of these. Suitability of a model depends on the extent to which the impacts of the reactions, and thermal and transport processes, are predicted for conditions outside of the database used in developing the model (1–4).

Batch Reactor. A batch reactor is one in which a feed material is treated as a whole for a fixed period of time. Batch reactors may be preferred for small-scale production of high priced products, particularly if many sequential operations are employed to obtain high product yields, eg, a process requiring a complex cycle of temperature–pressure–reactant additions. Batch reactors also may be justified when multiple, low volume products are produced in the same equipment or when continuous flow is difficult, as it is with highly viscous or sticky solids-laden liquids, eg, in the manufacture of polymer resins where molecular weight and product quality are markedly affected by increasing viscosity and heat removal demands. Because residence times can be more uniform in batch reactors, better yields and higher selectivity may be obtained than with continuous reactors. This advantage exists when undesired reaction products inhibit the reaction, side reactions are of lower order than that desired, or the product is an unstable or reactive intermediate.

Batch reactors often are used to develop continuous processes because of their suitability and convenient use in laboratory experimentation. Industrial practice generally favors processing continuously rather than in single batches, because overall investment and operating costs usually are less. Data obtained in batch reactors, except for very rapid reactions, can be well defined and used to predict performance of larger scale, continuous-flow reactors. Almost all batch reactors are well stirred; thus, ideally, compositions are uniform throughout and residence times of all contained reactants are constant.

Semibatch Reactor.   The semibatch reactor is similar to the batch reactor but has the additional feature of continuous addition or removal of one or more components. For example, gradual addition of chlorine to a stirred vessel containing benzene and catalyst results in higher yields of di- and trichlorobenzene than the inclusion of chlorine in the original batch. Similarly, thermal decomposition of organic liquids is enhanced by continuously removing gaseous products. Constant pressure can be maintained and chain-terminating reaction products removed from the system. In addition to better yields and selectivity, gradual addition or removal assists in controlling temperature particularly when the net reaction is highly exothermic. Thus, use of a semibatch reactor intrinsically permits more stable and safer operation than in a batch operation.

Continuous-Flow Stirred-Tank Reactor.   In a continuous-flow stirred-tank reactor (CSTR), reactants and products are continuously added and withdrawn. In practice, mechanical or hydraulic agitation is required to achieve uniform composition and temperature, a choice strongly influenced by process considerations, ie, multiple specialty product requirements and mechanical seal pressure limitations. The CSTR is the idealized opposite of the well-stirred batch and tubular plug-flow reactors. Analysis of selected combinations of these reactor types can be useful in quantitatively evaluating more complex gas-, liquid-, and solid-flow behaviors.

Because the compositions of mixtures leaving a CSTR are those within the reactor, the reaction driving forces, usually reactant concentrations, are necessarily low. Therefore, except for zero- and negative-order reactions, a CSTR requires the largest volume of the reactor types to obtain desired conversions. However, the low driving force makes possible better control of rapid exothermic and endothermic reactions. When high conversions of reactants are needed, several CSTRs in series can be used. Equally good results can be obtained by dividing a single vessel into compartments while minimizing back-mixing and short-circuiting. The larger the number of stages, the closer performance approaches that of a tubular plug-flow reactor.

Continuous-flow stirred-tank reactors in series are simpler and easier to design for isothermal operation than are tubular reactors. Reactions with narrow operating temperature ranges or those requiring close control of reactant concentrations for optimum selectivity benefit from series arrangements. If severe heat-transfer requirements are imposed, heating or cooling zones can be incorporated within or external to the CSTR. For example, impellers or centrally mounted draft tubes circulate liquid upward, then downward through vertical heat-exchanger tubes. In a similar fashion, reactor contents can be recycled through external heat exchangers.

Tubular Reactor. The tubular reactor is a vessel through which flow is continuous, usually at steady state, and configured so that conversion and other dependent variables are functions of position within the reactor rather than of time. In the ideal tubular reactor, the fluids flow as if they were solid plugs or pistons, and reaction time is the same for all flowing material at any given tube cross section; hence, position is analogous to time in the well-stirred batch reactor. Tubular reactors resemble batch reactors in providing initially high driving forces, which diminish as the reactions progress down the tubes.

Flow in tubular reactors can be laminar, as with viscous fluids in small-diameter tubes, and greatly deviate from ideal plug-flow behavior, or turbulent, as with gases, and consequently closer to the ideal (Fig. 2). Turbulent flow generally is preferred to laminar flow, because mixing and heat transfer when normal to flow are improved and less back-mixing is introduced in the direction of flow. For slow reactions and especially in small laboratory and pilot-plant reactors, establishing turbulent flow can result in inconveniently long reactors or may require unacceptably high feed rates. Depending on the consequences in process development and impact on process economics, compromises, though necessary, may not prove acceptable.

Multiphase Reactors. The overwhelming majority of industrial reactors are multiphase reactors. Some important reactor configurations are illustrated in Figures 3 and 4. The names presented are often employed, but are not the only ones used. The presence of more than one phase, whether or not it is flowing, confounds analyses of reactors and increases the multiplicity of reactor configurations. Gases, liquids, and solids each flow in characteristic fashions, either dispersed in other phases or separately. Flow patterns in these reactors are complex and phases rarely exhibit idealized plug-flow or well-stirred flow behavior.

A fixed-bed reactor is packed with catalyst. If a single phase is flowing, the reactor can be analyzed as a tubular plug-flow reactor or modified to account for axial diffusion. If both liquid and gas or vapor are injected downward through

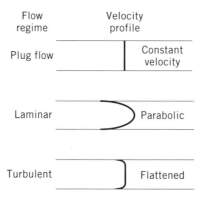

Fig. 2.   Flow characteristics for single-phase flows.

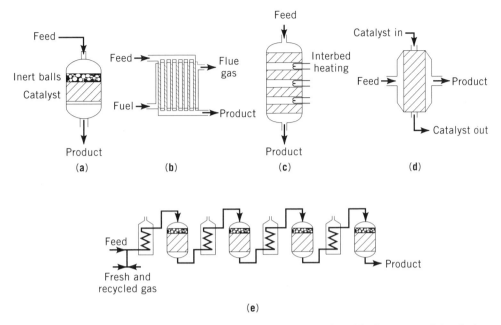

Fig. 3. Multiple fixed-bed configurations: (**a**) adiabatic fixed-bed reactor, (**b**) tubular fixed beds, (**c**) staged adiabatic reactor with interbed heating (cooling), (**d**) moving radial fixed-bed reactor, and (**e**) trickle beds in series.

the catalyst bed, or if substantial amounts of vapor are generated internally, the reactors are mixed-phase, downflow, and fixed-bed reactors. If the liquid and gas rates are so low that the liquid flows as a continuous film over the catalyst, the reactors are called trickle beds. Trickle beds have potential advantages of lower pressure drops and superior access for gaseous reactants to the catalyst; however, restricted access can also be a disadvantage, eg, where direct gas contact promotes undesired side reactions.

At higher total flow rates, particularly when the liquid is prone to foaming, the reactor is a pulsed column. This designation arises from the observation that the pressure drop within the catalyst bed cycles at a constant frequency as a result of liquid temporarily blocking gas or vapor pathways. The pulsed column is not to be confused with the pulse reactor used to obtain kinetic data in which a pulse of reactant is introduced into a tube containing a small amount of catalyst.

Downflow of reactants is preferred because reactors are more readily designed mechanically to hold a catalyst in place and are not prone to inadvertent excessive velocities, which upset the beds. Upflow is used less often but has the advantage of optimum contacting between gas, liquid, and catalyst over a wider range of conditions. Mixed-phase, upflow, and fixed-bed reactors offer higher liquid holdups and greater assurance of attaining uniform catalyst wetting and radial flow distribution, the consequences of which are more uniform temperature distribution and greater heat transfer.

At high liquid flow rates in these co-current fixed-bed reactors, gas becomes the dispersed phase and bubble flow develops; flow characteristics are similar

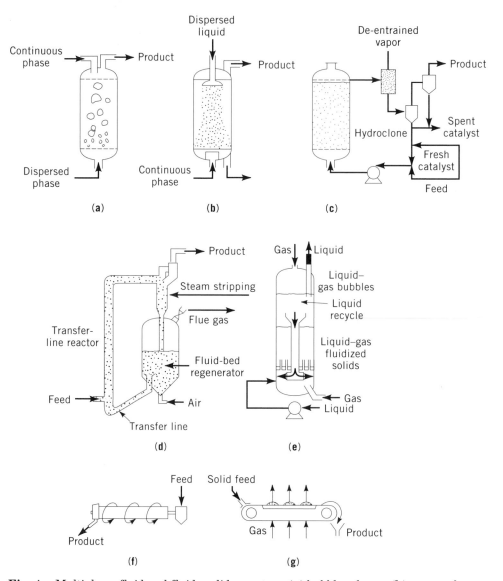

Fig. 4. Multiphase fluid and fluid−solids reactors: (**a**) bubble column, (**b**) spray column, (**c**) slurry reactor and auxiliaries, (**d**) fluidization unit, (**e**) gas−liquid−solid fluidized reactor, (**f**) rotary kiln, and (**g**) traveling grate or belt drier.

to those in countercurrent packed-column absorbers. At high gas rates, spray and slug flows can develop. Moving beds are fixed-bed reactors in which spent catalyst or reactive solids are slowly removed from the bottom and fresh material is added at the top. A fixed bed that collects solid impurities present in the feed or produced in the early reaction stages is a guard bed. If catalyst deposits are periodically burned or otherwise removed, the operation is cyclic, and the catalyst remaining behind the combustion front is regenerated.

In bubble column reactors, gas bubbles flow upward through a slower moving liquid. The bubbles, which rise in essentially plug flow, draw liquid in their wakes and thereby induce back-mixing in the liquid with which they have come in contact. Analogously, in spray columns, liquid as droplets descend through a fluid, usually a gas. Both bubble and spray columns are used for reactions where high interfacial areas between phases are desirable. Bubble column reactors are used for reactions where the rate-limiting step is in the liquid phase, or for slow reactions where contacting is not critical. An important variant of the bubble column reactor is the loop reactor, commonly used for both multiphase and highly viscous systems. Loop reactors are distinguishable by their hydraulically or mechanically driven fluid recirculation, which offers the benefits of the well-stirred behavior of CSTRs and high average reactant concentrations of tubular reactors.

Reactors are termed fluidized or fluid beds if upward gas or liquid flows, alone or in concert, are sufficiently high to suspend the solids and make them appear to behave as a liquid. This process is usually referred to as fluidization. The most common fluid bed is the gas-fluidized bed. With gas feeds, the excess gas over the minimum required for fluidization rises as discrete bubbles, through which the surrounding solids circulate. At higher gas rates, such beds lose their clearly defined surface, and the particles are fully suspended. Depending on the circumstances, these reactors are variously called riser, circulating-fluidized, fast-fluidized, or entrainment reactors. In ebullating-bed or gas–liquid–solid reactors, the solids are fluidized by liquid and gas, with gas primarily providing lifting power in the former, and liquid in the latter. These become slurry bubble column reactors (less precisely, slurry reactors) at higher rates when the beds begin to lose their defined surfaces. Slurry bubble column reactors that contain finely powdered solids are often termed and treated as bubble column reactors because such suspensions are homogeneous.

A reactor is termed a radial or panel-bed reactor when gas or vapor flow perpendicular to a catalyst-filled annulus or panel. These are used for rapid reactions to reduce stresses on the catalyst or to minimize pressure drops. Similar cross-flow configurations also are used for processing solids moving downward under gravity while a gas passes horizontally through them. Rotary kilns, belt dryers, and traveling grates are examples. Cross-flow reactors are not restricted to solids-containing systems. Venturis, in which atomized liquids are injected across the gas stream, are effective for fast reactions and similarly for generating small gas bubbles in downward-flowing liquids where mass transport across the gas–liquid interface is limiting.

Flow Regimes in Multiphase Reactors.   Reactant contacting, product separations, rates of mass and heat transport, and ultimately reaction conversion and product yields are strong functions of the gas and liquid flow patterns within the reactors. The nomenclature of commonly observed flow patterns or flow regimes reflects observed flow characteristics, ie, annular, bubbly, plug, slug, spray, stratified, and wavy.

Multiphase reactor flow regimes depend on both physical properties, eg, density, viscosity, and interfacial tension, and hydrodynamic and gravitational forces exerted on fluids. Mechanistic models, eg, those describing the annular mixed-phase flow between fixed-bed catalyst pellets as analogous to flow through two-dimensional slits, are emerging for predicting phase boundaries between regimes (5). However, flow maps, generally designed for specific reactor types, operating conditions, and fluid combinations are most commonly used to depict gas and liquid flow rates associated with various flow regimes and to delineate the phase (6).

Reactor Selection

Selection of a reactor, whether for a new application or a changing situation, is often determined by economics, reliability, or availability of a proven system that is amenable to extension in a new service. For example, fixed beds and slurry reactors are favored at high pressures over fluid beds, fluidized systems are less likely to develop hot spots or be subject to temperature runaways, and downflow vapor-phase fixed-bed technology originally developed for desulfurizing vaporized naphthas and light gas oils has been extended to mixed-phase operation with higher boiling gas oils and residua. Introduction of new catalysts, improvements in process and equipment design and operations, transient reactor behavior, heat generation and removal, reduced emissions and waste minimization, and safety-related hazards are other issues that influence the continued and extended use of a reactor configuration.

Catalyst Development. Traditional slurry polypropylene homopolymer processes suffered from formation of excessive amounts of low grade amorphous polymer and catalyst residues. Introduction of catalysts with up to 30-fold higher activity together with better temperature control have almost eliminated these problems (7). Although low reactor volume and available heat-transfer surfaces ultimately limit further productivity increases, these limitations are less restrictive with the introduction of more finely suspended metallocene catalysts and the emergence of industrial gas-phase fluid-bed polymerization processes (see METALLOCENE CATALYSTS (SUPPLEMENT)).

Design and Operations Improvement. Process improvements may permit older reactor types to remain competitive and slow displacement by other technological advances. Fluid catalytic cracking (FCC) reactors (8) evolved from long-residence time fluidized beds, designed for processing atmospheric gas oils, to short-contact time riser reactors capable of handling metals-laden vacuum gas oils and residua; feeds are more readily processed using hydrogen at high pressures, although at higher cost. Improvements in feed injectors and termination zone configurations (where product is separated from catalyst and further reactions quenched) heat and pressure balance control, solids retention, and mechanical reliability are essential in achieving higher conversion levels and increased yields of desired products made possible by the development of zeolite catalysts. These enhancements have been incorporated into both new and existing units. The introduction of highly active zeolites plays a dominant role, but catalyst superiority is not sufficient by itself.

Heat Generation and Removal. The impact of heat generation and removal on kinetics similarly affects reactor selection criteria. Problems of maximizing outputs from exothermic or endothermic reactions occur with reversible reactions, eg, ammonia synthesis or water gas shift reactions; temperature selective reactions, reforming or hydroformylation; or rapidly deactivating side reactions, eg, catalytic cracking of petroleum or various polymerization processes. With a reverse reaction, whose conversion rate increases with temperature more rapidly than that of the desired forward exothermic reaction, the temperature needed for maximum conversion decreases as the reaction progresses. Consequently, a batch reactor with a heat exchanger or small-diameter tubes jacketed in a coolant might be used. There are many options for limiting temperature rises or programming temperature distributions; even large-diameter fixed beds having poor heat transfer have been employed. Other options include using interstage coolers, injecting a volatile liquid (eg, water) to cool by evaporation, recycling unreacted feed or a portion of the product, and adding inert diluent to the feed or generating volatile liquid within catalyst beds. The advantage of complicating reactor design, by allowing generation and propagation of large axial and radial temperature gradients, is the lower cost of using these large, essentially adiabatic fixed-bed reactors.

Transient Reactor Behavior. Reactors can operate advantageously with moving thermal or concentration fronts that are either created by periodic flow reversal or are self-generated. Low level contaminants or waste products can be efficiently removed in adiabatic fixed beds with periodic flow reversals by taking advantage of higher outlet reaction temperatures generated in earlier cycles to accelerate exothermic reactions. Energy and cost savings are effected by this substitution of internal heat transfer for external exchange (9). When locally ignited, highly exothermic noncatalytic solid–solid and solid–gas reactions can proceed down tubular reactors as moving fronts in an analogous fashion to flame propagation in gases. Combustion of solid propellants, synthesis of intermetallic compounds (eg, Ni–Al), and manufacture of borides and carbides of transition- and rare-earth metals are examples (10). Transients also can have adverse effects as in causing local overheating of catalyst pellets in essentially steady-state operations.

Reduced Emissions and Waste Minimization. Reducing harmful emissions and minimizing wastes within a process by inclusion of additional reaction and separation steps and catalyst modification may be substantially better than end-of-pipe cleanup or even simply improving maintenance, housekeeping, and process control practices. SO_2 and NO reduction to their elemental products in fluid catalytic cracking units exemplifies the use of such a strategy (11).

Safety-Related Hazards. High temperature excursions and runaway are two of the many safety-related hazards that can affect reactor selection. Hazard assessments go beyond analyses of process variations and must consider system integration (eg, start-up and shutdown practices), system faults (eg, power failure, loss of cooling, leaks between streams), operational errors (eg, incorrect feed rates, under- or over-charging catalysts and solvents), and abnormal situations (eg, escaping contamination or fires and explosions emanating from neighboring process equipment). Continuously operated reactors, with their dedication to singular process objectives, generally lower inventories and more technically sophisticated control may be selected for being inherently safer than intermittent batch operations. However, continuously operated reactors are often coupled with other process functions and thus are subject to further incidents during the start-up and shutdown phases of operation (12).

Coupling Reaction Kinetics with Transport

The interacting effects of reaction kinetics with mass, momentum, and heat transport and the mixing and interchange of components are essential elements in establishing reactors as process systems. Reactor response to various design and operating variables largely depends on the relative kinetic and transport rates at the physical scales of the phenomena being addressed. Three levels of scale are normally addressed: micro-scale or molecular level, eg, molecular mean free path within a catalyst pore; meso-scale or catalyst particle or eddy size, eg, transport and reaction in and around catalyst pellets, droplets, and bubbles; and macro-scale or reactor scale, eg, integration of micro and meso-scale phenomena over an entire reactor (13).

Back-Mixing, Staging, and Recycle.   The extent of back-mixing and degree of staging or recycle directly affects overall reactor size and catalyst requirements for reactions with greater than zero-order dependencies on limiting reactants. Back-mixing does not necessarily imply reduced conversions. Complete back-mixing can decrease catalyst deactivation by diluting catalyst poisons, if these are limiting constituents. Because total conversion is a function of the kinetic behavior of the entire reactor system, the overall effects of reactor back-mixing must be considered. Back-mixing, staging, and recycle can markedly influence reaction pathways to alter product yields. In well-stirred batch polymerization, initiation and termination reactions are modified to establish widely differing molecular weight distributions (14). Tubular reactors can produce narrow molecular weight distributions, although throughput may be limited by the mixing rate and conductive heat transfer. Figure 5 illustrates the differing product distributions effected by changing individual recycle rates for a methane chlorination process (15).

The full range of flows from plug to well-mixed can be obtained by staging, whereby several reactor vessels are operated in sequence or by introducing recycle around the entire system. Thus the back-mixing character of individual reactors is only consequential to their specific designs. Three to six well-mixed stages in series often approach plug-flow behavior for first-order reactions that are carried out to 70% conversion. Greater staging is required for higher conversions and the actual number of stages depends on the specific reaction mechanisms and reaction rates for the system of interest. On the other hand, one or more plug-flow reactors in series with a large recycle stream around the reactor system behave as a well-mixed system. Thus, the degree of back-mixing sets criteria for system selection, but not necessarily for selecting the reactor type.

Reaction and Transport Interactions.   The importance of the various design and operating variables largely depends on relative rates of reaction and transport of reactants to the reaction sites. If transport rates to and from reaction sites are substantially greater than the specific reaction rate at meso-scale reactant concentrations, the overall reaction rate is uncoupled from the transport rates and increasing reactor size has no effect on the apparent reaction rate, the macro-scale reaction rate. When these rates are comparable, they are coupled, that is they affect each other. In these situations, increasing reactor size alters mass- and heat-transport rates and changes the apparent reaction rate. Conversions are underestimated in small reactors and selectivity is affected. Selectivity

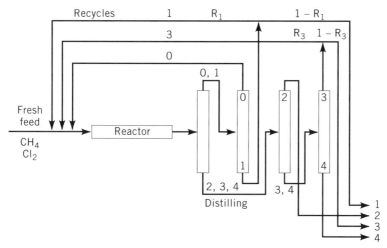

Recycle ratios		Fraction of CH$_4$ converted into			
R$_1$	R$_3$	CH$_3$Cl	CH$_2$Cl$_2$	CHCl$_3$	CCl$_4$
0	0	0.29	0.42	0.26	0.03
1	0	0	0.56	0.39	0.05
0	1	0.15	0.36	0	0.49
1	1	0	0.38	0	0.62

Fig. 5. Effects of product recycle on methane chlorination product distributions where 1 = methyl chloride, 2 = methylene chloride, 3 = chlorine, and 4 = carbon tetrachloride (15).

does not exhibit such consistent impacts and any effects of size on selectivity must be determined experimentally.

For well-defined reaction zones and irreversible, first-order reactions, the relative reaction and transport rates are expressed as the Hatta number, Ha (16). Ha equals $(k_1 D_A / k_L)^{-1}$, where k_1 = reaction rate constant, D_A = molecular diffusivity of reactant A, and k_L = mass-transfer coefficient. Reaction dominates with $Ha < 0.3$ and transport is limiting with $Ha > 2$. Well-defined zones encompass the reactor reaction volume elements (Fig. 6) for a fast irreversible homogeneous reaction, $A + B \rightarrow$ products.

For more complex systems, the degree of coupling requires evaluating radial or interfacial (in a multiphase reactor) concentration and temperature gradients within a reactor. If these gradients are ca 10% or less than the total concentration gradients of limiting reactants or inhibitor products, the reaction is uncoupled. At 10–90%, the reaction is coupled and above 90% the reaction is transport- or diffusion-controlled. Systematic experimentation is required to establish the reaction-induced gradients. Such work represents a significant commitment of time and effort, but is essential to predicting the performance characteristics of heterogeneous and homogeneous catalysts and determining radical and ionic reaction mechanisms.

Mass- and Heat-Transport. Phenomenological correlations describing the interrelated effects of fluid dynamic and geometric parameters on transport rate are commonly used to estimate mass- and heat-transport rates. Such generalized correlations have been developed from results obtained in a variety of systems.

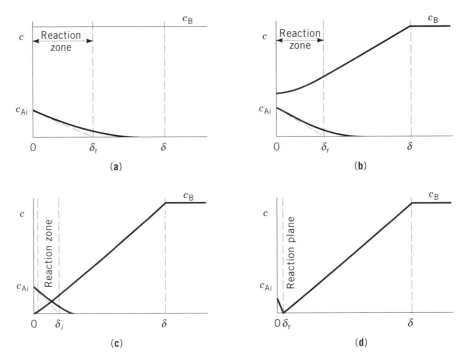

Fig. 6. Reaction zones for a first-order, fast irreversible homogeneous reaction, in reactants A and B with $Ha > 2$ and (**a**) $< D_B c_B / v_B D_A c_{Ai}$; (**b**) $\sim D_B c_B / v_B D_A c_{Ai}$; (**c**) $5 \times D_B c_B / v_B D_A c_{Ai}$; and (**d**) $10 \times D_B c_B / v_B D_A c_{Ai}$ (instantaneous) where c_{Ai} and c_B are bulk concentration of reactants A and B, and v_B is the kinematic viscosity of B, m^2/s. The initial slope of the curve (−−−), intercepted at δ_i, gives the size of the reaction zone. For component A reacting with B, δ is the end where B is fed.

These are based on gradients linear, across the resistance paths (film theory), or nonlinear because of insufficient time for reaching steady state (penetration theory). For example, in a vapor-phase fixed-bed reactor, the rate of mass transport from flowing reactant into a catalyst pellet can be determined from the correlations relating the mass-transfer factor, J_D; $\kappa \rho / G (Sc)^{2/3}$, where κ = mass-transfer coefficient, ρ = gas density, and Sc = the Schmidt number; and $\mu \rho D_v$, where D_v = molecular diffusivity to catalyst void volume ϵ and Reynolds number $D_p G / \mu$, where D_p = pellet diameter, G = mass velocity, and μ = viscosity (18). Equivalent relationships are available for heat transport. Mathematical methods for analyzing mass- and heat-transport phenomena in reacting systems are available (19). Because the correlations are highly dependent on and specific to the nature of the phases in the reactor, care is necessary to assure selecting only those correlations appropriate to the given system. Correlations based on air–water systems are particularly suspect when applied to nonaqueous systems.

Complex Flow Behavior

The concepts of well-mixed and plug-flow become inadequate when flow patterns deviate significantly from ideal behavior. Adsorption and subsequent desorption from within catalyst pellets in a trickle bed, solids recirculating downward to compensate for those entrained into rising bubbles in a fluid bed, and liquid held in the wake of a solid slurry exemplify back-mixing or bypassing in mixed-phase tubular reactors. Practical considerations, such as rapid mixing requirements, heating and reaction quenching demands, variations in flows through inlet distributors, the presence of internal baffles or structural obstructions, and inadequacies in locations of tank inlets and outlets, can result in substantial deviations from idealized flow behavior.

Residence Time Distributions. A range of residence times results from departures from ideal behavior. Residence time can be coupled with reaction kinetics and used in assessing reactor performance. The techniques for measuring residence times generally are used to define residence-time distributions (RTDs) of flowing and confined circulating elements. The RTD in a flow reactor can be determined by analyzing the effluent pattern resulting from a step function or pulsed change in an inlet flow property, eg, the concentration of one component or of an added tracer. Periodic functions, eg, sine waves, can also be used to measure time-averaged amplitude ratios and phase shifts. Thus, they provide inherently higher accuracy but require complex experimental technique. Comparable methods are applicable to analyze nonuniform batch reactors, but these require some means for introducing the tracer transient and for sampling or monitoring the resultant response.

Experimental Methodology. Perfect pulses and step functions are difficult experimental objectives. Pulses cannot be instantaneously injected nor step functions sharply outlined, and sinusoidal responses may take long times to reach steady states. Allowances are required for the solubility of gas tracers in adjacent liquid phases and for liquid tracer adsorption on solids and vessel walls. Difficulties exist in accurate determination of these concentrations and determination and extrapolation of concentrations in the long-tail portion of a distribution where concentrations are low. An effective approach to reduce the contribution of the tail is to use the corresponding Laplace transform integrals. The ready availability of high speed computers permits multiple least-square curve fitting of the data using a range of values for the arbitrary constant s in the Laplace transform integrals (20).

Figure 7 shows responses of outlet concentrations to step and pulse changes in inlet concentrations for various well-defined flow systems. For plug-flow, an inert tracer is not observed until a time that is equal to the mean residence time in the vessel has elapsed. With complete mixing, a response to a step change shows immediately and rises progressively; a tracer also responds instantaneously but then diminishes. Intermediate distributions are obtained for plug-flow with superimposed longitudinal mixing, as occurs with turbulent flow in an empty pipe or single-phase flow through a fixed-bed reactor. Laminar flow in pipes is also an intermediate distribution; a tracer is first observed in the effluent at half the mean residence time. When flow is restricted to part of the reactor vessel and stagnant fluid or recirculating eddies occupy the remainder, some tracer fluid does not appear in the effluent in a reasonable amount of time.

Interpretation of Residence Time Distributions. The residence time distributions for these models can be expressed mathematically (1–4). Similarly, other RTDs can be predicted for other flow models or combinations of models and many of these are also available. Response curves are not necessarily unique to

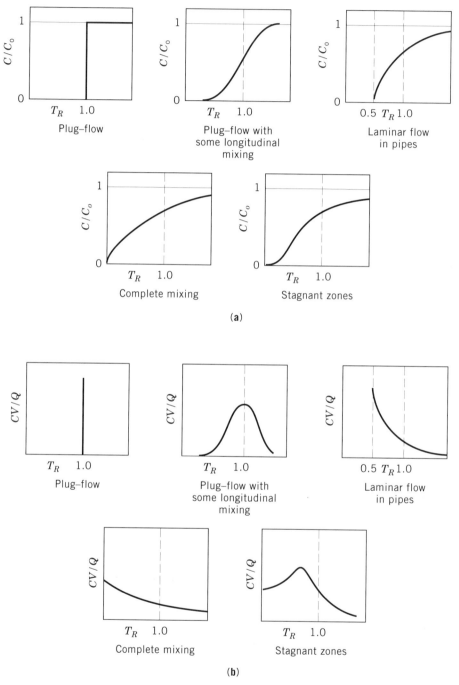

Fig. 7. Residence time distributions where U = velocity, V = reactor volume, t = time, $T_R = Ut/V$, C/C_o = tracer concentration to initial concentration, and Q = reactor volume: (**a**) output responses to step changes; (**b**) output responses to pulse inputs.

1020

specific flow patterns. Consequently, their validity depends on the degree to which the selected model adequately represents the effects of changing physical properties, operating variables, geometry, or size of the unit. On the other hand, these distributions can yield considerable insight in problem situations, when the responses do not meet expectations. The concept of residence time distribution is applicable to more than fluids. Varying catalyst deactivation in a reactor, as the result of loss of catalyst surface area by coking or other aging mechanisms, can be evaluated with particle population-balance models. Other applications include monitoring particle breakup and agglomeration, drop coalescence, other mixing phenomena, and biochemical and crystal growth (1,2).

Interpretation of the resultant response curves is accomplished in a manner somewhat analogous to determination of reaction orders. Combinations of plug-flow and well-mixed reactors of various size fractions of the total reactor vessel are arranged in series and parallel combinations. The number of possible combinations can exceed the number of variables used to differentiate independently among them, so that model selection depends on judgment. Once a model or combination of models is selected for further consideration, the quantitative expressions describing the models can be coupled with reaction kinetics into equations for predicting product yields. Many of the often used model combinations have been described (1–3,21). Figure 8 shows various reactor models.

For zero- and first-order reactions, the combining sequence has no effect on predicted conversion. This is not the case for higher or intermediate-order re-

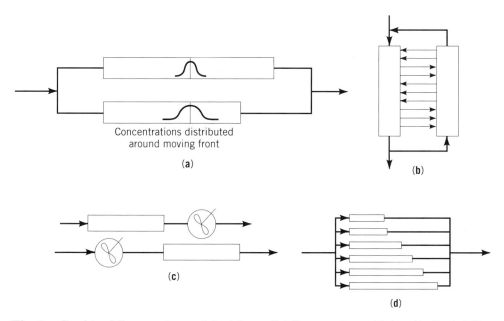

Fig. 8.   Combined flow reactor models: (**a**) parallel flow reactors with longitudinal diffusion (diffusivities can differ), (**b**) internal recycle–cross-flow reactor (the recycle can be in either direction), comprising two countercurrent plug-flow reactors with interconnecting distributed flows, (**c**) plug-flow and well-mixed reactors in series, and (**d**) zero-intermixing model, in which plug-flow reactors are parallel and a distribution of residence times duplicates that of a well-mixed reactor.

actions. Residence time distributions provide insight into the macromixed state of the system but provide no information about the mixing sequence, ie, the localized micromixed states within the reactor. The model combinations depicted in Figures 8c and 8d describe reactors that are well macromixed but not micromixed, or do not necessarily provide mixing sufficiently early in the process. In one case of a second-order reaction of premixed reactants and with well-stirred and plug-flow reactor volumes in a 4:1 ratio, the following conversions were calculated: plug-flow followed by complete mixing, 75%; complete mixing followed by plug-flow, 72%; and zero intermixing, 77%. These differences may represent either unacceptably high yield losses or the need to double the reactor volume (22). Despite these limitations, such models have been useful in developing, improving, and troubleshooting heterogeneous catalytic and noncatalytic thermal reactors, ion-exchange and chromatographic systems, and polymerization, leaching, and crystallization operations. More details on residence time distribution (RTD) methodologies and their application to chemical reactors are available (23) and additional discussion of micromixing and macromixing are provided (24,25).

Cold-Flow Models.   Use of tracers to obtain RTD characteristics followed by inferential analysis of flow patterns establishes conceptual but not actual detailed flow characteristics of a system. This limitation can be removed for reactions that are zero- or first-order or that can be so treated, eg, pseudo first-order reactions, with the aid of cold-flow models. These are replicas or portions of reactors that operate at ambient conditions and are designed for detailed observation of flow patterns. Often the model is constructed in part of glass or plastic to permit photographing phenomena. The use of cold-flow models offers many advantages, including potential for greater understanding, faster data collection because steady operation and accurate measurements are more easily obtained, and a greater range of geometries and flow conditions that can be studied. The development times should be shorter and research and development costs reduced as a result of adding this approach.

There are numerous applications of cold-flow models to all reactor types. Cold-flow models are used in studying hydrodynamic and kinetic fundamentals, extending reactor concepts, developing hardware, and improving operations. Fundamental studies include determining macro- and micromixing characteristics and contacting efficiencies, and establishing the effects of axial- and radial-flow distributions on reaction yields. Reactor studies embrace establishing limiting conditions of flow instabilities and extending operating ranges for acceptable performance, and providing generalized correlations for mixed-phase flow pressure drops and holdups and selecting optimum feed injection and product removal locations. Hardware development includes improving inlet distributor and quench hardware designs, defining impeller and tank baffle configurations, and determining mixing capabilities of jet nozzles. Operations improvement includes eliminating adverse flow distributions that promote erosion, attrition, misting, fouling, hot spotting, and temperature runaway.

Studies of individual bubbles rising in a two-dimensional gas–liquid–solid reactor provide detailed representations of bubble-wake interactions and projections of their impact on performance (Fig. 9). The details of flow, in this case bubble shapes, associated wake structures, and resultant bubble rise velocities and wake dynamics are important in characterizing reactor performance (26).

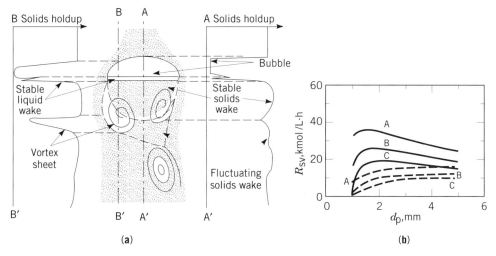

Fig. 9. Bubble-wake interactions in a gas–liquid–solid reactor: (**a**) solids concentration profile within bubble-wake domain, where A–A′ and B–B′ represent planes through the bubble, vortex, and wake; (**b**) projected impact of interactions on reaction rate as function of particle size and liquid velocity, where (—) is with wake effects and (– – –) is without, d_p is particle size, R_{SV} is reactor rate per unit volume of reactor (space velocity), and A = 3.5, B = 4.5, and C = 5.5 in cm/s (26).

The cold models are often large enough that little scaling is required. Fluids with known physical properties that are as similar as possible to those of the real system are best used to minimize fluid dynamic scaling. Because large cold-flow models require large quantities of feed, possibly in excess of what is reasonable for a pilot-plant operation, one approach to reducing these requirements without sacrificing the scale of the experiment is to use a reactor slice. In effect, a two-dimensional representation is employed. Such a unit was used, by immersing large-diameter tubes in a limestone fluid bed, to determine the effects on fluidization performance and heat-transfer characteristics (27). This represented the first phase of the use of coal-fired fluid-bed technology for refinery and petrochemical plant process heating. The transparent test unit was 2.3 m wide by 3.7 m high by 0.3 m deep. Horizontal bundles of tubes up to 15 cm inner diameter (ID) were used in experiments involving 2 t of limestone and ca 5 m^3/s air at 170 kPa (10,000 scf/min at 10 psig). Mixing characteristics were determined by photographs of color-tagged limestone as often as every second at three locations within the bed. Radial heat transfer from each tube was determined with electrically heated thin nichrome strips, which were embedded in the tubes, and with specially constructed thermocouple probes. The conditions for acceptable and impaired solids movement and heat transport within the system were established as functions of geometric configurations and operating variables.

Radial density gradients in FCC and other large-diameter pneumatic transfer risers reflect gas–solid maldistributions and reduce product yields. Cold-flow units are used to measure the transverse catalyst profiles as functions of gas velocity, catalyst flux, and inlet design. Impacts of measured flow distributions have been evaluated using a simple four lump kinetic model and assuming dispersed catalyst clusters where all the reactions are assumed to occur coupled with a continuous gas phase. A 3 wt % conversion advantage is determined for injection feed around the riser circumference as compared with an axial injection design (28).

Computer Simulation

Computer simulation of the reactor kinetic hydrodynamic and transport characteristics reduces dependence on phenomenological representations and idealized models and provides visual representations of reactor performance. Modern quantitative representations of laminar and turbulent flows are combined with finite difference algorithms and other advanced mathematical methods to solve coupled nonlinear differential equations. The speed and reduced cost of computation, and the increased cost of laboratory experimentation, make the former increasingly useful.

Computational Fluid Dynamics. Detailed quantitative predictions of free, wall, and recirculating flows, both turbulent and laminar, are possible in representations of fluctuations and other transient behavior. The decay patterns of three-dimensional jets from rectangular orifices have been modeled according to a finite-difference procedure, in which local turbulent energy and its dissipation are characterized by the $k-\epsilon$ model, where k is kinetic energy per unit mass of mean velocity field and ϵ is the dissipation rate (29). Flow separation caused by obstacles or other recirculating flows have also been simulated. The $k-\epsilon$ model can provide acceptable results for many situations and numerous industrial computational fluid dynamic (CFD) software packages are available. Broadened multiphase reactor applicability is foreseen with the growing understanding of turbulent energy transport mechanisms across phase boundaries and the introduction of new methodologies for simulating fluid interactions with physical boundaries and interfaces, eg, splitting the turbulence spectrum into multiple parts.

Validation and Application. Validated CFD examples are emerging (30) as are examples of limitations and misapplications (31). Realism depends on the adequacy of the physical and chemical representations, the scale of resolution for the application, numerical accuracy of the solution algorithms, and skills applied in execution. Data are available on performance characteristics of industrial furnaces and gas turbines; systems operating with turbulent diffusion flames have been studied for simple two-dimensional geometries and selected conditions (32). Turbulent diffusion flames are produced when fuel and air are injected separately into the reactor. Second-order and infinitely fast reactions coupled with mixing have been analyzed with the $k-\epsilon$ model to describe the macromixing process.

Various advanced two-dimensional models being developed show promise of predicting bubble dynamics (Fig. 10), wall heat-transfer rates, and combustion product distributions (33). A three-dimensional $k-\epsilon$ model has been formulated for fully baffled mixing tanks in which periodic flows produced by moving blades are time averaged (34). The model simulates both radial flow and axial recalculation patterns produced in these tanks (Fig. 11). The performance

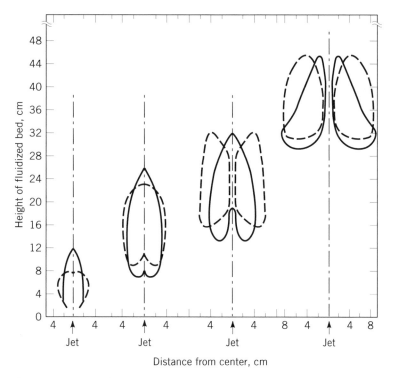

Fig. 10. Formation, growth, and splitting of bubbles in gas-fluidized beds, where (—) is theory ($\epsilon > 0.8$) and (---) is experiment.

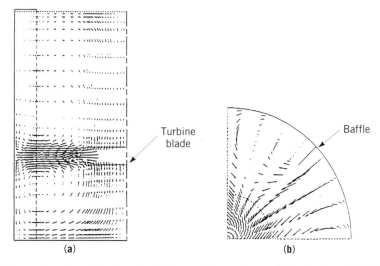

Fig. 11. Computer-simulated recirculating patterns in a mixing tank with full baffles: (**a**) elevation view shows circulation patterns generated by turbine blades; (**b**) plane view shows the effect of the baffle on the radial velocity vectors above the turbine blades.

with complex multiple reactions has been modeled by including micromixing through the addition of a coalescence–dispersion (c–d) model (35).

Axial and transverse velocity profiles and temperature distributions have been numerically determined for a vapor-phase epitaxy reactor, ie, a tubular reactor of rectangular cross section where chemical vapor deposition (CVD) is used to produce high purity crystals and dielectric films, and to treat material surfaces (36). Induction heating to 700–1200°C of a susceptor, eg, a carbon deposit placed on the bottom of the channel, generates a vigorous but laminar three-dimensional free convective flow pattern. Calculated spiral flow patterns and temperature distributions show good agreement with experimental measurements in a reactor 2.6 × 4.3 × 170 cm (Fig. 12). Calculated reactant concentrations also agree with measurements for experiments in which the induction-heated carbon deposit is burned and the local concentrations of carbon dioxide are measured. Growth and doping of GaAs from trimethylgallium and arsine by CVD has been simulated with particular attention to the distribution of impurities; kinetic rate constants are estimated using transition-state theory (37). Predicting film thickness and composition uniformity requires accurate simulation of the elementary gas-phase and surface reactions, thermal radiation, and the two- and three-dimensional mixed natural and forced convection patterns for the reaction and hydrodynamic interactions, particularly near the inlet (38).

Nonintrusive Instrumentation.   Essential to quantitatively enlarging fundamental descriptions of flow patterns and flow regimes are localized nonintru-

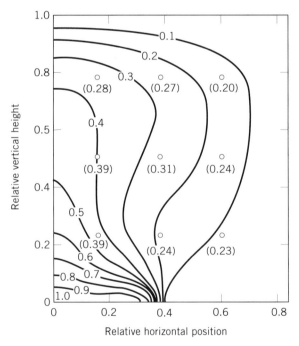

Fig. 12. Calculated reduced temperature, T_R, distributions for an epitaxial reactor cross section compared with experimental data (o). Gr (Grashof number) = 1.5×10^4, $L/H = 0.827$, Pr (Prandtl number) = 0.69; Re (Reynolds number) = 12.8 (29).

sive measurements. Early investigators used time-averaged pressure traverses for holdups, and pilot tubes for velocity measurements. In the 1990s investigators use laser-Doppler and hot film anemometers, conductivity probes, and optical fibers to capture time-averaged turbulent fluctuations (39).

Measurements with laser-Doppler and hot film techniques reveal the helical upward flow in the central region of bubble column and slurry bubble column reactors, the downward region near the wall, and an anisotropic turbulent region in the transition zone (40). Figure 13 shows these flow patterns and the split-film probe data processing system employed. A computer-aided radioactive particle tracking system (CARPT) is used in analyzing solids circulation in fluid beds (41). X-ray traverses and radioactive tracers (qv) have been used for radial- and axial-averaged phase concentrations in both pilot and large commercial units (42).

Hydrodynamics and Chemical Kinetics. Capabilities for specifying arrays of reactions and using these representations in analyzing reactor behavior have emerged from the union of high speed computing with modern analytical chemistry methods. Kinetic models generally contain minimal descriptions of hydrodynamics, usually as idealized reactors; similarly hydrodynamic models usually contain over-simplified kinetic descriptions. Whereas many reactants and products, particularly inorganic and specialty organic compounds, are readily processed as individual species, hydrocarbon mixtures, such as those derived from petroleum, must be divided into lumped groups and treated mathematically as invariant species (43,44).

Lumped species have evolved from utilizing distillation or chromatographic fractions for predicting FCC gasoline yields (43), to employing structural groups

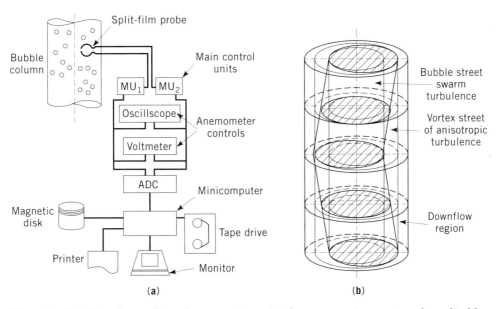

Fig. 13. Bubble column flow characteristics: (**a**) data processing system for split-film probe used to determine flow characteristics, where ADC = automated data center; (**b**) schematic representation of primary flow patterns.

based on nmr spectra and elemental analysis for catalytic hydrotreating to remove sulfur and nitrogen (45), then to working with Monte Carlo (integration methodology) (46) assemblies of 10,000 feed molecules and their reaction fragments divided among reactive moieties (eg, alkylaromatics, alkylnaphthalenes, and olefins) and using such techniques as nmr, elemental analysis, and molecular weight determinations for the pyrolysis of a Hondo (Mexico) crude asphaltene; total conversion and product selectivity have been predicted to within a few percent (47).

Laboratory Reactors

Fundamental reaction kinetics and chemical and physical properties, where necessary, are best obtained with experimental reactors that are different from those used during process development or as part of commercial operations. It is important that the reactor be similar to one of the basic types, have known flow patterns, and in the case of multiphase reactors, have known flow regimes, operate isothermally, and provide data over wide ranges of variables. The range of the conditions to be studied must be sufficiently large for adequately defining the effects of variables, but the rate equations need not accurately reflect true reaction mechanisms. However, extracting the intrinsic chemical kinetics leads to a better understanding of the process and more assured extrapolations of results to conditions that have not been studied. Recommended experimental reactor types are shown in Table 1. The best selection depends on careful analysis of the anticipated reactions, specific chemical and physical properties of the system, and appraisals of the objectives of the study (49).

Well-Stirred Batch and Continuous Reactors. Agitated batch and continuous fed vessels are particularly useful for liquid and predominantly liquid multiphase systems. These are especially convenient for low pressure and modest temperature operations. Well-mixed batch reactors are preferred for slow reactions, because compositions can be accurately measured with time. The

Table 1. Experimental Reactors[a]

Reaction mixture	Reactor	Remarks
gas	tubular	integral or differential
liquid	batch	slow reactions
	continuous-flow stirred-tank	moderately fast reactions
	tubular	fast reactions
solid	batch fluidized bed	inert gas or liquid for fluidization
gas–liquid	gas–liquid contactor[b]	slow reactions
	laminar film or jet	fast reactions
liquid–liquid	batch	good agitation
	continuous-flow stirred-tank	no settling of dispersed phase
	film	fast reactions
gas–solid	tubular	integral or differential with fixed bed
liquid–solid	batch	good dispersion of solids
	tubular	fixed or fluidized bed

[a]Ref. 48.
[b]Interfacial area must be known.

continuous-flow stirred-tank reactor (CSTR) is better for fast reactions because conversion and composition at a given condition is well defined. Interfacial areas in liquid–liquid and liquid–gas systems are not well defined and may need to be established by some independent means or by a different type of reactor; liquid films or laminar liquid jets, because of their known interfacial areas, are useful for this purpose. Mechanical stirring rather than shaking or rocking should be used to produce well-defined mixing characteristics. Flat or dish-bottomed vertical cylindrical tanks containing four equally spaced baffles which are vertically mounted on the circumference are preferred. These baffles, with widths of one-tenth the vessel diameter, prevent swirling and vortexing and produce good mixing with either marine or flat-blade turbines. All experimental stirred reactors should be equipped with a variable-speed drive to ensure sufficient agitation.

Tubular Reactors. The tubular reactor is excellent for obtaining data for fast thermal or catalytic reactions, especially for gaseous feeds. With sufficient volume or catalyst, high conversions, as would take place in a large-scale unit, are obtained; conversion represents the integral value of reaction over the length of the tube. Short tubes or pancake-shaped beds are used as differential reactors to obtain instantaneous reaction rates, which can be computed directly because composition changes can be treated as differential amounts. Initial reaction rates are obtained with a fresh feed. Reaction rates at higher conversions can be established either by using synthetic feeds, if the intermediates can be accurately specified, or by using a prereactor to supply feed of the correct concentration. In rare cases, recirculation around the differential reactor might be analyzed to follow the course of a reaction.

Maintenance of isothermal conditions requires special care. Temperature differences should be minimized and heat-transfer coefficients and surface areas maximized. Electric heaters, steam jackets, or molten salt baths are often used for such purposes. Separate heating or cooling circuits and controls are used with inlet and outlet lines to minimize end effects. Pressure or thermal transients can result in longer lived transients in the individual catalyst pellets, because concentration and temperature gradients within catalyst pores adjust slowly.

Catalyst Effectiveness. Even at steady-state, isothermal conditions, consideration must be given to the possible loss in catalyst activity resulting from gradients. The loss is usually calculated based on the effectiveness factor, which is the diffusion-limited reaction rate within catalyst pores divided by the reaction rate at catalyst surface conditions (50). The effectiveness factor E, in turn, is related to the Thiele modulus, ϕ $(R_p ka/D_e)^{-1}$, where R_p is an effective pellet radius, k the first-order rate constant, a the internal surface area, and D_e the effective diffusivity. It is desirable for E to be as close as possible to its maximum value of unity. Various formulas have been developed for E, which are particularly useful for analyzing reactors that are potentially subject to thermal instabilities, such as hot spots and temperature runaways (1,48,51).

Model Reactions. Independent measurements of interfacial areas are difficult to obtain in liquid–gas, liquid–liquid, and liquid–solid–gas systems. Correlations developed from studies of nonreacting systems may be satisfactory. Comparisons of reaction rates in reactors of known small interfacial areas, such as falling-film reactors, with the reaction rates in reactors of large but undefined areas can provide an effective measure of such surface areas. Another method is substitution of a model reaction whose kinetics are well established and where the physical and chemical properties of reactants are similar and limiting mechanisms are comparable. The main advantage of employing a model reaction is the use of easily processed reactants, less severe operating conditions, and simpler equipment.

Table 2. Laboratory Models for Industrial Reactors[a]

Industrial reactor		Laboratory model		
Type	Contact times, s	Model	Contact times, s	Interfacial area/vol, cm^{-1}
packed bed	0.03–1.0	falling film or	0.1–2	20–70
		stirred tank	0.06–10	0.002–0.54
venturi	0.004	laminar jet	0.001–0.1	20–80
spray column	0.03–0.1	laminar jet or	0.001–0.1	20–80
bubble column	0.01	rotating roller	0.002–0.1	500–1250
CSTR	0.01			

[a]Ref. 53.

A model reaction, charcoal-catalyzed decomposition of hydrogen peroxide to water and oxygen, has been used in establishing the efficiency of catalyst utilization in mixed-phase, downflow, fixed-bed reactors operating as trickle beds (52). Another technique employed for reactions occurring at near ambient conditions utilizes a laboratory-scale reactor and reactants with known interfacial properties (53). For example, a string of catalyst spheres could represent a falling-film reactor for methylstyrene hydrogenation or a falling film for the packed bed in propylene oxidation (Tables 2 and 3).

Table 3. Reactor Applications and Laboratory Models[a]

Reaction	Industrial reactor	Laboratory model
propene oxidation	packed bed	falling film
oxygen absorption		
benzene sulfonation	packed bed	stirred tank
propene oxidation		
Claus sulfur removal		
selective H$_2$S removal		
chlorine absorption	venturi scrubber	laminar jet
methylstyrene hydrogenation[b]	falling film	string of catalyst spheres

[a]Ref. 53.
[b]Mixture of isomers.

Calorimetry. Many processes, particularly in the manufacture of fine chemicals, specialty polymers, and pharmaceuticals, are progressed with insufficient information on molecular structures, by-products, or kinetics. State-of-the-art well-stirred calorimeters are available expressly designed for batch or continuous reactors to establish the potential for worst-case temperature excursions and permit rational specification of relief systems (54–56). This is important because many of the processes include exothermic reactions capable of temperature runaway, though at times disguised by overall reactions appearing thermally benign. In the manufacture of an osteoporosis drug, the determination that significant decomposition would begin within 25°C of the intended operating conditions resulted in substituting a CSTR for a batch reactor. This allowed the advantages of reduced reactant inventory, self-thermal regulation, and lessened relief and containment requirements at the expense of a minimal reduction in product yields (57).

Scale-Up

The objective of scale-up is to select designs and operating conditions so that similar responses, comparable yields, and product distributions are obtained in the different sized units. Scaling from laboratory units or pilot plants to industrial-sized reactors utilizes similar methodologies to those used in extrapolating to new situations for existing reactor operations, but with greater uncertainty and complexity. An evolving trend is the use of scale-independent unified models which incorporate reaction kinetics, thermodynamics, hydrodynamics, and hardware representations despite the source (42). The maximum advantages lie in assessing reaction paths and complex flow behavior and transport, and using computer simulation in projecting reactor performance. Effective scale-up requires disciplined assessment of micro-, meso-, and macro-responses of reactor configurations relative to process requirements.

Scale-Up Principles. Key factors affecting scale-up of reactor performance are nature of reaction zones, specific reaction rates, and mass- and heat-transport rates to and from reaction sites. Where considerable uncertainties exist or large quantities of products are needed for market evaluations, intermediate-sized demonstration units between pilot and industrial plants are useful. Matching overall fluid flow characteristics within the reactor might determine the operative criteria. Ideally, the smaller reactor acts as a volume segment of the larger one. Flow distributions are not markedly influenced by the wall, turbulent eddies are approximately the same size, and bubbles in bubble columns and fluid beds are the same size. Under these conditions, the residence time distributions of feeds, intermediates, and products should be comparable. In catalytic processes, catalyst particles should remain in the reactor for the same lengths of time.

In cases where a large reactor operates similarly to a CSTR, fluid dynamics sometimes can be established in a smaller reactor by external recycle of product. For example, the extent of solids back-mixing and liquid recirculation increases with reactor diameter in a gas–liquid–solids reactor. Consequently, if gas and liquid velocities are maintained constant when scaling and the same space velocities are used, then the smaller pilot unit should be of the same overall height. The net result is that the large-diameter reactor is well mixed and no temperature gradients occur even with a highly exothermic reaction. The smaller reactor approaches plug-flow behavior and exhibits a large temperature gradient. In this case, external recycle provides the same degree of back-mixing as is provided by internal circulation in the larger diameter reactor.

Dimensional Analysis. Dimensional analysis can be helpful in analyzing reactor performance and developing scale-up criteria. Seven dimensionless

Table 4. Dimensionless Groups in Chemical Reaction Systems

Name	Formula[a]	Proportional to
Arrhenius group	$\dfrac{\Delta E}{RT}$	$\dfrac{\text{activation energy}}{\text{potential energy of fluid}}$
Damköhler group I	$\dfrac{rL}{UC}$	$\dfrac{\text{chemical reaction rate}}{\text{bulk mass flow rate}}$
Damköhler group II	$\dfrac{rL^2}{D_v C}$	$\dfrac{\text{chemical reaction rate}}{\text{molecular diffusion rate}}$
Damköhler group III	$\dfrac{QrL}{C_p \rho U \Delta T}$	$\dfrac{\text{heat liberated}}{\text{bulk transport of heat}}$
Damköhler group IV	$\dfrac{QrL^2}{k \Delta T}$	$\dfrac{\text{heat liberated}}{\text{conductive heat transfer}}$
Reynolds number, Re	$\dfrac{LU\rho}{\mu}$	$\dfrac{\text{inertial force}}{\text{viscous force}}$
Thuring's radiation group	$\dfrac{\rho C_p U}{\epsilon S T^3}$	$\dfrac{\text{bulk transport of heat}}{\text{radiant heat transfer}}$

[a] ΔE = activation energy, J/mol; R = gas constant, J/(mol·K); T = absolute temperature, K; r = reaction rate, mol/(cm^3·s); L = characteristic length, m; U = fluid velocity, m/s; C = concentration, mol/cm^3; D_v = molecular diffusivity, cm^2/s; Q = heat generated, J/mol; C_p = specific heat, J/(g·K); ρ = density, g/cm^3; ΔT = temperature rise, K; k = thermal conductivity, W/(m·K); μ = viscosity, g/(m·s); ϵ = emissivity of surface, dimensionless; and S = Stefan-Boltzmann constant, W/(m^2·K^4).

groups used in generalized rate equations for continuous flow reaction systems are listed in Table 4. Other dimensionless groups apply in specific situations (58–61). Compromising assumptions are often necessary, and their validation must be established experimentally or by analogy to previously studied systems.

Improving batch manufacturing processes often employs dimensional analysis for assessing process requirements for suspending solids and catalysts, particle dissolution, reduction, and growth, and for enhancing heat transfer. Power requirements and pumping capacities for mechanically agitated reactors are related to viscous, gravitational, interfacial, and inertial forces through the Reynolds, Froude, and Weber dimensionless numbers, respectively, as defined in Table 5. Scale-up similitude of such well-mixed reactors requires maintaining constant power and pump capacity per unit volume, impeller tip speed, and Reynolds number. Because these conditions cannot all be maintained simultaneously and allowances must be made for non-Newtonian fluid behavior, constant

Table 5. Representative Dimensionless Groups for Agitated Reactors

Dimensionless group	Formula[a]	Important function
Weber, We	$\rho N^2 D^3/\sigma$	emulsification
Reynolds, Re	$\rho N D^2/\mu$	laminar, turbulent flows
Froude, Fr	$N^2 D/g$	free surfaces, vortices
Prandtl, Pr	$C_p \mu/k$	heat transfer

[a] N = agitator speed, s^{-1}; D = impeller diameter, cm; C_p = specific heat, J/(g·°C)[b]; k = thermal conductivity, J/(s·cm·°C); ρ = density, g/cm^3; μ = viscosity, g/(m·s); and σ = surface tension, g/s^2.
[b] To convert J to cal, divide by 4.184.

power per unit volume (P/V), tip speed, or a judicial compromise between these is used. These scaling factors do not account for such phenomena as shear dependence of particle integrity or additions of chemical stabilizers. In such cases, data from different reactor sizes and geometries extending over a substantial range of operating variables are needed for scale-up and extrapolation to unfamiliar conditions.

Fixed and Fluid Beds Compared. To understand scale-up principals, consideration of differences in approach used in scaling fixed- and fluid-bed reactors is useful. The markedly different fluidization characteristics in different-sized gas–fluid-bed units dictate process development based on a series of pilot and demonstration unit sizes. This is in marked contrast to fixed-bed process developments. Various fixed-bed hydrodesulfurization, hydrocracking, and catalytic reforming processes have been scaled based on data from pilot units with reactors only 13–44 mm dia; the industrial reactors have diameters of 2.5–9 m. In contrast, 300–600-mm dia vessels have been used in low conversion, fluid catalytic cracking pilot units; 380-mm and 2.4-m dia vessels have been used for high conversion fluid-bed catalytic reforming; vessels 300 mm and 2.4 m dia have been employed for a multiple fluid-bed iron ore reduction process development; and 75-mm and 2.9-m dia reactors have been used for a high temperature coke gasification process. Capacity scale-up factors of only one or two orders of magnitude have been employed in scaling to the 7.5–17-m dia vessels used industrially in these fluid-bed processes. For comparison, capacity scale-up factors for the fixed-bed developments exceeded 50,000. Typical scale-up factors employed in developing a range of past processes are given in Table 6 (62).

Fixed-Bed Scale-Up. Paramount with fixed-bed reactors, particularly with highly exothermic reactions or when expensive catalysts are used, is efficient catalyst utilization. Distributors and redistributors, when multiple beds are required for heat addition or removal, must not channel or segregate reactants and uniform flows must be maintained for the entire length of the reactors.

Table 6. Laboratory Unit, Pilot-Plant, and Commercial Scale-Up Ratios

		Scale of operation, kg/h		Scale-up ratio	
System	Applications	Laboratory	Pilot-plant	Lab to pilot-plant	Pilot-plant to commercial
gaseous	ammonia, methanol	0.01–0.10	10–100	500–1000	200–1500
gaseous reactants, liquid or solid products	sulfuric acid, urea, maleic anhydride	0.01–0.2	10–100	200–500	100–500
liquid and gaseous reactants, liquid products	benzene chlorination, oxidation	0.01–0.2	1–30	100–500	100–500
liquid reactants, solid or viscous liquid products	polymerizations, agricultural chemicals	0.005–0.2	1–20	20–200	20–250
solid reactants, solid products	phosphoric acid, cement, ore smelting	0.10–1.0	10–100	10–100	10–200

For trickle-bed reactors, this requires operating at high liquid mass velocities (1,63), preferably 3–8 kg/(m^2·s). Such feed rates are desirable and easily obtained in large industrial units, but a burden to pilot unit operations. Thus, despite the lower product yields and differing responses to operating variables compared with the larger industrial units, some process developments have been carried out at one-tenth of these rates in order to handle smaller quantities of reactants and reduce the overall height of the pilot units. However, more demanding process developments with trickle beds are best scaled in high mass velocity reactors.

Figure 14 illustrates the complex fluid dynamics of trickle-bed reactors as they are positioned on a generalized flow regime map (Baker plot). Contacting between the catalyst and the dispersed liquid film covering the catalyst, and the film's resistance to gas transport into the catalyst, particularly with vapor generation within the catalyst, are not simple functions of liquid and gas velocities. Crucial to efficient contacting and gas transport are extent of partial wetting of the catalyst pellets, rates of liquid renewal of the catalyst surface, and fraction of catalyst pores filled with liquid, factors which are largely determined empirically (64).

The complexities are exemplified by the respective vapor-phase and mixed-phase catalytic hydrodesulfurization of naphtha and heavy gas oils over pelletized solid catalysts composed of cobalt molybdate on an alumina base. These pellets are highly porous with large internal surface areas. In the vapor-phase process, sulfur-laden naphtha molecules and hydrogen reach the reaction sites by diffusion into the pores from the pellet surface. After reaction on the interior pore surfaces, the desulfurized products and hydrogen sulfide reverse the process and diffuse from the catalyst pores. For mixed-phase catalytic hydrodesulfurization, the countercurrent diffusion of hydrogen and hydrogen sulfide between vapor

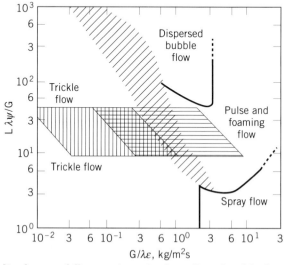

Fig. 14. Generalized map of flow regimes in downflow fixed-bed reactors using Baker plot parameters, $\lambda = (\delta_G/\delta_{air}\cdot\delta_L/\delta_W)^{0.5}$ and $\psi = \sigma_W/\sigma_L(\mu_L/\mu_W(\delta_W/\delta_L)^2)^{1/3}$, where ▤ represents pilot reactors; ▥, industrial reactors; and ▨, transition low–high interaction.

and liquid phases also is required. Capillary forces in the pores keep the catalyst wet and the active sites separate from the vapor. The exothermic heats of reaction must also pass from the reaction sites to the reaction media or to a heat-exchange surface. Heat generated at the reaction sites is conducted from the catalytic surface and the adsorbed reaction liquids to the pellet surfaces, then into the reacting fluids within the reactor. In a tubular fixed-bed reactor, heat conduction to the wall through the catalyst and fluids also is operative. The rates of these steps must be determined and compared with intrinsic reaction rates to establish overall catalyst effectiveness.

Fluid-Bed Scale-Up.   One difficulty in scaling up fluid-bed processes stems from the fact that fluid-bed behavior is as much a function of bed dimensions as of physical properties of the gas and solids. Thus, wall effects cannot be neglected and the various mechanisms that limit contacting must be taken into account. Scaling is possible when the effects of the important variables that influence the reactions are known and a means for predicting conditions can be specified. Particularly important to scaling is knowledge of the flow regimes governing gas–solids contacting. The various flow regimes, such as particulate fluidization, spouting, grid-produced slugging, and controlled and free bubbling, obey different physical laws. Slugging or slug-flow behavior occurs when the rising bubbles approach the size of the reactor diameter. In this case, scale-up requires catalyst content to vary proportionally to reactor diameter and superficial velocity to vary proportionally to its square root. Walls exert some effect on bubbles that are as small as one-tenth of the reactor's diameter. Careful attention must be paid to fundamental mechanisms and operating details described in the literature (65). All potentially operative factors must be evaluated and checked for consistency. A largely empirical scale-up methodology that has emerged from industrial practices (66) is available to assist in quantifying a scale-up strategy.

Analysis of a method of maximizing the usefulness of small pilot units in achieving similitude is described in Reference 67. The pilot unit should be designed to produce fully developed large bubbles or slugs as rapidly as possible above the inlet. Usually, the basic reaction conditions of feed composition, temperature, pressure, and catalyst activity are kept constant. Constant catalyst activity usually requires use of the same particle size distribution and therefore constant minimum fluidization velocity which is usually much less than the superficial gas velocity. Mass transport from the bubble by diffusion may be less than by convective exchange between the bubble and the surrounding emulsion phase.

These design fundamentals result in the requirement that space velocity, effective space–time, fraction of bubble gas exchanged with the emulsion gas, bubble residence time, bed expansion relative to settled bed height, and length-to-diameter ratio be held constant. Effective space–time, the product of bubble residence time and fraction of bubble gas exchanged, accounts for the reduction in gas residence time because of the rapid ascent of bubbles, and thereby for the lower conversions compared with a fixed bed with equal gas flow rates and catalyst weights.

These requirements are not all mutually compatible. In the case of noncatalytic reactions, bubble residence time may be the significant parameter. For rapid reactions, the fraction of gas exchanged between the rising gas bubbles and the continuous emulsion phase is the important factor. For a slow reaction with a large minimum fluidization velocity, space velocity is of primary concern. With slow reactions and poor contacting, effective space–time is most important as exemplified in a study of ozone reduction catalyzed by iron scale. Good agreement was found among the various fluid-bed reactors with diameters of 50, 200, and 760 mm, bed heights of 0.9–3 m, and gas velocities of 6–24 cm/s. Scale-up in this case requires catalyst inventory to vary proportionally to reactor diameter and superficial velocity to vary proportionally to its square root (68).

Additional Selection Factors. System characteristics and physical and chemical properties can be principal considerations in reactor selection. Catalyst size and strength, fouling and deactivation, and heat release and reactor stability exemplify the issues for catalytic processes.

Catalyst Size. Initial catalyst activity and catalyst life increase with decreasing size. Catalyst particles 20–40 μm in diameter can be used in fluidized reactors. Fixed beds require at least 1.5-mm diameter pellets, with smaller sizes precluded because of their high pressure drops and greater tendency to foul. Slurry and gas–liquid–solid fluidized-bed reactors can be operated with intermediate sized pellets. The higher activity and reduced rate of deactivation of a smaller catalyst may offset the larger reactor volumes required for bed expansion, which are inherent in these systems. If a single, fixed-bed reactor with 1.5-mm catalyst is considered as the basis of comparison, a multistage gas–liquid–solid bed with 0.8-mm catalyst requires an activity advantage of ca 200% to give a break-even reactor size; for a slurry containing 100-μm catalyst, ca 50% advantage is needed. For reactions controlled by diffusion of reactants into the pores of catalysts, such increases in activity and greater are achievable (69). Reactions are diffusion-limited when the characteristic diffusion time is larger than that of the chemical reaction. Reactions often are limited by diffusion when liquid viscosity is high, reactant molecules are large with low diffusivities, high catalyst interfacial areas exist in small pores, or reaction rates are fast. '

Catalyst Strength. A superior but physically weak catalyst may preclude the use of a fixed bed. Such a catalyst might be suitable for one of the other reactor configurations because catalyst strength can increase with decreasing size. For use in fixed beds, catalyst strengths of ca 3 kg per particle are desirable to prevent excessive breakage, particularly if pressure drop across the bed increases significantly with time. Lower strengths are required of catalysts that are used in gas–liquid–solid reactors. For a slurry reactor, catalyst attrition in auxiliary equipment, eg, pumps and hydroclones, can be limiting. Catalyst replacement cost must be considered for catalyst lost by attrition and from the inherent inefficiencies in containing inventory within a system. Inventory maintenance is much more complex in fluidized systems, particularly when catalyst properties change with time in the reactor.

Fouling and Deactivation. Fixed beds usually are favored if the catalyst deactivates in less than three months with economic factors dictating actual tolerable limits. If plugging is more rapid than deactivation but fouling only occurs in the first bed of a train, fixed-bed reactors might still be used because this bed is accessible and can be designed as a moving or guard bed. Similarly, if only a first bed deactivates rapidly, this configuration may be preferred. If a catalyst deactivates rapidly or plugging is widespread, the fluidized systems are preferred because the catalyst can be rapidly removed from the reactor, regenerated in other vessels, and returned.

Heat Release and Reactor Stability. Highly exothermic reactions, such as with phthalic anhydride manufacture or Fischer-Tropsch synthesis, compounded with the low thermal conductivity of catalyst pellets, make fixed-bed reactors vulnerable to temperature excursions and runaways. The larger fixed-bed reactors are more difficult to control and thus may limit the reactions to jacketed bundles of tubes with diameters under ~5 cm. The concerns may even be sufficiently large to favor the more complex but back-mixed slurry reactors.

Volumetric heat generation increases with temperature as a single or multiple S-shaped curves, whereas surface heat removal increases linearly. The shapes of these heat-generation curves and the slopes of the heat-removal lines depend on reaction kinetics, activation energies, reactant concentrations, flow rates, and the initial temperatures of reactants and coolants (70). The intersections of the heat-generation curves and heat-removal lines represent possible steady-state operations called stationary states (Fig. 15). Multiple stationary states are possible. Control is introduced to establish the desired steady-state operation, produce products at targeted rates, and provide safe start-up and shutdown. Control methods can affect overall performance by their way of adjusting temperature and concentration variations and upsets, and by the closeness to which critical variables are operated near their limits.

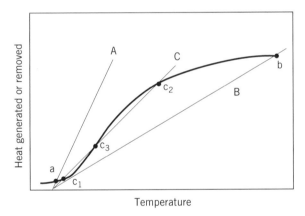

Fig. 15. Temperature vs heat generation or removal in establishing stationary states. The heavy line (—) shows the effect of reaction temperature on heat-generation rates for an exothermic first-order reaction. Curve A represents a high rate of heat removal resulting in the reactor operating at a low temperature with low conversion, ie, stationary state at a; B represents a low rate of heat removal and consequently both a high temperature and high conversion at its stationary state, b; and at intermediate heat removal rates, ie, C, multiple stationary states are attainable, c_1 and c_2. The stationary state at c_3 is unstable and obtainable only with close computer control.

Minimum Pilot-Plant Size

Economy of time and resources dictate using the smallest sized facility possible to assure that projected larger scale performance is within tolerable levels of risk and uncertainty. Minimum sizes of such laboratory and pilot units often are set by operability factors not directly involving internal reactor features. These include feed and product transfer line diameters, inventory control in feed and product separation systems, and preheat and temperature maintenance requirements. Most of these extraneous factors favor large units. Large industrial plants can be operated with high service factors for years, whereas it is not unusual for pilot units to operate at sustained conditions for only days or even hours.

Small Unit Characteristics. Small-diameter lines are prone to plugging and pressure drop buildup. Compensating with oversized lines, which reduce velocities, can be detrimental with mixed-phase flows. Reduced flow rates encourage solids deposition, impair heat transfer, and promote slug flow, and hence cause coking and fouling. Extraneous reactions occurring in feed and product inventories confound any data analysis. In smaller reactors, greater portions of the total system inventory tend to be heated or cooled and so are more likely to have thermal and catalytic reactions outside the reactor confines.

Small reactors usually require additional heat inputs to match the lower heat losses per unit of reactor inventory of larger industrial units. A large unit may require heat removal. The modes selected for redressing heat imbalances can affect results, particularly in systems of multiple reactions with different reaction orders, activation energies, or equilibria limits. Many different techniques are employed and all represent substantial compromise, including additional feed preheat or segmented heaters on the reactor. Alternatively, reactor variables, eg, temperature, pressure, catalyst activity, or feed compositions, can be modified to increase reaction rates of exothermic reactions. The resultant deviations may be substantial and may prevent the unit from operating as a fully integrated system; consequently, they may be the factors setting minimum reactor size.

Reactor Internals and Unit Hardware. Requirements for mixing feed components or separating products may determine minimum pilot unit size. If reactants cannot be premixed before they are passed into the reactor, the effectiveness of the inlet distributor in mixing the reactants can markedly affect reactor performance. This is especially true for gases, multiple phases, or liquid streams of greatly different kinematic viscosities. Distributors in industrial units typically have large numbers of injection points of quite diverse design characteristics, some of which are depicted in Figure 16 for fluidized-bed applications. Flow variations through these parallel paths can lead to poor flow distributions within a reactor, thus reducing product yields and selectivity. In some circumstances, undesirable side products can foul portions of the distributor and further upset flow patterns. Where this is important, or where the possibilities and consequences are insufficiently understood and independent means cannot be employed to assure adequate distribution, the pilot plant must be sized to accommodate such a distributor. Spacing should be comparable to those distributors that are anticipated to be used industrially.

Additional definition of the operative mechanisms can obviate the need for the larger unit. It may be possible to assess limitations in a smaller unit that has only a few injection points on the distributor. The unit could be used to evaluate distributor designs that permit a wide range of acceptable operating conditions. However, if the acceptable range proves smaller than desired, the larger pilot unit would then be needed to establish acceptable performance.

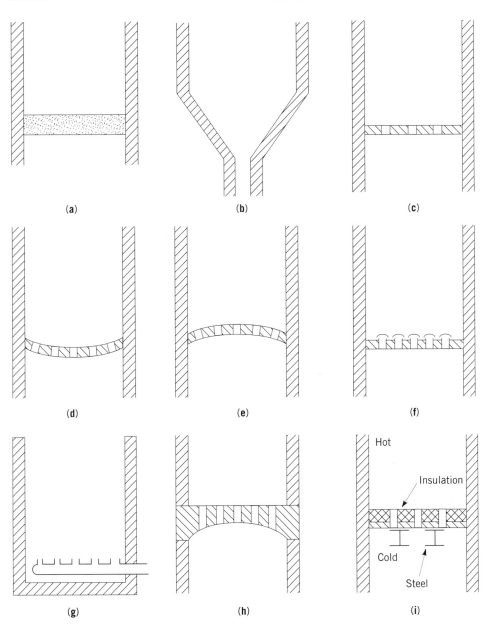

Fig. 16. Typical distributor grid designs for fluidized-bed applications: (**a**) porous plate, (**b**) cone, (**c**) perforated plate, (**d**) downward dish, (**e**) upward dish, (**f**) bubble caps, (**g**) pipe grid, (**h**) refractory arch, and (**i**) insulated grid.

Localized Reaction Phenomena. High temperature gasification of high ash content coals in a fluidized bed and the gas-phase production of linear low density polyethylene (LLDPE) are illustrative of the impacts of localized reaction phenomena. In high temperature coal gasification, particles may overheat and sinter. Successive contacting of coal with oxidizing and reducing atmospheres, a consequence of internal solids circulation, may alter particle surfaces and affect kinetics (71). Union Carbide's UNIPOL fluid bed requires conditions for preventing agglomeration of the polymer and controlling the particle sizes in the fluidized bed (72). Agglomeration occurs wherever the fluid-bed temperature, particularly near the gas distributor, reaches the polymer melting point. Reaction heat must be uniformly generated and removed even during start-up, shutdown, and other transient operating conditions. In addition, equilibrium particle size distributions must be established that balance the desire for low fluidizing gas rates and low concentrations of fine particles overhead against the need to maintain the high fines concentrations required for establishing fluid beds that are freely bubbling and the need to limit oversize particle production. For successful reactor performance, nonfouling operation must be demonstrated in a unit that is sufficiently large to include these influences.

Highly Integrated Processing. Integrated multiple process and multiple process functions within individual reactors are growing considerations and where employed must be taken into account when designing a pilot-plant program. These close couplings are driven by the economic and environmental requirements for thermally efficient utilization of raw materials, obtaining high yields of desired products, conserving air, water, and land resources, and minimizing health and safety related hazards.

One example is the interlaced effects of coal liquefaction, slurry preheating, and solvent hydrotreating on conversion and yield in the Exxon donor solvent (EDS) process (42). Crushed coal is contacted with molecular hydrogen and a hydrogen-donor diluent at ca 14–17 MPa (ca 140–170 atm) and 430–480°C in multiple upflow liquefaction reactors after being preheated in a two-phase (slurry–hydrogen) furnace. The dehydrogenated donor stream is regenerated in a conventional downflow, fixed-bed hydrotreater. In addition, a heavy bottoms fraction from vacuum distillation is processed in an integrated fluid-bed coking and gasification unit (Fig. 17). These process steps are highly interactive and lack of management results in a downward trend of increasingly lower product yields. The vicious cycle of downward spiraling performance and root causes that trigger such events, such as flow maldistributions, impaired heat transfer in the preheater, variations in feed reactivity, and off-spec hydrotreating catalyst activity, are shown in Figure 18 (73).

Industrial Reactor Development and Application

Most reactors have evolved from concentrated efforts focused on one type of reactor. Some processes have emerged from parallel developments using markedly different reactor types. In most cases, the reactor selected for laboratory study has become the reactor type used industrially because further development usually favors extending this technology. Descriptions of some industrially important petrochemical processes and their reactors are available (74–76). Following are illustrative examples of reactor usage, classified according to reactor type.

Batch Reactors. The batch reactor is frequently encountered in petrochemical, pharmaceutical, food, and mining processes, eg, alkylation, emulsion polymerization, hydrocarbon fermentation, glycerolysis of fats, ozonation, and metal chelation. The processes often require achieving uniform dispersions of micrometer-sized drops and providing adequate exothermic heat removal, es-

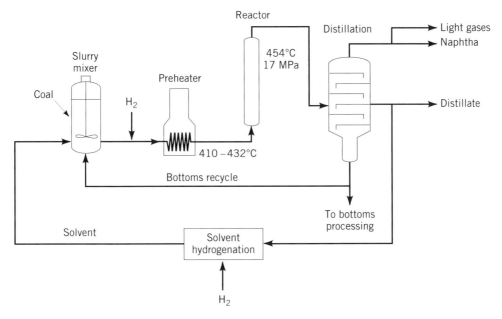

Fig. 17. Schematic diagram of the coal liquefaction section of the Exxon donor solvent (EDS) process. To convert MPa to psi, multiply by 145.

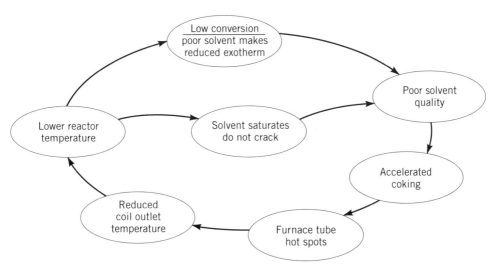

Fig. 18. Cycle of downward spiraling performance with unmanaged process interactions.

pecially where product quality is adversely affected by temperature. In the production of cobalt oleate from its acetate, an aqueous dispersion of cobalt hydroxide intermediate is contacted with an immiscible oleic acid–olefin mixture to form the oleate. Elimination of environmentally unacceptable nitrogen sparging practices and lower gummy sediment deposits have been achieved along with increased production and reduced power consumption through the installation of larger sized pitched blade turbines and substitution of partial baffles for full baffles (77).

Batch reactors are used in manufacturing plastic resins, eg, polyesters, phenolics, alkyds, urea–formaldehydes, acrylics, and furans. Such reactors generally are 6–40 m^3 (ca 200–1400 ft^3) baffled tanks, in which there are blades or impellers connected from above by long shafts, and heat is transferred either through jacketed walls or by internal coils. Finger-shaped baffles near the top are used instead of full-length baffles. Raw materials are held at temperatures of up to 275°C for ca 12 h until the polymerized liquid becomes sufficiently viscous. One plant has been designed to produce 200 different types and grades of synthetic resins in three 18-m^3 (ca 640-ft^3) reactors (78).

Thermoplastic molds can be considered batch reactors, with product uniformity affected by the transport of impurities within the molds and the shear stresses induced in the feed during injection. Chemical reactions take such forms as gas generation release from blowing agents, additional catalyzed polymerization of monomer–polymer adducts, and curing induced transformations. Reactions of urea, thiourea, or substituted ureas with excess formaldehyde and free methylol groups prevent excessive shrinking and cracking of urea–formaldehyde resins.

Laminate processing of thermosetting resin matrices (epoxies, silicones, or polyesters), eg, to fabricate large void-free high performance structural composites, are carried out in evacuated autoclave reactors. In autoclave-vacuum degassing, a commonly used procedure, successive layers of fibers coated with epoxy resin and containing a catalytic adduct are placed in a bag against a smooth tool surface and covered with glass or plastic bleeder fabrics to absorb excess resin (Fig. 19a). After vacuum is applied, temperature is increased in stages to 135°C. The pressure, raised to 586 kPa (85 psi) is only partly transmitted to the resin. Resin viscosity first decreases, but increases as the reactions accelerate with increasing temperature and pressure. As the fiber network is compressed, resin flows through the plies, and the laminate consolidates (79). A simplified reactor model is used in constructing stability maps for identifying resin pressures that are acceptable for limiting the growth of voids containing entrapped air and vaporized moisture.

Semibatch Reactors. Semibatch reactors are the most versatile of reactor types. Thermoplastic injection molds are semibatch reactors in which shaped plastic articles are produced from melts. In molding thermoplastics, large clamping forces of up to 5000 metric tons are needed to keep molds together, while highly viscous polymers are forced into their cavities. Heat transfer is critical. If the molds are too cold, polymers solidify before filling is completed; if they are too hot, the time required for cooling delays production.

Reaction injection molding (RIM) circumvents the problems with injection molds and provides the technology for fabricating large articles, such as polyurethane automobile fenders and bumpers (Fig. 19b) (80). In RIM production of polyurethanes, polyol (a 3000–6000 mol wt polyester or polyether), together with catalyst, cross-linking agents, chain extenders, and occasionally foaming agents, are rapidly mixed with a stoichiometric amount of isocyanate, eg, 4,4'-diphenyl methane diisocyanate, and injected into molds for polymerization. Impingement mixheads with nozzle velocities of 90–150 m/s provide the intense

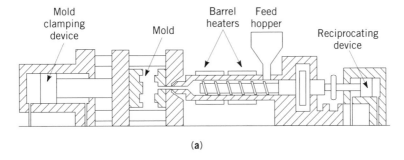

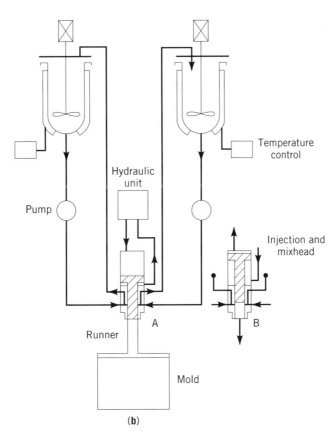

Fig. 19. Typical injection molding and reaction injection molding (RIM) machines: (**a**) injection molding machine for thermoplastics; (**b**) reaction injection molding machine, showing A, the closed position, where reagents circulate, and B, the open position, where reagents are mixed and dispensed to mold.

mixing needed to assure rapid uniform contact of reactants with the mold. Mold surfaces are maintained at 55–80°C, because the reactions are exothermic and polymer thermal conductivity is low. Molds of complex geometries are easily filled because of low viscosities of the starting materials. When surfactant-coated glass fibers are included to improve product flexibility and reduce thermal expansion, these semibatch processes are termed reinforced reaction injection molding (RRIM).

Continuous-Flow Stirred-Tank Reactors. The synthesis of p-tolualdehyde from toluene and carbon monoxide has been carried out using CSTR equipment (81). p-Tolualdehyde (PTAL) is an intermediate in the manufacture of terephthalic acid. Hydrogen fluoride–boron trifluoride catalyzes the carbonylation of toluene to PTAL. In the industrial process, separate stirred tanks are used for each process step. Toluene and recycle HF and BF_3 come in contact in a CSTR to form a toluene–catalyst complex; the toluene complex reacts with CO to form the PTAL–catalyst complex in another CSTR. Once the complex decomposes to the product and the catalyst is regenerated, hydrated HF–BF_3 is processed in a separate vessel, because by-product water deactivates the catalyst and promotes corrosion. High yields (95% of theoretical), high product purity (99.3% PTAL), and low catalyst consumption have been obtained.

The switch from the conventional cobalt complex catalyst to a new rhodium-based catalyst represents a technical advance for producing aldehydes by olefin hydroformylation with H_2 and CO, ie, by the oxo process (qv) (82). A 200 t/yr CSTR pilot plant provided scale-up data for the first industrial, 150,000 t/yr plant. A subsequent advance was based on using water-soluble HRh(CO)[P(m-sulfophenyl-Na)$_3$]$_3$ as the catalyst to produce butyraldehydes from propylene (83). Having both organic and water phases results in higher selectivity and activity achievable with an aqueous media, and complete catalyst–product separation, rendering easier waste product disposal. The system's flexibility permitted a 10,000-fold scale-up by Rhône-Poulenc and the later accommodation of a 100-fold increase in catalyst activity obtained using superior rhodium-based ligands (Fig. 20).

The physical nature of the low density polyethylene is a direct function of reactor type (84). The autoclave reactor produces resin with long-chain branches.

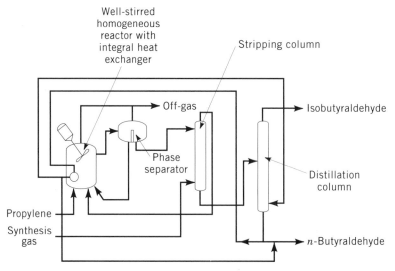

Fig. 20. Homogeneous rhodium-catalyzed oxo process in biphasic media developed by Ruhrchemie/Rhône-Poulenc (83).

However, a plug-flow tubular reactor can generate short-chain branches of the same molecular weight. High pressures (100–300 MPa (10–30 atm)) and moderate temperatures (150–200°C) are used in both processes. Peroxides or oxygen initiators are added to the preheated ethylene feed in the autoclave reactors, but can be injected at multiple points along the long (10,000–100,000 tube dia) tubular reactors. Because of these differences, the autoclave product is preferred for laminations and pipe moldings, whereas the tubular reactor (Fig. 21) is used to produce most films, blow and injection molding, and cable sheathing.

Thermal Tubular Reactors. Tubular reactors have been widely used for low temperature, liquid-phase noncatalytic oxidation, eg, butane to acetic acid and methyl ethyl ketone (MEK), p-xylene to terephthalic acid, cyclohexane to cyclohexanone and cyclohexanol, and n-alkanes to secondary alcohols, and high temperature pyrolysis, eg, thermal cracking of petroleum feeds to olefins, particularly ethylene. Generally, conversion and selectivity to any given product are low for these oxidations. Because runaway branch-chain reactions are possible, heat dissipation must be assured and oxygen concentrations controlled. These considerations often favor the use of a series of tubular reactors in plug flow with some back-mixing in each reactor to maintain sufficient radical concentrations to propagate the reactions. In butane oxidation, reaction rates are reduced as CO_2 and H_2O concentrations increase, so these products are best removed between reactor stages (86). In cyclohexane oxidation, selectivity to useful products decreases with conversion, as does the ease of separation.

Uncertainty in feed supply has encouraged the production of ethylene from a wide range of petroleum feeds, including high molecular weight gas oils in single large units, ca 250 t/h (87). Reactor configurations strongly depend on feed composition and the degree of flexibility must be weighed against the cost.

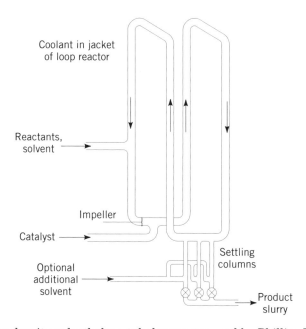

Fig. 21. A low density polyethylene tubular reactor used by Phillips Petroleum (85).

There are limits to the range of feeds that can be processed in tubular reactors. The residues are difficult to crack and thus must be processed in other reactor types, eg, fluidized beds (88).

The epitaxy reactor is a specialized variant of the tubular reactor in which gas-phase precursors are produced and transported to a heated surface where thin crystalline films and gaseous by-products are produced by further reaction on the surface. Similar to this chemical vapor deposition (CVD) are physical vapor depositions (PVD) and molecular beam generated deposits. Reactor details are critical to assuring uniform, impurity-free deposits and numerous designs have evolved (Fig. 22) (89).

Bubble Columns.   Bubble columns are finding increasing industrial application such as in ethylene dimerization, polymer manufacture, and liquid-phase oxidation. Bubble column processing offers the advantages of simplicity, favorable operating costs, and potentially superior product quality. Ethylene dimerization using homogeneous catalysts to produce 1-butene is favored compared to 1-butene produced by steam cracking, where butadiene must be separated and hydrogenated (90). Butadiene separation is complicated by the need to chemically remove the closely boiling isobutene, an impurity that later produces sticky polymers in some ethylene polymerizations.

Bubble columns are further employed with 1-butene as a comonomer in LLDPE manufacture with a Ziegler-Natta catalyst modified to prevent forming the titanium(III) complex responsible for producing high molecular weight polymers. Heat of reaction is removed by recycling liquid through a cooler. The

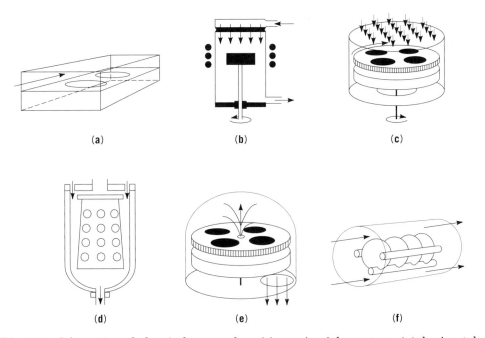

Fig. 22. Schematics of chemical vapor deposition epitaxial reactors: (**a**) horizontal reactor, (**b**) vertical pedestal reactor, (**c**) multisubstrate rotating disk reactor, (**d**) barrel reactor, (**e**) pancake reactor, and (**f**) multiple wafer-in-tube reactor (38).

catalyst is fed continuously to the reactor at temperatures of 50–60°C and low pressures, but sufficient to retain reactants in the liquid phase. The catalyst is incinerated after separation from the product. The Ziegler-Natta catalyst particles initially are 0.06–0.3 m dia and consist of aggregates of particles 10–100 mm in size. Suspended in solvent, the slurry aggregates grow with polymer formation to more than 8000 times their original volume and the particles become insignificant fractions of the total weight of the catalyst and polymer mixture of solids. The catalyst sites remain available for further reaction because the polymer-coated catalysts continuously break into smaller aggregates and are convected outward, where they are available to catalyze further reactions. Depending on the operating conditions, yields decrease or remain constant with time; reaction rates vary between first and second order. The differing responses depend on the intrinsic reaction rates and resistance to reactant transport from the bulk fluid to the available catalyst sites. These interactions between reaction rates and reactant transport are comparable to the impacts of the Thiele modulus on the loss in catalyst activity that results from the concentration gradients established in catalysts in fixed-bed reactors (91,92).

Bubble columns in series have been used to establish the same effective mix of plug-flow and back-mixing behavior required for liquid-phase oxidation of cyclohexane, as obtained with staged reactors in series. Well-mixed behavior has been established with both liquid and air recycle. The choice of one bubble column reactor was motivated by the need to minimize sticky by-products that accumulated on the walls (93). Here, high air rate also increased conversion by eliminating reaction water from the reactor, thus illustrating that the choice of a reactor system need not always be based on compromise, and solutions to production and maintenance problems are complementary. Unlike the liquid in most bubble columns, liquid in this reactor was intentionally well mixed.

Airlift Reactors. Airlift reactors are hydrodynamic variants of the bubble column in which the liquid or slurry circulates between two physically separated zones as a result of sparged gas in one zone inducing a density difference between the zones. The reactors have well-established uses in producing industrial and pharmaceutical chemicals, and in treating industrial and municipal wastes. Further applications stem from opportunities for waste minimization, eg, converting ethylene and chlorine to dichloroethane (94). Important features in an airlift reactor design are the means used for ensuring liquid circulation, establishing high gas–liquid interfacial areas, and efficiently separating gas and liquid. Numerous configurations are in use including physically separated internal loops and draft tubes, and external loops (Fig. 23). Also employed are vertical vessels with circulation induced by the downward injection of a bubbly liquid at velocities sufficient to force circulation (95).

Loop reactors are particularly suitable as bioreactors to produce, for example, single-cell protein (96). In this process, single yeast or bacteria cells feeding on methanol multiply in aqueous culture broths to form high value biomass at 35–40°C, 20 kg/m^3 cell concentrations, and specific growth rates of approximately 0.2 kg/h. Methanol substrate concentrations are maintained close to 1%, the cell mortality limit. Intense mixing and broth uniformity are essential, and must be accomplished without foaming, segregation of either bouyant or heavy components, or recourse to either chemical additives or mechanical agitation.

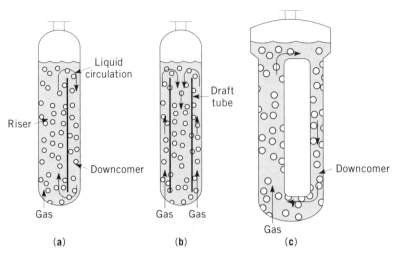

Fig. 23. Airlift reactors: (**a**) split cylinder internal loop, (**b**) draft tube internal loop, and (**c**) external loop (94).

Exothermic heat loads of 40 kw/m^3 must be efficiently removed internally with temperature differences of only 8°C available.

Spray Columns.   Spray columns have diverse specialized uses in biotechnology processing, catalyst manufacture, and minimization of waste products. Spray columns are used in the production of milk powder, cheese, and other fermentation products for direct heat-induced conversion of proteins, microorganisms, and enzymes, thus affecting color, flavor, nutritive value, and biological safety. Figure 24 shows the effect of column height (residence time) and drop size on minimizing protein denaturation, eg, β-lactoglobulin aggregates, while maximizing bacteria destruction, eg, *Micrococcus leteus* (97).

Thermal decomposition of spent acids, eg, sulfuric acid, is required as an intermediate step at temperatures sufficiently high to completely consume the organic contaminants by combustion; temperatures above 1000°C are required. Concentrated acid can be made from the sulfur oxides. Spent acid is sprayed into a vertical combustion chamber, where the energy required to heat and vaporize the feed and support these endothermic reactions is supplied by complete combustion of fuel oil plus added sulfur, if further acid production is desired. High feed rates of up to 30 t/d of uniform spent acid droplets are attained with a single rotary atomizer and decomposition rates of ca 400 t/ d are possible (98).

Similarly, incineration, pyrolysis, and partial oxidation are carried out in spray columns for the disposal of plastic wastes. Incinerators operate at about 1100°C with the feed injected as liquid slurries. Rotary kilns are often employed for solid feeds, with spraying of a solid slurry an essential design feature. Pyrolysis and partial oxidation are advantageous in that the polyolefins are chemically recycled. Oxygen-free pyrolysis of polymers at 600–800°C produces a range of products from methane to polycyclic aromatic compounds. Increasing the temperature to 1500°C by partial oxidation produces a synthesis gas containing carbon monoxide and hydrogen, as well as small quantities of methane and low molecular weight unsaturated compounds. In partial oxidation, molten polymer

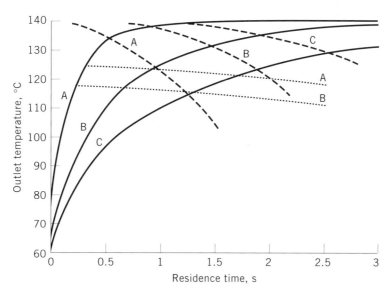

Fig. 24. Effect of residence time and outlet temperature on the chemical and biochemical conversions of skim milk in a spray column, where (——) represents the diameter of the droplet, A = 1 mm, B = 2 mm, and C = 3 mm; (– – –) represents aggregated β-lactoglobulin, A = 10%, B = 20%, and C = 30%; and (····) represents decimal reductions of *M. luteus* cells, A = >12, B = 3.

is sprayed through atomizing nozzles into the combustion chamber where it is mixed with oxygen (99).

Tubular Fixed-Bed Reactors. Bundles of downflow reactor tubes filled with catalyst and surrounded by heat-transfer media are tubular fixed-bed reactors. Such reactors are used most notably in steam reforming and phthalic anhydride manufacture. Steam reforming is the reaction of light hydrocarbons, preferably natural gas or naphthas, with steam over a nickel-supported catalyst to form synthesis gas, which is primarily H_2 and CO with some CO_2 and CH_4. Additional conversion to the primary products can be obtained by iron oxide-catalyzed water gas shift reactions, but these are carried out in large-diameter, fixed-bed reactors rather than in small-diameter tubes (65). The physical arrangement of a multitubular steam reformer in a box-shaped furnace has been described (1).

Approximately 45% of the world's phthalic anhydride production is by partial oxidation of *o*-xylene or naphthalene in tubular fixed-bed reactors. Approximately 15,000 tubes of 25-mm dia would be used in a 31,000 t/yr reactor. Nitrate salts at 375–410°C are circulated from steam generators to maintain reaction temperatures. The resultant steam can be used for gas compression and distillation as one step in reducing process energy requirements (100).

For fixed-bed reactors containing rapidly deactivating catalysts, the scheduled changes in operating variables to accommodate activity loss can have a marked effect on run length. This is exemplified by acetylene hydrochlorination to produce vinyl chloride in tubular fixed-bed reactors. Steel reactors, 2.4–3-m dia and 3–4.5 m high, with 150–500 tubes containing 6-mm $HgCl_2$-supported catalysts are used. Catalyst deactivation results from local overheating, which causes mercuric chloride to sublime from its carbon support. A variety of nearly optimal practices have been established, including changing both inlet and cooling-jacket temperatures to maintain conversion and decreasing mass flow rate (101).

Fixed-Bed Reactors. *Single-Phase Flow.* Fixed-bed reactors supplied with single-phase reactants are used extensively in the petrochemical industry for ammonia synthesis, catalytic reforming, other hydroprocesses, eg, hydrocracking and hydrodesulfurization, and oxidative dehydrogenation. The feeds in these processes are gases or vapors. The reactors generally are of large diameter, operate adiabatically, and often house multiple beds in individual pressure vessels. Bed geometries usually are determined by catalyst reactivity, so that the beds can be either tall and thin or short and squat. Variations of designs result from issues associated with the specific properties of the reactants or catalysts and differences in operating conditions. Either interbed cooling (or heating) or liquid quench additions are used to remove or supply reaction heat from the effluent of each bed. Multicomponent heterogeneous catalysts that require special activation treatments or unique handling requirements are used in most cases. Provisions must be made for restoring catalyst activity, either by replacement with fresh catalyst or by regeneration. Although reactants generally flow downward, air may be injected for catalyst regeneration at the bottom of the reactor.

Ammonia Synthesis. Reactor design for ammonia synthesis is subject to numerous constraints. Equilibrium favors high pressures (15–30 MPa (ca 150–300 atm)) and low temperatures (430–480°C) and kinetics favor the reverse conditions, ie, lower pressures and higher temperatures. The promoted iron catalyst tolerates a maximum temperature of 500°C. Cold synthesis gas is used to cool the pressure vessel walls to prevent hydrogen embrittlement. Consequently, many ammonia converters have been designed and are in use (102), eg, the radial H. Topsoe converter and the upflow ICI reactor. In one widely used design, several catalyst baskets are in series and cool synthesis gas is added between them. Radial reactors also have proven successful because smaller catalyst particles (1.5–3 mm) can be used without excessive pressure drop to increase catalyst utilization. Even horizontal converters containing three beds of catalysts have been designed, eg, the horizontal multibed Kellog reactor (1) (see AMMONIA).

Catalytic Reforming. Catalytic reforming is the primary refinery process for upgrading highly paraffinic, low octane gasoline-range feeds to higher octane products of about the same range of molecular weights. Supported bimetallic platinum catalysts, eg, Re or Ir as the second component, are generally used because of their superior activity maintenance characteristics compared with Pt alone. Because dehydrocyclization and dehydrogenation are principal reactions, catalytic reforming is a main source of aromatic chemicals, specifically benzene, toluene, and xylenes. For the same reason, the process is an important net producer of hydrogen. Isomerization, which also improves gasoline octane number, and hydrogenation also occur. Process definition involves compromises among operating conditions to assure product quality, eg, octane number, despite catalyst deactivation (103).

Consequent to these selection requirements, three basic reactor systems have evolved. The earliest involved three or four fixed beds in series and each was contained in a pressure vessel with reheat furnaces in between to compensate for the endothermic heat losses from dehydrogenation. Catalyst is regenerated after ca six months in this semiregenerative process. In cyclic reactors, there are four to five reactors in series with special valve arrangements so that one reactor is regenerated while the others remain in service. These units are more flexible to process requirements and can produce high yields of high octane products (Fig. 25). The latest in reactor system evolution is a series of slowly moving beds with radial gas flow. Essentially, continuous operation and the highest aromatic yield was achieved with steady catalyst addition to the top of three stacked reactors and subsequent removal from the bottom after about one to two months residence (104) (see CATALYSTS, REGENERATION–FCC UNITS).

Oligomerization. Dimerization and trimerization are used to produce supplemental quantities of 92+ octane-blending constituents of gasoline. In the UOP Hexall process, about 80% conversion of propylene to hexane and small amounts of nonene and dodecene are obtained in fixed-bed reactors. Low concentrations of the liquefied petroleum gas (LPG) feed are used because the reactions are highly exothermic. The reactor effluent is water cooled and unreacted LPG is recycled to adjust propylene feed concentrations. High purity propylene is a by-product. Existing tubular reactors used to produce polygasoline, largely nonene, can be retrofitted for this lower pressure and temperature process (105) (see LIQUEFIED PETROLEUM GAS).

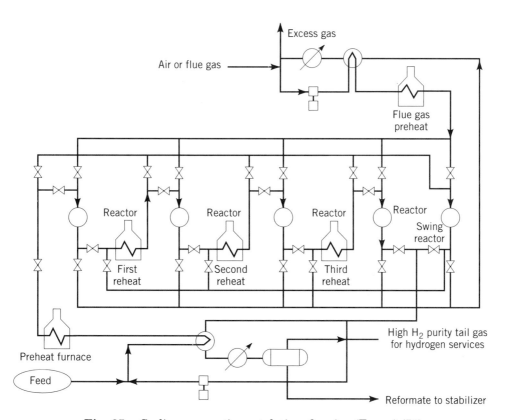

Fig. 25. Cyclic regenerative catalytic reforming (Exxon) (74).

Hydrocracking. Hydrocracking, an undesired reaction in catalytic reforming, is used to crack and hydrogenate at high pressures (1–10 MPa
(10–100 atm)) polycyclic aromatic feeds, eg, heavy gas oils, which are resistant to low pressure catalytic cracking. Hydrogen feed rate is the minimum
required to suppress coke formation, since higher rates increase deleterious aromatic hydrogenation to naphthenes. Reaction temperatures are ca 200–400°C
(106). Typically, four fixed beds are housed in a single pressure vessel, and hydrogen is injected between beds to control temperature. Temperature control is
essential to prevent large temperature excursions. With the advent of new noble
metal catalyst formulations, an additional set of fixed beds in a second pressure
vessel as a second stage can be used to provide more flexibility in the relative
amounts and qualities of gasoline and jet-diesel fuels produced (107). The main
impetus for such hydrotreating processes as hydrodesulfurization (HDS) was
the availability of hydrogen from catalytic reforming and the deleterious effects
of sulfur on downstream catalytic processing. Such processes minimize sulfur
and nitrogen oxide emissions, which are generated from these products when
they are burned as fuels. Relatively mild operating conditions (1300–4000°C
and 1–7 MPa (ca 10–70 atm)) and simple adiabatic downflow fixed beds are
used (108).

Catalytic Dehydrogenation. Similar reactor configurations are used for the
dehydrogenation of *n*-butene to butadiene. Conventional dehydrogenation with
steam is endothermic; the steam supplies heat and increases yields by reducing
n-butene partial pressures. Oxidative dehydrogenation is the primary butadiene
process. It is an exothermic, self-regenerative, heterogeneous catalyzed reaction,
in which the steam serves as a heat sink to moderate the temperature rise.
Typical temperatures in the fixed bed are 350–600°C, and most of the 0.6-m
deep beds operate at the higher temperatures (109).

Fixed-Bed Reactors. *Multi-Phase Flow.* Flow regimes and contacting
mechanisms in fixed-bed reactors that operate with mixtures of liquids and
gases are totally different from those with single-phase feeds. Nevertheless,
mixed-phase, downflow fixed-bed reactor designs are extensions of single-phase,
fixed-bed hydroprocessing technology and outwardly resemble such reactors. The
most generally used mixed-phase reactor is the trickle bed. Special distributors
are used to uniformly feed the two-phase mixtures (Fig. 26). Hydrodesulfurization, hydrocracking, hydrogenation, and oxidative dehydrogenation are carried
out with high boiling feeds in such reactors. Pressures generally are higher than
in their single-phase flow counterparts. Though some of these reactors may
operate in the pulsed-flow regime, these reactors retain their trickle bed name
because the same configurations are used. Furthermore, reactor instrumentation may not be suited for recording the pulsation, thus the flow regime would
not be noted. Where required, the temperature rise in trickle-bed reactors is
controlled as with single-phase reactors. Generally, liquid quench systems are
preferred.

Increased global urbanization provides the impetus to exploit improvements
in feed distribution and catalyst utilization to achieve very low (<500 ppm) sulfur levels in existing hydrodesulfurization reactors for a wide range of feeds. High
boiling feeds that have been successfully processed include virgin and cracked
heavy gas oils and residue. Such feeds contain metals, which are hydrodemetal-

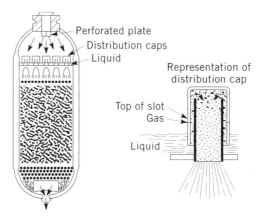

Fig. 26. Mixed-phase, trickle-bed distributor used for hydrodesulfurization in the Unicracking–HDS process (98).

lized and removed by deposition within the catalyst porous structures. Catalysts that are low in cost relative to noble metal supported catalysts slowly deactivate and are replaced after one to three years of use, depending on the severity of service (110,111). Vapor–liquid equilibria play an important role in catalyst deactivation. Moving trickle beds are also used in these applications. Catalysts can be fed as an oil slurry to the top and then flow downward through a series of beds. The conical bottom of each bed is designed to assure plug flow of catalyst (112).

Fluid-Bed Reactors.   The range of fluid-bed applications is large and diverse, as illustrated in Table 7 (65). One of the earliest and most well-known applications is fluid catalytic cracking (FCC). In FCC, gas oils are cracked at ca 500–525°C to produce gasoline and other light hydrocarbons; fine silica–alumina-based catalysts averaging ca 50–60-μm dia are used. The reactors are dense-phase fluid beds, dilute-phase transfer lines, or combinations of the two fluidization regimes. In the various configurations developed for this process, deactivated catalyst, which flows downward through a steam-stripping zone to remove residual products, is air lifted into a regenerator, where air at 600–700°C

Table 7. Fluidized-Bed Catalytic Reactors for Chemical Synthesis

Product or reaction	Process	Year first commercialized
phthalic anhydride	Sherwin-Williams-Badger	1945
Fischer-Tropsch synthesis[a]	Kellogg, Sasol	1955
vinyl acetate	Nihon Gosei	1956
acrylonitrile	Sohio	1960
ethylene dichloride	Monsanto	1961
chloromethane	Asahi Chemical	1965
maleic anhydride	Mitsubishi Chemical	1970
polyethylene (low density)[b]	Union Carbide	1977
polypropylene[b]	Mitsui Petrochemical	1984
o-cresol and 2,6-xylenol	Ashai Chemical	1984

[a]Reactors operate in the fast fluidization regime.
[b]Fluidized beds contain coarse particles.

is used to burn the coke. Regenerated catalyst is recalculated to the reactor (113). The advent of high activity catalysts containing zeolites widened the range of design configurations, feed properties, and operating conditions. These units are large and can process 6000 t/d of feed. Most feature short residence time, transfer line reactor sections.

Numerous other processes have followed the catalytic cracking lead, including fluid-bed coking of residue alone and combined with coke gasification (Fig. 27); catalytic reforming and catalytic oxidation of aromatics, eg, benzene, toluene, and naphthalene, to maleic and phthalic anhydrides; ammoxidation to acrylonitrile; and hydrogen chloride oxidation to chlorine. Process developments outside the petrochemical industry include iron ore reduction and spent nuclear fuel reprocessing. Thermal and catalytic coal gasification, oil-shale retorting, flue-gas desulfurization, and ethanol dehydration to produce ethylene represent new and renewed initiatives for fluidization technology.

A principal commercial development is the UNIPOL process for producing low density polyethylene and polypropylene. The Union Carbide process uses 50–200 μm silica-supported catalyst particles containing 0.4% chromium, 4.5% titanium, and 0.3% fluorine as oxides to produce polyethylene with melt indexes up to 1.5–2°C/min. Copolymerization of ethylene with α-olefins of three to six carbons promote branching and result in low density products (114). Operating pressures for these polyethylene and polypropylene fluidized-bed processes are dictated by economic considerations; higher pressure increases polymerization

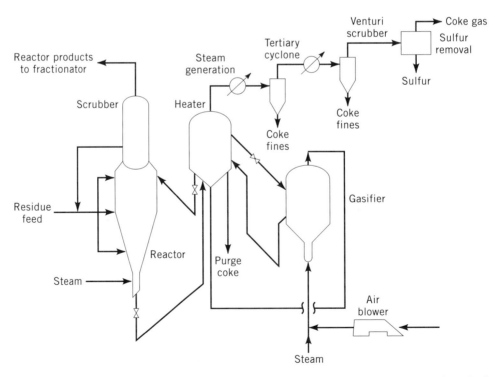

Fig. 27. A Flexicoking unit combining fluid coking and gasification using three-bed fluidized beds.

rates and improves fluidization, and lower pressures favor thinner reactor wall construction, lower compression costs, and less ethylene escape during product removal. Products are removed intermittently by lock hopper settling tanks (115). The modes of operation and contacting of reactants and catalysts can markedly affect reactor performance. For example, returning compressed and partially condensed unreacted feed to the fluid bed sometimes improves temperature control and increases productivity (116).

Gas−Liquid−Solid Fluidization Reactors. The logical extensions of gas-fluidization technology are the ebullating bed gas−liquid−solid and slurry reactors, where both liquid and gas fluidize the solids (19,117). Such reactors can be used for catalytically hydrodesulfurizing and hydrodemetalizing residual fuels, upgrading heavy gas oils and residue for further processing, and converting non-pumpable heavy crudes and bitumens into transportable lighter crudes suitable for conventional processing. Liquid recycle increases the total flow rate to that desired for good fluidization (118). Catalyst activity is maintained by periodic addition and removal through lock hoppers (119,120).

Indirect coal liquefaction is carried out in an ebullating bed reactor, in which synthesis gas together with inert hydrocarbon liquids fluidize the copper−zinc catalyst particles (see COAL CONVERSION PROCESSES). The $CO−CO_2−H_2$ mixture is converted to methanol at 230−250°C and 7.8 MPa (ca 77 atm) in this reactor. Catalyst activity increases as particle size decreases from 3 to 1 mm, which is one typical range for such reactors (121). Similarly, Fischer-Tropsch hydrocarbon synthesis is carried out in gas−liquid−solid reactors. The synthesis gas is generated from natural gas with both thermal partial oxidation and catalytic steam reforming carried out in a single fluid-bed reactor. The highly integrated combination produces clean products and exhibits high thermal efficiency, efficient heat removal, and economy of scale, features of increasing focus in the development and evolution of reactor technology (122).

BIBLIOGRAPHY

"Reactor Technology" in *ECT* 3rd ed., Vol. 19, pp. 880−914, by B. L. Tarmy, Exxon Research and Engineering Co.

1. G. F. Froment and K. B. Bischoff, *Chemical Reactor Analysis and Design*, John Wiley & Sons, Inc., New York, 1979.

2. K. R. Westerterp, W. P. M. van Swaaij, and A. A. C. M. Beenackers, *Chemical Reactor Design and Operation*, John Wiley & Sons, Inc., New York, 1984.

3. O. Levenspiel, *Chemical Reaction Engineering*, 2nd ed., John Wiley & Sons, Inc., New York, 1972.

4. H. S. Fogler, *Elements of Chemical Reaction Engineering*, 2nd ed., P.T.R. Prentice Hall, Englewood Cliffs, N.J., 1992.

5. R. A. Holub, *Hydrodynamics of Trickle Bed Reactors*, D.Sc. Thesis, Washington University, St. Louis, Mo., May 1990.

6. T. R. Melli, W. B. Kolb, J. M. deSantos, and L. E. Scriven, "Cocurrent Downflow in Packed Beds: Microscale Roots of Macroscale Flow Regimes," *AIChE National Meeting*, Houston, Tex., Apr. 1989.

7. H. Short and W. Price, *Chem. Eng.* **92**, 22 (1985).

8. P. K. Ladwig, T. R. Steffens, S. L. Laley, D. P. Leta, and R. D. Patel, "Resid Processing in Fluid Catalytic Crackers," *Foster Wheeler Heavy Oils Conference*, Orlando, Fla., June 7, 1993.

9. Y. S. Mantros, ed., *Unsteady State Processes in Catalysis*, VSP BV, the Netherlands, 1990, pp. 131–159.

10. H. Hlavacek, J. Puszynski, J. Degreve, and S. Kumar, *Chem. Eng. Sci.* **41**(4), 877–882 (1986).

11. A. Bhattacharyya and J. S. Yoo, in J. S. Magee and M. M. Mitchell, Jr., eds., *Fluid Catalytic Cracking: Science and Technology, Studies in Surface Science and Catalysis*, Vol. 76, Elsevier Science Publisher BV, Amsterdam, the Netherlands, pp. 531–562, 1993.

12. J. Barton and R. Rodgers, eds., *Chemical Reaction Hazards*, Institution of Chemical Engineers, Rugby, Warwickshire, U.K., 1993, pp. 7–12.

13. Ref. 2, p. 2.

14. Ref. 1, pp. 437–443.

15. Ref. 2, pp. 122–123.

16. Ref. 2, pp. 371–383.

17. Ref. 2, pp. 397.

18. P. N. Dwivedi and S. N. Upadhyay, *Ind. Eng. Chem. Process Des. Dev.* **16**(2), 157 (1977).

19. Y. T. Shah, *Gas–Liquid–Solid Reactor Design*, McGraw-Hill Book Co., Inc., New York, 1979; R. B. Bird, W. E. Stewart, and E. N. Lightfoot, *Transport Phenomena*, John Wiley & Sons, Inc., New York, 1960.

20. J. H. Seinfeld and L. Lapidus, *Mathematical Methods in Chemical Engineering*, Prentice Hall, Inc., Englewood Cliffs, N.J., 1974, pp. 372–382.

21. G. Astarita, *Mass Transfer With Chemical Reaction*, Elsevier Publishing Co., Amsterdam, the Netherlands, 1967; C. Y. Wen and L. T. Fan, *Models for Flow Systems and Chemical Reactors*, Marcel Dekker, Inc., New York, 1975.

22. Ref. 1, p. 612; Ref. 19, p. 84.

23. Ref. 7, pp. 708–803.

24. R. S. Brodkey, *Turbulence in Mixing Operations*, Academic Press, Inc., New York, 1975.

25. D. Hyman, in T. B. Drew, J. W. Hoopes, Jr., and T. Vermeulen, eds., *Advances in Chemical Engineering*, Academic Press, Inc., New York, 1962, pp. 119–202.

26. L. S. Fan and K. Tsuchiya, *Bubble Wake Dynamics In Liquids and Liquid–Solid Suspensions*, Butterworth-Heinemann, Boston, Mass., 1990.

27. D. C. Cherrington and L. P. Golan, *43rd Mid-Year Refining Meeting*, American Petroleum Institute, Toronto, Canada, May 9, 1978.

28. R. Korbee, O. C. Snip, J. C. Schouten, and C. M. van den Bleek, *Chem. Eng. Sci.* **49**(24B), 5819–5832 (1994).

29. F. Durst, B. E. Launder, F. W. Schmidt, and J. H. Whitelaw, *Turbulent Shear Flows*, Vol. 1, Springer-Verlag, Berlin, 1979.

30. B. E. Launder, *Int. J. Heat Fluid Flow* **10**(4), 282–300 (Dec. 1989).

31. B. E. Launder, *Appl. Sci. Res.* **48**, 247–269 (1991); R. V. A. Oliemans, ed., *Computational Fluid Dynamics for the Petrochemical Process Industry*, Kluwer Academic Publishers, the Netherlands, 1991.

32. S. E. Elghobashi and W. M. Pun, *15th Symposium (International) on Combustion*, Tokyo, Japan, 1974, pp. 1353–1365.

33. D. Gidaspow, Donald Q. Kern Award lecture, *23rd National Heat Transfer Conference, Denver, Colo., Aug. 5, 1985*, AIChE, New York.

34. J. C. Middleton, F. Pierce, and P. M. Lynch, *Inst. Chem. Eng. Symp. Ser.* **8**(87), 239–246 (1984).

35. G. K. Patterson, *70th Annual AIChE Meeting*, New York, Nov. 1977.
36. T. Hanzawa, Z. Sakauchi, U. Ito, K. Kato, and T. Tadaki, *J. Chem. Eng. (Japan)* **10**(4), 313 (1977).
37. H. Simka, P. Merchant, P. Futerko, and K. F. Jensen, *13th International Symposium on Chemical Reaction Engineering, Reaction Engineering: Science and Technology*, Baltimore, Md., Sept. 25–28, 1994, p. D11.
38. K. Jensen, *Handbook of Crystal Growth*, Vol. 3, D. T. J. Hurle, ed., Elsevier Science Publishers BV, Amsterdam, the Netherlands, 1994, pp. 541–598.
39. R. Torvik and H. F. Svendsen, *Chem. Eng. Sci.* **45**, 2325–2332 (1990).
40. K. Franz, T. Borner, H. J. Kantorek, and R. Buchholz, *Ger. Chem. Eng.* **7**, 365–374 (1984).
41. M. P. Dudukovic, N. Devanathan, and R. Holub, *Revue De L'Institut Francais Du Petrole*, **46**, 439–465 (1991).
42. B. L. Tarmy and C. A. Coulaloglou, *Chem. Eng. Sci.* **47**(13/14), 3231–3246 (1992).
43. J. Wei and J. C. W. Juo, *Ind. Eng. Chem. Fundam.* **8**, 114–123 (1969); V. W. Weekman, *AIChE Monograph Series 11*, Vol. 75, AIChE, New York, 1979.
44. D. K. Liguras and D. T. Allen, *Ind. Eng. Chem. Res.* **28**, 674–683 (1989).
45. M. R. Gray, *Ind. Eng. Chem. Res.* **29**, 505–512 (1990).
46. W. H. Press, S. A. Teukolsky, W. T. Vetterling, and B. P. Flannery, *Numerical Recipes in FORTRAN—The Art of Scientific Computing*, 2nd ed., Cambridge University Press, New York, 1992.
47. M. Neurock, A. Nigam, D. Trauth, and M. T. Klein, *Chem. Eng. Sci.* **49**(24A), 4153–4177 (1994).
48. H. Kramers and K. R. Westerterp, *Elements of Chemical Reactor Design and Operation*, Academic Press, Inc., New York, 1963, p. 228.
49. V. W. Weekman, Jr., *AICHE J.* **20**(5), 833 (1974).
50. Ref. 1, pp. 178–187.
51. N. G. Karanth and R. Hughes, *Catal. Rev.* **9**(2), 169 (1974).
52. R. M. Koros, in "Chemical Reaction Engineering," *4th International 16th European Symposium*, Heidelberg, Germany, Apr. 6–8, 1976, pp. IX 372–381.
53. A. Laurent and J. C. Charpentier, *Int. Chem. Eng.* **23**(2), 265 (1983).
54. S. Waldram, in Ref. 37, pp. PC-4.
55. R. N. Landau, D. G. Blackmond, and H.-H. Tung, *Ind. Eng. Chem. Res.* **33**, 814–820 (1994).
56. J. Barton and R. Rodgers, eds., *Chemical Reaction Hazards*, Institution of Chemical Engineers, Rugby, Warwickshire, U.K., 1993, pp. 63–99.
57. P. T. Woodrow, A. D. Epstein, M. Futran, A. L. Forman, and P. F. McKenzie, in Ref. 37, pp. B1-6.
58. Ref. 19, pp. 322,646.
59. J. Beek, in Ref. 25, pp. 203–271.
60. N. P. Cheremisinoff and R. Gupta, *Handbook of Fluids in Motion*, Ann Arbor Science Publishers, Ann Arbor, Mich., 1983, pp. 76–77.
61. A. Bisio and R. L. Kabel, *Scaleup of Chemical Processes*, John Wiley & Sons, Inc., New York, 1985, p. 6.
62. Ref. 61, p. 6.
63. C. N. Satterfield, *Mass Transfer in Heterogeneous Catalysis*, The MIT Press, Cambridge, Mass., 1970.
64. A. Gianetto and V. Specchia, *Chem. Eng. Sci.* **47**(13/14), 3197–3213 (1992).
65. D. Kunii and O. Levenspiel, *Fluidization Engineering*, 2nd ed., Butterworth-Heinemann, Boston, Mass., 1991.
66. J. M. Matsen, in A. Bisio, and R. L. Kabel, eds., *Scaleup of Chemical Processes*, John Wiley & Sons, Inc., New York, 1985, pp. 347–405.

67. J. M. Matsen and B. L. Tarmy, *Chem. Eng. Prog. Symp. Ser.* **66**(101), 1 (1970).
68. C. G. Frye, W. C. Lake, and N. C. Eckstrom, *AICHE J.* **4**, 403 (1958).
69. P. B. Weisz and J. S. Hicks, *Chem. Eng. Sci.* **17**, 265 (1962).
70. Ref. 1, p. 445.
71. R. Lewis, J. P. Strackey, W. P. Haynes, R. R. Santore, and D. Dubis, *Proceedings of the Tenth Synthetic Pipeline Gas Symposium*, Chicago, Ill., Oct. 30–Nov. 1, 1978, pp. 207–219.
72. *Chem. Eng.* **86**(26), 80 (1979).
73. Ref. 42, p. 3244.
74. B. C. Gates, J. R. Katzer, and G. C. A. Schuit, *Chemistry of Catalytic Processes*, McGraw-Hill Book Co., Inc., New York, 1979.
75. A. V. Hahn, *The Petrochemical Industry-Markets and Economics*, McGraw-Hill Book Co., Inc., New York, 1970.
76. C. N. Satterfield, *Heterogeneous Catalysis in Practice*, McGraw-Hill Book Co., Inc., New York, 1980.
77. R. R. Hemrajani, *5th European Conference on Mixing*, June 10–12, 1985, Wurzburg, Germany.
78. *Can. Chem. Process.*, 21 (Apr. 1978).
79. J. L. Kardos, M. P. Dudukovic, E. L. McKague, and M. W. Lehman, *Void Formation and Transport during Composite Laminate Processing: An Initial Model Framework*, STP 797, ASTM, Philadelphia, Pa., 1983, pp. 96–109.
80. M. F. Edwards, *Chemical Reaction Engineering of Polymer Processing, Reaction Injection Moulding, Inst. Chem. Eng. Symp. Ser.* **8**(87), 783–796 (1984).
81. S. Fujiyama and T. Kasahara, *Hydrocarbon Process.*, 147 (Nov. 1978).
82. *Chem. Eng.* **1**, 10 (Dec. 5, 1977).
83. J. Haggin, *Chem. Eng. News* **72**(41), 28–36 (Oct. 10, 1994).
84. Y. Tomura, N. Nagashima, and K. Shirayama, *Hydrocarbon Process.*, 151 (Nov. 1978).
85. J. L. Throne, *Plastics Process Engineering*, Marcel Dekker, Inc., New York, 1979, pp. 122, 126.
86. J. B. Saunby and B. W. Kill, *Hydrocarbon Process.*, 247 (Nov. 1976).
87. J. M. Dluzniewski and J. E. Wallace, *Oil Gas J.*, 64 (Oct. 16, 1978).
88. L. F. Hatch and S. Matar, *Hydrocarbon Process.*, 129 (Mar. 1978).
89. Ref. 38, p. 545.
90. D. Commereuc, Y. Chauvin, J. Gaillard, and J. Leonard, *Hydrocarbon Process.* **63**(11), 118 (1984).
91. L. K. Doraiswany and M. M. Sharma, *Heterogeneous Reactions, Analyses, Examples and Reactor Design*, Vol. 2, *Fluid–Fluid Solid Reactions*, John Wiley & Sons, Inc., New York, 1984, pp. 299–300.
92. Ref. 85, pp. 182–189.
93. J. Alagy, P. Trambouze, and H. Van Landeghem, *Advances in Chemistry Series, Evanston, Ill., Aug. 27–29, 1974*, American Chemical Society, Washington, D.C., 1974, No. 133, pp. 644–645.
94. Y. Chisti and M. Moo-Young, *Chem. Eng. Prog.* **89**(6), 38–45 (June 1993).
95. U.S. Pat. 5,190,733 (Mar. 2, 1993), M. B. Ajinkya, R. M. Koros, and B. L. Tarmy (to Exxon Research & Engineering Co.).
96. P. De Jong and H. J. L. J. van der Linden, *Chem. Eng. Sci. Special Issue* **47**(13/14), 3761–3768 (1992).
97. H. Blenke, in T. K. Ghose, A. Fiechter, and N. Blakebrough, eds., *Advances in Biochemical Engineering*, Vol. 13, Springer-Verlag, Berlin, 1979, pp. 120–214.
98. U. Sander and G. Daradimos, *Chem. Eng. Process.*, 57 (Sept. 1978).
99. H. Buckhorn, M. Burckschat, and H. Deusser, *J. Anal. Appl. Pyrolysis* **8**, 427 (1985).

100. A. Wiedmann and W. Gierer, *Chem. Eng.*, 62 (Jan. 29, 1979).
101. A. F. Ogunye and W. H. Ray, *Ind. Eng. Chem. Process Des. Dev.* **9**(4), 619 (1970).
102. Ref. 77, pp. 304–306.
103. Ref. 75, pp. 184–193; Ref. 77, pp. 247–256.
104. A. M. Atani, in G. J. Antos, A. M. Aitani, and J. M. Parera, eds., *Catalytic Naphtha Reforming: Science and Technology*, Marcel Dekker, Inc., New York, 1995, pp. 409–436.
105. D. J. Ward, R. H. Friedlander, R. Frame, and T. Imai, *Hydrocarbon Process.* **64**(5), 81 (1985).
106. Ref. 77, pp. 258–259.
107. A. Billon, J. P. Franck, J. P. Peries, E. Fehr, E. Gallei, and E. Lorenz, *Hydrocarbon Process.*, 122 (May 1978).
108. Ref. 77, pp. 259–265.
109. L. M. Welch, L. J. Croce, and H. F. Christmann, *Hydrocarbon Process.*, 131 (Nov. 1978).
110. R. L. Richardson, F. C. Riddick, and M. Ishikawa, *Oil Gas J.*, 80 (May 28, 1979).
111. "1978 Refining Process Handbook," *Hydrocarbon Process.*, 121 (Sept. 1978).
112. W. C. Van Zull Langhout, C. Ouwerkerk, and K. M. A. Pronk, *Oil Gas J.*, 120 (Dec. 1, 1980).
113. Ref. 1, pp. 662–666.
114. U.S. Pat. 4,011,382 (Mar. 8, 1977), I. J. Levine and F. J. Karol (to Union Carbide); Ger. Pat. 2,609,889 (Sept. 23, 1976), I. J. Levine and F. J. Karol (to Union Carbide).
115. Eur. Pat. 71,430 (Feb. 9, 1983), R. G. Aronson (to Union Carbide).
116. Eur. Pat. 89,691 (Sept. 28, 1983), J. M. Jenkins, III, T. M. Jones, R. L. Jones, and S. Beret (to Union Carbide).
117. Ref. 96, pp. 3181–3826.
118. K. H. Reichert and R. Michael, "Polymerization in Bubble Columns, Problems of Mass and Heat Transfer at High Solid Contents," *Inst. Chem. Eng. Symp. Ser.* **8**(87), 659–666 (1984).
119. R. P. Van Driesen, V. Caspers, A. R. Campbell, and G. Lunin, *Hydrocarbon Process.*, 107 (May 1979).
120. *Hydrocarbon Process.*, 133 (Sept. 1978).
121. M. B. Sherwin and M. E. Frank, *Hydrocarbon Process.*, 122 (Nov. 1976).
122. B. Eisenberg, G. C. Lahn, R. A. Fiato, and G. R. Say, presentation at *Alternate Energy '93*, Council on Alternate Fuels, Colorado Springs, Colo., Apr. 27–30, 1993.

General References

References 1–4, 9, 11, 17, 18, 65, 74–76, and 92 present further discussions of fundamental concepts, as well as additional perspective on some industrially important processes.

BARRY L. TARMY
TBD Technology

RECORDING DISKS. See INFORMATION STORAGE MATERIALS.

RECREATIONAL SURFACES

Recreational surfaces are synthetic, durable areas of consistent properties designed for various activities, including the high performance requirements of baseball, cricket, field hockey, football, golf, jumping, soccer, tennis, track, wrestling, and others. The category also includes indoor–outdoor carpets and similar materials designed for low maintenance in home or light recreational service. The characteristics of the artificial playing surface may be selected to match natural surfaces under ideal conditions or may have special features for specific sports purposes. In all cases, the intent is to provide appropriate functionality for the activity combined with good durability of the product. In most cases, the artificial surface permits greatly increased utilization compared with natural grass.

A grass-like artificial surface was installed for the first time in 1964, at Moses Brown School (Providence, Rhode Island) (1). In 1966, artificial turf was installed in the Houston Astrodome in Texas. These surfaces consisted of green pigmented, nylon-6,6 pile ribbon, with a cross-section resembling that of natural grass. Since that period, other fabrics of various pile ribbon and constructions have continued to become available commercially for indoor and outdoor facilities.

Resilient surfacing compositions for recreational use were introduced in tennis courts in the early 1950s. This led to the first all-weather, resilient athletic track installation at the University of Florida (Gainesville) in 1958, using a composition similar to that of tennis courts. The polyurethane running track used in the Olympic games at Mexico City in 1968 started a new era in this highly competitive sport. The uniformity and resilience of the synthetic track contributed to new speed records, and such tracks soon became standard throughout the world. In addition to the performance properties, the all-weather aspects of a synthetic track offered advantages for scheduled events and practice. The original system contained a rubber- and clay-filled polyurethane with polyurethane chips embedded in the surface. A wide variety of systems followed, including polyurethane with vinyl or rubber chips or sand finish for an all-weather skidproof surface, so-called sandwich tracks consisting of a bound rubber base with a polyurethane surface, bound rubber chips alone with a painted surface, and many others. A similar, smooth surface product was used for basketball. For tennis, a sand finish proved acceptable, but the polyurethanes were not competitive in cost with coated asphalt (qv) and molded vinyls and polyolefins.

Other more recent examples of recreational surfaces or components are artificial turf variations for golf tee mats and croquet, permanent resilient base layers replacing asphalt or asphalt and shock-absorbing underpad in artificial turf field installations, and sand-filled turf.

These grass-like and resilient installations require substantial amounts of synthetic materials. A typical sports field covered with artificial turf requires approximately 15,000 kg of fabric, 15,000–30,000 kg of shock-absorbing underpad, and 5,000–10,000 kg of adhesive and seaming materials. The artificial surface for a 0.40-km running track may require 50,000–70,000 kg of materials. Paint striping and marking of turf, tracks, and courts call for additional materials.

Types of Surfaces

Recreational surfaces must provide certain performance characteristics with acceptable costs, lifetimes, and appearance. Arbitrary but useful distinctions may be made for classification purposes, depending on the principal function: a covering intended primarily to provide an attractive surface for private leisure activities, eg, patio surfaces; a surface designed for service in a specific sport, eg, track surfaces; or a grass-like surface designed for a broad range of heavy-duty recreational activities, including professional athletics, eg, artificial turf for outdoor sports.

Light-Duty Recreational Surfaces. Artificial surfaces intended for incidental recreational use, eg, swimming pool decks, patios, and landscaping, are designed primarily to provide a practical, durable, and attractive surface. Minimum cost is a prime consideration and has driven the quality of some such products to a low level. Most surfaces in this category utilize polypropylene ribbon and a tufted fabric construction (see OLEFIN POLYMERS, POLYPROPYLENE).

Single-Use Athletic Surfaces. Included here are running tracks, tennis courts, golf tee mats, putting greens, and other installations designed for a particular sport or recreational use. Specific performance criteria are important and differ depending on the application. In the case of a tennis surface, for example, friction and resilience characteristics are critical because they affect footing and the behavior of the tennis ball. Another special application is a warning area or track adjacent to a sports surface, eg, the area between the fence and the playing portion of a baseball field. This frequently is covered with a special surface which must feel different from the main area to warn the player approaching the fence.

Multipurpose Recreational Surfaces. The performance demands control the design for artificial surfaces in this category, which include, for example, the playing surface for American football and soccer. The shock absorbency of the system affects player safety and long-term performance under very heavy, usually multipurpose use. The grass-like fabrics used for these applications are made from various pile materials, including polypropylene, nylon-6,6, nylon-6, and polyester (see FIBERS, OLEFIN; FIBERS, POLYESTER; POLYAMIDES). The fabric may be woven, knitted, or tufted. The underpad is derived from various materials, representing a compromise of properties. Because of the importance of safety and performance, fabric and installation costs are higher than those for the lighter duty surfaces.

Performance Characteristics

User-Related Properties. The most important element in the player's contact with the surface is traction. Shoe traction for light-duty consumer purposes need address only provision of reasonable footing. The frictional characteristics are obviously of much greater importance in surfaces designed for athletic use. For specialized surfaces such as a track, shoe traction is especially critical. With grass-like surfaces, traction is significantly affected by pile density and height, and other aspects of fabric construction.

The coefficient of static friction between the playing surface and the shoe determines traction. To test traction for grass-like surfaces, the force required to initiate movement in a weighted sports shoe resting on the artificial turf is measured (2). The coefficient of static friction is defined as the force pull in a direction parallel to the playing surface, divided by vertical force loading. An alternative method is simply to place the weighted shoe on an inclined plane of the surface of interest, and to determine the angle, θ, at which slippage is initiated. Because of resolution of the gravity forces involved, coefficient of friction is tan θ. By either method, vertical force loading must be sufficient to approximate actual penetration. The shoe characteristics significantly affect the traction.

Typical static friction coefficients are given in Table 1. These data demonstrate that the absolute traction values for synthetic surfaces are satisfactory in comparison with natural turf, provided that shoes with the appropriate surfaces are employed. Synthetic surfaces by virtue of their construction are to a degree directional, a characteristic which, when substantial, can significantly affect both player performance and ball roll. This effect is evident in a measurement of shoe traction in various directions with respect to the turf–pile angle. Some traction characteristics are directly affected by the materials. For example, nylon pile fabrics, exhibiting higher moisture regain, have different traction characteristics under wet and dry conditions than do polypropylene-based materials. Effects of artificial turf fabric construction on shoe traction are given in Table 2. Especially effective in aiding fabric surface uniformity is texturing of the pile ribbon, a process available for the two principal pile materials: nylon and polypropylene.

Abrasiveness of an artificial turf surface upon contact with bare skin is a performance criterion to be considered. Artificial surfaces are more abrasive than natural grass in good condition, although the latter typically contains much higher levels of bacteria (3). A suitable laboratory method for comparing

Table 1. Traction Characteristics of Surfaces[a]

Surface	Static friction coefficient[b]		Directionality index[c]
	Dry	Wet	
	Recreational surface		
tufted polypropylene	1.7–2.0	1.8–2.1	0.1–0.2
nylon-6,6			
knitted	1.8–2.0	1.6–1.8	0.10–0.25
tufted	1.9–2.1		0.05–0.15
woven	1.9–2.1		0.05–0.15
natural grass[d]	1.0–2.2	0.7–1.4	
	Indoor–outdoor carpeting		
tufted polypropylene	0.4–1.5	0.4–1.5	

[a]Ranges measured with appropriate sports shoes for the indicated surfaces.
[b]Defined as the average value measured in the four principal directions parallel to the fabric surface: two across the pile, one with, and one against the pile.
[c]Defined as the average absolute deviation of the four traction values from the mean.
[d]The range is determined by the type and condition of grass.

Table 2. Effect of Fabric Construction on Traction Characteristics

Fabric[a] pile density	Description	Static friction coefficient[b]	Directionality index[b]
standard	height at 1.27 cm	1.8–2.0	0.2
	increased curl[c]	1.8–2.0	0.15
	texturized[d]	1.8–2.2	0.02–0.05
high	height at 1.02 cm	1.8–2.2	0.15

[a]55.5 tex (500 den) nylon-6,6 pile ribbon.
[b]See Table 1 for definitions.
[c]Curl is an index of filament modification imparted to the ribbon during processing.
[d]Refers to a fiber-modification process.

abrasiveness among artificial turfs is to use the turf as the abrader on a Schiefer fabric tester (ASTM D1175) and determine the weight loss per cycle when applied to a stiff, friable foam or the like (4).

In general, user-related properties should encompass a good balance of traction, comfort, and safety. The right combination of energy recovery and shock absorption enables the athlete to perform at maximum potential in relative comfort. For running tracks, resilience minimizes energy dissipation on the surface. A particular range is optimal for the track modulus of elasticity (5). All of these considerations are dependent on the footgear selected. Spikes on running shoes should be long enough to provide adequate traction with easy separation. Ideally, dull spikes afford traction by depression of the surface and gain additional energy for the runner, because the resilient surface rebounds when the weight is lifted. With turf, a specific shoe design for the surface is important.

Game-Related Properties. For some activities, such as running and wrestling, the only consideration is the direct impact by the player. For others, eg, tennis, baseball, or soccer, the system must also provide acceptable ball-to-surface contact properties. Important ball-response properties on the artificial surface are coefficients of restitution and friction, because these directly determine the angle, speed, and spin of the ball.

The coefficient of restitution is defined as the ratio of the vertical components of the impact and rebound velocities resulting when a ball is dropped or thrown onto a playing surface. The velocities or related rebound heights may be measured photographically. Criteria such as ball inflation pressure, air temperature, and other details must be specified.

The coefficients of static friction between a ball and the playing surface are the ratios of the horizontal forces necessary to initiate a sliding or rolling motion across the surface to the normal forces (wt) perpendicular to the surface. The sliding and rolling coefficients of dynamic friction are similarly defined in terms of the forces necessary to sustain uniform motion across the playing surface. These friction coefficients determine slip or retention of inertial effects present upon impact. In golf, for example, the driven ball may bounce forward after the first impact with the surface, and bounce backward after the second. In this particular example, the combination of velocities and friction creates a slipping condition on the first bounce; on the second, the rotational backspin imparted to the ball when first hit is activated by sufficiently large friction. In soccer, on the other hand, the ball in play rarely slips because a coefficient of friction ≥ 0.4, which is almost always achieved, is sufficient to transfer momentum.

Values for ball-response parameters in various sports are given in Table 3 (6). Artificial surfaces can be designed to match certain desirable game-response parameters of natural grass surfaces and provide these properties consistently. Also, response is confined to a narrower range and is less affected by weather. As a general rule, artificial turf surfaces tend to be somewhat livelier in ball response, velocity, and distance of roll, with coefficients of friction lower than those for natural grass.

A more recent artificial turf product designed for a specific use is the golf driving mat, used for practice drives with conventional tees at commercial driving ranges. These products must be designed to withstand the considerable forces imparted to the surface by the golf club and the twisting motion of the golfer's heel. An additional accommodation is a means for inserting tees into the mat.

Impact Properties. Artificial playing surfaces for moderate to heavy use must provide shock absorbency for player comfort and safety. This is achieved by incorporation of a resilient layer, usually a shock-absorbing underpad.

An ideal shock-absorbing medium, eg, for football in the United States, would combine a reasonable softness in normal shoe contact with a high capacity for dissipation or distribution of kinetic energy involved in the impact of a player's fall. Various foamed elastomers are suitable for this purpose (see ELASTOMERS, SYNTHETIC). The design criterion is the ability to dissipate energy of motion by reducing impact deceleration through hysteresis losses in the material. A useful device for characterizing the required properties is a dynamic mechanical impact tester (ASTM F355). It employs an instrumented missile that is allowed to fall freely from a specific height onto the surface. Sensing components record electronically the force- and displacement-time profiles of the missile

Table 3. Physical Parameters^a for Ball Response from Sports Surfaces

Sports surface	Coefficient of restitution	Friction coefficients		
		Static	Sliding	Rolling^b
soccer				
nylon turf-pad	0.7	0.4	0.3–0.4	
polyester turf-pad	0.7	0.4	0.3–0.4	
polypropylene turf-pad	0.7	0.5	0.5	
baseball				
nylon turf-pad	0.6	0.5–0.6	0.6	0.1–0.2
natural grass	0.5	1.0	0.8	0.2
golf				
nylon turf-pad	0.5	0.3–0.5		0.1–0.2
natural grass	0.4	0.5		0.1–0.2
tennis				
nylon turf-pad	0.7	0.4–0.6		0.1–0.2
natural grass	0.8	0.7		0.1–0.2

^{a}Approximate values.
^{b}Rolling resistance increases markedly with rolling velocity.

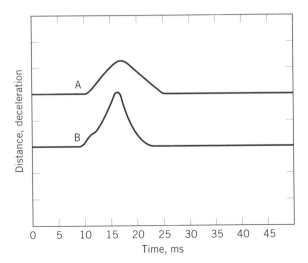

Fig. 1. Deceleration and penetration curves from dynamic impact tester. Time vs distance penetrated (A) and deceleration (B). Distance between vertical ticks for A = 12.7 mm; for B = 50 g.

throughout the interval of first penetration and rebound from the playing surface. The plots shown in Figure 1 illustrate the acceleration and displacement between initial and final contact of the missile with the playing surface, both vs time. The deceleration forces increase as the missile penetrates the surface, reach a maximum, and decrease as the missile rebounds from the surface. The effectiveness of the shock-absorbing medium is indicated by the height of the maximum or, more accurately, the integrated profile throughout the duration of impact, calculated according to the severity index, $\oint g^{5/2}dt$, where g = acceleration expressed in multiples of that due to gravity and dt is the time differential (7). The more effective the shock-absorbing material, the less sharply peaked the g_{max} curve and the broader and shallower the g_{max} profile. Effective performance would be achieved by a material displaying large hysteresis in which the impact of the falling weight is progressively absorbed without rebound. Clearly, because a useful system must also be reversible, this extreme example is not practical. Useful materials for shock absorbency have the ability to dissipate gradually the impact of the falling object, with a substantial conversion of the total kinetic energy to heat through hysteresis losses.

These shock-absorbing characteristics of underpad materials or resilient surfaces are functions of material selection, physical composition, thickness, and temperature. The sensitivity of performance to physical characteristics of the shock-absorbing medium is illustrated in Table 4 and Figure 2. Clearly, thicker materials offer better shock absorbency. However, in practice an excessively thick underpad may result in unsure footing as well as increased cost. Thermal effects must be considered because use temperatures can easily range from below freezing to 65°C in the sun. The ideal system would provide a relatively flat g_{max} response over this range. A compromise is a design for g_{max} peaks up to about 250 and a severity index below 1000 within a reasonable range of temperatures.

Table 4. Properties of Typical Underpad Materials

Property	Foam, % closed cells		Poured elastomer
	75–85	90–95	
thickness, cm	1.6	1.6	1.0
tensile strength, kPa[a]	620	655	2700
density, kg/m^3	96	256	1300–1400
g_{max}[b] at °C			
21	85	125	105[c]
−12	105	150	
49	120	150	

[a]To convert kPa to psi, multiply by 0.145.
[b]61-cm (2 ft) drop height of 9-kg (20 lb) flathead missile (ASTM F355), unless otherwise noted.
[c]22.8-cm (9 in.) drop height of hemispherical body.

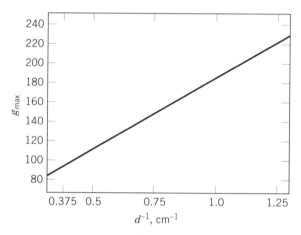

Fig. 2. Shock absorption g_{max} vs thickness d for typical polyurethane resilient surface, measured according to Procedure B of ASTM F355, using a 6.8-kg hemispherical missile dropped from a height of 30 cm.

By way of comparison, g_{max} for natural grass playing fields in late autumn ranges from about 75 for wet fields to 280 for frozen turf (8). The intermediate values observed depend on soil type, moisture, condition, and other variables.

Durability. Grass-like surfaces intended for heavy-duty athletic use should have a service life of at least eight years, a common warranty period provided by suppliers. Lifetime is more or less proportional to the ultraviolet (uv) exposure (sunlight) and to the amount of face ribbon available for wear, but pile density and height also have an effect. Color is a factor; generally uv absorption is highest with red fabrics and least with blue. In addition, different materials respond differently to abrasive wear. These effects cannot be measured except in simulated field use and controlled laboratory experiments, which do not necessarily reflect field conditions.

An instrument commonly in use for providing uv exposures, with intermittent condensation cycles, is the QUV Accelerated Weathering Tester (Q-Panel Company, Cleveland, Ohio). The grass-like ribbons on fabrics exposed may be monitored periodically for changes in physical properties. The Taber and Schiefer abrasion tests (ASTM D1175) evaluate fabrics and fabric constructions for potential wear properties. However, as the conflicting data in Table 5 indicate, any specific accelerated wear test to predict longevity of fabrics is suspect, unless the tests are applied to closely related fabrics for which actual wear-use data are available. Schiefer evaluation, performed periodically on grass-like fabrics exposed to outdoor weathering, is a useful accelerated wear test.

Other tests provide indexes to grab strength (ASTM D1682), tensile strength, and surface durability, including tuft bind (ASTM D1335), in which the force required to dislodge a surface element from its backing is measured. These and other tests (9) are indicators of physical endurance, as can best be simulated in the laboratory.

Artificial surfaces must be resistant to cigarette burns, vandalism, and other harm. Fire resistance is most critically evaluated by the NBS flooring radiant panel test (10). In this test, a gas-fired panel maintains a heat flux, impinging on the sample to be tested, between 1.1 W/cm^2 at one end and 0.1 W/cm^2 at the other. The result of the burn is reported as the flux needed to sustain flame propagation in the sample. Higher values denote greater resistance to burning; results depend on material and surface construction. Polypropylene turf materials are characterized by critical radiant flux indexes which are considerably lower than those for nylon and acrylic polymers (qv) (11).

Table 5. Abrasion Tests for Artificial Turf, Effective Pile Loss, %

Surface	Fiber size, tex[a]	Taber method[b]	Schiefer method[c]
tufted polypropylene	844	5	7
knitted nylon-6,6	55.6	21	0.2
knitted polypropylene	33.3		2.7

[a]To convert tex to den, divide by 0.1111.
[b]ASTM D1175, Rotary Platform, Double Head Method (5000 cycles).
[c]ASTM D1175, Uniform Abrasion Method (5000 cycles).

Materials and Components

A grass-like recreational surface system includes the top material directly available for use and observation, backing materials that serve to hold together or reinforce the system, fabric-backing finish, a shock-absorbing underpad system if any, and adhesives (qv) or other joining materials. The system is installed over a subbase, usually of asphalt or concrete.

Surface Materials. Pile materials used in grass-like surfaces may be selected from fiber-forming synthetic polymers, such as polyolefins, polyamides, polyesters, polyacrylates, vinyl polymers, and many others (see FIBERS, ELASTOMERIC). These polymers exhibit good mechanical strength in the necessary direction. The materials shown in Table 6 are thermoplastic polymers that may be suitably pigmented before or during extrusion. Utilization of uv and heat stabilizers (qv) is essential for outdoor use. An artificial turf of greater pile height and lower pile density is a suitable component of sand-filled turf fields.

Table 6. Typical Properties of Yarns Suitable for Pile Components of Artificial Surfaces

Property	Polypropylene	Poly(ethylene terephthalate)	Nylon-6,6
density, g/cm³	0.91	1.38	1.14
melting point, °C	170	250	265
tenacity, N/tex[a]	0.22	0.18–0.35	0.31
elongation, %	25	30–100	33
moisture regain[b] at 21°C and 65% rh, %	0.1	0.4	4

[a]To convert N/tex to g/den, multiply by 11.33.
[b]Equilibrium moisture or water content.

Backing Materials. Any fiber-forming polymer with reasonable tenacity may be used in backing materials, including polyamides, polyesters, and polypropylenes. The backing provides strength and offers a medium to which the pile fibers can be attached. It is usually not visible in the finished product, nor does its presence contribute much to the characteristics of the playing surface. However, it provides dimensional stability and prolongs service life. Some properties of fabric backing materials may be inferred from the data in Table 6; however, more highly drawn fiber equivalents of greater strength are employed for backing applications.

Backing Finish. The backing material must be consolidated with the pile ribbon. In tufting, for example, the tufts are locked to the backing medium by the primary finish. In weaving and knitting, the finish seals and stabilizes the product. Backing materials are usually applied as a coating which is subsequently heat-cured. For tufting, preferred choices are poly(vinyl acetate), poly(vinyl chloride), polyurethane resins, and latex formulations. For knitted fabrics, poly(vinylidene chloride), acrylics, or polystyrene–rubber latices are used.

Underpads. Shock-absorbing underpad material is usually made of foamed elastomer, which provides good energy absorption at reasonable cost (see FOAMED PLASTICS). The foamed materials may be poly(vinyl chloride), polyethylene, polyurethanes, or combinations of these and other materials. A newer material of interest is made from ground rubber particles, bonded with a polyurethane. Typical foam densities may range from 32 to 320 kg/m³. Important criteria include tensile strength, elongation, open-cell vs closed-cell construction, availability in continuous lengths, softness in energy-absorbing properties, resistance to chemicals, water absorption, compression-set resistance, and cost. Resistance to water absorption is very important, especially if the system is subjected to below-freezing temperatures. Closed-cell materials are most resistant to moisture.

Some recreational surfaces, in particular the lighter weight materials for patios and similar applications, are installed without shock-absorbing underpads. A light coating is usually joined directly to the turf during manufacture, providing a certain degree of softness and grip adequate for the service intended.

Coating, Adhesives, and Joining Materials. Grass-like surfaces are employed over substantial areas, and lengths of rolls must be joined, glued, or sewn together. A variety of adhesives, ranging from low cost poly(vinyl acetate) materials to cross-linked epoxy cements, is utilized. Sewing threads may be selected from the group of drawn, high tenacity yarns such as nylon-6,6 and polyester.

The common asphalt tennis courts have been improved significantly by synthetic all-weather coatings with superior appearance and characteristics. Typical coatings are vinyl or acrylic compositions in various colors. Poured-in-place and preformed systems of polyurethane, vinyl, and rubber, although often used, are more expensive than coated asphalt or concrete. More recently, open molded mats that are easily installed as interlocking tiles have been used to construct new or to repair tennis courts. These mats are less expensive than the poured-in-place and preformed systems that require gluing. They are easily removed if necessary, and thus are easily and quickly repaired. In addition, the open structure allows rapid drainage after rain. Such molded systems are usually made of polyolefins, vinyls, or polyolefin–rubber blends, with pigmentation and stabilizers. Properties of materials, typically polyurethanes, used for the smooth recreational surfaces of running tracks and tennis courts are shown in Table 7.

Table 7. Properties of Poured-in-Place Polyurethane Resilient Surfaces

Property	ASTM test method	Desired range	Typical values[a]		
			A	B	C
impact resilience, %	D2632	30–50	42	49	36
hardness, Shore A-2	D2240	45–65	60	48	50
breaking strength, kPa[b]	D412	>2,800	4,800	2,700	3,400
elongation to break, %	D412	100–300	210	298	168
tear strength, N/m[c]	D624	>11,000	17,200	15,200	11,200
10% compression, kPa[b]	D575	590–860	720	500	520
compression recovery, %	D395 (A)	>95	100	99.6	98.7

[a]Of laboratory samples prepared identically from different raw materials.
[b]To convert kPa to psi, multiply by 0.145.
[c]To convert N/m to lbf/in. (ppi) sample thickness, divide by 175.1.

Fabrication of Grass-Like Surfaces

Tufting. The tufting process is frequently employed in the construction of grass-like surfaces (12). The manufacturing techniques are essentially those developed for the carpet industry with characteristics of high speed and economy. Pile yarn is inserted into the back side of a woven or nonwoven fabric constituting the primary backing by a series of needles, each creating a loop or tuft as the yarn penetrates the backing, and forms the desired pattern on the other side. For artificial surfaces, the looped tufts that form in this process are cut to provide the desired individual blades in the playing surface. Cutting elements incorporated in the tufting machine sever the loops automatically in the process of forming the pile.

Depending on the fabric width and desired pile density, a tufting machine may incorporate 1000–2000 needles, which simultaneously insert the tufts across the fabric width. The needles may operate at speeds above 500 strokes per minute, contributing to a highly efficient output of fabric yardage. The primary backing for tufted surfaces is usually a woven, synthetic filament fabric.

After the tufts have been inserted, pile fiber and backing components are fused together by applying a backing finish and, optionally, a reinforcing secondary backing fabric (Fig. 3). In terms of total recreational surface production, tufting is by far the most utilized fabric construction method.

Knitting. The knitting process, as applied to manufacture of artificial turf and related products, provides a high strength, interlocked assembly of pile fibers and backing yarns (Fig. 3). Pile yarn, stitch yarn, and stuffer yarn are assembled in one operation. The pile and stitch yarns run in the machine or warp direction. The stuffer yarns interlock the rows (wales) formed by the pile and stitch yarns, knotting the system together in the width direction. Knitted fabrics typically possess high strength and high tuft bind.

A machine with approximately 1000 needles continuously produces a fabric 5-m wide. The assembly process is more complex, slower, and more expensive than tufting. The pile yarn and stitch yarn are inserted into the knitting needle, and the stuffer yarn is interlocked with the others through a separate feed mechanism of the machine. As with tufting, the loops of the pile fabric formed are slit, creating the desired individual blades.

The knitted fabric is subjected to a finishing operation in which a suitable backing material is applied to penetrate the yarn contact points and stabilize the structure. This process is usually accompanied by a heat treatment which stabilizes the fabric and conditions the pile.

Weaving. Weaving is a slower process than tufting or knitting. The process consists of a two- or three-dimensional meshing of warp, pile, and fill yarns that may be of different types. In contrast to knitted fabrics, yarns are not knotted together but interwoven at right angles. The pile yarns are cut by a series of wires that are continuously assembled into and withdrawn through the fabric

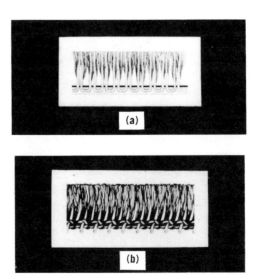

Fig. 3.   Construction of (**a**) pile fiber held in place by latex (tufted) and (**b**) pile fiber tied into backing (knitted) grass-like artificial surfaces.

loops. A suitable finish further stabilizes the fabric. The weaving technique is little used for recreational surfaces.

Finishing. In each of the processes discussed above, the artificial turf fabric is subjected to a finishing operation in which an adhesive, usually poly(styrene–butadiene), polyurethane, or poly(vinylidene chloride), is applied to the back side, with or without the optional reinforcing secondary backing, bonding the components and stabilizing the material. The finish may be applied with a knife or roll, in paste or foam form, followed by a heating and drying stage. The temperatures of application affect the pile ribbon properties.

Underlayment. An installed artificial turf system may or may not include components between the fabric and the subbase. Such components are not required for light-duty applications, but are essential in attaining the shock-absorbing properties required by heavy-duty surfaces. The foam underpads for shock-absorbing systems are made by incorporating a chemical blowing agent into the foam latex or plastisol. Voids of controlled size and number are uniformly distributed throughout the foam material. Closed-cell foam structures resist water absorption and are preferred for outdoor use. The underpad may be produced in slab or continuous roll form, usually up to five feet (1.5 m) in width.

Installation

Grass-like surfaces for heavy-duty athletic use usually are glued to or laid over a subbase of asphalt or other permanent foundation material. The shock-absorbing underpad component is in contact with the subbase layer, and the turf component is placed on top of the underpad. Turf panels are fastened together by sewing or gluing, and the entire perimeter of the grass-like surface is securely anchored to the subbase.

Newer variations of the normal asphaltic subbase include a permeable version (4) which allows rainwater to trickle through holes punched into the underpad and to drain away through the asphaltic subbase to a pipe grid system under the base (Fig. 4).

Bonding of the underpad to the subbase can vary from full glue-down or strip gluing, to loose-laying. The latter technique, without glue-bonding to asphalt, is employed in a float-drain construction in which rainwater drains laterally from the field between the pad and the asphalt. Glue is also omitted

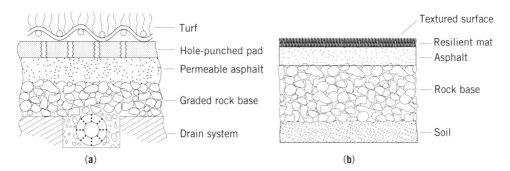

Fig. 4. Cross sections of (**a**) typical artificial turf and (**b**) resilient track surfaces.

at the subbase interface in constructions where the pad is permanently bonded to fabric in the factory, and the composite is installed in the field. Loose-laying again is employed on convertible fields, such as indoor stadiums, where the turf can be removed to permit other sports uses of the subbase.

The Magic Carpet system (AstroTurf Industries) (13) is a novel installation technique in which turf and adherent underpad are unwound from a core and floated onto the subbase of a playing field by air pressure supplied from ducts in the base. To expose the subbase for other sports uses, the turf–pad combination is rolled up and stored at the edge of the field or in a recessed storage compartment.

For other recreational surfaces, such as running tracks, the installation techniques are quite different. Most are poured-in-place. An interlocking tile technique may be employed for tennis courts. In all cases, adequate provision for weathering and water drainage is essential. In general, the resilient surfaces are installed over a hard base (see Fig. 4) that contains the necessary curbs to provide the finished level. Outdoors, asphalt is the most common base, and indoors, concrete. A poured-in-place polyurethane surface (14) is mixed on-site and cast from at least two components, an isocyanate and a filled polyol of the polyether or polyester type. The latter has traditionally contained an organic mercury catalyst (15), which provides a system with selective reaction toward organic hydroxyl groups, reducing moisture sensitivity (see URETHANE POLYMERS); however, amine-type catalysts, eg, Dabco 33 LV, and other nonmercury catalysts are finding increasing environmental acceptance. The isocyanate is of the polymeric type, a toluene diisocyanate of methylenebis(phenyl isocyanate)-based prepolymer. Similar systems are used as binders for scrap rubber granules. The surface properties can be varied by the type and amount of fillers and the size of the rubber granules.

The mixed liquid is pumped into the area, where it cures and forms a slab. It may be poured in two layers to eliminate imperfections in the base. The first layer may be a preformed rubber slab which is glued to the base, or a mixture of reground rubber and binders or rubber and polyurethane. A textured surface may be imparted to the second coat with sand or chips.

Outdoor running tracks, indoor basketball courts, field house surfaces, and stadium turf require line markings and decorations. These are painted on with two-component epoxy paints or water-based acrylic latex, depending on the permanence desired. In stadiums used for different sports the markings can be changed. Markings in different colors for different sports may be desired in community and school installations. Indoor paints are usually permanent and compatible with the surface, ie, they are flexible without cracking and accept finish and maintenance coatings. An innovation is use of knitted- or tufted-in (white) pile yarn to provide permanent line markings in grass-like surfaces. Numerals and emblems of white or other contrasting colors may be inlain on the green fabric.

Commercial Products

For the purpose of illustrating the recreational surface market, manufacturers and trade names of commercial products are provided in Tables 8 and 9.

Table 8. Manufacturers and Trade Names of Artificial Turf Surfaces for Multisport Use

Product	Description	Manufacturer	Country
AstroTurf[a]	knitted nylon-6,6 fabrics	Southwest Recreational Industries	U.S.
Edel Grass	knitted nylon-6,6 and tufted polypropylene fabrics	Edel Grass	Netherlands
Konygreen	knitted nylon-6 fabrics	Kolon Industries	S. Korea
Omniturf	tufted polypropylene sand-filled fabrics	Southwest Recreational Industries	U.S.
Playfield Turf	tufted polypropylene fabrics	Playfield Industries	U.S.
PolyPro	knitted monofilament polypropylene fabrics	Southwest Recreational Industries	U.S
Polytan	tufted polypropylene fabrics	various	Germany
Sportilux	tufted and knitted nylon and polypropylene fabrics	Desso/DLW AG	Netherlands/ Germany
Stadia Turf	tufted polypropylene fabrics	Southwest Recreational Industries	U.S.
Supergrasse	tufted polypropylene fabrics	Balsam Pacific	Australia
Tartan Turf[b]	tufted nylon-6,6 fabrics	3M	U.S.
Toray Turf	knitted nylon fabrics	Mitsubishi	Japan

[a]A subsidiary of Monsanto Company prior to 1988. [b]Discontinued in 1976, no fields extant; included for historical perspective.

Table 9. Manufacturers and Trade Names of Other Recreational Surfaces

Product	Description	Manufacturer[a]
AstroTurf	knitted nylon-6,6 for golf mats	AstroTurf/Southwest Industries
Iron Mat	woven polypropylene for golf mats	AGR[b]
Pro Turf	knitted nylon-6,6 for golf mats	Southwest Synthetic Turf
Turf King	knitted nylon-6,6 for golf mats	Wittek
UltiMat	woven polypropylene for golf mats	Easy Picker
Martin Surfaces	polyurethane surfaces for track	Martin
Mondo	polyurethane surfaces for track	Mondo[c]
Southwest Track Products	polyurethane surfaces for track	Southwest Recreational Industries
Deco Turf	acrylic latex surface for tennis	Koch Materials
Kramer Court	sand-filled tufted fabrics for tennis	Kramer Court
Laykold	acrylic latex surface for tennis	Advanced Polymer Technology
Nova Grass	sand-filled tufted fabrics for tennis	Nova Grass
OmniCourt	sand-filled tufted fabrics for tennis	Advanced Polymer Technology
Plexipave	acrylic latex surface for tennis	California Products

[a]Manufactured in the United States unless otherwise noted. [b]A division of Mohawk Carpet. [c]Manufactured in Italy.

In terms of surface area covered, artificial turf is the dominant commercial product in the category of artificial surfaces described in this article. Light-duty surfaces are the largest representative, followed by tennis court and multipurpose recreational surfaces.

BIBLIOGRAPHY

"Recreational Surfaces" in *ECT* 3rd ed., Vol. 19, pp. 922–936, by W. F. Hamner and T. A. Orofino, Monsanto Co.

1. *Sports Illustrated*, 40th Anniversary Issue, Sept. 19, 1994, p. 98.
2. T. A. Orofino and J. W. Leffingwell, *ASTM STP 1073 - 1990*, 166–175 (1990).
3. Technical data, Monsanto Co., Dalton, Ga., Dec. 1970.
4. T. A. Orofino, *Polym. News* **10**, 294 (1985).
5. T. A. McMahon and P. R. Greene, *J. Biomech.* **12**, 893 (1979); A. Chase, *Science* **81**, 90 (Apr. 1981).
6. J. J. Burke and F. B. Roghelia, technical data, Monsanto Co., Dalton, Ga., 1978; G. Raumann, technical data, Monsanto Co., Dalton, Ga., 1967; J. Vinicki, technical data, Monsanto Co., Dalton, Ga., 1966.
7. C. C. Chou and G. W. Nyquist, *Soc. Autom. Eng.* (740082), 398–410 (1974).
8. E. M. Milner and J. R. Gilliam, technical data, Monsanto Co., Dalton, Ga., 1977.
9. R. D. Breland, *ASTM STP 1073*, 176–182 (1990).
10. I. A. Benjamin and S. Davis, *Final Report No. NBSIR 78-1436*, National Bureau of Standards, U.S. Department of Commerce, Gaithersburg, Md., Apr. 1978; T. Kashiwagi, *J. Fire Flamm. Cons. Prod. Flamm.* **1**, 267 (1974).
11. I. A. Benjamin and C. H. Adams, *Fire J.* **63** (Mar. 1976).
12. G. Robinson, *Carpets and Other Textile Floor Coverings*, Trinsky Press, London, 1972.
13. U.S. Pat. 4,399,954 (Aug. 23, 1983), K. D. Arrant (to Monsanto Co.).
14. J. H. Saunders and K. C. Frisch, *Polyurethanes, Chemistry, and Technology*, Vol. 1, John Wiley & Sons, Inc., New York, 1963; Vol. 2, 1964.
15. U.S. Pat. 3,583,945 (June 8, 1971), J. Robins (to Minnesota Mining and Manufacturing Co.).

General Reference

T. A. Orofino and W. F. Hamner, "Recreational Surfaces," in J. I. Kroschwitz, ed., *Encyclopedia of Polymer Science and Engineering*, 2nd ed., Vol. 14, John Wiley & Sons, Inc., New York, pp. 245–260.

T. A. OROFINO
H. G. SWEENIE
AstroTurf Industries

RECYCLING

INTRODUCTION

Recycling is the process by which materials are separated from waste destined for disposal and remanufactured into usable or marketable materials. The amount of public attention given to recycling has increased noticeably since the mid-1980s, but recycling itself is an age-old process. For thousands of years, households and businesses have recycled goods to save materials or lower costs. Steel and paper mills, for example, have historically recovered their process waste for reuse because doing so makes economic sense. Likewise, archeologists have uncovered the use of recycling in early Mayan and Egyptian civilizations (1). It appears that even in ancient times people were aware of the economic value of reusing and recycling many household discards.

Nevertheless, there is little disputing that widespread public interest in recycling is largely a modern phenomenon. In a little over three decades (1960–1992), the amount of municipal solid waste (MSW) recycled in the United States increased from 6.7 to 15–20% (2). Between 1990 and 1993 alone, it is estimated that the number of curbside collection programs increased more than eightfold, from 600 to over 5000. Well over 78 million people are served by such programs (3). In short, the number of government supply-side recycling programs has skyrocketed.

More often than not, however, the demand for post-consumer materials has failed to keep pace with this boom in collection. In many regions of the United States and elsewhere, the supply of recyclable materials is so great that cities have been forced to either store the materials or curtail the number of items collected. Many principal cities worldwide have reported occasions when source-separated materials were actually sent to dumps or incinerators rather than being recycled (4).

As a result of this oversupply, scrap values for many recyclable materials have fallen noticeably over the past few years. Further complicating matters are new efforts from regulators and environmental activists to mandate the reuse of certain materials (rates and dates) and that products be made with specified amounts of recycled material (product content laws). Such demand-side measures distort market forces and do not appear to be justified on either economic or environmental grounds.

This article introduces the subject of recycling and discusses how and why recycling has become an important public policy issue. Specifically, the article discusses the amount of material collected and processed for recycling, the composition of this material, how collected material is handled, and the economic aspects of recycling. More thorough discussions of specific materials are found in subsequent articles.

Industrial Materials

Although more often associated with household and commercial waste, recycling has proven to be very successful in the industrial arena. Industrial recycling is the recovery for reuse or sale of materials from what otherwise would be wastes destined for disposal (5). Typically, the reclaimable materials employed in industrial recycling may consist of obsolete products, spent materials, industrial by-products or residues, or pollution control products. The recycling of many of these products is so well established that under standard commercial practices such materials are destined only for recovery, not for disposal.

The actual processing of industrial discards varies in complexity by material type. Recycling obsolete products, such as old or damaged automobiles, for example, may be quite simple. Typically, the hulk of the automobile is shredded and the pieces separated into ferrous and nonferrous metals (RECYCLING, FERROUS METALS; NONFERROUS METALS). The separated materials are then sent to be resmelted or are exported. In the United States, this form of recycling normally recovers 75% of the materials in obsolete automobiles (6). Alternatively, processing industrial by-products and pollution control products can be considerably more complicated. Because these materials often consist of complex mixtures of metals or chemicals, recycling must take place in several stages. Interestingly, the industrial recycling of these complex materials, many of which are considered hazardous, has both environmental and economic benefits. Not only does recycling separate valuable constituents, but in so doing it also removes hazardous materials. Thus, industrial recycling removes the threat that these materials pose to the environment and public health.

Determining the actual amount of industrial material that is recycled is difficult. Because much industrial recycling takes place at the plant level, few aggregate statistics are available. A 1992 survey of a number of commodity groups conducted by the U.S. Bureau of Mines, however, calculated that over 240 million metric tons of industrial material was being recycled yearly. This figure, which does not include an additional 2.8×10^3 t of organic material recycled, is very conservative and likely underestimates the extent of industrial recycling in the United States.

One illustration of the benefits afforded by industrial recycling is provided by reprocessing of dust collected from air pollution control equipment on steelmaking furnaces (7). Over 500×10^3 t of steelmaking dust, which contains mostly iron and constituents of slag, is collected annually in the United States. If sent to a landfill for disposal, the material would be classified as a hazardous waste and would have to be encased in three times its volume of concrete. Ironically, the metallic constituents of steelmaking dust which make it hazardous are the same constituents that also make the dust valuable. As a result, industrial processes have been developed that remove these valuable metals, leaving behind a slag that is not generally classified by U.S. law as hazardous.

Although industrial recycling has historically been very successful, there is significant debate about whether or not the reclamation of industrial material should be counted as recycling. Many environmental activists argue that the reuse of industrial material should not be regarded as true recycling because in many instances the material is pre-consumer rather than post-consumer.

This debate, however, appears to ignore the obvious environmental and economic benefits afforded by industrial recycling.

Municipal Solid Waste

Municipal solid waste (MSW) is most often defined as post-consumer solid waste generated by households (eg, single and multifamily units), commercial establishments (eg, retailers and offices), and institutions (eg, schools, hospitals, and government offices). Discards from each of these sectors account for approximately one-third of total MSW, respectively. Normally, MSW is classified as either material waste, ie, items such as paper, yard waste, metals, and glass, or product waste, which encompasses both durable and nondurable goods as well as packaging (qv). Beyond these simple classifications, defining MSW has been problematic because of disagreements regarding specific materials and the proper classification of pre- and post-consumer waste.

Interestingly, the difficulty defining MSW has led to many inaccurate policy conclusions. Most notably, it is often assumed that the United States generates far more waste than other (particularly European) countries. However, generally other countries define MSW as that which the municipality collects, ie, household waste. Given that household waste accounts for only about 45% of U.S. MSW, it is incorrect to conclude on the basis of aggregate figures that the United States is more wasteful than other industrialized countries.

Figure 1 schematically depicts the system that has developed to manage solid waste. Both materials recycling and energy recovery are viable options to either landfilling or nonrecuperative incineration. Composting, which is not present in Figure 1, is not widely used in the United States as a method of handling MSW (8). This is primarily because composted material contains relatively high concentrations of heavy metals and supplies very few plant nutrients. Thus, large-scale, commercially viable uses of compost have been limited (9). The problems associated with composting, however, do highlight an important feature of recycling: products must be recovered in quantities large enough and of sufficient quality to be acceptable to potential users or else these recovered materials cannot be recycled economically.

QUANTITY AND COMPOSITION

Because the actual quantity and composition of waste are highly dependent on local use habits, income, as well as the degree of urbanization, examining how much MSW Americans generate annually is difficult and confusing (10). The proportion of paper and packaging material in MSW, for example, may be significantly lower in rural communities than in cities due to a greater reliance on fresh foods and less access to newspapers and magazines. Similarly, research suggests that the quantity, and presumably the composition, of MSW has considerable seasonal variation (11). Understanding the factors that affect the local composition of waste is important in determining the actual amounts of recoverable materials and therefore potential revenues from recycling.

Estimates of per capita waste generation can be misleading. Although often reported in the popular press, the magnitude of these figures depends on the

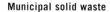

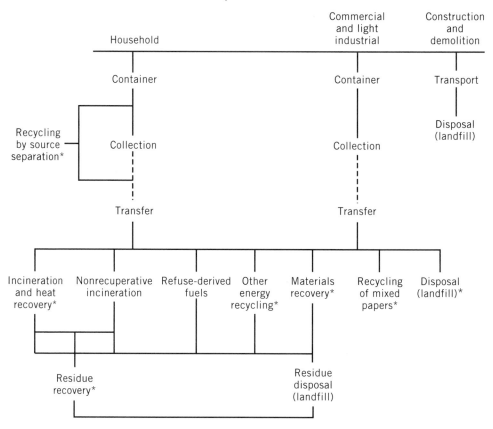

Fig. 1. Municipal solid waste management system where (*) indicates recycling options and (−−−), optional transfer.

size of the community, how the statistics are gathered, and the percentage of waste in a given residential region. One of 37 U.S. cities of varying sizes, for instance, found that daily per capita waste generation rates ranged from 0.9 to 4.3 kg (12). Such inconsistency underlines the importance of using locally gathered data when setting solid waste policy. The use of national averages in designing facilities to handle MSW can result in large economic mistakes due to inaccurate estimates regarding the amount and type of waste to be received.

According to the U.S. EPA's best estimates, Americans generated approximately 196 million metric tons of MSW in 1990. This figure is expected to rise to 208 million metric tons of waste in 1996, a projection which suggests a substantial slowing in the rate of increase in MSW generation. Figure 2 depicts the proportion of MSW by component materials since 1960 (13). A number of studies have examined waste generation rates in Europe (14). By far the largest contributors to MSW are paper and paperboard products (37.5% by weight) (see RECYCLING, PAPER). Yard waste, including leaves, grass clippings, weeds, and prunings, represents the second largest category of waste. The yard waste proportion of total discards has declined steadily, however, and this decline will

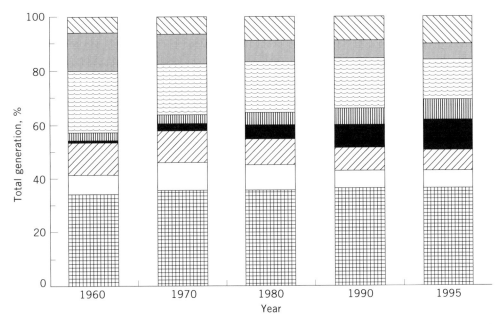

Fig. 2. Materials generated in the municipal waste stream from 1960 to 1995 (% of total generation), where (▦) represents paper and paperboard, (■) plastics, (▨) food waste, (□) glass, (▥) wood, (▨) metals, (▤) yard waste, and (◩) other.

likely accelerate because many individual states have banned yard wastes from municipal trash. The percentages of glass, metals, and food waste in MSW have likewise declined somewhat since the 1970s due in large measure to lightweighting and substitution by other materials.

On the other hand, the fraction of plastics has grown, increasing from <1% of MSW in 1960 to ~9.6% in 1995. This increase corresponds to the substitution of other materials with plastics and the greater reliance on plastics as a source of packaging material (see RECYCLING, PLASTICS). With regard to this increase in plastic packaging, it should be noted that the fraction of food residues in MSW is statistically related to plastic packaging. An analysis of 1990 data showed that a unit weight of plastic packaging has a high statistical probability of causing a reduction of 2.2 unit weights of food waste (15).

As with generation rates, the chemical composition of MSW varies significantly with local socioeconomic and demographic conditions. The average chemical composition of MSW in the United States is given in Table 1 (16).

There are numerous misconceptions about the sources of various chemical elements in waste, particularly those that are potential acid formers when the waste is incinerated or mechanically converted and used as a refuse-derived fuel. For example, it is often mistakenly stated that the source of chlorine in waste, hence a potential source of HCl emissions, is poly(vinyl chloride). The relative contents of selected, potentially acid-forming elements in the organic portion of a sample of waste collected from various households in one U.S. East Coast city is given in Table 2 (17). In this city, a chief source of chlorine in the waste is NaCl, probably from food waste.

Table 1. Analysis of MSW Composition, wt %[a]

Composition	Proximate analysis	Elemental analysis[b] As received[c]	Dry basis
moisture	19.7–31.3		
ash	9.4–26.8		
volatile	36.8–56.2		
fixed carbon	0.6–14.6		
carbon		23.45–33.47	48.7
hydrogen		3.38–4.72	
nitrogen		0.19–0.37	0.82
chlorine		0.13–0.32	0.66
sulfur		0.19–0.33	0.26
potassium			0.10
oxygen		15.37–31.90	

[a]Ref. 16.
[b]Magnetic metals removed.
[c]Having 19.7–31.3 wt % H_2O.

Table 2. Relative Contents of Selected Elements as Percentage of the Total[a]

Refuse category	Dry weight basis, organic portion only, wt % C	N	S	P	Cl
textiles	7.29	43.35	25.64	13.38	5.55
wood	4.74	0.80	1.75	1.07	0.73
garden waste	8.65	18.18	5.53	16.63	3.66
rubber and leather	5.81	4.10	17.14	1.05	14.17
food waste	9.21	29.31	8.23	49.62	17.04
paper	54.39	1.80	40.19	17.28	22.98
plastics	9.91	2.46	1.52	0.97	35.87
Total	*100.00*	*100.00*	*100.00*	*100.00*	*100.00*

[a]Ref. 17.

PROCESSING RECYCLABLE MATERIALS

Recovery Rates.   The rate at which MSW is recovered for recycling varies by region. The best estimate for the average recovery of MSW in the United States is approximately 15%. This figure underestimates total recovery because it excludes both domestic reuse, such as washing and reusing plastic containers, as well as the reuse of industrial scrap. As Figure 3 highlights, the amount of MSW that was separated for recycling was relatively small throughout the 1960s and 1970s; total recovery never exceeded 10% of generation (13). Increased public concern about the environment along with a massive proliferation of curbside collection programs, however, led to noticeable increases in recovery rates throughout the late 1980s and early 1990s (18). The U.S. EPA estimates that, given current trends, over 20% of MSW will be recovered for recycling by the year 2000.

Projected national recovery rates much beyond 20% are questionable. Although close to 85% (by weight) of MSW is composed of potentially recyclable

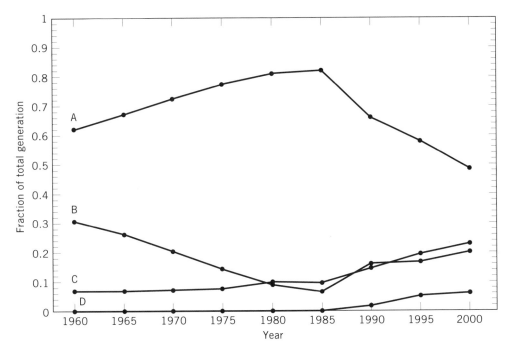

Fig. 3. Materials recovery, composting, combustion, and discards of municipal solid waste from 1969 to 2000 (1995–2000 estimated) as a fraction of total generation, where A represents landfill and other; B, combustion; C, recovery for recycling; and D, recovery for composting.

material (paper and paperboard, metals, glass, plastics, wood, and yard waste), far less is economically recyclable (19). Given market and technology constraints, one projection is that only 36 wt % of MSW is potentially recoverable for recycling (excluding yard waste) (19). A case study of Phoenix, Arizona was even more conservative, concluding that only 28 wt % of MSW was recoverable for recycling (20).

These potential recovery rates should not be equated with likely recovery rates. The cited recovery rate of 36% must be further reduced by the extent of participation, limits on what can be collected (eg, waste in sparsely populated regions is more difficult to collect than in densely populated communities), and recovered material discarded because it does not meet buyers' specifications for recycling. Given these constraints, the likely limit to the amount of MSW that can be recycled appears to be between 16 and 25%, depending on the amount of yard waste recycled, ie, if composting is considered to be recycling. Indeed, many of the more ambitious state recycling goals (some approaching 50%) seem well out of reach. A 1994 report concludes that reaching recovery rates beyond 25% will require doubling curbside collection by 2000, a logistically formidable and costly proposition (21).

Preparation of Collected Materials. The actual amount of recovered MSW that can be recycled to meet buyers' quality specifications is highly dependent upon how the material is collected and processed. There are primarily three methods available to collect MSW for recycling: mixed waste, waste with commingled recyclables, or waste with separated recyclables. Which method of collection is chosen, in turn, determines the amount of preparation that is needed prior to reclamation and reuse.

A large percentage of the MSW directed to recycling, particularly that collected through residential curbside collection programs, is processed at material recovery facilities (MRFs). In 1993 there were 172 MRFs operating in the United States, processing 18,644 tons a day for an average of 132 tons a day for each MRF (22). The amount of equipment required to process recyclables in these facilities varies significantly. At one extreme, MRFs can operate with only a tipping floor, where MSW is dumped for manual sorting, and a baler. Other MRFs are highly mechanized, employing automated sorting and processing equipment connected by a network of conveyor belts. All MRFs, regardless of the degree of automation, must rely on a good deal of manual labor for certain sorting and quality control functions. On average, human labor is the largest component of MRF operating expenses, accounting for >33% of overall processing costs.

Mixed MSW. The preparation of mixed MSW for recycling is essentially a four-step process. In the first stage, commonly referred to as previewing, the mixed waste is dumped on a tipping floor where oversized materials, potential explosives and readily flammable materials, and any other items that could damage processing equipment are removed. The waste is then crushed or shredded. Reducing the volume of mixed MSW makes for easier handling and more cost-effective shipping. The different components of the waste stream are separated from each other in the removal or segregation process. Depending on the composition of the mixed MSW, several different technologies may be employed at this stage including air classification to separate lighter materials in the stream from heavier ones, magnetic separation to remove ferrous metals, and screening to separate materials of different size. Manual labor is also used at this stage to separate materials such as newspaper, glass, and different types of plastic (12). After the segregation process, valuable materials are normally baled or otherwise prepared for transportation to market. Residue waste is either incinerated or landfilled. In many systems, the leftover material is shipped to a waste-to-energy facility where it is converted into refuse-derived fuel.

The principal advantage of handling mixed MSW is that it requires no change in the existing waste collection system. The manager of a processing facility can simply recover those materials for which market conditions are favorable, and dispose of the remaining waste. Unfortunately, mixed waste facilities are capital-intensive and operating them requires relatively high amounts of energy and maintenance. As a result, mixed waste processors must reap greater revenues from the sale of recyclable materials or charge higher tipping fees in order to cover operating expenses. In addition, some materials, especially paper, plastics, and corrugated materials, become too contaminated to be recycled.

Commingled Recyclables. The technology required to process commingled recyclables is dependent upon the types of materials collected. As a general rule, however, commingled materials are first dumped into a receiving pit. After initial inspection, they are loaded onto conveyors and separated using many of the same techniques described above. As with mixed waste processors, handlers of commingled materials normally employ a combination of automated and manual systems.

Relative to mixed MSW, commingled waste has at least two advantages. First, the risk of contaminating recyclable materials with foreign waste is significantly reduced. Thus, recyclables ultimately sent to market are often of higher quality, thereby enabling processors to obtain higher prices. Second, facility managers do not have to worry as much about hazardous waste materials threatening equipment or employees.

The primary disadvantage of commingled collection is that municipalities must operate a different collection system for recyclables than for other household and commercial waste. The success of these new collection programs is highly dependent on public participation. As a result, the amount of waste collected in commingled systems may be small.

Separated Recyclables. Even when initial separation of recyclables takes place at the household level, the separated materials still require some preparation before being sent to market. By visually inspecting materials, facility processors are able to remove any remaining contaminants. This ensures that recyclables will be of sufficient quality to meet buyer specifications. In addition, materials are often shredded, crushed, or baled to facilitate cost-effective shipping.

There are four key advantages to handling separated materials: (*1*) separated materials systems are far less labor-intensive than other collection schemes (as mentioned earlier, labor costs are the largest component of most recovery facilities' operating expenses); (*2*) the equipment needed to handle separated material is relatively simple and inexpensive; (*3*) source separation is often the only method of resource recovery suitable for small communities; and (*4*) separated programs can be designed and implemented quickly.

As with commingled recyclables, however, processing separated materials requires a different collection system, thereby increasing the cost of local solid waste programs. Moreover, the success of separation systems requires extraordinary public cooperation. The general experience is that only the higher socioeconomic groups are likely to participate. Many working-class communities may not desire to participate or may become easily disenchanted with the program. If overall participation is low, material collection (and sales) may not be sufficient to cover the costs of collection and other required activities.

Refuse-Derived Fuel. Many processing facilities divert a portion of the material that is not recovered for recycling to waste-to-energy plants, also referred to as resource recovery facilities, where the material is employed as fuel. The processes involved in the production of refuse-derived fuel (RDF) are outlined in Figure 4 (23). Nine different RDFs have been defined, as listed in Table 3 (24). There are several ways to prepare RDF-3, which is perhaps the most popular form and is the feed used in the preparation of densified refuse-derived fuel (d-RDF). All forms of RDF are part of the broader set of waste-derived fuels (WDF), which includes various waste biomass, eg, from silvaculture or agriculture (see FUELS FROM BIOMASS; FUELS FROM WASTE).

RDF-3 is intended for use as a supplement with coal for semisuspension or suspension firing or for use by itself in similar boilers. d-RDF is intended as a

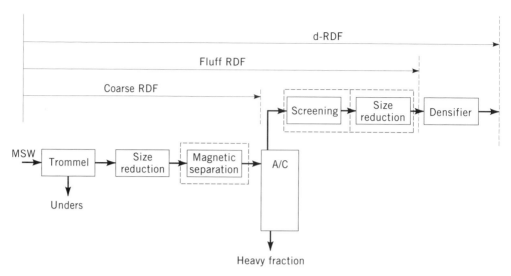

Fig. 4. Typical sequence of unit processes for RDF production where A/C = air classification. Optional locations are indicated by surrounding dashed lines.

Table 3. Definitions of Refuse-Derived Fuels[a]

RDF-1	wastes used as fuel in discarded form
RDF-2	wastes processed to coarse particle size with or without removal of magnetic metals
RDF-3	as MSW-derived shredded fuel which has been processed for the removal of metal, glass, and other entrained inorganic material; generally, this material has a particle size such that 95 wt % passes through a 5-cm mesh screen
RDF-4	combustible waste processed into powder form; 95 wt % passes through a 2.0-mm (10-mesh) screen
RDF-5	combustible waste compressed into pellets, slugettes, cubettes, or briquettes
RDF-6	combustible waste processed into gaseous fuel

[a]Ref. 24.

supplement with stoker coal or for use by itself in stoker boilers. Several methods and alternatives for producing RDF-3 or d-RDF have been described (25).

Because there is no single material called RDF and because the composition and therefore fuel properties depend on the composition of the starting MSW and the methods of processing, it is impossible to give what might be an average set of fuel properties. Table 4 gives the results of a typical RDF fuel analysis from a waste-to-energy plant in Maine (26). A number of analyses have examined how the presence of ash as well as the high moisture content of RDF affect its quality as a fuel. The thermodynamic balance for possible drying of the fuel has also been examined (27). It is unlikely that any RDF production process will be able to afford drying the material.

Mechanical and Chemical Recycling. The vast majority of recovered materials which are not burned for energy are simply remanufactured into second-generation products. Such mechanical recycling works primarily by applying heat (or in the case of paper, various chemicals) to the sorted and cleaned waste

Table 4. Typical RDF Fuel Analysis[a], %

Component	Percent[b]
Ultimate analysis	
moisture	29.2
carbon	32.2
oxygen	24.2
hydrogen	4.2
nitrogen	0.4
chlorine	0.1
sulfur	0.2
ash	9.5
Proximate analysis	
moisture	29.2
volatile matter	52.3
fixed carbon	9.0
ash	9.5
heating value, kJ/kg[c]	13,450

[a]Maine Energy Recovery Co. (computed on an as-received basis).
[b]Unless otherwise noted.
[c]To convert kJ/kg to Btu/lb, divide by 2.319.

and then refashioning the liquid material into new products. For many materials (especially plastics), however, mechanical recycling has several significant drawbacks: it is labor intensive and therefore quite costly to operate, it requires relatively clean streams of post-consumer materials, and in the case of plastics it requires separation by resin type and color to achieve high market value. As a result, a number of projects are underway to develop chemical technologies that can convert recovered wastes back into the higher value raw materials from which they were made (28,29).

Widespread interest in chemical recycling has thus far been confined largely to Europe (30). This is primarily because tough new recycling regulations, particularly in Germany, have made massive investments in these advanced technologies more economically attractive (2). Certain methods of chemical recycling, including the methanolysis and glycolysis of post-consumer plastics, have received attention in the United States, but commercial application of these techniques has been limited by the need for a clean, relatively pure feedstream. The plastics industry claims to be making improvements in these technologies which will reduce the need for cleaning and sorting (31). Serious impediments to the widespread use of chemical recycling still exist, however, including public opposition, the large capital expenditures for new chemical recycling facilities, and the present low prices for many virgin materials.

Economic Aspects of Recycling

Production. Several key components of MSW enjoy relatively high rates of recycling. Of all the aluminum produced in MSW in 1990, 38.1% was recovered for recycling. Over 60% of this recovered aluminum was comprised of aluminum beverage containers. The primary reason for the success of aluminum recycling is that collecting and reprocessing post-consumer aluminum is more cost effective than mining and processing bauxite. Similarly, paper and paperboard is recycled at approximately a 29% rate.

Typically, it takes decades to achieve such high recycling rates, eg, the case of aluminum. As Figure 5 illustrates, five years after beginning an industrywide push for recycling, the recycling rate for aluminum cans was less than 5% (32). A steady climb took place for the next 10 years with rates reaching approximately 25%. Only after 20 years did aluminum can recycling hit the nearly 50% rate. Recycling of poly(ethylene terephthalate) (PET) soft drink bottles is slightly ahead of this pace.

Other MSW discards such as glass (20% recycled), ferrous metals (13%), and plastics (2%) may begin to exhibit marked improvement in their recycling rates due to the explosion in curbside recycling programs and in construction of reprocessing facilities. Numerous private-sector initiatives to build recycling infrastructures are underway. The success of these efforts will ultimately depend on a variety of factors including the future composition of MSW, public participation rates, the ability to substitute capital for labor in the processing of materials, and the availability of competitively priced virgin materials.

Economic Analysis.   The economic success of recycling programs is subject to the following inequality where X = the cost to recover recyclable materials, Y = the cost of disposal, and Z = the value of the resource recovered.

$$X - Y \leq Z$$

Basic economic theory suggests that, in the earliest periods, society will rely primarily on virgin material because it is cheaper than collecting and recycling post-consumer goods. As the stock of virgin material is consumed over time, however, a point is reached when the costs of extraction and the price of this material will begin to rise. With the rise in virgin prices, consumer demand for alternative materials including recyclables will slowly increase as will Z.

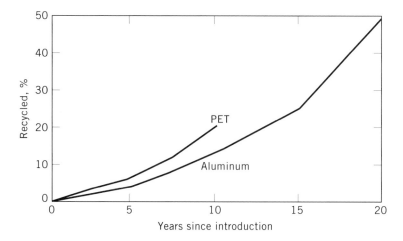

Fig. 5.   Nonrefillable container recycling.

Concurrently, increased investment in technology will likely lower the cost (X) to recover and recycle post-consumer materials. Eventually, the inequality above is satisfied for some materials and recycling becomes economically viable.

In practice, the massive increase in the supply of recyclable materials of late has distorted this process. Since 1985, the amount of material recovered for recycling from MSW has increased threefold, from less than 17 million tons a year in 1985 to over 50 million tons (estimated) in 1995. Unfortunately, this surge in supply has far outpaced the demand for recyclables, causing a precipitous decline in the price paid for these materials. As Figure 6 illustrates, the average scrap value for recyclables declined from \$97/t in 1988 to just over \$44/t in 1992 (19). Comparison of this price to the average cost of collecting and sorting recyclables (~\$175/t on average), makes the economic dilemma obvious. Assuming a buyer for recycling materials can be found, the market price paid to the recycler covers barely one-fourth of the collection and sorting costs, resulting in a large net loss.

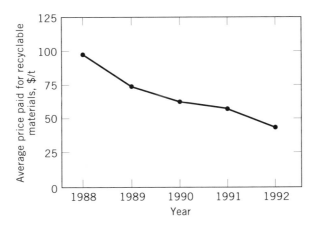

Fig. 6. Supply-side effects of U.S. curbside collection programs.

Table 5 shows the ratio of scrap values for a selected group of materials to the net recovery costs for these materials. The net recovery costs are calculated by subtracting average landfill dumping fees from the collection and processing costs ($X - Y$). If recyling a particular material is to make economic sense for a municipality, the ratio of scrap value to net recovery costs should equal or exceed 1. Only recycling aluminum cans is economically justified according to these national averages.

As mentioned earlier, using national averages is misleading because local conditions, such as high landfill dumping fees or a nearby reprocessing plant that can use the recovered materials, may make recycling other products economical. The comparisons made in Table 5, however, highlight the sort of economic analysis that can help municipalities determine which materials merit recycling.

Increasingly, however, municipalities are being denied the opportunities to make these sorts of economic comparisons by intrusive state and federal regulations. A variety of studies have examined government recycling programs in Europe (2,33). Although U.S. legislation has tended to be less stringent

Table 5. Ratio of Scrap Value to Net Recovery Costs[a],
$Z/X-Y$

Material	Ratio
newspaper	0.105
corrugated containers	0.279
mixed waste paper	0.057
aluminum cans	4.194
steel cans	0.403
clear glass	0.374
PET bottles and containers	0.496
HDPE[b] bottles and containers	0.495

[a]This ratio was calculated by the following method. The average cost to collect and transport recyclables ($125) was added to the processing costs for each of the materials listed. The average landfill tipping fee ($30) was then subtracted from this total, giving the denominator of the ratio. The numerator is simply the average scrap value for each of the recovered materials listed.
[b]Natural and mixed-color high density polyethylene.

than that in Europe, programs designed to increase the demand for recyclable materials have been considered or enacted at both the state and federal levels. Among the states, the demand-side programs that have been debated vary. In several instances, legislators have simply banned materials that could not easily be reused or recycled, in effect forcing manufacturers to utilize recycled or recyclable materials. A more common approach is to enact recycled-content mandates, or laws specifying what percentages of recycled materials must be used in manufacturing certain products.

As of January 1993, 24 states, which together consume over 60% of the newsprint in the United States, had minimum-content requirement for newspaper. Table 6 provides a listing of these and other state recycled-content measures (34). Idaho and Virginia require that recycled materials be used in road construction projects. California, Connecticut, Maryland, Oregon, and Wisconsin have content laws covering items such as telephone directories and glass (see RECYCLING, GLASS).

The motivation for measures such as these is simple. Government supply-side programs have led to the over-collection of certain recyclable materials, which has in turn driven down the price paid for these materials. By establishing stringent recycled-content laws, policymakers hope to force manufacturers to take financial responsibility for the waste they produce. The logic seems to be that as recycled-content and reuse requirements increase the demand for recyclable materials, the value of these materials will increase. Only by artificially elevating scrap values, it is argued, can recycling be made profitable and sustainable.

In fact, recycling was a valuable U.S. industry for decades (witness the success of industrial recycling), especially in metals, newsprint, and some types of glass and aluminum cans. Only when regulators, in response to public pressure, attempted to mandate the collection of materials that were in many instances expensive to collect and process did recycling begin to encounter problems.

Table 6. State Minimum-Content Laws

State	Material	Recycled-content requirement
Arizona	newsprint	50% by 2000
California	newsprint	50% by 2000
	glass containers	65% by 2005
	plastic bags	10% for 1.0 mil[a] bags by 1993; 30% for 0.75 mil[a] bags by 1995
	plastic containers	25% post-consumer material or 25% recycling rate or reusable/ refillable five times or reduced 10% in content by 1995
Connecticut	newsprint	50% by 2000
	telephone directories	40% by 2001
Florida	newsprint	50% by 1993
Illinois	newsprint	23% by 1993
Maryland	newsprint	40% by 1998
	telephone directories	40% by 2000
Missouri	newsprint	50% by 2000
North Carolina	newsprint	40% by 1997
Oregon	glass containers	50% by 2000
	newsprint	7.5% by 1995
	plastic containers	same as California law
Rhode Island	newsprint	40% by 2000
Texas	newsprint	30% by 2000
Wisconsin	newsprint	45% by 2001
	plastic containers	10% by 1995

[a] 1 mil = 25.4 μm.

The future success of recyling largely depends on how governments respond to the problems described above. There are serious questions as to whether it is wise for the government to interefere further in recycling markets. Indeed, in many instances, collection programs should be scaled back to collect only those materials for which there is demand. Moreover, recycling rates should not be the proxy for how well society utilizes its resources. The most successful waste management programs are those that consider the relative cost effectiveness and environmental impacts of all available options for handling waste: source reduction, recycling (including composting), incineration, and landfilling. Ignoring the economics of waste management, as many government programs do, has virtually no environmental benefits and may rob funds from other important social programs.

BIBLIOGRAPHY

"Introduction" under "Recycling" in *ECT* 3rd ed., Vol. 19, pp. 936–951, by H. Alter, Chamber of Commerce of the United States.

1. J. H. Alexander, *In Defense of Garbage*, Praeger Publishing, Westport, Conn., 1993; W. L. Rathje, "The History of Garbage," *Garbage Mag.* (Sept./Oct. 1990).
2. C. Boerner and K. Chilton, *Environment* **36**(1), 6–15, 32–33 (1994).

3. Franklin Associates, Ltd. and Keep America Beautiful, Inc., *The Role of Recycling in Integrated Solid Waste Management to the Year 2000: Summary*, Keep America Beautiful, Inc., Washington, D.C., Sept. 1994.

4. C. Boerner, in Ref. 3; S. Pattison, *Consumer's Res.*, 36 (Sept. 1993).

5. H. Alter, "The Recycling of Industrial Materials," *Design for Avoidance, Design for Recycling*, EMPA, Dübendorf, Switzerland, 1995.

6. H. Alter, in Ref. 5, p. 65.

7. H. Alter, in Ref. 5, pp. 62–63.

8. G. E. Stone, C. Wiles, and C. Clemons, *Composting at Johnson City*, Vols. 1 and 2, Final Report on Joint USEPA–TVA Composting Project, U.S. Environmental Protection Agency, Cincinnati, Ohio, 1975; Report EPA/530/SW-31r.2; *Guidelines for Local Government on Solid Waste Management*, Reprot SW-17c, U.S. Environmental Protection Agency, Washington, D.C., 1971. For a more general discussion of composting see R. T. Haug, *Composting Engineering: Principles and Practice*, Ann Arbor Science Publishers, Inc., Ann Arbor, Mich., 1980.

9. H. Alter, *The Greatly Growing Garbage Problem: A Guide to Municipal Solid Waste Management For Communities and Businesses*, U.S. Chamber of Commerce, Washington, D.C., 1988.

10. A. Porteous, *Refuse Derived Fuels*, Applied Science Publishers, London, 1981; H. Alter, *Resource Rec. Conserv.* **5**(1), 1 (1980); W. Rathje and C. Murphy, *Rubbish: The Archaeology of Garbage*, Harpers Collins, New York, 1992.

11. J. C. Even, P. Arberg, J. R. Parker, and H. Alter, *Res. Conserv.* **6**(3/4), 197 (1981).

12. U.S. Congress, Office of Technology Assessment, *Facing America's Trash: What Next for Municipal Solid Waste*, OTA-O-424, U.S. Government Printing Office, Washington, D.C., 1989.

13. Franklin Associates, *Characterization of Municipal Solid Waste in the United States: 1992 Update*, Report EPA/530-R-92-019 (PB 92-207-166), U.S. Environmental Protection Agency, Office of Solid Waste and Emergency Response, Washington, D.C., 1992.

14. World Resources Institute, *World Resources 1988–89*, Basic Books, Inc., New York, 1988.

15. H. Alter, *Waste Man. Res.* **11**, 319–332 (1993).

16. *Resource Recovery and Waste Reduction: Third Report to Congress*, Report SW-161, U.S. Environmental Protection Agency, Washington, D.C., 1975; R. A. Lowe, *Energy Recovery from Waste*, Report SW-36d.ii, U.S. Environmental Protection Agency, Washington, D.C., 1975.

17. H. Alter, G. Ingle, and E. R. Kaiser, *Solid Wastes Man. (England)* **64**(12), 706 (1974).

18. C. Boerner, in Ref. 2.

19. Franklin Associates, *Characterization of Municipal Solid Waste in the United States: 1990 Update*, Report EPA/530-SW-90-042 (PB 90-215112), U.S. Environmental Protection Agency, Office of Solid Waste and Emergency Response, Washington, D.C., 1990.

20. W. L. Rathje, D. C. Wilson, and W. W. Hughes, *The Phoenix Recyclables Report: Characterization of Recyclable Materials in Residential Solid Waste*, report to the City of Phoenix, Arizona, The Garbage Project, Bureau of Applied Research in Anthropology, University of Arizona, Tucson, 1989.

21. Ref. 3, p. iii.

22. *The Cost to Recycle at a Materials Recovery Facility*, National Solid Wastes Management Association, Washington, D.C., 1992.

23. L. Diaz, G. Savage, and C. G. Golueke, *Resource Recovery From Municipal Solid Wastes Volume II: Final Processing*, CRC Press, Boca Raton, Fla., 1982.

24. ASTM Standards E775, E776, E777, E778, E790, E791, and others in preparation by subcommittee E38.01, American Society for Testing and Materials, Philadelphia, Pa., 1981.

25. H. Alter and J. J. Dunn, Jr., *Solid Waste Conversion to Energy: Current European and U.S. Practices*, Marcel Dekker, Inc., New York, 1980, Chapt. 5; H. Alter and J. A. Campbell, in J. L. Jones and S. B. Radding, eds., *The Preparation and Properties of Densified Refuse-Derived Fuel, Thermal Conversion of Solid Wastes and Biomass*, American Chemical Society, Washington, D.C., 1980, pp. 127–142.

26. P. J. Knox ed., *Resource Recovery of Municipal Solid Waste*, American Institute of Chemical Engineers, New York, 1988.

27. H. P. Sheng and H. Alter, *Resource Rec. Conserv.* **1**(1), 85 (1975); Ref. 25, Chapt. 8.

28. D. Rothman and E. Chynoweth, *Chem. Week*, 20 (Mar. 2, 1994); V. Williams, "Chemical Recycling for Rubbers," *Eur. Rubber J.* (July 1993); R. Goddard, *Packaging Week (U.K.)*, **9**(4), 21 (June 1993).

29. A. Miller, *Chem. Ind. (U.K.)*, (1), 8 (Jan. 3, 1994).

30. P. Mapleston, "Chemical Recycling May be an Option to Meet Reclaim Levels," *Modern Plast.* (Nov. 1993), courtesy Lexis/Nexis Information Services.

31. Technical data, "The Evolution of Plastics Recycling Technology," American Plastics Council, Washington, D.C., 1993.

32. *The New Materials Society: Materials Shifts in the New Society*, Vol. 3, U.S. Department of the Interior, U.S. Bureau of Mines, Washington, D.C., 1990, p. 4.19.

33. C. Boerner and K. Chilton, "Who is Responsible for Garbage," *Environ. Eng.* **30**(4), 16 (Oct. 1994); J. W. Porter, *European Recycling Activities . . . And Lessons for the United States*, Waste Policy Center, Sterling, Va., May 1994.

34. *Waste Age*, 30 (Mar. 1993).

General References

J. Abert, ed., *Resource Recovery Guide*, Van Nostrand Reinhold Co., New York, 1983.

American Institute of Chemical Engineers (AICHE), *Status Report on Municipal Solid Waste Recycling*, The Government Programs Steering Committee of the AICHE, Washington, D.C., 1992.

C. Boerner and K. Chilton, *Environment* **36**(1), 7–15, 32–33 (Jan./Feb. 1994); *Am. Enterprise* **5**(2), 14–18 (Mar./Apr. 1994); *USA Today* **122**(2588), 78–81 (1994); *Environ. Eng.* **30**(4), 16–17, 23–25 (1994).

A. Brown and co-workers, *Energy Recovery Through Waste Combustion*, Elsevier Applied Science, London, 1988.

L. Diaz and co-workers, *Resource Recovery From Municipal Solid Waste*, Vols. I and II, CRC Press, Boca Raton, Fla., 1982.

J. Holmes, *Refuse Recycling and Recovery*, John Wiley & Sons, Inc., New York, 1981.

P. Knox, ed., *Resource Recovery of Municipal Solid Wastes*, American Institute of Chemical Engineers, New York, 1988.

W. Rathje and C. Murphy, *Rubbish: The Archaeology of Garbage*, Harpers Collins, New York, 1992.

T. Tietenberg, *Environmental and Natural Resource Economics*, 3rd ed., Harper Collins, New York, 1992, Chapt. 8.

U.S. Congress, Office of Technology Assessment, *Facing America's Trash: What Next for Municipal Solid Waste*, OTA-O-424, U.S. Government Printing Office, Washington, D.C., Oct. 1989.

Christopher Boerner
Kenneth Chilton
Washington University

METALS

Ferrous metals, **1092**
Nonferrous metals, **1106**

FERROUS METALS

Recycling of ferrous scrap is a principal worldwide activity where more than 300 million metric tons are consumed annually. Numerous grades of scrap and many technical and economic complexities are involved. The focus herein is on the United States, where scrap consumption in 1994 totaled 70 million t. The term ferrous metals refers to iron (qv) and steel (qv) scrap materials of three origins: home scrap, generated within the steel mill or foundry, and being a known material of high quality; prompt industrial scrap, produced from trimmings and discards during product manufacture; and obsolete scrap, ie, old scrap from discarded or rejected items. Ferrous scrap supply is roughly estimated at about 35% home scrap, 25% prompt industrial scrap, and 40% obsolete scrap (1,2).

Primary consumers for ferrous scrap are the iron and steel mills and foundries. Minor consumers include ferroalloy producers, copper producers for use in copper precipitation (see RECYCLING, NONFERROUS METALS), and the chemical industry. The steel industry consumes about three-fourths of the total. Scrap consumption for ferroalloy production, copper precipitation, and the chemical industry total less than one million t. The United States is the leading exporter of ferrous scrap, exporting almost nine million t in 1994, valued at about $1.3 billion. Total value of domestic scrap purchases and exports in 1994 was $8 billion (2).

History of Ferrous Scrap Recycling

A comprehensive history of ferrous scrap recycling in the United States beginning in 1646 has been published (3). Some selected events of interest include the first open-hearth steel produced (1868); the first commercial direct-arc electric furnace steel produced in the United States (1906); the first large induction furnace installed in the United States (1914); the first basic oxygen steel produced in the United States (1954); the rise of continuous casters and scrap-based minimills (1965); the first thin-slab continuous caster at a minimill (1989); U.S. exports and imports of ferrous scrap at record highs (1990); and the end of production of open-hearth steel (1991).

Supply and demand of scrap is affected by a variety of factors, including technological advances in iron- and steelmaking and scrap processing, war-time effects such as price controls and export restrictions, economic depressions, and foreign trade in scrap. The open-hearth furnace, which was once a primary part of steelmaking in an integrated plant, was capable of melting up to 100% scrap or any combination of hot metal (molten pig iron) and scrap. Open-hearth steelmaking declined rapidly in the 1980s and was phased out in 1991 in the United States. The lost demand for scrap by this process has been offset by advances in electric arc furnace (EAF) and basic oxygen furnace (BOF) processes. Developments impacting scrap include minimills, continuous casting, and minimill thin-slab casting.

Sources and Types of Ferrous Scrap

Home scrap, because it is generated within the plant during the production of steel (qv) or cast iron, has a known composition and is always recycled. Manufacturing facilities and steel service centers produce prompt industrial scrap during preparation to fabricator's specifications and during the fabrication of various industrial, commercial, or consumer products. Prompt industrial scrap is recycled because its chemical composition and physical characteristics are known or readily obtainable. Additional processing of the prompt industrial scrap may be required in some cases. For example, detinning and compacting may be required for new tin-plate scrap to be fully recyclable in the steel industry. Also, the increased use of galvanized steel in automobile manufacture and increasing environmental restrictions on EAF dust may lead to additional processing for galvanized scrap as this material becomes available in larger quantities.

The availability of prompt industrial scrap is directly related to the level of industrial activity. Producers generally do not accumulate prompt industrial scrap because of storage requirements and inventory control costs. Thus, it is rapidly available to the scrap consumer or the ferrous scrap industry. Prompt industrial scrap comes from imported steel as well as domestic steel mill products. Obsolete scrap, also known as old or post-consumer scrap, is widely used. Trends in new steelmaking capacity and the reduced proportions of premium scrap indicate that use of obsolete scrap should be expected to increase.

Descriptions of Primary Ferrous Scrap Grades. *Low Phosphorus Plate and Punchings.* This scrap, a prompt industrial scrap consisting of punchings or stampings, plate scrap, and bar croppings, has uniform known chemical analysis, is clean, and has high furnace recovery (yield) when melted. Maximum allowable phosphorus and sulfur contents are 0.05% each. Other unwanted residual elements are also uniformly low. Size limitations are specified.

No. 1 Busheling. This scrap is a clean prompt industrial scrap limited to about 30 cm (12 in.) in any dimension. It includes new factory busheling such as sheet clippings and stampings (free of old auto body and fender stock); metals which have been coated, limed, or vitreous enameled; and electrical steel containing more than 0.5% silicon.

No. 1 and Electric Furnace Bundles. No. 1 bundles produced at the plants where scrap is generated are designated No. 1 industrial or No. 1 factory bundles. This prompt industrial scrap consists of light-gauge sheet (no coated, rusted, or incinerated steel) tightly compacted to dense bales and having a bulk density of at least 1200 kg/m^3 (75 lb/ft^3). Bundles produced in a dealer's plant from prompt scrap collected at the scrap generating plant are designated No. 1 dealer bundles. Electric furnace bundles are smaller bales suitable for foundry use.

Cut Structural and Plate. A premium scrap at least 0.6 cm (1/4 in.) thick which is prepared in different lengths, this consists of clean structural shapes, crop ends, ship scrap, and shearings having maximum phosphorus and sulfur contents of 0.05% each. It is frequently marketed as prompt scrap.

No. 1 Heavy Melting Steel. Wrought iron and/or steel scrap ≥0.6 cm (1/4 in.) in thickness is available in three grades according to size limitations. Although not usually considered a premium grade, it is available in all primary scrap markets. Because of its wide use over many years, No. 1 heavy melting scrap is commonly used as a bellwether or reference grade for pricing purposes and various analyses of scrap use, price, and availability.

No. 2 Heavy Melting Steel.   Supplied in more than one grade form based on length and width requirements and a minimum thickness of about 0.3 cm (1/8 in.), this scrap is lower quality than No. 1 heavy melting. There is the possible inclusion of a wide variety of items such as alloy steels and coated materials.

No. 2 Bundles and Other Bundles.   This scrap is officially described as old black (uncoated) and galvanized steel. It is hydraulically compressed to a suitable size for furnace charging and weighs at least 1200 kg/m^3 (75 lb/ft^3). It is free of tin or lead coated or vitreous enameled material and may include household appliances and other obsolete light-gauge items. It may be designated as No. 3 bundles if a significant amount of coated materials are present. Steel can scrap obtained from a municipal waste incinerator and compressed into dense bales is referred to as an incinerator bundle. Other bundle designations include Terne plate bundles, bundled No. 2 steel, and silicon bundles, depending on the specific items contained.

Railroad Rails.   A relatively high quality scrap is generated by railroad rails. There are some variations in chemical analysis, however, depending on when the rail was produced.

Turnings and Borings.   This scrap grade is a broad description for the fragments generated when machining steel (turnings) and cast iron (borings). There are several distinct grades that include only steel turnings in the loose, crushed, or briquetted form; similar forms from mixtures of turnings and borings; and loose or briquetted borings. Turnings and borings are generated as prompt industrial scrap, but are commonly unsegregated and nonuniform in quality. Uniform higher quality turnings and borings are available at a higher price. Possible contaminants include elements from alloy steels; machinability additives such as bismuth, tellurium, lead, and sulfur; nonferrous machining fragments; and water and oil.

Slag Scrap.   Steelmaking slag scrap consists of irregular-sized nuggets of steel recovered magnetically from crushed steelmaking slag. Depending on the steels produced and scrap used during the steelmaking process, this scrap may have various alloying or tramp residuals. Ironmaking slag scrap is a high carbon iron magnetically recovered from crushed blast furnace slag. The metallic portion of this scrap has a uniform chemical analysis and is desirable for iron foundries.

Shredded or Fragmentized.   This scrap is predominantly shredded automobiles and light trucks. It may include scrap from other products such as appliances. The Institute of Scrap Recycling Industries, Inc. (ISRI) specifications describe two grades, one having a higher average density resulting from more effective shredding. This also allows the magnetic separators and air classifiers to achieve a better separation of the ferrous material from the nonmagnetic fractions. Shredded scrap is widely used in steel mills and iron foundries.

Machinery and Cupola Cast Iron.   This designation combines two cast-iron grades known as drop broken machinery cast and cupola cast. The former consists of clean, heavy machined cast iron that has been broken into pieces suitable for cupola charging. Cupola cast consists of a wide variety of cast-iron scrap, including automobile blocks and parts from agricultural and other machinery, in sizes suitable for the cupola. Machinery and cupola cast-iron scrap are primary sources of scrap for the iron foundries.

Cast-Iron Borings. Several grades of cast-iron borings and drillings that are free of steel turnings, lumps, and rusted material are called cast-iron borings. Some are supplied in hot or cold briquetted form to specified densities. Oil and water limits may be specified also.

Motor Blocks. Automobile and truck motors prepared to varying degrees of quality constitute a grade of scrap called motor blocks. The higher quality grades are stripped blocks from which most of the steel and nonferrous and nonmetallic parts are removed and the blocks are broken to cupola size. Degreasing the scrap further improves its quality.

There are more than 100 distinct grades or codes for ferrous scrap listed by the ISRI (4). Some have similar grade descriptions but differ in physical size, form, unacceptable items, and certain chemical limitations. The ISRI descriptions form the basis for many detailed specifications negotiated between the scrap supplier and scrap customer. These are typically more elaborate and definitive regarding acceptable quality for a particular user's plant. Using the ISRI and other sources, descriptions and comments for some of the most commonly used grades are given herein (5,6). These represent about 85% of the total scrap purchases made from 1991 through 1994.

Scrap Sources and Processing. More than 5000 scrap processing facilities in the United States purchase scrap from individuals and businesses and prepare it to specification for delivery to the mills, foundries, and other consumers of scrap (7). Most of the scrap preparation facilities are small, but there are also large, corporate-owned plants having numerous employees and the capability to collect and process large amounts of scrap. Total scrap processing capability is estimated at about 130 million t/yr (4). Processing of scrap involves sizing as required by the consumer and sorting to take advantage of alloy content and control residual elements.

The primary types of equipment used in scrap preparation are shredders, shears, and balers. Shredders capable of reducing automobiles, appliances, and other scrap to small fragments (fist-size) are common. Shears are used to cut large pieces of scrap into sizes as needed for convenient handling by the mill, foundry, or other scrap consumer. Balers are used to compact light-gauge material into rectangular bundles for easy handling and furnace charging. Some large automotive and appliance stamping plants have their own equipment to prepare scrap for direct shipment to steel mills. Others collect the scrap for removal by the scrap facility that buys the scrap. Considerable coordination is required to assure proper scrap segregation so that the highest scrap value and consistent supply is maintained.

Automobiles are the largest source of obsolete scrap. Other important sources of obsolete scrap include the demolition of steel structures and railroad companies. The latter provide a steady flow of scrap from their fabricating shops and from the recovery of worn out or abandoned track and railroad cars. All iron and steel products are recyclable if economically retrieved when scrapped.

Modern shredding and upgrading techniques are factors in the widespread acceptability of automobile scrap. Each vehicle contains about 70% or one ton of recoverable iron and steel scrap (8). Most scrap autos go through dismantling operations to remove discrete salvageable items such as the battery, radiator, alternator, and catalytic converter (see RECYCLING, NONFERROUS METALS). The dismantled autos are typically flattened to increase bulk density and the number of units per shipment to the shredding plant. Shredding has become the primary auto scrap recycling process. The scrap industry in the United States operates more than 180 shredders having a total annual shredded scrap output of over 10 million t (9). Shredders reduce the autos to several centimeters or fist-size fragments. The shredded material is processed through air classifiers, magnetic separators, liquid media operations, and sometimes manual sorting to upgrade scrap quality. Quality improvements have led to wide acceptance of shredded scrap by domestic and overseas consumers which is reflected in its price. The price is consistently higher than No. 1 heavy melting steel scrap, the bellwether grade used in large quantities. ISRI estimates that more scrap automobiles are recycled each year than were produced by the three principal U.S. automobile producers in that year. Nearly 11 million t of ferrous scrap were recovered from automobiles and light trucks in 1994 (8).

Shredding operations may also include appliances (white goods) which are being recycled in increasing quantities at least partly because of the banning of appliances from landfills in many states. The recycling rate for appliances in 1994 was 70% compared to 62% in 1993 when 1.4 million t of ferrous scrap was recovered from 36 million appliances. The recycling rate was 55% in 1992 (10,11).

Municipal solid waste (MSW) provides some forms of recyclable ferrous scrap, but the quality and markets are limited. The need to maximize recycling and minimize landfill disposal is forcing increased recovery of combustibles for energy and mixed ferrous scrap for recycling. This includes municipal source separation and curbside pickup programs, materials recovery facilities (MRFs), waste-to-energy (WTE) incinerators, and landfill mining. The industry and recovery of materials for recycling have been growing steadily in terms of the amount of MSW diverted from landfills for recycling and composting. The trend is expected to continue as many states and municipalities establish recycling laws and goals for recycling and waste reduction (qv) (12). The ferrous metal content in MSW ranges from about 5 to 10% (13) and consists mostly of steel beverage and food cans. Steel cans are recovered in increasing quantities in curbside programs, MRFs, and WTE operations. The recovery rate of steel cans has increased steadily to 34% in 1991, 41% in 1992, 48% in 1993, and 53% in 1994. A total of 1.4 million t of steel reportedly were recovered from 17.6 billion cans in 1994 (8,14). A 1993 survey of 39 operating WTE plants showed that over 250,000 t of ferrous scrap were recovered annually by these plants, mostly after incineration. The total ferrous scrap recovered annually from all MSW is estimated at 900,000 t (13). Landfill mining, or excavation of existing landfill sites, is being used to extend landfill life and reduce landfill closure costs. A small amount of ferrous scrap, mostly after incineration, has been recovered for recycling (see INCINERATORS; WASTES, INDUSTRIAL).

Automobile and industrial oil filters are another source of ferrous scrap being diverted from landfills. Processing equipment and melting procedures are being developed for routine melting in steel furnaces. Recycling of the filters requires thorough drainage of residual oil (for recycle) to avoid hazardous waste regulation and prevent dangerous overheating in the steel melting furnace. The number of oil filters discarded annually in the United States is estimated at more than 400 million auto oil filters containing over 155,000 t of steel. Some steel plants recycle oil filters commercially (15,16).

Construction and demolition (C&D) debris is a potentially large source of recyclables. However, as of 1995, generation rates and ferrous scrap content were not well established and estimates were highly variable. Ferrous materials in C&D debris are typically reinforcing bars, wire mesh, and structural steel. Some of the scrap is sold for recycling once concrete is effectively removed and the scrap is sized to specification (17).

Primary Uses of Ferrous Scrap

All wrought steel is produced by basic oxygen furnace (BOF) and electric-arc furnace (EAF) processes. The steel industry consists of two principal types of steelmaking facilities: integrated plants, which use blast furnaces, BOFs, and EAFs, and specialty plants and minimills, which use EAFs exclusively. The integrated steel plants begin the steelmaking process with iron ore, which is reduced in blast furnaces and converted to molten pig iron, also known as hot metal. Hot metal is transferred to a BOF where scrap is added and refining takes place. Final refining and alloy additions to produce the desired grade of steel follow. The proportion of scrap added to the BOF is typically 10–30% of the total metal charge and averaged about 23% in 1994 (1,7,18). Some integrated plants also have electric-arc furnaces for converting scrap to certain specialized grades of steel.

EAF-based plants were once used primarily to produce highly specialized grades of steel, such as stainless and tool steels. The EAF-based minimill steel industry has assumed a principal role in production of a variety of carbon-steel grades, once the exclusive domain of integrated steel producers. EAFs are used as rapid scrap melting devices producing steel from nearly 100% scrap charges. Refining is limited and obtaining the desired grade of steel product depends on the quality of scrap and other charge materials. Addition of pig iron or direct reduced iron (DRI) (see IRON BY DIRECT REDUCTION) during EAF steelmaking may be required to obtain the desired chemical compositions and limit residual elements. Pig iron is a high carbon metal produced in blast furnaces and, for a given plant, has uniform chemical composition. Carbon content is about 4% or higher and residual metal concentrations are low. DRI is produced from iron ore and is about 86–88% metallic iron. Unreduced iron oxides and other oxides from the original ore remain as a gangue which may require special slagmaking additions in some furnace processes. DRI and the hot briquetted form of DRI (HBI) is very uniform in quality with only trace levels of undesirable metal residual elements making it useful for blending with lower grade, lower priced scrap.

Ferrous foundries consist of two types: steel foundries in which electric furnaces (EAF and induction) are used, and iron foundries in which hot-blast cupolas and/or electric furnaces are used. Electric furnaces use virtually 100% scrap charges. Cupolas are shaft furnaces which use preheated air, coke, fluxes, and metallic charges. Scrap is over 90% of the metallic charge. Cupolas accounted for about 64% of total iron foundry scrap consumption in 1994 and electric furnaces accounted for about 34%. The balance was consumed by other furnaces, such as air furnaces. Iron foundry products have a high carbon content and the scrap charge usually contains a high percentage of cast iron or is used in combination with pig iron.

Table 1 shows the average percentages of scrap and pig iron used in the metallic charges for each of the three principal furnace types. DRI consumption averaged about 2% in electric furnaces and only a fraction of 1% in BOFs and cupolas. These percentages do not include the scrap consumed in blast furnaces and certain other special furnaces which amounted to 1.9 million t in 1994. DRI consumption in blast furnaces totaled 490,000 t in 1994.

Table 1. Ferrous Scrap and Pig Iron in the Steel Industry, 1994[a]

Type of furnace	Type of charge	
	Scrap, %	Pig iron, %
basic oxygen	23	76
electric-arc	94	4
cupola	94	5

[a]Ref. 1.

During the 1990–1994 period, the BOF proportion of steelmaking increased from 59.1% in 1990 to 61.0% in 1994; the EAF proportion increased from 37.3 to about 39.0%. Only a small amount of steel was produced in open-hearth furnaces (1990 and 1991). Raw steel production during this period ranged between 79.7 and 91.2 million t. Scrap consumption by steel mills averaged 60% of raw steel production. Iron foundry castings shipments ranged between 6.9 and 8.5 million t; steel foundry castings shipments ranged between 0.9 and 1.1 million t. A summary of scrap consumption by steel mills and foundries for this period is shown in Table 2.

Continuous casting increased steadily from about 67% of total raw steel in 1990 to almost 90% by the end of 1994. The increase in continuous casting and other process improvements reduced the amount of home scrap produced by steelmakers from 19 million t in 1990 to 14 million t in 1994. The reduced home scrap production and increasing need for low residual feed material resulted in increasing use of scrap alternatives such as pig iron and DRI. During this five-year period, DRI use in steel mills increased from 690,000 t to 1.5 million t. Steel

Table 2. Scrap Consumption in the United States, 10^6, t[a]

Industry	1990	1991	1992	1993	1994
steelmaking	54.0	49.0	50.0	53.0	54.0
steel foundries	1.9	1.6	1.6	1.9	2.0
iron foundries	13.0	11.0	11.0	13.0	14.0
Total	*68.9*	*61.6*	*62.6*	*67.9*	*70.0*

[a]Ref. 1.

foundries consumed little or no DRI and iron foundry consumption decreased from 18,000 t in 1990 to about 2,000 t in 1994. Total pig iron consumption ranged between 45 and 51 million t, peaking in 1994 at 50 million t consumed by steel mills, 1 million t consumed by iron foundries, and about 10,000 t consumed by steel foundries. Pig iron imports increased steadily from about 350,000 t in 1990 to 2.5 million t in 1994. Although the consumption of DRI and pig iron increased, most of the home scrap decrease was made up by purchased new scrap (4).

Residual Elements.   Residual elements, a principal concern to scrap users, have been the subject of many studies to identify sources, effects on product properties, and means of elimination or control. Steelmakers need to control several elements associated with scrap in order to meet the quality requirements of their final products. Copper and tin are of particular concern because these are not volatile, oxidizable, or otherwise refinable by normal furnace practice. Dilution using virgin materials or prime quality scrap may be required to obtain the desired chemical composition. These elements, in relatively small concentrations, degrade the hot workability and deep drawing quality of many steels. Other elements such as chromium, nickel, and molybdenum affect various mechanical properties. Cast irons are produced in several classes, such as gray iron, malleable iron, and ductile iron, each having specific microstructural characteristics that are critical for meeting product specifications. Low concentrations of lead, tin, or arsenic, for example, inhibit desirable graphite structure in ductile iron. Copper, which is useful in promoting a desirable pearlitic structure in gray iron, also promotes pearlite formation in ferritic ductile irons, causing serious loss of product ductility. Whereas the effects of individual elements may be known, interactions of two or more elements are difficult to determine. Volatile elements such as zinc and cadmium are of concern because these may significantly increase the costs of particulate control and disposal. Relatively large samples of scrap (eg, ≥20 t lots) have been melted to obtain homogenized melt analysis for the presence and concentrations of residual elements. Some mills have also devised back-calculation methods to determine scrap quality in use and as a means of monitoring scrap sources. Analyses of scrap have been published by various investigators over many years. These can be misleading, however, if used to indicate the expected analysis of a given grade of scrap from all suppliers and at different periods of time. Although these data can be useful in indicating the general level of some of the residual elements found in various types of scrap, in practice the actual compositions can vary widely, particularly in obsolete scrap. Table 3 shows concentrations for some residual elements obtained for scrap samples (5,6).

The steel mills and foundries consume many different grades of scrap. Scrap grade and price must be balanced with overall economics of use. Such factors as metal yield during melting and uniformity of quality regarding desired physical form and chemical analysis are important. The primary users' products must meet strict chemical and physical quality requirements and, although scrap generally offers the most economical form of iron units, proper selection and blending is essential to assure acceptable final product quality. Prompt industrial scrap has historically been very uniform having known or obtainable chemical analysis and is thus in high demand. Obsolete scrap is typically very heterogeneous. Chemical analysis of this scrap is not practical on a routine

Table 3. Residual Elements in Samples of Scrap, wt %

Scrap grade	Chromium	Copper	Nickel	Molybdenum	Tin
ironmaking slag scrap	0.05	0.05	0.02	0.01	0.005
No. 1 busheling	0.04	0.07	0.03	0.01	0.008
electric furnace bundles	0.10	0.07	0.07	0.02	0.010
plate and punchings	0.06	0.09	0.06	0.01	0.015
railroad, rails, etc	0.07	0.09	0.07	0.01	0.015
cut plate and structural	0.09	0.13	0.09	0.02	0.025
No. 1 heavy melting steel	0.10	0.25	0.09	0.03	0.025
shredded scrap	0.14	0.22	0.11	0.02	0.02
No. 2 heavy melting	0.18	0.55	0.20	0.04	0.04
No. 2 bundles	0.18	0.50	0.10	0.03	0.10
turnings and borings	0.40	0.20	0.40	0.15	0.015

basis. Chemical acceptability of many grades of obsolete scrap are based on user experience and relationship with scrap suppliers.

During the 1990–1994 period, four scrap grades dominated purchases by steelmakers having a combined average of about 60% of total scrap purchases. These are No. 1 and No. 2 heavy melting steel, No. 1 and electric furnace bundles, and shredded scrap. Three grades having a combined average of more than 50% represented the principal purchases by steel foundries: low phosphorus plate and punchings, cut structural and plate, and No. 1 heavy melting steel. Iron foundries made significant purchases of four grades averaging about 50% of the total: low phosphorus plate and punchings, cut structural and plate, shredded, and No. 1 busheling. All three industries purchased many other grades in smaller quantities. The primary obsolete grades such as No. 1 heavy melting steel and shredded have higher residual content than prompt industrial scrap, but tend to be relatively predictable in uniformity and productivity in the users' plants.

Trends in Scrap Recycling

Increased steel production in the scrap-based EAFs and the growth of continuous casting have significantly influenced scrap consumption. As the scrap-based minimills and specialty EAF plants increased their proportion of scrap consumption, the availability of prime quality home scrap declined. The yield of raw steel to finished steel in a continuous cast process can be more than 90% compared to ≤75% by the ingot casting process. The reduced supply of home scrap has caused an increasing reliance on purchased scrap, ie, prompt industrial and obsolete, to meet steelmaking needs. The proportion of home scrap compared to total scrap consumption in steel mills has decreased from about 35% in 1990 to 26% in 1994 (1). The use of continuous casting is widespread and future impacts on ferrous scrap supply can be expected to be less than in the early 1990s. Home scrap production in the foundries has been relatively stable.

Minimills and other EAF plants are expanding into flat-rolled steel products which, by some estimates, require 50–75% low residual scrap or alternative raw material. Up to 16 million t of new capacity are expected to be added in the

United States between 1994 and 2000 (18). Developments in other parts of the world also impact scrap use and supply. Possible scrap deficiencies of several million tons have been projected for EAFs in East Asia and in parts of Europe. This puts additional strains on the total scrap supply, particularly low residual scrap (19,20). The question of adequate supply of low residual scrap is always a controversial one. Some analysts see serious global shortages in the first decade of the twenty-first century; others are convinced that the scrap industry has the capability to produce scrap in the quantities and quality to meet foreseeable demand. This uncertainty in combination with high scrap prices has led to increased use of scrap alternatives where the latter is price competitive with premium scrap. Use of pig iron has increased in EAF plants and more capacity is being installed for DRI and HBI outside the United States.

Increased use of galvanized steel for corrosion protection and alloy steels for weight reduction in automobile manufacturing are changing the character of prompt industrial scrap and shredded scrap. Prompt industrial scrap from automobile stamping plants contain increasing amounts of zinc-coated scrap which may be as high as 40–50% of the scrap generated. Scrapped autos also have higher zinc content in the ferrous fraction. About two-thirds of new plant galvanizing capacity is for the automotive market; the rest is for the building and construction industries and appliances (21). Also, shipments of galvanized sheet by the domestic steel industry increased steadily since 1991 to about 13 million t in 1994. Imports of galvanized sheet have been substantial also, totaling about 1.7 million t in 1994 (22). Zinc content of galvanized sheet may vary from <1% to at least 10% depending on sheet thickness and coating weight used. A rough estimate based on total zinc used for galvanizing and total galvanized sheet shipments, without accounting for zinc losses during production, indicates an industry average of about 4%. Other zinc-containing coatings include Galvalume (Bethlehem Steel Corporation) and Zincrometal. Galvalume coating is an aluminum–zinc alloy having <50% zinc and, for a given coating weight, contains less total zinc than galvanized steel. Zincrometal is a zinc-based paint coating. The zinc content of the coated steel is about 0.5%. The use of Zincrometal declined in the early 1990s. Remelting of zinc-coated scrap results in high loadings of zinc and, to some extent, associated lead and cadmium, in the furnace dust. EAF furnace dust, listed as a hazardous waste, is receiving considerable research attention. Some of the dust is commercially processed for zinc recovery and safe disposal.

Weight-reduction efforts through use of high strength low alloy steels (HSLA) adds more residual elements to resulting scrap and results in less shredded scrap recovered from each auto. Whereas the percentage of total iron and steel in a typical car remained approximately the same from the mid-1980s to the mid-1990s, the total weight of iron and steel parts decreased about 5%. Projections to 2003, based on Corporate Average Fuel Economy (CAFE) standards, suggest that total weight of iron and steel in the typical car might decrease another 10%. This would represent a total decrease of about 160 kg from the mid-1980s to 2003. Industry and government are cooperating to improve recycling techniques and to utilize such concepts as Design for Recycling (registered trademark of ISRI) in automobile manufacturing and other consumer items. The objectives include development of designs and materials that permit easy disassembly of components, maximum recyclability, and prevention of pollution during recycling (23–26).

Factors Influencing Ferrous Scrap Recycling

Supply, Demand, and Prices. The economics of ferrous scrap recycling involve a complex variety of factors related to the demand of consuming industries and cost of supplying scrap. Cost of scrap is affected by such factors as collection, regional availability, processing and upgrading to acceptable quality, transportation, environmental controls, and export demand. Many grades of scrap are involved. The principal consumers exerting the strongest demand are the iron and steel mills and foundries, which must balance the price of various types of scrap with the internal cost of using the scrap. The availability and price of scrap are continually monitored by governmental and private groups because of the importance to the domestic economy. Adequate availability of low residual scrap is continually debated. Scrap consumers have increased use of pig iron and direct-reduced iron ore products as alternatives to low residual scrap, and possibly to permit greater use of lower price obsolete scrap by blending.

Thousands of scrap facilities purchase scrap from individuals and businesses and process it to specification or broker it for steel and foundry consumers. Scrap dealers, processors, and brokers employ more than 40,000 people in the United States (2). The overall scrap market consists of numerous regional U.S. markets and the international export market. Brokers help to even out various supply–demand imbalances that may exist within regions. The ferrous scrap market depends on the vitality of the iron and steel industry. When steelmaking and foundry operating rates and demand are low industrial scrap prices fall (27). Price volatility and price ranges for various grades of scrap can be extensive. Obsolete, dealer-prepared scrap may not be as volatile as industrial scrap because dealers can opt to hold scrap if prices are below their costs. The generators of prompt industrial scrap, however, generally must dispose of their scrap each month, selling at the highest bid prices.

The demand and prices for individual grades of scrap are also dependent on the various costs associated with converting the scrap to desired products in the mills and foundries. Scrap that has the appropriate cleanliness, chemical quality, and physical characteristics to provide the highest quality and yield at the lowest melting cost is the most desirable. Various grades of scrap may be combined by the consumer to optimize scrap price. Cleanliness, or the absence of nonmetallics, is important to maximize yields and minimize environmental loadings in gas scrubbers. Chemical quality is important to assure minimum tramp or residual elements. Volatile elements increase dust and sludge disposal costs; nonvolatile and nonoxidizable elements which may be deleterious to the final product must be diluted using higher price raw materials, including pig iron, DRI, and premium scrap.

Scrap metal prices are listed in several publications, including the *American Metal Market* (AMM) and *Iron Age Scrap Price Bulletin* (IA). The AMM and IA compile composite prices for No. 1 heavy melting steel based on prices in Chicago, Pittsburgh, and Philadelphia. In an analysis of scrap prices through 1991 (3), the AMM composite price for the No. 1 heavy melting steel scrap grade was tracked. Since 1974, there has been a significant increase in the proportion of purchased scrap consumed relative to home scrap. Other factors affecting the scrap industry include increased energy costs and the added costs in complying with various federal, state, and local environmental regulations. From 1992 to 1995, the average annual AMM composite price for No. 1 heavy melting steel scrap was $85/t in 1992, $112/t in 1993, $127/t in 1994, and $135/t through mid-1995. Other scrap grades also showed high prices in 1995. The AMM weekly composite price for shredded scrap ranged between about $143–$155/t. Regional consumer buying prices published for prompt industrial and other grades showed that in mid-1995 No. 1 bundles and cut structural and plate sold at levels of ~$155 and $185/t, respectively (28).

Export Considerations.   The United States is the largest exporting nation of iron and steel scrap. During the 1990–1994 period, exports ranged from 8.8 million to 11.6 million t and values ranged between $1.1 and $1.6 billion. Ferrous scrap is exported to more than 50 countries but most goes to just a few countries. The Republic of Korea, Canada, Japan, Turkey, India, and Mexico received more than 80% of the total in 1994. The trade surplus for all grades of scrap is about $1 billion and exports represent about 20% of the total value of the United States ferrous scrap market. Scrap trading operates in a free market, but because of its importance to the industrial sector and national security there have been periods when the federal U.S. Government found it necessary to impose price controls and export restrictions; for example, during the two World Wars and the Korean conflict. Price controls and export restrictions were also imposed in 1973–1974 because of high steel and pig iron production and high demand for scrap by the steel mills and ferrous foundries which caused unusual increases in scrap prices.

Environmental and Regulatory Aspects.   Ferrous scrap recycling provides many well-documented environmental benefits including reduced roadside litter, landfill requirements, and pollution and energy consumption compared to use of virgin materials. Use of scrap for steelmaking results in large reductions in air pollution, water use and pollution, mining wastes, and energy consumption while also conserving iron ore, coal, and limestone. The savings in landfill space is also considerable. Recycling operations do, however, generate certain emissions and waste streams that are being subjected to increasingly stricter environmental controls, thus increasing the cost of recycling in many cases.

Regulations affecting metal recycling are numerous, including comprehensive regulations resulting from the Clean Air Act Amendments of 1990 (CAA), the Clean Water Act, the Resource Conservation and Recovery Act (RCRA), and the Comprehensive Environmental Response, Compensation, and Liability Act (CERCLA/"Superfund"). RCRA and CERCLA have the most impact on metals recycling (29). Under RCRA, solid wastes are regulated separately as hazardous (Subtitle C) or nonhazardous waste (Subtitle D). One source of controversy is the disincentives to recycling that result when recyclable scrap is designated as waste. Such a designation results in higher costs for transportation, processing, storage, and disposal. Definition of waste is an issue also in international agreements on export and import of scrap and wastes. An example is the Basel Convention which was designed to prevent international shipments and improper management of hazardous wastes and prevent developing countries from becoming waste depositories for developed countries. The Basel Convention, negotiated under the auspices of the United Nations Environmental Program (UNEP), was signed by more than 100 international states and has been in force since May 5, 1992, but as of 1995 it had not been ratified by all the countries (30). The main concern to recyclers is that hazardous waste is not well defined. The criteria for waste used may not be realistic, and there is a lack of distinction between recyclables and waste for disposal. Ferrous scrap metals may fall outside Basel Convention rules. Considerable dialogue is in progress to assure proper understanding and definition of waste and recyclables and avoid unnecessary prohibition of scrap shipments.

EAF dust is regulated as hazardous waste because of the presence of unacceptable levels of contaminants such as lead, cadmium, and chromium. Furnaces that use large amounts of galvanized scrap generate dust from which zinc can be recovered and recycled. Regulations require that the dust be processed for recovery if it contains ≥15% zinc. Dust containing <15% zinc must either be processed for recovery or be stabilized for disposal. Dusts from BOF operations fall into the low zinc category and recovery economics for zinc from these dusts are not as favorable as for the higher zinc contents in EAF dusts. EAF furnaces in the United States generate about 16.5–18 kg of dust per ton of steel produced, or about 600,000 t/yr. The average zinc content is 19%. About 86% of the dust is processed for zinc recovery, 2–3% is used in fertilizers, and the balance is landfilled (31). An international assessment reported that ~30% of the dust generated worldwide is processed for the extraction of heavy metals and recovery of zinc. The estimated cost of treating the dust is more than $200/t, and various research projects are underway to develop more cost-effective technologies for treating furnace dust and for dezincing scrap before melting (32,33) (see ZINC AND ZINC ALLOYS).

The recycling of automobiles has eliminated a former blight and is a significant source of scrap. However, new regulations are requiring adjustments at scrap processing facilities. In order to control smoke from shredders, for example, costly filters and scrubbers have been installed. The shredder residue, or fluff, produced during shredding operations has received special attention. There is concern about contaminants such as PCBs, lead, and cadmium. Studies on fluff material and scrap industry interaction with the United States EPA have indicated that fluff does not warrant regulation (29) but it represents another cost to shredder operators. Another problem is the presence of as many as 30 different types of plastics that are nonrecyclable in the mixed state. The amount of fluff has been increasing owing to increased use of nonmetallics in automobile manufacturing. Without recovery and recycling the only viable disposition is disposal. As of 1995, landfill disposal costs ranged up to $100/t. Alternative methods of disposal are being researched, including recovery of recyclable materials, chemical fixation, and energy generation (see FUELS FROM WASTE). The amount of fluff generated per vehicle is about 250–340 kg (500–750 lbs) (31,34,35).

Radioactive scrap has the potential to become a serious problem domestically and internationally. Several accidental meltings of ferrous scrap containing discrete radioactive sources have occurred in the United States, including at least four in 1993 and two in 1994. Clean-up costs have averaged over $1 million per incident. The problem has been formally addressed by ISRI and by the United

States Nuclear-Regulatory Commission (36). Irradiated scrap has also received attention by the United States Department of Energy (DOE), which sponsored tests to develop technologies for decontaminating and recycling scrap metal from commercial and DOE nuclear facilities (see NUCLEAR REACTORS) (37).

Standards

The most comprehensive set of descriptions of ferrous scrap are published by ISRI. Individual steel mill and foundry consumers usually follow the ISRI specifications, although many also incorporate specific requirements tailored to the needs of the consuming facility.

Scrap from municipal refuse may be in the form of source-separated steel cans, a mixed ferrous fraction, metal magnetically separated from mixed waste or incinerator ash, and C&D debris. An ASTM specification (E1134-86) was developed in 1991 for source-separated steel cans. The Steel Recycling Institute has a descriptive steel can specification entitled "Steel Can Scrap Specifications". Published standards for municipal ferrous scrap also include ASTM E701-80, which defines chemical and physical test methods, and ASTM E702-85 which covers the chemical and physical requirements of ferrous scrap for several scrap-consuming industries.

Iron and steel products must meet increasingly strict quality standards, requiring the steel and foundry industries to have strict control over their raw materials. Iron and steel producers continually seek scrap of uniform consistent quality. Because most forms of obsolete scrap are heterogeneous, scrap consumers rely on scrap origin, reliability of the scrap supplier, and specifications tailored to their particular plants to assure acceptable quality.

BIBLIOGRAPHY

"Ferrous Metals" under "Recycling" in *ECT* 3rd ed., Vol. 19, pp. 952–962, by J. Early, National Bureau of Standards.

1. G. W. Houck, *Iron and Steel Scrap in June 1995*, U.S. Bureau of Mines Mineral Industry Surveys, Washington, D.C., Sept. 1995.
2. G. W. Houck, *U.S. Bureau of Mines Mineral Commodity Summaries 1995*, Washington, D.C., Jan. 1995, pp. 88–89.
3. R. E. Brown, *Metal Prices in the United States Through 1991*, U.S. Bureau of Mines Report, Washington, D.C., 1993, pp. 73–80.
4. Technical data, Institute of Scrap Recycling Industries, Inc. (ISRI), Washington, D.C., 1996.
5. A. J. Stone and P. H. Meyst, *Ferrous Scrap Materials Manual*, ICRI Report No. 517, Iron Casting Research Institute (ICRI), Columbus, Ohio, Mar. 31, 1988.
6. R. D. Burlingame, *Ferrous Scrap Explained*, Luria Brothers & Co., Inc., Cleveland, Ohio, 1981.
7. G. W. Houck, *Recycled Metals in the United States*, U.S. Bureau of Mines Special Publication, Washington, D.C., Oct. 1993, pp. 27–31.
8. *J. Met.* **47**(5), 4 (May 1995).
9. R. E. Brown, *Iron and Steel Scrap*, U.S. Bureau of Mines 1991 Annual Report, Washington, D.C., pp. 35–36.

10. *ASTM Stand. News* **22**(6), 8 (June 1994).

11. *Am. Metal Mark.* **102**(136), 4 (July 18, 1994).

12. C. L. McAdams, *Waste Age* **25**(4), 187–190 (Apr. 1994).

13. M. Rogoff, *Am. City County* **109**(5), 41 (Apr. 1994).

14. *Am. Met. Mark.* **102**(80), 10 (Apr. 27, 1994).

15. *Recycl. Magnet* **5**(2), 8 (Winter 1994).

16. K. D. Peaslee, *J. Met.* **46**(2), 44–46 (Feb. 1994).

17. *Construction Waste & Demolition Debris Recycling—A Primer*, Publication. No. GR-REC 300, The Solid Waste Association of North America, Silver Spring, Md., Oct. 1993.

18. P. F. Marcus and K. M, Kirsis, paper presented at the *Steel Survival Strategies X Conference*, New York, June 20, 1995, 72 pp.

19. *Am. Met. Mark.* **103**(104), 7A, 10A (May 31, 1995).

20. D. F. Barnett, *Am. Met. Mark.* **102**(140), 14 (July 22, 1994).

21. N. L. Samways, *Iron Steel Eng.*, D-19 (Feb. 1993).

22. Technical data, American Iron and Steel Institute (AISI), Washington, D.C., Dec. 1994 and Apr. 22, 1995.

23. *Am. Met. Mark.* **102**(45), 17 (Mar. 8, 1994).

24. *Am. Met. Mark.* **102**(104), 9 (June 1, 1994).

25. *J. Met.* **46**(11), 4 (Nov. 1994).

26. *Adv. Mater. Proc.* **146**(4), 7–8 (Oct. 1994).

27. R. R. Jordan and G. L. Crawford, *The McGraw-Hill Recycling Handbook*, McGraw-Hill Book Co., Inc., New York, 1993, pp. 15.1–15.27.

28. *Am. Met. Mark.* **103**(111), 9 (June 8, 1995).

29. R. J. Foster, *Information Circular 9397*, U.S. Bureau of Mines, Washington, D.C., 1994, pp. 10–14.

30. F. Veys, *Proceedings of the Second International Conference on the Recycling of Metals, Amsterdam, the Netherlands, Oct. 19–21, 1994*, ASM International, Materials Park, Ohio, pp. 31–42.

31. P. N. H. Bhakta, *J. Met.* **46**(2), 38 (Feb. 1994).

32. *Am. Met. Mark.* **102**(132), 6 (July 12, 1994).

33. J. H. Jolly, in Ref. 7, p. 70.

34. *Am. Met. Mark.* **102**(99), 10 (May 24, 1994).

35. B. J. Jody, E. J. Daniels, P. V. Bonsignore, and N. F. Brockmeier, in Ref. 31, pp. 40–43.

36. R. E. Brown, *Iron and Steel Scrap*, U.S. Bureau of Mines 1992 Annual Report, Washington, D.C., Oct. 1993, p. 1.

37. *Am. Met. Mark.* **102**(68), 9 (Apr. 11, 1994); **102**(116), 9 (June 17, 1994).

HARRY V. MAKAR
Consultant

NONFERROUS METALS

A nonferrous metal is any metal other than iron (qv) or one of the many iron alloys. For the nonferrous metals, recycling of aluminum, copper (qv), lead (qv), and zinc is of primary economic importance. Recycling of any material is dictated by economic factors and production from primary sources may be used as a reference point. The decision to recycle from secondary materials is a dynamic one and must be evaluated on an ongoing basis. The principal factors that determine recyclability include availability of both primary and secondary sources, purity, cost of energy, cost of waste generated, and regulatory environment. Several of these factors are interrelated. The emphasis on recycling has increased the availability of many materials. As natural ore bodies are depleted, the relative availability of waste materials is expected to increase. Two key advantages of recycling are the conservation of resources and protection of the environment (see ALUMINUM AND ALUMINUM ALLOYS; ZINC AND ZINC ALLOYS).

Nonferrous metals are primarily used as the metal or in an alloyed form, facilitating the recycle of the metal or alloy because much of the recycling process requires only that the scrap metal be sorted, melted, and cast (pyrometallurgical processing). The scrap is classified into two general categories, new and old. New scrap derives from material that never reaches the market, such as scrap generated during the manufacture of products. Old scrap is obtained from worn out or discarded products. Often, the recycling process consumes only 5–50% of the energy needed to produce from an ore and in most cases generates only a small portion of the waste. As of 1996, much of the aluminum and copper in use comes from scrap. The proportion of several nonferrous metals that are obtained through recycle is illustrated in Figure 1.

Concurrent with the implementation of the Resource Conservation and Recovery Act (RCRA) significant increases in the recycling of nonferrous metals from metal-bearing wastes occurred. The primary aim of RCRA was to establish

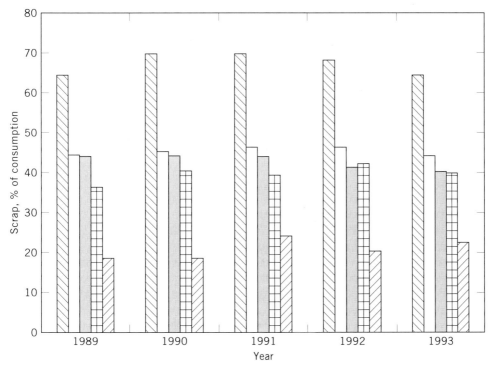

Fig. 1. Recycling of the nonferrous metals (◧) lead, (□) nickel (stainless steel), (■) copper, (▦) aluminum, and (▨) zinc from secondary sources from 1989 to 1993.

hazard definitions for wastes and to maintain an accountability for all generated wastes that were classified as hazardous. The term recycling has no regulatory significance. Some recycling is regulated, some is not. Regeneration, a regulatory subset of reclamation, which itself is a subset of recycling, is regulated (see REGULATORY AGENCIES, CHEMICAL PROCESS INDUSTRY). It has been estimated that there are 222 million t of metal-bearing hazardous wastes produced annually from metal-related industries (1) (see WASTES, HAZARDOUS WASTE TREATMENT).

Metal finishing (see METAL SURFACE TREATMENTS; METAL TREATMENTS) and plating (see ELECTROPLATING; ELECTROLESS PLATING) industries produce over 11 million t of wastewater treatment sludges (2). Although the metal content of these wastes has not been specified, it is expected that reclamation of metals from such sources could account for significant additional resource recovery. These wastes consist primarily of inorganic sludges and solids containing several metals, predominantly present as hydroxides and basic metal salts. A selective analytical survey of one specific waste type, ie, electroplating (qv) sludges (3), found an average metal content of 16.4% copper, 7.5% nickel, and 3.6% zinc where copper was the primary metal. For copper, nickel, and zinc electroplating sludges the metals averaged 8.4, 8.9, and 8.9%, respectively, plus having an average iron and calcium content of 4.4 and 5.4%. Not all metal-containing hazardous waste sludges contain from 8 to 16% copper.

Metal and metallic salt wastes are primarily generated by the electronic, electroplating, metal finishing, and machining industries, and as consumer waste. Significant quantities of metal-containing spent catalysts are also produced by the organic chemical industry. Much of the metal-bearing waste is classified as hazardous by the U.S. EPA under RCRA. This classification may result from characteristics such as corrosivity or toxicity. The classification, however, may also be arbitrarily assigned because of the specific nature of the waste source. The reclamation of metals from these types of wastes presents several regulatory concerns. Many of the concerns and disadvantages to recycling and reclamation of metals from hazardous waste have been presented (3). Often, wastes such as salts, basic salts, and hydroxides are more amenable to hydrometallurgical methods of reclamation.

Classification of Recycled Materials

The initial step in the recycle of metals is the physical segregation of the metals from other materials. For new scrap this process is straightforward; for old scrap it can be expensive. This classification and segregation of scrap is of importance to the producers of the metals from secondary materials. Historically, much of the classification has relied on hand sorting which can be reliable, but it is labor intensive. The recycling of automotive scrap is illustrative of the techniques that can be used to separate nonferrous metals into broad categories.

Automobile scrap generates almost 20% of old scrap copper yet only contains 10–35 kg per unit (4). The average mass of an automobile, which has a useful life expectancy of 10 years, is 1100 kg. The unit operations normally used in the recycling of automotive scrap are illustrated in Figure 2. Much of the copper, aluminum, and lead in the form of radiators, starters, generators, and

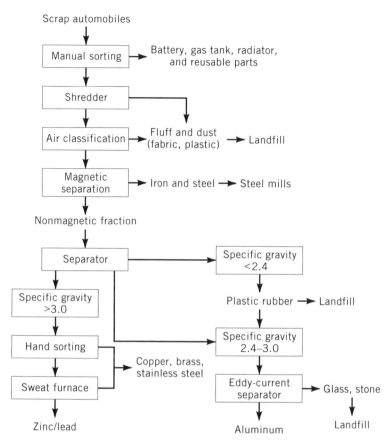

Fig. 2. Unit operations and industrial usage for the recycling of automotive scrap.

batteries is removed prior to shredding. The automobile is shredded into pieces of 10 × 20 cm and air blown to remove fabric and plastic. The iron and steel are removed magnetically. The use of screening and heavy media or eddy-current (or both) separates glass, rubber, and ceramic components from the nonferrous metals. Heavy media separation (HMS) has become quite sophisticated. It relies on flotation (qv) in order to minimize hand sorting. One such HMS system has resulted in a final aluminum content of 99.5% (5). A typical analysis of aluminum isolated with less sophisticated classification techniques from automotive scrap follows. The nonferrous components are collected, baled, and shipped to recyclers.

Metal	Quantity, %
aluminum after remelting	83–97
silicon	0.3–10.0
iron	0.5–0.9
copper	1.0–3.2
magnesium	<0.05–0.5
manganese	0.1–0.3
zinc	0.1–8.0

Pyrometallurgical Methods of Recycling

Much of the technology used in the reclamation of metals from metal-bearing wastes was developed by the mining industries. The primary means of recycling metal from metal and alloyed scrap is via pyrometallurgy (see METALLURGY).

Aluminum. The recycling of aluminum scrap offers significantly reduced energy requirements when compared to refining from ore. The energy requirements to produce aluminum from bauxite ore are 222 GJ/t (53,000,000 kcal/t), whereas the analogous requirements from scrap are only 10 GJ/t (2,400,000 kcal/t) (see ALUMINUM AND ALUMINUM ALLOYS). There are several melting techniques used, each having advantages and limitations. Aluminum is very reactive and has a strong tendency to form a dross, ie, an oxide containing entrained metallics, at the surface of the metal. Up to 5% of the molten metal forms dross during processing. The dross produced from the melting and oxidation of pure metal is known as white dross, whereas the dross containing fluxes and impurity metals is called black dross. In order to recover the aluminum from the dross, salts, usually chlorides, must be added. This treatment produces a secondary dross containing alumina, some aluminum metal, salts, and impurities. Disposal of this waste, undergoing increasing restrictions, is becoming a less attractive alternative. Therefore, new technologies are being pursued such as centrifugation (6), flotation, plasma, and arc melting which require no salt addition to recover white dross. Use of these techniques allows for the production of a salable or landfillable black dross (7). Alternatively, salt recovery plants are being installed.

There are several furnaces employed in the melting of scrap aluminum (8). It is common to melt scrap in a standard reverberatory furnace using oil or gas and air/oxygen heating. The flame is applied directly to the scrap. The round-top melter is a high capacity reverberatory furnace having a removable top for ease of charging (9). Up to 100 t of material can be contained in the furnace and over 20 t can be charged at once. A variation of the dry hearth melter is the tower melter where the scrap is preheated by the exhaust gases from the melter. The charge is arranged so that it is conveyed closer and closer to the hearth. This type of arrangement is thermally efficient relative to the round-top melters but requires the use of heavy-gauge aluminum scrap. For the melting of light-gauge scrap, a flotation melter (10) has been used. The flotation melter uses a conical vertical chamber where the exhaust gases from the furnace are recirculated. Light-gauge scrap is introduced to the conical chamber where it is heated. The high velocity of the gases at the narrow exit port of the chamber suspends the scrap whereupon melting falls into the hearth. One of the primary advantages of the flotation melter is that no salts are required. Oxidation of the aluminum is minimized by low retention times at the higher temperatures and by careful control of the burner oxygen content. There was only one commercial unit in use as of 1995.

Rotary-barrel or vertical cement mixer-type rotary gas- and oil-fired furnaces offer several advantages in the processing of scrap. These can be operated in batch or continuous mode. Salt, ie, a 1:1 mixture of potassium and sodium chloride containing fluoride as a wetter, is usually charged simultaneously to the scrap. The rotation of the furnace allows for the excellent mixing of the two.

A broad spectrum of scrap gauge and purity can be processed. These furnaces are simple to operate, require relatively little capital and floor space, are thermally quite efficient, and the throughput can be high. The primary disadvantage to the rotary-barrel furnace is production of relatively high quantities of dross and salts.

Other technologies that attempt to minimize the formation of white dross have been developed. Plasma heating of rotary furnaces has been used since about 1990. The higher attainable temperatures associated with the plasma have resulted in lower dross and salt waste production. More recently, the development of arc heating has resulted in further improvements. It is more energy efficient than the alternative rotary furnace techniques and results in less waste. A comparison of the energy and mass requirements and by-product generation from the fuel/air/oxygen, plasma, and arc processes are given in Table 1 (7).

The processing of used beverage containers represents a significant portion of the recycled aluminum. The material is separated by the consumer and compacted into bales for shipping to the secondary recycler. The bales are crushed and shredded followed by removal of magnetic materials and gravel (see SEPARATIONS, MAGNETIC SEPARATIONS). Delacquering or decoating operations are being used to provide additional heat input into the melting process. A typical used aluminum beverage can (UBC) recycling process is illustrated in Figure 3.

Copper.   Metals that cannot be separated from copper (qv) by traditional physical methods, eg, alloys of zinc, tin, and lead, can be eliminated by melting and oxidation. The following general conclusions can be made (3): zinc is removed by reduction to zinc metal followed by fume oxidation and collection; aluminum and iron are removed by air oxidation to slag; lead partitions between the alloy, slag, and vapor, as does tin (vapors are a mixture of lead(II) oxide, sulfide, and chloride, and tin(II) oxide and sulfide and tin chloride); nickel partitions between the slag and metal (outlet is the electrolyte tankhouse bleed); and precious metals are removed during electrorefining.

The common grades of scrap copper are shown in Table 2. No. 1 scrap is usually melted in a fuel-fired, tilting reverbatory furnace and cast or used in the direct production of alloys. No. 2 scrap (88–99% Cu) can be used to produce alloys but more often it is melted and fluxed in a reverbatory furnace. The scrap is also blended with blister copper in the anode furnace of secondary smelters to dilute the lead and nickel in preparation for electrowinning. Tankhouse maximums are 0.13% lead and 0.21% nickel. Before dilution the values often run upward of

Table 1. Aluminum Dross Treatment Using Three Rotary Furnace Technologies[a,b]

Treatment	Rotary furnace type			Centrifuge
	Air/O_2/fuel	Plasma	Arc	
energy, kWh				
input	630	475	473	350
loss	294	341	160	180
by-product residues, kg	625	566	525	520
recovered aluminum, kg (%)	475 (95)	453 (90.5)	475 (95)	480 (96)

[a]Refs. 6 and 7.
[b]For production from 1000 kg of 50 wt % free aluminum.

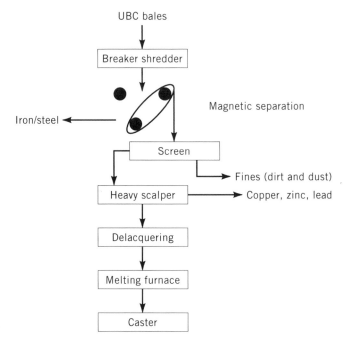

Fig. 3. Used aluminum beverage cans (UBC) recycling process.

0.4% lead and 0.7% nickel. The conditions in the reverbatory furnace are initially oxidizing to remove oxidizable and volatile impurities. Limestone flux is added to produce a slag. The slag is withdrawn and the flame is made reducing in character. The molten copper is poured into anodes for electrorefining. Brass and bronze (low iron) are melted in a rotary induction or gas-fired hearth. The composition is adjusted to specification, first by appropriate charge alloy selection, then by air (O_2) blowing to remove a portion of the zinc, followed by final composition adjustment by adding more alloy scrap, plus possibly tin ingot.

Low grade scrap (15–70% Cu) requires complete smelting, converting, and refining. A general flow for the refinement of low grade copper scrap is shown in Figure 4. The economic recycling of low grade copper scrap and residues requires complete smelting and converting operations. The smelting operation consists of a blast furnace operated under strongly reducing conditions. Alternatively, the top-blown rotary converter can be used to perform both the blast and converter functions to produce the blister copper. The blast furnace operation produces a slag that is low in copper and discardable. The converting furnace (Pierce-Smith, Hoboken, New Jersey) or top-blown rotary is distinctly more oxidizing, giving much lower levels of impurities by selective oxidation, but producing a high copper-containing slag (see COPPER). This slag is usually recycled to the blast furnace.

Blast furnaces are charged through the top with coke, flux (usually iron metal and silica), and scrap while air is injected through tuyeres continuously at the bottom just above the black copper. The coke (100 kg/t slag) burns to maintain furnace temperatures of 1200°C, provides the reductant, and maintains an open border. A charge of 10 t/h is typical. The furnace produces a molten black

Table 2. Purchased Scrap Categories and Treatment Methods[a]

Scrap type	Composition, %	U.S. quantity recycled in 1990, t $\times$ 10^3	Most common recycle treatment
		Copper scrap	
No. 1 copper wire and heavy scrap	99+ Cu	420	melted and fire refined, cast in shapes for rolling or extrusion or as ingots for sale
No. 2 wire, heavy and light scrap	88–99 Cu	340	melted, fire refined, and electrorefined; electro-refined cathodes used for all applications
		Alloy scrap	
leaded yellow brass scrap	~65 Cu, ~35 Zn, ~1 Pb	250	melted to form new brass, alloy composition adjusted with copper or zinc and lead
yellow and low brass including cartridges	65–80 Cu, 35–20 Zn	220	melted to form new brass, alloy composition adjus-ted with copper or zinc
low grade scrap and residues	10–88 Cu	140	smelted and converted to remove Al, Fe, Sn, and Zn then fire and electrorefined
automobile radiators		90	same as No. 2 scrap
red brass scrap	80–85 Cu, 10 Zn, 7 Pb, 3 Sn	70	melted and cast as ingot for subsequent alloymaking
bronze scrap, nickel–copper scrap, and other alloy scrap		70	melted to form similar alloy
Total		*1600*	

[a]Ref. 11.

copper that contains about 80% copper. The zinc, lead, and tin oxides are collected from the off-gas of the furnace in baghouses and a silicate slag containing about 1% of the copper is discarded. Iron reports to the slag and black copper and zinc to the slag and fume.

The top-blown rotary furnace can also be used for smelting, converting, and fire refining. The charge of low grade scrap, silica, limestone, and coke are melted with a burner located above the charge. The melted charge is continuously rotated and the oxygen fuel ratio of the burner adjusted to produce the required level of impurity. The slag is removed and the oxygen ratio is again adjusted to convert the black copper followed by fire refining.

The molten black copper is converted under more oxidizing conditions. Sufficient heat is generated during the oxidation of the iron and zinc to be

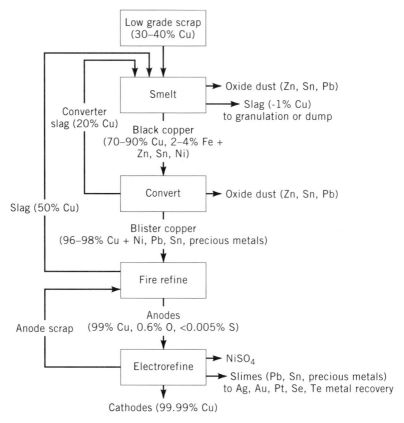

Fig. 4. General flow for the refinement of low grade copper scrap.

autothermal. The blister copper contains 96–98% copper with <1% nickel, lead, and tin; the remainder is oxygen and sulfur. The blister is fire refined by melting in a hearth furnace under reducing conditions to produce the low oxygen-containing 99% copper. Anodes are poured followed by electrorefining of the anodes to complete the process.

Norddeutsche Affinerie A-G, (Hamburg, Germany) uses an electric-arc furnace to melt low grade scrap and flux (12). Air is introduced for oxidation of impurities. The primary advantages of the electric-arc furnace are that it produces only about 20% of the expected off-gas from a blast furnace, and more completely reduces the charge producing a slag containing low quantities of heavy metals.

Zinc. Secondary zinc metallics are usually melted and selectively drossed or vaporized. Lead is removed from the bottom of the melt as an insoluble and iron forms the intermetallic $FeZn_3$ at the zinc–lead interface. Aluminum forms a top dross intermetallic with iron, $FeAl_3$. Further purification is obtained through distillation. A Pinto furnace is especially capable of removing iron from galvanizer grade top and bottom drosses (13). The Pinto furnace is a cylindrical vessel equipped with agitation that holds about 2 t of molten metal. The melt (650°C) is stirred for 20 minutes. Aluminum is added to form the high melting, low density intermetallic $FeAl_3$. The zinc is separated from the intermetallic by the effects of the agitation. Air is introduced to oxidize the surface of the intermetallic, further separating the zinc from the dross. The dross contains about 10% each of aluminum and iron. The zinc is advanced to a reverbatory furnace for alloying with aluminum.

The zinc retort is a large heated bottle made of silicon carbide and graphite containing about 2500 kg of material (13). One batch per day is common producing 50 t/month of zinc per retort. Oxides are usually not processed in the retorts. Zinc metal or oxide can be produced in the retorts but economics dictate that a specialty zinc dust or zinc oxide is obtained. Dust can be manufactured in a surface condenser at the rate of 1 t/m^3 of volume. For the production of French process oxide, the zinc vapor is burned in air. The residue from the retort is poured out while hot, and depending on conditions may have very little zinc remaining. Energy consumption for the retort is around 15.9 GJ/t (3,800,000 kcal/t).

Muffle furnaces are used to upgrade zinc metallics and dross. A typical muffle furnace is divided into two sections. One section is for melting and the other for vaporizing. Zinc scrap is added to the melting portion of the furnace. High melting point materials are skimmed and screened for metal recovery. Iron reacts with aluminum in the scrap to form the insoluble $FeAl_3$ which is skimmed. The recovered dross contains about 60% zinc and 2–3% lead. The melted zinc flows under a partition to the vaporizing chamber. The vaporizing chamber is usually heated in an upper chamber separated from the vaporizing metal by a muffle constructed of brick. The combustion gas exits into the melting unit. The vaporized zinc is either condensed or burned to the oxide. Aluminum and copper accumulate in the melt, the lead is tapped from under the melt, and the cadmium exits with the zinc. Up to 10 t of the 36-t melt is tapped weekly to produce a high aluminum, high zinc slab that is sold for the production of aluminum diecast alloys. The slab contains 10–30% copper, 15–50% zinc, <0.1% chromium, and <1.5% the remainder aluminum.

Larvik furnaces are electrically heated muffle furnaces that offer improved thermal efficiency. These are used to treat lower grade zinc residues containing significant quantities of iron such as $FeZn_3$ as bottom dross. Energy consumption is 5.4–7.9 GJ/t (1,300,000–1,900,000 kcal/t). Vapor is superheated by graphite resistors and moves countercurrent to the liquid. As the vapors cool, the melt is heated to near boiling (907°C) as both exit through a reflux column. Lead is refluxed into the melt where it falls to the bottom and is tapped. Materials that do not melt are skimmed. For the production of zinc, care is taken to avoid the addition of aluminum-containing materials which would form a $FeAl_3$ top dross containing high levels of entrained zinc. High boiling point impurities concentrate as they flow toward a tap at the far (high temperature) end of the furnace. A phosphorus-containing flux is added to lower the melting point for tapping. Most of the zinc is fumed off in this end of the furnace leaving an iron alloy containing 10% phosphorus and 2% zinc. A typical Larvik furnace treats 1800 to 2800 t/yr of zinc (13).

Sweating furnaces are used to process heavily contaminated scrap where the lead–zinc–aluminum–copper materials are not alloyed. Both reverbatory and muffle furnaces are used for this purpose. The metals are separated by melting points from organics, nonmetallic residues, and higher melting metals. Commonly, sweating furnaces are used to separate low melting point metals such as zinc, aluminum, and lead. The metals are separated sequentially by melting point, ie, lead (327°C), zinc (420°C), and aluminum (660°C). The energy requirements when sweating lead and zinc are around 753 MJ/t (180,000 kcal/t) of feed, and for aluminum, 1.3 GJ/t (300,000 kcal/t).

Hydrometallurgical Methods of Recycling

Changes in environmental regulations, notably RCRA, have brought about increased implementation of hydrometallurgical means for metals recycling. Because many of the wastes are classified as hazardous, there are economic incentives to recycle rather than discard such materials. Significant improvements have been realized in the efficiency of hydrometallurgical or aqueous solution-based processing (14). Some of the reasons (15) given by the mining industry for this shift in emphasis from pyrometallurgy to hydrometallurgy (see SUPPLEMENT) are increasing environmental legislation which has had a significant affect on the industry primarily owing to the reduced emission requirements for sulfur dioxide from the smelting of sulfide concentrates, increasing costs to produce suitable grades of concentrates for smelter purposes as the grades of ore continue to decrease, increasing costs of energy and labor, the amenability of complex sulfide ores to processing by chloride leaching, and the relative ease of separations of multimetal materials using hydrometallurgical techniques (16).

This shift in emphasis by the mining industry has led to the development and use of a variety of improved techniques, in particular the commercial availability of several metal specific extractants. These techniques are particularly useful in the separations and recycling of metals from metal sludges and metal salt solutions.

Concentrated waste solutions are obtained from spent metal plating baths and etchants. However, the majority of metal wastes are solids or sludges obtained from the hydrolysis of metal-bearing solutions and industrial process effluents. Most of these water-insoluble wastes are composed of hydroxides or basic salts of the contained metals. For processing by hydrometallurgical routes the materials must be brought into solution usually by acid or ammoniacal or alkaline digestion.

Precipitation. *Hydrolysis and Cementation.* Precipitation is one of the oldest techniques used for metal–metal and metal–solution separations. Precipitation can be illustrated by the following reactions:

$$Cu^{2+} + 2\,Fe^{3+} + 6\,OH^- \xrightarrow{\text{H}^+ \text{ to pH 3.5}} Fe_2O_3\,(s) + 3\,H_2O + Cu^{2+}$$

$$Cu^{2+} + Fe^0 \longrightarrow Cu^0\,(s) + Fe^{2+}$$

The first equation is an example of hydrolysis and is commonly referred to as chemical precipitation. The separation is effective because of the differences in solubility products of the copper(II) and iron(III) hydroxides. The second equation is known as reductive precipitation and is an example of an electrochemical reaction. The use of more electropositive metals to effect reductive precipitation is known as cementation. Precipitation is used to separate impurities from a metal in solution such as iron from copper (eq. 1), or it can be used to remove the primary metal, copper, from solution (eq. 2). Precipitation is commonly practiced for the separation of small quantities of metals from large volumes of water, such as from industrial waste processes.

For a sparingly soluble salt, the product of the total molecular concentrations of the ions is a constant for a given temperature. Thus for the dissolution of the electrolyte,

$$A_n B_m \rightleftharpoons n\,A^{m+} + m\,B^{n-} \qquad K_{sp} = [A^{m+}]^n[B^{n-}]^m$$

where K_{sp} is the solubility product constant and the brackets denote molar concentrations of the ionic constituents. Although activities of ions should be used in the strictest sense, molar concentrations are most often substituted in dilute solutions. The effects of pH on the solubility of certain metal hydroxides and sulfides are illustrated in Figure 5.

Hydrolysis is the interaction of the ion or ions of a salt with water to produce a weak acid, weak base, or both. When a metal hydroxide is brought into equilibrium with water the following occurs:

$$M(OH)_n \rightleftharpoons M^{n+} + n\,OH^-$$

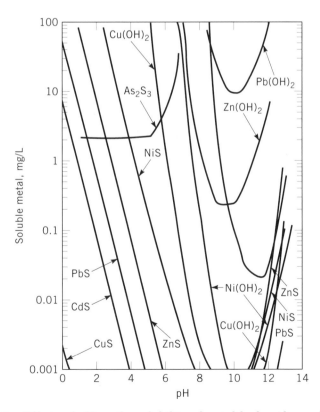

Fig. 5. Effects of pH on the solubility of metal hydroxides and sulfides.

From the solubility product data the thermodynamic tendency of a metal hydroxide to precipitate as a function of pH can be determined. The actual solubilities on complex samples must be determined empirically; however, general predictions can be made regarding metal–metal separations by hydrolysis.

Cementation, the removal of a metal ion from solution by reduction of the metal with a more electropositive material, is also known as reductive precipitation. For a simple case the following can be used:

$$M^{x+} + N^0 \longrightarrow M^0 + N^{x+}$$

where N^0 is a more electropositive metal than M^0, and N^{x+} and M^{x+} are the hydrated ions of those metals, respectively. Because the process in water is galvanic, the driving force of the process can be gauged using the standard reduction potentials for the metals in question.

A classic example of cementation is the removal of copper from solution by elemental iron.

$$Cu^{2+} + 2\,e^- \longrightarrow Cu^0 \qquad E^0 = +0.337 \text{ V}$$

$$Fe^{2+} + 2\,e^- \longrightarrow Fe^0 \qquad E^0 = -0.440 \text{ V}$$

Iron metal preferentially dissolves; the copper does not. Therefore, the anodic and cathodic reactions are as follows, to give the solid copper.

Anodic $\qquad\qquad\qquad Fe^0 \longrightarrow Fe^{2+} + 2\,e^-$

Cathodic $\qquad\qquad\qquad Cu^{2+} + 2\,e^- \longrightarrow Cu^0$

The potential of the reaction is given as E^0_{cell} = (cathodic − anodic reaction) = $0.337 - (-0.440) = +0.777$ V. The positive value of the standard cell potential indicates that the reaction is spontaneous as written (see ELECTROCHEMICAL PROCESSING). In other words, at thermodynamic equilibrium the concentration of copper ion in the solution is very small. The standard cell potentials are, of course, only guides to be used in practice, as rarely are conditions sufficiently controlled to be called standard. Other factors may alter the driving force of the reaction, eg, cementation using aluminum metal is usually quite anomalous. Aluminum tends to form a relatively inert oxide coating that can reduce actual cell potential.

Because copper is a relatively unreactive metal it is much more common to remove copper from solution by cementation than it is to cement impurities from solutions of copper. In the removal of copper as an impurity from a more electropositive metal, cementation can be exceedingly useful. It is common practice to cement copper from nickel and zinc electrowinning solutions using nickel and zinc metal, respectively (17). After a chemical or group of chemicals is precipitated from solution, a physical separation must be effected. This is usually accomplished by thickening followed by filtration and washing. Often, the precipitated product (cement) is leached again for purification.

Electrolytic Precipitation. In 1800, 31 years before Faraday's fundamental laws of electrolysis, Cruikshank observed that copper metal could be precipitated from its solutions by the current generated from Volta's pile (18). This technique forms the basis for the production of most of the copper and zinc metal worldwide.

Electrowinning is the deposition of metals from a leach solution by electrolysis where the metal is produced by cathodic reduction. Cementation relies on galvanic potential differences to effect metal reduction. Electrowinning can be considered a form of reductive precipitation, but it must rely on the superimposition of a more negative potential on the system than usually found at equilibrium if the metal is to be deposited at the cathode. Oxidation occurs at the insoluble anode. Cell potentials can be carefully controlled to allow the selective separation of copper from many metals. By contrast, cementation is quite nonselective. Electrorefining differs from electrowinning in that the anodic and cathodic reactions are the reverse of one another. Impure or blister copper metal from fire refining (pyrometallurgical operations) is the soluble anode. The copper ion formed by dissolution of the anode is plated at the cathode.

For many waste streams, electrical efficiencies are compromised owing to the corrosivity of the solution toward the precipitated metals and/or the low concentrations of metals that must be removed. The presence of chloride in the solution is particularly troublesome because of the formation of elemental chlorine at the anode. Several commercial cells have become available that attempt to address certain of these problems (19).

Solvent Extraction. The selective partitioning of metals by liquid–liquid solvent extraction is one of the most powerful methods of hydrometallurgical separation. Solvent extraction can also be used to concentrate dilute metal streams to facilitate further processing. The metal extraction process can be made highly selective or exclusive either by design of the extractant or by the chemical conditions imposed during extraction. An extractant's selectivity or affinity for groups of metals is determined by its electronic and/or steric properties. Alternatively, the aqueous environment supporting the metal ions can be altered by change in pH, electrolyte content, or by the incorporation of complexing agents in the aqueous solution that either compete with the extractant, owing to compound formation in the aqueous layer, or facilitate the separation process (ion pair formation).

In practice the extractant is carried by an aqueous immiscible solvent. The aqueous layer containing the metal ions is contacted (mixed) with the solvent. Depending on conditions, typically pH, certain of the metal ions are extracted into the organic layer. The mixed organic aqueous system is allowed to separate and the metal-laden organic is then contacted by an aqueous solution that strips the metal ion from the organic. Under normal operating conditions extractants do not give absolute metal separation in one stage of contact. Several stages of solvent–aqueous contact are usually required in order to achieve the degree of separations and efficiencies desired. This requires that continuous processing be utilized in order to be economically acceptable.

One of the most common contacting schemes uses the mixer–settler contactor. The organic and aqueous streams are pumped into the mixer continuously where the streams are contacted prior to exiting by overflow into the settler stage. The two phases are given sufficient time in the settler for phase separation. The organic layer, if lighter than the aqueous, overflows a weir at the exit end of the settler and is advanced to the next stage. The heavier aqueous exits the settler, eg, prior to the weir, and is advanced to the next stage or from the system. It should be noted that, in practice, the aqueous always contacts the organic in a countercurrent flow, ie, loaded aqueous contacts lean organic and lean aqueous contacts the stripped organic. This gives optimum multistage efficiency of extraction and stripping. Several other contacting schemes are available commercially and have been detailed (20). Commonly available commercial extractants are detailed in Table 3.

Commercial Hydrometallurgical Process for Multimetal Waste Separation. Many of the wastes that lend themselves to hydrometallurgical recovery are obtained as hydroxide sludges and solutions. Most are highly complex and contain several metals of economic interest. Therefore, a general scheme of processing needs to be developed. Often a combination of technologies is employed if efficient separations are to occur. One such process (21) that has been operated at the pilot-plant level is illustrative of a multistage process scheme for mixed metal hydroxide sludges from the electroplating and metal finishing industries. The process includes recovery stages for copper, nickel, zinc, and chromium. Consideration is given for iron, aluminum, and calcium removal so that the primary metals of interest are not contaminated. No efforts have been expended to recover these secondary materials. A simplified block flow sheet of the process is given in Figure 6.

The primary steps in the process are (*1*) leaching of sludges by sulfuric acid; (*2*) rejection of high iron materials by selective precipitation as jarosite; (*3*) solid–liquid separation of the leachate from the residues and jarosite and disposal of the iron-containing residues; (*4*) separation of the copper from the leach by solvent extraction (hydroxyoxime extractants) followed by electrowinning to metal or crystallization as copper sulfate pentahydrate; (*5*) raffinate from the copper circuit is advanced to a zinc solvent extraction circuit for selective removal of zinc by di-2-ethylhexylphosphoric acid followed by crystallization of the strip liquor to give zinc sulfate; (*6*) the raffinate from the zinc circuit is advanced to an electro- or chlorine oxidation of Cr(III) to Cr(VI) followed by selective precipitation of chromate as the lead salt from lead sulfate; (*7*) the lead chromate is leached in sulfuric acid to generate a chromic acid solution and regenerate the lead sulfate; and (*8*) the nickel remaining in the filtrate from the chromium precipitation is removed by precipitation as the sulfide or hydroxide.

The process and economics are detailed (21). Owing to the complex nature of the wastes, the process becomes economical only at high production volumes. Several alternative schemes could be developed based on available technologies. Of primary importance is a thorough understanding of the types and constituents of the wastes that feed the processes. Once this is defined, the process options must be considered and tested. A knowledge of what the process cannot do, ie, its limitations, is just as important as a clear understanding of process capabilities.

Economic Aspects

Aluminum. In the United States there are 22 primary aluminum reduction plants. Production of aluminum from scrap accounted for about 3,100,000 t

Table 3. Commercially Available Extractants

Type	Examples	Manufacturer	Uses
Acid complexes and chelating extractants			
alkylaryl sulfonamide	LIX34	Henkel Corp.	copper extraction from acidic leach liquors
alkyl phosphoric acids	di-2-ethylhexyl phosphoric acid (DEHPA)	Mobil, Daihachi, Albright Wilson	uranium and europium extraction, nickel–cobalt separation
alkyl phosphonic acids	2-ethylhexylphosphonic acid mono-2-ethylhexylester	Daihachi	nickel–cobalt separation
alkyl phosphinic acids	Cyanex 272	American Cyanamid	nickel–cobalt separation
alkyl thiophosphinic acids	Cyanex 301, 302	American Cyanamid	nickel–cobalt separation
aryl sulfonic acids	dinonyl naphthalene sulfonic acid	King Industries	magnesium extraction
carboxylic acids	naphthenic and versatic acids	Shell	copper–nickel separation, yttrium recovery
β-diketones	Hostarex DK16, LIX54	Hoechst, Henkel	copper extraction from ammoniacal solution
hydroxyloximes	LIX63, 84, 860	Henkel	copper and nickel extraction
	P5000 series	ICI-Acorga	copper extraction
	MOC-45, -55TD, -100TD	AlliedSignal	copper extraction
oxines	Kelex 100, 120	Sherex	copper extraction
Ion-pair extractants			
oxines	Kelex 100, 120	Sherex	copper extraction
secondary amines	LA-1 and LA-2	Rohm & Haas	uranium extraction
	Adogen 283	Sherex	proposed for vanadium and tungsten extraction
tertiary amines	various alamines, in particular Alamine 336	Henkel	widely used, uranium cobalt extraction, tungsten, vanadium, cobalt–nickel separation
quaternary amines	various adogens, in particular Adogen 381	Sherex	various uses, uranium cobalt, etc
	Aliquat 336	Henkel	vanadium, possibly chromium, tungsten, and uranium
	Adogen 464	Sherex	similar to Aliquat 336
phosphine oxides, alkyl	Cyanex 921, 923, 925	American Cyanamid	recovery of uranium, tin, arsenic from copper
pyridine dicarboxylate	CLX50	ICI-Acorga	copper extraction from chloride leach liquors

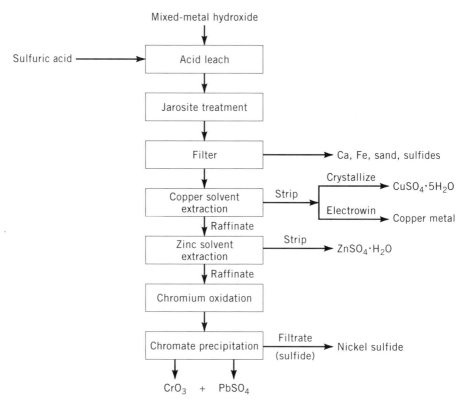

Fig. 6. Diagram showing a suggested commercial process for the separation of metal values from a multimetallic waste.

of the metal in 1994. Secondary aluminum accounts for over 40% of all production (22,23). Owing to increased imports and falling prices, aluminum production fell to its lowest level in many years in 1994. Domestic production capacity by the beginning of 1995 was only 75%. From January 1993 to January 1994 the spot price of aluminum had fallen from $1.23/kg to around $0.90/kg. By January 1995 aluminum had rebounded significantly reaching a spot price of over $2.19/kg. Used aluminum beverage cans (UBCs) generally sell for a minimum of $0.33/kg under spot prices. Of the secondary production approximately 45% comes from new manufacturing scrap and 55% from old, discarded metal. Used aluminum beverage cans have a recycle rate of almost 70% (24) and represent most of the tonnage of recycled aluminum. The aluminum disposition is as follows: packaging, 30%; transportation, 26%; construction, 17%; electrical, 9%; consumer durables, 8%; and other uses, 10%.

Cobalt. There is no U.S. mine production of cobalt. Refining of imported nickel–cobalt mattes has not occurred since the mid-1980s. About 1600 t of secondary cobalt was recycled from scrap by 13 facilities in the United States representing ~22% of total U.S. consumption. The price of the metal was around $44/kg. Most is imported from Zaire and Zambia. Increasing quantities are coming from Russia. Historically, the price of cobalt has been quite volatile and dependent on the political environment in those countries. Cobalt is used in superalloys, 40%; catalysts, 14%; paint driers, 11%; magnetic alloys, 10%; and cemented carbides and other uses, 16%.

Copper. Domestic mine production of copper metal in 1994 was over 1,800,000 t. Whereas U.S. copper production increased in the 1980s and 1990s, world supply declined in 1994. There are eight primary and five secondary smelters, nine electrolytic and six fire refiners, and fifteen solvent extraction–electrowinning (SX–EW) plants. Almost 540,000 t/yr of old scrap copper and alloy are recycled in the United States accounting for ~24% of total U.S. consumption (11). New scrap accounted for 825,000 t of contained copper. Almost 80% of the new scrap was consumed by brass mills. The ratio of new-to-old scrap is about 60:40% representing ~38% of U.S. supply.

Recovered copper in electronic scrap (old) is small in comparison representing about 14,000 t/yr of copper from 68,000 t/yr of waste (25). Electronic scrap accounts for iron (27,000 t), tin (14,000 t), nickel, lead, and aluminum (6,800 t each), and zinc (3,500 t). Precious metal value, which is the primary economic reason for the reclamation of electronic waste, accounts for 345 t of gold, twice that in silver, and some palladium.

The New York Commodity Exchange (Comex) prices for cathode copper in January 1993, 1994, and 1994 were $2.218/kg, $1.844/kg, and $3.084/kg, respectively. The primary uses for copper metal and alloy are construction, 42%; electrical/electronic, 24%; industrial machinery, 13%; transportation equipment, 11%; and consumer/general products, 10%. Copper compounds for use in agriculture and industry account for about 1% of total copper consumption.

Lead. Up to 75% of lead-containing products is derived from secondary materials. The largest single use of lead (qv) is in the manufacture of lead–acid batteries (see BATTERIES, SECONDARY BATTERIES) accounting for almost 80% of total lead consumption. Over 90% of all lead in batteries is recycled. The secondary lead smelter capacity at 26 sites in the United States is 1,000,000 t. Production of secondary lead approached 870,000 t. The price on the London Metals Exchange for lead was $0.489/kg in January 1994 and $0.666/kg in January 1995. In the United States in 1996 a premium is applied to the lead of around $0.176/kg. The principal use of lead is in transportation, accounting for about 70% of all lead products. The primary products are storage batteries. Smaller amounts of lead are used in fuel tanks, solders, and bearings. Almost 25% of lead is used in electrical, electronic, ammunition, construction, and protective coatings. The remainder is used in ceramics, crystal glass, in ballasts, foils, containers, and specialty chemicals.

Nickel. Around 56,000 t of nickel were recycled from scrap in 1994. This is just over one-third of the consumption of the metal. The only mining and smelter in the United States was idle in 1994, although INMETCO (Ellwood City, Pennsylvania) was a primary consumer of recyclable metal in the production of stainless steel. Almost one-half of all nickel went into the production of stainless steels with an additional 29% in superalloys and nonferrous alloys production, and 16% in electroplating. Uses for nickel were transportation, 29%; chemical industry (principally catalysts), 14%; electrical equipment, 11%; construction, 9%; and metal products, 8%. The remainder was used in machinery, household products, petroleum (qv), and other. Nickel prices have increased in spite of a global oversupply of the metal, primarily because of the increasing demands for stainless steel in developing countries.

Tin. There are no primary tin smelters in the United States. About 12,000 t of old and new tin scrap and alloy was recycled in 1994. Of this, 7,400 t was old scrap. The majority of recycled tin was recovered from detinning plants and secondary nonferrous metal processing plants. This recycled quantity was over one-sixth of the tin consumed in the United States in 1994. Of the primary tin consumed, 80% is by 25 companies in the following areas: cans and containers, 32%; electrical, 22%; transportation, 11%; construction, 10%; and other, 25%.

Zinc. Five mines produce over two-thirds of all domestic zinc. An additional 20 mines produce the remaining. There are 20 smelters that process secondary zinc. Six of them use electric-arc furnace (EAF) dusts as the source material. Zinc production from secondary sources accounted for almost 360,000 t, of which 140,000 t was in the form of old scrap. The average price of slab zinc metal was $0.9967/kg and $1.1565/kg in January 1994 and 1995, respectively (26). The primary uses of slab zinc are galvanizing, 47%; zinc-based alloy, 21%; brass and bronze, 13%; and other, 19%. Additionally, there were 126,000 t of zinc used as compounds or dust in the chemical, agriculture, rubber, and paint industries.

Waste Management

Total elimination of metal-bearing wastes is not a likely occurrence owing to process and product requirements and consumer disposal tendencies. On-site waste minimization efforts have been extensive in the industrial sector, however, owing to economic and regulatory driving forces. Local and state programs have had a significant impact on the residential recycling of metal, primarily aluminum, and on glass, plastics, and paper products (see WASTE REDUCTION).

In spite of these efforts, significant quantities of metal-bearing wastes, hazardous or not, continue to be earmarked for disposal rather than reclamation. There are several reasons. Often the wastes are mixed from several process streams, contain relatively low concentrations of many recoverable metals, and may contain high concentrations of water, silicates, and secondary metals, such as calcium and iron.

Although it is generally not practical to eliminate production of the wastes, it is often possible to produce a waste that is more attractive to a reclaimer. Some factors that can be considered are segregation of waste streams, maximization of recoverable metal content and quantity, and minimization of secondary metals, leach insolubles, organics, and chelating agents (see WASTES, INDUSTRIAL).

Maximization of metal content is usually achieved by removal of water, either through evaporation (qv) or drying, although dilute solutions are often concentrated by other techniques such as precipitation or ion exchange (qv). The choice of precipitant for sludge production can have a significant impact on the metal value. Sludges produced by overtreating with lime may contain very high levels of calcium and low levels of recyclable metals. The controlled addition of lime or the use of sodium carbonate or hydroxide may add significant value to the sludge. The addition of filter aids, sulfides, and carbon usually depresses sludge value and often such treatment can be accomplished in a separate stage to render all but a small portion of the waste recyclable.

Waste exchanges function as brokers for waste and often are able to match generators with recyclers of the waste. A number of public and private waste exchanges have been listed (1,18). There are several commercial operations that reclaim metals from wastes. Those companies located in the United States that have a significant interest in the hydrometallurgical processing of metal finishing, electroplating, catalyst, and electronic wastes are as follows:

Alpha Omega	Hackensack, New Jersey
American Chemet	Deerfield, Illinois
American Microtrace	Virginia Beach, Virginia
Bay Zinc	Moxee, Washington
Brush Wellman	Toledo, Ohio
Cozinco	Denver, Colorado
Elwood Chemical	Darlington, Pennsylvania
Encycle	Corpus Christi, Texas
Eticam	Warsick, Rhode Island
Gulf Chemical	Freeport, Texas
Hall Chemical	Wickliffe, Ohio
Heritage Environmental	Indianapolis, Indiana
Hydromet	Newman, Illinois
Madison Industries	Old Bridge, New Jersey
Mineral R&D	Freeport, Texas
NSSI	Houston, Texas
OMG	Cleveland, Ohio
Phibro-Tech	Fort Lee, New Jersey
Port Nickel	Braithwaite, Louisiana
Proler International	Cooleage, Arizona
Southern Water Treatment	Greenville, South Carolina
Tetra Resources	West Memphis, Arkansas
Trichem	Atlanta, Georgia
US Filter	Roseville, Minnesota
Zaclon	Cleveland, Ohio

BIBLIOGRAPHY

"Nonferrous Metals" under "Recycling" in *ECT* 3rd ed., Vol. 19, pp. 967–979, by G. Bourcier, Reynolds Metals Co.

1. C. S. Brooks, *Metal Recovery from Industrial Waste*, Lewis Publishing, Chelsea, Mich., 1991.
2. R. M. Bricka and M. J. Cullinane, *Proc. 42nd Purdue Indus. Waste Conf.* **42**, 809 (1987).
3. R. Odle, I. Martinez, and L. Deets, *J. Met.* **43**, 28 (1991).
4. A. K. Biswas, W. G. Davenport, and B. Betterton, *Extractive Metallurgy of Copper*, Pergamon, Oxford, U.K., 1994, pp. 414–433.
5. L. Ching-Hwa, C. Tei-Chih, and C. Sang-Teh, *JOM* **46**(5), 42 (1994).
6. B. Kos, "Waste Management and Recycling," *International Symposium*, CIM, Vancouver, B.C., Canada, Aug. 1995.

7. M. G. Drouet, M. Handfield, J. Meunier, and C. B. Laflamme, *JOM* **46**(5), 26 (1994).

8. R. D. Peterson, *JOM* **47**(2), 27 (1995).

9. J. C. Olive and J. L. Kirby, in E. R. Cutshall, ed., *Light Metals 1992*, TMS, Warrendale, Pa., 1992, pp. 863–868.

10. O. H. Perry, in U. Mannweiller, ed., *Light Metals 1994*, TMS, Warrendale, Pa., 1994, pp. 879–852.

11. J. L. W. Jolly and D. L. Edelstein, *Copper, Annual Report 1990*, U.S. Bureau of Mines, Washington, D.C., Apr. 1992.

12. K. Gochmann, *CIM Bull.* **85**(958), 150–156 (1992).

13. P. B. Queneau, B. J. Hansen, and D. E. Spiller, in "Recycling Lead and Zinc in the United States," *Proceedings of the Milton E. Wadsworth 4th International Symposium on Hydrometallurgy*, AIME SME/TMS, Salt Lake City, Utah, Aug. 1–5, 1993.

14. R. Winand, *J. Met.* **42**, 25 (1990).

15. R. Price, in J. Duprizac and A. Monhemius, eds., *Iron Control in Hydrometallurgy*, Ellis Horwood, Chichester, U.K., 1987, p. 143.

16. B. Kolodziej and Z. Adamski, *Hydrometal.* **12**, 117 (1984).

17. J. Cigan, T. McKay, and T. O'Keefe, eds., *Lead–Zinc–Tin '80, Proceedings of a World Symposium on Metallurgy and Environmental Control*, The Metallurgical Society, Warrendale, Pa., 1979.

18. C. L. Mantell, *Industrial Electrochemistry*, McGraw-Hill Book Co., Inc., New York, 1950, p. 268.

19. S. A. K. Palmer, M. A. Breton, T. J. Nunno, D. M. Sullivan, and N. F. Suprenant, *Metal/Cyanide Containing Wastes, Treatment Technologies, Pollution Technology*, Review No. 158, Noyes Data Corp., Park Ridge, N.J., 1988.

20. G. M. Ritcey and A. W. Ashbrook, *Solvent Extraction*, Part II, Elsevier, New York, 1984.

21. L. G. Twidwell, EPA/600/2-85/128, Oct. 1985, U.S. EPA, Washington, D.C., 1985.

22. M. C. Munson, *JOM* **47**(4), 54 (1995).

23. *Mineral Commodity Summaries*, U.S. Bureau of Mines, Washington, D.C., 1995.

24. I. K. Wernick, *JOM* **46**(4), 39 (1994).

25. E. Y. L. Sum, *J. Met.* **43**, 53 (1991).

26. *Mineral Industry Surveys*, U.S. Bureau of Mines, Washington, D.C., Apr. 3, 1995.

General References

Am. Met. Mark./Metalwork. News, Fairchild Publishing Co., New York.

Met. Prog., American Society of Metals, Metals Park, Ohio.

Scrap Age, Three Sons Publishing Co., Niles, Ill.

Solid Wastes Management, Communication Channels, Inc., Atlanta, Ga.

Waste Age, National Solid Waste Management Association, Inc., Washington, D.C.

H. WAYNE RICHARDSON
Phibro-Tech, Inc.

GLASS

Glass is a material having properties that provide attributes for many commercial products. As some of these products reach the end of their useful life and are discarded, there is often the opportunity to have the glass recycled into other useful products. In many respects, this alternative is preferred over the glass entering a municipal waste stream for landfill disposal.

Glass chemistry is typically categorized by its oxide components (see GLASS). Table 1 compares oxide components in the principal categories of glass products, which can potentially be recycled. Individual glass products usually have modest variations from these averages. Different colors are derived from minor ingredient additions. In most instances, color separation of recycled glass is necessary to avoid color quality concerns upon remelting. Chemical content differences between glass product categories limit the opportunity to recycle between glass categories.

Table 1. Oxide Components in Glass, %

Oxide	Glass containers	Flat for window and automotive	Fiber glass		Lighting
			Textile	Wool	
SiO_2	72.5	70.7	55.0	57.0	76.4
B_2O_3			7.0	5.2	14.6
Al_2O_3	1.5	0.7	14.8	8.0	2.0
CaO	10.0	9.5	20.5	8.1	
MgO	0.5	3.8	0.5	4.2	
Na_2O	14.5	13.3	1.0	14.5	5.4
K_2O	0.5	0.9	1.0	2.1	1.6

Levels of Recycling

There has been a recognition by the public and municipal governments to promote greater levels of recycling. U.S. consumers are recycling glass containers at a faster rate than ever before. In 1994, the glass container recycling rate reached 37% (1). The recycling rate has increased for seven consecutive years from 1988 to 1994, which indicates that the glass container industry's efforts to build a sustainable, national glass recycling infrastructure are proving successful.

Used and recovered, ie, post-consumer, commercial glass, as well as off-specification glass, suitable for remelting is referred to as cullet. The 37% glass recycling rate reflects the percentage of containers actually being recycled into salable products by manufacturers, not just the amount being collected. This percentage is based on the total number of all jars and bottles sold, not just a specific segment of the container market. Recycled glass (cullet) is not only made into new bottles and jars, but also used for secondary markets such as fiber glass and glasphalt, ie, paving asphalt utilizing crushed cullet as a grog constituent, replacing stone aggregate (2).

Commercial glass can be recycled when sufficient quantities can economically justify the development of a processing infrastructure. A variety of classification and separation issues must be addressed. There are chemical differences between the largest glass product categories, including glass containers; window and automotive glass; electronic glasses, eg, light bulbs, fluorescent tubes, and TV tubes; fiber glass, including insulating wool and textile types; and home cookware (see Table 1). Typically, only container and flat post-consumer glass is recycled commercially. A significant proportion of past cullet recycling has been in the glass container industry. Other segments, such as insulating fiber glass, are increasing the use of post-consumer cullet (3).

Approximately 10.6 million tons of glass containers (41 billion containers) are manufactured in the United States annually (4). In addition, an estimated 800,000 tons of glass containers are imported annually into the United States. Glass containers comprise approximately 4% of the municipal solid waste stream by weight. Normally, glass containers are the second largest contributor by weight to a recycling program, exceeded only by newspapers. It is estimated that 90% of U.S. communities having curbside recycling programs collect at least one color of container glass (5).

Cullet is one of the four principal ingredients in container glass. The other three are sand, limestone, and soda ash, which are all domestically plentiful and relatively inexpensive. Cullet, by melting at a lower temperature than glass-forming raw materials, allows for a reduction of energy input to melting furnaces. This prolongs furnace life. Substituting cullet for raw materials can also reduce emissions into the atmosphere. From a nonmanufacturing perspective, using cullet conserves landfill space for disposal of nonrecyclable materials (6).

Cullet for container production is a mixture of in-house rejects and post-consumer glass that has been beneficiated to remove contaminants, such as content residues, bottle tops, and labels that can cause imperfections in the finished containers. Cullet colors must also be segregated to meet manufacturer's specifications. The quality of the cullet is an important factor in determining both the quantity to be used and the quality of the finished containers. Production of glass from 100% cullet has been demonstrated, and continuous operation at 70–80% cullet is not uncommon. The addition of some virgin raw materials both allows the producer to adjust feed composition to meet product specifications, and aids the furnace's refining process, such as the elimination of gaseous bubbles, referred to as seeds.

Optimal recycling is properly defined as the remanufacturing or reprocessing of discarded materials into the same product. The definition of primary recycling was originally adopted by the American Society of Testing Materials in 1981 (7). In primary recycling, ie, closed-loop recycling, a material is fabricated back into the same material product. Making glass containers into new glass containers or newspapers into newspapers are examples of this. In 1992, the U.S. Department of Commerce declared that an effective recycling program should have a closed-loop system of reuse or a cycle of continuous reapplication (4).

Secondary recycling, or cascading, reprocesses discarded materials into other materials or products. Examples of secondary uses for glass include road aggregate, reflector beads, and fiber glass. Closed-loop glass recycling is preferred by the glass container industry. However, when this is not economically or technically feasible, secondary recycling is critical to avoid landfilling. A recovery or utilization rate includes tonnage from both primary and secondary recycling.

The State of New Jersey's Mandatory Recycling Act required counties and municipalities to recycle 60% of the municipal solid waste stream by 1995 (1). Although the law does not require curbside collection systems, this is by far the most popular recycling collection method in New Jersey. By 1993, New Jersey reached a 46% recycling rate and led the United States in recycling.

Economics

Glass manufacturers have typically compared the cost of batch raw materials to produce one ton of glass with the purchased price of cullet for recycling. The justification to pay more for cullet than for raw materials grew with the recognition of other attributes. For container glass, a relative 10% increase in cullet reduces the melting energy by 2.5%, particulate by 8%, NO_x by 4%, and SO_x by 10%.

Recycling of glass containers saves energy, but not as significant a quantity as compared to reuse. The primary energy saved includes energy required for the entire product life cycle, starting with raw materials in the ground and ending with either final waste disposition in a landfill or recycled material collection, processing, and return to the primary manufacturing process (8). Actual savings depend on many factors, including population density; locations of landfills, recovery facilities, and glass plants; and process efficiencies at the specific facilities available. They decrease if there is no local processing facility or glass plant, or if material losses in the recycling loop are high.

Regulatory motivations can be established in which the government institutes requirements for the recycling or refurbishment of recycled products. However, these kinds of regulations can prove costly, cumbersome, and difficult for all parties involved, whether individual, business, or government. Financial motivation is most effective in accelerating recycling and building an infrastructure to support effective recycling. To create a strong infrastructure, efficient, affordable, convenient collection systems, including transportation of the products to the processing site, are necessary (6).

Quality and Specifications

Container glass typically has a bulk density of 2500 kg/m^3 (155 lb/ft^3); unbroken containers average 350 to 435 kg/m^3 (22–27 lb/ft^3). Crushed into a furnace-ready state, typically less than 10 mm ($\sim 3/8$ in.), the bulk density averages 1300–1500 kg/m^3 (81–94 lb/ft^3). Cullet quality specifications are manufacturing company-specific. Typically, the specification includes a representative sampling technique, particle size gradation, color mix ratio, allowable organic and moisture content, and the specific absence of nonglass contaminants.

Glass manufacturers, as well as other industries such as insulating fiber glass, asphalt grog, structural clay and brick additives, and foam glass, all use cullet as a raw material. The recovery and recycled content requirements for each product demands that cullet meet specific quality requirements because it becomes a larger percentage of the raw material blend.

The glass industry is investing significant resources into improving glass processing and recycling technologies. These innovations include automated color separation, mechanized ceramic sorting, organic color coating, and cullet pulverization technologies (9). Because the glass container industry cannot use amber or green glass to make clear containers, glass used for recycling must be color-sorted. All colors of recycled glass must also be contaminant free.

Cullet processing in the 1990s relies on contaminant removal and separation of glass colors. When the cullet is crushed and contaminants are smaller in size, it is technically difficult or economically unfeasible to process to an acceptable quality level. Container manufacturers want to maximize recycling rates, but cullet must meet quality standards. There are significant quantities of contaminated cullet being generated which are not acceptable for recycling into new bottles and jars.

Many collection systems have not adequately prevented nonglass contaminants, such as ceramic tableware, metals, foils, and other foreign articles, from becoming incorporated in the glass stream destined for melting into new containers. Greater recycling rates have made the contamination more serious. This has resulted in glass manufacturers experiencing damage to furnaces and other equipment, higher production losses, more customer filling line breakage, and potential consumer complaints.

A number of processing schemes have been explored to separate various nonglass contaminants from crushed cullet. Some metals can be easily removed. Methods for nonmetal separation have been found expensive, complicated, marginally effective, and not readily adaptable to high tonnage requirements (9). In evaluating the nature of typical contaminants resulting in defects, it becomes apparent to the glass producers that many of the contaminant materials can be melted in a glass batch if they are of a smaller size. The smaller particle can be more effectively dissolved by other raw material constituents (3).

California legislation requiring recycled glass in insulating fiber glass created a need to produce a fine cullet compatible with their handling equipment. To meet this need, vertical shaft impact (VSI) crusher equipment has been installed by a number of cullet processing companies. The glass is pulverized to a particle size similar to other batch ingredients. This supplemental processing system has shown considerable promise in solving some of the container industry's problems in cullet contamination. These devices rely on receiving conventionally crushed cullet (less than $3/4$ in. in diameter). By accelerating the incoming material particles using a high speed (1500–2500 rpm) rotor or table, momentum forces from velocities up to 300 ft/s cause a breakdown of the glass particles upon impact. These units are capable of handling up to 100 t/h using approximately 2.2 kW (3 horsepower) per ton for crushing (3).

Metal component wear is minimized by employing crushing, grinding, and pulverizing mechanisms with material-to-material contact. Compared to ore minerals, glass can be crushed using less required force and has less equipment abrasion wear concerns with this type of equipment. This system has been proven to be cost-effective in the sand industry by lowering the initial capital and operating costs, including maintenance labor, down time, and energy requirements. The glass industry is realizing similar benefits.

The process of pulverized cullet reduction yields a product having near-batch equivalent sizing (−12 mesh (≤1.7 mm)) and in a furnace-ready condition. Foil-backed paper, lead and other metals, and some tableware ceramics can be removed in an oversized scalping operation after the first pass through the system. Other contaminants are reduced to a fine particle size that can be assimilated into the glass composition during melting.

All processed material is screened to return the coarse fraction for a second pass through the system. Process feed rates are matched to operating variables such as rpm speed and internal clearances, thus minimizing the level of excess fines (−200 mesh (≤0.075 mm)). At one installation (3) the following product size gradation of total smaller than mesh size (cumulative minus) was obtained:

Mesh	mm	%
12	1.7	99
16	1.2	87
20	0.84	70
40	0.42	52
70	0.21	33
100	0.14	16
140	0.10	12
200	0.075	6
325	0.04	2

By using this type of size gradation, nonglass contaminants in the cullet are significantly smaller than their normally crushed size of >1–3 mm. Also, a more uniform size for all batch ingredients ensures less particle size segregation and superior batch homogeneity. Most categories of typical contaminants have a significantly better chance of being assimilated in the batch melting process if the contaminant particle size is similar to the sand (3).

In the recycling of wine bottles, the presence of lead foil is a concern. Lead pieces sink to the furnace bottom, damage the refractory brick paving, and jeopardize furnace life. The pulverized cullet processing method allows some of the lead to be captured in the scalping operation. The remaining pieces are small enough to be incorporated into the glass structure (<50 ppm as PbO) or to exit the furnace in the form of a stone.

Some cullet contaminants continue to be melting concerns. Pyrocerams, especially certain clear glass-like cookingware articles, are difficult to identify after crushing and result in highly stressed defects in the finished glass. For these reasons, this processing method must be placed after existing primary glass processing circuits, and pulverizing must not be considered a substitute for satisfactory conventional sorting.

A number of insulating fiber glass facilities are using the pulverization method for purchased container cullet and expect to expand its use. In the future, this type of material may lead to preheating options using fluidized-bed technology. Batch agglomeration, including cullet, will also be a more feasible option.

Sources of Recoverable Glass

As practiced in the 1990s, cullet recyclers implement the following procedures: (*1*) purchase of scrap glass from a variety of sources; (*2*) supply of transportation to and from the beneficiation facility; (*3*) removal of foreign objects, ie, nonglass materials, including plastic, metal, and ceramic; (*4*) crushing of glass to a specified size; and (*5*) selling of cullet as raw material to a glass manufacturer. Recovery of post-consumer cullet is typically obtained from two categories: municipal solid waste (MSW) and direct resource recovery glass.

Generally, recyclables are either collected at curbside or deposited by consumers at various types of drop-off locations, such as local recycling centers, community service clubs, dealers, and commercial buyback centers. Curbside collections of recyclables can be accomplished either in conjunction with the pickup of all MSW or as a separate activity. Co-collection systems range from complete commingling of all waste for later separation at a mixed waste processing facility to transporting essentially source-separated recyclables in the same truck as MSW.

Separate collection programs generally use bins or toters into which consumers have deposited mixed recyclables. Contents are then sorted at curbside in conjunction with loading onto a separate collection vehicle. Curbside sorting is labor- and equipment-intensive. In suburban settings, collection rates can vary from 78–85 homes per hour using a two-way sort (recyclable versus nonrecyclable), to only 45 homes per hour using a full sort of newspapers, plastics, and glass. Thus, the trend is to two-way sorting at curbside, followed by separation of commingled recyclables at a material recovery facility (MRF) (6).

The variety of designs in use for MRFs in the United States results in different quality products. As recycling rates increase, new facilities are built. For glass separation, the following devices are typically utilized.

Screening Machine. This machine automatically screens out broken glass and other small particles of material. The pieces of broken glass are too small to be sorted by color. Instead, they are cleaned and processed together to produce a mixed-color glass aggregate that has been used to make champagne bottles, glass-based asphalt, and fiber glass products.

Inclined Sorting Table. Glass bottles, aluminum cans, and plastic bottles pass over an inclined sorting machine. Rotating chain curtains automatically divert the lightweight aluminum and plastic to each side of the table, whereas the heavier glass falls through the chains and brushes.

Glass Sorting Platform. Recycling center employees separate green, amber, and clear glass, dropping them into separate bins. Workers remove ceramic, mirror, or window glass because glass recycling plants typically refuse shipments containing these items.

Glass Beneficiation. Glass bottles and jars are crushed to the desired particle size. The material is screened for the removal of organic, eg, paper and plastic, labels, as well as closures, eg, caps and lids, not removed in earlier operations. Magnetic and nonferrous metal-detection devices remove these contaminants. Optical sorting equipment has not yet been adopted as of 1996 on an industrywide scale, on account of high capital and operating costs. The primary focus for avoiding contamination is precrushing activities.

Glass Storage.   Processed glass is stored on concrete pads or in silo bins until delivered to glass manufacturers.

Hand separation of glass bottles results in a much more usable product and therefore a higher recovery rate, ie, a tradeoff of labor for energy (8). If glass is broken before the colors are separated, the resulting product is of low value for container manufacture. Mixed-color glass may be salable for glasphalt, glass beads for reflective paint, or possibly fiber glass, but there is little or no market for it in containers.

The bulk of all cullet shipments are by over-the-road trucks. One cullet supplier reports mode shares of over 70% truck, 25% rail, and less than 5% barge (9). Another reports a 100% truck share (8). Although cullet suppliers often operate their own truck fleets, common carriers generally handle the longer hauls, often as return hauls. Given the mix of short one-way hauls versus long hauls with loaded returns, a reasonable average shipment distance appears to be on the order of 321.8 km (200 miles) (6).

There are about 65 glass container manufacturing facilities in the United States. Owens-Brockway supplies approximately 40% of all glass containers manufactured in this country; Anchor, Ball-Foster, and Gallo Glass account for most of the remaining 60%. Owens-Brockway and Gallo account for the largest internal or co-located beneficiation capacity ($\sim$12%), whereas Allwaste is the leading independent supplier of glass cullet ($\sim$42%). Allwaste currently operates 24 plants throughout the United States and owns the three oldest commercial suppliers: Advance Cullet, Circo, and Bassichis Company.

Recycling is not the total answer to the solid waste problem. However, efficiently operated recycling programs can easily divert 35% or more of municipal solid waste away from disposal. Curbside collection systems offer the opportunity to collect the greatest amount of recyclables in the most cost-efficient manner. It is imperative that communities and recyclers operate programs that are glass-friendly and which result in color-separated, contaminant-free material.

BIBLIOGRAPHY

"Recycling (Glass)" in *ECT* 3rd ed., Vol. 19, pp. 963–966, by P. Marsh, Marsh-Eco-Service Co., Inc.

1. *Americans Recycling Glass Containers at a Faster Rate Than Ever Before*, Glass Packaging Institute (GPI), Washington, D.C., Apr. 17, 1995.

2. *Resource Recycling Magazine* (1994).

3. C. P. Ross, *Glass Researcher*, **3**(2), 10–11 (1994).

4. *Glass Containers: Current Industrial Reports*, M32G(91)-13, U.S. Department of Commerce, Bureau of the Census, Washington, D.C., May 1992.

5. *Solid Waste Management Policy*, Glass Packaging Institute, Feb. 24, 1994.

6. L. L. Gaines and M. M. Mintz, *Energy Implications of Glass-Container Recycling*, Argonne National Laboratory, Argonne, Ill., Mar. 1994.

7. *Thesaurus on Resource Recovery Terminology*, American Society for Testing and Materials, Philadelphia, Pa., 1981.
8. E. Babcock and co-workers, *The Glass Industry: An Energy Perspective*, Energetics, Inc., Columbia, Md., prepared for Pacific Northwest Laboratory, Richland, Wash., Sept. 1988.
9. Technical data, C. Busey, Allwaste Recycling, Inc., Houston, Tex., Apr. 1995.

General Reference

Characterization of Municipal Solid Waste in the United States: 1992 Update, EPA/530-R-92-019, U.S. Environmental Protection Agency, Washington, D.C., July 1992.

C. Philip Ross
Creative Opportunities, Inc.